REFERENCE

Grzimek's
Animal Life Encyclopedia

Second Edition

●●●●

Grzimek's Animal Life Encyclopedia

Second Edition

●●●●

Volume 17
Cumulative Index

Melissa C. McDade, Project Editor

Synapse, the Knowledge Link Corporation, Indexing Services

Michael Hutchins, Series Editor

In association with the American Zoo and Aquarium Association

Detroit • New York • San Diego • San Francisco • Cleveland • New Haven, Conn. • Waterville, Maine • London • Munich

THOMSON
GALE

Grzimek's Animal Life Encyclopedia, Second Edition

Volume 17: Cumulative Index

Project Editor
Melissa C. McDade

Editorial
Madeline Harris, Christine Jeryan, Kate Kretschmann, Mark Springer

Indexing Services
Synapse, the Knowledge Link Corporation

Permissions
Margaret Chamberlain

Imaging and Multimedia
Randy Bassett, Mary K. Grimes, Lezlie Light, Christine O'Bryan, Barbara Yarrow, Robyn V. Young

Product Design
Tracey Rowens, Jennifer Wahi

Manufacturing
Wendy Blurton, Dorothy Maki, Evi Seoud, Mary Beth Trimper

For more information contact
The Gale Group, Inc.
27500 Drake Rd.
Farmington Hills, MI 48331-3535
Or you can visit our Internet site at
http://www.gale.com

Cover photo of chimpanzees (*Pan troglodytes*) by K & K Ammann, Bruce Coleman, Inc. Back cover photos of sea anemone by AP/Wide World Photos/University of Wisconsin-Superior; land snail, lionfish, golden frog, and green python by JLM Visuals; red-legged locust © 2001 Susan Sam; hornbill by Margaret F. Kinnaird; and tiger by Jeff Lepore/Photo Researchers. All reproduced by permission.

ISBN 0-7876-5362-4 (vols. 1–17 set)
0-7876-6570-3 (vol. 17)

This title is also available as an e-book.
ISBN 0-7876-7750-7 (17-vol set)

Contact your Gale sales representative for ordering information.

LIBRARY OF CONGRESS CATALOGING-IN-PUBLICATION DATA

Grzimek, Bernhard.
[Tierleben. English]
Grzimek's animal life encyclopedia.— 2nd ed.
v. cm.
Includes bibliographical references.
Contents: v. 1. Lower metazoans and lesser deuterosomes / Neil Schlager, editor — v. 2. Protostomes / Neil Schlager, editor — v. 3. Insects / Neil Schlager, editor — v. 4-5. Fishes I-II / Neil Schlager, editor — v. 6. Amphibians / Neil Schlager, editor — v. 7. Reptiles / Neil Schlager, editor — v. 8-11. Birds I-IV / Donna Olendorf, editor — v. 12-16. Mammals I-V / Melissa C. McDade, editor — v. 17. Cumulative index / Melissa C. McDade, editor.
ISBN 0-7876-5362-4 (set hardcover : alk. paper)
1. Zoology—Encyclopedias. I. Title: Animal life encyclopedia. II. Schlager, Neil, 1966- III. Olendorf, Donna IV. McDade, Melissa C. V. American Zoo and Aquarium Association. VI. Title.
QL7 .G7813 2004
590'.3—dc21
2002003351

Printed in Canada
10 9 8 7 6 5 4 3 2 1

Recommended citation: *Grzimek's Animal Life Encyclopedia,* 2nd edition. Edited by Michael Hutchins, Melissa C. McDade, Donna Olendorf, and Neil Schlager. Farmington Hills, MI: Gale Group, 2004.

Contents

Foreword vi
Cumulative Index 1

Foreword

Earth is teeming with life. No one knows exactly how many distinct organisms inhabit our planet, but more than 5 million different species of animals and plants could exist, ranging from microscopic algae and bacteria to gigantic elephants, redwood trees and blue whales. Yet, throughout this wonderful tapestry of living creatures, there runs a single thread: deoxyribonucleic acid or DNA. The existence of DNA, an elegant, twisted organic molecule that is the building block of all life, is perhaps the best evidence that all living organisms on this planet share a common ancestry. Our ancient connection to the living world may drive our curiosity, and perhaps also explain our seemingly insatiable desire for information about animals and nature. Noted zoologist, E. O. Wilson, recently coined the term "biophilia" to describe this phenomenon. The term is derived from the Greek *bios* meaning "life" and *philos* meaning "love." Wilson argues that we are human because of our innate affinity to and interest in the other organisms with which we share our planet. They are, as he says, "the matrix in which the human mind originated and is permanently rooted." To put it simply and metaphorically, our love for nature flows in our blood and is deeply engrained in both our psyche and cultural traditions.

Our own personal awakenings to the natural world are as diverse as humanity itself. I spent my early childhood in rural Iowa where nature was an integral part of my life. My father and I spent many hours collecting, identifying and studying local insects, amphibians and reptiles. These experiences had a significant impact on my early intellectual and even spiritual development. One event I can recall most vividly. I had collected a cocoon in a field near my home in early spring. The large, silky capsule was attached to a stick. I brought the cocoon back to my room and placed it in a jar on top of my dresser. I remember waking one morning and, there, perched on the tip of the stick was a large moth, slowly moving its delicate, light green wings in the early morning sunlight. It took my breath away. To my inexperienced eyes, it was one of the most beautiful things I had ever seen. I knew it was a moth, but did not know which species. Upon closer examination, I noticed two moon-like markings on the wings and also noted that the wings had long "tails," much like the ubiquitous tiger swallow-tail butterflies that visited the lilac bush in our backyard. Not wanting to suffer my ignorance any longer, I reached immediately for my *Golden Guide to North American Insects* and searched through the section on moths and butterflies. It was a luna moth! My heart was pounding with the excitement of new knowledge as I ran to share the discovery with my parents.

I consider myself very fortunate to have made a living as a professional biologist and conservationist for the past 20 years. I've traveled to over 30 countries and six continents to study and photograph wildlife or to attend related conferences and meetings. Yet, each time I encounter a new and unusual animal or habitat my heart still races with the same excitement of my youth. If this is biophilia, then I certainly possess it, and it is my hope that others will experience it too. I am therefore extremely proud to have served as the series editor for the Gale Group's rewrite of *Grzimek's Animal Life Encyclopedia*, one of the best known and widely used reference works on the animal world. *Grzimek's* is a celebration of animals, a snapshot of our current knowledge of the Earth's incredible range of biological diversity. Although many other animal encyclopedias exist, *Grzimek's Animal Life Encyclopedia* remains unparalleled in its size and in the breadth of topics and organisms it covers.

The revision of these volumes could not come at a more opportune time. In fact, there is a desperate need for a deeper understanding and appreciation of our natural world. Many species are classified as threatened or endangered, and the situation is expected to get much worse before it gets better. Species extinction has always been part of the evolutionary history of life; some organisms adapt to changing circumstances and some do not. However, the current rate of species loss is now estimated to be 1,000–10,000 times the normal "background" rate of extinction since life began on Earth some 4 billion years ago. The primary factor responsible for this decline in biological diversity is the exponential growth of human populations, combined with peoples' unsustainable appetite for natural resources, such as land, water, minerals, oil, and timber. The world's human population now exceeds 6 billion, and even though the average birth rate has begun to decline, most demographers believe that the global human population will reach 8–10 billion in the next 50 years. Much of this projected growth will occur in developing countries in Central and South America, Asia and Africa—regions that are rich in unique biological diversity.

Finding solutions to conservation challenges will not be easy in today's human-dominated world. A growing number of people live in urban settings and are becoming increasingly isolated from nature. They "hunt" in supermarkets and malls, live in apartments and houses, spend their time watching television and searching the World Wide Web. Children and adults must be taught to value biological diversity and the habitats that support it. Education is of prime importance now while we still have time to respond to the impending crisis. There still exist in many parts of the world large numbers of biological "hotspots"—places that are relatively unaffected by humans and which still contain a rich store of their original animal and plant life. These living repositories, along with selected populations of animals and plants held in professionally managed zoos, aquariums and botanical gardens, could provide the basis for restoring the planet's biological wealth and ecological health. This encyclopedia and the collective knowledge it represents can assist in educating people about animals and their ecological and cultural significance. Perhaps it will also assist others in making deeper connections to nature and spreading biophilia. Information on the conservation status, threats and efforts to preserve various species have been integrated into this revision. We have also included information on the cultural significance of animals, including their roles in art and religion.

It was over 30 years ago that Dr. Bernhard Grzimek, then director of the Frankfurt Zoo in Frankfurt, Germany, edited the first edition of *Grzimek's Animal Life Encyclopedia*. Dr. Grzimek was among the world's best known zoo directors and conservationists. He was a prolific author, publishing nine books. Among his contributions were: *Serengeti Shall Not Die*, *Rhinos Belong to Everybody* and *He and I and the Elephants*. Dr. Grzimek's career was remarkable. He was one of the first modern zoo or aquarium directors to understand the importance of zoo involvement in *in situ* conservation, that is, of their role in preserving wildlife in nature. During his tenure, Frankfurt Zoo became one of the leading western advocates and supporters of wildlife conservation in East Africa. Dr. Grzimek served as a Trustee of the National Parks Board of Uganda and Tanzania and assisted in the development of several protected areas. The film he made with his son Michael, *Serengeti Shall Not Die*, won the 1959 Oscar for best documentary.

Professor Grzimek has recently been criticized by some for his failure to consider the human element in wildlife conservation. He once wrote: "A national park must remain a primordial wilderness to be effective. No men, not even native ones, should live inside its borders." Such ideas, although considered politically incorrect by many, may in retrospect actually prove to be true. Human populations throughout Africa continue to grow exponentially, forcing wildlife into small islands of natural habitat surrounded by a sea of humanity. The illegal commercial bushmeat trade—the hunting of endangered wild animals for large scale human consumption—is pushing many species, including our closest relatives, the gorillas, bonobos and chimpanzees, to the brink of extinction. The trade is driven by widespread poverty and lack of economic alternatives. In order for some species to survive it will be necessary, as Grzimek suggested, to establish and enforce a system of protected areas where wildlife can roam free from exploitation of any kind.

While it is clear that modern conservation must take the needs of both wildlife and people into consideration, what will the quality of human life be if the collective impact of short-term economic decisions is allowed to drive wildlife populations into irreversible extinction? Many rural populations living in areas of high biodiversity are dependent on wild animals as their major source of protein. In addition, wildlife tourism is the primary source of foreign currency in many developing countries and is critical to their financial and social stability. When this source of protein and income is gone, what will become of the local people? The loss of species is not only a conservation disaster; it also has the potential to be a human tragedy of immense proportions. Protected areas, such as national parks, and regulated hunting in areas outside of parks are the only solutions. What critics do not realize is that the fate of wildlife and people in developing countries is closely intertwined. Forests and savannas emptied of wildlife will result in hungry, desperate people, and will, in the long-term lead to extreme poverty and social instability. Dr. Grzimek's early contributions to conservation should be recognized, not only as benefiting wildlife, but as benefiting local people as well.

Dr. Grzimek's hope in publishing his *Animal Life Encyclopedia* was that it would "...disseminate knowledge of the animals and love for them," so that future generations would "...have an opportunity to live together with the great diversity of these magnificent creatures." As stated above, our goals in producing this updated and revised edition are similar. However, our challenges in producing this encyclopedia were more formidable. The volume of knowledge to be summarized is certainly much greater in the twenty-first century than it was in the 1970's and 80's. Scientists, both professional and amateur, have learned and published a great deal about the animal kingdom in the past three decades, and our understanding of biological and ecological theory has also progressed. Perhaps our greatest hurdle in producing this revision was to include the new information, while at the same time retaining some of the characteristics that have made *Grzimek's Animal Life Encyclopedia* so popular. We have therefore strived to retain the series' narrative style, while giving the information more organizational structure. Unlike the original *Grzimek's*, this updated version organizes information under specific topic areas, such as reproduction, behavior, ecology and so forth. In addition, the basic organizational structure is generally consistent from one volume to the next, regardless of the animal groups covered. This should make it easier for users to locate information more quickly and efficiently. Like the original Grzimek's, we have done our best to avoid any overly technical language that would make the work difficult to understand by non-biologists. When certain technical expressions were necessary, we have included explanations or clarifications.

Considering the vast array of knowledge that such a work represents, it would be impossible for any one zoologist to have completed these volumes. We have therefore sought specialists from various disciplines to write the sections with which they are most familiar. As with the original *Grzimek's*, we have engaged the best scholars available to serve as topic editors, writers, and consultants. There were some complaints about inaccuracies in the original English version that may have been due to mistakes or misinterpretation during the complicated translation process. However, unlike the original *Grzimek's*, which was translated from German, this revision has been completely re-written by English-speaking scientists. This work was truly a cooperative endeavor, and I thank all of those dedicated individuals who have written, edited, consulted, drawn, photographed, or contributed to its production in any way. The names of the topic editors, authors, and illustrators are presented in the list of contributors in each individual volume.

The overall structure of this reference work is based on the classification of animals into naturally related groups, a discipline known as taxonomy or biosystematics. Taxonomy is the science through which various organisms are discovered, identified, described, named, classified and catalogued. It should be noted that in preparing this volume we adopted what might be termed a conservative approach, relying primarily on traditional animal classification schemes. Taxonomy has always been a volatile field, with frequent arguments over the naming of or evolutionary relationships between various organisms. The advent of DNA fingerprinting and other advanced biochemical techniques has revolutionized the field and, not unexpectedly, has produced both advances and confusion. In producing these volumes, we have consulted with specialists to obtain the most up-to-date information possible, but knowing that new findings may result in changes at any time. When scientific controversy over the classification of a particular animal or group of animals existed, we did our best to point this out in the text.

Readers should note that it was impossible to include as much detail on some animal groups as was provided on others. For example, the marine and freshwater fish, with vast numbers of orders, families, and species, did not receive as detailed a treatment as did the birds and mammals. Due to practical and financial considerations, the publishers could provide only so much space for each animal group. In such cases, it was impossible to provide more than a broad overview and to feature a few selected examples for the purposes of illustration. To help compensate, we have provided a few key bibliographic references in each section to aid those interested in learning more. This is a common limitation in all reference works, but *Grzimek's Animal Life Encyclopedia* is still the most comprehensive work of its kind.

I am indebted to the Gale Group, Inc. and Senior Editor Donna Olendorf for selecting me as Series Editor for this project. It was an honor to follow in the footsteps of Dr. Grzimek and to play a key role in the revision that still bears his name. *Grzimek's Animal Life Encyclopedia* is being published by the Gale Group, Inc. in affiliation with my employer, the American Zoo and Aquarium Association (AZA), and I would like to thank AZA Executive Director, Sydney J. Butler; AZA Past-President Ted Beattie (John G. Shedd Aquarium, Chicago, IL); and current AZA President, John Lewis (John Ball Zoological Garden, Grand Rapids, MI), for approving my participation. I would also like to thank AZA Conservation and Science Department Program Assistant, Michael Souza, for his assistance during the project. The AZA is a professional membership association, representing 215 accredited zoological parks and aquariums in North America. As Director/William Conway Chair, AZA Department of Conservation and Science, I feel that I am a philosophical descendant of Dr. Grzimek, whose many works I have collected and read. The zoo and aquarium profession has come a long way since the 1970s, due, in part, to innovative thinkers such as Dr. Grzimek. I hope this latest revision of his work will continue his extraordinary legacy.

Silver Spring, Maryland, 2001
Michael Hutchins
Series Editor

Cumulative Index

Bold page numbers indicate the primary discussion of a topic; page numbers in italics indicate illustrations; "t" indicates a table.

A

AAA (animal-assisted activities), 14:293
Aardvarks, 12:*48*, 12:129, 12:135, 15:131, 15:134, **15:155–159,** 15:*156*, 15:*157*, 15:*158*
Aardwolves, **14:359–367,** 14:*360*, 14:*361*, 14:*362*, 14:*363*, 14:*364*
  behavior, 14:259, 14:362
  conservation status, 14:362
  distribution, 14:362
  evolution, 14:359–360
  feeding ecology, 14:255, 14:260, 14:362
  habitats, 14:362
  physical characteristics, 14:360
  reproduction, 14:261, 14:362
  taxonomy, 14:359–360
AAT (animal-assisted therapy), 14:293
Aba-aba, 4:57, 4:231, 4:232–233, 4:*235*, 4:*237*
Abaco boas, 7:410
Abalone, 1:51, 2:431, 2:*432*, 2:*433*
Abatus cordatus. See Heart urchins
Abbott's boobies, 8:213–214, 8:*216*, 8:*217*
Abbott's duikers, 16:84*t*
Abbott's starlings, 11:410
Abdim's storks, 8:236, 8:267
Abedus spp., 3:62
Abeillia spp. *See* Hummingbirds
Aberdare shrews, 13:277*t*
Aberrant tooth shells, 2:470, 2:*472*, 2:*473–474*
Abert squirrels, 16:*167*, 16:*168*, 16:*171*
Abiotic environments, 1:24, 2:25–26
  See also Habitats
Ablabys taenianotus. See Cockatoo waspfishes
Able, K. W., 5:28, 5:39
Aborigines, 8:*23*
Abouts. *See* Rudds
Abramis spp., 4:298
Abramis brama. See Carp breams
Abrawayaomys spp., 16:264, 16:265
Abrocoma spp., 16:443–444, 16:445
Abrocoma bennettii. See Bennett's chinchilla rats
Abrocoma boliviensis. See Bolivian chinchilla rats
Abrocoma cinerea. See Ashy chinchilla rats
Abrocomidae. *See* Chinchilla rats
Abronia spp. *See* Alligator lizards
Abronia aurita. See Coban alligator lizards
Abroscopus spp., 11:6
Abrothricines, 16:263–264
Abrothrix longipilis, 16:266–267
Abrothrix olivaceus, 16:267, 16:269
Abtrichia antennata, 3:*378*, 3:379, 3:*380*
Aburria aburri. See Wattled guans
Abyssinian genets, 14:338
Abyssinian hyraxes, 15:177
Abyssinian longclaws, 10:375
Abyssinian sandgrouse. *See* Lichtenstein's sandgrouse
Abyssinian scimitarbills, 10:66
Abyssinian slaty flycatchers, 11:*28*, 11:*38–39*
Abyssinian wild asses. *See* African wild asses
Abyssinians, 14:291
Abyssobrotula galatheae, 5:15
Abyssocotidae, 4:57
Acacia spp., 3:49
Acacia rats. *See* Tree rats
Acadian sparrows. *See* Nelson's sharp-tailed sparrows
Acalypteratae, 3:357
Acanthagenys rufogularis. See Spiny-cheeked honeyeaters
Acanthaster planci. See Crown-of-thorns
Acanthemblemaria spp. *See* Barnacle blennies
Acanthemblemaria maria. See Secretary blennies
Acanthis flammea. See Common redpolls
Acanthis hornemanni. See Hoary redpolls
Acanthisitta chloris. See Riflemen
Acanthisittidae. *See* New Zealand wrens
Acanthixalus spp., 6:279, 6:283
Acanthixalus spinosus. See African wart frogs
Acanthiza spp. *See* Thornbills
Acanthiza chrysorrhoa. See Yellow-rumped thornbills
Acanthiza katherina. See Mountain thornbills
Acanthiza lineata. See Striated thornbills
Acanthiza pusilla. See Brown thornbills
Acanthiza reguloides. See Buff-rumped thornbills
Acanthizidae. *See* Australian warblers
Acanthocephala. *See* Thorny headed worms
Acanthochromis polyacanthus. See Marine damselfishes
Acanthocybium solandri. See Wahoos
Acanthodactylus spp., 7:297
Acanthodii, 4:10
Acanthodrilidae, 2:65
Acanthomorpha, 4:448
Acanthophis spp. *See* Death adders
Acanthophis antarcticus. See Death adders
Acanthophthalmus kuhli. See Coolie loaches
Acanthophthalmus semicinctus. See Coolie loaches
Acanthopidae, 3:177
Acanthopsis choirorhynchus. See Horseface loaches
Acanthopterygii, 4:12
Acanthorhynchus superciliosus. See Western spinebills
Acanthornis magnus. See Scrubtits
Acanthostega spp., 6:7
Acanthoxia spp., 3:203
Acanthuridae, 4:67–68, 5:391
Acanthuroidei, **5:391–404,** 5:*394*, 5:*395*
  behavior, 5:392–393
  conservation status, 5:393
  distribution, 5:392
  evolution, 5:391
  feeding ecology, 5:393
  habitats, 5:392
  humans and, 5:393
  physical characteristics, 5:391
  reproduction, 5:393
  species of, 5:*396*–403
  taxonomy, 5:391
Acanthurus spp., 5:396
Acanthurus chirurgus. See Doctorfishes
Acanthurus coeruleus. See Blue tangs
Acanthurus leucosternon. See Powder blue tangs
Acanthurus lineatus. See Lined surgeonfishes
Acari, 2:333
Acartia clausi, 2:*303*, 2:304
Accentors. *See* Hedge sparrows
Accipiter spp. *See* Goshawks; Sparrowhawks
Accipiter gentilis. See Northern goshawks
Accipiter henstii. See Henst's goshawks
Accipiter madagascariensis. See Madagascar sparrowhawks
Accipiter minullus. See African little sparrowhawks
Accipiter nisus. See Eurasian sparrowhawks
Accipiter novaehollandiae. See Variable goshawks
Accipitridae. *See* Eagles; Hawks
Acerentomidae, 3:*93*, 3:*94*
Acerentomoidea, 3:93
Acerodon jubatus. See Golden-crowned flying foxes
Aceros cassidix. See Sulawesi red-knobbed hornbills
Aceros waldeni. See Rufous-headed hornbills
Achatina spp., 2:414
Achatinella mustelina. See Agate snails

Achatinidae, 2:411
Acheilognathinae, 4:297
Acherontia atropos. See Death's head hawk moths
Acheta spp., 3:203
Acheta domesticus, 3:203
Achiridae, 5:450, 5:452, 5:453
Achiropsettidae, 5:450
Achistridae, 1:417
Achistrum spp., 1:417
Acid rain, 6:57
Acids, 3:21, 3:24
Acinonychinae. *See* Cheetahs
Acinonyx jubatus. See Cheetahs
Acipenser spp., 4:213
Acipenser brevirostrum. See Shortnose sturgeons
Acipenser fulvescens. See Lake sturgeons
Acipenser gueldenstaedtii, 4:*213*
Acipenser oxyrhinchus. See Atlantic sturgeons
Acipenser sturio. See Atlantic sturgeons
Acipenser transmontanus. See White sturgeons
Acipenseridae, 4:11, 4:58
Acipenseriformes, 4:18, **4:213–220,** 4:*216*
Acmaeiadae, 2:423
Acmaeina, 2:423
Acoelomorpha, 1:11, 1:179
Acoels, **1:179–183, 1:*179–183*,** 1:*180,* 2:35
 behavior, 1:180
 conservation status, 1:179–180
 distribution, 1:179–180
 evolution, 1:5, 1:7, 1:11, 1:179, 2:12, 2:13
 feeding ecology, 1:179–180
 habitats, 1:179–180
 humans and, 1:179–180
 physical characteristics, 1:179–180
 reproduction, 1:179–180
 species of, 1:*181*–182
 taxonomy, 1:5, 1:7, 1:11, 1:179, 2:12, 2:13
Acoleidae, 1:226
Acomyinae, 16:283
Acomys cahirinus. See Egyptian spiny mice
Acomys nesiotes. See Cyprus spiny mice
Acomys russatus. See Golden spiny mice
Aconaemys spp., 16:433
Aconaemys fuscus. See Rock rats
Aconaemys fuscus fuscus, 16:433
Aconaemys fuscus porteri, 16:433
Aconaemys sagei. See Sage's rock rats
Acontias spp., 7:327, 7:328
Acontinae, 7:327
Acontophiops spp., 7:327
Acorn barnacles. *See* Rock barnacles
Acorn-nesting ants, 3:68
Acorn shells. *See* Rock barnacles
Acorn woodpeckers, 10:*148*
Acorn worms, 1:13, 1:22, 1:*443,* 1:*444,* 1:445, 1:*446,* 1:*447,* 1:448–449
Acouchis, 16:124, 16:407, 16:408–409, 16:*411,* 16:*412,* 16:413–414
Acoustics. *See* Hearing
Acraniates. *See* Lancelets
Acrida spp., 3:203
Acridoidea, 3:201
Acridophaga spp., 7:445, 7:447
Acridotheres spp., 11:409
Acridotheres cristatellus. See Crested mynas
Acridotheres melanopterus. See Black-winged mynas
Acridotheres tristis. See Common mynas
Acris spp., 6:48, 6:66, 6:225, 6:228–229
Acris crepitans. See Northern cricket frogs
Acrobates spp., 13:140, 13:141, 13:142
Acrobates pygmaeus. See Pygmy gliders
Acrobatidae. *See* Feather-tailed possums
Acrobothriidae, 1:225
Acrocephalinae, 11:1, 11:2
Acrocephalus spp., 11:2–5
Acrocephalus arundinaceus. See Great reed warblers
Acrocephalus bistrigiceps. See Black-browed reed-warblers
Acrocephalus familiaris. See Millerbirds
Acrocephalus paludicola. See Aquatic warblers
Acrocephalus palustris. See Marsh warblers
Acrocephalus rufesecens, 11:5
Acrocephalus sechellensis. See Seychelles warblers
Acrocephalus stentoreus, 11:5
Acrocephalus vaughani, 11:6
Acrochordidae. *See* File snakes
Acrochordus spp. *See* File snakes
Acrochordus arafurae. See Arafura file snakes
Acrochordus granulatus. See Little file snakes
Acrochordus javanicus. See Java file snakes
Acrocinus longimanus. See Harlequin beetles
Acromis sparsa, 3:62
Acronurus, 5:396
Acropora millepora, 1:*110,* 1:*117*
Acropora palmata. See Elkhorn corals
Acropternis spp. *See* Tapaculos
Acropternis orthonyx. See Ocellated tapaculos
Acrothoracica, 2:273, 2:274
Acrotus willoughbyi. See Ragfishes
Acrulocercus bishopi. See Bishop's oo
Acryllium vulturinum. See Vulturine guineafowl
Actenoides concretus. See Rufous-collared kingfishers
Actiniaria. *See* Sea anemones
Actinodura sodangorum. See Black-crowned barwings
Actinopterygii, 4:10, 4:11–12, 6:7
Actinotrocha spp. *See* Actinotrochs
Actinotrocha branchiata, 2:491
Actinotrocha harmeri, 2:491, 2:494
Actinotrocha pallida, 2:491
Actinotrocha sabatieri, 2:491
Actinotrocha vancouverensis, 2:491, 2:494
Actinotrochs, 2:491, 2:492
Actinulidae, 1:123
Action Plan for Australian Birds 2000, 11:522
Actitis hypoleucos. See Common sandpipers
Actitis macularia. See Spotted sandpipers
Active mimicry, 2:38
 See also Behavior
Actophila, 2:412, 2:413, 2:414
Actophilornis africanus. See African jacanas
Actophilornis albinucha. See Malagasy jacanas
Actornithophilus spp., 3:250
Aculeata, 3:406
Acutotyphlops spp., 7:380, 7:381
Acutotyphlops subocularis. See Bismarck blindsnakes
Acyphoderes sexualis, 2:38
Acyrthosiphon pisum. See Pea aphids
Acyrtops spp., 5:355
Acyrtus spp., 5:355
Adamson's grunters, 5:222
Adanson, M., 11:379
Adapiformes, 14:1, 14:3
Adapis spp., 14:1
Adaptive radiation, 8:3, 11:341, 12:12
Addax nasomaculatus. See Addaxes
Addaxes, 16:2, 16:28–34, 16:*36,* 16:*37,* 16:40
Adder pikes. *See* Lesser weevers
Adders
 African puff, 7:*27,* 7:447
 common, 7:447, 7:448, 7:*449,* 7:*459*–460
 death, 7:39, **7:483–488,** 7:*489,* **7:493–494**
 dwarf puff, 7:446
 Gaboon, 7:446, 7:447, 7:*449,* 7:*451,* 7:457
 night, 7:445–448, 7:*486*
 Rhombic night, 7:*449,* 7:*451*–452
 See also Eastern hog-nosed snakes
Addisonia excentrica, 2:*435*
Adelbert bowerbirds, 11:481
Adelgidae, 3:54
Adelie penguins, 8:147, 8:*149,* 8:*152,* 8:153–154
Adelobasileus spp., 12:11
Adelogyrinids, 6:10, 6:11
Adelophryne spp., 6:156
Adelotus spp., 6:35, 6:141
Adelotus brevis. See Tusked frogs
Adelphorphagy, 4:133
Adenomera spp., 6:32, 6:34, 6:156, 6:158
Adenomus spp., 6:184
Adenophorea. *See* Roundworms
Adenorhinos spp., 7:445
Adephaga, 3:315, 3:316, 3:318
 See also Beetles
Aders's duikers, 16:84*t*
Adiheterothripidae, 3:281
Adipose. *See* Body fat
Admiralty flying foxes, 13:331*t*
Adolfus spp., 7:297
Adrenal glands, lissamphibian, 6:20–21
Adrianichthyidae. *See* Ricefishes
Adrianichthyoidei, 5:79
Adrianichthys kruyti. See Duckbilled buntingis
Aechmophorus occidentalis. See Western grebes
Aechmorphorus clarkii. See Clark's grebes
Aedes spp., 3:76
Aedes aegypti. See Yellow fever mosquitos
Aedes trisseriatus, 3:*78*
Aega spp., 2:252
Aegism, 2:32
 See also Behavior
Aegithalidae. *See* Long-tailed titmice
Aegithalos caudatus. See Long-tailed titmice
Aegithalos concinnus. See Black-throated tits
Aegithalos fuliginosus. See Sooty tits
Aegithalos niveogularis. See White-throated tits
Aegithina spp. *See* Ioras
Aegithina lafresnayei. See Great ioras
Aegithina nigrolutea. See Marshall's ioras
Aegithina tiphia. See Common ioras
Aegithina viridissima. See Green ioras
Aeglaius spp., 11:302
Aegolius acadicus. See Northern saw-whet owls
Aegolius funereus. See Tengmalm's owls
Aegotheles spp. *See* Owlet-nightjars
Aegotheles albertisi. See Mountain owlet-nightjars
Aegotheles archboldi. See Archbold's owlet-nightjars
Aegotheles bennettii. See Barred owlet-nightjars
Aegotheles crinifrons. See Moluccan owlet-nightjars

Aegotheles cristatus. *See* Australian owlet-nightjars
Aegotheles insignis. *See* Feline owlet-nightjars
Aegotheles novaezealandiae, 9:387
Aegotheles savesi. *See* New Caledonian owlet-nightjars
Aegotheles tatei. *See* Spangled owlet-nightjars
Aegotheles wallacii. *See* Wallace's owlet-nightjars
Aegothelidae. *See* Owlet-nightjars
Aegyptopithecus spp., 14:2
Aellops titan. *See* White-banded sphinxlets
Aenigmatolimnas marginalis. *See* Striped crakes
Aeolidiella sanguinea, 2:*406,* 2:*408*
Aeoliscus strigatus. *See* Common shrimpfishes
Aeolothripidae, 3:281
Aepophilidae, 3:54
Aepyceros melampus. *See* Impalas
Aepycerotinae, 16:1, 16:27
Aepyorithidae. *See* Elephant birds
Aepyornis gracilis. *See* Elephant birds
Aepyornis hildebrandti. *See* Elephant birds
Aepyornis maximus. *See* Elephant birds
Aepyornis medius. *See* Elephant birds
Aepyornithidae. *See* Elephant birds
Aepypodius bruijnii. *See* Bruijn's brush turkeys
Aepyprymnus spp., 13:73, 13:74
Aepyprymnus rufescens. *See* Rufous bettongs
Aequipecten opercularis. *See* Queen scallops
Aequorea spp., 1:127
Aequorea victoria, 1:49, 1:129, 1:*131,* 1:137
Aequorin, 1:49
Aeretes melanopterus. *See* North Chinese flying squirrels
Aerodramus spp., 9:422–423
Aerodramus fuciphagus. *See* Edible-nest swiftlets
Aerodramus maximus. *See* Black-nest swiftlets
Aerodramus spodiopyglius. *See* White-rumped swiftlets
Aeromys tephromelas. *See* Black flying squirrels
Aeromys thomasi. *See* Thomas's flying squirrels
Aeronautes spp. *See* Swifts
Aeronautes saxatalis. *See* White-throated swifts
Aerosols, 1:52
  See also Conservation status
Aesculapius, 7:54–55
Aesop's Fables, 2:42
Aethia spp. *See* Auks
Aethia psittacula. *See* Parakeet auklets
Aethomerus spp. *See* Longhorn beetles
Aethopyga spp. *See* Sunbirds
Aethopyga boltoni. *See* Apo sunbirds
Aethopyga christinae. *See* Fork-tailed sunbirds
Aethopyga duyvenbodei. *See* Elegant sunbirds
Aethopyga gouldiae. *See* Gould's sunbirds
Aethopyga linaraborae. *See* Lina's sunbirds
Aethopyga primigenius. *See* Gray-hooded sunbirds
Aetiocetus spp., 15:119
Aetobatus narinari. *See* Spotted eagle rays
Aetosaurs, 7:17–18
Afghan foxes. *See* Blanford's foxes
Afghan mouse-like hamsters, 16:243
Afghan pikas, 16:488, 16:*496,* 16:*500*
Afghans, 14:288
AFGP (antifreeze glycopeptide), 5:321–322
Afrana spp., 6:248
Afrauropodidae, 2:375
Africa. *See* Ethiopian region; Palaearctic region
African anomalurids, 12:14
African antelopes, 15:272, 16:4, 16:5
African arawanas, 4:232, 4:*233–234*
African banana bats, 13:311, 13:497, 13:498
African bateleurs, 8:319
African bats. *See* Butterfly bats
African bay owls, 9:338
African black-headed orioles. *See* Eastern black-headed orioles
African black oystercatchers, 9:126–128, 9:*129,* 9:130–*131*
African black tits, 11:155
African black vultures. *See* Lappet-faced vultures
African blue-flycatchers, 11:*99,* 11:*100*
African broadbills, 10:179, 10:*180,* 10:*181*
African brown nightjars. *See* Brown nightjars
African brush-tailed porcupines, 16:*355,* 16:*361,* 16:362–363
African buffaloes, 15:*265,* 16:*3,* 16:*6,* 16:*16,* 16:*19*
  behavior, 16:5
  evolution, 16:11
  feeding ecology, 15:142, 15:272, 16:*14*
  humans and, 15:273, 16:9
  physical characteristics, 15:139, 15:142, 16:12, 16:*13*
  taxonomy, 16:11
  See also Buffaloes
African bullfrogs, 6:247, 6:*249,* 6:*254,* 6:261
African burrowing snakes, **7:461–464**
African bush-warblers. *See* Little rush-warblers
African butterflies, 3:37
African chevrotains. *See* Water chevrotains
African civets, 14:*336,* 14:*338,* 14:*339,* 14:*340*
African clawed frogs. *See* Common plantanna
African climbing mice. *See* Gray climbing mice
African collared doves, 9:251
African Convention for the Conservation of Nature and Natural Resources, 15:195, 15:209, 15:280
African crakes, 9:48, 9:49
African darter tetras. *See* Striped African darters
African dassie rats, 16:121
African dormice, 16:317, 16:318, 16:319, 16:*321*
African dwarf kingfishers, 10:6
African egg-eaters. *See* Common egg-eaters
African Elephant Specialist Group, IUCN, 15:163
African elephantfishes, 4:232, 4:234
African elephants, 12:*8,* 15:*165,* 15:*173,* 15:*174*–175
  behavior, 15:*162,* 15:*164,* 15:*166,* 15:*168,* 15:*171*
  distribution, 12:*135*
  feeding ecology, 15:*167*
  hearing, 12:82
  humans and, 12:173, 15:172
  physical characteristics, 12:11–12
  reproduction, 12:107, 12:*108,* 12:209
  taxonomy, 12:*29*
  vibrations, 12:83
  vision, 12:79
  in zoos, 12:209
  See also Elephants
African epauletted bats, 13:312
African finfoot, 9:69, 9:*71*
African firefinches, 11:393
African flycatchers, 11:25, 11:26
African forest-flycatchers. *See* Fraser's forest-flycatchers
African gannets. *See* Cape gannets
African gazelles, 16:7
African golden cats, 14:391*t*
African golden moles, 12:13
African goliath beetles, 3:13
African grass owls, 9:336, 9:342
African grass rats, 16:261*t*
African gray-headed greenbuls, 10:396
African gray hornbills, 10:72
African gray parrots, 8:19, 8:22
  See also Gray parrots
African gray tits, 11:155
African gray treefrogs. *See* Gray treefrogs
African gray woodpeckers. *See* Gray woodpeckers
African Great Lakes, cichlids, 5:275–276
African green broadbills, 10:*180,* 10:185–*186*
African greenbuls, 10:395
African ground hornbills. *See* Southern ground-hornbills
African ground squirrels, 16:143
African hedgehogs, 13:203
African hillstream catfishes, 4:352
African jabirus. *See* Saddlebills
African jacanas, 9:*108,* 9:*109,* 9:110, 9:*111,* 9:113–*114*
African king vultures. *See* Lappet-faced vultures
African knifefishes, 4:232, 4:233
African lampeyes, 5:89
African land snails, giant. *See* Giant African land snails
African leaf-monkeys. *See* Colobus monkeys
African linsangs, 14:337, 14:*339,* 14:*341*
African little sparrowhawks, 8:319, 8:*325,* 8:*337*
African long-tongued fruit bats, 13:*339,* 13:*343,* 13:345–346
African lungfishes, 4:3, 4:57, 4:201–203, 4:*204,* 4:205–*206*
African manatees. *See* West African manatees
African mole-rats, **16:339–350,** 16:*346*
  behavior, 12:147, 16:124, 16:343
  conservation status, 16:345
  distribution, 16:123, 16:*339,* 16:342
  evolution, 16:123, 16:339
  feeding ecology, 16:343–344
  habitats, 16:342–343
  humans and, 16:345
  parasites and, 12:78
  photoperiodicity, 12:74
  physical characteristics, 12:72–75, 16:339–342
  reproduction, 16:126, 16:344–345
  seismic communication, 12:76
  species of, 16:347–349, 16:349*t*–350*t*
  taxonomy, 16:121, 16:339
  vision in, 12:76–77
African mudfishes, 4:*292,* 4:294–295
African native mice, 16:249
African nightjars, 9:368–369
African otter shrews, 12:14
African oystercatchers. *See* African black oystercatchers

African painted-snipes. *See* Greater painted-snipes
African pale white-eyes. *See* Cape white-eyes
African palm civets, 14:335, 14:*339*, 14:341–*342*
African palm swifts, 9:*425*, 9:*431*
African pancake tortoises. *See* Pancake tortoises
African paradise-flycatchers, 11:*98*, 11:*99*, 11:100–*101*
African penduline tits, 11:147, 11:*150*, 11:*151*–152
African penguins, 8:148, 8:151
African pied wagtails, 10:372
African pikes, 4:*335*, 4:*337*, 4:338
African pittas, 10:194, 10:*196*, 10:*199*–200
African polka-dot catfishes, 4:*353*
African pouched mice. *See* Gambian rats
African pouched rats. *See* Gambian rats
African puff adders, 7:*27*, 7:447
African pygmy falcons, 8:349, 11:383
African pygmy geese, 8:363, 8:369, 8:*374*, 8:*381*
African pygmy-kingfishers, 10:*11*, 10:*20*–21
African quailfinches, 11:*358*, 11:365–*366*
African red-billed hornbills. *See* Red-billed hornbills
African red-eyed bulbuls, 10:397
African red-winged starlings. *See* Red-winged starlings
African religions, snakes in, 7:55
African rhinoceroses, 15:218, 15:220, 15:249
African river blindness nematodes, 1:35–36, 1:*296*, 1:299
African river-martins, 10:361, 10:*362*, 10:*363*
African robin chats, 10:484
African rock pythons, 7:55, 7:420, 7:421
African savanna francolins, 8:435
African scarabs, 3:43
African sedge warblers. *See* Little rush-warblers
African shrike-flycatchers. *See* Shrike-flycatchers
African sideneck turtles, **7:129–134,** 7:*131*
African silverbills, 11:*354*, 11:*358*, 11:*370*
African skimmers, 9:205
African slender-snouted crocodiles, 7:180
African snakeheads, 5:*440*, 5:*446*
African snipes, 9:176, 9:*181*, 9:*182*–183
African snooks. *See* Nile perches
African spotted dikkops, 9:144
African thrushes. *See* Olive thrushes
African tree pangolins. *See* Tree pangolins
African treefrogs, 6:3–4, 6:6, **6:279–290,** 6:*284*
African wart frogs, 6:*284*, 6:*285*
African warthogs. *See* Common warthogs
African watersnakes, 7:467, 7:468
African wild asses, 12:176–177, 15:222, 15:226, 15:228, 15:*231*, 15:*232*
African wild dogs, 14:*259*, 14:*275*
  behavior, 14:258, 14:268, 14:269
  conservation status, 12:216, 14:272, 14:273
  evolution, 14:265
  feeding ecology, 14:260, 14:270
  habitats, 14:267
  physical characteristics, 14:266
  reproduction, 12:107, 14:261, 14:262, 14:271, 14:272
African worms, 2:68, 2:*69*, 2:*70*
Afrixalus spp., 6:32, 6:279, 6:280, 6:281, 6:282–283
Afrixalus brachycnemis, 6:283
Afrixalus delicatus, 6:283
Afrixalus fornasinii. See Greater leaf-folding frogs
Afro-American river turtles, **7:137–142,** 7:*139*
Afro-Australian fur seals, 14:393, 14:407*t*
Afromastacembelinae, 5:151
Afropavo congensis. See Congo peafowls
Afrotheria, 12:12, 12:26, 12:33, 15:134, 15:161, 15:177
Afrotis spp. *See* Bustards
Afrotis afraoides. See White-quilled bustards
Agacris insectivora, 3:204
Agalychnis spp., 6:49, 6:227
Agalychnis callidryas. See Red-eyed treefrogs
Agalychnis craspedopus, 6:227
Agalychnis moreletii, 6:228–229
Agama hispida. See Spiny agamas
Agami herons, 8:239, 8:241, 8:*247*
Agamia agami. See Agami herons
Agamidae, 7:204, **7:209–222,** 7:*212–213*
  behavior, 7:211
  conservation status, 7:211
  distribution, 7:*209*, 7:210
  evolution, 7:209
  feeding ecology, 7:211
  habitats, 7:210
  humans and, 7:211
  physical characteristics, 7:209–210
  reproduction, 7:211
  species of, 7:214–221
  taxonomy, 7:209
Agaminae, 7:209
Agamodon spp., 7:287, 7:288
Agamodon anguliceps, 7:*289*
Agapornis spp. *See* Budgerigars
Agapornis roseicollis. See Rosy-faced lovebirds
Agassiz, Louis, 5:89
Agassiz's dwarf cichlids, 5:*282*, 5:*284*
Agassiz's slickheads, 4:392
Agastoschizomus lucifer, 2:*338*, 2:*340*, 2:348–349
Agate snails, 2:*416*, 2:*418*
Agathiphagidae, 3:55
Agelaioides badius. See Baywings
Agelaius spp. *See* Icteridae
Agelaius chopi. See Chopi blackbirds
Agelaius icterocephalus. See Yellow-hooded blackbirds
Agelaius phoeniceus. See Red-winged blackbirds
Agelaius virescens. See Brown-and-yellow marshbirds
Agelaius xanthomus. See Yellow-shouldered blackbirds
Agelasida, 1:77, 1:79
Agelastes meleagrides. See White-breasted guineafowl
Agelastes niger. See Black guineafowl
Agema spp., 2:317
Aggressive mimicry, 4:68, 5:342
Aggressive vocalizations, 6:47–48
Agile gibbons, 14:207, 14:208–209, 14:210, 14:211, 14:*217*, 14:219–*220*
Agile mangabeys, 14:194, 14:204*t*
Agile wallabies, 13:*88*, 13:*91*, 13:*93*–94
Agkistrodon spp., 7:445, 7:446, 7:448
Agkistrodon bilineatus, 7:448
Agkistrodon contortrix. See Southern copperheads
Agkistrodon piscivorus. See Cottonmouths
Aglaeactis spp. *See* Hummingbirds
Aglaicercus spp. *See* Hummingbirds
Aglantha digitale, 1:*130*, 1:145
Aglaophenia picardi, 1:25
Aglaophenia pluma, 1:25, 1:*131*, 1:137–138
Aglaopheniids, 1:129
Aglyptodactylus spp., 6:281, 6:291, 6:292, 6:293
Agnathans, 4:27, 4:77, 4:83
Agonidae. *See* Poachers
Agonistic behavior, 4:62–63
Agonostomus spp., 5:59
Agonostomus monticola. See Mountain mullets
Agouti paca. See Pacas
Agouti taczanowskii. See Mountain pacas
Agoutidae. *See* Pacas
Agoutis, 12:132, **16:407–415,** 16:*408*, 16:*410*, 16:415*t*
Agreement on the Conservation of Albatrosses and Petrels, 8:126
Agricultural pests, 3:75
  citrus leaf miners, 3:394
  Diptera, 3:361
  Lepidoptera, 3:389
  Orthoptera, 3:207–208
  See also Pests
Agriculture, 3:78–79, 12:213, 12:215
  See also Domestic cattle
Agriochoeridae, 15:264
Agriocnemis femina. See Southeast Asian damselflies
Agrostichthys spp., 4:447
Agrostichthys parkeri, 4:449, 4:454
Agulhas long-billed larks, 10:346
Agulla spp. *See* Snakeflies
Agyrtria spp. *See* Hummingbirds
Agyrtria leucogaster, 9:443
Ahaetulla spp., 7:466
Ahaggar hyraxes, 15:177, 15:189*t*
Ahlquist, Jon, 10:2
  on Apodiformes, 9:415
  on babblers, 10:513
  on Charadriiformes, 9:101
  on Corvidae, 11:503
  on frogmouths, 9:377
  on Irenidae, 10:415
  on logrunners, 11:69
  on long-tailed titmice, 11:141
  on magpie geese, 8:363
  on Old World warblers, 11:1
  on Passeriformes, 10:169, 10:171
  on Pelecaniformes, 8:183
  on Piciformes, 10:85–86
  on toucans, 10:125
  on treecreepers, 11:177
  on trumpeters, 9:77
  on wrentits, 10:513
Aida-Wedo (Loa god), 7:57
Aidemedia. See Gapers
Ailao moustache toads, 6:*112*, 6:114–115
Ailao spiny toads. *See* Ailao moustache toads
Ailoscolecidae, 2:65
Ailuroedus spp., 11:477, 11:478, 11:480
Ailuroedus buccoides. See White-eared catbirds
Ailuroedus crassirostris. See Green catbirds
Ailuroedus melanotis. See Black-eared catbirds
Ailuropoda melanoleuca. See Giant pandas
Ailurops spp. *See* Bear cuscuses
Ailurops melanotis. See Yellow bear cuscuses
Ailurops ursinus. See Sulawesi bear cuscuses

Ailurus spp., 14:309
Ailurus fulgens. See Red pandas
Aimophila aestivalis. See Bachman's sparrows
Aipichthyidae, 5:1
Aipysurus laevis. See Olive seasnakes
Airbreathing catfishes, 4:352
Airbreathing mammals, 12:67–68
See also specific animals
Aïstopods, 6:10
Aix galericulata. See Mandarin ducks
Aix sponsa. See Wood ducks
Ajaia ajaja. See Roseate spoonbills
Akepas, 11:*345*, 11:*346*
Akialoas, 11:343
Akiapolaaus. *See* Lanai hookbills
Akikikis, 11:*345*, 11:*346*–347
Akodon azarae. See Azara's field mice
Akodon boliviensis, 16:270
Akodon cursor, 16:267
Akodon kofordi, 16:270
Akodon mollis, 16:270
Akodon montensis. See Forest mice
Akodon puer, 16:270
Akodon reigi, 16:266
Akodon subfuscus, 16:270
Akodon torques, 16:270
Akodon varius, 16:270
Akodontines, 16:263–264
Akohekohes, 11:342, 11:*345*, 11:*347*
Akysidae, 4:353
Alabama cavefishes, 5:6, 5:7, 5:*9*, 5:10
Alabama shads, 4:278
Alabes spp. *See* Singleslits
Alabes dorsalis. See Common shore-eels
Alabes parvulus. See Pygmy shore-eels
Alaemon spp. *See* Larks
Alaemon alaudipes. See Greater hoopoe-larks
Alagoas antwrens, 10:242
Alagoas curassows, 8:401, 8:414, 8:*417*, 8:*421*
Alagoas foliage-gleaners, 10:211
Alagoas tyrannulets, 10:275
Alajuela toads. *See* Golden toads
Alaotra grebes, 8:169, 8:171
Alaria spp., 1:199
Alaska blackcods. *See* Sablefishes
Alaska blackfishes, 4:379–380, 4:*382*, 4:*383*
Alaska king crabs. *See* Red king crabs
Alaska longspurs. *See* Lapland longspurs
Alaska pollocks, 5:27, 5:30, 5:*31*, 5:*35*–36
Alaskan brown bears. *See* Brown bears
Alaskan shore bugs, 3:263
Alatau salamanders. *See* Semirechensk salamanders
Alauahios, 11:343
Alauda spp. *See* Larks
Alauda arvensis. See Sky larks
Alauda gulgula. See Oriental skylarks
Alauda razae. See Raso larks
Alaudidae. *See* Larks
Alaus oculatus. See Eyed click beetles
Alaysia spp., 2:91
Albacore, 5:407, 5:*408*, 5:*410*, 5:414–415
Albanerpetontidae, 6:13
Albatrosses, 8:107–110, **8:113–122,** 8:*117*
behavior, 8:109, 8:*114*–115
conservation status, 8:116
distribution, 8:*113*, 8:114
evolution, 8:107, 8:113
feeding ecology, 8:109, 8:115
habitats, 8:108, 8:114
humans and, 8:110, 8:116
physical characteristics, 8:107, 8:113–114
reproduction, 8:109–110, 8:*114*, 8:115–116
species of, 8:*118*–122
taxonomy, 8:113
Albertina's starlings. *See* Bare-eyed mynas
Albert's lyrebirds, 10:*330*–331, 10:*331*, 10:*333*, 10:*334*
Albignac, 14:74–75
Albula spp., 4:249
Albula vulpes. See Bonefishes
Albulidae. *See* Bonefishes
Albuliformes, 4:11, **4:249–253,** 4:*251*
Alca impennis. See Great auks
Alca torda. See Razorbills
Alcae. *See* Auks
Alcedines, 10:1–4
Alcedinidae. *See* Kingfishers
Alcedininae, 10:5–6
Alcedo atthis. See Common kingfishers
Alcelaphinae, 16:1
Alcelaphini, 16:27, 16:28, 16:29, 16:31, 16:34
Alcelaphus buselaphus. See Red hartebeests
Alcelaphus buselaphus buselaphus. See Bubal hartebeests
Alcelaphus buselaphus cokii. See Coke's hartebeests
Alcelaphus buselaphus swaynei, 16:33–34
Alcelaphus buselaphus tora, 16:33–34
Alcelaphus lichtensteini. See Lichtenstein's hartebeests
Alces spp., 15:379, 15:383, 15:384
Alces alces. See Moose
Alces (Cervalces) latifrons, 15:*380*
Alces (Libralces) gallicus, 15:*380*
Alcidae, **9:219–229,** 9:*221*, 9:*223*
behavior, 9:*221*
conservation status, 9:222
distribution, 9:*219*, 9:221
evolution, 9:219
feeding ecology, 9:221–222
habitats, 9:221
humans and, 9:222
physical characteristics, 9:220
reproduction, 9:102, 9:222
species of, 9:*224*–*229*
taxonomy, 9:219–220
Alcids. *See* Alcidae
Alcippe spp. *See* Fulvettas
Alcippe chrysotis. See Golden-breasted fulvettas
Alcyonacea, 1:103
Alcyonaria. *See* Octocorallia
Alcyonidium spp., 2:25
Aldabra tortoises, 7:47, 7:70, 7:*72*
Alder flycatchers, 10:269, 10:275
Alderflies, 3:7, **3:289–295,** 3:*290*, 3:*292*, 3:*293*
Aldrichetta spp., 5:59
Aldrovandia spp., 4:249
Aleadryas rufinucha. See Rufous-naped whistlers
Alectoris spp. *See* Rock partridges
Alectoris graeca. See Rock partridges
Alectoris philbyi. See Philby's rock partridges
Alectura lathami. See Australian brush-turkeys
Alena (Aztekoraphidia) schremmeri. See Schremmer's snakeflies
Alepisauridae, 4:432
Alepisauroidei, 4:431
Alepisaurus ferox. See Longnose lancetfishes
Alepocephalidae. *See* Slickheads
Alepocephaloidea, 4:390
Alepocephalus agassizii. See Agassiz's slickheads
Alepocephalus rostratus. See Slickheads
Alepocephalus tenebrosus. See California slickheads
Alestiidae, 4:335
Alethe montana. See Usambara robin-chats
Alethinophidia, 7:198, 7:201
Aleurodicus dugesi. See Giant whiteflies
Aleurothrixus antidesmae, 3:*259*
Alewives, 4:*280*, 4:*281*–282
Alexteroon spp., 6:279, 6:283
Alexteroon obstetricans, 6:35
Alfalfa leaf cutter bees, 3:79, 3:*410*, 3:417, 3:*418*
Alfalfa springtails. *See* Lucerne fleas
Alfoncinos, 5:113, 5:115
Algae, 1:*9*, 1:32–33
Algae eaters, 4:57, 4:321, 4:*322*, 4:*326*, 4:*328*, 4:333
Algerian hedgehogs, 13:212*t*
Algerian nuthatches, 11:169
Ali, Salim
on babblers, 10:507
on fire-tailed myzornis, 10:519–520
on ioras, 10:416
Alice's Adventures in Wonderland (Carroll), 9:270
Alimentary canal, 3:*20*–21
Alkali bees, 3:79
Alkaloids, 6:197, 6:198, 6:199, 6:200, 6:208
Allactaga balicunica, 16:212, 16:216
Allactaga bullata, 16:212, 16:216
Allactaga elater. See Little five-toed jerboas
Allactaga euphratica, 16:216
Allactaga firouzi, 16:216
Allactaga major, 16:213–216
Allactaga severtzovi, 16:212–216
Allactaga sibirica, 16:215, 16:216
Allactaga tetradactyla. See Four-toed jerboas
Allactaga vinogradovi, 16:216
Allactaginae. *See* Five-toed jerboas
Allactodipus bobrinskii. See Bobrinski's jerboas
Allantois, 12:93
Alle alle. See Dovekies
Allen, Arthur A., 8:23
Allen, G. H., 5:353
Allen, G. R., 5:219
Allenbatrachus spp., 5:42
Allenopithecus spp., 14:191
Allenopithecus nigroviridis. See Allen's swamp monkeys
Allen's big-eared bats, 13:*505*, 13:*506*, 13:509
Allen's chipmunks, 16:160*t*
Allen's hummingbirds, 9:*440*
Allen's olingos, 14:316*t*
Allen's squirrels, 16:174*t*
Allen's swamp guenons. *See* Allen's swamp monkeys
Allen's swamp monkeys, 14:194, 14:*196*, 14:*197*
Allen's woodrats, 16:277*t*
Allied rock-wallabies, 13:102*t*
Allied rufous treecreepers. *See* Rufous treecreepers
Alligator gars, 4:221, 4:*222*, 4:*223*, 4:*224*
Alligator lizards, **7:339–344,** 7:*341*
Alligator mississippiensis. See American alligators
Alligator sinensis. See Chinese alligators

Alligator snappers. *See* Alligator snapping turtles
Alligator snapping turtles, 7:94, 7:*96*
  behavior, 7:*35*, 7:39
  as food, 7:47
  physical characteristics, 7:70
  See also Snapping turtles
Alligatoridae, 7:19, **7:171–178,** 7:*174*
  See also Alligators; Caimans
Alligatorinae. *See* Alligators
Alligators, 7:32, 7:60, 7:157–165, **7:171–178,** 7:*174*
  See also American alligators
Allocebus spp. *See* Hairy-eared mouse lemurs
Allocebus trichotis. See Hairy-eared mouse lemurs
Allocricetulus spp., 16:239
Allocricetulus curtatus. See Mongolian hamsters
Allocricetulus eversmanni. See Eversmann's hamsters
Allodontichthys spp., 5:92
Allograpta obliqua. See Chevroned hover flies
Allopauropus carolinensis, 2:*375*, 2:*377*
Allopauropus gracilis, 2:*376*
Alloperla roberti, 3:143
Allophryne ruthveni. See Ruthven's frogs
Allophrynidae. *See* Ruthven's frogs
Allopteridae, 3:305
Allotriognatha. *See* Lampridiformes
Alluroididae, 2:65
Almidae, 2:65
Alofia spp., 2:317, 2:320
Alopex lagopus. See Arctic foxes
Alopias vulpinus. See Thresher sharks
Alopochen aegyptiacus. See Egyptian geese
Alopoglossus spp., 7:304
Alopoglossus angulatus, 7:*304*
Alosa alabamae. See Alabama shads
Alosa pseudoharengus. See Alewives
Alosa sapidissima. See American shads
Alouatta spp. *See* Howler monkeys
Alouatta belzebul. See Red-handed howler monkeys
Alouatta caraya. See Black howler monkeys
Alouatta coibensis. See Coiba howler monkeys
Alouatta guariba. See Brown howler monkeys
Alouatta palliata. See Mantled howler monkeys
Alouatta palliata mexicana, 14:162
Alouatta pigra. See Mexican black howler monkeys
Alouatta sara. See Bolivian red howler monkeys
Alouatta seniculus. See Venezuelan red howler monkeys
Alouatta seniculus insulanus, 14:163
Alpacas, 12:180, **15:313–323,** 15:*318*, 15:*319*, 15:*321*
Alphaeidae, 2:200
Alpine accentors, 10:*461*, 10:*462–463*
Alpine chamois. *See* Northern chamois
Alpine charrs. *See* Charrs
Alpine choughs, 11:504
Alpine ibex, 12:139, 15:269, 16:87, 16:*89*, 16:91, 16:95, 16:104*t*
Alpine marmots, 12:76, 12:147, 12:148, 16:*144*, 16:*150*, 16:*155*, 16:157
Alpine musk deer. *See* Himalayan musk deer
Alpine pikas, 16:501*t*
Alpine pipits, 10:372
Alpine salamanders, 6:38, 6:367
Alpine shrews, 13:*254*, 13:*258*, 13:260
Alpine swifts, 9:415, 9:422, 9:*425*, 9:*430*–431
Alpine toads, 6:110, 6:111
Alponis brunneicapilla. See White-eyed starlings
Alsodes spp., 6:157
Alsophis sanctaecrucis, 7:470
Alston's woolly mouse opossums, 12:*256*, 12:*261–262*
Altai argalis, 16:90
Altamira orioles, 10:*170*
Altamira yellowthroats, 11:291
Alternation of generations. *See* Reproductive duality
Alticola spp., 16:225
Alticola argentatus. See Silvery mountain voles
Alticus spp., 5:342
Altiphrynoides spp., 6:184
Altiphrynoides malcomi. See Malcolm's Ethiopian toads
Altitude, in migration, 8:32
  See also specific species
Altricial offspring, 12:95–96, 12:97, 12:106, 12:108
  See also Reproduction
Aluterus monoceros, 5:469
Aluterus scriptus. See Scrawled filefishes
Alydidae, 3:264
Alymlestes spp., 16:480
Alymlestes kielanae, 16:480
Alytes spp. *See* Midwife toads
Alytes cisternasii. See Iberian midwife toads
Alytes dickhilleni, 6:90
Alytes muletensis. See Majorca midwife toads
Alytes obstetricans. See Midwife toads
Amabiliidae, 1:226
Amadina spp., 11:354
Amadina erythrocephala. See Red-headed finches
Amadina fasciata. See Cut-throat finches
Amadina gouldiae. See Gouldian finches
Amakihis, 11:342, 11:343
Amami jays, 11:508
Amami rabbits, 12:135, 16:482, 16:487, 16:505, 16:509, 16:516*t*
Amami thrushes, 10:485
Amami woodcocks, 9:180
Amandava amandava. See Red avadavats
Amandava formosa. See Green avadavats
Amandava subflava. See Zebra waxbills
Amani sunbirds, 11:210
Amarsipidae. *See Amarsipus carlsbergi*
Amarsipus carlsbergi, 5:421
Amathimysis trigibba, 2:*218*, 2:*220*
Amaurornis phoenicurus. See White-breasted waterhens
Amazilia spp., 9:443, 9:448
Amazilia castaneiventris. See Chestnut-bellied hummingbirds
Amazilia distans. See Táchira emeralds
Amazilia tzacatl. See Rufous-tailed hummingbirds
Amazon bamboo rats, 16:*452*, 16:*454*, 16:455–456
Amazon kingfishers, 10:*11*, 10:*21–22*
Amazon leaffishes, 5:*242*, 5:*246*, 5:251
Amazon River dolphins. *See* Botos
Amazon tree boas, 7:*410*, 7:411
Amazona autumnalis. See Red-lored Amazons
Amazona guildingii. See St. Vincent Amazons
Amazona ochrocephala. See Yellow-crowned Amazons
Amazona vittata. See Puerto Rican Amazon parrots
Amazonetta brasiliensis. See Brazilian teals
Amazonian manatees, 15:191–193, 15:*209*, 15:*210*, 15:*211*–212
Amazonian marmosets, 14:115, 14:116, 14:117, 14:120
Amazonian poison frogs, 6:*202*, 6:206, 6:208
Amazonian skittering frogs, 6:*229*, 6:*231*, 6:239
Amazonian snail-eaters, 7:*471*, 7:*479*, 7:480
Amazonian tapirs. *See* Lowland tapirs
Amazonian umbrellabirds, 10:305, 10:306, 10:308, 10:*309*, 10:*316*–317
Amba Mata (Indian goddess), 7:56
Ambassidae. *See* Glassfishes
Amber, fossil insects in, 3:8, 3:9
Amberjacks, 4:68
Amblonyx cinereus. See Asian small-clawed otters
Ambloplites rupestris. See Rock basses
Amblycera, 3:249, 3:250, 3:251
Amblycercus holosericeus, 11:301
Amblycipitidae, 4:353
Amblydoras hancockii. See Blue-eye catfishes
Amblyeleotris wheeleri. See Gorgeous prawn-gobies
Amblynura kleinschmidti. See Pink-billed parrotfinches
Amblyodipsas spp. *See* Purple-glossed snakes
Amblyopinae, 5:373, 5:375
Amblyopsidae, 4:57, 5:5, 5:6
Amblyopsis rosae. See Ozark cavefishes
Amblyopsis spelaea. See Northern cavefishes
Amblyornis spp., 11:477, 11:478, 11:480, 11:481
Amblyornis inornatus. See Vogelkops
Amblyornis macgregoriae. See Macgregor's bowerbirds
Amblyornis subalaris. See Streaked bowerbirds
Amblyospiza albifrons. See Grosbeak weavers
Amblypoda, 15:131
Amblypygi, 2:333
Amblyraja radiata, 4:177
Amblyramphus holosericeus. See Scarlet-headed blackbirds
Amblyrhiza spp., 16:469, 16:470
Amblyrhiza inundata. See Anguilla-St. Martin giant hutias
Amblyrhynchus cristatus. See Marine iguanas
Amblysominae, 13:215
Amblysomus spp. *See* Golden moles
Amblysomus gunningi. See Gunning's golden moles
Amblysomus hottentotus. See Hottentot golden moles
Amblysomus iris. See Zulu golden moles
Amblysomus julianae. See Juliana's golden moles
Ambon flying foxes, 13:331*t*
Ambrosia beetles, 3:63, 3:322
Ambulacralia, 2:4
Ambulocetidae, 15:2
Ambulocetus spp., 15:41
Ambystoma spp., 6:39, 6:42, 6:49, 6:355, 6:358
Ambystoma californiense, 6:356, 6:359, 6:360
Ambystoma cingulatum. See Flatwoods salamanders
Ambystoma gracile. See Northwestern salamanders
Ambystoma lermaense, 6:356

Ambystoma macrodactylum, 6:356
Ambystoma mavortium. *See* Tiger salamanders
Ambystoma mexicanum. *See* Mexican axolotl
Ambystoma opacum, 6:*31*, 6:35, 6:358
Ambystoma talpoideum, 6:356
Ambystoma tigrinum. *See* Tiger salamanders
Ambystomatidae. *See* Mole salamanders
Ameiva spp., 7:309, 7:310, 7:311, 7:312
Ameiva ameiva. *See* Giant ameivas
Ameiva polops. *See* St. Croix ground lizards
Ameletopsidae, 3:55
Amerana spp., 6:248
American alligators, 7:*106*, 7:*172*, 7:*174*, 7:175
distribution, 7:*175*
evolution, 7:171
farming, 7:48–49
as food, 7:48
humans and, 7:52
physical characteristics, 7:171
reproduction, 7:*4*, 7:*36*, 7:42, 7:157, 7:161, 7:163, 7:164, 7:*173*
American anhingas, 8:*10*, 8:*204*, 8:209
American avocets, 9:133, 9:*134*, 9:*136*, 9:139–*140*
American beetles, 3:322
American bison, 16:*12*, 16:*16*, 16:*22*–23
behavior, 16:5, 16:6, 16:*11*, 16:*13*–14
conservation status, 12:215, 16:*14*
distribution, 16:4
domestication, 15:145–146
evolution, 16:2
reproduction, 12:*82*
American black bears, 14:*296*, 14:*301*, 14:*302*–303
conservation status, 14:300
evolution, 14:295
habitats, 14:296–297
reproduction, 12:109, 12:110, 14:*299*
translocation of, 12:224
American black ducks, 8:371
American black oystercatchers, 9:126
American black rails, 9:1
American black swifts. *See* Black swifts
American black vultures, 8:234, 8:276, 8:278, 8:*279*, 8:281–*282*
American burbots. *See* Burbots
American burying beetles, 3:322, 3:*325*, 3:*332*–333
American butterfishes. *See* Butterfishes (Stromateidae)
American chameleons. *See* Green anoles
American cliff swallows, 10:358, 10:*359*, 10:*362*, 10:*364*–365
American cockroaches, 3:76, 3:149, 3:*150*, 3:151, 3:*152*, 3:*154*, 3:157–158
American congers, 4:*255*, 4:*259*, 4:263–*264*
American coots, 9:*46*, 9:*50*, 9:51
American crocodiles, 7:*162*, 7:179, 7:*183*, 7:*184*
American crows, 11:505, 11:506, 11:*508*, 11:*511*, 11:520–*521*
American cuckoos, 9:311
American dabchicks. *See* Least grebes
American darters. *See* American anhingas
American diamond girdle wearers, 1:345, 1:*346*, 1:*347*, 1:349
American dippers, 10:475, 10:*476*, 10:*477*, 10:*478*, 10:*480*
American eels, 4:*256*, 4:*257*, 4:*260*, 4:*261*, 4:262–263
American finfoot. *See* Sungrebes
American gizzard shads. *See* Gizzard shads
American golden-plovers, 8:29, 9:162, 9:*165*, 9:*166*–167
American goldfinches, 11:323–*324*, 11:*326*, 11:*331*–332
American graylings. *See* Arctic graylings
American horseshoe crabs, 2:*327*, 2:*328*, 2:*330*, 2:*331*–332
American hover flies, 3:*43*
American jabirus. *See* Jabirus
American jacanas. *See* Northern jacanas
American Kennel Club, 14:288
American Killifish Association, 5:94
American king crabs. *See* Red king crabs
American leaf-nosed bats, **13:413–434,** 13:*421*, 13:*422*
behavior, 13:416–417
conservation status, 13:420
distribution, 13:*413*, 13:415
evolution, 13:413
feeding ecology, 13:314, 13:417–418
habitats, 13:311, 13:415–416
humans and, 13:316–317, 13:420
physical characteristics, 13:413–415, 13:*414*
reproduction, 13:419–420
species of, 13:*423*–432, 13:433*t*–434*t*
taxonomy, 13:413
American least shrews, 13:*253*, 13:*257*–258
American leeches. *See* North American medicinal leeches
American lesser golden-plovers. *See* American golden-plovers
American Livebearer Association, 5:94
American Livestock Breed Conservancy, 15:282
American martens, 12:110, 12:132, 14:*320*, 14:*324*
American Medical Association, icon of, 7:55
American mink, 14:*326*, 14:*328*–329
conservation status, 14:324
distribution, 12:132
ecological niche, 12:117
feeding ecology, 14:323
physical characteristics, 14:321
reproduction, 12:105, 12:110, 14:324
in United Kingdom, 12:182–183
American mourning doves, 9:*244*, 9:*248*, 9:251, 9:*254*, 9:*259*–260
American mud turtles, **7:121–127,** 7:*124*
American oystercatchers, 9:*126*, 9:*127*, 9:*129*, 9:*130*
American oysters. *See* Eastern American oysters
American paddlefishes, 4:*216*, 4:*219*, 4:220
American painted-snipes. *See* South American painted-snipes
American perches. *See* Yellow perches
American Pet Products Manufacturers Association, 14:291
American pied oystercatchers. *See* American oystercatchers
American pikas, 12:*134*, 16:*494*, 16:*496*, 16:*497*
behavior, 16:482, 16:*484*, 16:492, 16:*493*
distribution, 16:491
feeding ecology, 16:*495*
reproduction, 16:486
American pipits, 10:372
American pocket gophers, 12:74
American primary screwworms. *See* New World primary screwworms
American pronghorns. *See* Pronghorns
American pygmy shrews, 13:199, 13:247, 13:*252*, 13:*254*, 13:*260*, 13:262–263
American redstarts, 11:*287*, 11:288, 11:289
American robins, 10:484–*485*, 10:*491*, 10:502–*503*
American saltwater crocodiles. *See* American crocodiles
American sand lances. *See* Inshore sand lances
American shads, 4:*280*, 4:*281*, 4:282
American shrew moles, 13:280, 13:*282*, 13:*283*, 13:*284*
American Sign Language (ASL), 12:160–162
American Society for the Prevention of Cruelty to Animals (ASPCA), 14:293
American stag-moose, 15:380
American striped cuckoos, 9:312, 9:313, 9:*316*, 9:*327*–328
American sturgeons, 4:*75*
American tadpole shrimps. *See* Longtail tadpole shrimps
American tailed caecilians, 6:5, 6:39, 6:411, 6:*412*, **6:415–418,** 6:*417*
American toads, 6:*45*, 6:*184*, 6:190
American treecreepers. *See* Brown creepers
American tube dwelling anemones, 1:*109*, 1:*111*, 1:115
American tufted titmice, 11:156, 11:157
American walkingsticks, common, 3:*226*, 3:*228*, 3:230
American water shrews, 13:195, 13:196, 13:247, 13:249, 13:*254*, 13:*261*, 13:263
American white pelicans, 8:186, 8:225, 8:226, 8:*229*, 8:230–*231*
American wigeons, 8:*375*, 8:*383*–*384*
American woodcocks, 9:177, 9:180
American Zoo and Aquarium Association (AZA), 12:203, 12:204, 12:209
Americans with Disabilities Act of 1990, 14:293
Americobdellidae, 2:76
Ameridelphia, 12:11
Amero-Australian treefrogs, **6:225–243,** 6:*231*–*232*
behavior, 6:228–229, 6:233–242
conservation status, 6:230, 6:233–242
defense mechanisms, 6:66
distribution, 6:5, 6:6, 6:*225*, 6:228, 6:*233*–*242*
evolution, 6:4, 6:225–226
feeding ecology, 6:229, 6:233–242
habitats, 6:228, 6:233–242
humans and, 6:230, 6:233–242
physical characteristics, 6:215, 6:226–228, 6:233–242, 6:281
reproduction, 6:229–230, 6:233–242
tadpoles, 6:227–228
taxonomy, 6:225–226, 6:233–242
Ametabolous, 3:*33*
Amethyst woodstars, 9:439
Amia spp., 4:230
Amia calva. *See* Bowfins
Amietia spp., 6:248
Amietia vertebralis. *See* Wide-mouthed frogs
Amiidae, 4:11, 4:229
Amiiformes. *See* Bowfins
Amiinae, 4:229
Amino acids, 3:24, 12:26–27, 12:30–31

Amiopsinae, 4:229
Amiskwia sagittiformis, 1:433
Amitermes spp., 3:*166*
Amitermes laurensis, 3:165
Amitermes medius, 3:165
Amitermes meridionalis, 3:165
Amitermes vitosus, 3:165
Amitermitinae, 3:165
Ammocrypta spp., 4:62
Ammodorcas clarkei. *See* Dibatags
Ammodramus leconteii. *See* Le Conte's sparrows
Ammodramus nelsoni. *See* Nelson's sharp-tailed sparrows
Ammodytes americanus. *See* Inshore sand lances
Ammodytidae. *See* Sand lances
Ammomanes spp. *See* Larks
Ammomanes cincturus. *See* Bar-tailed larks
Ammomanes grayi. *See* Gray's larks
Ammonia, 3:21
Ammonia toxicity, 7:3
Ammonites, 2:*10*
Ammophila spp., 3:*336*
Ammospermophilus spp., 16:143, 16:144
Ammospermophilus harrisii. *See* Harris's antelope squirrels
Ammospermophilus nelsoni. *See* Nelson's antelope squirrels
Ammospermophilus parryii. *See* Arctic ground squirrels
Ammotragus spp., 16:87
Ammotragus lervia. *See* Barbary sheep
Amniataba spp., 5:220
Amnion, 12:6–10, 12:10, 12:92–93
  See also Reproduction
Amniotes, 7:12–13, 7:*14*, 7:*15*
Amo spp., 6:247, 6:248
Amolopini, 6:247
Amolops spp., 6:247, 6:248, 6:249, 6:250
Amolops formosus. *See* Beautiful torrent frogs
Amorphochilus spp., 13:467–468
Amorphochilus schnablii. *See* Schnabeli's thumbless bats
Amorphoscelidae, 3:177, 3:178
Ampeliceps spp., 11:408
Ampeliceps coronatus. *See* Golden-crested mynas
Ampelion stresemanni. *See* White-cheeked cotingas
Ampheristus spp., 5:16
Amphibdella spp., 1:214
Amphibians
  art and, 6:*52*, 6:*54*
  behavior, **6:44–50**
  biogeography, 6:*4*–5
  communication, 6:44–48
  conservation status, 6:6, **6:56–60**
  defined, **6:3–6**
  deformities in, 6:56–57, 6:*59*
  diseases of, 6:57, 6:60
  distribution, 6:*4*, 6:5–6
  egg attendance, 6:34–35, 6:*216*, 6:*217*
  egg deposition, 6:31–32, 6:39, 6:230, 6:251, 6:*364*
  egg development and hatching, 6:33–34
  egg fertilization, 6:32–33
  egg transportation, 6:35–37
  evolution, 6:3, 6:4–5, **6:7–14**, 6:28
  feeding ecology, 6:6, 6:54
  as food, 6:54, 6:252, 6:256
  habitats, 6:6, 6:7
  humans and, **6:51–55**
  as introduced species, 6:54, 6:58, 6:191, 6:251, 6:262
  larvae, 6:28, 6:*36*, **6:39–43**
  literature and, 6:52, 6:*53*
  medicinal uses of, 6:53
  metamorphosis, 6:28, 6:39, 6:42–43
  migrating, 6:59, 6:356
  as pets, 6:54–55, 6:58
  physical characteristics, 6:3, **6:15–27**, 6:*16*
  population decline of, 6:56–59
  protection of, 6:59–60
  reproduction, 6:3, 6:18, **6:28–38**
  vs. reptiles, 7:3
  taxonomy, 6:3–*4*, 6:*11*
  See also specific topics and types of amphibians
Amphibolidae, 2:413
Amphibolus nebulosus, 2:117
Amphichthys spp., 5:41
Amphidromous fishes, 4:50
  See also specific fishes
Amphiglossus astrolabi, 7:329
Amphiliidae, 4:352
Amphilina foliacea, 1:231, 1:*232*, 1:*234–235*
Amphilinidae, 1:225
Amphilinidea, 1:225, 1:226, 1:227, 1:228, 1:230, 1:231
Amphinectomys spp., 16:268
Amphionides spp. *See Amphionides reynaudii*
Amphionides reynaudii, 2:*195*
Amphionididae. *See Amphionides reynaudii*
Amphionids, **2:195–196**
Amphioxiformes. *See* Lancelets
Amphioxus spp. *See* Lancelets
Amphioxus lanceolatus. *See* Lancelets
Amphipholis squamata. *See* Dwarf brittle stars
Amphipithecus spp., 14:2
Amphipods, **2:261–272**, 2:*263*, 2:*264*
  behavior, 2:261
  conservation status, 2:262
  distribution, 2:261
  evolution, 2:261
  feeding ecology, 2:29, 2:261
  habitats, 2:261
  humans and, 2:262
  physical characteristics, 2:261
  reproduction, 2:261–262
  species of, 2:265–271
  taxonomy, 2:261
Amphiporus spp., 1:253
Amphiprion spp. *See* Anemonefishes
Amphiprion ocellaris. *See* Clown anemonefishes
Amphipsocidae, 3:243
Amphisbaena alba. *See* White-bellied wormlizards
Amphisbaenidae. *See* Wormlizards
Amphispiza bilineata. *See* Black-throated sparrows
Amphiuma spp., 6:35, 6:405, 6:407
Amphiuma means. *See* Two-toed amphiumas
Amphiuma pholeter. *See* One-toed amphiumas
Amphiuma tridactylum. *See* Three-toed amphiumas
Amphiumas, 6:5, 6:13, 6:323, 6:325, **6:405–410**, 6:*408*
Amphiumidae. *See* Amphiumas
Amphiura filiformis, 1:*391*, 1:395–*396*
Amphiuridae, 1:387, 1:388
Amphizoidae, 3:54
Amplectic positions. *See* Amplexus
Amplexidiscus fenestrafer. *See* Elephant ear polyps
Amplexus, 6:*65*, 6:68, 6:304, 6:365–366
Ampulicidae, 3:148
Ampullaria canaliculata. *See* Apple snails
Ampullarioidea, 2:445, 2:446
Amsterdam albatrosses, 8:113
Amsterdam Island fur seals. *See* Subantarctic fur seals
Amur eastern red-footed falcons. *See* Amur falcons
Amur falcons, 8:353, 8:*358*
Amur sea stars. *See* Northern Pacific sea stars
Amur tigers, 12:*49*, 14:370
Amynthas corticis, 2:*69*, 2:71
Amytornis housei. *See* Black grasswrens
Amytornis striatus. *See* Striated grasswrens
Amytornithinae. *See* Grasswrens
Anabantidae, 5:427
Anabantoidei. *See* Labyrinth fishes
Anabas spp., 5:427, 5:428
Anabas testudineus. *See* Climbing perches
Anabathmis spp. *See* Sunbirds
Anabathmis reichenbachii. *See* Reichenbach's sunbirds
Anablepidae, 4:57, 5:89, 5:92, 5:93
Anableps spp., 5:89, 5:91, 5:93, 5:94
Anableps anableps. *See* Largescale foureyes
Anableps dowi. *See* Pacific foureyed fishes
Anabrus simplex. *See* Mormon crickets
Anacanthus barbatus, 5:467
Anacondas, 7:202, 7:409, 7:410, 7:411
  See also Green anacondas
Anadia spp., 7:303
Anaea spp., 2:38
Anaea butterfly caterpillars, 1:42
Anairetes alpinus. *See* Ash-breasted tit tyrants
Anal glands, 12:37
  See also Physical characteristics
Anambra waxbills, 11:357
Anamia, 4:*324*, 4:*325*
Anaora spp., 5:365
Anaplectes rubriceps. *See* Red-headed finches
Anapsida, 7:4–*5*, 7:12–14
Anarcroneuria spp., 3:142
Anareolatae, 3:221
Anarhichadidae. *See* Wolffishes
Anarhichas spp., 5:310, 5:311, 5:313
Anarhichas lupus, 5:310, 5:311
Anarhichas orientalis, 5:311
Anarhynchus frontalis. *See* Wrybills
Anarrhichthys ocellatus. *See* Wolf-eels
Anas acuta. *See* Northern pintails
Anas albogularis. *See* Andaman teals
Anas americana. *See* American wigeons
Anas aucklandica. *See* Brown teals
Anas castanea. *See* Chestnut teals
Anas chlorotis. *See* Brown teals
Anas clypeata. *See* Northern shovelers
Anas eatoni. *See* Eaton's pintails
Anas falcata. *See* Falcated teals
Anas formosa. *See* Baikal teals
Anas gibberifrons. *See* Sunda teals
Anas gracilis. *See* Gray teals
Anas laysanensis. *See* Laysan ducks
Anas melleri. *See* Meller's duck
Anas nesiotis. *See* Campbell Island teals
Anas platyrhynchos. *See* Mallards
Anas waigiuensis. *See* Salvadori's teals

Anas wyvilliana. See Hawaiian ducks
Anaspidaceans, **2:181–183,** 4:9
Anaspides tasmaniae, 2:*181,* 2:*183*
Anaspididae, 2:181
Anaspidinea, 2:181
Anastomus spp. *See* Storks
Anastomus oscitans. See Asian openbills
Anathana spp., 13:292
Anathana ellioti. See Indian tree shrews
Anatidae. *See* Ducks; Geese; Swans
Anatinae. *See* Dabbling ducks
Anatomy. *See* Physical characteristics
Anchieta, José de, 16:404
Anchieta's barbets, 10:115
Anchoa mitchilli. See Bay anchovies
Anchoveta, 4:55, 4:75, 4:279–*280,* 4:*285,* 4:287–288
Anchovies, 4:12, 4:44, 4:67, 4:277–*280,* 4:283, 4:*285*–287
Ancistrocerus inflictus. See Katydids
Ancistrona vagelli. See Wandering seabird lice
Ancistropsyllidae, 3:347, 3:348
Ancistrus triradiatus. See Branched bristlenose catfishes
Ancodonta, 15:264, 15:266
Ancylostoma caninum. See Dog hookworms
Ancylostoma duodenale, 1:35
Andalgalomys spp., 16:265
Andalusian hemipodes. *See* Small buttonquails
Andaman barn owls, 9:338, 9:341
Andaman dark serpent eagles. *See* Andaman serpent-eagles
Andaman horseshoe bats, 13:388–389, 13:392
Andaman red-whiskered bulbuls, 10:398
Andaman serpent-eagles, 8:*325,* 8:*336*
Andaman teals, 8:364
Andamia spp., 5:342
Andean avocets, 9:134
Andean cats, 14:391*t*
Andean cocks-of-the-rock, 10:305, 10:*310,* 10:*318*
Andean condors, 8:233, 8:236, 8:275, 8:276, 8:*277,* 8:278, 8:*279,* 8:*284*
Andean flamingos, 8:303, 8:304, 8:305, 8:306, 8:*307,* 8:*310*
Andean hairy armadillos, 13:184, 13:191*t*
Andean hillstars, 9:442, 9:444
Andean mice, 16:*271,* 16:*273,* 16:274–275
Andean mountain cavies, 16:399*t*
Andean night monkeys, 14:135, 14:138, 14:141*t*
Andean tapirs. *See* Mountain tapirs
Andean titis, 14:148
Anderson, Atholl, 8:98
Andigena spp. *See* Mountain toucans
Andigena hypoglauca. See Gray-breasted mountain toucans
Andigena laminirostris. See Plate-billed mountain toucans
Andinichthyidae, 4:351
Andinichthys spp., 4:351
Andinomys spp., 16:265
Andinomys edax. See Andean mice
Andinophryne spp., 6:184
Andrew's beaked whales, 15:69*t*
Andrews's frigatebirds. *See* Christmas frigatebirds
Andrews's three-toed jerboas, 16:212, 16:216, 16:223*t*
Andrias spp., 6:34, 6:49
Andrias davidianus. See Chinese giant salamanders
Andrias japonicus. See Japanese giant salamanders
Andriashevia aptera, 5:310
Androdon spp. *See* Hummingbirds
Andropadus spp., 10:396
Androphobus viridis. See Papuan whipbirds
Anecic earthworms, 2:66
Aneides spp., 6:391
Aneides ferreus, 6:391
Aneides lugubris. See Arboreal salamanders
Aneides vagrans, 6:391
Anelytropsis spp., 7:272
Anemonefishes, 2:*34,* 5:*299,* 5:*302*
  behavior, 4:64, 5:*294*–*295*
  habitats, 5:294
  reproduction, 5:297
  species of, 5:304
Anemones, **1:103–116,** 1:*109,* 1:*110,* 2:*34,* 2:36
  behavior, 1:29–30, 1:32, 1:40, 1:42, 1:106
  clownfish and, 1:33
  conservation status, 1:107
  distribution, 1:105
  evolution, 1:103
  feeding ecology, 1:27–28, 1:106–107
  habitats, 1:105
  humans and, 1:107–108
  physical characteristics, 1:103–*105*
  reproduction, 1:107
  species of, 1:111–116
  taxonomy, 1:103
Anemotaxis, 1:40
Angel insects. *See* Zorapterans
Angelfishes, **5:235–244,** 5:*245,* 5:*282*
  behavior, 5:239–240
  conservation status, 5:242–243
  distribution, 5:237–238
  evolution, 5:235
  feeding ecology, 4:67, 5:240–242
  habitats, 4:48, 5:238–239
  humans and, 5:243–244
  physical characteristics, 5:235–237
  reproduction, 4:65
  species of, 5:252–253, 5:287–288
  taxonomy, 5:235
Angelsharks, 4:11, **4:161–165,** 4:*163*
Angiostrongylus cantonensis. See Rat lungworms
Angleheads, **7:209–222**
Angler catfishes. *See* Squarehead catfishes
Anglerfishes, 4:14–15, **5:47–57,** 5:*48,* 5:*49,* 5:*52*
  behavior, 5:48–50
  conservation status, 5:51
  distribution, 5:47
  evolution, 5:47
  feeding ecology, 5:50
  habitats, 5:48
  humans and, 5:51
  physical characteristics, 4:14–15, 4:*22,* 5:47
  reproduction, 4:26, 4:33, 5:50–51
  species of, 5:*53*–*56*
  taxonomy, 5:47
Anglewings, 3:388
Angolan colobus, 14:184*t*
Angolan dwarf guenons. *See* Angolan talapoins
Angolan pittas. *See* African pittas
Angolan pythons, 7:420
Angolan talapoins, 14:191, 14:193, 14:*196,* 14:*198*–199
Angoni vlei rats, 16:*287,* 16:*292*–293
Angora goats, 16:*91*
Anguidae, 7:206, 7:207, **7:339–345,** 7:*341*
Anguilla spp., 4:14, 4:35
Anguilla anguilla. See European eels
Anguilla oceanica. See American congers
Anguilla rostrata. See American eels
Anguilla-St. Martin giant hutias, 16:471*t*
Anguillidae, 4:68, 4:255
Anguilliformes, **4:255–270,** 4:*259*–*260*
  behavior, 4:*256*
  conservation, 4:258
  distribution, 4:255
  evolution, 4:255
  feeding ecology, 4:256, 4:*257*
  habitats, 4:255–256
  humans and, 4:258
  physical characteristics, 4:255
  reproduction, 4:256–258
  species of, 4:261–269
  taxonomy, 4:11, 4:255
Anguimorpha, 7:196, 7:198
Anguinae, 7:339, 7:340
  See also Glass lizards
Anguis spp. *See* Slowworms
Anguis fragilis, 7:339, 7:340
Angwantibos, 12:116, 14:13, 14:14, 14:16, 14:*17,* 14:*19*–20
Anhima cornuta. See Horned screamers
Anhimidae. *See* Screamers
Anhinga anhinga. See American anhingas
Anhingas, 8:183–186, **8:201–210**
Anhingidae. *See* Anhingas
Anhydrophryne spp., 6:245, 6:251
Anianiaus, 11:*345,* 11:347–*348*
Aniliidae. *See* False coral snakes
Anilioidea, 7:198
Anilius scytale. See False coral snakes
Anilocra laticaudata, 2:*252*
Animal-assisted activities (AAA), 14:293
Animal-assisted therapy (AAT), 14:293
Animal husbandry, in zoos, 12:209–211
Animal Record Keeping System (ARKS), 12:206
Animal rights movement, 12:183, 12:212, 14:293–294
Animal Welfare and Conservation Organization, 12:*216*
Animalia, 2:8
  See also Taxonomy; specific species
Anis, 9:311–313, 9:*314,* 9:325–326
Anisomorpha buprestoides, 3:222–223
Anisopodidae, 3:55
Anisoptera. *See* Dragonflies
Anisozygoptera, 3:133
Anistominae, 3:54
Ankober serins, 11:324
Annam broad-headed toads, 6:111, 6:*112,* 6:*115*
Annam leaf turtles, 7:119
Annam pheasants. *See* Edwards' pheasants
Annam spadefoot toads. *See* Annam broad-headed toads
Annamia normani. See Anamia
Annamite striped rabbits, 16:487, 16:509, 16:*510,* 16:*512,* 16:513
Annandia spp., 6:251

Anna's hummingbirds, 8:*10*, 9:443, 9:*444*, 9:447, 9:*453*, 9:*467*
Annelida, 2:3, 2:*25*, 2:45, 2:59
evolution, 2:14
reproduction, 2:16, 2:17, 2:21, 2:22–23
taxonomy, 2:35
See also specific annelids
Anniella spp. *See* Legless lizards
Anniella pulchra. See California legless lizards
Anniellinae. *See* Legless lizards
Annobón paradise-flycatchers, 11:97
Annobón white-eyes, 11:227
Annulipalpia, 3:375, 3:376
Anoas, 12:137, 16:4, 16:11, 16:24*t*
Anobiid beetles. *See* Death watch beetles
Anodonthyla spp., 6:35, 6:304
Anodonthyla boulengerii. See Boulenger's climbing frogs
Anodorhynchus hyacinthinus. See Hyacinth macaws
Anoles, **7:243–257,** 7:*248*
behavior, 7:244–245
conservation status, 7:246–247
distribution, 7:*243*, 7:244
evolution, 7:243
feeding ecology, 7:245–246
habitats, 7:244
humans and, 7:247
physical characteristics, 7:243–244
reproduction, 7:207, 7:246
species of, 7:*250–257*
taxonomy, 7:243
See also Green anoles
Anolis spp. *See* Anoles
Anolis carolinensis. See Green anoles
Anomalepididae. *See* Early blindsnakes
Anomalepis spp., 7:369, 7:370–371
Anomalepis aspinosus, 7:371
Anomalocaris spp., 12:64
Anomalodesmata, 2:451
Anomalopidae. *See* Flashlightfishes
Anomalops katoptron. See Splitfin flashlightfishes
Anomalopteryx didiformis. See Moas
Anomalopus spp., 7:328
Anomalospiza imberbis. See Cuckoo finches
Anomaluridae. *See* Scaly-tailed squirrels
Anomalurus spp., 16:123
Anomalurus beecrofti. See Beecroft's anomalures
Anomalurus derbianus. See Lord Derby's anomalures
Anomalurus pelii. See Pel's anomalures
Anomalurus pusillus. See Lesser anomalures
Anomochilidae. *See* False blindsnakes
Anomochilus spp. *See* False blindsnakes
Anomochilus leonardi, 7:*387*, 7:*388*
Anomochilus weberi, 7:387, 7:*388*
Anomura, 2:197, 2:198, 2:200
Anopheles spp., 3:76
Anoplans, 1:11, **1:245–251,** 1:*248*
Anoplocephalidae, 1:226, 1:231
Anoplodactylus evansi, 2:*323*, 2:*324–325*
Anoplogaster cornuta. See Common fangtooths
Anoplogasteridae. *See* Fangtooth fishes
Anoplogastridae. *See* Fangtooth fishes
Anoplolepis gracilipes. See Yellow crazy ants
Anoplopoma fimbria. See Sablefishes
Anoplopomatidae, 5:179
Anoplotheriidae, 15:264
Anoplura. *See* Sucking lice
Anorrhinus tickelli. See Tickell's brown hornbills
Anostomidae, 4:335
Anostomus anostomus. See Striped headstanders
Anostraca. *See* Fairy shrimps
Anostracina, 2:135
Anotheca spinosa. See Spiny-headed treefrogs
Anotomys spp. *See* Aquatic rats
Anotomys leander, 16:266
Anotopteridae, 4:432
Anoura geoffroyi. See Geoffroy's tailless bats
Anourosorex spp., 13:248
Anourosorex squamipes. See Mole-shrews
Anourosoricini, 13:247
Anous spp. *See* Noddies
Anser anser. See Greylag geese
Anser caerulescens. See Snow geese
Anser canagica. See Emperor geese
Anser indicus. See Bar-headed geese
Anseranas semipalmata. See Magpie geese
Anseriformes, **8:363–368**
Anserinae. *See* Geese; Swans
Anseropoda placenta, 1:369
Ansonia spp., 6:184, 6:188–189
Ansonia longidigita. See Long-fingered slender toads
Ant bears. *See* Giant anteaters
Ant hill pythons. *See* Pygmy pythons
Ant thrushes, **10:239–256,** 10:*243–244*, 10:487
behavior, 10:240–241
conservation status, 10:241–242
distribution, 10:*239*, 10:240
evolution, 10:239
feeding ecology, 10:241
habitats, 10:240
humans and, 10:242
physical characteristics, 10:239–240
reproduction, 10:241
taxonomy, 10:239
Antalis entalis. See Tusk shells
Antarctic blue whales. *See* Blue whales
Antarctic bottlenosed whales. *See* Southern bottlenosed whales
Antarctic cods. *See* Notothens
Antarctic fur seals, 14:*393*, 14:*401*, 14:*402*, 14:403
distribution, 12:138, 14:394, 14:395
feeding ecology, 14:397
humans and, 12:*119*
physical characteristics, 12:*66*
reproduction, 14:398, 14:399
Antarctic giant petrels. *See* Southern giant petrels
Antarctic krill, 2:*185*, 2:*188*, 2:189–191, 2:*190*
Antarctic minke whales, 15:1, 15:120, 15:125, 15:130*t*
Antarctic petrels, 8:108, 8:*126*
Antarctic toothfishes, 5:323
Antarctic Treaty, Weddell seals, 14:431
Antarctica, 12:138
Antarctoperlaria, 3:141, 3:142
Antaresia spp., 7:420, 7:421
Antaresia perthensis. See Pygmy pythons
Antbirds, 10:170
Anteater chats, 10:*490*, 10:*498*
Anteaters, **13:171–179,** 13:*175*, 13:*176*
behavior, 13:150–151, 13:172–173, 13:177–179
conservation status, 13:152–153, 13:175, 13:177–179
distribution, 13:*171*, 13:172, 13:*177–179*
evolution, 13:147–149, 13:171
feeding ecology, 13:151–152, 13:173–174, 13:177–179
habitats, 13:150, 13:172, 13:177–179
humans and, 13:153, 13:175, 13:177–179
physical characteristics, 12:39, 12:46, 13:149, 13:171–172, 13:177–179
reproduction, 12:94, 13:174–175, 13:177–179
taxonomy, 13:147–149, 13:171, 13:177–179
Antechinomys laniger. See Kultarrs
Antechinus flavipes. See Yellow-footed antechinuses
Antechinus minimus, 12:279
Antechinus stuartii. See Brown antechinuses
Antechinus swainsonii. See Dusky antechinuses
Antechinuses, 12:279–283, 12:290
brown, 12:282, 12:*293*, 12:*296–297*
dusky, 12:279, 12:299*t*
yellow-footed, 12:277
Antedon bifida. See Rosy feather stars
Antelope ground squirrels. *See* Nelson's antelope squirrels
Antelope jackrabbits, 16:487, 16:*506*, 16:515*t*
Antelope squirrels, 12:131, 16:143
Harris', 16:*150*, 16:*154*, 16:156
Nelson's, 16:124, 16:145, 16:147, 16:160*t*
Antelopes, 15:263, 16:1–4
African, 15:272, 16:4, 16:5
blue, 16:28
domestication and, 15:146
feeding ecology, 15:142
giant sable, 16:33
humans and, 15:273
migrations, 12:87
royal, 16:60, 16:63, 16:71*t*
Tibetan, 12:134, 16:*9*
See also specific types of antelope
Antenna catfishes, 4:352
Antennae, 3:*18*, 3:27
See also Physical characteristics; specific species
Antennariidae. *See* Sargassumfishes
Antennarioidei. *See* Frogfishes
Antennarius spp. *See* Frogfishes
Antennarius maculatus. See Warty frogfishes
Antennarius multiocellatus. See Longlure frogfishes
Antennarius pictus. See Painted frogfishes
Antennarius radiosus. See Big-eye frogfishes
Antennarius striatus, 5:51
Anteroporidae, 1:226
Anthelidae, 3:386
Anthiinae, 5:255, 5:256, 5:260, 5:262
Anthobaphes spp. *See* Sunbirds
Anthobaphes violacea. See Orange-breasted sunbirds
Anthocephala spp. *See* Hummingbirds
Anthochaera carunculata. See Red wattlebirds
Anthocidaris crassispina. See Short-spined sea urchins
Anthomedusae, 1:123, 1:124–125
Anthony, J., 4:192
Anthophiloptera dryas. See Balsam beasts
Anthopleura xanthogrammica. See Giant green anemones
Anthops ornatus. See Flower-faced bats

Anthoscopus spp., 11:148, 11:149
Anthosigmella spp., 1:29
Anthozoa, **1:103–122**, 1:*109*, 1:*110*, 2:13
behavior, 1:106
conservation status, 1:107
distribution, 1:105
evolution, 1:103
feeding ecology, 1:106–107
habitats, 1:24, 1:25, 1:105
humans and, 1:107–108
physical characteristics, 1:103–*105*
reproduction, 1:107
species of, 1:*111*–121
taxonomy, 1:10, 1:103
Anthracocaridomorpha, 2:235
Anthracoceros marchei. *See* Palawan hornbills
Anthracoceros montani. *See* Sulu hornbills
Anthracomedusa turnbulli, 1:147
Anthracotheriidae, 15:136, 15:264
Anthracotheroidea, 15:302
Anthracothorax spp. *See* Hummingbirds
Anthreptes spp. *See* Sunbirds
Anthreptes fraseri. *See* Scarlet-tufted sunbirds
Anthreptes malacensis. *See* Plain-throated sunbirds
Anthreptes rectirostris. *See* Green sunbirds
Anthreptes reichenowi. *See* Plain-backed sunbirds
Anthreptes rhodolaema. *See* Red-throated sunbirds
Anthreptes rubritorques. *See* Banded sunbirds
Anthropoidea, 14:1, 14:3
Anthropoides spp., 9:23, 9:25
Anthropoides virgo. *See* Demoiselle cranes
Anthroscopus spp. *See* African penduline tits
Anthus spp. *See* Pipits
Anthus antarcticus. *See* South Georgia pipits
Anthus berthelotii. *See* Berthelot's pipits
Anthus caffer. *See* Bush pipits
Anthus campestris. *See* Tawny pipits
Anthus cervinus. *See* Red-throated pipits
Anthus chacoensis. *See* Chaco pipits
Anthus chloris. *See* Yellow-breasted pipits
Anthus correndera. *See* Correndera pipits
Anthus crenatus. *See* Yellow-tufted pipits
Anthus gustavi. *See* Pechora pipits
Anthus gutturalis. *See* Alpine pipits
Anthus hodgsoni. *See* Olive-backed pipits
Anthus lineiventris. *See* Striped pipits
Anthus longicaudatus. *See* Long-tailed pipits
Anthus lutescens. *See* Yellowish pipits
Anthus melindae. *See* Malindi pipits
Anthus nattereri. *See* Ochre-breasted pipits
Anthus nilghiriensis. *See* Nilgiri pipits
Anthus novaeseelandiae. *See* Australasian pipits
Anthus nyassae. *See* Woodland pipits
Anthus petrosus. *See* Rock pipits
Anthus pratensis. *See* Meadow pipits
Anthus richardi. *See* Richard's pipits
Anthus roseatus. *See* Rosy pipits
Anthus rubsecens. *See* American pipits
Anthus sokokensis. *See* Sokoke pipits
Anthus spragueii. *See* Sprague's pipits
Anthus trivialis. *See* Tree pipits
Antibyx spp. *See* Blacksmith plovers
Anticancer drugs, 1:44–46, 1:49
See also Humans
Anticoagulants, 2:76
Antidorcas marsupialis. *See* Springboks
Antifreeze, 4:49–50, 5:321–322
Antigomonidae, 1:275
Antigonia rubescens. *See* Red boarfishes
Antillean beaked whales. *See* Gervais' beaked whales
Antillean cloud swifts. *See* White-collared swifts
Antillean emeralds. *See* Puerto Rican emeralds
Antillean fruit-eating bats, 13:*422*, 13:*424*, 13:426
Antillean ghost-faced bats, 13:436, 13:442*t*
Antillean grackles, 11:302–303
Antillean manatees, 15:191, 15:211
See also West Indian manatees
Antillesoma antillarum, 2:*99*, 2:100
Antillothrix spp., 14:143
Antilocapra americana. *See* North American pronghorns
Antilocapra americana mexicana spp., 15:415–416
Antilocapridae. *See* Pronghorns
Antilocaprinae, 15:266–267, 15:411
Antilope cervicapra. *See* Blackbucks
Antilophia bokermanni. *See* Araripe manakins
Antilopinae, 16:1, **16:45–58**, 16:*49*, 16:56*t*–57*t*
Antilopine kangaroos. *See* Antilopine wallaroos
Antilopine wallaroos, 13:101*t*
Antimora microlepis, 5:27
Antioch dunes shieldbacks, 3:207
Antioquia bristle tyrants, 10:275
Antipaluria urichi, 3:*235*, 3:236–*237*
Antipatharia, 1:103, 1:104, 1:105, 1:106, 1:108
Antipathella fiordensis. *See* Black corals
Antipodean albatrosses, 8:113
Antipodean fur seals. *See* New Zealand fur seals
Antiponemertes allisonae, 1:255
Antlerless deer, 15:267
See also Chinese water deer
Antlers, 12:*5*, 12:10, 12:19–20, 12:22, 12:23, 12:99, 15:*132*
See also Physical characteristics
Antlions, 3:9, 3:63, 3:*306*, 3:*307*, 3:*310*, 3:*311*, 3:312–313
Antonina graminis. *See* Rhodesgrass mealybugs
Antrobathynella stammeri, 2:*179*
Antrozous spp. *See* Pallid bats
Antrozous pallidus. *See* Pallid bats
Ants, 2:23, 2:35, 2:36, 2:37, **3:405–418**, 3:*409*, 3:*410*
aphids and, 3:6, 3:49, 3:*61*, 3:406
behavior, 3:59, 3:65, 3:71, 3:406
biomes, 3:57
butterflies and, 3:72
conservation status, 3:87–88, 3:407
distribution, 3:67
ecosystem services of, 3:72
education and, 3:83
evolution, 3:405
feeding ecology, 3:5–6, 3:63, 3:406–407
habitats, 3:406
humans and, 3:407–408
medicinal uses, 3:81
as pests, 3:75, 3:76
physical characteristics, 3:27, 3:405
plants and, 3:49
reproduction, 3:36, 3:59, 3:*69*, 3:407
scale insects and, 3:49
social structure, 3:49, 3:67, 3:68–69, 3:70, 3:72, 3:406
species of, 3:418
taxonomy, 3:405
See also specific types of ants
Antsingy leaf chameleons. *See* Armored chameleons
Anura, 6:3, **6:61–68**
behavior, 6:64–67
defense mechanisms, 6:*64*, 6:66
distribution, 6:4–6, 6:63–64
egg transportation, 6:35–37
evolution, 6:4–5, 6:11–13, 6:15, 6:61–62
feeding ecology, 6:6, 6:26, 6:67
habitats, 6:26, 6:63–64
larvae, 6:39–43
physical characteristics, 6:*19*, 6:*22*, 6:25–26, 6:62–*63*, 6:66
predators of, 6:65–67
reproduction, 6:28–*29*, 6:*30*–34, 6:*32*, 6:38, 6:*65*, 6:68
taxonomy, 6:4, 6:61–62
vocalizations, 6:*22*, 6:26, 6:30
See also Frogs; Toads
Anurogryllus spp., 3:207
Anus, 2:3, 2:21
See also Physical characteristics
Aonyx spp., 14:321
Aotidae. *See* Night monkeys
Aotus spp. *See* Night monkeys
Aotus azarai. *See* Azari's night monkeys
Aotus dindinensis, 14:135
Aotus hershkovitzi. *See* Hershkovitz's night monkeys
Aotus lemurinus. *See* Gray-bellied night monkeys
Aotus lemurinus griseimembra, 14:138
Aotus miconax. *See* Andean night monkeys
Aotus nancymaae. *See* Nancy Ma's night monkeys
Aotus nigriceps. *See* Black-headed night monkeys
Aotus trivirgatus. *See* Three-striped night monkeys
Aotus vociferans. *See* Noisy night monkeys
Aoudads. *See* Barbary sheep
Apalharpactes narina. *See* Narina trogons
Apalharpactes reinwardtii. *See* Javan trogons
Apalis spp., 11:7
Apalis flavida. *See* Yellow-breasted apalis
Apaloderma vittatum. *See* Collared trogons
Apalone spp. *See* Snapping turtles
Apalone spinifera. *See* Spiny softshells
Apalopteron familiare. *See* Bonin white-eyes
Apapanes, 11:*342*, 11:*345*, 11:*348*
Aparallactinae, 7:461
Aparallactus spp. *See* Centipede eaters
Aparallactus jacksonii. *See* Jackson's centipede eaters
Aparasphenodon spp., 6:226
Apennine chamois. *See* Southern chamois
Apes
behavior, 14:7, 14:8
encephalization quotient, 12:149
enculturation of, 12:162
evolution, 14:2
feeding ecology, 14:9
habitats, 14:6
language, 12:160–161
memory, 12:152–153

numbers and, 12:155
taxonomy, 14:1, 14:3
theory of mind, 12:159–160
See also Gibbons; Great apes
Aphanius spp., 5:91
Aphanius fasciatus, 5:91
Aphantophryne spp., 6:36, 6:305
Aphelocephala spp. *See* Whitefaces
Aphelocephala leucopsis. See Southern whitefaces
Aphelocephala pectoralis. See Tasmanian thornbills
Aphelocoma californica. See Western scrub-jays
Aphelocoma coerulescens. See Florida scrub-jays
Aphelocoma ultramarina. See Gray-breasted jays
Aphia minuta. See Transparent gobies
Aphidae, 3:54
Aphididae, 3:264
Aphids, 2:25, 2:37, **3:259–260, 3:267–268**
ants and, 3:6, 3:49, 3:*61*
biological control of, 3:80
defense mechanisms, 3:66
eusocial, 3:68
feeding ecology, 3:63
humans and, 3:57
migrating, 3:66
pea, 3:*265*, 3:267–*268*
as pests, 3:75
reproduction, 3:*34*, 3:37–38, 3:39, 3:44, 3:59
See also Hemiptera; Sternorrhyncha
Aphos spp., 5:41
Aphragmophora, 1:433
Aphrastura spinicauda. See Thorn-tailed rayaditos
Aphredoderidae. *See* Pirate perches
Aphredoderoidei, 5:5
Aphredoderus sayanus. See Pirate perches
Aphriza virgata. See Surfbirds
Aphrocallistes vastus. See Cloud sponges
Aphrodite (Greek goddess), 7:54
Aphroditidae, 2:45
Aphrophoridae, 3:55
Aphyiine gobies, 5:375
Aphyocypris spp., 4:298
Aphyonidae, 5:15, 5:16, 5:18
Aphyosemion spp., 5:90
Aphyosemion australe. See Chocolate lyretails
Aphyosemion franzwerneri, 5:92
Apicomplexa, 2:11
Apicotermitinae, 3:168
Apiculture, 3:*75*, 3:79
Apidium spp., 14:2
Apis spp. *See* Bees
Apis dorsata. See Rock bees
Apis mellifera. See Honeybees
Apis nigrocincta, 3:68
Apistidae. *See* Longfinned waspfishes
Apistogramma agassizii. See Agassiz's dwarf cichlids
Apistus carinatus. See Ocellated waspfishes
Aplacophorans, 2:14, **2:379–385,** 2:*381*, 2:*382*
Apletodon spp., 5:355
Aplidium spp., 1:452
Aplidium albicans, 1:46
Aploactinidae. *See* Velvetfishes
Aplocheilichthys spp. *See* African lampeyes
Aplocheiloidei, 5:90, 5:91, 5:92
Aplodactylidae. *See* Seacarps
Aplodontia rufa. See Mountain beavers
Aplodontia rufa nigra, 16:133
Aplodontia rufa phaea, 16:133
Aplodontidae. *See* Mountain beavers
Aplonis spp., 11:408, 11:410
Aplonis brunneicapilla. See White-eyed starlings
Aplonis cinerascens. See Rarotonga starlings
Aplonis corvine. See Kosrae Mountain starlings
Aplonis crassa. See Tanimbar starlings
Aplonis feadensis. See Atoll starlings
Aplonis fusca bulliana. See Lord Howe Island starlings
Aplonis fusca fusca. See Norfolk Island starlings
Aplonis grandis. See Brown-winged starlings
Aplonis mystacea. See Yellow-eyed starlings
Aplonis pelzelni. See Pohnpei Mountain starlings
Aplonis santovestris. See Mountain starlings
Aplonis zelandica. See Tanimbar starlings
Aploparaksis spp., 1:229
Aplousobranchia, 1:452, 1:455
Aplydine, 1:46
Aplysina archeri. See Stove-pipe sponges
Aplysina cauliformis. See Row pore rope sponges
Aplysinellidae, 1:79
Apo mynas, 11:410
Apo sunbirds, 11:210
Apoanagyrus lopezi, 3:*409*, 3:*411*, 3:414–415
Apochela, 2:115
Apocrine sweat glands, 12:36, 12:37–38
See also Physical characteristics
Apocrita, 3:405–407
See also Ants; Bees; Wasps
Apoda. *See* Caecilians
Apodemes, 3:17
Apodemus agrarius. See Striped field mice
Apodemus sylvaticus. See Long-tailed field mice
Apodes. *See* Anguilliformes
Apodi, 9:415
Apodida, 1:417, 1:418, 1:420, 1:421
Apodidae. *See* Swifts
Apodiformes, **9:415–419**, 9:*417*
Apodinae. *See* Swifts
Apodora spp., 7:420, 7:421
Apodora papuana. See Papuan pythons
Apogonidae. *See* Cardinalfishes
Apolemia uvaria, 1:127, 1:*133*, 1:144
Apolinar's wrens, 10:529
Apolysis, 3:31
Apomorphies, 2:13
Aporrectodea caliginosa. See Common field worms
Aposematic coloration, 3:65
Apostlebirds, 11:453, 11:*454*, 11:*455*, 11:*457*–458
Apostolepis spp., 7:467, 7:468
Appalachian woodland salamanders, 6:*29*
Appendicular skeleton, 12:41
See also Physical characteristics
Appendicularia. *See* Larvaceans
Apple bugs. *See* Whirligig beetles
Apple snails, 2:*448*, 2:449
Apposition eyes, 3:27
Apristurus spp., 4:114
Aproteles bulmerae. See Bulmer's fruit bats
Apseudes intermedius, 2:*237*, 2:*238*
Apseudes spectabilis, 2:236
Apseudomorpha, 2:235
Aptenodytes spp. *See* Penguins
Aptenodytes forsteri. See Emperor penguins
Apteromantis aptera, 3:181
Apteronotid eels. *See* Black ghosts
Apteronotidae, 4:369, 4:370
Apteronotus spp., 4:369, 4:370, 4:371
Apteronotus albifrons. See Black ghosts
Apteropanorpidae, 3:341
Apterygidae. *See* Kiwis
Apterygota, 3:27, 3:31, 3:33, 3:38
Apteryx australis. See Brown kiwis
Apteryx haastii. See Great spotted kiwis
Apteryx owenii. See Little spotted kiwis
Aptoenodytes patavonicus. See King penguins
Apus spp. *See* Swifts
Apus apus. See Common swifts
Apus batesi. See Bates's swifts
Apus caffer. See White-rumped swiftlets
Apus horus. See Horus swifts
Apus melba. See Alpine swifts
Apus pacificus. See Pacific swifts
Apus pallidus. See Pallid swifts
Aquaculture, 2:41
for drug production, 1:45
vs. fisheries, 1:51–52
See also Humans
Aquarana spp., 6:248, 6:250, 6:261
Aquariums, 4:*72*, 4:74, 4:303, 5:94
Aquatic beetles. *See* Water beetles
Aquatic bugs. *See* Water bugs
Aquatic desmans, 13:196, 13:197, 13:198–199, 13:279, 13:280
Aquatic genets, 14:337, 14:338, 14:*339*, 14:*341*
Aquatic insects, 3:7, 3:22
See also specific aquatic insects
Aquatic leaf leeches, 2:*75*
Aquatic mammals, 12:14
adaptations, **12:62–68**
field studies, 12:201
hair, 12:3
hearing, 12:82
locomotion, 12:44
neonatal milk, 12:127
reproduction, 12:91
See also specific aquatic mammals
Aquatic moles, 13:197, 13:198–199
Aquatic pygmy backswimmers, 3:260
Aquatic pyralids, 3:387
Aquatic rats, 16:265
Aquatic shrews, 13:197, 13:198–199
Aquatic swamp toads, 6:*186*, 6:194
Aquatic tenrecs, 13:234*t*
Aquatic warblers, 11:6, 11:7
Aquila adalberti. See Spanish imperial eagles
Aquila audax. See Australian wedge-tailed eagles
Aquila chrysaetos. See Golden eagles
Aquila gurneyi. See Gurney's eagles
Aquila heliaca. See Imperial eagles
Aquila nipalensis. See Steppe eagles
Aquila verreauxii. See Verreaux's eagles
AR. *See* Aspect ratio
Ara-A (arabinosyl adenine), 1:44
Ara ararauna. See Blue and yellow macaws
Ara-C (arabinosyl cytosine), 1:44
Ara chloroptera. See Green-winged macaws
Ara macao. See Scarlet macaws
Arabia. *See* Ethiopian region
Arabian babblers, 10:505, 10:507, 10:*510*, 10:*514*
Arabian camels. *See* Dromedary camels
Arabian oryx, 16:*36*
behavior, 16:40
conservation status, 12:139, 16:33, 16:40

distribution, 16:29, 16:40, 16:*41*
evolution, 16:28
feeding ecology, 16:41
habitats, 16:29, 16:41
humans and, 12:139, 16:34, 16:41
physical characteristics, 16:40
reproduction, 16:32, 16:41
taxonomy, 16:28, 16:40
Arabian Peninsula. *See* Palaearctic region
Arabian rousettes. *See* Egyptian rousettes
Arabian sand gazelles, 12:139
Arabian sandboas, 7:409–411
Arabian tahrs, 16:90, 16:92, 16:94, 16:95, 16:97, 16:*100*, 16:101
Arabian woodpeckers, 10:150
Arabinosyl adenine (Ara-A), 1:44
Arabinosyl cytosine (Ara-C), 1:44
Arabuko-Sokoke Forest, 16:524
Aracaris, 10:125–128
Arachnida, 2:25, 2:*337*, 2:*338*
behavior, 2:37, 2:335
conservation status, 2:336
distribution, 2:335
evolution, 2:333
feeding ecology, 2:335
habitats, 2:335
humans and, 2:336
physical characteristics, 2:333–*334*, 2:*336*
reproduction, 2:23, 2:335
species of, 2:339–352
taxonomy, 2:333
See also specific arachnids
Arachnocampa luminosa. *See* New Zealand glowworms
Arachnothera spp. *See* Sunbirds
Arachnothera chrysogenys. *See* Yellow-eared spiderhunters
Arachnothera juliae. *See* Whitehead's spiderhunters
Aradid bugs, 3:260
Aradomorpha, 3:259
Arafura file snakes, 7:439–*442*, 7:*443*
Aramidae. *See* Limpkins
Aramides axillaris. *See* Rufous-necked wood-rails
Aramides ypecaha. *See* Giant wood-rails
Araneae, 2:333
Arapaima, 4:232, 4:233, 4:*235*, 4:*237*, 4:240
Arapaima gigas. *See* Arapaima
Araripe manakins, 10:297, 10:*298*, 10:299–*300*
Araripe's soldiers. *See* Araripe manakins
Aratinga pertinax. *See* Brown-throated parakeets
Aratinga wagleri. *See* Scarlet-fronted parrots
Arawanas, 4:232, 4:*233*–234, 4:*235*, 4:*237*, 4:240
Arboreal anteaters, 13:147
Arboreal dung beetles, 3:87–88
Arboreal mammals, 12:14
field studies of, 12:202
locomotion, 12:43
vision, 12:79
See also Habitats; specific arboreal mammals
Arboreal pangolins, 16:107–113
Arboreal salamanders, 6:*394*, 6:396–397
Arboreal spiny rats, white-faced, 16:450, 16:*452*, 16:*455*
Arborescent puyas, 9:442
Arborimus spp., 16:225
Arborimus albipes. *See* White-footed voles
Arborimus longicaudus. *See* Red tree voles
Arborophila spp. *See* Hill-partridges
Arborophila ardens. *See* Hainan hill-partridges
Arborophila cambodiana. *See* Chestnut-headed hill-partridges
Arborophila davidi. *See* Orange-necked hill-partridges
Arborophila rubirostris. *See* Red-billed hill-partridges
Arborophila rufipectus. *See* Sichuan hill-partridges
Arborophila torqueola. *See* Asian hill-partridges
Arch-beaked whales. *See* Hubb's beaked whales
Archaea, 1:49
Archaebacteria, 2:8
Archaeobatrachians, 6:4, 6:68
Archaeoceti, 15:2–3
Archaeocyathans, 1:10
Archaeoganga spp., 9:231
Archaeolemurinae. *See* Baboon lemurs
Archaeophiala spp., 2:387
Archaeopsittacus verreauxi, 9:275
Archaeopteryx, 8:3, 8:*8*, 8:*14*
Archaeotherium spp., 15:264
Archaeotraguludus krabiensis, 15:325
Archaoindris spp., 14:63
Archboldia spp., 11:477, 11:478
Archboldia papuensis. *See* Archbold's bowerbirds
Archbold's bowerbirds, 11:*478*, 11:481, 11:*482*, 11:*485*
Archbold's owlet-nightjars, 9:388
Archea. *See* Archaebacteria
Archenteron, 2:4, 2:21
Archeocytes, 1:21
Archeoindris fontoynonti, 14:63
Archeolemur edwardsi. *See* Baboon lemurs
Archeolepis mane, 3:383
Archerfishes, 4:50, 4:69, 5:235–239, 5:241–244
See also Banded archerfishes
Archer's larks, 10:346
Archey's frogs, 6:69, 6:70, 6:*71*, 6:*72*
Archiacanthocephala, 1:311
Archichauliodes diversus (adult). *See* Dobsonflies
Archichauliodes diversus (larvae). *See* Black creepers
Archidermaptera, 3:195
Archigetes iowensis, 1:230
Archigetes limnodrili, 1:230
Archigetes sieboldi, 1:230
Archilestes californica, 3:*135*
Archilochus colubris. *See* Ruby-throated hummingbirds
Archimandrita tessellata, 3:147
Archipsocopsis spp., 3:244
Archipsocus spp., 3:244
Archispirostreptus syriacus, 2:364
Architaenoglossa, 2:445, 2:446
Architettix compacta, 3:259
Architeuthis spp. *See* Giant squids
Archolipeurus nandu, 3:*251*
Archosargus. *See* Sheepsheads
Archosauria, 7:16, 7:17–19
Archostemata, 3:315, 3:316
See also Beetles
Arcos spp., 5:355
Arcovestia spp., 2:91
Arctic charrs. *See* Charrs
Arctic foxes, 12:132, 14:*260*, 14:267, 14:*270*, 14:272, 14:283*t*
Arctic fulmars. *See* Northern fulmars
Arctic graylings, 4:*408*, 4:*410*, 4:419
Arctic ground squirrels, 12:113, 16:144, 16:145, 16:*149*, 16:*151*–152
Arctic hares, 16:482, 16:*482*, 16:487, 16:*506*, 16:*508*, 16:515*t*
Arctic jaegers. *See* Arctic skuas
Arctic lemmings, 16:*229*
Arctic loons, 8:159, 8:*162*, 8:164–165
Arctic peregrine falcons, 8:349
Arctic skuas, 9:204, 9:*210*, 9:*211*
Arctic sousliks. *See* Arctic ground squirrels
Arctic terns, 8:29, 8:*31*, 9:102
Arctic warblers, 11:7, 11:*8*, 11:*20*–21
Arctictis binturong. *See* Binturongs
Arctiid moths, 2:37
Arctiids, 3:387, 3:388
Arctocebus spp. *See* Angwantibos
Arctocebus aureus. *See* Golden angwantibos
Arctocebus calabarensis. *See* Calabar angwantibos
Arctocephalinae. *See* Fur seals
Arctocephalus australis. *See* South American fur seals
Arctocephalus forsteri. *See* New Zealand fur seals
Arctocephalus galapagoensis. *See* Galápagos fur seals
Arctocephalus gazella. *See* Antarctic fur seals
Arctocephalus philipii. *See* Juan Fernández fur seals
Arctocephalus pusillus. *See* Afro-Australian fur seals
Arctocephalus pusillus doriferus. *See* Australian fur seals
Arctocephalus pusillus pusillus. *See* Cape fur seals
Arctocephalus townsendi. *See* Guadalupe fur seals
Arctocephalus tropicalis. *See* Subantarctic fur seals
Arctocyonids, 15:132, 15:133
Arctogalidia trivirgata. *See* Small-toothed palm civets
Arctoperlaria, 3:141, 3:142
Arctoscopus spp., 5:335
Arctoscopus japonicus. *See* Sailfin sandfishes
Arctosorex polaris, 13:251
Arctostylopids, 15:135
Ardea alba. *See* Great white egrets
Ardea americana. *See* Whooping cranes
Ardea antigone. *See* Sarus cranes
Ardea carunculata. *See* Wattled cranes
Ardea cinerea. *See* Gray herons
Ardea goliath. *See* Goliath herons
Ardea grus. *See* Eurasian cranes
Ardea herodias. *See* Great blue herons
Ardea humbloti, 8:245
Ardea idea, 8:245
Ardea insignis, 8:245
Ardea melanocephala. *See* Black-headed herons
Ardea purpurea. *See* Purple herons
Ardea virgo. *See* Demoiselle cranes
Ardeola spp. *See* Herons
Ardeola ralloides. *See* Squacco herons
Ardeotis spp. *See* Bustards
Ardeotis kori. *See* Kori bustards
Ardeotis nigriceps. *See* Great Indian bustards
Ardipithecus ramidus, 14:242

Arenaria interpres. See Ruddy turnstones
Arenariinae. *See* Turnstones
Arend's golden moles, 13:215
Arenicola grubei, 1:40
Arenicola marina. See Lugworms
Arenivaga spp., 3:148
Arenophryne spp., 6:35, 6:148, 6:149
Arenophryne rotunda. See Sandhill frogs
Areolatae, 3:221
Arfak berrypeckers, 11:189, 11:*190*
Arfak ringtails, 13:122*t*
See also Ringtail possums
Argalis, 12:178, 16:104*t*
Altai, 16:90
behavior, 16:92
distribution, 16:91, 16:92
evolution, 16:87
physical characteristics, 15:268, 16:89, 16:90
reproduction, 16:94
Argentina spp., 4:392
Argentina silus. See Greater argentines
Argentine ants, 3:69, 3:72
Argentine hemorrhagic fever, 16:270
Argentine Society of Mammalogists, 16:430
Argentines, 4:389, 4:390, 4:392
Argentinidae, 4:390, 4:391, 4:392
Argentinoidea, 4:390
Argentinoidei, 4:389, 4:391, 4:392, 4:393
Argonauta spp., 2:476, 2:478
Argonauta argo. See Greater argonauts
Argonauts, greater, 2:*481*, 2:483–*484*
Argonemertes hillii, 1:255
Argulidae, 2:289
Argulus spp., 2:*289*–290
Argulus foliaceus. See Fish lice
Argulus japonicus, 2:*291*, 2:*292*, 2:293
Argusianus argus. See Great argus pheasants
Argyropelecus aculeatus. See Silver hatchetfishes
Argyrotheca cistellula, 2:*524*, 2:*526*
Arhynchobdellida, 2:75
Ariidae. *See* Sea catfishes
Ariomma spp., 5:421
Ariomma indicum. See Indian ariommas
Ariommatidae. *See Ariomma* spp.
Arion lusitanicus. See Spanish slugs
Ariopsis bonillai. See New Granada sea catfishes
Aripuanã marmosets, 14:117, 14:118, 14:121, 14:*125*, 14:*127*, 14:130–131
Aristophanes, wasps and, 3:408
Aristotle, 1:417, 1:451
classification of animals, 12:149
on hoopoes, 10:63
on monkfishes, 5:56
zoos and, 12:203
Ariteus flavescens, 13:420
Arixenia esau, 3:*198*, 3:*199*
Arixeniina, 3:195, 3:196
Arizona brush lizards. *See* Common sagebrush lizards
Arizona gray squirrels, 16:167, 16:*168*, 16:*171*–172
Arkansas Department of Environmental Quality, 5:11
Arkansas kingbirds. *See* Western kingbirds
Arkhars. *See* Argalis
ARKS (Animal Record Keeping System), 12:206
Arlequinus spp., 6:279
Armadillidae, 2:250
Armadillididae, 2:250
Armadillidium spp., 2:251
Armadillidium vulgare. See Common pill woodlice
Armadillo lizards, 7:*320–321*
Armadillo officinalis, 2:252
Armadillos, 13:147–152, **13:181–192**, 13:*186*, 13:190*t*–192*t*
behavior, 13:150–151, 13:183–184
conservation status, 13:152–153, 13:185
distribution, 13:150, 13:*181*, 13:182–183
evolution, 13:147–149, 13:181–182
feeding ecology, 13:151–152, 13:184
habitats, 13:150, 13:183
humans and, 13:153, 13:185
physical characteristics, 13:149, 13:182
reproduction, 12:94, 12:103, 12:110–111, 13:152, 13:185
species of, 13:187–190, 13:190*t*–192*t*
See also specific types of armadillos
Armillifer spp., 2:318, 2:320
Armillifer agkistrodontis, 2:320
Armillifer armillatus, 2:*318*, 2:*319*, 2:320
Armillifer grandis, 2:320
Armillifer moniliformis, 2:320
Armilliferidae, 2:318
Armitage, K. B., 9:444
Armored catfishes, 4:352, 4:353
Armored chameleons, 7:232, 7:*234*, 7:235–*236*
Armored gurnards, 5:165
Armored katydids, 3:208
Armored rats, 16:*452*, 16:453–*454*
Armored sea cucumbers. *See* Slipper sea cucumbers
Armored sea robins, 5:163, 5:164, 5:165
Armored shrews, 13:196, 13:200, 13:*270*, 13:*273*–274
Armored spiny rats, 16:458*t*
Armored sticklebacks, 5:*138*, 5:*142*–143
Armored xenarthrans. *See* Cingulata
Armorhead catfishes, 4:353
Armorheads, 5:235, 5:236, 5:238, 5:240, 5:242, 5:243
See also Pentacerotidae
Arms race, 12:63–64
Army ants, 2:37, 3:66, 3:68, 3:72, 10:241
Armyworms, 3:56, 3:66, 3:359–360, 3:389
Árnason, Úlfur, 15:103
Arno gobies, 5:*379*, 5:*382*, 5:385–386
Arnoux's beaked whales, 15:69*t*
Aromobates spp., 6:197–199
Aromobates nocturnus. See Venezuelan skunk frogs
Arothron hispidus. See White-spotted puffers
Arothron meleagris. See Guinea fowl puffers
Arowanas, 4:4
Arraus. *See* South American river turtles
Arrector pili, 12:38
See also Physical characteristics
Arredondo's solenodons, 13:237–238
Arrow gobies, 5:375
Arrow loaches, 4:*324*, 4:*326*–327
Arrow worms, 1:9, **1:*433*–442**, 1:*435*, 1:*436*
behavior, 1:42, 1:434
conservation status, 1:435
distribution, 1:434
evolution, 1:433
feeding ecology, 1:434–435
habitats, 1:434
humans and, 1:435
physical characteristics, 1:433–434
reproduction, 1:22, 1:435
species of, 1:*437*–442
taxonomy, 1:9, 1:13, 1:433
Art
amphibians in, 6:*52*, 6:*54*
birds in, 8:26
fishes in, 4:72–73
insects in, 3:74
protostomes in, 2:42
See also Humans
Artamidae. *See* Woodswallows
Artamus cinereus. See Black-faced woodswallows
Artamus cyanopterus. See Dusky woodswallows
Artamus fuscus. See Ashy woodswallows
Artamus insignis. See Bismarck woodswallows
Artamus leucorhynchus. See White-breasted woodswallows
Artamus minor. See Little woodswallows
Artamus superciliosus. See White-browed woodswallows
Artedidraconidae. *See* Plunderfishes
Artedius spp., 5:180, 5:183
Artedius fenestralis. See Padded sculpins
Artedius harringtoni. See Scalyhead sculpins
Artedius lateralis. See Flathead sculpins
Artemia spp., 2:136, 2:137
Artemia franciscana, 2:*136*
Artemia salina, 2:*138*
Artemiidae, 2:135
Artemiina, 2:135, 2:136
Arthopoides regulorum. See Gray crowned cranes
Arthroleptella spp., 6:245, 6:248, 6:251
Arthroleptidae, 6:6, 6:35, 6:68, **6:265–271**, 6:*267*
Arthroleptides spp., 6:247
Arthroleptides dutoiti, 6:251
Arthroleptinae, 6:265, 6:266
Arthroleptis sechellensis. See Seychelles frogs
Arthroleptis stenodactylus. See Common squeakers
Arthroleptis tanneri. See Tanner's litter frogs
Arthroleptis wahlbergii. See Bush squeakers
Arthropleona, 3:99
Arthropleura spp., 2:363
Arthropoda, 2:3, 2:25, 2:333
evolution, 2:14
parasitism and, 2:33
reproduction, 2:17, 2:20–21, 2:23
taxonomy, 2:5, 2:35
See also specific arthropods
Arthrosaura spp., 7:303
Arthrotardigrada, 2:115, 2:117, 2:118
Artibeus harti, 13:*416*
Artibeus jamaicensis. See Jamaican fruit-eating bats
Artibeus lituratus. See Great fruit-eating bats
Artibeus watsoni, 13:416
Articulata. *See* Articulate lampshells
Articulate lampshells, **2:*521*–527**, 2:*524*
behavior, 2:522
conservation status, 2:523
distribution, 2:522
evolution, 2:521
feeding ecology, 2:522–523
habitats, 2:522
humans and, 2:523
physical characteristics, 2:521–522
reproduction, 2:523

species of, 2:*525*–527
taxonomy, 2:521
Artiocetus clavis, 15:266
Artiodactyla, **15:263–273**
behavior, 15:269–271
cetaceans and, 12:30, 15:2
conservation status, 15:272–273
distribution, 12:129, 12:132, 12:136, 15:269
evolution, 15:131–133, 15:135–138, 15:263–267
feeding ecology, 15:141–142, 15:271–272
habitats, 15:269
humans and, 15:273
physical characteristics, 12:40, 12:79, 15:138–140, 15:267–269
reproduction, 12:127, 15:272
ruminant, 12:10
taxonomy, 15:263–267
See also specific artiodactyls
Artisornis spp., 11:6
Artisornis moreaui, 11:7
Artrellias. *See* Crocodile monitors
Aruana. *See* Arawanas
Arvicanthis niloticus. *See* African grass rats
Arvicola spp. *See* Water voles
Arvicola terrestris. *See* Northern water voles
Arvicolinae, 16:124, **16:225–238,** 16:*231*, 16:*232*, 16:237*t*–238*t*, 16:270, 16:282, 16:283
behavior, 16:226–227, 16:233–237, 16:237*t*–238*t*
conservation status, 16:229, 16:233–237, 16:237*t*–238*t*
distribution, 16:*225*, 16:226, 16:*233–237*, 16:237*t*–238*t*
evolution, 16:225
feeding ecology, 16:227–228, 16:233–238, 16:237*t*–238*t*
habitats, 16:233–237
humans and, 16:229–230, 16:233–237
physical characteristics, 16:225–226, 16:233–237, 16:237*t*–238*t*
reproduction, 16:228–229, 16:233–237
taxonomy, 16:225, 16:233–237, 16:237*t*–238*t*
Asagi, 4:*299*
Asajirus spp., 1:479
Asajirus indicus, 1:479, 1:480, 1:*481*, 1:*482*
Asano, Toshio, 12:161–162
Asbestopluma spp., 1:10
Asbestopluma hypogea. *See* Carnivorous sponges
Ascalaphidae, 3:305, 3:306, 3:307, 3:308
Ascaphidae. *See* Tailed frogs
Ascaphus spp., 6:29, 6:33, 6:68
Ascaphus montanus. *See* Rocky Mountain tailed frogs
Ascaphus truei. *See* Coastal tailed frogs
Ascaphus truei montanus. *See* Rocky Mountain tailed frogs
Ascaris lumbricoides. *See* Maw-worms
Ascension frigatebirds, 8:193, 8:195, 8:*196*, 8:*198*
Ascension Island frigatebirds. *See* Ascension frigatebirds
Ascension rails, 9:48
Aschelminthes, 1:9
Aschiphasmatidae, 3:221, 3:222
Aschiza, 3:357
Ascidia spp., 1:29, 1:45–46, 1:48, 1:452
Ascidiacea. *See* Sea squirts
Ascotan Mountain killifishes, 5:*95*, 5:*98–99*
Ascothoracica, 2:273, 2:274, 2:275
Ascothorax ophioctenis, 2:277, 2:279
Asellia tridens. *See* Trident leaf-nosed bats
Aselliscus tricuspidatus. *See* Temminck's trident bats
Asellota, 2:249, 2:250
Asellus aquaticus. *See* Water lice
Asemichthys taylori, 5:182
Asexual reproduction, 1:15–16, 1:20–23, 2:22–24
See also Reproduction
Ash-breasted tit tyrants, 10:275
Ash-throated antwrens, 10:242
Ash-throated flycatchers, 10:*270*, 10:273
Ash-winged antwrens, 10:*244*, 10:249–*250*
Ashaninka rats, 16:443
Ashbyia lovensis. *See* Gibberbirds
Ash's larks, 10:346
Ashy bulbuls, 10:*400*, 10:410–*411*
Ashy chinchilla rats, 16:443, 16:*444*, 16:445, 16:*446*, 16:*447*
Ashy-crowned finch-larks, 10:*343*
Ashy drongos, 11:437, 11:438, 11:*440*, 11:*443*
Ashy-faced owls, 9:338, 9:341
Ashy flycatchers, 11:*29*, 11:*30*–31
Ashy-fronted bearded bulbuls. *See* White-throated bulbuls
Ashy minivets, 10:386
Ashy robins. *See* Gray-headed robins
Ashy roundleaf bats, 13:410*t*
Ashy storm-petrels, 8:137
Ashy woodpeckers. *See* Gray-faced woodpeckers
Ashy woodswallows, 11:459
Asia. *See* Oriental region; Palaearctic region
Asian asses. *See* Asiatic wild asses
Asian banded boars, 12:179
Asian bay owls. *See* Oriental bay owls
Asian blue quails. *See* King quails
Asian bony tongues, 4:232
Asian brown wood-owls, 9:*353*, 9:*359*–360
Asian brush-tailed porcupines. *See* Asiatic brush-tailed porcupines
Asian buffaloes. *See* Water buffaloes
Asian clams, 2:454
Asian cockroaches, 3:*152*, 3:*155*–156
Asian common toads. *See* Common Sunda toads
Asian dippers. *See* Brown dippers
Asian dowitchers, 9:178, 9:180
Asian drongo-cuckoos, 9:*316*, 9:*322*
Asian eared nightjars, 9:401
Asian elephants, 15:*173*, 15:*174*
conservation status, 15:171, 15:172
distribution, 15:166
evolution, 15:162–164
feeding ecology, 15:*169*, 15:*170*
humans and, 12:173
physical characteristics, 15:165
reproduction, 12:209
in zoos, 12:209, 12:*211*
See also Elephants
Asian fairy-blue-birds. *See* Fairy bluebirds
Asian garden dormice, 16:327*t*
Asian giant softshell turtles, 7:152
Asian giant tortoises, 7:71
Asian golden weavers, 11:379
Asian grass lizards, 7:298
Asian hamster mice. *See* Mouse-like hamsters
Asian hill-partridges, 8:399
Asian hillstream catfishes, 4:353
Asian horned frogs, 6:110, 6:*112*, 6:115–116
Asian house shrews. *See* Musk shrews
Asian knifefishes, 4:232
Asian koels. *See* Common koels
Asian kukrisnakes, 7:467, 7:469
Asian leaf-monkeys, 14:172
Asian leaffishes, 4:57
Asian loach catfishes, 4:353
Asian long-tailed shrikes, 10:426
Asian magpie-robins. *See* Magpie-robins
Asian medicine, reptiles in, 7:50–51
Asian mountain toads, 6:*112*, 6:116
Asian narrow-headed softshell turtles, 7:70, 7:*152*
Asian openbills, 8:*267*, 8:268, 8:*269*, 8:*271*
Asian pigs, 15:149
Asian rhinoceroses, 15:220, 15:221, 15:222
Asian river turtles, 7:66
Asian rosy-finches, 11:*324*
Asian sea basses. *See* Barramundis
Asian small-clawed otters, 14:325
Asian spadefoot toads. *See* Asian horned frogs
Asian tailed caecilians, **6:419–424,** 6:*421*
distribution, 6:5, 6:*419*, 6:*422–423*
physical characteristics, 6:416, 6:419, 6:422–423
reproduction, 6:39, 6:420, 6:422–423
taxonomy, 6:411, 6:*412*, 6:415, 6:419, 6:422–423, 6:425–426
Asian tapirs. *See* Malayan tapirs
Asian toadfrogs, 6:5, 6:64, **6:109–117,** 6:*112*
Asian toads. *See* Common Sunda toads
Asian treefrogs, **6:291–300,** 6:*294*
distribution, 6:6, 6:*291*, 6:292, 6:*295–299*
physical characteristics, 6:248, 6:281, 6:291–292, 6:295–299
reproduction, 6:*30*, 6:32, 6:292–293, 6:295–299
taxonomy, 6:4, 6:279, 6:291, 6:295–299
Asian two-horned rhinoceroses. *See* Sumatran rhinoceroses
Asian yellow weavers, 11:378
Asiatic black bears, 12:195, 14:295, 14:296, 14:297, 14:298, 14:300, 14:306*t*
Asiatic brown bears. *See* Brown bears
Asiatic brush-tailed porcupines, 16:*355*, 16:*358*, 16:363
Asiatic cave swiftlets, 12:53
Asiatic giant salamanders, **6:343–347,** 6:*344*
Asiatic golden cats, 14:390*t*
Asiatic ibex. *See* Siberian ibex
Asiatic long-tailed shrews. *See* Hodgson's brown-toothed shrews
Asiatic mouflons, 12:178
See also Urials
Asiatic needletails. *See* White-throated needletails
Asiatic rat fleas. *See* Oriental rat fleas
Asiatic salamanders, **6:335–342,** 6:*338*
distribution, 6:5, 6:*335*, 6:336, 6:*339–341*
evolution, 6:13, 6:291, 6:335–336
giant, **6:343–347,** 6:*344*
physical characteristics, 6:24, 6:336, 6:339–341
reproduction, 6:32, 6:34, 6:337, 6:339–342
taxonomy, 6:323, 6:335–336, 6:339–341, 6:343
Asiatic shrew-moles. *See* Chinese shrew-moles

Asiatic tapirs. *See* Malayan tapirs
Asiatic water shrews, 13:195
Asiatic white cranes. *See* Siberian cranes
Asiatic wild asses, 12:177, 15:*231*
behavior, 15:228, 15:235
conservation status, 15:222, 15:229, 15:235
distribution, 12:134, 15:*234*
feeding ecology, 15:235
habitats, 15:234
humans and, 15:235
physical characteristics, 15:217, 15:226, 15:234
reproduction, 15:235
taxonomy, 15:*234*
Asiatic wild dogs, 14:270
Asiatic wild horses. *See* Przewalski's horses
Asiatic wild sheep, 15:271
Asilids, 3:360
Asio spp., 9:347
Asio capensis. See Marsh owls
Asio flammeus. See Short-eared owls
Asio otus. See Flammulated owls
Asities, 10:170, **10:187–191**
ASL (American Sign Language), 12:160–162
Asoriculus maghrebensis, 13:247
Aspasma spp., 5:355
Aspasmichthys spp., 5:355
Aspasmodes spp., 5:355
Aspasmogaster spp., 5:355
Aspatha gularis. See Blue-throated motmots
ASPCA (American Society for the Prevention of Cruelty to Animals), 14:293
Aspect ratio (AR), bat wings, 12:57–58
Aspidites spp., 7:419, 7:420, 7:421
Aspidites melanocephalus. See Black-headed pythons
Aspidites ramsayi. See Womas
Aspidochirotida, 1:417, 1:418, 1:419, 1:421
Aspidogaster conchicola, 1:*202*, 1:*204*–205
Aspidogastrea, 1:197, 1:198, 1:199, 1:200
Aspidonotini. *See* Malagasy helmet katydids
Aspidontus spp., 4:69
Aspidontus taeniatus. See Mimic blennies
Aspidosiphon elegans, 2:98
Asplanchna priodonta, 1:*263*, 1:*264*
Aspredinichthys tibicen, 4:354
Aspredinidae, 4:352
Aspredo aspredo, 4:354
Assa spp., 6:149
Assa darlingtoni. See Hip pocket frogs
Assam rabbits. *See* Hispid hares
Assam sucker frogs. *See* Beautiful torrent frogs
Assassin bugs, 2:11, 2:38, 3:64, 3:76, 3:260, 3:261, 3:264
Assemblages, 4:42, 4:43
Asses, 12:176–177, **15:225–236,** 15:*231*
behavior, 15:218–219, 15:226–228
conservation status, 15:221–222, 15:229–230
distribution, 15:217, 15:226
evolution, 15:215–216, 15:225
feeding ecology, 15:219–220, 15:228
habitats, 15:217–218, 15:226
humans and, 15:222–223, 15:229–230
physical characteristics, 15:139, 15:216–217, 15:225–226
reproduction, 15:220–221, 15:228–229
species of, 15:232, 15:234–235
taxonomy, 15:215–216, 15:225
Association France Vivipare, 5:94
Associative learning, 1:41, 2:39–40
See also Behavior
Astacidea, 2:197, 2:198
Astacus spp. *See* Crayfish
Asterias spp., 1:368, 1:370
Asterias amurensis. See Northern Pacific sea stars
Asterias forbesi, 1:370
Asterias vulgaris. See Northern sea stars
Asterina gibbosa, 1:370
Asterina phylactica, 1:370
Asteriomyzostomatidae, 2:59
Asteroidea. *See* Sea stars
Asteronychidae, 1:387
Asterophryinae, 6:35, 6:301–302
Asterophrys turpicola. See New Guinea bush frogs
Asteroschematidae, 1:387
Asthenes spp. *See* Canasteros
Asthenes luizae. See Cipo castaneros
Asthenosoma varium. See Fire urchins
Astley's leiothrix, 10:507
Astrapia spp. *See* Birds of paradise
Astrapia mayeri. See Ribbon-tailed astrapias
Astrapotheria, 12:11, 15:131, 15:133–136
Astrapotherium spp., 15:133–134
Astrid's rotifers, 1:*263*, 1:*265–266*
Astroblepidae, 4:352
Astrobrachion constrictum. See Snake stars
Astropecten irregularis. See Sand stars
Astrophorida, 1:77, 1:79
Astropyga magnifica. See Magnificent urchins
Astroscopus spp., 5:334
Astroscopus guttatus. See Northern stargazers
Astyanax mexicanus jordani. See Blind cavefishes (Characins)
Astylosterninae, 6:265, 6:266
Astylosternus diadematus. See Crowned forest frogs
Atalaye nesophontes, 13:243
Atelerix spp. *See* African hedgehogs
Atelerix albiventris. See Central African hedgehogs
Atelerix algirus. See Algerian hedgehogs
Atelerix frontalis. See Southern African hedgehogs
Atelerix sclateri. See Somalian hedgehogs
Ateles spp. *See* Spider monkeys
Ateles belzebuth. See White-bellied spider monkeys
Ateles belzebuth hybridsus. See Variegated spider monkeys
Ateles chamek. See Peruvian spider monkeys
Ateles fusciceps. See Brown-headed spider monkeys
Ateles fusciceps robustus. See Colombian black spider monkeys
Ateles geoffroyi. See Geoffroy's spider monkeys
Ateles hybridus, 14:160
Ateles marginatus. See White-whiskered spider monkeys
Ateles paniscus. See Black spider monkeys
Atelidae, **14:155–169,** 14:*161*
behavior, 14:157
conservation status, 14:159–160
distribution, 14:*155*, 14:156
evolution, 14:155
feeding ecology, 14:158
habitats, 14:156–157
humans and, 14:160
physical characteristics, 14:155–156
reproduction, 14:158–159
species of, 14:*162*–166, 14:166*t*–167*t*
taxonomy, 14:155
Atelins, 14:155, 14:157, 14:158, 14:159
Atelocynus microtis. See Short-eared dogs
Atelognathus spp., 6:157
Atelognathus patagonicus. See Patagonia frogs
Atelophryniscus spp., 6:184
Atelopus spp., 6:49, 6:184, 6:189, 6:190
Atelopus rubriventris. See Yungus redbelly toads
Atelopus varius. See Harlequin frogs
Atelopus zeteki, 6:189
Athena (Greek goddess), 7:54
Athenarian burrowing anemones. *See* Starlet sea anemones
Athene blewitti. See Forest owlets
Athene cunicularia. See Burrowing owls
Athene noctua. See Little owls
Atherinidae. *See* Silversides
Atherinidarum spp., 5:67
Atheriniformes, **5:67–77,** 5:*72*
behavior, 5:69
conservation status, 5:70–71
distribution, 5:68
evolution, 4:12, 5:67
feeding ecology, 5:69
habitats, 5:68–69
humans and, 5:71
physical characteristics, 5:68
reproduction, 4:35, 5:69–70
species of, 5:*73–76*
taxonomy, 5:67
Atherinomorpha, 4:12
Atherinomorus capricornensis, 5:69
Atherinomorus endrachtensis. See Eendracht land silversides
Atheris spp., 7:445, 7:447
Atheris squamigera. See Green bush vipers
Atherton scrubwrens, 11:57
Atherurus africanus. See African brush-tailed porcupines
Atherurus macrourus. See Asiatic brush-tailed porcupines
Atilax spp., 14:347
Atilax paludinosus. See Marsh mongooses
Atitlán grebes, 8:169, 8:171
Atius, 9:418
Atlantic albacores, 5:407
Atlantic argentines. *See* Greater argentines
Atlantic bluefin tunas, 5:407, 5:*408*, 5:415, 5:*417*
Atlantic bottlenosed dolphins. *See* Common bottlenosed dolphins
Atlantic bottlenosed whales. *See* Northern bottlenosed whales
Atlantic butterfishes, 5:311–312
Atlantic cods, 4:75, 5:*27*–30, 5:*31*, 5:*32*–33
Atlantic corals. *See* Sea pansies
Atlantic cutlassfishes. *See* Largehead hairtails
Atlantic flashlightfishes, 5:*114*
Atlantic gray seals. *See* Gray seals
Atlantic guitarfishes, 4:*180*, 4:*186*, 4:187–188
Atlantic hagfishes, 4:77, 4:78, 4:79, 4:*80*
Atlantic halibuts, 5:453
Atlantic herrings, 4:279, 4:*280*, 4:*281*, 4:283–284
Atlantic horseshoe crabs. *See* American horseshoe crabs
Atlantic humpbacked dolphins, 15:57*t*
Atlantic mackerels, 5:*408*, 5:*409*, 5:413–414

Atlantic manta rays, 4:*61*, 4:177, 4:*181*, 4:*182*, 4:183–184
Atlantic menhadens, 4:277, 4:*280*, 4:282–*283*
Atlantic midshipmen, 5:*43*, 5:*45*
Atlantic midwater eelpouts, 5:312
Atlantic mudskippers, 5:375, 5:*378*, 5:*382*, 5:386–387
Atlantic murres. *See* Common murres; Marbled murrelets
Atlantic needlefishes. *See* Atlantic sauries
Atlantic oysters. *See* Eastern American oysters
Atlantic puffins. *See* Puffins
Atlantic ridleys. *See* Kemp's ridley turtles
Atlantic royal flycatchers, 10:275
Atlantic salmons, 1:52, 4:*408*, 4:415–*416*
Atlantic sand fiddler crabs. *See* Sand fiddler crabs
Atlantic sauries, 5:*82*, 5:*84*
Atlantic seals. *See* Gray seals
Atlantic shads. *See* American shads
Atlantic snipe eels. *See* Slender snipe eels
Atlantic spotted dolphins, 15:*2*, 15:*43*, 15:*44*, 15:46, 15:47
Atlantic States Marine Fisheries Planning Commission, 4:261
Atlantic sturgeons, 4:215, 4:*216*, 4:*217*, 4:218
Atlantic tarpons, 4:243, 4:*244*, 4:*245*, 4:*246*–247
Atlantic tomcods, 5:*31*, 5:*32*, 5:34
Atlantic torpedos, 4:*180*, 4:*183*, 4:186–187
Atlantic walruses, 14:409, 14:*410*, 14:*412*, 14:415
Atlantic white-sided dolphins, 15:9, 15:57*t*
Atlantic whiting. *See* Silver hakes
Atlantisia elpenor. See Ascension rails
Atlantisia rogersi. See Inaccessible rails
Atlantoxerus spp., 16:143, 16:144
Atlantoxerus getulus. See Barbary ground squirrels
Atlapetes brunneinucha. See Chestnut-capped brush-finches
Atlapetes pallidiceps. See Pale-headed brush finches
Atlas beetles, 3:59
Atlas moths, 3:43, 3:47, 3:387–388, 3:*392*, 3:*400*, 3:402
Atmosphere, subterranean, 12:72–73
Atoll fruit doves, 9:243, 9:250
Atoll starlings, 11:410
Atopophrynus spp., 6:156
Atractaspididae. *See* African burrowing snakes
Atractaspidinae. *See Atractaspis* spp.
Atractaspis spp., 7:461, 7:483
Atractaspis bibronii. See Southern burrowing asps
Atractaspis microlepidota. See Small-scaled burrowing asps
Atractosteus spp., 4:221
Atractosteus spatula. See Alligator gars
Atractosteus tristoechus. See Cuban gars
Atractosteus tropicus. See Tropical gars
Atrato glass frogs, 6:*218*, 6:221–*222*
Atrichornis clamosus. See Noisy scrub-birds
Atrichornis rufescens. See Rufous scrub-birds
Atrichornithidae. *See* Scrub-birds
Atropoides spp., 7:445
Atropoides nummifer, 7:447, 7:448
Atrytone logan. See Delaware skippers
Atta spp. *See* Leaf cutter ants
Atta cephalotes. See Leaf cutter ants
Atta sexdens. See Leaf cutter ants
Attacus atlas. See Atlas moths
Attagis spp. *See* Seedsnipes
Attagis gayi. See Rufous-bellied seedsnipes
Attagis malouinus. See White-bellied seedsnipes
Attenborough, David, 11:507, 13:217
Attenuation, bird songs and, 8:39
Atthis spp. *See* Hummingbirds
Attrition, landscape, 3:86–87
Atypus affinis. See Purse web spiders
Atz, Jim, 4:192
Aubria spp., 6:247
Auchenipteridae, 4:30, 4:66, 4:352
Auchenoglanididae, 4:352
Auchenorrhyncha, 3:259, 3:260
See also Cicadas
Auckland Island teals. *See* Brown teals
Auckland sea lions. *See* Hooker's sea lions
Audition. *See* Ears
Auditory Neuroethology Lab, 12:54
Auditory system, amphibian, 6:23, 6:26
Audubon, John James
on Eastern phoebes, 10:285
on hummingbirds, 9:452
on New World warblers, 11:286
on passenger pigeons, 9:252
Audubon Screen Tours, 8:27
Audubon Society, 8:22, 8:26, 8:27
Audubon's cottontails. *See* Desert cottontails
Audubon's warblers. *See* Yellow-rumped warblers
Auffenberg, Walter, 7:53, 7:361
The Auk (journal), 8:27
Auks, 9:101, 9:104–105, **9:219–229,** 9:*223*
Aulacorhynchus spp. *See* Green toucanets; Toucanets
Aulacorhynchus huallagae. See Yellow-browed toucanets
Aulacorhynchus prasinus. See Emerald toucanets
Aulacothrips dictyotus. See Ectoparasitic thrips
Auliscomys boliviensis, 16:267
Aulopidae, 4:431, 4:432
Aulopiformes, 4:45, **4:431–440,** 4:*435*
Aulopus purpurissatus. See Sergeant bakers
Aulorhynchidae. *See* Tubesnouts
Aulostomidae. *See* Trumpetfishes
Aulostomus spp. *See* Trumpetfishes
Aulostomus chinensis, 5:133
Aulostomus maculatus. See West Atlantic trumpetfishes
Aurelia aurita. See Moon jellies
Auricillocoris spp., 3:55
Auricles, 12:9, 12:72
See also Physical characteristics
Auriparus flaviceps. See Verdins
Aurochs, 12:177, 16:*17*, 16:21–*22*
Austin, Oliver, 10:415, 10:506
Austral blackbirds, 11:302
Austral conures, 9:276
Austral kingdom, 3:55
Austral perches, **5:219–222,** 5:*223*, 5:*227*, 5:*229*, **5:230–232**
Australasian bitterns, 8:237, 8:245
Australasian bushlarks, 10:343, 10:*347*
Australasian carnivorous marsupials. *See* Dasyuromorphia
Australasian catbirds. *See* Green catbirds
Australasian eared nightjars, 9:401
Australasian figbirds, 11:*429*, 11:*430*, 11:*431*
Australasian gannets, 8:*216*, 8:*219*
Australasian kites, 8:318
Australasian larks, 10:*348*
Australasian Monotreme/Marsupial Action Plan, 13:71
Australasian pipits, 10:372
Australasian pseudo babblers, 11:127
Australembia rileyi, 3:*235*, 3:*236*
Australembiidae, 3:55
Australes. *See* Chocolate lyretails
Australian acacia gall thrips, 3:*286*–287
Australian basses, 5:221
Australian black-breasted buzzards, 8:319
Australian blue headed koels. *See* Common koels
Australian boobooks. *See* Southern boobook owls
Australian bronze cuckoos. *See* Horsfield's bronze-cuckoos
Australian brush-turkeys, 8:404, 8:405, 8:*406*, 8:*409*, 11:60
Australian cassowaries. *See* Southern cassowaries
Australian cattle dogs, 14:292
Australian chats, **11:65–68,** 11:*67*
Australian congollis, 5:321, 5:322
Australian coursers. *See* Australian pratincoles
Australian creepers, **11:133–139,** 11:*136*
Australian cycad thrips, 3:*286*, 3:287
Australian diamond doves, 9:241
Australian dragonflies, 3:133
Australian Environment Protection and Biodiversity Conservation Act of 1999, 9:14
Australian fairy-wrens, **11:45–53,** 11:*47*
Australian false vampire bats, 13:313, 13:*380*, 13:*381*, 13:*382*, 13:*383*, 13:*384*–385
Australian fantails, 11:83–84
Australian flower swallows. *See* Mistletoebirds
Australian flowerpeckers. *See* Mistletoebirds
Australian freshwater crocodiles. *See* Johnstone's crocodiles
Australian fur seals, 14:394
Australian gannets, 8:*184*
Australian gastric-brooding frogs, 6:37
Australian grassfinches, 11:354
Australian graylings, 4:391
Australian greater painted-snipes, 9:115, 9:117
Australian ground frogs, **6:139–146,** 6:*142*
distribution, 6:5–6, 6:*139*, 6:140, 6:*143–145*
reproduction, 6:32, 6:34, 6:141, 6:143–145
taxonomy, 6:139–140, 6:143–145, 6:155
Australian honeyeaters, **11:235–253,** 11:*239–240*
behavior, 11:236–237
conservation status, 11:237–238
distribution, 11:*235*, 11:236
evolution, 11:235
feeding ecology, 11:237
habitats, 11:236
humans and, 11:238
physical characteristics, 11:235–236
reproduction, 11:237
species of, 11:*241–253*
taxonomy, 11:55, 11:235
Australian hopping mice. *See* Australian jumping mice
Australian jumping mice, 16:*250*, 16:*254*, 16:*260*
Australian kestrels, 8:348
Australian Koala Foundation, 13:49

Australian lampreys. *See* Short-headed lampreys
Australian little buttonquails, 9:20
Australian lungfishes, 4:6, 4:57, **4:197–200,** 4:*198*, 4:*199*
Australian magpie-larks, 10:341, 11:453, 11:*454*, 11:*455*, 11:*456*
Australian magpies, 11:467, 11:468, 11:*470*, 11:*473*–474
Australian marsupial moles, 12:13
Australian masked owls, 9:336–338, 9:*339*, 9:340–*341*
Australian owlet-nightjars, 9:370, 9:*388*, 9:*390*, 9:*392*–393
Australian passerines, 8:11, 8:16
Australian pelicans, 8:183, 8:*226*
Australian peregrine falcons, 8:349
Australian phasmid, 3:43
Australian pied oystercatchers, 9:126
Australian pratincoles, 9:151, 9:152, 9:*154*, 9:*157*
Australian prowfishes, 5:163, 5:164, 5:165
Australian pygmy monitors. *See* Short-tailed monitors
Australian ravens, 11:505
Australian regent bowerbirds. *See* Regent bowerbirds
Australian region, 12:136–138
  feral cats, 12:185–186
  rabbit control, 12:186–188, 12:193
Australian robins, **11:105–113,** 11:*107*
Australian sand goannas, 7:37, 7:*360*, 7:361, 7:362
Australian sea lions, 14:*401*, 14:404
  behavior, 12:*145*
  distribution, 14:394, 14:*403*
  habitats, 14:*395*
  reproduction, 14:*396*, 14:398, 14:*399*
Australian skinks, 7:202
Australian smelts, 4:57, 4:*389*, 4:391, 4:*394*, 4:*400*, 4:401
Australian Society for Fish Biology, 4:393, 4:398
Australian songlarks, 10:341
Australian spotted barramundis, 4:232, 4:233, 4:234
Australian stick nest rats. *See* Greater stick-nest rats
Australian stone-plovers. *See* Beach thick-knees
Australian termites, giant, 3:161–162, 3:164, 3:166, 3:*169*, 3:*170*, 3:172
Australian toadlets, **6:147–154,** 6:*150*
Australian tupongs. *See* Australian congollis
Australian warblers, **11:55–63,** 11:*58*
Australian watering pots. *See* Watering pot shells
Australian wedge-tailed eagles, 8:321
Australian white ibises, 8:*294*
Australian Wildlife Protection Act, 3:322
Australian yellow white-eyes, 11:227
Australo-American sideneck turtles, **7:77–84**
Australo-Papuan elapids, 7:483, 7:484, 7:486
Australopithecines, 12:20, 14:247
Australopithecus spp., 14:242
Australopithecus afarensis, 14:*242*, 14:*243*
Australopithecus africanus, 14:242
Australopithecus bahrelghazali, 14:242
Australopithecus boisei, 14:242
Australopithecus gahri, 14:242
Australopithecus robustus, 14:242, 14:247, 14:250
Austrelaps labialis, 7:486
Austrobatrachus spp., 5:42
Austrochaperina fryi. *See* Fry's whistling frogs
Austrochaperina robusta. *See* Fry's whistling frogs
Austroclaudius spp., 3:55
Austrocynipidae, 3:55
Austroglanididae, 4:352
Austrognathia australiensis, 1:*333*, 1:*334*, 1:335
Austrolebias nigripinnis. *See* Blackfin pearl killifishes
Austroniidae, 3:55
Austroperlidae, 3:55, 3:141
Austropetaliidae, 3:55
Austrophasma spp., 3:217
Austrophasma redelinghuisensis, 3:*219*
Austrosimulium spp., 3:55
Austrosimulium pestilens. *See* Dawson River black flies
Autapomorphies, 12:7
Autarchoglossa, 7:197–198, 7:200, 7:202, 7:203–204, 7:206, 7:207
Autochthonous preservation, 3:7
Automaris spp., 3:*385*
Automedusa, 1:123, 1:127, 1:129
Automeris cecrops. *See* Silkmoth caterpillars
Autotrophy, 2:11, 2:26
  See also Feeding ecology
Avadavats, 11:354, 11:356, 11:365
Avahi spp. *See* Avahis
Avahi laniger. *See* Eastern woolly lemurs
Avahi occidentalis, 14:63
Avahis, **14:63–69,** 14:*68*
  behavior, 14:6, 14:8, 14:65–66
  conservation status, 14:67
  distribution, 14:65
  evolution, 14:63
  feeding ecology, 14:8, 14:66
  habitats, 14:65
  physical characteristics, 14:63–65
  species of, 14:*69*
  taxonomy, 14:63
Avian flight. *See* Flight
Avian malaria, 11:237–238, 11:343, 11:348
Avian pox, 11:343, 11:348
Avian songs. *See* Songs, bird
Aviceda madagascariensis. *See* Madagascar cuckoo-hawks
Avila-Pires, T. C. S., 7:307
Avitabatrachus spp., 6:12
Avocets, 9:104, **9:133–136, 9:139–141**
Ax, Peter, 1:12
Axis spp., 12:19, 15:269, 15:357, 15:358
Axis axis. *See* Chitals
Axis calamianensis. *See* Calamaian deer
Axis deer. *See Axis* spp.
Axis kuhlii. *See* Bawean hog deer
Axis porcinus. *See* Hog deer
Axolotl salamanders. *See* Mexican axolotl
Aye-ayes, **14:85–89,** 14:*86*, 14:*87*, 14:*88*
  behavior, 14:86–87
  conservation status, 14:88
  distribution, 12:136, 14:86
  evolution, 14:4, 14:85
  feeding ecology, 14:9, 14:87–88
  habitats, 14:86
  humans and, 14:88
  physical characteristics, 12:46, 12:83, 14:85
  reproduction, 14:10–11, 14:88
  taxonomy, 14:4, 14:85
Aye-ayes (extinct). *See Daubentonia robusta*
Aysheaia pedunculata, 2:109
Aythya americana. *See* Red-headed ducks
Aythya innotata. *See* Madagascar pochards
Ayu, 4:390, 4:391, 4:392, 4:*394*, 4:*400*
AZA. *See* American Zoo and Aquarium Association
Azara's agoutis, 16:407, 16:409, 16:415*t*
Azara's field mice, 16:266–270
Azara's tuco-tucos, 16:*426*, 16:*427*
Azari's night monkeys, 14:135, 14:*136*, 14:138, 14:141*t*
Azemiopinae. *See* Fea's vipers
Azemiops feae. *See* Fea's vipers
Azorella spp., 9:190
Azores bullfinches, 11:324
Aztecs, reptiles and, 7:55
Azure-breasted pittas, 10:195
Azure jays, 11:504
Azure rollers, 10:52, 10:53
Azure tits, 11:156
Azure-winged magpies, 11:*510*, 11:*515*–516
Azygiidae, 1:199

B

Babakotia spp., 14:63
Babblers, **10:505–524,** 10:*509*–*510*
  behavior, 10:507
  conservation status, 10:507
  distribution, 10:*505*, 10:506
  evolution, 10:505
  feeding ecology, 10:507
  habitats, 10:506–507
  humans and, 10:508
  physical characteristics, 10:506
  reproduction, 10:507
  species of, 10:*511*–*523*
  taxonomy, 10:505
Babbling starlings, 11:*409*, 11:*413*, 11:*424*–425
Babies. *See* Neonates
Babina spp., 6:248, 6:251
Babirousinae. *See* Babirusas
Babirusas, 15:*281*, 15:*283*
  behavior, 15:278, 15:287
  conservation status, 15:280–281, 15:288
  distribution, 12:137, 15:*287*
  evolution, 15:275
  feeding ecology, 15:278, 15:287–288
  habitats, 15:287
  humans and, 15:288
  physical characteristics, 15:*276*, 15:287
  reproduction, 15:279, 15:288
  taxonomy, 15:275, 15:287
Baboon lemurs, 14:63, 14:*68*, 14:*69*, 14:72
Baboons, 14:6, 14:188, 14:189, 14:191, 14:193, 14:194
Babyrousa spp. *See* Babirusas
Babyrousa babyrussa. *See* Babirusas
Bacardi bats, 13:317
Bachia spp., 7:303
Bachia bresslaui, 7:*305*, 7:*306*
Bachman's sparrows, 11:*267*, 11:*280*–281
Bachman's warblers, 11:290, 11:291
Bacillidae, 3:221, 3:222

Bacillus spp., 3:223
Bacillus rossius. *See* Mediterranean stick insects
Backswimmers, 3:260, 3:*266*, 3:*276–277*
Bacteria
drugs and, 1:48–49
origins of life and, 2:8–9
sponge metabolites and, 1:45
symbiosis and, 1:31–32
Bacterial infections, 3:39
Bactrian camels, 15:*320*
behavior, 15:317
conservation status, 15:317, 15:318
distribution, 12:134, 15:*315*
domestication of, 12:179, 15:150
evolution, 15:313
habitats, 15:316
humans and, 15:*319*
physical characteristics, 15:268, 15:314–*315*
taxonomy, 15:313
Bactrocera dorsalis, 3:60
Baculum, 12:91, 12:111
See also Physical characteristics
Badgers, **14:319–325**, 14:*322*, 14:*326*, **14:329–334**
behavior, 12:145, 14:258–259, 14:321–323
distribution, 14:321
Eurasian, 12:77, 14:324
evolution, 14:319
feeding ecology, 14:260, 14:323
habitats, 14:321
humans and, 14:324
physical characteristics, 14:319–321
reproduction, 14:324
species of, 14:*329*, 14:331, 14:332*t*–333*t*
taxonomy, 14:256, 14:319
torpor, 12:113
Badis badis, 5:238
Baeolophus spp. *See* American tufted titmice
Baeolophus wollweberi. *See* Bridled titmice
Baeopogon clamans. *See* Sjöstedt's honeyguide greenbuls
Baeriida, 1:58
Baetis spp., 3:125–126
Bagre graso, 4:354
Bagrid catfishes, 4:66, 4:352
Bagridae. *See* Bagrid catfishes
Bagworms, 3:386, 3:*391*, 3:*395*, 3:400–401
Bahama lancelets, 1:*489*, 1:*496–497*
Bahama swallows, 10:360
Bahama woodstars, 9:444
Bahaman raccoons, 14:309, 14:310, 14:315*t*
Bahamian funnel-eared bats, 13:461, 13:*462*, 13:*463*, 13:464
Bahamian hutias, 16:461, 16:*462*, 16:467*t*
Bahia antwrens. *See* Black-capped antwrens
Bahia hairy dwarf porcupines, 16:366, 16:368, 16:*370*, 16:*371*, 16:373
Bahia spinetails, 10:211
Bahia tyrannulets, 10:275
Baijis, 15:5, 15:10, **15:19–22**, 15:*20*, 15:*21*
Baikal oilfishes, 4:57, 5:179, 5:182
Baikal sculpins, 5:179, 5:180
Baikal seals, 12:138, 14:422, 14:*426*, 14:*427*, 14:430
Baikal teals, 8:365
Bailey's pocket mice, 16:208*t*
Baillonius bailloni. *See* Saffron toucanets
Baillon's crakes, 9:47
Baiomys taylori. *See* Pygmy mice
Baird's beaked whales, 15:*59*, 15:*61*, 15:62, 15:*63*, 15:*64*, 15:65
Baird's cormorants. *See* Pelagic cormorants
Baird's sandpipers, 9:179
Baird's tapirs. *See* Central American tapirs
Baird's trogons, 9:479
Baitfishes, 4:44
Bajaichthys spp., 4:448
Baker, A. N., 1:381
Balaena spp. *See* Bowhead whales
Balaena mysticetus. *See* Bowhead whales
Balaeniceps rex. *See* Shoebills
Balaenicipitidae. *See* Shoebills
Balaenidae, 15:3, 15:4, 15:103, **15:107–118**, *15:114*
Balaenoptera acutorostrata. *See* Northern minke whales
Balaenoptera bonaerensis. *See* Antarctic minke whales
Balaenoptera borealis. *See* Sei whales
Balaenoptera brydei, 15:120
Balaenoptera edeni. *See* Bryde's whales
Balaenoptera musculus. *See* Blue whales
Balaenoptera physalus. *See* Fin whales
Balaenopteridae. *See* Rorquals
Balaenula spp., 15:107
Balanoglossus spp., 1:445
Balanoglossus australiensis, 1:*444*
Balanoglossus clavigerus, 1:444–445
Balanoglossus gigas, 1:443
Balanophis spp., 7:468
Balanotaeniidae, 1:225
Balantiopteryx infusca, 13:358
Balantiopteryx plicata. *See* Gray sac-winged bats
Balanus balanoides. *See* Northern rock barnacles
Balanus psittacus, 2:276
Bald eagles, 8:27, 8:*49*, 8:*315*, 8:*321*, 8:323, 12:192
See also Eagles
Bald-headed uakaris. *See* Bald uakaris
Bald-headed woodshrikes. *See* Bornean bristleheads
Bald ibises, 8:293
Bald uakaris, 14:144, 14:145, 14:146, 14:*147*, 14:148, 14:*149*, 14:*150*, 14:151
Baldpates. *See* American wigeons
Baldwin, Carole C., 4:431
Baleaphryne spp. *See* Majorca midwife toads
Balearica regulorum. *See* Gray crowned cranes
Balearicinae cranes, 9:23
Baleen whales, 12:14, 15:1–2
behavior, 15:7
conservation status, 12:215
digestive system, 12:120–121
echolocation, 12:86
evolution, 15:3–4
feeding ecology, 15:8
migrations, 12:169
physical characteristics, 12:67
reproduction, 15:9
See also specific types of baleen whales
Bali cattle, 12:178
Bali mynas, 11:407, 11:408, 11:410, 11:*412*, 11:*420*
Bali starlings. *See* Bali mynas
baliscassius-hottentotottus, 11:437
Balistapus undulatus. *See* Orange-striped triggerfishes
Balistes vetula. *See* Queen triggerfishes
Balistidae. *See* Triggerfishes
Balistoides conspicillum. *See* Clown triggerfishes
Balitoridae, 4:321, 4:322
Balitorinae, 4:321
Ball pythons, 7:*23*, 7:420, 7:*423*, 7:*427*
Ball sea-hedgehogs. *See* Tuxedo pincushion urchins
Balloon animals. *See* Balloon water bears
Balloon water bears, 2:*119*, 2:120–121
Ballou, J. D., 12:207
Balon, E. L., 4:191, 4:193
Balsam beasts, 3:*209*, 3:*213*–214
Baltic amber, 3:8, 3:9
Baltic gray seals. *See* Gray seals
Baltimore orioles, 11:*303*, 11:304, 11:*308*, 11:310–*311*
Baluchitherium spp., 12:12
Baluchitherium grangeri. *See* *Indricotherium transouralicum*
Bamboo antshrikes. *See* Fasciated antshrikes
Bamboo bats, 13:311, 13:497, 13:*504*, 13:*513*, 13:514
Bamboo rats, 16:123
Amazon, 16:*452*, 16:*454*, 16:455–456
East African, 12:74
large, 16:*283*, 16:*287*, 16:*288*, 16:294–295
montane, 16:450–451, 16:*452*, 16:*455*, 16:457
southern, 16:450, 16:*452*, 16:*454*, 16:456–457
Bana leaf litter frogs, 6:*112*, 6:*113*
Banana bats, African, 13:497, 13:498
Banana boas. *See* Southern bromeliad woodsnakes
Banana fishes. *See* Horseface loaches
Banana frogs. *See* Greater leaf-folding frogs
Banana pipistrelles, 13:311
Banana slugs, North American, 2:413
Bananaquites, 11:285
Band cusk-eels, 5:*19*, 5:*21–22*
Band-rumped storm-petrels, 8:138
Band-tailed barbthroats, 9:*454*, 9:*457*
Band-tailed earthcreepers, 10:*212*, 10:*216–217*
Band-winged grasshoppers, 3:205, 3:206
Band-winged nightjars, 9:402, 9:403
Bandanna prawns. *See* Banded coral shrimps
Banded anteaters. *See* Numbats
Banded archerfishes, 5:*240*, 5:*246*, 5:*248*, 5:253
Banded bay cuckoos, 9:*316*, 9:*321*
Banded boxer shrimps. *See* Banded coral shrimps
Banded caecilians, 6:*433*–434
Banded cat sharks, 4:*105*
Banded coral shrimps, 2:34, 2:*202*, 2:*204*, 2:213–214
Banded cotingas, 10:305, 10:308, 10:*310*, 10:*313*–314
Banded duikers. *See* Zebra duikers
Banded dwarf boas. *See* Banded woodsnakes
Banded finches. *See* Double-barred finches
Banded green sunbirds. *See* Green sunbirds
Banded hare-wallabies, 13:*91*
Banded kingfishers, 10:6, 10:*12*, 10:16–*17*
Banded knifefishes, 4:12, 4:369–371, 4:*372*, 4:*374*–375
Banded leaf-monkeys, 14:*173*, 14:175, 14:*176*, 14:179–*180*
Banded linsangs, 14:*336*, 14:345*t*
Banded mantids. *See* Boxer mantids

Banded mongooses, 14:259, 14:*349*, 14:350, 14:*351*, 14:357*t*
Banded palm civets, 14:*337*, 14:344*t*
Banded pipefishes. *See* Ringed pipefishes
Banded pittas, 10:195
Banded pygmy sunfishes, 5:196, 5:199, 5:*201*, 5:*205*–206
Banded ratsnakes, 7:466, 7:470
Banded rubber frogs, 6:*307*, 6:315–316
Banded seasnakes. *See* Sea kraits
Banded stilts, 9:102, 9:*134*, 9:*135*
Banded sunbirds, 11:210
Banded wattle-eyes, 11:27
Banded woodsnakes, 7:434, 7:*436*
Bandfishes, 4:369–371, 4:*372*, 4:*375*–*376*
Bandicoots, 12:93, 12:137, **13:1–7**, 13:*10*, 13:19
  dry-country, 13:1, **13:9–18**
  rainforest, **13:9–18**
Bandicota spp., 16:249–250
Bands, human, 14:251
Bandtailed earless lizards. *See* Common lesser earless lizards
Bandy-bandy snakes, 7:486, 7:*489*, 7:*494*, 7:498
Banggai cardinalfishes, 5:262
Bangos. *See* Milkfishes
Banjo catfishes, 4:352, 4:353, 4:354
Bank diamonds. *See* Spotted pardalotes
Bank herrings. *See* Atlantic herrings
Bank swallows. *See* Sand martins
Bank voles, 16:*228*, 16:*232*, 16:*234*, 16:236
Banka snakeheads, 5:438
Bankia setacea. *See* Shipworms
Banks flying foxes, 13:331*t*
Banksianus fulgidus. *See* Pesquet's parrots
Bannan caecilians, 6:*421*, 6:*422*, 6:423
Banner-tailed kangaroo rats, 16:200, 16:*201*–202, 16:203, 16:*204*, 16:*207*
Bannerfishes, 5:*241*
Bannerman's turacos, 9:300, 9:301–302
Bannerman's weavers, 11:379
Bantengs, 12:178, 16:4, 16:24*t*
Baphetids, 6:8
Bar-breasted mousebirds, 9:*470*, 9:471, 9:*472*, 9:*473*
Bar-headed geese, 8:32, 8:365, 8:371
Bar jacks, 5:*257*
Bar-tailed godwits, 9:177
Bar-tailed larks, 10:345
Bar-tailed treecreepers, 11:177
Bar-tailed trogons. *See* Collared trogons
Bar-winged cinclodes, 10:*212*, 10:*217*–218
Bar-winged wood wrens, 10:529
Baramundi, 4:68
Barasinghas, 12:19, 15:357, 15:*358*, 15:361, 15:*363*, 15:367–*368*, 15:384
Barathronus spp., 5:18
Barbados raccoons, 14:262, 14:309, 14:310
Barbary apes. *See* Barbary macaques
Barbary falcons. *See* Peregrine falcons
Barbary ground squirrels, 16:159*t*
Barbary lions, 14:380
Barbary macaques, 14:*195*, 14:200–201
  behavior, 12:*150*
  distribution, 14:5, 14:191, 14:*198*
  evolution, 14:189
  habitats, 14:191
  reproduction, 12:*49*, 14:10
Barbary sheep, 16:4, 16:87, 16:90–92, 16:94, 16:*96*, 16:*98*, 16:100
Barbastella barbastellus. *See* Western barbastelles
Barbatula barbatula. *See* Stone loaches
Barbeled hound sharks, 4:113
Barbeled leaf fishes. *See* Amazon leaffishes
Barbels, 4:*306*, 4:*307*
Barber pole shrimps. *See* Banded coral shrimps
Barber's pole worms, 1:*296*, 1:300–301
Barbets, 10:85–88, **10:113–123**, 10:*118*
  behavior, 10:115–116
  conservation status, 10:117
  distribution, 10:*113*, 10:114
  evolution, 10:113
  feeding ecology, 10:116
  habitats, 10:114–115
  humans and, 10:117
  physical characteristics, 10:113–114
  reproduction, 10:116–117
  species of, 10:*119*–*122*
  taxonomy, 10:113
Barbinae, 4:297
Barbini, 4:303
Barbitistes constrictus, 3:206
Barbourisia rufa. *See* Red whalefishes
Barbourisiidae. *See* Red whalefishes
Barbourula busuangensis. *See* Philippine barbourulas
Barbourulas, **6:83–88**, 6:*85*
Barbus spp., 4:298
Barbus barbus. *See* Barbels
Barbus esocinus, 4:298
Barca snakeheads, 5:438
Barchatus spp., 5:42
Bardach, J. E., 5:38
Bare-eared curassows. *See* Alagoas curassows
Bare-eyed mynas, 11:410, 11:*412*, 11:*415*–416
Bare-headed parrots. *See* Pesquet's parrots
Bare-legged scops-owls. *See* Seychelles scops-owls
Bare-necked fruitcrows, 10:307, 10:312
Bare-necked umbrellabirds, 10:*311*, 10:315–*316*
Bare-tailed woolly opossums, 12:*250*, 12:252, 12:*255*, 12:*257*
Bare-throated bellbirds, 10:305, 10:306, 10:*311*, 10:*321*
Bare-throated francolins. *See* Red-necked francolins
Bare-throated whistlers, 11:116
Barentsia spp., 1:319
Barentsia discreta. *See* Marine colonial entoprocts
Barentsiidae, 1:319
Barisia spp. *See* Alligator lizards
Barita gymnocephala. *See* Bornean bristleheads
Bark beetles
  behavior, 3:59–60, 3:319
  feeding ecology, 3:63
  habitats, 3:56
  humans and, 3:323
  reproduction, 3:62, 3:322
  temperature and, 3:37
  wrinkled, 3:317
Bark-feeding insects, 3:56
Bark-gnawing beetles, 3:321
Barking deer. *See* Indian muntjacs
Barking treefrogs, 6:*41*
Barklice, 3:243
Barkudia spp., 7:328
Barn owls, 8:40, 8:262, **9:335–344**, 9:*337*, 9:*338*, 9:*339*
Barn Owls (Taylor), 9:335
Barn swallows, 10:*358*, 10:*362*, 10:363–*364*
Barnacle blennies, 5:341, 5:*342*
Barnacles, **2:273–281**, 2:*277*
  behavior, 2:28, 2:43, 2:275
  conservation status, 2:276
  distribution, 2:274–275
  evolution, 2:273
  feeding ecology, 2:275
  habitats, 2:275
  humans and, 2:276
  physical characteristics, 2:273–275
  reproduction, 2:5, 2:275–276
  species of, 2:*278*–*281*
  taxonomy, 2:273
Barokinesis, 1:38
Barracudas, 4:48, 4:68, 4:69, 5:*405*, **5:405–407**, 5:*408*, **5:415–416**
Barracudinas, 4:432, 4:433
Barramundi perches. *See* Barramundis
Barramundis, 5:*221*, 5:*224*, 5:*225*–*226*
Barred antshrikes, 10:*244*, 10:*246*–*247*
Barred bandicoots. *See* Eastern barred bandicoots
Barred buttonquails, 9:14, 9:*15*, 9:17–*18*
Barred cuckoo-doves, 9:*253*, 9:*257*
Barred eagle-owls, 9:*352*, 9:*357*–*358*
Barred flagtails. *See* Flagtail kuhlias
Barred fruiteaters, 10:305
Barred galaxias, 4:391
Barred munias. *See* Spotted munias
Barred owlet-nightjars, 9:388, 9:*390*, 9:*392*
Barred quails, 8:456
Barred sorubims. *See* Tiger shovelnose catfishes
Barred warblers, 11:5
Barred waxbills. *See* Common waxbills
Barrel sponges, 1:*81*, 1:*82*, 1:83
Barreleyes, 4:*389*, 4:390–392, 4:*394*, 4:398–*399*
Barrelfishes, 5:422
Barrington iguanas, 7:*244*
Bartail flatheads. *See* Indian flatheads
Bartel's wood-owls. *See* Asian brown wood-owls
Bártók, Bela, 8:19
Bartramia longicauda. *See* Upland sandpipers
Bary, Anton de. *See* de Bary, Anton
Barycholos spp., 6:156
Barylamda spp., 15:135
Baryphthengus martii. *See* Rufous motmots
Baryphthengus ruficapillus. *See* Rufous-capped motmots
Baseodiscus delineatus, 1:*248*, 1:*249*
Basileuterus griseiceps. *See* Gray-headed warblers
Basileuterus ignotus. *See* Pirre warblers
Basilia falcozi. *See* Spider bat flies
Basilinna spp. *See* Hummingbirds
Basilinna leucotis. *See* White-eared hummingbirds
Basiliscus basiliscus. *See* Common basilisks
Basiliscus plumifrons. *See* Green basilisks
Basilornis spp., 11:408
Basilornis galeatus. *See* Helmeted mynas
Basilornis miranda. *See* Apo mynas

Basilosauridae, 15:2
Basilosaurus spp., 15:3, 15:119
Basipodellidae, 2:284
Basket stars, **1:387–394,** 1:*391*
  behavior, 1:388–389
  conservation status, 1:390
  distribution, 1:387
  evolution, 1:387
  feeding ecology, 1:389
  habitats, 1:388
  humans and, 1:390
  physical characteristics, 1:387, 1:*388*
  reproduction, 1:389
  species of, 1:*393*, 1:394
  taxonomy, 1:387
Basking sharks, 4:11, 4:67, 4:*131*, 4:133–*134*, 4:*136*–137
Basommatophora, 2:412, 2:413, 2:414, 2:415
Bass bait (larvae). *See* Eastern dobsonflies
Bassaricyon spp. *See* Olingos
Bassaricyon alleni. See Allen's olingos
Bassaricyon beddardi. See Beddard's olingos
Bassaricyon gabbii. See Olingos
Bassaricyon lasius. See Harris's olingos
Bassaricyon pauli, 14:310
Bassariscus spp., 14:309, 14:310
Bassariscus astutus. See Ringtails (Procyonidae)
Bassariscus sumichrasti. See Cacomistles
Basses, 4:68
  See also specific types of basses
Basslets, 5:255, 5:256, 5:258, 5:259, 5:261, 5:263
Bastard soldierfishes. *See* Blackbar soldierfishes
Bat Conservation International, 13:420
Bat-eared foxes, 14:*275*, 14:281
  behavior, 14:*260*
  canine distemper and, 12:216
  distribution, 14:*280*
  evolution, 14:265
  physical characteristics, 14:266–267, 14:*267*
  reproduction, 14:271, 14:272
Bat falcons, 8:349, 8:352
Bat fleas, 3:349, 3:*351*, 3:*352*
Bat flies, 3:47, 3:*363*, 3:*369*–370
Bat parasites, 3:263
  See also specific parasites
Bat rays, 4:128
Bat stars, 1:*369*
Batara cinerea. See Giant antshrikes
Bate, D. M. A., 16:333
Bate's dwarf antelopes. *See* Dwarf antelopes
Bate's pygmy antelopes. *See* Dwarf antelopes
Bate's slit-faced bats, 13:376*t*
Batesian mimicry, 3:65, 3:320, 5:342
Bates's swifts, 9:424
Bates's weavers, 11:379
Batfishes (Ephippidae), 4:69, 5:391, 5:392, 5:393
Batfishes (*Ogcocephalus* spp.), 5:*47*, 5:50
Bath sponges, 1:79, 1:80, 1:*81*, 1:*83*, 1:85
Bathmocercus spp., 11:4, 11:7
Bathochordaeus spp., 1:473
Bathochordaeus charon. See Giant larvaceans
Bathyblennius antholops, 5:342
Bathycallionymus spp., 5:365, 5:366
Bathycrinids, 1:356
Bathydraconidae. *See* Dragonfishes
Bathyergidae. *See* African mole-rats
Bathyerginae. *See* Dune mole-rats
Bathyergus spp. *See* Dune mole-rats
Bathyergus janetta. See Namaqua dune mole-rats
Bathyergus suillus. See Cape dune mole-rats
Bathylagidae, 4:390
Bathylaginae, 4:391
Bathylagus nigrigenys, 4:392
Bathymasteridae. *See* Ronquils
Bathynellaceans, **2:177–179**
Bathynellidae, 2:177–178
Bathynomus giganteus, 2:249
Bathypera ovoida, 1:453
Bathypterois quadrifilis. See Tripodfishes
Bathysauridae, 4:432
Bathysauroides gigas, 4:431
Bathysauropsis gracilis, 4:431
Bathysauropsis malayanus, 4:431
Bathysciadiidae, 2:435, 2:436
Bathysphyraenops simplex, 5:220
Bathysquilloidea, 2:168, 2:171
Batillipes spp., 2:117
Batillipes noerrevangi, 2:117
Batis capensis. See Cape batis
Batodon spp., 13:193
Batonodoides, 13:194
Batrachemys spp., 7:77
Batrachemys dahli, 7:77
Batrachoides spp., 5:41, 5:42
Batrachoides manglae. See Cotuero toadfishes
Batrachoididae. *See* Toadfishes
Batrachoidiformes. *See* Toadfishes
Batrachoidinae, 5:41
Batrachomoeus spp., 5:42
Batrachophrynus spp., 6:157, 6:158, 6:159
Batrachophrynus macrostomus, 6:159
Batrachophrynus patagonicus. See Patagonia frogs
Batrachosauroids, 6:13
Batrachoseps spp., 6:35, 6:325, 6:391
Batrachoseps campi. See Inyo Mountains salamanders
Batrachostominae. *See* Frogmouths
Batrachostomus spp. *See* Frogmouths
Batrachostomus auritus. See Large frogmouths
Batrachostomus cornutus. See Sunda frogmouths
Batrachostomus harterti. See Dulit frogmouths
Batrachostomus hodgsoni. See Hodgson's frogmouths
Batrachostomus mixtus. See Bornean frogmouths
Batrachostomus moniliger. See Sri Lanka frogmouths
Batrachostomus stellatus. See Gould's frogmouths
Batrachotoxin, 6:53, 6:197, 6:198, 6:209
Batrachuperus spp., 6:335, 6:336, 6:337
Batrachuperus gorganensis, 6:337
Batrachuperus londongensis, 6:336
Batrachuperus mustersi, 6:336–337
Batrachuperus persicus, 6:337
Batrachuperus tibetanus. See Tibetan stream salamanders
Batrachyla spp., 6:156, 6:157
Batrachylodes spp., 6:247, 6:248, 6:251
Batrichthys spp., 5:42
Bats, 12:12–13, **13:307–318**
  behavior, 13:312–313
  vs. birds, 8:*46–47*, 8:*48*
  brains, 12:49
  conservation status, 13:315–316
  distribution, 12:129, 12:132, 12:136, 12:137, 12:138, 13:310–311
  evolution, 12:11, 13:308–309
  feeding ecology, 13:313–314
  field studies, 12:200–201
  flight adaptations in, 12:52–60, 12:*53–59*
  habitats, 13:311
  hearing, 12:82
  hearts, 12:45
  humans and, 13:316–317
  as keystone species, 12:216–217
  locomotion, 12:14, 12:43–44
  navigation, 12:87
  physical characteristics, 13:307–308, 13:310
  reproduction, 12:90–91, 12:94, 12:95, 12:103, 12:105–106, 12:110, 13:314–315
  taxonomy, 13:309
  See also Vespertilionidae; specific types of bats
Baudó oropendolas, 11:305
Baüplan, 2:8, 2:13
Baurubatrachus spp., 6:13
Baw Baw frogs, 6:141, 6:*142*, 6:145
Bawean hog deer, 15:360
Bay anchovies, 4:*280*, 4:*283*, 4:286–287
Bay-breasted warblers, 11:*287*
Bay cats, 14:390*t*
Bay duikers, 16:*76*, 16:*78*, 16:*83*
Bay owls. *See* Oriental bay owls
Bay-winged cowbirds. *See* Baywings
Bay-winged hawks. *See* Harris' hawks
Bay wrens, 10:*530*, 10:534–*535*
Baya weavers, 11:*381*, 11:*387*
Baylis, Jeffrey R., 8:40
Baywings, 11:304, 11:*307*, 11:*320*
Bdelloidea, 1:259, 1:260, 1:261–262
Bdellonemertidae, 1:255
Bdellophis vittatus. See Banded caecilians
Bdellostoma stoutii. See Pacific hagfishes
Bdelloura candida, 1:*189*, 1:*193*–194, 1:194
Bdeogale spp., 14:347
Bdeogale crassicauda. See Bushy-tailed mongooses
Beach curlews. *See* Beach thick-knees
Beach fleas. *See* Beach hoppers
Beach hoppers, 2:*264*, 2:*267*, 2:268
Beach stone-curlews. *See* Beach thick-knees
Beach thick-knees, 9:144, 9:145, 9:147–*148*
Beacham, T. D., 5:33
Beaded lacewings, 3:307, 3:308, 3:*310*, 3:*311*
Beaded lizards, Mexican. *See* Mexican beaded lizards
Beagles, 14:288
Beaked sandfishes. *See* Sandfishes
Beaked whales, **15:59–71,** 15:*63*, 15:*69t–70t*
  behavior, 15:6, 15:7, 15:61
  conservation status, 15:10, 15:61–62
  distribution, 15:*59*, 15:61
  evolution, 15:3, 15:59
  feeding ecology, 15:8, 15:61
  habitats, 15:61
  humans and, 15:62
  physical characteristics, 15:4, 15:59, 15:61
  reproduction, 15:61
  species of, 15:64–68, 15:*69t–70t*
  taxonomy, 15:3, 15:59
Beaks. *See* Bills and beaks
Bean, Tarleton H., 5:351
Bean bugs. *See* Southern green stink bugs
Bean clams. *See* Coquina clams

Bear cats. *See* Binturongs
Bear cuscuses, 13:57
Beard, K. Christopher, 14:91
Beard worms, **2:85–89,** 2:*87*
Bearded barbets, 10:115
Bearded bellbirds, 10:305, 10:*311*, 10:*320*–321
Bearded dragons, 7:*41*, 7:*205*, 7:*212*, 7:*218*, 7:219
Bearded ghouls, 5:166, 5:*167*, 5:*172*, 5:176
Bearded partridges. *See* Bearded wood-partridges
Bearded pigs, 15:*279*, 15:280, 15:289*t*
Bearded pygmy chameleons. *See* Short-tailed chameleons
Bearded reedlings, 10:505, 10:506, 10:507, 10:*509*, 10:520–*521*
Bearded sakis, 14:143–148, 14:*147*, 14:*149*, 14:*150*–151
Bearded seals, 14:435*t*
Bearded tits. *See* Bearded reedlings
Bearded toad heads. *See* Toad-headed agamas
Bearded tree-quails. *See* Bearded wood-partridges
Bearded vultures, 8:321
Bearded waspfishes. *See* Ocellated waspfishes
Bearded wood-partridges, 8:458, 8:*459*, 8:*460*
Beardfishes, 4:448, 5:*1*, **5:1–3,** 5:*3*
Bears, **14:295–307,** 14:*301*
behavior, 14:258, 14:297–298
conservation status, 14:262, 14:300
distribution, 12:131, 12:136, 14:*295*, 14:296
evolution, 14:295–296
feeding ecology, 14:255, 14:260–261, 14:298–299
field studies of, 12:*196*, 12:200
gaits of, 12:*39*
habitats, 14:296–297
hibernation, 12:113
humans and, 14:300
Ice Age, 12:24
physical characteristics, 14:296
reproduction, 12:109, 12:110, 14:299
species of, 14:*302*–305, 14:306*t*
taxonomy, 14:256, 14:295
in zoos, 12:*210*
Beautiful nuthatches, 11:169
Beautiful sunbirds, 11:209
Beautiful torrent frogs, 6:*254*, 6:261
Beauty lizards. *See* Brown garden lizards
Beavers, 12:14, **16:177–184,** 16:*182*
behavior, 16:179
conservation status, 16:180
distribution, 16:124, 16:*177*, 16:179
evolution, 16:122, 16:177
feeding ecology, 16:179–180
habitats, 16:179
humans and, 16:127, 16:180
Ice Age, 12:24
North American, 12:111
physical characteristics, 16:123, 16:177–179
reproduction, 12:107, 16:126, 16:180
species of, 16:183–184
taxonomy, 16:122, 16:177
See also specific types of beavers
Becards. *See* Tyrant flycatchers
Beccari's mastiff bats, 13:*484*, 13:494*t*
Bechstein's bats, 12:85
Becket, Thomas à, 3:85
Bed bugs, 3:64, 3:*266*, 3:270–*271*
Beddard's olingos, 14:310, 14:316*t*
Beddome's Indian frogs, 6:*254*, 6:263–264
Bedotiidae. *See* Malagasy rainbowfishes
Bee-eaters, 10:1–4, 10:*3*, **10:39–49,** 10:*40*, 10:*43*
behavior, 10:39–40
evolution, 10:39
feeding ecology, 10:40–41, 10:*42*
physical characteristics, 10:39
reproduction, 10:*41*–42
species of, 10:*44–48*
taxonomy, 10:39
Bee flies, 3:360
Bee hummingbirds, 9:438, 9:448
Bee robbers. *See* Death's head hawk moths
Bee stings, 3:81
Beecroft, John, 16:303
Beecroft's anomalures, 16:*301*, 16:303, 16:*305*
Beef tapeworms, 1:*227*, 1:*233*, 1:239
distribution, 1:*235*
humans and, 1:35, 1:231
physical characteristics, 1:228, 1:*229*
Beehives, 3:79
Beekeeping. *See* Apiculture
Bees, 2:25, 3:35, 3:79, **3:405–418,** 3:*410*
behavior, 3:59, 3:65, 3:406
conservation status, 3:407
evolution, 3:10, 3:405
feeding ecology, 3:20, 3:42, 3:63, 3:406–407
habitats, 3:406
humans and, 3:407–408
metamorphosis, 3:34
as pests, 3:50
physical characteristics, 3:405
pollination by, 3:14, 3:78, 3:79
reproduction, 3:36, 3:38, 3:39, 3:59, 3:407
social, 3:68, 3:69
species of, 3:*411*–413
symbolism, 3:74, 3:407–408
taxonomy, 3:405
vision, 3:30
Beeswax, 3:79, 3:81
Beethoven, Ludwig, 8:19
Beetle crickets, 3:*210*, 3:*211*, 3:212
Beetles, 2:38, **3:315–334,** 3:*324*, 3:*325*
behavior, 3:47, 3:60, 3:64, 3:65, 3:319–320
biological control and, 3:80
conservation status, 3:82, 3:88, 3:322
distribution, 3:319
evolution, 3:8, 3:315
feeding ecology, 3:42, 3:63, 3:320–321
habitats, 3:319
humans and, 3:74, 3:319
jewelry, 3:74
logging and, 3:87
medicinal uses, 3:81
metamorphosis, 3:34
as pests, 3:50, 3:75, 3:76, 3:323
physical characteristics, 3:10, 3:18, 3:*24*, 3:43, 3:317–319
pollination by, 3:14, 3:79
reproduction, 3:29, 3:38, 3:45, 3:59, 3:62, 3:321–322
species of, 3:*327*–333
symbolism, 3:74
taxonomy, 3:315–316
See also specific types of beetles
Begall, S., 16:435
Begle, D. P., 4:389
Behavior
amphibians, **6:44–50**
anurans, 6:64–67
African treefrogs, 6:282, 6:285–289
Amero-Australian treefrogs, 6:228–229, 6:233–242
Arthroleptidae, 6:266, 6:268–270
Asian toadfrogs, 6:110–111, 6:113–117
Asian treefrogs, 6:292, 6:295–299
Australian ground frogs, 6:140, 6:143–145
Bombinatoridae, 6:83–84, 6:86–88
Bufonidae, 6:185, 6:188–194
Discoglossidae, 6:90, 6:92–94
Eleutherodactylus spp., 6:66, 6:158
ghost frogs, 6:132, 6:133–134
glass frogs, 6:216, 6:219–223
leptodactylid frogs, 6:158, 6:162–171
Madagascan toadlets, 6:318, 6:319–320
marine toads, 6:*46*, 6:191
Mesoamerican burrowing toads, 6:96–97
Myobatrachidae, 6:148, 6:151–153
narrow-mouthed frogs, 6:303, 6:*304*, 6:308–316
New Zealand frogs, 6:70, 6:72–74
parsley frogs, 6:127, 6:129
Pipidae, 6:100, 6:103–106
poison frogs, 6:66–67, 6:199, 6:203–209
Ruthven's frogs, 6:212
Seychelles frogs, 6:136, 6:137–138
shovel-nosed frogs, 6:273–*274*, 6:277
spadefoot toads, 6:119–120, 6:124–125
tadpoles, 6:41–42, 6:44
tailed frogs, 6:78, 6:80–81
three-toed toadlets, 6:179, 6:181–182
toads, 6:46–48
true frogs, 6:250–251, 6:255–263
vocal sac-brooding frogs, 6:*174*, 6:175–176
Xenopus spp., 6:47, 6:100
caecilians, 6:44, 6:412
American tailed, 6:416, 6:417–418
Asian tailed, 6:420, 6:422–423
buried-eye, 6:432, 6:433–434
Kerala, 6:426, 6:428
tailless, 6:436, 6:439–441
salamanders, 6:45, 6:325
amphiumas, 6:406, 6:409–410
Asiatic, 6:336, 6:339–341
Cryptobranchidae, 6:345
lungless, 6:44, 6:392, 6:395–403
mole, 6:44, 6:356, 6:358–360
Pacific giant, 6:350, 6:352
Proteidae, 6:378, 6:381–383
Salamandridae, 6:45, 6:*365*–366, 6:370–375
sirens, 6:328, 6:331–332
torrent, 6:385, 6:387
birds, 9:182–188
albatrosses, 8:109, 8:114–115, 8:118–122
Alcidae, 9:*221*, 9:224–228
anhingas, 8:202, 8:207–209
Anseriformes, 8:365–366
ant thrushes, 10:240–241, 10:245–256
Apodiformes, 9:417
asities, 10:188, 10:190
Australian chats, 11:66, 11:67
Australian creepers, 11:134, 11:137–139
Australian fairy-wrens, 11:45, 11:48–53
Australian honeyeaters, 11:236–237, 11:241–253
Australian robins, 11:105, 11:108–112

Australian warblers, 11:56, 11:59–63
babblers, 10:507, 10:511–523
barbets, 10:115–116, 10:119–122
barn owls, 9:336–337, 9:340–343
bee-eaters, 10:39–40
bird songs and, 8:38–39
birds of paradise, 11:491, 11:495–501
bitterns, 8:242–243
Bombycillidae, 10:448, 10:451–454
boobies, 8:212–213
bowerbirds, 11:479, 11:483–488
broadbills, 10:178, 10:181–186
bulbuls, 10:397, 10:402–413
bustards, 9:*92*, 9:*93*, 9:96–99
buttonquails, 9:12–13, 9:16–21
caracaras, 8:349
cassowaries, 8:76
Charadriidae, 9:162, 9:166–172
Charadriiformes, 9:102
chats, 10:486–487
chickadees, 11:157, 11:160–166
chowchillas, 11:70, 11:72–73
Ciconiiformes, 8:234–235
Columbidae, 9:249, 9:255–266
Columbiformes, 9:243
condors, 8:234, 8:235, 8:277–278
Coraciiformes, 10:1–4
cormorants, 8:202, 8:205–206, 8:207–209
Corvidae, 11:505–506, 11:512–522
cotingas, 10:306–307, 10:312–322
crab plovers, 9:122
Cracidae, 8:414–415
cranes, 9:*24–26*, 9:31–35
cuckoo-shrikes, 10:386
cuckoos, 9:312, 9:318–329
dippers, 10:476, 10:479–482
diversity of, 8:17–18
diving-petrels, 8:143, 8:145–146
doves, 9:243, 9:249, 9:255–266
drongos, 11:438, 11:441–445
ducks, 8:371–372
eagles, 8:320–321
elephant birds, 8:104
emus, 8:84
Eupetidae, 11:76, 11:78–80
fairy bluebirds, 10:416, 10:423–424
Falconiformes, 8:314
falcons, 8:349
false sunbirds, 10:188, 10:191
fantails, 11:85–86, 11:89–95
finches, 11:323–324, 11:328–339
flamingos, 8:305
flowerpeckers, 11:190
fowls, 8:435
frigatebirds, 8:194, 8:197–198
frogmouths, 9:368–369, 9:378, 9:381–385
Galliformes, 8:401
gannets, 8:212–213
geese, 8:371–372
Glareolidae, 9:152, 9:155–160
grebes, 8:170–171, 8:174–180
Gruiformes, 9:*2–3*
guineafowl, 8:427
gulls, 9:211–217
hammerheads, 8:262
Hawaiian honeycreepers, 11:342, 11:346–351
hawks, 8:320–321
hedge sparrows, 10:459, 10:462–464
herons, 8:234–235, 8:242–243
hoatzins, 8:466
honeyguides, 10:138, 10:141–144
hoopoes, 10:62
hornbills, 10:73, 10:78–84
hummingbirds, 9:443–445, 9:455–467
ibises, 8:234–235, 8:292
Icteridae, 11:293, 11:309–322
ioras, 10:416, 10:419–420
jacamars, 10:92
jacanas, 9:108–109, 9:112–114
kagus, 9:42
kingfishers, 10:8, 10:13–23
kiwis, 8:90
lapwings, 9:166–172
Laridae, 9:206, 9:211–217
larks, 10:343–344, 10:348–354
leafbirds, 10:416, 10:420–423
limpkins, 9:37–38
logrunners, 11:70, 11:72–73
long-tailed titmice, 11:142, 11:145–146
lyrebirds, 10:330–331, 10:334–335
magpie-shrikes, 11:468, 11:471–475
manakins, 10:296, 10:299–303
mesites, 9:6, 9:8–9
moas, 8:97
monarch flycatchers, 11:97, 11:100–103
motmots, 10:32, 10:35–37
moundbuilders, 8:404–405
mousebirds, 9:469, 9:473–476
mudnest builders, 11:453, 11:456–458
New World blackbirds, 11:293, 11:309–322
New World finches, 11:263–264, 11:269–282
New World quails, 8:457
New World vultures, 8:234–235, 8:277–278
New World warblers, 11:289, 11:293–299
New Zealand wattle birds, 11:447, 11:449–450
New Zealand wrens, 10:204, 10:206–207
nightjars, 9:368–370, 9:403, 9:408–414
nuthatches, 11:168, 11:171–175
oilbirds, 9:374
Old World flycatchers, 11:26, 11:30–43
Old World warblers, 11:4–5, 11:10–23
Oriolidae, 11:*428*, 11:431–434
ostriches, 8:100–101
ovenbirds, 10:210, 10:214–228
owlet-nightjars, 9:388, 9:391–393
owls, 9:332, 9:336–337, 9:340–343, 9:348–349, 9:354–365
oystercatchers, 9:126, 9:130–131
painted-snipes, 9:116, 9:118–119
palmchats, 10:455
pardalotes, 11:201, 11:204–206
parrots, 9:276–277, 9:283–297
Passeriformes, 10:172–173
Pelecaniformes, 8:185
pelicans, 8:185, 8:226
penduline titmice, 11:147, 11:151–152
penguins, 8:149–150, 8:153–157
Phasianidae, 8:435
pheasants, 8:435
Philippine creepers, 11:184, 11:186–187
Picidae, 10:149, 10:154–167
Piciformes, 10:86–87
pigeons, 9:243, 9:249, 9:255–266
pipits, 10:373–374, 10:378–383
pittas, 10:194, 10:197–201
plantcutters, 10:326, 10:327–328
plovers, 9:166–172
potoos, 9:396, 9:399–400
pratincoles, 9:155–160
Procellariidae, 8:124, 8:128–132
Procellariiformes, 8:109
pseudo babblers, 11:128
puffbirds, 10:102, 10:106–111
rails, 9:48–49, 9:57–66
Raphidae, 9:270, 9:272–273
Recurvirostridae, 9:134, 9:137–140
rheas, 8:70
rollers, 10:52–53, 10:55–58
sandgrouse, 9:231–232, 9:235–238
sandpipers, 9:178, 9:182–188
screamers, 8:394
scrub-birds, 10:337–338, 10:339–340
secretary birds, 8:344
seedsnipes, 9:190, 9:193–195
seriemas, 9:85–86, 9:88–89
sharpbills, 10:292
sheathbills, 9:198, 9:200–201
shoebills, 8:287
shrikes, 10:427–428, 10:432–438
sparrows, 11:398, 11:401–406
spoonbills, 8:235, 8:292
storks, 8:234–235, 8:265–266
storm-petrels, 8:136, 8:140–141
Sturnidae, 11:408, 11:414–425
sunbirds, 11:209, 11:213–225
sungrebes, 9:70, 9:71–72
swallows, 10:358–359, 10:363–369
swans, 8:371–372
swifts, 9:422–423, 9:426–431
tapaculos, 10:258, 10:262–267
terns, 9:211–217
thick-knees, 9:144–145, 9:147–148
thrushes, 10:486–487
tinamous, 8:58
titmice, 11:157, 11:160–166
todies, 10:26, 10:29–30
toucans, 10:126–127, 10:130–134
tree swifts, 9:434, 9:435–436
treecreepers, 11:178, 11:180–181
trogons, 9:478, 9:481–485
tropicbirds, 8:185, 8:187, 8:190–191
trumpeters, 9:78, 9:81–82
turacos, 9:300–301, 9:304–310
typical owls, 9:348–349, 9:354–365
tyrant flycatchers, 10:271–273, 10:278–288
vanga shrikes, 10:440, 10:443–444
Vireonidae, 11:256, 11:258–262
wagtails, 10:373–374, 10:378–383
weaverfinches, 11:354–355, 11:360–372
weavers, 11:377, 11:382–394
whistlers, 11:117, 11:120–126
white-eyes, 11:228–229, 11:232–233
woodcreepers, 10:230, 10:232–236
woodhoopoes, 10:66, 10:68–69
woodswallows, 11:460, 11:462–465
wrens, 10:527–528, 10:531–538

fishes, **4:60–71**
Acanthuroidei, 5:392–393, 5:396–403
Acipenseriformes, 4:213–214, 4:217–220
Albuliformes, 4:250, 4:252–253
angelsharks, 4:162, 4:164–165
anglerfishes, 5:48–50, 5:53–56
Anguilliformes, 4:*256*, 4:260–268
Atheriniformes, 5:69, 5:73–76
Aulopiformes, 4:433, 4:436–439

Australian lungfishes, 4:198
beardfishes, 5:1, 5:3
Beloniformes, 5:80, 5:83–86
Beryciformes, 5:115, 5:118–121
bichirs, 4:210, 4:211–212
blennies, 5:342, 5:345–348
bowfins, 4:230
bullhead sharks, 4:98, 4:101–102
butterflyfishes, 4:62–63
Callionymoidei, 5:366, 5:369–371
carps, 4:298–299, 4:307–319
catfishes, 4:353, 4:357–366
characins, 4:337, 4:342–349
chimaeras, 4:92, 4:94–95
coelacanths, 4:191–192
Cypriniformes, 4:298–299, 4:307–319, 4:322, 4:325–333
Cyprinodontiformes, 5:92, 5:96–102
damselfishes, 4:62, 4:70
dogfish sharks, 4:152, 4:155–157
dories, 5:124, 5:126–128, 5:130
eels, 4:*256*, 4:260–268
electric eels, 4:370–371, 4:373–377
elephantfishes, 4:60, 4:62, 4:233, 4:238
Elopiformes, 4:243, 4:246–247
Esociformes, 4:380, 4:383–386
flatfishes, 5:452, 5:455–463
Gadiformes, 5:29, 5:32–39
gars, 4:222, 4:224–227
Gasterosteiformes, 5:134–135, 5:139–147
gobies, 5:375, 5:380–388
Gobiesocoidei, 5:356, 5:359–363
Gonorynchiformes, 4:290, 4:293–295
ground sharks, 4:114–115, 4:119–128
Gymnotiformes, 4:370–371, 4:373–377
hagfishes, 4:79, 4:80–81
herrings, 4:277–278, 4:281–287
Hexanchiformes, 4:144, 4:146–148
Labroidei, 5:277, 5:284–290, 5:294–295, 5:301–307
labyrinth fishes, 5:428, 5:431–434
lampreys, 4:84–85, 4:88–90
Lampridiformes, 4:449, 4:452–454
lanternfishes, 4:442, 4:444–446
lungfishes, 4:*202*, 4:203, 4:205–206
mackerel sharks, 4:132–133, 4:135–141
minnows, 4:298–299, 4:314–315
morays, 4:*256*, 4:264–266
mudminnows, 4:380, 4:383–386
mullets, 5:60, 5:62–65
Ophidiiformes, 5:17, 5:20–22
Orectolobiformes, 4:105, 4:108–110
Osmeriformes, 4:392, 4:395–401
Osteoglossiformes, 4:232–233, 4:237–241
Percoidei, 5:197, 5:202–208, 5:212, 5:214–217, 5:221, 5:225–226, 5:228, 5:230–233, 5:239–240, 5:247–253, 5:261, 5:266–268, 5:270–273
Percopsiformes, 5:5, 5:8–12
pikes, 4:380, 4:383–386
ragfishes, 5:352
Rajiformes, 4:177, 4:182–188
rays, 4:177, 4:182–188
Saccopharyngiformes, 4:272, 4:274–276
salmons, 4:*63*, 4:406–407, 4:410–419
sawsharks, 4:168, 4:170–171
Scombroidei, 5:406, 5:409–418
Scorpaeniformes, 5:158, 5:161–162, 5:165, 5:169–178, 5:181–182, 5:186–193
Scorpaenoidei, 5:165
skates, 4:177, 4:187
snakeheads, 5:438, 5:442–446
South American knifefishes, 4:370–371, 4:373–377
southern cod-icefishes, 5:322, 5:325–328
Stephanoberyciformes, 5:106, 5:108–110
Stomiiformes, 4:422–423, 4:426–429
Synbranchiformes, 5:151, 5:154–156
Tetraodontiformes, 5:469–470, 5:474–484
toadfishes, 5:42, 5:44–45
Trachinoidei, 5:333, 5:337–339
trouts, 4:62, 4:406–407
Zoarcoidei, 5:311, 5:315–318
insects, 3:5–6, **3:59–66**
ants, 3:59, 3:65, 3:71, 3:406
bees, 3:59, 3:65, 3:406
beetles, 3:47, 3:60, 3:64, 3:65, 3:319–320
book lice, 3:243–244, 3:246–247
bristletails, 3:114, 3:116–117
caddisflies, 3:376, 3:379–381
cockroaches, 3:65, 3:148–149, 3:153–158
Coleoptera, 3:319–320, 3:327–333
crickets, 3:204–206
diplurans, 3:107, 3:110–111
Diptera, 3:359–360, 3:364–367, 3:369–374
dragonflies, 3:46, 3:133–134, 3:*135*
earwigs, 3:196, 3:199–200
fleas, 3:349, 3:352–354
flies, 3:60, 3:359–360
grasshoppers, 3:59–60
Hemiptera, 3:261, 3:267–279
Hymenoptera, 3:406, 3:411–424
lacewings, 3:307, 3:311–314
Lepidoptera, 3:*385*–386, 3:393–404
mantids, 3:41, 3:65, 3:180, 3:183–186
Mantophasmatodea, 3:219
mayflies, 3:60, 3:126, 3:129–130
Mecoptera, 3:343, 3:345–346
Megaloptera, 3:290, 3:293–294
mosquitos, 3:59, 3:359
moths, 3:60, 3:65, 3:385–386
Odonata, 3:133–134, 3:*135*, 3:137–138
Orthoptera, 3:204–206, 3:211–216
Phasmida, 3:222–223, 3:228–232
Phthiraptera, 3:251, 3:254–256
proturans, 3:93, 3:96–97
rock-crawlers, 3:189, 3:192–193
snakeflies, 3:298, 3:300–302
springtails, 3:100, 3:103–104
stoneflies, 3:142, 3:145–146
strepsipterans, 3:336, 3:338–339
termites, 3:164–166, 3:*166*, 3:170–175
thrips, 3:282, 3:285–287
Thysanura, 3:119, 3:122–123
wasps, 3:59, 3:65
webspinners, 3:233–234, 3:236–237
zorapterans, 3:240, 3:241
lower metazoans and lesser deuterostomes, 1:27–30, **1:37–43**
acoels, 1:180, 1:181–182
anoplans, 1:246, 1:249–251
arrow worms, 1:42, 1:434, 1:437–441
box jellies, 1:148, 1:151–152
calcareous sponges, 1:58, 1:61–64
cnidarians, 1:37
Anthozoa, 1:29, 1:30, 1:32, 1:40, 1:42, 1:106, 1:111–120
Hydrozoa, 1:26, 1:127–128, 1:134–145
comb jellies, 1:42, 1:170, 1:173–177
demosponges, 1:79, 1:82–86
echinoderms
Crinoidea, 1:357, 1:361–364
Echinoidea, 1:40, 1:403–404, 1:408–415
Ophiuroidea, 1:388–389, 1:393–398
sea cucumbers, 1:42, 1:421, 1:425–431
sea daisies, 1:382, 1:385–386
sea stars, 1:369, 1:373–379
enoplans, 1:255, 1:256–257
entoprocts, 1:319, 1:323–325
flatworms, 1:42, 1:43
monogeneans, 1:38–40, 1:214, 1:218, 1:220–223
tapeworms, 1:230, 1:234–242
free-living flatworms, 1:40, 1:187, 1:190–195
gastrotrichs, 1:270, 1:272–273
girdle wearers, 1:345, 1:347–349
glass sponges, 1:68, 1:71–75
gnathostomulids, 1:332, 1:334–335
hair worms, 1:305–306, 1:308–309
hemichordates, 1:444, 1:447–450
jaw animals, 1:328–329
jellyfish, 1:24, 1:30, 1:32, 1:38–39, 1:42, 1:155, 1:160–167
kinorhynchs, 1:276, 1:279–280
lancelets, 1:487–488, 1:490–497
larvaceans, 1:473–474, 1:476–478
nematodes, 1:43
roundworms, 1:285, 1:287–291
secernenteans, 1:294, 1:297, 1:299–304
Orthonectida, 1:99, 1:101–102
placozoans, 1:88
priapulans, 1:337, 1:340–341
rhombozoans, 1:93, 1:96–98
rotifers, 1:38, 1:43, 1:261, 1:264–267
Salinella salve, 1:91
salps, 1:467, 1:471–472
sea squirts, 1:454, 1:458–466
sorberaceans, 1:480, 1:482–483
thorny headed worms, 1:312, 1:314–316
Trematoda, 1:37, 1:41, 1:199–200, 1:203–211
wheel wearers, 1:352
mammals, 12:9–10, **12:140–148**
aardvarks, 15:156–157
agoutis, 16:408, 16:411–414, 16:415*t*
anteaters, 13:172–173, 13:177–179
armadillos, 12:71, 13:183–184, 13:187–190, 13:190*t*–192*t*
Artiodactyla, 15:269–271
aye-ayes, 14:86–87
baijis, 15:20
bandicoots, 13:3
dry-country, 13:10–11, 13:14–16, 13:16*t*–17*t*
rainforest, 13:10–11, 13:14–16, 13:16*t*–17*t*
bats, 13:312–313
American leaf-nosed bats, 13:416–417, 13:423–432, 13:433*t*–434*t*
bulldog bats, 13:446, 13:449–450
disk-winged, 13:474, 13:476–477
Emballonuridae, 13:356–357, 13:360–363, 13:363*t*–364*t*
false vampire, 13:381, 13:384–385
funnel-eared, 13:460, 13:463–465
horseshoe, 13:390, 13:396–400
Kitti's hog-nosed, 13:368
Molossidae, 13:484–485, 13:490–493, 13:493*t*–495*t*

mouse-tailed, 13:351, 13:353
moustached, 13:435–436, 13:440–441, 13:442*t*
New Zealand short-tailed, 13:454, 13:457–458
Old World fruit, 13:312, 13:319–320, 13:325–330, 13:331*t*–332*t*, 13:335, 13:340–347, 13:348*t*–349*t*
Old World leaf-nosed, 13:402, 13:405, 13:407–409, 13:409*t*–410*t*
Old World sucker-footed, 13:480
slit-faced, 13:371, 13:373, 13:375, 13:376*t*–377*t*
smoky, 13:468, 13:470–471
Vespertilionidae, 13:498–500, 13:506–514, 13:515*t*–516*t*, 13:521, 13:524–525, 13:526*t*
bears, 14:258, 14:297–298, 14:302–305, 14:306*t*
beavers, 16:179, 16:183–184
bilbies, 13:20–21
botos, 15:28–29
Bovidae, 16:5–6
Antilopinae, 16:46, 16:48, 16:50–55, 16:56*t*–57*t*
Bovinae, 16:13–14, 16:18–23, 16:24*t*–25*t*
Caprinae, 16:92, 16:98–103, 16:103*t*–104*t*
duikers, 16:74, 16:80–84, 16:84*t*–85*t*
Hippotraginae, 16:30–31, 16:37–42, 16:42*t*–43*t*
Neotraginae, 16:60–62, 16:66–71, 16:71*t*
bushbabies, 14:25–26, 14:28–32, 14:32*t*–33*t*
Camelidae, 15:316–317, 15:320–323
Canidae, 12:148, 14:267–269, 14:276–283, 14:283*t*–284*t*
capybaras, 16:402–*404*
Carnivora, 12:117, 12:141, 12:145, 14:257–260
cats, 12:145, 14:371–372, 14:379–389, 14:390*t*–391*t*
Caviidae, 16:391–392, 16:395–398, 16:399*t*
Cetacea, 12:86, 15:6–7
Balaenidae, 15:109–110, 15:115–117
beaked whales, 15:61, 15:64–68, 15:69*t*–70*t*
dolphins, 15:5, 15:44–46, 15:53–55, 15:56*t*–57*t*
franciscana dolphins, 15:24
Ganges and Indus dolphins, 15:14
gray whales, 15:95–97
Monodontidae, 15:84–85, 15:90
porpoises, 15:35, 15:38–39, 15:39*t*
pygmy right whales, 15:104
rorquals, 15:6, 15:7, 15:122–123, 15:127–130, 15:130*t*
sperm whales, 15:74–75, 15:79–80
chevrotains, 15:327–328, 15:331–333
Chinchillidae, 16:378–379, 16:382–383, 16:383*t*
colugos, 13:302, 13:304
coypus, 16:474–476
cultural, 12:157–159
dasyurids, 12:289, 12:294–298, 12:299*t*–301*t*
Dasyuromorphia, 12:280–281
deer
Chinese water, 15:374–375
muntjacs, 15:344–345, 15:349–354
musk, 15:336–337, 15:340–341
New World, 15:383, 15:388–396*t*
Old World, 15:359, 15:364–371*t*
Dipodidae, 16:213–215, 16:219–222, 16:223*t*–224*t*
Diprotodontia, 13:35–36
domestication and, 12:174–175
dormice, 16:318–319, 16:323–326, 16:327*t*–328*t*
duck-billed platypuses, 12:230–231, 12:244–245
Dugongidae, 15:200, 15:203
echidnas, 12:237, 12:240*t*
elephants, 15:167–169, 15:174–175
Equidae, 15:226–228, 15:226*t*, 15:232–235
Erinaceidae, 13:204–205, 13:209–213, 13:212*t*–213*t*
giant hutias, 16:469, 16:471*t*
gibbons, 14:210–211, 14:218–222
Giraffidae, 12:167, 15:401–403, 15:408
great apes, 14:228–232, 14:237–240
gundis, 16:312–313, 16:314–315
Herpestidae, 14:349–351, 14:354–356, 14:357*t*
Heteromyidae, 16:200–202, 16:205–209, 16:208*t*–209*t*
hippopotamuses, 15:306, 15:310–311
humans, 14:249–252, 14:*251*, 14:*252*
hutias, 16:461–462, 16:467*t*
Hyaenidae, 14:362, 14:364–367
hyraxes, 15:180–181, 15:186–188, 15:189*t*
Indriidae, 14:*64*, 14:65–66, 14:69–72
Insectivora, 13:197–198
koalas, 13:45–46
Lagomorpha, 16:482–483
lemurs, 14:6, 14:7, 14:8, 14:49, 14:50–51, 14:55–60
Cheirogaleidae, 14:36–37, 14:41–44
sportive, 14:8, 14:75–76, 14:79–83
Leporidae, 16:506–507, 16:511–515, 16:515*t*–*516*t
Lorisidae, 14:14–15, 14:18–20, 14:21*t*
luxury organs and, 12:22
Macropodidae, 12:145, 13:36, 13:86–87, 13:93–100, 13:101*t*–102*t*
manatees, 15:206, 15:211–212
Megalonychidae, 13:156–157, 13:159
moles
golden, 13:198, 13:216–217, 13:220–222, 13:222*t*
marsupial, 13:26–27, 13:28
monitos del monte, 12:274
monkeys
Atelidae, 14:157, 14:162–166, 14:166*t*–167*t*
Callitrichidae, 14:116–120, 14:*123*, 14:127–131, 14:132*t*
Cebidae, 14:103–104, 14:108–112
cheek-pouched, 14:191–192, 14:197–204, 14:204*t*–205*t*
leaf-monkeys, 14:7, 14:174, 14:178–183, 14:184*t*–185*t*
night, 14:137, 14:141*t*
Pitheciidae, 14:145–146, 14:150–153, 14:153*t*
monotremes, 12:230–231
mountain beavers, 16:132
Muridae, 16:284, 16:288–295, 16:296*t*–297*t*
Arvicolinae, 16:226–227, 16:233–238, 16:237*t*–238*t*
hamsters, 16:240–241, 16:245–246, 16:247*t*
Murinae, 16:251, 16:255–260, 16:261*t*–262*t*
Sigmodontinae, 16:267–268, 16:272–276, 16:277*t*–278*t*
musky rat-kangaroos, 13:70–71
Mustelidae, 12:145, 12:*186*, 14:258–259, 14:321–323, 14:327–331, 14:332*t*–333*t*
numbats, 12:304
octodonts, 16:435, 16:438–440, 16:440*t*
opossums
New World, 12:251–252, 12:257–263, 12:264*t*–*12:265*t
shrew, 12:268, 12:270
Otariidae, 14:396–397, 14:402–406
pacaranas, 16:387
pacas, 16:418, 16:*419*, 16:423–424
pangolins, 16:110–112, 16:115–120
peccaries, 15:293–294, 15:298–300
Perissodactyla, 15:218–219
Petauridae, 13:126–127, 13:131–133, 13:133*t*
Phalangeridae, 13:60, 13:*63*–*66t*
pigs, 15:278, 15:284–288, 15:289*t*–290*t*
pikas, 16:491–493, 16:497–501, 16:501*t*–502*t*
pocket gophers, 16:188–189, 16:192–194, 16:195*t*–197*t*
porcupines
New World, 16:367–368, 16:371–373, 16:374*t*
Old World, 16:352–353, 16:357–364
possums
feather-tailed, 13:141, 13:144
honey, 13:136–137
pygmy, 13:106–107, 13:110–111
primates, 14:6–8
Procyonidae, 14:310, 14:313–315, 14:315*t*–316*t*
pronghorns, 15:412–414
Pseudocheiridae, 13:115–116, 13:119–123, 13:122*t*–123*t*
rat-kangaroos, 13:74–75, 13:78–81
rats
African mole-rats, 12:71, 12:74, 12:76, 12:83, 12:147, 16:*340*, 16:343, 16:347–349, 16:349*t*–350*t*
cane, 16:334, 16:337–338
chinchilla, 16:444, 16:447–448
dassie, 16:*331*
spiny, 16:450, 16:453–457, 16:458*t*
rhinoceroses, 15:253, 15:257–261
rodents, 16:124–125
sengis, 16:519–521, 16:524–530, 16:531*t*
shrews
red-toothed, 13:248–250, 13:255–263, 13:263*t*
West Indian, 13:244
white-toothed, 13:267, 13:271–275, 13:276*t*–277*t*
Sirenia, 15:192–193
solenodons, 13:238–239, 13:240–241
springhares, 16:308–309
squirrels
flying, 16:136, 16:139–140, 16:141*t*–142*t*
ground, 12:*71*, 12:141, 16:124, 16:144–145, 16:151–158, 16:158*t*–160*t*

scaly-tailed, 16:300, 16:302–305
tree, 16:165–166, 16:169–174, 16:174t–175t
Talpidae, 13:280–281, 13:284–287, 13:287t–288t
tapirs, 15:239–240, 15:245–247
tarsiers, 14:94, 14:97–99
tenrecs, 13:227–228, 13:232–234, 13:234t
three-toed tree sloths, 13:163, 13:167–169
tree shrews, 13:291, 13:294–296, 13:297t–298t
true seals, 12:85, 12:87, 14:418–419, 14:427–428, 14:430–434, 14:435t
tuco-tucos, 16:426–427, 16:429–430, 16:430t–431t
ungulates, 15:142
Viverridae, 14:337, 14:340–343, 14:344t–345t
walruses, 14:412–414
wombats, 13:52, 13:55–56
Xenartha, 13:150–151
protostomes, 2:33, **2:35–40**
amphionids, 2:196
amphipods, 2:261, 2:265–271
anaspidaceans, 2:182, 2:183
aplacophorans, 2:380, 2:382–385
Arachnida, 2:37, 2:335, 2:339–352
articulate lampshells, 2:522, 2:525–527
bathynellaceans, 2:178, 2:179
beard worms, 2:86, 2:88
bivalves, 2:452, 2:457–466
caenogastropods, 2:446, 2:449
centipedes, 2:354–355, 2:358–362
cephalocarids, 2:132, 2:133
Cephalopoda, 2:477–478, 2:482–488
chitons, 2:395, 2:397–400
clam shrimps, 2:148, 2:150–151
copepods, 2:300, 2:304–310
Crustacea, 2:32
cumaceans, 2:229–230, 2:232
Decapoda, 2:32, 2:33–34, 2:199–200, 2:204–213
deep-sea limpets, 2:436, 2:437
earthworms, 2:43, 2:66, 2:70–73
echiurans, 2:104, 2:106–107
fairy shrimps, 2:136–137, 2:139
fish lice, 2:289–290, 2:292–293
freshwater bryozoans, 2:497, 2:500–502
horseshoe crabs, 2:*328*, 2:331–332
Isopoda, 2:250–251, 2:254–259
krill, 2:186, 2:189–192
leeches, 2:76, 2:80–82
leptostracans, 2:162, 2:164–165
lophogastrids, 2:226, 2:227
mantis shrimps, 2:169–170, 2:173–175
marine bryozoans, 2:504, 2:506–508, 2:509, 2:513–514
mictaceans, 2:241, 2:242
millipedes, 2:364–365, 2:367–370
monoplacophorans, 2:388, 2:390–391
mussel shrimps, 2:312, 2:314–315
mysids, 2:216, 2:219–222
mystacocarids, 2:295, 2:297
myzostomids, 2:60, 2:62
Neritopsina, 2:440, 2:442–443
nonarticulate lampshells, 2:516, 2:518–519
Onychophora, 2:37, 2:110, 2:113–114
pauropods, 2:376, 2:377
peanut worms, 2:98, 2:100–101
phoronids, 2:492, 2:494
Polychaeta, 2:32, 2:46, 2:50–56
Pulmonata, 2:*5*, 2:414, 2:417–421
remipedes, 2:125–126, 2:128
sea slugs, 2:403–404, 2:407–409
sea spiders, 2:321, 2:324–325
spelaeogriphaceans, 2:243, 2:244
symphylans, 2:371–372, 2:373
tadpole shrimps, 2:142, 2:145–146
tanaids, 2:236, 2:238–239
tantulocaridans, 2:284, 2:286
Thecostraca, 2:28, 2:275, 2:278–281
thermosbaenaceans, 2:245, 2:247
tongue worms, 2:33, 2:319–320
true limpets, 2:424, 2:426–427
tusk shells, 2:470, 2:473–474
Vestimentifera, 2:92, 2:94
Vetigastropoda, 2:430, 2:433–434
water bears, 2:117, 2:120–123
water fleas, 2:153–154, 2:157–158
reptiles, **7:34–46**
activity patterns, 7:43–45
African burrowing snakes, 7:462–464
African sideneck turtles, 7:129, 7:132–133
Afro-American river turtles, 7:137, 7:140–141
Agamidae, 7:211, 7:214–221
Alligatoridae, 7:172–173, 7:175–177
Anguidae, 7:340, 7:342–344
Australo-American sideneck turtles, 7:78, 7:81–83
big-headed turtles, 7:135
blindskinks, 7:272
blindsnakes, 7:381–385
boas, 7:411, 7:413–417
Central American river turtles, 7:100
chameleons, 7:228–230, 7:235–241
chemosensory systems and, 7:34–38
colubrids, 7:468–469, 7:473–481
Cordylidae, 7:320–321, 7:324–325
Crocodylidae, 7:180, 7:184–187
early blindsnakes, 7:371
Elapidae, 7:485, 7:491–498
false blindsnakes, 7:388
false coral snakes, 7:400
file snakes, 7:441
Florida wormlizards, 7:284
Gekkonidae, 7:261, 7:265–269
Geoemydidae, 7:116, 7:118–119
gharials, 7:168
Helodermatidae, 7:354–355
Iguanidae, 7:244–*245*, 7:250–257
Kinosternidae, 7:121–122, 7:125–126
knob-scaled lizards, 7:348, 7:351
Lacertidae, 7:298, 7:301–302
learning, 7:39–40
leatherback seaturtles, 7:101
Microteiids, 7:304, 7:306–307
mole-limbed wormlizards, 7:280, 7:281
Neotropical sunbeam snakes, 7:406
New World pond turtles, 7:105–106, 7:109–113
night lizards, 7:292, 7:294–295
pig-nose turtles, 7:76
pipe snakes, 7:396
play, 7:39
pythons, 7:420–421, 7:424–428
seaturtles, 7:*86*, 7:89–91
shieldtail snakes, 7:392–394
skinks, 7:329, 7:332–337
slender blindsnakes, 7:375–377
snapping turtles, 7:93, 7:95–96
softshell turtles, 7:152, 7:154–155
spade-headed wormlizards, 7:288, 7:289
splitjaw snakes, 7:430
Squamata, 7:204–206
sunbeam snakes, 7:402
tactile cues, 7:40–42
Teiidae, 7:311, 7:314–316
tortoises, 7:143, 7:146–148
Tropidophiidae, 7:434, 7:436–437
tuatara, 7:190
Varanidae, 7:360–361, 7:365–368
Viperidae, 7:447, 7:451–459
vision and, 7:38–39
wormlizards, 7:274, 7:276
See also Social behavior
Behavior modification. *See* Learning
Behavioral ecology, 12:148
Behavioral fevers, 7:45
Being John Malkovich (movie), 8:22
Beintema, J. J., 16:311
Beiras, **16:59–72,** 16:63, 16:71t
Beisa oryx, 16:32
Belalangs, 5:*168*, 5:*172*–173
Belanger's tree shrews, 13:292, 13:297t
Belding's ground squirrels, 16:124, 16:*147*, 16:160t
Belding's sparrows. *See* Savannah sparrows
Belding's yellowthroats, 11:291
Belford's honeyeaters. *See* Belford's melidectes
Belford's melidectes, 11:*239*, 11:251–*252*
Belidae, 3:55
Belkin's dune tabanid flies, 3:360
Bell geckos, 7:262
Bell-Jarman principle, 12:119
Bell miners, 11:238, 11:*240*, 11:*243*–244
Bellbirds. *See* Bell miners; Cotingas; Kokakos
Bell's dab lizards. *See* Spiny-tailed agamas
Bell's false brook salamanders. *See* Bell's salamanders
Bell's salamanders, 6:*393*, 6:402–*403*
Bell's vireos, 11:*257*, 11:*258*–*259*
Belomys pearsonii. See Hairy-footed flying squirrels
Belonidae. *See* Needlefishes
Beloniformes, 4:12, **5:79–87,** 5:*82*
Belonion apodion, 5:80
Belonogaster rufipennis, 9:470
Belontia spp. *See* Belontiinae
Belontia signata, 5:427, 5:429
Belontiinae, 5:427
Belostoma spp., 3:*62*
Belostomatidae, 3:44
Belted kingfishers, 10:2, 10:8, 10:*10*, 10:*11*, 10:*22*–*23*
Beluga sturgeons, 4:*216*, 4:*219*
Belugas, **15:81–91,** 15:*82*, 15:*83*, 15:*86*, 15:*89*, 15:*90*
behavior, 15:84–85
conservation status, 15:9, 15:86–87
distribution, 15:*81*, 15:82–83
echolocation, 12:*85*, 12:87
evolution, 15:81
feeding ecology, 15:85
habitats, 15:83–84, 15:*87*
humans and, 15:10–11, 15:87–88
physical characteristics, 15:81–82, 15:*85*, 15:*88*
reproduction, 15:85–86
taxonomy, 15:81

Bembridae. *See* Deepwater flatheads
Bemis, W. E., 4:229
Beneden, Édouard van, 1:93
Beneden, P. J. van, 1:31
Beneficial insects, 3:78–82
Bengal cats, 14:*372*
Bengal floricans, 9:3, 9:94, 9:*95*, 9:98–*99*
Bengal monitor lizards, 7:52, 7:*360*, 7:361
Bengal pittas. *See* Indian pittas
Bengal slow lorises, 14:16, 14:21*t*
Bengali water mongooses, 14:349, 14:352
Beni titis, 14:148
Bennett's cassowaries, 8:*78*, 8:79–*80*
Bennett's chinchilla rats, 16:443, 16:444, 16:*446*, 16:*447*
Bennett's owlet-nightjars. *See* Barred owlet-nightjars
Bennett's tree kangaroos, 13:*86*, 13:*91*, 13:*99*–100
Bennett's wallabies. *See* Red-necked wallabies
Bennett's woodpeckers, 10:*153*, 10:*158*
Bensch's rails. *See* Subdesert mesites
Bensonomys spp., 16:264
Bentham, Jeremy, 1:47
Bentheuphausiidae, 2:185
Benthic fishes, 4:47, 4:48, 4:49, 4:62, 4:64, 4:65
See also specific fish
Benthobatis spp., 4:176
Benthomisophria palliata, 2:*302*, 2:308–309
Benthopectinidae, 1:368
Benthopelagic fishes, 4:47–49
See also specific fish
Benthosema spp., 4:443
Benthosema fibulatum. *See* Brooch lanternfishes
Benthosema pterotum. *See* Skinnycheek lanternfishes
Berardius spp., 15:59
Berardius arnouxii. *See* Arnoux's beaked whales
Berardius bairdii. *See* Baird's beaked whales
Berberentulus huisunensis, 3:*95*, 3:96, 3:*97*
Bereis' treefrogs. *See* Hourglass treefrogs
Beremendiini, 13:247
Bergh, Henry, 14:293
Bergmann's rule, 8:5, 12:21, 14:245
Bering Sea beaked whales. *See* Stejneger's beaked whales
Berlese-type funnels, 3:94
Berlioz's silver pheasants, 8:433
Bermuda blackbirds. *See* Gray catbirds
Bermuda petrels, 8:131, 8:*131*
Bermudamysis speluncola, 2:217
Bern Convention of European Wildlife and Natural Habitats. *See* Convention on the Conservation of European Wildlife and Natural Habitats
Bernaks. *See* Emerald notothens
Bernard's hermit crabs. *See* Common hermit crabs
Bernard's wallaroos. *See* Black wallaroos
Bernatzik, H. A., 8:226
Bernier's vangas, 10:439, 10:441
Bernoulli lift, 12:56
Beroe spp., 1:28, 1:171
Beroe cucumis, 1:28
Beroe forskalii, 1:*172*, 1:*173*
Beroe ovata, 1:171
Beroida, 1:169
Berothidae, 3:305
Berra, Tim M., 4:57, 4:59
Berrypeckers, 11:189–191, 11:*192*
Berthelot's pipits, 10:374, 10:375, 10:*377*, 10:*382*
Berycidae, 5:113
Beryciformes, **5:113–122,** 5:*117*
Bess beetles, 3:319, 3:322
Betsileo golden frogs, 6:*294*, 6:295–*296*
Betsileo poison frogs. *See* Betsileo golden frogs
Betsileo reed frogs, 6:*284*, 6:286
Betta spp., 5:427, 5:428, 5:429
Betta burdigala, 5:427
Betta chloropharynx, 5:427
Betta livida, 5:429
Betta macrostoma. *See* Brunei beauties
Betta miniopinna, 5:427, 5:429
Betta persephone, 5:429
Betta schalleri, 5:427
Betta spilotogena, 5:429
Betta splendens. *See* Siamese fighting fishes
Bettas. *See* Siamese fighting fishes
Bettongia spp. *See* Bettongs
Bettongia gaimardi. *See* Tasmanian bettongs
Bettongia lesueur. *See* Boodies
Bettongia penicillata. *See* Brush-tailed bettongs
Bettongia tropica. *See* Northern bettongs
Bettongs, 13:34, 13:73
brush-tailed, 13:39, 13:74, 13:*75*, 13:76, 13:77, 13:*78*–79
northern, 13:74, 13:76, 13:77, 13:*79*
rufous, 13:*74*, 13:*75*, 13:*76*, 13:77, 13:79, 13:80–81
Tasmanian, 13:*73*, 13:74, 13:76, 13:77, 13:*78*
Bewick's wrens, 10:527, 10:528, 10:529
Bezoar goats, 12:179
Bharals. *See* Blue sheep
Biak monarchs, 11:*99*, 11:*101*–102
Bias musicus. *See* Black-and-white flycatchers
Biases, in subterranean mammal research, 12:70–71
Biatas nigropectus. *See* White-bearded antshrikes
Bibimys spp., 16:266
Bibionidae, 3:359–360
Bibron, G., 7:379
Bibron's burrowing asps. *See* Southern burrowing asps
Bibron's geckos, 7:261
Bicheno finches. *See* Double-barred finches
Bichirs, 4:11, 4:15, 4:24, 4:57, **4:209–212,** 4:*210*, 4:*211*
Bichon frises, 14:289
Bicolor anthias, 4:*33*
Bicolor parrotfishes, 4:*4*
Bicolor-spined porcupines, 16:365, 16:374*t*
Bicolored antpittas, 10:242
Bicolored mouse-warblers, 11:*58*, 11:*62*–63
Bicolored shrews, 13:*266*
Bicornis hornbills, 8:*10*
Bicornuate uterus, 12:90–91
Bidder's organ, 6:183
Bidyanus spp., 5:220
Bidyanus bidyanus. *See* Silver perches
Bifasciated larks. *See* Greater hoopoe-larks
Bifax spp., 5:42
Bifax lacinia, 5:41
Bifurcating flatworms, 1:*20*
Big-bellied seahorses, 5:*133*
Big black horse flies, 3:*363*, 3:*372*
Big brown bats, 13:*499*, 13:*505*, 13:507–*508*
distribution, 13:311, 13:498
echolocation, 12:87
reproduction, 13:315, 13:500–501
smell, 12:85
Big cats, 12:136
See also Cheetahs; Leopards; Lions
Big crested mastiff bats, 13:495*t*
Big cypress fox squirrels, 16:*164*
Big-eared bats
Allen's, 13:*505*, 13:*506*, 13:509
common, 13:433*t*
Big-eared climbing rats, 16:277*t*
Big-eared flying foxes, 13:*324*, 13:*329*, 13:330
Big-eared flying mice, 16:300, 16:*301*, 16:*304*–305
Big-eared forest treefrogs, 6:*284*, 6:288–289
Big-eared free-tailed bats. *See* Giant mastiff bats
Big-eared mastiff bats, 13:495*t*
Big-eye barenoses. *See* Humpnose big-eye breams
Big-eye breams. *See* Humpnose big-eye breams
Big-eye frogfishes, 5:*52*, 5:*53*
Big-footed bats, Rickett's, 13:313
Big-headed ants, 3:72
Big-headed turtles, 7:32, **7:135–*136*,** 7:*139*, 7:*141*–142
Big mouth gulper eels. *See* Pelican eels
Big naked-backed bats, 13:435, 13:442*t*
Big pocket gophers, 16:196*t*
Big-thumbed treefrogs. *See* Eiffinger's Asian treefrogs
Big white cranes. *See* Whooping cranes
Bigelow, H. B., 5:56
Bigeye herrings. *See* Alewives
Bigeye tunas, 4:*26*, 5:407
Bigeyed sixgill sharks, 4:144
Bigeyes, 4:48, 4:67, 5:255, 5:256, 5:258–263
Bighead carps, 4:299–300, 4:303
Bighorn sheep, 12:178, 16:97, 16:99, 16:102
behavior, 12:*141*, 16:5, 16:*92*
conservation status, 16:95
distribution, 12:131, 16:91, 16:*92*
evolution, 16:87
feeding ecology, 16:93
habitats, 15:269
Bigmouth buffaloes, 4:*322*, 4:*330*
Bigscales, 5:105, 5:106
Bilateria, 1:3–6, 1:7–11, 2:4, 2:10–11, 2:13, 2:36
Bilbies, 12:137, 13:2, 13:3, 13:5, **13:19–23,** 13:*20*
Bilenca, D. N, 16:267
Billfishes. *See* Longnose gars
Bills and beaks, 8:3, 8:*5*, 8:*10*
larks, 10:342, 10:*345*
pelicans, 8:183
toucan, 10:126, 10:*127*
See also Physical characteristics
Binaural hearing, 12:81
Bini free-tailed bats, 13:494*t*
Binocularity, 12:79
See also Physical characteristics
Binturongs, 14:335, 14:336, 14:*339*, 14:*342*–343
Biochemical analysis, 8:27
Biochemistry, 2:4
Biodiversity, 1:48–53, 2:29, 12:11–15
conservation biology and, 12:216–225
indicators, 1:49, 3:82

of protostomes, 2:25
surveys, 12:194–198
threats to, 12:213
zoo management and, 12:206–209
See also Conservation status
Biodiversity Action Plan (U.K.), 2:56
Biogenetic law, 1:7
Biogeography, 3:84
amphibian, 6:*4–5*
birds, 8:*17*–18
fishes, **4:52–59,** 4:55*t*
insects, **3:52–58**
mammals, **12:129–139,** 12:*131,* 12:220
Squamata, 7:203–204
Bioindicators. *See* Biodiversity, indicators
Bioinformatics, 2:4–5
See also Taxonomy
Bioko Allen's bushbabies, 14:25, 14:26, 14:32*t*
Biological control, 3:77, 3:80, 12:193
Apoanagyrus lopezi and, 3:415
beetles and, 3:323
lacewings and, 3:309
mongooses for, 12:190–191
rabbits, 12:187–188
Biological diversity. *See* Biodiversity
Biological hotspots, 12:218
Biology, conservation, 1:48, 12:216–225
The Biology of Millipedes, 2:364
Bioluminescence, 1:24, 1:31, 4:392, 4:422, 4:441
Biomes, 3:56–57
See also Habitats
Bioparks, 12:205
Biotic environments, 1:24
Bioturbation, 2:43
See also Behavior
Bipalium pennsylvanicum. See Land planarians
Bipedidae. *See* Mole-limbed wormlizards
Bipes spp. *See* Mole-limbed wormlizards
Bipes biporus. See Two-legged wormlizards
Bipes canaliculatus. See Tropical wormlizards
Bipes tridactylus, 7:280
Bipulvina spp., 2:387
Birch mice, **16:211–224,** 16:*218,* 16:223*t*–224*t*
Bird fleas, 3:*347,* 3:348
Bird flight. *See* Flight
Bird snakes, 7:466
Bird songs. *See* Songs, bird
Bird Watcher's Digest, 8:27
Birdbeak burrfishes, 4:*468*
Birder's World (journal), 8:27
Birding, 8:22–23, 8:27
Birding (journal), 8:27
BirdLife International
on Abbott's booby, 8:217
on Apodiformes, 9:418
on babblers, 10:507
on biak monarchs, 11:101
on fantails, 11:87
on introduced species, 11:508
on owls, 9:333, 9:351
on pelicans, 8:230
on pittas, 10:195
on rail-babblers, 11:79
on storks, 8:267
on weavers, 11:378
Birds
vs. bats, 12:58–59
behavior, 8:7, 8:17–18
biogeography, 8:*17*–18
defined, **8:3–18**
digestion, 8:10–11
distribution, 8:*17*–18
ecology, 8:15–17
evolution, 8:3
excretion, 8:10–11
feathers, 8:*12,* 8:*14,* 8:*16*
fledgling care, 8:14–15
flight, 8:*11,* **8:45–57**
humans and, **8:19–28,** 8:*22,* 8:*23,* 8:*25,* 12:184
lice and, 3:249, 3:250, 3:251
life history, 8:11
mammal predation and, 12:189–191
vs. mammals, 12:6
migration, 8:23, **8:29–35,** 8:42, 9:450, 11:4–5
physical characteristics, 8:1–9, 8:*5,* 8:*6,* 8:*15*
physiology, 8:9–11, 8:*13*
reproduction, 8:*4,* 8:7, 8:11–14
See also specific birds and bird groups
The Birds (Aristotle), 10:63
The Birds (movie), 8:22
Bird's nest corals. *See* Cauliflower corals
Bird's nest sponges, 1:*70,* 1:72–73
Birds of Africa from Seabirds to Seed-Eaters (Stuarts), 8:128
Birds of Hawaii (Munro), 11:350
Birds of paradise, **11:489–502,** 11:*493–494*
behavior, 11:491
conservation status, 11:492
distribution, 11:*489,* 11:490
evolution, 11:489
feeding ecology, 11:*490,* 11:491
habitats, 11:491
humans and, 11:492
physical characteristics, 11:489–490
reproduction, 11:491–492
species of, 11:*495–502*
taxonomy, 11:477, 11:489
Birds of the World (Austin), 10:506
Birdwatching, 8:37–38
Birdwing butterflies, 3:84, 3:88, 3:90, 3:389
giant, 3:388
Queen Alexandra's, 3:389, 3:*390,* 3:*396,* 3:398–399
Birth control, in zoos, 12:207
The Birth of Venus, 2:42
Biryong rock-crawlers, 3:*191,* 3:*192,* 3:193
Bisexual reproduction, 4:29
See also Reproduction
Bishop-birds, 11:376, 11:*381*
Bishop's fossorial spiny rats, 16:450, 16:458*t*
Bishop's oo, 11:*239,* 11:*250*
Bismarck blindsnakes, 7:380, 7:382, 7:*384*–385
Bismarck woodswallows, 11:459
Bison, 12:*50,* 12:*104,* 12:*130,* 15:263, **16:11–25,** 16:*17*
behavior, 12:*146,* 15:271, 16:5, 16:13–14
conservation status, 16:14
distribution, 12:131, 16:*11,* 16:13
evolution, 15:138, 16:11
feeding ecology, 16:14, 16:24*t*–25*t*
habitats, 16:13, 16:24*t*–25*t*
humans and, 12:139, 16:8, 16:15
migrations, 12:164, 12:168
physical characteristics, 15:140, 15:267, 16:11–13
reproduction, 12:*104,* 15:272, 16:14
species of, 16:18–*25t*
taxonomy, 16:11
in zoos, 12:210
See also specific types of bison
Bison spp. *See* Bison
Bison bison. See American bison
Bison bonasus. See Wisents
Biswamoyopterus biswasi. See Namdapha flying squirrels
Bitia spp. *See* Keel-bellied watersnakes
Bitia hydroides. See Keel-bellied watersnakes
Biting flies, 3:43
Biting lice, 3:249
See also Chewing lice
Bitis spp., 7:445, 7:446, 7:448
Bitis arietans. See African puff adders
Bitis gabonica. See Gaboon adders
Bitis nasicornis, 7:447
Bitis peringueyi, 7:447
Bittacidae. *See* Hangingflies
Bitterlings, 4:298–301, 4:303, 4:*305,* 4:*311,* 4:317
Bitterns, 8:233–236, **8:239–260**
behavior, 8:242–243
conservation status, 8:244–245
distribution, 8:*239,* 8:241
evolution, 8:239
feeding ecology, 8:243–244
habitats, 8:241–242
humans and, 8:245
physical characteristics, 8:233–234, 8:239–241
reproduction, 8:244
species of, 8:*256–257*
taxonomy, 8:233, 8:239
Bivalves, **2:451–467,** 2:*455,* 2:*456*
behavior, 2:452
conservation status, 2:454
distribution, 2:452
evolution, 2:451
feeding ecology, 2:452–453
habitats, 2:452
humans and, 2:454
physical characteristics, 2:451–452, 2:*453*
reproduction, 2:24, 2:453–454
species of, 2:*457*–466
taxonomy, 2:451
Bivalvia. *See* Bivalves
Biwadrilidae, 2:65
Biziura lobata. See Musk ducks
Blaberidae, 3:147, 3:148, 3:149, 3:150
Blaberoidea, 3:147
Blaberus giganteus. See Death's head cockroaches
Black agoutis, 16:415*t*
See also Mexican black agoutis
Black-and-cinnamon fantails, 11:84
Black-and-crimson orioles. *See* Crimson-bellied orioles
Black-and-red broadbills, 10:*178,* 10:*180,* 10:*182*
Black-and-rufous sengis, 16:522, 16:524, 16:531*t*
Black-and-scarlet pittas. *See* Graceful pittas
Black-and-white bears. *See* Giant pandas
Black-and-white becards, 10:*273*
Black-and-white colobus, 14:7
Black-and-white fantails. *See* Willie wagtails
Black-and-white flycatchers, 11:*28,* 11:*41*

Black-and-white nuthatches. *See* Black-and-white warblers
Black-and-white shrike-flycatchers. *See* Black-and-white flycatchers
Black-and-white tegus, 7:*311*, 7:312
Black-and-white warblers, 11:*289*, 11:*292*, 11:*295*–296
Black-and-white–striped soapfishes. *See* Sixline soapfishes
Black-and-yellow broadbills, 10:179, 10:*180*, 10:*184*
Black-and-yellow silky flycatchers, 10:447
Black angelfishes. *See* Freshwater angelfishes
Black-backed butcherbirds, 11:472
Black-backed geese. *See* Comb ducks
Black-backed gulls, 9:*207*
Black-backed jackals. *See* Silverback jackals
Black-backed mouse-warblers. *See* Bicolored mouse-warblers
Black-backed orioles, 11:304
See also Baltimore orioles
Black-backed pittas. *See* Superb pittas
Black-backed squirrels. *See* Peters's squirrels
Black-bearded flying foxes. *See* Big-eared flying foxes
Black-bearded sakis. *See* Bearded sakis
Black-bearded tomb bats, 13:*357*, 13:364*t*
Black bears, 14:297, 14:298
See also American black bears; Asiatic black bears
Black bee-eaters, 10:*43*, 10:*45*
Black beetles. *See* Oriental cockroaches
Black-bellied bustards, 9:*92*
Black-bellied hamsters, 16:*244*
behavior, 16:245
conservation status, 16:245–246
distribution, 16:*245*
evolution, 16:239
feeding ecology, 16:*241*, 16:*242*, 16:245
habitats, 16:*240*, 16:245
humans and, 16:127, 16:243, 16:246
physical characteristics, 16:239, 16:245
reproduction, 16:245
taxonomy, 16:239, 16:245
Black-bellied plovers, 9:161, 9:162, 9:164
Black-bellied sandgrouse, 9:*234*, 9:*237*
Black-bellied storm-petrels, 8:137
Black-belted flowerpeckers, 11:191
Black-billed koels. *See* Common koels
Black-billed magpies, 11:507, 11:509
Black-billed scythebills. *See* Red-billed scythebills
Black-billed turacos, 9:300
Black-billed woodhoopoes, 10:65
Black breams. *See* Mountain perches; Sooty grunters
Black-breasted barbets, 10:115
Black-breasted buttonquails, 9:14, 9:*15*, 9:*18*–19
Black-breasted buzzard-kites. *See* Black-breasted buzzards
Black-breasted buzzards, 8:*328*, 8:*331–332*
Black-breasted hazel grouse. *See* Chinese grouse
Black-breasted kites. *See* Black-breasted buzzards
Black-breasted leaf turtles, 7:7, 7:*116*
Black-breasted pittas. *See* Gurney's pittas; Rainbow pittas
Black-breasted pufflegs, 9:450
Black-breasted sunbirds. *See* Olive-backed sunbirds (*Cinnyris* spp.)
Black-breasted trillers, 10:386, 10:387
Black-browed albatrosses. *See* black-browed mollymawks
Black-browed mollymawks, 8:114, 8:*115*, 8:116, 8:*117*, 8:120–121
Black-browed reed-warblers, 11:*6*
Black bulbuls, 10:397, 10:*400*, 10:*410*
Black butcherbirds, 11:468, 11:469
Black caimans, 7:157, 7:171, 7:173
Black-capped antwrens, 10:*244*, 10:248–*249*
Black-capped capuchins, 14:101, 14:102, 14:104–105, 14:106, 14:*107*, 14:110–*111*
Black-capped chickadees, 11:156, 11:*157*, 11:*159*, 11:*160*–161
Black-capped donacobius, 10:*530*, 10:537–*538*
Black-capped honeyeaters. *See* Strong-billed honeyeaters
Black-capped manakins, 10:297
Black-capped marmots, 16:*146*, 16:159*t*
Black-capped mockingthrushes. *See* Black-capped donacobius
Black-capped red siskins, 11:325, 11:339
Black-capped terns, 9:203
Black-capped tits. *See* Black-capped chickadees
Black-capped vireos, 11:*256*, 11:*257*, 11:*259*–260
Black caracaras, 8:348, 8:350
Black carps, 4:*303*
Black catbirds, 10:468
Black-cheeked robins. *See* Gray-headed robins
Black-chinned weavers, 11:379
Black cockatoos, 9:277–278, 9:279
Black cods. *See* Maori chiefs
Black-collared barbets, 10:*118*, 10:*120*–121
Black-collared bulbuls, 10:395, 10:*400*, 10:*412*
Black colobus, 14:174, 14:175, 14:184*t*
Black corals, 1:103, 1:107–108, 1:*109*, 1:115, 1:*116*
See also Antipatharia
Black cormorants. *See* Great cormorants
Black crappies, 5:*200*, 5:*202*, 5:205
Black creepers, 3:291, 3:*292*, 3:*293*, 3:294
Black crested gibbons, 14:207, 14:210, 14:*216*, 14:*220*, 14:221
Black-crowned barwings, 10:*509*, 10:*518*
Black-crowned cranes, 9:24–25
Black-crowned finch-larks. *See* Black-crowned sparrow-larks
Black-crowned night herons, 8:236, 8:*244*, 8:*246*, 8:254–*255*
Black-crowned pittas. *See* Graceful pittas
Black-crowned sparrow-larks, 10:*344*, 10:*345*, 10:346, 10:*346*, 10:*349*
Black-crowned tchagras, 10:426, 10:*431*, 10:*433*–434
Black cuckoo-shrikes, 10:386
Black currawongs. *See* Pied currawongs
Black dolphins, 15:56*t*
Black dragonfishes, 4:421, 4:*425*, 4:429
Black drongos, 11:*438*, 11:439, 11:*440*, 11:*444*–445
Black duikers, 16:*75*, 16:*78*, 16:80–*81*
Black-eared bushtits. *See* Bushtits
Black-eared catbirds, 11:478
Black-eared flying foxes. *See* Blyth's flying foxes
Black-eared miners, 11:238
Black-eared sparrow-larks, 10:342, 10:345
Black emus. *See* King Island emus
Black-eyed bulbuls. *See* Common bulbuls
Black-faced antbirds, 10:*251–252*
Black-faced antthrushes, 10:*243*, 10:*254*
Black-faced black spider monkeys. *See* Peruvian spider monkeys
Black-faced cuckoo-shrikes, 10:*387*
Black-faced honeycreepers. *See* Po'oulis
Black-faced honeyeaters. *See* Yellow-tufted honeyeaters
Black-faced impalas, 16:34
Black-faced kangaroos. *See* Western gray kangaroos
Black-faced lion tamarins, 14:11, 14:124, 14:132*t*
Black-faced parrotfinches. *See* Pink-billed parrotfinches
Black-faced sheathbills, 9:197, 9:198, 9:*200*
Black-faced spoonbills, 8:236, 8:293
Black-faced woodswallows, 11:*461*, 11:*462*
Black falcons, 8:347, 8:348
Black fantails, 11:84
Black fellows. *See* Black creepers
Black fins. *See* Lesser weevers
Black flies, 3:56, 3:76, 3:88, 3:360
See also Dawson River black flies
Black flying foxes, 12:*5*, 13:*320*, 13:*324*, 13:328–*329*
Black flying squirrels, 16:141*t*
Black-footed albatrosses, 8:113, 8:115, 8:116
Black-footed cats, 14:370, 14:390*t*
Black-footed ferrets, 12:214, 12:223
conservation status, 12:204, 14:262, 14:325
feeding ecology, 14:*261*, 14:323
habitats, 12:*115*, 14:319–320
Black-footed penguins. *See* Penguins
Black-footed squirrels. *See* Yucatán squirrels
Black four-eyed opossums, 12:265*t*
Black-fronted dotterels, 9:161
Black-fronted duikers, 16:84*t*
Black-fronted nunbirds, 10:103, 10:*105*, 10:109–*110*
Black-fronted titis, 14:148
Black ghosts, 4:369, 4:370, 4:*372*, 4:*373*
Black gibbons. *See* Black crested gibbons
Black grasswrens, 11:*47*, 11:*51–52*
Black grenadiers. *See* Roundnose grenadiers
Black grouse, 8:401
Black grubs. *See* Black-spot flatworms
Black guans, 8:*417*, 8:419–420
Black guillemots, 9:*223*, 9:*226*
Black guineafowl, 8:426–428
Black-handed spider monkeys. *See* Geoffroy's spider monkeys
Black-head minnows. *See* Fathead minnows
Black-headed bushmasters, 7:*450*, 7:*452*, 7:456
Black-headed ducks, 8:367, 8:372, 8:*375*, 8:*391*
Black-headed gulls. *See* Common black-headed gulls
Black-headed herons, 8:242
Black-headed ibises, 8:293
Black-headed nasute termites, 3:*169*, 3:173–*174*
Black-headed night monkeys, 14:135, 14:141*t*
Black-headed orioles. *See* Eastern black-headed orioles

Black-headed pardalotes. *See* Striated pardalotes
Black-headed penduline titmice, 11:147
Black-headed pittas. *See* Hooded pittas
Black-headed pythons, 7:*423*, 7:*424*, 7:425
Black-headed shrikes. *See* Long-tailed shrikes
Black-headed snakes, 7:466
Black-headed squirrel monkeys. *See* Bolivian squirrel monkeys
Black-headed tits, 11:149
Black-headed uacaris. *See* Black uakaris
Black-headed weavers. *See* Village weavers
Black herons, 8:*240*, 8:*247*, 8:*253*
Black honeyeaters, 11:237
Black-hooded antwrens, 10:242
Black-hooded magpie-shrikes, 11:468
Black-hooded orioles, 11:427
Black howler monkeys, 14:*159*, 14:166*t*
Black incas, 9:418
Black jays. *See* White-winged choughs
Black kangaroos. *See* Black wallaroos
Black katy chitons, 2:394, 2:*396*, 2:*399*–400
Black kites, 8:18, 8:320, 8:*328*, 8:*333*
Black lampshells, 2:*524*, 2:*525*
Black larks, 10:342, 10:*344*, 10:345
Black lechwes, 16:34
Black-legged kittiwakes, 9:*204*, 9:*210*, 9:*213*–214
Black-legged seriemas, 9:*87*, 9:88–*89*
Black lemurs, 14:*50*, 14:*53*, 14:*57*–58
Black-lipped pearl oysters, 2:*43*, 2:*455*, 2:*461*–*462*
Black-lipped pikas. *See* Plateau pikas
Black-lipped toads. *See* Common Sunda toads
Black long-spined urchins. *See* Long-spined sea urchins
Black macaws. *See* Palm cockatoos
Black macrotermes, 3:*169*, 3:*170*, 3:173
Black magpies. *See* White-winged choughs
Black mambas, 7:*490*, 7:*491*
Black mangabeys, 14:194, 14:205*t*
Black-mantled fairy bluebirds. *See* Philippine fairy bluebirds
Black-mantled tamarins, 14:118
Black marsh turtles, 7:72
Black-masked/white-collared kingfishers. *See* Collared kingfishers
Black microhylids, 6:305
Black miniature donkeys, 12:*172*
Black morays. *See* Green morays
Black mullets. *See* Flathead mullets
Black muntjacs, 15:*344*, 15:346, 15:*347*, 15:*350*–351
Black-naped monarchs, 11:*99*, 11:*102*
Black-naped woodpeckers. *See* Gray-faced woodpeckers
Black-necked cranes, 9:3, 9:24, 9:25, 9:26
Black-necked red cotingas, 10:*307*
Black-necked rock hyraxes, 15:189*t*
Black-necked screamers. *See* Northern screamers
Black-necked spitting cobras, 7:*490*, 7:*492*–493
Black-necked stilts, 9:*101*, 9:*134*
Black-nest swiftlets, 9:424
Black nunbirds, 10:103
Black oligobrachias, 2:*87*, 2:*88*
Black orioles, 11:428
Black owls. *See* Sooty owls
Black oystercatchers. *See* African black oystercatchers
Black phoebes, 10:273
Black-polled yellowthroats, 11:291
Black porpoises. *See* Burmeister's porpoises
Black pricklebacks, 5:*314*, 5:*317*
Black pythons, 7:420
Black rails, 9:51–52, 9:*55*, 9:*58*
Black rats, 12:190, 16:*123*, 16:253, 16:*254*
behavior, 12:*186*, 16:*126*, 16:256
conservation status, 16:256, 16:270
distribution, 16:123, 16:*256*
feeding ecology, 16:256
habitats, 16:256
humans and, 16:127, 16:256
physical characteristics, 16:256
reproduction, 16:256
taxonomy, 16:256
Black ratsnakes, 7:43
Black redstarts, 10:486, 10:*491*, 10:*498*–499
Black rhinoceroses, 15:*256*, 15:259–261
behavior, 15:218, 15:*250*, 15:*251*, 15:*253*
conservation status, 15:*222*, 15:254
distribution, 15:*260*
feeding ecology, 12:*118*, 15:220
habitats, 15:*252*
physical characteristics, 12:*47*
reproduction, 15:*252*
Black ribbon eels. *See* Ribbon morays
Black-ringed finches. *See* Double-barred finches
Black ruffs, 5:*423*, 5:*424*
Black-rumped agoutis, 16:407, 16:415*t*
Black-rumped buttonquails, 9:*15*, 9:*16*–17
Black-rumped flamebacks. *See* Lesser flame-backed woodpeckers
Black-rumped golden-backed woodpeckers. *See* Lesser flame-backed woodpeckers
Black-rumped silverbills. *See* African silverbills
Black saddled tobies, 4:70
Black salmons. *See* Atlantic salmons
Black scimitarbills. *See* Common scimitarbills
Black sea lions. *See* California sea lions
Black seedeaters. *See* Variable seedeaters
Black-shouldered bats. *See* Evening bats
Black-shouldered nightjars, 9:401
Black-shouldered opossums, 12:249–253, 12:264*t*
Black sickle-winged guans. *See* Black guans
Black sicklebills, 11:492
Black-sided flowerpeckers. *See* Gray-sided flowerpeckers
Black sittellas, 11:*170*, 11:173–*174*
Black skimmers, 9:205, 9:*210*, 9:*216*–217
Black snakes, 7:462, 7:485
Black spider monkeys, 14:*157*, 14:167*t*
Black-spined toads. *See* Common Sunda toads
Black-spot flatworms, 1:198, 1:199, 1:*202*, 1:*205*, 1:210
Black-spotted barbets, 10:115, 10:116, 10:*118*, 10:*121*–*122*
Black-spotted cuscuses, 13:39, 13:60, 13:*63*, 13:64
Black-spotted salamanders, 6:372
Black squirrel monkeys. *See* Blackish squirrel monkeys
Black sticks, 4:*306*, 4:*310*, 4:312
Black stilts, 9:103, 9:133–*134*, 9:137, 9:138–*139*
Black storks, 8:265, 8:266, 8:*269*, 8:*271*–*272*
Black storm-petrels, 8:136, 8:137
Black-streaked puffbirds, 10:102
Black-stripe minnows, 4:*389*, 4:390, 4:392, 4:*394*, 4:*396*, 4:397
Black-striped duikers. *See* Bay duikers
Black-striped wallabies, 13:101*t*
Black suckers. *See* Longnose suckers
Black swans, 8:365, 8:367
Black swifts, 9:422, 9:*425*, 9:426–*427*
Black-tailed deer, 15:*380*, 15:*386*, 15:*390*–391
evolution, 15:379
guilds of, 12:118
habitats, 15:383
humans and, 15:385
migrations, 12:165
physical characteristics, 12:38, 15:*381*
reproduction, 15:*383*, 15:384
Black-tailed godwits, 9:177, 9:178, 9:180
Black-tailed hairy dwarf porcupines, 16:365, 16:374*t*
Black-tailed hutias, 16:467*t*
Black-tailed jackrabbits, 16:484, 16:*485*, 16:487, 16:*508*, 16:515*t*
Black-tailed native-hens, 9:48
Black-tailed phascogales. *See* Brush-tailed phascogales
Black-tailed prairie dogs, 16:*144*, 16:*147*, 16:*149*, 16:154–155
behavior, 16:*145*, 16:*146*
conservation status, 16:148
distribution, 16:*153*
habitats, 12:*75*
Black-tailed treecreepers, 11:134
Black terns, 9:*204*, 9:*210*, 9:*216*
Black-thighed falconet, 8:313
Black threadsnakes, 7:*374*, 7:*485*
Black thrips, 3:*282*
Black-throated antbirds, 10:*243*, 10:*253*–254
Black-throated blue warblers, 11:*286*
Black-throated divers. *See* Arctic loons
Black-throated finches, 10:*172*
Black-throated honeyguides. *See* Greater honeyguides
Black-throated huet-huets, 10:*261*, 10:*262*
Black-throated sparrows, 11:*265*
Black-throated sunbirds. *See* Olive-backed sunbirds (*Cinnyris* spp.)
Black-throated tits, 11:*145*–*146*
Black-throated wattle-eyes, 11:*28*, 11:*42*
Black tiger shrimps. *See* Giant tiger prawns
Black tinamous, 8:59, 8:60, 8:*60*, 8:*61*
Black-tipped hangingflies, 3:*344*, 3:*345*
Black toucanets, 10:125, 10:126
Black tree kangaroos, 13:32
Black tufted-ear marmosets, 14:117
Black turtles. *See* Pacific green turtles
Black uakaris, 14:*144*, 14:145, 14:153*t*
Black urchins. *See* Short-spined sea urchins
Black vultures. *See* American black vultures
Black wallabies. *See* Swamp wallabies
Black wallaroos, 13:101*t*
Black widow finches. *See* Dusky indigobirds
Black widow spiders, 2:37, 2:336
Black wildebeests, 16:27, 16:28, 16:29, 16:32, 16:33, 16:*35*, 16:38–*39*
Black-winged flycatcher-shrikes, 10:*388*, 10:*393*
Black-winged ioras. *See* Common ioras
Black-winged mynas, 11:410, 11:*412*, 11:419–*420*

Black-winged petrels, 8:109
Black-winged starlings. *See* Black-winged mynas
Black-winged stilts, 9:133–135, 9:*136*, 9:137–*138*
Black woodhoopoes, 10:66
Black woodpeckers, 9:251, 10:149, 10:*152*, 10:*163*
Blackback butterflyfishes, 5:*237*
Blackbacks. *See* Winter flounders
Blackbar soldierfishes, 5:*117*, 5:*120*–121
Blackbar triggerfishes, 5:*473*, 5:*476*–477
Blackbarred garfishes. *See* Blackbarred halfbeaks
Blackbarred halfbeaks, 5:*82*, 5:*84*, 5:85
Blackbirds, 8:25, 10:*484*–485, 10:*491*, 10:*502*
See also New World blackbirds
Blackbucks, 12:136, 16:2, 16:5, 16:48, 16:56*t*
Blackcaps, 8:33, 11:*8*, 11:*22*
Blackcheek blennies, 5:*344*, 5:*346*, 5:348
Blackchins, 4:441, 4:442
Blackcods. *See* Sablefishes
Blackfin icefishes, 5:*321*
Blackfin pearl killifishes, 5:*95*, 5:*98*, 5:101–102
Blackfin wolf herrings. *See* Dorab wolf herrings
Blackfishes, 5:219–222, 15:2, 15:44
See also Alaska blackfishes; Black ruffs; Bowfins
Blackflies, 2:29
Blackish blindsnakes, 7:382, 7:383–*384*
Blackish-headed spinetails, 10:211
Blackish oystercatchers, 9:126
Blackish squirrel monkeys, 14:101, 14:102, 14:105–106, 14:*107*, 14:*109*–110
Blackpoll warblers, 8:*30*
Black's graylings. *See* Arctic graylings
Blacksaddled coral groupers, 5:*265*, 5:271–*272*
Blacksaddled coral trouts. *See* Blacksaddled coral groupers
Blackscoters, 8:370
Blacksmith plovers, 9:161
Blackspot green knifefishes, 4:*369*
Blackspotted wrasses, 4:*4*, 4:*33*
Blackstarts, 10:486, 10:*490*, 10:*497*–498
Blacktails. *See* Indigo snakes
Blackwater fever, 2:12
Bladder grasshoppers, 3:201, 3:203
Bladdernose seals. *See* Hooded seals
Blainville's beaked whales, 15:*60*, 15:61, 15:*62*, 15:*63*, 15:*66*–67
Blakiston's eagle-owls, 9:333, 9:345, 9:347, 9:351–*352*, 9:*358*–359
Blakiston's fish-owls. *See* Blakiston's eagle-owls
Blanding's turtles, 7:106
Blanford's foxes, 14:*275*, 14:*280*
Blanford's fruit bats, 13:349*t*
Blanus cinereus, 7:274
Blarina spp., 13:248
Blarina brevicauda. *See* Northern short-tailed shrews
Blarina carolinensis. *See* Southern short-tailed shrews
Blarina hylophaga. *See* Elliot's short-tailed shrews
Blarinella spp., 13:248
Blarinella quadraticauda. *See* Chinese short-tailed shrews
Blarinellini, 13:247
Blarinini, 13:247
Blarinomys spp. *See* Shrew mice
Blarinomys breviceps. *See* Brazilian shrew mice
Blasius's horseshoe bats, 13:393, 13:*394*, 13:*398*–399
Blastaea, 1:7
Blastobasids, 3:389
Blastocerus spp., 15:382
Blastocerus campestris. *See* Pampas deer
Blastocerus dichotomus. *See* Marsh deer
Blastocoel, 2:4
Blastopores, 1:19, 2:3, 2:15–16, 2:21
Blastulation, 1:18–19, 2:20–21, 2:23–24
See also Reproduction
Blatta orientalis. *See* Oriental cockroaches
Blatta surinamensis. *See* Suriname cockroaches
Blattella asahinai. *See* Asian cockroaches
Blattella beybienkoi. *See* Asian cockroaches
Blattella germanica. *See* German cockroaches
Blattella vaga, 3:149
Blattellidae, 3:147, 3:149
Blattidae, 3:149
Blattodea. *See* Cockroaches
Blattoidea, 3:147
Blaxter, K. L., 15:145
Bleached earless lizards. *See* Common lesser earless lizards
Bleda canicapilla. *See* Gray-headed bristlebills
Bledius spp., 3:322
Bleeding-heart pigeons, 9:*249*
Bleeding-heart tetras, 4:*340*, 4:*343*–344
Bleeker, P., 5:427
Blennies, 4:47, 4:48, 4:50, 4:67, **5:341–348,** 5:*344*
Blenniidae. *See* Blennies
Blennioidei. *See* Blennies
Blephariceriini, 3:54
Blesboks, 16:27, 16:29, 16:33, 16:*35*, 16:*39*–40
Blind cave characins. *See* Blind cavefishes (Characins)
Blind cave eels, 5:151, 5:152, 5:*153*, 5:*154*, 5:155
Blind cave gudgeons, 5:375, 5:*379*, 5:*381*
Blind cave loach of Xiaao, 4:*324*, 4:*328*, 4:332–333
Blind cave salamanders. *See* Olms
Blind cave tetras. *See* Blind cavefishes (Characins)
Blind cavefishes (Characins), 4:335, 4:337, 4:*340*, 4:343, 4:*344*
Blind cavefishes (Cypriniformes), 4:*324*, 4:329, 4:332–333
Blind gobies, 5:375
Blind loaches, 4:*324*, 4:*326*, 4:327
Blind mole-rats, 12:70, 12:73, 12:74, 12:115, 12:116, 16:123, 16:124
circadian rhythms and, 12:74–75
orientation, 12:75
vision, 12:76–77
Blind river dolphins. *See* Ganges and Indus dolphins
Blind scolopenders, 2:*357*, 2:*358*
Blind sharks, 4:105
Blind swamp cave eels, 5:151, 5:152
Blind tree mice. *See* Malabar spiny dormice
Blindskinks, **7:271–272**
Blindsnakes, **7:379–385,** 7:*380*, 7:*381*
early, 7:198, 7:201, 7:202, **7:369–372**
false, **7:387–388**
skulls of, 7:201
slender, 7:371, **7:373–377**
spotted, 7:*197*
taxonomy, 7:198
Blister beetles, 3:81, 3:316, 3:320
Blood, crocodilian, 7:161
Blood-feeding leeches, 2:75, 2:76
See also Medicinal leeches; Tiger leeches
Blood-feeding pests, 3:76
Blood flukes, 1:*198*, 1:199
See also Human blood flukes
Blood platelets, 12:8
See also Physical characteristics
Blood pythons, 7:*423*, 7:*426*–427
Blood-sucking flies, 3:20
Bloodsuckers. *See* Brown garden lizards
Bloodsucking conenoses. *See* Kissing bugs
Bloodworms, 3:80
Bloody bellies, 1:*172*, 1:175, 1:*176*
Blossom bats, 13:333, 13:*336*, 13:*339*, 13:*346*–347
Blotched genets, 14:*336*, 14:344*t*
Blotched pipe snakes, 7:*395*, 7:*396*
Blotched snakeheads, 5:438
Blotched upsidedown catfishes, 4:*356*, 4:*362*, 4:363
Blotcheye soldierfishes, 5:*116*
Blow lugs. *See* Lugworms
Blowflies, 3:76
Blowholes, 12:67
Blubber, 12:66
See also Physical characteristics; Whales
Blue-and-white bearded wildebeests. *See* Blue wildebeests
Blue-and-white flycatchers, 11:*26*
Blue-and-white kingfishers. *See* White-rumped kingfishers
Blue-and-white mockingbirds, 10:*469*, 10:*470*–471
Blue-and-yellow macaws, 9:290
Blue angelfishes, 5:*243*
Blue antelopes, 16:28
Blue banded sea perches, 5:*255*
Blue-bearded bee-eaters, 10:*43*, 10:*44*
Blue-bellied broadbills. *See* Hose's broadbills
Blue-bellied poison frogs, 6:*201*, 6:*205*, 6:206–207
Blue-bellied rollers, 10:*1*, 10:51–52, 10:*54*, 10:*55*–56
Blue-billed dove-petrels. *See* Broad-billed prions
Blue-billed malimbes, 11:*380*, 11:*384*
Blue birds of paradise, 11:491, 11:492
Blue-black grassquits, 11:*267*, 11:*276*–277
Blue bottles. *See* Portuguese men of war
Blue-breasted quails. *See* King quails
Blue bustards, 9:93, 9:*95*, 9:97–*98*
Blue butterflies, 3:88, 3:90
Blue-capped ifrits, 11:75, 11:77, 11:*80*–81
Blue chaffinches, 11:325, 11:*326*, 11:328–*329*
Blue channel catfishes. *See* Channel catfishes
Blue-cheeked jacamars. *See* Yellow-billed jacamars
Blue chromis, 5:*299*, 5:304–*305*
Blue cods, 5:*332*
Blue corals, 1:103, 1:104, 1:107
Blue couas, 9:*317*, 9:324–*325*
Blue cranes, 9:27
Blue-crested plantain-eaters. *See* Hartlaub's turacos

Blue-crowned motmots, 10:31–32, 10:*33*, 10:*34*, 10:*37*
Blue-crowned pigeons, 9:*241*
Blue cuckoo-shrikes, 10:*388*, 10:*389*
Blue-diademed motmots. *See* Blue-crowned motmots
Blue discus, 5:277, 5:*283*, 5:*287*, 5:288–289
Blue ducks, 8:365
Blue duikers, 15:272, 16:*74*, 16:84*t*
Blue-eared barbets, 10:116
Blue-eared kingfishers. *See* Lilac-cheeked kingfishers
Blue-eye catfishes, 4:*356*, 4:360–*361*
Blue-eye urchins. *See* Long-spined sea urchins
Blue-eyed imperial shags, 8:*185*
Blue-eyed lemurs, 14:*2*
Blue-eyes, 4:57, 5:67, 5:68
Blue-faced boobies. *See* Masked boobies
Blue fantails, 11:83
Blue fliers. *See* Red kangaroos
Blue flycatchers. *See* African blue-flycatchers
Blue-footed boobies, 8:*213*, 8:219–220
Blue-gray carpet sharks, 4:105
Blue-gray flycatchers. *See* Ashy flycatchers
Blue-gray gnatcatchers, 11:*2*, 11:*4*, 11:*8*, 11:*10*–11
Blue grouse, 8:434
Blue hares. *See* Mountain hares
Blue-headed fantails, 11:83, 11:*88*, 11:*89*–90
Blue-headed picathartes. *See* Gray-necked picathartes
Blue-headed pittas, 10:195
Blue jays, 11:504, 11:*506*, 11:*510*, 11:*512*–513
Blue jewel-babblers, 11:77, 11:*78*–79
Blue korhaans. *See* Blue bustards
Blue Madagascar coucals. *See* Blue couas
Blue magpies, 11:504
Blue-mantled fairy bluebirds. *See* Fairy bluebirds
Blue marlins, 5:*408*, 5:*411*
Blue-masked leafbirds, 10:417, 10:*418*, 10:*422*–423
Blue mockingbirds, 10:467
Blue monkeys, 12:38, 14:205*t*
Blue morphos, 3:*392*, 3:397, 3:*398*
Blue Mountain vireos, 11:256
Blue mussels. *See* Common blue mussels
Blue-naped mousebirds, 9:*472*, 9:*475*
Blue-naped sunbirds. *See* Purple-naped sunbirds
Blue-necked cranes. *See* Gray crowned cranes
Blue-necked jacamars. *See* Yellow-billed jacamars
Blue neon gobies. *See* Neon gobies
Blue pikes. *See* Walleyes
Blue plantain-eaters. *See* Great blue turacos
Blue platys, 5:*89*
Blue ringed octopods, 2:*475*, 2:479
Blue sharks, 4:116, 4:*118*, 4:123–*124*
Blue sheep, 16:87, 16:92–93, 16:94, 16:*96*, 16:*99*, 16:100–101
Blue-spotted cornetfishes, 5:*138*, 5:*140*–141
Blue-spotted mudskippers, 5:*376*
Blue-spotted stingrays, 4:*177*
Blue-spotted sunfishes, 5:*200*, 5:202–*203*
Blue starfishes, 1:370, 1:*372*, 1:*375*, 1:378–379
Blue striped grunts, 4:*62*
Blue swallows, 10:360
Blue-tailed emeralds, 9:438
Blue-tailed trogons. *See* Javan trogons
Blue tangs, 5:*393*
See also Palette tangs
Blue-throated hummingbirds, 9:416
Blue-throated motmots, 10:32, 10:*34*, 10:*35*
Blue-throated rollers, 10:51–52
Blue tiger prawns. *See* Giant tiger prawns
Blue titmice, 11:*156*, 11:158
Blue-toed rocket frogs, 6:*201*, 6:203–*204*
Blue-tongued skinks, 7:*328*
Blue trevallys, 4:*26*
Blue vangas, 10:439, 10:440, 10:441, 10:442
Blue whales, 12:63, 15:130*t*
behavior, 15:6
conservation status, 15:9, 15:125
distribution, 15:5, 15:*121*
feeding ecology, 15:8, 15:123
humans and, 15:125
physical characteristics, 12:12, 12:63, 12:79, 12:82, 12:213, 15:4, 15:120
reproduction, 15:124
Blue wildebeests, 16:*29*, 16:42*t*
behavior, 16:5, 16:31
coevolution and, 12:216
conservation status, 12:220–221
distribution, 16:29
feeding ecology, 16:32
habitats, 16:29–30
humans and, 16:34
migrations, 12:*114*, 12:*166*, 12:*169*, 16:6
reproduction, 16:*28*
taxonomy, 16:27
Blue-winged leafbirds, 10:417, 10:*418*, 10:*421*
Blue-winged parrots, 9:277
Blue-winged pittas, 10:194, 10:*195*
Blue-winged warblers, 11:285, 11:286, 11:*292*, 11:*294*
Blue-winged wasps. *See* Digger wasps
Blue-winged yellow warblers. *See* Blue-winged warblers
Blueback salmons. *See* Sockeye salmons
Bluebacks. *See* Coho salmons
Bluebills, 11:353, 11:354
Bluebirds, 10:485, 10:*487*, 10:488, 10:489, 10:*494*
Bluebottle fishes. *See* Man-of-war fishes
Bluebucks, 16:27, 16:33, 16:41*t*
Bluefishes, 4:49, 4:68, **5:211–212**, 5:*213*, **5:216–217**
Bluegill sunfishes. *See* Bluegills
Bluegills, 4:66, 5:196, 5:197, 5:*198*, 5:*201*, 5:*203*–204
Bluehead wrasses. *See* Blueheads
Blueheads, 5:296, 5:*299*, 5:*303*–304
Blueleg mantis shrimps. *See* Giant mantis shrimps
Blues
large, 3:*390*, 3:394–*395*
Western pygmy, 3:384
Bluestreak cleaner wrasses, 4:69, 5:*293*, 5:*300*, 5:301–*302*
Bluestriped grunts, 5:*264*, 5:*267*, 5:268
Bluethroats, 10:483, 10:484, 10:486
Bluish slate antshrikes. *See* Cinereous antshrikes
Blunt-eared bats. *See* Peruvian crevice-dwelling bats
Blunt-headed vinesnakes, 7:468
Blunt-nosed vipers. *See* Levantine vipers
Blunthead cichlids, 5:*279*, 5:*282*, 5:*288*, 5:289
Blunthead puffers, 5:471
Bluntnose sixgill sharks, 4:143, 4:144, 4:*145*, 4:*147*–148
Blyth's flying foxes, 13:*323*, 13:*326*
Blyth's horseshoe bats, 13:*388*, 13:389, 13:390, 13:391
Blyth's reed warblers, 9:314
Boa spp., 7:410, 7:411
Boa constrictors, 7:409–411, 7:*412*, 7:*413*
Boa dumerili. *See* Dumeril's boas
Boa madagascariensis. *See* Madagascar boas
Boa mandrita. *See* Madagascar tree boas
Boarfishes, 5:235, 5:236, 5:238, 5:240, 5:242
See also Pentacerotidae
Boas, 7:26, 7:40, 7:42, 7:*197*, 7:205, **7:409–417**, 7:*412*
See also specific types of boas
Boat-billed herons, 8:239, 8:241, 8:*246*, 8:258–259
Bobcats, 12:*50*, 14:370, 14:*373*, 14:*378*, 14:*384*, 14:389
Bobolinks, 11:301–303, 11:*307*, 11:*321*–*322*
Bobrinski's jerboas, 16:212, 16:213, 16:216, 16:*218*, 16:*219*, 16:222
Bobtail eels. *See* Bobtail snipe eels
Bobtail snipe eels, 4:*4*, 4:271, 4:*273*, 4:*274*
Bobtail squids, 2:475, 2:478
See also Butterfly bobtail squids
Bobtails, 7:*329*, 7:*331*, 7:*332*, 7:336
Bobwhites. *See* Northern bobwhite quails
Bocaccios, 5:166, 5:*168*, 5:174–*175*
Bocages mole-rats, 16:349*t*
Bock, W. J., 10:329
Bockadams. *See* Dog-faced watersnakes
Bodenheimer's pipistrelles, 13:313
Bodotriidae, 2:229
Body fat, 12:120
in hominids, 12:23–24
in neonates, 12:125
storage of, 12:19, 12:174
See also Physical characteristics
Body lice, human, 3:252, 3:*253*, 3:*255*–*256*
Body mass, 12:11–12, 12:24–25
Body organization. *See* Baüplan
Body size, 12:41
bats, 12:58–59
blue whales, 12:213
competition and, 12:117
domestication and, 12:174
field studies and, 12:200–202
Ice Ages and, 12:17–25
life histories and, 12:97–98
nocturnal mammals and, 12:115
reproduction and, 12:15
sexual dimorphism and, 12:99
tropics and, 12:21
See also Physical characteristics
Body temperature, 2:26, 4:37, 5:*406*, 8:9–10
body size and, 12:21
sperm and, 12:91, 12:103
subterranean mammals, 12:73–74
See also Physical characteristics; Thermoregulation
Boeseman's rainbowfishes, 5:*69*–*72*, 5:*75*
Bog turtles, 7:105
Bogart, J. P., 4:190
Bogdanowicz, W., 13:387
Bogert, Charles M., 7:43
Boggi. *See* Bobtails
Bogotá sunangels, 9:449, 9:450–451
Bohadschia argus, 1:*417*

Bohadschia graeffei. See Sea cucumbers
Bohemian waxwings, 10:447, 10:448, 10:*450*, 10:451–*452*
Böhme, Wolfgang, 7:223–224
Bohor reedbucks, 16:27, 16:29, 16:30, 16:34
Boidae. *See* Boas
Boiga irregularis. See Brown treesnakes
Boinae, 7:206, 7:409–411
Boinski, S., 14:104
Boissonneaua spp. *See* Hummingbirds
Boissonneaua jardini. See Velvet-purple coronets
Bojophlebia prokopi, 3:13, 3:125
Boland, Chris, 11:457
Boleophthalmus boddarti. See Blue-spotted mudskippers
Boles, W. E., 10:385, 11:459
Bolinopsis spp., 1:171
Bolinopsis gracilis, 1:28
Bolinopsis vitrea, 1:28
Bolitoglossa spp., 6:35, 6:325, 6:391
Bolitoglossa pesrubra. See Talamancan web-footed salamanders
Bolitoglossa rostrata, 6:35
Bolitoglossa subpalmata, 6:397–398
Bolitoglossini, 6:390, 6:391
Bolivian blackbirds, 11:293, 11:304
Bolivian bleating frogs, 6:*307*, 6:313
Bolivian chinchilla rats, 16:443, 16:444, 16:445, 16:*446*, 16:*447*, 16:448
Bolivian earthcreepers, 10:*212*, 10:*216*
Bolivian hemorrhagic fever, 16:270
Bolivian pygmy blue characins, 4:336
Bolivian red howler monkeys, 14:167*t*
Bolivian spinetails, 10:211
Bolivian squirrel monkeys, 14:101, 14:102, 14:103, 14:*107*, 14:*108*
Bolivian tuco-tucos, 16:430*t*
Boll weevils, 3:60
Bolma rugosa, 2:*430*
Boltenia echinata, 1:453
Bolyeria multocarinata. See Smooth-scaled splitjaws
Bolyeriidae. *See* Splitjaw snakes
Bombardier beetles, 3:320
Bombay canaries. *See* American cockroaches
Bombay ducks, 4:431, 4:432, 4:433, 4:434
Bombay swamp eels, 5:152
Bombina spp. *See* Fire-bellied toads
Bombina bombina. See Fire-bellied toads
Bombina orientalis. See Oriental fire-bellied toads
Bombina variegata. See Yellow-bellied toads
Bombinatoridae, **6:83–88**, 6:*85*
  distribution, 6:5, 6:*83*, 6:*86–88*
  evolution, 6:4, 6:64, 6:84, 6:86–88
  larvae, 6:41, 6:84, 6:86–88
  physical characteristics, 6:26, 6:83, 6:86–88
  taxonomy, 6:83, 6:86–88
Bombus spp. *See* Bumblebees
Bombus lapidarius. See Large red-tailed bumblebees
Bombycilla cedrorum. See Cedar waxwings
Bombycilla garrulus. See Bohemian waxwings
Bombycilla japonica. See Japanese waxwings
Bombycillidae, **10:447–454**, 10:*450*
Bombycillinae. *See* Waxwings
Bombyliid, 3:359
Bombyx mori. See Silkworms
Bon Dieu Loa, 7:57
Bonaparte's gulls, 9:205
Bonaparte's nightjars, 9:405
Bonaparte's tinamous. *See* Highland tinamous
Bonasa bonasia. See Hazel grouse
Bonasa sewerzowi. See Chinese grouse
Bonasa umbellus. See Ruffed grouse
Bonefishes, 4:11, **4:249–253**, 4:*250*
Bonellia spp. *See* Spoon worms
Bonellia viridis. See Green bonellias
Bonellidae, 2:103, 2:104
Bones, 12:41
  bats, 12:58–59
  growth of, 12:10
  See also Physical characteristics
Bongos, 15:138, 15:*141*, 16:4, 16:*5*, 16:11
Bonin Island honeyeaters. *See* Bonin white-eyes
Bonin siskins, 11:324
Bonin white-eyes, 11:227, 11:235
Bonitos, 5:406, 5:407
Bonn Convention for the Conservation of Migratory Animals
  Brazilian free-tailed bats, 13:493
  free-tailed bats, 13:487
Bonnet macaques, 14:194
Bonnethead sharks, 4:115, 4:*117*, 4:*127*–128
Bonobos, 14:*236*, 14:*239*–240
  behavior, 14:228, 14:231–232
  conservation status, 14:235
  distribution, 14:227
  evolution, 12:33, 14:225
  feeding ecology, 14:232–233
  habitats, 14:227
  humans and, 14:235
  language, 12:162
  physical characteristics, 14:226–227
  reproduction, 14:235
  taxonomy, 14:225
Bonteboks, 16:27, 16:29, 16:34, 16:*35*, 16:*39*–40
Bony fishes. *See* Atlantic menhadens
Bony-headed treefrogs. *See* Manaus long-legged treefrogs
Bony tongues, 4:11, **4:231–241**, 4:*235–236*
Boobies, 8:183, 8:184, **8:211–223**
  behavior, 8:212–213
  conservation status, 8:214–215
  distribution, 8:*211*, 8:212
  evolution, 8:211
  feeding ecology, 8:213
  habitats, 8:212
  humans and, 8:215
  physical characteristics, 8:211
  reproduction, 8:213–214
  species of, 8:*217*–222
  taxonomy, 8:211
Boobook owls. *See* Southern boobook owls
Boocercus euryceros, 15:141
Boodies, 13:74, 13:77, 13:79–*80*
Booidea, 7:198
Book lice, **3:243–248**, 3:*244*, 3:*245*
Book scorpions, 2:*337*, 2:*345*, 2:347–348
Boomslangs, 7:466, 7:467, 7:*472*, 7:*474*–475
Boophis spp., 6:281, 6:291, 6:292, 6:293
Boophis erythrodactylus. See Forest bright-eyed frogs
Boot sponges. *See* Sharp-lipped boot sponges
Booted eagles, 8:321
Bootettix argentatus, 3:206
Boraker, D. K., 16:436
Borchiellini, C., 1:57
Border collies, 14:288–289
Boreal chickadees, 11:156
Boreal owls, 9:350
Boreal peewees. *See* Olive-sided flycatchers
Boreal tits, 11:157
Borean long-tailed porcupines. *See* Long-tailed porcupines
Boreidae. *See* Snow scorpionflies
Borers, 3:389
Boreus brumalis. See Snow scorpionflies
Bornean bay cats. *See* Bay cats
Bornean bristleheads, 10:425, 10:429, 11:469, 11:*470*, 11:*471*
Bornean frogmouths, 9:379
Bornean gibbons. *See* Mueller's gibbons
Bornean orangutans, 12:*156*, 12:222, 14:225, 14:*236*, 14:*237*
Bornean peacock-pheasants, 8:437
Bornean smooth-tailed tree shrews, 13:292, 13:*293*, 13:*294*–295
Bornean tree shrews, 13:292, 13:297*t*
Bornean yellow muntjacs, 15:346, 15:*347*, 15:*349*, 15:352
Borneo crocodiles, 7:179
Borneo frogs, 6:110
Borneo short-tailed porcupines. *See* Thick-spined porcupines
Bos spp., 16:11, 16:13, 16:14
Bos domestica. See Bali cattle
Bos frontalis. See Gayals
Bos gaurus. See Gaurs
Bos grunniens. See Yaks
Bos indicus. See Brahma cattle
Bos javanicus. See Bantengs
Bos mutus. See Yaks
Bos primigenius. See Aurochs
Bos sauveli. See Koupreys
Bos taurus. See Aurochs; Domestic cattle
Boselaphini, 16:1, 16:11
Boselaphus tragocamelus. See Nilgais
Bostockia spp. *See* Nightfishes
Bostockia porosa. See Nightfishes
Bostrichid beetles. *See* False powderpost beetles
Bostrychia bocagei. See Dwarf olive ibises
Bostrychia hagedash. See Hadada ibises
Bot flies, 3:64, 3:361, 3:*368*, 3:370
  See also Horse bot flies
Botaurinae. *See* Bitterns
Botaurus poiciloptilus. See Australasian bitterns
Botaurus stellaris. See Eurasian bitterns
Botha's larks, 10:346
Bothidae, 4:70, 5:450, 5:453
Bothma, J. P., 15:171
Bothragonus swani. See Rockheads
Bothriechis spp., 7:445, 7:447
Bothriechis aurifer. See Yellow-blotched palm-pitvipers
Bothriechis schlegelii. See Gold morph of eyelash vipers
Bothriocephalidae, 1:225
Bothriopsis spp., 7:445, 7:447
Bothriopsis bilineata, 7:447
Bothrochilus spp., 7:420, 7:421
Bothrochilus boa. See Ringed pythons
Bothrocophias spp., 7:445
Bothrocophias hyoprora, 7:447
Bothrops spp., 7:445
Bothrops alcatraz, 7:448

Bothrops ammodytoides. See Patagonian lanceheads
Bothrops asper, 7:448
Bothrops insularis, 7:448
Bothus leopardinus. See Pacific leopard flounders
Bothus lunatus. See Peacock flounders
Bothus ocellatus, 5:*451*
Botia dario. See Tiger loaches
Botia macracantha. See Clown loaches
Boticelli, Sandro, 2:42
Botiinae, 4:321
Botos, 15:5, 15:*8,* **15:27–31,** 15:*28,* 15:*29,* 15:*30*
Botryllus schlosseri, 1:*456,* 1:*461,* 1:466
Botta's pocket gophers. *See* Valley pocket gophers
Bottle flies, 3:357, 3:361
Bottle squids, 2:475
Bottleheads. *See* Northern bottlenosed whales
Bottleneck species, 12:221–222
Bottlenosed dolphins, 12:27, 12:*67,* 12:68, 15:2
  behavior, 12:*143,* 15:44, 15:45, 15:46
  chemoreception, 12:84
  conservation status, 15:49–50
  distribution, 15:5
  echolocation, 12:*87*
  feeding ecology, 15:*47*
  habitats, 15:44
  humans and, 15:10–11
  language, 12:*161*
  physical characteristics, 12:67–68
  reproduction, 15:48
  See also Common bottlenosed dolphins
Bottlenosed whales, 15:7
  See also Northern bottlenosed whales; Southern bottlenosed whales
Boubous. *See* Bush-shrikes
Boucard, Adolphe, 9:452
Boucard tinamous. *See* Slaty-breasted tinamous
Bougainvillia principis, 1:28
Boulengerella lucius. See Golden pike characins
Boulengerochromis microlepis. See Giant cichlids
Boulenger's Asian tree toads. *See* Brown tree toads
Boulenger's callulops frogs, 6:*306,* 6:*308*–309
Boulenger's climbing frogs, 6:*306,* 6:310
Boulenger's earless toads. *See* Chirinda toads
Boulengerula spp., 6:436
Boulton's hill-partridges. *See* Sichuan hill-partridges
Bourgueticrinida, 1:355
Bourletiella hortensis. See Garden springtails
Bovichtidae. *See* Thornfishes
Bovidae, **16:1–106**
  Antilopinae, 16:1, **16:45–58,** 16:*49,* 16:*56t–57t*
  behavior, 12:145, 12:146–147, 16:5–6
  Bovinae, 16:1, **16:11–25,** 16:*16,* 16:*17,* 16:*24t–25t*
  conservation status, 16:8
  digestive systems of, 12:14
  distribution, 16:4
  evolution, 15:137, 15:138, 15:265, 15:266, 15:267, 16:1–2
  feeding ecology, 15:142, 15:271, 16:6–7
  habitats, 16:4–5
  Hippotraginae, 16:1, **16:27–43,** 16:*35,* 16:*41t–42t*
  horns, 12:40
  humans and, 16:8–9
  Neotraginae, 16:1, **16:59–72,** 16:*65,* 16:*71t*
  physical characteristics, 15:142, 15:267, 15:268, 16:2–4
  reproduction, 15:143, 16:7–8
  taxonomy, 16:1–2
  See also Caprinae; Duikers
Bovinae, 16:1, **16:11–25,** 16:*16,* 16:*17,* 16:*24t–25t*
Bovini, 16:1, 16:11
Bow flies, 3:64
Bowdoin's beaked whales. *See* Andrew's beaked whales
Bowerbankia spp., 2:25
Bowerbirds, 10:174, **11:477–488,** 11:*482*
  behavior, 11:479
  conservation status, 11:481
  distribution, 11:*477,* 11:478
  evolution, 11:477
  feeding ecology, 11:479
  habitats, 11:478–479
  humans and, 11:481
  physical characteristics, 11:477–478
  reproduction, 11:479–480
  species of, 11:*483–488*
  taxonomy, 11:477
Bowfins, 4:6–7, 4:11, 4:14, 4:23, 4:*26,* 4:57, **4:229–230,** 6:7
Bowhead whales, **15:107–118,** 15:*114,* 15:*115*–116
  behavior, 15:109–110
  conservation status, 15:9–10, 15:111–112
  distribution, 15:5, 15:*107,* 15:109
  evolution, 12:*65,* 15:107
  feeding ecology, 15:110
  habitats, 15:109
  humans and, 15:112
  migrations, 12:87, 12:170
  physical characteristics, 15:107, 15:109
  reproduction, 15:110–111
  species of, 15:*115*–118
  taxonomy, 15:107
Box jellies, 1:10, **1:147–152,** 1:*148,* 1:*150*
Box turtles, 7:*106*
  Chinese three-striped, 7:119
  eastern, 7:*68,* 7:106, 7:*108,* 7:112–*113*
  Malayan, 7:*12,* 7:*115*
  ornate, 7:72
  yellow-margined, 7:*117,* 7:*118*–119
Boxer mantids, 3:*182,* 3:*183,* 3:186–187
Boxer shrimps, 2:197, 2:198
Boxfishes, 5:467–471
Brabant cichlids. *See* Blunthead cichlids
Brace's emeralds, 9:450
Brachaeluridae. *See* Blind sharks
Brachionichthys hirsutus. See Spotted handfishes
Brachionus spp., 1:260
Brachionus plicatilis, 1:261
Brachiopoda, 1:9, 1:13, 1:22
Brachycarida, 2:245
Brachycephalidae. *See* Three-toed toadlets
Brachycephalus spp., 6:180
Brachycephalus didactyla, 6:180
Brachycephalus ephippium. See Pumpkin toadlets
Brachycephalus pernix. See Southern three-toed toadlets
Brachycera, 3:357, 3:358, 3:359
Brachydelphis spp., 15:23
Brachygalba goeringi. See Pale-headed jacamars
Brachygalba lugubris. See Brown jacamars
Brachyhypopomus spp., 4:370
Brachyhypopomus pinnicaudatus, 4:370, 4:*372,* 4:*373,* 4:375
Brachyistius frenatus. See Kelp perches
Brachylagus spp. *See* Pygmy rabbits
Brachylagus idahoensis. See Pygmy rabbits
Brachylophus spp., 7:244
Brachymeles spp., 7:327, 7:328
Brachymerus bifasciatus. See Banded rubber frogs
Brachymetroides atra, 3:259
Brachyophidium rhodogaster, 7:391
Brachypauropodidae, 2:375, 2:376
Brachypelma smithi. See Mygalomorph spiders
Brachyphylla cavernarum. See Antillean fruit-eating bats
Brachyphyllinae, 13:413, 13:415
Brachyplatystoma filamentosum. See Lau-lau
Brachyplatystoma rousseauxii, 4:371
Brachypoda, 2:131
Brachypteracias leptosomus. See Short-legged ground-rollers
Brachypteracias squamigera. See Scaly ground-rollers
Brachypteraciidae. *See* Ground-rollers
Brachyramphus spp. *See* Murrelets
Brachyramphus marmoratus. See Marbled murrelets
Brachytarsophrys spp., 6:110, 6:111
Brachytarsophrys intermedia. See Annam broad-headed toads
Brachyteles spp. *See* Muriquis
Brachyteles arachnoides. See Southern muriquis
Brachyteles hypoxanthus. See Northern muriquis
Brachytrupinae, 3:204
Brachyura. *See* True crabs
Brachyuromy betsileonensis. See Malagasy reed rats
Braconidae, 3:298
Bradbury, J. W., 13:342
Bradley, James Chester, 3:221
Bradypodidae. *See* Three-toed tree sloths
Bradypodion spp., 7:*37,* 7:224, 7:227, 7:232
Bradypodion thamnobates. See KwaZulu-Natal Midlands dwarf chameleons
Bradypterus spp., 11:3, 11:4, 11:7
Bradypterus baboecala. See Little rush-warblers
Bradypterus sylvaticus, 11:5
Bradypus spp. *See* Three-toed tree sloths
Bradypus infuscatus. See Three-toed tree sloths
Bradypus pygmaeus. See Monk sloths
Bradypus torquatus. See Maned sloths
Bradypus tridactylus. See Pale-throated three-toed sloths
Bradypus variegatus. See Brown-throated three-toed sloths
Bradytriton spp., 6:392
Brahma cattle, 16:*4*
Brahminy ducks. *See* Ruddy shelducks
Brahminy kites, 8:320, 8:323
Brahminy starlings, 11:*408*
Braincase, 12:8, 12:9
Brains, 7:8, 12:6, 12:36, 12:49
  enlargement of, 12:9
  evolution of, 12:19
  learning and, 12:141
  life spans and, 12:97–98
  Neanderthals, 12:24
  placentation and, 12:94

size of, 12:149–150
See also Physical characteristics
Bramble sharks, 4:151, 4:152
Bramblings, 11:*326*, 11:*329–330*
Bramidae. *See* Pomfrets
Branch herrings. *See* Alewives
Branched bristlenose catfishes, 4:*4*, 4:*356*, 4:*358*, 4:361–362
Branchinecta spp., 2:136
Branchinecta ferox, 2:137
Branchinecta gigas, 2:135, 2:137
Branchinectidae, 2:135
Branchinella spp., 2:137
Branchioceriantbus imperator, 1:127
Branchiopoda, 2:135, 2:141, 2:147
Branchiostegus spp., 5:256
Branchiostoma spp. *See* Lancelets
Branchiostoma belcheri. *See* Smalltail lancelets
Branchiostoma floridae. *See* Florida lancelets
Branchiostoma lanceolatum. *See* European lancelets
Branchiostoma nigeriense, 1:25
Branchiostomatidae. *See* Lancelets
Branchipodidae, 2:135
Branchipodopsis spp., 2:136
Branchiura. *See* Fish lice
Brandt's cormorants, 8:*203*, 8:*204*, 8:*206*
Brandt's hamsters, 16:247*t*
Brandt's hedgehogs, 13:213*t*
Branick's giant rats. *See* Pacaranas
Branta canadensis. *See* Canada geese
Branta sandvicensis. *See* Hawaiian geese
Braswell, A. L., 7:315
Braude, S., 12:71
Brauer's bristlemouths, 4:*425*, 4:*426*
Brauer's inocelliid snakeflies, 3:*299*, 3:300–*301*
Brazil nut poison frogs, 6:199, 6:*202*, 6:*205*
Brazilian agoutis. *See* Red-rumped agoutis
Brazilian free-tailed bats, 13:*484*, 13:*485*, 13:*486*, 13:*489*, 13:*492*, 13:493
behavior, 13:312, 13:484–485
distribution, 13:310
feeding ecology, 13:485
humans and, 13:487–488
maternal recognition, 12:85
reproduction, 13:315, 13:486–487
Brazilian guinea pigs, 16:*392*
Brazilian mergansers, 8:365, 8:367, 8:*375*, 8:389–390
Brazilian poison frogs, 6:*202*, 6:207–208
Brazilian shrew mice, 16:*271*, 16:*273*, 16:274
Brazilian spiny tree rats, 16:451, 16:458*t*
Brazilian tapirs. *See* Lowland tapirs
Brazilian teals, 8:371
Brazilian three-banded armadillos, 13:192*t*
Brazilozoros spp., 3:239
Brazza's swallows, 10:360
Brechites vaginiferus. *See* Watering pot shells
Breeding. *See* Reproduction
Breeding Bird Survey, 8:23, 10:448, 10:453
Breeding seasons. *See* Reproduction
Breeds, dog, 14:288–289
Bregmaceros mcclellandi, 5:27
Bregmacerotidae, 5:25–29, 5:27
Brennania belkini. *See* Belkin's dune tabanid flies
Brenowitz, E. A., 8:40
Brentidae, 3:317
Brevicepines, 6:35
Breviceps spp., 6:35, 6:68, 6:303, 6:304–305
Breviceps adspersus. *See* Bushveld rain frogs
Breviceps gibbosus. *See* Cape rain frogs
Breviceps macrops, 6:305
Breviciptinae, 6:302
Brevoortia tyrannus. *See* Atlantic menhadens
Brewer's blackbirds, 11:302–303, 11:304
Brewer's moles. *See* Hairy-tailed moles
Brewster's warblers, 11:285–286, 11:294
Bridges's degus, 16:433, 16:440*t*
Bridled beauties. *See* Bluestreak cleaner wrasses
Bridled nail-tailed wallabies, 13:*40*, 13:*92*, 13:*98–99*
Bridled parrotfishes, 4:*26*
Bridled titmice, 11:*158*, 11:*159*, 11:*162*
Brienomyrus spp., 4:231
Brienomyrus brachyistius, 4:233
Briggs, J. C., 5:355
Bright-headed cisticolas. *See* Golden-headed cisticolas
Brills. *See* Windowpane flounders
Brindlebasses, 5:*293*
Brindled bandicoots. *See* Northern brown bandicoots
Brine flies, 3:359
Brine shrimps, 2:36, 2:*136*, 2:*138*, 2:139, 2:*140*
Bristle-nosed barbets, 10:115
Bristle-spined porcupines. *See* Thin-spined porcupines
Bristle-thighed curlews, 9:180
Bristlebirds, 11:55–57, 11:*58*, 11:*59*
Bristleheads, 11:467–469, 11:*470*, 11:471
Bristlemouths, 4:421, 4:422, 4:423, 4:*425*, 4:*426*
Bristletails, 3:10–13, 3:19, 3:57, **3:*113*–117**, 3:*115*
See also Silverfish
Bristleworms, 2:50–56
Bristly millipedes, 2:363, 2:364, 2:365, 2:*366*, 2:369–*370*
British Red Data Book, on wart biter katydids, 3:85
Brits. *See* Atlantic herrings
Brittle stars, **1:387–399**, 1:*389*, 1:*391*, 2:32
behavior, 1:388–389
conservation status, 1:390
evolution, 1:387
feeding ecology, 1:389
habitats, 1:388
humans and, 1:390
physical characteristics, 1:387, 1:*388*
reproduction, 1:389–390
species of, 1:*393–398*
symbiosis and, 1:*35*
taxonomy, 1:387
Britz, R., 5:427
Broad-banded sand swimmers, 7:*331*, 7:*333*, 7:334–335
See also Half-girdled snakes
Broad-billed dove-petrels. *See* Broad-billed prions
Broad-billed hummingbirds, 9:*439*
Broad-billed motmots, 10:31–32, 10:*34*, 10:*35*–36
Broad-billed prions, 8:*127*, 8:130–131
Broad-billed todies, 10:26
Broad-faced potoroos, 13:39
Broad fish tapeworms, 1:35, 1:231, 1:*232*, 1:240–*241*
Broad-footed moles, 13:287*t*
Broad-headed skinks, 7:*328*, 7:*330*, 7:*336–337*
Broad-headed snakes, 7:485, 7:486
Broad-headed toads, 6:110
Broad sea fans, 1:107
Broad-snouted caimans, 7:164, 7:171
Broad-striped dasyures, 12:299*t*
Broad-tailed hummingbirds, 9:443, 9:445
Broad-winged hawks, 8:321
Broadbills, 10:170, **10:177–186**, 10:*180*
Broadcast spawning, 1:18, 1:21–22, 2:17–18, 2:23–24, 2:27
See also Reproduction
Broadnose sevengill sharks, 4:*143*–144, 4:*145*, 4:*146*, 4:148
Brocket deer, 15:379, 15:382, 15:384, 15:396*t*
Broken-wing display, 9:74
Brolga cranes, 9:25
Bromeliad boas. *See* Southern bromeliad woodsnakes
Bromeliad dwarf boas. *See* Southern bromeliad woodsnakes
Bromeliad woodsnakes, 7:*436–437*
Brontë, Emily, 10:462
Brontotheres, 15:136
Bronx Zoo, 12:205
Bronze-cuckoos, 11:57
Bronze quolls, 12:279
Bronze sunbirds, 11:209
Bronze-tailed sicklebills. *See* White-tipped sicklebills
Bronzed drongos, 11:438
Bronzed shags. *See* New Zealand king shags
Brooch lanternfishes, 4:*441*
Brood parasitism, 4:354
cowbirds, 11:*256*, 11:291, 11:321
cuckoos, 9:313–314, 11:445
honeyguides as, 10:139
indigobirds, 11:356, 11:378
Viduinae, 11:378
weaver-finches, 11:356
whydahs, 11:356, 11:378
See also Reproduction
Brook charrs. *See* Brook trouts
Brook trouts, 4:*32*, 4:*405*, 4:*408*, 4:417–*418*, 5:421, 5:422
Brookesia spp., 7:224–225, 7:227
Brookesia perarmata. *See* Armored chameleons
Brooks, D., 10:127
Broom, R., 16:329
Brophy, B., 15:151
Brosme spp., 5:26
Brosme brosme, 5:27
Brosmophysis spp., 5:18
Brotogeris jugularis. *See* Orange-chinned parakeets
Brotula barbata, 5:18
Brotulinae, 5:16
Brotulotaeniinae, 5:16
Brow-antlered deer. *See* Eld's deer
Brown, J. H., 16:200
Brown, William L., 10:130
Brown agoutis, 16:*408*
Brown-and-yellow marshbirds, 11:*307*, 11:316–*317*
Brown antechinuses, 12:282, 12:*293*, 12:*296*–297
Brown-backed flowerpeckers, 11:191
Brown-backed honeyeaters, 11:*236*
Brown-bearded sheath-tail bats, 13:*356*
Brown bears, 12:*23*, 14:*301*, 14:*303*

behavior, 12:145, 14:258, 14:297, 14:298
conservation status, 14:300
distribution, 14:296
evolution, 14:*295–296*
feeding ecology, 12:*124*, 14:260, 14:299
habitats, 14:297
physical characteristics, 14:296
taxonomy, 14:295
Brown birds. *See* Brown tremblers
Brown boobies, 8:*216*, 8:*222*
Brown brocket deer. *See* Gray brocket deer
Brown butterflies, 3:57
Brown caimans. *See* Common caimans
Brown-capped tits, 11:157
Brown-capped vireos. *See* Warbling vireos
Brown capuchins. *See* Black-capped capuchins
Brown-cheeked hornbills, 10:75
Brown-chested flycatchers. *See* Brown-chested jungle-flycatchers
Brown-chested jungle-flycatchers, 11:*29*, 11:*33*–34
Brown creepers, 11:118, 11:177, 11:*179*, 11:*180*–181
Brown dippers, 10:475, 10:*478*, 10:*479*–480
Brown discus. *See* Blue discus
Brown dorcopsises, 13:102*t*
Brown drakes. *See* Brown mayflies
Brown ducks. *See* Brown teals
Brown-eared bulbuls. *See* Ashy bulbuls
Brown eared-pheasants, 8:*438*, 8:*451*
Brown falcons, 8:353, 8:358–359
Brown fat, 12:113
Brown flycatchers. *See* Jacky winters
Brown four-eyed opossums, 12:250–253, 12:264*t*
Brown frogs, 6:*254*, 6:263
conservation status, 6:57, 6:263
distribution, 6:63, 6:263
humans and, 6:53, 6:252, 6:263
reproduction, 6:*31*, 6:49, 6:263
Brown garden lizards, 7:*213*, 7:214–*215*
Brown greater bushbabies, 14:*24*, 14:25, 14:26, 14:*27*, 14:*30*, 14:31
Brown hairy dwarf porcupines, 16:366, 16:374*t*
Brown hares. *See* European hares
Brown hawk-owls, 9:347, 9:349
Brown hawks. *See* Brown falcons
Brown-headed cowbirds, 11:291, 11:293, 11:304, 11:*307*, 11:*320–321*
Brown-headed nuthatches, 11:*169*, 11:*170*, 11:*172*
Brown-headed spider monkeys, 14:167*t*
Brown-headed stork-billed kingfishers. *See* Stork-billed kingfishers
Brown honeyeaters, 11:*240*, 11:246–*247*
Brown howler monkeys, 14:159, 14:160, 14:166*t*
Brown hyenas, 14:*361*, 14:*362*, 14:*363*, 14:366
behavior, 14:259, 14:260, 14:362
distribution, 14:*365*
evolution, 14:359
feeding ecology, 14:260–261, 14:*360*, 14:362
taxonomy, 14:359–360
Brown jacamars, 10:91
Brown jays, 11:504
Brown kingfishers. *See* Laughing kookaburras
Brown kiwis, 8:*10*, 8:89, 8:90, 8:*90*, 8:*91*, 8:92, 8:*92–93*
Brown lemmings, 16:*228*
Brown lemurs, 14:8, 14:52, 14:*54*, 14:*55*, 14:59
Brown long-eared bats, 13:*502*, 13:*504*, 13:*513*–514
Brown mayflies, 3:*128*, 3:*129*–130
Brown mesites, 9:5, 9:*6*, 9:7, 9:*8–9*
Brown mouse lemurs. *See* Red mouse lemurs
Brown murine bats. *See* Brown tube-nosed bats
Brown nightjars, 9:368, 9:401, 9:404
Brown owls. *See* Tawny owls
Brown pelicans, 8:*183*, 8:185, 8:186, 8:225, 8:*226*, 8:*227*, 8:*229*, 8:*231–232*
Brown ragfishes. *See* Ragfishes
Brown rats, 12:*184*, 12:*187*, 12:190, 16:*253*, 16:*254*
behavior, 16:251, 16:257–258
conservation status, 16:253, 16:258
distribution, 16:*256*, 16:257
feeding ecology, 16:258
habitats, 16:250–251, 16:257
humans and, 16:258
physical characteristics, 16:257
reproduction, 12:105–106, 16:258
taxonomy, 16:257
Brown recluse spiders, 2:336
Brown roatelos. *See* Brown mesites
Brown roofed turtles, 7:72
Brown sandboas, 7:410
Brown shrike-thrushes. *See* Gray shrike-thrushes
Brown shrikes, 10:430
Brown skuas, 9:163, 9:206
Brown smoothhound sharks, 4:128
Brown snakes, 7:483, 7:485, 7:*489*, 7:*496*, 7:497–498
Brown-spotted yellow wing dragonflies, 3:*22*
Brown-tailed apalis. *See* Yellow-breasted apalis
Brown-tailed sicklebills. *See* White-tipped sicklebills
Brown teals, 8:363, 8:364, 8:367, 8:*374*, 8:*384*
Brown thornbills, 11:56, 11:57
Brown thrashers, 8:39, 10:465, 10:*466*, 10:*468*, 10:472–*473*
Brown-throated conures. *See* Brown-throated parakeets
Brown-throated parakeets, 9:*281*, 9:*290*–291
Brown-throated sunbirds. *See* Plain-throated sunbirds
Brown-throated three-toed sloths, 13:*149*, 13:*161*, 13:*163*, 13:*164*, 13:*166*, 13:*167*
Brown-throated treecreepers, 11:178
Brown titis, 14:*145*, 14:153*t*
Brown tree toads, 6:*187*, 6:193–194
Brown treecreepers, 11:*134*
Brown treesnakes, 7:470, 7:*472*, 7:*473*, 11:98, 11:230, 11:508
Brown tremblers, 10:*469*, 10:*473*–474
Brown trouts, 4:50, 4:*407*, 4:*408*, 4:*416*–417
Brown tube-nosed bats, 13:*523*, 13:524
Brown University, 12:57
Brown waxbills. *See* Common waxbills
Brown-winged starlings, 11:408
Brown woodpeckers. *See* Smoky-brown woodpeckers
Brown wren-babblers. *See* Pygmy wren-babblers
Brownbanded cockroaches, 3:76, 3:*152*, 3:*155*, 3:158
Brown's cormorants. *See* Brandt's cormorants
Brownsnout spookfishes, 4:392, 4:399
Brubrus, 10:425–430
Bruce, Murray, 9:338
Bruce's green pigeons, 9:*254*, 9:263–*264*
Brueelia spp., 3:251, 3:*252*
Brugia malayi, 1:35
Bruijn's brush turkeys, 8:401
Brumbies, 12:176
Brundin, L., 3:55
Brunei beauties, 5:429
Brunnich's guillemots. *See* Thick-billed murres
Brusca, G. J., 1:13
Brusca, R. C., 1:13
Brush lizards. *See* Common sagebrush lizards
Brush mice, 16:266
Brush-tailed bettongs, 13:39, 13:74, 13:*75*, 13:76, 13:77, 13:*78*–*79*
Brush-tailed phascogales, 12:*290*, 12:*292*, 12:297
Brush-tailed possums, 10:331
Brush-tailed rock wallabies, 13:85, 13:*92*, 13:*96*–*97*
Brush-turkeys. *See* Moundbuilders
Brushland tinamous, 8:58, 8:59, 8:*60*, 8:*65*
Brushtailed possums. *See* Common brushtail possums
Bruton, M. N., 4:190, 4:193
Bryaninops yongei. *See* Whip coral gobies
Bryant's sparrows. *See* Savannah sparrows
Brycon petrosus, 4:338
Bryde's whales, 15:94, 15:120, 15:122–*126*, 15:*128*, 15:129
Brygoo, E. R., 7:235
Bryobatrachus nimbus. *See* Moss frogs
Bryonia spp., 1:153
Bryoniida, 1:153
Bryopsis spp., 1:45
Bryozoans, 1:13, 1:22, 2:5, 2:25, 2:29
freshwater, **2:497–502,** 2:*499*
marine, **2:503–508,** 2:*505*, **2:509–514,** 2:*512*
See also Ectoprocta
Bubal hartebeests, 16:33
Bubalornis niger. *See* Red-billed buffalo weavers
Bubalornithinae. *See* Buffalo weavers
Bubalus spp., 16:11
Bubalus bubalis. *See* Water buffaloes
Bubalus depressicornis. *See* Anoas
Bubalus mephistopheles. *See* Short-horned water buffaloes
Bubble coral shrimps, 2:*197*
Bubble corals, 1:*106*
Bubbling kassina, 6:*284*, 6:287–*288*
Bubo spp. *See* Eagle-owls
Bubo ascalaphus. *See* Savigny's eagle-owls
Bubo blakistoni. *See* Blakiston's eagle-owls
Bubo bubo. *See* Eurasian eagle-owls
Bubo lacteus. *See* Verreaux's eagle owls
Bubo poirrieri, 9:331, 9:345
Bubo sumatranus. *See* Barred eagle-owls
Bubo virginianus. *See* Great horned owls
Bucco capensis. *See* Collared puffbirds
Bucconidae. *See* Puffbirds
Bucephala spp. *See* Golden-eye ducks
Buceros bicornis. *See* Great hornbills
Buceros hydrocorax. *See* Rufous hornbills
Buceros rhinoceros. *See* Rhinoceros hornbills
Bucerotes, 10:1–4
Bucerotidae. *See* Hornbills
Bucerotinae. *See* Hornbills

Bucket-tailed loriciferans, 1:345, 1:*346*, 1:*348*, 1:349
Buckler dories, 5:124, 5:*125*, 5:128–*129*
Bucorvinae. *See* Hornbills
Bucorvus abyssinicus. See Northern ground-hornbills
Bucorvus leadbeateri. See Southern ground-hornbills
Budgerigars, 8:24, 8:40, 9:276, 9:278–280, 9:*282*, 9:285–*286*
Budgett's frogs, 6:*161*, 6:162–*163*
Budgies. *See* Budgerigars
Budorcas taxicolor. See Takins
Budorcas taxicolor bedfordi. See Golden takins
Buergeria spp., 6:291, 6:292, 6:293
Buergeria buergeri. See Buerger's frogs
Buergerinae, 6:291
Buerger's frogs, 6:*294*, 6:*295*
Buettner-Janusch, 14:75
Buff-backed herons. *See* Cattle egrets
Buff-banded rails, 9:*49*, 9:52
Buff-breasted buttonquails, 9:14, 9:19
Buff-breasted flowerpeckers. *See* Fire-breasted flowerpeckers
Buff-breasted sandpipers, 9:177–178, 9:179
Buff-breasted scrubwrens. *See* White-browed scrubwrens
Buff-breasted warblers. *See* Mangrove gerygones
Buff-cheeked gibbons. *See* Golden-cheeked gibbons
Buff-rumped thornbills, 11:56
Buff-spotted flufftails, 9:48, 9:52, 9:*55*, 9:*57*
Buff-spotted woodpeckers, 10:144
Buffalo sculpins, 5:181, 5:182
Buffalo weavers, 11:375, 11:376, 11:377–378, 11:*380*, 11:382
Buffaloes, **16:11–25,** 16:*17*, 16:24*t*–*25t*
  See also African buffaloes; Water buffaloes
Buffenstein, Rochelle, 12:74
Buffy flower bats, 13:*422*, 13:*425*, 13:426–427
Buffy-headed marmosets, 14:124, 14:*125*, 14:*130*
Buffy tufted-ear marmosets, 14:117, 14:121
Bufo spp.
  defense mechanisms, 6:66
  evolution, 6:183
  feeding ecology, 6:67
  larvae, 6:42
  medicinal uses of, 6:53
  reproduction, 6:31–32
Bufo americanus. See American toads
Bufo anotis. See Chirinda toads
Bufo boreas, 6:67
Bufo bufo. See Common European toads
Bufo calamita. See Natterjack toads
Bufo ephippium. See Pumpkin toadlets
Bufo fuscus. See Common spadefoot
Bufo houstonensis. See Houston toads
Bufo laevis. See Common plantanna
Bufo marinus. See Marine toads
Bufo melanostictus. See Common Sunda toads
Bufo obstetricans. See Midwife toads
Bufo periglenes. See Golden toads
Bufo quercicus. See Oak toads
Bufo siachinensis. See Pakistani toads
Bufo viridis. See Green toads
Bufocephala spp., 7:77
Bufoides spp., 6:184
Bufonidae, **6:183–195,** 6:*186–187*
  behavior, 6:185, 6:188–194
  conservation status, 6:185, 6:188–194
  distribution, 6:5, 6:*183*, 6:184, 6:*188–194*
  evolution, 6:183
  feeding ecology, 6:185, 6:188–194
  habitats, 6:184, 6:188–194
  humans and, 6:185, 6:188–194
  physical characteristics, 6:183–184, 6:188–194
  reproduction, 6:35, 6:*185*, 6:188–194
  species of, 6:188–194
  taxonomy, 6:188–194, 6:197
Bugeranus spp., 9:23
Bugeranus carunculatus. See Wattled cranes
Bugfishes. *See* Atlantic menhadens
Bugs, 3:8, 3:9, 3:10, 3:68, 3:261
  aquatic, 3:260, 3:261
  palm, 3:9
  terrestrial scutellerid, 3:260
  See also Hemiptera; specific types of bugs
Bugtilemur spp., 14:35
Bugula turbinata, 2:*512*, 2:*513*
Bukovin mole rats, 16:297*t*
Bulb-tip anemones. *See* Magnificent sea anemones
Bulbonaricus spp., 5:133
Bulbuls, **10:395–413,** 10:*400–401*
  behavior, 10:397
  conservation status, 10:398–399
  distribution, 10:*395*, 10:396
  evolution, 10:395
  feeding ecology, 10:398
  habitats, 10:396–397
  humans and, 10:399
  physical characteristics, 10:395–396
  reproduction, 10:398
  species of, 10:*402–413*
  taxonomy, 10:395
Bulky kakapos, 9:275
Bull sharks, 4:*30*, 4:*117*, 4:*120*–121, 4:123
Bullbirds. *See* Umbrellabirds
Bulldog bats, 12:*53*, **13:443–451,** 13:*447*, 13:*448*
Bulldog plecos, 4:*353*
Buller, Lawry, 10:207
Buller's mollymawks, 8:114, 8:115
Buller's pocket gophers, 16:*191*, 16:*192*, 16:194–195
Bullfinches, 11:323, 11:324, 11:*327*, 11:335
Bullfrogs
  African, 6:247, 6:*249*, 6:*254*, 6:261
  South American, 6:*157*, 6:*160*, 6:166–167
  See also Indian tiger frogs; North American bullfrogs
Bullhead buffaloes. *See* Bigmouth buffaloes
Bullhead catfishes, 4:57, 4:352
Bullhead sharks, 4:11, **4:97–103,** 4:*100*
Bullock's orioles. *See* Baltimore orioles
Bullrouts, 5:164
Bullseye snakeheads, 5:438, 5:441, 5:443–444
Bullsnakes, in Hopi snake dance, 7:57
Bulmer's fruit bats, 13:348*t*
Bulo Berti boubous, 10:427, 10:429
Bulweria bulwerii. See Bulwer's petrels
Bulweria fallax. See Joaquin's petrels
Bulwer's petrels, 8:*132*
Bulwer's pheasants, 8:434
Bumblebee bats. *See* Kitti's hog-nosed bats
Bumblebee catfishes. *See* Coral catfishes
Bumblebees, 3:*21*, 3:60, 3:68, 3:70–71, 3:72, 3:405
  large red-tailed, 3:*410*, 3:*412*
  long-tongued, 3:78
  pollination by, 3:78
  radar tagged, 3:*86*
Bungarus fasciatus, 7:51
Bunkers. *See* Atlantic menhadens
Bunker's earless lizards. *See* Common lesser earless lizards
Bunocephalus spp., 4:357
Bunolagus monticularis. See Riverine rabbits
Bunopithecus spp., 14:207
Buntings. *See* New World finches
Bunyoro rabbits, 16:482, 16:505
Buoyancy, 4:24, 12:68
Buphaginae. *See* Oxpeckers
Buphagus erythrorhynchus. See Red-billed oxpeckers
Burbots, 5:26, 5:*31*, 5:*35*, 5:36
  See also Lotidae
Burch, J. B., 2:412
Burchell's coursers, 9:*154*, 9:*158*
Burchell's sandgrouse, 9:*232*
Burchell's zebras. *See* Plains zebras
Burgess shale, 12:64
Burhinidae. *See* Thick-knees
Burhinus spp. *See* Thick-knees
Burhinus capensis. See Spotted dikkops
Burhinus grallarius. See Bush thick-knees
Burhinus oedicnemus. See Stone-curlews
Burhinus senegalensis. See Senegal thick-knees
Burhinus vermiculatus. See Water dikkops
Buried-eye caecilians, 6:5, 6:411–*412*, **6:431–434,** 6:*432*, 6:*433*
Burmeister's porpoises, 15:33–36, 15:*37*, 15:*38*, 15:39
Burmeister's seriemas. *See* Black-legged seriemas
Burmese hornbills. *See* Plain-pouched hornbills
Burmese pythons, 7:*420*
Burmese spadefoot toads, 6:*110*, 6:*112*, 6:116–117
Burramyidae. *See* Pygmy possums
Burramys spp. *See* Mountain pygmy possums
Burramys parvus. See Mountain pygmy possums
Burridge, C. P., 4:59
Burros, 12:182
Burrowing asps, 7:461
Burrowing bettongs. *See* Boodies
Burrowing boas. *See* Neotropical sunbeam snakes
Burrowing gobies, 4:70
Burrowing mud anemones. *See* American tube dwelling anemones
Burrowing owls, 9:333, 9:346, 9:349, 9:350, 9:*353*, 9:362–*363*
Burrowing parakeets, 9:278
Burrowing pythons. *See* Calabar boas
Burrowing snakes
  African, **7:461–464**
  colubrids, 7:466
  Neotropical, 7:468
  Nilgiri, 7:*393*
  yellow-and-black, 7:462
Burrowing starfishes. *See* Sand stars
Burrowing techniques, 7:274, 7:280, 7:288
  See also Behavior
Burrowing toads. *See* Mesoamerican burrowing toads
Burrowing wombats, 13:32

Burrows, 4:62, 4:64
Bursa spp., 2:447
Bursaphaloenchus exylophilus, 1:36
Bursovaginoidea, 1:331, 1:332
Burton, T. C., 6:301
Burton's legless lizards. *See* Burton's snake lizards
Burton's snake lizards, 7:262, 7:*264*, 7:*265*, 7:269
Buru Island monarchs. *See* White-tipped monarchs
Burunduks. *See* Siberian chipmunks
Burying beetles, 3:321, 3:322
See also American burying beetles
Bush canaries. *See* Yellowheads
Bush cows. *See* Lowland tapirs
Bush-crickets, 3:31, 3:38, 3:39, 3:201
See also Speckled bush-crickets
Bush dassies. *See* Bush hyraxes
Bush dogs, 14:266, 14:267, 14:269, 14:270, 14:*274*, 14:284*t*
Bush duikers, 16:*75*, 16:*76*, 16:77
Bush grasshoppers, 3:201
Bush hyraxes, 15:177–*179*, 15:180–*185*, 15:*186*, 15:187–188
Bush pigs, 12:136, 15:278, 15:280, 15:*283*, 15:*284*–285
Bush pipits, 10:373, 10:*377*, 10:*382*–383
Bush-quails, 8:434
Bush rats, 16:251
Bush-shrikes, 10:425–430, 10:*431*, 10:432–*433*, 10:*433*, 10:*434*
Bush squeakers, 6:*267*, 6:269
Bush stone-curlews, 9:*144*
Bush thick-knees, 9:144, 9:145–146
Bush turkeys. *See* Australian brush-turkeys
Bush wrens, 10:203
Bushbabies, 14:1, **14:23–34**, 14:*27*
behavior, 14:25–26
conservation status, 14:26
distribution, 14:5, 14:*23*, 14:24
evolution, 14:2, 14:3, 14:23
feeding ecology, 14:9, 14:26
habitats, 14:6, 14:24
humans and, 14:26
physical characteristics, 14:23–24
reproduction, 14:26
species of, 14:28–32, 14:32*t*–33*t*
taxonomy, 14:4, 14:23
Bushbucks, 16:11, 16:14, 16:24*t*
Bushcreepers, 10:209
Bushdogs. *See* Tayras
Bushhoppers, 3:208
Bushmasters, 7:*450*, 7:*452*, 7:456
Bushtits, 10:169, 11:141, 11:142, 11:*146*
Bushveld gerbils, 16:296*t*
Bushveld horseshoe bats, 13:390
Bushveld pipits. *See* Bush pipits
Bushveld rain frogs, 6:35, 6:*306*, 6:*309*
Bushveld sengis, 16:*518*, 16:519, 16:531*t*
Bushy-tailed gundis. *See* Speke's pectinators
Bushy-tailed hutias. *See* Black-tailed hutias
Bushy-tailed mongooses, 14:357*t*
Bushy-tailed opossums, 12:249–253, 12:*255*, 12:*259*
Bushy-tailed rats, western Malagasy, 16:*284*
Bushy-tailed woodrats, 16:*124*, 16:277*t*
Bustard quail. *See* Barred buttonquails
Bustards, 9:1–4, **9:91–100**, 9:*95*
Bustards, Hemipodes and Sandgrouse (Johnsgard), 9:1, 9:11
Busuanga jungle toads. *See* Philippine barbourulas
Butcherbirds, 11:467–468, 11:*470*, 11:472–473
Buteo jamaicencis. See Red-tailed hawks
Buteo lagopus. See Rough-legged buzzards
Buteo magnirostris. See Roadside hawks
Buteo platypterus. See Broad-winged hawks
Buteo swainsoni. See Swainson's hawks
Buteogallus aequinoctialis. See Crab hawks
Butis koilomatodon, 5:377
Butler, Percy M., 13:289
Butterfishes, 5:295, 5:298, 5:*300*, 5:*301*, 5:304
See also Rock whitings
Butterfishes (Stromateidae), 5:*421*, **5:421–426**, 5:*423*
Butterflies, 1:42, 2:37–38, **3:383–399**, 3:*390*, 3:*391*, 3:*392*
ants and, 3:72
behavior, 3:65, 3:*385*–386
as bioindicators, 3:82
conservation status, 3:82, 3:87, 3:88, 3:90, 3:388–389
distribution, 3:385
evolution, 3:383, 3:385
farming of, 3:80
feeding ecology, 3:42, 3:46, 3:48, 3:63, 3:386–387
flowers and, 3:14
habitats, 3:385
humans and, 3:74, 3:389
migration, 3:37
as pests, 3:50, 3:75
physical characteristics, 3:10, 3:*24*, 3:384–385
puddling, 3:38
reproduction, 3:34, 3:36, 3:38, 3:387–388
species of, 3:394–399
See also specific types of butterflies
Butterfly agamas, 7:*213*, 7:*215*, 7:220
Butterfly bats, 13:515*t*
See also Gervais' funnel-eared bats
Butterfly bobtail squids, 2:*481*, 2:*484*, 2:487–488
Butterfly clams. *See* Coquina clams
Butterfly cods. *See* Red lionfishes
Butterfly gardening, 3:91
Butterfly houses, 3:*90*
Butterflyfishes, 5:49, **5:235–247**, 5:*237*, 5:*245*
behavior, 4:62–63, 4:232–233, 5:239–240
conservation status, 4:234, 5:242–243
distribution, 4:57, 4:232, 5:235–237, 5:237–238
evolution, 5:235
feeding ecology, 4:67, 4:233, 5:240–242
habitats, 4:48, 4:50, 4:232, 5:238–239
humans and, 4:234, 5:243–244
physical characteristics, 4:231–232, 5:235–237
reproduction, 4:233–234, 5:242
species of, 4:240–*241*, 5:*247*
taxonomy, 4:231, 5:235
Buttonquails, **9:11–22**, 9:*12*, 9:*15*
behavior, 9:12–13
conservation status, 9:14
distribution, 9:*11*, 9:12
evolution, 9:11
feeding ecology, 9:13
habitats, 9:12–13
humans and, 9:14
physical characteristics, 9:12
reproduction, 9:13–14
species of, 9:*16–21*
taxonomy, 9:1, 9:11–12
Buzzards, 8:318
By the wind sailors, 1:10, 1:127, 1:*132*, 1:136–137
Bycanistes brevis. See Silvery-cheeked hornbills
Bycanistes cylindricus. See Brown-cheeked hornbills
Byrrhoidea, 3:316
Byrsotria fumigata, 3:150
Bythitidae, 5:15–18
Bythitoidei, 5:16, 5:18

C

Cabanis's greenbuls, 10:397
Cabassous spp. *See* Naked-tailed armadillos
Cabassous centralis. See Northern naked-tailed armadillos
Cabassous chacoensis. See Chacoan naked-tail armadillos
Cabassous tatouay. See Greater naked-tailed armadillos
Cabassous unicinctus. See Southern naked-tailed armadillos
Cabbage birds. *See* Spotted bowerbirds
Cabbage white butterflies, 3:37, 3:39, 3:*391*, 3:*399*
Cabbagehead jellyfish. *See* Cannonball jellyfish
Cabezons, 5:183, 5:*185*, 5:*186*, 5:189–190
Cabot, John, 5:30
Cabral, Pedro, 15:242
Cacajao spp. *See* Uakaris
Cacajao calvus. See Bald uakaris
Cacajao calvus calvus. See White bald uakaris
Cacajao calvus novaesi. See Novae's bald uakaris
Cacajao calvus rubicundus. See Red bald uakaris
Cacajao calvus ucayalii. See Ucayali bald uakaris
Cacajao melanocephalus. See Black uakaris
Cacatua spp. *See* Cockatoos
Cacatua banksii. See Black cockatoos
Cacatua galerita. See Sulphur-crested cockatoos
Cacatua pastinator. See Western corella parrots
Cacatua tenuirostris. See Slender-billed parrots
Cachamas. *See* Pirapitingas
Cacicus spp. *See* Caciques
Cacicus cela. See Yellow-rumped caciques
Cacicus koepckeae. See Selva caciques
Cacicus melanicterus. See Yellow-winged caciques
Cacicus montezuma. See Montezuma's oropendolas
Cacicus sclateri. See Ecuadorian caciques
Caciques, 11:301–304, 11:306, 11:*308*, 11:309–310
Cacklers. *See* Gray-crowned babblers
Cacomistles, 14:310, 14:316*t*
Cacosterninae, 6:245–246, 6:251
Cacosternum spp., 6:245
Cactospiza pallida. See Woodpecker finches
Cactus ground-finches, 11:*265*
Cactus mealybugs, 3:81

Cactus mice, 16:*264*
Cactus wrens, 10:174, 10:526, 10:527, 10:528, 10:*530*, 10:*531*
Caddisflies, **3:375–381,** 3:*376*, 3:*377*, 3:*378*
behavior, 3:64, 3:376
evolution, 3:7, 3:375
feeding ecology, 3:63, 3:376
reproduction, 3:34, 3:38, 3:60, 3:376–377
Tobias's, 3:57
Caduceus, 7:54–55
Caecilia spp., 6:436
Caecilia bivittatum. See Two-lined caecilians
Caecilia compressicauda. See Cayenne caecilians
Caecilia glutinosa. See Ceylon caecilians
Caecilia oxyura. See Red caecilians
Caecilia thompsoni, 6:412
Caecilians, **6:411–413**
American tailed, 6:5, 6:39, 6:411, 6:*412*, **6:415–418,** 6:*417*
behavior, 6:44, 6:412
buried-eye, 6:5, 6:*411–412*, **6:431–434,** 6:*432*, 6:*433*
distribution, 6:5, 6:412
evolution, 6:4, 6:9–10, 6:11, 6:13, 6:15, 6:411
feeding ecology, 6:6, 6:412
habitats, 6:412
Kerala, 6:5, 6:411, 6:*412*, **6:425–429,** 6:*427*
as pets, 6:55
physical characteristics, 6:15, 6:26–27, 6:411–412, 6:415–416
reproduction, 6:29–30, 6:32, 6:35, 6:38, 6:39, 6:412
taxonomy, 6:3, 6:411, 6:*412*
See also Asian tailed caecilians; Tailless caecilians
Caeciliidae. *See* Tailless caecilians
Caeciliusidae, 3:243
Caelifera. *See* Short-horned grasshoppers
Caenis spp., 3:126
Caenogastropods, **2:445–449,** 2:*448*
Caenolestes spp., 12:267–268
Caenolestes fuliginosus. See Silky shrew opossums
Caenolestidae. *See* Shrew opossums
Caenotheriidae, 15:264
Caerulean paradise-flycatchers, 11:97
Caerulin, 6:240
Caesionidae. *See* Fusiliers
Cagebirds, 8:22, 8:24
bulbuls as, 10:399
canaries as, 11:325, 11:339
parrots as, 9:280
red-billed leiothrix as, 10:517
turacos as, 9:302
weaverfinches as, 11:356, 11:357
See also Humans
Cahows. *See* Bermuda petrels
Caiman crocodilus. See Common caimans
Caiman latirostris. See Broad-snouted caimans
Caiman lizards, 7:201, 7:309, 7:310, 7:311, 7:312, 7:*313*, 7:*316*
Caiman yacare. See Yacaré caimans
Caimaninae. *See* Caimans
Caimans, 7:157–165, **7:171–178,** 7:*174*
Caipora bambuiorum, 14:155
Cairina moschata. See Muscovy ducks
Cairina scutulata. See White-winged ducks
Calabar angwantibos, 14:16, 14:*17*, 14:*19*–20
Calabar boas, 7:409, 7:411, 7:*412*, 7:415–*416*
Calabar ground pythons. *See* Calabar boas
Calamaian deer, 15:360
Calamanostraca, 2:141
Calamanthus spp. *See* Fieldwrens
Calamaria spp. *See* Reedsnakes
Calamonastes spp., 11:6
Calamospiza melanocorys. See Lark buntings
Calandra larks, 10:345, 10:346, 10:*347*, 10:*351*
Calandrella spp. *See* Larks
Calandrella brachydactyla. See Greater short-toed larks
Calandrella rufescens, 10:*345*
Calanoida, 2:299, 2:300
Calanus finmarchicus, 2:*303*, 2:304–305
Calcarea. *See* Calcareous sponges
Calcareous sponges, **1:57–65,** 1:*58*, 1:*59*, 1:*60*
behavior, 1:58
conservation status, 1:59
distribution, 1:58
evolution, 1:57
feeding ecology, 1:58
habitats, 1:24, 1:58
humans and, 1:59
physical characteristics, 1:*58*
reproduction, 1:20, 1:21, 1:59
species of, 1:*61–65*
taxonomy, 1:57–58
Calcarius lapponicus. See Lapland longspurs
Calcaronea, 1:57, 1:58
Calcichordates, 1:13
Calcinea, 1:57, 1:58
Calcium, subterranean mammals and, 12:74
Calcochloris spp., 13:216
Calcochloris obtusirostris. See Yellow golden moles
Caldon turtles. *See* Leatherback seaturtles
Caldwell, R. L., 2:168
Calechidna spp., 7:445
Calicalicus spp. *See* Vanga shrikes
Calicalicus madagascariensis. See Red-tailed vangas
Calicalicus rufocarpalis. See Red-shouldered vangas
Calicivirus, 12:187–188, 12:223
Calico basses. *See* Black crappies
Calidrinae, 9:175–180
Calidris spp., 9:169
Calidris acuminata. See Sharp-tailed sandpipers
Calidris alba. See Sanderlings
Calidris alpina. See Dunlins
Calidris bairdii. See Baird's sandpipers
Calidris canutus. See Red knots
Calidris ferruginea. See Curlews
Calidris fuscicollis. See White-rumped sandpipers
Calidris maritima. See Purple sandpipers
Calidris mauri. See Western sandpipers
Calidris minuta. See Little stints
Calidris ptilocnemis. See Rock sandpipers
Calidris temminckii. See Temminck's stints
Calidris tenuirostris. See Great knots
California anchovies. *See* Northern anchovies
California angelsharks. *See* Pacific angelsharks
California bullhead sharks, 4:*29*, 4:*97*, 4:98, 4:*100*, 4:*101*
California condors, 8:27, 8:234, 8:236, 8:275, 8:276, 8:278, 8:*279*, 8:*283–284*
California Department of Fish and Games Committee on Threatened Trout, 4:411
California flyingfishes, 5:*82*, 5:*85*–86
California giant salamanders, 6:350, 6:*352*
California gray squirrels. *See* Western gray squirrels
California grunions, 5:69, 5:72, 5:*73*–74
California horned lizards. *See* Texas horned lizards
California jays. *See* Western scrub-jays
California lampshells, 2:*524*, 2:*525*, 2:527
California leaf-nosed bats, 12:79, 13:315, 13:*417*, 13:420, 13:*422*, 13:*423*
California legless lizards, 7:*341*, 7:342–*343*
California market squids, 2:*22*, 2:477
California meadow voles, 16:125–126
California mice, 16:125–126, 16:278*t*
California newts, 6:*368*, 6:373–374
California quails, 8:456, 8:457
California scorpionfishes, 5:*168*, 5:*171*, 5:174
California sea lions, 12:124, 14:394, 14:396, 14:*397*, 14:399, 14:*401*, 14:*404*, 14:405
California sheepheads, 5:*300*, 5:302–*303*
California slickheads, 4:*394*, 4:*395*
California smoothtongues, 4:*394*, 4:*396*, 4:398
California thrashers, 10:467
California tonguefishes, 5:*454*, 5:*456*–457
California whiptail lizards, 7:*310*
California woodpeckers, 10:149
Californian needlefishes, 5:*82*, 5:84–*85*
Caligo spp., 3:*386*
Caligo memnon. See Owl butterflies
Caliroa cerasi. See Pear and cherry slugs
Callaeas wilsoni, 11:449
Callaeidae. *See* New Zealand wattle birds
Callagur borneoensis. See Painted terrapins
Callanthiidae. *See* Splendid perches
Callibaetis spp., 3:125–126
Callicebinae. *See* Titis
Callicebus spp. *See* Titis
Callicebus barbarabrownae. See Northern Bahian blond titis
Callicebus brunneus. See Brown titis
Callicebus coimbrai. See Coimbra's titis
Callicebus cupreus. See Red titis
Callicebus donacophilus, 14:143, 14:144
Callicebus medemi. See Medem's collared titis
Callicebus melanochir. See Southern Bahian masked titis
Callicebus modestus, 14:143
Callicebus moloch. See Dusky titi monkeys
Callicebus nigrifrons. See Black-fronted titis
Callicebus oenathe. See Andean titis
Callicebus olallae. See Beni titis
Callicebus ornatus. See Ornate titis
Callicebus personatus. See Masked titis
Callicebus torquatus. See Collared titis
Callichthyidae, 4:351, 4:352
Callimico goeldii. See Goeldi's monkeys
Callionymidae. *See* Dragonets
Callionymoidei, **5:365–372,** 5:*368*
Callionymus spp., 5:365, 5:366
Callionymus sanctaehelenae. See St. Helena dragonets
Callipepla californica. See California quails
Callipepla gambelii. See Gambel's quail
Calliphlox spp. *See* Hummingbirds
Calliphlox amethystinus. See Amethyst woodstars
Calliphlox evelynae. See Bahama woodstars
Calliphoridae. *See* Bottle flies
Callipodida, 2:365
Callipogon armillatus, 3:*318*

Callisaurus draconoides. See Zebra-tailed lizards
Callithrix spp. *See* Eastern Brazilian marmosets
Callithrix argentata. See Silvery marmosets
Callithrix aurita. See Buffy tufted-ear marmosets
Callithrix flaviceps. See Buffy-headed marmosets
Callithrix humeralifera. See Tassel-eared marmosets
Callithrix jacchus. See Common marmosets
Callithrix penicillata. See Black tufted-ear marmosets
Callitrichidae, **14:115–133,** 14:*125*, 14:*126*
  behavior, 12:147, 14:116–120, 14:*123*
  claws, 12:39
  conservation status, 14:124
  distribution, 14:*115*, 14:116
  evolution, 14:115
  feeding ecology, 14:120–121
  habitats, 14:116
  humans and, 14:124
  physical characteristics, 14:115–116
  reproduction, 14:121–124
  species of, 14:*127–131*, 14:132*t*
  taxonomy, 14:4, 14:115
Calliurichthys spp., 5:365, 5:366
Callixalus spp., 6:280
Callobatrachus spp., 6:12
Callopistes spp., 7:309, 7:312
Callopistes flavipunctatus, 7:310
Callopistes maculatus, 7:310
Calloplesiops spp., 5:256
Callorhinchidae, 4:91
Callorhinchus milii. See Ghost sharks
Callorhinus spp. *See* Northern fur seals
Callorhinus ursinus. See Northern fur seals
Callosciurus spp., 16:163
Callosciurus erythraeus. See Pallas's squirrels
Callosciurus pygerythrus, 16:167
Callosciurus quinquestriatus, 16:167
Calloselasma spp., 7:445
Calloselasma rhodostoma. See Malaysian pitvipers
Calls. *See* Behavior; Songs, bird; Vocalizations
Callulops robustus. See Boulenger's callulops frogs
Calocitta formosa. See White-throated magpie-jays
Calodactylodes aureus. See Indian golden geckos
Calomys spp., 16:269, 16:270
Calomys laucha, 16:267, 16:270
Calomys musculinus, 16:269, 16:270
Calomys venustus, 16:269
Calomyscinae, 16:282
Calomyscus bailwardi. See Mouse-like hamsters
Calomyscus hotsoni. See Hotson's mouse-like hamsters
Calomyscus mystax. See Afghan mouse-like hamsters
Calomyscus tsolovi. See Tsolov's mouse-like hamsters
Calomyscus urartensis. See Urartsk mouse-like hamsters
Calonectris diomedea. See Cory's shearwaters
Calotes, **7:209–211,** 7:214–215
Calotes liocephalus, 7:211
Calotes versicolor. See Brown garden lizards
Calothorax spp. *See* Hummingbirds
Calothorax lucifer. See Lucifer hummingbirds
Calpe eustrigiata, 3:387
Caluella spp., 6:302
Calumia spp., 5:375
Calumma spp., 7:*37*, 7:224, 7:227, 7:232
Calumma boettgeri, 7:224
Calumma nasuta, 7:224
Calumma parsonii. See Parson's chameleons
Caluromyinae, 12:249
Caluromys spp. *See* Woolly opossums
Caluromys philander. See Bare-tailed woolly opossums
Caluromysiops spp. *See* Black-shouldered opossums
Caluromysiops irrupta. See Black-shouldered opossums
Calymmodesmus montanus, 2:365
Calyptahyla crucialis, 6:37
Calypte spp. *See* Hummingbirds
Calypte anna. See Anna's hummingbirds
Calypte costae. See Costa hummingbirds
Calypteratae, 3:357
Calyptogena magnifica. See Giant vent clams
Calyptomena spp. *See* Broadbills
Calyptomena hosii. See Hose's broadbills
Calyptomena viridis. See Green broadbills
Calyptomena whiteheadi. See Whitehead's broadbills
Calyptommatus spp., 7:303
Calyptura cristata. See Kinglet calypturas
Camaroptera spp., 11:6
Camaroptera brachyura, 11:6
Camas pocket gophers, 16:195*t*
Cambrian explosion, 1:8–9, 2:10–11
  See also Evolution
Camel crickets, 3:201, 3:203, 3:204, 3:207, 3:*210*, 3:212–*213*
Camelidae, **15:313–323,** 15:*319*
  behavior, 15:316–317
  conservation status, 15:317–318
  distribution, 12:132, 12:134, 15:*313*, 15:315
  domestication of, 15:150–151
  evolution, 15:313–314
  feeding ecology, 15:317
  habitats, 15:315–316
  humans and, 15:318
  physical characteristics, 15:314–315
  reproduction, 15:317
  species of, 15:*320–323*
  taxonomy, 15:267, 15:313–314
Camelini, 15:313
Camels, 15:*151*, 15:263, **15:313–323**
  behavior, 15:316–317
  conservation status, 15:317–318
  distribution, 15:*313*, 15:315
  evolution, 15:136, 15:137, 15:138, 15:265, 15:313–314
  feeding ecology, 15:271, 15:317
  habitats, 15:315–316
  humans and, 15:273, 15:318
  neonatal milk, 12:127
  physical characteristics, 15:314–315
  reproduction, 15:143, 15:317
  species of, 15:320–323
  taxonomy, 15:313–314
  thermoregulation, 12:113
Camelus spp. *See* Camels
Camelus bactrianus. See Bactrian camels
Camelus dromedarius. See Dromedary camels
Camelus ferus. See Bactrian camels
Camelus thomasi, 12:179
Cameras, in field studies, 12:194–200
Cameron, Elizabeth, 11:86
Cameroon picathartes. *See* Gray-necked picathartes
Cameroon scaly-tails, 16:300, 16:*301*, 16:*305*
Camigiun narrow-mouthed frogs, 6:305
Camouflaged catfishes, 4:367
Campan's chameleons. *See* Jeweled chameleons
Campbell, J., 7:340
Campbell black-browed mollymawks, 8:114
Campbell Island teals, 8:363, 8:364, 8:367
Campbell's monkeys, 14:204*t*
Campecopea hirsuta, 2:250
Campellolebias spp., 4:66, 5:93
Campephaga spp. *See* Cuckoo-shrikes
Campephaga flava. See Black cuckoo-shrikes
Campephaga lobata. See Western wattled cuckoo-shrikes
Campephaga phoenicea. See Red-shouldered cuckoo-shrikes
Campephagidae. *See* Cuckoo-shrikes
Campephilus imperialis. See Imperial woodpeckers
Campephilus principalis. See Ivory-billed woodpeckers
Campethera bennettii. See Bennett's woodpeckers
Campethera nivosa. See Buff-spotted woodpeckers
Campfire program (Zimbabwe), 12:219
Campo miners, 10:*212*, 10:*214*
Campochaera spp. *See* Cuckoo-shrikes
Campochaera sloetii. See Golden cuckoo-shrikes
Campodea fragilis, 3:*109*, 3:*110*
Campodeidae, 3:107, 3:*107*
Camponotus spp., 3:408
Campostoma anomalum. See Stonerollers
Campsicnemus spp., 3:358
Campsicnemus mirabilis. See Ko'okay spurwing long-legged flies
Campsurus spp., 3:125
Camptostoma imberbe. See Northern beardless-tyrannulets
Campylomormyrus cassaicus, 4:233
Campylomormyrus phantasticus, 4:233
Campylopterus rufus. See Rufous sabrewings
Campylorhamphus pucherani. See Greater scythebills
Campylorhamphus trochilirostris. See Red-billed scythebills
Campylorhynchus spp., 10:526–527, 10:528
Campylorhynchus brunneicapillus. See Cactus wrens
Campylorhynchus nuchalis. See Stripe-backed wrens
Campylorhynchus yucatanicus. See Yucatan wrens
Canada birds. *See* White-throated sparrows
Canada geese, 8:*48*, 8:364, 8:373, 8:*374*, 8:*378*–379
Canada jays. *See* Gray jays
Canada lynx, 14:371, 14:*378*, 14:*383*, 14:388–389
Canadia spp., 2:45
Canadian beavers. *See* North American beavers
Canadian caribou, 12:180
Canadian cranes. *See* Sandhill cranes
Canadian Parks and Wilderness Society cloud sponges, 1:74

Farrea occa, 1:74
Canadian porcupines. *See* North American porcupines
Canal turtles. *See* Leatherback seaturtles
Canaries, 8:39, 8:42
See also Island canaries
Canary bird fishes. *See* Plainfin midshipmen
Canary Islands chaffinches. *See* Blue chaffinches
Canary Islands finches. *See* Blue chaffinches
Canary Islands giant lizards, 7:59
Canary Islands oystercatchers, 9:125
Canary shrews, 13:*276t*
Canasteros, 10:210
Cancer, 1:44–46, 1:49
Candelabrum penola, 1:127
Candiru, 4:*352*, 4:*356*, 4:*358*, 4:366–367
Candlefishes. *See* Eulachons
Candoia spp., 7:409, 7:410
See also Viper boas
Candoia aspera. *See* Viper boas
Candoia bibroni. *See* Fiji Island boas
Candy cane sea cucumbers, 1:*423*, 1:426–*427*
Candy-striped catfishes. *See* Iridescent shark-catfishes
Cane mice, 16:270
Cane rats, 16:121, **16:333–338**, 16:*336*
Cane toads. *See* Marine toads
Canebrakes. *See* Timber rattlesnakes
Canidae, **14:265–285**, 14:*275*
behavior, 12:145, 12:147, 12:148, 14:258, 14:267–269
conservation status, 14:262, 14:272–273
distribution, 12:131, 14:*265*, 14:267
evolution, 14:265–266
feeding ecology, 14:261, 14:269–271
habitats, 14:267
humans and, 14:262, 14:263, 14:273–274
as keystone species, 12:217
physical characteristics, 14:266–267
reproduction, 14:261, 14:271–272
species of, 14:*276–283*, 14:*283t–284t*
taxonomy, 14:256, 14:265–266
Canine heartworms, 1:*296*, 1:298
Canines (teeth), 12:46, 12:99
See also Teeth
Canirallus oculeus. *See* Gray-throated rails
Canis spp., 14:265
Canis adustus. *See* Side-striped jackals
Canis aureus. *See* Golden jackals
Canis dirus. *See* Dire wolves
Canis familiaris. *See* Domestic dogs
Canis familiaris dingo. *See* Dingos
Canis latrans. *See* Coyotes
Canis lupus. *See* Gray wolves
Canis lupus lupus. *See* Eurasian wolves
Canis lupus pallipes, Indian wolves
Canis mesomelas. *See* Silverback jackals
Canis rufus. *See* Red wolves
Canis simensis. *See* Ethiopian wolves
Cannatella, David, 6:215
Cannibalism, intrauterine, 4:133
See also Reproduction
Cannonball jellyfish, 1:157, 1:*158*, 1:163–*164*
Cansumys spp. *See* Gansu hamsters
Cantharidae, 3:81, 3:320
Cantherhines macrocerus. *See* Whitespotted filefishes
Canthigaster spp., 4:70
Canthigaster coronata. *See* Crowned tobies
Canthigaster ocellicincta. *See* Shy tobies
Canthigaster papua. *See* False-eye puffers
Canthigaster rapaensis. *See* Rapa Island tobies
Canthigaster solandri. *See* Spotted tobies
Canthigaster valentini. *See* Black saddled tobies
Canthumeryx spp., 15:265
Canyon wrens, 10:526, 10:527, 10:*530*, 10:531–*532*
Caouana turtles. *See* Leatherback seaturtles
Cape Barren geese, 8:366, 8:371, 8:*373*, 8:*374*, 8:*379*
Cape batis, 11:*26*, 11:*28*, 11:42–*43*
Cape buffaloes. *See* African buffaloes
Cape bulbuls, 10:*397*
Cape crombecs. *See* Long-billed crombecs
Cape dikkops. *See* Spotted dikkops
Cape dune mole-rats, 16:345, 16:*346*, 16:*347*
Cape filesnakes, 7:*471*, 7:*477*, 7:478
Cape flat lizards, 7:*321*, 7:*323*, 7:*324*–325
Cape fur seals, 12:*106*, 14:394
Cape gannets, 8:214, 8:*216*, 8:*218–219*
Cape golden moles, 13:197, 13:215, 13:217, 13:*222t*
Cape gray mongooses, 14:350
Cape griffons, 8:323
Cape ground squirrels. *See* South African ground squirrels
Cape grysboks, 16:63, 16:*71t*
Cape grysbucks. *See* Cape grysboks
Cape hares, 16:*483*
Cape horseshoe bats, 13:389, 13:392, 13:*395*, 13:*398*
Cape hunting dogs. *See* African wild dogs
Cape hyraxes. *See* Rock hyraxes
Cape long-billed larks, 10:342, 10:345, 10:*347*, 10:*348*–349
Cape Lopez lyretails. *See* Chocolate lyretails
Cape mole-rats, 16:*340*, 16:*346*
behavior, 16:343, 16:347
conservation status, 16:345, 16:347
distribution, 16:342, 16:*347*
evolution, 16:339
feeding ecology, 16:347
habitats, 16:342, 16:347
humans and, 16:347
photoperiodicity, 12:74
physical characteristics, 16:347
reproduction, 16:344–345
taxonomy, 16:339, 16:347
Cape mountain zebras, 15:222
Cape pangolins. *See* Ground pangolins
Cape porcupines. *See* South African porcupines
Cape puffbacks. *See* Cape batis
Cape rain frogs, 6:305
Cape rock sengis. *See* Cape sengis
Cape rough-scaled lizards, 7:298
Cape rousettes. *See* Egyptian rousettes
Cape seahorses, 5:136
Cape sengis, 16:522, 16:*531t*
Cape spinytail iguanas, 7:47–48, 7:51, 7:*248*, 7:*251*–252
Cape sugarbirds, 11:*239*, 11:252–*253*
Cape terrapins. *See* Helmeted turtles
Cape thick-knees. *See* Spotted dikkops
Cape turtledoves, 9:*251*
Cape wagtails, 10:372, 10:374, 10:376
Cape weavers, 11:377
Cape white-eyes, 11:*231*, 11:*232*
Cape York cockatoos. *See* Palm cockatoos
Capelins, 4:391, 4:392, 4:*394*, 4:*399*–400
Capensibufo spp., 6:184
Capercaillies, 8:399, 8:401, 8:434, 8:437, 8:*439*, 8:*442*–443
Caperea marginata. *See* Pygmy right whales
Capingentidae, 1:225
Capitella capitata, 2:*48*, 2:*52*
Capitellids, 2:47
Capito aurovirens. *See* Scarlet-crowned barbets
Capito maculicoronatus. *See* Spot-crowned barbets
Capito niger. *See* Black-spotted barbets
Capniidae, 3:54, 3:141
Capoeta tetrazona. *See* Tiger barbs
Capped gibbons. *See* Pileated gibbons
Capped herons, 8:234
Capped langurs, 12:*119*, 14:175
Capped terns, 9:203–204
Capped wheatears, 10:*491*, 10:*497*
Capra spp., 16:87, 16:88, 16:90, 16:91, 16:92, 16:95
Capra aegagrus. *See* Bezoar goats; Wild goats
Capra aegagrus hircus. *See* Valais goats
Capra caucasica. *See* West Caucasian turs
Capra cylindricornis. *See* East Caucasian turs
Capra falconeri. *See* Markhors
Capra falconeri heptneri. *See* Tajik markhors
Capra hircus. *See* Domestic goats
Capra ibex. *See* Alpine ibex; Nubian ibex
Capra prisca. *See* Angora goats
Capra pyrenaica. *See* Spanish ibex
Capra sibirica. *See* Siberian ibex
Capra walie. *See* Walia ibex
Caprella spp., 2:*261*
Caprella californica. *See* Skeleton shrimps
Caprellidea, 2:261
Capreolus spp. *See* Roe deer
Capreolus capreolus. *See* European roe deer
Capreolus pygargus. *See* Siberian roe deer
Capricorn white-eyes, 11:228
Capricornis spp., 16:2, 16:87
Capricornis crispus. *See* Japanese serows
Capricornis sumatraensis. *See* Serows
Capricornis swinhoei. *See* Formosan serows
Caprimulgidae. *See* Nightjars
Caprimulgiformes, **9:367–371**
Caprimulgus affinis. *See* Madagascar nightjars
Caprimulgus candicans. *See* White-winged nightjars
Caprimulgus carolinensis. *See* Chuck-will's-widows
Caprimulgus concretus. *See* Bonaparte's nightjars
Caprimulgus cristatus. *See* Australian owlet-nightjars
Caprimulgus europaeus. *See* European nightjars
Caprimulgus eximius. *See* Golden nightjars
Caprimulgus grandis. *See* Great potoos
Caprimulgus griseus. *See* Gray potoos
Caprimulgus indicus. *See* Gray nightjars
Caprimulgus longipennis. *See* Standard-winged nightjars
Caprimulgus longirostris. *See* Band-winged nightjars
Caprimulgus macrurus. *See* Large-tailed nightjars
Caprimulgus minor. *See* Common nighthawks
Caprimulgus nigriscapularis. *See* Black-shouldered nightjars

Caprimulgus noctitherus. See Puerto Rican nightjars
Caprimulgus nuttallii. See Common poorwills
Caprimulgus pectoralis. See Fiery-necked nightjars
Caprimulgus prigoginei. See Itombwe nightjars
Caprimulgus ruficollis. See Red-necked nightjars
Caprimulgus rufigena. See Rufous-cheeked nightjars
Caprimulgus solala. See Nechisar nightjars
Caprimulgus strigoides. See Tawny frogmouths
Caprimulgus tristigma. See Freckled nightjars
Caprimulgus vociferus. See Whip-poor-wills
Caprinae, 12:134, **16:87–105,** 16:*96,* 16:*97*
  behavior, 16:92–93
  conservation status, 16:94–95
  distribution, 16:*87,* 16:90–91
  evolution, 16:1, 16:87–89
  feeding ecology, 16:93–94
  habitats, 15:269, 16:91–92, 16:98–102
  humans and, 16:95
  physical characteristics, 16:89–90
  reproduction, 16:7, 16:8, 16:94
  species of, 16:98–104*t*
  taxonomy, 16:1, 16:87–89
Caprinae Specialist Group (CSG), IUCN, 16:87
Caprini, 16:87, 16:88–89, 16:90
Capriolinae. *See* New World deer
Caproidae, 5:123
Caprolagus spp. *See* Hispid hares
Caprolagus hispidus. See Hispid hares
Capromimus spp., 5:123
Capromyidae. *See* Hutias
Capromys spp., 16:462, 16:463
Capromys pilorides. See Cuban hutias
Captive breeding programs, 6:59, 6:190, 6:347, 12:223–224
  Australian honeyeaters, 11:238
  Ciconiiformes and, 8:236
  cranes, 9:27–28, 9:29
  Guam rails, 9:3
  kokakos, 11:449
  mynas, 11:420
  parrots, 9:280
  rails, 9:59, 9:66
  Recurvirostridae, 9:135
  regent honeyeaters, 11:243
  shore plovers, 9:163
  typical owls, 9:351
  See also Conservation status; Domestication; Zoos
Captivity, reptiles in, 7:40
  See also Humans
Captured birds. *See* Cagebirds
Capuchin birds, 10:305, 10:*306,* 10:307
Capuchins, 14:4, **14:101–107,** 14:*107,* **14:110–112**
  behavior, 14:7, 14:103–104, 14:250
  conservation status, 14:11, 14:105, 14:106
  distribution, 14:*101,* 14:103
  evolution, 14:101
  feeding ecology, 14:9, 14:104–105
  habitats, 14:103
  humans and, 12:*174,* 14:106
  memory, 12:153
  physical characteristics, 14:101–102
  reproduction, 14:105
  species of, 14:*110*–112
  taxonomy, 14:101
Capybaras, 9:109, 12:14, 16:123, 16:126, **16:401–406,** 16:*402,* 16:*403,* 16:*404,* 16:*405*
Carabidae, 3:54, 3:62, 3:322
Caracal (Felis) caracal. See Caracals
Caracals, 14:372, 14:*377,* 14:*387*–388
Caracanthidae. *See* Orbicular velvetfishes
Caracanthus maculatus. See Spotted coral crouchers
Caracaras, **8:347–362**
  behavior, 8:349
  conservation status, 8:351
  evolution, 8:347
  extinct species, 8:351
  feeding ecology, 8:349–350
  habitats, 8:348–349
  humans and, 8:352
  pesticides, 8:351
  reproduction, 8:350–351
  species of, 8:*354*–361
  taxonomy, 8:347
Carangidae, 4:14, 4:48, 4:68, 5:255–263
Carangoides ferdau. See Blue trevallys
Caranx hippos. See Crevalle jacks
Caranx ruber. See Bar jacks
Carapidae. *See* Pearlfishes
Carapus spp. *See* Pearlfishes
Carapus bermudensis, 5:*15,* 5:*17,* 5:*19,* 5:*20,* 5:21
Carapus chavesi. See Carapus bermudensis
Carapus recifensis. See Carapus bermudensis
Carassius auratus. See Crucian carps
Carausius morosus. See Indian stick insects
Carbohydrates, 12:120
  See also Feeding ecology
Carbon cycle, 12:213
Carbon dioxide, subterranean mammals and, 12:72–73, 12:114
Carbon dioxide emissions, 1:52–53
Carcharhinidae. *See* Requiem sharks
Carcharhiniformes. *See* Ground sharks
Carcharhinus amblyrhynchoides, 4:116
Carcharhinus amblyrhynchos. See Gray reef sharks
Carcharhinus amboinensis, 4:116
Carcharhinus borneensis, 4:116
Carcharhinus brachyurus. See Cooper sharks
Carcharhinus brevipinna, 4:116
Carcharhinus hemiodon, 4:116
Carcharhinus leucas, 4:116
Carcharhinus limbatus, 4:116
Carcharhinus longimanus. See Oceanic whitetip sharks
Carcharhinus melanopterus, 4:116
Carcharhinus obscurus, 4:116
Carcharhinus plumbeus, 4:116
Carcharias kamoharai. See Crocodile sharks
Carcharias taurus. See Sand tiger sharks
Carcharodon spp., 4:131
Carcharodon carcharias. See White sharks
Carcharodon megalodon. See Megalodon sharks
Carcinonemertidae, 1:255
Carcinoscorpius rotundicauda, 2:328
Carcinus spp. *See* Shore crabs
Carcinus maenas. See European green crabs
Cardiidae, 2:452, 2:453
Cardim, Fernão, 16:404
Cardinal tetras, 4:*340,* 4:*344*–345
Cardinalfishes, 4:46, 4:67, 4:69, 5:255, 5:256, 5:258–263, 5:345–346
Cardiocraniinae, 16:211, 16:212
Cardiocranius paradoxus. See Five-toed pygmy jerboas
Cardioderma cor. See Heart-nosed bats
Cardiovascular system. *See* Circulatory system
Carditidae, 2:453
Cardon turtles. *See* Leatherback seaturtles
Carduelinae. *See* Finches
Carduelis atriceps. See Black-capped red siskins
Carduelis carduelis. See European goldfinches
Carduelis chloris. See Greenfinches
Carduelis cucullata. See Red siskins
Carduelis johannis. See Warsangli linnets
Carduelis monguilloti. See Vietnam greenfinches
Carduelis siemiradzkii. See Saffron siskins
Carduelis spinus. See Eurasian siskins
Carduelis tristis. See American goldfinches
Carduelis yarrellii. See Yellow-faced siskins
Caretta caretta. See Loggerhead turtles
Carettochelyidae. *See* Pig-nose turtles
Carettochelys insculpta. See Pig-nose turtles
Cariama cristata. See Red-legged seriemas
Cariamidae. *See* Seriemas
Caribbean flamingos. *See* Greater flamingos
Caribbean land iguanas, 7:207
Caribbean manatees. *See* West Indian manatees
Caribbean monk seals. *See* West Indian monk seals
Caribbean motmots. *See* Blue-crowned motmots
Caribbean spiny lobster, 2:199
Caribbean trembling thrushes. *See* Brown tremblers
Caribou. *See* Reindeer
Caridea. *See* True shrimps
Caridonax fulgidus. See White-rumped kingfishers
Carleton, M. D., 16:282
Carlocebus spp., 14:143
Carmine, 3:264
Carmine bee-eaters, 10:*3,* 10:39, 10:*41,* 10:*43,* 10:*48*
Carnivora, 2:25, 3:46, 3:63, 12:14, **14:255–263**
  behavior, 12:141, 12:145, 14:257–260
  color vision, 12:79
  conservation status, 14:262
  digestive system, 12:120, 12:123–124
  distribution, 12:129, 12:131, 12:132, 12:136, 14:257
  evolution, 14:255–256
  feeding ecology, 14:260–261
  guilds, 12:117
  habitats, 14:257
  hearing, 12:82
  humans and, 14:262–263
  locomotion, 12:14
  neonatal milk, 12:126–127
  physical characteristics, 14:256–257
  reproduction, 12:90, 12:92, 12:94, 12:95, 12:98, 12:109, 12:110, 14:261–262
  taxonomy, 14:255, 14:256
  teeth of, 12:14
  water balance, 12:113
  See also Feeding ecology; specific carnivores
Carnivorous caterpillars, 3:387
Carnivorous diets. *See* Feeding ecology
Carnivorous fishes, 4:50
  See also specific fish

Carnivorous sponges, 1:28, 1:80, 1:*81*, 1:*82*, 1:85–86
Carnivorous tadpoles, 6:39
Carolina anoles. *See* Green anoles
Carolina chuck-wills. *See* Chuck-will's-widows
Carolina doves. *See* American mourning doves
Carolina mantids, 3:*178*
Carolina parakeets, 9:276
Carolina pygmy sunfishes, 5:196
Carolina waterdogs. *See* Dwarf waterdogs
Carolina wrens, 8:22, 10:*528*
Carollia castanea. *See* Chestnut short-tailed bats
Carollia perspicillata. *See* Seba's short-tailed bats
Carolliinae, 13:413, 13:417–418, 13:420
Carp breams, 4:*300*
Carpal glands, 12:37
See also Physical characteristics
Carpenter ants, 3:63, 3:64, 3:75
Carpenter bees, large, 3:*410*, 3:*413*
Carpenter frogs. *See* Woodworker frogs
Carpet beetles, 3:75
Carpet sharks, **4:105–112**, 4:*106*
Carpet vipers. *See* Saw-scaled vipers
Carphophis spp. *See* Wormsnakes
Carpiodes asiaticus. *See* Chinese suckers
Carpitalpa spp. *See* Arend's golden moles
Carpitalpa arendsi. *See* Arend's golden moles
Carpococcyx viridis, 9:315
Carpodacus mexicanus. *See* House finches
Carpodectes antoniae, 10:308
Carpoids, 1:13
Carpornis melanocephalus, 10:308
Carpospiza brachydactyla. *See* Pale rock sparrows
Carps, **4:297–320**, 4:*303*, 4:*305–306*
behavior, 4:298–299
conservation, 4:303
distribution, 4:298
evolution, 4:297
feeding ecology, 4:299–300
habitats, 4:298
humans and, 4:303
physical characteristics, 4:297–298, 4:*299*, 4:*301*
reproduction, 4:67, 4:69, 4:*300*–303
species of, 4:307–313
taxonomy, 4:297
Carr, Archie, 12:218, 12:223
Carr, Mark H., 1:51
Carrier pigeons, 9:255–256
Carrion beetles, 3:62, 3:64, 3:321
Carrion crows, 11:505, 11:506, 11:*511*, 11:*521*
Carroll, Lewis, 9:270
Carroll, Robert
on amniote eggs, 7:12–13
on archosauria, 7:17
on pterodactyloids, 7:19
Carrying capacity, of zoos, 12:207
Carson-Ewart, B. M., 5:365
Carteriospongia foliascens, 1:*81*, 1:*83*, 1:85
Carterodon sulcidens. *See* Cerrado rats
Carukia barnesi. *See* Irukandjis
Carybdea marsupialis, 1:147
Carybdea rastoni. *See* Southern sea wasps
Carybdea sivickisi, 1:147–148
Carybdeidae, 1:147
Caryophyllaeidae, 1:225
Caryophyllaeus laticeps, 1:*232*, 1:*235*
Caryophyllidea, 1:225, 1:227, 1:228, 1:231
Casarea dussumieri. *See* Splitjaw snakes
Cascade torrent salamanders, 6:*387*
Cascades salamanders. *See* Cascade torrent salamanders
Case, J., 5:38
Caseasauria, 12:10
Casinycteris argynnis. *See* Short-palate fruit bats
Casmerodius albus. *See* Great white egrets
Caspian seals, 12:138, 14:422, 14:435*t*
Caspian terns, 9:203, 9:*210*, 9:*214*
Caspiomyzon spp., 4:83
Casques, hornbill, 10:72
Cassava mealybugs, 3:415
Cassin's alseonax. *See* Cassin's flycatchers
Cassin's flycatchers, 11:*29*, 11:*31*
Cassin's gray flycatchers. *See* Cassin's flycatchers
Cassin's honeyguides, 11:222
Cassiopea xamachana. *See* Upside-down jellyfish
Cassis spp., 2:447
Cassowaries, 8:53–56, **8:75–81**
Caste, in social insects, 3:69–70
See also Behavior
Castellarini, F., 16:269
Castor spp. *See* Beavers
Castor canadensis. *See* North American beavers
Castor fiber. *See* Eurasian beavers
Castoridae. *See* Beavers
Castoroides spp., 16:177
Castro-Vaszuez, A., 16:269
Casuariidae. *See* Cassowaries
Casuarius bennettii. *See* Bennett's cassowaries
Casuarius casuarius. *See* Southern cassowaries
Casuarius unappendiculatus. *See* One-wattled cassowaries
Cat bears. *See* Giant pandas
Cat-birds. *See* Blue-gray gnatcatchers
Cat-eyed frogs, 6:110
Cat-eyed snakes, 7:469, 7:481
Cat Fanciers' Association, 14:291
Cat fleas, 3:57, 3:349, 3:350
Cat squirrels. *See* Fox squirrels
Catagonus spp., 15:291
Catagonus wagneri. *See* Chacoan peccaries
Catamounts. *See* Pumas
Catarrhini, 14:1
Catbirds, 11:477, 11:478–480, 11:*482*, 11:483
See also Mimids
Catch tentacles, 1:30
See also Physical characteristics
Catenotaeniidae, 1:226
Catenula spp., 1:188
Catenulida, 1:11, 1:185
Cateridae, 1:275
Caterodon spp., 16:450
Caterodon sulcidens. *See* Cerrado rats
Caterpillars, 1:42, 3:32, 3:34, 3:389
carnivorous, 3:387
defense mechanisms, 3:65
hemolymph, 3:25
life histories, 3:36
mechanoreception, 3:28
moth, 3:80
as pests, 3:75
physical characteristics, 3:18
silk moth, 3:*388*
swallowtail, 3:65
vision, 3:28
Catesby, Mark, 10:155, 10:157, 10:164, 10:166
Catfishes, **4:351–367**, 4:*355–356*
behavior, 4:353
conservation, 4:354
distribution, 4:57, 4:352–353
evolution, 4:351
feeding ecology, 4:68, 4:69, 4:353
habitats, 4:353
humans and, 4:354
physical characteristics, 4:*4*, 4:6, 4:*22*, 4:24, 4:60, 4:62, 4:351–*352*
reproduction, 4:30, 4:66, 4:354
species of, 4:357–367
taxonomy, 4:12, 4:351
Catharacta spp., 9:204, 9:207
Catharacta chilensis. *See* Chilean skuas
Catharacta lonnbergi. *See* Brown skuas
Catharacta maccormicki. *See* South Polar skuas
Catharacta skua. *See* Great skuas
Catharopeza bishopi. *See* Whistling warblers
Cathartes spp. *See* New World vultures
Cathartes aura. *See* Turkey vultures
Cathartes burrovianus. *See* Lesser yellow-headed vultures
Cathartes melambrotus. *See* Greater yellow-headed vultures
Cathartidae. *See* New World vultures
Catharus guttatus. *See* Hermit thrushes
Catharus minimus. *See* Gray-cheeked thrushes
Catharus mustelina. *See* Wood thrushes
Catharus ustulatus. *See* Swainson's thrushes
Cathedral-mound building termites, 3:165
Cathemeral lemurs, 14:8
Cathemerality, 14:8
Catherpes spp., 10:526
Cathetocephalidae, 1:226
Catlocarpio siamensis, 4:298
Catopinae, 3:54
Catopras, 5:242
Catoprion spp., 4:69
Catoptrophorus semipalmatus. *See* Willets
Catopuma badia. *See* Bay cats
Catopuma temminckii. *See* Asiatic golden cats
Catostomidae. *See* Suckers
Catostominae, 4:321
Catostomus catostomus. *See* Longnose suckers
Catreus wallichi. *See* Cheers (*Catreus* spp.)
Cats, **14:369–392**, 14:*377*, 14:*378*
behavior, 12:145, 14:371–372
conservation status, 14:262, 14:374–375
digestive system, 12:120
distribution, 12:131, 12:136, 14:257, 14:*369*, 14:370–371
evolution, 14:369
feeding ecology, 14:260, 14:372
feral, 12:185–186, 14:257
habitats, 14:371
humans and, 14:374–376
learning set, 12:152
neonatal requirements, 12:126
physical characteristics, 14:370
reproduction, 12:103, 12:106, 12:109, 12:111, 14:372–373
self recognition, 12:159
species of, 14:*379*–389, 14:390*t*–391*t*
taxonomy, 14:256, 14:369
vision, 12:84
in zoos, 12:210
See also Domestic cats
Catsharks, 4:11, 4:*30*, 4:114, 4:115

Cattle, 14:*247*
aurochs, 12:177, 16:*17*, 16:21–*22*
behavior, 16:6
Brahma, 16:4
conservation status, 12:215
digestive system, 12:122–123
distribution, 16:4
evolution, 15:265, 16:1
feeding ecology, 12:118, 15:271
humans and, 16:8
physical characteristics, 16:2
vision, 12:79
Zebu, 16:6
See also Bovidae; Domestic cattle; Ungulates; specific types of cattle
Cattle dogs, Australian, 14:292
Cattle egrets, 8:242, 8:*247*, 8:*251–252*
Cattle lice, 3:76
Caturidae, 4:229
Catworms, 2:*49*, 2:*53*
Caucasian salamanders, 6:38, 6:367
Caucasilacerta spp., 7:297
Caudacaecilia spp., 6:419
Caudal autonomy. *See* Tails, regeneration of
Caudata, 6:3, 6:4, 6:15, **6:323–326,** 6:327
See also Newts; Salamanders
Caudinidae, 1:420
Caudiverbera spp., 6:156–159
Caudiverbera caudiverbera. See Helmeted water toads
Caudofoveata. *See* Chaetodermomorpha
Cauliflower corals, 1:*109*, 1:*120*–121
Caulolatilus spp., 5:256
Causinae. *See* Night adders
Causus spp. *See* Night adders
Causus rhombeatus. See Rhombic night adders
Cavalli-Sforza, L. L., 14:250
Cave basses, 5:255, 5:258–263
Cave bears, Ice Age, 12:24
Cave beetles, 3:319
Cave-birds. *See* Rockwarblers
Cave crickets, 3:203, 3:204, 3:207
Cave-dwelling bats. *See* Long-fingered bats
Cave dwelling psocids, 3:243, 3:244
Cave fruit bats. *See* Dawn fruit bats
Cave gudgeons. *See* Blind cave gudgeons
Cave-hyenas, Ice Age, 12:24
Cave-lions, Ice Age, 12:24
Cave owls. *See* Common barn owls
Cavefishes (Amblyopsidae), 4:57, 5:5, 5:6, 5:*8*–10
See also Blind cavefishes (Characins)
Cavefishes (Cypriniformes), 4:*324*, 4:*328*, 4:329, 4:332–333
Cavefishes of Nam Lang, 4:*324*, 4:*328*
Cavia spp. *See* Guinea pigs
Cavia aperea. See Guinea pigs
Cavia fulgida. See Shiny guinea pigs
Cavia magna. See Greater guinea pigs
Cavia porcellus. See Domestic guinea pigs
Cavia tschudii. See Montane guinea pigs
Caviar, 4:*75*, 4:215
Cavibelonia, 2:379
Cavies, 12:72, **16:389–400,** 16:*394*, 16:399*t*
Caviidae, 12:72, 16:123, 16:125, **16:389–400,** 16:*394*, 16:399*t*
Caviinae, 16:389–390
Caviomorphs, 16:121–126
Cayenne caecilians, 6:*438*, 6:440–441
CCAMLR. *See* Convention on the Conservation of Antarctic Marine Living Resources
Cebidae, 14:4, **14:101–112,** 14:*107*
behavior, 14:7, 14:103–104
conservation status, 14:11, 14:105–106
distribution, 14:*101*, 14:102–103
evolution, 14:101
feeding ecology, 14:9, 14:104–105
habitats, 14:103
humans and, 12:*174*, 14:106
physical characteristics, 14:101–102
reproduction, 14:105
species of, 14:*108*–112
taxonomy, 14:101
Cebinae, 14:101–102, 14:103
Cebu bearded pigs, 15:276, 15:279, 15:280, 15:281, 15:289*t*
Cebu flowerpeckers, 11:191
Cebuella pygmaea. See Pygmy marmosets
Cebupithecia spp., 14:143
Cebus spp. *See* Capuchins
Cebus albifrons. See White-fronted capuchins
Cebus apella. See Black-capped capuchins
Cebus apella robustus, 14:106
Cebus capucinus. See White-throated capuchins
Cebus libidinosus, 14:101
Cebus nigritus, 14:101
Cebus olivaceus. See Weeper capuchins
Cebus xanthosternos. See Yellow-breasted capuchins
Cedar waxwings, 10:447, 10:*448*, 10:*450*, 10:*451*
Cedarberg katydids, 3:203
Cedarbergeniana imperfecta. See Cedarberg katydids
Cediopsylla spp., 3:350
Celebes bee-eaters. *See* Purple-bearded bee-eaters
Celebes cuscuses. *See* Sulawesi bear cuscuses
Celebes hornbills. *See* Sulawesi red-knobbed hornbills
Celebes macaques, 14:*188*, 14:193–194
Celebes pied imperial pigeons. *See* White imperial pigeons
Celebes pigs, 15:280, 15:282, 15:289*t*
Celebes rainbowfishes, 4:57, 5:*72*, 5:*75*, 5:76
Celestus spp., 7:339
Celestus hylaius, 7:*341*, 7:*343*
Celestus macrotus, 7:339
Celeus brachyurus. See Rufous woodpeckers
Celithemis eponina. See Brown-spotted yellow wing dragonflies
Cell determination, 2:19
Cell differentiation, 1:18
Cellana spp., 2:425
Cellularization theory, 1:7
Cellulase, 12:122
Cemophora spp. *See* Scarletsnakes
Cenocrinus asterius. See Great West Indian sea lilies
Centinelan extinction, 3:87
Centipede eaters, 7:461
Centipedes, **2:353–362,** 2:*357*
behavior, 2:354–355
conservation status, 2:356
distribution, 2:354
evolution, 2:353
feeding ecology, 2:*353*, 2:*355*
habitats, 2:354
humans and, 2:356
locomotion, 2:*19*
physical characteristics, 2:353–*354*
reproduction, 2:355–356
species of, 2:358–362
taxonomy, 2:353
Central African hedgehogs, 13:*196*, 13:*205*, 13:*206*, 13:*207*, 13:*213t*
Central America. *See* Neotropical region
Central American agoutis, 16:*408*, 16:*410*, 16:*411*
Central American river turtles, **7:99–100**
Central American tapirs, 15:218, 15:238, 15:*239*, 15:*241*, 15:242, 15:*244*, 15:*245*–246
Central Asian gazelles, 12:134
Central Asian ground squirrels, 16:143
Central mudminnows, 4:*379*, 4:*382*, 4:*383*, 4:386
Central nervous system, 3:*25–26*, 6:21–22
Central netted dragons, 7:*212*, 7:*216*
Central Panay Mountains National Park, 10:75
Central Park Zoo, New York, 12:203
Central stonerollers. *See* Stonerollers
Central valley grasshoppers, 3:207
Centrarchidae. *See* Sunfishes
Centriscidae. *See* Shrimpfishes
Centrocercus minimus. See Gunnison sage grouse
Centrocercus urophasianus. See Sage grouse
Centroderes eisigii, 1:*278*, 1:*279*
Centroderidae, 1:275
Centrodraco spp., 5:365
Centrodraco insolitus. See Draconetts
Centrolene spp., 6:215, 6:216, 6:217
Centrolene geckoideum. See Pacific giant glass frogs
Centrolene heloderma. See Pichincha glass frogs
Centrolene prosoblepon. See Nicaragua glass frogs
Centrolene valerioi. See La Palma glass frogs
Centrolenella fleischmanni. See Fleischmann's glass frogs
Centrolenella ignota. See Lynch's Cochran frogs
Centrolenidae. *See* Glass frogs
Centrolophidae. *See* Medusafishes
Centrolophus niger. See Black ruffs
Centronycteris centralis. See Shaggy-haired bats
Centrophoridae. *See* Gulper sharks
Centrophorus spp., 4:152
Centrophorus granulosus. See Gulper sharks
Centropodinae. *See* Coucals
Centropomidae, 4:68, 5:219, 5:220, 5:222
Centropomus spp., 5:220, 5:221
Centropomus undecimalis. See Common snooks
Centropus chlororhynchus. See Green-billed coucals
Centropus grillii, 9:313
Centropus nigrorufus, 9:315
Centropus sinensis. See Greater coucals
Centropus steerii, 9:315
Centropyge spp. *See* Pygmy angelfishes
Centropyge resplendens, 5:242–243
Centrozoros spp., 3:239
Centruroides vittatus. See Striped scorpions
Centurio senex. See Wrinkle-faced bats
Cephalobaena spp., 2:317
Cephalobaenida, 2:317, 2:318, 2:319, 2:320
Cephalobaenidae, 2:317
Cephalocarids, **2:131–133**

Cephalochlamydidae, 1:225
Cephalochordata. *See* Lancelets
Cephalodella gibba, 1:*263*, 1:*266–267*
Cephalodiscus gilchristi, 1:46
Cephalodiscus gracilis, 1:*446*, 1:*447*
Cephalofovea tomahmontis, 2:*112*, 2:*113*, 2:114
Cephalophinae. *See* Duikers
Cephalopholis spp., 5:262
Cephalophus spp., 16:5, 16:73
Cephalophus adersi. *See* Aders's duikers
Cephalophus callipygus. *See* Peters's duikers
Cephalophus dorsalis. *See* Bay duikers
Cephalophus harveyi. *See* Harvey's duikers
Cephalophus jentinki. *See* Jentink's duikers
Cephalophus leucogaster. *See* White-bellied duikers
Cephalophus maxwelli. *See* Maxwell's duikers
Cephalophus monticola. *See* Blue duikers
Cephalophus natalensis. *See* Natal duikers
Cephalophus niger. *See* Black duikers
Cephalophus nigrifrons. *See* Black-fronted duikers
Cephalophus ogilbyi. *See* Ogilby's duikers
Cephalophus rufilatus. *See* Red-flanked duikers
Cephalophus silvicultor. *See* Yellow-backed duikers
Cephalophus spadix. *See* Abbott's duikers
Cephalophus weynsi. *See* Weyns's duikers
Cephalophus zebra. *See* Zebra duikers
Cephalopoda, 1:37, **2:475–489**, 2:*481*
behavior, 2:477–478
conservation status, 2:479
distribution, 2:477
evolution, 2:475
feeding ecology, 2:478
habitats, 2:477
humans and, 2:479–480
physical characteristics, 2:475–477, 2:*476*
reproduction, 2:478–479
species of, 2:*482*–489
taxonomy, 2:475
Cephalopterus spp. *See* Cotingas
Cephalopterus glabricollis. *See* Bare-necked umbrellabirds
Cephalopterus ornatus. *See* Amazonian umbrellabirds
Cephalopterus penduliger. *See* Long-wattled umbrellabirds
Cephalopyrus flammiceps. *See* Fire-capped tits
Cephalorhyncha, 1:9, 1:13
Cephalorhyncha asiatica, 1:*278*, 1:*279*
Cephalorhynchidae, 1:275
Cephalorhynchus spp., 15:43
Cephalorhynchus commersonii. *See* Commerson's dolphins
Cephalorhynchus eutropia. *See* Black dolphins
Cephalorhynchus hectori. *See* Hector's dolphins
Cephaloscyllium ventriosum. *See* Swellsharks
Cephalostatin 1, 1:46
Cephalurus spp., 4:115
Cephisus siccifolius. *See* Spittle bugs
Cephoidea, 3:54
Cepphus spp. *See* Guillemots
Cepphus columba. *See* Pigeon guillemots
Cepphus grylle. *See* Black guillemots
Ceractinomorpha, 1:77
Cerambycid beetles, 2:37
Cerambycidae, 3:53, 3:320
Ceraplectana spp., 1:421
Cerastes spp., 7:445
Cerastes cerastes. *See* Horned vipers
Cerathyla proboscidea. *See* Sumaco horned treefrogs
Ceratias holboelli. *See* Krøyers deep sea anglerfishes
Ceratioidei, 5:47, 5:48, 5:50, 5:51, 5:55
Ceratitis capitata. *See* Mediterranean fruit flies
Ceratobactrachini, 6:246, 6:251
Ceratobatrachus spp., 6:66, 6:246
Ceratobatrachus guentheri, 6:251
Ceratocumatidae, 2:229
Ceratodontidae, 4:57
Ceratodontiformes. *See* Australian lungfishes
Ceratodus forsteri. *See* Australian lungfishes
Ceratogymna elata. *See* Yellow-casqued hornbills
Ceratophora tennentii. *See* Leaf-horned agamas
Ceratophryinae, 6:155, 6:158
Ceratophrys spp., 6:39, 6:65, 6:67, 6:155, 6:157–159, 6:308
Ceratophrys aurita, 6:157
Ceratophrys cornuta. *See* Suriname horned frogs
Ceratophrys occidentalis. *See* Cururu lesser escuerzo
Ceratophrys turpicola. *See* New Guinea bush frogs
Ceratophyllidae, 3:347, 3:348
Ceratophylloidea, 3:347
Ceratopogonids, 3:358
Ceratoscopelus maderensis, 4:443
Ceratoscopelus warmingii, 4:443
Ceratotherium spp., 15:216, 15:249, 15:251
Ceratotherium simum. *See* White rhinoceroses
Ceraurus pleurexanthemus. *See* Trilobites
Cerberus rynchops. *See* Dog-faced watersnakes
Cercartetus spp. *See* Pygmy possums
Cercartetus caudatus. *See* Long-tailed pygmy possums
Cercartetus concinnus. *See* Western pygmy possums
Cercartetus lepidus. *See* Little pygmy possums
Cercartetus nanus. *See* Eastern pygmy possums
Cercocebus spp., 14:188, 14:191
Cercocebus albigena. *See* Gray-cheeked mangabeys
Cercocebus atys. *See* Sooty mangabeys
Cercocebus galeritus. *See* Agile mangabeys
Cercocebus torquatus. *See* Collared mangabeys
Cercococcyx spp., 9:312
Cercomacra cinerascens. *See* Gray antbirds
Cercomela melanura. *See* Blackstarts
Cercopagis pengoi. *See* Fishhook water fleas
Cercopithecidae. *See* Old World monkeys
Cercopithecinae. *See* Cheek-pouched monkeys
Cercopithecini. *See* Guenons
Cercopithecoides spp., 14:172
Cercopithecus spp. *See* Guenons
Cercopithecus ascanius. *See* Red-tailed monkeys
Cercopithecus campbelli. *See* Campbell's monkeys
Cercopithecus cephus. *See* Moustached guenons
Cercopithecus diana. *See* Diana monkeys
Cercopithecus dryas, 14:194
Cercopithecus erythrogaster, 14:193–194
Cercopithecus erythrotis, 14:194
Cercopithecus hamlyni, 14:194
Cercopithecus lhoesti. *See* l'Hoest's guenons
Cercopithecus mitis. *See* Blue monkeys
Cercopithecus petaurista. *See* Lesser white-nosed monkeys
Cercopithecus preussi, 14:193–194
Cercopithecus sclateri, 14:193–194
Cercopithecus solatus, 14:194
Cercotrichas galactotes. *See* Rufous bush chats
Cercyonis sthenele sthenele, 3:388
Cerdocyon thous. *See* Crab-eating foxes
Cerebral cortex, 12:24, 12:49
See also Physical characteristics
Cerebratulus spp., 1:245
Cerebratulus lacteus, 1:247
Cereopsis geese. *See* Cape Barren geese
Cereopsis novaehollandiae. *See* Cape Barren geese
Ceriantharia. *See* Tube anemones
Ceriantheopsis americanus. *See* American tube dwelling anemones
Cerianthids, 1:104
Cerithiomorpha, 2:445
Cerorhinca monocerata. *See* Rhinoceros auklets
Cerrado rats, 16:450–451, 16:*452*, 16:454, 16:*455*
Cerrophidion spp., 7:445
Certhia spp. *See* Treecreepers
Certhia afra. *See* Greater double-collared sunbirds
Certhia americana. *See* Brown creepers
Certhia asiatica. *See* Purple sunbirds
Certhia brachydactyla. *See* Short-toed treecreepers
Certhia cruentata. *See* Scarlet-backed flowerpeckers
Certhia familiaris. *See* Eurasian treecreepers
Certhia famosa. *See* Malachite sunbirds
Certhia leucophaea. *See* White-throated treecreepers
Certhia malacensis. *See* Plain-throated sunbirds
Certhia nipalensis. *See* Rusty-flanked treecreepers
Certhia novaehollandiae. *See* New Holland honeyeaters
Certhia rectirostris. *See* Green sunbirds
Certhia senegalensis. *See* Scarlet-chested sunbirds
Certhia verticalis. *See* Green-headed sunbirds
Certhia violacea. *See* Orange-breasted sunbirds
Certhiidae. *See* Treecreepers
Certhilauda spp. *See* Larks
Certhilauda brevirostris. *See* Agulhas long-billed larks
Certhilauda burra. *See* Ferruginous larks
Certhilauda chuana. *See* Short-clawed larks
Certhilauda curvirostris. *See* Cape long-billed larks
Certhionyx niger. *See* Black honeyeaters
Certhionyx variegatus. *See* Pied honeyeaters
Cerulean warblers, 11:288, 11:*292*, 11:*298–299*
Ceruminous glands, 12:37
Cervalces scotti. *See* American stag-moose
Cervera's wrens. *See* Zapata wrens
Cervidae. *See* Deer
Cervinae. *See* Old World deer
Cervoidea, 15:266
Cervus spp., *15:358*
Cervus albirostris. *See* White-lipped deer
Cervus alfredi. *See* Visayan spotted deer
Cervus duvaucelii. *See* Barasinghas
Cervus elaphus. *See* Red deer

Cervus elaphus canadensis. See Elk
Cervus elaphus nannodes. See Tule elk
Cervus elaphus siamensis, 15:361
Cervus elaphus sibiricus. See Siberian marals
Cervus elaphus xanthopygus. See Izubrs
Cervus eldi. See Eld's deer
Cervus nippon. See Sika deer
Cervus (Rusa) mariannus. See Philippine brown deer
Cervus (Rusa) timorensis. See Timor deer
Cervus schomburgki. See Schomburgk's deer
Cervus timorensis rusa. See Javan rusa deer
Cervus unicolor. See Sambars
Ceryle alcyon. See Belted kingfishers
Ceryle rudis. See Pied kingfishers
Ceryle torquata. See Ringed kingfishers
Cerylinae, 10:5–7
Cestida, 1:169
Cestoda. *See* Tapeworms
Cestodaria, 1:226
Cestracion francisci. See California bullhead sharks
Cestraeus spp., 5:59
Cestum veneris. See Venus's girdles
Cetacea, **15:1–11**
behavior, 15:6–7
conservation status, 15:9–10
distribution, 12:129, 12:138, 15:5
echolocation, 12:86
evolution, 15:2–4
feeding ecology, 15:8
geomagnetic fields, 12:84
habitats, 15:5–6
hippopotamuses and, 12:30
humans and, 15:10–11
locomotion, 12:44
physical characteristics, 12:67, 15:4–5
reproduction, 12:92, 12:94, 12:103–104, 15:9
taxonomy, 12:14, 15:2–4, 15:133, 15:266
See also Whales
Cetartiodactyla, 12:14, 15:266–267
Cetomimidae. *See* Flabby whalefishes
Cetomimoidea, 5:105, 5:106
Cetopsidae, 4:352
Cetorhinidae, 4:67
Cetorhinus maximus. See Basking sharks
Cetoscarus bicolor. See Bicolor parrotfishes
Cetotheriidae, 15:3
Cettia spp., 11:3, 11:5, 11:7
Cettia cetti. See Cetti's warblers
Cetti's warblers, 11:4, 11:*9*, 11:14–*15*
Ceuthophilus spp., 3:207
Ceylon caecilians, 6:31, 6:*421*, 6:*422*–423
Ceylon frogmouths. *See* Large frogmouths
Ceylon hawk cuckoos. *See* Common hawk-cuckoos
Ceylon mynas. *See* Sri Lanka mynas
Ceyx pictus. See African pygmy-kingfishers
Ceyx rufidorsum. See Rufous-backed kingfishers
CFCs (chlorofluorocarbons), 1:52
Chabert's vangas, 10:439, 10:440, 10:441
Chaca. *See* Squarehead catfishes
Chaca chaca. See Squarehead catfishes
Chachalacas, **8:413–423**, 8:*415*
behavior, 8:401, 8:414–415
conservation status, 8:416
distribution, 8:*413*, 8:414
evolution, 8:413
feeding ecology, 8:415–416
habitats, 8:414
humans and, 8:416
physical characteristics, 8:413–414
reproduction, 8:416
species of, 8:*418–419*
taxonomy, 8:416
Chacidae, 4:353
Chacma baboons, 14:*189*, 14:*190*, 14:191, 14:193
Chaco pipits, 10:372
Chacoan hairy armadillos, 13:191*t*
Chacoan naked-tail armadillos, 13:190*t*
Chacoan peccaries, 15:291, 15:*294*, 15:*295*, 15:*297*, 15:*299*–300
Chacophrys spp., 6:155
Chaenocephalus aceratus. See Blackfin icefishes
Chaenopsidae, 5:341, 5:343
Chaenotonotida, 1:270
Chaerephon spp., 13:483
Chaerephon gallagheri, 13:487
Chaerephon jobensis. See Northern mastiff bats
Chaerephon plicata. See Wrinkle-lipped free-tailed bats
Chaerephon pumila. See Lesser-crested mastiff bats
Chaeropus spp., 13:10
Chaeropus ecaudatus. See Pig-footed bandicoots
Chaeteessidae, 3:177
Chaetetids, 1:10
Chaetocercus spp., 9:442, 9:449
Chaetocercus berlepschi. See Esmeraldas woodstars
Chaetocercus bombus. See Little woodstars
Chaetoderma argenteum, 2:*381*, 2:383–*384*
Chaetodermomorpha, 2:*379*, 2:380
Chaetodipus spp. *See* Coarse-haired pocket mice
Chaetodipus baileyi. See Bailey's pocket mice
Chaetodipus hispidus. See Hispid pocket mice
Chaetodipus spinatus. See Spiny pocket mice
Chaetodon collare. See Redtail butterflyfishes
Chaetodon flavocoronatus, 5:242–243
Chaetodon litus, 5:242–243
Chaetodon lunula, 5:239
Chaetodon marleyi, 5:242–243
Chaetodon melanotus. See Blackback butterflyfishes
Chaetodon multicinctus. See Pebbled butterflyfishes
Chaetodon obliquus, 5:242–243
Chaetodon polylepis. See Pyramid butterflyfishes
Chaetodon punctatofasciatus, 5:241
Chaetodon robustus, 5:242–243
Chaetodon trifascialis, 5:241
Chaetodontidae. *See* Butterflyfishes
Chaetognatha. *See* Arrow worms
Chaetomys spp., 16:365, 16:366, 16:367
Chaetomys subspinosus. See Thin-spined porcupines
Chaetonotida, 1:269–270
Chaetophractus nationi. See Andean hairy armadillos
Chaetophractus vellerosus. See Small hairy armadillos
Chaetophractus villosus. See Large hairy armadillos
Chaetops spp. *See* Rock jumpers
Chaetopsylla spp., 3:350
Chaetopterus variopedatus. See Parchment worms
Chaetorhynchus spp. *See* Pygmy drongos
Chaetorhynchus papuensis. See Pygmy drongos
Chaetostoma spp. *See* Bulldog plecos
Chaetura spp., 9:418, 9:424
Chaetura pelagica. See Chimney swifts
Chaetura vauxi. See Vaux's swifts
Chaeturinae. *See* Swifts
Chaffinches, 11:323, 11:324, 11:*326*, 11:*328*
Chagas' disease, 2:11, 12:254
Chai, Peng, 9:440, 10:92
Chain catsharks, 4:*118*, 4:*125*, 4:126
Chain salps, 1:*468*
Chain vipers. *See* Russel's vipers
Chaisognathus granti. See Chilean stag beetles
Chalcides spp., 7:327, 7:329
Chalcidoid wasps, 3:64
Chalcites basalisx. See Horsfield's bronze-cuckoos
Chalcomitra spp. *See* Sunbirds
Chalcomitra balfouri. See Socotra sunbirds
Chalcomitra senegalensis. See Scarlet-chested sunbirds
Chalcoparia spp. *See* Sunbirds
Chalcoparia singalensis. See Ruby-cheeked sunbirds
Chalcophaps indica. See Emerald doves
Chalcorana spp., 6:247
Chalcosoma caucasus. See Atlas beetles
Chalcostigma spp., 9:442
Chalcostigma olivaceum. See Olivaceous thornbills
Chalinolobus spp., 13:497
Chalinolobus gouldii. See Gould's wattled bats
Chalinolobus variegatus. See Butterfly bats
Chamaea fasciata. See Wrentits
Chamaeleo spp., 7:*37*, 7:224, 7:227, 7:232
Chamaeleo calyptratus. See Veiled chameleons
Chamaeleo caroliquarti, 7:223
Chamaeleo chamaeleon. See Common chameleons
Chamaeleo dilepis. See Flap-necked chameleons
Chamaeleo jacksonii. See Jackson's chameleons
Chamaeleo namaquensis, 7:228
Chamaeleo parsonii. See Parson's chameleons
Chamaeleonidae. *See* Chameleons
Chamaepelia inca. See Inca doves
Chamaepetes spp. *See* Sickle-winged guans
Chamaepetes unicolor. See Black guans
Chamaesaura spp. *See* Grass lizards
Chamaesaura anguina. See Common grass lizards
Chambered nautilus. *See* Pearly nautilus
Chambius kasserinensis, 16:517
Chameleons, **7:223–242**, 7:*233–234*
behavior, 7:38, 7:228–230
conservation status, 7:232
distribution, 7:*223*, 7:227–228
evolution, 7:223–224
eyes, 7:32
feeding ecology, 7:230
habitats, 7:228
humans and, 7:232
physical characteristics, 7:198, 7:224–227, 7:*225*, 7:*226*, 7:*231*
reproduction, 7:*230*–232
species of, 7:235–241
superstitions and, 7:57
taxonomy, 7:223–224

See also specific types of chameleons
Chamois, 12:137, 15:271, 16:87–90, 16:*91*, 16:92, 16:94
northern, 16:6, 16:87, 16:*96*, 16:*100*
southern, 16:87, 16:103*t*
Championica montana, 3:202
Champsocephalus gunnari. *See* Mackerel icefishes
Champsochromis caeruleus. *See* Trout cichlids
Champsodontidae. *See* Gapers
Chancelière, Gennes del la, 9:273
Chandidae. *See* Glassfishes
Chanidae, 4:289, 4:290
Channa spp., 5:437
Channa argus. *See* Northern snakeheads
Channa aurantimaculata. *See* Orange-spotted snakeheads
Channa bankanensis. *See* Banka snakeheads
Channa barca. *See* Barca snakeheads
Channa bleheri. *See* Rainbow snakeheads
Channa gachua. *See* Dwarf snakeheads
Channa maculata. *See* Blotched snakeheads
Channa marulius. *See* Bullseye snakeheads
Channa micropeltes. *See* Giant snakeheads
Channa orientalis. *See* Walking snakeheads
Channa pleurophthalmus. *See* Ocellated snakeheads
Channa striata. *See* Striped snakeheads
Channel basses. *See* Red drums
Channel-billed cuckoos, 9:*316*, 9:*323*–324
Channel-billed toucans, 10:126
Channel catfishes, 4:*355*, 4:*361*
Channeled rockfishes. *See* Deepwater scorpionfishes
Channichthyidae. *See* Icefishes
Channidae. *See* Snakeheads
Channoidei. *See* Snakeheads
Chanos spp., 4:290, 4:291
Chanos chanos. *See* Milkfishes
Chaparana spp., 6:246, 6:249
Chaperina fusca. *See* Saffron-bellied frogs
Chapman, C. A., 14:163
Chapman, Graeme, 11:458
Chapman's antshrikes. *See* Barred antshrikes
Chappell Island tiger snakes, 7:496
Characidae. *See* Characins
Characiformes. *See* Characins
Characins, 4:36, 4:69, **4:335–350**, 4:*340–341*
behavior, 4:337
conservation, 4:338–339
distribution, 4:337
evolution, 4:335
feeding ecology, 4:337–338
habitats, 4:337
humans and, 4:339
physical characteristics, 4:336, 4:*337*
reproduction, 4:66, 4:67, 4:*336*, 4:*337*, 4:*338*
species of, 4:342–350
taxonomy, 4:335
Charadrii, 9:101, 9:104, 9:107
Charadriidae, **9:161–173,** 9:*165*
behavior, 9:162
conservation status, 9:163–164
distribution, 9:*161*, 9:162
evolution, 9:161
feeding ecology, 9:162
habitats, 9:162
humans and, 9:164
physical characteristics, 9:161–162
reproduction, 9:162–163
species of, 9:*166–173*
taxonomy, 9:161
See also Lapwings; Plovers
Charadriiformes, 9:45, **9:101–105,** 9:107
Charadrius spp. *See* Plovers
Charadrius alexandrinus. *See* Snowy plovers
Charadrius cinctus. *See* Red-kneed dotterels
Charadrius dubius. *See* Little-ringed plovers
Charadrius hiaticula. *See* Ringed plovers
Charadrius himantopus. *See* Black-winged stilts
Charadrius javanicus. *See* Javan plovers
Charadrius marginatus. *See* White-fronted plovers
Charadrius melanops. *See* Black-fronted dotterels
Charadrius melodius. *See* Piping plovers
Charadrius montanus. *See* Mountain plovers
Charadrius obscurus. *See* New Zealand dotterels
Charadrius pallidus. *See* Chestnut-banded sandplovers
Charadrius pecuarius. *See* Kittlitz's plovers
Charadrius peronii. *See* Malaysian plovers
Charadrius placidus. *See* Long-billed plovers
Charadrius rubricollis. *See* Hooded plovers
Charadrius sanctaehelenae. *See* St. Helena plovers
Charadrius semipalmatus. *See* Semipalmated plovers
Charadrius thoracicus. *See* Madagascar plovers
Charadrius vociferus. *See* Killdeer
Charadrius wilsonia. *See* Wilson's plovers
Charilaidae, 3:203
Charina spp., 7:410–411
Charina reinhardtii. *See* Calabar boas
Charina trivirgata. *See* Rosy boas
Charitornis albertinae. *See* Bare-eyed mynas
Charles mockingbirds, 10:468
Charlotte's Web, 2:41
Charmosyna papou. *See* Papuan lories
Charniodiscus arboreus. *See* Sea pens
Charniodiscus oppositus. *See* Pink paddles
Charrs, 4:62, 4:405, 4:*408*, 4:*413*, 4:417
Chasmistes spp., 4:322
Chasmistes muriei, 4:322
Chatham Islands albatrosses. *See* Chatham mollymawks
Chatham Islands black robins, 11:106, 11:*107*, 11:*110*–111
Chatham Islands oystercatchers, 9:126
Chatham Islands robins. *See* Chatham Islands black robins
Chatham mollymawks, 8:114, 8:115, 8:*117*, 8:*120*
Chatham robins. *See* Chatham Islands black robins
Chatham snipes, 9:180
Chatrabus spp., 5:42
Chats, **10:483–490, 10:492–499**
Australian, **11:65–68,** 11:*67*
behavior, 10:486–487
conservation status, 10:488–489
distribution, 10:*483*, 10:485, 10:*492–499*
evolution, 10:483
feeding ecology, 10:487
habitats, 10:485–486
humans and, 10:489
physical characteristics, 10:483–485
reproduction, 10:487–488
species of, 10:*492–499*
taxonomy, 10:483
Chatterers. *See* Gray-crowned babblers
Chattering kingfishers, 10:6
Chaudhuriidae, 4:57, 5:151
Chauliodes spp., 3:*289*
Chauliodinae. *See* Fishflies
Chauliodus sloani. *See* Viperfishes
Chauna chavaria. *See* Northern horned screamers; Northern screamers
Chauna torquata. *See* Southern horned screamers; Southern screamers
Chaunacidae. *See* Sea toads
Chaunacoidei. *See* Sea toads
Chaunoproctus ferreorostris. *See* Bonin siskins
Checkered beetles, 3:320, 3:321
Checkered sengis, 16:*519*, 16:*521*, 16:522, 16:*523*, 16:*524*, 16:525
Cheek-pouched monkeys, **14:187–206,** 14:*195*, 14:*196*
behavior, 14:191–192
conservation status, 14:194
distribution, 14:*187*, 14:191
evolution, 14:187–189
feeding ecology, 14:192–193
habitats, 14:191
humans and, 14:194
physical characteristics, 14:189–191
reproduction, 14:193
species of, 14:*197*–204, 14:204*t*–205*t*
taxonomy, 14:187–188
Cheek teeth. *See* Molar teeth; Premolars
Cheers (*Catreus* spp.), 8:434
Cheetahs, 12:*10*, 12:*93*, 12:*206*, 14:*374*, 14:*377*
behavior, 14:258, 14:371
bottleneck species, 12:221
conservation status, 12:216
distribution, 14:370
feeding ecology, 14:372
humans and, 14:262, 14:290
Ice Age, 12:24
physical characteristics, 14:370
running speed, 12:42, 12:*45*
species of, 14:*381*–382
taxonomy, 14:369
Chega. *See* Squarehead catfishes
Cheilinus undulatus. *See* Humphead wrasses
Cheilobranchinae. *See* Singleslits
Cheilodactylidae. *See* Morwongs
Cheilodactylus spp., 5:238
Cheilodactylus vittatus. *See* Hawaiian morwongs
Cheilodipterus spp., 5:260
Cheilodipterus nigrotaeniatus, 5:345–346
Cheilopogon pinnatibarbatus. *See* Four-winged flyingfishes
Cheilopogon pinnatibarbatus californicus. *See* California flyingfishes
Cheilostomatida, 2:509
Cheimarrichthyidae. *See* Torrentfishes
Cheimarrichthys fosteri. *See* Torrentfishes
Cheirogaleidae. *See* Dwarf lemurs; Mouse lemurs
Cheirogaleus spp. *See* Dwarf lemurs
Cheirogaleus major. *See* Greater dwarf lemurs
Cheirogaleus medius. *See* Western fat-tailed dwarf lemurs
Cheiromeles spp., 13:483, 13:484, 13:486
Cheiromeles parvidens. *See* Lesser naked bats
Cheiromeles torquatus. *See* Naked bats
Chelemys macronyx, 16:268
Cheliceates, 2:5

Cheliceraformes, 2:333
Chelictinia spp. *See* White-tailed kites
Chelidae. *See* Australo-American sideneck turtles
Chelidinae, 7:77
Chelidonichthys lucerna. *See* Tub gurnards
Chelidonichthys spinosus. *See* Red gurnards
Chelidoptera tenebrosa. *See* Swallow-winged puffbirds
Chelidorhynx spp., 11:83
Chelifer cancroides. *See* Book scorpions
Chelodina longicollis. *See* Common snakeneck turtles
Chelodina rugosa. *See* Northern snakenecks
Chelodina siebenrocki. *See* New Guinea snakeneck turtles
Chelodina steindachneri. *See* Steindachner's turtles
Chelodininae, 7:77
Chelomophrynus bayi, 6:95
Chelonia mydas. *See* Green seaturtles
Chelonians, 7:47, 7:53
  See also Tortoises; Turtles
Cheloniidae. *See* Seaturtles
Chelus fimbriatus. *See* Matamatas
Chelydra spp. *See* Snapping turtles
Chelydra serpentina. *See* Snapping turtles
Chelydridae. *See* Snapping turtles
Chelyosoma orientale, 1:*457*, 1:460–*461*
Chemical compounds, 12:120
Chemical defense, 2:37, 3:65
  See also Behavior
Chemical insecticides, 3:77
Chemical pollution, 7:60–62
Chemokinesis, 1:39
Chemoreception, 12:84
  See also Physical characteristics
Chemosensory cues
  Anura, 6:45–46
  beetles, 3:320
  lacewings, 3:307
  lissamphibians, 6:22–23
  Orthoptera, 3:207
  salamanders, 6:24, 6:44–45
  See also Behavior; specific species
Chemosensory systems, 4:21–22, 4:60, 4:62, 7:31
  behavior and, 7:34–38
  Squamata, 7:195, 7:198, 7:200, 7:204–205
  See also Physical characteristics
Chemotaxis, 1:40
Chen caerulescens. *See* Snow geese
Chenonetta jubata. *See* Maned ducks
Chernoff, B., 5:67
Cherrie's pocket gophers, 16:196*t*
Cherry salmons, 4:*409*, 4:412–*413*
Cherry sawfly pear slugs. *See* Pear and cherry slugs
Cherry slugworms. *See* Pear and cherry slugs
Cherrystone clams. *See* Northern quahogs
Chersomanes albofasciata. *See* Spike-heeled larks
Chersophilus duponti. *See* Dupont's larks
Chestnut-backed buttonquails, 9:14, 9:19
Chestnut-backed mousebirds, 9:*472*, 9:*473*–474
Chestnut-backed scimitar-babblers, 10:*510*, 10:*511*–512
Chestnut-backed sparrow-larks, 10:345–346
Chestnut-banded sandplovers, 9:163
Chestnut-bellied herons. *See* Agami herons
Chestnut-bellied hummingbirds, 9:418, 9:450
Chestnut-bellied sandgrouse, 9:*233*, 9:*234*, 9:*236*–237
Chestnut-bellied whitefaces. *See* Southern whitefaces
Chestnut-breasted sandgrouse. *See* Chestnut-bellied sandgrouse
Chestnut-breasted whitefaces, 11:57
Chestnut bulbuls. *See* Ashy bulbuls
Chestnut-capped brush-finches, 11:*267*, 11:*278*–279
Chestnut-collared kingfishers. *See* Rufous-collared kingfishers
Chestnut-collared swifts, 9:422
Chestnut-eared aracaris, 10:127, 10:*129*, 10:*130*
Chestnut-eared finches. *See* Zebra finches
Chestnut-headed hill-partridges, 8:437
Chestnut-mandibled toucans, 10:*87*
Chestnut rails, 9:49, 9:*56*, 9:*65*
Chestnut sac-winged bats, 13:356, 13:364*t*
Chestnut short-tailed bats, 13:433*t*
Chestnut-sided silvereyes, 11:229
Chestnut-sided warblers, 11:*290*
Chestnut-sided white-eyes, 11:233
Chestnut teals, 8:364
Chestnut wattle-eyes, 11:*28*, 11:*41*–42
Chettusia spp., 9:161
Chevroderma turnerae, 2:*381*, 2:*384*–385
Chevron snakeheads. *See* Striped snakeheads
Chevroned hover flies, 3:*363*, 3:371–*372*
Chevrotains, **15:325–334**, 15:*330*
  behavior, 15:327–328
  conservation status, 15:329
  distribution, 15:*325*, 15:327
  evolution, 15:137, 15:265–266, 15:325–326
  feeding ecology, 15:328
  habitats, 15:327
  humans and, 15:329
  physical characteristics, 15:138, 15:267–268, 15:271, 15:326–327
  reproduction, 15:328–329
  species of, 15:*331*–333
  taxonomy, 15:267
Chewing lice, **3:249–257**
Chiapas catfishes, 4:352
Chiasmocleis boliviana. *See* Bolivian bleating frogs
Chiasmocleis ventrimaculatus. *See* Dotted humming frogs
Chiasmodontidae. *See* Deep-sea swallowers
Chibchanomys orcesi. *See* Water mice
Chickadees, **11:155–166**, 11:*159*
  behavior, 11:157
  conservation status, 11:158
  distribution, 11:*155*, 11:156
  evolution, 11:155
  feeding ecology, 11:157
  habitats, 11:156–157
  humans and, 11:158
  physical characteristics, 11:155–156
  reproduction, 11:157–158
  taxonomy, 11:155
Chickarees. *See* North American red squirrels
Chicken turtles, 7:106
Chickens
  disease, 8:25
  greater prairie, 8:434, 8:435, 8:437, 8:*439*, 8:*444*–445
  meat, 8:21
Chicks. *See* Reproduction
Chiefdoms, human, 14:251–252
Chiffchaffs, 11:*8*, 11:*20*
Chiggers. *See* Chigoes
Chigoes, 3:*351*, 3:353–*354*
Chihuahuan collared lizards. *See* Common collared lizards
Chilatherina sentaniensis, 5:70
Chile Darwin's frogs, 6:36, 6:173, 6:174, 6:*176*
Chilean coruros. *See* Coruros
Chilean dolphins. *See* Black dolphins
Chilean flamingos, 8:303, 8:304, 8:305, 8:306, 8:*307*, 8:*308*–*309*
Chilean guemals. *See* Chilean huemuls
Chilean huemuls, 15:379, 15:384, 15:*387*, 15:*390*, 15:392
Chilean mouse opossums. *See* Elegant fat-tailed opossums
Chilean perches, 5:219, 5:221, 5:222
Chilean pilchards. *See* South American pilchards
Chilean plantcutters. *See* Rufous-tailed plantcutters
Chilean pudus. *See* Southern pudus
Chilean Red List, on *Spalacopus cyanus maulinus*, 16:439
Chilean sardines. *See* South American pilchards
Chilean seedsnipes. *See* Least seedsnipes
Chilean shrew opossums, 12:267, 12:268, 12:*269*, 12:*270*
Chilean skuas, 9:206
Chilean stag beetles, 3:59
Chilean tree mice, 16:266, 16:*271*, 16:*273*, 16:274
Chilean woodstars, 9:450
Chilognatha, 2:363, 2:365
Chilomycterus spp., 5:471
Chilopoda. *See* Centipedes
Chiloporter eatoni, 3:*128*, 3:*129*
Chilorhinophis spp. *See* Yellow-and-black burrowing snakes
Chiloscyllium punctatum. *See* Banded cat sharks
Chimaeras, 4:10, **4:91–95**, 4:*93*
Chimaeridae, 4:91
Chimaeriformes. *See* Chimaeras
Chimaerocestidae, 1:226
Chimaeropsyllidae, 3:347, 3:348–349
Chimango caracaras, 8:348
Chimara albomaculata, 3:*375*
Chimarrogale spp., 13:247, 13:248, 13:249
Chimarrogale himalayica. *See* Himalayan water shrews
Chimarrogale phaeura. *See* Sunda water shrews
Chimney swallows. *See* Barn swallows
Chimney swifts, 9:419, 9:*422*, 9:*423*, 9:*425*, 9:*427*–428
Chimpanzees, 12:*196*, 14:*226*, 14:*236*
  behavior, 12:145, 12:*151*, 12:157, 14:7, 14:*229*–231
  conservation status, 12:215, 12:*220*, 14:235
  distribution, 14:5, 14:227
  enculturation of, 12:162
  evolution, 12:33
  feeding ecology, 14:9, 14:233–234
  habitats, 14:227
  humans and, 14:235
  language, 12:160–162
  memory, 12:153–154
  numbers and, 12:155–156, 12:162

physical characteristics, 14:226, 14:227
reproduction, 14:10, 14:235
self recognition, 12:159
social cognition, 12:160
species of, 14:*239*
taxonomy, 14:225
theory of mind, 12:159–160
tool use, 12:157
touch, 12:83
China. *See* Oriental region
China alligators. *See* Chinese alligators
Chinchilla spp., 16:377
Chinchilla brevicaudata. *See* Short-tailed chinchillas
Chinchilla lanigera. *See* Long-tailed chinchillas
Chinchilla mice, 16:265, 16:270
Chinchilla rats, **16:443–448**, 16:*446*
Chinchillas, 12:132, **16:377–384**, 16:*381*, 16:383*t*
Chinchillidae. *See* Chinchillas; Viscachas
Chinchillones. *See* Chinchilla rats
Chinchillula spp. *See* Chinchilla mice
Chinese algae eaters, 4:*322*, 4:*328*, 4:333
Chinese Alligator Fund, 7:173
Chinese alligators, 7:59, 7:161, 7:171–*176*, 7:*174*
Chinese black-headed gulls. *See* Saunder's gulls
Chinese bulbuls, 10:398
Chinese carps, 4:298
Chinese catfishes, 4:353
Chinese crested mynas. *See* Crested mynas
Chinese crested terns, 9:209
Chinese desert cats, 14:390*t*
Chinese dormice, 16:*322*, 16:*324*
Chinese egrets, 8:241
Chinese ferret badgers, 14:332*t*
Chinese giant salamanders, 6:49, 6:54, 6:343–347, 6:*344*, 6:*346*
Chinese goldfishes, 4:303
Chinese gray shrikes, 10:426
Chinese grouse, 8:*440*, 8:*443*
Chinese gymnures. *See* Hainan gymnures
Chinese hamsters, 16:243
Chinese hedgehogs, 13:204, 13:212*t*
Chinese hillstream loaches, 4:*324*, 4:*328*
Chinese jumping mice, 16:213, 16:216, 16:223*t*
Chinese jungle mynas. *See* Crested mynas
Chinese lake dolphins. *See* Baijis
Chinese liver flukes. *See* Oriental liver flukes
Chinese mantids, 3:179, 3:*181*, 3:*182*, 3:185–*186*
Chinese Materia Medica, 7:50
Chinese medicine. *See* Asian medicine
Chinese Meishan pigs, 15:149
Chinese mitten crabs, 2:*203*, 2:*206*
Chinese monal pheasants. *See* Chinese monals
Chinese monals, 8:434, 8:*438*, 8:*449–450*
Chinese muntjacs. *See* Reeve's muntjacs
Chinese olive flycatchers. *See* Brown-chested jungle-flycatchers
Chinese pangolins, 16:110, 16:113, 16:*114*, 16:*115*
Chinese penduline tits, 11:149
Chinese pygmy dormice, 16:297*t*
Chinese quails. *See* King quails
Chinese red pikas, 16:501*t*
Chinese river crabs. *See* Chinese mitten crabs
Chinese rock squirrels, 16:143, 16:144
Chinese schemers. *See* Silver carps
Chinese short-tailed shrews, 13:*253*, 13:*255–256*
Chinese shrew-moles, 13:282, 13:288*t*
Chinese softshell turtles, 7:48, 7:151, 7:152, 7:*153*, 7:*154*, 7:155
Chinese starlings. *See* Crested mynas
Chinese stripe-necked turtles, 7:*117*, 7:*118*, 7:119
Chinese sturgeons, 4:*214*
Chinese suckers, 4:*4*, 4:*319*, 4:*322*, 4:330–*331*
Chinese three-striped box turtles, 7:119
Chinese thrushes. *See* Hwameis
Chinese water deer, 15:343, **15:373–377**, 15:*374*, 15:*375*, 15:*376*
Chinese water pheasants. *See* Pheasant-tailed jacanas
Chinook salmons, 4:405, 4:*409*, 4:*412*, 4:415
Chinstrap penguins, 8:147, 8:149, 8:*150*
Chioglossa spp., 6:17, 6:24
Chioglossa lusitanica. *See* Golden-striped salamanders
Chionactis spp. *See* Shovel-nosed snakes
Chionea spp., 3:358
Chionidae. *See* Sheathbills
Chionis alba. *See* Pale-faced sheathbills
Chionis minor. *See* Black-faced sheathbills
Chionomys spp., 16:228
Chipmunks, 12:131, 16:127, 16:143, 16:144, 16:145, 16:147
Allen's, 16:160*t*
eastern, 16:126, 16:*148*, 16:*149*, 16:153–*154*
least, 16:*149*, 16:*152*
red-tailed, 16:159*t*
Siberian, 16:144, 16:158*t*
Tamias palmeri, 16:147–148
Townsend's, 16:*146*
Chipping sparrows, 11:*267*, 11:*281*
Chipping squirrels. *See* Eastern chipmunks
Chippy. *See* Chipping sparrows
Chiribiquete emeralds, 9:438
Chiridota hydrothermica. *See* Hydrothermal vent sea cucumbers
Chiridotea caeca. *See* Sand isopods
Chiridotidae, 1:420
Chirinda forest toads. *See* Chirinda toads
Chirinda toads, 6:*187*, 6:194
Chirindia spp., 7:273
Chiriquá pocket gophers, 16:196*t*
Chirixalus spp., 6:291, 6:292
Chirixalus eiffingeri. *See* Eiffinger's Asian treefrogs
Chirocentridae. *See* Wolf herrings
Chirocentrus spp., 4:277
Chirocentrus dorab. *See* Dorab wolf herrings
Chirocephalidae, 2:135
Chirodactylus grandis, 5:242
Chiroderma improvisum, 13:420
Chiroderma villosum. *See* Hairy big-eyed bats
Chirodropidae, 1:147
Chirolophis spp., 5:310
Chirolophis nugator. *See* Mosshead warbonnets
Chiromantis spp., 6:49, 6:281, 6:282, 6:291, 6:292
Chiromantis xerampelina. *See* Gray treefrogs
Chironectes spp. *See* Water opossums
Chironectes minimus. *See* Water opossums
Chironemidae. *See* Kelpfishes
Chironex fleckeri. *See* Sea wasps
Chironomidae, 3:55, 3:359
Chiropotes spp. *See* Bearded sakis
Chiropotes albinasus. *See* White-nosed bearded sakis
Chiropotes satanas. *See* Bearded sakis
Chiropotes satanas satanas. *See* Southern bearded sakis
Chiropotes satanas utahicki. *See* Uta Hick's bearded sakis
Chiroptera. *See* Bats
Chiroptera Specialist Group, IUCN, 13:358, 13:420
Chiropterotriton spp., 6:325, 6:391
Chiroteuthids, 2:476
Chiroxiphia linearis. *See* Long-tailed manakins
Chirugus scopoli. *See* Pheasant-tailed jacanas
Chirus. *See* Tibetan antelopes
Chisel-toothed kangaroo rats, 16:202, 16:203, 16:209*t*
Chitala blanci, 4:234
Chitala chitala. *See* Clown knifefishes
Chitala lopis, 4:232
Chitala ornata. *See* Clown featherbacks
Chitals, 15:269, 15:358, 15:*362*, 15:*365–366*
Chitin, 2:42, 3:8, 3:17–18, 3:27, 3:31
Chitons, **2:393–401**, 2:*394*, 2:*396*
behavior, 2:395
conservation status, 2:395
distribution, 2:395
evolution, 2:393
feeding ecology, 2:395
habitats, 2:395
humans and, 2:395
physical characteristics, 2:393–395
reproduction, 2:395
species of, 2:*397*–401
taxonomy, 2:393
Chitra chitra. *See* Asian narrow-headed softshell turtles
Chitra indica. *See* Asian narrow-headed softshell turtles
Chitwood, B. G., 1:283, 1:293
Chizaerhis leucogaster. *See* White-bellied go-away-birds
Chlamydera spp., 11:477, 11:478–481
Chlamydera cerviniventris. *See* Fawn-breasted bowerbirds
Chlamydera lauterbachi. *See* Yellow-breasted bowerbirds
Chlamydera maculata. *See* Spotted bowerbirds
Chlamydera nuchalis. *See* Great bowerbirds
Chlamydochaera spp. *See* Cuckoo-shrikes
Chlamydochaera jeffreyi. *See* Black-breasted trillers
Chlamydogobius spp., 5:375, 5:376
Chlamydogobius eremius. *See* Desert gobies
Chlamydogobius gloveri. *See* Dalhousie gobies
Chlamydogobius micropterus, 5:377
Chlamydogobius squamigenus. *See* Edgbaston gobies
Chlamydosaurus kingi. *See* Frilled lizards
Chlamydoselachidae, 4:143
Chlamydoselachus anguineus. *See* Frilled sharks
Chlamydotis spp. *See* Bustards
Chlamydotis macqueenii. *See* Macqueen's bustards
Chlamydotis undulata. *See* Houbara bustards
Chlamyphorus spp., 13:149, 13:182
Chlamyphorus retusus. *See* Chacoan hairy armadillos

Chlamyphorus truncatus. *See* Pink fairy armadillos
Chlamys opercularis. *See* Queen scallops
Chlamytheres. *See* Giant armadillos
Chlidonias spp. *See* Marsh terns
Chlidonias niger. *See* Black terns
Chloebia spp., 11:354
Chloebia gouldiae. *See* Gouldian finches
Chlopsidae, 4:255
Chlorella spp., 2:32
Chloridops spp., 11:343
Chloridops kona. *See* Kona grosbeaks
Chloridops regiskongi. *See* King Kong finches
Chlorion lobatum. *See* Metallic hunting wasps
Chlorocebus spp., 14:188, 14:191
Chlorocebus aethiops. *See* Grivets
Chloroceyle amazona. *See* Amazon kingfishers
Chlorocichla spp., 10:395–399
Chlorocichla flavicollis. *See* Yellow-throated leaf-loves
Chlorocichla flaviventris. *See* Yellow-bellied greenbuls
Chlorocichla laetissima. *See* Joyful greenbuls
Chlorofluorocarbons (CFCs), 1:52
Chlorolius spp., 6:280
Chloroperlidae, 3:54, 3:141
Chloropeta spp., 11:7
Chlorophthalmidae, 4:432
Chlorophthalmoidei, 4:431
Chlorophthalmus acutifrons. *See* Greeneyes
Chloropipo flavicapilla. *See* Yellow-headed manakins
Chloropsis spp. *See* Leafbirds
Chloropsis aurifrons. *See* Golden-fronted leafbirds
Chloropsis cochinchinensis. *See* Blue-winged leafbirds
Chloropsis cyanopogon. *See* Lesser green leafbirds
Chloropsis flavipennis. *See* Philippine leafbirds
Chloropsis hardwickii. *See* Orange-bellied leafbirds
Chloropsis palawanensis. *See* Yellow-throated leafbirds
Chloropsis venusta. *See* Blue-masked leafbirds
Chlorostilbon spp. *See* Hummingbirds
Chlorostilbon bracei. *See* Brace's emeralds
Chlorostilbon maugaeus. *See* Puerto Rican emeralds
Chlorostilbon mellisugus. *See* Blue-tailed emeralds
Chlorostilbon olivaresi. *See* Chiribiquete emeralds
Chlorostilbon ricordii. *See* Cuban emeralds
Chlorotalpa spp., 13:215
Chlorotalpa duthieae. *See* Congo golden moles
Chlorotalpa sclateri. *See* Sclater's golden moles
Choanoflagellates, 1:7–8, 2:12
Chocó Endemic Bird Area (Colombia), 10:103
Chocó poorwills, 9:405
Choco tinamous, 8:59
Choco vireos, 11:256
Chocolate australes. *See* Chocolate lyretails
Chocolate-chip sea stars, 1:*367*
Chocolate flycatchers. *See* Abyssinian slaty flycatchers
Chocolate lyretails, 5:*95*, 5:96–97
Choeradodinae, 3:179
Choeradodis rhomboidea, 3:*182*, 3:*184*–185
Choeronycteris mexicana. *See* Mexican hog-nosed bats
Choeropsis liberiensis. *See* Pygmy hippopotamuses
Cholevinae, 3:54
Choloepus spp., 13:149, 13:151, 13:153, 13:155, 13:156, 13:157
Choloepus didactylus. *See* Southern two-toed sloths
Choloepus hoffmanni. *See* Hoffman's two-toed sloths
Chologaster agassizii. *See* Spring cavefishes
Chologaster cornuta. *See* Swampfishes
Chondrichthyes, 4:10–11, 4:15, 4:23, 4:26, 4:91, 4:173
Chondrohierax uncinatus. *See* Hook-billed kites
Chondrohierax wilsonii. *See* Cuban kites
Chondrosida, 1:77, 1:79
Chonopeltis spp., 2:289
Chopi blackbirds, 11:305, 11:*307*, 11:*317*–318
Chopi-grackles. *See* Chopi blackbirds
Chordates, 1:13, 1:22, 2:3, 2:11
See also specific chordate species
Chordeiles acutipennis. *See* Lesser nighthawks
Chordeiles minor. *See* Common nighthawks
Chordeiles rupestris. *See* Sand-colored nighthawks
Chordeilinae. *See* Nighthawks
Chordonia, 2:4
Chorioallantoic placenta, 12:7, 12:12, 12:93
See also Physical characteristics; Reproduction
Chorion, 12:92–94
See also Physical characteristics; Reproduction
Chorisochismini, 5:355, 5:356
Chorisochismus dentex. *See* Rocksuckers
Choristida. *See* Astrophorida
Choristidae, 3:341
Chorthippus brunneus. *See* Field grasshoppers
Chorus frogs, 6:*231*, 6:238–239
Choruses. *See* Vocalizations
Choughs, 11:505
alpine, 11:504
red-billed, 11:504, 11:*511*, 11:518–519
white-winged, 11:*453*, 11:456–457, 11:504
Chousinghas, 16:*17*
behavior, 16:23
conservation status, 16:24
distribution, 16:*18*, 16:23
evolution, 15:265, 16:11
feeding ecology, 16:23
habitats, 12:136, 16:23
humans and, 16:24
physical characteristics, 16:2, 16:12, 16:23
reproduction, 16:14, 16:23
taxonomy, 15:265, 16:11, 16:23
See also Antelopes
Chow chows, 14:293
Chowchillas, **11:69–74**, 11:*73*
Chowder clams. *See* Northern quahogs
Chozchozes, 16:433, 16:435, 16:*437*, 16:*438*, 16:439
Christianity, snakes in, 7:57
Christidis, L., 10:385, 11:459
Christinus marmoratus. *See* Marbled geckos
Christmas Bird Counts, 8:22, 8:23
Christmas frigatebirds, 8:193, 8:195, 8:*196*, 8:*197*–198
Christmas Island blindsnakes, 7:381, 7:382
Christmas Island frigatebirds. *See* Christmas frigatebirds
Christmas Island white-eyes, 11:227, 11:228
Chroicocephalus saundersi. *See* Saunder's gulls
Chromadoria, 1:283
Chromatophores, 7:*226*
Chromis spp., 5:294
Chromis cyanea. *See* Blue chromis
Chromis sanctaehelenae, 5:298
Chromodoris willani, 2:*405*
Chromosome multiplication, 12:30–31
Chrossarchus spp. *See* Cusimanses
Chrotogale owstoni. *See* Owston's palm civets
Chrotopterus auritus. *See* Woolly false vampire bats
Chrysalis, monarch butterfly, 3:*389*
Chrysaora spp., 1:157
Chrysaora dactylometre, 1:*153*
Chrysaora quinquecirrha. *See* Sea nettles
Chrysemys picta. *See* Painted turtles
Chrysemys picta picta. *See* Eastern painted turtles
Chrysiptera spp., 5:294
Chrysis spp., 3:406
Chrysis coerulans. *See* Cuckoo wasps
Chrysobatrachus spp., 6:280
Chrysochloridae. *See* Golden moles
Chrysochlorinae, 13:215
Chrysochloris spp., 13:197, 13:215
Chrysochloris asiatica. *See* Cape golden moles
Chrysochloris stuhlmanni. *See* Stuhlmann's golden moles
Chrysococcyx spp., 9:313
Chrysococcyx caprius. *See* Dideric cuckoos
Chrysococcyx cupreus. *See* Emerald cuckoos
Chrysococcyx klaas. *See* Klaas's cuckoos
Chrysococcyx russatus. *See* Gould's bronze cuckoos
Chrysocyon brachyurus. *See* Maned wolves
Chrysolampis spp. *See* Hummingbirds
Chrysolampis mosquitus. *See* Ruby topaz
Chrysolophus amherstiae. *See* Lady Amherst's pheasants
Chrysomelid beetles, 3:57
Chrysomelidae, 3:65
Chrysomyia spp. *See* Latrine flies
Chrysopa spp., 3:*308*, 3:309
Chrysopelea spp. *See* Flyingsnakes
Chrysoperla spp., 3:307, 3:309
Chrysopidae, 3:305, 3:306, 3:307, 3:308, 3:309
Chrysospalax spp., 13:215
Chrysospalax trevelyani. *See* Large golden moles
Chrysospalax villosus. *See* Rough-haired golden moles
Chrysuronia spp. *See* Hummingbirds
Chthamalus stellatus, 2:28
Chthonicola sagittatus. *See* Speckled warblers
Chubby frogs. *See* Malaysian painted frogs
Chubs. *See* Sea chubs
Chucao tapaculos, 10:260, 10:*261*, 10:*263*
Chuck-will's-widows, 9:*368*, 9:370, 9:403–404, 9:*407*, 9:*410*
Chuckawallas. *See* Common chuckwallas
Chucks. *See* Chuck-will's-widows; Common chuckwallas
Chuditches, 12:*280*, 12:283, 12:288, 12:290–291, 12:*293*, 12:*294*–295
Chunga burmeisteri. *See* Black-legged seriemas
Chytridiomycosis, 6:57, 6:60
Cibolacris parviceps, 3:204

Cicadabirds, 10:386
Cicadas, **3:259–264**
behavior, 2:36, 3:59–60, 3:261, 3:270
conservation status, 3:270
distribution, 3:*269*, 3:270
evolution, 3:9, 3:259
feeding ecology, 3:63, 3:262, 3:270
habitats, 3:270
humans and, 3:264, 3:270
physical characteristics, 3:260, 3:270
reproduction, 3:263, 3:270
symbolism, 3:74
taxonomy, 3:259, 3:270
See also Seventeen-year cicadas
Cicadomorpha, 3:259
Cichla spp., 5:276
Cichla temensis. *See* Speckled pavons
Cichladusa guttata. *See* Spotted palm-thrushes
Cichlasoma dovii. *See* Guapotes
Cichlasoma nicaraguense. *See* Mogas
Cichlidae. *See* Cichlids
Cichlids, 4:*34*, 4:36, 4:52, 4:62, 4:66, **5:275–289**, 5:*282*, 5:*283*
Cicinnurus spp. *See* Birds of paradise
Cicinnurus regius. *See* King birds of paradise
Cicinnurus regius cocineifrons, 11:500
Cicinnurus regius regius, 11:500
Cicinnurus respublica. *See* Wilson's birds of paradise
Ciconia abdimii. *See* Abdim's storks
Ciconia boyciana. *See* Oriental white storks
Ciconia ciconia. *See* European white storks
Ciconia episcopus. *See* Woolly-necked storks
Ciconia maguari. *See* Maguari storks
Ciconia nigra. *See* Black storks
Ciconiidae. *See* Storks
Ciconiiformes, **8:233–237**, 8:*234*
See also specific Ciconiiformes
Ciconiini. *See* Storks
Cigarette beetles, 3:76
Ciguatera poisoning, 5:215–216, 5:399
Ciguatoxin, 4:265
Cihuacoatl (Mayan goddess), 7:55
Ciliated animals, 1:9, 2:5, 2:11, 2:12
Ciliocincta spp., 1:99
Ciliocincta sabellariae, 1:*100*, 1:101, 1:*102*
Cimex lectularius. *See* Bed bugs
Cimicomorpha, 3:259
Cimmarons, 12:176
Cimolestes spp., 14:255–256
Cinachyra antarctica, 1:29
Cincinnati Zoo, 12:203
Cinclidae. *See* Dippers
Cinclocerthia ruficauda. *See* Brown tremblers
Cinclodes aricomae. *See* Royal cinclodes
Cinclodes fuscus. *See* Bar-winged cinclodes
Cinclosma spp. *See* Quail thrushes
Cinclosomatidae. *See* Eupetidae
Cinclus cinclus. *See* Eurasian dippers
Cinclus leucocephalus. *See* White-capped dippers
Cinclus mexicanus. *See* American dippers
Cinclus pallasii. *See* Brown dippers
Cinclus schulzi. *See* Rufous-throated dippers
Cinereous antshrikes, 10:*244*, 10:*247–248*
Cinereous cockroaches, 3:149, 3:*152*, 3:*154*
Cinereous flycatchers. *See* Ashy flycatchers
Cingulata, 13:147, 13:148, 13:182
Cinnamon bears. *See* American black bears
Cinnamon-breasted tits. *See* Rufous-bellied tits
Cinnamon-rumped foliage-gleaners, 10:*213*, 10:*226–227*
Cinnamon-rumped trogons, 9:479
Cinnamon-tailed fantails, 11:87
Cinnamon treefrogs. *See* Painted Indonesian treefrogs
Cinnycerthia spp., 10:527
Cinnyricinclus fermoralis. *See* Abbott's starlings
Cinnyris spp. *See* Sunbirds
Cinnyris afra. *See* Greater double-collared sunbirds
Cinnyris asiaticus. *See* Purple sunbirds
Cinnyris chloropygius. *See* Olive-bellied sunbirds
Cinnyris coccinigaster. *See* Splendid sunbirds
Cinnyris collaris. *See* Collared sunbirds
Cinnyris cupreus. *See* Copper sunbirds
Cinnyris dussumieri. *See* Seychelles sunbirds
Cinnyris fuscus. *See* Dusky sunbirds
Cinnyris gouldiae. *See* Gould's sunbirds
Cinnyris jugularis. *See* Olive-backed sunbirds (*Cinnyris* spp.)
Cinnyris minima. *See* Crimson-backed sunbirds
Cinnyris moreaui. *See* Moreau's sunbirds
Cinnyris neergaardi. *See* Neergaard's sunbirds
Cinnyris oseus decorsei. *See* Palestine sunbirds
Cinnyris pulchellus. *See* Beautiful sunbirds
Cinnyris rockefelleri. *See* Rockefeller's sunbirds
Cinnyris rufipennis. *See* Rufous-winged sunbirds
Cinnyris ursulae. *See* Ursula's sunbirds
Ciona spp., 1:452
Ciona intestinalis, 1:*457*, 1:*458*
Cipo castaneros, 10:211
Circadian rhythms, 4:60, 12:74–75
See also Physical characteristics
Circannual cycles, subterranean mammals, 12:75
See also Physical characteristics
Circannual rhythms, migration and, 8:32
Circellium bacchus. *See* Dung beetles
Circulatory system, 3:24–*25*, 4:23–24, 6:17, 7:28–29, 7:67, 7:69–70, 8:9–10, 12:45–46, 12:60
See also Physical characteristics
Circus aeruginosus. *See* Western marsh harriers
Circus approximans. *See* Pacific marsh harriers
Circus cyaneus. *See* Hen harriers
Circus macrosceles. *See* Madagascar harriers
Circus macrourus. *See* Pallid harriers
Circus maillardi. *See* Reunion harriers
Circus pygargus. *See* Montagu's harriers
Cirolana spp., 2:251
Cirolanidae, 2:249
Cirrata. *See* Finned octopods
Cirratulids, 2:46, 2:47
Cirrhitichthys aprinus, 5:242
Cirrhitichthys falco, 4:63
Cirrhitidae. *See* Hawkfishes
Cirrhitoidea, 5:235, 5:237
Cirrhitops fasciatus. *See* Red-barred hawkfishes
Cirrhitus pinnulatus, 5:236
Cirripedes. *See* Barnacles
Cissa spp., 11:504
Cisticola spp., 11:1, 11:3–5, 11:7, 11:394
Cisticola anonymus, 11:5
Cisticola chubbi, 11:4
Cisticola exilis. *See* Golden-headed cisticolas
Cisticola hunteri, 11:4
Cisticola juncidis. *See* Zitting cisticolas
Cisticola nigriloris, 11:4, 11:5, 11:6
Cistoclemmys flavomarginata. *See* Yellow-margined box turtles
Cistothorus spp., 10:527, 10:528
Cistothorus apolinari. *See* Apolinar's wrens
Cistothorus palustris. *See* Marsh wrens
Cistothorus platensis. *See* Sedge wrens
Citellus spp. *See* Ground squirrels
CITES. *See* Convention on International Trade in Endangered Species
Citharichthys sordidus. *See* Pacific sanddabs
Citharidae, 5:450
Citharinidae, 4:69, 4:335, 4:339
Citrils, 11:323
Citrus leaf miners, 3:*391*, 3:*394*
Citrus spine nematodes, 1:*296*, 1:301–302
Cittura cyanotis. *See* Lilac-cheeked kingfishers
Civet cats. *See* African civets
Civet oil, 14:338
Civets, **14:335–345**, 14:*339*
behavior, 14:258, 14:337
conservation status, 14:338
distribution, 12:136, 14:336
evolution, 14:335
feeding ecology, 14:337
humans and, 14:338
physical characteristics, 14:335–336
reproduction, 14:337
species of, 14:*340–342*, 14:344*t*–345*t*
taxonomy, 14:256, 14:335
Civettictis spp., 14:335, 14:338
Civettictis civetta. *See* African civets
Cladistic biogeography, 3:53–54
Cladistics. *See* Taxonomy
Cladocarpus lignosus, 1:127
Cladocera. *See* Water fleas
Cladorhizidae, 1:28
Cladorhynchus spp. *See* Stilts
Cladorhynchus leucocephalus. *See* Banded stilts
Cladoselache, 4:10
Clam shrimps, **2:147–151**, 2:*149*
Clamator spp. *See* Cuckoos
Clams, 2:*36*, 2:451, 2:454
Asian, 2:454
giant vent, 2:*456*, 2:*462*, 2:466
skeletons, 2:*18*
yo-yo, 2:*455*, 2:*461*, 2:465
See also Bivalves; specific types of clams
Clamworms, 2:25, **2:45–57**, 2:*48*, 2:*49*
behavior, 2:46
conservation status, 2:47
distribution, 2:46
evolution, 2:45
feeding ecology, 2:46–47
habitats, 2:46
humans and, 2:47
physical characteristics, 2:45–46
reproduction, 2:47
species of, 2:*50–56*
taxonomy, 2:45
See also Ragworms
Clangula hyemalis. *See* Oldsquaws
Clapp, W. F., 5:56
Clapper larks, 10:344
Clapper rails, 9:49, 9:51, 9:52
Clarence galaxias, 4:391
Clarias batrachus. *See* Walking catfishes
Clarias gariepinus. *See* Sharptooth catfishes
Clariidae, 4:352
Clarion wrens, 10:529

Clarius spp., 4:24
Clarke, Bryan, 2:411
Clarke's gazelles. *See* Dibatags
Clarke's weavers, 11:379
Clark's grebes, 8:171
Clark's lacerta, 7:299
Clark's nutcrackers, 11:*504*, 11:506
Claroglanididae, 4:353
Claroteidae, 4:352
Classic theory, of origin of life, 2:8
Classical conditioning, 1:41, 2:40
See also Behavior
Classification. *See* Taxonomy
Classification of species. *See* Taxonomy
Clathrina spp., 1:*59*
Clathrina heronensis, 1:*60*, 1:*61*
Clathrinida, 1:58
Claudius spp., 7:121
Clavagellidae, 2:451
Clavelina spp., 1:454
Clavelina dellavallei, 1:*456*, 1:*459*
Clavelina lepadiformis, 1:*456*, 1:459–*460*
Clavelina picta. *See* Painted tunicates
Clavopsella michaeli, 1:26
Clawed frogs, **6:99–107**, 6:*103–106*
Clawed lobsters, 2:197
Clawed toads. *See* Common plantanna
Claws, 12:39
See also Physical characteristics
Clay-licks, 9:290
Cleaner gobies, 4:69, 5:376, 5:384
Cleaner organisms, 2:33–34
Cleaner wrasses, 5:295–*296*
See also Bluestreak cleaner wrasses
Cleanerfishes, 4:50, 4:69
Clearance rates, 1:29
Clearfin lizardfishes, 4:433, 4:*435*, 4:*438*, 4:439
Clearnose skates, 4:*180*, 4:*186*, 4:187
Cleavage, 1:18–19, 1:21–23, 2:3, 2:4, 2:15–16, 2:18–21, 2:23–24, 2:35
See also Reproduction
Cleidopus gloriamaris, 5:*116*
Clelia spp. *See* Mussuranas
Clelia clelia. *See* Mussuranas
Clemens, W. A., 5:351
Clemmys guttata. *See* Spotted turtles
Clench, M. H., 10:329
Cleptoparasites, 3:406
Cleptornis marchei. *See* Golden white-eyes
Clethrionomys spp. *See* Red-backed voles
Clethrionomys gapperi. *See* Southern red-backed voles
Clethrionomys glareolus. *See* Bank voles
Clevelandia ios. *See* Arrow gobies
Clever Hans (horse), 12:155
Click beetles, 3:316, 3:*325*, 3:*327*, 3:329–330
Clicking crickets, Suriname, 3:*210*, 3:*211*–212
Clicking peltops, 11:*470*, 11:471–*472*
Clidomys spp., 16:469, 16:470
Cliff swallows. *See* American cliff swallows
Climaciella spp., 3:307
Climaciella brunnea, 3:307
Climacoceras spp., 15:265
Climacteridae. *See* Australian creepers
Climacteris spp. *See* Australian creepers
Climacteris affinis. *See* White-browed treecreepers
Climacteris erythrops. *See* Red-browed treecreepers
Climacteris leucophaea. *See* White-throated treecreepers
Climacteris melanura. *See* Black-tailed treecreepers
Climacteris picumnus. *See* Brown treecreepers
Climacteris rufa. *See* Rufous treecreepers
Climate
changes in, 6:57, 7:61
glaciers and, 12:17–19
mammalian evolution and, 12:11, 12:17–19
Climbing catfishes, 4:352
Climbing galaxias, 4:391
Climbing mice, gray, 16:*286*, 16:*290*–291
Climbing perches, 5:428, 5:*430*, 5:*431*
Climbing shrews. *See* Forest musk shrews
Clingfishes, 4:47, 4:48, 4:50, 5:*355*, **5:355–363**, 5:*358*
Clinidae, 5:341, 5:343
Clinocottus spp., 5:180
Clinocottus globiceps. *See* Mosshead sculpins
Clinoid sponges, 1:29
Clinostomus elongatus. *See* Redside daces
Clinotarsus spp., 6:249
Clinton, Bill, 12:184
Clinus spatulatus, 5:343
Cliona spp., 1:29
Cliona celata. *See* Yellow boring sponges
Clitellata, 2:23, 2:65
Cloaca, 6:32–33
Cloacotaenia megalops, 1:230
Cloaked lizards. *See* Frilled lizards
Cloeotis percivali. *See* Percival's trident bats
Clonorchis sinensis. *See* Oriental liver flukes
Close-barred sandgrouse. *See* Lichtenstein's sandgrouse
Clotbey larks. *See* Thick-billed larks
Clothes moths, 3:75, 3:*391*, 3:*403*–404
Clothoda urichi, 3:*233*
Cloud sponges, 1:*67*, 1:*68*, 1:*70*, 1:*73*–74
Cloud swifts. *See* White-collared swifts
Clouded leopards, 14:370, 14:372, 14:*377*, 14:*385*, 14:387
Clover springtails. *See* Lucerne fleas
Clown anemonefishes, 5:*299*, 5:*302*, 5:304
Clown beetles, 3:318, 3:321, 3:323
Clown featherbacks, 4:*26*, 4:232
See also Clown knifefishes
Clown fishes, 2:34
Clown knifefishes, 4:*231*–234, 4:*235*, 4:239–240, 4:*241*
Clown loaches, 4:*322*, 4:*326*, 4:331–332
Clown shrimps. *See* Harlequin shrimps
Clown triggerfishes, 4:*4*, 4:*31*, 5:*473*, 5:*475*–476
Clownfish-anemone symbiosis, 1:33
Club-footed bats. *See* Bamboo bats
Club urchins. *See* Slate-pencil urchins
Clupea harengus. *See* Atlantic herrings
Clupea harengus pallasii. *See* Pacific herrings
Clupea pallasii. *See* Pacific herrings
Clupeidae, 4:34, 4:44, 4:48, 4:277
Clupeiformes. *See* Herrings
Clupeoidei, 4:277
Clupeomorpha, 4:11–12
Clydesdale horses, 15:*216*
Clyomys spp., 16:450
Clyomys bishopi. *See* Bishop's fossorial spiny rats
Clyptraeidae, 2:446
Clytia spp., 1:129
Clytia gracilis, 1:129
Clytoceyx rex. *See* Shovel-billed kookaburras
Clytoctantes alixii. *See* Recurve-billed bushbirds
Clytolaima spp. *See* Hummingbirds
Clytomyias spp. *See* Orange-crowned wrens
Clytomyias insignis. *See* Orange-crowned wrens
Cnemidocarpa spp., 1:452
Cnemidophorus spp., 7:309, 7:310, 7:311, 7:312
Cnemidophorus gularis. *See* Texas spotted whiptails
Cnemidophorus inornatus. *See* Little striped whiptails
Cnemidophorus laredoensis. *See* Laredo striped whiptails
Cnemidophorus sexlineatus. *See* Six-lined racerunners
Cnemidophorus tigris, 7:312
Cnemidophorus tigris mundus. *See* California whiptail lizards
Cnemidophorus uniparens. *See* Desert grassland whiptails
Cnemidophorus vanzoi. *See* St. Lucia whiptails
Cnemophilinae. *See* Birds of paradise
Cnemophilus spp. *See* Birds of paradise
Cnemophilus macgregorii, 11:490
Cnidarians, 2:9, 2:12–13, 2:15, 2:32
behavior, 1:37
drugs and, 1:45
feeding ecology, 1:27–30
habitats, 1:24
reproduction, 1:19–20, 1:21
taxonomy, 1:10
See also Corals; Hydrozoa; Jellyfish
Coachella sand-lizards. *See* Coachella Valley fringe-toed lizards
Coachella uma. *See* Coachella Valley fringe-toed lizards
Coachella Valley fringe-toed lizards, 7:207, 7:*248*, 7:*253*, 7:255
Coalfishes. *See* Pollocks; Sablefishes
Coarse-filter conservation, 3:84, 3:86, 3:88–89
Coarse-haired pocket mice, 16:199, 16:200, 16:203
Coarse-haired wombats. *See* Common wombats
Coastal brown bears. *See* Brown bears
Coastal giant salamanders, 6:*350*, 6:*352*
Coastal miners, 10:*212*, 10:214–*215*
Coastal tailed frogs, 4:191, 6:42, 6:77, 6:*78*, 6:*79*, 6:80–*81*
Coat coloration, 12:99
Coatimundis. *See* White-nosed coatis
Coatis, 14:256, 14:310, 14:311, 14:316*t*, 15:239
Coban alligator lizards, 7:*341*, 7:*343*–344
Cobias, 5:255, 5:256, 5:258–263
Cobitidae. *See* Loaches
Cobitinae, 4:321
Cobitis barbatula. *See* Stone loaches
Cobitis choirorhynchus. *See* Horseface loaches
Cobitis fossilis. *See* Weatherfishes
Cobitis kuhlii. *See* Coolie loaches
Cobitis macracanthus. *See* Clown loaches
Cobitoidea, 4:321–322
Cobras, 7:42, 7:56, 7:205–206, **7:483–488**, 7:*490*, **7:492–493**
Cobweb spiders. *See* Long-bodied cellar spiders
Coccidae, 3:264
Coccinelid beetles, 3:65

Coccothraustes vesperitnus. See Evening grosbeaks
Cocculina japonica. See Japanese deep-sea limpets
Cocculinidae, 2:435
Cocculiniformia. *See* Deep-sea limpets
Cocculinoidea, 2:435, 2:436
Coccyzinae. *See* American cuckoos
Coccyzus spp. *See* Cuckoos
Coccyzus americanus. See Yellow-billed cuckoos
Coccyzus melacoryphus. See Dark-billed cuckoos
Coccyzus pumilis, 9:313
Cochineal insects, 3:81–82
Cochlearius cochlearius. See Boat-billed herons
Cochleoceps spp., 5:355, 5:357
Cochleoceps orientalis. See Eastern cleaner-clingfishes
Cochliomyia hominivorax. See New World primary screwworms
Cochranella spp., 6:215, 6:216
Cochranella griffithsi. See Ecuador Cochran frogs
Cochranella igonta. See Lynch's Cochran frogs
Cochranella ocellata. See Spotted Cochran frogs
Cociella crocodila, 5:158
Cockatiels, 9:276, 9:280, 9:*282,* 9:*284–285*
Cockatoo-parrots. *See* Cockatiels
Cockatoo waspfishes, *5:167,* 5:*169,* 5:177
Cockatoos, 9:275–279, 9:*282,* 9:283–284
Cockerell's fantails, 11:87
Cockles, 2:454
See also Bivalves
Cockroaches, 2:25, **3:147–159,** 3:*152*
behavior, 3:65, 3:148–149
conservation status, 3:150
distribution, 3:148
education and, 3:83
evolution, 3:8, 3:9, 3:10, 3:12, 3:13, 3:147
farming of, 3:80
feeding ecology, 3:64, 3:149
habitats, 3:148
humans and, 3:150–151
as pests, 3:76, 3:151
physical characteristics, 3:19, 3:22–24, 3:27, 3:147–148
reproduction, 3:30, 3:33–34, 3:39, 3:62, 3:149–150
species of, 3:*153*–158
taxonomy, 3:147
vs. termites, 3:161–162
See also specific types of cockroaches
Cocks-of-the-rock. *See* Cotingas
Coconut lories. *See* Rainbow lorikeets
Coconut stick insects, 3:*227,* 3:*230,* 3:231
Cocoons, parrotfishes and, 5:*295,* 5:*297*
Cod worms, 1:*296,* 1:302
Codlike fishes, 4:448, 5:15
Cods, 4:*30,* 4:75, **5:25–30,** 5:*31,* **5:32–34**
Cody, M. L., 9:444
Coecilia rostratus. See Frigate Island caecilians
Coelacanth Conservation Council, 4:193
Coelacanthiformes. *See* Coelacanths
Coelacanthimorpha, 4:12
Coelacanths, 4:12, 4:32, **4:189–196,** 4:*190,* 4:*194,* 6:7, 12:62
Coeligena spp. *See* Hummingbirds
Coeligena prunellei, 9:418
Coeloblastula, 2:21, 2:23
Coelodonta antiquitatis. See Woolly rhinoceroses
Coelom, 1:3–6, 1:19, 2:3–4, 2:15–16, 2:21, 2:35
Coeloplanids, 2:32
Coelops frithi. See East Asian tailless leaf-nosed bats
Coendou spp. *See* Tree porcupines
Coendou bicolor. See Bicolor-spined porcupines
Coendou ichillus, 16:365
Coendou koopmani. See Koopman's porcupines
Coendou melanurus. See Black-tailed hairy dwarf porcupines
Coendou pallidus, 16:365
Coendou paragayensis, 16:365
Coendou prehensilis. See Prehensile-tailed porcupines
Coendou pruinosus. See Frosted hairy dwarf porcupines
Coendou roosmalenorum, 16:365
Coendou rothschildi. See Rothschild's porcupines
Coendou sneiderni. See White-fronted hairy dwarf porcupines
Coenocorypha pusilla. See Chatham snipes
Coeranoscincus spp., 7:328
Coerebinae. *See* Bananaquites
Coevolution, 3:14, 3:48–49, 12:216–217
See also Evolution
Coffeesnakes, 7:468
Coffin flies, 3:357, 3:358
Coffinfishes. *See* Sea toads
Cogger, H. G., 7:216
Cognition, 12:140–142, **12:149–164**
See also Behavior
Cohen, D. M., 5:25, 5:26, 5:27
Coho salmons, 4:*409,* 4:*412*
Coiba howler monkeys, 14:160, 14:167*t*
Coiba Island agoutis, 16:409, 16:*410,* 16:*412,* 16:413
Coimbra's titis, 14:11, 14:148
Coke's hartebeests, 16:*31*
Colaptes auratus. See Northern flickers
Colaptes auratus rufipileus. See Guadeloupe flickers
Colaptes chrysoides. See Gilded flickers
Colasisi. *See* Philippine hanging-parrots
Colburn's tuco-tucos, 16:430*t*
Cold seep worms, **2:91–95,** 2:*93*
Cole, Charles J., 7:306
Coleman's shrimps, 1:*34*
Coleoidea, 2:475
Coleonyx variegatus. See Western banded geckos
Coleonyx variegatus abbotti. See San Diego banded geckos
Coleoptera, 3:35, **3:315–334,** 3:*324,* 3:*325*
behavior, 3:319–320
circulatory system, 3:24
conservation status, 3:322
distribution, 3:319
evolution, 3:315
feeding ecology, 3:320–321
habitats, 3:319
humans and, 3:322–323
physical characteristics, 3:19, 3:22, 3:317–319
reproduction, 3:44, 3:321–322
species of, 3:*327*–333
taxonomy, 3:315–316
See also Beetles; Weevils
Coleorrhyncha, 3:259, 3:260
See also Moss bugs
Coleridge, Samuel Taylor, 8:116
Coleura seychellensis. See Seychelles sheath-tailed bats
Colias spp., 3:54
Colibri spp. *See* Violet-ears
Colibri coruscans. See Sparking violet-ears
Colibri thalassinus. See Green violet-ears
Coliidae. *See* Mousebirds
Colinus virginanus ridgwayi. See Masked bobwhites
Colinus virginianus. See Northern bobwhite quails
Colisa spp., 5:427, 5:428
Colisa labiosa. See Thick-lipped gouramies
Colisa lalia. See Dwarf gouramies
Colius castanotus. See Chestnut-backed mousebirds
Colius colius. See White-backed mousebirds
Colius indicus. See Red-faced mousebirds
Colius leucocephalus. See White-headed mousebirds
Colius striatus. See Bar-breasted mousebirds
Collar complex, 1:7
Collared anteaters. *See* Northern tamanduas; Southern tamanduas
Collared aracaris, 10:*128*
Collared broadbills. *See* Silver-breasted broadbills
Collared carpet sharks, 4:105
Collared crescentchests, 10:*261,* 10:*265*
Collared falconets, 8:348
Collared flycatchers, 11:*29,* 11:*32*–33
Collared hedgehogs, 13:213*t*
Collared kingfishers, 10:6, 10:*11,* 10:*19*
Collared lemmings, 16:226, 16:230
Collared lizards, 7:243, 7:244, 7:*249,* 7:*250*
Collared mangabeys, 14:194, 14:*196,* 14:*198,* 14:202
Collared owlet-nightjars. *See* Barred owlet-nightjars
Collared palm-thrushes. *See* Spotted palm-thrushes
Collared peccaries, 15:267, 15:291, 15:*292,* 15:*293,* 15:*294,* 15:*296,* 15:*297,* 15:*298*
Collared pikas, 16:*480,* 16:484, 16:491, 16:*493,* 16:*495,* 16:501*t*
Collared plains-wanderers. *See* Plains-wanderers
Collared pratincoles, 9:*152,* 9:153, 9:*154,* 9:*155*
Collared puffbirds, 10:*105,* 10:106–*107*
Collared sunbirds, 11:208, 11:*211,* 11:*215*–216
Collared titis, 14:143, 14:144, 14:145, 14:*149,* 14:*152*
Collared tits. *See* White-naped tits
Collared trogons, 9:478, 9:*480,* 9:482–*483*
Collared tuco-tucos, 16:431*t*
Collembola. *See* Springtails
Collette, B. B., 5:405
Colli, Guarino R., 7:306
Collias, Elsie, 11:377
Collias, Nicholas, 10:173, 11:377
Collies, 14:288–289, 14:292
Colliformes. *See* Mousebirds
Collocalia bartschi. See Guam swiftlets
Collocalia elaphra. See Seychelles swiftlets
Collocalia francica. See Mascarene swiftlets
Collocalia leucophaeus. See Tahiti swiftlets
Collocalia sawtelli. See Atius

Collocalia troglodytes. *See* Pygmy swiftlets
Collocalia whiteheadi. *See* Whitehead's swiftlets
Colluricincla spp., 11:116
Colluricincla harmonica. *See* Gray shrike-thrushes
Colluricincla sanghirensis. *See* Sangihe shrike-thrushes
Colluricincla woodwardi. *See* Sandstone shrike-thrushes
Colluricinda megahyncha. *See* Little shrike-thrushes
Colobidae, 14:4
Colobinae. *See* Leaf-monkeys
Colobus spp., 14:172, 14:173
Colobus angolensis. *See* Angolan colobus
Colobus guereza. *See* Mantled guerezas
Colobus monkeys, 12:14, 14:5, 14:11, 14:172, 14:173
  Angolan, 14:184*t*
  black, 14:174, 14:175, 14:184*t*
  black-and-white, 14:7
  king, 14:175, 14:184*t*
  olive, 14:173, 14:175, 14:*176*, 14:*178*, 14:179
  Penant's red, 14:185*t*
  red, 14:7, 14:172–175
Colobus polykomos. *See* King colobus
Colobus satanas. *See* Black colobus
Colobus vellerosus, 14:175
Coloconger raniceps. *See* Froghead eels
Colocongridae, 4:255
Cololabis adocetus, 5:79
Colombian black spider monkeys, 14:*159*
Colombian grebes, 8:169, 8:171
Colombian potoos. *See* Rufous potoos
Colombian tinamous, 8:59
Colombian weasels, 14:324
Colombian woolly monkeys, 14:160, 14:*161*, 14:*165*, 14:166
Colonial entoprocts, 1:319, 1:321
  freshwater, 1:319, 1:*320*, 1:*324*
  marine, 1:*320*, 1:*322*, 1:*323*
Colonial theory, 1:7–8, 2:12
  See also Evolution
Coloninae, 3:54
Colophon spp., 3:322
Coloptychon spp. *See* Alligator lizards
Color vision, 3:27, 12:79
  See also Physical characteristics; Vision
Colorado beetles, 3:*317*
Colorado pikeminnows, 4:298, 4:299–300, 4:*303*, 4:*307*, 4:315
Colorado squawfishes. *See* Colorado pikeminnows
Coloration, 4:61, 4:69–70, 7:30, 12:38, 12:173
  aposematic, 3:65
  beetles, 3:318–319
  birds, 8:5–6, 8:8, 8:9
  chameleons, 7:224–225, 7:*226*, 7:*227*
  crocodilians, 7:158–159
  flash, 1:42, 2:38–39 (*See also* Behavior)
  Lepidoptera and, 3:386
  Orthoptera, 3:203
  parrotfishes, 5:296–297
  Squamata, 7:203
  wrasses, 5:296–297
  See also Behavior; Physical characteristics
Colorful pufflegs, 9:438
Colossendeis megalonyx, 2:323, 2:*324*
Colosteids, 6:8–9
Colostethus spp., 6:34, 6:66, 6:197–200
Colostethus caeruleodactylus. *See* Blue-toed rocket frogs
Colostethus stepheni. *See* Stephen's rocket frogs
Colostethus subpunctatus, 6:*32*
Colostrum, 12:126
  See also Reproduction
Coluber spp. *See* Racers
Coluber constrictor. *See* Eastern racers
Colubridae. *See* Colubrids
Colubrids, **7:465–482,** 7:*471*, 7:*472*
  behavior, 7:468–469
  conservation status, 7:470
  distribution, 7:*465*, 7:467
  evolution, 7:465–467
  feeding ecology, 7:469
  habitats, 7:468
  humans and, 7:470
  physical characteristics, 7:467
  reproduction, 7:469–470
  species of, 7:*473*–481
  taxonomy, 7:198, 7:465–467
Colubrinae, 7:466, 7:468
Colubroidea. *See* Colubrids
Colugos, 12:14, 12:43, 12:136, **13:299–305,** 13:*300*, 13:*303*
Columba spp., 9:243
Columba calcaria, 9:241
Columba capensis. *See* Namaqua doves
Columba cristata. *See* Western crowned pigeons
Columba leuconota. *See* Snow pigeons
Columba livia. *See* Rock pigeons
Columba luctuosa. *See* White imperial pigeons
Columba luzonica. *See* Luzon bleeding hearts
Columba macroura. *See* American mourning doves
Columba magnfica. *See* Wompoo fruit doves
Columba palumbus. *See* Wood pigeons
Columba turtur. *See* European turtledoves
Columba unchall. *See* Barred cuckoo-doves
Columba waalia. *See* Bruce's green pigeons
Columbellidae, 2:446
Columbia sturgeons. *See* White sturgeons
Columbian ground squirrels, 12:*71*, 16:*150*, 16:*154*, 16:157
Columbian mammoths, 12:139
Columbicola columbae. *See* Slender pigeon lice
Columbidae. *See* Pigeons
Columbiformes, **9:241–246,** 9:247
Columbina passerina. *See* Common ground doves
Columbinae, 9:247
Columbus, Christopher, 15:242
Colymbus fulica. *See* Sungrebes
Colymbus immer. *See* Common loons
Colymbus stellatus. *See* Red-throated loons
Comactinia echinoptera, 1:*360*, 1:*362*, 1:363
Comasteridae, 1:355
Comatulid crinoids, 2:32, 2:*33*
Comatulida. *See* Feather stars
Comb bearers. *See* Comb jellies
Comb-crested jacanas, 9:108
Comb ducks, 8:364, 8:*374*, 8:*380*–381
Comb jellies, **1:169–177,** 1:*172*, 2:12–13
  behavior, 1:42, 1:170
  conservation status, 1:51, 1:171
  distribution, 1:169
  evolution, 1:169
  feeding ecology, 1:28–29, 1:170–171
  habitats, 1:24, 1:170
  humans and, 1:171
  physical characteristics, 1:169
  reproduction, 1:19, 1:21, 1:171
  species of, 1:*173*–177
  taxonomy, 1:9, 1:10–11, 1:169
Comb-toed jerboas, 16:212–215, 16:*218*, 16:220, 16:*221*
Combat behaviors. *See* Behavior
Combfishes, 5:179
Comephoridae. *See* Baikal oilfishes
Comicus spp., 3:203
Commensalism, 1:31, 1:32, 2:32, 2:*33*, **12:171–181**
  See also Behavior
Commerson's dolphins, 15:49, 15:56*t*
Commerson's leaf-nosed bats, 13:401, 13:*402*, 13:*404*, 13:410*t*
Commission for the Conservation of Antarctic Marine Resources
  mackerel icefishes, 5:327
  toothfishes, 5:323
Committee on the Status of Endangered Wildlife in Canada, 4:217
  black-tailed prairie dogs, 16:155
  harbor porpoises, 15:35, 15:38
  Vancouver Island marmots, 16:156
Common adders, 7:447, 7:448, 7:*449*, 7:*459*–460
Common American walkingsticks, 3:*227*, 3:*228*, 3:230
Common anchovies. *See* Bay anchovies
Common angelsharks. *See* Angelsharks
Common Asian toads. *See* Common Sunda toads
Common barn owls, 9:*332*, 9:335–*336*, 9:*337*, 9:*338*, 9:*339*, 9:*341*–342, 9:351
Common barnacles. *See* Rock barnacles
Common basilisks, 7:25
Common bentwing bats, 13:310, 13:315, 13:*520*, 13:*521*, 13:*523*, 13:524–*525*
Common big-eared bats, 13:433*t*
Common bitterns. *See* Eurasian bitterns
Common black-headed gulls, 9:173, 9:*210*, 9:*212*
Common blackbirds. *See* Blackbirds
Common blossom bats. *See* Southern blossom bats
Common blue mussels, 2:*455*, 2:*458*–459
Common bottlenosed dolphins, 15:1, 15:*47*, 15:*52*
  behavior, 15:7, 15:46, 15:54
  conservation status, 15:54
  distribution, 15:*54*
  feeding ecology, 15:54
  habitats, 15:*5*, 15:54
  humans and, 15:11, 15:54
  neonatal requirements, 12:126
  physical characteristics, 15:54
  reproduction, 15:54
  taxonomy, 15:54
Common brittle stars, 1:*392*, 1:*397*
Common bronzewings, 9:248
Common brushtail possums, 12:215, 13:*58*, 13:*62*, 13:*65*
  behavior, 13:37
  conservation status, 13:61
  distribution, 13:34, 13:59
  feeding ecology, 13:38, 13:60
  habitats, 13:34, 13:59
  humans and, 13:61
  translocation of, 12:224

Common buffalofishes. *See* Bigmouth buffaloes
Common bulbuls, 10:396, 10:*401*, 10:*402*
Common bush vipers. *See* Green bush vipers
Common bushtits. *See* Bushtits
Common buttonquails. *See* Barred buttonquails
Common caimans, 7:161, 7:163, 7:164, 7:171, 7:173, 7:*174*, 7:*176*–177
Common canaries. *See* Island canaries
Common caracaras. *See* Crested caracaras
Common carps, 4:69, 4:*299*, 4:303, 4:*306*, 4:*310*–311
Common chachalacas. *See* Plain chachalacas
Common chaffinches. *See* Chaffinches
Common chameleons, 7:228, 7:*229*, 7:*233*, 7:*237*
Common chiffchaffs. *See* Chiffchaffs
Common chimpanzees. *See* Chimpanzees
Common chuckwallas, 7:245, 7:*248*, 7:*250*, 7:252–253
Common clothes moths. *See* Webbing clothes moths
Common clownfishes. *See* Clown anemonefishes
Common collared lizards, 7:*249*, 7:*250*
Common coots, 9:*47*
Common coucals. *See* Greater coucals
Common cranes. *See* Eurasian cranes
Common crossbills. *See* Red crossbills
Common crow-pheasants. *See* Greater coucals
Common crows. *See* American crows
Common cuckoos, 9:312–315, 9:*316*, 9:319–*320*
Common cuscuses, 13:60, 13:*66t*
Common cuttlefishes, 2:*481*, 2:486–*487*
Common daces, 4:*306*, 4:*307*, 4:314
Common diving-petrels, 8:143, 8:*144*, 8:*145*
Common dolphinfishes, 4:339, 5:*211*, 5:*213*, 5:*215*–216
Common dolphins. *See* Short-beaked saddleback dolphins
Common dormice. *See* Hazel dormice
Common eagle-owls. *See* Eurasian eagle-owls
Common earwigs. *See* European earwigs
Common eels. *See* American eels; European eels
Common egg-eaters, 7:*469*, 7:*472*, 7:*474*
Common eiders, 8:363, 8:372
Common Eurasian moles. *See* European moles
Common European octopods. *See* Common octopods
Common European snakeflies, 3:*299*, 3:*301*, 3:302
Common European toads, 6:*37*
Common European white-toothed shrews, 13:*266*, 13:267, 13:*268*, 13:*270*, 13:*271*
Common fairy bluebirds. *See* Fairy bluebirds
Common fangtooths, 5:*117*, 5:*118*, 5:119
Common field worms, 2:*69*, 2:70–71
Common fiscals, 10:427
Common flickers. *See* Northern flickers
Common forest treefrogs. *See* Luzon bubble-nest frogs
Common foxes. *See* Crab-eating foxes
Common gar pikes. *See* Longnose gars
Common garter snakes, 7:469, 7:*471*, 7:478–*479*
Common genets, 14:*335*, 14:*337*, 14:*339*, 14:*340*–341
Common gibbons. *See* Lar gibbons
Common gobies, 5:375, 5:376–377
Common goose barnacles, 2:32, 2:*274*, 2:*277*, 2:280
Common grackles, 11:302–303
See also Hill mynas
Common grass lizards, 7:*320*
Common grebes. *See* Little grebes
Common ground doves, 9:*254*, 9:*261*
Common guillemots. *See* Common murres
Common hamsters. *See* Black-bellied hamsters
Common harvestmen, 2:*338*, 2:*345*–346
See also Long-bodied cellar spiders
Common hatchetfishes. *See* River hatchetfishes
Common hawk-cuckoos, 9:*317*, 9:*319*
Common hermit crabs, 2:*203*, 2:*207*, 2:210–211
Common hippopotamuses, 15:*303*, 15:*308*, 15:*309*
behavior, 15:*301*, 15:*304*, 15:*305*, 15:*307*, 15:310
conservation status, 15:311
distribution, 15:*310*
feeding ecology, 15:310
habitats, 15:269, 15:306, 15:310
humans and, 15:311
physical characteristics, 15:267, 15:310
reproduction, 15:*306*, 15:310–311
taxonomy, 15:310
Common honeybees. *See* Honeybees
Common house-martins. *See* House martins
Common house rats. *See* Black rats
Common iguanas. *See* Green iguanas
Common Indian sandgrouse. *See* Chestnut-bellied sandgrouse
Common Indian toads. *See* Common Sunda toads
Common ioras, 10:416, 10:417, 10:*418*, 10:*419*
Common jackals. *See* Golden jackals
Common jacks. *See* Crevalle jacks
Common jollytails, 4:*54*, 4:58–59, 4:391, 4:392, 4:393
Common kingfishers, 10:6, 10:*8*, 10:*11*, 10:*21*
Common kingsnakes, 7:470
Common kiwis. *See* Brown kiwis
Common koels, 9:312, 9:313, 9:*316*, 9:*322*–323, 11:508
Common langurs. *See* Northern plains gray langurs
Common lesser earless lizards, 7:*248*, 7:*252*, 7:253–254
Common limpets, 2:*426*–*427*
Common long-nosed armadillos. *See* Nine-banded armadillos
Common loons, 8:159, 8:*161*, 8:*162*, 8:*165*–166
Common magpies. *See* Eurasian magpies
Common marmosets, 14:*116*, 14:117, 14:118, 14:121–124, 14:*125*, 14:129–*130*
See also Marmosets
Common melipotes. *See* Common smoky honeyeaters
Common minnows. *See* Eurasian minnows
Common mole-rats, 16:124, 16:*341*, 16:*346*
behavior, 12:74, 16:348
conservation status, 16:348
distribution, 16:*347*, 16:348
feeding ecology, 16:348
habitats, 16:348
humans and, 16:348
physical characteristics, 12:*70*, 16:*342*, 16:348
reproduction, 16:344
taxonomy, 16:348
See also African mole-rats
Common moles. *See* European moles
Common moorhens, 9:*3*, 9:47, 9:*49*, 9:50
Common mountain viscachas. *See* Mountain viscachas
Common mudpuppies. *See* Mudpuppies
Common munias. *See* Spotted munias
Common murres, 9:219, 9:*223*, 9:*224*
Common musk turtles. *See* Stinkpots
Common mussels. *See* Common blue mussels
Common mynahs. *See* Common mynas
Common mynas, 9:163, 11:*412*, 11:*419*
Common nautilus. *See* Pearly nautilus
Common nectar-feeding fruit bats. *See* Dawn fruit bats
Common needleflies, 3:*144*, 3:*145*
Common newts. *See* Smooth newts
Common night adders. *See* Rhombic night adders
Common nighthawks, 9:369–370, 9:402–404, 9:*405*, 9:*407*, 9:*408*
Common nightingales. *See* Nightingales
Common octopods, 1:42, 2:37, 2:475, 2:477, 2:478, 2:479, 2:*481*, 2:484–*485*
Common opossums, 12:249–254
Common paradise kingfishers, 10:*12*, 10:14–*15*
Common parsley frogs. *See* Parsley frogs
Common partridges. *See* Gray partridges
Common pheasants. *See* Ring-necked pheasants
Common pill woodlice, 2:*249*, 2:250, 2:*253*, 2:*254*–255
Common plantanna, 6:41, 6:47, 6:53, 6:65, 6:100–*101*, 6:*103*–104
Common plate-tailed geckos, 7:*264*, 7:*267*, 7:268
Common poorwills, 9:370, 9:402–404, 9:*407*, 9:*409*
Common porcupines, 16:353, 16:*356*, 16:*358*
Common porpoises. *See* Harbor porpoises
Common potoos. *See* Gray potoos
Common pratincoles. *See* Collared pratincoles
Common prions. *See* Broad-billed prions
Common pygmy woodlice, 2:*253*, 2:*257*, 2:259–260
Common rat fleas. *See* Oriental rat fleas
Common redpolls, 11:*326*, 11:330–*331*
Common redshanks, 9:177, 9:178
Common reedbucks. *See* Southern reedbucks
Common remoras, 5:*213*, 5:*214*, 5:216
Common rheas. *See* Greater rheas
Common ringed plovers. *See* Ringed plovers
Common ringtails, 13:113, 13:*114*–115, 13:116, 13:117, 13:*118*, 13:120–*121*
See also Ringtail possums
Common rosellas. *See* Eastern rosellas
Common rough-scaled lizards, 7:298
Common rough woodlice, 2:*253*, 2:*258*–259
Common sabertooths. *See* Common fangtooths

Common sagebrush lizards, 7:*249*, 7:*255–256*
Common sandgrouse. *See* Chestnut-bellied sandgrouse
Common sandpipers, 9:178
Common sapsuckers. *See* Yellow-bellied sapsuckers
Common sawsharks, 4:*167*, 4:*169*, 4:*170*
Common scimitarbills, 10:66, 10:*67*, 10:68–*69*
Common scops-owls. *See* Eurasian scops-owls
Common screech owls. *See* Eastern screech owls
Common seadragons. *See* Weedy seadragons
Common seals. *See* Harbor seals
Common shads. *See* American shads
Common shiny woodlice, 2:*253*, 2:*254*, 2:257–258
Common shipworms, 2:*49*, 2:*53*, 2:54, 2:*456*, 2:*457*
Common shore-eels, 5:356, 5:*358*, 5:*359*
Common short-tailed porcupines. *See* Common porcupines
Common shrews, 13:*248*, 13:*254*, 13:*261*
See also Shrews
Common shrimpfishes, 5:*138*, 5:139–*140*
Common sicklebills. *See* White-tipped sicklebills
Common side-blotched lizards, 7:246, 7:*249*, 7:*255*, 7:256
Common silverfish. *See* Silverfish
Common slug-eaters, 7:*471*, 7:*477*
Common smoky honeyeaters, 11:*239*, 11:*252*
Common snakeneck turtles, 7:*78*, 7:*80*, 7:*82*–83
Common snapping turtles. *See* Snapping turtles
Common snipes, 9:176, 9:180
Common snooks, 5:*219*, 5:*224*, 5:*225*
Common soles, 5:*454*, 5:*460*, 5:462–463
Common spadefoot, 6:122, 6:*123*, 6:*124*
Common spoonbills. *See* Spoonbills
Common spotted cuscuses, 13:*62*, 13:*63*, 13:64
Common squeakers, 6:*267*, 6:*268*
Common squirrel monkeys, 14:101, 14:*102*, 14:103, 14:*105*, 14:*107*, 14:*109*
Common squirrelfishes. *See* Squirrelfishes
Common starlings. *See* European starlings
Common stilts. *See* Black-winged stilts
Common stonechats. *See* Stonechats
Common striped scorpions. *See* Striped scorpions
Common sturgeons. *See* Atlantic sturgeons
Common sunbeam snakes, 7:*401*, 7:*402*
Common sunbird-asities, 10:188, 10:*189*, 10:*190*–191
Common Sunda toads, 6:*187*, 6:191
Common swifts, 9:417, 9:419, 9:422, 9:424, 9:*425*, 9:429–*430*
Common tailorbirds, 11:*2*, 11:5, 11:*9*, 11:*18*
Common tenrecs, 13:*231*
behavior, 13:228, 13:232
conservation status, 13:232
distribution, 13:*232*
feeding ecology, 13:232
habitats, 13:*229*, 13:232
humans and, 13:200, 13:232
physical characteristics, 12:*84*, 13:226, 13:*227*, 13:232
reproduction, 13:232
taxonomy, 13:232
See also Tenrecs
Common terns, 9:*210*, 9:214–*215*
Common tree shrews, 13:*289*, 13:*290*, 13:*291*, 13:*293*, 13:*294*, 13:295
Common tree toads. *See* Brown tree toads
Common treecreepers. *See* Eurasian treecreepers
Common trumpeters, 9:78, 9:79, 9:*80*, 9:*81*
Common turtledoves. *See* European turtledoves
Common vampires. *See* Vampire bats
Common wallaroos, 13:101*t*
Common wambengers. *See* Brush-tailed phascogales
Common warthogs, 15:264, 15:275, 15:280, 15:*283*, 15:*285*–286
See also Warthogs
Common water fleas, 2:*156*, 2:*158*
Common waxbills, 11:*358*, 11:*364*–365, 11:393
Common whitefishes. *See* Lake whitefishes
Common whitethroats. *See* Whitethroats
Common wombats, 13:*35*, 13:*54*
behavior, 13:37, 13:*52*, 13:*53*, 13:55
conservation status, 13:40, 13:55
distribution, 13:*55*
feeding ecology, 13:55
habitats, 13:*52*, 13:55
humans and, 13:53, 13:55
physical characteristics, 13:51, 13:55
reproduction, 13:38, 13:55
taxonomy, 13:*55*
See also Wombats
Common yellowthroats, 11:288
Common zebras. *See* Plains zebras
Commons, tragedy of the, 1:47–48
Commonwealth Biological Control Act (Australia), 12:188
Communication, 1:37, 1:41–43, 4:19–23, 4:60, 4:61–62
amphibians, 6:44–48
avian, 8:18
caddisflies, 3:376
chemosensory cues for, 6:22–23, 6:24, 6:44–46
echolocation, 12:53–55, 12:*59*, 12:*85*–87
electroreceptors for, 6:15
lacewings, 3:307
in mating, 3:59–60
mechanoreceptors for, 6:15
Orthoptera, 3:206, 3:207
seismic, 12:76
senses and, 12:84
social insect, 3:72
stoneflies, 3:142
subterranean mammals, 12:114
symbolic, 12:161–162
vibrations in, 12:83
See also Behavior; Songs, bird; Vocalizations
Communities, 4:42
See also Behavior; specific types of communities
Comobatrachus spp., 6:12
Comoro black bulbuls. *See* Black bulbuls
Comoro black flying foxes. *See* Livingstone's fruit bats
Comoro bulbuls. *See* Black bulbuls
Comoro drongos, 11:439
Comoro thrushes, 10:489
Comparative psychology, 1:37–38
Compasses, avian, 8:33–34
Competition, 1:26–27, 2:28–29, 3:45–46, 4:43
deer, 12:118
inter- and intra-specific, 12:114–115
scramble, 6:48–49
sperm, 12:98–99, 12:105–106
See also Behavior; Feeding ecology
Competitive exclusion, 2:28
Complex-toothed flying squirrels, 16:136
Compound eyes, 3:27–*28*
Computers, bird songs and, 8:38
Conant, Roger, 7:455
Concave-casqued hornbills. *See* Great hornbills
Concave-crowned horned toads. *See* Burmese spadefoot toads
Concentricycloidea. *See* Sea daisies
Concertina locomotion, 7:26, 7:*29*
See also Behavior
Conchostraca. *See* Clam shrimps
Conchs, 2:43
Concolor gibbons. *See* Black crested gibbons
Conditioned stimuli, 1:41, 2:40
See also Behavior
Conditioning, 1:41
See also Behavior
Condonts, 4:9
The Condor (journal), 8:27
Condoro Island flying foxes. *See* Island flying foxes
Condors, **8:275–285**
behavior, 8:234, 8:277–278
conservation status, 8:278
distribution, 8:*275*, 8:276–277
evolution, 8:275
feeding ecology, 8:277, 8:278
habitats, 8:277
humans and, 8:278
physical characteristics, 8:276
reproduction, 8:278
species of, 8:*280*–284
taxonomy, 8:275
Condrostoma spp., 4:299–300
Condylactis gigantea, 1:28
Condylarthra, 15:131–132, 15:133, 15:135, 15:136, 15:263, 15:266, 15:343
Condylura spp., 13:281
Condylura cristata. *See* Star-nosed moles
Condylurinae, 13:279
Cone-headed chameleons. *See* Veiled chameleons
Cone shells, 2:29, 2:41, 2:446, 2:447, 2:*448*, 2:449
Cone snails, 1:45
Conebills, 11:285, 11:291
Conehead katydids, 3:203, 3:204
Conenoses, 3:259
Conepatus spp., 14:319, 14:321
Conepatus mesoleucus. *See* Western hog-nosed skunks
Confused flour beetles, 3:76
Conger eels. *See* American congers
Conger oceanicus. *See* American congers
Congiopodidae. *See* Pigfishes
Congiopodus peruvianus. *See* South American pigfishes
Congo eels. *See* Wrymouths
Congo golden moles, 13:222*t*
Congo peafowls, 8:433
Congo swallows. *See* African river-martins

Congo water civets. *See* Aquatic genets
Congosorex polli. *See* Poll's shrews
Congridae, 4:48, 4:67, 4:70, 4:255
Congrogadinae, 5:256
Conicera spp., 3:358
Conidae, 2:446
Conidens spp., 5:355
Conies. *See* American pikas
Coniopterygidae, 3:305, 3:306, 3:307, 3:309
Conirostrum spp. *See* Conebills
Connecticut river shads. *See* American shads
Connecticut warblers, 11:*288*
Connochaetes spp. *See* Wildebeests
Connochaetes gnou. *See* Black wildebeests
Connochaetes taurinus. *See* Blue wildebeests
Conocephalinae, 3:206
Conocephalus spp., 3:203
Conocephalus discolor. *See* Long-winged coneheads
Conocyemidae, 1:93
Conolophus pallidus. *See* Barrington iguanas
Conophoralia, 1:331, 1:332
Conopophila albogularis. *See* Rufous-banded honeyeaters
Conostoma spp. *See* Parrotbills
Conotoxins, 1:45, 2:29
Conozoa hyalina. *See* Central valley grasshoppers
Conraua spp., 6:246, 6:248, 6:249, 6:256
Conraua crassipes, 6:246
Conraua goliath. *See* Goliath frogs
Conrauini, 6:246
Conservation Action Plan for Eurasian Insectivores
  alpine shrews, 13:260
  Chinese short-tailed shrews, 13:256
  common shrews, 13:261
  elegant water shrews, 13:259
  Eurasian pygmy shrews, 13:262
  Eurasian water shrews, 13:258
  giant shrews, 13:262
  Himalayan water shrews, 13:259
  Hodgson's brown-toothed shrews, 13:259
  mole-shrews, 13:255
Conservation biology, 1:48, 12:216–225
Conservation Endowment Fund, 12:204
Conservation International, 12:218
Conservation medicine, 12:222–223
Conservation status
  amphibians, 6:6, **6:56–60**
    anurans
      African treefrogs, 6:283, 6:285–289
      Amero-Australian treefrogs, 6:230, 6:233–242
      Arthroleptidae, 6:266, 6:268–271
      Asian toadfrogs, 6:111, 6:113–117
      Asian treefrogs, 6:293, 6:295–299
      Australian ground frogs, 6:141, 6:143–145
      Bombinatoridae, 6:84, 6:86–88
      brown frogs, 6:57, 6:263
      Bufonidae, 6:185, 6:188–194
      Discoglossidae, 6:90, 6:92–94
      ghost frogs, 6:132, 6:133–134
      glass frogs, 6:57, 6:217, 6:219–223
      leptodactylid frogs, 6:159, 6:162–171
      Madagascan toadlets, 6:318, 6:319–320
      Mesoamerican burrowing toads, 6:97
      Myobatrachidae, 6:149, 6:151–153
      narrow-mouthed frogs, 6:305, 6:308–316
      New Zealand frogs, 6:70, 6:72–74
      parsley frogs, 6:128, 6:129
      Pipidae, 6:102, 6:104–106
      poison frogs, 6:200, 6:203–209
      Ruthven's frogs, 6:212
      Salamandridae, 6:367, 6:370–375
      Seychelles frogs, 6:136, 6:137–138
      shovel-nosed frogs, 6:275, 6:277
      sirens, 6:329, 6:331–332
      spadefoot toads, 6:122, 6:124–125
      tailed frogs, 6:78, 6:80–81
      tailless caecilians, 6:437, 6:439–441
      three-toed toadlets, 6:180, 6:181–182
      torrent salamanders, 6:386, 6:387
      true frogs, 6:251, 6:255–263
      vocal sac-brooding frogs, 6:174, 6:176
      Xenopus spp., 6:102, 6:251
    caecilians
      American tailed, 6:416, 6:417–418
      Asian tailed, 6:420, 6:422–423
      buried-eye, 6:432, 6:433–434
      Kerala, 6:426, 6:428
    salamanders
      amphiumas, 6:407, 6:409–410
      Asiatic, 6:337, 6:339–342
      Cryptobranchidae, 6:346–347
      lungless, 6:392, 6:395–403
      mole, 6:356, 6:358–360
      Pacific giant, 6:350, 6:352
      Proteidae, 6:379, 6:381–383
  birds
    albatrosses, 8:110, 8:116, 8:118–122
    Alcidae, 9:219, 9:224–229
    anhingas, 8:202–203, 8:207–209
    Anseriformes, 8:367–368
    ant thrushes, 10:241–242, 10:245–256
    Apodiformes, 9:418
    asities, 10:188, 10:190
    Australian chats, 11:66, 11:67–68
    Australian creepers, 11:135, 11:137–139
    Australian fairy-wrens, 11:46, 11:48–53
    Australian honeyeaters, 11:237–238, 11:241–253
    Australian robins, 11:106, 11:108–112
    Australian warblers, 11:57, 11:59–63
    babblers, 10:507, 10:511–523
    barbets, 10:117, 10:119–123
    barn owls, 9:338, 9:340–343
    bee-eaters, 10:42
    birds of paradise, 11:492, 11:495–502
    Bombycillidae, 10:448, 10:451–454
    boobies, 8:214–215
    bowerbirds, 11:481, 11:483–488
    broadbills, 10:179, 10:181–186
    bulbuls, 10:398–399, 10:402–413
    bustards, 9:91, 9:93–94, 9:96–99
    buttonquails, 9:14, 9:16–21
    Caprimulgiformes, 9:370
    caracaras, 8:351
    cassowaries, 8:77
    Charadriidae, 9:163–164, 9:167–173
    Charadriiformes, 9:103–104
    chats, 10:488–489
    chickadees, 11:158, 11:160–166
    chowchillas, 11:71, 11:72–74
    Ciconiiformes, 8:236
    Columbidae, 9:251–252, 9:255–266
    condors, 8:278
    cormorants, 8:186, 8:202–203, 8:205–206, 8:207–209
    Corvidae, 11:507–508, 11:512–522
    cotingas, 10:308, 10:312–322
    crab plovers, 9:123
    Cracidae, 8:416
    cranes, 9:27–29, 9:31–36
    cuckoo-shrikes, 10:387
    cuckoos, 9:315, 9:318–329
    dippers, 10:477, 10:479–482
    diving-petrels, 8:144, 8:145–146
    doves, 9:251–252, 9:255–266
    drongos, 11:439, 11:441–445
    ducks, 8:372–373
    eagles, 8:323
    elephant birds, 8:104
    emus, 8:85–86
    Eupetidae, 11:76, 11:78–81
    fairy bluebirds, 10:417, 10:423–424
    Falconiformes, 8:315–316
    falcons, 8:351
    false sunbirds, 10:188, 10:191
    fantails, 11:86–87, 11:89–95
    finches, 11:324–325, 11:328–339
    flamingos, 8:306
    flowerpeckers, 11:191, 11:193–198
    fowls, 8:436–437
    frigatebirds, 8:195, 8:197–198
    frogmouths, 9:379, 9:381–385
    Galliformes, 8:401–402
    gannets, 8:214–215
    geese, 8:372–373
    Glareolidae, 9:153, 9:155–160
    grebes, 8:171, 8:174–181
    Gruiformes, 9:3
    guineafowl, 8:429
    gulls, 9:211–217
    hammerheads, 8:263
    Hawaiian honeycreepers, 11:343–344, 11:346–351
    hawks, 8:323
    hedge sparrows, 10:460, 10:462–464
    herons, 8:236, 8:244–245
    hoatzins, 8:467
    honeyguides, 10:139, 10:141–144
    hoopoes, 10:63
    hornbills, 10:75, 10:78–84
    hummingbirds, 9:443, 9:449–451, 9:455–467
    ibises, 8:236, 8:293
    Icteridae, 11:305–306, 11:309–312
    ioras, 10:417, 10:419–420
    jacamars, 10:93
    jacanas, 9:110, 9:112–114
    kagus, 9:43
    kingfishers, 10:10, 10:13–23
    kiwis, 8:90
    lapwings, 9:167–173
    Laridae, 9:208–209, 9:211–217
    larks, 10:346, 10:348–354
    leafbirds, 10:417, 10:420–423
    limpkins, 9:37, 9:38–39
    logrunners, 11:71, 11:72–74
    long-tailed titmice, 11:143, 11:145–146
    lyrebirds, 10:332, 10:334–335
    magpie-shrikes, 11:469, 11:471–475
    manakins, 10:297, 10:299–303
    mesites, 9:6, 9:8–9
    moas, 8:98
    monarch flycatchers, 11:97, 11:100–103
    motmots, 10:33, 10:35–37
    moundbuilders, 8:407

mousebirds, 9:469, 9:470
mudnest builders, 11:449–452, 11:456–458
New World blackbirds, 11:309–322
New World finches, 11:266, 11:269–283
New World quails, 8:458
New World vultures, 8:278
New World warblers, 11:290–291, 11:293–299
New Zealand wattle birds, 11:448, 11:449–451
New Zealand wrens, 10:204, 10:206–207
nightjars, 9:405, 9:408–414
nuthatches, 11:169, 11:171–175
oilbirds, 9:375
Old World flycatchers, 11:27, 11:30–43
Old World warblers, 11:6–7, 11:10–23
Oriolidae, 11:428, 11:431–434
ostriches, 8:101
ovenbirds, 10:210–211, 10:214–228
owlet-nightjars, 9:389, 9:391–393
owls, 9:333–334, 9:338, 9:340–343, 9:351, 9:354–365
oystercatchers, 9:127–128, 9:130–131
painted-snipes, 9:117, 9:118–119
palmchats, 10:456
pardalotes, 11:202, 11:204–206
parrots, 9:279–280, 9:283–298
Pelecaniformes, 8:186
pelicans, 8:186, 8:227
penduline titmice, 11:149, 11:151–153
penguins, 8:151, 8:153–157
Phasianidae, 8:436–437
pheasants, 8:436–437
Philippine creepers, 11:184, 11:186–187
Picidae, 10:150, 10:154–167
Piciformes, 10:88
pigeons, 9:251–252, 9:255–266
pipits, 10:375–376, 10:378–383
pittas, 10:195, 10:197–201
plantcutters, 10:326, 10:327–328
plovers, 9:167–173
potoos, 9:397, 9:399–400
pratincoles, 9:155–160
Procellariidae, 8:126
Procellariiformes, 8:110
pseudo babblers, 11:128, 11:130–131
puffbirds, 10:103–104, 10:106–111
rails, 9:51–52, 9:57–67
Raphidae, 9:270–271, 9:272–273
Recurvirostridae, 9:135, 9:137–141
rheas, 8:71
rollers, 10:53, 10:55–58
sandgrouse, 9:233, 9:235–238
sandpipers, 9:103, 9:179–180, 9:182–188
screamers, 8:394
scrub-birds, 10:338, 10:339–340
secretary birds, 8:345
seedsnipes, 9:191, 9:193–195
seriemas, 9:86, 9:88–89
sharpbills, 10:293
sheathbills, 9:199, 9:200–201
shoebills, 8:288
shrikes, 10:429, 10:432–438
sparrows, 11:398, 11:401–406
spoonbills, 8:293
storks, 8:236, 8:267
storm-petrels, 8:138, 8:140–141
Sturnidae, 11:410, 11:414–425
sunbirds, 11:210, 11:213–225
sunbitterns, 9:73, 9:75
sungrebes, 9:70, 9:71–72
swallows, 10:360–361, 10:363–369
swans, 8:372–373
swifts, 9:424, 9:426–431
tapaculos, 10:259, 10:262–267
terns, 9:211–217
thick-knees, 9:145–146, 9:147–148
thrushes, 10:488–489
titmice, 11:158, 11:160–166
todies, 10:27–28, 10:29–30
toucans, 10:127–128, 10:130–135
tree swifts, 9:434, 9:435–436
treecreepers, 11:178, 11:180–181
trogons, 9:479, 9:481–485
tropicbirds, 8:189, 8:190–191
trumpeters, 9:79, 9:81–82
turacos, 9:301–302, 9:304–310
typical owls, 9:351, 9:354–365
tyrant flycatchers, 10:274–275, 10:278–288
vanga shrikes, 10:441–442, 10:443–444
Vireonidae, 11:256, 11:258–262
wagtails, 10:375–376, 10:378–383
weaverfinches, 11:356–357, 11:360–372
weavers, 11:378–379, 11:382–394
whistlers, 11:118, 11:120–126
white-eyes, 11:229–230, 11:232–233
woodcreepers, 10:230, 10:232–236
woodhoopoes, 10:66, 10:68–69
woodswallows, 11:460, 11:462–465
wrens, 10:529, 10:531–538
fishes, 4:**74–76**
Acanthuroidei, 5:393, 5:397–403
Acipenseriformes, 4:214–215, 4:217–220
Albuliformes, 4:250, 4:252–253
angelsharks, 4:162, 4:164–165
anglerfishes, 5:51, 5:53–56
Anguilliformes, 4:258, 4:262–269
Atheriniformes, 5:70–71, 5:73–76
Aulopiformes, 4:434, 4:436–439
Australian lungfishes, 4:199
beardfishes, 5:2, 5:3
Beloniformes, 5:81, 5:83–86
Beryciformes, 5:116, 5:118–121
bichirs, 4:210, 4:211–212
blennies, 5:343, 5:345–348
bowfins, 4:230
bullhead sharks, 4:98, 4:101–102
Callionymoidei, 5:366–367, 5:370–371
carps, 4:303, 4:308–313
catfishes, 4:354, 4:357–367
characins, 4:338–339, 4:342–350
chimaeras, 4:92, 4:94–95
coelacanths, 4:192–193
Cypriniformes, 4:303, 4:307–319, 4:322, 4:325–333
Cyprinodontiformes, 5:94, 5:96–102
dogfish sharks, 4:152–153, 4:156–158
dories, 5:124, 5:126–127, 5:129–130
eels, 4:258, 4:262–269
electric eels, 4:371, 4:373–377
Elopiformes, 4:244, 4:247
Esociformes, 4:380, 4:383–386
flatfishes, 5:453, 5:455–463
freshwater fishes, 4:76
Gadiformes, 5:29–30, 5:33–40
gars, 4:222, 4:224–227
Gasterosteiformes, 5:136, 5:139–141, 5:143–147
gobies, 5:377, 5:380–388
Gobiesocoidei, 5:357, 5:359–363
Gonorynchiformes, 4:291, 4:293–295
ground sharks, 4:116, 4:119–128
Gymnotiformes, 4:371, 4:373–377
hagfishes, 4:79, 4:80–81
herrings, 4:278, 4:281–288
Hexanchiformes, 4:144, 4:146–148
Labroidei, 5:280–281, 5:284–290, 5:298, 5:301–307
labyrinth fishes, 5:429, 5:431–434
lampreys, 4:85, 4:88–90
Lampridiformes, 4:450, 4:453–454
lanternfishes, 4:443, 4:445–446
lungfishes, 4:203, 4:205–206
mackerel sharks, 4:133, 4:135–141
minnows, 4:303, 4:314–315
morays, 4:258, 4:265–267
mudminnows, 4:380, 4:383–386
mullets, 5:60, 5:62–65
Ophidiiformes, 5:18, 5:20–22
Orectolobiformes, 4:105, 4:109–111
Osmeriformes, 4:393, 4:395–401
Osteoglossiformes, 4:234, 4:237–241
Percoidei, 5:199, 5:202–208, 5:212, 5:214–217, 5:222, 5:225–226, 5:228–233, 5:242–243, 5:247–253, 5:262, 5:267–268, 5:270–273
Percopsiformes, 5:6, 5:8–12
pikes, 4:380, 4:383–386
ragfishes, 5:353
Rajiformes, 4:178, 4:182–188
rays, 4:178, 4:182–188
Saccopharyngiformes, 4:272, 4:274–276
salmons, 4:407, 4:410–419
sawsharks, 4:168, 4:170–171
Scombroidei, 5:407, 5:409–418
Scorpaeniformes, 5:159, 5:161–162, 5:166, 5:169–178, 5:183, 5:186–193
Scorpaenoidei, 5:166
skates, 4:178, 4:187
snakeheads, 5:439, 5:442–446
South American knifefishes, 4:371, 4:373–377
southern cod-icefishes, 5:322–323, 5:325–328
Stephanoberyciformes, 5:106, 5:108–110
Stomiiformes, 4:424, 4:426–429
Stromateoidei, 5:422
Synbranchiformes, 5:152, 5:154–156
Tetraodontiformes, 5:471, 5:475, 5:477–484
toadfishes, 5:42, 5:44–46
Trachinoidei, 5:335, 5:337–339
whale sharks, 4:105, 4:110
Zoarcoidei, 5:313, 5:315–318
insects, 3:82, **3:84–91**
book lice, 3:244, 3:246–247
bristletails, 3:114, 3:116–117
caddisflies, 3:377, 3:379–381
cockroaches, 3:150, 3:153–158
Coleoptera, 3:322, 3:327–333
diplurans, 3:108, 3:110–111
Diptera, 3:360–361, 3:364–374
earwigs, 3:197, 3:199–200
fleas, 3:350, 3:352–355
Hemiptera, 3:263, 3:267–280
Hymenoptera, 3:407, 3:412–424
lacewings, 3:308–309, 3:311–314
Lepidoptera, 3:388–389, 3:393–404
mantids, 3:181, 3:183–187
Mantophasmatodea, 3:219

mayflies, 3:127, 3:129–130
Mecoptera, 3:343, 3:345–346
Megaloptera, 3:291, 3:293–294
Odonata, 3:134–135, 3:137–138
Orthoptera, 3:207, 3:211–216
Phasmida, 3:223–224, 3:228–232
Phthiraptera, 3:252, 3:254–256
proturans, 3:94, 3:96–97
rock-crawlers, 3:190, 3:192–193
snakeflies, 3:298, 3:300–302
springtails, 3:101, 3:103–104
stoneflies, 3:143, 3:145–146
strepsipterans, 3:336, 3:338–339
termites, 3:168, 3:171–175
thrips, 3:282–283, 3:285–287
Thysanura, 3:120, 3:122–123
webspinners, 3:234, 3:236–237
zorapterans, 3:240, 3:241
lower metazoans and lesser deuterostomes, **1:47–55**, 1:*52*
acoels, 1:180, 1:182
anoplans, 1:247, 1:249–251
arrow worms, 1:435, 1:437–442
box jellies, 1:149, 1:151–152
calcareous sponges, 1:59, 1:61–65
cnidarians
Anthozoa, 1:107, 1:111–121
Hydrozoa, 1:129, 1:134–145
comb jellies, 1:51, 1:171, 1:173–177
demosponges, 1:80, 1:82–86
echinoderms
Crinoidea, 1:359, 1:361–364
Echinoidea, 1:405, 1:409–415
Ophiuroidea, 1:48, 1:390, 1:394–398
sea cucumbers, 1:48, 1:422, 1:425–431
sea daisies, 1:383, 1:385–386
sea stars, 1:48, 1:370, 1:373–379
enoplans, 1:255, 1:256–257
entoprocts, 1:321, 1:323–325
flatworms
free-living flatworms, 1:188, 1:190–195
monogeneans, 1:215, 1:219–223
tapeworms, 1:231, 1:234–242
gastrotrichs, 1:270, 1:272–273
girdle wearers, 1:48, 1:345, 1:348–349
glass sponges, 1:69, 1:71–75
gnathostomulids, 1:332, 1:334–335
hair worms, 1:306, 1:308–309
hemichordates, 1:445, 1:447–450
jaw animals, 1:329
jellyfish, 1:51, 1:156, 1:160, 1:162–167
kinorhynchs, 1:277, 1:279–280
lancelets, 1:488, 1:490–497
larvaceans, 1:474, 1:476–478
nematodes
roundworms, 1:285, 1:287–289, 1:291–292
secernenteans, 1:295, 1:298–299, 1:301–304
Orthonectida, 1:99, 1:101–102
placozoans, 1:89
priapulans, 1:338, 1:340–341
rhombozoans, 1:94, 1:96–98
rotifers, 1:262, 1:264–267
Salinella salve, 1:92
salps, 1:468, 1:471–472
sea squirts, 1:455, 1:458–466
sorberaceans, 1:480, 1:482–483
thorny headed worms, 1:313, 1:315–316
Trematoda, 1:200, 1:203–211
wheel wearers, 1:354
mammals, 12:15, **12:213–225**
aardvarks, 15:158
agoutis, 16:409, 16:411–414, 16:415*t*
anteaters, 13:175, 13:177–179
armadillos, 13:185, 13:187–190, 13:190*t*–192*t*
Artiodactyla, 15:272–273
aye-ayes, 14:88
baijis, 15:21
bandicoots, 13:4–6
dry-country, 13:11–12, 13:14–16, 13:16*t*–17*t*
rainforest, 13:11–12, 13:14–17*t*
bats, 13:315–316
American leaf-nosed bats, 13:420, 13:423–432, 13:433*t*–434*t*
bulldog, 13:447, 13:449–450
disk-winged, 13:474, 13:476–477
false vampire, 13:382, 13:384–385
funnel-eared, 13:461, 13:463–465
horseshoe, 13:392–393, 13:396–400
Kitti's hog-nosed, 13:369
Molossidae, 13:487, 13:490–493, 13:493*t*–495*t*
mouse-tailed, 13:352, 13:353
moustached, 13:436, 13:440–441, 13:442*t*
New Zealand short-tailed, 13:455, 13:457–458
Old World fruit, 13:321–322, 13:325–330, 13:331*t*–332*t*, 13:337, 13:340–347, 13:348*t*–349*t*
Old World leaf-nosed, 13:405, 13:407–409, 13:409*t*–410*t*
Old World sucker-footed, 13:480
slit-faced, 13:373, 13:375–376, 13:376*t*–377*t*
smoky, 13:468, 13:470–471
Vespertilionidae, 13:501–502, 13:506–510, 13:512–514, 13:515*t*–516*t*, 13:522, 13:524–525, 13:526*t*
bears, 14:262, 14:300, 14:302–305, 14:306*t*
beavers, 16:180, 16:183–184
bilbies, 13:22
botos, 15:29–30
Bovidae, 16:8
Antilopinae, 16:48, 16:50–56, 16:56*t*–57*t*
Bovinae, 16:14–15, 16:18–23, 16:24*t*–25*t*
Caprinae, 16:94–95, 16:98–103, 16:103*t*–104*t*
duikers, 16:77, 16:80–84, 16:84*t*–85*t*
Hippotraginae, 16:33–34, 16:37–42, 16:42*t*–43*t*
Neotraginae, 16:63, 16:66–71, 16:71*t*
bushbabies, 14:26, 14:28–32, 14:32*t*–33*t*
Camelidae, 15:317–318, 15:320–323
Canidae, 14:272–273, 14:277–283, 14:283*t*–284*t*
capybaras, 16:405
Carnivora, 14:262
cats, 14:374–375, 14:380–389, 14:390*t*–391*t*
Caviidae, 16:393, 16:395–398, 16:399*t*
Cetacea, 15:9–10
Balaenidae, 15:9–10, 15:111–112, 15:115–118
beaked whales, 15:61–62, 15:69*t*–70*t*
dolphins, 15:48–50, 15:53–55, 15:56*t*–57*t*
Emballonuridae, 13:358, 13:360–363, 13:363*t*–364*t*
franciscana dolphins, 15:25
Ganges and Indus dolphins, 15:15–16
gray whales, 15:99–100
Monodontidae, 15:86–87, 15:90–91
porpoises, 15:35–36, 15:38–39, 15:39*t*
pygmy right whales, 15:105
rorquals, 15:9, 15:127–130, 15:130*t*
sperm whales, 15:77–78, 15:79–80
chevrotains, 15:329, 15:331–333
Chinchillidae, 16:380, 16:382–383, 16:383*t*
cognition and, 12:150–151
colugos, 13:302, 13:304
coypus, 16:477
dasyurids, 12:290–291, 12:294–298, 12:299*t*–301*t*
Dasyuromorphia, 12:283
deer
Chinese water, 15:376
muntjacs, 15:346, 15:349–354
musk, 15:338, 15:340–341
New World, 15:384, 15:388–396, 15:396*t*
Old World, 15:360–361, 15:364–371, 15:371*t*
Dipodidae, 16:216–217, 16:219–222, 16:223*t*–224*t*
Diprotodontia, 13:39–40
domestication and, 12:175
dormice, 16:321, 16:323–326, 16:327*t*–328*t*
duck-billed platypuses, 12:233–234, 12:246–247
Dugongidae, 15:201, 15:203–204
echidnas, 12:238, 12:240*t*
elephants, 15:171–172, 15:174–175
Equidae, 15:229–230, 15:232–235, 15:236*t*
Erinaceidae, 13:207, 13:209–213, 13:212*t*–213*t*
giant hutias, 16:470, 16:471*t*
gibbons, 14:214–215, 14:218–222
Giraffidae, 15:405, 15:408–409
great apes, 14:235, 14:237–240
gundis, 16:313, 16:314–315
Herpestidae, 14:352, 14:354–356, 14:357*t*
Heteromyidae, 16:203, 16:205–209, 16:208*t*–209*t*, 16:209*t*
hippopotamuses, 15:308, 15:311
hutias, 16:463, 16:467*t*
Hyaenidae, 14:362, 14:365–367
hyraxes, 15:184, 15:186–188, 15:189*t*
Indriidae, 14:67, 14:69–72
Insectivora, 13:199–200
koalas, 13:48–49
Lagomorpha, 16:487
lemurs, 14:52, 14:56–60
Cheirogaleidae, 14:39, 14:41–44
sportive, 14:76, 14:79–83
Leporidae, 16:509, 16:511–515, 16:515*t*–516*t*
Lorisidae, 14:16, 14:18–20, 14:21*t*
Macropodidae, 13:89, 13:93–100, 13:101*t*–102*t*
manatees, 15:208–209, 15:211–212
Megalonychidae, 13:158, 13:159
moles

golden, 13:217, 13:220–222, 13:222*t*
marsupial, 13:27, 13:28
monitos del monte, 12:274
monkeys
Atelidae, 14:159–160, 14:162–166, 14:166*t*–167*t*
Callitrichidae, 14:124, 14:127–131, 14:132*t*
Cebidae, 14:105–106, 14:108–112
cheek-pouched, 14:194, 14:197–204, 14:204*t*–205*t*
leaf-monkeys, 14:175, 14:178–183, 14:184*t*–185*t*
night, 14:138, 14:141*t*
Pitheciidae, 14:148, 14:150–153, 14:153*t*
monotremes, 12:233–234
mountain beavers, 16:133
Muridae, 16:285, 16:288–295, 16:296*t*–297*t*
Arvicolinae, 16:229, 16:233–238, 16:237*t*–238*t*
hamsters, 16:243, 16:245–246, 16:247*t*
Murinae, 16:252–253, 16:256–260, 16:261*t*–262*t*
Sigmodontinae, 16:270, 16:272–276, 16:277*t*–278*t*
musky rat-kangaroos, 13:71
Mustelidae, 14:324–325, 14:327–331, 14:332*t*–333*t*
numbats, 12:305
octodonts, 16:436, 16:438–440, 16:440*t*
Otariidae, 14:399–400, 14:402–406
pacaranas, 16:388
pacas, 16:420, 16:423–424
pangolins, 16:113, 16:115–120
peccaries, 15:295–296, 15:298–300
Perissodactyla, 15:221–222
Petauridae, 13:129, 13:131–133, 13:133*t*
Phalangeridae, 13:60, 13:*63–66t*
pigs, 15:280–281, 15:284–288, 15:289*t*–290*t*
pikas, 16:495, 16:497–501, 16:501*t*–502*t*
pocket gophers, 16:190, 16:192–195*t*, 16:196*t*–197*t*
porcupines
New World, 16:369, 16:372–373, 16:374*t*
Old World, 16:353–354, 16:357–364
possums
feather-tailed, 13:142, 13:144
honey, 13:138
pygmy, 13:108, 13:110–111
primates, 14:11
Procyonidae, 14:310–311, 14:313–315, 14:315*t*–316*t*
pronghorns, 15:415–416
Pseudocheiridae, 13:117, 13:119–123, 13:122*t*–123*t*
rat-kangaroos, 13:76, 13:78–81
rats
African mole-rats, 16:345, 16:347–349, 16:349*t*–350*t*
cane, 16:335, 16:337–338
chinchilla, 16:445, 16:447–448
dassie, 16:331
spiny, 16:450–451, 16:453–457, 16:458*t*
rhinoceroses, 15:254, 15:257–262
rodents, 16:126
sengis, 16:522, 16:524–530, 16:531*t*
shrews
red-toothed, 13:252, 13:255–263, 13:263*t*
West Indian, 13:244
white-toothed, 13:268, 13:271–275, 13:276*t*–277*t*
Sirenia, 15:191, 15:194–195
solenodons, 13:239, 13:241
springhares, 16:310
squirrels
flying, 16:136, 16:137, 16:141*t*–142*t*
ground, 16:147–148, 16:151–158, 16:158*t*–160*t*
scaly-tailed, 16:300, 16:303–305
tree, 16:167, 16:169–174*t*, 16:175*t*
Talpidae, 13:282, 13:284–287, 13:287*t*–288*t*
tapirs, 15:241–242, 15:245–247
tarsiers, 14:95, 14:97–99
tenrecs, 13:229, 13:232–234, 13:234*t*
three-toed tree sloths, 13:165, 13:167–169
tree shrews, 13:292, 13:294–296, 13:297*t*–298*t*
true seals, 14:262, 14:422, 14:427, 14:429–434, 14:435*t*
tuco-tucos, 16:427, 16:429–430, 16:430*t*–431*t*
Viverridae, 14:338, 14:340–343, 14:344*t*–345*t*
walruses, 14:415
wombats, 13:40, 13:53, 13:55–56
Xenartha, 13:152–153
protostomes, 2:43
amphionids, 2:196
amphipods, 2:262, 2:265–271
anaspidaceans, 2:182, 2:183
aplacophorans, 2:380, 2:382–385
Arachnida, 2:336, 2:340–352
articulate lampshells, 2:523, 2:525–527
bathynellaceans, 2:178, 2:179
beard worms, 2:86, 2:88
bivalves, 2:454, 2:458–466
caenogastropods, 2:447, 2:449
centipedes, 2:356, 2:358–362
cephalocarids, 2:132, 2:133
Cephalopoda, 2:479, 2:483, 2:485–489
chitons, 2:395, 2:398–401
clam shrimps, 2:148, 2:150–151
copepods, 2:300, 2:304–310
cumaceans, 2:230, 2:232
Decapoda, 2:201, 2:204–214
deep-sea limpets, 2:436, 2:437
earthworms, 2:67, 2:70–73
echiurans, 2:104, 2:107
fairy shrimps, 2:137, 2:139–140
fish lice, 2:290, 2:292–293
freshwater bryozoans, 2:498, 2:500–502
horseshoe crabs, 2:328, 2:331–332
Isopoda, 2:252, 2:255–260
krill, 2:187, 2:189–192
leeches, 2:77–78, 2:80–83
leptostracans, 2:162, 2:164–165
lophogastrids, 2:226, 2:227
mantis shrimps, 2:171, 2:173–175
marine bryozoans, 2:504, 2:506–508, 2:511, 2:513–514
mictaceans, 2:241, 2:242
millipedes, 2:365, 2:367–370
monoplacophorans, 2:388, 2:390–391
mussel shrimps, 2:312, 2:314–315
mysids, 2:217, 2:219–222
mystacocarids, 2:296, 2:297
myzostomids, 2:60, 2:62
Neritopsina, 2:441, 2:442–443
nonarticulate lampshells, 2:516, 2:518–519
Onychophora, 2:111, 2:113–114
pauropods, 2:376, 2:377
peanut worms, 2:98, 2:100–101
phoronids, 2:492, 2:494
Polychaeta, 2:47, 2:50–56
Pulmonata, 2:414, 2:417–421
remipedes, 2:126, 2:128
sea slugs, 2:405, 2:407–410
sea spiders, 2:322, 2:324–325
spelaeogriphaceans, 2:243, 2:244
symphylans, 2:372, 2:373
tadpole shrimps, 2:142, 2:145–146
tanaids, 2:236, 2:238–239
tantulocaridans, 2:284, 2:286
Thecostraca, 2:276, 2:278–281
thermosbaenaceans, 2:246, 2:247
tongue worms, 2:320
true limpets, 2:425, 2:426–427
tusk shells, 2:470, 2:473–474
Vestimentifera, 2:92, 2:94–95
Vetigastropoda, 2:431, 2:433–434
water bears, 2:118, 2:121–123
water fleas, 2:155, 2:157–158
reptiles, 7:10–11, **7:59–63**
African burrowing snakes, 7:462–464
African sideneck turtles, 7:130, 7:132–133
Afro-American river turtles, 7:138, 7:140–142
Agamidae, 7:211, 7:214–221
Alligatoridae, 7:173, 7:175–177
American alligators, 7:48–49
Anguidae, 7:340, 7:342–344
Australo-American sideneck turtles, 7:78, 7:81–83
big-headed turtles, 7:136
blindskinks, 7:272
blindsnakes, 7:382–385
boas, 7:411, 7:413–417
Central American river turtles, 7:100
chameleons, 7:232, 7:235–241
colubrids, 7:470, 7:473–481
Cordylidae, 7:322, 7:324–325
crocodilians, 7:10–11
Crocodylidae, 7:181–182, 7:184–187
early blindsnakes, 7:371
Elapidae, 7:486–487, 7:491–498
false blindsnakes, 7:388
false coral snakes, 7:400
file snakes, 7:442
Florida wormlizards, 7:285
Gekkonidae, 7:262–263, 7:265–269
Geoemydidae, 7:116, 7:118–119
gharials, 7:169
Helodermatidae, 7:357
Iguanidae, 7:246–247, 7:250–257
Kinosternidae, 7:122–123, 7:125–126
knob-scaled lizards, 7:349, 7:351
Lacertidae, 7:298–299, 7:301–302
leatherback seaturtles, 7:102
Microteiids, 7:304, 7:306–307
mole-limbed wormlizards, 7:280, 7:281
Neotropical sunbeam snakes, 7:406
New World pond turtles, 7:106–107, 7:109–113
night lizards, 7:292, 7:294–295
pig-nose turtles, 7:76

pipe snakes, 7:397
pythons, 7:421–422, 7:424–428
seaturtles, 7:87, 7:89–91
shieldtail snakes, 7:392–394
skinks, 7:329, 7:332–337
slender blindsnakes, 7:375–377
snapping turtles, 7:94, 7:95–96
softshell turtles, 7:152, 7:154–155
spade-headed wormlizards, 7:288, 7:289
splitjaw snakes, 7:430–431
Squamata, 7:207
sunbeam snakes, 7:403
Teiidae, 7:312, 7:314–316
Testudines, 7:72–73
tortoises, 7:10, 7:72–73, 7:144, 7:146–149
Tropidophiidae, 7:435–437
tuatara, 7:191
turtles, 7:10–11, 7:72–73
Varanidae, 7:363, 7:365–368
Viperidae, 7:448, 7:451–460
wormlizards, 7:274, 7:276
zoos and, 12:204, 12:*208–209*
See also Extinct species; Humans
Contact chemoreception, 3:26–27
Contaminants, environnmental. *See* Pollution
Continental drift, 4:58, 7:196
Contopus spp. *See* Pewees
Contopus borealis. *See* Olive-sided flycatchers
Contopus pertinax. *See* Greater pewees
Contopus sordidulus. *See* Western wood-pewees
Contopus virens. *See* Eastern wood-pewees
Contreras, L. C.
on coruros, 16:435, 16:436
on octodonts, 16:433
Conuropsis carolinensis. *See* Carolina parakeets
Conuropsis fratercula, 9:275
Conus geographus. *See* Geography cone shells
Conus magnus, 1:45
Conus marmoreus, 2:447
Conus textile, 2:447
Conus tulipus, 2:447
Convention for Migratory Species, 9:29
Convention for the Conservation of Anadromous Stocks in the North Pacific, 4:413
Convention for the Conservation of Nature and Natural Resources (Africa), 15:195, 15:209, 15:280
Convention on International Trade in Endangered Species, 3:82, 3:90, 8:27, 14:374
Acipenseriformes, 4:215, 4:217
addaxes, 16:39
African elephants, 15:175
American crocodiles, 7:184
Anthozoa, 1:107
Arabian oryx, 16:40
Asian elephants, 15:174
Asian treefrogs, 6:293
Australian lungfishes, 4:199
babblers, 10:507, 10:516, 10:517
babirusas, 15:288
bald uakaris, 14:151
banded cotingas, 10:314
barn owls, 9:342
on barn owls, 9:341
Beecroft's anomalures, 16:303
big-eared flying mice, 16:305
bivalves, 2:454
black-capped capuchins, 14:111
black corals, 1:115
Bolivian squirrel monkeys, 14:108
brown-throated three-toed sloths, 13:167
bulbuls, 10:405
buttonquails, 9:19
capuchins, 14:105, 14:106
Central American river turtles, 7:100
Chacoan peccaries, 15:300
chameleons, 7:232
coelacanths, 4:193
Coleoptera, 3:322
collared peccaries, 15:298
collared titis, 14:152
colubrids, 7:470
common chameleons, 7:238
common hippopotamuses, 15:311
common squirrel monkeys, 14:109
Cordylidae, 7:322
cranes, 9:32–36
Cryptobranchidae, 6:347
dugongs, 15:201
dusky titi monkeys, 14:152
elephants, 15:171
emperor scorpions, 2:351
European medicinal leeches, 2:81
fairy bluebirds, 10:417
Garrulax canorus taewanus, 10:516
gharials, 7:169
giant clams, 2:464
gray-backed sportive lemurs, 14:79
gray-necked picathartes, 10:523
Gruiformes, 9:3
guanacos, 15:322
Helodermatidae, 7:357
hippopotamuses, 15:308
hornbills, 10:80, 10:81
hummingbirds, 9:449
Hydrozoa, 1:129
Indian pangolins, 16:116
Jackson's chameleons, 7:239
jaguars and, 14:386
jeweled chameleons, 7:239
Johnstone's crocodiles, 7:185
kagus, 9:43
leafbirds, 10:417
leatherback seaturtles, 7:102
leeches, 2:78
Lepidoptera, 3:388
lesser Malay mouse deer, 15:332
Lord Derby's anomalures, 16:303
Madagascar day geckos, 7:268
manatees, 15:208–209
masked titis, 14:153
Milne-Edwards's sportive lemurs, 14:81
minor chameleons, 7:240
mole salamanders, 6:359
Mount Omei babblers, 10:517
mouse lemurs, 14:39
mugger crocodiles, 7:185
mussuranas, 7:480
mynas, 11:417, 11:420
Neotropical sunbeam snakes, 7:406
Nile crocodiles, 7:186
northern sportive lemurs, 14:83
olms, 6:381
Orectolobiformes, 4:105
Osteoglossiformes, 4:234, 4:240
pacas, 16:423
panther chameleons, 7:240
parrots, 9:280, 9:295
Parson's chameleons, 7:236–237
peccaries, 15:295
Pel's anomalures, 16:303
Perissodactyla, 15:221–222
Phelsuma spp., 7:263
pigs, 15:280–281
Pitheciidae, 14:148
pittas, 10:195
plains-wanderers, 9:21
poison frogs, 6:200
prehensile-tailed skinks, 7:333
pygmy hippopotamuses, 15:311
pygmy hogs, 15:287
pygmy right whales, 15:105
pythons, 7:421
Queen Alexandra's birdwings, 3:398–399
red-billed hill tits, 10:517
red mouse lemurs, 14:43
red-tailed sportive lemurs, 14:80
Russel's vipers, 7:458
saltwater crocodiles, 7:187
Scleractinia, 1:117, 1:118, 1:119, 1:120
Sirenia, 15:194–195
sloths, 13:153
small-toothed sportive lemurs, 14:83
southern African hedgehogs, 13:207, 13:210
sperm whales, 15:78
sportive lemurs, 14:76
squirrel monkeys, 14:105
straw-headed bulbuls, 10:399
sunbird-asities, 10:188
tapirs, 15:243
Teiidae, 7:312
turacos, 9:301
Varanidae, 7:363
veiled chameleons, 7:237
vicuñas, 15:321
Viperidae, 7:448
vocal sac-brooding frogs, 6:174, 6:176
volcano rabbits, 16:514
water chevrotains, 15:331
weasel sportive lemurs, 14:82
weaverfinches, 11:361, 11:364, 11:365, 11:366, 11:372
weeper capuchins, 14:112
white-faced sakis, 14:150
white-footed sportive lemurs, 14:82
white-fronted capuchins, 14:110
white-lipped peccaries, 15:299
white-necked picathartes, 10:523
white-nosed bearded sakis, 14:153*t*
white-throated capuchins, 14:111
See also Conservation status
Convention on the Conservation of Antarctic Marine Living Resources, 2:190
Convention on the Conservation of European Wildlife and Natural Habitats
alpine shrews, 13:260
bath sponges, 1:85
carnivorous sponges, 1:86
demosponges, 1:80
Discoglossidae, 6:90
Eurasian beavers, 16:180, 16:183
European wels, 4:366
Lacertidae, 7:299
parsley frogs, 6:128
Pyrenean brook salamanders, 6:371
on Roman snails, 2:420
salmons, 4:416
smooth snakes, 7:474
Convention on the Law of the Sea (1982)
common dolphinfishes, 5:215

pompano dolphinfishes, 5:214
Convergent evolution, 2:7, 12:27
of birds, 8:3
of marine mammals, 12:68
of subterranean mammals, 12:78
See also Evolution
Convict blennies, 5:331–335
Convoluta spp., 1:33
Convoluta roscoffensis. See Green flatworms
Convolutriloba longifissura, 1:*181*, 1:*182*
Convolutriloba retrogemma, 1:179
Conway-Morris, S., 1:433
Cook, Captain James, 12:191
Cook, W. M., 16:268
Cookie-cutter sharks, 4:139, 4:152, 4:*154*, 4:*155*–156
Coolie loaches, 4:*324*, 4:*325*, 4:332
Cooloola monsters, 3:203
Cooloolidae. *See* Cooloola monsters
Coons. *See* Northern raccoons
Cooper, William E. Jr., 7:348
Cooper sharks, 4:*115*
Cooperation, 1:31–32
See also Behavior
Cooperative breeding
Australian fairy-wrens, 11:46
in bee-eaters, 10:41–42
nuthatches, 11:174
Old World warblers, 11:6
in todies, 10:27
trumpeters, 9:77, 9:79
white-winged choughs, 11:457
in wrens, 10:528–529
See also Reproduction
Coopers of the sea, 2:*263*, 2:*269*, 2:270
Cooties. *See* Human head/body lice
Coots, 9:1–4, **9:45–67,** 9:*46*
Copadichromis eucinostomus, 5:279
Cope, E. D., 7:353, 7:379, 15:132
Copeina. *See* Splash tetras
Copella spp., 4:66
Copella arnoldi. See Splash tetras
Copepods, 1:38, **2:299–310,** 2:*302*, 2:*303*
behavior, 2:300
conservation status, 2:300
distribution, 2:300
evolution, 2:299
feeding ecology, 2:36, 2:300
habitats, 2:300
humans and, 2:300–301
physical characteristics, 2:299–300
reproduction, 2:300
species of, 2:*304*–310
taxonomy, 2:299
Cope's gray treefrogs, 6:47, 6:*232*, 6:*236*
Cope's lizards. *See* Long-nosed leopard lizards
Cophixalus riparius. See Wilhelm rainforest frogs
Cophotis spp., 7:211
Cophylinae, 6:301, 6:302
Copiphora rhinoceros. See Rhinoceros katydids
Copiphorinae, 2:36
Copper sunbirds, 11:208
Copper-tailed glossy-starlings, 11:410, 11:*413*, 11:*422*
Copperheads (snake)
in religions, 7:57
reproduction, 7:*36*
southern, 7:52, 7:447, 7:448
Copperheads (squirrel). *See* Golden-mantled ground squirrels
Coppersmith barbets, 10:114–*115*, 10:*118*, 10:*119*
Coppery-chested jacamars, 10:92, 10:*94*, 10:97–98
Coppery ringtail possums, 13:*115*, 13:123*t*
Coprophagy, 12:48, 12:121, 16:485
See also Behavior
Copsychus saularis. See Magpie-robins
Copsychus sechellarum. See Seychelles magpie-robins
Coptopsyllidae, 3:347, 3:348
Coptotermes acinaciformis, 3:165
Coptotermes formosanus, 3:72
Copulation, 1:18, 1:21–23, 2:17, 2:23–24, 12:103–104
See also Reproduction
Copulatory plugs, 12:105–106
Coquerel's mouse lemurs, 14:36, 14:*37*, 14:39, 14:*40*, 14:*41*, 14:43–44
Coquerel's sifakas, 14:*65*
Coquina clams, 2:*451*, 2:*456*, 2:*463*, 2:464
Coracias cyanogaster. See Blue-bellied rollers
Coracias galbula. See Baltimore orioles
Coracias garrulus. See European rollers
Coracias sagittata. See Olive-backed orioles
Coracias spatulata. See Racket-tailed rollers
Coracias temminckii. See Temminck's rollers
Coracias tibicen. See Australian magpies
Coracii, 10:3–4
Coraciidae. *See* Rollers
Coraciiformes, **10:1–4**
Coracina spp. *See* Cuckoo-shrikes
Coracina azurea. See Blue cuckoo-shrikes
Coracina lineata. See Yellow-eyed cuckoo-shrikes
Coracina newtoni. See Reunion cuckoo-shrikes
Coracina novaehollandiae. See Black-faced cuckoo-shrikes
Coracina ostenta. See White-winged cuckoo-shrikes
Coracina papuensis. See White-bellied cuckoo-shrikes
Coracina pectoralis. See White-breasted cuckoo-shrikes
Coracina tenuirostris. See Cicadabirds
Coracina typica. See Mauritius cuckoo-shrikes
Coracornis raveni. See Maroon-backed whistlers
Coragyps atratus. See American black vultures
Coral bleaching, 1:33, 1:53
See also Conservation status
Coral cardinalfishes. *See* Pajama cardinalfishes
Coral catfishes, 4:*355*, 4:*360*, 4:365
Coral crouchers, 4:42, 4:43, 4:48
Coral dragonets. *See* Lancer dragonets
Coral-dwelling hawkfishes, 5:240
Coral reef habitats, 4:35, 4:42, 4:43, 4:*43*, 4:44, 4:47–48, 4:*54*, 4:61
Coral Reef Initiative (U.S.), 1:107
Coral reefs, 1:26–*27*, 1:33, 1:*51*
See also Corals
Coral reefs, conservation of, 2:30
Coral snakes
behavior, 7:468
coloration, 7:203
distribution, 7:484
feeding ecology, 7:486
North American, 7:*490*, 7:*491*–492
taxonomy, 7:483
venom and, 7:206
Coral trouts, 5:255, 5:271–*272*
See also Epinephelini
Corallimorpharia. *See* Mushroom corals
Coralliozetus tayrona, 5:343
Corallium rubrum. See Red corals
Corallivores, 4:50
Corallus spp., 7:410, 7:411
Corallus caninus. See Emerald tree boas
Corallus cropanii, 7:411
Corallus hortulanus. See Amazon tree boas
Corals, **1:103–122,** 1:*109*, 1:*110*
behavior, 1:29, 1:30, 1:42, 1:106
conservation status, 1:107
distribution, 1:105
evolution, 1:3–4, 1:6, 1:103
feeding ecology, 1:27, 1:106–107
habitats, 1:24, 1:105
humans and, 1:107–108
lace, 2:*509*, 2:*510*
physical characteristics, 1:103–*105*
reproduction, 1:19, 1:107
species of, 1:114–121
symbiosis, 1:33
taxonomy, 1:9, 1:103
Corbet, G. B., 16:199, 16:211
Corbicula spp. *See* Asian clams
Corbus graculinus. See Pied currawongs
Corby messengers, 11:509
Corcoracinae. *See* White-winged choughs
Corcorax melanorhamphos. See White-winged choughs
Cordicephalus spp., 6:12
Cordon-bleu, 11:354, 11:*358*, 11:363–364
Cordulegastridae, 3:54
Cordylidae, 7:204, 7:206, **7:319–325,** 7:*323*
Cordylinae. *See* Girdled lizards
Cordylobia anthropophaga, 3:64
Cordylophora caspia, 1:25
Cordylosaurus subtessellatus. See Dwarf plated lizards
Cordylus spp. *See* Girdled lizards
Cordylus cataphractus. See Armadillo lizards
Cordylus giganteus. See Giant girdled lizards
Coregonus alpenae, 4:407
Coregonus clupeaformis. See Lake whitefishes
Coregonus johannae, 4:407
Coregonus lavaretus, 4:410
Coregonus nigripinnis, 4:407
Coreidae. *See* Flag-legged insects
Coreids, 3:262
Corella parallelogramma, 1:*457*, 1:461, 1:*462*
Corematodus spp., 4:69
Coreoperca spp., 5:220, 5:221
Coreosiniperca spp., 5:220
Corgis, 14:292
Corixidae, 3:260
Cormobates leucophaea. See White-throated treecreepers
Cormobates placens. See White-throated treecreepers
Cormorants, 8:21, 8:183–186, **8:201–210**
Cormura brevirostris. See Chestnut sac-winged bats
Corn buntings, 11:*264*, 11:*268*, 11:*275*–276
Corn earworms, 3:*392*, 3:*393*, 3:397
Cornbirds. *See* Golden-headed cisticolas
Corncrakes, 9:50, 9:52, 9:*53*, 9:*60*
Cornetfishes, 5:131–132, 5:133, 5:135
blue-spotted, 5:*138*, 5:*140*–141

red, 5:134
Corneum, 12:36
See also Physical characteristics
Cornitermes spp., 3:165
Cornsnakes, 7:*39*, 7:470
Coroidea, 11:477
Corolla spectabilis, 2:*406*, 2:*408*, 2:409–410
Coronarctidae, 2:117
Coronatae, 1:153–154, 1:155, 1:156
Coronella austriaca. *See* Smooth snakes
Cororos, 16:123
Corphaena equiselis. *See* Pompano dolphinfishes
Corpora quadrigemina, 12:6
See also Physical characteristics
Corpus spp., 1:32
Corpus luteum, 12:92
See also Physical characteristics
Correndera pipits, 10:372
Corroboree toadlets, 6:59, 6:148
Corrugated water frogs, 6:246, 6:*253*, 6:*258*
Corsac foxes, 14:284*t*
Corsican stick insects. *See* Mediterranean stick insects
Cortés, Hernando, 3:81, 9:451, 12:203, 15:216
Cortez chubs, 5:*246*, 5:*249*
Cortisole, 12:148
Corucia spp., 7:329
Corucia zebrata. *See* Prehensile-tailed skinks
Coruros, 12:74, 12:75, 16:434, 16:435, 16:436, 16:*437*, 16:*438*, 16:439
Corvidae, 10:171, 10:173, **11:503–524,** 11:*510–511*
behavior, 11:505–506
conservation status, 11:507–508
distribution, 11:*503*, 11:504–505
evolution, 11:503
feeding ecology, 11:*504*, 11:506–507, 11:*508*
habitats, 11:505
humans and, 11:509
physical characteristics, 11:503–504
reproduction, 11:507
species of, 11:*512–522*
taxonomy, 11:503
Corviis corax. *See* Ravens
Corvinella spp. *See* Shrikes
Corvinella corvina. *See* Yellow-billed shrikes
Corvinella melanoleuca. *See* Magpie-shrikes
Corvus spp. *See* Corvidae
Corvus albus. *See* Pied crows
Corvus brachyrhynchos. *See* American crows
Corvus caurinus. *See* Northwestern crows
Corvus corax. *See* Northern ravens; Ravens
Corvus corone. *See* Carrion crows
Corvus corone cornix. *See* Hooded crows
Corvus coronoides. *See* Australian ravens
Corvus florensis. *See* Flores crows
Corvus frugilegus. *See* Eurasian rooks
Corvus hawaiiensis. *See* Hawaiian crows
Corvus kubaryi. *See* Mariana crows
Corvus leucognaphalus. *See* White-necked crows
Corvus macrorhynchos. *See* Large-billed crows
Corvus mexicanus. *See* Great-tailed grackles
Corvus minutus. *See* Cuban palm crows
Corvus monedula. *See* Western jackdaws
Corvus nasicus. *See* Cuban crows
Corvus olivaceus. *See* Eastern whipbirds
Corvus orru. *See* Torresian crows
Corvus palmarum. *See* Hispaniolan palm crows
Corvus splendens. *See* House crows
Corydalidae, 3:289, 3:290, 3:291
Corydalinae, 3:289
Corydalus spp., 3:291
Corydalus cornutus. *See* Eastern dobsonflies
Corydendrium parasiticum, 1:30
Corydon spp. *See* Broadbills
Corydon sumatranus. *See* Dusky broadbills
Corydoras spp. *See* Plated catfishes
Corydoras aeneus. *See* Catfishes
Corydoras hastatus. *See* Dwarf corydoras
Corydoras revelatus, 4:351
Corymorpha nutans, 1:127, 1:*132*, 1:134
Corynactis californica. *See* Jeweled anenomes
Coryne tubulosa, 1:25
Coryphaena spp. *See* Dolphinfishes
Coryphaena hippurus. *See* Common dolphinfishes
Coryphaenidae. *See* Dolphinfishes
Coryphaenoides spp., 5:27
Coryphaenoides rupestris. *See* Roundnose grenadiers
Cory's shearwaters, 8:123, 8:*127*, 8:*128*
Corystes cassivelaunus. *See* Masted crabs
Corythaeola spp. *See* Turacos
Corythaeola cristata. *See* Great blue turacos
Corythaeolinae. *See* Turacos
Corythaix hartlaubi. *See* Hartlaub's turacos
Corythaix porphyreolophus. *See* Purple-crested turacos
Corythaixoides concolor. *See* Gray go-away-birds
Corythoichthys isigakius, 5:136
Corythomantis spp., 6:226
Corytophanes percarinatus, 7:246
Corytophaninae, 7:243–244
COSEWIC. *See* Committee on the Status of Endangered Wildlife in Canada
Cossid moths, 3:81
Cossidae, 3:389
Cossypha spp. *See* African robin chats
Cossypha heuglini. *See* White-browed robin chats
Costa hummingbirds, 9:444
Costa Rican heliconias, 9:445
Costa Rican quetzals. *See* Resplendent quetzals
Costa Rican worm salamanders, 6:*394*, 6:401–402
Cotinga cayana. *See* Spangled cotingas
Cotinga cotinga. *See* Purple-breasted cotingas
Cotinga maculata. *See* Banded cotingas
Cotinga maynana. *See* Plum-throated cotingas
Cotinga ridgwayi. *See* Turquoise cotingas
Cotingas, 10:170–171, **10:305–323,** 10:*309–311*
behavior, 10:306–307
conservation status, 10:308
distribution, 10:*305*, 10:306
feeding ecology, 10:307
habitats, 10:306
humans and, 10:308
physical characteristics, 10:305–306
reproduction, 10:307
species of, 10:*312–322*
Cotingidae. *See* Cotingas
Cottidae. *See* Sculpins
Cottocomephoridae. *See* Baikal sculpins
Cottoidei. *See* Sculpins
Cotton bollworms. *See* Corn earworms
Cotton bugs. *See* Staining bugs
Cotton rats, 16:127, 16:128, 16:263, 16:268–270, 16:275–276
Cotton stainers. *See* Staining bugs
Cotton-top tamarins, 14:119, 14:*122*, 14:123, 14:124, 14:*126*, 14:*127*
Cottonfishes. *See* Bowfins
Cottonmouths, 7:*35*, 7:37, 7:57, 7:447, 7:*450*, 7:*452*–453
Cottontails, 16:*507*, 16:*510*, 16:514–515
behavior, 16:482
evolution, 16:505
habitats, 16:481, 16:506
humans and, 16:488, 16:509
Cottunculidae, 5:179
Cottus spp., 5:180, 5:183
Cottus bairdii, 5:180
Cottus cognatus, 5:180
Cottus echinatus, 5:183
Cotuero toadfishes, 5:42
Coturnicops noveboracensis. *See* Yellow rails
Coturnix chinensis. *See* King quails
Coturnix coturnix. *See* European common quail; Quails
Coturnix novaezelandiae. *See* New Zealand quails
Cotylorhiza tuberculata, 1:40
Cotylosauria, 7:4–5
Couas, 9:311, 9:*317*, 9:324–325
Coucals, 9:311–312, 9:314, 9:315, 9:*317*, 9:325
Couch's spadefoot toads, 6:*121*, 6:*123*, 6:*124*–125
Coues' flycatchers. *See* Greater pewees
Cougars. *See* Pumas
Coumba cuneata. *See* Diamond doves
Coumba oenas. *See* Stock pigeons
Countershading, 12:38
See also Physical characteristics
Counting, 12:155
See also Cognition
Counting birds, amateur. *See* Christmas Bird Counts
Coura caerulea. *See* Blue couas
Courols, 10:3–4, 10:51–53, 10:*54*, 10:57–*58*
Coursers, **9:151–154, 9:158–160**
Courtship behavior, 6:29–31, 6:48–50
pheromones for, 6:44–45, 6:45–46
salamanders, 6:44–45, 6:49
vocalizations and, 6:35, 6:47, 6:49
See also Behavior; Reproduction
Couvalli jacks. *See* Crevalle jacks
Cove oysters. *See* Eastern American oysters
Cow-killers. *See* Velvet ants
Cowan, I. M., 5:353
Cowbirds, 11:*256*, 11:291, 11:302, 11:303, 11:304, 11:320, 11:321
See also Brown-headed cowbirds
Cownose rays, 4:176, 4:177
Cowries, 2:41, 2:447
Cows. *See* Cattle
Cowsharks, 4:11, 4:143
Coxal bone, 12:41
See also Physical characteristics
Coyotes, 12:*136*, 12:*143*, 14:*258*, **14:265–273,** 14:*266*, 14:*273*
behavior, 12:148, 14:268
conservation status, 14:272
evolution, 14:265
feeding ecology, 14:270, 14:271
habitats, 14:267
humans and, 14:262, 14:273
physical characteristics, 14:266–267
taxonomy, 14:265

Coypus, 16:*122*, **16:473–478,** 16:*474*, 16:*475*, 16:*476*, 16:*478*
distribution, 16:123, 16:*473*, 16:474
habitats, 16:124, 16:474
humans and, 12:184, 16:126–127, 16:477
physical characteristics, 16:123, 16:473–474
Cozumel Island coatis, 14:316*t*
Cozumel Island raccoons, 14:310, 14:311, 14:315*t*
Cozumel thrashers, 10:468
Cozumel wrens, 10:529
Crab-eater seals, 12:138, 14:260, 14:420, 14:*425*, 14:*433*–434
Crab-eating foxes, 14:*268*, 14:271, 14:*275*, 14:*279*, 14:282–283
Crab-eating kingfishers. *See* Shovel-billed kookaburras
Crab-eating raccoons, 14:315*t*
Crab hawks, 8:321
Crab lice, 3:*249*
Crab plovers, 9:104, **9:121–123,** 9:*122*
Crab spiders, 1:42, 2:38
Crabs, 2:25, **2:197–214,** 2:*202*, 2:*203*
behavior, 2:32, 2:35, 2:43, 2:199–200
conservation status, 2:43, 2:201
distribution, 2:199
evolution, 2:197–198
feeding ecology, 2:200
habitats, 2:199
humans and, 2:*42*, 2:201
physical characteristics, 2:198
reproduction, 2:*16*, 2:201
species of, 2:*204*–214
taxonomy, 2:197–198
Cracidae, **8:413–423**
behavior, 8:414–415
conservation status, 8:416
distribution, 8:*413*, 8:414
evolution, 8:413
feeding ecology, 8:415–416
habitats, 8:414
humans and, 8:416
physical characteristics, 8:413–414
reproduction, 8:416
species of, 8:*419–421*
taxonomy, 8:416
Cracraft, Joel, 9:41, 10:52
Cracticidae. *See* Magpie-shrikes
Cracticus louisiadensis. See Tagula butcherbirds
Cracticus mentalis. See Black-backed butcherbirds
Cracticus nigrogularis. See Black-hooded magpie-shrikes
Cracticus quoyi. See Black butcherbirds
Cracticus torquatus. See Gray butcherbirds
Crag lizards, 7:319
Crag martins, 10:*362*, 10:*368*
Crakes, 9:46–50, 9:52, 9:*54*, 9:63–64
Cranchiids, 2:476
Crane flies, 3:357, 3:358
European marsh, 3:*363*, 3:*369*, 3:373–374
phantom, 3:*360*
Cranes, 8:26, 9:1–4, **9:23–36,** 9:*30*
behavior, 9:*24–26*
captive breeding, 9:27–28, 9:29
conservation status, 9:27–29
distribution, 9:*23*, 9:24
evolution, 9:23
feeding ecology, 9:26–27
habitats, 9:24
humans and, 9:29
migration, 9:24, 9:28–29, 9:35
physical characteristics, 9:24
reproduction, 9:24, 9:27, 9:28
species of, 9:*31–36*
taxonomy, 9:23–24
vocalizations, 9:24–25
Crangon septemspinosa. See Sevenspine bay shrimps
Craniata, 4:9
Craniidae, 2:515, 2:516
Craniiformea. *See* Craniidae
Cranioleuca hellmayri. See Streak-capped spinetails
Cranioleuca henricae. See Bolivian spinetails
Craniscus spp., 2:515
Craseonycteris thonglongyai. See Kitti's hog-nosed bats
Craspedacusta sowerbyi, 1:*133*, 1:142
Crassiclitellata, 2:65
Crassostrea virginica. See Eastern American oysters
Craterocephalus fluviatilis, 5:70
Crateropus gutturalis. See Babbling starlings
Crateroscelis nigrorufa. See Bicolored mouse-warblers
Craterostigmomorpha, 2:353, 2:356
Craterostigmus tasmanianus. See Tasmanian remarkables
Cratogeomys spp., 16:185
Cratogeomys castanops. See Yellow-faced pocket gophers
Cratogeomys fumosus. See Smoky pocket gophers
Cratogeomys merriami. See Merriam's pocket gophers
Cratogeomys neglectus. See Querétaro pocket gophers
Cratogeomys zinseri. See Zinser's pocket gophers
Crawfish frogs, 6:*247*
Crax daubentoni. See Yellow-knobbed curassows
Crax globulosa. See Wattled curassows
Crax rubra. See Great curassows
Crayfish, 2:25, 2:197, 2:199, 2:200, 2:201
Creagus furcatus. See Swallow-tailed gulls
Cream-colored coursers, 9:*152*
Cream-colored pratincoles. *See* Gray pratincoles
Creamy-bellied antwrens. *See* Black-capped antwrens
Creamy-bellied gnatcatchers, 11:7
Creatophora spp., 11:409
Creediidae. *See* Sandburrowers
Creek perches. *See* Longear sunfishes
Creek rats, 16:*254*, 16:*255*, 16:258–259
Creepers
Australian, **11:133–139,** 11:*136*
brown, 11:118, 11:177, 11:*179*, 11:180–181
bush, 10:209
earth, 10:*212*, 10:*215*, 10:*216–217*
honey, 10:172
Philippine, 10:505, **11:183–188,** 11:*185*
wall, 11:167, 11:*170*, 11:174–175
wood, 10:170–171, **10:229–237,** 10:*231*
See also Treecreepers
Creeping eruptions. *See* Dog hookworms
Creeping water bugs, 3:*266*, 3:*275–276*
Cremastocheilus spp., 3:321
Crematogaster scutellaris, 3:*61*, 3:*69*
Crenicara spp., 5:279
Crenichthys spp., 5:89
Crenicichla alta. See Millets
Crenimugil crenilabis. See Fringelip mullets
Creocele spp., 5:355
Creodonta, 12:11
Crepidophryne spp., 6:184
Crepuscular activity cycle, 12:55
See also Behavior
Crest-tailed marsupial mice. *See* Mulgaras
Crested agoutis, 16:409, 16:415*t*
Crested barbets, 10:114
Crested bellbirds, 11:115–117, 11:*119*, 11:*124*–125
Crested berrypeckers, 11:189, 11:190, 11:191
Crested birds of paradise, 11:*494*, 11:*495*
Crested bulbuls. *See* Red-whiskered bulbuls
Crested buntings, 11:*268*, 11:*276*
Crested caracaras, 8:*313*, 8:348, 8:*349*, 8:350, 8:353, 8:*354*, 8:*417*
Crested cuckoo-doves, 9:*253*, 9:*257–258*
Crested finchbills, 10:396, 10:*400*, 10:*411*
Crested firebacks, 8:434
Crested free-tailed bats. *See* Lesser-crested mastiff bats
Crested gallitos, 10:259, 10:*261*, 10:*263*–264
Crested gardener birds. *See* Macgregor's bowerbirds
Crested gibbons, 14:207, 14:210, 14:215, 14:*216*, 14:*220*, 14:221
Crested guans, 8:417, 8:*419*
Crested guineafowl, 8:426–428
Crested honeycreepers. *See* Akohekohes
Crested ibises. *See* Japanese ibises
Crested jays, 11:504
Crested larks, 10:341, 10:342, 10:343, 10:*344*, 10:*345*, 10:346, 10:*347*, 10:*352*
Crested lizards. *See* Desert iguanas
Crested mynas, 8:24, 11:*412*, 11:*418*–419
See also Helmeted mynas
Crested oarfishes. *See* Crestfishes
Crested oropendolas, 11:305
Crested owlet-nightjars. *See* Australian owlet-nightjars
Crested parrots. *See* Cockatiels
Crested partridges. *See* Crested wood-partridges
Crested penguins. *See* Penguins
Crested pitohuis, 11:116
Crested rats, 16:*287*, 16:*289*, 16:293
Crested scorpionfishes, 5:*167*, 5:169, 5:*170*
Crested seriemas. *See* Red-legged seriemas
Crested shelducks, 8:367, 8:372–373
Crested shrike-tits. *See* Eastern shrike-tits
Crested terns, 9:203
Crested tits, 11:157
Crested tree swifts, 9:433, 9:434, 9:*435*
Crested wood-partridges, 8:434, 8:*440*, 8:*448*–449
Crestfishes, 4:*451*, 4:*453*–454
Crevalle jacks, 5:*264*, 5:*266*, 5:267
Crevice bats. *See* Peruvian crevice-dwelling bats
Crex crex. See Corncrakes
Crex egregia. See African crakes
Cribinopsis fernaldi. See Crimson anemones
Cribos. *See* Indigo snakes
Cricetinae. *See* Hamsters
Cricetomyinae, 16:282

Cricetomys gambianus. See Gambian rats
Cricetulus spp. *See* Rat-like hamsters
Cricetulus barabensis. See Striped dwarf hamsters
Cricetulus migratorius. See Gray dwarf hamsters
Cricetus spp. *See* Black-bellied hamsters
Cricetus cricetus. See Black-bellied hamsters
Cricket frogs, 6:48, 6:66, **6:265–271,** 6:*267*
Crickets, **3:201–208,** 3:*210,* **3:211–214**
behavior, 3:59–60, 3:204–206
education and, 3:83
farming of, 3:80
mechanoreception, 3:26
reproduction, 3:31, 3:38, 3:207
See also specific types of crickets
Criconema civellae. See Citrus spine nematodes
Cricosaura spp., 7:291
Cricosaura typica. See Cuban night lizards
Crimson anemones, 1:*106*
Crimson-backed sunbirds, 11:208, 11:*212,* 11:*220*
Crimson-bellied orioles, 11:*430,* 11:432–*433*
Crimson-breasted barbets. *See* Coppersmith barbets
Crimson chats, 11:65, 11:*66,* 11:*67*
Crimson finches, 11:354
Crimson-hooded manakins, 10:*296*
Crimson horned pheasants. *See* Satyr tragopans
Crimson rosellas, 9:279
Crimson seedcrackers, 11:*355,* 11:*358,* 11:*362*
Crimson tang. *See* Crimson chats
Crimson-throated barbets, 10:115
Crimson topaz, 9:444, 9:*453,* 9:*461*–462
Crimson tragopans. *See* Satyr tragopans
Crimson-wings, 11:353, 11:354
Crinia darlingtoni. See Hip pocket frogs
Crinia georgiana, 6:147, 6:148
Crinia nimbus. See Moss frogs
Crinia presignifera, 6:147
Crinia remota, 6:147, 6:148
Crinia signifera, 6:147
Crinia tasmaniensis, 6:147
Crinifer piscator. See Western gray plantain-eaters
Criniferinae. *See* Turacos
Criniferoides leucogaster. See White-bellied go-away-birds
Criniger spp., 10:395–399
Criniger calurus. See Red-tailed greenbuls
Criniger chloronotus. See Eastern bearded greenbuls
Criniger finschii. See Finsch's bulbuls
Criniger flaveolus. See White-throated bulbuls
Criniger pallidus. See Puff-throated bulbuls
Crinkle-breasted manucodes. *See* Crinkle-collared manucodes
Crinkle-collared manucodes, 11:*494,* 11:495–*496*
Crinoid clingfishes, 5:355, 5:*356,* 5:*358,* 5:*359*–360
Crinoidea, **1:355–366,** 1:*360–361*
behavior, 1:357
conservation status, 1:359
distribution, 1:356
evolution, 1:*12,* 1:355
feeding ecology, 1:357–358
habitats, 1:357
humans and, 1:359
physical characteristics, 1:355–*356*
reproduction, 1:358–359
species of, 1:*362–365*
taxonomy, 1:355
Crinoids, 2:32, 2:*33*
Crisia eburna. See Joint-tubed bryozoans
Crissal thrashers, 10:466
Cristatellidae, 2:497
Crisulipora occidentalis, 2:*503*
Crithagra imberbis. See Cuckoo finches
Croakers, 5:255
See also Sciaenidae
Croaking gouramies, 5:427, 5:428
Crocias langbianis. See Gray-crowned crocias
Crocidura spp., 13:265, 13:266–267, 13:268
Crocidura attenuata. See Gray shrews
Crocidura canariensis. See Canary shrews
Crocidura dsinezumi. See Dsinezumi shrews
Crocidura fuliginosa. See Southeast Asian shrews
Crocidura horsfieldii. See Horsfield's shrews
Crocidura leucodon. See Bicolored shrews
Crocidura monticola. See Sunda shrews
Crocidura negrina. See Negros shrews
Crocidura russula. See Common European white-toothed shrews
Crocidura suaveolens. See Lesser white-toothed shrews
Crocidurinae. *See* White-toothed shrews
Crocodile icefishes. *See* Mackerel icefishes
Crocodile monitors, 7:*364,* 7:*365,* 7:368
Crocodile newts. *See* Mandarin salamanders
Crocodile sharks, 4:131–133, 4:*134,* 4:*139,* 4:141
Crocodile tegus, 7:309, 7:311, 7:*313,* 7:315–*316*
Crocodilefishes, 5:159
Crocodiles, 7:59, 7:157–165, **7:179–188,** 7:*183*
See also Nile crocodiles; Saltwater crocodiles
Crocodilians, **7:157–165**
conservation status, 7:10–11, 7:59, 7:61, 7:164
evolution, 7:18–19, 7:157
farming, 7:48
as food, 7:48
heart, 7:10
humans and, 7:52–53
integumentary system, 7:29
limbs, 7:25
in mythology, 7:55
reproduction, 7:6–7, 7:42, 7:163–164
salt glands, 7:30
skeleton, 7:23–24
skulls, 7:5, 7:26
superstitions and, 7:57
teeth, 7:9, 7:26
territoriality, 7:44
water balance, 7:30
Crocodilurus sp. *See* Crocodile tegus
Crocodilurus lacertinus. See Crocodile tegus
Crocodylidae, **7:179–188,** 7:*183*
See also Crocodiles
Crocodylinae. *See* Crocodiles
Crocodylomorpha, 7:157
Crocodylus acutus. See American crocodiles
Crocodylus cataphractus. See African slender-snouted crocodiles
Crocodylus johnstonii. See Johnstone's crocodiles
Crocodylus niloticus. See Nile crocodiles
Crocodylus novaeguineae. See New Guinea crocodiles
Crocodylus palustris. See Mugger crocodiles
Crocodylus porosus. See Saltwater crocodiles
Crocodylus raninus. See Borneo crocodiles
Crocodylus rhombifer. See Cuban crocodiles
Crocuta crocuta. See Spotted hyenas
Croizat, Leon, 3:53
Cromeria spp., 4:289, 4:290
Crook, J. J., 11:377
Crook-bill plovers. *See* Wrybills
Crop milk, 9:244, 9:251
Cross-fostering, 9:424, 11:110–111
Cross foxes. *See* Red foxes
Cross River Allen's bushbabies, 14:32*t*
Crossarchus spp., 14:347
Crossarchus obscurus. See Western cusimanses
Crossaster papposus, 1:369
Crossbills, 11:323–324, 11:*325,* 11:*326,* 11:336–337
Crossing over (genetics), 1:15
Crossley's babblers, 10:*509,* 10:*520*
Crossodactylodes spp., 6:156
Crossodactylus spp., 6:156
Crossoptilon spp. *See* Eared pheasants
Crossoptilon harmani. See Tibetan eared-pheasants
Crossoptilon mantchuricum. See Brown eared-pheasants
Crotalinae. *See* Pitvipers
Crotalus spp., 7:445, 7:446, 7:448
Crotalus adamanteus. See Eastern diamondback rattlesnakes
Crotalus atrox. See Western diamondback rattlesnakes
Crotalus cerastes. See Sidewinding
Crotalus durissus, 7:55, 7:448
Crotalus horridus. See Timber rattlesnakes
Crotalus viridis. See Prairie rattlesnakes
Crotalus willardi, 7:447
Crotaphatrema spp., 6:431, 6:432
Crotaphytinae, 7:243, 7:244
Crotaphytus spp., 7:246
Crotaphytus collaris. See Common collared lizards
Croton bugs. *See* German cockroaches
Crotophaga glandarius. See Great spotted cuckoos
Crotophaga major. See Greater anis
Crotophaga sulcirostris. See Groove-billed anis
Crotophaginae. *See* Anis
Crown jellyfish, 1:25, 1:153, 1:*159,* 1:*161*–162
See also Coronatae
Crown-of-thorns, 1:29, 1:367, 1:369, 1:370, 1:*371,* 1:*376,* 1:377
Crowned babblers, 11:127
Crowned cranes, 9:1, 9:23–27
Crowned eagles, 8:321
Crowned forest frogs, 6:*267,* 6:*270*
Crowned gibbons. *See* Pileated gibbons
Crowned lemurs, 14:*9,* 14:*51,* 14:*54,* 14:*57,* 14:59–60
Crowned pigeons, 9:247
Crowned sandgrouse, 9:235
Crowned thrips, 3:*284,* 3:*286*
Crowned tobies, 4:70
Crows, 8:25, 10:171, **11:503–509,** 11:*511,* **11:519–522**
Crows of the World (Goodwin), 11:504

Crucian carps, 4:*5*, 4:22, 4:*301*, 4:303, 4:*306*, 4:*308*–309
Crustacea, 2:25
behavior, 2:*19*, 2:32, 2:36
feeding ecology, 2:36
as food, 2:41
insects and, 3:10, 3:11
physical characteristics, 2:*18*
reproduction, 2:17, 2:21, 2:23
See also specific crustaceans
Crypseloides senex. See Great dusky swifts
Cryptacanthodes aleutensis. See Dwarf wrymouths
Cryptacanthodes bergi, 5:310, 5:313
Cryptacanthodes giganteus, 5:310
Cryptacanthodes maculatus. See Wrymouths
Cryptacanthodidae. *See* Wrymouths
Cryptic golden moles, 13:215
Cryptobatrachus spp., 6:37, 6:230
Cryptobiosis, 3:37
Cryptobranchidae, **6:343–347,** 6:*344*
distribution, 6:5, 6:*343*, 6:*345*
evolution, 6:13, 6:343
reproduction, 6:29, 6:32, 6:34, 6:345–346
taxonomy, 6:323, 6:343
Cryptobranchoidea, 6:323, 6:343
Cryptobranchus spp., 6:13, 6:34, 6:347
Cryptobranchus alleganiensis. See Hellbenders
Cryptobranchus guiday, 6:343
Cryptobranchus matthewi, 6:343
Cryptobranchus saskatchewanensis, 6:343
Cryptobranchus scheuchzeri, 6:343
Cryptocercidae. *See Cryptocercus* spp.
Cryptocercus spp., 3:150, 3:162
Cryptochiton stelleri. See Gumboot chitons
Cryptochitons, 2:395
Cryptochloris spp., 13:215
Cryptochloris wintoni. See De Winton's golden moles
Cryptochloris zyli. See Van Zyl's golden moles
Cryptocotyle lingua, 1:200
Cryptodira, 7:13–14, 7:65
See also Turtles
Cryptomys spp., 16:339, 16:342–343, 16:344, 16:345
See also African mole-rats
Cryptomys amatus, 16:345
Cryptomys anselli. See Zambian mole-rats
Cryptomys bocagei. See Bocages mole-rats
Cryptomys damarensis. See Damaraland mole-rats; Kalahari mole-rats
Cryptomys darlingi. See Mashona mole-rats
Cryptomys foxi. See Nigerian mole-rats
Cryptomys hottentotus. See Common mole-rats
Cryptomys hottentotus hottentotus. See Common mole-rats
Cryptomys hottentotus pretoriae. See Highveld mole-rats
Cryptomys mechowi. See Giant Zambian mole-rats
Cryptomys ochraceocinereus. See Ochre mole-rats
Cryptomys zechi. See Togo mole-rats
Cryptoprocta spp., 14:347, 14:348
Cryptoprocta ferox. See Fossa
Cryptops hortensis. See Blind scolopenders
Cryptopsaras couesii. See Seadevils
Cryptospiza spp. *See* Crimson-wings
Cryptotermes brevis. See West Indian powderpost drywood termites
Cryptothylax spp., 6:280
Cryptotis spp., 13:198, 13:248, 13:250
Cryptotis brevis. See Tusked frogs
Cryptotis meridensis. See Mérida small-eared shrews
Cryptotis parva. See American least shrews
Cryptotriton spp., 6:392
Crypturellus boucardi. See Slaty-breasted tinamous
Crypturellus cinnamomeus. See Thicket tinamous
Crypturellus columbianus. See Colombian tinamous
Crypturellus kerriae. See Choco tinamous
Crypturellus noctivagus. See Yellow-legged tinamous
Crypturellus saltuarius. See Magdalena tinamous
Crypturellus tataupa. See Tataupa tinamous
Crypturellus transfasciatus. See Pale-browed tinamous
Crypturellus variegatus. See Variegated tinamous
Crystal gobies, 5:375
Crystallogobius linearis. See Crystal gobies
Crystals, fossil insects in, 3:8
Ctenocephalides spp., 3:349
Ctenocephalides canis. See Dog fleas
Ctenocephalides felis felis. See Cat fleas
Ctenochaetus spp., 5:396
Ctenochaetus striatus. See Striped bristletooths
Ctenocheilocaris spp., 2:295
Ctenodactylidae. *See* Gundis
Ctenodactylus spp., 16:311, 16:312, 16:313
Ctenodactylus gundi. See North African gundis
Ctenodactylus vali. See Desert gundis
Ctenoluciidae, 4:335, 4:337
Ctenomyidae. *See* Tuco-tucos
Ctenomys spp. *See* Tuco-tucos
Ctenomys australis. See Southern tuco-tucos
Ctenomys azarae. See Azara's tuco-tucos
Ctenomys boliviensis. See Bolivian tuco-tucos
Ctenomys colburni. See Colburn's tuco-tucos
Ctenomys conoveri, 16:425
Ctenomys emilianus. See Emily's tuco-tucos
Ctenomys latro. See Mottled tuco-tucos
Ctenomys magellanicus. See Magellanic tuco-tucos
Ctenomys mattereri, 16:427
Ctenomys mendocinus, 16:427
Ctenomys nattereri. See Natterer's tuco-tucos
Ctenomys opimus, 16:427
Ctenomys pearsoni. See Pearson's tuco-tucos
Ctenomys perrensis. See Goya tuco-tucos
Ctenomys pundti, 16:425
Ctenomys rionegrensis. See Rio Negro tuco-tucos
Ctenomys saltarius. See Salta tuco-tucos
Ctenomys sociabilis. See Social tuco-tucos
Ctenomys talarum. See Talas tuco-tucos
Ctenomys torquatus. See Collared tuco-tucos
Ctenopharyngodon idellus. See Grass carps
Ctenophora. *See* Comb jellies
Ctenophorus inermis. See Central netted dragons
Ctenophorus isolepis. See Military dragons
Ctenophthalmidae, 3:347, 3:349
Ctenopoma spp., 5:427, 5:428
Ctenopoma multispinis, 5:428
Ctenops spp., 5:427, 5:428, 5:433
Ctenosaura hemilopha. See Cape spinytail iguanas
Ctenostomata, 2:509
Ctenotus spp., 7:329
Ctenotus quattuordecimlineatus. See Fourteen-lined comb eared skinks
Cubacubana spelaea, 3:*121*, 3:*123*
Cuban boas, 7:*412*, 7:*413*, 7:415
Cuban crocodiles, 7:160
Cuban crows, 11:508
Cuban emeralds, 9:450
Cuban flickers. *See* Northern flickers
Cuban flower bats, 13:433*t*
Cuban flycatchers. *See* Giant kingbirds
Cuban funnel-eared bats. *See* Small-footed funnel-eared bats
Cuban gars, 4:*223*, 4:*224–225*
Cuban ground boas, 7:*434*–435
Cuban hutias, 16:461, 16:*462*, 16:463, 16:*464*, 16:*465*
Cuban kites, 8:323
Cuban marsh wrens. *See* Zapata wrens
Cuban night lizards, 7:291, 7:*293*, 7:*294*, 7:295
Cuban palm crows, 11:508
Cuban red-winged blackbirds. *See* Red-winged blackbirds
Cuban solenodons, 13:237, 13:*238*
Cuban sparrows, 11:266
Cuban todies, 10:*26*, 10:*29*
Cuban treefrogs, 6:*231*, 6:*238*
Cuban trogons, 9:*480*, 9:*483*
Cubanichthys spp., 5:91
Cubirea spp., 2:318, 2:320
Cubitermes spp., 3:167
Cubomedusae. *See* Box jellies
Cubozoa. *See* Box jellies
Cuckoo bees, 3:406
Cuckoo finches, 11:375–376, 11:378, 11:*381*, 11:393–*394*
Cuckoo rollers. *See* Courols
Cuckoo-shrikes, **10:385–394,** 10:*388*
Cuckoo wasps, 3:*410*, 3:413–*414*
Cuckoos, **9:311–330,** 9:315, 9:*316–317*, 9:443
behavior, 9:312
conservation status, 9:315
distribution, 9:*311*, 9:312
evolution, 9:311
feeding ecology, 9:313
habitats, 9:312
humans and, 9:315
physical characteristics, 9:311–312
reproduction, 9:*313*–315
songs, 8:19, 8:40–41
species of, 9:*318–329*
taxonomy, 9:311
Cuckoo's spittles. *See* Spittle bugs
Cucujoidea, 3:316
Cuculidae. *See* Cuckoos
Cuculinae. *See* Old World cuckoos
Cuculus americanus. See Yellow-billed cuckoos
Cuculus audeberti. See Thick-billed cuckoos
Cuculus basalis. See Horsfield's bronze-cuckoos
Cuculus caeruleus. See Blue couas
Cuculus canorus. See Cuckoos; Single nesting cuckoos
Cuculus caprius. See Dideric cuckoos
Cuculus glandarius. See Great spotted cuckoos
Cuculus lugubris. See Asian drongo-cuckoos
Cuculus pallidus. See Pallid cuckoos

Cuculus paradiseus. See Greater racket-tailed drongos
Cuculus scolopaceus. See Common koels
Cuculus solitarius. See Red-chested cuckoos
Cuculus sonneratii. See Banded bay cuckoos
Cuculus taitensis. See Long-tailed koels
Cuculus varius. See Common hawk-cuckoos
Cucumberfishes, 4:391
Cuis, 16:391–392, 16:*394*, 16:*395*, 16:396
Culeolus spp., 1:454
Culeolus likae, 1:*457*, 1:*458*, 1:464–465
Culicicapa spp., 11:84, 11:105
Culicicapa ceylonensis. See Gray-headed flycatchers
Culpeos, 14:265, 14:284*t*
Culter spp., 4:298
Culter alburnus. See Upper mouths
Cultrinae, 4:297
Cultural entomology, 3:74
Cultural pest controls, 3:77
Culture, 12:157–159
  bird significance, 8:19
  bird songs and, 8:19–20, 8:42
  See also Behavior; Literature
Cultus cods. *See* Lingcods
Cumaceans, **2:229–233,** 2:*231*
Cuon alpinus. See Dholes
Cuora amboinensis. See Malayan box turtles
Cuora trifasciata. See Chinese three-striped box turtles
Cup-nesting caciques, 11:301
Cupedid beetles, 3:*324*, 3:*328–329*
Cupesid beetles. *See* Cupedid beetles
Curaeus curaeus. See Austral blackbirds
Curaeus forbesi. See Forbes's blackbirds
Curassows, **8:413–423**
  behavior, 8:414–415
  conservation status, 8:416
  distribution, 8:*413*, 8:414
  evolution, 8:413
  feeding ecology, 8:415–416
  habitats, 8:414
  humans and, 8:416
  physical characteristics, 8:413–414
  reproduction, 8:416
  species of, 8:*421*–423
  taxonomy, 8:416
Curculio nucum. See Nut weevils
Curculionidae, 3:322
Curimatidae, 4:335, 4:337
Curio, Eberhard, 9:250
Curl-crested manucodes, 11:489
Curlews, 9:175–178, 9:180
Curly-crested helmet-shrikes. *See* White helmet-shrikes
Curly-headed pelicans. *See* Dalmatian pelicans
Currawongs, 11:86, 11:467–469, 11:*470*, 11:474–475
Cursorial locomotion. *See* Running
Cursorius spp., 9:152
Cursorius africanus. See Double-banded coursers
Cursorius cursor. See Cream-colored coursers
Cursorius rufus. See Burchell's coursers
Curtis, Edward, 2:471
Cururu lesser escuerzo, 6:*160*, 6:169
Curve-billed honeycreepers, 11:343
Cuscomys spp., 16:443
Cuscomys ashaninka. See Ashaninka rats
Cuscomys oblativa, 16:443
Cuscuses, **13:57–67,** 13:*62*, 13:*66t*
  behavior, 13:35, 13:60
  conservation status, 13:39, 13:61
  distribution, 13:34, 13:59
  evolution, 13:31, 13:32, 13:57–58
  feeding ecology, 13:60
  habitats, 12:137, 13:34–35, 13:59–60
  humans and, 13:62
  physical characteristics, 13:33, 13:58
  reproduction, 13:39, 13:60–61
  species of, 13:64–65, 13:66*t*
  taxonomy, 13:31, 13:32, 13:57–58
Cushion stars, 1:*371*, 1:*372*, 1:377–378, 1:379
  conservation status, 1:370
  distribution, 1:368, 1:*373*, 1:*374*
  physical characteristics, 1:367
  reproduction, 1:370
Cusimanses, 14:349, 14:350
Cusk-eels, 4:49, **5:15–18,** 5:*16*, 5:*19*, **5:21–22**
Cusks. *See* Tusks
Cuspidariidae, 2:453
Cut-throat finches, 11:372
Cuticles, 3:17–19, 3:*26*, 3:*27*
  See also Molting; Physical characteristics
Cutlass fishes. *See* Banded knifefishes
Cuttlefishes, **2:475–481,** 2:*481*, **2:486–487**
  behavior, 2:37, 2:477–478
  conservation status, 2:479
  distribution, 2:477
  evolution, 2:475
  feeding ecology, 2:478
  habitats, 2:477
  humans and, 2:479–480
  physical characteristics, 2:475–477
  reproduction, 2:478–479
  species of, 2:486–*487*
  taxonomy, 2:475
Cutworms, 3:389
Cuvier, Georges, 5:163, 5:427
Cuvier's beaked whales, 15:*60*, 15:61, 15:62, 15:*63*, 15:*65*, 15:66
Cuvier's dwarf caimans, 7:157, 7:171
Cuvier's pike characins. *See* Golden pike characins
Cuvier's toucans. *See* White-throated toucans
Cyamids, 2:261
Cyamus scammoni. See Gray whale lice
Cyanea spp., 1:157
Cyanea capillata. See Lion's mane jellyfish
Cyanocitta spp., 11:504
Cyanocitta cristata. See Blue jays
Cyanocitta stelleri. See Steller's jays
Cyanocola spp., 11:504
Cyanocorax caeruleus. See Azure jays
Cyanocorax chrysops. See Plush-crested jays
Cyanocorax morio. See Brown jays
Cyanolanius madagascarinus. See Electrified blue vangas
Cyanoliseus patagonus. See Burrowing parakeets
Cyanolyca spp., 11:505
Cyanomitra spp. *See* Sunbirds
Cyanomitra obscura. See Western olive sunbirds
Cyanomitra verticalis. See Green-headed sunbirds
Cyanopica cyana. See Azure-winged magpies
Cyanoramphus cookii. See Norfolk Island parakeets
Cyanthus latirostris. See Broad-billed hummingbirds
Cyathocephalus truncatus, 1:230
Cyathopharynx furcifer, 5:*278*
Cycad thrips, Australian, 3:*284*, 3:*286*, 3:287
Cycadothrips albrechti. See Australian cycad thrips
Cyclanorbinae, 7:151
Cyclarhidinae. *See* Peppershrikes
Cyclarhis spp. *See* Peppershrikes
Cyclarhis gujanensis. See Rufous-browed peppershrikes
Cyclaspis longicaudata, 2:*231*, 2:*232*
Cycleptinae, 4:321
Cyclestheria hislopi, 2:*149*, 2:*150*
Cyclestheriidae, 2:147, 2:148
Cyclichthys orbicularis. See Birdbeak burrfishes
Cycliophora. *See* Wheel wearers
Cyclodomorphus branchialis, 7:*331*, 7:*333*–334
Cyclomya, 2:387
Cycloneuralia, 1:9, 1:12–13
Cyclopes spp., 13:151
Cyclopes didactylus. See Silky anteaters
Cyclophoroidea, 2:445, 2:446
Cyclophyllidea, 1:226, 1:227–228, 1:229, 1:230, 1:231
Cyclopidae, 2:299
Cyclopoida, 2:299
Cyclops spp., 2:*300*
Cyclops viridis, 2:*299*
Cyclopteridae. *See* Lumpfishes
Cyclopterus lumpus. See Lumpfishes
Cycloramphinae, 6:155–156, 6:158
Cycloramphus spp., 6:156, 6:157
Cycloramphus culeus. See Titicaca water frogs
Cyclorana spp., 6:155, 6:227, 6:228–229
Cyclorana platycephala. See Water-holding frogs
Cycloraninae, 6:139
Cyclorhagida, 1:275, 1:276, 1:277
Cyclorrhapha, 3:357, 3:358, 3:359, 3:360
Cyclorrhynchus spp. *See* Auks
Cyclosalpa affinis, 1:*469*, 1:472
Cyclostomata. *See* Stenolaemata
Cyclothone spp., 4:424
Cyclothone braueri. See Brauer's bristlemouths
Cyclotyphlops spp. *See Cyclotyphlops deharvengi*
Cyclotyphlops deharvengi, 7:380, 7:381
Cyclura spp. *See* Caribbean land iguanas
Cyclura carinata, 7:245
Cyclura carinata carinata. See Turks and Caicos iguanas
Cyclura collei. See Jamaican iguanas
Cyclura cornuta cornuta. See Rhinoceros iguanas
Cyclurus spp., 4:230
Cydippida, 1:169
Cyema atrum. See Bobtail snipe eels
Cyematidae, 4:271, 4:272
Cyematoidei, 4:271
Cygnus atratus. See Black swans
Cygnus buccinator. See Trumpeter swans
Cygnus olor. See Mute swans
Cylicobdellidae, 2:76
Cylindrachetidae. *See* False mole crickets
Cylindrophis aruensis, 7:396
Cylindrophis engkariensis, 7:395–396
Cylindrophis lineatus, 7:395
Cylindrophis maculatus. See Blotched pipe snakes
Cylindrophis ruffus. See Red-tailed pipe snakes
Cymatogaster aggregata. See Shiner perches
Cymbacephalus beauforti. See Crocodilefishes
Cymbilaimus lineatus. See Fasciated antshrikes

Cymbirhynchus macrorhynchos. *See* Black-and-red broadbills
Cynictis spp., 14:347
Cynictis penicillata. *See* Yellow mongooses
Cynocephalidae. *See* Colugos
Cynocephalus spp. *See* Colugos
Cynocephalus variegatus. *See* Malayan colugos
Cynocephalus volans. *See* Philippine colugos
Cynodontia, 12:10, 12:11
Cynogale bennettii. *See* Otter civets
Cynoglossidae. *See* Tonguefishes
Cynolebias spp., 4:29
Cynolebias bellottii, 5:90
Cynolebias maculatus, 5:90
Cynolebiatinae, 5:90
Cynomys spp. *See* Prairie dogs
Cynomys gunnisoni. *See* Gunnison's prairie dogs
Cynomys ludovicianus. *See* Black-tailed prairie dogs
Cynomys mexicanus, 16:147
Cynopoecilus spp., 5:93
Cynops spp., 6:371
Cynops pyrrhogaster. *See* Japanese fire-bellied newts
Cynopterus spp. *See* Short-nosed fruit bats
Cynopterus brachyotis. *See* Lesser short-nosed fruit bats
Cynopterus sphinx. *See* Indian fruit bats
Cyomys ludovicianus. *See* Black-tailed prairie dogs
Cyornis caerulatus. *See* Large-billed blue-flycatchers
Cyornis tickelliae. *See* Orange-breasted blue flycatchers
Cyphoderris spp. *See* Grigs
Cyphorhinus spp., 10:527
Cyphorhinus aradus. *See* Song wrens
Cypraeidae, 2:446
Cypress cockroaches, 3:149, 3:150
Cypress trouts. *See* Bowfins
Cyprinella spp., 4:301
Cyprinidae. *See* Minnows
Cypriniformes, **4:297–320,** 4:*303–306*, **4:321–334,** 4:*322–323*
  behavior, 4:298–299, 4:322
  conservation, 4:303, 4:322
  distribution, 4:298, 4:321
  evolution, 4:321
  feeding ecology, 4:299–300, 4:*322*
  habitats, 4:298, 4:321–322
  humans and, 4:303, 4:322
  physical characteristics, 4:297–298, 4:*301*, 4:*302*, 4:321
  reproduction, 4:297, 4:*300*–303, 4:*302*, 4:322
  species of, 4:307–319, 4:325–333
  taxonomy, 4:36, 4:297, 4:321
Cyprininae, 4:297
Cyprinocirrhites polyactis, 5:238–239, 5:240, 5:242
Cyprinodon spp., 5:91, 5:92
Cyprinodon diabolis. *See* Devils Hole pupfishes
Cyprinodon inmemoriam, 5:94
Cyprinodon tularosa, 5:90
Cyprinodon variegatus. *See* Sheepsheads
Cyprinodontidae. *See* Killifishes
Cyprinodontiformes, 4:12, 4:29, 4:35, **5:89–104,** 5:*95*
  behavior, 5:92
  conservation status, 5:94
  distribution, 5:91
  evolution, 5:89–90
  feeding ecology, 5:92
  habitats, 5:91–92
  humans and, 5:94
  physical characteristics, 5:90–91
  reproduction, 5:92–93
  species of, 5:*96*–102
  taxonomy, 5:89–90
Cyprinodontoidei, 5:90, 5:91, 5:92
Cyprinus acutidorsalis. *See* Chinese carps
Cyprinus auratus. *See* Crucian carps
Cyprinus barbus. *See* Barbels
Cyprinus carpio. *See* Common carps
Cyprinus catostomus. *See* Longnose suckers
Cyprinus erythrophthalmus. *See* Rudds
Cyprinus gobio. *See* Gudgeons
Cyprinus leuciscus. *See* Common daces
Cyprinus phoxinus. *See* Eurasian minnows
Cyprinus rerio. *See* Zebrafishes
Cyprinus tinca. *See* Tenches
Cyprus spiny mice, 16:261*t*
Cypseloides cherriei. *See* Spot-fronted swifts
Cypseloides niger. *See* Black swifts
Cypseloides phelpsi. *See* Chestnut-collared swifts
Cypseloides rutilus. *See* Chestnut-collared swifts
Cypseloides senex. *See* Great dusky swifts
Cypseloidinae. *See* Swifts
Cypselus comatus. *See* Whiskered tree swifts
Cypsilurus spp., 5:80
Cypsiurus parvus. *See* African palm swifts
Cyrotreta, 1:10
Cyrtocrinida, 1:355
Cyrtodiopsis dalmanni, 3:*362*, 3:366–367, 3:*373*
Cyst nematodes, 1:36
Cystignathus senegalensis. *See* Bubbling kassina
Cystisoma fabricii, 2:*264*, 2:*269*–270
Cystisoma pellucidum, 2:*264*
Cystodytes spp., 1:451
Cystodytes incrassatus, 1:451
Cystophora cristata. *See* Hooded seals
Cytarabine, 1:49
Cytoplasmic incompatibility, 3:39
Cyttarops spp., 13:355
Cyttomimus spp., 5:123
Cyzicidae, 2:147

D

Dabbling ducks, 8:363, 8:364, 8:369
Dabchicks. *See* Little grebes
Dabelsteen, T., 8:41
Daboia spp., 7:445
Daboia russelii. *See* Russel's vipers
Dacelo. *See* Kookaburras
Dacelo novaeguineae. *See* Laughing kookaburras
Daces. *See* Common daces
Dactylethra mülleri. *See* Müller's plantanna
Dactylochirotida, 1:417–421
Dactylogyrus vastator, 1:*216*, 1:*219*–220
Dactylomys dactylinus. *See* Amazon bamboo rats
Dactylomys peruanus, 16:451
Dactylopodola baltica, 1:270, 1:*271*, 1:*272*, 1:273
Dactylopodolidae, 1:269
Dactylopsila spp., 13:127, 13:128
Dactylopsila megalura. *See* Great-tailed trioks
Dactylopsila palpator. *See* Long-fingered trioks
Dactylopsila tatei. *See* Tate's trioks
Dactylopsila trivirgata. *See* Striped possums
Dactylopsilinae, 13:125, 13:126
Dactyloptena spp., 5:157
Dactyloptena orientalis. *See* Oriental helmet gurnards
Dactylopteridae. *See* Flying gurnards
Dactylopteroidei. *See* Flying gurnards
Dactylopterus spp., 5:157
Dactylopterus volitans. *See* Sea robins
Dactylopus spp. *See* Young fingered dragonets
Dactylopus coccus. *See* Cochineal insects
Dactylopus dactylopus. *See* Young fingered dragonets
Dactyloscopidae, 5:341
Dactyloscopus tridigitatus. *See* Sand stargazers
Daddy longlegs spiders. *See* Long-bodied cellar spiders
Daector spp., 5:41–42
Dagaa, 4:*303*, 4:*313*, 4:316–317, 5:281
Dagestan turs. *See* East Caucasian turs
Dagetella spp., 4:209
Dagger moths, 3:36
Daggertooths, 4:432, 4:433
Dahlella spp., 2:161
Dahlella caldariensis, 2:*163*, 2:*164*
Dairy cattle. *See* Domestic cattle
Dairy goats. *See* Domestic goats
Dalatias licha. *See* Kitefin sharks
Dalatiidae. *See* Kitefin sharks
d'Albertis's ringtail possums, 13:122*t*
  See also Ringtail possums
Dalhousie gobies, 5:375
Dallia spp., 4:379
Dallia pectoralis. *See* Alaska blackfishes
Dall's porpoises, 15:33–*34*, 15:35, 15:36, 15:39*t*
Dall's sheep, 12:*103*, 12:178
Dalmatian pelicans, 8:186, 8:225, 8:226, 8:228, 8:*229*, 8:*230*
Dalmatians, 14:289, 14:292
Dalpiazinidae, 15:13
Dalyellioida, 1:186
Dama dama. *See* Fallow deer
Dama gazelles, 16:48, 16:57*t*
Dama wallabies. *See* Tammar wallabies
Damaliscus hunteri. *See* Hunter's hartebeests
Damaliscus lunatus. *See* Topis
Damaliscus lunatus jimela, 16:33
Damaliscus lunatus korrigum. *See* Korrigums
Damaliscus lunatus lunatus. *See* Tsessebes
Damaliscus pygargus. *See* Blesboks; Bonteboks
Damara mole-rats. *See* Damaraland mole-rats
Damaraland dikdiks. *See* Kirk's dikdiks
Damaraland mole-rats, 16:124, 16:*342*, 16:344–345, 16:*346*, 16:*348*
Damaraland red-billed hornbills. *See* Red-billed hornbills
Damballah-Wed (Loa god), 7:57
Damophila spp. *See* Hummingbirds
Dampwood termites, 3:164
Damselfishes, 4:26, 4:64, **5:293–298,** 5:*299*, 5:*303*, **5:304–306**
  behavior, 4:62, 4:70, 5:294–295
  distribution, 4:*58*, 5:294
  feeding ecology, 4:50, 4:67, 5:295–296
  habitats, 4:42, 4:47, 4:48, 5:294
  physical characteristics, 4:26, 5:293–294
  reproduction, 4:35, 5:296–298
Damselflies, 3:*54*, ***3:133–139,*** 3:*135*
  behavior, 3:63, 3:65, 3:133–134

as bioindicators, 3:82
evolution, 3:8, 3:133
reproduction, 3:*60*, 3:61, 3:62, 3:134
threats to, 3:88
Danaines, 3:388
Danaus spp. *See* Monarch butterflies
Danaus plexippus. *See* Monarch butterflies
Dance-flies, 3:360
Dance language, honeybee, 3:72
Dancing shrimps. *See* Harlequin shrimps
Danh-gbi (African god), 7:55
Daniels, M., 9:275
Danio rerio. *See* Zebrafishes
Danionella translucida, 4:298
Danioninae, 4:297
Danube catfishes. *See* European wels
Daphnia magna, 2:*153*, 2:*154*
Daphnia pulex. *See* Common water fleas
Daphnids, 2:36
Daphoenositta spp. *See* Sittellas
Daphoenositta miranda. *See* Black sittellas
Daptrius spp. *See* Caracaras
Daptrius americanus. *See* Red-throated caracaras
Daptrius ater. *See* Black caracaras
Darevskia spp., 7:297, 7:298
Darevskia clarkorum. *See* Clark's lacerta
Darién pocket gophers, 16:196*t*
Dark Annamite muntjacs. *See* Truong Son muntjacs
Dark-backed weavers, 11:*380*, 11:*385*
Dark-billed cuckoos, 9:312
Dark-capped bulbuls. *See* Common bulbuls
Dark-eyed juncos, 11:*268*, 11:*271*
Dark-handed gibbons. *See* Agile gibbons
Dark kangaroo mice, 16:*201*, 16:209*t*
Dark red boobooks. *See* Southern boobook owls
Dark-throated orioles, 11:428
Dark-throated thrushes, 10:485
Dark-winged trumpeters, 9:78, 9:*80*, 9:*82*
Darker chowchillas, 11:69
Darkling beetles, 3:316, 3:317, 3:319, 3:321
D'Arnaud's barbets, 10:114, 10:*118*, 10:120–*121*
Darters, 4:62, 4:*340*, 4:*342*, 4:347, **5:195–199**, 5:*200*, 5:*203*, **5:206**, 8:183–186
Darwin, Charles, 2:7, 8:42
on animal weapons, 15:268
on armadillos, 13:181
on Bovidae, 16:2
on *Calomys laucha*, 16:267
on domestic pigs, 15:149
on mind and behavior, 12:149, 12:150
on ungulates, 15:143
See also Evolution; Galápagos finches
Darwin stick insects, 3:*225*, 3:*230*–231
Darwin's foxes, 14:273
Darwin's frogs, 6:36, 6:*45*, 6:173, 6:*174*, 6:*175*–176
Darwin's rheas. *See* Lesser rheas
Dascyllus spp., 2:34
Dascyllus aruanus. *See* Humbug damsels
Dascyllus reticulatus, 4:42
Dascyllus trimaculatus. *See* Threespot dascyllus
Dassie rats, 16:121, **16:*329*–332**, 16:*330*
Dasyatidae. *See* Stingrays
Dasyatis americana. *See* Southern stingrays
Dasyatis centroura. *See* Roughtail stingrays
Dasycercus cristicauda. *See* Mulgaras
Dasydytes spp., 1:269
Dasydytidae, 1:269
Dasykaluta rosamondae. *See* Little red kalutas
Dasyornis spp. *See* Bristlebirds
Dasyornis brachypterus. *See* Bristlebirds
Dasyornis broadbenti. *See* Rufous bristlebirds
Dasyornis longirostris. *See* Western bristlebirds
Dasypeltis spp. *See* Egg-eaters
Dasypeltis scabra. *See* Common egg-eaters
Dasypodidae. *See* Armadillos
Dasypodinae, 13:182
Dasypogon diadema, 3:*362*, 3:*364*
Dasyprocta spp., 16:407, 16:*408*
Dasyprocta azarae. *See* Azara's agoutis
Dasyprocta coibae. *See* Coiba Island agoutis
Dasyprocta cristata. *See* Crested agoutis
Dasyprocta fuliginosa. *See* Black agoutis
Dasyprocta guamara. *See* Onnoco agoutis
Dasyprocta kalinowskii. *See* Kalinowski's agoutis
Dasyprocta leporina. *See* Red-rumped agoutis
Dasyprocta mexicana. *See* Mexican black agoutis
Dasyprocta prymnolopha. *See* Black-rumped agoutis
Dasyprocta punctata. *See* Central American agoutis
Dasyprocta ruatanica. *See* Roatán Island agoutis
Dasyprocta variegata. *See* Brown agoutis
Dasyproctidae. *See* Agoutis
Dasypsyllus spp., 3:349
Dasypus spp., 13:152, 13:*183*, 13:185
Dasypus hybridus. *See* Southern long-nosed armadillos
Dasypus kappleri. *See* Great long-nosed armadillos
Dasypus novemcinctus. *See* Nine-banded armadillos
Dasypus pilosus. *See* Hairy long-nosed armadillos
Dasypus sabanicola. *See* Llanos long-nosed armadillos
Dasypus septemcinctus. *See* Seven-banded armadillos
Dasyrhynchidae, 1:226
Dasyuridae. *See* Dasyurids
Dasyurids, 12:114, 12:277–284, **12:287–301**, 12:*292*, 12:*293*
Dasyuroidea. *See* Dasyuromorphia
Dasyuromorphia, 12:137, **12:277–285**
Dasyurus albopunctatus. *See* New Guinean quolls
Dasyurus geoffroii. *See* Chuditches
Dasyurus hallucatus. *See* Northern quolls
Dasyurus maculatus. *See* Spotted-tailed quolls
Dasyurus spartacus. *See* Bronze quolls
Dasyurus viverrinus. *See* Eastern quolls
Data Deficient status, 3:85
See also Conservation status
Daubentonia spp. *See* Aye-ayes
Daubentonia madagascariensis. *See* Aye-ayes
Daubentonia robusta, 14:85, 14:86, 14:88
Daubentoniidae. *See* Aye-ayes
Daubenton's bats, 13:310–311, 13:313, 13:315, 13:500, 13:*504*, 13:*511*, 13:514
Daurian hedgehogs, 13:*208*, 13:211
Daurian pikas, 16:484, 16:488, 16:*492*
Davainea proglottina, 1:*233*, 1:*236*, 1:237
Davaineidae, 1:226, 1:231
David's echymiperas, 13:12
Davy's naked-backed bats, 13:435, 13:*437*, 13:442*t*
Dawn, bird songs and, 8:37
Dawn fruit bats, 13:307, 13:*339*, 13:*344*, 13:345
Dawson River black flies, 3:*363*, 3:*364*, 3:371
Dayherons, 8:233
Dayton, Paul, 1:48
DDT. *See* Pesticides
de Bary, Anton, 1:31, 2:31, 2:33
de Pinna, Mario C. C., 4:351
De Sá, Rafael, 6:301
De Winton's golden moles, 13:217, 13:222*t*
Dead-leaf mantids, 3:*48*, 3:181, 3:*182*, 3:*183*, 3:185
Dead leaf mimeticas, 3:*209*, 3:214–*215*
"Dear enemy" phenomenon, 7:36–37
Death adders, 7:39, **7:483–488**, 7:*489*, **7:493–494**
Death watch beetles, 3:75, 3:319–320
Death's head cockroaches, 3:147, 3:*149*
Death's head hawk moths, 3:*390*, 3:*398*, 3:402–403
Decapoda, 1:42, **2:197–214**, 2:199, 2:*202*, 2:*203*, 2:475–476
behavior, 2:199–200
conservation status, 2:201
distribution, 2:199
evolution, 2:197–198
feeding ecology, 2:200
habitats, 2:199
humans and, 2:201
physical characteristics, 2:198, 2:*199*
reproduction, 2:18, 2:201
species of, 2:*204*–214
taxonomy, 2:197–198
Decapodiformes. *See* Decapoda
Decapterus spp., 4:67, 5:260
Deception, intentional, 12:159–160
See also Cognition
Declarative memory, 12:154
See also Cognition
Declining Amphibian Populations Task Force, 6:60
Decomposition, 2:*26*
Deconychura longicauda. *See* Long-tailed woodcreepers
Decorator urchins. *See* Short-spined sea urchins
Decticus verrucivorus. *See* Wart biter katydids
Deep-pitted poachers. *See* Rockheads
Deep-pitted sea-poachers. *See* Rockheads
Deep-sea fishes, 4:56, 4:392
See also specific fishes
Deep-sea girdle wearers, Japanese, 1:*346*, 1:*348*–349
Deep-sea gulpers. *See* Pelican eels
Deep-sea limpets, **2:435–437**
Deep-sea mysids, 2:215, 2:216
Deep-sea perches. *See* Orange roughies
Deep-sea slickheads. *See* Slickheads
Deep-sea smelts, 4:389, 4:390, 4:391, 4:392
Deep-sea swallowers, 5:331–335
Deep-sea tubeworms. *See* Hydrothermal vent worms
Deep slope and wall habitats, 4:48
See also Habitats
Deep water giant mysids. *See* Giant red mysids
Deep water habitats, 4:56
See also Habitats
Deep water reef corals, 1:*110*, 1:*117*, 1:118
Deepscale dories. *See* Thorny tinselfishes

Deepwater dragonets, 5:365–367, 5:*368*
Deepwater eels. *See* Bobtail snipe eels
Deepwater flatheads, 5:157–159
Deepwater jacks, 5:166
Deepwater scorpionfishes, 5:*167*, 5:*174*, 5:175–176
Deer, 15:145–146, 15:263, 15:*362*, 15:*363*
 antlers, 12:40
 behavior, 15:270, 15:271
 digestive systems, 12:14
 distribution, 12:131, 12:134, 12:136, 12:137
 evolution, 15:137, 15:265–266
 gigantism and, 12:19–20
 Ice Age evolution, 12:22–23
 mate selection, 12:102
 physical characteristics, 15:267, 15:268, 15:269
 population indexes, 12:195
 reproduction, 12:103, 15:143
 sexual dimorphism, 12:99
 See also specific types of deer
Deer flies, 3:361
Deer mice, 12:*91*, 16:124, 16:128, 16:*265*, 16:266
Defassa waterbucks. *See* Waterbucks
Defense mechanisms, 1:29–30, 1:41–42, 2:37–39
 amphibians, 6:*45*, 6:*46*, 6:49–50
 bird songs and, 8:38–40, 8:41
 evolution of, 12:18–19
 insects, 3:64–66
 senses and, 12:84
 See also Behavior
Deforestation, 1:52–53, 8:26, 12:214
 See also Conservation status
Deformities, amphibian, 6:56–57, 6:*59*
Degeeriella spp., 3:251
Degodi larks, 10:346
Degus, 16:*122*, 16:*434*, 16:*435*, 16:436, 16:*437*, 16:*438*, 16:440
 Bridges's, 16:433, 16:440*t*
 moon-toothed, 16:433, 16:440*t*
Deimatidae, 1:419
Deinacrida heteracantha. *See* New Zealand giant wetas
Deinagkistrodon spp., 7:445, 7:448
Deinagkistrodon acutus. *See* Hundred-pace pitvipers
Deinogalerix spp., 13:194
Deinosuchus spp., 7:157, 7:158
Deirochelyinae, 7:105
Deirochelys reticularia. *See* Chicken turtles
DeKay's brown snakes. *See* Brown snakes
Delacour, Jean
 on Anatidae, 8:369
 on babblers, 10:505, 10:507, 10:513
 on bulbuls, 10:395
 on Irenidae, 10:415
 on sunbirds, 11:207
 on white-necked picathartes, 10:522
Delanymys brooksi. *See* Delany's swamp mice
Delany's swamp mice, 16:297*t*
Delaware skippers, 3:*389*
Delayed fertilization, 12:109–110
 See also Reproduction
Delayed implantation, 12:109–110, 12:111
 See also Reproduction
Delcourt's giant geckos, 7:259, 7:262–263
Deleornis. *See* Sunbirds
Deleornis fraseri. *See* Scarlet-tufted sunbirds
Delhi sands flower-loving flies, 3:361
Delichon urbica. *See* House martins
Delitzschala bitterfeldensis, 3:12
Dellichthys spp., 5:355
Delmarva fox squirrels, 16:167
Delomys spp., 16:264
Delphacid treehoppers, 3:*265*, 3:*269*, 3:273
Delphacidae, 3:264
Delphacodes kuscheli. *See* Delphacid treehoppers
Delphinapterus leucas. *See* Belugas
Delphinidae. *See* Dolphins
Delphininae, 15:41
Delphinoidea, 15:81
Delphinus spp., 15:2, 15:43–44
Delphinus delphis. *See* Short-beaked saddleback dolphins
Deltamys kempi. *See* Kemp's grass mice
Delusory parasitosis, 3:77
Demand and consumption economies, 12:223
Demansia spp. *See* Whipsnakes
Demidoff's bushbabies, 14:7, 14:*24*, 14:*25*, 14:*27*, 14:*29*, 14:31
Demodex folliculorum. *See* Demodicids
Demodicids, 2:*337*, 2:*339*–340
Demoiselle cranes, 9:*30*, 9:*31*–32
Demon adders. *See* Rhombic night adders
Demon stingers. *See* Bearded ghouls
Demosponges, **1:77–86,** 1:*78*, 1:*81*
 behavior, 1:79
 conservation status, 1:80
 distribution, 1:79
 evolution, 1:77–78
 feeding ecology, 1:27, 1:80
 habitats, 1:24, 1:79
 humans and, 1:80
 physical characteristics, 1:*78*–79
 reproduction, 1:80
 species of, 1:*82*–86
 taxonomy, 1:77–78
Demospongiae. *See* Demosponges
Dendragapus obscurus. *See* Blue grouse
Dendraster excentricus. *See* Western sand dollars
Dendroaspis spp. *See* Mambas
Dendroaspis polylepis. *See* Black mambas
Dendrobates spp., 6:29, 6:37–38, 6:66, 6:197–200
Dendrobates arboreus, 6:199
Dendrobates auratus. *See* Green poison frogs
Dendrobates betsileo. *See* Betsileo golden frogs
Dendrobates castaneoticus. *See* Brazil nut poison frogs
Dendrobates fantasticus, 6:206
Dendrobates histrionicus, 6:*202*
Dendrobates imitator. *See* Imitating poison frogs
Dendrobates minutus. *See* Blue-bellied poison frogs
Dendrobates mysteriosus, 6:198
Dendrobates pumilio. *See* Strawberry poison frogs
Dendrobates reticulatus. *See* Red-backed poison dart frogs
Dendrobates speciosus, 6:198
Dendrobates vanzolinii, 6:38, 6:199, 6:200, 6:*202*
Dendrobates variabilis, 6:206
Dendrobates ventrimaculatus. *See* Amazonian poison frogs
Dendrobatidae. *See* Poison frogs
Dendrobranchiata, 2:197, 2:198, 2:201
Dendrocephalus spp., 2:136
Dendroceratida, 1:77, 1:79
Dendrochirotida, 1:417–421
Dendrochirus spp., 5:166
Dendrochirus zebra. *See* Zebra turkeyfishes
Dendrocincla anabatina. *See* Tawny-winged woodcreepers
Dendrocincla fuliginosa. *See* Plain-brown woodcreepers
Dendrocincla homochroa. *See* Ruddy woodcreepers
Dendrocitta spp., 11:504
Dendrocitta vagabunda. *See* Rufous treepies
Dendrocolaptidae. *See* Woodcreepers
Dendrocopos dorae. *See* Arabian woodpeckers
Dendrocopos leucotos. *See* White-backed woodpeckers
Dendrocopos major. *See* Great spotted woodpeckers
Dendrocygna eytoni. *See* Plumed whistling ducks
Dendrocygna javanica. *See* Lesser whistling-ducks
Dendrocygna viduata. *See* White-faced whistling ducks
Dendrogale spp., 13:292
Dendrogale melanura. *See* Bornean smooth-tailed tree shrews
Dendrogale murina. *See* Northern smooth-tailed tree shrews
Dendrohyrax spp., 15:178, 15:183
Dendrohyrax arboreus. *See* Southern tree hyraxes
Dendrohyrax dorsalis. *See* Western tree hyraxes
Dendrohyrax validus. *See* Eastern tree hyraxes
Dendroica spp., 11:287
Dendroica angelae. *See* Elfin-wood warblers
Dendroica caerulescens. *See* Black-throated blue warblers
Dendroica castanea. *See* Bay-breasted warblers
Dendroica cerulea. *See* Cerulean warblers
Dendroica coronata. *See* Yellow-rumped warblers
Dendroica discolor. *See* Prairie warblers
Dendroica graciae. *See* Grace's warblers
Dendroica kirtlandii. *See* Kirtland's warblers
Dendroica occidentalis. *See* Hermit warblers
Dendroica pensylvanica. *See* Chestnut-sided warblers
Dendroica petichia. *See* Yellow warblers
Dendroica striata. *See* Blackpoll warblers
Dendroica townsendi. *See* Townsend's warblers
Dendrolagus spp., 13:83, 13:85, 13:86
Dendrolagus bennettianus. *See* Bennett's tree kangaroos
Dendrolagus dorianus. *See* Doria's tree kangaroos
Dendrolagus goodfellowi. *See* Goodfellow's tree kangaroos
Dendrolagus lumholtzi. *See* Lumholtz's tree kangaroos
Dendrolagus matschiei. *See* Matschie's tree kangaroos
Dendrolagus mbaiso. *See* Forbidden tree kangaroos
Dendrolagus scottae. *See* Black tree kangaroos
Dendromurinae, 16:282
Dendromus melanotis. *See* Gray climbing mice
Dendromus vernayi, 16:291
Dendronanthus indicus. *See* Forest wagtails

Dendronephthya hemprichi. See Red soft tree corals
Dendrophryniscus spp., 6:184
Dendropicos goertae. See Gray woodpeckers
Dendrortyx barbatus. See Bearded wood-partridges
Dendrortyx macroura. See Long-tailed wood-partridges
Dendroscansor decurvirostris, 10:203
Dendrotriton spp., 6:392
Denham's bustards, 9:*92*, 9:93
Denisonia maculata, 7:486
Dense beaked whales. *See* Blainville's beaked whales
Density estimates, 12:195, 12:196–198, 12:200–202
Dental formulas, 12:46
See also Teeth
Dentaliida, 2:469, 2:470
Dentalium priseum, 2:*469*
Dentalium sexangulare, 2:*469*
Dentalium striatum, 2:*469*
Dentary-squamosal joint, 12:9
Dentatherina merceri. See Pygmy silversides
Dentatherinidae. *See* Pygmy silversides
Denticipitidae. *See* Denticle herrings
Denticipitoidei, 4:277
Denticle herrings, 4:57, 4:277
Dentition. *See* Teeth
Dent's horseshoe bats, 13:*394*, 13:*398*
Deoterthridae, 2:284
Deposit feeding, 2:29, 2:36
See also Feeding ecology
Deppe's squirrels, 16:175*t*
Depth perception, 12:79
See also Vision
Derichthydae, 4:255
Derilissus spp., 5:355
Derived similarities, 12:28
Dermacentor andersoni. See Rocky Mountain wood ticks
Dermaptera. *See* Earwigs
Dermatemys mawii. See Central American river turtles
Dermis, 12:36
See also Physical characteristics
Dermochelys coriacea. See Leatherback seaturtles
Dermogenys spp., 5:81
Dermogenys pusillus, 5:81
Dermophis spp., 6:437, 6:439
Dermophis mexicanus. See Mexican caecilians
Dermoptera. *See* Colugos
Derocheilocaris spp., 2:295
Derocheilocaris typicus, 2:*295*, 2:*297*
Derodontids, 3:317
Deroplatyinae, 3:179
Deroplatys spp. *See* Dead-leaf mantids
Deroplatys lobata. See Dead-leaf mantids
Deroptyus accipitrinus. See Red-fan parrots
Des Murs's wiretails, 10:*212*, 10:218–*219*
Desert agamas. *See* Spiny agamas
Desert ants, 3:68
Desert bandicoots, 13:2, 13:5, 13:11, 13:17*t*
Desert bighorn sheep, 16:92
Desert birds. *See* Gibberbirds
Desert camel spiders. *See* Egyptian giant solpugids
Desert cavies. *See* Mountain cavies
Desert chats. *See* Gibberbirds
Desert cottontails, 16:*507*, 16:*510*, 16:*514*
Desert darkling beetles, 3:317, 3:319
Desert dormice, 16:317, 16:318, 16:320, 16:*322*, 16:*324*, 16:325
Desert gobies, 5:375
Desert golden moles. *See* Grant's desert golden moles
Desert grasshoppers, 3:201, 3:202, 3:203
Desert grassland whiptails, 7:312, 7:*313*, 7:*314*, 7:315
Desert gundis, 16:312, 16:313, 16:*314*, 16:315
Desert horned vipers. *See* Horned vipers
Desert iguanas, 7:36, 7:*248*, 7:*252*
Desert kangaroo rats, 16:200, 16:*201*–202, 16:*203*
Desert larks, 10:345
Desert lions, 14:372
Desert lizards. *See* Desert iguanas
Desert locusts, 3:63, 3:66, 3:204, 3:208
Desert lynx. *See* Caracals
Desert night lizards, 7:291, 7:*293*, 7:*294*
Desert oryx, 16:30
Desert plated lizards, 7:319–321
Desert pocket gophers, 16:195*t*
Desert pocket mice, 16:199, 16:200
Desert rain frogs, 6:305
Desert shrews, 13:*248*, 13:*254*, 13:259–*260*
Desert tortoises, 7:*145*, 7:*146*, 7:147
Desert warthogs. *See* Warthogs
Desert whales. *See* Gray whales
Desert whip scorpions. *See* Giant whip scorpions
Deserts, 12:20–21, 12:113, 12:*117*
See also Habitats
Desmana spp., 13:279
Desmana moschata. See Russian desmans
Desmaninae. *See* Aquatic desmans
Desmans, **13:279–288**, 13:*283*, 13:287*t*
aquatic, 12:14, 13:196–199, 13:279, 13:280
behavior, 13:198
distribution, 12:134
feeding ecology, 13:198–199
habitats, 13:197
physical characteristics, 13:195–196
taxonomy, 13:193, 13:194
Desmarest's spiny pocket mice, 16:*204*, 16:*206*
Desmatolagus gobiensis, 16:480
Desmatosuchus spp., 7:157
Desmodema spp., 4:447
Desmodontinae, 13:413, 13:415
Desmodus rotundus. See Vampire bats
Desmognathinae, 6:389–390
Desmognathus spp., 6:34, 6:35, 6:44–45, 6:389
Desmognathus aeneus, 6:34, 6:35, 6:45
Desmognathus fuscus. See Dusky salamanders
Desmognathus marmoratus, 6:35
Desmognathus wrighti, 6:45
Desmoscolex squamosus, 1:*286*, 1:287
Desmostylia, 15:131, 15:133
Determinate cleavage, 1:19
See also Reproduction; Spiral cleavage
Determination, cellular, 2:19
Detritivores, 2:26, 3:64, 3:166, 4:50
Deuterostomes
defined, 2:3–6
evolution, 2:13–14
reproduction, 2:15–16, 2:19–20
Deuterostomia, 1:6, 1:*8*, 1:9–10, 1:13, 1:19, 1:*25*
Deutsche Killifisch Gemeinschaft, 5:94
Development. *See* Life histories; Reproduction
Devil-fishes (whale). *See* Gray whales
Devilfishes (Cephalopoda). *See* Common octopods; Giant Pacific octopods
Devil's coach-horses, 3:*325*, 3:*332*, 3:333
Devil's darning needles. *See* Odonata
Devils Hole pupfishes, 5:*95*, 5:*97*–98
Devonobiomorpha, 2:353
Devonohexapodus bocksbergensis, 3:10–11
Dexteria spp., 2:136
Dexteria floridana, 2:137
Dhabb lizards, as food, 7:48
Dholes, 14:258, 14:265, 14:266, 14:268, 14:270, 14:273, 14:283*t*
Diacodexeidae, 15:263
Diacodexis spp., 15:263, 15:264
Diactor bilineatus. See Leaf-footed bugs
Diadectomorphs, 6:10
Diadem roundleaf bats, 13:*402*, 13:*406*, 13:407–*408*
Diadema spp. *See* Long-spined sea urchins
Diadema antillarum, 1:401
Diadema savignyi. See Long-spined sea urchins
Diadema steosum, 1:402
Diademed sifakas, 14:63, 14:*66*, 14:67
Diademichthyini, 5:355, 5:356
Diademichthys spp. *See* Urchin clingfishes
Diademichthys lineatus. See Urchin clingfishes
Diadophis spp. *See* Ring-necked snakes
Diadophis punctatus. See Ring-necked snakes
Diaemus youngii. See White-winged vampires
Diamond, J., 15:146, 15:151
Diamond-backed watersnakes, 7:470
Diamond birds. *See* Pardalotes
Diamond dories. *See* Thorny tinselfishes; Tinselfishes
Diamond doves, 9:*254*, 9:*259*
Diamond filefishes, 5:469
Diamond firetails, 11:356, 11:*359*, 11:*366*–367
Diamond girdle wearers. *See* American diamond girdle wearers
Diamond gouramies. *See* Pearl gouramies
Diamond Java sparrows. *See* Diamond firetails
Diamond pythons, 7:420–421
Diamond sparrows. *See* Diamond firetails
Diamondback rattlesnakes
eastern, 7:40, 7:50, 7:475
western, 7:38, 7:40, 7:42, 7:44, 7:50, 7:*197*
Diamondback terrapins, 7:107, 7:*108*, 7:109–*110*
Diamondfishes. *See* Monos
Diamphipnoidae, 3:141
Diana monkeys, 14:193, 14:204*t*
Dian's tarsiers, 14:93, 14:95, 14:*96*, 14:*97*, 14:98–99
Diapause, 3:36–37, 3:49–50, 12:109
Diaphanopterodea, 3:52
Diapheromera arizonensis, 3:*224*
Diapheromera femorata. See Common American walkingsticks
Diapheromeridae, 3:221, 3:222
Diaphorus spp., 3:360
Diaphragm, 12:8, 12:36, 12:45–46
See also Physical characteristics
Diaphus spp., 4:442, 4:443
Diapsida, 7:*5*, 7:15–17
Diard's trogons, 9:479
Diarthrognathus spp., 12:11
Diaspididae, 3:264

Diastylidae, 2:229
Diastylis sculpta, 2:*231*, 2:*232*
Dibamidae. *See* Blindskinks
Dibamus spp., 7:271, 7:*272*
Dibamus bourreti, 7:*271*
Dibatags, 16:48, 16:56*t*
Dibblers, 12:288
Dibranchiates. *See* Neocoleoids
Dicaeidae. *See* Flowerpeckers
Dicaeum spp. *See* Flowerpeckers
Dicaeum aeneum. *See* Midget flowerpeckers
Dicaeum agile. *See* Thick-billed flowerpeckers
Dicaeum anthonyi. *See* Flame-crowned flowerpeckers
Dicaeum celebicum. *See* Gray-sided flowerpeckers
Dicaeum chrysorrheum. *See* Yellow-vented flowerpeckers
Dicaeum concolor. *See* Plain flowerpeckers
Dicaeum cruentatum. *See* Scarlet-backed flowerpeckers
Dicaeum everetti. *See* Brown-backed flowerpeckers
Dicaeum geelvinkianum. *See* Red-capped flowerpeckers
Dicaeum haematostictum. *See* Black-belted flowerpeckers
Dicaeum hirundinaceum. *See* Mistletoebirds
Dicaeum ignipectus. *See* Fire-breasted flowerpeckers
Dicaeum proprium. *See* Whiskered flowerpeckers
Dicaeum quadricolor. *See* Cebu flowerpeckers
Dicaeum vincens. *See* White-throated flowerpeckers
Dicamptodon spp., 6:33, 6:34, 6:349
Dicamptodon aterrimus, 6:350
Dicamptodon copei, 6:350
Dicamptodon ensatus. *See* California giant salamanders
Dicamptodon tenebrosus. *See* Coastal giant salamanders
Dicamptodontidae. *See* Pacific giant salamanders
Dicentrarchus labrax. *See* European sea basses; Sea basses
Diceratheres spp., 15:249
Dicerorhinus spp., 15:216, 15:217
Dicerorhinus hemitoechus. *See* Steppe rhinoceroses
Dicerorhinus kirchbergensis. *See* Merck's rhinoceroses
Dicerorhinus sumatrensis. *See* Sumatran rhinoceroses
Dicerorhinus tagicus, 15:249
Diceros spp., 15:216, 15:251
Diceros bicornis. *See* Black rhinoceroses
Diceros bicornis longipes, 15:260
Dicerotinae, 15:249
Dichaeteuridae, 1:269
Dichistiidae, 5:237, 5:244
Dichobunidae, 15:263, 15:264
Dichrostigma flavipes. *See* Yellow-footed snakeflies
Dick bridegroom fishes. *See* Pineconefishes
Dickensonia costata. *See* Flatworms
Dickinsonia, 2:10
Diclidophora merlangi, 1:*217*, 1:*220*, 1:221–222
Diclidurinae, 13:355
Diclidurus spp. *See* Ghost bats
Diclidurus albus. *See* Northern ghost bats
Dicosmoecus gilvipes. *See* October caddisflies
Dicotyles spp., 15:291
Dicrocerus spp., 15:266, 15:343
Dicrocoelium dendriticum. *See* Lancet flukes
Dicrodon spp., 7:309, 7:310
Dicroglossinae, 6:246, 6:248, 6:251
Dicroglossini, 6:246
Dicromatism, sexual, 7:41
See also Reproduction
Dicrostonyx spp., 16:226
Dicrostonyx groenlandicus. *See* Northern collared lemmings
Dicrostonyx torquatus. *See* Arctic lemmings
Dicrostonyx vinogradovi, 16:229
Dicruridae. *See* Drongos
Dicrurus spp. *See* Drongos
Dicrurus adsimilis. *See* Fork-tailed drongos
Dicrurus aldabranus, 11:439
Dicrurus andamanensis, 11:439
Dicrurus bracteatus. *See* Spangled drongos
Dicrurus fuscipennis. *See* Comoro drongos
Dicrurus leucophaeus. *See* Ashy drongos
Dicrurus ludwigii. *See* Square-tailed drongos
Dicrurus macrocercus. *See* Black drongos
Dicrurus megarhynchus. *See* Ribbon-tailed drongos
Dicrurus modestus. *See* Velvet-mantled drongos
Dicrurus paradiseus. *See* Greater racket-tailed drongos
Dicrurus sumatranus. *See* Sumatran drongos
Dicrurus waldenii. *See* Mayotte drongos
The Dictionary of Birds, 11:505
Dictyoceratida, 1:77, 1:79
Dictyodendrillidae, 1:79
Dicyema acuticephalum, 1:*94*, 1:*95*, 1:*96*
Dicyemennea antarcticensis, 1:*95*, 1:*97*
Dicyemidae, 1:21, 1:93
Dicyemodeca deca, 1:*93*, 1:*95*, 1:*96*, 1:97
Dicynodontia, 12:10
Didelphidae. *See* New World opossums
Didelphimorphia. *See* New World opossums
Didelphinae, 12:249
Didelphis spp. *See* Southern opossums
Didelphis aurita, 12:251
Didelphis marsupialis. *See* Southern opossums
Didelphis virginiana. *See* Virginia opossums
Didemnids, 1:454
Didemnum spp., 1:452
Didemnum commune, 1:*456*, 1:462, 1:*463*
Didemnum studeri, 1:*456*, 1:462–*463*
Dideric cuckoos, 9:*317*, 9:*320*, 11:385
Didodinae, 16:211
Didric cuckoos. *See* Dideric cuckoos
Didunculines. *See* Tooth-billed pigeons
Didunculus strigirostris. *See* Tooth-billed pigeons
Didus solitaria. *See* Rodrigues solitaires
Didynamipus spp., 6:184
Diederick cuckoos. *See* Dideric cuckoos
Diederik cuckoos. *See* Dideric cuckoos
Diel migrations, 1:24
See also Behavior
Dieldrin. *See* Pesticides
Diesingia spp., 2:318
Diet. *See* Feeding ecology
Differentiation, cellular, 2:19
Digenea simplex. *See* Red algae
Digenetic trematodes, 1:197–199, 1:200
Digestive system, 12:46–48
avian, 8:10–11
carnivores, 12:120
herbivores, 12:14–15, 12:120–124
insects, 3:19–21, 3:31
lissamphibians, 6:17
reptiles, 7:9–10, 7:26–28
See also Feeding ecology; Physical characteristics
Digger wasps, 3:62, 3:*409*, 3:420–*421*
Digitigrades, 12:*39*
Dijkgraf, Sven, 12:54
Dikdiks, **16:59–72**
behavior, 16:5, 16:60–62
conservation status, 16:63
distribution, 16:*59*–60, 16:71*t*
evolution, 16:59
feeding ecology, 15:142, 15:272, 16:62
habitats, 15:269, 16:60
humans and, 16:64, 16:66–70
physical characteristics, 15:142, 16:59, 16:66–70
reproduction, 16:7, 16:62–63
species of, 16:66–71*t*
taxonomy, 16:59
Dilaridae, 3:305, 3:306, 3:308
Dilepididae, 1:226
Dilododontids, 15:135
Dimorphic fantails, 11:84, 11:*88*, 11:*91*–*92*
Dimorphic rufous fantails. *See* Dimorphic fantails
Dimorphism, 1:18, 3:59, 7:41, 8:54
See also Reproduction
Dimorphognathus spp., 6:247
Dinagat gymnures, 13:207, 13:213*t*
Dinagat moonrats. *See* Dinagat gymnures
Dinaromys spp., 16:228
Dineutus discolor. *See* Whirligig beetles
Dingisos. *See* Forbidden tree kangaroos
Dingos, 12:136, 12:180, 14:257, 14:266, 14:*270*, 14:273, 14:292
Dinilysia spp., 7:16
Dinocerata, 12:11, 15:131
Dinoflagellates, 1:33, 2:11
Dinomyidae. *See* Pacaranas
Dinomys branickii. *See* Pacaranas
Dinopercidae. *See* Cave basses
Dinopithecus spp., 14:189
Dinopium benghalense. *See* Lesser flame-backed woodpeckers
Dinoponera spp., 3:408
Dinoponera quadriceps. *See* Ponerine ants
Dinornis giganteus. *See* Moas
Dinornis novaezealandiae. *See* Moas
Dinornis struthoides. *See* Moas
Dinornithdae. *See* Moas
Dinosauria, 7:19–22
Dinosaurs, mammals and, 12:33
Dioclerus spp., 3:55
Diodon holacanthus, 5:471
Diodon hystrix. *See* Spot-fin porcupinefishes
Diodontidae. *See* Porcupinefishes
Diodorus Siculus, 14:290
Dioecocestidae, 1:226, 1:230
Dioecotaeniidae, 1:226, 1:230
Diomedea spp. *See* Great albatrosses
Diomedea albatrus. *See* Short-tailed albatrosses
Diomedea amsterdamensis. *See* Amsterdam albatrosses
Diomedea antipodensis. *See* Antipodean albatrosses
Diomedea bulleri. *See* Buller's mollymawks

Diomedea cauta. See Shy albatrosses
Diomedea cauta cauta. See White-capped albatrosses
Diomedea cauta eremita. See Chatham mollymawks
Diomedea chlororhynchos. See Yellow-nosed mollymawks
Diomedea chrysostoma. See Gray-headed mollymawks
Diomedea dabbenena. See Tristan albatrosses
Diomedea epomophora. See Southern royal albatrosses
Diomedea epomophora sanfordi. See Northern royal albatrosses
Diomedea exulans. See Wandering albatrosses; Wandering royal albatrosses
Diomedea gibsoni. See Gibson's albatrosses
Diomedea immutabilis. See Laysan albatrosses
Diomedea impavida. See Campbell black-browed mollymawks
Diomedea irrorata. See Waved albatrosses
Diomedea melanophris. See Black-browed mollymawks
Diomedea nigripes. See Black-footed albatrosses
Diomedea platei. See Pacific mollymawks
Diomedea salvini. See Salvin's mollymawks
Diomedeidae. *See* Albatrosses
Diomedes (Greek hero), 8:116
Diopsiulus regressus, 2:365
Diphascon spp., 2:117
Diphascon recamieri, 2:117
Diphylla ecaudata. See Hairy-legged vampire bats
Diphyllidea, 1:225, 1:227
Diphyllobothriidae, 1:225
Diphyllobothrium latum. See Broad fish tapeworms
Diphyodonty, 12:9, 12:89
Diplecogaster spp., 5:355
Diplecogaster bimaculata bimaculata. See Two-spotted clingfishes
Diploblastic phyla, 1:19, 1:20, 2:12–13
  See also specific species
Diplocardia missippiensis, 2:68
Diplocardia riparia. See River worms
Diplocotyle olriki, 1:230
Diplocrepini, 5:355, 5:356
Diplodactylinae, 7:259, 7:261
Diplogale hosei. See Hose's palm civets
Diploglossinae, 7:339, 7:340
Diploglossus spp., 7:339, 7:340
Diploglossus fasciatus, 7:340
Diplogrammus spp., 5:365
Diplomesodon spp., 13:265, 13:266–267
Diplomesodon pulchellum. See Piebald shrews
Diplometopon spp., 7:287
Diplomys caniceps, 16:450–451
Diplomys labilis. See Rufous tree rats
Diplomys rufodorsalis, 16:450
Diplomystidae, 4:352
Diplopoda, 2:374
Diploprionini, 5:255
Diplopterinae, 3:150
Diplosoma spp., 1:455
Diplostomum flexicaudum, 1:200
Diplostraca, 2:147
Diplurans, 3:4, 3:10, 3:11, **3:107–111,** 3:*109*
Dipnoi, 4:12
Dipodidae, 12:134, 16:121, 16:123, **16:211–224,** 16:*218*, 16:*223t–224t*, 16:283
Dipodinae, 16:211, 16:212
Dipodoidea, 16:211
Dipodomyinae, 16:199
Dipodomys spp., 16:124, 16:199, 16:200
Dipodomys deserti. See Desert kangaroo rats
Dipodomys elator. See Texas kangaroo rats
Dipodomys elephantinus. See Elephant-eared kangaroo rats
Dipodomys heermanni. See Heermann's kangaroo rats
Dipodomys heermanni morroensis, 16:203
Dipodomys ingens. See Giant kangaroo rats
Dipodomys merriami, 16:200, 16:203
Dipodomys merriami parvus, 16:203
Dipodomys microps. See Chisel-toothed kangaroo rats
Dipodomys nelsoni. See Nelson's kangaroo rats
Dipodomys nitratoides. See San Joaquin Valley kangaroo rats
Dipodomys nitratoides exilis, 16:203
Dipodomys nitratoides tipton, 16:203
Dipodomys ordii. See Ord's kangaroo rats
Dipodomys spectabilis. See Banner-tailed kangaroo rats
Dipodomys stephensi. See Stephen's kangaroo rats
Diporochaetidae, 2:65
Dippers, 10:173, **10:475–482,** 10:*478*
Diprotodontia, 12:137, **13:31–41,** 13:83
Dipsas spp. *See* Snail-suckers
Dipsas indica. See Amazonian snail-eaters
Dipsocoromorpha, 3:259
Dipsosaurus spp., 7:246
Dipsosaurus dorsalis. See Desert iguanas
Diptera, 3:35, **3:357–374,** 3:*362*, 3:*363*
  behavior, 3:359–360
  circulatory system, 3:24
  conservation status, 3:360–361
  distribution, 3:359
  evolution, 3:357
  feeding ecology, 3:360
  habitats, 3:359
  humans and, 3:361
  physical characteristics, 3:22, 3:358–359
  reproduction, 3:360
  species of, 3:*364*–374
  taxonomy, 3:357
  wing-beat frequencies, 3:23
Diptera spp. *See* Blackflies
Dipteropeltis spp., 2:289
Dipturus spp., 4:176
Dipus sagitta. See Hairy-footed jerboas
Dipylidiidae, 1:226, 1:231
Dire wolves, 14:265
Direct development, 1:19, 2:21, 2:23
  See also Reproduction
Direct exploitation, 12:215
  See also Humans
Direct observations, 12:198–199, 12:200–202
Directional hearing, owls, 9:337
The Directory of Australian Birds (Schodde and Mason), 10:329
Diretmidae. *See* Spinyfins
Dirofilaria immitis. See Canine heartworms
Dischidodactylus spp., 6:156
Dischistiidae. *See* Galjoens
Discina spp., 2:515
Discinidae, 2:515, 2:516
Discinisca spp., 2:515
Discodeles spp., 6:246, 6:248, 6:251
Discodeles opisthodon. See Faro webbed frogs
Discodermia dissolute, 1:45
Discodermolide, 1:45
Discoglossidae, **6:89–94,** 6:*91*
  distribution, 6:5, 6:*89*, 6:*92–94*
  evolution, 6:64, 6:89
  larvae, 6:41
  taxonomy, 6:89, 6:92–94
Discoglossus montalenti, 6:90
Discoglossus nigriventer, 6:90
Discoglossus pictus. See Painted frogs (Discoglossidae)
Discoglossus sardus. See Tyrrhenian painted frogs
Discosura spp. *See* Hummingbirds
Discosura longicauda, 9:440
Discotrema spp. *See* Crinoid clingfishes
Discotrema crinophila. See Crinoid clingfishes
Discradisca spp., 2:515
Disculicipitidae, 1:226
Discus rays. *See* Freshwater stingrays
Diseases, 2:31
  amphibian, 6:57, 6:60
  Argentine hemorrhagic fever, 16:270
  bacterial infections, 3:39
  bird, 8:16, 11:237–238, 11:343, 11:348
  bird carriers, 8:25–26
  Bolivian hemorrhagic fever, 16:270
  calicivirus, 12:187–188, 12:223
  cancer, 1:44–46, 1:49
  Chagas', 2:11, 12:254
  Giarda, 16:127
  Gilchrist's disease, 16:127
  hantaviruses, 16:128, 16:270
  HIV, 12:222
  infectious, 1:34–36, 12:215–216, 12:222–223
  from insects, 3:76
  Lyme disease, 16:127
  myxoma virus, 12:187
  parasitic, 1:34–36, 2:28, 2:33
  from protists, 2:11–12
  rabbit calicivirus disease, 12:187–188, 12:223
  reptilian, 7:61, 7:62
  rickets, 12:74
  tetanus, 12:77–78
  viral resistance, 12:187
  See also Humans
Disk anemones. *See* Elephant ear polyps
Disk-winged bats, 13:311, **13:473–477,** 13:*475*
Disney's Animal Kingdom, 12:204
Dispersal
  distribution and, 12:130–131
  for inbreeding avoidance, 12:110
  See also Distribution
Dispersed social systems, 12:145–148
Dispholidus spp. *See* Boomslangs
Dispholidus typus. See Boomslangs
Disporella hispida, 2:*505*, 2:*506*, 2:507
Disruptive coloration, 12:38
  See also Physical characteristics
Dissorophids, 6:9, 6:11
Dissostichus eleginoides. See Patagonian toothfishes
Dissostichus mawsoni. See Antarctic toothfishes
Dissotrocha aculeata, 1:*263*, 1:*266*, 1:267
Distance sampling, 12:198, 12:200–202
Distaplia cylindrica, 1:454, 1:*456*, 1:*459*, 1:463–464

Distichopora violacea, 1:*127*, 1:*130*, 1:141, 2:*510*
Distoechurus spp., 13:140, 13:141, 13:142
Distoechurus pennatus. *See* Feather-tailed possums
Distoleon perjerus, 3:309
Distribution
amphibians, 6:*4*, 6:5–6
anurans, 6:4–6, 6:63–64
African treefrogs, 6:*279*, 6:281, 6:*285–289*
Amero-Australian treefrogs, 6:5, 6:6, 6:*225*, 6:228, 6:*233–242*
Arthroleptidae, 6:*265*, 6:*268–270*
Asian toadfrogs, 6:*109*, 6:110, 6:*113–116*
Asian treefrogs, 6:6, 6:*291*, 6:292, 6:*295–299*
Australian ground frogs, 6:5–6, 6:*139*, 6:140, 6:*143–145*
Bombinatoridae, 6:5, 6:*83*, 6:*86–88*
brown frogs, 6:63, 6:263
Bufonidae, 6:5, 6:*183*, 6:184, 6:*188–194*
Discoglossidae, 6:5, 6:*89*, 6:*92–94*
Eleutherodactylus spp., 6:6
frogs, 6:4–5, 6:63–64
ghost frogs, 6:*131*, 6:*133*
glass frogs, 6:6, 6:216, 6:*219–223*
leptodactylid frogs, 6:6, 6:*155*, 6:157–158, 6:*162–171*
Madagascan toadlets, 6:*317*–318, 6:*319*–320
Mesoamerican burrowing toads, 6:*95*, 6:96
Myobatrachidae, 6:*147*, 6:148, 6:*151–153*
narrow-mouthed frogs, 6:4, 6:5, 6:6, 6:*301*, 6:303, 6:*308–315*
New Zealand frogs, 6:*69*, 6:*72–74*
parsley frogs, 6:*127*, 6:*129*
Pipidae, 6:6, 6:*99*, 6:100, 6:*103–106*
poison frogs, 6:6, 6:*197*, 6:198, 6:*203–209*
Ruthven's frogs, 6:*211*
Seychelles frogs, 6:*135*, 6:136, 6:*137*
shovel-nosed frogs, 6:*273*
spadefoot toads, 6:*119*, 6:*124–125*
tailed frogs, 6:5, 6:77, 6:*80–81*
three-toed toadlets, 6:*179*, 6:*181–182*
toads, 6:5, 6:63–64
true frogs, 6:4–5, 6:6, 6:*245*, 6:250, 6:*255–263*
vocal sac-brooding frogs, 6:*173*, 6:*175–176*
caecilians, 6:5, 6:412
American tailed, 6:*415*, 6:416, 6:*417–418*
Asian tailed, 6:5, 6:*419*, 6:*422–423*
buried-eye, 6:432, 6:*433*
Kerala, 6:426, 6:*428*
tailless, 6:*435*, 6:*439–440*
salamanders, 6:5, 6:325
amphiumas, 6:*405*, 6:406, 6:*409–410*
Asiatic, 6:5, 6:*335*, 6:336, 6:*339–341*
Caudata, 6:325
Cryptobranchidae, 6:5, 6:*343*, 6:*345*
lungless, 6:5, 6:*389*, 6:392, 6:*395–403*
mole, 6:5, 6:*355*, 6:356, 6:*358–360*
Pacific giant, 6:5, 6:*349*, 6:350, 6:*352*
Proteidae, 6:*377*, 6:378, 6:*381–382*
Salamandridae, 6:325, 6:*363*, 6:364, 6:*370–375*
sirens, 6:5, 6:*327*, 6:328, 6:*331–332*
torrent, 6:5, 6:*385*, 6:*387*
birds
albatrosses, 8:*113*, 8:114, 8:*118–121*
Alcidae, 9:*219*, 9:221, 9:*224–228*
anhingas, 8:*201*, 8:202, 8:*207–209*
Anseriformes, 8:*363*, 8:364
ant thrushes, 10:*239*, 10:240, 10:*245–256*
Apodiformes, 9:417
asities, 10:*187*, 10:188, 10:*190*
Australian chats, 11:*65*, 11:*67–68*
Australian creepers, 11:*133*, 11:134, 11:*137–139*
Australian fairy-wrens, 11:*45*, 11:*48–53*
Australian honeyeaters, 11:*235*, 11:236, 11:*241–253*
Australian robins, 11:*105*, 11:*108–112*
Australian warblers, 11:*55*, 11:*59–63*
babblers, 10:*505*, 10:506, 10:*511–523*
barbets, 10:*113*, 10:114, 10:119–122
barn owls, 9:*335*–356, 9:*340–343*
bee-eaters, 10:*39*
birds of paradise, 11:*489*, 11:490, 11:*495–501*
bitterns, 8:*239*, 8:241
Bombycillidae, 10:447–448, 10:*451–454*
boobies, 8:*211*, 8:212
bowerbirds, 11:*477*, 11:478, 11:*483–488*
bristletails, 3:113, 3:*116*–117
broadbills, 10:*177*, 10:178, 10:*181*–186
bulbuls, 10:*395*, 10:396, 10:402–412
bustards, 9:*91*–92, 9:*96–99*
butterflyfishes, 4:240
buttonquails, 9:*11*, 9:12, 9:*16–21*
caracaras, 8:*347*, 8:348
cassowaries, 8:*75*, 8:76
Charadriidae, 9:*161*, 9:162, 9:*166–172*
chats, 10:*483*, 10:485, 10:*492–499*
chickadees, 11:*155*, 11:156, 11:*160–165*
chowchillas, 11:*69*, 11:70
Ciconiiformes, 8:234
Columbidae, 9:*247*–248, 9:*255–266*
Columbiformes, 9:243
condors, 8:*275*, 8:276–277
Coraciiformes, 10:*1*, 10:3–4
Cordylidae, 7:*319*, 7:320, 7:*324–325*
cormorants, 8:*201*, 8:202, 8:205–206, 8:*207–209*
Corvidae, 11:*503*, 11:504–505, 11:*512–523*
cotingas, 10:*305*, 10:306, 10:*312–322*
crab plovers, 9:*121*
Cracidae, 8:*413*, 8:414
cranes, 9:*23*, 9:24, 9:*31–35*
cuckoo-shrikes, 10:*385*–386, 10:*389–393*
cuckoos, 9:*311*, 9:312, 9:*318–329*
dippers, 10:*475*–476, 10:*479–481*
diving-petrels, 8:*143*, 8:*145–146*
doves, 9:*243*, 9:247–248, 9:*255–266*
drongos, 11:*437*, 11:438, 11:*441–445*
ducks, 8:*369*, 8:370–371
eagles, 8:*317*, 8:319–320
elephant birds, 8:*103*, 8:104
emus, 8:*83*, 8:84
Eupetidae, 11:*75*, 11:76, 11:*78–80*
fairy bluebirds, 10:*415*, 10:416, 10:*423–424*
Falconiformes, 8:*313*, 8:314
falcons, 8:*347*, 8:348
false sunbirds, 10:*187*, 10:*190*–191
fantails, 11:*83*, 11:84–85, 11:*89–94*
finches, 11:*323*, 11:*328–339*
flamingos, 8:*303*, 8:304
flowerpeckers, 11:*189*, 11:*193–198*
frigatebirds, 8:*193*–194, 8:*197–198*
frogmouths, 9:368, 9:*377*, 9:378, 9:*381–385*
Galliformes, 8:*399*, 8:400
gannets, 8:*211*, 8:212
geese, 8:*369*, 8:370–371
Glareolidae, 9:*151*, 9:*155–160*
grebes, 8:*169*, 8:170, 8:*174–180*
Gruiformes, 9:2
guineafowl, 8:*425*, 8:427
hammerheads, 8:261
Hawaiian honeycreepers, 11:*341*–342, 11:*346–351*
hawks, 8:*317*, 8:319–320
hedge sparrows, 10:*459*, 10:*462–464*
herons, 8:*239*, 8:241
hoatzins, 8:*465*, 8:466
honeyguides, 10:*137*, 10:*141–144*
hoopoes, 10:*61–62*
hornbills, 10:*71*, 10:72–73, 10:*78–83*
hummingbirds, 9:*437*, 9:442, 9:*455–467*
ibises, 8:*291–292*
Icteridae, 11:*301*, 11:302, 11:*309–321*
ioras, 10:*415*, 10:416, 10:*419–420*
jacamars, 10:*91*, 10:92, 10:*95–98*, 10:99
jacanas, 9:*107*, 9:108, 9:*112–114*
kagus, 9:*41*
kingfishers, 10:*5*, 10:7–8, 10:*13–23*
kiwis, 8:*89*
lapwings, 9:*167–172*
Laridae, 9:*203*, 9:205, 9:*211–217*
larks, 10:*341*, 10:342–343, 10:*348–354*
leafbirds, 10:*415*, 10:416, 10:*420–423*
limpkins, 9:*37*
logrunners, 11:*69*, 11:70
long-tailed titmice, 11:*141*, 11:*145–146*
loons, 8:*159*, 8:160, 8:163–166
lyrebirds, 10:*329*, 10:330, 10:*334–335*
magpie-shrikes, 11:*467*, 11:468, 11:*471–475*
major populations, 8:*17*–18
mesites, 9:*5*
mimids, 10:*465*, 10:466, 10:*470–473*
moas, 8:*95*, 8:96
monarch flycatchers, 11:*97*, 11:*100–103*
motmots, 10:*31*, 10:32, 10:*35–37*
moundbuilders, 8:*403*, 8:404
mousebirds, 9:*469*
New World blackbirds, 11:*309–321*
New World finches, 11:*263*, 11:*269–282*
New World quails, 8:400, 8:*455*, 8:456
New World vultures, 8:234, 8:*275*, 8:276–277
New World warblers, 11:*285*, 11:288, 11:*293–299*
New Zealand wattle birds, 11:*447*, 11:*449–450*
New Zealand wrens, 10:*203*, 10:204, 10:*206–207*
nightjars, 9:368, 9:*401*, 9:402, 9:*408–413*
nuthatches, 11:*167*, 11:168, 11:*171–174*
oilbirds, 9:*373*, 9:374
Old World flycatchers, 11:*25*, 11:26
Old World warblers, 11:*1*, 11:3, 11:*10–23*
Oriolidae, 11:*427*, 11:428, 11:*431–434*
ostriches, 8:*99*, 8:100
ovenbirds, 10:*209*–210, 10:*214*–215
owlet-nightjars, 9:*387*, 9:388, 9:*391–392*

owls, 9:332, 9:*340–343*, 9:*346*–347, 9:*354–365*
oystercatchers, 9:*125*–126, 9:*130–131*
painted-snipes, 9:*115*, 9:116, 9:*118–119*
palmchats, 10:*455*
pardalotes, 11:*201*, 11:*204–206*
parrots, 9:*275*, 9:276, 9:*283–297*
Passeriformes, 10:172
Pelecaniformes, 8:184
pelicans, 8:184, 8:225
penduline titmice, 11:*147*, 11:*151–152*
penguins, 8:*147*, 8:149, 8:*153–157*
Phasianidae, 8:*433*, 8:434
pheasants, 8:400, 8:*433*, 8:434
Philippine creepers, 11:*183*, 11:*186–187*
Picidae, 10:*147*, 10:148, 10:*154–167*
Piciformes, 10:86
pigeons, 9:243, 9:*247*–248, 9:*255–266*
pipits, 10:*371*, 10:372–373, 10:*378–383*
pittas, 10:*193*, 10:*197–201*
plantcutters, 10:*325*, 10:326, 10:*327–328*
plovers, 9:*166–172*
potoos, 9:*395*, 9:*399–400*
pratincoles, 9:*155–160*
prions, 8:*123*
Procellariidae, 8:*123*, 8:*129–130*
Procellariiformes, 8:*107*
pseudo babblers, 11:*127*, 11:*130*
puffbirds, 10:*101*, 10:102
rails, 9:*45*, 9:47, 9:*57–67*
Raphidae, 9:*269*, 9:270, 9:*272–273*
Recurvirostridae, 9:*133*, 9:134
rheas, 8:*69*
roatelos, 9:*5*
rollers, 10:*51*, 10:52, 10:*55–58*
sandgrouse, 9:*231*, 9:*235–238*
sandpipers, 9:*175*, 9:176–177, 9:*182–188*
screamers, 8:*393*, 8:394
scrub-birds, 10:*337*, 10:*339–340*
secretary birds, 8:*343*, 8:344
seedsnipes, 9:*189*–190, 9:*193–195*
seriemas, 9:*85*, 9:*88–89*
sharpbills, 10:*291*, 10:292
sheathbills, 9:*198*, 9:*200–201*
shoebills, 8:287
shrikes, 10:*425*, 10:426–427, 10:*432*–438
sparrows, 11:*397*–398, 11:*401–405*
spoonbills, 8:*291–292*
storks, 8:265
storm-petrels, 8:*135*, 8:*139–141*
Sturnidae, 11:*407*, 11:408, 11:*414–425*
sunbirds, 11:*207*, 11:208, 11:*213–225*
sunbitterns, 9:*73*
sungrebes, 9:*69*, 9:*71*–72
swallows, 10:*357*, 10:358, 10:*363*–364
swans, 8:*369*, 8:370–371
swifts, 9:*421*, 9:422, 9:*426–431*
tapaculos, 10:*257*, 10:258, 10:*262–267*
terns, 9:*211–217*
thick-knees, 9:*143*–144, 9:*147–148*
thrushes, 10:*483*, 10:485, 10:*499–504*
tinamous, 8:*57*
titmice, 11:*155*, 11:156, 11:*160–165*
todies, 10:*25*, 10:26, 10:29
toucans, 10:*125*, 10:126, 10:130–134
tree swifts, 9:*433*, 9:*435–436*
treecreepers, 11:*177*, 11:*180–181*
trogons, 9:*477*, 9:*481–485*
tropicbirds, 8:*187*, 8:*190–191*
trumpeters, 9:*77*, 9:78, 9:*81–82*
turacos, 9:*299*, 9:300, 9:*304–310*
typical owls, 9:*345*, 9:346–347, 9:*354–365*
tyrant flycatchers, 10:*269*, 10:271, 10:*278–288*
vanga shrikes, 10:*439*, 10:440, 10:*443–444*
Vireonidae, 11:*255*, 11:*258–262*
wagtails, 10:*371*, 10:372–373, 10:*378–383*
weaverfinches, 11:*353*, 11:354, 11:*360–372*
weavers, 11:*375*, 11:376, 11:*382–394*
whistlers, 11:*115*, 11:116, 11:*120–125*
white-eyes, 11:*227*, 11:228, 11:232–233
woodcreepers, 10:*229*, 10:*232–236*
woodhoopoes, 10:*65*, 10:66, 10:68, 10:*68–69*
woodswallows, 11:*459*, 11:*462–464*
wrens, 10:*525*, 10:526–527, 10:*531–538*
fishes, **4:52–59**, 4:*53*, 4:*55t*, 4:*56*
Acanthuroidei, 5:392, 5:*396*–403
Acipenseriformes, 4:213, 4:*217*–220, 4:*219*
Albuliformes, 4:249, 4:*252–253*
angelsharks, 4:162, 4:*164–165*
anglerfishes, 5:47, 5:*53*–56
Anguilliformes, 4:255, 4:*260–268*
Atheriniformes, 5:68, 5:*73*–76
Aulopiformes, 4:432, 4:*436–438*, 4:439
Australian lungfishes, 4:57, 4:*198*
beardfishes, 5:1, 5:*3*
Beloniformes, 5:80, 5:*83–85*
Beryciformes, 5:114–115, 5:*118*–121
bichirs, 4:57, 4:210, 4:*211*
blennies, 5:342, 5:*345*–348
bowfins, 4:57, 4:*229*, 4:230
bullhead sharks, 4:98, 4:*101–102*
butterflyfishes, 4:57
Callionymoidei, 5:366, 5:*369*–371
carps, 4:298, 4:*307–319*
catfishes, 4:57, 4:352–353, 4:*357–362*, 4:363–366
characins, 4:337, 4:*342–349*
chimaeras, 4:91, 4:*94–95*
coelacanths, 4:*191*
Cypriniformes, 4:298, 4:*307–319*, 4:321, 4:*325*–333, 4:*326*
Cyprinodontiformes, 5:91, 5:*96*–102
damselfishes, 4:*58*
dispersal and, 4:58–59
dogfish sharks, 4:151, 4:*155*–157
dories, 5:124, 5:*126*–130
eels, 4:255, 4:*260–268*
electric eels, 4:370, 4:*373–376*
elephantfishes, 4:57, 4:*238*
Elopiformes, 4:243, 4:246–247
Esociformes, 4:380, 4:*383–386*, 4:*384*
flatfishes, 5:452, 5:455–463
freshwater fishes, 4:36–37, 4:56–58, 4:*57*
Gadiformes, 5:26–28, 5:*32*–34, 5:*35*, 5:36–39
gars, 4:221, 4:*224–226*, 4:227
Gasterosteiformes, 5:134, 5:*139*–147
gobies, 5:374, 5:*380*–388
Gobiesocoidei, 5:356, 5:*359*–363
Gonorynchiformes, 4:290, 4:*293*–295
ground sharks, 4:114, 4:*119–128*
Gymnotiformes, 4:370, 4:*373–376*
hagfishes, 4:78, 4:*80–81*
herrings, 4:277, 4:*281–287*
Hexanchiformes, 4:144, 4:*146–147*, 4:148
killifishes, 4:52, 4:57
Labroidei, 5:276–277, 5:*284*–290, 5:294, 5:*301*–306
labyrinth fishes, 5:427–428, 5:*431*–434
lampreys, 4:84, 4:*88–89*
Lampridiformes, 4:449, 4:*452*–454, 4:*453*
lanternfishes, 4:442, 4:*444*–446
lungfishes, 4:202, 4:*205–206*
mackerel sharks, 4:132, 4:*135–141*
marine fishes, 4:52, 4:53–56
minnows, 4:52, 4:58, 4:298, 4:314–315
morays, 4:255, 4:264–266
mudminnows, 4:380, 4:*383*–386, 4:*384*
mullets, 5:59, 5:*62–65*
Ophidiiformes, 5:16, 5:*20–22*
Orectolobiformes, 4:105, 4:*108–109*
Osmeriformes, 4:390–391, 4:*395*–401, 4:*396*, 4:*399–400*
Osteoglossiformes, 4:232, 4:*237–241*
Percoidei, 5:196, 5:*202*–208, 5:211, 5:*214–217*, 5:220, 5:*225*–233, 5:237–238, 5:247–253, 5:258–259, 5:*266*–272
Percopsiformes, 5:5, 5:*8*–12
pikes, 4:58, 4:380, 4:*383*–386, 4:*384*
ragfishes, 5:*352*
Rajiformes, 4:176, 4:*182–188*
rays, 4:176, 4:*182–188*
Saccopharyngiformes, 4:271, 4:*274*–276, 4:*275*
salmons, 4:52, 4:58, 4:405–*406*, 4:*411–413*, 4:*414*–419, 4:*416*, 4:*418*
sawsharks, 4:168, 4:*170–171*
Scombroidei, 5:406, 5:*409*–418
Scorpaeniformes, 5:158, 5:*161*–162, 5:164, 5:*169*–178, 5:180, 5:*186*–193
Scorpaenoidei, 5:164
skates, 4:176, 4:187
snakeheads, 5:437–438, 5:*442–446*
South American knifefishes, 4:370, 4:*373–376*
southern cod-icefishes, 5:322, 5:*325*–328
Stephanoberyciformes, 5:106, 5:*108*–110
stingrays, 4:*52*, 4:176
Stomiiformes, 4:422, 4:*426*–429, 4:*427*, 4:*428*
Stromateoidei, 5:422
sturgeons, 4:58, 4:213, 4:*217–219*
Synbranchiformes, 5:151, 5:*154*–156
Tetraodontiformes, 5:468, 5:*474*–484
toadfishes, 5:41–42, 5:*44–45*
Trachinoidei, 5:333, 5:*337*–339
trouts, 4:58, 4:405
whale sharks, 4:*108*
Zoarcoidei, 5:310, 5:*315–317*
insects, **3:52–58**
book lice, 3:243, 3:*246*–247
caddisflies, 3:375, 3:*379*–381
cockroaches, 3:148, 3:*153*–158
Coleoptera, 3:319, 3:*327*–333
diplurans, 3:107, 3:*110*–111
Diptera, 3:359, 3:*364*–374
earwigs, 3:196, 3:*199*–200
fleas, 3:348–349, 3:*352–354*
Hemiptera, 3:261, 3:*267*–279
Hymenoptera, 3:406, 3:*411*–424
lacewings, 3:306, 3:*311*–314
Lepidoptera, 3:385, 3:*393*–404
mantids, 3:179, 3:*183–186*
Mantophasmatodea, 3:218, 3:*219*
mayflies, 3:125, 3:*129*–130
Mecoptera, 3:342, 3:*345*–346
Megaloptera, 3:290, 3:*293*–294
Odonata, 3:133, 3:*137–138*

  - Old World flycatchers, 11:*30–43*
  - Orthoptera, 3:203, 3:*211–215*
  - Phasmida, 3:222, 3:*228–232*
  - Phthiraptera, 3:250, 3:*254–256*
  - proturans, 3:93, 3:*96–97*
  - rock-crawlers, 3:189, 3:*192*–193
  - snakeflies, 3:297, 3:*300–302*
  - springtails, 3:99, 3:*103–104*
  - stoneflies, 3:142, 3:*145*–146
  - strepsipterans, 3:336, 3:*338*–339
  - termites, 3:163, 3:*170–174*
  - thrips, 3:281, 3:*285*–287
  - Thysanura, 3:119, 3:122–*123*
  - webspinners, 3:233, 3:*236–237*
  - zorapterans, 3:240, 3:*241*
- lower metazoans and lesser deuterostomes
  - acoels, 1:180, 1:*181*–182
  - anoplans, 1:245, 1:*249–250*
  - arrow worms, 1:434, 1:*437–441*
  - box jellies, 1:147, 1:*151–152*
  - calcareous sponges, 1:58, 1:*61*–64
  - cnidarians
    - Anthozoa, 1:105, 1:*111–120*
    - Hydrozoa, 1:127, 1:134–145
  - comb jellies, 1:169, 1:*173*–177
  - demosponges, 1:79, 1:*82*–86
  - echinoderms
    - Crinoidea, 1:356, 1:*361*–364
    - Echinoidea, 1:402, 1:*408*–415
    - Ophiuroidea, 1:387, 1:*393*–398
    - sea cucumbers, 1:420–421, 1:425–431
    - sea daisies, 1:382, 1:*385*–386
    - sea stars, 1:368, 1:*373*–379
  - enoplans, 1:253, 1:*256*–257
  - entoprocts, 1:319, 1:*323*–325
  - flatworms
    - free-living flatworms, 1:187, 1:*190*–195
    - monogeneans, 1:214, 1:*218–223*
    - tapeworms, 1:229, 1:*234*–241
  - gastrotrichs, 1:270, 1:*272*–273
  - girdle wearers, 1:344, 1:*347*–349
  - glass sponges, 1:68, 1:*71*–75
  - gnathostomulids, 1:331, 1:*334*–335
  - hair worms, 1:305, 1:*308*–309
  - hemichordates, 1:444, 1:*447*–449
  - jaw animals, 1:328, 1:*329*
  - jellyfish, 1:154, 1:*160*–167
  - kinorhynchs, 1:276, 1:*279–280*
  - lancelets, 1:487, 1:*490–496*
  - larvaceans, 1:473, 1:476–478
  - nematodes
    - roundworms, 1:284, 1:287–291
    - secernenteans, 1:294, 1:297–304
  - Orthonectida, 1:99, 1:101–*102*
  - placozoans, 1:88
  - priapulans, 1:337, 1:*340*–341
  - rhombozoans, 1:93, 1:*96*–98
  - rotifers, 1:260, 1:*264*–267
  - *Salinella salve*, 1:*91*
  - salps, 1:467, 1:471–472
  - sea squirts, 1:454, 1:*458*–466
  - sorberaceans, 1:480, 1:*482*–483
  - thorny headed worms, 1:312, 1:*314*–315
  - Trematoda, 1:198, 1:*203*–211
  - wheel wearers, 1:*352*
- mammals, **12:129–139**
  - aardvarks, 15:*155*, 15:156
  - agoutis, 16:*407*, 16:*411*–414, 16:415*t*
  - anteaters, 13:*171*, 13:172, 13:*177*–179
  - armadillos, 13:*181*, 13:182–183, 13:*187*–190, 13:190*t*–192*t*
  - Artiodactyla, 12:129, 12:132, 12:136, 15:269
  - aye-ayes, 14:*85*, 14:86
  - baijis, 15:*19*, 15:20
  - bandicoots, 13:2
    - dry-country, 13:*9*, 13:10, 13:*14*–16, 13:16*t*–17*t*
    - rainforest, 13:*9*, 13:10, 13:*14*–17*t*
  - bats, 13:310–311
    - American leaf-nosed, 13:*413*, 13:415, 13:*423*–432, 13:433*t*–434*t*
    - bulldog, 13:*443*, 13:445, 13:*449–450*
    - disk-winged, 13:*473*, 13:*476*–477
    - false vampire, 13:*379*, 13:*384*–385
    - funnel-eared, 13:*459*, 13:460, 13:*463*–465
    - horseshoe, 13:*387*, 13:388–389, 13:*396*–400
    - Kitti's hog-nosed, 13:*367*, 13:368
    - Molossidae, 13:*483*, 13:*490*–493, 13:493*t*–495*t*
    - mouse-tailed, 13:351, 13:*351*, 13:*353*
    - moustached, 13:435, 13:*435*, 13:*440*–441, 13:442*t*
    - mudnest builders, 11:*453*, 11:456–457
    - New Zealand short-tailed, 13:*453*, 13:454, 13:457
    - Old World fruit, 13:*319*, 13:*325*–330, 13:331*t*–332*t*, 13:*333*, 13:334, 13:*340*–347, 13:348*t*–349*t*
    - Old World leaf-nosed, 13:*401*, 13:402, 13:*407*–409, 13:409*t*–410*t*
    - Old World sucker-footed, 13:*479*, 13:480
    - slit-faced, 13:*371*, 13:*375*, 13:376*t*–377*t*
    - smoky, 13:*467*, 13:*470*
    - Vespertilionidae, 13:*497*, 13:498, 13:*506*–514, 13:515*t*–516*t*, 13:*519*, 13:520–521, 13:*524–525*, 13:526*t*
  - bears, 14:*295*, 14:296, 14:*302*–305, 14:306*t*
  - beavers, 16:*177*, 16:179, 16:*183*
  - bilbies, 13:*19*–20
  - botos, 15:*27*, 15:28
  - Bovidae, 16:4
    - Antilopinae, 16:*45*, 16:46, 16:*50*–55, 16:56*t*–57*t*
    - Bovinae, 16:*11*, 16:13, 16:*18*–23, 16:24*t*–25*t*
    - Caprinae, 16:*87*, 16:90–91, 16:*98*–103, 16:103*t*–104*t*
    - duikers, 16:*73*, 16:*80*–84, 16:84*t*–85*t*, 16:85*t*
    - Hippotraginae, 16:*27*, 16:29, 16:*37*–42*t*, 16:43*t*
    - Neotraginae, 16:*59*–60, 16:*66*–71*t*
  - bushbabies, 14:5, 14:*23*, 14:24, 14:*28*–32, 14:32*t*–33*t*
  - Camelidae, 15:*313*, 15:315, 15:*320*–323
  - cane rats, 16:*333*, 16:334, 16:*337–338*
  - Canidae, 14:*265*, 14:267, 14:*276*–283, 14:283*t*–284*t*
  - capybaras, 16:*401*, 16:402
  - Carnivora, 12:129, 12:131, 12:132, 12:136, 14:257
  - cats, 12:131, 12:136, 14:257, 14:*369*, 14:370–371, 14:*379*–389, 14:390*t*–391*t*
  - Caviidae, 16:*389*, 16:390, 16:*395*–398, 16:399*t*
  - Cetacea, 12:129, 12:138, 15:5
    - Balaenidae, 15:5, 15:*107*, 15:109, 15:*115*–117
    - beaked whales, 15:*59*, 15:61, 15:*64*–68, 15:69*t*–70*t*
    - dolphins, 15:*41*, 15:43, 15:*53*–55, 15:56*t*–57*t*
    - Emballonuridae, 13:*355*, 13:*360*–363, 13:363*t*–364*t*
    - franciscana dolphins, 15:*23*, 15:24
    - Ganges and Indus dolphins, 15:*13*, 15:14
    - gray whales, 15:*93*, 15:94–95
    - Monodontidae, 15:*81*, 15:82–83, 15:*90–91*
    - porpoises, 15:*33*, 15:34, 15:*38*–39, 15:39*t*
    - pygmy right whales, 15:*103*, 15:104
    - rorquals, 15:5, 15:*119*, 15:120–121, 15:*127*–130, 15:130*t*
    - sperm whales, 15:*73*, 15:74, 15:*79*–80
  - chevrotains, 15:*325*, 15:327, 15:*331*–333
  - Chinchillidae, 16:*377*, 16:*382*, 16:383, 16:383*t*
  - colugos, 13:*299*, 13:302, 13:*304*
  - coypus, 16:123, 16:*473*, 16:474
  - dasyurids, 12:*287*, 12:288, 12:*294*–298, 12:299*t*–301*t*
  - Dasyuromorphia, 12:279
  - deer
    - Chinese water, 15:*373*, 15:374
    - muntjacs, 15:*343*, 15:344, 15:*349*–354
    - musk, 15:*335*, 15:336, 15:*340*–341
    - New World, 15:*379*, 15:382, 15:*388*–396, 15:396*t*:
    - Old World, 15:*357*, 15:359, 15:*364*–371, 15:371*t*
  - Dipodidae, 16:*211*, 16:212, 16:*219*–222, 16:223*t*–224*t*
  - Diprotodontia, 13:34
  - dormice, 16:*317*–318, 16:*323*–326, 16:327*t*–328*t*
  - duck-billed platypuses, 12:137, 12:230, 12:*243*, 12:244
  - Dugongidae, 15:*199*, 15:*203*
  - echidnas, 12:*235*, 12:236, 12:240*t*
  - elephants, 12:*135*, 15:*161*, 15:166, 15:*174*–175
  - Equidae, 15:*225*, 15:226, 15:*232*–235, 15:236*t*
  - Erinaceidae, 13:*203*, 13:204, 13:*209*–213, 13:212*t*–213*t*
  - giant hutias, 16:469, 16:*469*, 16:471*t*
  - gibbons, 14:*207*, 14:210, 14:*218*–222
  - Giraffidae, 15:*399*, 15:401, 15:*408*
  - great apes, 14:*225*, 14:227, 14:*237*–240
  - gundis, 16:*311*, 16:312, 16:*314*–315
  - Herpestidae, 14:*347*, 14:348–349, 14:*354*–356, 14:357*t*
  - Heteromyidae, 16:*199*, 16:200, 16:*205*–209, 16:208*t*–209*t*
  - hippopotamuses, 15:*301*, 15:306, 15:*310*–311
  - humans, 14:*241*, 14:247
  - hutias, 16:*461*, 16:467*t*
  - Hyaenidae, 14:*359*, 14:362, 14:*364*–366
  - hyraxes, 15:*177*, 15:179, 15:*186*–188, 15:189*t*
  - Indriidae, 14:*63*, 14:65, 14:*69*–72
  - Insectivora, 13:197
  - koalas, 13:*43*, 13:44
  - Lagomorpha, 16:481

lemurs, 12:135–136, 14:5, 14:*47*, 14:49, 14:*55*–60
Cheirogaleidae, 14:*35*, 14:36, 14:*41*–44
sportive, 14:*73*, 14:75, 14:*79–83*
Leporidae, 16:*505*–506, 16:*511*–515, 16:515*t*–516*t*
Lorisidae, 14:*13*, 14:14, 14:*18*–20, 14:21*t*
Macropodidae, 12:137, 13:*83*, 13:84–85, 13:*93*–100, 13:101*t*–102*t*
manakins, 10:*299–303*
manatees, 15:*205*, 15:*211*–212
Megalonchidae, 13:*159*
Megalonychidae, 13:*155*–156
mesites, 9:*8–9*
moles
golden, 13:197, 13:*215*, 13:216, 13:*220–222*, 13:222*t*
marsupial, 13:*25*, 13:26, 13:*28*
monitos del monte, 12:*273*, 12:274
monkeys
Atelidae, 14:*155*, 14:156, 14:*162*–166, 14:166*t*–167*t*
Callitrichidae, 14:*115*, 14:116, 14:*127*–131, 14:132*t*
Cebidae, 14:*101*, 14:102–103, 14:*108*–112
cheek-pouched, 14:*187*, 14:191, 14:*197*–204, 14:204*t*–205*t*
leaf-monkeys, 14:5, 14:*171*, 14:173, 14:*178*–183, 14:184*t*–185*t*
night, 14:*135*, 14:137, 14:141*t*
Pitheciidae, 14:*143*, 14:144–145, 14:*150–152*, 14:153*t*
monotremes, 12:136–137, 12:230
mountain beavers, 16:*131*
Muridae, 12:134, 16:123, 16:*281*, 16:283, 16:*288–295*, 16:296*t*–297*t*
Arvicolinae, 16:*225*, 16:226, 16:*233*–238, 16:237*t*–238*t*
hamsters, 16:*239*–240, 16:*245*–247*t*
Murinae, 16:*249*, 16:250, 16:*255*–260
Sigmodontinae, 16:*263*, 16:265–266, 16:*272*–276, 16:277*t*–278*t*
musky rat-kangaroos, 13:*69*, 13:70
Mustelidae, 14:*319*, 14:*327*–331, 14:332*t*–333*t*
numbats, 12:*303*
octodonts, 16:*433*, 16:434, 16:*438*–440, 16:440*t*
opossums
New World, 12:*249*, 12:250–251, 12:*257–63*, 12:264*t*–265*t*
shrew, 12:*267*, 12:*270*
Otariidae, 14:*393*, 14:394–395, 14:*402*–406
pacaranas, 16:*385*, 16:386
pacas, 16:*417*, 16:418, 16:*423*–424
pangolins, 16:*107*, 16:110, 16:*115*–119
peccaries, 15:*291*, 15:292, 15:*298–299*
Perissodactyla, 15:217
Petauridae, 13:*125*, 13:*131*–133, 13:133*t*
Phalangeridae, 13:*57*, 13:59, 13:*63–66t*
pigs, 15:*275*, 15:276–277, 15:*284*–288, 15:289*t*–290*t*
pikas, 16:*491*, 16:*497*–501, 16:501*t*–502*t*
pocket gophers, 16:*185*, 16:187, 16:*192*–194, 16:195*t*–197*t*
porcupines
New World, 16:*365*, 16:367, 16:*371*–373, 16:374*t*
Old World, 16:*351*, 16:352, 16:*357*–364
possums
feather-tailed, 13:*139*, 13:140, 13:*144*
honey, 13:*135*, 13:136
pygmy, 13:*105*, 13:106, 13:*110*–111
primates, 14:5–6
Procyonidae, 14:*309*, 14:309–310, 14:*313–315*, 14:315*t*–316*t*
pronghorns, 15:*411*, 15:412
Pseudocheiridae, 13:*113*, 13:114, 13:*119*–123, 13:122*t*–123*t*
rat-kangaroos, 13:*73*, 13:74, 13:*78–80*
rats
African mole-rats, 16:*339*, 16:342, 16:*347*–349, 16:349*t*–350*t*
chinchilla, 16:*443*, 16:444, 16:*447*–448
dassie, 16:*329*, 16:330
spiny, 16:*449*, 16:450, 16:*453*–457, 16:458*t*
rhinoceroses, 15:*249*, 15:252, 15:*257*–261
rodents, 16:123
sengis, 16:*517*, 16:519, 16:*524*–530, 16:531*t*
shrews
red-toothed, 13:*247*, 13:248, 13:*255*–263, 13:263*t*
West Indian, 13:*243*, 13:244
white-toothed, 13:*265*, 13:266, 13:*271*–275, 13:276*t*–277*t*
Sirenia, 15:192
solenodons, 13:*237*, 13:238, 13:*240*
springhares, 16:*307*, 16:308
squirrels
flying, 16:*135*, 16:*139–140*, 16:141*t*–142*t*
ground, 12:131, 16:*143*, 16:144, 16:*151*–158, 16:158*t*–160*t*
scaly-tailed, 16:*299*, 16:300, 16:*302–305*
tree, 16:*163*–164, 16:*169*–174*t*, 16:175*t*
Talpidae, 13:*279*, 13:280, 13:*284*–287, 13:287*t*–288*t*
tapirs, 15:*237*, 15:238, 15:*245*–247
tarsiers, 14:*91*, 14:93, 14:*97*–99
tenrecs, 13:*225*, 13:226, 13:*232*–234, 13:234*t*
three-toed tree sloths, 13:*161*, 13:162, 13:*167*–169
tree shrews, 13:*289*, 13:291, 13:*294–296*, 13:297*t*–298*t*
true seals, 12:129, 12:138, 14:*417*–418, 14:*427*–434, 14:435*t*
tuco-tucos, 16:*425*, 16:*429*–430, 16:430*t*–431*t*
Viverridae, 14:*335*, 14:336, 14:*340*–343, 14:344*t*–345*t*
walruses, 12:138, 14:*409*, 14:411–412
wombats, 13:*51*, 13:*55*–56
Xenartha, 13:150
protostomes
amphionids, 2:196
amphipods, 2:261, 2:*265*–271
anaspidaceans, 2:181, 2:*183*
aplacophorans, 2:380, 2:*382–385*
Arachnida, 2:335, 2:*339*–352
articulate lampshells, 2:522, 2:*525*–527
Australo-American sideneck turtles, 7:*81*–83, 7:*82*
bathynellaceans, 2:178, 2:*179*
beard worms, 2:86, 2:*88*
bivalves, 2:452, 2:*457*–466
caenogastropods, 2:446, 2:449
centipedes, 2:354, 2:*358*–362
cephalocarids, 2:131, 2:*133*
Cephalopoda, 2:477, 2:*482*–488
chitons, 2:395, 2:*397*–400
clam shrimps, 2:147, 2:*150*–151
copepods, 2:300, 2:304–309
cumaceans, 2:229, 2:*232*
Decapoda, 2:199, 2:*204*–213
deep-sea limpets, 2:435, 2:*437*
earthworms, 2:66, 2:*70*–73
echiurans, 2:103, 2:*106*–107
fairy shrimps, 2:136, 2:139–*140*
fish lice, 2:289, 2:*292*–293
freshwater bryozoans, 2:497, 2:*500*–502
horseshoe crabs, 2:328, 2:*331*–332
Isopoda, 2:250, 2:*254*–259
krill, 2:185, 2:*189*–192
leeches, 2:76, 2:*80*–82
leptostracans, 2:161, 2:*164*–165
lophogastrids, 2:225, 2:*227*
mantis shrimps, 2:168, 2:*173*–175
marine bryozoans, 2:503–504, 2:*506*–508, 2:509, 2:*513*–514
mictaceans, 2:241, 2:*242*
millipedes, 2:364, 2:*367–370*
monoplacophorans, 2:388, 2:*390*–391
mussel shrimps, 2:312, 2:*314*
mysids, 2:216, 2:*219*–221
mystacocarids, 2:295, 2:*297*
myzostomids, 2:59, 2:*62*
Neritopsina, 2:440, 2:*442*–443
nonarticulate lampshells, 2:515, 2:*518*–519
Onychophora, 2:110, 2:*113*–114
pauropods, 2:375, 2:377
peanut worms, 2:98, 2:100–101
phoronids, 2:491, 2:494, 2:*495*
Polychaeta, 2:46, 2:*50*–56
Pulmonata, 2:413, 2:417–421
remipedes, 2:125, 2:*128*
sea slugs, 2:403, 2:*407*–409
sea spiders, 2:321, 2:*324*–325
spelaeogriphaceans, 2:243, 2:*244*
symphylans, 2:371, 2:*373*
tadpole shrimps, 2:141–142, 2:*145*–146
tanaids, 2:235, 2:*238*–239
tantulocaridans, 2:283–284, 2:*286*
Thecostraca, 2:274, 2:*278–281*
thermosbaenaceans, 2:245, 2:*247*
tongue worms, 2:319
true limpets, 2:423–424, 2:*426*–427
tusk shells, 2:470, 2:*473*–474
Vestimentifera, 2:91–92, 2:*94*
Vetigastropoda, 2:430, 2:*433*–434
water bears, 2:116, 2:120–123
water fleas, 2:153, 2:*157*–158
reptiles, 7:434, 7:*436*–437
African burrowing snakes, 7:*461*, 7:462, 7:*463*
African sideneck turtles, 7:*129*, 7:*132–133*
Afro-American river turtles, 7:*137*, 7:*140*—142
Agamidae, 7:*209*, 7:210, 7:*214*–221
Alligatoridae, 7:*171*, 7:172, 7:*175*–177
American alligators, 7:*175*
Anguidae, 7:*339*, 7:340, 7:*342*–344
Australo-American sideneck turtles, 7:77
big-headed turtles, 7:*135*
blindskinks, 7:*271*, 7:272
blindsnakes, 7:*379*, 7:381, 7:*383*–385
boas, 7:*409*, 7:410–411, 7:*413*–417

Central American river turtles, 7:*99*, 7:100
chameleons, 7:*223*, 7:227–228, 7:*235*–240
colubrids, 7:*465*, 7:467, 7:*473*–481
crocodilians, 7:161–162
Crocodylidae, 7:*179*, 7:*184*–187
early blindsnakes, 7:*369*, 7:370–*371*
Elapidae, 7:*483*, 7:484, 7:*491*–498
false blindsnakes, 7:*387*–*388*
false coral snakes, 7:*399*, 7:400
file snakes, 7:*439*, 7:*441*, 7:*442*
Florida wormlizards, 7:*283*, 7:284
Gekkonidae, 7:*259*, 7:260, 7:*265*–*269*
Geoemydidae, 7:*115*, 7:*118*–119
gharials, 7:*167*, 7:168
Helodermatidae, 7:*353*, 7:*354*
Iguanidae, 7:*243*, 7:244, 7:*250*–*256*
Kinosternidae, 7:*121*, 7:*125*–126
knob-scaled lizards, 7:*347*, 7:348, 7:*351*
Lacertidae, 7:*297*, 7:298, 7:*301*–302
leatherback seaturtles, 7:*101*
Microteiids, 7:*303*, 7:*306*–307
mole-limbed wormlizards, 7:*279*, 7:280, 7:*281*
Neotropical sunbeam snakes, 7:*405*–406
New World pond turtles, 7:*105*, 7:109–*113*
night lizards, 7:*291*, 7:*294*–*295*
pig-nose turtles, 7:*75*
pipe snakes, 7:*395*, 7:*396*
pythons, 7:*419*, 7:420, 7:*424*–428
seaturtles, 7:*85*, 7:*89*–91
shieldtail snakes, 7:*391*, 7:392, 7:*393*
skinks, 7:*327*, 7:328, 7:*332*–337
slender blindsnakes, 7:*373*, 7:374, 7:*376*–*377*
snapping turtles, 7:*93*, 7:*95*–96
softshell turtles, 7:*151*, 7:*154*–155
spade-headed wormlizards, 7:*287*, 7:288, 7:*289*
splitjaw snakes, 7:*429*, 7:430
Squamata, 7:203–204
sunbeam snakes, 7:*401*–402
Teiidae, 7:*309*, 7:310, 7:*314*–*316*
tortoises, 7:*143*, 7:*146*–*148*
Tropidophiidae, 7:*433*, 7:434, 7:*436*–437
tuatara, 7:*189*–*190*
turtles, 7:70
Varanidae, 7:*359*, 7:360, 7:*365*–368
Viperidae, 7:*445*, 7:447, 7:*451*–*459*
wormlizards, 7:*273*, 7:274, 7:*276*
Diterpenoid glycosides, 1:45
Ditrachybothriidae, 1:225
Ditrysian lepidopterans, 3:384
Diurnal birds of prey. *See* Falconiformes
Diurnal mammals, 12:79, 12:80
See also specific species
Divariscintilla yoyo. See Yo-yo clams
Diversity. *See* Biodiversity
Dives dives. See Melodious blackbirds
Diving beetles, 3:318, 3:319, 3:323
Diving mammals, 12:67–68
See also specific diving mammals
Diving-petrels, 8:107–110, **8:143–146**
Diving shrews, 13:*249*
Djakonov's rock-crawlers, 3:*191*, 3:*192*, 3:193
Djarthia murgonensis, 12:277
Djibouti francolins, 8:437
DNA, 2:8, 2:11
phylogenetics and, 12:26–33
in reproduction, 1:15
See also Reproduction
DNA analysis, 8:3, 8:27
See also Taxonomy
Dob lizards. *See* Spiny-tailed agamas
Doberman pinschers, 14:289
Dobsonflies, 2:29, **3:289–295,** 3:*292*, 3:*293*, 3:294
Dobsonia spp. *See* Naked-backed fruit bats
Dobsonia magna. See Naked-backed fruit bats
Dobson's horseshoe bats, 13:*393*
Dobson's long-tongued dawn bats. *See* Dawn fruit bats
Dobson's shrew tenrecs, 13:*231*, 13:*233*
Doctorfishes, 4:*176*
Dodos, 9:241, 9:244–245, 9:247, **9:269–274,** 9:*270*, 9:*272*
Dog-faced fruit bats. *See* Indian fruit bats
Dog-faced watersnakes, 7:465–466, 7:*472*, 7:*473*, 7:476–477
Dog family. *See* Canidae
Dog fleas, 3:*348*, 3:350
Dog heartworms. *See* Canine heartworms
Dog hookworms, 1:*296*, 1:299–300
Dog tapeworms, 1:231, 1:*233*, 1:*238*–239
Dogania subplana. See Malayan softshell turtles
Dogfish sharks, 4:11, **4:151–158,** 4:*154*
Dogfishes. *See* Alaska blackfishes; Central mudminnows; Lesser spotted dogfishes; Piked dogfishes
Doglike bats. *See* Lesser dog-faced bats
Dogs, 12:*215*, **14:265–285,** 14:*275*
behavior, 14:258, 14:268–269
conservation status, 14:272–273
distribution, 12:137, 14:267
evolution, 12:29, 14:265–266
feeding ecology, 14:260, 14:269–271
gaits of, 12:*39*
habitats, 14:267
humans and, 14:273
neonatal requirements, 12:126
physical characteristics, 14:266–267
reproduction, 12:106, 12:107, 14:271–272
self recognition, 12:159
smell, 12:*81*
species of, 14:*278*, 14:281–282, 14:283*t*, 14:284*t*
taxonomy, 14:256, 14:265–266
See also Domestic dogs
Dolabella auricularia. See Sea hares
Dolastatin 10, 1:45
Dolicholagus longirostris. See Longsnout blacksmelts
Dolichonyx oryzivorus. See Bobolinks
Dolichopithecus spp., 14:172, 14:189
Dolichopodid flies, 3:358
Dolichopteryx longipes. See Brownsnout spookfishes
Dolichotinae, 16:389
Dolichotis spp., 16:389, 16:390
Dolichotis patagonum. See Maras
Dolioletta gegenbauri. See Doliolida
Doliolida, 1:467, 1:468, 1:*469*, 1:470
Doliornis remseni, 10:308
Dollarbirds, 10:*52*, 10:*54*, 10:*56*
Dollfusilepis hoploporus, 1:230
Dologale spp., 14:347
Dolops spp., 2:289, 2:290
Dolops ranarum, 2:*291*, 2:*292*, 2:293
Dolphin gulls, 9:203
Dolphinfishes, 4:49, 4:61, **5:211–212,** 5:*213*, **5:214–216**
Dolphins, 12:14, **15:41–58,** 15:*42*, 15:*43*, 15:*52*
behavior, 15:5, 15:44–46
conservation status, 15:9, 15:48–50
distribution, 12:138, 15:*41*, 15:43
echolocation, 12:86, 12:87
encephalization quotient, 12:149
evolution, 15:3, 15:41–42
feeding ecology, 15:46–47
field studies of, 12:201
habitats, 15:5, 15:43–44
hearing, 12:82
humans and, 12:173, 15:10, 15:51
physical characteristics, 12:63, 12:66, 12:67, 15:4, 15:42–43
reproduction, 12:91, 12:94, 12:103, 15:47–48
self recognition, 12:159
smell, 12:80
species of, 15:53–55, 15:*56t*–*57t*
taxonomy, 15:41–42
touch, 12:83
See also Common dolphinfishes; specific types of dolphins
Dome pressure receptors (DPRs), 7:160
Domestic animals, pests of, 3:76
Domestic camels, 12:179–180, 15:150–151
See also Camels
Domestic cats, 12:180, **14:289–294,** 14:*290*
animal rights movement and, 14:293–294
behavior, 12:145
distribution, 12:137, 14:370
evolution, 14:290–291
feeding ecology, 12:*175*, 14:372
history of, 12:172
humans and, 14:291, 14:292, 14:375
introduced species, 12:215
popularity of, 14:291
reproduction, 14:291–292
vision, 12:79
wildlife and, 14:291
See also Cats
Domestic cattle, 12:177, 15:*148*, 15:263, 16:8
behavior, 15:271, 16:6
cloning, 15:151
dairy, 12:122, 15:151
domestication of, 12:172, 15:145
humans and, 15:273, 16:8, 16:15
neonatal milk, 12:127
stomach, 12:*124*
See also Cattle; Ungulates; specific types of cattle
Domestic dogs, 12:180, **14:287–294**
animal rights movement and, 14:293–294
behavior, 12:143, 12:*180*, 14:268, 14:269
breeds, 14:288–289
distribution, 14:267, 14:288
evolution, 12:29, 14:287
history of, 12:172
humans and, 14:262, 14:288, 14:292–293
popularity of, 14:291
reproduction, 14:292
See also Dogs
Domestic donkeys, 12:176–177, 15:*219*
Domestic elephants, 15:172
See also Elephants
Domestic goats, 12:*177*, 12:179, 15:145, 15:*147*, 15:263
distribution, 16:91

domestication of, 12:172, 15:145, 15:146–148
humans and, 15:273, 16:8, 16:9, 16:95
See also Goats
Domestic guinea pigs, 12:180–181, 16:*391*, 16:399*t*
See also Guinea pigs
Domestic horses, 12:175–176, 15:*151*, 15:236*t*
behavior, 15:*230*
distribution, 15:217
domestication of, 15:149–150
humans and, 15:223
reproduction, 15:229
smell, 12:80
See also Horses
Domestic llamas, 12:180, 15:151
Domestic pigs, 12:179, 15:263
conservation status, 15:281
distribution, 15:*277*
domestication of, 12:172, 15:145, 15:148–149
humans and, 15:*150*, 15:273, 15:281, 15:282
pot-bellied, 12:191, 15:*278*
reproduction, 15:272, 15:279, 15:280
Yorkshire, 12:*172*
See also Pigs
Domestic rabbits, 12:180
See also Rabbits
Domestic sheep, 12:*174*, 12:178–179, 15:263, 16:*94*
distribution, 16:91
domestication of, 15:145, 15:146–148
humans and, 15:*147*, 15:273, 16:8, 16:95
physical characteristics, 15:268
reproduction, 12:*90*
See also Sheep
Domesticated asses. *See* Domestic donkeys
Domestication, 8:21–22, **12:171–181, 15:145–153**
See also Zoos; specific domestic animals
Dominance, 12:*141*
See also Behavior
Dominance hierarchies, 4:63
Dominican amber, 3:8
Donacidae, 2:451, 2:452
Donacobius spp., 10:525
Donacobius atricapillus. See Black-capped donacobius
Donald Duck, 8:22
Donax clams. *See* Coquina clams
Donax variabilis. See Coquina clams
Donkeys, 12:176–177, 15:*219*, 15:*229*
conservation status, 15:230
domestication of, 12:*172*, 15:145, 15:*146*
feeding ecology, 15:229
humans and, 15:223
Donnelly, Maureen, 6:301
Doodlebugs. *See* Antlions
Dorab wolf herrings, 4:*280*, 4:*281*
Dorabs. *See* Dorab wolf herrings
Doradidae, 4:352
Dorados. *See* Common dolphinfishes
Doras hancockii. See Blue-eye catfishes
D'Orbigny's seedsnipes. *See* Gray-breasted seedsnipes
Dorcadia spp., 3:350
Dorcadia ioffi. See Sheep and goat fleas
Dorcas gazelles, 12:222, 16:48, 16:*49*, 16:*50*–51
Dorcatragus megalotis. See Beiras
Dorcopsis spp., 13:83, 13:85, 13:86
Dorcopsis hageni. See White-striped dorcopsises
Dorcopsis luctuosa. See Gray dorcopsises
Dorcopsis macleayi. See Papuan forest wallabies
Dorcopsis veterum. See Brown dorcopsises
Dorcopsises
brown, 16:102*t*
gray, 13:*91*, 13:*94*, 13:100
white-striped, 16:102*t*
Dorcopsulus spp., 13:83, 13:85, 13:86
Dorcopsulus vanheurni. See Lesser forest wallabies
Doria's tree kangaroos, 13:36, 13:102*t*
Doricha spp. *See* Hummingbirds
Dorichthys spp., 5:134
Dories, **5:123–130,** 5:*125*
See also Walleyes
Dormancy. *See* Energy conservation; Hibernation
Dormice, 12:134, 16:122, 16:125, **16:317–328,** 16:*322*, 16:327*t*–328*t*
Dormitator maculatus. See Fat sleepers
Dormouse possums, 13:35
Dornorns. *See* Reef stonefishes
Dorosoma cepedianum. See Gizzard shads
Dorosoma petenense. See Threadfin shads
Dorst, J., 9:444
Doryfera spp. *See* Hummingbirds
Doryfera ludovicae. See Green-fronted lancebills
Dorylus spp., 3:68
Doryphallophoridae, 2:284
Doryrhamphus spp., 5:135
Doryrhamphus dactyliophorus. See Ringed pipefishes
Dosidicus gigas. See Humboldt squids
Dotted humming frogs, 6:303
Dottybacks, 4:48, 5:255
Double-banded coursers, 9:153, 9:*154*, 9:*159*
Double-banded sandgrouse, 9:231
Double-barred finches, 11:*355*, 11:*359*, 11:367–*368*
Double-crested cormorants, 8:27, 8:186, 8:*202*, 8:*203*, 8:*204*, 8:*205–206*
Double helix, 12:26
Double-toothed barbets, 10:115
Double-wattled cassowaries. *See* Southern cassowaries
Douc langurs. *See* Red-shanked douc langurs
Douglas's squirrels, 16:*164*
Dovekies, 9:219, 9:220, 9:*223*, 9:*227*
Dover soles. *See* Common soles
Doves, 8:20, 8:25, 9:241–246, **9:247–267,** 9:*253–254*
behavior, 9:243, 9:249
conservation status, 9:251–252
distribution, 9:243, 9:*247*–248
evolution, 9:241, 9:247
feeding ecology, 9:243–244, 9:249–250
habitats, 9:248–249
humans and, 9:252
physical characteristics, 9:241–243, 9:247
reproduction, 9:244, 9:250–251
species of, 9:*255–266*
taxonomy, 9:241, 9:247
Dowitchers, 9:177
Dowler, R. C., 16:270
DPRs. *See* Dome pressure receptors
Dracaena spp. *See* Caiman lizards
Dracaena guianensis, 7:310
Dracaena paraguayensis. See Paraguayan caiman lizards
Draco volans. See Flying lizards
Dracoderidae, 1:275
Draconetta spp. *See* Draconettas
Draconetta xenica. See Draconettas
Draconettas, 5:365, 5:*368*, 5:*369*, 5:371
Draconettidae. *See* Deepwater dragonets
Draconetts, 5:*368*, 5:*370*, 5:371
Draculo spp., 5:365
Draft dogs, 14:293
Draft guards, 14:288
Draft horses, 12:*176*
Draft mammals, 12:15
See also specific draft mammals
Dragon lizards, **7:209–211,** 7:*212*, 7:*213*, **7:216–220**
Dragonets, **5:365–372,** 5:*368*
Dragonfishes, **4:421–425,** 4:*422*, 4:*425*, **4:428–429,** 5:321, 5:322
Dragonflies, 3:52, **3:133–139,** 3:*134*
behavior, 3:46, 3:133–134, 3:*135*
as bioindicators, 3:82
conservation of, 3:88, 3:89, 3:90–91, 3:135
feeding ecology, 3:63, 3:134
fossil, 3:7, 3:*8*, 3:9, 3:10, 3:12, 3:13
jewelry, 3:74
physical characteristics, 3:5, 3:10, 3:22, 3:27, 3:133
reproduction, 3:34, 3:61, 3:62, 3:134
threats to, 3:88
Drahomira spp., 2:387
Dreissena spp. *See* Zebra mussels
Dreissena polymorpha. See Zebra mussels
Dreissenidae, 2:452
Drepaneidae. *See* Sicklefishes
Drepanididae. *See* Hawaiian honeycreepers
Drepanidinae, 11:342
Drepanis spp. *See* Sicklebills
Drepanis coccinea. See Iiwis
Drepanoplectes jacksoni. See Jackson's widow-birds
Drepanorhynchus spp. *See* Sunbirds
Drepanorhynchus reichenowi. See Golden-winged sunbirds
Drepanornis spp. *See* Birds of paradise
Drepanornis bruijnii. See Pale-billed sicklebills
Drepanosticta anscephala. See Forest-dwelling damselflies
Dreptes spp. *See* Sunbirds
Dreptes thomensis. See São Tomé sunbirds
Driftfishes, 5:421
Driftwood catfishes, 4:352
Drills, 14:188, 14:191, 14:193–194, 14:205*t*
Driloleirus macelfreshi, 2:67
Driver ants, 3:68, 3:72
Dromadidae. *See* Crab plovers
Dromaius ater. See King Island emus
Dromaius baudinianus. See Kangaroo Island emus
Dromaius novaehollandiae. See Emus
Dromas ardeola. See Crab plovers
Dromedary camels, 12:*122*, 15:*315*, 15:*319*, 15:*320*–321
behavior, 15:316
domestication of, 12:179, 15:150–151
habitats, 15:316
humans and, 15:318
physical characteristics, 15:268, 15:314
taxonomy, 15:313

Dromiciops australis. *See* Monitos del monte
Dromiciops gliroides. *See* Monitos del monte
Dromococcyz phasianellus. *See* Pheasant cuckoos
Dromomerycids, 15:137
Drone carpenter bees. *See* Large carpenter bees
Drongo cuckoos. *See* Asian drongo-cuckoos
Drongos, **11:437–445,** 11:*440*
Drosophila spp., 3:61, 3:360
Drosophila lanaiensis. *See* Lanai pomace flies
Drosophila melanogaster. *See* Fruit flies
Drugs. *See* Pharmaceuticals
"Drugs from the Sea," 1:44
Drugstore beetles, 3:76
Drums, 5:255
  See also Sciaenidae
Dry-country bandicoots, 13:1, **13:9–18,** 13:*11*, 13:16*t*–17*t*
Dryland mouse opossums. *See* Orange mouse opossums
Drymarchon spp. *See* Indigo snakes
Drymarchon corais. *See* Indigo snakes
Drymarchon corais couperi. *See* Eastern indigo snakes
Drymocichla spp., 11:4
Drymodes brunneopygia. *See* Southern scrub robins
Drymophila squamata. *See* Scaled antbirds
Drymornis bridgesii. *See* Scimitar-billed woodcreepers
Dryococelus australis. *See* Lord Howe Island stick insects
Dryocopus galeatus. *See* Helmeted woodpeckers
Dryocopus lineatus. *See* Lineated woodpeckers
Dryocopus martius. *See* Black woodpeckers
Dryocopus pileatus. *See* Pileated woodpeckers
Dryodora spp., 1:171
Dryolimnas cuvieri. *See* White-throated rails
Dryomys spp. *See* Forest dormice
Dryomys laniger. *See* Woolly dormice
Dryomys nitedula. *See* Forest dormice
Dryomys sichuanensis. *See* Chinese dormice
Dryoscopus spp. *See* Bush-shrikes
Dryoscopus gambensis. *See* Northern puffbacks
Drywood termites, 3:164
  Giant Sonoran, 3:*169*, 3:*171*
  West Indian powderpost, 3:*169*, 3:*170*–171
Dsinezumi shrews, 13:276*t*
Duality, reproductive, 1:16
  See also Reproduction
Duberria spp. *See* Slug-eaters
Duberria lutrix. *See* Common slug-eaters
DuBois, Sieur, 9:269
Duck-billed platypuses, 12:*231*, **12:243–248,** 12:*244*
  behavior, 12:230–231, 12:244–*245*, 12:*246*
  conservation status, 12:233–234, 12:246–247
  distribution, 12:137, 12:230, 12:*243*, 12:244
  evolution, 12:228, 12:243
  feeding ecology, 12:231, 12:245
  habitats, 12:*229*, 12:230, 12:244, 12:*247*
  humans and, 12:234, 12:247
  physical characteristics, 12:228–230, 12:243–*244*, 12:*246*
  reproduction, 12:101, 12:106, 12:108, 12:*228*, 12:231–233, 12:*232*, 12:245–246
  taxonomy, 12:243
Duck hawks. *See* Peregrine falcons
Duckbill garfishes. *See* Shortnose gars
Duckbill Poso minnows. *See* Duckbilled buntingis
Duckbilled buntingis, 5:80, 5:81, 5:*82*, 5:*83*–84
Duckbills, 5:331–335
Ducks, 8:20, 8:21, **8:369–392**
  behavior, 8:371–372
  conservation status, 8:372–373
  distribution, 8:*369*, 8:370–371
  evolution, 8:369
  feeding ecology, 8:372
  habitats, 8:371
  humans and, 8:373
  physical characteristics, 8:369–370
  reproduction, 8:372
  species of, 8:*376*–392
  taxonomy, 8:369
Ducula spp. *See* Fruit pigeons
Ducula luctuosa. *See* White imperial pigeons
Ducula spilorrhoa. *See* Torresian imperial pigeons
Duellman, W. E., 6:425
Duellmanohyla spp., 6:227
Duets. *See* Songs, bird
Dugesia spp., 1:187, 1:188
Dugesia dorotocephala, 1:38
Dugesia tigrina. *See* Freshwater planarians
Dugong dugon. *See* Dugongs
Dugongidae, 15:*199*, **15:199–204,** 15:*202*
  behavior, 15:192–193
  conservation status, 15:194–195
  distribution, 15:192
  evolution, 15:191–192
  feeding ecology, 15:193
  habitats, 15:192
  humans and, 15:195–196
  physical characteristics, 12:44, 15:192
  reproduction, 12:91, 12:94, 15:193–194
  species of, 15:203–204
  taxonomy, 15:191–192
  See also Dugongs; Steller's sea cows
Dugongids. *See* Dugongidae
Dugonginae. *See* Dugongs
Dugongs, 15:*194*, **15:199–204,** 15:*200*, 15:*202*, 15:*203*
  behavior, 15:192–193, 15:200
  conservation status, 15:195, 15:201
  distribution, 15:*191*, 15:192, 15:*199*
  evolution, 15:191, 15:199
  feeding ecology, 15:193, 15:*195*, 15:200
  habitats, 15:192
  humans and, 15:195–196, 15:201
  physical characteristics, 15:*192*, 15:199
  reproduction, 12:103, 15:194, 15:201
  taxonomy, 15:191
Duikers, 16:1, **16:73–85,** 16:*78*, 16:*79*
  behavior, 16:5, 16:74
  conservation status, 16:80–84, 16:84*t*–85*t*
  distribution, 16:*73*
  evolution, 15:265, 16:73
  feeding ecology, 16:74, 16:76
  habitats, 16:74
  humans and, 16:77
  physical characteristics, 15:138, 15:267, 16:2, 16:73
  reproduction, 16:76–77
  species of, 16:80–83, 16:84*t*–85*t*
  taxonomy, 16:73
Dulidae. *See* Palmchats
Dulit frogmouths, 9:378, 9:379
Dull-blue flycatchers, 11:*29*, 11:*35*–36
Dulus dominicus. *See* Palmchats
Dumbacher, Jack, 11:76, 11:118
Duméril, A., 7:379
Dumeril's boas, 7:411
Dumetella carolinensis. *See* Gray catbirds
Dunce hinnies, 12:177
Dune mole-rats, 16:339, 16:342, 16:343, 16:344
  Cape, 16:345, 16:*346*, 16:*347*
  Namaqua, 16:*344*, 16:345, 16:349*t*
Dune squeakers. *See* Common squeakers
Dung, insects and, 3:42
Dung beetles, 3:*44*, 3:64, 3:88, 3:322, 3:323
  arboreal, 3:87–88
  conservation, 3:*85*
  feeding ecology, 3:321
  sacred scarab, 3:*316*
Dung flies, 3:61
Dung scarabs. *See* Dung beetles
Dungeness crabs, 2:*42*
Dunlins, 9:175
Dunnarts, 12:280–282, 12:289
Dunnocks, 10:*460*, 10:*461*, 10:*462*
Duplicendentata. *See* Lagomorpha
Dupont's larks, 10:342, 10:343, 10:*345*, 10:346
Duprat, Hubert, 3:377
Dürer, Albrecht, 3:323, 9:451, 15:254
Durrell, Gerald, 7:430
Durrell Wildlife Conservation Trust, 15:287
d'Urville, Dumont, 8:147
Dusicyon australis. *See* Falkland Island wolves
Dusky antechinuses, 12:279, 12:299*t*
Dusky Asian orioles, 11:427–428
Dusky barn owls. *See* Sooty owls
Dusky-blue flycatchers. *See* Dull-blue flycatchers
Dusky broadbills, 10:177, 10:179, 10:*180*, 10:*185*
Dusky brown orioles, 11:*428*
Dusky bushbabies, 14:26, 14:32*t*
Dusky buttonquails. *See* Barred buttonquails
Dusky dolphins, 12:124, 15:*46*, 15:47
Dusky fantails, 11:87
Dusky flycatchers, 10:271
Dusky flying foxes, 13:331*t*
Dusky-footed sengis, 16:531*t*
Dusky-footed woodrats, 16:277*t*
Dusky friarbirds, 11:*428*
Dusky hopping mice, 16:261*t*
Dusky indigobirds, 11:*381*, 11:*393*
Dusky leaf-nosed bats, 13:*403*, 13:*404*
Dusky marsupials. *See* Dusky antechinuses
Dusky moorhens, 9:47, 9:50
Dusky redshanks. *See* Spotted redshanks
Dusky salamanders, 6:*391*, 6:*393*, 6:*395*
Dusky São Tomé sunbirds. *See* São Tomé sunbirds
Dusky sengis, 16:531*t*
Dusky sleepers, 5:374
Dusky sunbirds, 11:208
Dusky thrushes, 10:485
Dusky titi monkeys, 14:143, 14:144, 14:145, 14:*149*, 14:151–*152*
Dusky woodswallows, 11:*460*, 11:*461*, 11:*463*
Dusty tree kangaroos. *See* Bennett's tree kangaroos
Duvernoy's glands, 7:467
Dvorák, Antonín, 8:19
Dwarf African clawed frogs, 6:45–46

Dwarf antelopes, 16:2, 16:5, 16:7, 16:*60*, 16:63, 16:*65*, 16:*66*, 16:68–69
Dwarf armadillos. *See* Pichi armadillos
Dwarf bitterns, 8:233
Dwarf blue sheep, 16:87, 16:91, 16:94, 16:103*t*
Dwarf boas, 7:*434*
Dwarf brittle stars, 1:390, 1:*392*, 1:394–*395*
Dwarf brocket deer, 15:396*t*
Dwarf bushbabies. *See* Demidoff's bushbabies
Dwarf buttonquails. *See* Black-rumped buttonquails
Dwarf cassowaries. *See* Bennett's cassowaries
Dwarf chimpanzees. *See* Bonobos
Dwarf cichlids, Agassiz's, 5:*282*, 5:*284*
Dwarf corydoras, 4:*356*, 4:*358*–359
Dwarf crocodiles, 7:157, 7:179
Dwarf deer, 15:381, 15:384
Dwarf dories, 5:123, 5:124
Dwarf dormice. *See* African dormice
Dwarf duikers, 16:73, 16:77
Dwarf epauletted fruit bats, 13:*338*, 13:*343*–344
Dwarf flying mice. *See* Zenker's flying mice
Dwarf flying squirrels, 16:136, 16:*139*
Dwarf geese. *See* African pygmy geese
Dwarf gibbons. *See* Kloss gibbons
Dwarf gouramies, 5:*428*
Dwarf hamsters, 16:*125*, 16:239, 16:240
Dwarf herrings, 4:*278*
Dwarf hippopotamuses, 15:301
Dwarf honeyguides, 10:139
Dwarf hutias, 16:467*t*
Dwarf lemurs, 14:*4*, **14:35–45,** 14:*40*
  behavior, 14:36–37
  conservation status, 14:39
  distribution, 14:*35*, 14:36
  evolution, 14:35
  feeding ecology, 14:37–38
  habitats, 14:36
  humans and, 14:39
  physical characteristics, 14:35
  reproduction, 14:38
  species of, 14:*41*–42
  taxonomy, 14:3, 14:35
Dwarf little fruit bats, 13:417, 13:*421*, 13:*430*
Dwarf minke whales, 12:*200*
Dwarf mongooses, 12:146, 12:148, 14:259, 14:261, 14:*350*, 14:*353*, 14:*354*, 14:356
Dwarf mud anemones. *See* Starlet sea anemones
Dwarf olive ibises, 8:293
Dwarf plated lizards, 7:320
Dwarf ponies, 15:*218*
Dwarf porcupines. *See* Hairy dwarf porcupines
Dwarf puff adders, 7:446
Dwarf pygmy gobies, 5:*378*, 5:*381*, 5:386
Dwarf rabbits, 16:*507*
Dwarf scalytails. *See* Lesser anomalures
Dwarf sea urchins. *See* Pea urchins
Dwarf sirens, **6:327–333,** 6:*330*
Dwarf sloths. *See* Monk sloths
Dwarf snakeheads, 5:438, 5:439
Dwarf sperm whales, 15:1, 15:6, 15:73, 15:74, 15:76, 15:77, 15:78
Dwarf threadworms. *See* Threadworms
Dwarf tinamous, 8:59
Dwarf waterdogs, 6:377, 6:378, 6:*380*, 6:382–383
Dwarf wrymouths, 5:310
Dyacopterus spadiceus. *See* Dyak fruit bats
Dyak fruit bats, 12:106, 13:*339*, 13:*341*, 13:344
Dyaphorophyia castanea. *See* Chestnut wattle-eyes
Dyar's law, 3:31
Dyck, Jan, 9:276
Dycrostonix torquatus, 16:270
Dyer, B. S., 5:67
Dyes, 1:44, 2:41, 3:81–82
  See also Humans
Dynamic lift, 12:56
Dynastes hercules. *See* Hercules beetles
Dyscophinae, 6:301, 6:302
Dyscophus spp., 6:301, 6:302, 6:304
Dyscophus antongilii. *See* Tomato frogs
Dyscophus guineti. *See* Sambava tomato frogs
Dysdercus albofasciatus. *See* Staining bugs
Dysichthys amaurus. *See* Camouflaged catfishes
Dysichthys coracoideus. *See* Guitarrita
Dysithamnus mentalis. *See* Plain antvireos
Dysithamnus puncticeps. *See* Spot-crowned antvireos
Dysmorodrepanis, 11:343
Dysmorodrepanis munroi. *See* Lanai hookbills
Dysomma brevirostre. *See* Pignosed arrowtooth eels
Dytiscidae, 3:322
Dytiscus marginalis. *See* Great water beetles
Dzhungarian hamsters, 16:*240*, 16:*242*, 16:243, 16:247*t*
Dzigettais. *See* Asiatic wild asses

E

Eagle-owls, 9:331–333, 9:345–347, 9:349, 9:351–*352*, 9:357–359
Eagle rays, 4:176
Eagles, **8:317–341,** 8:318, 12:184, 12:192
  behavior, 8:320–321
  conservation status, 8:323
  distribution, 8:*317*, 8:319–320
  evolution, 8:317–318
  feeding ecology, 8:321–322
  habitats, 8:320
  humans and, 8:323–324
  pesticides, 8:26
  physical characteristics, 8:318–319
  reproduction, 8:321, 8:322–323
  species of, 8:*329–341*
  symbolism, 8:20
  taxonomy, 8:317–318
Ealey's ningauis. *See* Pilbara ningauis
Ear tufts, 9:331–332
Eared doves, 9:243
Eared grebes, 8:169, 8:172, 8:*173*, 8:178–*179*
Eared hutias, 16:461, 16:467*t*
Eared pheasants, 8:434
Eared quetzals, 9:479, 9:*480*, 9:*484*
Eared seals, **14:393–408,** 14:*401*
  behavior, 14:396–397
  conservation status, 14:262, 14:399–400
  distribution, 14:394–395
  evolution, 14:393
  feeding ecology, 14:397–398
  habitats, 14:395–396
  humans and, 14:400
  physical characteristics, 14:393–394
  reproduction, 14:262, 14:398–399
  species of, 14:*402*–406, 14:407*t*–408*t*
  taxonomy, 14:256, 14:393
Eared trogons. *See* Eared quetzals
Earless dragons, 7:*213*, 7:*218*, 7:220
Earless monitors, **7:359–368,** 7:*364*, 7:*365*
Early blindsnakes, 7:198, 7:201, 7:202, **7:369–372**
Ears, 7:9, 7:32, 7:159–160, 12:8–9, 12:*41*
  avian, 8:6, 8:39–40
  bats, 12:54
  subterranean mammals, 12:77
  See also Physical characteristics
Earth-boring beetles, 3:322
Earth centipedes, 2:*357*, 2:*358*–359
Earthcreepers, 10:*212*, 10:*215*, 10:*216–217*
Earthworms, 2:25, **2:65–74,** 2:*69*
  behavior, 2:35, 2:36, 2:43, 2:66
  conservation status, 2:67
  distribution, 2:66
  evolution, 2:65
  feeding ecology, 2:36, 2:66
  habitats, 2:66
  humans and, 2:68
  physical characteristics, 2:*18*, 2:65–*66*, 2:*67*
  reproduction, 2:16, 2:22, 2:67
  species of, 2:*70–73*
  taxonomy, 2:65
Earwigs, 3:62, 3:81, 3:*195*, **3:195–200,** 3:*196*, 3:*198*
  See also specific types of earwigs
East African black mud turtles, 7:*130*, 7:*131*, 7:*133*
East African hedgehogs, 13:*198*
East African long-nosed sengis. *See* Rufous sengis
East African mole rats, 12:74, 16:297*t*
East African sandboas, 7:411, 7:*412*, 7:*416*–417
East African serrated mud turtles, 7:*131*, 7:132–*133*
East Asian tailless leaf-nosed bats, 13:409*t*
East Caucasian turs, 16:91, 16:92, 16:104*t*
East Fresian sheep, 15:147
East Mediterranean blind mole-rats, 12:74, 12:76–77
Eastern American oysters, 2:*455*, 2:*457*, 2:459
Eastern Australian galaxiids, 4:390
Eastern barred bandicoots, 13:4, 13:5, 13:10, 13:*11*–12, 13:*14*
Eastern bearded greenbuls, 10:*400*, 10:*408*–409
Eastern black-headed orioles, 11:*430*, 11:*433*–434
Eastern blindsnakes. *See* Blackish blindsnakes
Eastern bluebirds, 10:485, 10:*487*, 10:488, 10:*490*, 10:*494*
Eastern box turtles, 7:*68*, 7:106, 7:*108*, 7:112–*113*
Eastern Brazilian marmosets, 14:115, 14:116, 14:117, 14:120
Eastern bristlebirds. *See* Bristlebirds
Eastern broad-billed rollers. *See* Dollarbirds
Eastern chipmunks, 16:126, 16:*148*, 16:*149*, 16:153–*154*
Eastern cleaner-clingfishes, 5:*358*, 5:*359*, 5:360
Eastern collared lizards. *See* Common collared lizards
Eastern cottontails, 16:481, 16:487, 16:*510*, 16:*512*, 16:514–515

Eastern diamondback rattlesnakes, 7:40, 7:50, 7:475
Eastern dobsonflies, 3:*292*, 3:*293*–294
Eastern earless lizards. *See* Common lesser earless lizards
Eastern European hedgehogs, 13:212*t*
Eastern fox squirrels, 16:*164*, 16:*166*, 16:*168*, 16:*169*, 16:171
Eastern freshwater cods, 5:222
Eastern gizzard shads. *See* Gizzard shads
Eastern gorillas, 14:225, 14:*228*, 14:232, 14:*236*, 14:*238*–239
Eastern grass owls, 9:336, 9:337, 9:338, 9:*339*, 9:*342*–343
Eastern gray kangaroos, 12:108–109, 12:148, 12:185, 13:*32*, 13:38, 13:*87*, 13:*91*, 13:*94*
Eastern gray squirrels. *See* Gray squirrels
Eastern hog-nosed snakes, 7:*197*, 7:*470*, 7:*471*, 7:*475*, 7:480–481
Eastern horseshoe bats, 13:*388*, 13:390, 13:392, 13:*395*, 13:*396*
Eastern horseshoe crabs. *See* Japanese horseshoe crabs
Eastern indigo snakes, 7:470, 7:475
Eastern kingbirds, 10:272, 10:273, 10:275
Eastern lowland gorillas, 14:225, 14:227, 14:*230*, 14:232
Eastern lubber grasshoppers, 3:*79*
Eastern Malagasy ring-tailed mongooses, 14:*348*
Eastern meadowlarks, 11:*304*
Eastern moles, 12:*71*, 12:*187*, 13:*195*, 13:198, 13:280, 13:*281*, 13:*283*, 13:*284*–285
Eastern mud turtles, 7:121, 7:*122*
Eastern mudminnows, 4:379
Eastern narrow-mouthed toads, 6:*307*, 6:*312*–313
Eastern newts, 6:*29*, 6:44, 6:*45*, 6:365, 6:*368*, 6:*371*–372
Eastern oysters. *See* Eastern American oysters
Eastern painted turtles, 7:*65*
Eastern pearl shells. *See* European pearly mussels
Eastern phoebes, 10:273, 10:*276*, 10:284–*285*
Eastern pipistrelles, 13:*504*, 13:*507*, 13:513
Eastern pocket gophers. *See* Plains pocket gophers
Eastern pygmy possums, 13:*106*, 13:107, 13:*108*, 13:*109*, 13:*110*
Eastern quolls, 12:*279*, 12:280, 12:283, 12:*287*, 12:299*t*
Eastern racers, 7:470
Eastern red-footed falcons. *See* Amur falcons
Eastern rock sengis, 16:*518*, 16:*523*, 16:*528*–529
Eastern rosellas, 9:*282*, 9:*285*
Eastern screech owls, 9:*331*, 9:*352*, 9:*355*
Eastern scrub-birds. *See* Rufous scrub-birds
Eastern sharp-snouted frogs. *See* Spotted snout-burrowers
Eastern short-tailed opossums. *See* Southern short-tailed opossums
Eastern shrike-tits, 11:*119*, 11:*120*
Eastern side-blotched lizards. *See* Common side-blotched lizards
Eastern spadefoot toads, 6:*121*, 6:122
Eastern spiny softshell turtles, 7:*152*
Eastern spotted skunks, 14:*321*, 14:324
Eastern subterranean termites, 3:*169*, 3:*171*, 3:172–173
Eastern tarsiers. *See* Spectral tarsiers
Eastern towhees, 11:*267*, 11:*279*
Eastern tree hyraxes, 15:177, 15:179, 15:180, 15:184, 15:*185*, 15:*187*
Eastern tube-nosed fruit bats. *See* Queensland tube-nosed bats
Eastern warbling vireos. *See* Warbling vireos
Eastern whipbirds, 10:331, 11:70, 11:75–76, 11:77, 11:*79*–*80*
Eastern whitefaces. *See* Southern whitefaces
Eastern whitefishes. *See* Lake whitefishes
Eastern wood-pewees, 10:275
Eastern woodrats, 16:125, 16:277*t*
Eastern woolly lemurs, 14:63–64, 14:*65*, 14:*68*, 14:*69*
Eastern yellow robins, 11:*106*, 11:*107*, 11:*109*
Eastern zebra-tailed lizards. *See* Zebra-tailed lizards
Eastwood's seps, 7:322
Eating habits. *See* Feeding ecology
Eaton's pintails, 8:364
Eburana spp., 6:247
Ecacanthothrips tibialis. *See* Crowned thrips
Eccentric sand dollars. *See* Western sand dollars
Eccrine sweat glands, 12:36
Ecdysis, 2:14, 3:31–32
Ecdysozoa, 1:6, 1:9, 2:5, 2:13
Echeneidae. *See* Remoras
Echeneis spp., 5:211
Echidnas, **12:235–241,** 12:*239*, 12:240*t*
behavior, 12:230–231, 12:*237*
conservation status, 12:233–234, 12:238
distribution, 12:137, 12:230, 12:*235*, 12:236
evolution, 12:227–228, 12:235
feeding ecology, 12:231, 12:237
habitats, 12:230, 12:237
humans and, 12:234, 12:238
physical characteristics, 12:228–230, 12:235–236
reproduction, 12:106, 12:*107*, 12:108, 12:231–233, 12:237–238
taxonomy, 12:227, 12:235
Echidnophaga gallinacea, 3:349
Echigo Plain moles, 13:282
Echiichthys vipera. *See* Lesser weevers
Echimyidae. *See* Spiny rats
Echimys spp. *See* Spiny rats
Echimys blainvillei, 16:450–451
Echimys chrysurus. *See* White-faced arboreal spiny rats
Echimys pictus, 16:451
Echimys rhipidurus, 16:451
Echimys thomasi, 16:450
Echinarachnius parma. *See* Sand dollars
Echiniscoidea, 2:115
Echiniscoides hoepneri, 2:117
Echiniscoides sigismundi, 2:116
Echiniscoides sigismundi polynesiensis, 2:116
Echiniscoides sigismundi sigismundi. *See* Tidal water bears
Echiniscus testudo. *See* Turtle water bears
Echinobothriidae, 1:225
Echinococcus granulosus. *See* Dog tapeworms
Echinococcus multilocularis. *See* Fox tapeworms
Echinocyamus pusillus. *See* Pea urchins
Echinocyamus scaber, 1:401
Echinoderes spp., 1:*275*, 1:276
Echinoderes sensibilis, 1:*278*, 1:*280*
Echinoderidae, 1:275–276, 1:277
Echinoderms, 1:13, 1:39, 1:42, 1:*402*
Crinoidea, 1:*12*, **1:355–366**
Echinoidea, **1:401–416**
evolution, 2:11
Ophiuroidea, 1:48, **1:387–399**
sea cucumbers, **1:*417*–431**
sea daisies, **1:381–386**
sea stars, **1:367–380**
symbiosis, 2:32, 2:33
taxonomy, 2:3
Echinoidea, **1:401–416,** 1:*402*, 1:*406*, 1:*407*
behavior, 1:403–404
conservation status, 1:405
distribution, 1:402
evolution, 1:401
feeding ecology, 1:*403*, 1:404–405
habitats, 1:402
humans and, 1:405
physical characteristics, 1:401–402, 1:*403*
reproduction, 1:22, 1:405
species of, 1:*408*–415
taxonomy, 1:401
Echinophallidae, 1:225
Echinoprocta rufescens. *See* Short-tailed porcupines
Echinops spp., 13:225, 13:226, 13:227, 13:230
Echinops telfairi. *See* Lesser hedgehog tenrecs
Echinorhinidae. *See* Bramble sharks
Echinorhinus brucus. *See* Bramble sharks
Echinosorex gymnura. *See* Malayan moonrats
Echinosorinae, 13:203
Echinosquilla guerini, 2:169
Echinostoma revolutum. *See* Echinostomes
Echinostomes, 1:*202*, 1:205–*206*
Echinostrephus aciculatus. *See* Rock boring urchins
Echinothrips americanus. *See* Black thrips
Echinotriton andersoni, 6:*46*
Echiodontines, 5:17
Echiopsis atriceps, 7:486
Echiopsis curta, 7:486
Echiostoma barbatum, 4:*422*
Echis spp., 7:445
Echis carinatus. *See* Saw-scaled vipers
Echiurans, 2:9, 2:13, 2:21, 2:24, **2:103–107,** 2:*105*
Echiuridae, 2:103
Echiuridea, 2:103
Echolocation, 8:6, 12:53–55, 12:*59*
oilbirds, 9:374
swiftlets, 9:422–423
toothed whales, 12:*85*–87
See also Behavior; Physical characteristics
Echymipera spp., 13:10
Echymipera clara. *See* Large-toothed bandicoots
Echymipera davidi. *See* David's echymiperas
Echymipera echinista. *See* Menzies' echymiperas
Echymipera kalubu. *See* Spiny bandicoots
Echymipera rufescens. *See* Rufous spiny bandicoots
Eciton spp., 3:68
Eciton burchelli. *See* Army ants
Eciton hamatum. *See* Army ants
Ekloniaichthys spp. *See* Weedsuckers
Ekloniaichthys scylliorhiniceps. *See* Weedsuckers
Eclairs sur L'Au-Dela (Messiaen), 10:332
Eclectus parrots, 9:276, 9:*282*, 9:*287*–288
Eclectus roratus. *See* Eclectus parrots

Ecological niches, 2:27–28, 3:41, 12:114–115, 12:117, 12:149–150, 12:183–184
See also Habitats
Ecology, **1:24–30,** 1:*25*, 1:49, **2:25–30,** 2:43, **3:41–51,** 3:87, **12:113–119**
animal intelligence and, 12:151
behavioral, 12:148
birds and, 8:15–17
classic themes in, 3:45–47
conservation biology and, 12:216–218
domestication and, 12:175
freshwater, **4:36–41**
historical, 4:70
importance of insects, 3:42–43
introduced species and, 12:182–193
landscaping and, 3:91
learning behavior and, 12:156
marine, **4:42–51,** 4:*43*, 4:*44*
See also Behavior; Conservation status; Feeding ecology; Habitats
Economic entomologists, 3:78
Economics, in conservation biology, 12:219–220, 12:223
Economidichthys trichonis, 5:377
Ecosystems, 2:*26*, 3:72, 3:85–86, 3:88
See also Conservation status
Ecotourism. *See* Humans
Ecsenius pictus. *See* White-lined comb-tooth blennies
Ecteinascidia turbinate, 1:46
Ecteinascidin 743, 1:46
Ectobiinae, 3:149
Ectobius duskei. *See* Russian steppe cockroaches
Ectocommensals, 2:31–32
See also Behavior
Ectoderm, 2:15, 2:20, 2:35
Ectoparasites, 3:64, 4:50
Ectoparasitic thrips, 3:*285*–286
Ectophylla alba. *See* White bats
Ectopistes migratorius. *See* Passenger pigeons
Ectoprocta, 1:9, 1:13, 1:22, 1:29
See also Bryozoans
Ectosymbionts, 2:31
Ectosymbiosis, 1:32
Ectotherms, 6:3
Ectothermy, 7:28, 7:198
See also Physical characteristics
Ecuador Cochran frogs, 6:217, 6:*218*, 6:*220–221*
Ecuador shrew opossums. *See* Silky shrew opossums
Ecuadorian caciques, 11:304
Ecuadorian hillstars, 9:444
Ecuadorian marsupial frogs. *See* Riobamba marsupial frogs
Edalorhina spp., 6:66, 6:156, 6:158
Edalorhina perezi. *See* Perez's snouted frogs
Edelia spp. *See* Western pygmy perches
Edelia vittata. *See* Western pygmy perches
Edentates, 12:11
locomotion, 12:14
reproduction, 12:90
See also Armadillos; Sloths
Edestoid sharks, 4:10
Edgbaston gobies, 5:377
Ediacaran fauna, 2:9–10
See also Evolution
Ediacaria flindersi. *See* Jellyfish
Edible dormice, 16:317, 16:*318*, 16:*319*, 16:*320*, 16:321, 16:*322*, 16:*323*, 16:326
Edible frogs. *See* Roesel's green frogs
Edible-nest swiftlets, 9:424, 9:*425*, 9:*428*
Edmunds, G. F., 3:55
Edolius ludwigii. *See* Square-tailed drongos
Edolius megarhynchus. *See* Ribbon-tailed drongos
Education
insects and, 3:82–83
zoos and, 12:204, 12:*205*
Edward lyrebirds. *See* Superb lyrebirds
Edwards, H. Milne, 2:195
Edwards' pheasants, 8:433, 8:437, 8:*438*, 8:*450–451*
Edwardsina gigantea. *See* Giant torrent midges
Edwardsina tasmaniensis. *See* Tasmanian torrent midges
Eel blennies, 5:255, 5:259
Eel catfishes. *See* Channel catfishes
Eel knifefishes. *See* Banded knifefishes
Eel suckers. *See* Sea lampreys
Eelpouts, **5:309–313,** 5:*314*, **5:317–318**
See also Burbots
Eels, **4:255–270,** 4:*259–260*
band cusk-eels, 5:*19*, 5:*21–22*
behavior, 4:*256*
conservation, 4:258
distribution, 4:255
evolution, 4:255
feeding ecology, 4:17, 4:256
habitats, 4:50, 4:255–256
humans and, 4:258
physical characteristics, 4:255
reproduction, 4:47, 4:256–258
species of, 4:261–269
taxonomy, 4:11, 4:255, 4:261–269
See also specific types of eels
Eeltail catfishes, 4:352, 4:353
Eendracht land silversides, 5:*72*, 5:*73*
Efts, 6:323, 6:*368*, 6:372
Egernia spp., 7:327, 7:329
Egernia striata. *See* Night skinks
Egg-eaters, common, 7:*469*, 7:*472*, 7:*474*
Egg laying, 3:61–62
See also Reproduction
Egg-manipulation programs, Chatham Islands black robins, 11:110–111
Egg mimicry, 9:314
Egg predation
wrens, 10:527–528
Eggs
amphibian, 6:31–37, 6:39
avian, 8:7
elephant birds, 8:104
incubation, 8:12–14
kiwis, 8:90
ostriches, 8:22, 8:26, 8:101
fishes, 4:*29–33*, 4:*31*, 4:*34*, 4:46, 4:65, 4:66
mammalian
delayed implantation, 12:110, 12:111
mammalian, 12:91–92, 12:101, 12:102–104, 12:106
marsupials, 12:108
placentals, 12:109
reptiles, 7:*6–7*
See also Larvae; Oocytes; Ovulation; Reproduction
Eggs (food), 7:47–48
Egrets, 8:26, 8:233–236, 8:241, 8:244, 8:245
Egretta spp. *See* Egrets
Egretta ardesiaca. *See* Black herons
Egretta eulophotes, 8:245
Egretta garzetta. *See* Little egrets
Egretta ibis. *See* Cattle egrets
Egretta thula. *See* Snowy egrets
Egretta tricolor. *See* Tricolor herons
Egretta vinaceigula, 8:245
Egyptian banded cobras, 7:*484*
Egyptian culture, 6:51
Egyptian free-tailed bats, 13:495*t*
Egyptian fruit bats. *See* Egyptian rousettes
Egyptian geese, 8:262
Egyptian giant solpugids, 2:*338*, 2:*348*
Egyptian maus, 14:291
Egyptian mongooses, 14:351–352, 14:357*t*
Egyptian mythology, reptiles in, 7:55
Egyptian plovers, 9:151, 9:152, 9:*154*, 9:157–*158*
Egyptian rat fleas. *See* Oriental rat fleas
Egyptian rousettes, 12:46, 13:*309*, 13:*338*, 13:*340*
Egyptian slit-faced bats, 13:*371*, 13:*372*, 13:*373*, 13:*374*, 13:*375*–376
Egyptian spiny mice, 16:*251*, 16:*254*, 16:*257*, 16:259
Egyptian tomb bats, 13:364*t*
Egyptian vultures, 8:*328*, 8:*334*
Ehrlich, Paul, 1:48, 3:48, 12:217
Eidolon dupreanum. *See* Madagascan fruit bats
Eidolon helvum. *See* Straw-colored fruit bats
EIDs (emerging infectious diseases), 12:222–223
Eiffinger's Asian treefrogs, 6:292, 6:293, 6:*294*, 6:297
Eigenmann, Carl H., 4:351
Eigenmann, Rosa Smith, 4:351
Eigenmannia spp., 4:*369*
Eigenmannia lineata. *See* Glass knifefishes
Eigenmannia virescens, 4:376
Eira barbara. *See* Tayras
Eisenberg, John F., 16:201, 16:434, 16:486
Eisenia andrei, 2:68
Eisenia fetida, 2:68
Eisentraut, M., 13:341
Ekaltadeta sina, 13:70
Ekbom's syndrome, 3:77
El Niño, 6:57
El Segundo flower-loving flies, 3:361
Elacatinus oceanops. *See* Neon gobies
Elachyglossa spp., 6:249, 6:252
Elaenia chiriquensis. *See* Lesser elaenias
Elaenia flavogaster. *See* Yellow-bellied elaenias
Elaenias. *See* Tyrant flycatchers
Elaeniinae. *See* Tyrant flycatchers
Elands, 15:145–146, 16:6, 16:*9*, 16:11, 16:13
Elanus spp. *See* White-tailed kites
Elanus scriptus. *See* Letter-winged kites
Elaphe spp. *See* Ratsnakes
Elaphe guttata. *See* Cornsnakes
Elaphe guttata guttata. *See* Cornsnakes
Elaphe obsoleta. *See* Black ratsnakes
Elaphodus cephalophus. *See* Tufted deer
Elaphrini, 3:54
Elaphurus davidianus. *See* Pere David's deer
Elapidae, 7:206, **7:483–499,** 7:*489*, 7:*490*
behavior, 7:485
conservation status, 7:486–487
distribution, 7:*483*, 7:484
evolution, 7:483–484
feeding ecology, 7:485–486
habitats, 7:484–485

humans and, 7:487–488
physical characteristics, 7:484
reproduction, 7:486
species accounts, 7:*491*–498
taxonomy, 7:483–484
Elapinae, 7:484
Elapognathus minor, 7:486
Elapomorphus spp., 7:467
Elapotinus spp., 7:461
Elasipodida, 1:417, 1:418, 1:419, 1:420, 1:421
Elasmobranchs, 4:34
Elasmodontomys spp., 16:469
Elasmodontomys obliquus. See Puerto Rican giant hutias
Elasmotherium sibiricum, 15:249
Elassoma boehlkei. See Carolina pygmy sunfishes
Elassoma okefenokee, 5:*196*
Elassoma zonatum. See Banded pygmy sunfishes
Elassomatidae. *See* Pygmy sunfishes
Eld's deer, 12:19, 12:218, 15:361, 15:*363*, 15:*367*, 15:368
Electra pilosa. See Sea mats
Electric catfishes, 4:69, 4:352, 4:*355*, 4:*362*
Electric discharges, 4:19–20, 4:69, 4:175, 4:177, 4:233, 4:362, 4:370–371
Electric eels, 4:6, 4:20, 4:66, 4:69, **4:369–377,** 4:*370*, 4:*372*, 4:*374*
Electric fishes, 4:57, 4:60
See also specific fishes
Electric rays, 4:11, 4:173–179, 4:*181*
Electrified blue vangas, 10:439
Electrolocation, 4:232
Electron carinatum. See Keel-billed motmots
Electron microscopy, 2:4
Electron platyrhynchum. See Broad-billed motmots
Electrona antarctica, 4:443, 4:*444*, 4:*445*
Electrophoridae, 4:69
Electrophorus spp., 4:20
Electrophorus electricus. See Electric eels
Electroreceptive systems, 4:19–20, 4:23, 4:60, 4:62, 4:68, 4:232
Electroreceptors, 6:15
Eledone spp. *See* Lesser octopods
Elegant crescentchests, 10:*258*
Elegant crested-tinamous, 8:*55*, 8:58, 8:*60*, 8:*66*
Elegant-eyed lizards, 7:*203*
Elegant fat-tailed opossums, 12:253, 12:264*t*
Elegant pittas, 10:194
Elegant sunbirds, 11:210
Elegant trogons, 9:*478*, 9:*479*
Elegant water shrews, 13:*253*, 13:*258*, 13:259
Eleginopidae. *See Eleginops maclovinus*
Eleginops maclovinus, 5:321, 5:322
Elenia spp., 2:318, 2:319, 2:320
Eleonora's falcons, 8:314, 8:349, 8:350
Eleothreptus anomalus. See Sickle-winged nightjars
Eleotridae, 4:50, 5:373, 5:374, 5:375
Eleotrididae. *See* Eleotridae
Eleotris amblyopsis. See Large-scale spinycheek sleepers
Eleotris fusca. See Dusky sleepers
Elephant birds, 8:26, **8:103–105,** 8:*104*
Elephant ear polyps, 1:*110*, 1:*116*
Elephant-eared kangaroo rats, 16:209*t*
Elephant lice, 3:249, 3:251, 3:*253*, 3:*254*
Elephant seals
distribution, 14:418
feeding ecology, 14:420
habitats, 14:257
physical characteristics, 14:256, 14:417
reproduction, 14:261
sexual dimorphism, 12:99
See also Northern elephant seals; Southern elephant seals
Elephant sharks. *See* Ghost sharks
Elephant shrews. *See* Sengis
Elephant trunk snakes. *See* Java file snakes
Elephantfishes, 4:*236*, 4:239
behavior, 4:60, 4:62, 4:94, 4:233, 4:238
distribution, 4:57, 4:94, 4:*238*
feeding ecology, 4:68, 4:94, 4:238
habitats, 4:232
humans and, 4:94, 4:234
physical characteristics, 4:94, 4:231, 4:238
taxonomy, 4:11, 4:94, 4:238
Elephantidae. *See* Elephants
Elephantinae, 15:162
Elephantnose fishes, 4:19, 4:*232*, 4:234, 4:*236*, 4:*237*, 4:238
Elephants, 2:59, 12:*18*, 12:*48*, 12:*221*, 15:146, **15:161–175,** 15:*163*, 15:*173*
behavior, 15:167–169
conservation status, 15:171–172
digestive systems of, 12:14–15
distribution, 12:*135*, 12:136, 15:*161*, 15:166
evolution, 15:131, 15:133, 15:134, 15:161–165
feeding ecology, 15:169–170
feet, 12:*38*
field studies of, 12:*199*, 12:*200*
habitats, 15:166–167
humans and, 12:173, 15:172
locomotion, 12:42
metabolism, 12:21
migrations, 12:167
neonatal milk, 12:127
physical characteristics, 12:11–12, 12:79, 12:82, 12:83, 15:165–166
reproduction, 12:90, 12:91, 12:94, 12:95, 12:98, 12:103, 12:106–108, 12:209, 15:170–171
species of, 15:174–175
taxonomy, 12:*29*, 15:131, 15:133, 15:134, 15:161–165
teeth, 12:46
vibrations, 12:83
in zoos, 12:209
Elephantulus spp., 16:518–519, 16:521
Elephantulus brachyrhynchus. See Short-snouted sengis
Elephantulus edwardii. See Cape sengis
Elephantulus fuscipes. See Dusky-footed sengis
Elephantulus fuscus. See Dusky sengis
Elephantulus intufi. See Bushveld sengis
Elephantulus myurus. See Eastern rock sengis
Elephantulus revoili. See Somali sengis
Elephantulus rozeti. See North African sengis
Elephantulus rufescens. See Rufous sengis
Elephantulus rupestris. See Western rock sengis
Elephas spp., 15:162, 15:164–165
Elephas ekorensis, 15:163
Elephas hysudricus, 15:163–164
Elephas hysudrindicus, 15:164
Elephas maximus. See Asian elephants
Elephas maximus indicus, 15:164
Elephas maximus maximus, 15:164, 15:165
Elephas maximus sumatrensis, 15:164
Eleutheria spp., 1:129
Eleutherochir spp., 5:365
Eleutherodactylinae, 6:156, 6:179
Eleutherodactylus spp.
behavior, 6:66, 6:158
distribution, 6:6
evolution, 6:155
habitats, 6:158
physical characteristics, 6:*157*
reproduction, 6:*30*, 6:33, 6:35, 6:68, 6:158
taxonomy, 6:156
Eleutherodactylus coqui. See Puerto Rican coqui
Eleutherodactylus curtipes, 6:*46*
Eleutherodactylus iberia. See Iberian rain frogs
Eleutherodactylus jasperi. See Golden coqui
Eleutherodactylus johnstonei, 6:159
Eleutherodactylus karlschmidti, 6:159
Eleutherodactylus planirostris, 6:34
Elf owls, 9:331, 9:345, 9:*346*, 9:*348*
Elfin-wood warblers, 11:291
Elgaria spp. *See* Alligator lizards
Elgaria kingii, 7:340
Elgaria parva, 7:339
Eligmodontia morgani, 16:267
Eligmodontia typus, 16:268, 16:269
Eliomys spp. *See* Garden dormice
Eliomys melanurus. See Asian garden dormice
Eliomys quercinus. See Garden dormice
Eliurus myoxinus. See Western Malagasy bushy-tailed rats
Eliurus tanala. See Greater Malagasy bushy-tailed rats
Elizabeth II (Queen of England), 16:423
Elk, 12:80, 12:*115*, 12:164–165, 15:145–146, 15:358
Elkhorn corals, 1:*110*, 1:*117*, 1:118
Elliot's short-tailed shrews, 13:198
Elliot's storm-petrels, 8:135, 8:138
Ellisella kirschbaumi, 4:369
Ellis's sandpipers, 9:179–180
Ellobiidae, 2:411, 2:413
Ellobius spp., 16:225
Elminia spp., 11:97
Elminia longicauda. See African blue-flycatchers
Elopichthys bambusa, 4:298, 4:299–300
Elopidae, 4:69, 4:243
Elopiformes, 4:11, **4:243–248,** 4:*245*
Elopomorpha, 4:11, 4:255
Elops spp., 4:243
Elops saurus. See Ladyfishes
Elosia aspera. See Warty tree toads
Elpidia glacialis, 1:421
Elpidiidae, 1:419, 1:420
Elpistostege spp., 6:7
Elseya dentata. See Victoria river snappers
Elseya novaeguineae. See New Guinea snapping turtles
Elseyornis spp., 9:161
Elusor spp., 7:78
Elvers, 4:260–261
Elvira spp. *See* Hummingbirds
Elysia rufescens, 1:45
Elysia viridis, 2:*406*, 2:*407*, 2:409
Emballonura alecto. See Small Asian sheath-tailed bats
Emballonura monticola. See Lesser sheath-tailed bats

Emballonura semicaudata. See Pacific sheath-tailed bats
Emballonuridae, **13:355–365,** 13:*359,* 13:363*t*–364*t*
Emballonurinae, 13:355
Embaneura vachrameevi, 3:305
Emberiza calandra. See Corn buntings
Emberiza cia. See Rock buntings
Emberiza citrinella. See Yellowhammers
Emberiza lathami. See Crested buntings
Emberiza nivalis. See Snow buntings
Emberiza quelea. See Red-billed queleas
Emberiza sandwichensis. See Savannah sparrows
Emberiza schoeniclus. See Reed buntings
Emberiza unicolor. See Plumbeous sierra-finches
Emberizidae. *See* New World finches
Emberizinae, 11:263
Embernagra brunnei-nucha. See Chestnut-capped brush-finches
Embiidina. *See* Webspinners
Embioptera. *See* Webspinners
Embiotocidae. *See* Surfperches
Emblema spp. *See* Firetails
Embolomeres, 6:10
Embrithopoda, 15:131, 15:133
Embryogenesis, 1:18–19, 1:21–23, 2:18–19
See also Reproduction
Embryology, 1:7, 2:7–8, 2:15–16, 2:35
See also Reproduction
Embryonic development, 12:12
See also Reproduction
Embryonic membranes, 12:92–94
Embryos, 4:34, 4:109, 12:92–94, 12:*94,* 14:*248*
avian, 8:*4,* 8:13–14
crocodilians, 7:163–164
development of, 7:6–7, 7:13
retention of, 7:7
See also Reproduction
Emerald cuckoos, 11:219
Emerald doves, 9:*254,* 9:*258–259*
Emerald forest frogs, 6:*161,* 6:170–171
Emerald notothens, 5:*324,* 5:*326,* 5:327–328
Emerald rockcods. *See* Emerald notothens
Emerald toucanets, 10:*126*
Emerald tree boas, 7:410, 7:*412,* 7:*414*
Emerging infectious diseases (EIDs), 12:222–223
Emerita analoga. See Pacific sand crabs
Emeus crassus. See Moas
Emily's tuco-tucos, 16:430*t*
Emin's shrikes, 10:425, 10:426
Emlen, Stephen, 10:40
Emmons, L., 16:165, 16:266, 16:457
Emotional fevers, 7:45
Emperor angelfishes, 5:240, 5:*242,* 5:*245,* 5:*252–253*
Emperor birds of paradise, 11:491, 11:492
Emperor geese, 8:*370*
Emperor kingfishers. *See* Shovel-billed kookaburras
Emperor nautilus. *See* Pearly nautilus
Emperor penguins, 8:*16,* 8:147–151, 8:*152,* 8:*153*
Emperor scorpions, 2:*337,* 2:*348,* 2:350–351
Emperor tamarins, 14:*117,* 14:119, 14:121, 14:132*t*
Emperors, **5:255–263,** 5:*264,* **5:268,** 5:*269*
Empetrichthyinae, 5:89
Empetrichthys spp., 5:89
Empididae, 3:358, 3:359
Empidonax spp. *See* Flycatchers
Empidonax alnorum. See Alder flycatchers
Empidonax flaviventris. See Yellow-bellied flycatchers
Empidonax hammondii. See Hammond's flycatchers
Empidonax oberholseri. See Dusky flycatchers
Empidonax traillii. See Willow flycatchers
Empidornis semipartitus. See Silverbirds
Empire gudgeons, 5:377
Empusidae, 3:177
Emulation, 12:158
See also Behavior
Emus, 8:21, 8:53–56, 8:*54,* **8:83–88,** 8:*84,* 8:*85,* 8:*86,* 8:*87–88*
Emydidae. *See* New World pond turtles
Emydinae, 7:105
Emydocephalus annulatus. See Turtle-headed seasnakes
Emydura spp., 7:78
Emys orbicularis. See European pond turtles
Enamelled birds. *See* King of Saxony birds of paradise
Enamelled birds of paradise. *See* King of Saxony birds of paradise
Enantiornithes, 8:3
Encentrum astridae. See Astrid's rotifers
Encephalitis, from birds, 8:25
Encephalization quotient (EQ), 12:149
Encheliophis spp., 5:17
Enchelyopus cimbrius, 5:28
Encoding information, 12:152
Enculturation, of apes, 12:162
Endangered species, in zoos, 12:204, 12:208, 12:212
See also Conservation status
Endangered Species Act (U.S.), 1:118, 8:27, 12:184
Acipenseriformes, 4:217
American alligators, 7:48–49
beetles, 3:322
black-tailed prairie dogs, 16:155
dugongs, 15:201
Fea's muntjacs, 15:351
flatwoods salamanders, 6:358
flycatchers, 10:275
gray-backed sportive lemurs, 14:79
gray whales, 15:99
gray wolves, 14:277
Idaho ground squirrels, 16:158
insects, 3:83
island foxes, 12:192
lesser Malay mouse deer, 15:332
manatees, 15:209
Milne-Edwards's sportive lemurs, 14:81
northern sportive lemurs, 14:83
pythons, 7:421
red mouse lemurs, 14:43
red-tailed sportive lemurs, 14:80
salmons, 4:414
Sirenia, 15:195
small-toothed sportive lemurs, 14:83
sperm whales, 15:78
sportive lemurs, 14:76
spruce-fir moss spiders, 2:343
tiger salamanders, 6:360
tree squirrels, 16:167
vernal pool tadpole shrimps, 2:146
Vespertilioninae, 13:502
weasel sportive lemurs, 14:82
white-footed sportive lemurs, 14:82
woodpeckers, 10:150, 10:160
Endeiolepis, 4:9
Endemic species, 1:49–50
See also Conservation status
Endoconidae, 2:446
Endocoprids, 3:322
Endocrine disrupters, amphibians and, 6:56
Endocrine system, 3:28, 6:18–20
Endocytobiosis, 1:32
Endoderm, 2:15, 2:20, 2:35
See also Physical characteristics
Endodontoid snails, Pacific, 2:414
Endogeic worms, 2:66
Endometrium, 12:106
See also Physical characteristics
Endomyzostomatidae, 2:59
Endoparasites, 3:64
Endopterygota, 3:27, 3:315–316
Endosymbiosis, 1:32, 2:8, 2:11, 2:31
Endotheliodothelium placentation, 12:93–94
Endothermy, 12:3, 12:40–41, 12:50, 12:113
See also Physical characteristics
Endoxocrinus parrae. See West Atlantic stalked crinoids
Endracht hardyheads. *See* Eendracht land silversides
Eneoptera spp., 3:203
Eneoptera surinamensis. See Suriname clicking crickets
Energy, 7:10, 7:28, 7:43–45, 12:59, 12:71–72, 12:120
See also Physical characteristics
Energy conservation
Apodiformes, 9:416–417, 9:422
Caprimulgiformes, 9:370
Ciconiiformes, 8:235
common poorwills, 9:403
hummingbirds, 9:416, 9:440–442, 9:445–446
in migration, 8:32
nightjars, 9:403
Engleman's lizardfishes, 4:*431*
English dippers. *See* Eurasian dippers
English sparrows. *See* House sparrows
English starlings. *See* European starlings
Engraulidae. *See* Anchovies
Engraulis mordax. See Northern anchovies
Engraulis ringens. See Anchoveta
Engystoma carolinense. See Eastern narrow-mouthed toads
Engystoma ornatum. See Ornate narrow-mouthed frogs
Enhydra lutris. See Sea otters
Enhydris spp., 7:468
Enicocephalomorpha, 3:259
Enicognathus ferrugineus. See Austral conures
Enicurinae. *See* Forktails
Enicurus leschenaulti. See White-crowned forktails
Enigmonia spp., 2:452
Enneabatrachus spp., 6:12
Enneacanthus spp., 5:195
Enneacanthus gloriosus. See Blue-spotted sunfishes
Enneapterygius mirabilis. See Miracle triplefins
Enoicyla pusilla, 3:376
Enophrys spp., 5:183
Enophrys bison. See Buffalo sculpins
Enoplans, 1:11, **1:253–257,** 1:*254*

Enoplia, 1:283
Enoplosidae. *See* Oldwives
Enoplosus armatus. See Oldwives
Enrichment programs, in zoos, 12:209–210
Ensatina spp., 6:35, 6:*390*, 6:391, 6:*393*, 6:*398*–399
Ensatina eschscholtzii. See Ensatina spp.
Ensifera. *See* Long-horned grasshoppers
Ensifera spp. *See* Hummingbirds
Ensifera ensifera. See Sword-billed hummingbirds
Entale tusks. *See* Tusk shells
Entedononecremnus krauteri, 3:62
Entelodontidae, 15:264, 15:275
Enterobius vermicularis. See Pinworms
Enterocoely, 2:3, 2:21
Enterogona, 1:452
Enteropneusta. *See* Acorn worms
Entertainment
  birds and, 8:22
  zoos and, 12:203–204
Entobdella soleae, 1:213, 1:*214*, 1:*216*, 1:*218*–219
Entoconchidae, 2:445
Entocytherids, 2:312
Entognatha, 3:4
Entomacrodus cadenati, 5:343
Entomology, 1:37–38, 3:74, 3:77–78
  See also Behavior
Entomophagy, 3:80–81
Entomophila albogularis. See Rufous-banded honeyeaters
Entomophobia, 3:77
Entoprocts, 1:9–10, 1:13, **1:319–325,** 1:*321*, 1:*322*
Entotheonella palauensis, 1:45
Enulius spp., 7:469
Environment. *See* Habitats
Environment Australia, on Southern gastric brooding frogs, 6:153
Environment Protection and Biodiversity Conservation Act 1999 (Commonwealth of Australia), 13:76
Environmental Protection Agency (U.S.), on fertilizers, 6:57
Eoacanthocephala, 1:311
Eobatrachus spp., 6:12
Eobothus minimus, 5:449
Eobuglossus eocenicus, 5:449
Eocacecilia spp., 6:*8*, 6:11, 6:13
Eochanna chorlakkiensis, 5:437
Eodendrogale spp., 13:291
Eodiscoglossus spp., 6:11–12, 6:89
Eogaleus spp., 4:113
Eohypsibius nadjae, 2:116
Eolophus roseicapillus. See Galahs
Eomeropidae, 3:341, 3:342
Eonycteris spp. *See* Dawn fruit bats
Eonycteris spelaea. See Dawn fruit bats
Eopelobatinae, 6:119
Eopsaltria australis. See Eastern yellow robins
Eopterum devonicum, 3:12
Eorhinophrynus septentrionalis, 6:95
Eosentomidae, 3:93, 3:94
Eosentomoidea, 3:93
Eosentomon palustre, 3:*95*, 3:*96*
Eosimias spp., 14:2
Eosuchia, 7:5
Eosynanceja brabantica, 5:164
Eotragus spp., 15:265, 16:1
Eoxenopoides reuingi, 6:99
Eoxenos laboulbenei, 3:*337*, 3:*338*–339
Eozapus spp., 16:212
Eozapus setchuanus. See Chinese jumping mice
Eparctocyon, 15:343
Epauletted fruit bats, 13:312
  dwarf, 13:*338*, 13:*343*–344
  Wahlberg's, 13:*313*, 13:*338*, 13:*343*
Epeira spp. *See* Web-spinning spiders
Ephemera danica, 3:*126*
Ephemera vulgata. See Brown mayflies
Ephemeroptera. *See* Mayflies
Ephippidae, 4:69, 5:391
Ephippigerinae, 3:205
Ephippiorhynchus senegalensis. See Saddlebills
Ephydra brucei, 3:359
Ephydra cinera. See Brine flies
Ephydrids, 3:358
Epibatidine, 6:208
Epibulus insidiator. See Slingjaw wrasses
Epicaridea, 2:249, 2:250
Epicrates spp., 7:410, 7:411
Epicrates angulifer. See Cuban boas
Epicrates cenchria. See Rainbow boas
Epicrates exsul. See Abaco boas
Epicrates gracilis. See Haitian vine boas
Epicrates inornatus. See Puerto Rican boas
Epicrates monensis. See Mona boas
Epicrates subflavus. See Jamaican boas
Epicrionops spp., 6:416
Epicrionops marmoratus. See Marbled caecilians
Epicuticle, 3:18–19
Epidermis, 3:19, 12:36
  See also Physical characteristics
Epigeic worms, 2:66
Epigonichthys spp., 1:485, 1:487
Epigonichthys cultellus, 1:*489*, 1:495–*496*
Epigonichthys lucayanus. See Bahama lancelets
Epilamprinae, 3:148
Epimachus spp. *See* Birds of paradise
Epimachus fastuosus. See Black sicklebills
Epimenia australis, 2:380, 2:*381*, 2:*382*–*383*
Epimyrma spp., 3:69
Epinephelinae, 5:255, 5:261, 5:262
Epinephelini, 5:255
Epinephelus spp., 5:262
Epinephelus lanceoletus. See Brindlebasses
Epinephelus striatus. See Nassau groupers
Epineuralia, 2:4
Epiophlebia laidlawi. See Living fossils
Epipedobates spp., 6:66, 6:197–200, 6:*199*
Epipedobates tricolor. See Phantasmal poison frogs
Epipelagic species, 1:24
Epipelagic zone, 4:49, 4:56
Epiperipatus biolleyi, 2:*112*, 2:*113*
Epiplatys spp., 5:93
Epiplatys bifasciatus, 5:93
Epipyropids, 3:389
Episcada salvinia rufocincta. See South American butterflies
Episodic memory, 12:154–155
  See also Cognition
Epitheliochorial placentation, 12:93–94
Epitrachys rugosus, 2:75
Epixerus spp., 16:163
Epixerus ebii, 16:167
Epixerus wilsoni, 16:167
Epomophorus spp. *See* Epauletted fruit bats
Epomophorus gambianus. See Gambian epauletted fruit bats
Epomophorus wahlbergi. See Wahlberg's epauletted fruit bats
Epomops spp. *See* African epauletted bats
Epomops franqueti. See Singing fruit bats
Eponymous bat falcons, 8:314
Eptatretinae, 4:*78*
Eptatretus spp., 4:77
Eptatretus burgeri. See Japanese hagfishes
Eptatretus stoutii. See Pacific hagfishes
Eptesicus spp., 13:497, 13:498
Eptesicus fuscus. See Big brown bats
Eptesicus nilssoni. See Northern bats
Eptesicus serotinus. See Serotine bats
Epthianura albifrons. See White-fronted chats
Epthianura aurifrons. See Orange chats
Epthianura crocea. See Yellow chats
Epthianura lovensis. See Gibberbirds
Epthianura tricolor. See Crimson chats
Epthianuridae. *See* Australian chats
EQ (encephalization quotient), 12:149
Equal holoblastic cleavage, 2:20
  See also Reproduction
Equetus spp., 5:258
Equidae, **15:225–236,** 15:*231*
  behavior, 15:142, 15:219, 15:226–228
  conservation status, 15:221, 15:222, 15:229–230
  distribution, 12:130–131, 12:134, 15:217, 15:*225*, 15:226
  evolution, 15:136, 15:137, 15:215, 15:216, 15:225
  feeding ecology, 12:118–119, 15:220, 15:228
  habitats, 15:218, 15:226
  humans and, 15:222, 15:223, 15:230
  physical characteristics, 15:139, 15:140, 15:216, 15:217, 15:225–226
  reproduction, 15:221, 15:228–229
  species of, 15:232–235, 15:236*t*
  taxonomy, 15:225
  See also Asses; Horses; Zebras
Equilibrium theory of island biogeography, 3:56
Equus spp. *See* Equidae
Equus africanus. See African wild asses
Equus africanus africanus. See Nubian wild asses
Equus africanus somaliensis. See Somali wild asses
Equus asinus. See Domestic donkeys
Equus burchellii. See Plains zebras
Equus burchellii burchelli. See True plains zebras
Equus caballus. See Tarpans
Equus caballus caballus. See Domestic horses
Equus caballus przewalskii. See Przewalski's horses
Equus caballus silvestris. See Forest tarpans
Equus grevyi. See Grevy's zebras
Equus hemionus. See Asiatic wild asses
Equus hemionus hemippus. See Syrian onagers
Equus hemionus khur. See Khurs
Equus hemionus kulan. See Kulans
Equus hemionus onager. See Persian onagers
Equus kiang. See Kiangs
Equus quagga. See Quaggas
Equus zebra. See Mountain zebras
Equus zebra hartmanni. See Hartmann's mountain zebras
Equus zebra zebra. See Cape mountain zebras
Erdmann, M. V., 4:189

Erect-crested penguins, 8:148, 8:150
Eremalauda starki. See Stark's larks
Eremiainae, 7:297
Eremiaphilidae, 3:177
Eremiascincus spp., 7:329
Eremiascincus richardsonii. See Broad-banded sand swimmers
Eremitalpa spp. *See* Grant's desert golden moles
Eremitalpa granti. See Grant's desert golden moles
Eremobius phoenicurus. See Band-tailed earthcreepers
Eremodipus lichtensteini. See Lichtenstein's jerboas
Eremomela spp., 11:4, 11:7
Eremomela icteropygialis. See Yellow-bellied eremomelas
Eremophila alpestris. See Horned larks
Eremophila bilopha. See Temminck's larks
Eremopterix spp. *See* Sparrow-larks
Eremopterix australis. See Black-eared sparrow-larks
Eremopterix grisea. See Ashy-crowned finch-larks
Eremopterix leucotis. See Chestnut-backed sparrow-larks
Eremopterix nigriceps. See Black-crowned sparrow-larks
Eremopterix verticalis. See Gray-backed sparrow-larks
Eremotheres, 13:148
Erethistidae, 4:353
Erethizon spp., 16:366
Erethizon dorsatum. See North American porcupines
Erethizontidae. *See* New World porcupines
Eretmochelys imbricata. See Hawksbills
Ereuniidae, 5:179
Ergasilus sieboldi, 2:*302*, 2:307
Ergaticus versicolor. See Pink-headed warblers
Ericabatrachus spp., 6:247
Eridacnis radcliffei, 4:114
Erignathus barbaus. See Bearded seals
Erilepis zonifer. See Skilfishes
Erinaceidae, 13:193, 13:194, **13:203–214,** 13:*208*, 13:212*t*–213*t*
Erinaceinae, 13:203, 13:204
Erinaceus albiventris. See East African hedgehogs
Erinaceus amurensis. See Chinese hedgehogs
Erinaceus concolor. See Eastern European hedgehogs
Erinaceus europaeus. See Western European hedgehogs
Eriocheir sinensis. See Chinese mitten crabs
Eriocnemis spp. *See* Pufflegs
Eriocnemis luciani. See Sapphire-vented pufflegs
Eriocnemis mirabilis. See Colorful pufflegs
Eriocnemis nigrivestis. See Black-breasted pufflegs
Eriocraniidae, 3:54
Eristalis spp., 3:359
Eristicophis spp., 7:445
Erithacus calliope. See Siberian rubythroats
Ermia spp., 7:445
Ermines, 9:171, 12:116, 14:*324*, 14:*326*, 14:*327*–328
Ernest, K. A., 16:269
Erophylla sezekorni. See Buffy flower bats
Erpetoichthys spp., 4:209
Erpetoichthys calabaricus. See Reedfishes
Erpeton tentaculatus. See Tentacled snakes
Erpobdellidae, 2:76
Erpobdelliformes, 2:75
Erritzoe, Helga Boullet, 10:194
Erritzoe, Johannes, 10:194
Erycinae, 7:409–411
Erymnochelys madagascariensis. See Madagascan big-headed turtles
Erythrinidae, 4:335, 4:338
Erythrocebus spp. *See* Patas monkeys
Erythrocebus patas. See Patas monkeys
Erythrocercus spp., 11:97
Erythrocytes, 12:8, 12:36, 12:45, 12:60
See also Physical characteristics
Erythrogonys spp., 9:161
Erythrosquilloidea, 2:168, 2:171
Erythrura spp. *See* Parrotfinches
Erythrura cyaneovirens. See Red-headed parrotfinches
Erythrura kleinschmidti. See Pink-billed parrotfinches
Erythrura prasina. See Pin-tailed parrotfinches
Erythrura regia. See Royal parrotfinches
Erythrura viridifacies. See Green-faced parrotfinches
Eryx spp., 7:411
Eryx colubrinus. See East African sandboas
Eryx jayakari. See Arabian sandboas
Eryx johnii. See Brown sandboas
Eryx muelleri. See Sahara sandboas
Eryx somalicus, 7:411
Eryx tataricus, 7:411
ESA. *See* Endangered Species Act
Esacus spp. *See* Thick-knees
Esacus magnirostris. See Beach thick-knees
Esacus recurvirostris. See Great thick-knees
Escarpia spp., 2:91
Eschmeyer, W. N., 4:390
Eschrichtiidae. *See* Gray whales
Eschrichtius robustus. See Gray whales
Eskimo curlews, 9:103–104, 9:178, 9:180
Esmeraldas woodstars, 9:450
Esocidae, 4:58, 4:68
Esociformes, **4:379–387,** 4:*382*
Esox spp., 4:379, 4:380, 4:381
Esox lucius. See Northern pikes
Esox masquinongy. See Muskellunges
Esperlans. *See* California smoothtongues
Estesius spp., 6:13
Estivation cycle, 4:*202*, 4:203, 4:206
Estrelda rhodopareia. See Jameson's firefinches
Estrilda astrild. See Common waxbills
Estrilda poliopareia. See Anambra waxbills
Estrildidae. *See* Weaverfinches
Estrous periods, 12:109
See also Reproduction
Estuarine crocodiles. *See* Saltwater crocodiles
ESUs (evolutionary significant units), 3:85
Etheostoma caeruleum. See Rainbow darters
Etheostoma nigrum. See Johnny darters
Etheostoma olmstedi. See Tessellated darters
Ethics
in conservation biology, 3:84, 12:217–218
zoos and, 12:212
Ethiopian hedgehogs, 13:213*t*
Ethiopian region, 12:135–136
Ethiopian wolves, 14:262, 14:269, 14:273, 14:*275*, 14:*276*, 14:277–278
Ethology, 1:37–38
See also Behavior
Etmopteridae. *See* Lantern sharks
Etmopterus spp., 4:151, 4:152
Etroplus spp., 5:279
EU Birds Directive, on larks, 10:346
Euarchontoglires, 12:26, 12:*31*, 12:33
Eubacteria. *See* Bacteria
Eubalaena spp. *See* Right whales
Eubalaena australis. See Southern right whales
Eubalaena glacialis. See North Atlantic right whales
Eubalaena japonica. See North Pacific right whales
Eublepharinae, 7:259, 7:262
Eublepharis macularius. See Leopard geckos
Eubucco bourcierii. See Red-headed barbets
Eubucco tucinkae. See Scarlet-hooded barbets
Eucestoda, 1:226–227
Euchirus longimanus. See Long-armed chafers
Euchnemis fornasini. See Greater leaf-folding frogs
Euchoreutes spp., 16:211
Euchoreutes naso. See Long-eared jerboas
Euchoreutes setchuanus. See Chinese jumping mice
Euchoreutinae. *See* Long-eared jerboas
Euchroma gigantea. See Giant metallic ceiba borers
Eucidaris tribuloides. See Slate-pencil urchins
Eucla cods, 5:25–26, 5:28, 5:29
Euclichthyidae. *See* Eucla cods
Euclichthys polynemus. See Eucla cods
Euclimacia spp., 3:307
Euclimacia torquata. See Mantid lacewings
Eucnemis betsileo. See Betsileo reed frogs
Eucnemis seychellensis. See Seychelles treefrogs
Eucnemis viridiflavus. See Painted reed frogs
Eucopia spp., 2:225
Eucopiidae. *See Eucopia* spp.
Eucrossorhinus dasypogon. See Tasseled wobbegongs
Eucycliophora. *See* Wheel wearers
Eudendrium carneum, 1:30
Eudendrium glomeratum, 1:*130*, 1:139–140
Eudendrium racemosum, 1:25
Eudendrium rameum, 1:25
Euderma maculatum. See Spotted bats
Eudimorphodon spp., 7:*18*
Eudistoma spp., 1:452
Eudocimus ruber. See Scarlet ibises
Eudontomyzon spp., 4:83
Eudrilidae, 2:65, 2:66
Eudrilus eugeniae. See African worms
Eudromia elegans. See Elegant crested-tinamous
Eudromias spp., 9:161
Eudromias morinellus. See Eurasian dotterels
Eudynamys scolopacea. See Common koels
Eudyptes spp. *See* Penguins
Eudyptes chrysocome. See Rockhopper penguins
Eudyptes chrysolophus. See Macaroni penguins
Eudyptes pachyrhynchus. See Fiordland penguins
Eudyptes robustus. See Snares penguins
Eudyptes sclateri. See Erect-crested penguins
Eudyptula minor. See Little penguins
Euechinoidea, 1:401

Eugenes fulgens. See Magnificent hummingbirds
Euglandina rosea. See Rosy wolfsnails
Eugongylus spp., 7:327
Eugralla spp. *See* Tapaculos
Eugralla paradoxa. See Ochre-flanked tapaculos
Euholognatha, 3:142–143
Eukaryotes, 2:8–9
See also specific eukaryotes
Eukoenenia draco, 2:*338*, 2:346–347
Eukrohnia fowleri, 1:434, 1:435, 1:*436*, 1:440–*441*
Eukrohniidae, 1:433
Eulabeornis castaneoventris. See Chestnut rails
Eulacestoma nigropectus. See Ploughbills
Eulachons, 4:68, 4:392–*395*, 4:*394*, 4:400–401
Eulampis spp. *See* Hummingbirds
Eulampis jugularis. See Purple-throated caribs
Eulamprus quoyi, 7:329
Eulemur spp., 14:8
Eulemur coronatus. See Crowned lemurs
Eulemur macaco flavifrons. See Blue-eyed lemurs
Eulemur mongoz. See Mongoose lemurs
Euleptorhamphus viridis, 5:80
Eulichadidae, 3:54
Eulimnadia texana. See Texan clam shrimps
Eumalacostraca, 2:181, 2:185
Eumastacidae, 3:201
Eumastacoidea. *See* Monkey grasshoppers
Eumastax spp., 3:204
Eumeces spp. *See* Skinks
Eumeces laticeps. See Broad-headed skinks
Eumeces obsoletus. See Great Plains skinks
Eumecichthys spp., 4:447
Eumecichthys fiski. See Unicornfishes
Eumenes spp., 3:406
Eumenes fraternus. See Potter wasps
Eumeryx spp., 15:265, 15:266
Eumetazoans, 1:10
Eumetopias jubatus. See Steller sea lions
Eumomota superciliosa. See Turquoise-browed motmots
Eumops glaucinus. See Wagner's bonneted bats
Eumops perotis. See Western bonneted bats
Eumyias sordida. See Dull-blue flycatchers
Eunectes spp., 7:410
Eunectes murinus. See Green anacondas
Eungella day frogs. *See* Eungella torrent frogs
Eungella torrent frogs, 6:46, 6:*150*, 6:152–153
Eunice aphroditois, 2:*46*
Eunice viridis. See Palolo worms
Eunicella verrucosa. See Broad sea fans
Eunicids, 2:47
Euoticus spp., 14:23, 14:26
Euoticus elegantulus. See Southern needle-clawed bushbabies
Euoticus pallidus. See Northern needle-clawed bushbabies
Euparkerella spp., 6:156
Eupelycosauria, 12:10
Eupetaurus cinereus. See Woolly flying squirrels
Eupetes spp. *See* Rail babblers
Eupetes caerulescens. See Blue jewel-babblers
Eupetes macrocerus. See Malaysian rail-babblers
Eupetidae, **11:75–81**
Euphagus spp. *See* Icteridae
Euphagus carolinus. See Rusty blackbirds
Euphagus cyanocephalus. See Brewer's blackbirds
Euphanerops, 4:9
Euphausia spp., 2:185
Euphausia pacifica. See North Pacific krill
Euphausia superba. See Antarctic krill
Euphausiacea. *See* Krill
Euphausiidae, 2:185
Eupherusa spp. *See* Hummingbirds
Eupherusa cyanophrys. See Oaxaca hummingbirds
Eupherusa poliocerca. See White-tailed hummingbirds
Euphlyctis spp., 6:249
Euphlyctis cyanophlyctis, 6:250
Eupholus bennetti, 3:*315*
Euphractinae, 13:182
Euphractus sexcinctus. See Yellow armadillos
Euphydryas spp., 3:54
Euphyllia cristata. See Frogspawn corals
Euplectella aspergillum. See Venus's flower baskets
Euplectes spp., 11:375, 11:376
Euplectes ardens. See Red-collared widow-birds
Euplectes jacksoni. See Jackson's widow-birds
Euplectes orix. See Southern red bishops
Euplectes progne. See Long-tailed widows
Eupleres spp., 14:347
Eupleres goudotii. See Falanoucs
Euplerinae, 14:335
Eupodotis spp. *See* Bustards
Eupodotis humilis. See Little brown bustards
Eupodotis caerulescens. See Blue bustards
Eupolyphaga everestinia, 3:148
Euporismites balli, 3:305
Eupotmicroides spp., 4:151
Euprimates, 14:1
Euproctus asper. See Pyrenean brook salamanders
Euproctus platycephalus, 6:367
Euprotomicrus bispinatus. See Pygmy sharks
Eupsophus spp., 6:157
Euptilotis neoxenus. See Eared quetzals
Eupyrgidae, 1:419
Eurarchintoglires, 15:134
Eurasian avocets. *See* Pied avocets
Eurasian badgers. *See* European badgers
Eurasian beavers, 16:177, 16:*182*, 16:*183*
Eurasian bitterns, 8:*246*, 8:*256–257*
Eurasian blackbirds. *See* Blackbirds
Eurasian bullfinches, 11:*327*, 11:*335*
Eurasian chiffchaffs. *See* Chiffchaffs
Eurasian collared doves, 8:24
Eurasian coots, 9:52
Eurasian crag martins. *See* Crag martins
Eurasian cranes, 9:24–27, 9:29, 9:*30*, 9:*34*
Eurasian crows. *See* Carrion crows
Eurasian dippers, 10:475, 10:*476*, 10:477, 10:*478*, 10:*479*
Eurasian dotterels, 9:161, 9:163
Eurasian eagle-owls, 9:331, 9:345, 9:347, 9:*348*, 9:*352*, 9:*357*
Eurasian golden orioles, 11:301–302, 11:*429*, 11:*430*, 11:*434*
Eurasian green woodpeckers, 10:*148*
Eurasian griffons, 8:323
Eurasian hobbies, 8:315, 8:349
Eurasian jays, 11:504, 11:506–507, 11:*510*, 11:513–*514*
Eurasian kingfishers. *See* Common kingfishers
Eurasian lynx, 14:*378*, 14:*379*, 14:388
Eurasian magpies, 11:506, 11:*510*, 11:516–*517*
Eurasian minnows, 4:*306*, 4:*309*, 4:314
Eurasian nutcrackers. *See* Spotted nutcrackers
Eurasian nuthatches. *See* Nuthatches
Eurasian oystercatchers, 9:126, 9:127
Eurasian penduline tits. *See* European penduline tits
Eurasian plovers. *See* Northern lapwings
Eurasian pond turtles, **7:115–120**
Eurasian pratincoles, 9:152
Eurasian pygmy-owls, 9:347
Eurasian pygmy shrews, 13:*254*, 13:*256*, 13:261–262
Eurasian red squirrels. *See* Red squirrels
Eurasian river turtles, **7:115–120**
Eurasian rooks, 11:505
Eurasian scops-owls, 9:*352*, 9:*354*
Eurasian siskins, 11:*327*, 11:*333*
Eurasian sparrowhawks, 8:*319*, 8:321, 8:323
Eurasian spoonbills. *See* Spoonbills
Eurasian tawny owls. *See* Tawny owls
Eurasian thick-knees, 9:*144*
Eurasian tree sparrows. *See* Tree sparrows
Eurasian treecreepers, 11:177, 11:178, 11:*179*, 11:*180*
Eurasian turtledoves, 9:*253*
Eurasian water shrews, 13:197, 13:*249*, 13:*250*, 13:*253*, 13:*258*
Eurasian watersnakes, 7:468
Eurasian wild boars. *See* Eurasian wild pigs
Eurasian wild pigs, 12:179, 15:148, 15:149, 15:*283*, 15:*285*, 15:288
behavior, 12:146
conservation status, 15:*280*, 15:281
evolution, 15:*264*, 15:275
habitats, 15:277
humans and, 15:281–*282*
reproduction, 15:279
smell, 12:85
Eurasian wolves, 12:180
Eurasian wood-owls. *See* Tawny owls
Eurasian woodcocks, 9:177, 9:179–180, 9:*181*, 9:*182*
Eurasian wrynecks. *See* Northern wrynecks
Eurocephalus spp. *See* Helmet-shrikes
Europe. *See* Palaearctic region
European badgers, 12:77, 14:324, 14:*326*, 14:*329*, 14:331
behavior, 12:145, 14:258–259, 14:*319*, 14:321–323
feeding ecology, 14:260
physical characteristics, 14:321
European beaked whales. *See* Gervais' beaked whales
European beavers. *See* Eurasian beavers
European bee-eaters, 10:*42*, 10:*43*, 10:47–48
European bison, 16:24*t*
European brown bears. *See* Brown bears
European cabbage whites, 3:39, 3:*390*, 3:*399*
European carps. *See* Common carps
European catfishes. *See* European wels
European chameleons. *See* Common chameleons
European common frogs. *See* Brown frogs
European common quail, 8:25
European Community, pufferfish products and, 5:471
European Community Habitats Directive, 7:238, 16:246
European earwigs, 3:57, 3:*195*, 3:*198*, 3:*199*–200

European eels, 4:6, 4:*256*, 4:257, 4:*260*, 4:*261–262*
European field crickets, 3:91
European fire-bellied toads, 6:193
European fire salamanders, 6:*369*, 6:*373*
European free-tailed bats, 13:*486*, 13:495*t*
European genets. *See* Common genets
European goldfinches, 11:*327*, 11:*332–333*
European goshawks. *See* Northern goshawks
European green crabs, 2:33, 2:201
European green toads. *See* Green toads
European greenfinches. *See* Greenfinches
European ground squirrels, 16:147, 16:*150*, 16:*155*–156
European hamsters. *See* Black-bellied hamsters
European hares, 16:*487*, 16:*508*, 16:*510*, 16:*511*–512
  behavior, 16:482, 16:483, 16:*486*
  conservation status, 12:222
  distribution, 16:481, 16:505
  humans and, 16:488
  reproduction, 16:485–486, 16:508
European hedgehogs, 13:206
European honeybees. *See* Honeybees
European kingfishers. *See* Common kingfishers
European lancelets, 1:488, 1:*489*, 1:*490*, 1:495
European mantids, 3:*177*, 3:179, 3:*182*, 3:*184*, 3:185
European marsh crane flies, 3:*363*, 3:*369*, 3:373–374
European medicinal leeches, 2:75, 2:77–*78*, 2:79, 2:*80*–81
European mink, 12:117, 12:132, 14:324
European moles, 12:70–71, 12:*73*, 13:200, 13:*280*, 13:*281*, 13:*283*, 13:*285*–286
  communication, 12:76
European mouflons, 12:178
European mudminnows, 4:379, 4:380, 4:*382*, 4:*384*, 4:385–386
European newts, 6:365–366
European night crawlers, 2:66, 2:68
European nightjars, 9:368, 9:370, 9:402–404, 9:*407*, 9:411–*412*
European otters, 12:132, 14:*326*, 14:*330*
European pearly mussels, 2:*455*, 2:*462*–463
European penduline tits, 11:147, 11:*148*, 11:149, 11:*150*, 11:*151*
European perches, 5:49, 5:198
European pied flycatchers. *See* Pied flycatchers
European pigs, 12:191–192
European pilchards, 4:*280*, 4:285–*286*
European pine martens, 12:132, 14:332*t*
European plaice. *See* Plaice
European pond turtles, 7:*71*, 7:*107*, 7:*108*, 7:*112*
European rabbits, 16:*507*, 16:*510*, 16:513, 16:*514*
  behavior, 16:482, 16:483
  distribution, 16:481, 16:505–506
  habitats, 16:482
  humans and, 16:488, 16:509
  integrated rabbit control, 12:186–188, 12:193
  physical characteristics, 16:507
  reproduction, 16:485, 16:486–487, 16:507–508
  taxonomy, 16:505
European red deer, 12:19
European red squirrels. *See* Red squirrels
European Register of Marine Species, on Acoela, 1:179
European rock-thrushes. *See* Rock thrushes
European roe deer, 15:384, 15:*386*, 15:*388*
European rollers, 10:52, 10:*53*, 10:*54*, 10:*55*
European salamanders, **6:363–375,** 6:*368–369*
European sea basses, 4:*31*, 5:198
European serins, 11:323, 11:*327*, 11:*334*
European smelts, 4:392, 4:393
European snakeflies, common, 3:*299*, 3:*301*, 3:302
European sousliks. *See* European ground squirrels
European stag beetles, 3:323, 3:*326*, 3:*329*, 3:330–331
European starlings, 8:19, 8:24, 11:407–409, 11:*410*, 11:*413*, 11:*421–422*
European stone loaches. *See* Stone loaches
European storm-petrels, 8:136, 8:137, 8:138
European sturgeons. *See* Beluga sturgeons
European swallows. *See* Barn swallows
European treefrogs, 6:*232*, 6:*235*
European turtledoves, 9:*253*, 9:*256*–257
European Union Birds Directive. *See* EU Birds Directive
European Union Habitat and Species Directive, on smooth snakes, 7:474
European wasps, 3:72
European water shrews, 13:195
European wels, 4:*355*, 4:*360*, 4:366
European white storks, 8:20, 8:236, 8:265, 8:266, 8:*267*, 8:*269*, 8:272
European wild boars, 15:149
European wood wasps, 3:*410*, 3:*421*–422
European wrynecks. *See* Northern wrynecks
Euros. *See* Common wallaroos
Euroscaptor mizura. See Japanese mountain moles
Euroscaptor parvidens. See Small-toothed moles
Eurostopodus spp., 9:401
Eurostopodus diabolicus. See Heinrich's nightjars
Eurostopodus guttatus. See Spotted nightjars
Eurostopodus mystacalis. See White-throated nightjars
Euryalidae, 1:387
Euryapsida, 7:*5*, 7:14–15
Euryapteryx curtus. See Moas
Euryapteryx geranoides. See Moas
Eurycantha spp., 3:224
Eurycea spp., 6:34, 6:390–391
Eurycea bislineata. See Two-lined salamanders
Eurycea multiplicata, 6:29
Eurycea quadridigitata, 6:390
Eurycea rathbuni. See Texas blind salamanders
Euryceros prevostii. See Helmet vangas
Eurychelidon sirintarae. See White-eyed river martins
Eurycnema spp., 3:222
Eurycnema osiris. See Darwin stick insects
Eurydactylodes spp., 7:261
Eurydice pulchra, 2:251
Euryhaline fishes, 4:47
Eurylaimidae. *See* Broadbills
Eurylaimus blainvillii. See Clicking peltops
Eurylaimus ochromalus. See Black-and-yellow broadbills
Eurylaimus samarensis. See Visayan wattled broadbills
Eurylaimus steerii. See Mindanao broadbills
Eurymylids. *See* Rodents
Eurynorhynchus pygmeus. See Spoon-billed sandpipers
Eurypauropodidae, 2:375, 2:376
Eurypharyngidae, 4:271, 4:272
Eurypharynx pelecanoides. See Pelican eels
Eurypyga helias. See Sunbitterns
Eurypyga helias meridionalis, 9:73–74
Eurypygidae. *See* Sunbitterns
Eurystomus azureus. See Azure rollers
Eurystomus gularis. See Blue-throated rollers
Eurystomus orientalis. See Dollarbirds
Eurythoe complanata. See Fire worms
Euryzygomatomys spinosus. See Guiaras
Eusociality, 3:68, 12:146–147
Eusthenia nothofagi, 3:143
Eustheniidae, 3:141
Eusthenopteron spp., 6:*10*
Eustis, Dorothy Harrison, 14:293
Eustomias spp., 4:421, 4:422
Eustomias schmidti. See Scaleless dragonfishes
Eutaeniophorus spp., 5:106
Eutamias spp., 16:143
Eutardigrada, 2:115, 2:116, 2:117, 2:118
Euteleostei, 4:11
Eutetrarhynchidae, 1:226
Eutherians. *See* Placentals
Euthypoda spp., 3:203
Eutoxeres spp., 9:443
Eutoxeres aquila, 9:446–447, 9:456
Eutriorchis astur. See Serpent-eagles
Eutrophication, haplochromine cichlids and, 5:281
Evaporite lenses, 12:18
Even-toed ungulates. *See* Artiodactyla
Evening bats, 13:*504*, 13:*508*, 13:512–513
Evening grosbeaks, 8:24, 11:323, 11:*325*, 11:*327*, 11:*336*
Everett's hornbills. *See* Sumba hornbills
Everett's monarchs. *See* White-tipped monarchs
Evergreen bagworm moths. *See* Bagworms
Evermannellidae. *See* Sabertooth fishes
Evermannichthys spp., 5:374
Eversmann's hamsters, 16:247*t*
Eviota spp., 5:376
Evistias acutirostris. See Striped boarfishes
Evolution, **1:3–6**
  adaptive radiation and, 11:342, 11:350
  amphibians, 6:3, 6:4–5, **6:7–14,** 6:28
    anurans, 6:4–5, 6:11–13, 6:15, 6:61–62
      African treefrogs, 6:279–281
      Amero-Australian treefrogs, 6:4, 6:225–226
      Arthroleptidae, 6:265
      Asian toadfrogs, 6:109–110
      Asian treefrogs, 6:13, 6:291
      Australian ground frogs, 6:139–140
      Bombinatoridae, 6:83
      Bufonidae, 6:183
      Discoglossidae, 6:64, 6:89
      Eleutherodactylus spp., 6:155
      frogs, 6:4–5, 6:11–*13*, 6:15, 6:61–62
      glass frogs, 6:215
      leptodactylid frogs, 6:12–13, 6:155–157
      Madagascan toadlets, 6:317
      Mesoamerican burrowing toads, 6:95
      Myobatrachidae, 6:147
      narrow-mouthed frogs, 6:301–302

New Zealand frogs, 6:69
parsley frogs, 6:127
Pipidae, 6:99
poison frogs, 6:197
Ruthven's frogs, 6:211
Seychelles frogs, 6:135
shovel-nosed frogs, 6:273
spadefoot toads, 6:119
tailed frogs, 6:4, 6:64, 6:77
three-toed toadlets, 6:179
toads, 6:15, 6:61–62
true frogs, 6:245–248
vocal sac-brooding frogs, 6:173
caecilians, 6:4, 6:9–10, 6:11, 6:13, 6:15, 6:411
American tailed, 6:415–416
Asian tailed, 6:419
buried-eye, 6:431
gymnophionans, 6:13
Kerala, 6:425–426
tailless, 6:435
lissamphibians, 6:9, 6:11–13
salamanders, 6:4, 6:9–10, 6:11, 6:13, 6:15, 6:323–324
amphiumas, 6:405
Asiatic, 6:13, 6:291, 6:335–336
Caudata, 6:323–324
Cryptobranchidae, 6:13, 6:343
lungless, 6:13, 6:389–392
mole, 6:13, 6:355
Pacific giant, 6:13, 6:349
Proteidae, 6:377
Salamandridae, 6:13, 6:323–324, 6:363
sirens, 6:13, 6:327
torrent, 6:385
tetrapods, 6:7–10
birds, 8:3
albatrosses, 8:107, 8:113
Alcidae, 9:219
anhingas, 8:201, 8:207–209
Anseriformes, 8:363
ant thrushes, 10:239
Apodiformes, 9:415
asities, 10:187
Australian chats, 11:65
Australian creepers, 11:133
Australian fairy-wrens, 11:45
Australian honeyeaters, 11:235
Australian robins, 11:105
Australian warblers, 11:55
babblers, 10:505
barbets, 10:113
barn owls, 9:335
bee-eaters, 10:39
bird songs and, 8:42
birds of paradise, 11:489
bitterns, 8:239
Bombycillidae, 10:447
boobies, 8:211
bowerbirds, 11:477
broadbills, 10:177
bulbuls, 10:395
bustards, 9:23, 9:91
buttonquails, 9:11
Caprimulgiformes, 9:367
caracaras, 8:347
cassowaries, 8:75
Charadriidae, 9:161
Charadriiformes, 9:101, 9:107
chats, 10:483
chickadees, 11:155
chowchillas, 11:69
Ciconiiformes, 8:233
Columbidae, 9:247
Columbiformes, 9:241
condors, 8:275
Coraciiformes, 10:3–4
cormorants, 8:183, 8:201
Corvidae, 11:503
cotingas, 10:305
crab plovers, 9:121
Cracidae, 8:413
cranes, 9:23
cuckoo-shrikes, 10:385
cuckoos, 9:311
dippers, 10:475
diving-petrels, 8:143
doves, 9:247
drongos, 11:437
ducks, 8:369
eagles, 8:317–318
elephant birds, 8:103
emus, 8:83
Eupetidae, 11:75
fairy bluebirds, 10:415
Falconiformes, 8:313
falcons, 8:347
false sunbirds, 10:187
fantails, 11:83
finches, 11:323
flamingos, 8:303
flowerpeckers, 11:189
frigatebirds, 8:183, 8:193
frogmouths, 9:377
Galliformes, 8:399
gannets, 8:211
geese, 8:369
Glareolidae, 9:151
grebes, 8:169
Gruiformes, 9:1, 9:107
hammerheads, 8:261
Hawaiian honeycreepers, 11:341
hawks, 8:317–318
hedge sparrows, 10:459
herons, 8:233, 8:239
hoatzins, 8:465
honeyguides, 10:137
hoopoes, 10:61
hornbills, 10:71
hummingbirds, 9:437
ibises, 8:291
Icteridae, 11:301
ioras, 10:415
jacamars, 10:91
jacanas, 9:107
kagus, 9:41
kingfishers, 10:5–6
kiwis, 8:89
Laridae, 9:203–204
larks, 10:341
leafbirds, 10:415
logrunners, 11:69
long-tailed titmice, 11:141
lyrebirds, 10:329
magpie-shrikes, 11:467
manakins, 10:295
mesites, 9:5
moas, 8:95
monarch flycatchers, 11:97
motmots, 10:31–32
moundbuilders, 8:403–404
mousebirds, 9:469
mudnest builders, 11:453
New World finches, 11:263
New World quails, 8:455
New World vultures, 8:233, 8:275
New World warblers, 11:285–287
New Zealand wattle birds, 11:447
New Zealand wrens, 10:203
nightjars, 9:401
nuthatches, 11:167
oilbirds, 9:373–374
Old World flycatchers, 11:25
Old World warblers, 11:1–2
Oriolidae, 11:427
ostriches, 8:99
ovenbirds, 10:209
owlet-nightjars, 9:387
owls, 9:331
oystercatchers, 9:125
painted-snipes, 9:115
palmchats, 10:455
pardalotes, 11:201
parrots, 9:275
Passeriformes, 10:170
Pelecaniformes, 8:183
pelicans, 8:183, 8:225
penduline titmice, 11:147
penguins, 8:147–148
Phasianidae, 8:433
pheasants, 8:433
Philippine creepers, 11:183
Picidae, 10:147
Piciformes, 10:85–86
pigeons, 9:245, 9:247
pipits, 10:371
pittas, 10:193
plantcutters, 10:325
potoos, 9:395
Procellariidae, 8:123
Procellariiformes, 8:107
pseudo babblers, 11:127
puffbirds, 10:101
rails, 9:45
Raphidae, 9:269
Recurvirostridae, 9:133
rheas, 8:69
rollers, 10:51
sandgrouse, 9:231
sandpipers, 9:175
screamers, 8:393
scrub-birds, 10:337
secretary birds, 8:343
seedsnipes, 9:189
seriemas, 9:85
sharpbills, 10:291
sheathbills, 9:197
shoebills, 8:233, 8:287
shrikes, 10:425
sparrows, 11:397
spoonbills, 8:291
storks, 8:233, 8:265
storm-petrels, 8:135
Struthioniformes, 8:53–54
Sturnidae, 11:407
sunbirds, 11:207
sunbitterns, 9:73
sungrebes, 9:69
swallows, 10:357
swans, 8:369

swifts, 9:421
tapaculos, 10:257
thick-knees, 9:143
thrushes, 10:483
tinamous, 8:57
titmice, 11:155
todies, 10:25
toucans, 10:125
tree swifts, 9:433
treecreepers, 11:177
trogons, 9:477
tropicbirds, 8:183, 8:187
trumpeters, 9:77
turacos, 9:299
typical owls, 9:345
tyrant flycatchers, 10:269–270
vanga shrikes, 10:439
Vireonidae, 11:255
wagtails, 10:371
weaverfinches, 11:353
weavers, 11:375–376
whistlers, 11:115
white-eyes, 11:227
woodcreepers, 10:229
woodhoopoes, 10:65
woodswallows, 11:459
wrens, 10:525
fishes, **4:9–13,** 6:7
Acanthuroidei, 5:391
Acipenseriformes, 4:213
Albuliformes, 4:249
angelsharks, 4:161
anglerfishes, 5:47
Anguilliformes, 4:255
Atheriniformes, 5:67
Aulopiformes, 4:431
Australian lungfishes, 4:197
beardfishes, 5:1
behavior and, 4:70
Beloniformes, 5:79
Beryciformes, 5:113
bichirs, 4:209
blennies, 5:341
bowfins, 4:229
bullhead sharks, 4:97
Callionymoidei, 5:365
carps, 4:297
catfishes, 4:351
characins, 4:335
chimaeras, 4:91
coelacanths, 4:189–190
Cypriniformes, 4:297, 4:321
Cyprinodontiformes, 5:89–90
dogfish sharks, 4:151
dories, 5:123
eels, 4:255
electric eels, 4:369
Elopiformes, 4:243
Esociformes, 4:379
flatfishes, 5:449–450
Gadiformes, 5:25
gars, 4:221
Gasterosteiformes, 5:131–132
gobies, 5:373–374
Gobiesocoidei, 5:355
Gonorynchiformes, 4:289–290
ground sharks, 4:113
Gymnotiformes, 4:369
hagfishes, 4:9–10, 4:77
herrings, 4:277
Hexanchiformes, 4:143
Labroidei, 5:275–276, 5:293
labyrinth fishes, 5:427
lampreys, 4:9–10, 4:83
Lampridiformes, 4:447–448
lanternfishes, 4:441
lungfishes, 4:201
mackerel sharks, 4:131–132
minnows, 4:297
morays, 4:255
mudminnows, 4:379
mullets, 5:59
Ophidiiformes, 5:15–16
Orectolobiformes, 4:105
Osmeriformes, 4:389–390
Osteoglossiformes, 4:231
Percoidei, 5:195, 5:211, 5:219, 5:235, 5:255
Percopsiformes, 5:5
pikes, 4:379
ragfishes, 5:351
Rajiformes, 4:173–174
rays, 4:10, 4:173–174
Saccopharyngiformes, 4:271
salmons, 4:405
sawsharks, 4:167
Scombroidei, 5:405
Scorpaeniformes, 5:157, 5:163–164, 5:179
Scorpaenoidei, 5:163–164
sharks, 4:10
skates, 4:10, 4:11, 4:173–174
snakeheads, 5:437
South American knifefishes, 4:369
southern cod-icefishes, 5:321
Stephanoberyciformes, 5:105
stingrays, 4:11, 4:173
Stomiiformes, 4:421
Stromateoidei, 5:421
Synbranchiformes, 5:151
Tetraodontiformes, 5:467
toadfishes, 5:41
Trachinoidei, 5:331
Zoarcoidei, 5:309–310
insects, **3:7–16,** 3:44–45
book lice, 3:243
bristletails, 3:113
caddisflies, 3:375
cockroaches, 3:147
Coleoptera, 3:315
diplurans, 3:107
Diptera, 3:357
earwigs, 3:195
fleas, 3:347
Hemiptera, 3:259
Hymenoptera, 3:405
lacewings, 3:305
Lepidoptera, 3:383
major transitions in, 3:67
mantids, 3:177
Mantophasmatodea, 3:217
mayflies, 3:125
Mecoptera, 3:341
Megaloptera, 3:289
Odonata, 3:133
Orthoptera, 3:201
Phasmida, 3:221
Phthiraptera, 3:249
proturans, 3:93
rock-crawlers, 3:189
snakeflies, 3:297
social behavior, 3:70–71
springtails, 3:99
stoneflies, 3:141
strepsipterans, 3:335
termites, 3:161–163
thrips, 3:281
Thysanura, 3:119
webspinners, 3:233
zorapterans, 3:239
limbed vertebrates, 6:7–11
lower metazoans and lesser deuterostomes, **1:7–14,** 1:*8*
acoels, 1:5, 1:7, 1:179
anoplans, 1:245
arrow worms, 1:433
biodiversity and, 1:49
box jellies, 1:147
calcareous sponges, 1:57
cnidarians
Anthozoa, 1:103
Hydrozoa, 1:123
comb jellies, 1:169
demosponges, 1:77–78
echinoderms
Crinoidea, 1:*12*, 1:355
Echinoidea, 1:401
Ophiuroidea, 1:387
sea cucumbers, 1:417, 1:418–419
sea daisies, 1:381
sea stars, 1:367
enoplans, 1:253
entoprocts, 1:319
flatworms
free-living flatworms, 1:185
monogeneans, 1:213
tapeworms, 1:226–227
gastrotrichs, 1:269
glass sponges, 1:67
gnathostomulids, 1:331
hair worms, 1:305
hemichordates, 1:443
jaw animals, 1:327
jellyfish, 1:3–4, 1:6, 1:*9*, 1:153
kinorhynchs, 1:275
lancelets, 1:485
larvaceans, 1:473
nematodes
roundworms, 1:283
secernenteans, 1:293
Orthonectida, 1:99
placozoans, 1:87
priapulans, 1:337
rhombozoans, 1:93
rotifers, 1:259
Salinella salve, 1:91
salps, 1:467
sea squirts, 1:451–452
sorberaceans, 1:479
thorny headed worms, 1:311
Trematoda, 1:197
wheel wearers, 1:351
mammals, 12:10–11
aardvarks, 15:155
agoutis, 16:407
anteaters, 13:171
armadillos, 13:181–182
Artiodactyla, 15:131–133, 15:135–138, 15:263–267
aye-ayes, 14:85
baijis, 15:19

bandicoots, 13:1
dry-country, 13:9
rainforest, 13:9
bats, 13:308–309
American leaf-nosed, 13:413
bulldog, 13:443
disk-winged, 13:473
Emballonuridae, 13:355
false vampire, 13:379
funnel-eared, 13:459
horseshoe, 13:387
Kitti's hog-nosed, 13:367
Molossidae, 13:483
mouse-tailed, 13:351
moustached, 13:435
New Zealand short-tailed, 13:453
Old World fruit, 13:319, 13:333
Old World leaf-nosed, 13:401
Old World sucker-footed, 13:479
slit-faced, 13:371
smoky, 13:467
Vespertilionidae, 13:497, 13:519
bears, 14:295–296
beavers, 16:177
bilbies, 13:19
botos, 15:27
Bovidae, 16:1–2
Antilopinae, 16:45
Bovinae, 16:11
Caprinae, 16:87–89
duikers, 16:73
Hippotraginae, 16:27–28
Neotraginae, 16:59
bushbabies, 14:2, 14:3, 14:23
Camelidae, 15:313–314
Canidae, 14:265–266
capybaras, 16:401
Carnivora, 14:255–256
cats, 14:369
Caviidae, 16:389
Cetacea, 15:2–4
Balaenidae, 15:3, 15:107
beaked whales, 15:59
dolphins, 15:41–42
franciscana dolphins, 15:23
Ganges and Indus dolphins, 15:13
gray whales, 15:93
Monodontidae, 15:81
porpoises, 15:33
pygmy right whales, 15:103
rorquals, 12:*66*, 15:119
sperm whales, 15:73
chevrotains, 15:325–326
Chinchillidae, 16:377
coevolution, 12:216–217
colugos, 13:299–300
conservation biology and, 12:218
convergent, 12:27
coypus, 16:473
dasyurids, 12:287
Dasyuromorphia, 12:277
deer
Chinese water, 15:373
muntjacs, 15:343
musk, 15:335
New World, 15:379–382
Old World, 15:357–358
Dipodidae, 16:211
Diprotodontia, 13:31–32
dormice, 16:317
duck-billed platypuses, 12:228, 12:243
Dugongidae, 15: 199
echidnas, 12:235
elephants, 15:161–165
Equidae, 15:225
Erinaceidae, 13:203
giant hutias, 16:469
gibbons, 14:207–209
Giraffidae, 15:399
great apes, 12:33, 14:225
gundis, 16:311
Herpestidae, 14:347
Heteromyidae, 16:199
hippopotamuses, 15:301–302, 15:304
humans, 14:241–244, 14:*243*
hutias, 16:461
Hyaenidae, 14:359–360
hyraxes, 15:177–178
Ice Ages, 12:17–25, **12:17–25**
Indriidae, 14:63
Insectivora, 13:193–194
intelligence and, 12:149–150
koalas, 13:43
Lagomorpha, 16:479–480
lemurs, 14:47
Cheirogaleidae, 14:35
sportive, 14:73–75
Leporidae, 16:505
Lorisidae, 14:13
Macropodidae, 13:31–32, 13:83
manatees, 15: 205
marine mammals, **12:62–68**
Megalonychidae, 13:155
molecular genetics and phylogenetics, **12:26–35**
moles
golden, 13:215–216
marsupial, 13:25
monitos del monte, 12:273
monkeys
Atelidae, 14:155
Callitrichidae, 14:115
Cebidae, 14:101
cheek-pouched, 14:187–189
leaf-monkeys, 14:171–172
night, 14:135
Pitheciidae, 14:143
monotremes, 12:11, 12:33, 12:227–228
mountain beavers, 16:131
Muridae, 16:122, 16:281–282
Arvicolinae, 16:225
hamsters, 16:239
Murinae, 16:249
Sigmodontinae, 16:263–264
musky rat-kangaroos, 13:69–70
Mustelidae, 14:319
numbats, 12:303
nutritional adaptations, **12:120–128**
octodonts, 16:433
opossums
New World, 12:249–250
shrew, 12:267
Otariidae, 14:393
pacaranas, 16:385–386
pacas, 16:417
pangolins, 16:107
peccaries, 15:291
Perissodactyla, 15:215–216
Petauridae, 13:125
Phalangeridae, 13:57
pigs, 15:275
pikas, 16:491
pocket gophers, 16:185
porcupines
New World, 16:365–366
Old World, 16:351
possums
feather-tailed, 13:139–140
honey, 13:135
pygmy, 13:105–106
primates, 14:1–3
Procyonidae, 14:309
pronghorns, 15:411
Pseudocheiridae, 13:113
rat-kangaroos, 13:73
rats
African mole-rats, 16:339
cane, 16:333
chinchilla, 16:443
dassie, 16:329
spiny, 16:449
rhinoceroses, 15:249, 15:251
rodents, 16:121–122
sengis, 16:517–518
sexual selection and, 12:101–102
shrews
red-toothed, 13:247
West Indian, 13:243
white-toothed, 13:265
Sirenia, 15: 191–192
solenodons, 13:237
springhares, 16:307
squirrels
flying, 16:135
ground, 16:143
scaly-tailed, 16:299
tree, 16:163
subterranean adaptive, **12:69–78**
Talpidae, 13:279
tapirs, 15:237–238
tarsiers, 14:91
tenrecs, 13:225
three-toed tree sloths, 13:161
tree shrews, 13:289–291
true seals, 14:417
tuco-tucos, 16:425
ungulates, 15:131–138
Viverridae, 14:335
walruses, 14:409
wombats, 13:51
Xenartha, 13:147–149
See also Convergent evolution
of migration, 8:31–32
protostomes, **2:7–14**, 2:*9*, 2:*10*
amphionids, 2:195
amphipods, 2:261
anaspidaceans, 2:181
aplacophorans, 2:14, 2:379
Arachnida, 2:333
articulate lampshells, 2:521
bathynellaceans, 2:177
beard worms, 2:85
bivalves, 2:451
caenogastropods, 2:445
centipedes, 2:353
cephalocarids, 2:131
Cephalopoda, 2:475
chitons, 2:393
clam shrimps, 2:147
copepods, 2:299

cumaceans, 2:229
Decapoda, 2:197–198
deep-sea limpets, 2:435
earthworms, 2:65
echiurans, 2:103
fairy shrimps, 2:135
fish lice, 2:289
horseshoe crabs, 2:327
Isopoda, 2:249
krill, 2:185
leeches, 2:75
leptostracans, 2:161
lophogastrids, 2:225
mantis shrimps, 2:167
marine bryozoans, 2:503, 2:509
mictaceans, 2:241
millipedes, 2:363
monoplacophorans, 2:387
mussel shrimps, 2:311
mysids, 2:215
mystacocarids, 2:295
myzostomids, 2:59
Neritopsina, 2:439
nonarticulate lampshells, 2:515
onychophorans, 2:9, 2:109
pauropods, 2:375
peanut worms, 2:97
phoronids, 2:491
Polychaeta, 2:45
Pulmonata, 2:411
remipedes, 2:125
sea slugs, 2:403
sea spiders, 2:321
spelaeogriphaceans, 2:243
symphylans, 2:371
tadpole shrimps, 2:141
tanaids, 2:235
tantulocaridans, 2:283
Thecostraca, 2:273
thermosbaenaceans, 2:245
tongue worms, 2:317–318
true limpets, 2:423
tusk shells, 2:469
Vestimentifera, 2:91
Vetigastropoda, 2:429
water bears, 2:115
water fleas, 2:153
reptiles, **7:12–22**, 7:*13*
African burrowing snakes, 7:461
African sideneck turtles, 7:129
Afro-American river turtles, 7:137
Agamidae, 7:209
Alligatoridae, 7:171
Anguidae, 7:339
Australo-American sideneck turtles, 7:77, 7:81–83
big-headed turtles, 7:135
blindskinks, 7:271
blindsnakes, 7:379
boas, 7:409
Central American river turtles, 7:99
chameleons, 7:223–224
colubrids, 7:465–467
Cordylidae, 7:319
crocodilians, 7:18–19, 7:157
Crocodylidae, 7:179
early blindsnakes, 7:369
Elapidae, 7:483–484
false blindsnakes, 7:387
false coral snakes, 7:399
file snakes, 7:439
Florida wormlizards, 7:283–284
Gekkonidae, 7:259
Geoemydidae, 7:115
gharials, 7:167
Helodermatidae, 7:353
Iguanidae, 7:243
Kinosternidae, 7:121
knob-scaled lizards, 7:347
Lacertidae, 7:297
leatherback seaturtles, 7:101
lizards, 7:16
mesosaurs, 7:14
Microteiids, 7:303
mole-limbed wormlizards, 7:279
Neotropical sunbeam snakes, 7:405
New World pond turtles, 7:105
night lizards, 7:291
pareiasaurs, 7:14
pig-nose turtles, 7:75
pipe snakes, 7:395
pythons, 7:419
seaturtles, 7:14, 7:85, 7:89–91
shieldtail snakes, 7:391
skinks, 7:327–328
slender blindsnakes, 7:373
snakes, 7:16–17
snapping turtles, 7:93
softshell turtles, 7:151
spade-headed wormlizards, 7:287
splitjaw snakes, 7:429–430
squamata, 7:195–198
sunbeam snakes, 7:401
Teiidae, 7:309
tortoises, 7:143
Tropidophiidae, 7:433
tuatara, 7:189, 7:*190*
turtles, 7:13–14, 7:65
Varanidae, 7:359
Viperidae, 7:445–446
wormlizards, 7:273
Evolutionary significant units (ESUs), 3:85
Ex situ conservation, 12:223–224
See also Conservation status
Exaetoleon obtabilis, 3:309
Excited coloration, 7:9
See also Physical characteristics
Excretion
avian, 8:10–11
insects, 3:21
Excretory system
amphibian, 6:17–18, 6:298
fishes, 4:24–25
Exercise, 7:10
Exhibits, zoo, 12:204–205, 12:*210*
Exiliboa spp., 7:434
Exiliboa placata, 7:434
Exilisciurus spp., 16:163
Exite theory, 3:12
Exocoetidae. *See* Flyingfishes
Exocoetoidei, 5:79
Exoskeletons, 2:*4*, 2:14, 3:8, 3:17, 3:*27*
See also Physical characteristics
Exotics, introduction of, 12:215
Exploitation, of wildlife, 12:215
See also Humans
Exports. *See* Imports and exports
Extatosoma tiaratum, 3:222
Extatosoma tiaratum tiaratum. *See* Macleay's spectres
External brooding, 2:23
See also Reproduction
External fertilization, 2:17–18, 2:23
See also Reproduction
External skeletons, 2:*18*
See also Physical characteristics
Extinct species
amphibians, 6:56–59
Cynops wolterstorffi, 6:371
Discoglossus nigriventer, 6:90
golden coqui, 6:165
leptodactylid frogs, 6:159
Southern gastric brooding frogs, 6:153
true frogs, 6:251
birds, 4:75–76
babblers, 10:507
bonin siskins, 11:324
crested shelducks, 8:367
dodos, 9:241, 9:245, 9:270–271, 9:272
elephant birds, 8:103–104
emus, 8:87
Eurasian dippers, 10:477
gerygones, 11:57
great auks, 9:219–220, 9:228
grebes, 8:171
Gruiformes, 9:1, 9:3
Guadalupe storm-petrels, 8:110, 8:138
Guadeloupe flickers, 10:156
Hawaiian honeycreepers, 11:343, 11:349, 11:350
huias, 11:448
hummingbirds, 9:443
Kangaroo Island emus, 8:87
kokakos, 11:449
loons, 8:159
Madagascar pochards, 8:367
moas, 8:98
monarch flycatchers, 11:98
New Zealand quails, 8:437
New Zealand wrens, 10:203, 10:204, 10:207
Panurus biarmicus kosswigi, 10:521
parrots, 9:276
passenger pigeons, 9:245, 9:251–252
pink-headed ducks, 8:367
pipios, 11:118
rails, 9:51
Raphidae, 9:270–271, 9:272–273
Regulus calendula obscura, 11:7
solitaires, 10:489
spectacled cormorant, 8:186
starlings, 11:410
Struthiones, 8:99
tiekes, 11:450
white-eyes, 11:229
fishes
Aipichthyidae, 5:1
catfishes, 4:354
cichlids, 5:280
Cottus echinatus, 5:183
Cypriniformes, 4:303, 4:322
Cyprinodontiformes, 5:94
duckbilled buntingis, 5:81
ground sharks, 4:113
Osmeriformes, 4:393
Percoidei, 5:199
Rheocles sikorae, 5:70
salmons, 4:407
insects, 3:52, 3:53, 3:87
Alloperla roberti, 3:143

Antioch dunes shieldbacks, 3:207
Bojophlebia prokopi, 3:125
central valley grasshoppers, 3:207
Cercyonis sthenele sthenele, 3:388
Glaucopsyche lygdamus xerces, 3:388
Hydropsyche tobiasi, 3:377
Ko'okay spurwing long-legged flies, 3:360
Lanai pomace flies, 3:360
Necrotaulidae, 3:375
Oahu deceptor bush crickets, 3:207
Rhyacophila amabilis, 3:377
Rocky Mountain grasshoppers, 3:57
Speyeria adiaste atossa, 3:388
Tobias's caddisflies, 3:57
Triaenodes phalacris, 3:377
Triaenodes tridonata, 3:377
Tshekardocoleidae, 3:315
Tshekardoleidae, 3:329
Volutine stoneyian tabanid flies, 3:360
lower metazoans and lesser deuterostomes, 1:48, 1:50–53
Bryoniida, 1:153
Graptolithina, 1:443
Hemidasys agaso, 1:270
Heteractinida, 1:57
mammals, 12:129, 15:138, 16:126
animal intelligence and, 12:151
Archaoindris spp., 14:63
Archeoindris fontoynonti, 14:63
Archeolemur spp., 14:63
Babakotia spp., 14:63
baboon lemurs, 14:63, 14:72
Barbados raccoons, 14:262, 14:309, 14:310
Barbary lions, 14:380
black-footed ferrets, 14:262, 14:324
broad-faced potorros, 13:39
bubal hartebeests, 16:33
Camelus thomasi, 12:179
Caribbean monk seals, 14:422
causes of, 12:214–216
Columbian mammoths, 12:139
Cuscomys oblativa, 16:443
Daubentonia robusta, 14:85, 14:86, 14:88
desert bandicoots, 13:5, 13:11, 13:17*t*
dusky flying foxes, 13:331*t*
Falkland Island wolves, 14:262, 14:272
giant hutias, 16:469–471, 16:*470*, 16:471*t*
gray whales, 15:99
greater New Zealand short-tailed bats, 13:454, 13:455, 13:458
greater sloth lemurs, 14:71
Guadalupe storm-petrels, 14:291
Guam flying foxes, 13:332*t*
Hadropithecus spp., 14:63
Hippopotamus lemerlei, 12:136
Hypnomys spp., 16:317–318
imposter hutias, 16:467*t*
island biogeography and, 12:220–221
Jamaican monkeys, 12:129
koala lemurs, 14:73–74, 14:75
lesser bilbies, 13:22
lesser Haitian ground sloths, 13:159
Megaladapis edwardsi, 14:73
Megaladapis grandidieri, 14:73
Megaladapis madagascariensis, 14:73
Megaroyzomys spp., 16:270
Mesopropithecus spp., 14:63
Miss Waldron's red colobus, 12:214
Natalus stramineus primus, 13:463
Nesophontes spp., 12:133
Nesoryzomys darwini, 16:270
Nesoryzomys indefessus, 16:270
Omomyidae, 14:91
Oryzomys galapagoensis, 16:270
Oryzomys nelsoni, 16:270
Paleopropithecus spp., 14:63
pig-footed bandicoots, 13:*5*, 13:11, 13:15
Piliocolobus badius waldronae, 14:179
quaggas, 15:221, 15:*236t*
red deer, 15:370
red gazelles, 12:129
Robert's lechwes, 16:33
saber-toothed tigers, 14:369
Samana hutias, 16:467*t*
Schomburgk's deer, 15:360
sea minks, 14:262
sloth lemurs, 14:63
Steiromys spp., 16:366
Steller's sea cows, 12:215, 15:191, 15:195–196, 15:201, 15:203
Syrian onagers, 15:222
tarpans, 15:222
Tasmanian wolves, 12:137, 12:307
Toolache wallabies, 13:39
true plains zebras, 15:222
West Indian monk seals, 12:133, 14:435*t*
West Indian shrews, 13:244
Zanzibar leopards, 14:385
protostomes
Lottia alveus, 2:425
Rostroconchia, 2:469
Thermosphaeroma thermophilum, 2:252
reptiles
Alsophis sanctaecrucis, 7:470
Ameiva spp., 7:312
Delcourt's giant geckos, 7:259, 7:262–263
Eastwood's seps, 7:322
file snakes, 7:439
Galápagos tortoises, 7:147
Leiocephalus eremitus, 7:246–247
Leiocephalus herminieri, 7:246–247
Megalania prisca, 7:359
Navassa woodsnakes, 7:435
Phelsuma edwardnewtoni, 7:262
polyglyphanodontines, 7:309
reptilian skulls and, 7:4–5
smooth-scaled splitjaws, 7:429
Typhlops cariei, 7:382
Exxon-Valdez oil spill, 8:161
Eye gnats, 3:361
Eyed click beetles, 3:*325*, 3:*327*, 3:329–330
Eyed finger sponges, 1:*81*, 1:*82*
Eyed lizards, 7:298
Eyelight fishes, 5:115
Eyes
chameleons, 7:32, 7:225–226
crocodilians, 7:159
insects, 3:27–*28*
mammals, 12:5–6, 12:50, 12:77, 12:*80*
reptiles, 7:8–9
See also Physical characteristics; Vision
Eyespots, 6:*64*, 6:66
Eysacoris fabricii. *See* Shield bugs

F

Face flies, 3:361
Face mites. *See* Demodicids
Faces, 12:4–6
See also Physical characteristics
Facetotecta. *See Hansenocaris* spp.
Facial discs, 9:346, 9:*347*
Faciola spp., 1:200
Faciolidae, 1:199
Facultative symbiosis, 1:32, 2:31
See also Behavior
Fahay, M. P.
on Gadiformes, 5:25, 5:29
on red hakes, 5:28
on white hakes, 5:28, 5:39
Fairy armadillos, 13:149, 13:*185*, 13:*186*, 13:*189*, 13:190
Fairy basslets, 4:44, 4:48, 4:67, 5:255, 5:259, 5:261
Fairy bluebirds, **10:415–424,** 10:*416*, 10:*418*, 10:*423*
Fairy penguins. *See* Little penguins
Fairy pittas, 10:194, 10:195
Fairy shrimps, 2:*135*, **2:135–140,** 2:*138*
behavior, 2:136–137
conservation status, 2:137
distribution, 2:136
evolution, 2:135
feeding ecology, 2:137
habitats, 2:136
humans and, 2:137
physical characteristics, 2:135–136
reproduction, 2:137
species of, 2:139–140
taxonomy, 2:135
Fairy terns, 9:*103*
Fairy-wrens, Australian, **11:45–53,** 11:*47*
Falabella horses, 12:176
Falanoucs, 14:335, 14:336, 14:338, 14:*339*, 14:*342*, 14:343
Falcated teals, 8:370
Falcipennis africanus. *See* Gray-winged fracolins
Falcipennis canadensis. *See* Spruce grouse
Falco alopex. *See* Fox kestrels
Falco amurensis. *See* Amur falcons
Falco araea. *See* Seychelles kestrels
Falco berigora. *See* Brown falcons
Falco cenchroides. *See* Australian kestrels
Falco cherrug. *See* Saker falcons
Falco columbarius. *See* Merlin falcons
Falco eleonorae. *See* Eleonora's falcons
Falco fasciinucha. *See* Taita falcons
Falco hypoleucos. *See* Gray falcons
Falco naumanni. *See* Lesser kestrels
Falco novaeseelandiae. *See* New Zealand falcons
Falco peregrinus. *See* Peregrine falcons
Falco piscator. *See* Western gray plantain-eaters
Falco punctatus. *See* Mauritius kestrels
Falco rufigularis. *See* Bat falcons; Eponymous bat falcons
Falco rusticolus. *See* Gray morph gyrfalcons; Gyrfalcons
Falco subbuteo. *See* Eurasian hobbies
Falco subniger. *See* Black falcons
Falco vespertinus. *See* Red-footed falcons
Falconidae. *See* Caracaras; Falcons
Falconiformes, **8:313–316**
evolution, 8:275
feeding ecology, 8:277, 8:278
habitats, 8:277
humans and, 8:278
physical characteristics, 8:276

reproduction, 8:278
species of, 8:*280–284*
taxonomy, 8:275
See also Eagles; Falcons; Hawks
Falconry, 8:316, 8:324
Falcons, 8:21, 8:*22*, **8:347–362,** 9:94, 9:97
behavior, 8:349
conservation status, 8:351
distribution, 8:*347*, 8:348
evolution, 8:347
feeding ecology, 8:349–350
habitats, 8:348–349
humans and, 8:352
pesticides, 8:26, 8:351
reproduction, 8:350–351
species of, 8:*354*–361
taxonomy, 8:347
Falculea palliata. See Sicklebilled vangas
Falcunculus spp. *See* Shrike-tits
Falcunculus frontatus. See Eastern shrike-tits
Falcunculus gutturalis. See Crested bellbirds
Falkland Island wolves, 14:262, 14:272, 14:284*t*
Fall herrings. *See* Atlantic herrings
Fallow deer, 15:*362*, 15:*365*, 15:366
competition, 12:118
evolution, 12:19, 15:357
habitats, 15:358
humans and, 12:173
physical characteristics, 15:*357*
reproduction, 15:272
False blindsnakes, **7:387–388**
False catsharks, 4:113, 4:114, 4:*118*, 4:*124*, 4:125
False clown anemonefishes. *See* Clown anemonefishes
False coral snakes, 7:*399*, **7:399–400,** 7:*400*
False corals. *See* Mushroom corals
False-eye puffers, 5:*470*
False gharials, 7:59, 7:*157*, **7:179–188,** 7:*183*
False-head mimicry, 1:42, 2:37–38
See also Behavior
False killer whales, 12:87, 15:2, 15:*6*, 15:43, 15:46, 15:49
False mole crickets, 3:55, 3:205
False pacas. *See* Pacaranas
False pitvipers, 7:467, 7:468
False pottos, 14:*17*, 14:20
False powderpost beetles, 3:75
False rednose tetras. *See* False rummynose tetras
False rummynose tetras, 4:*340*, 4:*345*
False sunbirds. *See* Sunbird-asities
False talking catfishes. *See* Blue-eye catfishes
False vampire bats, **13:379–385,** 13:*383*
feeding ecology, 12:13, 12:*84*, 12:*85*, 13:313–314
humans and, 13:317
physical characteristics, 13:310
woolly, 13:419, 13:433*t*
False water-cobras, 7:468
False water rats, 16:262*t*
Fan shells. *See* Noble pen shells
Fan-tailed berrypeckers, 11:*190*, 11:*192*, 11:*198*
Fan-tailed cisticolas. *See* Zitting cisticolas
Fan-tailed ragfishes. *See* Ragfishes
Fan-tailed warblers. *See* Zitting cisticolas
Fang blennies, 4:69
Fangs, 7:27
See also Venom
Fangtooth fishes, 5:113, 5:115, 5:*117*, 5:*118*, 5:119
Fannia spp. *See* Little-house flies
Fanovana newtonias. *See* Red-tailed newtonias
Fanovana warblers. *See* Red-tailed newtonias
Fantail warblers. *See* Zitting cisticolas
Fantails, **11:83–95,** 11:*84*, 11:*88*
behavior, 11:85–86, 11:89–95
conservation status, 11:86–87, 11:89–95
distribution, 11:*83*, 11:*89–94*
evolution, 11:83
feeding ecology, 11:86, 11:89–95
habitats, 11:85, 11:89–94
humans and, 11:87, 11:89–95
physical characteristics, 11:84, 11:89–94
reproduction, 11:86, 11:89–95
taxonomy, 11:83–84, 11:89–94
FAO. *See* Food and Agriculture Organization of the United Nations
Far Eastern violet sea urchins. *See* Short-spined sea urchins
Farancia spp. *See* Mudsnakes
Farancia abacura. See Red-bellied mudsnakes
Farancia erytrogramma. See Rainbow snakes
Farming
insect, 3:80
reptiles, 7:48–*49*
Faro webbed frogs, 6:251, 6:*253*, 6:*255–256*
Farrea occa, 1:74
Fasciated antshrikes, 10:*244*, 10:*245*
Fasciola hepatica. See Liver flukes
Fat, migratory, 8:32
See also Body fat
Fat dormice. *See* Edible dormice
Fat innkeeper worms. *See* Innkeeper worms
Fat mice, 16:296*t*
Fat sleepers, 5:374
Fat-tailed dunnarts, 12:280, 12:288, 12:*289*, 12:300*t*
Fat-tailed gerbils, 16:296*t*
Fat-tailed mouse opossums, 12:250–253
Fat-tailed pseudantechinuses, 12:*283*, 12:288, 12:300*t*
Fatbacks. *See* Atlantic menhadens
Fathead minnows, 4:*33*, 4:301, 4:*302*, 4:*303*, 4:*307*, 4:314–315
Fathead sculpins, 5:179, 5:180, 5:182
Fathom fishes. *See* Eulachons
Fattening, seasonal, 12:168–169
See also Feeding ecology
Fatula cockroaches. *See* Madeira cockroaches
Fauna and Flora International (FFI), on black crested gibbons, 14:221
Fauriellidae, 3:281
Fawn-breasted bowerbirds, 11:*478*
Fawn-colored larks, 10:342
Fawn-eyed diamond birds. *See* Red-browed pardalotes
Fawn-flanked silvereyes, 11:229
Faxon, Walter, 11:286
Fea's muntjacs, 15:343, 15:346, 15:*347*, 15:*351*
Fea's rib-faced deer. *See* Fea's muntjacs
Fea's vipers, 7:445–448, 7:*449*, 7:*451*
Feather-footed jerboas. *See* Hairy-footed jerboas
Feather harvests. *See* Humans; specific species
Feather stars, **1:355–365,** 1:*357*, 1:*358*, 1:*359*
behavior, 1:356–357
conservation status, 1:358
distribution, 1:356
evolution, 1:355
feeding ecology, 1:357–358
habitats, 1:356
humans and, 1:358
physical characteristics, 1:355–*356*
reproduction, 1:358
species of, 1:*361–363*
taxonomy, 1:355
Feather-tailed possums, **13:139–145,** 13:*143*, 13:*144*
behavior, 13:35, 13:37, 13:141
conservation status, 13:142
distribution, 13:*139*, 13:140
evolution, 13:31, 13:139–140
feeding ecology, 13:141
habitats, 13:34, 13:141
humans and, 13:142
physical characteristics, 13:140
reproduction, 13:39, 13:142
species of, 13:144
taxonomy, 13:139–140
Featherfin knifefishes, 4:11
Feathertail gliders. *See* Pygmy gliders
Featherwing beetles, 3:317
Fecundity, 2:27, 6:31–32
See also Reproduction
Feduccia, Alan, 10:51
Feeding ecology
amphibians, 6:6, 6:54
anurans, 6:6, 6:26, 6:67
African treefrogs, 6:282, 6:285–289
Amero-Australian treefrogs, 6:229, 6:233–242
Arthroleptidae, 6:266, 6:268–271
Asian toadfrogs, 6:111, 6:113–117
Asian treefrogs, 6:292, 6:295–299
Australian ground frogs, 6:140–141, 6:143–145
Bombinatoridae, 6:4, 6:64, 6:84, 6:86–88
Bufonidae, 6:185, 6:188–194
Discoglossidae, 6:90, 6:92–94
frogs, 6:6, 6:26, 6:67
ghost frogs, 6:132, 6:133–134
glass frogs, 6:216, 6:219–223
leptodactylid frogs, 6:158, 6:162–171
Madagascan toadlets, 6:318, 6:319–320
marine toads, 6:67, 6:185, 6:191
Mesoamerican burrowing toads, 6:97
Myobatrachidae, 6:148, 6:151–153
narrow-mouthed frogs, 6:303–304, 6:308–316
New Zealand frogs, 6:70, 6:72–74
parsley frogs, 6:127, 6:129
Pipidae, 6:67, 6:100–101, 6:104–106
poison frogs, 6:199, 6:203–209
Ruthven's frogs, 6:212
Seychelles frogs, 6:136, 6:137–138
shovel-nosed frogs, 6:275
spadefoot toads, 6:120–121, 6:124–125
Suriname toads, 6:100–101, 6:103–106
tadpoles, 6:6, 6:37–38, 6:39, 6:40–41
tailed frogs, 6:78, 6:80–81
three-toed toadlets, 6:179, 6:181–182
true frogs, 6:251, 6:255–263
vocal sac-brooding frogs, 6:174, 6:175–176
Xenopus spp., 6:100–101
caecilians, 6:6, 6:412
American tailed, 6:416, 6:417–418
Asian tailed, 6:420, 6:422–423

buried-eye, 6:432, 6:433–434
Kerala, 6:426, 6:428
tailless, 6:436, 6:439–441
salamanders, 6:6, 6:24, 6:65, 6:326
amphiumas, 6:406–407, 6:409–410
Asiatic, 6:336–337, 6:339–341
Caudata, 6:326
Cryptobranchidae, 6:345, 6:*346*
lungless, 6:392, 6:395–403
mole, 6:356, 6:358–360
Pacific giant, 6:350, 6:352
Proteidae, 6:378, 6:381–383
Salamandridae, 6:326, 6:366–367, 6:370–375
sirens, 6:328–329, 6:331–332
torrent, 6:385, 6:387
birds, 8:15–17
Abbott's booby, 8:217
albatrosses, 8:109, 8:115, 8:118–122
Alcidae, 9:221–222, 9:224–228
anhingas, 8:185, 8:202
Anseriformes, 8:366
ant thrushes, 10:241, 10:245–256
Apodiformes, 9:417
asities, 10:188, 10:190
Australian chats, 11:66, 11:67
Australian creepers, 11:134–135, 11:137–139
Australian fairy-wrens, 11:46, 11:48–53
Australian honeyeaters, 11:237, 11:241–253
Australian robins, 11:106, 11:108–112
Australian warblers, 11:56, 11:59–63
babblers, 10:507, 10:511–523
barbets, 10:116, 10:119–122
barn owls, 9:337, 9:340–343
bee-eaters, 10:40–41, 10:*42*
birds of paradise, 11:*490*, 11:491, 11:495–501
bitterns, 8:243–244
Bombycillidae, 10:*448*, 10:451–454
boobies, 8:185, 8:213
bowerbirds, 11:479, 11:483–488
broadbills, 10:178, 10:181–186
bulbuls, 10:398, 10:402–413
bustards, 9:92–93, 9:96–99
buttonquails, 9:13, 9:16–21
Caprimulgiformes, 9:370
caracaras, 8:349–350
cassowaries, 8:76
Charadriidae, 9:162, 9:166–173
Charadriiformes, 9:102–103
chats, 10:487
chickadees, 11:157, 11:160–166
chowchillas, 11:70, 11:72–73
Ciconiiformes, 8:235
Columbidae, 9:249–250, 9:255–266
Columbiformes, 9:243–244
condors, 8:277, 8:278
Coraciiformes, 10:2–3
cormorants, 8:185, 8:202, 8:205–206, 8:207–209
Corvidae, 11:*504*, 11:506–507, 11:*508*, 11:512–522
cotingas, 10:307, 10:312–322
crab plovers, 9:122
Cracidae, 8:415–416
cranes, 9:26–27, 9:31–36
cuckoo-shrikes, 10:387
cuckoos, 9:313, 9:318–329
digestion, 8:10–11
dippers, 10:476, 10:479–482
diving-petrels, 8:143–144, 8:145–146
doves, 9:243–244, 9:249–250, 9:255–266
drongos, 11:438–439, 11:441–445
ducks, 8:372
eagles, 8:321–322
elephant birds, 8:104
emus, 8:85
Eupetidae, 11:76, 11:78–80
excretion, 8:10–11
fairy bluebirds, 10:416–417, 10:423–424
Falconiformes, 8:314–315
falcons, 8:349–350
false sunbirds, 10:188, 10:191
fantails, 11:86, 11:89–95
finches, 11:*324*, 11:*325*, 11:328–339
flamingos, 8:305
flight, 8:16
flowerpeckers, 11:190, 11:193–198
frigatebirds, 8:185, 8:194, 8:197–198
frogmouths, 9:379, 9:381–385
Galliformes, 8:399–400
gannets, 8:185, 8:*212*, 8:213
geese, 8:372
Glareolidae, 9:152, 9:155–160
grebes, 8:171, 8:174–180
Gruiformes, 9:2
guineafowl, 8:428
gulls, 9:211–217
hammerheads, 8:262
Hawaiian honeycreepers, 11:*342*–343, 11:346–351
hawks, 8:321–322
hedge sparrows, 10:459, 10:462–464
herons, 8:235, 8:242, 8:243–244
hoatzins, 8:466–467
honeyguides, 10:138–139, 10:141–144
hoopoes, 10:62
hornbills, 10:73–74, 10:78–84
human bird feeding, 8:23–24
hummingbirds, 9:*439*, 9:*441*, 9:445–447, 9:455–467
ibises, 8:292
Icteridae, 11:293–294, 11:309–322
ioras, 10:416, 10:419–420
jacamars, 10:92
jacanas, 9:109, 9:112–114
kagus, 9:42
kingfishers, 10:8–9, 10:13–23
kiwis, 8:90
lapwings, 9:167–173
Laridae, 9:206–207, 9:211–217
larks, 10:344–345, 10:348–354
leafbirds, 10:416, 10:420–423
limpkins, 9:38
logrunners, 11:70, 11:72–73
long-tailed titmice, 11:142, 11:145–146
lyrebirds, 10:331, 10:334–335
magpie-shrikes, 11:468, 11:471–475
manakins, 10:296, 10:299–303
mesites, 9:6, 9:8–9
moas, 8:97–98
monarch flycatchers, 11:97, 11:100–103
motmots, 10:32, 10:35–37
moundbuilders, 8:405
mousebirds, 9:470, 9:473–476
mudnest builders, 11:453, 11:456–458
New World blackbirds, 11:293–294, 11:309–322
New World finches, 11:264, 11:*269–282*
New World quails, 8:457
New World vultures, 8:235, 8:277, 8:278
New World warblers, 11:*287*, 11:289–290, 11:293–299
New Zealand wattle birds, 11:447, 11:449–450
New Zealand wrens, 10:204, 10:206–207
nightjars, 9:403–404, 9:408–414
nuthatches, 11:168, 11:171–175
oilbirds, 9:374–375
Old World flycatchers, 11:26–27, 11:30–43
Old World warblers, 11:5, 11:10–23
Oriolidae, 11:428–*429*, 11:431–434
ovenbirds, 10:210, 10:214–228
owlet-nightjars, 9:388–389, 9:391–393
owls, 9:333, 9:337, 9:340–343, 9:349–350, 9:354–355
oystercatchers, 9:126–127, 9:130–131
painted-snipes, 9:116, 9:118–119
palmchats, 10:455–*456*
pardalotes, 11:201, 11:204–206
parrots, 9:277–278, 9:283–297
Passeriformes, 10:173
Pelecaniformes, 8:185
pelicans, 8:185, 8:226
penduline titmice, 11:148, 11:151–152
penguins, 8:150, 8:153–157
Phasianidae, 8:435–436
pheasants, 8:435–436
Philippine creepers, 11:184, 11:186–187
Picidae, 10:149, 10:154–167
Piciformes, 10:87
pigeons, 9:243–244, 9:249–250, 9:255–266
pipits, 10:374, 10:378–383
pittas, 10:194, 10:197–201
plantcutters, 10:326, 10:327–328
plovers, 9:166–173
potoos, 9:396, 9:399–400
pratincoles, 9:155–160
Procellariidae, 8:124, 8:129–130
Procellariiformes, 8:108, 8:109
pseudo babblers, 11:128, 11:130–131
puffbirds, 10:103, 10:106–111
rails, 9:48, 9:49–50, 9:57–67
Raphidae, 9:270, 9:272–273
Recurvirostridae, 9:134–135
rheas, 8:70
rollers, 10:53, 10:55–58
sandgrouse, 9:232, 9:235–238
sandpipers, 9:178–179, 9:182–188
screamers, 8:394
scrub-birds, 10:338, 10:339–340
secretary birds, 8:344
seedsnipes, 9:190–191, 9:193–195
seriemas, 9:86, 9:88–89
sharpbills, 10:292
sheathbills, 9:198–199, 9:200–201
shoebills, 8:287–288
shrikes, 10:428, 10:*429*, 10:432–438
sparrows, 11:398, 11:401–406
spoonbills, 8:292
storks, 8:235, 8:266
storm-petrels, 8:136–137, 8:140–141
Struthioniformes, 8:54–55
Sturnidae, 11:408–409, 11:*410*, 11:414–425
sunbirds, 11:209, 11:213–225
sunbitterns, 9:74

sungrebes, 9:70, 9:71–72
swallows, 10:359, 10:363–369
swans, 8:372
swifts, 9:423, 9:426–431
tacto-location, 8:*266*
tapaculos, 10:259, 10:262–267
terns, 9:211–217
thick-knees, 9:145, 9:147–148
thrushes, 10:487
tinamous, 8:58
titmice, 11:*156*, 11:157, 11:160–166
todies, 10:26–27, 10:29–30
toucans, 10:127, 10:130–135
tree swifts, 9:434, 9:435
treecreepers, 11:178, 11:180–181
trogons, 9:*478–479*, 9:481–485
tropicbirds, 8:185, 8:187–188
trumpeters, 9:77, 9:78, 9:81–82
turacos, 9:301, 9:304–310
typical owls, 9:349–350, 9:354–365
tyrant flycatchers, 10:273, 10:278–288
vanga shrikes, 10:440–*441*, 10:443–444
Vireonidae, 11:256, 11:258–262
vultures, 8:277, 8:278
wagtails, 10:374, 10:378–383
weaverfinches, 11:*355*, 11:360–372
weavers, 11:377, 11:382–394
whistlers, 11:117, 11:120–126
white-eyes, 11:229, 11:232–233
woodcreepers, 10:*230*, 10:232–236
woodhoopoes, 10:66, 10:68–69
woodswallows, 11:460, 11:462–465
wrens, 10:528, 10:531–538
fishes, 4:17, 4:39, 4:40, 4:67–69
Acanthuroidei, 5:393, 5:397–403
Acipenseriformes, 4:214, 4:217–220
Albuliformes, 4:250, 4:252–253
angelsharks, 4:162, 4:164–165
anglerfishes, 5:50, 5:53–56
Anguilliformes, 4:256, 4:*257*, 4:261–269
Atheriniformes, 5:69, 5:73–76
Aulopiformes, 4:433, 4:436–439
Australian lungfishes, 4:198
beardfishes, 5:1, 5:3
Beloniformes, 5:80–81, 5:83–86
Beryciformes, 5:115–116, 5:118–121
bichirs, 4:210, 4:211–212
blennies, 5:342–343, 5:345–348
bowfins, 4:230
bullhead sharks, 4:98, 4:101–102
Callionymoidei, 5:366, 5:369–371
carps, 4:299–300, 4:308–313
catfishes, 4:68, 4:69, 4:353, 4:357–366
characins, 4:337–338, 4:342–350
chimaeras, 4:92, 4:94–95
coelacanths, 4:192
Cypriniformes, 4:299–300, 4:307–319, 4:*322*, 4:325–333
Cyprinodontiformes, 5:92, 5:96–102
damselfishes, 4:50, 4:67
dogfish sharks, 4:152, 4:156–158
dories, 5:124, 5:126–127, 5:129–130
eels, 4:17, 4:256, 4:*257*, 4:261–269
electric eels, 4:69, 4:371, 4:373–377
elephantfishes, 4:68, 4:238
Elopiformes, 4:243, 4:246–247
Esociformes, 4:380, 4:383–386
flatfishes, 5:452–453, 5:455–463
Gadiformes, 5:29, 5:33–39
gars, 4:222, 4:224–227
Gasterosteiformes, 5:135, 5:139–147
gobies, 5:375–376, 5:380–388
Gobiesocoidei, 5:356–357, 5:359–363
Gonorynchiformes, 4:290, 4:293–295
ground sharks, 4:115, 4:119–128
Gymnotiformes, 4:68, 4:371, 4:373–377
hagfishes, 4:79, 4:80–81
herrings, 4:67, 4:69, 4:278, 4:281–287
Hexanchiformes, 4:144, 4:146–148
Labroidei, 5:277–279, 5:284–290, 5:295–296, 5:301–307
labyrinth fishes, 5:428, 5:431–434
lampreys, 4:69, 4:*84*, 4:85, 4:88–90
Lampridiformes, 4:449–*450*, 4:452–454
lanternfishes, 4:443, 4:445–446
lungfishes, 4:203, 4:205–206
mackerel sharks, 4:133, 4:135–141
marine fishes, 4:44, 4:50–51
minnows, 4:69, 4:299–300, 4:314–315, 4:322
morays, 4:68, 4:256, 4:265–267
mudminnows, 4:380, 4:383–386
mullets, 4:17, 4:69, 5:60, 5:62–65
Ophidiiformes, 5:17, 5:20–22
Orectolobiformes, 4:105, 4:109–111
Osmeriformes, 4:392–393, 4:395–401
Osteoglossiformes, 4:233, 4:237–241
parrotfishes, 4:17, 4:50, 4:67
Percoidei, 5:197–198, 5:202–208, 5:212, 5:214–217, 5:221, 5:225–226, 5:228, 5:230–233, 5:240–242, 5:247–253, 5:260–261, 5:266–268, 5:270–273
Percopsiformes, 5:5, 5:8–12
pikes, 4:380, 4:383–386
ragfishes, 5:352–353
Rajiformes, 4:177–178, 4:182–188
rays, 4:177–178, 4:182–188
Saccopharyngiformes, 4:272, 4:274–276
salmons, 4:68, 4:407, 4:410–419
sawsharks, 4:168, 4:170–171
Scombroidei, 5:406, 5:409–418
Scorpaeniformes, 5:158, 5:161–162, 5:165, 5:169–178, 5:182, 5:186–193
Scorpaenoidei, 5:165
sea basses, 4:44, 4:68
sharks, 4:69
skates, 4:177–178, 4:187
snakeheads, 5:438–439, 5:442–446
South American knifefishes, 4:371, 4:373–377
southern cod-icefishes, 5:322, 5:325–328
Stephanoberyciformes, 5:106, 5:108–110
stingrays, 4:68, 4:177–178
Stomiiformes, 4:*423*–424, 4:426–429
Stromateoidei, 5:422
Synbranchiformes, 5:151, 5:154–156
Tetraodontiformes, 5:470, 5:474–475, 5:477–484
toadfishes, 5:42, 5:44–45
Trachinoidei, 5:333–334, 5:337–339
trouts, 4:68, 4:407
whale sharks, 4:67, 4:110
wrasses, 4:50, 4:68, 4:69
Zoarcoidei, 5:311–312, 5:315–318
insects, 3:5–6, 3:19–21, 3:41–42, 3:46, 3:62–64
ants, 3:5–6, 3:63, 3:406–407
aphids, 3:63
bees, 3:20, 3:42, 3:63, 3:406–407
beetles, 3:42, 3:63, 3:320–321
book lice, 3:244, 3:246–247
bristletails, 3:114, 3:116–117
butterflies, 3:42, 3:46, 3:48, 3:63, 3:386–387
caddisflies, 3:376, 3:379–381
cockroaches, 3:64, 3:149, 3:153–158
Coleoptera, 3:320–321, 3:327–333
diplurans, 3:107–108, 3:110–111
Diptera, 3:360, 3:364–374
dragonflies, 3:63, 3:134
earwigs, 3:196, 3:199–200
fleas, 3:47, 3:64, 3:349, 3:352–354
flies, 3:20, 3:35, 3:42, 3:47, 3:63, 3:64, 3:360
Hemiptera, 3:262, 3:267–279
Hymenoptera, 3:35, 3:406–407, 3:411–424
lacewings, 3:307–308, 3:311–314
Lepidoptera, 3:20, 3:46, 3:48, 3:*384*, 3:386–387, 3:393–404
mantids, 3:63, 3:180, 3:183–186
Mantophasmatodea, 3:219
mayflies, 3:126–127, 3:129–130
Mecoptera, 3:343, 3:345–346
Megaloptera, 3:290, 3:293–294
mosquitos, 3:64, 3:360
moths, 3:46, 3:48, 3:63, 3:386–387
Odonata, 3:134, 3:137–138
Orthoptera, 3:206–207, 3:211–216
Phasmida, 3:223, 3:228–232
Phthiraptera, 3:251, 3:254–256
proturans, 3:94, 3:96–97
rock-crawlers, 3:189–190, 3:192–193
sea skaters, 3:262
snakeflies, 3:298, 3:300–302
springtails, 3:100, 3:103–104
stoneflies, 3:142–143, 3:145–146
strepsipterans, 3:336, 3:338–339
termites, 3:166–167, 3:170–175
thrips, 3:282, 3:285–287
Thysanura, 3:119, 3:122–123
wasps, 3:63
water bugs, 3:63, 3:262
webspinners, 3:234, 3:236–237
zorapterans, 3:240, 3:241
lower metazoans and lesser deuterostomes, **1:27–30**
acoels, 1:180, 1:181–182
anoplans, 1:246, 1:249–251
arrow worms, 1:434–435, 1:437–442
box jellies, 1:148, 1:151–152
calcareous sponges, 1:58, 1:61–65
cnidarians, 1:28–30
Anthozoa, 1:28, 1:106–107, 1:111–120
Hydrozoa, 1:*125*, 1:128–129, 1:134–145
comb jellies, 1:28–29, 1:170–171, 1:173–177
demosponges, 1:27, 1:80, 1:82–86
echinoderms
Crinoidea, 1:357–358, 1:362–365
Echinoidea, 1:404–405, 1:408–415
Ophiuroidea, 1:389, 1:393–398
sea cucumbers, 1:421, 1:425–431
sea daisies, 1:382, 1:385–386
sea stars, 1:369–370, 1:373–379
enoplans, 1:254–255, 1:256–257
entoprocts, 1:321, 1:323–325
flatworms
free-living flatworms, 1:28, 1:187–188, 1:190–195
monogeneans, 1:214, 1:218, 1:220–223

tapeworms, 1:230, 1:234–242
gastrotrichs, 1:270, 1:272–273
girdle wearers, 1:345, 1:347–349
glass sponges, 1:68–69, 1:71–75
gnathostomulids, 1:332, 1:334–335
hair worms, 1:306, 1:308–309
hemichordates, 1:444–445, 1:447–450
jaw animals, 1:329
jellyfish, 1:29, 1:155–156, 1:160–167
kinorhynchs, 1:276, 1:279–280
lancelets, 1:488, 1:490–497
larvaceans, 1:*474*, 1:476–478
nematodes
roundworms, 1:285, 1:287–291
secernenteans, 1:294, 1:297, 1:299–304
Orthonectida, 1:99, 1:101–102
placozoans, 1:89
priapulans, 1:337–338, 1:340–341
rhombozoans, 1:93, 1:96–98
rotifers, 1:261, 1:264–267
Salinella salve, 1:91
salps, 1:467, 1:471–472
sea squirts, 1:454, 1:458–466
sorberaceans, 1:480, 1:482–483
thorny headed worms, 1:312, 1:314–316
Trematoda, 1:200, 1:203–211
wheel wearers, 1:352
mammals, 12:12–15
aardvarks, 15:157–158
agoutis, 16:408–409, 16:411–414, 16:415*t*
anteaters, 13:173–174, 13:177–179
armadillos, 13:184, 13:187–190, 13:190*t*–192*t*
Artiodactyla, 15:141–142, 15:271–272
aye-ayes, 14:*86*, 14:*87*–88
baijis, 15:20
bandicoots, 13:3–4
dry-country, 13:11, 13:14–16, 13:16*t*–17*t*
rainforest, 13:11, 13:14–17*t*
bats, 12:12–13, 12:54–55, 12:58, 12:59, 13:313–314
American leaf-nosed, 13:314, 13:417–418, 13:423–432, 13:433*t*–434*t*
bulldog, 13:446–447, 13:449–450
disk-winged, 13:474, 13:476–477
Emballonuridae, 13:357, 13:360–363, 13:363*t*–364*t*
false vampire, 13:313, 13:381–382, 13:384–385
funnel-eared, 13:461, 13:463–465
horseshoe, 13:390–391, 13:396–400
Kitti's hog-nosed, 13:368
Molossidae, 13:485, 13:490–493, 13:493*t*–495*t*
mouse-tailed, 13:351–352, 13:353
moustached, 13:436, 13:440–441, 13:442*t*
New Zealand short-tailed, 13:454, 13:457–458
Old World fruit, 13:320, 13:325–330, 13:331*t*–332*t*, 13:335–336, 13:340–347, 13:348*t*–349*t*
Old World leaf-nosed, 13:405, 13:407–409, 13:409*t*–410*t*
Old World sucker-footed, 13:480
slit-faced, 13:373, 13:375–376, 13:376*t*–377*t*
smoky, 13:468, 13:470–471
Vespertilionidae, 13:500, 13:506–514, 13:515*t*–516*t*, 13:521, 13:524–525, 13:526*t*
bears, 14:255, 14:298–299, 14:302–305, 14:306*t*
beavers, 16:179–180, 16:183–184
bilbies, 13:21–22
botos, 15:29
Bovidae, 16:6–7
Antilopinae, 16:48, 16:50–56, 16:56*t*–57*t*
Bovinae, 16:14, 16:18–23, 16:24*t*–25*t*
Caprinae, 16:92–93, 16:98–103*t*, 16:104*t*
duikers, 16:74, 16:76, 16:80–84, 16:84*t*–85*t*
Hippotraginae, 16:31–32, 16:37–42, 16:42*t*–43*t*
Neotraginae, 16:62, 16:66–71, 16:71*t*
bushbabies, 14:9, 14:26, 14:28–32, 14:32*t*–33*t*
Camelidae, 15:317, 15:320–323
Canidae, 14:269–271, 14:277–283, 14:283*t*–284*t*
capybaras, 16:404, 16:*405*
Carnivora, 14:260–261
cats, 14:372, 14:379–389, 14:390*t*–391*t*
Caviidae, 16:392–393, 16:395–398, 16:399*t*
Cetacea, 15:8
Balaenidae, 15:110, 15:115–117
beaked whales, 15:61, 15:64–68, 15:69*t*–70*t*
dolphins, 15:46–47, 15:53–55, 15:56*t*–57*t*
franciscana dolphins, 15:24
Ganges and Indus dolphins, 15:14–15
gray whales, 15:97–98
Monodontidae, 15:85, 15:90–91
porpoises, 15:35, 15:38–39, 15:39*t*
pygmy right whales, 15:104–105
rorquals, 15:8, 15:123–124, 15:127–130, 15:130*t*
sperm whales, 15:75–76, 15:79–80
chevrotains, 15:328, 15:331–333
Chinchillidae, 16:379–380, 16:382–383, 16:383*t*
colugos, 13:302, 13:304
coypus, 16:476–477
dasyurids, 12:289, 12:294–298, 12:299*t*–301*t*
Dasyuromorphia, 12:281
deer
Chinese water, 15:375
muntjacs, 15:345, 15:349–354
musk, 15:337, 15:340–341
New World, 15:383–384, 15:388–396, 15:396*t*
Old World, 15:359, 15:364–371, 15:371*t*
Dipodidae, 16:215, 16:219–222, 16:223*t*–224*t*
Diprotodontia, 13:37–38
dormice, 16:319–320, 16:323–326, 16:327*t*–328*t*
duck-billed platypuses, 12:231, 12:245
Dugongidae, 15:200, 15:203
echidnas, 12:237, 12:240*t*
elephants, 15:*167*, 15:169–170, 15:174–175
Equidae, 12:118–119, 15:228, 15:232–235, 15:236*t*
Erinaceidae, 13:206, 13:209–213, 13:212*t*–213*t*
giant hutias, 16:469, 16:471*t*
gibbons, 14:211–213, 14:218–222
Giraffidae, 15:403–404, 15:408–409
great apes, 14:232–234, 14:237–240
gundis, 16:313, 16:314–315
Herpestidae, 14:351–352, 14:354–356, 14:357*t*
Heteromyidae, 16:202, 16:205–209, 16:208*t*–209*t*
hippopotamuses, 15:306, 15:308, 15:310–311
humans, 14:*245*, 14:247–248
hutias, 16:462, 16:467*t*
Hyaenidae, 14:362, 14:364–367
hyraxes, 15:182, 15:186–188, 15:189*t*
Indriidae, 14:66, 14:69–72
Insectivora, 13:198–199
koalas, 13:46–47
Lagomorpha, 16:483–485
lemurs, 14:9, 14:51, 14:56–60
Cheirogaleidae, 14:37–38, 14:41–44
sportive, 14:8, 14:76, 14:79–83
Leporidae, 16:507, 16:511–515, 16:515*t*–516*t*
Lorisidae, 14:15, 14:18–20, 14:21*t*
Macropodidae, 13:87–88, 13:93–100, 13:101*t*–102*t*
manatees, 15:206, 15:208, 15:211–212
Megalonychidae, 13:157, 13:159
migrations and, 12:167–170
moles
golden, 12:84, 13:217, 13:220–222, 13:222*t*
marsupial, 13:27, 13:28
monitos del monte, 12:274
monkeys
Atelidae, 14:158, 14:162–166, 14:166*t*–167*t*
Callitrichidae, 14:120–121, 14:127–131, 14:132*t*
Cebidae, 14:104–105, 14:108–112
cheek-pouched, 14:192–193, 14:197–204, 14:204*t*–205*t*
leaf-monkeys, 14:8, 14:9, 14:174, 14:178–183, 14:184*t*–185*t*
night, 14:138, 14:141*t*
Pitheciidae, 14:146–147, 14:150–153, 14:153*t*
monotremes, 12:231
mountain beavers, 16:132–133
Muridae, 16:284, 16:288–295, 16:296*t*–297*t*
Arvicolinae, 16:227–228, 16:233–238, 16:237*t*–238*t*
hamsters, 16:241–242, 16:245–246, 16:247*t*
Murinae, 16:251–252, 16:255–260, 16:261*t*–262*t*
Sigmodontinae, 16:268–269, 16:272–276, 16:277*t*–278*t*
musky rat-kangaroos, 13:71
Mustelidae, 12:192, 14:260, 14:323, 14:327–331, 14:332*t*–333*t*
numbats, 12:304
octodonts, 16:435–436, 16:438–440, 16:440*t*
opossums

New World, 12:252–253, 12:257–263, 12:264*t*–265*t*
shrew, 12:268, 12:270
Otariidae, 14:397–398, 14:402–406
pacaranas, 16:387
pacas, 16:*418*–419, 16:423–424
pangolins, 16:112–113, 16:115–120
peccaries, 15:294–295, 15:298–300
Perissodactyla, 15:219–220
Petauridae, 13:127–128, 13:131–133, 13:133*t*
Phalangeridae, 13:60, 13:*63–66t*
pigs, 15:278–279, 15:284–288, 15:289*t*–290*t*
pikas, 16:493–494, 16:497–501, 16:501*t*–502*t*
pocket gophers, 16:189–190, 16:192–195*t*, 16:196*t*–197*t*
porcupines
New World, 16:368, 16:371–373, 16:374*t*
Old World, 16:353, 16:357–364
possums
feather-tailed, 13:141, 13:144
honey, 13:137
pygmy, 13:107, 13:110–111
primates, 14:8–9
Procyonidae, 14:310, 14:313–315, 14:315*t*–316*t*
pronghorns, 15:414
Pseudocheiridae, 13:116, 13:119–123, 13:122*t*–123*t*
rat-kangaroos, 13:75, 13:78–81
rats
African mole-rats, 16:343–344, 16:347–349, 16:349*t*–350*t*
cane, 16:334, 16:337–338
chinchilla, 16:445, 16:447–448
dassie, 16:331
spiny, 16:450, 16:453–457, 16:458*t*
rhinoceroses, 15:253, 15:257–261
rodents, 16:125
sengis, 16:521, 16:524–530, 16:531*t*
shrews
red-toothed, 13:250–251, 13:255–263, 13:263*t*
West Indian s, 13:244
white-toothed, 13:267, 13:271–275, 13:276*t*–277*t*
Sirenia, 15:193
solenodons, 13:239, 13:241
springhares, 16:309
squirrels
flying, 16:136, 16:139–140, 16:141*t*–142*t*
ground, 16:145, 16:147, 16:151–158, 16:158*t*–160*t*
scaly-tailed, 16:300, 16:302–305
tree, 16:166, 16:169–174*t*, 16:175*t*
Talpidae, 13:281–282, 13:284–287, 13:287*t*–288*t*
tapirs, 15:240–241, 15:245–247
tarsiers, 14:*94*–95, 14:97–99
tenrecs, 13:228–229, 13:232–234, 13:234*t*
three-toed tree sloths, 13:164–165, 13:167–169
tree shrews, 13:291–292, 13:294–296, 13:297*t*–298*t*
true seals, 14:260, 14:*261*, 14:419–420, 14:427–434, 14:435*t*
tuco-tucos, 16:427, 16:429–430, 16:430*t*–431*t*
ungulates, 15:141–142
Viverridae, 14:337, 14:340–343, 14:344*t*–345*t*
walruses, 14:260, 14:414
wombats, 13:52–53, 13:55–56
Xenartha, 13:151–152
protostomes, 2:29, 2:36–37
amphionids, 2:196
amphipods, 2:29, 2:261, 2:265–271
anaspidaceans, 2:182, 2:183
aplacophorans, 2:380, 2:382–385
Arachnida, 2:335, 2:339–352
articulate lampshells, 2:522–523, 2:525–527
bathynellaceans, 2:178, 2:179
beard worms, 2:86, 2:88
bivalves, 2:452–453, 2:457–466
caenogastropods, 2:446, 2:449
centipedes, 2:*353*, 2:*355*, 2:358–362
cephalocarids, 2:132, 2:133
Cephalopoda, 2:478, 2:483–489
chitons, 2:395, 2:397, 2:399–401
clam shrimps, 2:148, 2:150–151
copepods, 2:36, 2:300, 2:304–310
cumaceans, 2:230, 2:232
Decapoda, 2:200, 2:204–214
deep-sea limpets, 2:436, 2:437
earthworms, 2:36, 2:66, 2:70–73
echiurans, 2:104, 2:106–107
fairy shrimps, 2:137, 2:139
fish lice, 2:290, 2:292–293
freshwater bryozoans, 2:497, 2:500–502
horseshoe crabs, 2:328, 2:331–332
Isopoda, 2:29, 2:251–252, 2:254–259
krill, 2:186–187, 2:189–192
lampshells, 2:29
leeches, 2:29, 2:76, 2:80–82
leptostracans, 2:162, 2:164–165
lophogastrids, 2:226, 2:227
mantis shrimps, 2:170, 2:173–175
marine bryozoans, 2:504, 2:506–508, 2:510, 2:513–514
mictaceans, 2:241, 2:242
millipedes, 2:365, 2:367–370
monoplacophorans, 2:388
mussel shrimps, 2:312, 2:314–315
mysids, 2:216–217, 2:219–222
mystacocarids, 2:295, 2:297
myzostomids, 2:60, 2:62
Neritopsina, 2:440–441, 2:442–443
nonarticulate lampshells, 2:516, 2:518–519
Onychophora, 2:110, 2:113–114
pauropods, 2:376, 2:377
peanut worms, 2:98, 2:100–101
phoronids, 2:492, 2:494
Polychaeta, 2:36, 2:46–47, 2:50–56
Pulmonata, 2:414, 2:417–421
remipedes, 2:126, 2:128
sea slugs, 2:404–405, 2:407–409
sea spiders, 2:322, 2:324–325
spelaeogriphaceans, 2:243, 2:244
symphylans, 2:372, 2:373
tadpole shrimps, 2:142, 2:145–146
tanaids, 2:236, 2:238–239
tantulocaridans, 2:284, 2:286
Thecostraca, 2:275, 2:278–281
thermosbaenaceans, 2:245, 2:247
tongue worms, 2:320
true limpets, 2:424, 2:426–427
tusk shells, 2:470, 2:473–474
Vestimentifera, 2:92, 2:94–95
Vetigastropoda, 2:430, 2:433–434
water bears, 2:117, 2:120–123
water fleas, 2:154, 2:157–158
reptiles, 7:3–4, 7:26–28
African burrowing snakes, 7:462–464
African sideneck turtles, 7:130, 7:132–133
Afro-American river turtles, 7:137, 7:140–141
Agamidae, 7:211, 7:214–221
Alligatoridae, 7:173, 7:175–177
Anguidae, 7:340, 7:342–344
Australo-American sideneck turtles, 7:78, 7:81–83
big-headed turtles, 7:135–136
blindskinks, 7:272
blindsnakes, 7:382–385
boas, 7:411, 7:413–417
Central American river turtles, 7:100
chameleons, 7:230, 7:235–241
colubrids, 7:469, 7:473–481
Cordylidae, 7:321, 7:324–325
crocodilians, 7:162–163
Crocodylidae, 7:180, 7:184–187
early blindsnakes, 7:371
Elapidae, 7:485–486, 7:491–498
false blindsnakes, 7:388
false coral snakes, 7:400
file snakes, 7:441–442
Florida wormlizards, 7:284
Gekkonidae, 7:261–262, 7:265–269
Geoemydidae, 7:116, 7:118–119
gharials, 7:168
Helodermatidae, 7:355, 7:357
Iguanidae, 7:245–246, 7:250–257
Kinosternidae, 7:122, 7:125–126
knob-scaled lizards, 7:348, 7:351
Lacertidae, 7:298, 7:301–302
leatherback seaturtles, 7:101–102
Microteiids, 7:304, 7:306–307
mole-limbed wormlizards, 7:280, 7:281
Neotropical sunbeam snakes, 7:406
New World pond turtles, 7:106, 7:109–113
night lizards, 7:292, 7:294–295
pig-nose turtles, 7:76
pipe snakes, 7:396
plasticity and, 7:39–40
preferences and plasticity, 7:39–40
pythons, 7:421, 7:424–428
reptiles, 7:3–4
seaturtles, 7:86, 7:89–91
shieldtail snakes, 7:392–394
skinks, 7:329, 7:332–337
slender blindsnakes, 7:375–377
snapping turtles, 7:94, 7:95–96
softshell turtles, 7:152, 7:154–155
spade-headed wormlizards, 7:288, 7:289
splitjaw snakes, 7:430
Squamata, 7:204–*205*
suction, 7:3–4
sunbeam snakes, 7:402–403
Teiidae, 7:311–312, 7:314–316
tortoises, 7:143, 7:146–148
Tropidophiidae, 7:434–437
tuatara, 7:190
turtles, 7:67
Varanidae, 7:361–362, 7:365–368
Viperidae, 7:447–448, 7:451–459

wormlizards, 7:274, 7:276
Feet, 12:*38*
See also Physical characteristics
Feick's dwarf boas. *See* Banded woodsnakes
Feinsinger, P., 9:445
Fejervarya spp., 6:246, 6:250
Felid Taxonomic Advisory Group (TAG), 14:369
Felidae. *See* Cats
Felinae, 14:369
Feline owlet-nightjars, 9:387, 9:*388*, 9:*390*, 9:*391*
Felines. *See* Cats
Felis spp., 14:369
Felis bengalensis euptilura. *See* Bengal cats
Felis bieti. *See* Chinese desert cats
Felis catus. *See* Domestic cats; Feral cats
Felis chaus. *See* Jungle cats
Felis concolor. *See* Pumas
Felis concolor coryi. *See* Florida panthers
Felis margarita. *See* Sand cats
Felis nigripes. *See* Black-footed cats
Felis rufus. *See* Bobcats
Felis silvestris. *See* Wild cats
Felis silvestris libyca. *See* Libyan wild cats
Felis viverrina. *See* Fishing cats
Felou gundis, 16:312, 16:313
Felovia spp., 16:311
Felovia vae. *See* Felou gundis
Felted grass coccids. *See* Rhodesgrass mealybugs
Female reproductive system, 12:90–91
See also Physical characteristics; Reproduction
Fennec foxes, 14:*258*, 14:273
Fennecus zerda. *See* Fennec foxes
Feral Cat Coalition of San Diego, California, 14:292
Feral cats, 12:185–186, 14:257
See also Domestic cats; Wild cats
Feral Celebes pigs. *See* Timor wild boars
Feral cockroaches, 3:148, 3:150
Feral muntjacs, 12:118
Feral pigs, 12:191–192
See also Domestic pigs
Ferecetotherium spp., 15:73
Feresa attenuata. *See* Pygmy killer whales
Ferminia cerverai. *See* Zapata wrens
Feroculus spp., 13:265
Feroculus feroculus. *See* Kelaart's long-clawed shrews
Ferrets, 12:141, 12:321, 14:323, 14:324, 14:325
See also Black-footed ferrets
Ferruginous larks, 10:346
Ferruginous pygmy-owls, 9:333
Fertility control agents, 12:188
See also Reproduction
Fertilization, 1:18, 1:20–23, 2:17–18
of amphibian eggs, 6:32–33
delayed, 12:109–110
insects, 3:39
mollusks, 2:23–24
of reptilian eggs, 7:6
See also Reproduction; specific species
Fertilizers, amphibians and, 6:56, 6:57
Festucalex cinctus, 5:134
Feylinia spp., 7:327, 7:328
Feylininae. *See Feylinia* spp.
FFI (Fauna and Flora International), on black crested gibbons, 14:221
Fiber duikers, 16:73
Ficedula albicollis. *See* Collared flycatchers
Ficedula basilancia. *See* Little slaty flycatchers
Ficedula bonthaina. *See* Lampobattang flycatchers
Ficedula hypoleuca. *See* Pied flycatchers
Fiddler crabs, 2:201
Fiddlers. *See* Channel catfishes
Field crickets, 3:91, 3:207
Field grasshoppers, 3:*210*, 3:*211*
Field hamsters. *See* Black-bellied hamsters
Field mice, 16:270
Azara's, 16:266–270
long-tailed, 16:*250*, 16:261*t*
old, 16:125
striped, 16:261*t*
Field studies, **12:194–202**, 12:*195–201*
Field voles, 16:227, 16:*228*
Fieldfares, 10:486–487
Fieldwrens, 11:56
Fierasfer dubius. *See Carapus bermudensis*
Fierce snakes. *See* Taipans
Fiery-breasted bush-shrikes, 10:425
Fiery minivets, 10:*388*, 10:*392–393*
Fiery-necked nightjars, 9:369, 9:401, 9:404, 9:*407*, 9:*413*
Fiery-throated hummingbirds, 9:444, 9:447
Fifebirds. *See* Amazonian umbrellabirds
Fig wasps, 3:49
Figbirds, **11:427–432,** 11:*430*
Fighting behavior. *See* Behavior
Fiji Island boas, 7:410, 7:411
File snakes, 7:198, **7:439–444,** 7:*443*
See also Cape filesnakes
Filefishes, 4:48, 4:67, 4:69, 4:70, 5:467–471
See also Longnose filefishes; Scrawled filefishes
Filleul, A., 4:255
Filospermoidea, 1:331, 1:332
Filter feeding. *See* Suspension feeding
Fin whales, 12:67, 15:*120*, 15:*126*, 15:*127*–128
behavior, 15:6, 15:122
conservation status, 15:9, 15:125
distribution, 15:5, 15:121
feeding ecology, 15:8
Finback catsharks, 4:113, 4:114
Finbacks. *See* Fin whales
Finch-billed bulbuls. *See* Crested finchbills
Finches, 10:*172*, 10:173, **11:323–339,** 11:*326–327*
behavior, 11:323–324
conservation status, 11:324–325
distribution, 11:*323*
evolution, 11:323
feeding ecology, 11:*324*, 11:*325*
firefinches, 11:354, 11:*358*, 11:363
Galápagos, 8:3, 8:18, 8:42
habitats, 11:323
humans and, 11:325
physical characteristics, 11:323
quailfinches, 11:354, 11:*358*, 11:365–366
reproduction, 11:*324*
snow, 11:397, 11:*400*, 11:405
species of, 11:*328–339*
taxonomy, 11:286, 11:323
weaverfinches, 11:353–357
zebra, 11:354, 11:355, 11:357, 11:*359*, 11:367
See also specific types of finches
Fine-filter conservation, 3:84–85, 3:89–90
Fine-scaled yellowfins. *See* Smallscale yellowfins
Finfoot, 9:1–4
See also specific types of finfoot
Fingerfishes. *See* Monos
Fingerlings, 5:183
Fink, S. V., 4:289
Fink, W. L., 4:289, 4:389, 4:390, 4:421
Finless porpoises, 15:6, 15:33, 15:34, 15:*35*, 15:*36*, 15:39*t*
Finned octopods, 2:475, 2:477, 2:478
Finners. *See* Northern minke whales
Fins, 4:14–15, 4:17–18, 4:*22*
See also Physical characteristics
Finschia spp., 11:115
Finsch's bulbuls, 10:396
Fiordland penguins, 8:148, 8:151
Fire, salamanders and, 6:51–52
Fire ant decapitating flies, 3:*363*, 3:*369*, 3:370–371
Fire ants, red imported, 3:68, 3:70
Fire-bellied toads, 6:67, **6:83–88,** 6:*85*, 6:*86*, 6:193
Fire brats, **3:119–124**
Fire-breasted flowerpeckers, 11:*192*, 11:*196*
Fire-capped tits, 11:147, 11:*150*, 11:*152*
Fire control, 14:250
Fire corals, 1:*132*, 1:135
Fire dartfishes. *See* Fire gobies
Fire eels, 5:152, 5:*153*, 5:*154*
Fire gobies, 5:*375*, 5:*378*, 5:*383*, 5:387–388
Fire salamanders, 6:*18*, 6:38, 6:51, 6:367
Fire-tailed myzornis, 10:507, 10:508, 10:*509*, 10:519–*520*
Fire urchins, 1:*34*
Fire worms, 2:*5*, 2:*45*, 2:*49*, 2:*50*–51
Firebelly toads. *See* Fire-bellied toads
Firefinches, 11:353, 11:354, 11:*358*, 11:363
Firefishes. *See* Fire gobies
Fireflies, 3:60, 3:316, 3:317, 3:320, 3:321–322, 3:322–323
Firetails, 11:353, 11:*359*, 11:366–367
First International Conference on Conservation Biology, 1:48
Fischer's pygmy fruit bats, 13:348*t*
Fischer's turacos, 9:300
Fish hawks. *See* Ospreys
Fish ladders, 4:*73*
Fish leeches, 2:75, 2:77
Fish lice, **2:289–293,** 2:*291*, 2:*292*
Fish-owls, 9:332, 9:333
Fisheries, 1:50–52, 4:73–*74*, 5:280–281
anchoveta, 4:288
capelins, 4:400
channel catfishes, 4:361
herrings, 4:279
Osmeriformes, 4:393
protostomes, 2:41
See also Humans
Fishers, 12:132, 14:321, 14:323, 14:324
Fishery Management Plan for Sharks of the Atlantic Ocean, 4:*106*
Fishes, 2:32, 2:33, 4:3–8
amphibians and, 6:7, 6:58
behavior, **4:60–71**
benthic, 4:47, 4:48, 4:49
benthopelagic, 4:47, 4:48, 4:49
biogeography, **4:52–59**

communication signals, 4:19–23, 4:60, 4:61–62
conservation status, 4:**74–76**
coral reef, 4:35, 4:42, 4:43, 4:61
deep-sea, 4:56, 4:392
defined, **4:3–8**
distribution, **4:52–59,** 4:*55t,* 4:*56*
eggs, 4:*29*–33, 4:*30,* 4:*31,* 4:*34,* 4:46, 4:65, 4:66
evolution, **4:9–13**
feeding ecology, 4:17, 4:39, 4:40, 4:67–69
as food, 4:73–74, 4:*75*
freshwater, **4:36–41,** 4:50, 4:52
humans and, **4:72–76,** 4:*73,* 4:*74*
jawed *vs.* jawless, 4:10
marine, **4:42–51,** 4:*43,* 4:*44,* 4:50, 4:52
movement of, 4:14, 4:17–20, 4:*26*
pelagic, 4:39, 4:47, 4:48, 4:49
physical characteristics, 4:3–6, 4:*4*–7, **4:14–28,** 4:*15–19,* 4:*21,* 4:*22,* 4:*24–27*
reproduction, 4:25–27, **4:29–35,** 4:*31*–33, 4:*32,* 4:64–67
respiration, 4:6, 4:*19,* 4:23–24, 4:*25*
schools of, 4:63–64, 4:277–*278*
symbiosis and, 1:32
taxonomy, **4:9–13**
See also specific fish
Fishflies, **3:*289*–295**
Fishhook water fleas, 2:*156,* 2:*157*–158
Fishing, birds for, 8:21
Fishing bats, 13:313–314
Fishing cats, 14:*370*
Fishing genets. *See* Aquatic genets
Fishing-owls, 9:332, 9:333, 9:346–347, 9:349
Fishing rights, 4:73
See also Humans
Fishmoths. *See* Silverfish
Fission, 2:22
See also Reproduction
Fissurella spp., 2:431
Fissurelloidea, 2:429, 2:440
Fistularia spp. *See* Cornetfishes
Fistularia petimba. See Red cornetfishes
Fistularia tabacaria. See Blue-spotted cornetfishes
Fistulariidae. *See* Cornetfishes
Fitch, J. E., 5:352, 5:353
Fitzroy River turtles, 7:69
Five-toed dwarf jerboas. *See* Five-toed pygmy jerboas
Five-toed jerboas, 16:211, 16:212, 16:213, 16:214
Five-toed pygmy jerboas, 16:215, 16:216, 16:*218,* 16:*220*
Fjord seals. *See* Harp seals
Flabby whalefishes, 5:105, 5:106
Flabellifera, 2:249
Flacourt, Étienne de, 8:104
Flag-legged insects, 2:39
Flagella, sperm, 1:17
See also Reproduction
Flagship species, 1:49
See also Conservation status
Flagtail aholeholes. *See* Flagtail kuhlias
Flagtail kuhlias, 5:220, 5:*224,* 5:*227,* 5:228
Flagtail prochilodus, 4:*341,* 4:*345,* 4:348
Flagtails. *See* Kuhlias
Flame bowerbirds, 11:*478*
Flame-crowned flowerpeckers, 11:191
Flame nautilus. *See* Pearly nautilus
Flame robins, 11:105
Flame-templed babblers, 10:507, 10:*510,* 10:512–*513*
Flamingolepis liguloides, 1:230
Flamingos, **8:303–311,** 8:*307*
Flammulated owls, 9:333
Flannery, Tim, 13:9, 13:90
Flap-necked chameleons, 7:*230*
Flapped larks, 10:343
Flash coloration, 1:42, 2:38–39
See also Behavior
Flashjacks. *See* Bridled nail-tailed wallabies
Flashlightfishes, **5:113–116,** 5:*115,* 5:*117,* **5:118**
Flask-shaped sea cucumbers, 1:*423,* 1:427–*428*
Flat-backed millipedes, 2:*366,* 2:369
Flat lizards, 7:319, 7:320, 7:321, 7:322
Flat-tailed horned lizards, 7:207
Flatbills. *See* Tyrant flycatchers
Flatbottom sea stars. *See* Northern Pacific sea stars
Flatfishes, **5:449–465,** 5:*450,* 5:*454*
behavior, 5:452
conservation status, 5:453
distribution, 5:452
evolution, 5:449–450
feeding ecology, 5:452–453
habitats, 5:452
humans and, 5:453
physical characteristics, 4:14, 5:*450*–452
reproduction, 4:34, 4:35, 5:*451,* 5:453
species of, 5:455–463
taxonomy, 5:449–450
Flathead catfishes, 4:*351*
Flathead locust lobsters, 2:*203,* 2:*210,* 2:213
Flathead mullets, 5:59, 5:*61,* 5:*63,* 5:64
Flathead sculpins, 5:181
Flatheads, 4:68, **5:157–162,** 5:*160*
See also Duckbills; Northern bottlenosed whales; Southern bottlenosed whales
Flatwoods salamanders, 6:356, 6:*357,* 6:*358*
Flatworms, 1:*185,* 1:*187,* 2:13, 2:14, 2:15
behavior, 1:38, 1:42, 1:43
evolution, 1:*9*
free-living flatworms, **1:185–195,** 1:*189*
monogeneans, **1:213–224,** 1:*216,* 1:*217*
parasitic, **1:197–205,** 1:*202,* **1:210**
physical characteristics, 1:*186*
reproduction, 1:*16,* 1:20, 1:*20*
symbiosis, 1:33
tapeworms, **1:225–243,** 1:*232,* 1:*233*
three striped, 1:*3*
Fleas, **3:347–355,** 3:*351*
bat, 3:349, 3:*351,* 3:*352*
cat, 3:57, 3:349, 3:350
feeding ecology, 3:20, 3:47, 3:64
metamorphosis, 3:34
as pests, 3:76
physical characteristics, 3:18
Flectonotus spp., 6:37, 6:225, 6:230
Fledglings. *See* Reproduction
Flegler-Balon, Christine, 4:193
Flehman response, 12:80, 12:*103*
Fleischmann's glass frogs, 6:*216,* 6:*218,* 6:*222–223*
Fleming, T. H., 16:200
Flesh flies, 3:37, 3:64, 3:357, 3:361
Flesh-footed shearwaters, 8:*126*
Flickers, northern, 10:88, 10:149, 10:150, 10:*152,* 10:155–*156*
Flickertails. *See* Richardson's ground squirrels
Flies, 3:47, **3:357–374,** 3:*362,* 3:*363*
behavior, 3:60, 3:359–360
biting, 3:43
conservation status, 3:360–361
distribution, 3:359
evolution, 3:357
feeding ecology, 3:20, 3:35, 3:42, 3:47, 3:63, 3:64, 3:360
habitats, 3:359
humans and, 3:361
medicinal, 3:81
metamorphosis, 3:34
as pests, 3:75, 3:76, 3:361
physical characteristics, 3:18, 3:358–359
pollination by, 3:14, 3:78
reproduction, 3:38, 3:39, 3:360
species of, 3:*364*–374
taxonomy, 3:357
See also specific types of flies
Flight
bats, 12:12–13, 12:14, 12:43–44, **12:52–61,** 12:*53–59*
birds, 8:*11,* **8:45–57,** 12:52
advantages of, 8:45
aerobic capacity, 8:49
altitude, 8:50–51
body anatomy, 8:45, 8:*46*
feathers, 8:7, 8:8–9, 8:*11,* 8:*14*
food for, 8:16
hawking, 9:403, 9:446–447, 9:478–479, 11:438
heat dispersion, 8:48
landing, 8:49–50
metabolic power, 8:49
migration, 8:50
muscle anatomy, 8:47–49
origins of, 8:51
over land, 8:50
over sea, 8:50
sallying, 9:403
silent, 9:337
spoonbills, 8:292
takeoff, 8:49–50
water birds, 8:49
wing anatomy, 8:45–46, 8:*46,* 8:*47*
insects, 3:12–13, 3:*19,* 3:22–24, 3:45
See also Behavior; specific species
Flight membranes, of bats, 12:43, 12:56–58
Flight speed, migratory, 8:32
Flightless birds
dodos, 9:244
Gruiformes, 9:3
rails, 9:46, 9:47, 9:51
Flightless dwarf anomalures. *See* Cameroon scaly-tails
Flightless grebes. *See* Titicaca flightless grebes
Flint, Vladimir, 9:29
Floating habitats, 4:62
Flores crows, 11:508
Flores monarchs, 11:102
Florida alligators. *See* American alligators
Florida gars, 4:222, 4:*223,* 4:*226,* 4:227
Florida lancelets, 1:*490,* 1:495
Florida Manatee Sanctuary Act, 15:195, 15:209
Florida manatees, 15:191, 15:193, 15:195, 15:209, 15:211–212
See also West Indian manatees
Florida panthers, 12:*222*

Florida scrub-jays, 11:506, 11:507, 11:508
Florida spotted gars. *See* Florida gars
Florida water rats. *See* Round-tailed muskrats
Florida wormlizards, 7:*283*, **7:283–285**, 7:*284*
Floridazoros spp., 3:239
Florisuga spp., 9:447
Florisuga mellivora. *See* White-necked jacobins
Floscularia ringens, 1:*263*, 1:*265*, 1:266
Flounders, 4:70, 5:453
Flour beetles, confused, 3:76
Flour weevils, 3:57
Flower baskets, Venus's. *See* Venus's flower baskets
Flower bats, 13:314
  See also specific types of flower bats
Flower-faced bats, 13:409*t*
Flower flies, 3:357
Flower-loving flies, 3:361
Flower mantids. *See* Orchid mantids
Flower thrips, Western, 3:283, 3:*284*, 3:*285*
Flowerpecker weaver-finches, 11:354, 11:360
Flowerpeckers, **11:189–199**, 11:*192*
  behavior, 11:190
  conservation status, 11:191
  distribution, 11:*189*
  evolution, 11:189
  feeding ecology, 11:190
  habitats, 11:190
  humans and, 11:191
  physical characteristics, 11:189
  reproduction, 11:190–191
  species of, 11:*193–198*
  taxonomy, 11:189
Flowers of the wave, 5:*72*, 5:*73*, 5:*75–76*
Flufftails, 9:45, 9:46, 9:47
Fluffy gliders. *See* Yellow-bellied gliders
Fluid dynamics, powered flight and, 12:56
Fluid feeders, 3:20
Flukes, **1:197–211**, 1:*202*
  behavior, 1:38, 1:199–200
  conservation status, 1:200
  evolution, 1:197
  feeding ecology, 1:200
  habitats, 1:198–199
  humans and, 1:201
  as parasites, 1:34–35
  physical characteristics, 1:197–198
  reproduction, 1:20, 1:21, 1:200
  species of, 1:*204*–211
  taxonomy, 1:197
Fluorescent microscopy, 2:4
Fluvicolinae. *See* Tyrant flycatchers
Fluvicoline flycatchers. *See* Tyrant flycatchers
Fluviphylax spp., 5:90
Fly maggot farming, 3:80
Fly-trapping, passive, 9:379
Flycatcher-shrikes, 10:385–387
Flycatcher-thrushes, 10:487
Flycatcher warblers, 11:5
Flycatchers, 10:170–171, 10:173, 10:270
  gray-headed, 11:105, 11:*107*, 11:111
  monarch, **11:97–103**, 11:*99*
  shrike-flycatchers, 11:*28*, 11:40
  See also specific types of flycatchers
Flying ants, 3:68
Flying foxes, **13:319–333**, 13:*323*, 13:*324*
  behavior, 13:319–320
  conservation status, 13:321–322
  crepuscular activity cycle, 12:54
  distribution, 13:*319*
  evolution, 13:319
  feeding ecology, 13:320
  golden-crowned, 13:*338*
  gray-headed, 13:312, 13:320
  habitats, 13:319
  humans and, 13:322
  physical characteristics, 12:*54*, 12:58, 13:307, 13:319
  reproduction, 13:320–321
  species of, 13:*325*–330, 13:331*t*–332*t*
  taxonomy, 13:309, 13:319
Flying frogs, 6:291
Flying geckos, 7:*261*
Flying gurnards, **5:157–162**, 5:*160*
Flying jacks. *See* White-faced sakis
Flying lemurs. *See* Colugos
Flying lizards, 7:*209*, 7:*211*, 7:*213*, 7:*217*
Flying mice, 16:300
  big-eared, 16:300, 16:*301*, 16:*304*–305
  Zenker's, 16:300, 16:*301*, 16:*304*
  See also Pygmy gliders
Flying squids. *See Onychoteuthis* spp.
Flying squirrels, 12:14, 12:131, **16:135–142**, 16:*138*, 16:141*t*–142*t*
Flyingfishes, 4:4, 4:12, 4:*30*, 4:49, 4:67, 4:70, 5:79–81
Flyingsnakes, 7:468
Foam nest frogs. *See* Gray treefrogs
Fodiator spp., 5:79–80
Fodies, 11:376, 11:*381*, 11:388–389
Foetorepus spp., 5:365, 5:366
Folivores, 14:8
Folk medicine, reptiles in, 7:50–52, 7:*54*
  See also Humans; specific species
Folklore
  insects in, 3:74
  reptiles in, 7:54
  See also Humans; specific species
Food
  amphibians as, 6:54, 6:252, 6:256
  beetles as, 3:323
  fish as, 4:73–74, 4:*75*
  Hemiptera as, 3:264
  Hymenoptera as, 3:408
  insects as, 3:80–81
  Orthoptera as, 3:208
  protostomes as, 2:41
  reptiles as, 7:47–48
  See also Feeding ecology; Humans
Food and Agriculture Organization of the United Nations
  Alaska pollocks, 5:36
  albacore, 5:415
  Atlantic bluefin tunas, 5:415
  Atlantic mackerels, 5:414
  blue marlins, 5:411
  flathead mullets, 5:64
  gobies, 5:377
  king mackerels, 5:414
  largehead hairtails, 5:417
  livestock, 12:175
  mullets, 5:60
  sailfishes, 5:410
  skipjack tunas, 5:412
  swordfishes, 5:418
  wahoos, 5:412
Food web, 2:25
  See also Behavior; Feeding ecology
Foose, Thomas J., 12:206, 12:207
Foot drumming, 12:83
Foot-flagging displays, 6:46, 6:*49*
Foothill yellow-legged frogs, 6:*5*
Foraging
  memory and, 12:152–153
  subterranean mammals, 12:75–76
  See also Feeding ecology
Forbesella agassizi. *See* Spring cavefishes
Forbesichthys agassizii. *See* Spring cavefishes
Forbes's blackbirds, 11:305
Forbes's chestnut rails. *See* Forbes's forest-rails
Forbes's forest-rails, 9:*55*, 9:*57–58*
Forbidden tree kangaroos, 13:32
Forcepsfishes, 5:241–242
Forcipiger flavissimus. *See* Forcepsfishes
Forcipiger longirostris. *See* Longnose butterflyfishes
Forcipulatida, 1:368
Ford, Julian, 11:83
Ford, Linda, 6:215
Fordonia leucobalia. *See* White-bellied mangrove snakes
Fordyce, Ewan, 8:147
Forebrain, 12:6
  See also Physical characteristics
Foregut, 3:20–21
Foregut fermentation, 12:14–15, 12:47–48
Forelimbs, bat wings and, 12:56
Forensic entomologists, 3:78
Forest bright-eyed frogs, 6:*294*, 6:296–297
Forest cobras, 7:*490*, 7:*492*
Forest defoliators, 3:389
Forest Department (Government of Assam), on pygmy hogs, 15:287
Forest dingos, 12:180
Forest dormice, 16:317–320, 16:*322*, 16:*323–324*
Forest-dwelling bandicoots, 13:3, 13:4
Forest-dwelling damselflies, 3:*54*
Forest elephants, 15:162–163, 15:*166*
Forest entomologists, 3:78
Forest foxes. *See* Crab-eating foxes
Forest giants, 3:*136*, 3:*138*
Forest giraffes. *See* Okapis
Forest green treefrogs. *See* Kinugasa flying frogs
Forest hogs, 15:277, 15:*278*, 15:*283*, 15:*284*
Forest-living guenons, 14:188, 14:192, 14:193
Forest marmots. *See* Woodchucks
Forest mice, 16:266, 16:267, 16:270
  See also Gray climbing mice
Forest musk shrews, 13:197, 13:265, 13:266, 13:*269*, 13:*272*
Forest owlets, 9:351
Forest pests, 3:75
Forest rats, red, 16:*281*
Forest shrews, 13:*270*, 13:*273*
Forest spiny pocket mice, 16:199
Forest tarpans, 12:175
Forest vultures. *See* Greater yellow-headed vultures
Forest wagtails, 10:*377*, 10:*378*
Forest wallabies, 13:34, 13:35, 13:83, 13:86
Forest weavers. *See* Dark-backed weavers
Forest woodhoopoes, 10:66
Foresters. *See* Eastern gray kangaroos
Forests, 12:218–219
Forests, tropical. *See* Tropical forests
Forey, P., 4:189
Forficula auricularia. *See* European earwigs
Forficulina, 3:195–197

Fork-crowned lemurs. *See* Masoala fork-crowned lemurs
Fork-tailed drongos, 11:*439*
Fork-tailed flycatchers, 10:270
Fork-tailed pygmy tyrants, 10:275
Fork-tailed storm-petrels, 8:137, 8:138
Fork-tailed sunbirds, 11:208
Fork-tailed swifts, 9:422
Forkbeards. *See* Hakes
Forked-tail cats. *See* Channel catfishes
Forktails, 10:483
Formica spp. *See* Wood ants
Formicariidae. *See* Ant thrushes
Formicarius analis. See Black-faced antthrushes
Formicarius rufifrons. See Rufous-fronted ant thrushes
Formicivora erythronotos. See Black-hooded antwrens
Formicivora rufa. See Rusty-backed antwrens
Formosan serows, 16:87, 16:90
Formosan subterranean termites, 3:57
Forshaw, J. M., 9:275
Forty-spotted pardalotes, 11:202, 11:*203*, 11:*204*
Fossa, 12:*98*, 12:*198*, **14:347–355,** 14:*348*, 14:*353*, 14:*354*
Fossa fossana. See Malagasy civets
Fossil evidence. *See* Evolution
Fossil fuels, 12:213
See also Humans
Fossil insects, 3:7–9, 3:*8*, 3:*9*
Devonian, 3:10, 3:11
Mazon Creek, 3:11
winged, 3:12
Fossils. *See* Evolution
Fossorial mammals, 12:13, 12:69, 12:82
See also specific fossorial mammals
Fossorial spiny rats, Bishop's, 16:450, 16:458*t*
Foster, Mercedes S., 10:299
Foudia spp. *See* Fodies
Foudia flavicans, 11:378
Foudia madagascariensis. See Madagascar fodies
Foudia rubra, 11:378
Foudia sechellarum, 11:378
Founder species
bottlenecks and, 12:221–222
in zoo populations, 12:207–208
See also Conservation status
Four-banded sandgrouse, 9:231
Four-eyed opossums, 12:38, 12:254
Four-horned antelopes. *See* Chousinghas
Four-o'clocks. *See* Noisy friarbirds
Four-spined sandperches, 5:*331*
Four-striped grass mice, 12:*117*, 16:*252*
Four-toed hedgehogs. *See* Central African hedgehogs
Four-toed jerboas, 16:212, 16:216, 16:223*t*
Four-toed salamanders, 6:390, 6:*394*, 6:400
Four-toed sengis, 16:518, 16:519, 16:521, 16:*523*, 16:*526*–527
Four-winged flies. *See* Unique-headed bugs
Four-winged flyingfishes, 5:*80*
4d cells. *See* Mesentoblasts
Foureyed fishes, 4:57
See also Largescale foureyes
Foureyes. *See* Largescale foureyes
Fourteen-lined comb eared skinks, 7:*331*, 7:*332*, 7:333
Fowls, **8:433–453,** 8:*436*
behavior, 8:435
conservation status, 8:436–437
distribution, 8:*433*, 8:434
evolution, 8:433
feeding ecology, 8:435–436
habitats, 8:434–435
humans and, 8:437
physical characteristics, 8:433–434
reproduction, 8:436
species of, 8:*441–452*
taxonomy, 8:433
Fox kestrels, 8:*357*–358
Fox squirrels
big cypress, 16:*164*
Delmarva, 16:167
eastern, 16:*164*, 16:*166*, 16:*168*, 16:*169*, 16:171
Mexican, 16:175*t*
Fox tapeworms, 1:*228*, 1:231
Foxes, **14:265–273,** 14:*275*, **14:278–285**
behavior, 14:267–268
conservation status, 12:192, 14:272, 14:273
distribution, 14:267
evolution, 14:265
feeding ecology, 14:255, 14:269–270, 14:271
feral cats and, 12:185
habitats, 14:267
humans and, 14:273
as pests, 12:188
physical characteristics, 14:266–267
reproduction, 12:107, 14:271, 14:272
species of, 14:278–283, 14:283*t*, 14:284*t*
taxonomy, 14:265
Foxface rabbitfishes, 5:*395*, 5:*398*, 5:402–403
Foxfaces. *See* Foxface rabbitfishes
Fox's weavers, 11:379
Fragmentation, habitat, 12:214, 12:222–223
See also Conservation status
Franciscana dolphins, 15:6, 15:9, **15:23–26,** 15:*24*, 15:*25*
Francolinus afer. See Red-necked francolins
Francolinus camerunensis. See Mount Cameroon francolins
Francolinus gularis. See Swamp francolins
Francolinus lathami. See Latham's francolins
Francolinus leucoscepus. See Yellow-necked francolins
Francolinus nahani. See Nahan's francolins
Francolinus ochropectus. See Djibouti francolins
Francolinus pictus. See Painted francolins
Frank, Morris, 14:293
Frankfurt Zoo, Germany, 12:*210*
Frankham, R., 12:206, 12:208
Franklin Island house-building rats. *See* Greater stick-nest rats
Franklin Island stick nest rats. *See* Greater stick-nest rats
Frankliniella occidentalis. See Western flower thrips
Franquets epauletted bats. *See* Singing fruit bats
Franz Josef (Emperor of Austria), 12:203
Fraseria ocreata. See Fraser's forest-flycatchers
Fraser's dolphins, 15:57*t*
Fraser's forest-flycatchers, 11:*28*, 11:37–*38*
Fraser's sunbirds. *See* Scarlet-tufted sunbirds
Fratercula spp. *See* Puffins
Fratercula arctica. See Puffins
Fratercula corniculata. See Horned puffins
Freckled antechinuses. *See* Southern dibblers
Freckled frogmouths. *See* Tawny frogmouths
Freckled nightjars, 9:404
Fredericella sultana, 2:*499*, 2:*500*
Fredericellidae, 2:497
Frederickena unduligera. See Undulated antshrikes
Free-living barnacles, 2:275, 2:276
Free-living flatworms, 1:11, **1:185–195,** 1:*187*, 1:189
behavior, 1:40, 1:187
conservation status, 1:188
distribution, 1:187
evolution, 1:185
feeding ecology, 1:28, 1:187–188
habitats, 1:187
humans and, 1:188
physical characteristics, 1:*186–187*
reproduction, 1:18, 1:20, 1:21, 1:188
species of, 1:*190–195*
taxonomy, 1:185–186
Free Madagascar frogs, 6:*294*, 6:*296*
Free-tailed bats, **13:483–496,** 13:*487*, 13:*489*
behavior, 13:312, 13:484–485
conservation status, 13:487
distribution, 13:*483*
evolution, 13:483
feeding ecology, 13:485
habitats, 13:311
humans and, 13:487–488
physical characteristics, 13:307, 13:308, 13:310, 13:483
reproduction, 13:486–487
species of, 13:491–493, 13:493*t*–495*t*
taxonomy, 13:483
Freeze-susceptible insects, 3:37
Freeze-tolerant insects, 3:37
Fregata andrewsi. See Christmas frigatebirds
Fregata aquila. See Ascension frigatebirds
Fregata ariel. See Lesser frigatebirds
Fregata magnificens. See Magnificent frigatebirds
Fregata minor. See Great frigatebirds
Fregatidae. *See* Frigatebirds
Fregetta spp. *See* Storm-petrels
Fregetta grallaria. See White-bellied storm-petrels
Fregetta tropica. See Black-bellied storm-petrels
French girdle wearers, 1:*344*, 1:*346*, 1:*347*–348
French Island, Australia, 12:*223*
Frenulata, 2:85, 2:86
Frenzel, Johannes, 1:91
Freshies. *See* Johnstone's crocodiles
Freshwater aholeholes. *See* Jungle perches
Freshwater angelfishes, 5:*282*, 5:*287*–288
Freshwater batfishes. *See* Chinese suckers
Freshwater blennies, 5:342
Freshwater bryozoans, **2:497–502,** 2:*499*
Freshwater butterflyfishes, 4:232, 4:233, 4:*236*, 4:240–*241*, 5:49
Freshwater colonial entoprocts, 1:319, 1:*320*, 1:*322*, 1:*324*
Freshwater crabs, 2:199, 2:201
Freshwater decapods, 2:199, 2:201
Freshwater dogfishes. *See* Bowfins
Freshwater fairy shrimps. *See* Sudanese fairy shrimps
Freshwater fishes, **4:36–41,** 4:50, 4:52, 4:56–58, 4:*57*, 4:76
See also specific fishes

Freshwater habitats, 4:36, 4:*37–41*, 4:*38*
See also Habitats
Freshwater herrings. *See* Alewives
Freshwater kuhlias. *See* Jungle perches
Freshwater limpets, New Zealand, 2:*416*, 2:417
Freshwater mussels, 2:451, 2:453–454
Freshwater pearl mussels, 2:454
See also European pearly mussels
Freshwater planarians, 1:38, 1:*189*, 1:*191*, 1:194–195
Freshwater pompanos. *See* Pirapitingas
Freshwater sponges, 1:21, 1:78, 1:79, 1:*81*, 1:*84*, 1:86
Freshwater stingrays, 4:57, 4:*173*, 4:174, 4:176–177, 4:*180*, 4:*184*–186
Freshwater sturgeons. *See* Lake sturgeons
Freussaciidae, 2:411
Friarbirds, 11:*428*
Fricke, Hans, 4:189, 4:191, 4:193
Fricke, R., 5:365
Friedmann's larks, 10:346
Friendly whales. *See* Gray whales
Friends of the Arabuko-Sokoke Forest, 16:525
Fries' gobies, 5:377
Frigate Island caecilians, 6:*438*, 6:*440*
Frigatebirds, 8:183–186, **8:193–199,** 8:*196*
Frill-necked lizards. *See* Frilled lizards
Frilled anemones, 1:*110*, 1:112–*113*
Frilled coquettes, 9:*453*, 9:462–*463*
Frilled lizards, 7:*35*, 7:*210*, 7:*212*, 7:215–216, 7:*217*
Frilled sharks, 4:11, 4:143, 4:144, 4:*145*, 4:*146*–147
Fringe-lipped bats, 13:314, 13:*417*, 13:*422*, 13:*423*–424
Fringe-toed lizards. *See* Coachella Valley fringe-toed lizards
Fringelip mullets, 5:*61*, 5:*62*–63
Fringetails. *See* Silverfish
Fringilla aestivalis. *See* Bachman's sparrows
Fringilla albicilla. *See* Whiteheads
Fringilla albicollis. *See* White-throated sparrows
Fringilla algilis. *See* Thick-billed flowerpeckers
Fringilla amandava. *See* Red avadavats
Fringilla ardens. *See* Red-collared widow-birds
Fringilla atricollis. *See* African quailfinches
Fringilla bengalus. *See* Red-cheeked cordon-bleu
Fringilla bichenovii. *See* Double-barred finches
Fringilla bicolor. *See* Lark buntings
Fringilla domestica. *See* House sparrows
Fringilla erythrophthalmus. *See* Eastern towhees
Fringilla funerea. *See* Dusky indigobirds
Fringilla guttata. *See* Zebra finches
Fringilla hyemalis. *See* Dark-eyed juncos
Fringilla lapponica. *See* Lapland longspurs
Fringilla macroura. *See* Pin-tailed whydahs
Fringilla melba. *See* Green-winged pytilias
Fringilla melodia. *See* Song sparrows
Fringilla montana. *See* Tree sparrows
Fringilla montifringilla. *See* Bramblings
Fringilla nivalis. *See* Snow finches
Fringilla oryzivora. *See* Bobolinks
Fringilla passerina. *See* Chipping sparrows
Fringilla teydea. *See* Blue chaffinches
Fringillidae. *See* Finches
Fringillinae. *See* Finches
Frith, Cliff, 11:71
Fritillaria spp., 1:473
Fritillaria borealis, 1:*475*, 1:476
Fritzinger, Leopold J. F. J., 4:201
Frog-eyed geckos. *See* Common plate-tailed geckos
Frog-hoppers. *See* Spittle bugs
Frogfishes, 4:*64*, 5:*47–51*, 5:54
Froghead eels, 4:*259*, 4:*263*
Froglets, 6:36, 6:68, 6:*174*
Frogmouth catfishes, 4:353
Frogmouths, 9:367–369, **9:377–385,** 9:*380*
Frogs, 6:3, **6:61–68**
art and, 6:52, 6:*54*
chemosensory cues, 6:45–46
clawed, **6:99–107,** 6:*103–106*
cricket, 6:48, 6:66, **6:265–271,** 6:*267*
distribution, 6:4–5, 6:63–64
egg transportation, 6:36–37
evolution, 6:4–5, 6:11–*13*, 6:15, 6:61–62
feeding ecology, 6:6, 6:26, 6:67
flying, 6:291
as food, 6:54
ghost, 6:6, **6:131–134,** 6:*133*, 6:155
habitats, 6:26, 6:63–64
harlequin, **6:183–195,** 6:*184*, 6:*186–187*
jumping ability, 6:25–26, 6:*53*, 6:66, 6:*66*
larvae, 6:39–43
mythology and, 6:51–52
nest construction, 6:50
New Zealand, 6:4–5, 6:41, 6:64, **6:69–75,** 6:*71*
painted (Discoglossidae), **6:89–94,** 6:*91*
painted (Limnodynastidae), 6:*141*, 6:144–145
parsley, 6:5, 6:64, **6:127–130,** 6:*128*, 6:*129*
physical characteristics, 6:*20*, 6:*23*, 6:25–26, 6:62–63
predators of, 6:65–67
psychoactive drugs from, 6:53
reproduction, 6:28–38, 6:*31*, 6:*34*, 6:*35*, 6:*36*, 6:68
Ruthven's, 6:6, **6:211–213,** 6:*212*, 6:225
Seychelles, 6:6, 6:36, 6:68, **6:135–138,** 6:*137*
shovel-nosed, 6:6, **6:273–278,** 6:*274*, 6:*276*
vocal sac-brooding, 6:6, 6:36, **6:173–177,** 6:*175–176*
vocalizations, 6:*22*, 6:26, 6:30, 6:46–48, 6:304
water, **6:147–154,** 6:*150*
See also True frogs; specific types of frogs
Frogspawn corals, 1:*104*
Frost, D., 7:340
Frosted hairy dwarf porcupines, 16:365, 16:374*t*
Frostfishes. *See* Atlantic tomcods
Frostius spp., 6:184
Froth-bugs. *See* Spittle bugs
Frugivores, 14:8
Fruit bats, 12:13, 13:*310*, 13:312
Fruit doves, 9:243, 9:244, 9:247, 9:251
Fruit flies, 3:57, 3:60, 3:80, 3:357, 3:360, 3:361
education and, 3:83
eyes, 3:27
Mediterranean, 3:59, 3:*363*, 3:372–*373*
reproduction, 3:4, 3:29
Fruit pigeons, 9:242–243
Fruiteaters. *See* Cotingas
Fry, C. Hilary, 10:5–6, 10:39
Fry's whistling frogs, 6:*306*, 6:*310*–311
Fuelleborn's longclaws, 10:373
Fugus, 5:471, 5:*472*, 5:*476*, 5:482
Fulgoromorpha, 3:259
Fulica spp. *See* Coots
Fulica americana. *See* American coots
Fulica atra. *See* Eurasian coots
Fulica cornuta. *See* Horned coots
Fulica cristata. *See* Red-knobbed coots
Fulica gigantea. *See* Giant coots
Fulk, G. W., 16:435
Fulmars, 8:*108*, **8:123–127**
Fulmarus glacialis. *See* Northern fulmars
Fulmarus glacialoides. *See* Southern fulmars
Fulvettas, 10:506, 10:*509*
Fulvous-bellied ant-pittas, 10:*243*, 10:255–*256*
Fulvous-chested jungle-flycatchers, 11:*29*, 11:*34*
Fulvous leaf-nosed bats. *See* Fulvous roundleaf bats
Fulvous lemurs. *See* Brown lemurs
Fulvous roundleaf bats, 13:405, 13:*406*, 13:*407*, 13:408
Funambulus spp., 16:163
Funambulus tristriatus, 16:167
Funch, P., 1:12, 1:13, 1:327, 1:351
Function, **3:17–30**
Functional diseases, 2:33
Fundulidae, 5:92, 5:94
Fundulus spp., 5:89
Fundulus catenatus, 5:92, 5:93
Fundulus diaphanus, 5:92
Fundulus fuscus, 5:91
Fundulus heteroclitus, 5:94
Fundulus heteroclitus macrolepidotus. *See* Northern mummichogs
Fungia scutaria. *See* Mushroom corals
Fungus beetles, 3:316
Fungus gardens, 3:63
Funisciurus spp., 16:163
Funisciurus anerythrus, 16:165
Funisciurus carruthersi, 16:167
Funisciurus congicus. *See* Striped tree squirrels
Funisciurus isabella, 16:167
Funnel-eared bats, **13:459–465,** 13:*460*, 13:*461*, 13:*462*, 13:*463*
Funnels, Berlese-type, 3:94
Fur. *See* Hairs
Fur seals, 12:*106*, **14:393–403,** 14:*400*, 14:*401*, **14:407–408**
behavior, 14:396–397
conservation status, 12:*219*, 14:399
distribution, 12:138, 14:394–395
evolution, 14:393
feeding ecology, 14:397–398
habitats, 14:395–396
humans and, 14:400
physical characteristics, 14:393–394
reproduction, 12:*92*, 14:398–399
species of, 14:*402*–403, 14:407*t*–408*t*
taxonomy, 14:256
Fur trade, 12:173, 14:374, 14:376
See also Humans
Furbringer, M., 9:345
Furcifer spp., 7:*37*, 7:224, 7:227, 7:232
Furcifer campani. *See* Jeweled chameleons
Furcifer labordi, 7:232
Furcifer minor. *See* Minor chameleons
Furcifer pardalis. *See* Panther chameleons
Furgaleus macki, 4:116
Furina dunmalli, 7:486
Furipteridae. *See* Smoky bats

Furipterus spp., 13:467, 13:468
Furipterus horrens. *See* Smoky bats
Furnariidae. *See* Ovenbirds
Furnariinae. *See* Ovenbirds
Furnarius figulus. *See* Wing-banded horneros
Furnarius rufus. *See* Rufous horneros
Furniture bugs. *See* Silverfish
Furniture cockroaches. *See* Brownbanded cockroaches
Furniture termites. *See* West Indian powderpost drywood termites
Fusiliers, 4:67, 5:255, 5:257–263
Fussion-fission communities, 14:229

G

Gabela bush-shrikes, 10:429
Gabela helmet-shrikes, 10:429
Gabianus pacificus. *See* Pacific gulls
Gabon Allen's bushbabies, 14:*27*, 14:28–*29*
Gaboon adders, 7:446, 7:447, 7:*449*, 7:*451*, 7:457
Gaboon caecilians, 6:31, 6:436, 6:*436*
Gaboon vipers. *See* Gaboon adders
Gadella maraldi, 5:27
Gadfly petrels, 8:123–124
Gadiculus argenteus, 5:28
Gadidae, 5:25–30, 5:27
Gadiformes, **5:25–40**, 5:*27*, 5:*31*
  behavior, 5:29
  conservation status, 5:29–30
  distribution, 5:26–28
  evolution, 5:25
  feeding ecology, 5:29
  habitats, 5:28–29
  humans and, 5:30
  physical characteristics, 5:25–26
  reproduction, 5:29
  species of, 5:*32*–40
  taxonomy, 5:25
Gadila aberrans. *See* Aberrant tooth shells
Gadilida, 2:469, 2:470
Gadinae, 5:25
Gadomus spp., 5:27
Gadopsidae. *See* Blackfishes
Gadopsis spp. *See* Blackfishes
Gadopsis marmoratus. *See* River blackfishes
Gadus spp., 5:27
Gadus morhua. *See* Atlantic cods
Gaia: An Atlas of Planet Management, 12:218
Gaidropsarus spp., 5:28
Gaige, Helen, 6:211
Gail, Françoise, 1:479
Gaindatherium browni, 15:249
Gaits, 12:*39*
  See also Physical characteristics
Galadictis spp., 14:347, 14:348
Galagidae. *See* Bushbabies
Galago spp., 14:23
Galago alleni. *See* Bioko Allen's bushbabies
Galago cameronensis. *See* Cross River Allen's bushbabies
Galago demidoff. *See* Demidoff's bushbabies
Galago gabonensis. *See* Gabon Allen's bushbabies
Galago gallarum. *See* Somali bushbabies
Galago granti. *See* Grant's bushbabies
Galago matschiei. *See* Dusky bushbabies
Galago moholi. *See* Moholi bushbabies
Galago nyasae. *See* Malawi bushbabies
Galago orinus. *See* Uluguru bushbabies
Galago rondoensis. *See* Rondo bushbabies
Galago senegalensis. *See* Senegal bushbabies
Galago udzungwensis. *See* Uzungwa bushbabies
Galago zanzibaricus, 14:26
Galagoides spp., 14:23
Galagoides demidoff. *See* Demidoff's bushbabies
Galagoides thomasi. *See* Thomas's bushbabies
Galagoides zanzibaricus. *See* Zanzibar bushbabies
Galagonidae. *See* Bushbabies
Galagos, vision, 12:79
Galahs, 9:277
Galápagos boobies, 8:212, 8:220
Galápagos cormorants, 8:*202*, 8:*204*, 8:208–209
Galápagos doves, 9:248
Galápagos finches, 8:3, 8:18, 8:42, 11:323
Galápagos flamingos, 8:306
Galápagos fur seals, 12:*92*, 14:394, 14:395, 14:399, 14:*400*, 14:408*t*
Galápagos mockingbirds, 10:466, 10:*468*
Galápagos penguins, 8:148, 8:149, 8:150
Galápagos sea lions, 12:*64*, 12:*65*, 12:*131*, 14:*394*, 14:*397*, 14:399–400, 14:*401*, 14:*404*, 14:405–406
Galápagos tortoises, 7:47, 7:59, 7:*71*, 7:*145*, 7:*146*–147
Galatea racquet-tails. *See* Common paradise kingfishers
Galaxias brevipinnis. *See* Climbing galaxias
Galaxias cleaveri. *See* Tasmanian mudfishes
Galaxias fontanus. *See* Swan galaxias
Galaxias fuscus. *See* Barred galaxias
Galaxias johnstoni. *See* Clarence galaxias
Galaxias maculatus. *See* Common jollytails
Galaxias occidentalis. *See* Western minnows
Galaxias olidus. *See* Marbled galaxias
Galaxias parvus. *See* Swamp galaxias
Galaxias pedderensis. *See* Pedder galaxias
Galaxias pusillus nigrostriatus. *See* Black-stripe minnows
Galaxias truttaceus. *See* Trout minnows
Galaxiella munda. *See* Mudminnows
Galaxiella nigrostriata. *See* Black-stripe minnows
Galaxiella pusilla. *See* Eastern Australian galaxiids
Galaxiids, 4:58, **4:389–393**, 4:*394*, **4:396–397**
Galaxioidea, 4:390, 4:391
Galbula albirostris. *See* Yellow-billed jacamars
Galbula dea. *See* Paradise jacamars
Galbula galbula. *See* Green-tailed jacamars
Galbula pastazae. *See* Coppery-chested jacamars
Galbula ruficauda. *See* Rufous-tailed jacamars
Galbulae, 10:101
Galbulidae. *See* Jacamars
Galea spp., 16:389, 16:390
Galea flavidens. *See* Yellow-toothed cavies
Galea musteloides. *See* Cuis
Galeichthys bonillai. *See* Sea catfishes
Galemys spp., 13:279
Galemys pyrenaicus. *See* Pyrenean desmans
Galeocerdo spp., 4:115
Galeocerdo cuvier. *See* Tiger sharks
Galeodes arabs. *See* Egyptian giant solpugids
Galeommatidae, 2:452
Galeommatoidea, 2:452, 2:453
Galeomorphii, 4:11, 4:97, 4:113, 4:131
Galeorhinus, 4:113
Galeorhinus galeus, 4:116
Galericinae, 13:203
Galerida spp. *See* Larks
Galerida cristata. *See* Crested larks
Galerida theklae, seeThekla larks
Galidia spp., 14:347, 14:348
Galidia elegans. *See* Ring-tailed mongooses
Galidia elegans elegans. *See* Eastern Malagasy ring-tailed mongooses
Galidictis grandidieri. *See* Western Malagasy broad striped mongooses
Galidiinae, 14:347, 14:349, 14:352
Galil, Bella S., 3:221
Galjoens, 5:235, 5:236, 5:238, 5:239, 5:241, 5:242, 5:243
Gall gnats, 3:361
Gallardo, M. H., 16:433
Gallicolumba keayi. *See* Negros bleeding hearts
Gallicolumba luzonica. *See* Bleeding-heart pigeons
Gallicolumba rufigula. *See* Golden-heart pigeons
Gallicrex cinerea. *See* Watercocks
Gallinago spp., 9:176
Gallinago gallinago. *See* Common snipes
Gallinago imperialis. *See* Imperial snipes
Gallinago macrodactyla. *See* Madagascar snipes
Gallinago media. *See* Great snipes
Gallinago nemoricola. *See* Wood snipes
Gallinago nigripennis. *See* African snipes
Gallinago nobilis. *See* Noble snipes
Gallinula spp., 9:48, 9:50
Gallinula chloropus. *See* Common moorhens
Gallinula elegans. *See* Buff-spotted flufftails
Gallinula mortierii. *See* Tasmanian native hens
Gallinula pacifica. *See* Samoan moorhens
Gallinula phoenicurus. *See* White-breasted waterhens
Gallinula silvestris. *See* San Cristobal moorhens
Gallinula tenebrosa. *See* Dusky moorhens
Gallinula ventralis. *See* Black-tailed native-hens
Gallinules, 9:1–4, 9:46–52
Gallioformes, **8:399–402**
Gallirallus owstoni. *See* Guam rails
Gallirallus philippensis. *See* Buff-banded rails
Gallirex johnstoni. *See* Ruwenzori turacos
Gallirex porphyreolophus. *See* Purple-crested turacos
Galliwasps, **7:339–340**, 7:*341*, **7:343**
Galloisiana nipponensis. *See* Japanese rock-crawlers
Galloperdix spp. *See* Spurfowl
Gallotia spp. *See* Giant lizards
Gallotia gomerana. *See* Gomeran giant lizards
Gallotia simonyi. *See* Simony's giant lizards
Gallotiinae, 7:297
Galls, plant, 3:63
Gallus spp. *See* Junglefowls
Gallus gallus. *See* Red jungle fowl
Gambarian, P. P., 16:211
Gambelia spp., 7:246
Gambelia wislizenii. *See* Long-nosed leopard lizards
Gambel's quail, 8:*457*
Gambian epauletted fruit bats, 13:*336*
Gambian puffback-shrikes. *See* Northern puffbacks

Gambian rats, 16:*286*, 16:*289*–290
Gambian slit-faced bats, 13:376*t*
Gambusia affinis. See Mosquitofishes
Gamete exchange, 1:17–18
See also Reproduction
Gametes, 12:101, 12:105
See also Reproduction
Gametogenesis, 1:16–17, 1:25, 2:16–17, 4:32–33
See also Reproduction
Gammaridea, 2:261
Gammarus lacustris, 2:*264*, 2:*267*
Gamostolus subantarcticus. See Unique-headed bugs
Gampsocleis glabra. See Heath bush-crickets
Gampsonyx spp. *See* White-tailed kites
Gampsonyx swainsonii. See South American pearl kites
Ganeshida, 1:169
Ganges and Indus dolphins, 15:4, 15:5, **15:13–17**, 15:*14*, 15:*15*, 15:*16*
Ganges river dolphins. *See* Ganges and Indus dolphins
Ganges sharks, 4:116, 4:*117*, 4:*119*, 4:122–123
Gangetic leaffishes, 5:*246*, 5:*250*, 5:251
Gangetic morays. *See* Slender giant morays
Gannets, 8:183, 8:184, **8:211–219**, 8:*212*
Gans, Carl, 7:391, 7:392
Gansu hamsters, 16:239
Gansu moles, 13:280, 13:288*t*
Gansu pikas, 16:491, 16:*496*, 16:*499*, 16:500
Gaoligong pikas, 16:501*t*
Gapers, 5:163, 5:331–333, 5:335, 11:342
Garden ants, 3:68
Garden bulbuls. *See* Common bulbuls
Garden centipedes. *See* Garden symphylans
Garden dormice, 16:317, 16:318, 16:319, 16:*321*, 16:*322*, 16:*324*–325
Garden eels, 4:48, 4:67, 4:70
Garden slugs, 1:40, 2:25, 2:*413*
Garden snails, 2:*411*
Garden springtails, 3:101
Garden symphylans, 2:*373*
Garden warblers, 8:33, 11:*5*
Gardener bowerbirds. *See* Macgregor's bowerbirds
Gardening, butterfly, 3:90–91
Gardiner's frogs, 6:136, 6:*137*
Gardner, Allen, 12:160
Gardner, Beatrice, 12:160
Gardon rouges. *See* Rudds
Garefowls. *See* Great auks
Gargariscus prionocephalus. See Belalangs
Garibaldi damselfishes, 5:294, 5:*299*, 5:*305*
Garibaldis. *See* Garibaldi damselfishes
Garman, S., 4:143, 5:89
Garnet pittas, 10:195
Garnett's bushbabies. *See* Northern greater bushbabies
Garra spp., 4:298
Garra orientalis, 4:*302*
Garra pingi. See Black sticks
Garritornis spp. *See* Rufous babblers
Garritornis isidorei. See Rufous babblers
Garrulax spp. *See* Laughing thrushes
Garrulax cachinnans. See Nilgiri laughing thrushes
Garrulax canorus. See Hwameis
Garrulax canorus taewanus, 10:516
Garrulax galbanus. See Yellow-throated laughing thrushes
Garrulax leucolophus. See White-crested laughing thrushes
Garrulus spp., 11:504, 11:505, 11:507
Garrulus glandarius. See Eurasian jays
Garrulus lanceolatus. See Lanceolated jays
Garrulus lidthi. See Amami jays
Gars, 4:11, 4:23, **4:221–227**, 4:*223*
Garstang, W., 1:14
Garter snakes
behavior, 7:44
chemosensory behavior, 7:34–36
common, 7:469, 7:*471*, 7:*478–479*
food ecology, 7:39–40
reproduction, 4:470, 7:42
tactile cues, 7:40, 7:42
Gas exchange. *See* Respiratory system
Gasterochisma spp., 5:406
Gasteropelecidae, 4:335
Gasteropelecus sternicla. See River hatchetfishes
Gasterophilus spp. *See* Bot flies
Gasterophilus intestinalis. See Horse bot flies
Gasterorhamphosus zuppichinii, 5:131
Gasterosteidae, 4:35, 5:131, 5:133
Gasterosteiformes, **5:131–149**, 5:*137*, 5:*138*
behavior, 5:134–135
conservation status, 5:136
distribution, 5:134
evolution, 5:131–132
feeding ecology, 5:135
habitats, 5:134
humans and, 5:136
physical characteristics, 5:132–133
reproduction, 5:135–136
species of, 5:*139*–147
taxonomy, 5:131
Gasterosteoidei, 5:131, 5:133
Gasterosteus aculeatus. See Threespine sticklebacks
Gastric brooding frogs, 6:37, 6:149, 6:*150*, 6:153
Gastrocotyle trachuri, 1:*217*, 1:*219*, 1:222
Gastrocyathus spp., 5:355
Gastrocymba spp., 5:355
Gastroneuralia, 2:4
Gastropholis spp., 7:297
Gastrophryne spp., 6:301, 6:304
Gastrophryne carolinensis. See Eastern narrow-mouthed toads
Gastrophryne olivacea. See Great Plains narrow-mouthed toads
Gastropods, 1:44, 2:33, 2:41
Gastrosaccus spp., 2:216
Gastroscyphus spp., 5:355
Gastrotaenia cygni, 1:230
Gastrotheca spp. *See* Marsupial frogs
Gastrotheca ceratophrys, 6:37
Gastrotheca cornuta, 6:*32*
Gastrotheca guentheri, 6:37, 6:226–227
Gastrotheca plumbea, 6:37
Gastrotheca riobambae. See Riobamba marsupial frogs
Gastrotheca walkeri, 6:*23*
Gastrotheca weinlandii, 6:227
Gastrotrichs, 1:9, 1:12, 1:19, 1:22, 1:43, **1:269–273**, 1:*271*
Gastrulation, 1:19, 2:21, 2:23
See also Reproduction
Gators. *See* American alligators
Gaulins. *See* Least bitterns
Gaurs, 12:178, 15:267, 15:271, 16:2, 16:4, 16:6, 16:24*t*
Gause, C. F., 2:28
Gause's principle. *See* Competitive exclusion
Gavia adamsii. See Yellow-billed loons
Gavia arctica. See Arctic loons
Gavia immer. See Common loons
Gavia pacifica. See Pacific loons
Gavia stellata. See Red-throated loons
Gavialidae. *See* Gharials
Gavialis gangeticus. See Gharials
Gaviidae. *See* Loons
Gaviiformes. *See* Loons
Gayals, 12:180
Gayet, M., 4:289, 4:369
Gay's seedsnipes. *See* Rufous-bellied seedsnipes
Gazella spp., 15:265, 16:1, 16:7
Gazella bennettii, 16:48
Gazella cuvieri, 16:48
Gazella dama. See Dama gazelles
Gazella dama mhorr. See Mhorr gazelles
Gazella dorcas. See Dorcas gazelles
Gazella gazella. See Mountain gazelles
Gazella granti. See Grant's gazelles
Gazella leptoceros. See Slender-horned gazelles
Gazella rufifrons. See Red-fronted gazelles
Gazella rufina. See Red gazelles
Gazella soemmerringii. See Persian gazelles
Gazella spekei, 16:48
Gazella subgutturosa, 16:48
Gazella subgutturosa marica. See Arabian sand gazelles
Gazella thomsonii. See Thomson's gazelles
Gazelles, 15:146, **16:45–58**, 16:*49*
behavior, 16:46, 16:48
conservation status, 16:48
distribution, 16:4, 16:*45*, 16:46
evolution, 15:265, 16:1, 16:45
feeding ecology, 16:48
habitats, 16:46
humans and, 12:139, 16:48
physical characteristics, 16:2, 16:45
reproduction, 16:48, 16:50–55
species of, 16:50–57
taxonomy, 16:1, 16:45
Gebe cuscuses, 13:32, 13:60
Geckos, **7:259–269**, 7:*264*
behavior, 7:261
conservation status, 7:262–263
distribution, 7:*259*, 7:260
evolution, 7:259
feeding ecology, 7:261–262
in folk medicine, 7:51
habitats, 7:260–261
hearing, 7:32
humans and, 7:263
in mythology and religion, 7:55
physical characteristics, 7:259–*260*
reproduction, 7:262
species of, 7:*265*–269
taxonomy, 7:259
Geese, 8:20, **8:369–392**
behavior, 8:371–372
conservation status, 8:372–373
distribution, 8:*369*, 8:370–371
evolution, 8:369
feeding ecology, 8:372
habitats, 8:371

humans and, 8:373
physical characteristics, 8:369–370
reproduction, 8:372
species of, 8:*376–392*
taxonomy, 8:369
Gehyra vorax. See Voracious geckos
Geist, V., 16:89
Gekko gecko. See Tokay geckos
Gekko gecko azhari. See Tokay geckos
Gekko gecko gecko. See Tokay geckos
Gekko vittatus. See White-striped geckos
Gekkonidae, **7:259–269,** 7:*264*
behavior, 7:261
conservation status, 7:262–263
distribution, 7:*259*, 7:260
evolution, 7:259
feeding ecology, 7:261–262
habitats, 7:260–261
humans and, 7:263
physical characteristics, 7:259–260
reproduction, 7:262
species of, 7:*265–269*
taxonomy, 7:259
Gekkoninae, 7:259, 7:262
Gekkota, 7:196, 7:197–198, 7:200, 7:203–204, 7:207
Gelada baboons. *See* Geladas
Geladas, 14:*195*, 14:*201*, 14:203
behavior, 14:7, 14:192
conservation status, 14:194
distribution, 14:*192*
evolution, 14:189
feeding ecology, 14:193
habitats, 14:191
reproduction, 14:193
taxonomy, 14:188
Gelatinous species, 1:29
Gelechiid moths, 3:389
Gelechiidae, 3:387, 3:389
Gelochelidon spp. *See* Black-capped terns
Gelochelidon nilotica. See Gull-billed terns
Gelyelloida, 2:299
Gempylidae, 4:68, 5:405
Gempylus serpens. See Snake mackerels
Gemsboks, 12:*95*, 16:2, 16:4, **16:27–43,** 16:*29*, 16:*31*, 16:*32*, 16:*33*, 16:*34*, 16:41*t*
GenBank, 2:8
Gender changes. *See* Sex changes
Gene duplication, 12:30–31
Gene sequencing, 2:7–8, 2:11
Gene transfer, 1:31
See also Reproduction
Generation length, 12:207
Genetic coding, for migration, 8:32
Genetic diversity, 2:29
See also Biodiversity; Conservation status
Genetic drift, 12:222
Genetic sex determination (GSD), 7:72
See also Reproduction
Genetic variability, 1:15, 1:49
See also Conservation status
Genetics
evolution and, 12:26–35
small populations and, 12:221–222
zoo populations and, 12:206–209
Genets, **14:335–345,** 14:*339*
behavior, 14:258, 14:337
conservation status, 14:338
distribution, 14:336
evolution, 14:335
feeding ecology, 14:337
habitats, 14:337
physical characteristics, 14:335–336
reproduction, 14:338
species of, 14:*340–341*, 14:344*t*
taxonomy, 14:256, 14:335
Genetta spp., 14:335
Genetta abyssinica. See Abyssinian genets
Genetta genetta. See Common genets
Genetta johnstoni. See Johnston's genets
Genetta servalina, 14:336
Genetta tigrina. See Blotched genets
Genicanthus spp., 5:238, 5:242
Genital displays, squirrel monkeys, 14:104
Genomes, 12:26–27
Genomic conflict, 1:31
Gentle jirds. *See* Sundevall's jirds
Gentle lemurs, 14:3, 14:8, 14:10–11, 14:*52*
Gentoo penguins, 8:147, 8:*148*, 8:149, 8:150, 8:151
Genyochromis spp., 4:69
Genyophryninae, 6:35, 6:302, 6:305
Genypterus spp., 5:18
Genypterus blacodes, 5:16, 5:18
Geobates poecilopterus. See Campo miners
Geobatrachus spp., 6:156
Geocapromys spp., 16:461, 16:462
Geocapromys brownii. See Jamaican hutias
Geocapromys ingrahami. See Bahamian hutias
Geochelone denticulata. See South American yellow-footed tortoises
Geochelone gigantea. See Aldabra tortoises
Geochelone nigra. See Galápagos tortoises
Geochelone nigra vandenburgh. See Giant tortoises
Geochelone radiata. See Radiated tortoises
Geochelone sulcata. See Great African tortoises
Geococcyx californianus. See Greater roadrunners
Geoemyda japonica, 7:119
Geoemyda spengleri. See Black-breasted leaf turtles
Geoemydidae, **7:115–120,** 7:*117*
Geoffroy's bats, 12:54
Geoffroy's cats, 14:374, 14:*378*, 14:*384*–385
Geoffroy's horseshoe bats, 13:390, 13:392
Geoffroy's ocelots. *See* Geoffroy's cats
Geoffroy's spider monkeys, 12:7, 14:*160*, 14:*161*, 14:*163*
Geoffroy's tailless bats, 13:*422*, 13:427–*428*
Geoffroy's tamarins, 14:*123*, 14:132*t*
Geoffroy's woolly monkeys. *See* Gray woolly monkeys
Geoffroyus geoffroyi. See Red-cheeked parrots
Geogale spp. *See* Large-eared tenrecs
Geogale aurita. See Large-eared tenrecs
Geogalinae. *See* Large-eared tenrecs
Geography cone shells, 1:45, 2:29, 2:447, 2:*448*, 2:449
Geokinesis, 1:38–39
See also Behavior
Geologic periods, 3:52–53
Geological Society of America, 12:65
Geomagnetic fields, 12:84
Geomalia heinrichi. See Geomalias
Geomalias, 10:485
Geometrids, Hawaiian, 3:387
Geometriidae, 3:389
Geomyidae. *See* Pocket gophers
Geomyoidea, 16:121
Geomys spp., 16:185
Geomys arenarius. See Desert pocket gophers
Geomys bursarius. See Plains pocket gophers
Geomys personatus. See Texas pocket gophers
Geomys pinetis. See Southeastern pocket gophers
Geomys tropicalis. See Tropical pocket gophers
Geopelia cuneata. See Diamond doves
Geophilomorpha, 2:353, 2:354–355, 2:356
Geophis spp. *See* Neotropical burrowing snakes
Geophytes, 12:69, 12:76
Georges Bank flounders. *See* Winter flounders
Georychinae, 16:339
Georychus spp. *See* Cape mole-rats
Georychus capensis. See Cape mole-rats
Geoscapheinae, 3:150
Geoscapheus spp., 3:150
Geoscapheus robustus, 3:150
Geositta spp. *See* Miners
Geositta peruviana. See Coastal miners
Geospiza conirostris. See Cactus ground-finches
Geospiza fuliginosa. See Small ground-finches
Geotaxis, 1:39–40
See also Behavior
Geothlypis beldingi. See Belding's yellowthroats
Geothlypis speciosa. See Black-polled yellowthroats
Geothlypis trichas. See Common yellowthroats
Geotria spp., 4:83
Geotria australis. See Pouched lampreys
Geotriidae, 4:83
Geotrygon chrysia. See Key West quail doves
Geotrygon montana. See South American ruddy quail doves
Geotrypetes spp., 6:437
Geotrypetes seraphini. See Gaboon caecilians
Geoxus spp. *See* Mole mice
Geoxus valdivianus. See Long-clawed mole mice
Gephyrothuriidae, 1:419, 1:420
Geranopterus alatus, 10:51
Gerardia spp., 1:105
Gerbil mice. *See* Gray climbing mice
Gerbillinae, 16:127, 16:283–284, 16:285
Gerbillurus setzeri. See Setzer's hairy-footed gerbils
Gerbillus spp., 12:114
Gerbils, 16:123, 16:127, 16:128, 16:*282*, 16:*283*, 16:*284*, 16:296*t*
See also Sundevall's jirds
Gerenuks, 12:*154*, 15:140, 16:7, 16:*46*, 16:*47*, 16:48, 16:*49*, 16:*55–56*
Germ cells, 1:15, 1:18, 1:20–21
See also Reproduction
Germ layers, 1:19, 2:15, 2:20–21
Germ theory, of disease, 2:33
German bass. *See* Common carps
German brown trouts. *See* Brown trouts
German carps. *See* Common carps
German cockroaches, 3:57, 3:76, 3:*147*, 3:148, 3:151, 3:*152*, 3:*156*–157
German shepherds, 14:288–289
German wasps. *See* Yellow jackets
Geronticus calvus. See Bald ibises
Geronticus eremita. See Hermit ibises
Gerreidae. *See* Mojarras
Gerrhonotinae. *See* Alligator lizards
Gerrhonotus spp. *See* Alligator lizards
Gerrhonotus liocephalus. See Texas alligator lizards

Gerrhosaurinae. *See* Plated lizards
Gerrhosaurus spp. *See* Plated lizards
Gerrhosaurus skoogi. *See* Desert plated lizards
Gerridae, 3:54
Gerromorpha, 3:259
Gersemia antarctica, 1:106
Gervais' beaked whales, 15:69*t*
Gervais' funnel-eared bats, 13:461, 13:*462*, 13:*463*, 13:465
Gerygone spp. *See* Gerygones
Gerygone hyoxantha, 11:57
Gerygone insularis, 11:57
Gerygone levigaster. *See* Mangrove gerygones
Gerygone modesta, 11:57
Gerygone olivacea. *See* White-throated gerygones
Gerygones, 11:55, 11:56
Gestation, 12:95–96
  See also Reproduction
Ghana cuckoo-shrikes. *See* Western wattled cuckoo-shrikes
Gharials, 7:*167*, **7:167–170**, 7:*168*
  behavior, 7:168
  conservation status, 7:59, 7:169
  distribution, 7:167
  evolution, 7:19, 7:157–158, 7:167
  feeding ecology, 7:162, 7:168
  habitats, 7:161, 7:168
  physical characteristics, 7:167
  reproduction, 7:168–169
  taxonomy, 7:19, 7:157–158, 7:167, 7:179
  See also False gharials
Ghost bats, 13:308, **13:355–365**, 13:357, 13:*359*
  See also Australian false vampire bats
Ghost-faced bats, 13:316–317, 13:435, 13:436, 13:*439*, 13:*440*, 13:441, 13:442*t*
Ghost flatheads, 5:157, 5:158, 5:159
Ghost frogs, 6:6, **6:131–134**, 6:*133*
Ghost knifefishes. *See* Longtail knifefishes
Ghost moths, 3:385
Ghost owls. *See* Common barn owls
Ghost pipefishes, 5:131–132, 5:133, 5:135
Ghost sharks, 4:*93*, 4:94
Ghost shrimps, 2:197, 2:198, 2:*261*
Ghostfishes. *See* Wrymouths
Giant African land snails, 2:413, 2:414, 2:415, 2:*416*, 2:418–*419*
Giant Amazonian leeches, 2:77, 2:*79*, 2:*81*, 2:82
Giant ameivas, 7:*313*, 7:*314*
Giant ant-pittas, 10:242, 10:*243*, 10:*254*–255
Giant anteaters, 12:*29*, 13:*176*, 13:*177*, 13:178–179
  behavior, 13:*173*
  conservation status, 13:153
  feeding ecology, 13:151, 13:*174*
  habitats, 13:150, 13:172
  physical characteristics, 13:149
Giant antshrikes, 10:*244*, 10:*246*
Giant armadillos, 13:148, 13:150, 13:152, 13:*181*, 13:182, 13:*186*, 13:*189*–190
Giant Australian termites, 3:161–162, 3:164, 3:166, 3:*169*, 3:*170*, 3:172
Giant bandicoots, 13:17*t*
Giant barred frogs, 6:*142*, 6:*144*
Giant barrel sponges, 1:79
Giant beavers, 16:177
Giant birdwing butterflies, 3:388
Giant blindsnakes. *See* Schlegel's blindsnakes
Giant bottlenosed whales. *See* Baird's beaked whales
Giant cichlids, 5:*283*, 5:284–*285*
Giant clams, 2:454, 2:*456*, 2:*463*–464
Giant comet moths. *See* Atlas moths
Giant coots, 9:50, 9:*51*, 9:*56*, 9:*66*–*67*
Giant coral trouts. *See* Blacksaddled coral groupers
Giant damselfishes, 5:294
Giant duikers, 16:73
Giant earthworms, 2:*65*
Giant electric light bugs. *See* Giant water bugs
Giant elephant ear mushroom anemones. *See* Elephant ear polyps
Giant estuarine morays. *See* Slender giant morays
Giant fishkillers. *See* Giant water bugs
Giant flying squirrels, 16:135, 16:136, 16:137
Giant frogmouths. *See* Papuan frogmouths
Giant fulmars. *See* Southern giant petrels
Giant geckos, 7:262
Giant girdled lizards, 7:*323*, 7:*324*
Giant golden moles. *See* Large golden moles
Giant gouramies, 4:57, 5:427, 5:*430*, 5:*433*–434
Giant green anemones, 1:33, 1:*110*, 1:*111*
Giant haplochromises, 5:277–278
Giant hawkfishes, 5:237–238
Giant helmet katydids, 3:203, 3:205
Giant herbivorous armadillos, 13:181
Giant hornbills. *See* Great hornbills
Giant hummingbirds, 9:438, 9:439, 9:442, 9:448
Giant hutias, **16:469–471**, 16:*470*, 16:471*t*
Giant ibises, 8:293
Giant Indian fruit bats. *See* Indian flying foxes
Giant insects, 3:13–14
  See also specific insects
Giant Japanese spider crabs, 2:198
Giant kangaroo rats, 16:200, 16:201–*202*, 16:203, 16:*204*, 16:*207*
Giant kingbirds, 10:275
Giant kingfishers. *See* Laughing kookaburras
Giant land iguanas, 7:59
Giant larvaceans, 1:473, 1:*475*, 1:476
Giant leaf-nosed bats. *See* Commerson's leaf-nosed bats
Giant lizards, 7:59, 7:297, 7:298–299
Giant mantas. *See* Atlantic manta rays
Giant mantis shrimps, 2:170, 2:*172*, 2:*173*
Giant marbled sculpins. *See* Cabezons
Giant mastiff bats, 13:*489*, 13:491–*492*
Giant medusan worms, 1:419, 1:*423*, 1:425, 1:*426*
Giant metallic ceiba borers, 3:323, 3:*326*, 3:*327*–328
Giant muntjacs, 15:346, 15:*348*, 15:*353*–354
Giant musk turtles, 7:72
Giant noctules. *See* Greater noctules
Giant nuthatches, 11:169, 11:*170*, 11:*173*
Giant otter shrews, 13:196, 13:234*t*
Giant otters, 14:321, 14:324
Giant Pacific octopods, 2:476
Giant pandas, 12:*118*, 12:*208*, 14:*301*, 14:303–*304*
  behavior, 14:297
  captive breeding, 12:224
  conservation status, 14:300
  distribution, 12:136, 14:296
  feeding ecology, 14:255, 14:260, 14:*297*, 14:299
  habitats, 14:297
  physical characteristics, 14:296
  reproduction, 12:110, 14:299
  taxonomy, 14:295
Giant pangolins, 16:110, 16:112, 16:*114*, 16:117
Giant perches, **5:219–222**, 5:*224*, **5:225–227**
Giant petrels. *See* Southern giant petrels
Giant pittas, 10:195
Giant pixie. *See* African bullfrogs
Giant pocket gophers. *See* Large pocket gophers
Giant pouched bats, 13:364*t*
Giant prickly stick insects. *See* Macleay's spectres
Giant ragfishes. *See* Ragfishes
Giant rat-kangaroos, 13:70
Giant rats, 16:265, 16:*286*, 16:*289*, 16:291
Giant red mysids, 2:*225*, 2:*227*
Giant sable antelopes, 16:33
Giant salamanders, Asiatic, **6:343–347**, 6:*344*
  See also Pacific giant salamanders
Giant salmonflies, 3:*144*, 3:*145*
Giant schnauzers, 14:292
Giant sculpins. *See* Cabezons
Giant sea cows, 12:15
Giant sengis, 16:517–521
Giant shrews, 13:*254*, 13:*255*, 13:262
Giant silkmoths. *See* Atlas moths
Giant snakeheads, 5:437–439, 5:*440*, 5:*444*
Giant Sonoran drywood termites, 3:*169*, 3:*171*
Giant South American river turtles. *See* South American river turtles
Giant squids, 2:476, 2:479
Giant squirrels, 16:163, 16:166, 16:174
Giant stag. *See* Irish elk
Giant stoneflies, 3:*143*
Giant sunbirds. *See* São Tomé sunbirds
Giant thorny headed worms, 1:*315*–316
Giant thrips, 3:*281*
Giant tiger prawns, 2:*203*, 2:*210*, 2:211
Giant tigerfishes, 4:336, 4:*341*, 4:*342*
Giant toebiters. *See* Giant water bugs
Giant torrent midges, 3:360
Giant tortoises, 7:71, 7:*144*
Giant treefrogs. *See* White-lipped treefrogs
Giant tubeworms. *See* Hydrothermal vent worms
Giant vent clams, 2:*456*, 2:*462*, 2:466
Giant vinegaroons. *See* Giant whip scorpions
Giant water bugs, 3:260, 3:262, 3:263, 3:264, 3:*265*, 3:*268*–*269*
Giant wetas, New Zealand, 3:202, 3:207
Giant whip scorpions, 2:*337*, 2:*349*, 2:351–352
Giant white-eyes, 11:228
Giant whiteflies, 3:62
Giant wood-rails, 9:52, 9:*53*, 9:*61*–*62*
Giant wrens, 10:529
Giant yellow water bears, 2:118, 2:*119*, 2:122
Giant Zambian mole-rats, 12:73, 16:340, 16:345, 16:349*t*
Giarda, beavers and, 16:127
Gibba turtles, 7:78, 7:*80*, 7:*81*
Gibber chats. *See* Gibberbirds
Gibberbirds, 11:65, 11:*67*–*68*
Gibberfishes, 5:105, 5:106
Gibberichthyidae. *See* Gibberfishes
Gibberichthys spp. *See* Gibberfishes

Gibberichthys pumilus. See Gibberfishes
Gibbons, **14:207–223,** 14:*216,* 14:*217*
behavior, 12:145, 14:6–7, 14:8, 14:210–211
conservation status, 14:11, 14:214–215
distribution, 12:136, 14:5, 14:*207,* 14:210
evolution, 14:207–209
feeding ecology, 14:211–213
habitats, 14:210
humans and, 14:215
locomotion, 12:43
physical characteristics, 14:209–210
reproduction, 12:107, 14:213–214
self recognition, 12:159
sexual dimorphism, 12:99
species of, 14:*218–222*
taxonomy, 14:4, 14:207–208
in zoos, 12:210
Gibb's shrew-moles. *See* American shrew moles
Gibnuts. *See* Pacas
Gibson's albatrosses, 8:113
Gidley, J. W., 16:479
Gigantism, 9:245
Giganturidae. *See* Telescope fishes
Giganturoidei, 4:431
Gigas clams. *See* Giant clams
Gigliolella spp., 2:318
Gila monsters, 7:10, 7:37, **7:353–358,** 7:*354,* 7:*355,* 7:*356*
Gilbert's potoroos, 13:34, 13:39, 13:74, 13:76, 13:77, 13:*78*
Gilchrist's disease, beavers and, 16:127
Gilded flickers, 10:156
Gile's planigales, 12:300*t*
Gill, F. B., 10:174
Gill-net fisheries, 5:280–281
See also Fisheries
Gillbirds. *See* Red wattlebirds
Gilliard's flying foxes, 13:332*t*
Gillichthys mirabilis. See North American mudsucker gobies
Gills, 4:6, 4:*19,* 4:23–24, 4:*25,* 6:40, 6:41, 6:42
See also Physical characteristics
Gill's plantanna, 6:102
Gilquiniidae, 1:226
Ginglymostoma cirratum. See Orectolobiformes
Ginglymostomatidae. *See* Orectolobiformes
Ginkgo-toothed beaked whales, 15:69*t*
Ginsburgellus novemlineatus. See Nineline gobies
Gippsland giant worms, 2:67, 2:*69,* 2:*70,* 2:72
Giraffa spp., 15:265
Giraffa camelopardalis. See Giraffes
Giraffa camelopardalis reticulata. See Reticulated giraffes
Giraffa camelopardalis rothschildi. See Rothschild's giraffes
Giraffa camelopardalis tippelskirchi. See Masai giraffes
Giraffe gazelles. *See* Gerenuks
Giraffe-necked weevils, 3:*325,* 3:*327*
Giraffes, 12:6, 12:*82,* 12:*103,* 12:116, **15:399–408,** 15:*401,* 15:*403,* 15:*407*
behavior, 15:401–403, 15:*402*
conservation status, 15:405
distribution, 15:*399,* 15:401
evolution, 15:138, 15:265, 15:399
feeding ecology, 15:403–404, 15:*404*
habitats, 15:401
horns, 12:40, 12:102
humans and, 15:405–406
migrations, 12:167
physical characteristics, 15:138, 15:140, 15:399–401, 15:*402*
reproduction, 15:*135,* 15:143, 15:*400,* 15:404–405
species of, 15:408
taxonomy, 15:138, 15:263, 15:265, 15:399
Giraffidae, 12:135, **15:399–409,** 15:*407*
behavior, 12:167, 15:401–403, 15:408
conservation status, 15:405, 15:408–409
distribution, 15:*399,* 15:401, 15:*408*
evolution, 15:137, 15:266, 15:267, 15:399
feeding ecology, 15:403–404, 15:408–409
habitats, 15:401, 15:408
humans and, 15:405–406, 15:408–409
physical characteristics, 12:40, 12:102, 15:268, 15:399–401, 15:408
reproduction, 15:404–405, 15:408–409
taxonomy, 15:399, 15:408
Girdle wearers, **1:343–350,** 1:*346,* 2:5
behavior, 1:345
conservation status, 1:48, 1:345
distribution, 1:344
feeding ecology, 1:345
habitats, 1:344
humans and, 1:345
physical characteristics, 1:343–344
reproduction, 1:22, 1:345
species of, 1:*347–349*
taxonomy, 1:9, 1:12–13, 1:343
Girdled lizards, **7:319–325,** 7:*323*
Girdles, Venus's. *See* Venus's girdles
Girellinae, 5:241
Gizzard muds. *See* Gizzard shads
Gizzard nanny shads. *See* Gizzard shads
Gizzard shads, 4:*280,* 4:284–*285*
Glacidorbidae, 2:413, 2:414
Glacidorbis hedleyi, 2:*416,* 2:417
Glacier bears. *See* American black bears
Glaciers, 12:17–18, 12:*20*
Glaciopsyllus antarcticus, 3:349
Gladiator frogs, 6:50, 6:227, 6:228–229
Gladiators, **3:217–220,** 3:*218*
Glander, K. E., 14:162
Glandirana spp., 6:248
Glands
adrenal, 6:20–21
anal, 12:37
carpal, 12:37
ceruminous, 12:37
Duvernoy's, 7:467
parotid, 6:53, 6:184
pituitary, 6:19
poison, 6:66–67
prostate, 12:104
prothoracic, 3:28
salivary, 7:9–10
salt, 7:30
scent, 12:37, 12:81
sebaceous, 12:37
slime, 4:77
sternal, 12:37
sweat, 12:7, 12:36–38, 12:89
thyroid, 6:19
uterine, 12:94
See also Mammary glands; Physical characteristics
Glandulocaudine characins, 4:66
Glareola spp., 9:151
Glareola cinerea. See Gray pratincoles
Glareola isabella. See Australian pratincoles
Glareola nuchalis. See Rock pratincoles
Glareola pratincola. See Collared pratincoles
Glareolidae, **9:151–160,** 9:*154*
Glass catfishes, 4:352, 4:353, 4:*356,* 4:*359,* 4:365–366
Glass eyes. *See* Walleyes
Glass frogs, **6:215–223,** 6:*218*
conservation status, 6:*57,* 6:217, 6:219–223
distribution, 6:6, 6:*215,* 6:216, 6:*219–223*
physical characteristics, 6:215–216, 6:218, 6:219–223
reproduction, 6:*30,* 6:32, 6:35, 6:*36,* 6:*216*–217, 6:219–223
taxonomy, 6:215, 6:219–223
Glass gobies, 5:375
Glass knifefishes, 4:370, 4:*371,* 4:*372,* 4:*376*
Glass lizards, **7:339–342,** 7:*341*
Glass plants. *See* Glass-rope sponges
Glass-rope sponges, 1:*70,* 1:*71*
Glass shrimps. *See Mysis relicta*
Glass snakes. *See* Slender snipe eels
Glass sponges, **1:67–76,** 1:*70*
behavior, 1:68
conservation status, 1:69
distribution, 1:68
evolution, 1:67
feeding ecology, 1:68–69
habitats, 1:24, 1:68
humans and, 1:69
physical characteristics, 1:67–68, 1:*69*
reproduction, 1:69
species of, 1:*71–75*
taxonomy, 1:67
Glassfishes, 4:69, 5:255, 5:258–261, 5:263
Glasshouse whiteflies. *See* Greenhouse whiteflies
Glaucidium spp., 9:332, 9:333, 9:349
Glaucidium brasilianum. See Ferruginous pygmy-owls
Glaucidium californicum. See Northern pygmy owls
Glaucidium gnoma. See Northern pygmy owls
Glaucidium minutissimum, 9:345
Glaucidium perlatum. See Pearl-spotted owlets
Glaucis spp., 9:443, 9:448
Glaucis dohrnii. See Hook-billed hermits
Glaucis hirsuta. See Hairy hermits
Glaucomys spp., 16:136–137
Glaucomys sabrinus. See Northern flying squirrels
Glaucomys volans. See Southern flying squirrels
Glaucopsyche lygdamus xerces, 3:388
Glaucosoma hebraicum, 5:256
Glaucosomatidae, 5:255, 5:256, 5:258–263
Gleaning bats, 13:313
See also specific gleaning bats
Glen canyon chuckwallas. *See* Common chuckwallas
Gliders, 13:31, 13:35, 13:39
See also specific types of gliders
Gliding possums, 13:33, **13:125–133,** 13:133*t*
Glires, 16:479–480
See also Lagomorpha; Rodents
Gliridae, 16:283
Glironia spp. *See* Bushy-tailed opossums
Glironia venusta. See Bushy-tailed opossums
Glirulus japonicus. See Japanese dormice

Glittering kingfishers. *See* White-rumped kingfishers
Glittering-throated emeralds, 9:443, 9:449
Global positioning systems (GPS), 12:199
Global Strategy for the Management of Farm Animal Genetic Resources, 12:175
Global warming, 1:52–53
  See also Conservation status
Globe skimmers. *See* Wandering gliders
Globicephala spp. *See* Pilot whales
Globicephala macrorhynchus. *See* Short-finned pilot whales
Globicephala melas. *See* Long-finned pilot whales
Globicephalinae, 15:41
Globin genes, 12:30
Globodera spp., 1:36
Globular sea urchins. *See* Tuxedo pincushion urchins
Gloger's rule, 8:9
Glomerida, 2:363
Glomeridesmida, 2:363
Glomeris marginata. *See* Pill millipedes
Glossanodon struhsakeri, 4:392
Glossata, 3:384
Glossina palpalis. *See* Tsetse flies
Glossiphonia spp., 2:76
Glossiphoniidae, 2:75, 2:76, 2:77
Glossodoris atromarginata, 2:*406*, 2:*408*–409
Glossodoris stellatus, 2:*403*
Glossogobius ankaranensis, 5:375
Glossogobius giuris, 5:377
Glossolepis incisus. *See* Salmon rainbowfishes
Glossolepis wanamensis, 5:70
Glossophaga longirostris. *See* Southern long-tongued bats
Glossophaga soricina. *See* Pallas's long-tongued bats
Glossophaginae, 13:413, 13:414, 13:415, 13:417, 13:418, 13:420
Glossoscolecidae, 2:65, 2:66
Glossosoma nigrior, 3:*378*, 3:*379*
Glossosomatidae. *See* Turtle case makers
Glossosomatoidea. *See* Turtle case makers
Glossy-mantled manucodes, 11:491
Glossy wormsnakes. *See* Peters' wormsnakes
Glottidia spp., 2:515
Glottidia pyramidata, 2:*517*, 2:*518*, 2:519
Glover's pikas, 16:491
Glowing spider bugs. *See* New Zealand glowworms
Glowworms, 3:321
  New Zealand, 3:*362*, 3:*366*, 3:368–369
  pink, 3:*324*, 3:*330*
Gloyd, Howard, 7:455
Gloydius spp., 7:445, 7:446
Gloydius caliginosus, 7:447
Gloydius halys, 7:447
Gloydius monticola, 7:447
Gloydius strauchi. *See* Tibetan pitvipers
Glycerids, 2:47
Glycosides, 1:45
Glyphis gangeticus. *See* Ganges sharks
Glyphis glyphis, 4:116
Glyphotes spp., 16:163
Glyptemys muhlenbergii. *See* Bog turtles
Glyptodonts, 12:11, 13:147–150, 13:152, 13:181, 15:137, 15:138
Gnatcatchers, 11:1–2, 11:*3*, 11:*4*, 11:7–*8*, 11:10–11
Gnathanacanthidae. *See* Red velvetfishes
Gnathifera, 1:9, 1:11–12
Gnathodon strigirostris. *See* Tooth-billed pigeons
Gnathonemus petersii. *See* Elephantnose fishes
Gnathophausia spp., 2:226
Gnathophausia ingens. *See* Giant red mysids
Gnathostomata, 4:9, 4:10, 4:190
Gnathostomula armata, 1:*332*
Gnathostomula paradoxa, 1:*333*, 1:*334*, 1:335
Gnathostomulids, 1:9, 1:12, 1:19, 1:22, **1:331–335,** 1:*333*, 2:*5*
Gnats
  eye, 3:361
  gall, 3:361
  mourning, 3:66
  See also New Zealand glowworms
Gnatwrens, 11:1, 11:*8*, 11:10
Gnorimopsar chopi. *See* Chopi blackbirds
Gnus, 12:116, 15:140
Gnypetoscincus queenslandiae. *See* Prickly forest skinks
Go-away-birds, 9:300, 9:*301*, 9:*303*, 9:308–309
Goannas, 7:57, **7:359–368**
Goas. *See* Tibetan gazelles
Goatfishes, 4:48, 4:68, **5:235–244,** 5:*241*, 5:*246*, 5:*249*, **5:250–251**
Goats, 12:*5*, 15:146, 15:263, 16:1, **16:87–105,** 16:*96*, 16:*97*
  behavior, 12:*151*, 15:271, 16:92–93
  bezoar, 12:179
  conservation status, 16:94–95
  digestive system, 12:122
  distribution, 16:4, 16:*87*, 16:90–91
  evolution, 16:87–89
  feeding ecology, 16:93–94
  habitats, 16:91–92
  humans and, 15:273, 16:8, 16:9, 16:95
  migrations, 12:165, 15:138
  neonatal milk, 12:127
  physical characteristics, 16:5, 16:89–90
  reproduction, 16:94
  species of, 16:98–104*t*
  taxonomy, 16:87–89
  See also Domestic goats
Gobies, **5:373–389,** 5:*378*, 5:*379*
  behavior, 5:375
  conservation status, 5:377
  distribution, 5:374
  ecology, 4:42, 4:44, 4:45
  evolution, 5:373–374
  feeding ecology, 5:375–376
  habitats, 4:47–48, 4:50, 5:374–375
  humans and, 5:377
  physical characteristics, 5:374
  reproduction, 4:*30*, 5:376–377
  species of, 5:*380*–388
  taxonomy, 5:373
Gobiesocidae, 4:47, 5:355
Gobiesocinae. *See* Clingfishes
Gobiesocini, 5:355, 5:356
Gobiesocoidei, **5:355–363,** 5:*358*
Gobiesox spp., 5:355, 5:356
Gobiesox maeandricus. *See* Northern clingfishes
Gobiidae, 4:50, 4:64, 5:373, 5:374, 5:375, 5:377
Gobiinae, 5:373
Gobio spp., 4:298
Gobio gobio. *See* Gudgeons
Gobiobotinae, 4:297
Gobiocichla spp., 5:276
Gobiocypris rarus, 4:301
Gobiodon spp., 5:374
Gobiodontines, 5:375
Gobioidei. *See* Gobies
Gobioides broussoneti. *See* Violet gobies
Gobionellinae, 5:373
Gobioninae, 4:297
Gobiopterus spp., 5:375
Gobiopterus chuno. *See* Glass gobies
Gobiosoma spp., 4:69, 5:374
Gobiosoma genie, 5:376
Gobiosoma oceanops. *See* Neon gobies
Gobiosomine gobies, 5:374, 5:376
Gobius spp., 5:375
Goblin sharks, 4:124, 4:131–132, 4:*134*, 4:*139*, 4:140
Godwits, 9:175–178, 9:180
Godzilliognomus frondosus, 2:125
Godzillius robustus, 2:125
Goeldi's monkeys, **14:115–129,** 14:*121*, 14:*123*, 14:*126*
  behavior, 14:6–7, 14:117–120
  distribution, 14:116
  evolution, 14:115
  feeding ecology, 14:120, 14:121
  habitats, 14:116
  physical characteristics, 14:115, 14:116
  reproduction, 14:10, 14:121–124
  species of, 14:*127*, 14:129
  taxonomy, 14:4, 14:115
Goera spp., 3:377
Goethalsia spp. *See* Hummingbirds
Goggle eyes, 5:*265*, 5:*269*, 5:270
  See also Rock basses
Goitered gazelles. *See* Persian gazelles
Gola malimbes, 11:379
Gold-crested black bowerbirds. *See* Archbold's bowerbirds
Gold-fronted chloropsis. *See* Golden-fronted leafbirds
Gold-fronted fruitsuckers. *See* Golden-fronted leafbirds
Gold-mantled chloropsis. *See* Blue-winged leafbirds
Gold morph of eyelash vipers, 7:*447*, 7:448
Gold-striped frogs, 6:157, 6:*160*, 6:167
Gold-striped salamanders. *See* Golden-striped salamanders
Gold-striped soapfishes. *See* Sixline soapfishes
Golden angwantibos, 14:16, 14:21*t*
Golden anteaters. *See* Silky anteaters
Golden apple snails, 2:*28*
Golden arboreal alligator lizards. *See* Coban alligator lizards
Golden bandicoots, 13:*3*, 13:4, 13:12, 13:16*t*
Golden bats. *See* Old World sucker-footed bats
Golden-bellied tree shrews, 13:297*t*
Golden-bellied water rats, 16:261*t*
Golden bosunbirds. *See* White-tailed tropicbirds
Golden bowerbirds, 11:477, 11:*480*, 11:*482*, 11:*485*–486
Golden-breasted fulvettas, 10:*509*, 10:*518*–519
Golden-breasted tit babblers. *See* Golden-breasted fulvettas

Golden-breasted whistlers. *See* Golden whistlers
Golden-capped fruit bats. *See* Golden-crowned flying foxes
Golden carps. *See* Crucian carps
Golden cats, 14:390*t*–391*t*
Golden-cheeked gibbons, 14:*216*, 14:*220*, 14:221–*222*
Golden-cheeked warblers, 11:291
Golden coqui, 6:68, 6:156, 6:159, 6:*161*, 6:164–*165*
Golden-crested mynas, 11:*412*, 11:416–*417*
Golden-crowned flying foxes, 13:*338*, 13:*341*–342
Golden-crowned kinglets, 11:*8*, 11:*11*
Golden-crowned manakins, 10:297
Golden-crowned sifakas, 14:11, 14:63, 14:*67*
Golden cuckoo-shrikes, 10:386, 10:*388*, 10:*391*
Golden dart-poison frogs, 6:*202*, 6:209
Golden doves, 9:242
Golden eagles, 8:324, 12:192
Golden-eye ducks, 8:365, 8:370
Golden-faced sakis, 14:*146*
Golden-fronted leafbirds, 10:417, 10:*418*, 10:*421*–422
Golden grunts. *See* Bluestriped grunts
Golden hamsters, 16:239–240, 16:*241*, 16:243, 16:*244*, 16:*245*, 16:246–247
Golden-headed cisticolas, 11:*9*, 11:*12*–13
Golden-headed lion tamarins, 14:118, 14:*126*, 14:*128*–129
Golden-headed manakins, 10:*298*, 10:*302*
Golden-headed quetzels, 9:478
Golden-heart pigeons, 9:*250*
Golden horseshoe bats, 13:*406*, 13:*408*, 13:409
Golden jackals, 14:*268*, 14:271, 14:272
Golden jellyfish, 1:*159*, 1:162–*163*
Golden lion tamarins, 14:*120*, 14:*126*
  behavior, 14:117, 14:118
  conservation status, 12:204, 12:208, 12:224, 14:11
  humans and, 14:124
  physical characteristics, 14:115–116
  reproduction, 14:121–122, 14:124
Golden mantellas, 6:292, 6:293
Golden-mantled ground squirrels, 16:143, 16:159*t*
Golden-mantled rosellas. *See* Eastern rosellas
Golden masked owls, 9:340–341
Golden mice, 16:277*t*
Golden moles, 12:79–80, **13:215–223,** 13:*216*, 13:*219*
  behavior, 13:198
  distribution, 12:135, 13:197
  evolution, 12:29
  feeding ecology, 12:84
  habitats, 13:197
  physical characteristics, 12:29, 13:195–197
  species of, 13:220–222, 13:222*t*
  taxonomy, 13:193–194
Golden-naped weavers, 11:379
Golden nightjars, 9:402
Golden orioles. *See* Eurasian golden orioles
Golden paper wasps, 3:*409*, 3:*416*, 3:423
Golden perches, 5:221, 5:*223*, 5:*229*, 5:231–232
Golden phoronoids, 2:*491*, 2:*492*
Golden pike characins, 4:*341*, 4:*347*
Golden pipits, 10:372, 10:*377*, 10:379–*380*
Golden-plovers, 9:164
Golden poison frogs. *See* Golden dart-poison frogs
Golden pottos. *See* Calabar angwantibos
Golden regents. *See* Regent bowerbirds
Golden ringtail possums, 13:123*t*
Golden-rumped diamondbirds. *See* Forty-spotted pardalotes
Golden-rumped lion tamarins, 14:11, 14:117, 14:132*t*
Golden-rumped sengis, 16:521, 16:522, 16:*523*, 16:*524*–525
Golden shads. *See* Alewives
Golden snub-nosed monkeys, 14:*5*, 14:175, 14:*177*, 14:*182*–183
Golden sparrows, 11:398, 11:*400*, 11:402–*403*
Golden-spectacled warblers, 11:*8*, 11:*21*
Golden spiny mice, 16:*251*, 16:261*t*
Golden squirrels. *See* Persian squirrels
Golden stones, 3:*144*, 3:*145*–146
Golden-striped salamanders, 6:*370*
Golden swallows, 10:360
Golden takins, 16:2
Golden thorius, 6:*394*, 6:403
Golden-tipped bats, 13:*519*, 13:*520*, 13:521, 13:522
Golden toads, 6:56, 6:*185*, 6:*186*, 6:*192*
Golden trillers. *See* Golden cuckoo-shrikes
Golden trouts, 4:*409*, 4:*410*–411
Golden-tufted honeyeaters. *See* Yellow-tufted honeyeaters
Golden whistlers, 11:116, 11:*117*, 11:*119*, 11:121–*122*
Golden white-eyes, 11:227, 11:230
Golden whitefishes, 4:*75*
Golden-winged sunbirds, 11:207, 11:*208*, 11:209, 11:*212*, 11:*221*
Golden-winged warblers, 11:285, 11:286, 11:*292*, 11:*293*–294
Goldenfaces, 11:115
Goldeyes, 4:232, 4:234, 4:238
Goldfinches. *See* American goldfinches; European goldfinches
Goldfishes. *See* Crucian carps
Goldie's birds of paradise, 11:492
Goldmania spp. *See* Hummingbirds
Goliath beetles, 3:13, 3:43
Goliath cockatoos. *See* Palm cockatoos
Goliath frogs, 6:62, 6:246, 6:*253*, 6:*256*
Goliath herons, 8:233, 8:235, 8:241, 8:*247*, 8:*250*–*251*
Goliatus goliatus. *See* African goliath beetles
Gombessa. *See* Coelacanths
Gomeran giant lizards, 7:299
Gomphocerinae. *See* Slant-faced grasshoppers
Gompotheres, 15:137
Gonads, 1:20–23
  See also Reproduction
Gonatid squids, 2:477
Gonatodes albogularis. *See* Yellow-headed geckos
Gonatodes albogularis albogularis. *See* Yellow-headed geckos
Gonatodes albogularis bodinii. *See* Yellow-headed geckos
Gonatodes albogularis fuscus. *See* Yellow-headed geckos
Gonatodes albogularis notatus. *See* Yellow-headed geckos
Gongshan muntjacs, 15:346, 15:*348*, 15:*350*, 15:352
Gongylus gongylodes. *See* Wandering violin mantids
Goniastrea aspera, 1:*109*, 1:118–*119*
Goniistius spp., 5:238
Gonionemus spp., 1:129
Gonochorism, 1:17, 1:21–23, 2:16, 2:22–24, 4:433
  See also Reproduction
Gonodactyloidea, 2:168, 2:170, 2:171
Gonodactylus smithii, 2:170
Gonoleks, 10:425–430, 10:*431*, 10:*434*
Gonorynchidae, 4:289, 4:290
Gonorynchiformes, **4:289–296,** 4:*292*
Gonorynchus spp., 4:290, 4:291
Gonorynchus gonorynchus. *See* Sandfishes
Gonostomatidae. *See* Bristlemouths
Goodall, Jane, 12:157
Goodeidae. *See* Splitfins
Goodeinae, 5:89
Goodfellow's tree kangaroos, 13:33
Goodo. *See* Murray cods
Goodrich, Edwin S., 4:209
Goodwin, Derek, 11:504
Goose barnacles. *See* Common goose barnacles
Goose-beaked whales. *See* Cuvier's beaked whales
Goose-neck turtles. *See* Spiny softshells
Goosefishes. *See* Monkfishes
Gooseneck barnacles. *See* Common goose barnacles
Gopherus agassizii. *See* Desert tortoises
Gorakhnath, 7:56
Gorals, 16:87, 16:88, 16:89, 16:90, 16:92, 16:95
Gordian worms. *See* Gordiida
Gordiida, 1:305–306
Gordius aquaticus, 1:*307*, 1:*308*–309
Gorgasia preclara. *See* Splendid garden eels
Gorgeous gobies. *See* Gorgeous prawn-gobies
Gorgeous prawn-gobies, 5:375, 5:*379*, 5:*382*
Gorgeted wood-quails, 8:437
Gorgonian octocorals, 1:106
Gorgonocephalidae, 1:387
Gorgonocephalus arcticus. *See* Northern basket stars
Gorgonocephalus stimpsoni, 1:387
Gorgopithecus spp., 14:189
Gorilla spp. *See* Gorillas
Gorilla beringei. *See* Eastern gorillas
Gorilla beringei beringei. *See* Mountain gorillas
Gorilla beringei graueri. *See* Eastern lowland gorillas
Gorilla gorilla. *See* Western gorillas
Gorillas
  behavior, 12:*141*, 14:7, 14:228, 14:232
  conservation status, 12:215, 14:235
  distribution, 14:5, 14:227
  evolution, 12:33, 14:225
  feeding ecology, 14:232, 14:233
  field studies of, 12:200
  habitats, 14:6, 14:227
  humans and, 14:11
  memory, 12:153
  physical characteristics, 14:4, 14:226
  population indexes, 12:195
  reproduction, 14:234–235
  taxonomy, 14:225

in zoos, 12:208, 12:*210*
See also specific types of gorillas
Gorsachius goisagi. See Japanese night herons
Gorsachius magnificus, 8:245
Goshawks, 8:318, 8:320
Gosline, W. A., 4:351, 5:365
Gouania spp., 5:355
Gould, John, 9:415, 10:330, 10:332
Gouldian finches, 11:353, 11:*354,* 11:356–357, 11:*359,* 11:*369*–370
Gould's broadbills. *See* Silver-breasted broadbills
Gould's bronze cuckoos, 11:224
Gould's frogmouths, 9:378, 9:379, 9:*380,* 9:*383*
Gould's sunbirds, 11:207, 11:*212,* 11:*224–225*
Gould's violet-ears. *See* Sparking violet-ears
Gould's wattled bats, 12:*59*
Goura spp. *See* Columbiformes
Goura cristata. See Blue-crowned pigeons
Goura victoria. See Victoria crowned-pigeons
Gouramies, 4:24, 5:427, 5:428
See also specific types of gouramies
Gourinae. *See* Crowned pigeons
Government cats. *See* Channel catfishes
Goya tuco-tucos, 16:431*t*
GPS (global positioning systems), 12:199
Graceful clam shrimps, 2:*149,* 2:*150,* 2:151
Graceful lizardfishes. *See* Slender lizardfishes
Graceful pittas, 10:195, 10:*196,* 10:198–*199*
Grace's warblers, 11:290
Gracile chimpanzees. *See* Bonobos
Gracile opossums, 12:249
Gracilinanus spp. *See* Gracile opossums
Gracilinanus aceramarcae, 12:251, 12:254
Gracilinanus agilis, 12:251
Gracilinanus emiliae, 12:251
Gracilinanus marica. See Northern gracile mouse opossums
Gracilinanus microtarsus, 12:251
Gracilisuchus spp., 7:18
Gracillariidae, 3:389
Grackles, 11:301–306, 11:*307,* 11:318–319
See also Hill mynas; Sri Lanka mynas
Gracula cristatella. See Crested mynas
Gracula ptilogenys. See Sri Lanka mynas
Gracula religiosa. See Hill mynas
Gracupica spp., 11:409
Graeffea crouanii. See Coconut stick insects
Graell's black-mantled tamarins, 14:116
Grahame, Kenneth, 13:200
Grain beetles, 3:57, 3:76, 3:316
Grain moths, 3:44
Grain weevils, 3:57, 3:76
Grains, 8:25–26
Grallaria gigantea. See Giant ant-pittas
Grallaria rufocinerea. See Bicolored antpittas
Grallina spp. *See* Mudnest builders
Grallina bruijni. See Torrent-larks
Grallina cyanoleuca. See Australian magpie-larks
Grallinidae. *See* Mudnest builders
Grallininae. *See* Mudnest builders
Grammar. *See* Syntax
Grammatidae. *See* Basslets
Grammatostomias dentatus, 4:*422*
Grammicolepidae, 5:123, 5:124
Grammicolepis brachiusculus. See Thorny tinselfishes
Grammistes sexlineatus. See Sixline soapfishes
Grammistini, 5:255
Grampus griseus. See Risso's dolphins
Grampuses. *See* Giant whip scorpions
Grand Comoro flycatchers, 11:27, 11:34–*35*
Grand eclectus parrots. *See* Eclectus parrots
Grand potoos. *See* Great potoos
Grande, L., 4:229
Grande, T., 4:289
Grandisonia spp., 6:436
Graneledone spp., 2:478
Granite night lizards, 7:*292*
Grantiopsis heroni, 1:*60,* 1:*61,* 1:64
Grant's bushbabies, 14:26, 14:32*t*
Grant's desert golden moles, 12:71–72, 12:76, 13:*197,* 13:215, 13:*217,* 13:*219,* 13:*220*–221
Grant's gazelles, 15:271, 16:48, 16:57*t*
Graphidostreptus gigas, 2:363
Graphiurinae. *See* African dormice
Graphiurus spp. *See* African dormice
Graphiurus crassicaudatus. See Jentink's dormice
Graphiurus kelleni. See Kellen's dormice
Graphiurus murinus. See Woodland dormice
Graphiurus ocularis. See Spectacled dormice
Graphiurus parvus. See Savanna dormice
Graphiurus rupicola. See Stone dormice
Graphiurus surdus. See Silent dormice
Grapsus grapsus. See Red rock crabs
Graptemys spp. *See* Map turtles
Graptolithina, 1:443
Grass basses. *See* Black crappies
Grass bugs, 3:56
Grass carps, 4:301, 4:303, 4:*305,* 4:*309*
Grass-crown mealybugs. *See* Rhodesgrass mealybugs
Grass frogs. *See* Brown frogs
Grass lizards, 7:298, 7:*300,* 7:*301,* 7:*302,* 7:319–322
Grass owls. *See* Eastern grass owls
Grass snakes, 7:466
Grass worms, 3:56
Grassbirds, 11:1, 11:2
Grasscutters. *See* Cane rats
Grasseichthys spp., 4:289, 4:290
Grassfinches, 11:353–357
See also Weaverfinches
Grasshopper mice, 16:125
Grasshopper warblers, 11:*3,* 11:*9,* 11:*16*
Grasshoppers, 3:59–60, 3:65, **3:201–208,** 3:*210,* **3:211–212**
farming of, 3:80
as food, 3:80
habitats, 3:56, 3:57, 3:203–204
physical characteristics, 3:22, 3:27, 3:202–203
reproduction, 3:29–30, 3:31, 3:39, 3:207
symbolism, 3:74
See also specific types of grasshoppers
Grassquits, 11:264
Grasswrens, 11:45–*46,* 11:52
Grauer's broadbills. *See* African green broadbills
Grauer's shrews, 13:277*t*
Graveldivers, 5:309, 5:310, 5:312, 5:313
Gravestockia pharetroniensis, 1:57
Gravid aphids, 3:86
Gray, John E., 7:189, 7:223
Gray antbirds, 10:*243,* 10:*251*
Gray antwrens, 10:*244,* 10:*249*
Gray-backed fiscals, 10:428, 10:429
Gray-backed sparrow-larks, 10:346
Gray-backed sportive lemurs, 14:75, 14:*78,* 14:*79*
Gray-backed storm-petrels, 8:136, 8:137, 8:138
Gray-bellied douroucoulis. *See* Gray-bellied night monkeys
Gray-bellied night monkeys, 14:135, 14:138, 14:*139,* 14:*140*
Gray-bellied owl monkeys. *See* Gray-bellied night monkeys
Gray-breasted jays, 11:*505,* 11:506
Gray-breasted mountain toucans, 10:128, 10:*129,* 10:132–*133*
Gray-breasted robins. *See* Eastern yellow robins
Gray-breasted seedsnipes, 9:*190,* 9:191, 9:*192,* 9:*194*
Gray-breasted wood wrens, 10:*530,* 10:*535*–536
Gray brocket deer, 15:396*t*
Gray butcherbirds, 11:242, 11:*470,* 11:*472*–473
Gray catbirds, 8:39, 10:*467,* 10:*470*
Gray-cheeked flying squirrels, 16:141*t*
Gray-cheeked mangabeys, 14:*196,* 14:*201*–202, 14:204*t*
Gray-cheeked thrushes, 10:484, 10:*485*
Gray-chinned sunbirds. *See* Green sunbirds
Gray chubs. *See* Striped parrotfishes
Gray climbing mice, 16:*286,* 16:*290*–291
Gray-crowned babblers, 11:127, 11:128, 11:*129,* 11:*130*–131
Gray crowned cranes, 9:*26,* 9:27, 9:*30,* 9:*31*
Gray-crowned crocias, 10:507
Gray-crowned rosy finches, 11:*326,* 11:*337*–338
Gray dorcopsises, 13:*91,* 13:*94,* 13:100
Gray dwarf hamsters, 16:247*t*
Gray-eyed frogs, 6:*282*
Gray-faced woodpeckers, 10:*152,* 10:*163*–164
Gray falcons, 8:347
Gray fantails, 11:83, 11:*86,* 11:*88,* 11:*92*–93
Gray flycatchers. *See* Gray hypocolius
Gray four-eyed frogs, 6:63, 6:157, 6:*160,* 6:168
Gray four-eyed opossums, 12:250, 12:*256,* 12:263
Gray foxes, 14:265, 14:272, 14:284*t*
Gray fruit bats. *See* Black flying foxes
Gray gentle lemurs, 14:*50*
Gray gibbons. *See* Mueller's gibbons
Gray go-away-birds, 9:*301,* 9:*303,* 9:307–*308*
Gray gulls, 9:205
Gray hamsters, 16:243
Gray-headed bristlebills, 10:396
Gray-headed bush-shrikes, 10:427, 10:428, 10:*431,* 10:*435*
Gray-headed canary flycatchers. *See* Gray-headed flycatchers
Gray-headed chickadees. *See* Siberian tits
Gray-headed doves, 9:*254,* 9:*261*–*262*
Gray-headed flycatchers, 11:105, 11:*107,* 11:*111*
Gray-headed flying foxes, 13:312, 13:320
Gray-headed green woodpeckers. *See* Gray-faced woodpeckers
Gray-headed juncos, 11:271
Gray-headed kingfishers, 10:8
Gray-headed mollymawks, 8:114
Gray-headed robins, 11:*107,* 11:109–*110*

Gray-headed warblers, 11:291
Gray-headed woodpeckers. *See* Gray-faced woodpeckers
Gray herons, 8:*247*, 8:*248*
Gray herrings. *See* Alewives
Gray-hooded sunbirds, 11:210
Gray hypocolius, 10:447–*450*, 10:454
Gray jays, 11:504, 11:507, 11:*510*, 11:*514*–515
Gray jumpers. *See* Apostlebirds
Gray kangaroos, 13:34, 13:84
Gray kingbirds, 10:*270*
Gray long-eared bats, 13:516*t*
Gray louries. *See* Gray go-away-birds
Gray-mantled albatrosses. *See* Light-mantled albatrosses
Gray morph gyrfalcons, 8:314
Gray mouse lemurs, 14:*37*, 14:*38*, 14:39, 14:*40*, 14:*42*–43
Gray myotis, 12:*5*, 13:310–311, 13:316
Gray-neck night monkeys, 14:135
Gray-necked bald crows. *See* Gray-necked picathartes
Gray-necked picathartes, 10:507, 10:*509*, 10:*523*
Gray-necked wood-rails, 9:*48*
Gray nightjars, 9:*407*, 9:*411*
Gray nurse sharks. *See* Sand tiger sharks
Gray-olive bulbuls, 10:397
Gray parrots, 8:19, 8:22, 9:275, 9:*282*, 9:*288*
Gray partridges, 8:402, 8:434, 8:*440*, 8:*446*
Gray pelicans. *See* Spot-billed pelicans
Gray penduline tits. *See* African penduline tits
Gray petrels, 8:110
Gray phalaropes, 9:*178*
Gray pikes. *See* Walleyes
Gray plantain-eaters. *See* Western gray plantain-eaters
Gray potoos, 9:395, 9:*396*, 9:397, 9:*398*, 9:*399*–400
Gray pratincoles, 9:*154*, 9:*156*–157
Gray reef sharks, 4:115, 4:*116*, 4:*117*, 4:*119*–120
Gray rheboks, 16:27, 16:28, 16:29, 16:30, 16:31, 16:32
Gray-rumped tree swifts, 9:433
Gray sac-winged bats, 13:*359*, 13:360–*361*
Gray seals, 14:417, 14:418, 14:421–422, 14:*426*, 14:*427*–428
Gray sharks. *See* Greenland sharks
Gray shrews, 13:276*t*
Gray shrike-thrushes, 11:116–118, 11:*119*, 11:123–*124*
Gray-sided flowerpeckers, 11:*192*, 11:*196*–197
Gray silkies. *See* Gray silky flycatchers
Gray silky flycatchers, 10:447, 10:*450*, 10:*453*–454
Gray slender lorises, 14:*17*, 14:*18*
Gray slender mouse opossums, 12:251, 12:*256*, 12:261
Gray snub-nosed monkeys, 14:175, 14:185*t*
Gray squirrel fleas, 3:*349*
Gray squirrels, 12:*93*, 12:139, 16:163, 16:*164*, 16:*165*, 16:*168*, 16:*170*–171
  Arizona, 16:167, 16:*168*, 16:*171*–172
  vs. red squirrels, 12:183–184
  western, 16:175*t*
Gray swiftlets. *See* White-rumped swiftlets
Gray teals, 8:364
Gray-throated rails, 9:47–48
Gray-throated silvereyes, 11:229
Gray-throated sunbirds. *See* Plain-throated sunbirds
Gray tree kangaroos. *See* Bennett's tree kangaroos
Gray treefrogs, 6:*33*, 6:*34*, 6:47, 6:285, 6:*294*, 6:297–298
Gray tremblers. *See* Brown tremblers
Gray wagtails, 10:372, 10:*374*, 10:*377*, 10:*378*–379
Gray whale lice, 2:*263*, 2:265–*266*
Gray whales, **15:93–101,** 15:*94*, 15:*96*, 15:*97*, 15:*98*, 15:*99*, 15:*100*
  behavior, 15:95–97
  conservation status, 15:9, 15:99–100
  distribution, 15:5, 15:*93*, 15:94–95
  evolution, 15:3, 15:93
  feeding ecology, 15:97–98
  habitats, 15:6, 15:95
  humans and, 15:100
  migrations, 12:87
  physical characteristics, 15:4, 15:93–94
  reproduction, 15:*95*, 15:98–99
  taxonomy, 15:93
Gray-winged fracolins, 8:435
Gray-winged trumpeters. *See* Common trumpeters
Gray wolves, 12:*30*, 14:*271*, 14:*275*, 14:*276*–*277*
  behavior, 14:258, 14:*266*
  conservation status, 14:273
  distribution, 12:129, 12:131–132
  habitats, 14:267
  reproduction, 14:*255*, 14:*259*
  smell, 12:*83*
Gray woodpeckers, 10:*153*, 10:158–*159*
Gray woolly monkeys, 14:160, 14:*161*, 14:*165*–166
Graybacks. *See* Alewives
Grayfishes. *See* Piked dogfishes
Grayia spp., 7:467
Grayish mouse opossums, 12:264*t*
Gray's beaked whales, 15:70*t*
Gray's brush-turkeys. *See* Maleos
Gray's larks, 10:345
Gray's leaf insects. *See* Javan leaf insects
Gray's malimbes. *See* Blue-billed malimbes
Gray's sliders, 7:105
Graz Zoological Institute, 1:87
Great African gray treefrogs. *See* Gray treefrogs
Great African tortoises, 7:*66*
Great African wattle cranes. *See* Wattled cranes
Great albatrosses, 8:113
Great apes, **14:225–240,** 14:*236*
  behavior, 14:228–232
  conservation status, 12:215, 14:235
  distribution, 12:135, 12:136, 14:*225*, 14:227
  evolution, 14:225
  feeding ecology, 14:232–234
  habitats, 14:227–228
  humans and, 14:235
  imitation, 12:158
  language, 12:162
  physical characteristics, 14:225–227
  reproduction, 14:234–235
  self recognition, 12:159
  species of, 14:*237*–240
  taxonomy, 14:4, 14:225
  tool use, 12:157
  See also Apes
Great argus pheasants, 8:434, 8:*436*
Great auks, 9:104, 9:219–*220*, 9:222, 9:*223*, 9:*228*–*229*
Great barracudas, 5:*408*, 5:415–*416*
Great Basin kangaroo rats. *See* Chisel-toothed kangaroo rats
Great Basin pocket mice, 16:*200*
Great birds of paradise. *See* Greater birds of paradise
Great bitterns. *See* Eurasian bitterns
Great black cockatoos. *See* Palm cockatoos
Great blue herons, 8:234, 8:235, 8:241, 8:*242*, 8:*247*, 8:*248*–*249*
Great blue turacos, 9:299–300, 9:*303*, 9:*304*
Great bowerbirds, 11:477, 11:478
Great bustards, 9:1, 9:92, 9:*95*, 9:*96*
  See also Great Indian bustards
Great cormorants, 8:*204*, 8:*205*
Great crested flycatchers, 10:273, 10:*274*, 10:*276*, 10:*287*
Great crested grebes, 8:171, 8:172, 8:*173*, 8:*178*
Great crested newts, 6:45, 6:49, 6:59, 6:*369*, 6:374
Great curassows, 8:*400*
Great Danes, 14:289
Great dusky swifts, 9:*416*, 9:*423*
Great eagle-owls. *See* Eurasian eagle-owls
Great egrets. *See* Great white egrets
Great flashlightfishes. *See* Splitfin flashlightfishes
Great frigatebirds, 8:193, 8:*194*, 8:*195*
Great fruit-eating bats, 13:434*t*
Great gray kangaroos. *See* Eastern gray kangaroos
Great gray kiwis. *See* Great spotted kiwis
Great gray owls, 9:346, 9:*348*
Great gray slugs, 2:*416*, 2:*419*, 2:420
Great grebes, 8:171, 8:*173*, 8:*177*–*178*
Great hammerhead sharks, 4:*117*, 4:126–*127*
Great-helmeted hornbills. *See* Helmeted hornbills
Great hornbills, 10:73, 10:*75*, 10:76, 10:*77*, 10:79–*80*
Great horned owls, 9:347, 9:349, 9:*352*, 9:*356*–367
Great Indian bustards, 9:3, 9:*95*, 9:*96*–97
Great Indian hornbills. *See* Great hornbills
Great ioras, 10:415
Great jacamars, 10:91, 10:92, 10:*94*, 10:*96*
Great kiskadees, 10:270, 10:*271*, 10:*276*, 10:*280*
Great knots, 9:178, 9:*181*, 9:185–*186*
Great Lakes Fishery Commission, 4:86
Great Lakes longears. *See* Longear sunfishes
Great Lakes sturgeons. *See* Lake sturgeons
Great Lakes trouts. *See* Lake trouts
Great long-nosed armadillos, 13:191*t*
Great niltavas. *See* Large niltavas
Great northern divers. *See* Common loons
Great northern sea cows. *See* Steller's sea cows
Great palm cockatoos. *See* Palm cockatoos
Great Papuan frogmouths. *See* Papuan frogmouths
Great Philippine eagles, 8:319, 8:322
Great pied hornbills. *See* Great hornbills
Great Plains narrow-mouthed toads, 6:303, 6:312
Great Plains skinks, 7:42

Great potoos, 9:*369*, 9:395, 9:*396*, 9:*398*, 9:*400*
Great red-footed urchins. *See* Sea biscuits
Great reed warblers, 8:42, 11:*2*, 11:5, 11:*9*, 11:*17*
Great rufous woodcreepers, 10:*231*, 10:*235*
Great sapphirewings, 9:443
Great skuas, 9:204, 9:206
Great slaty woodpeckers, 10:*152*, 10:*166*
Great snakeheads. *See* Bullseye snakeheads
Great snipes, 9:179
Great sparrows. *See* Southern rufous sparrows
Great spinetails, 10:*213*, 10:*222*
Great spotted cuckoos, 9:312–314, 9:*316*, 9:*318*
Great spotted kiwis, 8:90, 8:*91*, 8:*93–94*
Great spotted woodpeckers, 10:*149*
Great sturgeons. *See* Beluga sturgeons
Great-tailed grackles, 11:*307*, 11:318–*319*
Great-tailed trioks, 13:133*t*
Great thick-knees, 9:144, 9:*145*
Great tinamous, 8:*60*, 8:*61–62*
Great tits, 11:155, 11:158, 11:*159*, 11:163–*164*
Great vase sponges. *See* Barrel sponges
Great water beetles, 3:*324*, 3:*329*
Great water diving beetles. *See* Great water beetles
Great West Indian sea lilies, 1:*361*, 1:*363*, 1:364–365
Great whales. *See* specific great whales
Great white cranes. *See* Siberian cranes
Great white egrets, 8:*233*, 8:*247*, 8:*249–250*
Great white herons. *See* Great blue herons
Great white pelicans, 8:226, 9:272
Great white sharks. *See* White sharks
Great yellow-headed vultures, 8:276
Greater adjutants, 8:235, 8:236
Greater Andean flamingos. *See* Andean flamingos
Greater anis, 9:312, 9:*316*, 9:*325–326*
Greater argentines, 4:*394*, 4:395–*396*
Greater argonauts, 2:*481*, 2:483–*484*
Greater bamboo lemurs, 14:11, 14:*48*
Greater bandicoot rats, 16:249–250
Greater bent-winged bats. *See* Common bentwing bats
Greater bilbies, 13:*2*, 13:4, 13:5, 13:6, 13:19–*20*, 13:*21*, 13:*22*
Greater birds of paradise, 11:*491*, 11:*494*, 11:*501*–502
Greater blue ringed octopods, 2:*475*
Greater bulldog bats, 13:313, 13:443, 13:*444*, 13:*445*, 13:*446*, 13:447, 13:*448*, 13:*449*
Greater bushbabies. *See* Brown greater bushbabies; Northern greater bushbabies
Greater cane rats, 16:333, 16:*334*, 16:*336*, 16:*337*
Greater coucals, 9:*317*, 9:*325*
Greater crested mynas. *See* Helmeted mynas
Greater Cuban nesophontes, 13:243
Greater dog-faced bats, 13:*356*, 13:*359*, 13:361–*362*
Greater doglike bats. *See* Greater dog-faced bats
Greater double-collared sunbirds, 11:*212*, 11:*222*–223
Greater dwarf lemurs, 14:*4*, 14:*36*, 14:*40*, 14:*41*–42
Greater Egyptian jerboas, 16:212, 16:216, 16:223*t*
Greater false vampire bats, 13:313–314, 13:*379*, 13:*381*, 13:*382*
Greater fat-tailed jerboas, 16:212, 16:214, 16:216, 16:223*t*
Greater five-lined skinks. *See* Broad-headed skinks
Greater flamingos, 8:*10*, 8:*46*, 8:303, 8:304, 8:*306*, 8:*307*, 8:*308*
Greater gliders, 13:33, 13:113–120, 13:*115*, 13:*118*, 13:*119*
Greater gliding possums, **13:113–123**, 13:122*t*–123*t*
Greater green-billed malkohas. *See* Green-billed malkohas
Greater guinea pigs, 16:399*t*
Greater hedgehog tenrecs, 13:200, 13:*227*, 13:234*t*
Greater honeyguides, 10:68, 10:138, 10:*140*, 10:*141*, 11:219
Greater hoopoe-larks, 10:342, 10:*345*, 10:*347*, 10:349–*350*
Greater horseshoe bats, 13:*389*, 13:*391*, 13:*395*, 13:*397*–398
behavior, 13:390
conservation status, 13:392–393
distribution, 13:388
reproduction, 13:314, 13:372
Greater house bats, 13:*484*, 13:486, 13:*489*, 13:*490*, 13:491
Greater Japanese shrew moles, 13:280, 13:282, 13:288*t*
Greater king starlings. *See* Helmeted mynas
Greater koa finches, 11:343, 11:*345*, 11:*349*
Greater kudus, 16:6, 16:11, 16:*12*, 16:*13*, 16:*16*, 16:*19*, 16:20, 16:25*t*
Greater leaf-folding frogs, 6:*280*, 6:282, 6:*284*, 6:285–286
Greater long-tailed hamsters, 16:239, 16:247*t*
Greater long-tongued fruit bats, 13:*339*, 13:*344*, 13:346
Greater long-tongued nectar bats. *See* Greater long-tongued fruit bats
Greater Malagasy bushy-tailed rats, 16:*281*
Greater Malay mouse deer, 15:325, 15:*326*, 15:*327*, 15:*328*, 15:*330*, 15:*332*–333
Greater marsupial moles. *See* Marsupial moles
Greater Mascarene flying foxes, 13:331*t*
Greater melampittas, 11:76
Greater mouse-eared bats, 13:497, 13:*503*, 13:515*t*
Greater mouse-tailed bats, 13:352
Greater mynas. *See* Helmeted mynas
Greater naked-backed bats. *See* Naked-backed fruit bats
Greater naked-tailed armadillos, 13:185, 13:*186*, 13:*189*
Greater New Zealand short-tailed bats, 13:454, 13:455, 13:*456*, 13:457–458
Greater noctules, 13:313, 13:516*t*
Greater one-horned rhinoceroses. *See* Indian rhinoceroses
Greater painted-snipes, 9:115, 9:*116*, 9:117, 9:*118*
Greater pewees, 10:*276*, 10:*286*
Greater prairie chickens, 8:434, 8:435, 8:437, 8:*439*, 8:*444*–445
Greater rabbit-eared bandicoots. *See* Greater bilbies
Greater racket-tailed drongos, 11:438, 11:*440*, 11:443–*444*
Greater rhabdornis, 11:183, 11:184, 11:*185*, 11:*186*
Greater rheas, 8:69–71, 8:*71*, 8:*72*, 8:*73*
Greater roadrunners, 9:*312*, 9:*313*, 9:*317*, 9:*327*
Greater sac-winged bats, 12:81, 13:*356–358*, 13:*359*, 13:*360*
Greater scaly-throated honeyguides, 10:137
Greater scimitarbills. *See* Common scimitarbills
Greater scythebills, 10:230
Greater sheathbills. *See* Pale-faced sheathbills
Greater short-tailed bats. *See* Greater New Zealand short-tailed bats
Greater short-toed larks, 10:346, 10:*347*, 10:351–*352*
Greater shrews, 13:276*t*
Greater sirens, 6:31, 6:327, 6:*328*, 6:*330*, 6:*332*
Greater sloth lemurs, 14:*68*, 14:*70*, 14:71
Greater sooty owls. *See* Sooty owls
Greater spear-nosed bats, 13:*422*, 13:*424*
behavior, 13:312, 13:416, 13:419
echolocation, 12:87
feeding ecology, 13:418
vocalizations, 12:85
Greater stick-nest rats, 16:*254*, 16:*259–260*
Greater streaked honeyeaters. *See* Greater Sulawesi honeyeaters
Greater Sulawesi honeyeaters, 11:*239*, 11:*250*–251
Greater sulphur crested cockatoos. *See* Sulphur-crested cockatoos
Greater swallow-tailed swifts, 9:423
Greater thornbirds, 10:*213*, 10:*224*
Greater thumbless bats. *See* Schnabeli's thumbless bats
Greater two-lined bats. *See* Greater sac-winged bats
Greater weevers, 5:335
Greater white-lined bats. *See* Greater sac-winged bats
Greater whitethroats. *See* Whitethroats
Greater yellow-headed vultures, 8:*279*, 8:281, 8:*281*
Greater yellowlegs, 9:179
Grebes, **8:169–181**, 8:*173*, 9:69
behavior, 8:170–171
conservation status, 8:171
distribution, 8:*169*, 8:170
evolution, 8:169
feeding ecology, 8:171
habitats, 8:170
humans and, 8:171–172
physical characteristics, 8:169–170
reproduction, 8:*171*
species of, 8:*174–181*
taxonomy, 8:169
Greeffiella minutum, 1:*286*, 1:287–288
Greek mythology, reptiles in, 7:54–55
Greek tortoises, 7:*67*, 7:*143*
Green acouchis, 16:*408*, 16:*410*, 16:*412*, 16:414
Green alfalfa aphids. *See* Pea aphids
Green anacondas, 7:202, 7:410, 7:411, 7:*412*, 7:*414*, 7:415
Green and gold bell frogs, 6:*58*
Green anemones, giant, 1:*110*, 1:*111*
Green anoles, 7:*35*, 7:38–39, 7:*244*, 7:246, 7:*248*, 7:*250*, 7:256–257

Green aracaris, 10:126
Green avadavats, 11:356
Green-backed firecrowns, 9:443, 9:449–450
Green-backed honeyguides, 10:139
Green-backed whipbirds. *See* Eastern whipbirds
Green basilisks, 7:*245*
Green basses. *See* Largemouth basses
Green-billed coucals, 9:312
Green-billed malkohas, 9:*317*, 9:*324*
Green bonellias, 2:*105*, 2:*106*–107
Green-breasted bush-shrikes, 10:429
Green-breasted manucodes. *See* Crinkle-collared manucodes
Green-breasted pittas. *See* Hooded pittas
Green broadbills, 10:*178*, 10:179, 10:*180*, 10:181–*182*, 10:185–*186*
Green bulbuls. *See* Golden-fronted leafbirds
Green bush vipers, 7:*449*, 7:456–*457*
Green catbirds, 11:*482*, 11:*483*
Green collared pigeons. *See* Pheasant pigeons
Green colobus. *See* Olive colobus
Green discus. *See* Blue discus
Green-faced parrotfinches, 11:356
Green figbirds. *See* Australasian figbirds; Timor figbirds
Green flatworms, 1:33, 1:40, 1:179, 1:*181*–182
Green frogs, 6:41, 6:46, 6:*47*, 6:49, 6:*67*, 6:252
See also Roesel's green frogs
Green-fronted lancebills, 9:*454*, 9:*458*–459
Green-headed olive sunbirds. *See* Green-headed sunbirds
Green-headed sunbirds, 11:*211*, 11:*218*–219
Green heads. *See* Big black horse flies
Green hermits, 9:*439*, 9:442, 9:*454*, 9:457–*458*
Green hunting crows. *See* Green magpies
Green hydra-*Chlorella* symbiosis, 1:32–33
Green iguanas, 7:47, 7:243, 7:245, 7:*246*
Green ioras, 10:417, 10:*418*, 10:*419*–420
Green lacewings, 3:*308*, 3:*310*, 3:*311*–312
Green magpies, 11:504, 11:509, 11:*510*, 11:*515*
Green manucodes. *See* Crinkle-collared manucodes
Green monkeys. *See* Grivets
Green morays, 4:*4*, 4:*26*, 4:*260*, 4:*265*–266
Green orioles. *See* Olive-backed orioles
Green peafowls, 8:399
Green pigeons, 9:244
Green pikes. *See* Walleyes
Green plovers. *See* Northern lapwings
Green poison frogs, 6:54, 6:198, 6:*202*, 6:204–*205*
Green pythons, 7:*206*, 7:420, 7:*421*, 7:*423*, 7:*425*, 7:426
Green ringtails, 13:*114*, 13:*116*, 13:*118*, 13:*119*, 13:120
Green sandpipers, 9:179
Green sea lampreys. *See* Sea lampreys
Green sea urchins, 1:*404*, 1:405
Green seaturtles, 7:*38*, 7:43, 7:*85*, 7:86, 7:*88*, 7:*90*–91
Green shads. *See* Alewives
Green springtails. *See* Varied springtails
Green stick insects. *See* Thunberg's stick insects
Green sugarbirds. *See* Malachite sunbirds
Green sunbirds, 11:*211*, 11:214–*215*
Green sunfishes, 5:198
Green swordtails, 4:32, 4:*33*, 5:94, 5:*95*, 5:*100*, 5:101
Green-tailed jacamars, 10:92, 10:*94*, 10:*95*–96
Green tiger barbs. *See* Tiger barbs
Green toads, 6:*187*, 6:192–193
Green toucanets, 10:125
Green tree pythons. *See* Green pythons
Green treefrogs, 6:53, 6:*231*, 6:240
Green turacos, 9:300
Green turtles. *See* Green seaturtles
Green vegetable bugs. *See* Southern green stink bugs
Green violet-ears, 9:439, 9:*446*
Green-winged macaws, 9:*278*
Green-winged pytilias, 11:*355*, 11:*358*, 11:*361*–362
Green-winged trumpeters. *See* Dark-winged trumpeters
Green woodhoopoes, 8:*10*, 10:65, 10:*66*, 10:*67*, 10:*68*
Greenbones. *See* Butterfishes
Greenbuls. *See* specific types of greenbuls
Greenewalt, C., 9:440
Greeneyes, 4:*435*, 4:*437*
Greenface sandsifters, 5:278
Greenfinches, 11:*326*, 11:*330*
Greenflies. *See* Aphids
Greenhalli's dog-faced bats, 13:494*t*
Greenheads. *See* Mallards
Greenhouse camel crickets, 3:*210*, 3:212–*213*
Greenhouse gases, 1:52–53
See also Conservation status
Greenhouse whiteflies, 3:*265*, 3:*267*
Greenland. *See* Nearctic region
Greenland collared lemmings. *See* Northern collared lemmings
Greenland sharks, 4:48, 4:151, 4:152, 4:*154*, 4:*155*, 4:156–157
Greenlets, 11:255, 11:*257*
Greenling sea trouts. *See* Kelp greenlings
Greenlings, **5:179–192,** 5:*184*
Greenshanks, 9:178, 9:179
Greenwood, H., 4:289, 5:123, 5:309, 5:421
Gregarious behavior, 12:145–147
See also Behavior
Grenadier weavers. *See* Southern red bishops
Grenadiers, **5:25–30,** 5:*31*, **5:36–37**
Grevy's zebras, 15:*215*, 15:*221*, 15:*227*, 15:*231*, 15:*232*–233
behavior, 15:219, 15:*228*
conservation status, 15:222
evolution, 15:216
physical characteristics, 15:*225*
reproduction, 15:221
Greylag geese, 8:21, 8:369, 8:373
Gribbles, 2:*253*
See also Common shipworms
Gridiron-tailed lizards. *See* Zebra-tailed lizards
Griffin, Donald, 12:54
Griffon vultures. *See* Himalayan griffon vultures
Grigs, 3:201, 3:207
Grillotiidae, 1:226
Grimwood's longclaws, 10:375
Grindles. *See* Bowfins
Grinnels. *See* Bowfins
Gripopterygidae, 3:141
Grivets, 14:*187*, 14:*196*, 14:199
behavior, 12:143, 14:7
communication, 12:84, 12:160
distribution, 14:*197*
evolution, 14:188
Grizzlies. *See* Brown bears
Grobben, Karl, 2:3–5
Grobecker, D. B., 5:47, 5:49
Gromphadorhina spp., 3:149
Gromphadorhina portentosa. *See* Madagascan hissing cockroaches
Grooming behavior. *See* Behavior
Groove-billed anis, 9:*314*, 11:*8*
Groove-billed barbets, 10:*85*
Groove-toothed rats. *See* Angoni vlei rats
Groove-toothed swamp rats. *See* Creek rats
Grooved jellyfish, 1:153
See also Coronatae
Grosbeak weavers, 11:377
Grosbeaks, 11:323–*324*, 11:327, 11:338, 11:343, 11:349, 11:377
Ground agamas. *See* Spiny agamas
Ground antbirds. *See* Ant thrushes
Ground beetles, 3:63, 3:90, 3:317–323
Ground cuckoo-shrikes, 10:*386*, 10:387
Ground cuscuses, 13:*60*, 13:*62*, 13:*64*–*65*
Ground finches. *See* African quailfinches
Ground hornbills. *See* Southern ground-hornbills
Ground-jays, 11:504
Ground pangolins, 16:*109*, 16:*110*, 16:*112*, 16:113, 16:*114*, 16:117–*118*
Ground parrots, 9:276
Ground pythons. *See* Neotropical sunbeam snakes
Ground-rollers, 10:1–4, 10:51–53
Ground sharks, **4:113–130,** 4:*117*–*118*
behavior, 4:114–115
conservation, 4:116
distribution, 4:114
evolution, 4:113
feeding ecology, 4:115
habitats, 4:114
humans and, 4:116
physical characteristics, 4:113–114
reproduction, 4:115–116
species of, 4:119–128
taxonomy, 4:11, 4:113
See also Greenland sharks
Ground sloths, 12:132, 15:137, 15:138
Ground sparrows, 11:397, 11:398, 11:405–406
Ground squirrels, **16:143–161,** 16:*149*–*150*
behavior, 12:*71*, 12:141, 16:124, 16:144–145
conservation status, 16:147–148
distribution, 12:131, 16:*143*, 16:144
evolution, 16:143
feeding ecology, 16:145, 16:147
habitats, 16:144
humans and, 16:126, 16:148
physical characteristics, 12:113, 16:143–144
reproduction, 16:147
species of, 16:151–158, 16:158*t*–160*t*
taxonomy, 16:143
Groundhogs. *See* Woodchucks
Groundlings. *See* Stone loaches
Groundsnakes, 7:466, 7:468, 7:469
Groupers, 4:44, 4:46–48, 4:61, 4:68, **5:255–263,** 5:*265*, 5:*269*, **5:270–271**
Grouse, 3:201, 8:400–401, 8:433–437, 8:*440*
Groves, C. P.
on capuchins, 14:101
on tarsiers, 14:91

Growth. *See* Reproduction
Grubs
scarabaeid, 3:56
underground grass, 3:389
white, 3:63
wichety, 3:81
Gruidae. *See* Cranes
Gruiformes, **9:1–4,** 9:*2,* 9:107
Grunions, 5:*69*
Grunt-fishes. *See* Grunt sculpins
Grunt sculpins, 5:*185,* 5:*189*
Grunters, **5:219–222,** 5:*223,* 5:*227,* 5:*229,* **5:232–233**
Grunts, 4:*62,* 5:255, 5:*264,* 5:*267,* 5:268
See also Haemulidae
Grus spp., 9:23
Grus americana. See Whooping cranes
Grus antigone. See Sarus cranes
Grus canadensis. See Sandhill cranes
Grus carunculatus. See Wattled cranes
Grus grus. See Eurasian cranes
Grus japonensis. See Red-crowned cranes
Grus leucogeranus. See Siberian cranes
Grus monacha. See Hooded cranes
Grus nigricollis. See Black-necked cranes
Grus vipio. See White-naped cranes
Gryllacrididae. *See* Raspy crickets
Gryllidae, 3:44, 3:203
Grylloblatodea, 3:54
Grylloblatta spp., 3:189–190
Grylloblatta campodeiformis. See Northern rock-crawlers
Grylloblattella pravdini. See Pravdin's rock-crawlers
Grylloblattidae, 3:*189*
Grylloblattina, 3:189
Grylloblattina djakonovi. See Djakonov's rock-crawlers
Grylloblattodea. *See* Rock-crawlers
Gryllodes sigillatus, 3:203
Grylloidea. *See* Crickets
Gryllotalpa spp., 3:205, 3:207
Gryllotalpinae, 3:44
Gryllotalpoidea. *See* Mole crickets
Gryllus campestris. See European field crickets
Grypania, 2:9
Grysboks, **16:59–72,** 16:71*t*
GSD. *See* Genetic sex determination
Guadalupe, Virgin of, 7:55
Guadalupe fur seals, 14:394, 14:395, 14:398, 14:399, 14:408*t*
Guadalupe storm-petrels, 8:110, 8:138, 14:291
Guadeloupe caracaras. *See* Crested caracaras
Guadeloupe flickers, 10:156
Guadeloupe raccoons, 14:309, 14:310, 14:316*t*
Guadeloupe woodpeckers, 10:*153,* 10:161–*162*
Guam flycatchers, 11:98
Guam flying foxes, 12:138, 13:332*t*
Guam rails, 9:3, 9:51, 9:52, 9:*53,* 9:*59*
Guam swiftlets, 9:418
Guam swifts, 9:424
Guanacaste squirrels. *See* Deppe's squirrels
Guanacos, 15:*134,* **15:313–323,** 15:*315,* 15:*319,* 15:*322*
behavior, 15:316–317, 15:*318*
distribution, 12:132, 15:*313,* 15:315
domestication of, 12:180
evolution, 15:151, 15:313–314
feeding ecology, 15:317
habitats, 15:316
humans and, 15:318
physical characteristics, 15:314
reproduction, 15:*316,* 15:317
taxonomy, 15:313
Guanay shags, 8:185–186
Guano bats. *See* Brazilian free-tailed bats
Guans, **8:413–423**
behavior, 8:414–415
conservation status, 8:416
distribution, 8:*413,* 8:414
evolution, 8:413
feeding ecology, 8:415–416
habitats, 8:414
humans and, 8:416
physical characteristics, 8:413–414
reproduction, 8:416
species of, 8:*419–421*
taxonomy, 8:416
Guapotes, 5:277
Guard dogs, 14:288, 14:289, 14:292–293
Guatemalan flickers. *See* Northern flickers
Gudgeons, 4:*306,* 4:*312,* 5:375, 5:377, 5:*379,* 5:*381*
Guenons, 14:5
behavior, 12:146, 14:7, 14:192
evolution, 14:188
guilds of, 12:116
habitats, 14:191
humans and, 14:194
physical characteristics, 14:189
reproduction, 14:10, 14:193
taxonomy, 14:188
Guenther's dikdiks, 16:71*t*
Guenther's long-snouted dikdiks. *See* Guenther's dikdiks
Guerezas, 14:174, 14:*175,* 14:*176,* 14:*178*
Guerrilla behavior, 1:26
See also Behavior
Guguftos. *See* Gerenuks
Guianan cocks-of-the-rock, 10:*308,* 10:*310,* 10:318–*319*
Guianan mastiff bats. *See* Greater house bats
Guianan sakis. *See* White-faced sakis
Guiaras, 16:458*t*
Guide dogs, 14:*293*
Guiffra, E., 15:149
Guilds, 4:42, 12:116–119
Guillemots, 9:*219,* 9:220, 9:*223,* 9:225–226
Guinea baboons, 14:*188,* 14:194, 14:205*t*
Guinea fowl puffers, 5:*469*
Guinea picathartes. *See* White-necked picathartes
Guinea pigs, 16:*392,* 16:*394,* 16:*395*–396
behavior, 12:148
distribution, 12:132
domestic, 12:180–181, 16:*391,* 16:399*t*
greater, 16:399*t*
humans and, 16:128, 16:393
montane, 16:399*t*
neonatal requirements, 12:126
neonates, 12:125
physical characteristics, 16:390
reproduction, 12:106
shiny, 16:399*t*
taxonomy, 16:121, 16:389
Guineafowl, 8:21, 8:399, 8:400, **8:425–431**
Guira cuckoos, 9:*315*
Guira guira. See Guira cuckoos
Guitarfishes, 4:11, 4:128, 4:173–179, 4:*180,* 4:*186,* 4:187–188
Guitarrero. *See* Guitarrita
Guitarrita, 4:*356,* 4:357–*358*
Gular sac, 9:24, 9:102
Gulf porpoises. *See* Vaquitas
Gulf ridleys. *See* Kemp's ridley turtles
Gulf Stream beaked whales. *See* Gervais' beaked whales
Gull-billed terns, 9:203
Gulls, 9:101–105, **9:203–217,** 9:*210*
behavior, 9:102, 9:206
conservation status, 9:208–209
distribution, 9:*203,* 9:205
evolution, 9:203–204
feeding ecology, 9:102–103, 9:206–207
habitats, 9:205–206
humans and, 9:209
physical characteristics, 9:203, 9:204–205
reproduction, 9:207–208
species of, 9:*211–217*
taxonomy, 9:101, 9:104, 9:203
Gulo spp., 14:321
Gulo gulo. See Wolverines
Gulper eels, 4:11, 4:*271,* 4:*273,* 4:*275,* 4:276
Gulper sharks, 4:151, 4:152
Gulpers, **4:271–276,** 4:*273*
Gumboot chitons, 2:394, 2:395, 2:*396,* 2:*397*–398
Gumdrops. *See* Spotted coral crouchers
Gummivory, 14:8–9
Gumshoe chitons. *See* Gumboot chitons
Gundis, **16:311–315**
Gundlach's hawks, 8:323
Gunnels, 5:309–313
Gunning's golden moles, 13:222*t*
Gunnison sage grouse, 8:433, 8:437
Gunnison's prairie dogs, 16:159*t*
Gunn's bandicoots. *See* Eastern barred bandicoots
Günther, A., 5:351, 7:189
Guppies, 4:32
Gurnard perches, 5:163
Gurnards
armored, 5:165
flying, **5:157–162,** 5:*160*
red, 5:*168,* 5:*171,* 5:177
tub, 5:*165*
Gurney's eagles, 8:*325,* 8:*340*
Gurney's pittas, 10:195, 10:*196,* 10:*197*
Gurry sharks. *See* Greenland sharks
Gursky, S., 14:98
Gustation, 3:26–27
See also Physical characteristics
Guttera plumifera. See Plumed guineafowl
Guttera pucherani. See Crested guineafowl
Guy's hermits. *See* Green hermits
Guyu spp., 5:220
Gygis spp. *See* Noddies
Gymnarchidae, 4:57, 4:231
Gymnarchus niloticus. See Aba-aba
Gymnelus spp., 5:310
Gymnobelideus spp., 13:128, 13:129
Gymnobelideus leadbeateri. See Leadbeater's possums
Gymnobucco peli. See Bristle-nosed barbets
Gymnobucco sladeni. See Sladen's barbets
Gymnocephalus cernuus. See Ruffes
Gymnocharacinus bergii. See Naked characins
Gymnocrex talaudensis. See Talaud rails

Gymnodoerus spp. *See* Cotingas
Gymnodoerus foetidus. *See* Bare-necked fruitcrows
Gymnodraco acuticeps. *See* Naked dragonfishes
Gymnogyps californianus. *See* California condors
Gymnolaemata, **2:509–514,** 2:*512*
Gymnomystax mexicanus. *See* Oriole blackbirds
Gymnophaps albertisii. *See* Papua mountain pigeons
Gymnophiona. *See* Caecilians
Gymnophthalmidae. *See* Microteiids
Gymnophthalmus spp., 7:303, 7:304
Gymnophthalmus cryptus, 7:306
Gymnophthalmus underwoodi, 7:*305*, 7:*306*–307
Gymnopis spp., 6:437
Gymnopis multiplicata, 6:*436*
Gymnorhamphichthys spp., 4:370
Gymnorhina tibicen. *See* Australian magpies
Gymnorhinus cyanocephalus. *See* Pinyon jays
Gymnorhynchidae, 1:226
Gymnorrex spp., 9:47
Gymnoscops insularis. *See* Seychelles scops-owls
Gymnothorax funebris. *See* Green morays
Gymnothorax nudivomer. *See* Starry moray eels
Gymnothorax prasinus. *See* Yellow morays
Gymnotid eels. *See* Banded knifefishes
Gymnotidae, 4:14, 4:60, 4:62, 4:369
Gymnotiformes, 4:68, **4:369–377,** 4:*372*
Gymnotus spp., 4:369
Gymnotus carapo. *See* Banded knifefishes
Gymnura altavela. *See* Spiny butterfly rays
Gymnures, 13:193, 13:195, 13:197, 13:199, **13:203–214,** 13:212*t*–213*t*
Gynaephora groenlandica, 3:36, 3:37, 3:385
Gynodiastylidae, 2:229
Gynogenesis, 3:59, 4:66
Gypaetus barbatus. *See* Bearded vultures
Gypohierax angolensis. *See* Palm nut vultures
Gyps bengalensis. *See* White-rumped vultures
Gyps coprotheres. *See* Cape griffons
Gyps fulvus. *See* Eurasian griffons
Gyps himalayensis. *See* Himalayan griffon vultures
Gyps indicus. *See* Long-billed vultures
Gyps rueppellii. *See* Rüppell's vultures
Gypsy moths, 3:60, 3:85, 3:388, 3:*392*, 3:395–*396*
Gyrfalcons, 8:21, 8:*314*, 8:348, 8:350, 8:353, 8:359–360
Gyrinids. *See* Whirligig beetles
Gyrinocheilidae. *See* Algae eaters
Gyrinocheilus aymonieri. *See* Chinese algae eaters
Gyrinocheilus pennocki. *See* Spotted algae eaters
Gyrinophilus spp., 6:34, 6:390, 6:391
Gyrinophilus prophyriticus, 6:*391*
Gyrinus mexicanus. *See* Mexican axolotl
Gyrocotylidea, 1:225, 1:226, 1:227, 1:228, 1:231
Gyrodactylids, 1:214–215
Gyrodactylus pungitii, 1:*216*, 1:*220*–221
Gyuyer, C., 6:301

H

Ha Shi Ma Yu oil, 6:53
Haast's eagles, 8:323
Habitat destruction and alteration, 1:49–50, 7:60
  See also Conservation status
Habitat loss, 12:214
  See also Conservation status
Habitats
  amphibians, 6:6, 6:7
    anurans, 6:26, 6:63–64
      African treefrogs, 6:281–282, 6:285–289
      Amero-Australian treefrogs, 6:228, 6:233–242
      Arthroleptidae, 6:265–266, 6:268–270
      Asian toadfrogs, 6:110, 6:113–116
      Asian treefrogs, 6:292, 6:295–299
      Australian ground frogs, 6:140, 6:143–145
      Bombinatoridae, 6:83, 6:86–88
      Bufonidae, 6:184, 6:188–194
      Discoglossidae, 6:89–90, 6:92–94
      Eleutherodactylus spp., 6:158
      frogs, 6:26, 6:63–64
      ghost frogs, 6:131, 6:133–134
      glass frogs, 6:216, 6:219–223
      leptodactylid frogs, 6:158, 6:162–171
      Madagascan toadlets, 6:318, 6:319–320
      Mesoamerican burrowing toads, 6:96
      Myobatrachidae, 6:148, 6:151–153
      narrow-mouthed frogs, 6:303, 6:308–315
      New Zealand frogs, 6:69, 6:72–74
      parsley frogs, 6:127, 6:129
      Pipidae, 6:100, 6:103–106
      poison frogs, 6:198–199, 6:203–209
      Ruthven's frogs, 6:211
      Salamandridae, 6:364–365, 6:370–375
      Seychelles frogs, 6:136, 6:137–138
      shovel-nosed frogs, 6:273
      sirens, 6:328, 6:331–332
      spadefoot toads, 6:119, 6:*120*, 6:124–125
      tailed frogs, 6:78, 6:80–81
      tailless caecilians, 6:436, 6:439–441
      three-toed toadlets, 6:179, 6:181–182
      toads, 6:26, 6:63–64
      torrent salamanders, 6:385, 6:387
      true frogs, 6:250, 6:255–263
      vocal sac-brooding frogs, 6:173, 6:175–176
    caecilians, 6:412
      American tailed, 6:416, 6:417–418
      Asian tailed, 6:419, 6:422–423
      buried-eye, 6:432, 6:433–434
      Kerala, 6:426, 6:428
    salamanders
      amphiumas, 6:406, 6:409–410
      Asiatic, 6:336, 6:339–341
      Cryptobranchidae, 6:345
      lungless, 6:392, 6:395–403
      mole, 6:356, 6:358–360
      newts, 6:*63*, 6:364–365, 6:370–375
      Pacific giant, 6:350, 6:352
      Proteidae, 6:378, 6:381–383
  birds
    albatrosses, 8:114, 8:118–122
    Alcidae, 9:219, 9:221, 9:224–228
    anhingas, 8:184, 8:202, 8:207–209
    Anseriformes, 8:364–365
    ant thrushes, 10:240, 10:245–256
    Apodiformes, 9:417
    asities, 10:188, 10:190
    Australian chats, 11:65–66, 11:67
    Australian creepers, 11:134, 11:137–139
    Australian fairy-wrens, 11:45, 11:48–53
    Australian honeyeaters, 11:236, 11:241–253
    Australian robins, 11:105, 11:108–112
    Australian warblers, 11:55–56, 11:57, 11:59–63
    babblers, 10:506–507, 10:511–523
    barbets, 10:114–115, 10:119–122
    barn owls, 9:335, 9:336, 9:340–343
    bee-eaters, 10:39
    birds of paradise, 11:491, 11:495–501
    bitterns, 8:241–242
    Bombycillidae, 10:448, 10:451–454
    boobies, 8:184, 8:212
    bowerbirds, 11:478–479, 11:483–488
    broadbills, 10:178, 10:181–186
    bulbuls, 10:396–397, 10:402–413
    bustards, 9:92, 9:96–99
    buttonquails, 9:12–13, 9:16–21
    caracaras, 8:348–349
    cassowaries, 8:76
    Charadriidae, 9:162, 9:166–172
    Charadriiformes, 9:101–102, 9:104
    chats, 10:485–486
    chickadees, 11:156–157, 11:160–166
    chowchillas, 11:69, 11:70, 11:72–73
    Ciconiiformes, 8:234, 8:236
    Columbidae, 9:248–249, 9:255–266
    condors, 8:277
    cormorants, 8:184, 8:202, 8:205–206, 8:207–209
    Corvidae, 11:505, 11:512–522
    cotingas, 10:306, 10:312–322
    crab plovers, 9:121–122
    Cracidae, 8:414
    cranes, 9:24, 9:31–36
    cuckoo-shrikes, 10:386
    cuckoos, 9:311, 9:312, 9:318–329
    dippers, 10:476, 10:479–482
    diving-petrels, 8:143, 8:145–146
    doves, 9:248–249, 9:255–266
    drongos, 11:438, 11:441–445
    ducks, 8:371
    eagles, 8:320
    elephant birds, 8:104
    emus, 8:84
    Eupetidae, 11:75, 11:76, 11:78–80
    fairy bluebirds, 10:416, 10:423–424
    Falconiformes, 8:314
    falcons, 8:348–349
    false sunbirds, 10:188, 10:191
    fantails, 11:85, 11:89–94
    finches, 11:323, 11:328–339
    flamingos, 8:304–305
    flowerpeckers, 11:190, 11:193–198
    frigatebirds and, 8:184, 8:194, 8:197–198
    frogmouths, 9:368, 9:378, 9:381–385
    Galliformes, 8:399
    gannets, 8:184, 8:212
    geese, 8:371
    Glareolidae, 9:152, 9:155–160
    grebes, 8:170, 8:174–180
    Gruiformes, 9:2
    guineafowl, 8:427
    gulls, 9:102–103, 9:211–217
    hammerheads, 8:262
    Hawaiian honeycreepers, 11:342, 11:346–351
    hawks, 8:320
    hedge sparrows, 10:459, 10:462–464

herons, 8:241–242
hoatzins, 8:466
honeyguides, 10:138, 10:141–144
hoopoes, 10:62
hornbills, 10:73, 10:78–84
hummingbirds, 9:442–443, 9:455–467
ibises, 8:292
Icteridae, 11:302–303, 11:309–322
ioras, 10:416, 10:419–420
jacamars, 10:92
jacanas, 9:108, 9:112–114
kagus, 9:41–42
kingfishers, 10:8, 10:13–23
kiwis, 8:90
lapwings, 9:167–172
Laridae, 9:205–206, 9:211–217
larks, 10:343, 10:348–354
leafbirds, 10:416, 10:420–423
limpkins, 9:37
logrunners, 11:69, 11:70, 11:72–73
long-tailed titmice, 11:142, 11:145–146
lyrebirds, 10:330, 10:334–335
magpie-shrikes, 11:468, 11:471–475
manakins, 10:295, 10:299–303
mesites, 9:5, 9:8–9
moas, 8:96–97
monarch flycatchers, 11:97, 11:100–103
motmots, 10:32, 10:35–37
moundbuilders, 8:404
mousebirds, 9:469, 9:473–476
mudnest builders, 11:453, 11:456–457
New World blackbirds, 11:309–322
New World finches, 11:263, 11:269–282
New World quails, 8:456–457
New World vultures, 8:277
New World warblers, 11:288–289, 11:293–299
New Zealand wattle birds, 11:447, 11:449–450
New Zealand wrens, 10:204, 10:206–207
nightjars, 9:368, 9:402–403, 9:408–414
nuthatches, 11:167, 11:168, 11:171–174
oilbirds, 9:373, 9:374
Old World flycatchers, 11:26, 11:30–43
Old World warblers, 11:3–4, 11:10–23
Oriolidae, 11:428, 11:431–434
ostriches, 8:100
ovenbirds, 10:210, 10:214–228
owlet-nightjars, 9:388, 9:391–392
owls, 9:332, 9:336, 9:340–343, 9:347–348, 9:354–365
oystercatchers, 9:126, 9:130–131
painted-snipes, 9:116, 9:118–119
palmchats, 10:455
pardalotes, 11:201, 11:204–206
parrots, 9:276, 9:283–297
partitioning, 11:4
Passeriformes, 10:172
Pelecaniformes, 8:184–185
pelicans, 8:184–185, 8:225–226
penduline titmice, 11:147, 11:151–152
penguins, 8:149, 8:153–157
Phasianidae, 8:434–435
pheasants, 8:434–435
Philippine creepers, 11:183, 11:186–187
Picidae, 10:148–149, 10:154–167
Piciformes, 10:86
pigeons, 9:248–249, 9:255–266
pipits, 10:373, 10:378–383
pittas, 10:193–194, 10:197–200
plantcutters, 10:326, 10:327–328
plovers, 9:102, 9:166–172
potoos, 9:395–396, 9:399–400
pratincoles, 9:155–160
Procellariidae, 8:124, 8:129–130
Procellariiformes, 8:*108*–109
pseudo babblers, 11:128, 11:130–131
puffbirds, 10:102, 10:106–110
rails, 9:47–48, 9:51–52, 9:57–66, 9:63
Raphidae, 9:269, 9:270, 9:272–273
Recurvirostridae, 9:134, 9:137–140
rheas, 8:69–70
rollers, 10:52, 10:55–58
sandgrouse, 9:231, 9:235–238
sandpipers, 9:175, 9:182–188
screamers, 8:394
scrub-birds, 10:337, 10:339
secretary birds, 8:344
seedsnipes, 9:189, 9:193–195
seriemas, 9:85, 9:88–89
sharpbills, 10:292
sheathbills, 9:198, 9:200–201
shoebills, 8:287
shrikes, 10:427, 10:432–438
sparrows, 11:398, 11:401–406
spoonbills, 8:292
storks, 8:265
storm-petrels, 8:135, 8:140–141
Sturnidae, 11:408, 11:414–425
sunbirds, 11:208–209, 11:213–225
sunbitterns, 9:74
sungrebes, 9:69–70, 9:71–72
swallows, 10:358, 10:363–369
swans, 8:371
swifts, 9:422, 9:426–431
tapaculos, 10:258, 10:262–267
terns, 9:211–217
thick-knees, 9:144, 9:147–148
thrushes, 10:485–486
tinamous, 8:57–58
titmice, 11:156–157, 11:160–166
todies, 10:26, 10:29
toucans, 10:126, 10:130–134
tree swifts, 9:433–434, 9:435–436
treecreepers, 11:177, 11:180–181
trogons, 9:478, 9:481–485
tropicbirds, 8:184, 8:187, 8:*190–191*
trumpeters, 9:78, 9:79, 9:81–82
turacos, 9:300, 9:304–310
typical owls, 9:347–348, 9:354–365
tyrant flycatchers, 10:271, 10:278–288
vanga shrikes, 10:440, 10:443–444
Vireonidae, 11:255–256, 11:258–262
wagtails, 10:373, 10:378–383
weaverfinches, 11:354, 11:360–372
weavers, 11:376–377, 11:382–394
whistlers, 11:116, 11:120–125
white-eyes, 11:228, 11:232–233
woodcreepers, 10:229, 10:232–236
woodhoopoes, 10:66, 10:68–69
woodswallows, 11:459, 11:462–464
wrens, 10:527, 10:531–538
fishes
Acanthuroidei, 5:392, 5:396–403
Acipenseriformes, 4:213, 4:217–220
Albuliformes, 4:249–250, 4:252–253
alteration of, 4:74–75
angelsharks, 4:162, 4:164–165
anglerfishes, 5:48, 5:53–56
Anguilliformes, 4:255–256, 4:260–268
Atheriniformes, 5:68–69, 5:73–76
Aulopiformes, 4:432–433, 4:436–439
Australian lungfishes, 4:198
beardfishes, 5:1, 5:3
behavior and, 4:62
Beloniformes, 5:80, 5:83–86
Beryciformes, 5:115, 5:118–121
bichirs, 4:210, 4:211–212
blennies, 4:48, 4:50, 5:342, 5:345–348
bowfins, 4:230
bullhead sharks, 4:98, 4:101–102
Callionymoidei, 5:366, 5:369–371
carps, 4:298, 4:308–313
catfishes, 4:353, 4:357–366
characins, 4:337, 4:342–349
chimaeras, 4:91–92, 4:94–95
coelacanths, 4:191
Cypriniformes, 4:298, 4:307–319, 4:321–322, 4:325–333
Cyprinodontiformes, 5:91–92, 5:96–102
damselfishes, 4:42, 4:47, 4:48
deep slope and wall, 4:48
deep water, 4:56
dogfish sharks, 4:151, 4:155–157
dories, 5:124, 5:126–128, 5:130
eels, 4:50, 4:255–256, 4:260–268
electric eels, 4:370, 4:373–377
Elopiformes, 4:243, 4:246–247
Esociformes, 4:380, 4:383–386
flatfishes, 5:452, 5:455–463
floating, 4:62
freshwater, 4:36, 4:*37*–41, 4:*38*, 4:52
Gadiformes, 5:28–29, 5:32–39
gars, 4:221–222, 4:224–227
Gasterosteiformes, 5:134, 5:139, 5:141–147
gobies, 4:42, 4:47, 4:48, 4:50, 5:374–375, 5:380–388
Gobiesocoidei, 5:356, 5:359–363
Gonorynchiformes, 4:290, 4:293–295
ground sharks, 4:114, 4:119–128
groupers, 4:48
Gymnotiformes, 4:370, 4:373–377
hagfishes, 4:78–79, 4:80–81
hawkfishes, 4:42, 4:43, 4:48
herrings, 4:277, 4:281–287
Hexanchiformes, 4:144, 4:146–148
Labroidei, 5:277, 5:284–290, 5:294, 5:301–306
labyrinth fishes, 5:428, 5:431–434
lake, 4:38–39, 4:40, 4:52
lampreys, 4:84, 4:88–90
Lampridiformes, 4:449, 4:452–454
lanternfishes, 4:442, 4:444–446
lionfishes, 4:48
lungfishes, 4:202–203, 4:205–206
mackerel sharks, 4:132, 4:135–141
marine, 4:47–50
minnows, 4:314–315
morays, 4:47, 4:48, 4:255–256, 4:264–266
mudminnows, 4:380, 4:383–386
mullets, 4:47, 5:59–60, 5:62–65
Ophidiiformes, 5:17, 5:20–22
Orectolobiformes, 4:105, 4:108–110
Osmeriformes, 4:391–392, 4:395–401
Osteoglossiformes, 4:232, 4:237–241
parrotfishes, 4:48
Percoidei, 5:196–197, 5:202–208, 5:211–212, 5:214–217, 5:220–221, 5:225–226, 5:228, 5:230–233, 5:238–239,

5:247–253, 5:259–260, 5:266–268, 5:270–273
Percopsiformes, 5:5, 5:8–12
pikes, 4:380, 4:383–386
ragfishes, 5:352
Rajiformes, 4:176–177, 4:182–188
rays, 4:176–177, 4:182–188
river, 4:39–40, 4:52
Saccopharyngiformes, 4:272, 4:274–276
salmons, 4:405–406, 4:410–419
sawsharks, 4:168, 4:170–171
Scombroidei, 5:406, 5:409–418
Scorpaeniformes, 5:158, 5:161–162, 5:164–165, 5:169–178, 5:180–181, 5:186–193
Scorpaenoidei, 5:164–165
seahorses, 4:48
shallow, 4:53–56
skates, 4:176–177, 4:187
snakeheads, 5:438, 5:442–446
South American knifefishes, 4:370, 4:373–377
southern cod-icefishes, 5:322, 5:325–328
squirrelfishes, 4:48
Stephanoberyciformes, 5:106, 5:108–110
Stomiiformes, 4:422, 4:426–429
streams, 4:39–40, 4:62
Stromateoidei, 5:422
subtital, 4:47
Synbranchiformes, 5:151, 5:154–156
Tetraodontiformes, 5:468–469, 5:474–484
toadfishes, 5:42, 5:44–45
Trachinoidei, 5:333, 5:337–339
tropical, 4:53
tunas, 4:49
wrasses, 4:48
Zoarcoidei, 5:310–311, 5:315–317
insects, 3:5, 3:86–88
book lice, 3:243, 3:246–247
bristletails, 3:57, 3:114, 3:116–117
caddisflies, 3:376, 3:379–381
cockroaches, 3:148, 3:153–158
Coleoptera, 3:319, 3:327–333
diplurans, 3:107, 3:110–111
Diptera, 3:359, 3:364–368, 3:370–374
earwigs, 3:196, 3:199–200
fleas, 3:349, 3:352–354
grasshoppers, 3:56, 3:57, 3:203–204
Hemiptera, 3:261, 3:267–268, 3:270–279
Hymenoptera, 3:406, 3:411–424
lacewings, 3:306–307, 3:311–314
Lepidoptera, 3:385, 3:393–404
mantids, 3:180, 3:183–186
Mantophasmatodea, 3:218–219
mayflies, 3:125–126, 3:129–130
Mecoptera, 3:342, 3:345–346
Megaloptera, 3:290, 3:293–294
Odonata, 3:133, 3:137–138
Orthoptera, 3:203–204, 3:211–215
Phasmida, 3:222, 3:228–232
Phthiraptera, 3:250, 3:254–256
proturans, 3:93, 3:96–97
rock-crawlers, 3:189, 3:192–193
snakeflies, 3:297, 3:300–302
springtails, 3:99–100, 3:103–104
stoneflies, 3:142, 3:145–146
strepsipterans, 3:336, 3:338–339
termites, 3:163–164, 3:170–175
thrips, 3:281, 3:285–287
Thysanura, 3:119, 3:122–123
webspinners, 3:233, 3:236–237
zorapterans, 3:240, 3:241
lower metazoans and lesser deuterostomes, 1:24–30
acoels, 1:180, 1:181–182
anoplans, 1:246, 1:249–250
arrow worms, 1:434, 1:437–441
box jellies, 1:147–148, 1:151–152
calcareous sponges, 1:24, 1:58, 1:61–64
cnidarians, 1:24
Anthozoa, 1:24, 1:25, 1:105, 1:111–120
Hydrozoa, 1:24, 1:127–128, 1:134–145
comb jellies, 1:170, 1:173–177
conservation of, 1:49–50
demosponges, 1:24, 1:79, 1:82–86
echinoderms
Crinoidea, 1:357, 1:362–365
Echinoidea, 1:402, 1:408–415
Ophiuroidea, 1:388, 1:393–398
sea cucumbers, 1:421, 1:425–431
sea daisies, 1:382, 1:385–386
sea stars, 1:369, 1:373–379
enoplans, 1:253–254, 1:256–257
entoprocts, 1:319, 1:323–325
flatworms
free-living flatworms, 1:187, 1:190–195
monogeneans, 1:214, 1:218, 1:220–223
tapeworms, 1:229–230, 1:234–240, 1:242
gastrotrichs, 1:270, 1:272–273
girdle wearers, 1:344, 1:347–349
glass sponges, 1:24, 1:68, 1:71–75
gnathostomulids, 1:331, 1:334–335
hair worms, 1:305, 1:308–309
hemichordates, 1:444, 1:447–450
jaw animals, 1:328
jellyfish, 1:24, 1:154–155, 1:160–162, 1:164–167
kinorhynchs, 1:276, 1:279–280
lancelets, 1:487, 1:490–496
larvaceans, 1:473, 1:476–478
migration and, 1:42–43
nematodes
roundworms, 1:284, 1:287–291
secernenteans, 1:294, 1:297, 1:299–304
Orthonectida, 1:99, 1:101–102
placozoans, 1:88
priapulans, 1:337, 1:340–341
rhombozoans, 1:93, 1:96–98
rotifers, 1:260–261, 1:264–267
Salinella salve, 1:91
salps, 1:467
sea squirts, 1:454, 1:458–466
sorberaceans, 1:480, 1:482–483
thorny headed worms, 1:312, 1:314–316
Trematoda, 1:198–199, 1:203–207, 1:209–211
wheel wearers, 1:352
mammals
aardvarks, 15:156
agoutis, 16:407, 16:411–414, 16:415*t*
anteaters, 13:172, 13:177–179
armadillos, 13:183, 13:187–190, 13:190*t*–192*t*
Artiodactyla, 15:269
aye-ayes, 14:86
baijis, 15:20
bandicoots, 13:2–3
dry-country, 13:10, 13:14–16, 13:16*t*–17*t*
rainforest, 13:10, 13:14–17*t*
bats, 13:311
bulldog, 13:445, 13:449–450
disk-winged, 13:474, 13:476–477
Emballonuridae, 13:355, 13:360–363, 13:363*t*–364*t*
false vampire, 13:379, 13:384–385
funnel-eared, 13:460, 13:463–465
horseshoe, 13:389–390, 13:396–400
Kitti's hog-nosed, 13:368
Molossidae, 13:484, 13:490–493, 13:493*t*–495*t*
mouse-tailed, 13:351, 13:353
moustached, 13:435, 13:440–441, 13:442*t*
New Zealand short-tailed, 13:454, 13:457
Old World fruit, 13:319, 13:325–330, 13:331*t*–332*t*, 13:334, 13:340–347, 13:348*t*–349*t*
Old World leaf-nosed, 13:402, 13:407–409, 13:409*t*–410*t*
Old World sucker-footed, 13:480
slit-faced, 13:371, 13:375, 13:376*t*–377*t*
smoky, 13:467–468, 13:470
Vespertilionidae, 13:498, 13:506, 13:508–514, 13:515*t*–516*t*, 13:521, 13:524–525, 13:526*t*
bears, 14:296–297, 14:302–305, 14:306*t*
beavers, 16:179, 16:183–184
bilbies, 13:20
botos, 15:28
Bovidae, 16:4–5
Antilopinae, 16:46, 16:50–55, 16:56*t*–57*t*
Bovinae, 16:13, 16:18–23, 16:24*t*–25*t*
Caprinae, 16:91–92, 16:98–103, 16:103*t*–104*t*
duikers, 16:74, 16:80–84, 16:84*t*–85*t*
Hippotraginae, 16:29–30, 16:37–42, 16:42*t*–43*t*
Neotraginae, 16:60, 16:66–71, 16:71*t*
bushbabies, 14:6, 14:24, 14:28–32, 14:32*t*–33*t*
Camelidae, 15:315–316, 15:320–323
Canidae, 14:267, 14:276–283, 14:283*t*–284*t*
capybaras, 16:402
Carnivora, 14:257
cats, 14:371, 14:379–389, 14:390*t*–391*t*
Caviidae, 16:390–391, 16:395–398, 16:399*t*
Cetacea, 15:5–6
Balaenidae, 15:109, 15:115–117
beaked whales, 15:61, 15:64–68, 15:69*t*–70*t*
dolphins, 15:43–44, 15:53–55, 15:56*t*–57*t*
franciscana dolphins, 15:24
Ganges and Indus dolphins, 15:14
gray whales, 15:95
Monodontidae, 15:83–84, 15:90
porpoises, 15:34–35, 15:38–39, 15:39*t*
pygmy right whales, 15:104
rorquals, 15:6, 15:121, 15:127–130, 15:130*t*
sperm whales, 15:74, 15:79–80
chevrotains, 15:327, 15:331–333
Chinchillidae, 16:378, 16:382–383, 16:383*t*
colugos, 13:302, 13:304
coypus, 16:124, 16:474

dasyurids, 12:288–289, 12:294–298, 12:299t–301t
Dasyuromorphia, 12:279–280
deer
Chinese water, 15:374
muntjacs, 15:344, 15:349–354
musk, 15:336, 15:340–341
New World, 15:382–384, 15:388–396, 15:396t
Old World, 15:359, 15:364–371, 15:371t
Dipodidae, 16:213, 16:219–222, 16:223t–224t
Diprotodontia, 13:34–35
dormice, 16:318, 16:323–326, 16:327t–328t
duck-billed platypuses, 12:*229*, 12:230, 12:244, 12:*247*
Dugongidae, 15:200, 15:203
echidnas, 12:237, 12:240t
elephants, 15:166–167, 15:174–175
Equidae, 15:226, 15:232–235, 15:236t
Erinaceidae, 13:204, 13:209–213, 13:212t–213t
giant hutias, 16:469, 16:471t
gibbons, 14:210, 14:218–222
Giraffidae, 15:401, 15:408
great apes, 14:227–228, 14:237–240
gundis, 16:312, 16:314–315
Herpestidae, 14:349, 14:354–356, 14:357t
Heteromyidae, 16:200, 16:205–208t, 16:209t
hippopotamuses, 15:306, 15:310–311
hot spots, 12:218–219
humans, 14:247
hutias, 16:461, 16:467t
Hyaenidae, 14:362, 14:364–367
hyraxes, 15:179–180, 15:186–188, 15:189t
Indriidae, 14:65, 14:69–72
Insectivora, 13:197
koalas, 13:44–45
Lagomorpha, 16:481–482
lemurs, 14:6, 14:49, 14:55, 14:57–60
Cheirogaleidae, 14:36, 14:41–44
sportive, 14:75, 14:79–83
Leporidae, 16:506, 16:511–5135, 16:515t–516t
Lorisidae, 14:14, 14:18–20, 14:21t
Macropodidae, 13:34, 13:35, 13:85–86, 13:93–100, 13:101t–102t
manatees, 15:206, 15:211–212
Megalonychidae, 13:156, 13:159
moles
golden, 13:197, 13:216, 13:220–222, 13:222t
marsupial, 13:26, 13:28
monitos del monte, 12:274
monkeys
Atelidae, 14:156–157, 14:162–166, 14:166t–167t
Callitrichidae, 14:116, 14:127–131, 14:132t
Cebidae, 14:103, 14:108–112
cheek-pouched, 14:191, 14:197–204, 14:204t–205t
leaf-monkeys, 14:174, 14:178–183, 14:184t–185t
night, 14:137, 14:141t
Pitheciidae, 14:145, 14:150–152, 14:153t
monotremes, 12:230
mountain beavers, 16:132
Muridae, 16:283–284, 16:288–295, 16:296t–297t
Arvicolinae, 16:226, 16:233–238, 16:237t–238t
hamsters, 16:240, 16:245–246, 16:247t
Murinae, 16:250–251, 16:255–260, 16:261t–262t
Sigmodontinae, 16:266–267, 16:272–276, 16:277t–278t
musky rat-kangaroos, 13:70
Mustelidae, 14:321, 14:327–331, 14:332t–333t
numbats, 12:304
octodonts, 16:434, 16:438–440, 16:440t
opossums
New World, 12:251, 12:257–263, 12:264t–265t
shrew, 12:268, 12:270
Otariidae, 14:395–396, 14:402–406
pacaranas, 16:386–387
pacas, 16:418, 16:423–424
pangolins, 16:110, 16:115–119
peccaries, 15:292–293, 15:298–300
Perissodactyla, 15:217–218
Petauridae, 13:125–126, 13:131–133, 13:133t
Phalangeridae, 13:59–60, 13:*63–66t*
pigs, 15:277, 15:284–288, 15:289t–290t
pikas, 16:491, 16:497–501, 16:501t–502t
pocket gophers, 16:187, 16:192–194, 16:195t–197t
porcupines
New World, 16:367, 16:371–373, 16:374t
Old World, 16:352, 16:357–364
possums
feather-tailed, 13:141, 13:144
honey, 13:136
pygmy, 13:106, 13:110–111
primates, 14:6
Procyonidae, 14:310, 14:313–315, 14:315t–316t
pronghorns, 15:412
Pseudocheiridae, 13:114–115, 13:119–123, 13:122t–123t
rat-kangaroos, 13:74, 13:78–81
rats
African mole-rats, 16:342–343, 16:347–349, 16:349t–350t
American leaf-nosed, 13:311, 13:415–416, 13:423–432, 13:433t–434t
cane, 16:334, 16:337–338
chinchilla, 16:444, 16:447–448
dassie, 16:330
spiny, 16:450, 16:453–457, 16:458t
rhinoceroses, 15:252–253, 15:257–261
rodents, 16:124
sengis, 16:519, 16:524–530, 16:531t
shrews
red-toothed, 13:248, 13:255–263, 13:263t
West Indian, 13:244
white-toothed, 13:266–267, 13:271–275, 13:276t–277t
Sirenia, 15:192
solenodons, 13:238, 13:240
springhares, 16:308
squirrels
flying, 16:135–136, 16:139–140, 16:141t–142t
ground, 16:144, 16:151–158, 16:158t–160t
scaly-tailed, 16:300, 16:302–305
tree, 16:164–165, 16:169–174t, 16:175t
Talpidae, 13:280, 13:284–287, 13:287t–288t
tapirs, 15:238, 15:245–247
tarsiers, 14:93–94, 14:97–99
tenrecs, 13:226–227, 13:232–234, 13:234t
three-toed tree sloths, 13:162, 13:167–169
tree shrews, 13:291, 13:294–296, 13:297t–298t
true seals, 14:257, 14:418, 14:427–428, 14:430–434, 14:435t
tuco-tucos, 16:425–426, 16:429–430, 16:430t–431t
Viverridae, 14:337, 14:340–343, 14:344t–345t
walruses, 12:14, 14:412
wombats, 13:52, 13:55–56
Xenartha, 13:150
protostomes
amphionids, 2:196
amphipods, 2:261, 2:265–271
anaspidaceans, 2:181–182, 2:183
aplacophorans, 2:380, 2:382–385
Arachnida, 2:335, 2:339–352
articulate lampshells, 2:522, 2:525–527
bathynellaceans, 2:178, 2:179
beard worms, 2:86, 2:88
bivalves, 2:452, 2:457–466
caenogastropods, 2:446, 2:449
centipedes, 2:354, 2:358–362
cephalocarids, 2:132, 2:133
Cephalopoda, 2:477, 2:482–488
chitons, 2:395, 2:397–400
clam shrimps, 2:147–148, 2:150–151
copepods, 2:300, 2:304–310
cumaceans, 2:229, 2:232
Decapoda, 2:199, 2:204–213
deep-sea limpets, 2:435, 2:437
earthworms, 2:66, 2:70–73
echiurans, 2:103–104, 2:106–107
fairy shrimps, 2:136, 2:139
fish lice, 2:289, 2:292–293
freshwater bryozoans, 2:497, 2:500–502
horseshoe crabs, 2:328, 2:331–332
Isopoda, 2:250, 2:254–259
krill, 2:185, 2:189–192
leeches, 2:76, 2:80–82
leptostracans, 2:161, 2:164–165
lophogastrids, 2:226, 2:227
mantis shrimps, 2:168, 2:173–175
marine bryozoans, 2:504, 2:506–508, 2:509, 2:513–514
mictaceans, 2:241, 2:242
millipedes, 2:364, 2:367–370
monoplacophorans, 2:388, 2:390–391
mussel shrimps, 2:312, 2:314
mysids, 2:216, 2:219–222
mystacocarids, 2:295, 2:297
myzostomids, 2:59, 2:62
Neritopsina, 2:440, 2:442–443
nonarticulate lampshells, 2:515–516, 2:518–519
Onychophora, 2:110, 2:113–114
pauropods, 2:375, 2:377
peanut worms, 2:98, 2:100–101
phoronids, 2:491, 2:494
Polychaeta, 2:46, 2:50–56

Pulmonata, 2:413–414, 2:417–421
remipedes, 2:125, 2:128
sea slugs, 2:403, 2:407–409
sea spiders, 2:321, 2:324–325
spelaeogriphaceans, 2:243, 2:244
symphylans, 2:371, 2:373
tadpole shrimps, 2:142, 2:145–146
tanaids, 2:235–236, 2:238–239
tantulocaridans, 2:284, 2:286
Thecostraca, 2:275, 2:278–280
thermosbaenaceans, 2:245, 2:247
tongue worms, 2:319
true limpets, 2:424, 2:426–427
tusk shells, 2:470, 2:473–474
Vestimentifera, 2:92, 2:94
Vetigastropoda, 2:430, 2:433–434
water bears, 2:116–117, 2:120–123
water fleas, 2:153, 2:157–158
reptiles, 7:436–437
African burrowing snakes, 7:462, 7:463
African sideneck turtles, 7:129, 7:132–133
Afro-American river turtles, 7:137, 7:140—141
Agamidae, 7:210, 7:214–221
Alligatoridae, 7:172, 7:175–177
Anguidae, 7:340, 7:342–344
Australo-American sideneck turtles, 7:77, 7:81–83
big-headed turtles, 7:135
blindskinks, 7:272
blindsnakes, 7:381, 7:383–385
boas, 7:411, 7:413–417
Central American river turtles, 7:100
chameleons, 7:228, 7:235–241
colubrids, 7:468, 7:473–481
Cordylidae, 7:320, 7:324–325
Crocodylidae, 7:179–180, 7:184–187
early blindsnakes, 7:371
Elapidae, 7:484–485, 7:491–498
false blindsnakes, 7:388
false coral snakes, 7:400
file snakes, 7:441
Florida wormlizards, 7:284
Gekkonidae, 7:260–261, 7:265–269
Geoemydidae, 7:115, 7:118–119
gharials, 7:168
Helodermatidae, 7:354
Iguanidae, 7:244, 7:250–256
Kinosternidae, 7:121, 7:125–126
knob-scaled lizards, 7:348, 7:351
Lacertidae, 7:298, 7:301–302
leatherback seaturtles, 7:101
Microteiids, 7:304, 7:306–307
mole-limbed wormlizards, 7:280, 7:281
Neotropical sunbeam snakes, 7:405–406
New World pond turtles, 7:105, 7:109–112
night lizards, 7:292, 7:294–295
pig-nose turtles, 7:76
pipe snakes, 7:396
pythons, 7:420, 7:424–428
seaturtles, 7:86, 7:89–91
shieldtail snakes, 7:392–394
skinks, 7:329, 7:332–337
slender blindsnakes, 7:374–377
snapping turtles, 7:93, 7:95–96
softshell turtles, 7:152, 7:154–155
spade-headed wormlizards, 7:288, 7:289
splitjaw snakes, 7:430
Squamata, 7:204
sunbeam snakes, 7:402
Teiidae, 7:310–311, 7:314–316
tortoises, 7:143, 7:146–148
Tropidophiidae, 7:434, 7:436–437
tuatara, 7:190
Varanidae, 7:360, 7:365–368
Viperidae, 7:447, 7:451–459
wormlizards, 7:274, 7:276
restoration of, 6:59
See also Biomes; Coral reef habitats
Habituation, 1:40–41, 2:39
See also Behavior
Hachisuka, Masauji, 9:269
Hadada ibises, 8:*295*, 8:*298*
Haddocks, 5:26, 5:27, 5:29, 5:30, 5:*31*, 5:*33*–34
Hadedahs. *See* Hadada ibises
Hadedas. *See* Hadada ibises
Hadromerida, 1:29, 1:77, 1:79
Hadropithecus spp., 14:63
Haeckel, Ernst, 1:7, 1:129, 1:488
Haeckelia rubra, 1:169
Haemadipsa picta. *See* Tiger leeches
Haemadipsa sumatrana, 2:78
Haemadipsidae, 2:75, 2:76
Haematomyzus elephantis. *See* Elephant lice
Haematopidae. *See* Oystercatchers
Haematopinus oliveri, 3:252
Haematops validirostris. *See* Strong-billed honeyeaters
Haematopus ater. *See* Blackish oystercatchers
Haematopus bachmani. *See* American black oystercatchers
Haematopus chathamensis. *See* Chatham Islands oystercatchers
Haematopus fuliginosus. *See* Sooty oystercatchers
Haematopus leucopodis. *See* Magellanic oystercatchers
Haematopus longirostris. *See* Australian pied oystercatchers
Haematopus meadewaldoi. *See* Canary Islands oystercatchers
Haematopus moquini. *See* African black oystercatchers
Haematopus ostralegus. *See* Eurasian oystercatchers
Haematopus palliatus. *See* American oystercatchers
Haematopus sulcatus, 9:125
Haematopus unicolor. *See* Variable oystercatchers
Haementeria ghilianii. *See* Giant Amazonian leeches
Haemochorial placentation, 12:94
Haemonchus contortus. *See* Barber's pole worms
Haemopidae, 2:75, 2:76
Haemotoderus spp. *See* Cotingas
Haemulidae, 5:255, 5:258–263
Haemulon sciurus. *See* Blue striped grunts; Bluestriped grunts
Haffer, J., 10:91, 10:125
Hafner, John C., 16:199
Hagenbeck, Carl, 12:204
Hagenbeck exhibits, 12:204–205
Hagfishes, **4:77–81,** 4:*80–81*
evolution, 4:9–10, 4:77
physical characteristics, 4:4, 4:6, 4:25, 4:77, 4:*78*, 4:80–81
reproduction, 4:27, 4:*30*, 4:34, 4:79, 4:80–81
Hagloidea, 3:201
Haideotriton spp., 6:391
Haikouichthys, 4:9
Haimara. *See* Trahiras
Hainan gymnures, 13:207, 13:213*t*
Hainan hares, 16:509
Hainan hill-partridges, 8:434
Hainan moonrats. *See* Hainan gymnures
Hair-crested drongos, 11:438
Hair follicle mites. *See* Demodicids
Hair worms, 1:9, 1:12, 1:21–22, **1:305–310,** 1:*307*, 2:5
Hairless bats. *See* Naked bats
Hairs, 12:3, 12:7, 12:38–39, 12:*43*
bats, 12:59
domestication and, 12:174
evolution of, 12:19, 12:22
for heat loss prevention, 12:51
longtail weasels, 12:*47*
marine mammals, 12:63, 12:*64*
in sexual selection, 12:24
See also Physical characteristics
Hairtail blennies, 5:*344*, 5:345
Hairy-armed bats. *See* Leisler's bats
Hairy bees, 3:81
Hairy big-eyed bats, 13:434*t*
Hairy dwarf porcupines
Bahía, 16:366, 16:*370*, 16:*371*, 16:373
black-tailed, 16:365, 16:374*t*
brown, 16:366, 16:374*t*
frosted, 16:365, 16:374*t*
orange-spined, 16:366, 16:374*t*
Paraguay, 16:366, 16:367, 16:368, 16:374*t*
white-fronted, 16:365, 16:374*t*
Hairy-eared mouse lemurs, 14:36, 14:38, 14:39, 14:*40*, 14:*41*
Hairy-fisted crabs. *See* Chinese mitten crabs
Hairy-footed flying squirrels, 16:141*t*
Hairy-footed jerboas, 16:214, 16:215, 16:216, 16:*218*, 16:*221*
Hairy frogs, 6:35, 6:*266*, 6:*267*, 6:270–271
Hairy-fronted muntjacs. *See* Black muntjacs
Hairy hermits, 9:*454*, 9:*456*–457
Hairy-legged vampire bats, 13:314, 13:*414*, 13:435*t*
Hairy long-nosed armadillos, 13:185, 13:191*t*
Hairy-nosed otters, 14:332*t*
Hairy-nosed wombats, 13:37, 13:51, 13:52
Hairy slit-faced bats, 13:376*t*
Hairy-tailed moles, 13:*282*, 13:*283*, 13:*284*, 13:286
Hairy-winged bats. *See* Harpy-headed bats
Hairyfishes, 5:*105*, 5:106, 5:*107*, 5:*109*
Haitian nesophontes, 13:243, 13:244
Haitian solenodons. *See* Hispaniolan solenodons
Haitian vine boas, 7:410
Hakes, 5:*25*, **5:25–30,** 5:*31*, **5:37–40**
Halaelurus spp., 4:115
Halammohydra schulzei, 1:*130*, 1:134
Halcyon chelicuti. *See* Striped kingfishers
Halcyon chloris. *See* Collared kingfishers
Halcyon godeffroyi. *See* Marquesas kingfishers
Halcyon leucocephala. *See* Gray-headed kingfishers
Halcyon sancta. *See* Sacred kingfishers
Halcyon tuta. *See* Chattering kingfishers
Halcyoninae, 10:5–7
Haldeman's black nasutes. *See* Black-headed nasute termites

Half-collared flycatchers. *See* Collared flycatchers
Half-girdled snakes, 7:486, 7:*489*, 7:*497*, 7:498
See also Broad-banded sand swimmers
Half-toed geckos. *See* House geckos
Halfbeaks, 4:12, 4:47, 4:67, 4:70, 5:79, 5:80–81
See also Blackbarred halfbeaks
Haliaeetus albicilla. *See* White-tailed eagles
Haliaeetus leucocephalus. *See* Bald eagles
Haliaeetus leucogaster. *See* White-bellied sea-eagles
Haliaeetus leucoryphus. *See* Pallas's sea-eagles
Haliaeetus pelagicus. *See* Steller's sea-eagles
Haliaeetus sanfordi. *See* Sanford's sea-eagles
Haliaeetus vociferoides. *See* Madagascar fish-eagles
Haliastur indus. *See* Brahminy kites
Haliastur sphenurus. *See* Whistling kites
Halibuts, 5:451, 5:452, 5:453
See also specific species
Halichoerus grypus. *See* Gray seals
Halichondria okadai, 1:45
Halichondrida, 1:77, 1:79
Halichondrin B, 1:45
Haliclona oculata. *See* Eyed finger sponges
Haliclystus spp., 1:10, 1:154
Haliclystus auricula. *See* Stalked jellyfish
Halicryptus spinulosus, 1:13
Halictinae, 3:68, 3:70
Halictophagus naulti, 3:*337*, 3:*338*
Halimochirugus alcocki. *See* Spikefishes
Haliotis spp., 2:431
Haliotis rufescens. *See* Red abalone
Haliotis tuberculata lamellosa, 2:*429*, 2:*431*
Haliotoidea, 2:429, 2:430, 2:440
Halipegus occidualis, 1:198
Halisarcida, 1:77, 1:79
Hall, Michelle, 11:456
Hall's babblers, 11:127, 11:128
Hallucigenia spp., 2:109
Hallucinogenic drugs, 6:53
Halmahera pythons, 7:*423*, 7:*425–426*
Halobates spp. *See* Sea skaters
Halobates micans. *See* Sea skaters
Halobatrachus spp., 5:42
Halobiotus spp., 2:117
Halobiotus crispae, 2:117
Halocordyle disticha, 1:30
Halocoryne epizoica, 1:128–129
Halocynthia aurantium, 1:453
Halocynthia pyriformis. *See* Sea pigs
Halocyptena microsoma. *See* Least storm-petrels
Halophryne spp., 5:42
Halosauridae, 4:249, 4:250
Halosauropsis spp., 4:249
Halosauropsis macrochir. *See* Halosaurs
Halosaurs, 4:*251*, 4:*252*
Halosaurus spp., 4:249
Halosbaena spp., 2:245
Halovelia electrodominica, 3:259
Halterina pulchella, 3:309
Halterina purcelli, 3:309
Hamadryas baboons, 12:146, 12:*209*, 14:7, 14:192, 14:194, 14:*195*, 14:*202–203*
Hambly, Wilfred, 7:55
Hamerkops. *See* Hammerheads
Hamilton, W. D., 3:70
Hamilton's frogs, 6:69, 6:70, 6:*71*, 6:*72–73*
Hamirostra melanosternon. *See* Australian black-breasted buzzards; Black-breasted buzzards
Hamlets, 4:30, 4:45, 4:68, 5:255, 5:262
Hammaticherus spp. *See* Cerambycid beetles
Hammer-headed fruit bats, 12:84, 13:312, 13:334, 13:*338*, 13:*341*, 13:342
Hammerhead sharks, 4:11, 4:27, 4:113, 4:114, 4:115
Hammerheads, 8:183, 8:225, 8:233, 8:234, 8:239, **8:261–263,** 8:*262*, 8:272
Hammerjaws, 4:432
Hammond's flycatchers, 10:271, 10:*276*, 10:*285*–286
Hamptophryne boliviana. *See* Bolivian bleating frogs
Hamsters, 16:123, 16:128, **16:239–248,** 16:*240*, 16:*241*, 16:*244*, 16:282, 16:283
Hancock's amblydoras. *See* Blue-eye catfishes
Hanging parrots, 9:279
Hangingflies, **3:341–346,** 3:*344*, 3:*345*
Hanlon, R. T., 2:477–478
Hannia spp., 5:220
Hansen, P. M., 4:156
Hansenocaris spp., 2:273, 2:274, 2:275
Hantaviruses
mice and, 16:128
Sigmodontinae, 16:270
Hanuman langurs. *See* Northern plains gray langurs
Hapalemur spp., 14:8
Hapalemur aureus, 14:11
Hapalemur griseus griseus. *See* Gray gentle lemurs
Hapalemur simus. *See* Greater bamboo lemurs
Hapalochlaena spp. *See* Blue ringed octopods
Hapalochlaena lunulata. *See* Greater blue ringed octopods
Hapaloptila castanea. *See* White-faced nunbirds
Haplobothriidae. *See* Haplobothriidea
Haplobothriidea, 1:225, 1:229
Haplochromine cichlids, 4:36, 5:281
Haplochromis spp., 5:281
Haplocylicini, 5:355, 5:356
Haplocylix spp., 5:355
Haplodiploidy, 3:70
Haplognathia ruberrima. *See* Red haplognathias
Haplognathiidae, 1:331
Haplonycteris fischeri. *See* Fischer's pygmy fruit bats
Haplophaedia spp. *See* Hummingbirds
Haplopharyngida, 1:185
Haplorhine primates
evolution, 12:11
reproduction, 12:94
See also Apes; Humans; Monkeys; Tarsiers
Haplosclerida, 1:29, 1:77, 1:78–79
Haplotaxida spp. *See* Giant earthworms
Haplotaxodon, 4:*34*
Happy jacks. *See* Gray-crowned babblers
Harbor porpoises, 15:7, 15:9, 15:33, 15:*34*, 15:35, 15:*36*, 15:*37*, 15:*38*
Harbor seals, 14:*261*, 14:418–420, 14:*424*, 14:*426*, 14:*428*–429
Hard clams. *See* Northern quahogs
Hard head whales. *See* Gray whales
Hardening, cuticle, 3:17, 3:32–33
Hardin, Garret, 1:47–48
Hardshell clams. *See* Northern quahogs
Hardwicke's hedgehogs. *See* Collared hedgehogs
Hardwicke's lesser mouse-tailed bats, 13:*352*, 13:*353*
Hardy, Laurence M., 7:306
Hare, Brian, 14:287
Hare-wallabies, 13:83, 13:84, 13:85, 13:86, 13:88
Harem groups. *See* One-male groups
Harems. *See* Polygyny
Hares, **16:505–516,** 16:*510*
behavior, 16:482–483, 16:506–507
conservation status, 12:222, 16:509
distribution, 12:129, 12:135, 16:*505*–506
evolution, 16:480, 16:505
feeding ecology, 16:484–485, 16:507
habitats, 16:481, 16:506
humans and, 16:488, 16:509
milk of, 12:126
physical characteristics, 12:38, 16:505
predation and, 12:115
reproduction, 16:485–487, 16:507–509
species of, 16:*511*–513, 16:515*t*–516*t*
taxonomy, 16:505
Harlequin bats, 13:*501*
Harlequin beetles, 3:317
Harlequin ducks, 8:364, 8:*375*, 8:388, 8:*389*
Harlequin filefishes. *See* Longnose filefishes
Harlequin frogs, **6:183–195,** 6:*184*, 6:*186–187*, 6:*189*
Harlequin ghost pipefishes. *See* Ornate ghost pipefishes
Harlequin poison frogs, 6:*202*, 6:205–206
Harlequin shrimps, 2:*202*, 2:*207*–208
Harlequin stink bugs, 3:*261*
Harlequin toads. *See* Harlequin frogs
Harlequins, 4:*303*, 4:*316*
Harlow, Harry, 12:151–152
Harney, B. A., 16:200
Harp seals, 14:*421*, 14:*422*, 14:424, 14:*426*, 14:*429*–430
Harpactes diardii. *See* Diard's trogons
Harpactes duvaucelii. *See* Scarlet-rumped trogons
Harpactes kasumba. *See* Red-naped trogons
Harpactes oreskios. *See* Orange-breasted trogons
Harpactes orrhophaeus. *See* Cinnamon-rumped trogons
Harpactes wardi. *See* Ward's trogons
Harpactes whiteheadi. *See* Whitehead's trogons
Harpacticoida, 2:299, 2:300
Harpadontidae, 4:431
Harpagiferidae. *See* Spiny plunderfishes
Harpagornis moorei. *See* Eagles; Haast's eagles
Harpedona spp., 3:55
Harpia harpyja. *See* Harpy eagles
Harpiocephalus spp., 13:519
Harpiocephalus harpia. *See* Harpy-headed bats
Harpy eagles, 8:319, 8:*326*, 8:*340–341*
Harpy fruit bats, 13:*339*, 13:*341*, 13:345
Harpy-headed bats, 13:*523*, 13:524
Harpy-winged bats. *See* Harpy-headed bats
Harpyionycteris whiteheadi. *See* Harpy fruit bats
Harrier hawks, 8:318, 8:319, 8:320
Harris, T., 10:425
Harrison, Ian J., 5:83, 5:94
Harris's antelope squirrels, 16:*150*, 16:*154*, 16:156
Harris's hawks, 8:322, 8:*325*, 8:338–339
Harris's olingos, 14:310, 14:316*t*
Harrovia longipes, 2:32, 2:*33*

Hartebeests, 15:139, **16:27–43,** 16:*30,* 16:*42t*
Hartenberger, J. L., 16:311
Hartert, Ernst, 10:505
Hartlaub's turacos, 9:300, 9:302, 9:*303,* 9:*307*
Hartman, Daniel, 15:191
Hartmann's mountain zebras, 15:222
Hartmeyeria triangularis, 1:*457,* 1:*462,* 1:465
Hartweg's spike-thumb frogs, 6:*231,* 6:238
Harvest mice, 16:*254,* 16:*255,* 16:258
 plains, 16:*264*
 saltmarsh, 16:*265*
 Sumichrast's, 16:*269*
Harvest spiders. *See* Common harvestmen; Long-bodied cellar spiders
Harvester ants, 3:60, 3:72, 3:408
Harvey's duikers, 16:85*t*
Hatchetfishes, 4:336, 4:*339–340,* 4:*349,* 4:421–423, 4:*425,* 4:426–427
Hatching. *See* Reproduction
Hatshepsut (Queen of Egypt), 12:203
Hawaii amakihis, 11:*343*
Hawaiian acorn worms, 1:*444,* 1:*446,* 1:*447,* 1:448–449
Hawaiian anthias, 4:*5*
Hawaiian crows, 8:27, 11:508
Hawaiian ducks, 8:364, 8:367
Hawaiian geese, 8:366, 8:367, 8:370, 8:371
Hawaiian geometrids, 3:387
Hawaiian hoary bats, 12:188–192
Hawaiian honeycreepers, 10:172, **11:341–352,** 11:*345,* 12:188
 behavior, 11:342
 conservation status, 11:343–344
 distribution, 11:*341*–342
 evolution, 11:341
 feeding ecology, 11:*342*–343
 habitats, 11:342
 humans and, 11:344
 physical characteristics, 11:341
 reproduction, 11:*343*
 species of, 11:*346–351*
 taxonomy, 11:341
Hawaiian Islands, introduced species, 12:188–192
Hawaiian monk seals, 12:138, 14:*420,* 14:422, 14:*425,* 14:*429,* 14:431–432
Hawaiian morwongs, 5:*245,* 5:*247*–248
Hawaiian squirrelfishes, 5:*115,* 5:116
Hawfinches, 11:324, 11:*327,* 11:*334*–335
Hawk-eagles, 8:321
Hawk moths, 9:445
  death's head, 3:*390,* 3:*398,* 3:402–403
Hawk owls. *See* Northern hawk-owls
Hawkfishes, 4:42, 4:43, 4:45–48, 4:63, 5:235–240, 5:242–244
Hawks, **8:317–341**
 behavior, 8:320–321
 buzzard-like, 8:318
 conservation status, 8:323
 distribution, 8:*317,* 8:319–320
 evolution, 8:317–318
 feeding ecology, 8:321–322
 habitats, 8:320
 humans and, 8:323–324
 pesticides, 8:26
 physical characteristics, 8:318–319
 reproduction, 8:321, 8:322–323
 species of, 8:*329–341*
 taxonomy, 8:317–318
Hawksbills, 7:*66*
Hayes, Cathy, 12:160
Hayes, Keith, 12:160
Hazel dormice, 16:318, 16:*319,* 16:*320,* 16:321, 16:*322,* 16:*323,* 16:326
Hazel grouse, 8:434
Head lice, human, 3:252, 3:*253,* 3:*255*–256
Headlands, conservation, 3:90
Health care, in zoos, 12:209–211
Hearing
 avian, 8:6, 8:39–40, 9:337
 fishes, 4:61
 mammals, 12:5–6, 12:8–9, 12:50, 12:72, 12:77, 12:*81,* 12:81–82
 See also Ears; Physical characteristics
Heart-nosed bats, 13:314, 13:*380,* 13:381–382
Heart urchins, 1:401, 1:403, 1:404, 1:405, 1:*406,* 1:*409,* 1:413
 See also Sea biscuits
Hearts, 7:160–161, 12:8, 12:36, 12:45, 12:*46,* 12:60
 See also Physical characteristics
Heartworms, canine, 1:*296,* 1:299
Heat loss, 12:51, 12:59–60
 See also Thermoregulation
Heath bush-crickets, 3:207
Heathwrens, 11:56
Heavy metals, pollution from, 1:25
Hecht, Max K., 7:16
Hectopsylla spp., 3:350
Hector's beaked whales, 15:70*t*
Hector's dolphins, 15:43, 15:*45,* 15:48
Hedge accentors. *See* Dunnocks
Hedge sparrows, **10:459–464,** 10:*461*
Hedgehog tenrecs, 13:199
Hedgehogs, 12:*185,* 13:193, 13:195–198, 13:200, **13:203–214,** 13:212*t*–213*t*
Hediste diversicolor. See Ragworms
Hedonic receptors, in reptiles, 7:42
Hedydipna spp. *See* Sunbirds
Hedydipna collaris. See Collared sunbirds
Hedydipna metallica. See Nile Valley sunbirds
Hedydipna pallidigaster. See Amani sunbirds
Hedydipna platura. See Pygmy sunbirds
Heel flies, 3:76
Heel-walkers, **3:217–220**
Heemstra, P. C., 5:123
Heermann's kangaroo rats, 16:201, 16:*204,* 16:*207,* 16:208
Hegner, Robert, 10:40
Heinrich's nightjars, 9:370, 9:405
Heinroth, Katharina, 9:255
Heinroth, Oskar, 9:255
Helaeomyia petrolei. See Petroleum flies
Helarctos malayanus. See Malayan sun bears
Heleioporus spp., 6:140, 6:141
Heleioporus australiacus, 6:141
Helena's parotias. *See* Lawes's parotias
Helenodora inopinata, 2:109
Heleophryne hewitti, 6:132
Heleophryne natalensis. See Natal ghost frogs
Heleophryne orientalis, 6:131
Heleophryne purcelli. See Purcell's ghost frogs
Heleophryne rosei. See Rose's ghost frogs
Heleophrynidae. *See* Ghost frogs
Helfman, G. S., 4:390, 5:365
Heliangelus spp. *See* Hummingbirds
Heliangelus regalis. See Royal sunangels
Heliangelus zusii. See Bogotá sunangels
Helicellid land snails, 2:414–415
Helicinidae, 2:439, 2:440, 2:441, 2:447
Heliconia psittacorum. See Costa Rican heliconias
Heliconines, 3:388
Heliconius charitonia. See Zebra butterflies
Helicophagus hypophthalmus. See Iridescent shark-catfishes
Helicoplacoids, 1:13
Helicops spp. *See* Neotropical watersnakes
Helicoradomenia juani, 2:*381,* 2:*382,* 2:383
Helicoverpa zea. See Corn earworms
Heliobolus spp., 7:297
Heliobolus lugubris. See Kalahari sand lizards
Heliodoxa spp. *See* Hummingbirds
Heliolais erythroptera. See Red-winged warblers
Heliomaster spp. *See* Hummingbirds
Heliomaster longirostris. See Long-billed starthroats
Heliopais personata. See Masked finfoot
Heliophobius spp. *See* Silvery mole-rats
Heliophobius argenteocinereus. See Silvery mole-rats
Helioporacea. *See* Blue corals
Heliornis fulica. See Sungrebes
Heliornis senegalensis. See African finfoot
Heliornithidae, 9:1–4, 9:45
Heliosciurus spp., 16:163
Heliosciurus rufobrachium, 16:165
Heliothryx spp., 9:443, 9:446
Heliothryx barroti. See Purple-crowned fairies
Heliozelidae, 3:384
Helix spp. *See* Lung-bearing snails
Helix aspersa, 2:*411*
Helix pomatia. See Roman snails
Hellbenders, 6:31, 6:49, 6:323, **6:343–347,** 6:*344*
Hellgrammites, 3:289
 See also Eastern dobsonflies
Helmet fleas, 3:*351,* 3:353, 3:*354*
Helmet gurnards. *See* Flying gurnards
Helmet-shrikes, 10:425–430, 10:*432*
Helmet vangas, 10:439, 10:440, 10:441
Helmetcrests, 9:442
Helmeted curassows. *See* Northern helmeted curassows
Helmeted guineafowl, 8:*426,* 8:426–428, 8:*429,* 8:*430*
Helmeted honeyeaters, 11:244
Helmeted hornbills, 10:72, 10:75–76, 10:77, 10:*80*–81
Helmeted mynas, 11:410, 11:*412,* 11:*415*
Helmeted terrapins. *See* Helmeted turtles
Helmeted turtles, 7:*131,* 7:*132*
Helmeted water toads, 6:*161,* 6:170
Helmeted woodpeckers, 10:150
Helminthophis spp., 7:369, 7:370, 7:371
Helminthophis flavoterminatus, 7:369–370, 7:371
Helminths, 1:34
Helmitheros vermivorus. See Worm-eating warblers
Helobdella spp., 2:76
Heloderma spp., 7:353
Heloderma horridum. See Mexican beaded lizards
Heloderma suspectum. See Gila monsters
Helodermatidae, 7:206, **7:353–358,** 7:*356*
Heloecius cordiformis. See Semaphore crabs
Helogale spp., 14:347
Helogale parvula. See Dwarf mongooses
Helohyus spp., 15:265

Helostoma spp., 5:427, 5:428
Helostoma temminckii. See Kissing gouramies
Helostomatidae, 4:57, 5:427
Helpers, in social systems, 12:146–147
  See also Behavior
Hemachatus haemachatus. See Rinkhal's cobras
Hemerobiidae, 3:305, 3:306, 3:307, 3:309
Hemerobiiformia, 3:305
Hemerodromus cinctus. See Three-banded coursers
Hemibelideus spp., 13:113
Hemibelideus lemuroides. See Lemuroid ringtails
Hemicentetes spp., 13:225, 13:226, 13:227–228
Hemicentetes nigriceps. See White-streaked tenrecs
Hemicentetes semispinosus. See Yellow streaked tenrecs
Hemichordates, **1:443–450**, 1:*446*
  behavior, 1:444
  conservation status, 1:445
  distribution, 1:444
  evolution, 1:443
  feeding ecology, 1:444–445
  habitats, 1:444
  humans and, 1:46, 1:445
  physical characteristics, 1:443–444
  reproduction, 1:22, 1:445
  species of, 1:*447*–450
  taxonomy, 1:13, 1:443
Hemichromis spp., 5:279
Hemidactyliini, 6:390
Hemidactylium spp., 6:35
Hemidactylium scutatum. See Four-toed salamanders
Hemidactylus frenatus. See House geckos
Hemidactylus turcicus. See Mediterranean geckos
Hemidasys agaso, 1:270
Hemiechinus auritus. See Long-eared hedgehogs
Hemiechinus collaris. See Collared hedgehogs
Hemiechinus nudiventris, 13:207
Hemieuryalidae, 1:387
Hemigaleidae. *See* Weasel sharks
Hemigalinae, 14:335
Hemigalus derbyanus. See Banded palm civets
Hemignathus spp. *See* Amakihis
Hemignathus lucidus. See Maui nukupuus
Hemignathus virens. See Hawaii amakihis
Hemilepidotus spp., 5:183
Hemilepidotus hemilepidotus. See Red Irish lords
Hemilepistus reaumuri, 2:250, 2:251
Hemilobophasma spp., 3:217
Hemimerina, 3:195, 3:196
Hemimetabolous insects, 3:34
Hemimysis margalefi, 2:*218*, 2:*219*, 2:220–221
Heminectes rufus. See Chile Darwin's frogs
Hemiodontidae, 4:335
Hemiphaga spp., 9:248
Hemiphractinae, 6:225, 6:227
Hemiphractus spp., 6:37, 6:66, 6:226–227, 6:229, 6:230, 6:308
Hemiphractus helioi. See Horned treefrogs
Hemiphractus johnsoni, 6:32
Hemiphractus proboscideus. See Sumaco horned treefrogs
Hemipodius melanogaster. See Black-breasted buttonquails
Hemipodius pyrrhothorax. See Red-chested buttonquails
Hemiprocne comata. See Whiskered tree swifts
Hemiprocne coronata. See Crested tree swifts
Hemiprocne longipennis. See Gray-rumped tree swifts
Hemiprocne mystacea. See Moustached tree swifts
Hemiprocnidae. *See* Tree swifts
Hemiptera, 3:31, 3:33–34, **3:259–280**, 3:*265*, 3:*266*
  behavior, 3:261
  conservation status, 3:263
  distribution, 3:261
  eusocial, 3:68
  evolution, 3:259
  feeding ecology, 3:20, 3:262
  habitats, 3:261
  humans and, 3:263–264
  physical characteristics, 3:260–261
  reproduction, 3:31, 3:44, 3:262–263
  species of, 3:*267*–280
  taxonomy, 3:259–260
  wing-beat frequencies, 3:23
Hemipus spp. *See* Flycatcher-shrikes
Hemipus hirundinaceus. See Black-winged flycatcher-shrikes
Hemiramphidae. *See* Halfbeaks
Hemiramphus far. See Blackbarred halfbeaks
Hemirhamphidae, 4:47, 4:67
Hemirhamphodon spp., 5:81
Hemiscyllidae. *See* Longtail carpet sharks
Hemisotidae. *See* Shovel-nosed frogs
Hemisquilla ensigera. See Giant mantis shrimps
Hemistragus spp., 16:87
Hemistragus hylocrius. See Nilgiri tahrs
Hemistragus jayakari. See Arabian tahrs
Hemisus spp., 6:35, 6:67
Hemisus guttatus. See Spotted snout-burrowers
Hemisus sudanensis. See Marbled snout-burrowers
Hemitaurichthys polycanthus, 5:238
Hemitaurichthys polylepis, 5:241
Hemithyris psittacea. See Black lampshells
Hemitilapia oxyrhynchus. See Giant haplochromises
Hemitriccus furcatus. See Fork-tailed pygmy tyrants
Hemitriccus kaempferi. See Kaempfer's tody tyrants
Hemitripteridae, 5:179
Hemiuridae, 1:200
Hemocoel, 3:24
Hemocytes, 3:24
Hemoglobin, 5:326, 12:30
  See also Physical characteristics
Hemolymph, 3:24
Hemotoxins, 7:52
Hemprich's hornbills, 10:73
Hen harriers, 8:*318*, 8:*326*, 8:336–337
Henderson, Robert W., 7:410
Henfishes. *See* Lumpfishes
Henicopernis longicauda. See Long-tailed buzzards
Henicorhina spp., 10:527
Henicorhina leucophrys. See Gray-breasted wood wrens
Henicorhina levcoptera. See Bar-winged wood wrens
Heniochus diphreutes. See Bannerfishes
Henry, J. Q., 1:180
Henst's goshawks, 8:323
Hepatoxylidae, 1:226
Hephaestus spp., 5:220
Hephaestus adamsoni. See Adamson's grunters
Hephaestus fuliginosus. See Sooty grunters
Hepialidae, 3:389
Hepsetidae, 4:335, 4:338
Hepsetus odoe. See African pikes
Heptapteridae, 4:352
Heptaxodontidae. *See* Giant hutias
Heptranchias perlo. See Sharpnose sevengill sharks
Heraldia spp., 5:135
Herbert, Thomas, 9:270
Herbert River ringtails, 13:117, 13:*118*, 13:*121*–122
Herbivores, 2:25, 3:46, 3:62–63
  digestive system, 12:120–124
  distribution, 12:135, 12:136
  in guilds, 12:118–119
  migrations, 12:164–166, 12:164–168
  periglacial environments and, 12:25
  subterranean, 12:69, 12:76
  teeth, 12:14–15, 12:37, 12:46
  See also Feeding ecology; Ungulates; specific herbivores
Herbivorous fishes, 4:67–68
  See also specific fishes
Herbivory, 7:27, 7:204
  See also Feeding ecology
Hercules beetles, 3:317, 3:*326*, 3:*330*, 3:331
Herding dogs, 14:288–289, 14:292
Herichthys carpintis. See Tampico cichlids
Herlings. *See* Brown trouts
Hermann's tortoises, 7:*145*, 7:*148*–149
Hermaphroditism, 1:17, 1:21–23, 2:16, 2:23, 4:25–27, 4:29–30, 4:45
  See also Reproduction; specific species
Hermit crabs, 2:197, 2:198, 2:201
Hermit hummingbirds, 9:446
Hermit ibises, 8:293, 8:*295*, 8:*297*–*298*
Hermit thrushes, 10:484, 10:485, 10:*491*, 10:*501*
Hermit warblers, 11:288
Hermits, 9:415, 9:438–*439*, 9:*441*–442, 9:446–449, 9:*454*–455, 11:286
Hernings. *See* Atlantic herrings
Herns. *See* Atlantic herrings
Hero shrews. *See* Armored shrews
Herodotius pattersoni, 16:517
Herodotus, 14:290
Herons, 8:233–236, 8:*234*, **8:239–260**
  behavior, 8:234–235, 8:242–243
  conservation status, 8:236, 8:244–245
  distribution, 8:*239*, 8:241
  evolution, 8:239
  feeding ecology, 8:235, 8:242, 8:243–244
  flight, 8:*234*
  habitats, 8:241–242
  humans and, 8:245
  pesticides, 8:26
  physical characteristics, 8:233–234, 8:239–241
  reproduction, 8:234, 8:236, 8:244
  species of, 8:*248*–*259*
  taxonomy, 8:239
Herpailurus yaguarondi. See Jaguarundis
Herpestes spp., 14:347, 14:348–349
Herpestes auropunctatus. See Indian mongooses
Herpestes edwardsii. See Indian gray mongooses
Herpestes ichneumon. See Egyptian mongooses

Herpestes javanicus. See Small Indian mongooses
Herpestes nyula. See Indian mongooses
Herpestes palustris. See Bengali water mongooses
Herpestes pulverulentus. See Cape gray mongooses
Herpestes sanguinus. See Slender mongooses
Herpestidae. *See* Fossa; Mongooses
Herpestinae, 14:347, 14:349
Herpetocypris reptans, 2:*311*
Herpetotheres cachinnans. See Laughing falcons
Herpsilochmus atricapillus. See Black-capped antwrens
Herpsilochmus parkeri. See Ash-throated antwrens
Herrel, Anthony, 7:347
Herring gulls, 9:*206*, 9:209, 9:*210*, 9:*211*–212
Herring jacks. *See* American shads
Herring smelts, 4:390
Herring whales. *See* Fin whales
Herrings, 4:57, **4:277–288,** 4:*278*, 4:*280*
  behavior, 4:277–278
  conservation, 4:278
  distribution, 4:277
  evolution, 4:277
  feeding ecology, 4:67, 4:69, 4:278
  habitats, 4:277
  humans and, 4:278–279
  physical characteristics, 4:277
  reproduction, 4:278
  species of, 4:281–288
  taxonomy, 4:12, 4:277
  See also Reef herrings
Herrons. *See* Atlantic herrings
Hershkovitz, P.
  on *Callicebus* spp., 14:143
  on capuchins, 14:101
  on squirrel monkeys, 14:101
Hershkovitz's night monkeys, 14:135, 14:138, 14:141*t*
Hesperids, 3:386
Hesperoperla pacifica. See Golden stones
Heteractinida, 1:57
Heteractis spp. *See* Sea anemones
Heteractis magnifica. See Magnificent sea anemones
Heteralocha acutirostris. See Huias
Heterenchelyidae, 4:255
Heterixalus spp., 6:280, 6:281
Heterixalus betsileo. See Betsileo reed frogs
Heterobathmiidae, 3:55
Heterocephalus spp. *See* Naked mole-rats
Heterocephalus glaber. See Naked mole-rats
Heteroclinus roseus. See Rosy weedfishes
Heterocypris salinus, 2:*313*, 2:314–315
Heterodera spp., 1:36
Heterodon spp. *See* Hog-nosed snakes
Heterodon platirhinos. See Eastern hog-nosed snakes
Heterodonta, 2:451
Heterodontiformes. *See* Bullhead sharks
Heterodontus spp., 4:97
Heterodontus francisci. See California bullhead sharks
Heterodontus galeatus, 4:98
Heterodontus japonicus, 4:98, 4:101, 4:102
Heterodontus mexicanus, 4:98
Heterodontus portusjacksoni. See Port Jackson sharks
Heterodontus quoyi, 4:98
Heterodontus ramalheira, 4:97, 4:98
Heterodontus zebra, 4:98, 4:101, 4:102
Heterodonty, 12:46
Heterodoxus spiniger, 3:250
Heterohyrax spp., 15:177
Heterohyrax antineae. See Ahaggar hyraxes
Heterohyrax brucei. See Bush hyraxes
Heterohyrax brucei antineae, 15:179
Heterohyrax chapini. See Yellow-spotted hyraxes
Heterojapyx gallardi, 3:*109*, 3:*110*–111
Heterokrohniidae, 1:433
Heteromirafra spp. *See* Larks
Heteromirafra archeri. See Archer's larks
Heteromirafra ruddi. See Rudd's larks
Heteromirafra sidamoensis. See Sidamo bushlarks
Heteromyias albispecularis. See Gray-headed robins
Heteromyid rodents, 12:131
  See also specific rodents
Heteromyidae, 16:121, 16:123, **16:199–210,** 16:*204*, 16:208*t*–209*t*
Heteromys spp., 16:199, 16:200, 16:201, 16:203
Heteromys desmarestianus. See Desmarest's spiny pocket mice
Heteromys nelsoni. See Nelson's spiny pocket mice
Heteronemertea, 1:245, 1:246–247
Heteronemiidae, 3:221, 3:222
Heteronetta atricapilla. See Black-headed ducks
Heterophyes heterophyes, 1:200
Heteropoda, 2:446
Heteroptera. *See* True bugs
Heteropteryx dilatata. See Jungle nymphs
Heterorhabditis spp., 1:36, 1:*45*
Heteroscyllium colcloughi. See Blue-gray carpet sharks
Heterostraci, 4:9
Heterotardigrada, 2:115, 2:116, 2:117
Heterothermy, 12:50, 12:59–60, 12:73, 12:113
  See also Thermoregulation
Heterothripidae, 3:281
Heterotis niloticus. See African arawanas
Heterotrophy, 2:11, 2:26
  See also Feeding ecology
Heterurethra. *See* Succineidae
Heuglin's coursers. *See* Three-banded coursers
Heuglin's robins. *See* White-browed robin chats
Hexabranchus sanguineus. See Spanish dancers
Hexacorallia, 1:103
Hexacrobylidae, 1:479, 1:480
Hexacrobylus spp. *See Asajirus* spp.
Hexactinellida. *See* Glass sponges
Hexagenia spp., 3:*127*
Hexagonoporus physeteris, 1:227
Hexagrammidae, 5:179
Hexagrammoidei. *See* Greenlings
Hexagrammos spp., 5:180
Hexagrammos decagrammus. See Kelp greenlings
Hexagrammos stelleri. See Whitespotted greenlings
Hexamerocerata, 2:375, 2:376
Hexanchidae, 4:143–144
Hexanchiformes, 4:11, **4:143–149,** 4:*145*
Hexanchiforms, 4:143
Hexanchus griseus. See Bluntnose sixgill sharks
Hexanchus nakamurai, 4:143
Hexanchus vitulus. See Bigeyed sixgill sharks
Hexapleomera robusta, 2:236
Hexapods. *See* Insects
Hexaprotodon spp., 15:301
Hexaprotodon liberiensis. See Pygmy hippopotamuses
Hexatrygon bickelli. See Sixgill stingrays
Hexatrygonidae, 4:176
Hexolobodon phenax. See Imposter hutias
Hi utsuri, 4:*299*
Hibernation, 12:113, 12:120, 12:169
  See also Behavior; Energy conservation
Hickety turtles. *See* Central American river turtles
Hickory shads. *See* Gizzard shads
Hidden-necked turtles. *See* Cryptodira
Hide beetles, 3:317, 3:321
Hieraaetus pennatus. See Booted eagles
Hieremys annandalii. See Yellow-headed temple turtles
Higgins, Robert P., 1:344
High altitudes, 12:113
High fin banded sharks. *See* Chinese suckers
High mountain voles, 16:225, 16:228
Highbrow crestfishes. *See* Crestfishes
Higher primates, 14:1, 14:2, 14:3, 14:7
  See also specific species
Higher vocal center, 8:39
Highland guans, 8:413
Highland motmots, 10:31, 10:32
Highland tinamous, 8:58, 8:59, 8:*59*, 8:*60*, 8:*62*
Highland wood wrens. *See* Gray-breasted wood wrens
Highland yellow-shouldered bats, 13:433*t*
Highveld mole-rats, 16:350*t*
Hihis. *See* Stitchbirds
Hildebrandtia spp., 6:247, 6:248, 6:249
Hildebrandt's horseshoe bats, 13:387, 13:389, 13:391, 13:*393*
Hildegarde's tomb bats, 13:*357*
Hill, J. E., 16:199, 16:211
Hill long-tongued fruit bats. *See* Greater long-tongued fruit bats
Hill mynas, 11:*412*, 11:*417*–418
Hill-partridges, 8:399, 8:434
Hill swallows. *See* House swallows
Hill wallaroos. *See* Common wallaroos
Hillstars, 9:438, 9:442, 9:444, 9:447, 9:450
Hillstream loaches, 4:322, 4:*324*, 4:*326*, 4:*328*
Hilty, Steven L., 10:130
Himalayan brown owls. *See* Asian brown wood-owls
Himalayan gorals, 16:87, 16:*96*, 16:98–*99*
Himalayan griffon vultures, 8:313, 8:316, 8:319
Himalayan marmots, 16:*146*
Himalayan monals, 8:435
Himalayan musk deer, 15:335, 15:*339*, 15:*340*
Himalayan pikas, 16:*494*, 16:501*t*
Himalayan pipistrelles, 13:*503*
Himalayan quails, 8:437
Himalayan snow bears. *See* Brown bears
Himalayan tahrs, 12:137, 16:90, 16:91, 16:92, 16:93, 16:*94*
Himalayan water shrews, 13:*253*, 13:*256*, 13:258–259

Himalayan wood-owls. *See* Asian brown wood-owls
Himantopus spp. *See* Stilts
Himantopus himantopus. See Black-winged stilts
Himantopus mexicanus. See Black-necked stilts
Himantopus novaezelandiae. See Black stilts
Himantornis haematopus. See Nkulengu rails
Himantornithinae, 9:45
Himantura chaophraya, 4:178
Himatione sanguinea. See Apapanes
Hind limbs, bats, 12:57
See also Bats
Hindgut, 3:20–21
See also Physical characteristics
Hindgut fermentation, 12:14–15, 12:48
Hinduism, 2:42, 7:56
See also Humans
Hingemouths. *See* African mudfishes
Hinnies, 12:177, 15:223
Hiodon spp., 4:232, 4:233
Hiodon alosoides. See Goldeyes
Hiodon tergisus. See Mooneyes
Hiodontidae. *See* Mooneyes
Hip pocket frogs, 6:36, 6:*150*, 6:151–152
Hipparionid horses, 15:137
Hippelates spp. *See* Eye gnats
Hippichthys spp., 5:134
Hippocamelus spp., 15:382
Hippocamelus antisensis. See Peruvian huemuls
Hippocamelus bisulcus. See Chilean huemuls
Hippocampus spp., 4:14
Hippocampus abdominalis. See Big-bellied seahorses
Hippocampus barbouri, 5:134
Hippocampus breviceps. See Short-nosed seahorses
Hippocampus capensis. See Cape seahorses
Hippocampus erectus. See Lined seahorses
Hippocampus kuda. See Yellow seahorses
Hippoglossus hippoglossus, 5:451
Hippoglossus stenolepis. See Pacific halibuts
Hippolais spp., 11:3, 11:4
Hippolais icterina, 11:5
Hippolais languida, 11:4
Hippolais polyglotta. See Melodious warblers
Hippopodius hippopus, 1:28
Hippopotamidae. *See* Hippopotamuses
Hippopotamus spp., 15:301
Hippopotamus amphibius. See Common hippopotamuses
Hippopotamus leeches, 2:76, 2:*79*, 2:*80*, 2:82–83
Hippopotamus lemerlei, 12:136, 15:302
Hippopotamuses, 12:*62*, 12:*116*, **15:301–312,** 15:*302*, 15:*304*, 15:*309*
behavior, 15:306
cetaceans and, 12:30
conservation status, 15:308
distribution, 12:135, 15:*301*, 15:306
evolution, 15:136, 15:137, 15:264, 15:267, 15:304
feeding ecology, 15:271, 15:306, 15:308
habitats, 15:269, 15:306
jacanas and, 9:109
physical characteristics, 12:68, 15:138, 15:139, 15:267, 15:304–306
reproduction, 12:91, 15:308
species of, 15:310–311
taxonomy, 15:301–302, 15:304
Hippopotamyrus pictus, 4:233
Hippopus spp., 2:454
Hippos. *See* Palette tangs
Hipposideridae. *See* Old World leaf-nosed bats
Hipposideros spp., 13:401, 13:402, 13:405
Hipposideros ater. See Dusky leaf-nosed bats
Hipposideros bicolor, 13:*404*
Hipposideros caffer. See Sundevall's roundleaf bats
Hipposideros cineraceus. See Ashy roundleaf bats
Hipposideros commersoni. See Commerson's leaf-nosed bats
Hipposideros diadema. See Diadem roundleaf bats
Hipposideros fulvus. See Fulvous roundleaf bats
Hipposideros larvatus. See Roundleaf horseshoe bats
Hipposideros ridleyi. See Ridley's roundleaf bats
Hipposideros ruber. See Noack's roundleaf bats
Hipposideros stenotis. See Northern leaf-nosed bats
Hipposideros terasensis, 13:405
Hippotraginae, 16:1, **16:27–44,** 16:*35–36*, 16:42*t*–43*t*
Hippotragini, 16:27, 16:28, 16:29, 16:31, 16:34
Hippotragus spp., 16:27, 16:28
Hippotragus equinus. See Roan antelopes
Hippotragus leucophaeus. See Bluebucks
Hippotragus niger. See Sable antelopes
Hippotragus niger variani. See Giant sable antelopes
Hiranetis braconiformis. See Assassin bugs
Hirolas. *See* Hunter's hartebeests
Hirudinea. *See* Leeches
Hirudinidae. *See* Medicinal leeches
Hirudo spp. *See* Leeches
Hirudo medicinalis. See European medicinal leeches
Hirundapus spp., 9:418, 9:424
Hirundapus caudacutus. See White-throated needletails
Hirundapus celebensis. See Purple needletails
Hirundinidae. *See* Martins; Swallows
Hirundo apus. See Common swifts
Hirundo atrocaerulea. See Blue swallows
Hirundo caudacuta. See White-throated needletails
Hirundo coronata. See Crested tree swifts
Hirundo fuciphaga. See Edible-nest swiftlets
Hirundo megaensis. See White-tailed swallows
Hirundo melba. See Alpine swifts
Hirundo nigra. See Black swifts
Hirundo pelagica. See Chimney swifts
Hirundo perdita. See Red Sea cliff-swallows
Hirundo pratinocola. See Collared pratincoles
Hirundo pyrrhonota. See American cliff swallows
Hirundo rustica. See Barn swallows
Hirundo tahitica. See House swallows
Hirundo zonaris. See White-collared swifts
Hispaniola hooded katydids, 3:*209*, 3:*215*
Hispaniolan crossbills, 11:324
Hispaniolan giant hutias, 16:471*t*
Hispaniolan hutias, 16:*463*, 16:*464*, 16:465–*466*
Hispaniolan palm crows, 11:508
Hispaniolan solenodons, 13:195, 13:198, 13:237, 13:*238*, 13:*239*, 13:240–241
Hispaniolan trogons, 9:479, 9:*480*, 9:483–*484*
Hispid cotton rats, 16:268, 16:*269*, 16:270, 16:*271*, 16:*272*, 16:*275–276*
Hispid hares, 16:482, 16:487, 16:505, 16:506, 16:509, 16:515*t*
Hispid pocket gophers, 16:196*t*
Hispid pocket mice, 16:*204*, 16:*205*
Hispidoberycidae, 5:105
Hispidoberyx ambagiosus. See Hispidoberycidae
Hissing adders. *See* Eastern hog-nosed snakes
Histeridae, 3:318
Histiodraco velifer. See Sailfin plunderfishes
Histiogamphelus spp., 5:133
Histiogamphelus cristatus, 5:133
Histoplasma capsulatum, 8:25
Histoplasmosis, 13:316
Historical ecology, 4:70
The History of Animals (Aristotle), 12:203
Histrio histrio. See Sargassumfishes
Histriobdella homari, 2:46
Histrionicus histrionicus. See Harlequin ducks
Hitchcock, Alfred, 8:22
HIV (human immunodeficiency virus), 12:222
Hoary bats, 13:310, 13:312, 13:497, 13:499, 13:500, 13:501, 13:*505*, 13:*510*–511
Hoary-headed dabchicks. *See* Hoary-headed grebes
Hoary-headed grebes, 8:170, 8:171, 8:*173*, 8:176–177
Hoary marmots, 16:*144*, 16:145, 16:158*t*
Hoary redpolls, 11:331
Hoary-throated spinetails, 10:211
Hoatzins, **8:465–468,** 8:*466*, 8:*467*, 10:326
Hoburogecko suchanovi, 7:259
Hochstetter's frogs, 6:69, 6:*70*, 6:*71*, 6:73–74
Hodgson's broadbills. *See* Silver-breasted broadbills
Hodgson's brown-toothed shrews, 12:*134*, 13:*254*, 13:*256*, 13:259
Hodgson's frogmouths, 9:*380*, 9:*384*
Hodomys alleni. See Allen's woodrats
Hodotermitidae, 3:164, 3:166
Hoeck, H. N., 16:391
l'Hoest's guenons, 14:188, 14:194
Hoffman's two-toed sloths, 13:*148*, 13:*156*, 13:*157*, 13:*158*
Hoffstetterichthys spp., 4:351
Hog deer, 12:19, 15:358, 15:*362*, 15:*364*–365
Hog lice, 3:76
Hog-nosed bats. *See* Kitti's hog-nosed bats; Mexican hog-nosed bats
Hog-nosed snakes, 7:205, 7:467
See also Eastern hog-nosed snakes
Hog slaters. *See* Water lice
Hogchokers, 5:*452*, 5:*454*, 5:455, 5:*456*
Hoge's sideneck turtles, 7:79
Hogfishes, 5:298, 5:*300*, 5:302, 5:*303*
Hoggar hyraxes. *See* Ahaggar hyraxes
Hoglice. *See* Water lice
Hokkaido salamanders, 6:*336*, 6:*338*, 6:340
Holacanthus bermudensis. See Blue angelfishes
Holacanthus ciliaris. See Queen angelfishes
Holarctic kingdom, 3:54, 3:55
Holarctic rails, 9:47
Holarctic wrens. *See* Winter wrens
Holaspis spp., 7:297
Holbrookia maculata. See Common lesser earless lizards
Holdaway, Richard, 11:83–84
Holjapyx diversiunguis, 3:*109*, 3:*110*, 3:111
Holoaden spp., 6:156

Holoarctic true moles, 12:13
Holoblastic cleavage, 2:20, 2:23–24
  See also Reproduction
Holocentridae, 4:48, 4:67, 5:113–116
Holocentrus spp., 5:113
Holocentrus ascensionis. See Squirrelfishes
Holocephalans, 4:10
Holocephali, 4:10, 4:91
Holochilus spp., 16:268
Holochilus chacarius, 16:270
Holometabola, 3:31, 3:34, 3:315–316
Holophagus gulo, 4:192
Holoptychius spp., 4:197
Holothuria (Thymiosycia) thomasi. See Tiger's tail sea cucumbers
Holothuriidae, 1:420, 1:421
Holothuroidea. *See* Sea cucumbers
Holotropical kingdom, 3:54–55
Holotypes, 2:8
Homalopsinae, 7:465–466, 7:468, 7:470
Homalopsis spp. *See* Puff-faced watersnakes
Homalopteridae, 4:321
Homalorhagida, 1:275, 1:276, 1:277
Homeostasis, 4:*23*
  See also Physical characteristics
Homeothermy, 12:50
  See also Thermoregulation
Homing behavior, 5:181–182, 7:43, 8:32–35
  See also Behavior
Homing pigeons, 8:22, 9:439
Hominidae, **14:225–253,** 14:*236*
  behavior, 14:228–232, 14:249–252
  conservation status, 14:235
  distribution, 14:*225*, 14:227, 14:*241*, 14:247
  evolution, 14:225, 14:241–244
  feeding ecology, 14:232–234, 14:247–248
  habitats, 14:227–228, 14:247
  physical characteristics, 12:23–24, 14:225–227, 14:244–247
  reproduction, 14:234–235, 14:248–249
  species of, 14:*237*–240
  taxonomy, 14:4, 14:225, 14:241–243
Homininae, 14:225, 14:241–243, 14:247, 14:248
Hominoids, 14:4, 14:5
Homo spp., 14:242, 14:243
Homo erectus, 12:20, 14:*243*, 14:244, 14:*249*
Homo habilis, 14:*243*
Homo neanderthalensis, 14:*243*–244
Homo rudolfensis, 14:243
Homo sapiens, **14:241–253,** 14:*245*, 14:*246*, 14:*247*, 14:*248*, 14:*250*
  animal intelligence and, 12:150
  behavior, 14:249–252, 14:*251*, 14:*252*
  body size of, 12:19
  desert adaptation of, 12:21
  domestication and commensals, **12:171–181**
  ecosystems and, 12:213
  evolution, 14:241–244, 14:*243*
  feeding ecology, 14:*245*, 14:247–248
  field studies of mammals, **12:194–202**
  habitats, 14:247
  mammalian invasives and pests, **12:182–193**
  physical characteristics, 14:244–247
  reproduction, 12:94, 12:104, 12:106, 12:109, 14:248–249
  taxonomy, 14:1, 14:3, 14:4, 14:241–243
  zoos, **12:203–212**
  See also Humans
Homo sapiens neanderthalensis, 14:243–244
Homoiodorcas spp., 16:1
Homology, 2:7, 12:27–28
Homoptera, 3:65, 3:75, 3:259, 3:260, 3:262–263
  See also Aphids; Cicadas; Scale insects
Homoroselaps spp., 7:483
Homoroselaps lacteus. See Spotted harlequin snakes
Homoscleromorpha, 1:77, 1:79
Homosclerophorida. *See* Homoscleromorpha
Homunculus spp., 14:143
Honey, 3:*75*, 3:79, 3:81
Honey badgers, 10:139, 14:259, 14:319, 14:*322*, 14:332*t*
Honey bears. *See* Kinkajous
Honey buzzards, 8:314
Honey possums, **13:135–138,** 13:*136*, 13:*137*, 13:*138*
  behavior, 13:35, 13:37, 13:136–137
  conservation status, 13:138
  distribution, 13:*135*, 13:136
  evolution, 13:31, 13:135
  feeding ecology, 13:38, 13:137
  habitats, 13:34, 13:35, 13:136
  humans and, 13:138
  physical characteristics, 13:135–136
  reproduction, 13:39, 13:137–138
  taxonomy, 13:135
Honey-pot ants, 3:49, 3:81, 3:408
Honeybees, 2:28, 2:35, 2:37, 3:*64*, 3:405, 3:408, 3:*410*, 3:*411*–412
  honey frame of, 3:*75*
  humans and, 3:79
  physical characteristics, 3:27
  pollination by, 3:72, 3:78, 3:79
  reproduction, 3:61, 3:67
  social structure, 3:49, 3:68, 3:70, 3:72
  workers, 3:*68*
Honeycomb cowfishes, 5:*469*
Honeycomb worms, 2:*48*, 2:*52*, 2:56
Honeycreepers, Hawaiian. *See* Hawaiian honeycreepers
Honeycutt, R. L.
  on Caviidae, 16:389
  on octodonts, 16:433
Honeyeaters, Australian. *See* Australian honeyeaters
Honeyguides, 10:85–88, **10:137–145,** 10:*140*, 11:219
Hood mockingbirds, 10:*467*, 10:*469*, 10:*472*
Hooded cranes, 9:3, 9:24–27
Hooded crows, 11:506
Hooded grebes, 8:171, 8:*173*, 8:179–180
Hooded katydids, Hispaniola, 3:*209*, 3:*215*
Hooded mergansers, 8:364
Hooded orioles, 11:*306*
Hooded pittas, 10:194, 10:195, 10:*196*, 10:*197*–198
Hooded plovers, 9:164
Hooded robins, 11:106
Hooded seals, 12:126, 12:127, 14:417, 14:420, 14:424, 14:*425*, 14:*432*–*433*
Hooded skunks, 14:333*t*
Hooded vultures, 8:320
Hooded warblers, 11:*292*, 11:*297*
Hoofed mammals. *See* Ungulates
Hook-billed bulbuls, 10:399
Hook-billed hermits, 9:443, 9:449
Hook-billed kingfishers, 10:6, 10:*12*, 10:*14*
Hook-billed kites, 8:322, 8:*327*, 8:*330*
Hook-billed vangas, 10:439, 10:440, 10:441, 10:*444*
Hook-lipped rhinoceroses. *See* Black rhinoceroses
Hooker's sea lions, 14:394, 14:395, 14:396, 14:399, 14:400, 14:*401*, 14:*403*–404
Hookworms, 1:34, 1:35, 1:*296*, 1:301
Hooligans. *See* Eulachons
Hoolock gibbons, 14:207–208, 14:209, 14:210, 14:211, 14:212, 14:213, 14:*216*, 14:*218*
Hooper, J. N. A., 1:57
Hoopoe larks. *See* Greater hoopoe-larks
Hoopoes, 8:26, 10:1–*2*, **10:61–63,** 10:*62*
Hooves, 12:39–40, 12:42
  See also Physical characteristics; Ungulates
Hopi snake dance, 7:56–57
Hoplias malabaricus. See Trahiras
Hoplichthyidae. *See* Ghost flatheads
Hoplobatrachus spp., 6:246, 6:248–251
Hoplobatrachus tigerinus. See Indian tiger frogs
Hoplobrotula spp., 5:16
Hoplocarida, 2:170
Hoplocephalus bungaroides. See Broad-headed snakes
Hoplocercinae, 7:243, 7:244
Hoplodactylus delcourti. See Delcourt's giant geckos
Hoplolatilus spp., 5:256
Hoplomys spp., 16:449
Hoplomys gymnurus. See Armored rats
Hoplonemertea. *See* Enoplans
Hoplophryne spp., 6:304
Hoplophryne rogersi, 6:37
Hoplopleura spp., 3:249
Hoplopsyllus spp., 3:349
Hoplosternum spp., 4:351
Hoplostethus atlanticus. See Orange roughies
Horaichthys spp., 5:81
Horizontal gene transfer, 1:31
  See also Reproduction
Hormaphidinae, 3:68
Hormogastridae, 2:65
Hormones, 3:28, 8:39
  See also Physical characteristics; specific species
Horn-faced bees, Japanese, 3:78–79
Horn flies, 3:76
Horn sharks. *See* Bullhead sharks; California bullhead sharks
Hornbill Specialist Group (IUCN), 10:75
Hornbills, 10:1–4, **10:71–84,** 10:77
  behavior, 10:73
  conservation status, 10:75
  evolution, 10:71
  feeding ecology, 10:73–74
  habitats, 10:73
  humans and, 10:75–76
  physical characteristics, 10:71–72
  reproduction, 10:74–75
  species of, 10:*78*–*84*
  taxonomy, 10:71
The Hornbills (Kemp), 10:71
Hornby's storm-petrels, 8:138
Horned beetles, 3:81
Horned coots, 9:50
Horned frogmouths. *See* Sunda frogmouths
Horned grebes, 8:170, 8:171
Horned guans, 8:401, 8:413, 8:*417*, 8:*420*–421
Horned land frogs, 6:*306*, 6:311

Horned larks, 10:342, 10:343, 10:*345*, 10:*347*, 10:*354*, 11:272, 11:273
Horned lizards, 7:204, 7:207, 7:243–247, 7:*249*, 7:*251*, 7:254
Horned puffins, 9:*102*
Horned screamers, 8:393, 8:394, 8:*395*, 8:*396*
Horned toads. *See* Texas horned lizards
Horned treefrogs, 6:*226*
Horned vipers, 7:*449*, 7:*457*–458
Hornelliellidae, 1:226
Horneros. *See* Ovenbirds
Hornets, 3:68, 3:74
Horns, 12:*5*, 12:10, 12:40, 12:*43*, 12:*47*, 12:102, 15:*132*
  See also Antlers
Horntail wasps, 3:75
Horny sponges, 1:80
Horny toads. *See* Texas horned lizards
Horse bot flies, 3:*363*, 3:*368*, 3:370
Horse-faced bats. *See* Hammer-headed fruit bats
Horse flies, 3:*29*, 3:76, 3:357, 3:360, 3:361, 3:*363*, 3:*372*
Horse-like antelopes. *See* Hippotragini
Horse mackerels, symbiosis and, 1:33
Horse racing, 12:*173*, 12:174
Horse-stingers. *See* Odonata
Horseface loaches, 4:*322*, 4:*328*, 4:331
Horsehair worms, 1:21, 1:*306*
Horsehead seals. *See* Gray seals
Horses, **15:225–236,** 15:*230*, 15:*231*
  behavior, 15:142, 15:219, 15:226–228
  conservation status, 15:221, 15:222, 15:229–230
  distribution, 15:217, 15:*225*, 15:226
  domestication, 15:145, 15:146, 15:149
  evolution, 15:136, 15:137, 15:138, 15:215, 15:216, 15:225
  feeding ecology, 15:141, 15:220, 15:228
  gaits of, 12:*39*
  habitats, 15:218, 15:226
  humans and, 15:222, 15:223, 15:230
  locomotion, 12:42, 12:*45*
  neonatal milk, 12:126
  numbers and, 12:155
  as pests, 12:182
  physical characteristics, 15:139, 15:140, 15:217
  reproduction, 12:95, 12:106, 15:221, 15:228–229
  species of, 15:232–236*t*
  taxonomy, 15:136, 15:137, 15:138, 15:215, 15:216, 15:225
  vision, 12:79
Horseshoe bats, 13:310, **13:387–400,** 13:*394*, 13:*395*
  behavior, 13:312, 13:390
  conservation status, 13:392–393
  distribution, 13:*387*, 13:388–389
  echolocation, 12:86
  evolution, 13:387
  feeding ecology, 13:390–391
  humans and, 13:393
  physical characteristics, 13:387–388
  reproduction, 13:315, 13:392
  species of, 13:*396*–400
  taxonomy, 13:387
Horseshoe crabs, **2:327–332,** 2:*330*
  behavior, 2:328
  conservation status, 2:43, 2:328
  distribution, 2:328
  evolution, 2:327
  feeding ecology, 2:328
  habitats, 2:328
  humans and, 2:41, 2:42, 2:328–329
  physical characteristics, 2:327–328, 2:*329*
  reproduction, 2:328
  species of, 2:*331–332*
  taxonomy, 2:327
Horsfield's bronze-cuckoos, 9:*317*, 9:*321*, 11:66
Horsfield's shrews, 13:276*t*
Horsfield's tarsiers. *See* Western tarsiers
Horus (Egyptian god), 8:*22*
Horus swifts, 9:424
Horwich, Robert, 9:28
Hose's broadbills, 10:178, 10:179, 10:*180*, 10:181–*182*
Hose's leaf-monkeys, 14:175, 14:184*t*
Hose's palm civets, 14:344*t*
Hose's pygmy flying squirrels, 16:*138*, 16:*139*, 16:140
Host fishes, 4:50
Host organisms, 2:31, 2:33
  See also Symbiosis
Host species. *See* Brood parasitism
Hot cave bats, 13:415–416
Hotson's mouse-like hamsters, 16:243
Hotspots, 1:50, 12:218
Hottentot buttonquails. *See* Black-rumped buttonquails
Hottentot golden moles, 13:215–216, 13:217, 13:*219*, 13:*220*
Houbara bustards, 9:92, 9:93, 9:*95*, 9:*97*
Houbaropsis spp. *See* Bustards
Houbaropsis bengalensis. *See* Bengal floricans
Houde, P., 9:41
Hound sharks, 4:11, 4:113, 4:115
Hounds, 14:288
Hourglass treefrogs, 6:229, 6:*232*, 6:*236*–237
House bats, 13:516*t*
House-building rats. *See* Greater stick-nest rats
House centipedes, 2:*357*, 2:*361*, 2:362
House crows, 11:505, 11:509, 11:*511*, 11:*519*–520
House finches, 8:31, 11:*325*
House flies, 3:25, 3:57, 3:76, 3:357, 3:361
House geckos, 7:260, 7:263, 7:*264*, 7:*267*
House martins, 10:*362*, 10:*365*–366
House mice, 16:*254*, 16:*255*–*256*
  behavior, 12:*192*
  distribution, 12:137, 16:123
  feral cats and, 12:185
  humans and, 12:173, 16:126, 16:128
  reproduction, 16:125, 16:*251*, 16:*252*
  smell, 12:80
House mynas. *See* Common mynas
House rats. *See* Black rats
House scorpions. *See* Book scorpions
House sparrows, 8:24, 9:440, 10:361, 11:397, 11:398, 11:*400*, 11:*401*
House swallows, 10:*362*, 10:*365*
House wrens, 10:174, 10:525–529, 10:*527*, 10:*530*, 10:*536*–*537*
Household pests, 3:75–76
Houston toads, 6:*187*, 6:*190*
Hover wasps, 3:68
Howella spp., 5:220–221
Howler monkeys, **14:155–169,** 14:*161*
  behavior, 12:*142*, 14:7, 14:157
  conservation status, 14:159–160
  distribution, 14:156
  evolution, 14:155
  feeding ecology, 14:158
  habitats, 14:156–157
  humans and, 14:160
  physical characteristics, 14:*160*
  population indexes, 12:195
  reproduction, 14:158–159
  species of, 14:*162*–163, 14:166*t*–167*t*
  taxonomy, 14:155
  translocation of, 12:224
Hox genes, 2:4–5, 7:24
Huachuca earless lizards. *See* Common lesser earless lizards
Hubbard's angel insects. *See* Hubbard's zorapterans
Hubbard's zorapterans, 3:*239*, 3:240, 3:*241*
Hubb's beaked whales, 15:69*t*
Huchens, 4:405
Hucho hucho. *See* Huchens
Huebner, S., 10:71
Huemuls. *See* Chilean huemuls; Peruvian huemuls
Huet huets. *See* Black-throated huet-huets
Hughes M. S., 8:40
Hugh's hedgehogs, 13:207, 13:213*t*
Huia spp., 6:247
Huias, 8:26, 11:*448*, 11:*450*–*451*
Huitzilopochtli, 9:451
Human blood flukes, 1:35, 1:*197*, 1:199, 1:*202*, 1:*204*, 1:210–211
Human diseases. *See* Diseases
Human fleas, 3:350
Human head/body lice, 3:252, 3:*253*, 3:*255*–256
Human immunodeficiency virus (HIV), 12:222
Human whipworms, 1:*286*, 1:290–291
Humane Society of the United States, 14:291, 14:294
Humans
  amphibians and, **6:51–55**
    anurans
      African treefrogs, 6:283, 6:285–289
      Amero-Australian treefrogs, 6:230, 6:233–242
      Arthroleptidae, 6:266, 6:268–271
      Asian toadfrogs, 6:111, 6:113–117
      Asian treefrogs, 6:293, 6:295–299
      Australian ground frogs, 6:141, 6:143–145
      Bombinatoridae, 6:84, 6:86–88
      brown frogs, 6:53, 6:252, 6:263
      Bufonidae, 6:185, 6:188–194
      Discoglossidae, 6:90, 6:92–94
      ghost frogs, 6:132, 6:133–134
      glass frogs, 6:217, 6:219–223
      leptodactylid frogs, 6:159, 6:162–171
      Madagascan toadlets, 6:318, 6:319–320
      marine toads, 6:51, 6:191
      Mesoamerican burrowing toads, 6:97
      Myobatrachidae, 6:149, 6:151–153
      narrow-mouthed frogs, 6:305, 6:308–316
      New Zealand frogs, 6:70, 6:72–74
      parsley frogs, 6:128, 6:129
      Pipidae, 6:102, 6:104–106
      poison frogs, 6:200, 6:203–209
      Ruthven's frogs, 6:212
      Seychelles frogs, 6:136, 6:137–138

shovel-nosed frogs, 6:275
spadefoot toads, 6:122, 6:124–125
tailed frogs, 6:78, 6:80–81
tailless caecilians, 6:437
three-toed toadlets, 6:180, 6:181–182
true frogs, 6:252, 6:255–263
vocal sac-brooding frogs, 6:174, 6:176
caecilians
American tailed, 6:416, 6:417–418
Asian tailed, 6:420, 6:422–423
buried-eye, 6:432, 6:433–434
Kerala, 6:426, 6:428
tailless, 6:437, 6:439–441
salamanders
amphiumas, 6:407, 6:409–410
Asiatic, 6:337, 6:339–342
Cryptobranchidae, 6:347
lungless, 6:392, 6:395–403
mole, 6:356, 6:358–360
Pacific giant, 6:351, 6:352
Proteidae, 6:379, 6:381–383
Salamandridae, 6:367, 6:370–375
sirens, 6:329, 6:331–332
torrent, 6:386, 6:387
birds and, **8:19–28**, 8:*22*, 8:*23*, 8:*25*
albatrosses, 8:110, 8:116, 8:118–122
Alcidae, 9:222, 9:224–229
anhingas, 8:203, 8:207–209
Anseriformes, 8:368
ant thrushes, 10:242, 10:245–256
Apodiformes, 9:418–419
art, 8:26
asities, 10:188, 10:190
Australian chats, 11:66, 11:67–68
Australian creepers, 11:135, 11:137–139
Australian fairy-wrens, 11:46, 11:48–53
Australian honeyeaters, 11:238, 11:241–253
Australian robins, 11:106, 11:108–112
Australian warblers, 11:57, 11:59–63
babblers, 10:508, 10:511–523
barbets, 10:117, 10:119–122
barn owls, 9:338, 9:340–343
bee-eaters, 10:42
bird domestication, 8:21–22
bird endangerment, 8:26–27
bird songs, 8:19–20
bird species, 8:24–25
birding, 8:22–23
birds of paradise, 11:492, 11:495–502
bitterns, 8:245
Bombycillidae, 10:449, 10:451–454
boobies, 8:215
bowerbirds, 11:481, 11:483–488
broadbills, 10:179, 10:181–186
bulbuls, 10:399, 10:402–413
bustards, 9:94, 9:96–99
buttonquails, 9:14, 9:16–21
Caprimulgiformes, 9:370–371
caracaras, 8:352
cassowaries, 8:77
Charadriidae, 9:164, 9:167–173
Charadriiformes, 9:103–104
chats, 10:489
chickadees, 11:158, 11:160–166
chowchillas, 11:71, 11:72–74
Ciconiiformes, 8:236–237
Columbidae, 9:252, 9:255–266
condors, 8:278
cormorants, 8:186, 8:203, 8:205–206, 8:207–209
Corvidae, 11:509, 11:512–522
cotingas, 10:308, 10:312–322
crab plovers, 9:123
Cracidae, 8:416
cranes, 9:29, 9:31–36
cuckoo-shrikes, 10:387
cuckoos, 9:315, 9:318–329
culture, 8:19
dippers, 10:477, 10:479–482
diseases, 8:25–26
diving-petrels, 8:144, 8:145–146
doves, 9:255–266
drongos, 11:439, 11:441–445
ducks, 8:373
eagles, 8:323–324
elephant birds, 8:104
emus, 8:86
Eupetidae, 11:76, 11:78–81
fairy bluebirds, 10:417, 10:423–424
Falconiformes, 8:316
falcons, 8:352
false sunbirds, 10:188, 10:191
fantails, 11:87, 11:89–95
feather use, 8:26
feeding birds, 8:23–24
finches, 11:325, 11:328–339
fishing, 8:21
flamingos, 8:306
flowerpeckers, 11:191, 11:193–198
frigatebirds, 8:195, 8:197–198
frogmouths, 9:379, 9:381–385
Galliformes, 8:402
gannets, 8:215
geese, 8:373
Glareolidae, 9:153, 9:155–160
grebes, 8:171–172, 8:174–180
guineafowl, 8:429
gulls, 9:211–217
hammerheads, 8:263
Hawaiian honeycreepers, 11:344, 11:346–351
hawks, 8:323–324
hedge sparrows, 10:460, 10:462–464
herons, 8:245
hoatzins, 8:468
honeyguides, 10:138–139, 10:141–144
hoopoes, 10:63
hornbills, 10:75–76, 10:78–84
hummingbirds, 9:451–452, 9:455–467
hunting, 8:21
ibises, 8:293
Icteridae, 11:306, 11:309–322
ioras, 10:417, 10:419–420
jacamars, 10:93
jacanas, 9:110, 9:112–114
kagus, 9:43
kingfishers, 10:10, 10:13–23
kiwis, 8:90
lapwings, 9:164, 9:167–173
Laridae, 9:209, 9:211–217
larks, 10:346, 10:348–354
leafbirds, 10:417, 10:420–423
limpkins, 9:39
literature, 8:26
logrunners, 11:71, 11:72–74
long-tailed titmice, 11:143, 11:145–146
lyrebirds, 10:332, 10:334–335
magpie-shrikes, 11:469, 11:471–475
manakins, 10:297, 10:299–303
mesites, 9:6, 9:8–9
moas, 8:98
monarch flycatchers, 11:98, 11:100–103
motmots, 10:33, 10:35–37
moundbuilders, 8:407
mousebirds, 9:471, 9:473–476
mudnest builders, 11:454, 11:456–458
New World blackbirds, 11:309–322
New World finches, 11:266, 11:269–283
New World quails, 8:458
New World vultures, 8:278
New World warblers, 11:291, 11:293–299
New Zealand wattle birds, 11:448, 11:449–451
New Zealand wrens, 10:204, 10:206–207
nightjars, 9:405–406, 9:408–414
nuthatches, 11:169, 11:171–175
oilbirds, 9:375
Old World flycatchers, 11:27, 11:30–43
Old World warblers, 11:7, 11:10–23
Oriolidae, 11:428, 11:431–434
ostriches, 8:101
ovenbirds, 10:211, 10:214–228
owlet-nightjars, 9:389, 9:391–393
owls, 9:334, 9:338, 9:340–343, 9:351, 9:354–365
oystercatchers, 9:128, 9:130–131
painted-snipes, 9:117, 9:118–119
palmchats, 10:456
pardalotes, 11:202, 11:204–206
parrots, 9:280, 9:283–298
Pelecaniformes, 8:186
pelicans, 8:186, 8:227–228
penduline titmice, 11:149, 11:151–153
penguins, 8:153–157
pet birds, 8:22
Phasianidae, 8:437
pheasants, 8:437
Philippine creepers, 11:184, 11:186–187
Picidae, 10:150–151, 10:154–167
Piciformes, 10:88
pigeons, 9:252, 9:255–266
pipits, 10:376, 10:378–383
pittas, 10:195, 10:197–201
plantcutters, 10:326, 10:327–328
plovers, 9:167–173
potoos, 9:397, 9:399–400
pratincoles, 9:155–160
Procellariidae, 8:123, 8:129–130
Procellariiformes, 8:110
pseudo babblers, 11:128
puffbirds, 10:104, 10:106–111
rails, 9:52, 9:57–67
Raphidae, 9:271, 9:272–273
Recurvirostridae, 9:135, 9:137–141
religious symbols, 8:26
rheas, 8:71
rollers, 10:53, 10:55–58
sandgrouse, 9:233, 9:235–238
sandpipers, 9:180, 9:182–188
screamers, 8:394
scrub-birds, 10:338, 10:339–340
secretary birds, 8:345
seedsnipes, 9:191, 9:193–195
seriemas, 9:86, 9:88–89
sharpbills, 10:293
sheathbills, 9:199, 9:200–201
shoebills, 8:288
shrikes, 10:429–430, 10:432–438

sparrows, 11:399, 11:401–406
spoonbills, 8:293
storks, 8:268
storm-petrels, 8:138, 8:140–141
Struthioniformes, 8:55
Sturnidae, 11:411, 11:414–425
sunbirds, 11:210, 11:213–225
sunbitterns, 9:75
sungrebes, 9:70, 9:71–72
swallows, 10:361, 10:363–369
swans, 8:373
swifts, 9:424, 9:426–431
tapaculos, 10:259–260, 10:262–267
terns, 9:211–217
thick-knees, 9:146, 9:147–148
thrushes, 10:489
tinamous, 8:59
titmice, 11:158, 11:160–166
todies, 10:28, 10:29–30
toucans, 10:128, 10:130–135
tree swifts, 9:434, 9:435–436
treecreepers, 11:178, 11:180–181
trogons, 9:479, 9:481–485
tropicbirds, 8:186, 8:189, 8:190–191
trumpeters, 9:79, 9:81–82
turacos, 9:302, 9:304–310
typical owls, 9:351, 9:354–365
tyrant flycatchers, 10:275, 10:278–288
vanga shrikes, 10:442, 10:443–444
Vireonidae, 11:256, 11:258–262
wagtails, 10:376, 10:378–383
weaverfinches, 11:357, 11:360–372
weavers, 11:379, 11:382–394
whistlers, 11:118, 11:120–126
white-eyes, 11:230, 11:232–233
woodcreepers, 10:230, 10:232–236
woodhoopoes, 10:66, 10:68–69
woodswallows, 11:460, 11:462–465
wrens, 10:529, 10:531–538
fishes and, **4:72–76**, 4:*73*, 4:*74*
Acanthuroidei, 5:393, 5:397–403
Acipenseriformes, 4:215, 4:217–220
Albuliformes, 4:250, 4:252–253
angelsharks, 4:162, 4:164–165
anglerfishes, 5:51, 5:53–56
Anguilliformes, 4:258, 4:262–269
Atheriniformes, 5:71, 5:73–76
Aulopiformes, 4:434, 4:436–439
Australian lungfishes, 4:199
beardfishes, 5:2, 5:3
Beloniformes, 5:81, 5:83–86
Beryciformes, 5:116, 5:118–121
bichirs, 4:210, 4:211–212
blennies, 5:343, 5:345–348
bowfins, 4:230
bullhead sharks, 4:99, 4:101–102
Callionymoidei, 5:367, 5:370–371
carps, 4:303, 4:308–313
catfishes, 4:354, 4:357–367
characins, 4:339, 4:342–350
chimaeras, 4:92, 4:94–95
coelacanths, 4:193
Cypriniformes, 4:303, 4:307–319, 4:322, 4:325–333
Cyprinodontiformes, 5:94, 5:96–102
dogfish sharks, 4:153, 4:156–158
dories, 5:124, 5:126–130
eels, 4:258, 4:262–269
electric eels, 4:371, 4:373–377
elephantfishes, 4:234
Elopiformes, 4:244, 4:247
Esociformes, 4:381, 4:383–386
flatfishes, 5:453, 5:455–460, 5:462–463
Gadiformes, 5:30, 5:33–40
gars, 4:222, 4:224–227
Gasterosteiformes, 5:136, 5:139–141, 5:143–147
gobies, 5:377, 5:380–388
Gobiesocoidei, 5:357, 5:359–363
Gonorynchiformes, 4:291, 4:293–295
ground sharks, 4:116, 4:120–128
Gymnotiformes, 4:371, 4:373–377
hagfishes, 4:79, 4:80–81
herrings, 4:278–279, 4:281–288
Hexanchiformes, 4:144, 4:146–148
Labroidei, 5:281, 5:284–290, 5:298, 5:301–307
labyrinth fishes, 5:429, 5:431–434
lampreys, 4:85–86, 4:88–90
Lampridiformes, 4:450, 4:453–454
lanternfishes, 4:443, 4:445–446
lungfishes, 4:203, 4:205–206
mackerel sharks, 4:133, 4:136–141
minnows, 4:303, 4:314–315
morays, 4:258, 4:265–267
mudminnows, 4:381, 4:383–386
mullets, 5:60, 5:62–65
Ophidiiformes, 5:18, 5:21–22
Orectolobiformes, 4:105–106, 4:109–111
Osmeriformes, 4:393, 4:395–401
Osteoglossiformes, 4:234, 4:237–241
Percoidei, 5:199, 5:202–208, 5:212, 5:214–217, 5:222, 5:225–233, 5:243–244, 5:247–253, 5:263, 5:267–268, 5:270–273
Percopsiformes, 5:6, 5:8–12
pikes, 4:381, 4:383–386
ragfishes, 5:353
Rajiformes, 4:178–179, 4:182–188
rays, 4:178–179, 4:182–188
Saccopharyngiformes, 4:272, 4:274–276
salmons, 4:407, 4:410–419
sawsharks, 4:168, 4:170–171
Scombroidei, 5:407, 5:409–412, 5:414–418
Scorpaeniformes, 5:159, 5:161–162, 5:166, 5:169–178, 5:183, 5:186–193
Scorpaenoidei, 5:166
skates, 4:178–179, 4:187
snakeheads, 5:439, 5:442–446
South American knifefishes, 4:371, 4:373–377
southern cod-icefishes, 5:323, 5:325–328
Stephanoberyciformes, 5:106, 5:108–110
Stomiiformes, 4:424, 4:426–430
Stromateoidei, 5:422
Synbranchiformes, 5:152, 5:154–156
Tetraodontiformes, 5:471, 5:475–479, 5:481–484
toadfishes, 5:42, 5:44–46
Trachinoidei, 5:335, 5:337–339
Zoarcoidei, 5:313, 5:315–318
insects and, **3:74–83**, 13:200
aphids, 3:57
book lice, 3:244, 3:246–247
bristletails, 3:114, 3:116–117
caddisflies, 3:377, 3:379–381
cockroaches, 3:150–151, 3:153–158
Coleoptera, 3:322–323, 3:327–333
diplurans, 3:108, 3:110–111
Diptera, 3:361, 3:364–374
earwigs, 3:197, 3:199–200
fleas, 3:350, 3:352–355
Hemiptera, 3:263–264, 3:267–280
Hymenoptera, 3:407–408, 3:412–424
insect distribution, 3:57
lacewings, 3:309, 3:311–314
Lepidoptera, 3:389, 3:393–397, 3:399–404
locusts, 3:207–208
mantids, 3:181, 3:183–187
Mantophasmatodea, 3:220
mayflies, 3:127, 3:129–130
Mecoptera, 3:343, 3:345–346
Megaloptera, 3:291, 3:294
Odonata, 3:135, 3:137–138
Orthoptera, 3:207–208, 3:211–216
Phasmida, 3:224, 3:228–232
Phthiraptera, 3:252, 3:254–256
proturans, 3:94, 3:96–97
rock-crawlers, 3:190, 3:192–193
snakeflies, 3:298, 3:300–302
springtails, 3:101, 3:103–104
stoneflies, 3:143, 3:145–146
strepsipterans, 3:336, 3:338–339
termites, 3:168, 3:171–175
thrips, 3:283, 3:285–287
Thysanura, 3:120, 3:122–123
webspinners, 3:234, 3:236–237
zorapterans, 3:240, 3:241
lower metazoans, lesser deuterostomes and, **1:44–46**
acoels, 1:180, 1:182
anoplans, 1:247, 1:249–251
arrow worms, 1:435, 1:437–442
biodiversity, 1:49
box jellies, 1:149, 1:151–152
calcareous sponges, 1:59, 1:61–65
cnidarians
Anthozoa, 1:107–108, 1:111–121
Hydrozoa, 1:129, 1:135–140, 1:142–145
comb jellies, 1:171, 1:173–177
demosponges, 1:80, 1:82–86
echinoderms
Crinoidea, 1:359, 1:362–365
Echinoidea, 1:405, 1:409–415
Ophiuroidea, 1:390, 1:394–398
sea cucumbers, 1:422, 1:425–431
sea daisies, 1:383, 1:385–386
sea stars, 1:370, 1:373–379
enoplans, 1:255, 1:257
entoprocts, 1:321, 1:323–325
flatworms
free-living flatworms, 1:188, 1:190–195
monogeneans, 1:215, 1:219–223
tapeworms, 1:231, 1:235–242
gastrotrichs, 1:270, 1:272–273
girdle wearers, 1:345, 1:348–349
glass sponges, 1:69, 1:71–75
gnathostomulids, 1:332, 1:334–335
hair worms, 1:306
hemichordates, 1:46, 1:445, 1:447–450
jaw animals, 1:329
jellyfish, 1:156–157, 1:161–167
kinorhynchs, 1:277, 1:279–280
lancelets, 1:488, 1:490–497
larvaceans, 1:474, 1:476–478
nematodes
roundworms, 1:285, 1:287–289, 1:291–292
secernenteans, 1:295, 1:298–299, 1:301–304
Orthonectida, 1:99, 1:101–102

placozoans, 1:89
priapulans, 1:338, 1:340–341
rhombozoans, 1:94, 1:96–98
rotifers, 1:262, 1:264–267
Salinella salve, 1:92
salps, 1:468, 1:471–472
sea squirts, 1:455, 1:458–466
sorberaceans, 1:480, 1:482–483
thorny headed worms, 1:313, 1:315–316
Trematoda, 1:201, 1:203, 1:205–211
wheel wearers, 1:354
mammals and, 12:15
aardvarks, 15:159
agoutis, 16:409, 16:411–414
anteaters, 13:175, 13:177–179
ape enculturation by humans, 12:162
armadillos, 13:185, 13:187–190
Artiodactyla, 15:273
aye-ayes, 14:88
baijis, 15:21–22
bandicoots, 13:6
dry-country, 13:12, 13:14–16
rainforest, 13:12, 13:14–16
bats, 13:316–317
American leaf-nosed, 13:316–317, 13:420, 13:423–432
bulldog, 13:447, 13:449–450
disk-winged, 13:474, 13:476–477
Emballonuridae, 13:358, 13:360–363
false vampire, 13:317, 13:382, 13:384–385
funnel-eared, 13:461, 13:463–465
horseshoe, 13:393, 13:396–400
Kitti's hog-nosed, 13:369
Molossidae, 13:487–488, 13:490–493
mouse-tailed, 13:352, 13:353
moustached, 13:438, 13:440–441
New Zealand short-tailed, 13:455, 13:457–458
Old World fruit, 13:322, 13:325–330, 13:337, 13:340–347
Old World leaf-nosed, 13:405, 13:407–409
Old World sucker-footed, 13:480
slit-faced, 13:373, 13:375–376
smoky, 13:468, 13:470–471
Vespertilionidae, 13:502, 13:506–514, 13:522, 13:524–525
bears, 14:300, 14:303–305
beavers, 16:180–181, 16:183–184
bilbies, 13:22
botos, 15:30
Bovidae, 16:8–9
Antilopinae, 16:48, 16:50–56
Bovinae, 16:15, 16:18–23
Caprinae, 16:95, 16:98–103
duikers, 16:77, 16:80–84
Hippotraginae, 16:34, 16:37–42
Neotraginae, 16:64, 16:66–70
bushbabies, 14:26, 14:28–32
Camelidae, 15:318, 15:320–323
Canidae, 14:262, 14:273–274, 14:277–278, 14:280–283
capybaras, 16:405–406
Carnivora, 14:262–263
cats, 14:290, 14:291, 14:292–293, 14:375–376, 14:380–389
Caviidae, 16:393, 16:395–398
Cetacea, 15:10–11
Balaenidae, 15:112, 15:115–118
beaked whales, 15:62, 15:64–68
dolphins, 15:51
franciscana dolphins, 15:25
Ganges and Indus dolphins, 15:16
gray whales, 15:100
Monodontidae, 15:87–88, 15:90–91
porpoises, 15:36, 15:38–39
pygmy right whales, 15:105
rorquals, 15:125, 15:128–130
sperm whales, 15:78, 15:79–80
chevrotains, 15:329, 15:331–333
Chinchillidae, 16:380, 16:382–383
colugos, 13:302, 13:304
coypus, 12:184, 16:126–127, 16:477
cross-species diseases, 12:222–223
dasyurids, 12:291, 12:294–298
Dasyuromorphia, 12:283–284
deer
Chinese water, 15:376
muntjacs, 15:346, 15:349–354
musk, 15:338, 15:340–341
New World, 15:384–385, 15:388–396, 15:388–396*t*, 15:396*t*
Old World, 15:361, 15:364–371, 15:371*t*
Dipodidae, 16:219–222
Diprotodontia, 13:40
distribution, 14:5, 14:*241*, 14:247
dogs, 14:288, 14:*289*, 14:292–293
dormice, 16:321, 16:323–326
duck-billed platypuses, 12:234, 12:247
Dugongidae, 15:191, 15:201, 15:203–204
echidnas, 12:238
elephants, 12:173, 15:172, 15:174–175
encephalization quotient, 12:149
Equidae, 15:230, 15:232–235
Erinaceidae, 13:207, 13:209–212
field studies of mammals, 12:*195–201*
giant hutias, 16:470
gibbons, 14:215, 14:218–222
Giraffidae, 15:405–406, 15:408–409
great apes, 14:235, 14:237–240
gundis, 16:313, 16:314–315
hearing, 12:82
Herpestidae, 14:352, 14:354–356
Heteromyidae, 16:203, 16:205–208
hippopotamuses, 15:308, 15:311
hutias, 16:463
Hyaenidae, 14:362, 14:365–367
hyraxes, 15:184, 15:186–188
Ice Age evolution of, 12:20–21
Indriidae, 14:67, 14:69–72
as invasive species, 12:184–185
koalas, 13:49
Lagomorpha, 16:487–488
lemurs, 14:52, 14:56–60
Cheirogaleidae, 14:39, 14:41–44
sportive, 14:77, 14:79–83
Leporidae, 16:509, 16:511–515
Lorisidae, 14:16, 14:18–20
Macropodidae, 13:89–90, 13:93–100
mammal distribution, 12:139
manatees, 15:209, 15:212
Megalonychidae, 13:158, 13:159
memory, 12:152, 12:154
milk of, 12:97
mirror self-recognition, 12:159
moles
golden, 13:218, 13:220–221
marsupial, 13:27, 13:28
monitos del monte, 12:275
monkeys
Atelidae, 14:160, 14:162–166
Callitrichidae, 14:124, 14:127–131
Cebidae, 14:106, 14:108–112
cheek-pouched, 14:194, 14:197–204
leaf-monkeys, 14:175, 14:178–183
night, 14:138
Pitheciidae, 14:148, 14:150–153
monotremes, 12:234
mountain beavers, 16:133
Muridae, 16:127–128, 16:285, 16:288–295
Arvicolinae, 16:229–230, 16:233–237
hamsters, 16:243, 16:245–247
Murinae, 16:253, 16:256–260
Sigmodontinae, 16:270, 16:272–276
musky rat-kangaroos, 13:71
Mustelidae, 14:324–325, 14:327–331
neonates, 12:125
numbats, 12:306
octodonts, 16:436, 16:438–440
opossums
New World, 12:254, 12:257–263
shrew, 12:268, 12:270
Otariidae, 14:400, 14:403–406
pacaranas, 16:388
pacas, 16:420–421, 16:423–424
pangolins, 16:113, 16:115–120
peccaries, 15:296, 15:298–300
Perissodactyla, 15:222–223
Petauridae, 13:129, 13:131–132
Phalangeridae, 13:60–61, 13:*63–66t*
pigs, 15:*150*, 15:273, 15:281–282, 15:282, 15:284–288, 15:289*t*–290*t*
pikas, 16:495, 16:497–501
pocket gophers, 16:190, 16:192–195
porcupines
New World, 16:369, 16:372–373
Old World, 16:354, 16:357–364
possums
feather-tailed, 13:142, 13:144
honey, 13:138
pygmy, 13:108, 13:110–111
primates, 14:11
Procyonidae, 14:311, 14:313–315
pronghorns, 15:416–417
Pseudocheiridae, 13:113–123, 13:117
rat-kangaroos, 13:76, 13:78–81
rats
African mole-rats, 16:345, 16:347–349
cane, 16:335, 16:337–338
chinchilla, 16:445, 16:447–448
dassie, 16:331
spiny, 16:451, 16:453–457
rhinoceroses, 15:254–255, 15:257–262
rodents, 16:126–128
sengis, 16:522, 16:524–530
shrews
red-toothed, 13:252, 13:255–263
West Indian, 13:244
white-toothed, 13:268, 13:271–275
Sirenia, 15:191, 15:195–196
social cognition, 12:160
solenodons, 13:239, 13:241
spermatogenesis, 12:103
springhares, 16:310
squirrels
flying, 16:137
ground, 16:126, 16:148, 16:151–158
scaly-tailed, 16:300, 16:303–305
tree, 16:167, 16:169–174

Talpidae, 13:282, 13:284–287
tapirs, 15:242–243, 15:245–247
tarsiers, 14:95, 14:97–99
tenrecs, 13:230, 13:232–233
three-toed tree sloths, 13:165, 13:167–169
tree shrews, 13:292, 13:294–296
true seals, 14:423–424, 14:428–434
tuco-tucos, 16:427, 16:429–430
Viverridae, 14:338, 14:340–343
walruses, 14:415
wombats, 13:53, 13:55–56
Xenartha, 13:153
See also Conservation status; Domestication; *Homo sapiens*
parasitism and, 2:33
protostomes and, **2:41–44**
amphionids, 2:196
amphipods, 2:262, 2:265–271
anaspidaceans, 2:182, 2:183
aplacophorans, 2:380, 2:383–385
Arachnida, 2:41, 2:336, 2:340–352
articulate lampshells, 2:523, 2:525–527
bathynellaceans, 2:178, 2:179
beard worms, 2:86, 2:88
bivalves, 2:454, 2:458–466
caenogastropods, 2:447, 2:449
centipedes, 2:356, 2:358–362
cephalocarids, 2:132, 2:133
Cephalopoda, 2:479–480, 2:483–489
chitons, 2:395, 2:398–401
clam shrimps, 2:148, 2:150–151
copepods, 2:300–301, 2:304–310
cumaceans, 2:230, 2:232
Decapoda, 2:201, 2:204–214
deep-sea limpets, 2:436, 2:437
earthworms, 2:68, 2:70–73
echiurans, 2:104, 2:107
fairy shrimps, 2:137, 2:139–140
fish lice, 2:290, 2:292–293
freshwater bryozoans, 2:498, 2:500–502
horseshoe crabs, 2:41, 2:42, 2:328–329, 2:332
Isopoda, 2:252, 2:255–260
krill, 2:187, 2:189–192
leeches, 2:42, 2:78, 2:80–83
leptostracans, 2:162, 2:164–165
lophogastrids, 2:226, 2:227
mantis shrimps, 2:171, 2:173–175
marine bryozoans, 2:504, 2:506–508, 2:511, 2:513–514
mictaceans, 2:241, 2:242
millipedes, 2:365, 2:367–370
monoplacophorans, 2:388, 2:390–391
mussel shrimps, 2:312, 2:314–315
mysids, 2:217, 2:220–222
mystacocarids, 2:296, 2:297
myzostomids, 2:60, 2:62
Neritopsina, 2:441, 2:442–443
nonarticulate lampshells, 2:516, 2:518–519
Onychophora, 2:111, 2:113–114
pauropods, 2:376, 2:377
peanut worms, 2:98, 2:100–101
phoronids, 2:492, 2:494
Polychaeta, 2:47, 2:50–56
Pulmonata, 2:414–415, 2:417–421
remipedes, 2:126, 2:128
sea slugs, 2:43, 2:405, 2:407–410
sea spiders, 2:322, 2:324–325
spelaeogriphaceans, 2:243, 2:244
symphylans, 2:372, 2:373
tadpole shrimps, 2:143, 2:146
tanaids, 2:236, 2:238–239
tantulocaridans, 2:284, 2:286
Thecostraca, 2:276, 2:279–281
thermosbaenaceans, 2:246, 2:247
tongue worms, 2:320
true limpets, 2:425, 2:426–427
tusk shells, 2:471, 2:473–474
Vestimentifera, 2:92, 2:94–95
Vetigastropoda, 2:431, 2:433–434
water bears, 2:118, 2:121–123
water fleas, 2:155, 2:157–158
reptiles and, **7:47–58**
African burrowing snakes, 7:462–464
African sideneck turtles, 7:130, 7:132–133
Afro-American river turtles, 7:138, 7:140–142
Agamidae, 7:211, 7:214–221
Alligatoridae, 7:173, 7:175–177
Anguidae, 7:340, 7:342–344
Australo-American sideneck turtles, 7:79, 7:81–83
big-headed turtles, 7:136
blindskinks, 7:272
blindsnakes, 7:382–385
boas, 7:411, 7:413–417
Central American river turtles, 7:100
colubrids, 7:470, 7:473–481
Cordylidae, 7:322, 7:324–325
Crocodylidae, 7:182, 7:184–187
early blindsnakes, 7:371
Elapidae, 7:487–488, 7:491–498
false blindsnakes, 7:388
false coral snakes, 7:400
feeding ecology, 7:232, 7:235–241
file snakes, 7:442
Florida wormlizards, 7:285
folk medicine, 7:50–52
Gekkonidae, 7:263, 7:265–269
Geoemydidae, 7:116, 7:118–119
gharials, 7:169
Helodermatidae, 7:357
Iguanidae, 7:247, 7:250–257
Kinosternidae, 7:123, 7:125–126
knob-scaled lizards, 7:349, 7:351
Lacertidae, 7:299, 7:301–302
leatherback seaturtles, 7:102
Microteiids, 7:304, 7:306–307
mole-limbed wormlizards, 7:280, 7:281
mythology and religions, 7:54–57
Neotropical sunbeam snakes, 7:406
New World pond turtles, 7:107, 7:109–113
night lizards, 7:292, 7:294–295
pig-nose turtles, 7:76
pipe snakes, 7:397
pythons, 7:421, 7:422, 7:424–428
reptile farming and ranching, 7:48–*49*
seaturtles, 7:87, 7:89–91
shieldtail snakes, 7:392–394
skinks, 7:329, 7:332–337
slender blindsnakes, 7:375–377
snapping turtles, 7:94, 7:95–96
softshell turtles, 7:152, 7:154–155
spade-headed wormlizards, 7:288, 7:289
splitjaw snakes, 7:431
sunbeam snakes, 7:403
Teiidae, 7:312, 7:314–316
tortoises, 7:144, 7:146–149
Tropidophiidae, 7:435–437
tuatara, 7:191
Varanidae, 7:363, 7:365–368
Viperidae, 7:448, 7:451–460
wormlizards, 7:275, 7:276
See also Conservation status
Humblotia flavirostris. See Grand Comoro flycatchers
Humblot's flycatchers. *See* Grand Comoro flycatchers
Humboldt penguins, 8:148, 8:150, 8:151
Humboldt squids, 2:479
Humboldt's woolly monkeys, 14:*156*, 14:167*t*
Humbug damsels, 5:*295*
Hume, Rob, 9:331
Hume's ground-jays, 11:507, 11:*510*, 11:*517*
Hume's groundpeckers. *See* Hume's ground-jays
Hume's owls, 9:347, 9:*348*
Humidity
behavior and, 1:38
subterranean mammals and, 12:73–74
Hummingbirds, 9:415–419, **9:437–468,** 9:*453–454*
behavior, 9:417, 9:443–445
conservation status, 9:418, 9:443, 9:449–451
distribution, 9:417, 9:*437*, 9:442
evolution, 9:415, 9:437
feeding ecology, 9:417, 9:*439*, 9:441, 9:*441*, 9:445–447
habitats, 9:417, 9:442–443
humans and, 9:418–419, 9:451–452
nectar and, 9:438–439, 9:441, 9:443, 9:445–446
nest construction, 9:447–448
physical characteristics, 9:415–417, 9:437, 9:438–442
physiology of, 9:416–417, 9:441–442, 9:445–446
reproduction, 9:417–418, 9:418, 9:447–449
sounds of, 8:*10*, 8:37
species of, 9:*455–467*
taxonomy, 9:415, 9:437–438
Humpback dolphins, 15:5, 15:6, 15:57*t*
Humpback salmons. *See* Pink salmons
Humpback whales, 15:*9*, 15:*125*, 15:130*t*
behavior, 15:7, 15:123
distribution, 15:5, 15:121
evolution, 12:*66*, 15:119
feeding ecology, 15:8, 15:*123*, 15:*124*
migrations, 12:87, 15:6
physical characteristics, 15:4, 15:120
taxonomy, 12:*66*, 15:119
vocalizations, 12:85
Humphead wrasses, 4:45, 4:50, 5:293, 5:298, 5:*299*, 5:*301*
Humpnose big-eye breams, 5:*264*, 5:268, 5:*269*
Humpwinged crickets, 3:201
Humu humus. *See* Blackbar triggerfishes
Hundred-pace pitvipers, 7:*450*, 7:*454–455*
Hundred-pacers. *See* Hundred-pace pitvipers
Hunter's hartebeests, 16:27, 16:29, 16:33, 16:42*t*
Hunting, 12:24–25, 14:248
See also Humans
Hunting birds, 8:21
Hunting cissas. *See* Green magpies
Hunting dogs, 12:146, 14:288, 14:292
Huon tree kangaroos. *See* Matschie's tree kangaroos
Hupotaenidia owstoni. See Guam rails
Huskies, Siberian, 14:289, 14:*292*, 14:293

Huso spp., 4:213
Huso huso. See Beluga sturgeons
Hussar monkeys. *See* Patas monkeys
Hutchinsoniella macracantha, 2:*131*, 2:*133*
Hutchinsoniellidae, 2:131
Hutias, 16:123, **16:461–467**, 16:*464*, 16:*467t*
See also Giant hutias
Hutterer, R., 16:433
Hutton's tube-nosed bats, 13:*526t*
HVC. *See* Higher vocal center
Hwameis, 10:*510*, 10:*516*
Hyacinth macaws, 9:275, 9:*282*, 9:*288–289*
Hyacinthine macaws. *See* Hyacinth macaws
Hyaena brunnea. See Brown hyenas
Hyaena hyaena. See Striped hyenas
Hyaena hyaena barbara, 14:360
Hyaena hyaena dubbah, 14:360
Hyaena hyaena hyaena, 14:360
Hyaena hyaena sultana, 14:360
Hyaena hyaena syriaca, 14:360
Hyaenidae. *See* Aardwolves; Hyenas
Hyalinobatrachium spp., 6:215, 6:216
Hyalinobatrachium aureoguttatum, 6:*217*
Hyalinobatrachium valerioi. See La Palma glass frogs
Hyalonema sieboldi. See Glass-rope sponges
Hybrid gibbons, 14:208–209, 14:210
Hybridization, 6:44, 6:263
Belford's melidectes, 11:251
red canaries, 11:339
redpolls, 11:331
turtledoves, 9:250
Hybridogenesis, 4:66
Hydra spp., 1:*16*, 1:32–33, 1:37, 1:125, 1:129, 2:32
Hydra vulgaris, 1:*132*, 1:143
Hydractinia spp., 1:27, 1:129
Hydractinia echinata, 1:26–27, 1:*130*, 1:140
Hydractinia symbiologicarpus, 1:30
Hydrallmania spp., 1:129
Hydrargira limi. See Central mudminnows
Hydrichthys mirus, 1:*130*, 1:140–141
Hydrobates pelagicus. See European storm-petrels
Hydrobatidae. *See* Storm-petrels
Hydrobatinae. *See* Storm-petrels
Hydrobiosidae, 3:375
Hydrocarbons, pollution from, 1:25
Hydrocenidae, 2:439, 2:440, 2:441, 2:447
Hydrochaeridae. *See* Capybaras
Hydrochaeris hydrochaeris. See Capybaras
Hydrocoryne miurensis, 1:127
Hydrocynus goliath. See Giant tigerfishes
Hydrodamalinae. *See* Steller's sea cows
Hydrodamalis gigas. See Steller's sea cows
Hydrodamalis stelleri. See Giant sea cows
Hydrodynastes spp. *See* False water-cobras
Hydroidomedusae, 1:29, 1:123, 1:124, 1:125, 1:127, 1:129
Hydroids. *See* Hydrozoa
Hydrolaetare spp., 6:156, 6:157, 6:158
Hydrolagus colliei. See Spotted ratfishes
Hydromantes spp., 6:5, 6:391
Hydromantes italicus. See Italian cave salamanders
Hydromantes platycephalus. See Mt. Lyell salamanders
Hydromedusinae, 7:77
Hydrometra argentina. See Water measurers
Hydromys chrysogaster. See Golden-bellied water rats
Hydrophasianus chirurgus. See Pheasant-tailed jacanas
Hydrophiidae, 7:483
Hydrophiinae, 7:196–197, 7:484
Hydrophiloidea, 3:316
Hydropotes spp. *See* Water deer
Hydropotes inermis. See Chinese water deer
Hydropotes inermis argyropus. See Siberian water deer
Hydropotes inermis inermis, 15:373
Hydropotinae. *See* Chinese water deer
Hydroprogne spp. *See* Crested terns
Hydroprogne caspia. See Caspian terns
Hydrops spp., 7:467, 7:468
Hydropsyche tobiasi, 3:377
Hydropsychidae, 3:376
Hydroptilidae. *See* Purse case makers
Hydroptiloidea. *See* Purse case makers
Hydrosaurus amboinensis. See Sailfin lizards
Hydrostatic skeletons, 2:*18*
See also Physical characteristics
Hydrothermal vent sea cucumbers, 1:*424*, 1:425, 1:*427*
Hydrothermal vent worms, 2:*91*, **2:91–95**, 2:*93*, 2:*94*
Hydrozoa, **1:123–148**, 1:*130*, 1:*131*, 1:*132*, 2:13
behavior, 1:26, 1:30, 1:127–128
conservation status, 1:129
distribution, 1:127
evolution, 1:123
feeding ecology, 1:*125*, 1:128–129
habitats, 1:24, 1:25, 1:127–128
humans and, 1:129
physical characteristics, 1:123–127, 1:*124*
pollution and, 1:25
reproduction, 1:16, 1:20, 1:21, 1:*126*, 1:129
species of, 1:134–145
taxonomy, 1:10, 1:123
Hydrurga leptonyx. See Leopard seals
Hyemoschus spp. *See* Water chevrotains
Hyemoschus aquaticus. See Water chevrotains
Hyenas, 12:*142*, 14:256, 14:*259*, 14:260, **14:359–367**, 14:*361*, 14:*363*
Hygophum spp., 4:443
Hygrokinesis, 1:38
Hyla spp., 6:5, 6:40, 6:42, 6:225, 6:228–229, 6:230
Hyla arborea. See European treefrogs
Hyla armata, 6:227
Hyla boans, 6:50, 6:227, 6:230
Hyla bürgeri. See Buerger's frogs
Hyla callidryas. See Red-eyed treefrogs
Hyla calypsa, 6:230
Hyla chrysoscelis. See Cope's gray treefrogs
Hyla ebraccata, 6:48
Hyla faber, 6:50
Hyla femoralis chrysoscelis. See Cope's gray treefrogs
Hyla geographica, 6:229
Hyla goinorum. See Amazonian skittering frogs
Hyla gratiosa. See Barking treefrogs
Hyla labialis, 6:228–229
Hyla leucophyllata. See Hourglass treefrogs
Hyla leucopygia, 6:230
Hyla marianae, 6:37
Hyla microcephala, 6:40
Hyla miliaria, 6:227
Hyla parviceps, 6:46
Hyla picadoi, 6:37
Hyla picturata, 6:227
Hyla prosoblepon. See Nicaragua glass frogs
Hyla riobambae. See Riobamba marsupial frogs
Hyla rosenbergi. See Rosenberg's treefrogs
Hyla septentrionalis. See Cuban treefrogs
Hyla spinosa. See Spiny-headed treefrogs
Hyla thorectes, 6:230
Hyla triseriata. See Chorus frogs
Hyla vasta, 6:227
Hyla versicolor, 6:47, 6:228–229, 6:236
Hyla wilderi, 6:37
Hyla xanthosticta, 6:230
Hyla zeteki, 6:37
Hylacola spp. *See* Heathwrens
Hylarana spp., 6:248, 6:250
Hylella fleischmanni. See Fleischmann's glass frogs
Hylella ocellata. See Spotted Cochran frogs
Hylexetastes uniformis. See Uniform woodcreepers
Hylidae. *See* Amero-Australian treefrogs
Hylinae, 6:225, 6:230
Hylobates spp., 14:207–208, 14:209, 14:211
Hylobates agilis. See Agile gibbons
Hylobates agilis albibarbis, 14:208
Hylobates concolor, 14:207
Hylobates gabriellae, 14:207
Hylobates hoolock. See Hoolock gibbons
Hylobates klossi. See Kloss gibbons
Hylobates lar. See Lar gibbons
Hylobates leucogenys, 14:207
Hylobates moloch. See Moloch gibbons
Hylobates muelleri. See Mueller's gibbons
Hylobates pileatus. See Pileated gibbons
Hylobates syndactylus. See Siamangs
Hylobatidae. *See* Gibbons
Hylobittacus apicalis. See Black-tipped hangingflies
Hylocharis spp. *See* Hummingbirds
Hylocharis cyanus. See White-chinned sapphires
Hylocharis eliciae, 9:443
Hylocharis grayi, 9:443
Hylocharis sapphirina, 9:443
Hylochoerus spp., 15:275, 15:276
Hylochoerus meinertzhageni. See Forest hogs
Hylocitrea bonensis. See Yellow-flanked whistlers
Hylodes spp., 6:156
Hylodes asper. See Warty tree toads
Hylodinae, 6:156
Hylomanes momotula. See Tody motmots
Hylomyinae, 13:203, 13:204
Hylomys spp., 13:203
Hylomys hainanensis. See Hainan gymnures
Hylomys megalotis. See Long-eared lesser gymnures
Hylomys sinens, 13:207
Hylomys sinensis. See Shrew gymnures
Hylomys suillus. See Lesser gymnures
Hylomys suillus parvus, 13:207
Hylonomus lyelli, 7:*14*
Hylonympha spp. *See* Hummingbirds
Hylopetes spp., 16:136
Hylopetes lepidus. See Gray-cheeked flying squirrels
Hylopezus dives. See Fulvous-bellied ant-pittas
Hylophilus spp. *See* Greenlets

Hylophilus throacicus. See Lemon-chested greenlets
Hylophylax naevia. See Spot-backed antbirds
Hylorchilus spp., 10:526, 10:527, 10:529
Hylorchilus navai. See Nava's wrens
Hylorchilus sumichrasti. See Slender-billed wrens
Hylorina spp., 6:157
Hylorina sylvatica. See Emerald forest frogs
Hyman, L. H., 1:7, 1:283
Hymenocera picta. See Harlequin shrimps
Hymenochirus spp., 6:45–46, 6:100, 6:102, 6:317
Hymenolaimus malacorhynchos. See Blue ducks
Hymenolepididae, 1:226, 1:231
Hymenolepis diminuta. See Rat tapeworms
Hymenolepis nana, 1:231
Hymenopodidae, 3:177
Hymenoptera, 3:35, 3:49, 3:68, **3:405–425**, 3:*409*, 3:*410*
  behavior, 3:406
  conservation status, 3:407
  distribution, 3:406
  evolution, 3:405
  feeding ecology, 3:35, 3:406–407
  habitats, 3:406
  humans and, 3:407–408
  kin selection theory and, 3:70
  physical characteristics, 3:19, 3:22, 3:24, 3:405
  reproduction, 3:34, 3:38, 3:39, 3:44, 3:71, 3:407
  social, 3:70, 3:71, 3:72
  species of, 3:*411*–424
  taxonomy, 3:405
Hymenopus coronatus. See Orchid mantids
Hynobiidae. *See* Asiatic salamanders
Hynobiinae, 6:335
Hynobius spp., 6:34, 6:335, 6:336, 6:337, 6:340
Hynobius abei, 6:337
Hynobius chinensis, 6:336, 6:337
Hynobius dunni, 6:337
Hynobius hidamontanus, 6:337
Hynobius leechii, 6:336
Hynobius naevius. See Japanese marbled salamanders
Hynobius nebulosus, 6:336
Hynobius retardatus. See Hokkaido salamanders
Hynobius stejnegeri, 6:337
Hynobius takedai, 6:337
Hynobius tokyoensis. See Tokyo salamanders
Hynobius tsuensis. See Tsushima salamanders
Hyocrinids, 1:356
Hyolitha, 2:*97*
Hyotherium spp., 15:265
Hypargos niveoguttatus. See Peter's twinspots
Hypercapnia, 12:73
Hyperia galba, 2:*263*, 2:*266*, 2:270
Hyperiidea, 2:261
Hyperoglyphe perciformis. See Barrelfishes
Hyperoliidae. *See* African treefrogs
Hyperoliinae, 6:279–280
Hyperolius spp., 6:32, 6:279, 6:282, 6:285
Hyperolius benguellensis, 6:286
Hyperolius erythrodactylus. See Forest bright-eyed frogs
Hyperolius marginatus, 6:287
Hyperolius marmoratus, 6:287
Hyperolius minutissimus, 6:281
Hyperolius nasutus. See Sharp-nosed reed frogs
Hyperolius parallelus, 6:287
Hyperolius pickersgilli, 6:283
Hyperolius spinigularis, 6:35
Hyperolius spinosus. See African wart frogs
Hyperolius tuberculatus, 6:287
Hyperolius viridiflavus. See Painted reed frogs
Hyperoodon spp. *See* Bottlenosed whales
Hyperoodon ampullatus. See Northern bottlenosed whales
Hyperoodon planifrons. See Southern bottlenosed whales
Hypertragulidae, 15:265
Hyphantornis rubriceps. See Red-headed weavers
Hyphessobrycon erythrostigma. See Bleeding-heart tetras
Hypnale spp., 7:445
Hypnale hypnale, 7:447
Hypnale nepa. See Sri Lankan hump-nosed pitvipers
Hypnomys spp., 16:317–318
Hypoboscidae, 3:358
Hypocnemis cantator. See Warbling antbirds
Hypocolius ampelinus. See Gray hypocolius
Hypoechinorhynchus thermaceri, 1:*311*
Hypogastrura armata, 3:101
Hypogeomys antimena. See Malagasy giant rats
Hypogeophis spp., 6:436, 6:440
Hypogeophis rostratus. See Frigate Island caecilians
Hypogramma spp. *See* Sunbirds
Hypogramma hypogrammicum. See Purple-naped sunbirds
Hypolectus spp., 4:45
Hypomesus pretiosus. See Surf smelts
Hyponeuralia, 2:4
Hypophthalmichthyinae, 4:297
Hypophthalmichthys molitrix. See Silver carps
Hypoplectrus spp., 4:30, 5:262
Hypopomidae, 4:369
Hypoprion brevirostris. See Lemon sharks
Hypopterus spp., 5:221
Hypoptychidae. *See* Sand eels
Hypoptychus dybowskii. See Sand eels
Hypopygus lepturus, 4:370
Hypopyrrhus pyrohypogaster. See Red-bellied grackles
Hypositta spp. *See* Vanga shrikes
Hypositta corallirostris. See Nuthatch vangas
Hyposmocoma spp., 3:388–389
Hypoterus macropterus, 5:220, 5:222
Hypothymis azurea. See Black-naped monarchs
Hypothymis coelestis. See Caerulean paradise-flycatchers
Hypoxia, 12:73, 12:114
Hypseleotris agilis, 5:377
Hypseleotris compressa. See Empire gudgeons
Hypselodoris bullocki, 2:*405*
Hypsidoridae, 4:351
Hypsidoris farsonensis, 4:351
Hypsignathus monstrosus. See Hammer-headed fruit bats
Hypsipetes spp., 10:395–399, 10:396
Hypsipetes flavala. See Ashy bulbuls
Hypsipetes indicus. See Yellow-browed bulbuls
Hypsipetes madagascariensis. See Black bulbuls
Hypsipetes mcclellandii. See Mountain bulbuls
Hypsipetes nicobariensis. See Nicobar bulbuls
Hypsipetes siquijorensis. See Streak-breasted bulbuls
Hypsiprimnodontidae. *See* Musky rat-kangaroos
Hypsiprymnodon moschatus. See Musky rat-kangaroos
Hypsurus caryi. See Rainbow seaperches
Hypsypops rubicundus. See Garibaldi damselfishes
Hyracoidea. *See* Hyraxes
Hyracotherium spp., 15:132
Hyraxes, **15:177–190**, 15:*178*, 15:*185*
  behavior, 15:180–181
  conservation status, 15:184
  digestive system, 12:14–15
  distribution, 15:*177*, 15:179
  evolution, 15:131, 15:133, 15:177
  feeding ecology, 15:182
  habitats, 15:179–180
  humans and, 15:184
  physical characteristics, 15:178–179
  reproduction, 12:94, 12:95, 15:182–183
  species of, 15:186–188, 15:189*t*
  taxonomy, 15:134, 15:177–178
Hyspiprymnodon bartholomai, 13:69–70
Hysterocarpus traskii. See Tule perches
Hystrichopsyllidae, 3:347, 3:348
Hystrichopsylloidea, 3:347
Hystricidae. *See* Old World porcupines
Hystricognathi, 12:11, 12:71, 16:121, 16:122, 16:125
Hystricomorph rodents, 12:92, 12:94
Hystrix spp., 16:351
Hystrix africaeaustralis. See South African porcupines
Hystrix brachyura. See Common porcupines
Hystrix crassispinis. See Thick-spined porcupines
Hystrix cristata. See North African porcupines
Hystrix indica. See Indian crested porcupines
Hystrix javanica. See Javan short-tailed porcupines
Hystrix macroura, 16:*358*
Hystrix pumila. See Indonesian porcupines
Hystrix sumatrae. See Sumatran porcupines
Hysudricus maximus, 15:164

I

Ibadan malimbes, 11:379
Ibalia leucospoides, 3:*409*, 3:415–*416*
Iberian desmans. *See* Pyrenean desmans
Iberian lynx, 14:262, 14:370–371, 14:374, 14:*378*, 14:*387*, 14:389
Iberian midwife toads, 6:*91*, 6:*92*
Iberian rain frogs, 6:62
Ibex, 16:4, 16:8, 16:9, 16:90, 16:92–93, 16:95
Ibidorhynchina struthersii. See Ibisbills
Ibidorhynchinae, 9:133
The Ibis (journal), 8:27
Ibisbills, 9:133, 9:*134*, 9:135, 9:*136*, 9:*137*
Ibises, 8:233–236, **8:291–301**
  behavior, 8:234–235, 8:292
  conservation status, 8:236, 8:293
  distribution, 8:*291*–292
  evolution, 8:291
  feeding ecology, 8:292
  habitats, 8:292
  humans and, 8:293
  physical characteristics, 8:233–234, 8:291

  reproduction, 8:236, 8:292–293
  species of, 8:*296–301*
  taxonomy, 8:233, 8:291
Ice Age, **12:17–25**
Ice-crawlers, 3:189
Ice rats, 16:*283*
Icebirds. *See* Broad-billed prions
Icefishes, 4:49, 4:391, 5:321, 5:322, 5:326
  See also Southern cod-icefishes
Icelidae, 5:179
Ichneumia spp., 14:347
Ichneumia albicauda. See White-tailed mongooses
Ichneumonidae, 3:39, 3:64, 3:298
Ichnotropis spp., 7:297
Ichnotropis capensis. See Cape rough-scaled lizards
Ichnotropis squamulosa. See Common rough-scaled lizards
Ichnumeons. *See* Egyptian mongooses
Ichthyoborus spp., 4:69
Ichthyomyines, 16:264, 16:268, 16:269
Ichthyomys spp., 16:266, 16:269
Ichthyomys pittieri, 16:269
Ichthyomyzon spp., 4:83
Ichthyomyzon unicuspis. See Silver lampreys
Ichthyophiidae. *See* Asian tailed caecilians
Ichthyophis spp., 6:35, 6:419, 6:422
Ichthyophis bannanicus. See Bannan caecilians
Ichthyophis glandulosus, 6:420
Ichthyophis glutinosus. See Ceylon caecilians
Ichthyophis kohtaoensis. See Koh Tao Island caecilians
Ichthyophis mindanaoensis, 6:420
Ichthyophis orthoplicatus. See Pattipola caecilians
Ichthyosaurs, 2:*10*, 7:5, 7:15
Ichthyostega spp., 6:7, 6:*10*
Icosteoidei. *See* Ragfishes
Icosteus aenigmaticus. See Ragfishes
Ictaluridae, 4:6, 4:57, 4:60, 4:62, 4:352
Ictalurus punctatus. See Channel catfishes
Icteria virens. See Yellow-breasted chats
Icteridae, **11:301–322,** 11:*307–308*
  behavior, 11:303
  conservation status, 11:305–306
  distribution, 11:*301*, 11:302
  evolution, 11:301
  feeding ecology, 11:293–294
  habitats, 11:302–303
  humans and, 11:306
  physical characteristics, 11:301–302
  reproduction, 11:304–305
  species of, 11:*309–322*
  taxonomy, 11:301
Icterine greenbuls, 10:*400*, 10:407–*408*
Icterine warblers, 11:*9*, 11:17–*18*
Icterus spp. *See* Orioles
Icterus abeillei. See Black-backed orioles
Icterus bonana. See Martinique orioles
Icterus chrysater. See Yellow-backed orioles
Icterus dives. See Melodious blackbirds
Icterus galbula. See Baltimore orioles
Icterus gularis. See Altamira orioles
Icterus icterocephalus. See Yellow-headed blackbirds
Icterus nigerrimus. See Jamaican blackbirds
Icterus nigrogularis. See Yellow orioles
Icterus oriole. See Hooded orioles
Icterus parisorum. See Scott's orioles
Icterus spurius. See Orchard orioles
Ictinia plumbea. See Plumbeous kites
Ictiobinae, 4:321
Ictiobus cyprinellus. See Bigmouth buffaloes
Ictonyx spp. *See* Zorillas
Ictonyx striatus. See Zorillas
Idaho ground squirrels, 16:147, 16:*150*, 16:*153*, 16:157–158
Idiacanthus fasciola. See Black dragonfishes
Idiocranium spp., 6:35, 6:436
Idiocranium russeli, 6:412, 6:436
Idionycteris phyllotis. See Allen's big-eared bats
Idiosepius spp., 2:476
Idiurus macrotis. See Big-eared flying mice
Idiurus zenkeri. See Zenker's flying mice
Idmidronea atlantica, 2:*505*, 2:*507*–508
Idodictyum spp., 2:*509*
Ifrita kowaldi. See Blue-capped ifrits
Ifrits, 11:75, 11:76, 11:77, 11:80–81
Iguana spp. *See* Iguanas
Iguana iguana. See Green iguanas
Iguanas, **7:243–257,** 7:*248*
  behavior, 7:244–245
  conservation status, 7:246–247
  distribution, 7:*243*, 7:244
  evolution, 7:243
  farming, 7:48
  feeding ecology, 7:245–246
  in folk medicine, 7:51
  as food, 7:48, 7:*51*
  habitats, 7:244
  humans and, 7:247
  physical characteristics, 7:243–244
  reproduction, 7:246
  species of, 7:*250–257*
  taxonomy, 7:243
  See also specific types of iguanas
Iguanidae, **7:243–257,** 7:*248*, 7:*249*
  behavior, 7:204, 7:244–*245*
  conservation status, 7:246–247
  distribution, 7:203–204, 7:*243*, 7:244
  evolution, 7:195, 7:196, 7:243
  feeding ecology, 7:245–246
  habitats, 7:244
  humans and, 7:247
  physical characteristics, 7:200, 7:243–244
  reproduction, 7:207, 7:246
  species of, 7:*250–257*
  taxonomy, 7:197–198, 7:243
Iguaninae, 7:204, 7:243–246
Iiwis, 11:342
Ikaite rotifers, 1:*263*, 1:*264–265*
Ikedidae, 2:103
Ili pikas, 16:487, 16:495, 16:502*t*
Illacme plenipes, 2:363
Illinois mud turtles. *See* Yellow mud turtles
Ilyocoris cimicoides. See Creeping water bugs
Imago, 3:125, 3:*126*
Imamura, Hisashi, 5:157, 5:163
Imantodes spp. *See* Blunt-headed vinesnakes
Imbrasia belina. See Mopane worms
Imitating poison frogs, 6:*202*, 6:*206*
Imitation, 12:158
  See also Behavior
Immersion exhibits, 12:205
Immigration, 2:27
  See also Behavior
Immortal jellyfish, 1:*123*, 1:*128*, 1:*130*, 1:139
Immunocontraception, 12:188
  See also Reproduction
Immunoglobins, neonatal, 12:126
Impalas, 16:*36*, 16:41–42
  behavior, 16:5, 16:31, 16:*39*, 16:41–42
  black-faced, 16:34
  conservation status, 16:31
  distribution, 16:29, 16:*39*
  evolution, 16:27
  habitats, 16:30
  physical characteristics, 16:28
  predation and, 12:116
  reproduction, 16:32
  smell, 12:80
  taxonomy, 16:27
Imperial eagles, 8:324
Imperial pheasants, 8:433
Imperial sandgrouse. *See* Black-bellied sandgrouse
Imperial shags, 8:184, 8:*185*
Imperial snipes, 9:177
Imperial woodpeckers, 10:88, 10:150
Implantation, delayed, 12:109–110
  See also Reproduction
Imports and exports, of reptiles, 7:53*t*
Imposter hutias, 16:467*t*
Impressions (fossils), 3:7–8
Inaccessible rails, 9:1, 9:*55*, 9:*61*
Inarticulata. *See* Nonarticulate lampshells
Inarticulated brachiopods. *See* Nonarticulate lampshells
Inbreeding, 12:110, 12:222
  See also Captive breeding; Domestication
Inca doves, 9:*254*, 9:*260–261*
Inca terns, 9:203
Incaichthys spp., 4:351
Incertametra santanensis, 3:259
Inchworms, 3:65
Incirrata. *See* Common octopods
Incisors, 12:46
  See also Teeth
Incompatibility, cytoplasmic, 3:39
Incrustations (fossils), 3:8
Incurvariidae, 3:389
Incus, 12:36
Indeterminate cleavage, 1:19
  See also Radial cleavage; Reproduction
Indexes
  indirect observations and, 12:199–202
  population, 12:195–196
Indian ariommas, 5:422
Indian blackbuck antelopes, 15:271
Indian bullfrogs. *See* Indian tiger frogs
Indian bustards. *See* Great Indian bustards
Indian buttonquails. *See* Barred buttonquails
Indian civets, 14:336, 14:345*t*
Indian cobras, 7:56
Indian crested porcupines, 16:*352*, 16:*356*, 16:*358*, 16:359–360
Indian desert hedgehogs, 13:204, 13:206, 13:*208*, 13:*210*
Indian elephants. *See* Asian elephants
Indian flapshell turtles, 7:*153*, 7:*154*
Indian flashlightfishes. *See* Splitfin flashlightfishes
Indian flatheads, 5:*160*, 5:*161*, 5:162
Indian flying foxes, 13:307, 13:*320*, 13:322, 13:*324*, 13:*327*, 13:328
Indian fruit bats, 13:*307*, 13:*338*, 13:*344*
Indian garden lizards. *See* Brown garden lizards
Indian gharials. *See* Gharials
Indian giant squirrels, 16:167, 16:174*t*

Indian golden geckos, 7:261
Indian gray hornbills, 10:72, 10:73, 10:75
Indian gray mongooses, 14:351–352
Indian hill mynas. *See* Hill mynas
Indian honeyguides. *See* Yellow-rumped honeyguides
Indian house crickets, 3:57
Indian mealmoths, 3:76, 3:*392*, 3:401–*402*
Indian mongooses, 12:190–191, 14:*350*, 14:352
Indian muntjacs, 15:343, 15:*344*, 15:*346*, 15:*347*, 15:*350*, 15:351–352
Indian mynas. *See* Common mynas
Indian mythology, snakes in, 7:*52*, 7:55–56
Indian pangolins, 16:110, 16:113, 16:*114*, 16:*115*, 16:116
Indian peafowls, 8:399
Indian pied hornbills, 10:75
Indian pied kingfishers. *See* Pied kingfishers
Indian pittas, 10:194, 10:*196*, 10:*200*
Indian porcupines. *See* Indian crested porcupines
Indian pythons, 7:42, 7:56, 7:420–422
Indian rabbits. *See* Central American agoutis
Indian rhinoceroses, 15:*223*, 15:*251*, 15:254, 15:*256*, 15:*258–259*
  behavior, 15:218
  conservation status, 12:224, 15:222, 15:254
  habitats, 15:252–253
  humans and, 15:254–255
Indian ringneck parakeets. *See* Rose-ringed parakeets
Indian sandgrouse. *See* Chestnut-bellied sandgrouse
Indian skimmers, 9:205
Indian spotted chevrotains. *See* Spotted mouse deer
Indian stick insects, 3:*227*, 3:*229–230*
Indian tapirs. *See* Malayan tapirs
Indian tent turtles, 7:115
Indian tiger frogs, 6:251, 6:*253*, 6:257
Indian toads. *See* Common Sunda toads
Indian tragopans. *See* Satyr tragopans
Indian tree shrews, 13:197, 13:292, 13:*293*, 13:*294*
Indian tree swifts. *See* Crested tree swifts
Indian treepies. *See* Rufous treepies
Indian white-backed vultures. *See* White-rumped vultures
Indian wild asses. *See* Khurs
Indian Wildlife Protection Act of 1972, 15:280
Indian wolves, 12:180
Indian wood ibis. *See* Painted storks
Indiana bats, 13:310–311, 13:316, 13:502
Indicator spp., 10:137–144
Indicator archipelagicus. See Malaysian honeyguides
Indicator indicator. See Greater honeyguides
Indicator maculatus. See Spotted honeyguides
Indicator minor. See Lesser honeyguides
Indicator pumilio. See Dwarf honeyguides
Indicator variegatus. See Greater scaly-throated honeyguides; Scaly-throated honeyguides
Indicator xanthonotus. See Yellow-rumped honeyguides
Indicatoridae. *See* Honeyguides
Indigo snakes, 7:466, 7:470, 7:*472*, 7:*475–476*
Indigobirds, 8:40, 11:356, 11:363, 11:375–378, 11:*381*
Indirana spp., 6:246, 6:249
Indirana beddomii. See Beddome's Indian frogs
Indirect development, 1:19, 2:21
  See also Reproduction
Indirect observations, 12:199–202
Indo-Burma hotspot, 12:218
Indo-Pacific bottlenosed dolphins, 15:7
Indo-Pacific crabs, 2:201
Indo-Pacific crocodiles. *See* Saltwater crocodiles
Indo-Pacific swamp crabs. *See* Mangrove crabs
Indochinese brush-tailed porcupines. *See* Asiatic brush-tailed porcupines
Indochinese gibbons. *See* Black crested gibbons
Indochinese shrews. *See* Gray shrews
Indonesian coelacanths, 4:189, 4:*191*, 4:193, 4:*194*, 4:199
Indonesian cuscuses, 13:34
Indonesian Komodo dragons. *See* Komodo dragons
Indonesian mountain weasels, 14:324
Indonesian porcupines, 16:*356*, 16:*357*
Indonesian stink badgers, 14:333*t*
Indopacetus spp., 15:59
Indopacetus pacificus. See Longman's beaked whales
Indopacific beaked whales. *See* Longman's beaked whales
Indoreonectes spp. *See* Hillstream loaches
Indostomidae, 4:57, 5:131, 5:133, 5:134
Indostomus spp. *See* Indostomidae
Indostomus paradoxus. See Armored sticklebacks
Indri indri. See Indris
Indricotherium asiaticum, 15:249
Indricotherium transouralicum, 15:216
Indriidae, 12:136, 14:4, **14:63–72,** 14:*68*
  behavior, 14:*64*, 14:65–66
  conservation status, 14:67
  distribution, 14:*63*, 14:65
  evolution, 14:63
  feeding ecology, 14:66
  habitats, 14:65
  humans and, 14:67
  physical characteristics, 14:63–65
  reproduction, 14:66–67
  species of, 14:*69–72*
  taxonomy, 14:63
Indris, **14:63–72,** 14:*65*, 14:*66*, 14:*68*
  behavior, 14:6, 14:8, 14:65, 14:66
  conservation status, 14:67
  distribution, 14:65
  evolution, 14:63
  feeding ecology, 14:66
  habitats, 14:65
  humans and, 14:67
  physical characteristics, 14:63–65
  reproduction, 14:66
  taxonomy, 14:4, 14:63
Induced luteinization, 12:92
Induced ovulation, 12:109
  See also Reproduction
Indus dolphins. *See* Ganges and Indus dolphins
Inermicapsiferidae, 1:226
Infections, parasitic, 1:34–36, 2:28, 2:33
Infectious diseases, 1:34–36, 12:215–216, 12:222–223
Influenza, birds and, 8:25
Infrared energy, 12:83–84
Ingerana spp., 6:246, 6:248, 6:249, 6:250, 6:251
Ingolfiellidea, 2:261
Inia geoffrensis. See Botos
Inia geoffrensis boliviensis, 15:27
Inia geoffrensis geoffrensis, 15:27
Inia geoffrensis humboldtiana, 15:27
Iniidae. *See* Botos
Inimicus spp. *See* Bearded ghouls
Inimicus didactylus. See Bearded ghouls
Inioidea, 15:23
Inkfishes, 4:447, 4:448
  See also Crestfishes
Inland dotterels, 9:161
Inland silversides, 5:70, 5:*72*, 5:*74*
Innkeeper worms, 2:104, 2:*105*, 2:*106*, 2:107
Inocellia crassicornis. See Schummel's inocelliid snakeflies
Inocelliid snakeflies
  Brauer's, 3:*299*, 3:300–*301*
  Schummel's, 3:*299*, 3:*300*
Inocelliidae, 3:297, 3:298
Inquilism, 2:31, 2:33
  See also Behavior
Insecta. *See* Insects
Insecticides, 3:77
Insectivora, **13:193–201,** 14:8
  field studies of, 12:201
  Iguanidae as, 7:246
  phylogenetic trees and, 12:33
  reproduction, 12:90, 12:91, 12:92, 12:94, 12:95
  subterranean, 12:71
  See also specific insectivores
Insectivorous bats, 13:313
  See also specific insectivorous bats
Insects, 2:25, 2:36, 2:37
  behavior, 3:5–6, **3:59–66**
  beneficial, 3:78–82
  biogeography, **3:52–58**
  as bioindicators, 3:82
  as biological controls, 3:80
  conservation status, 3:82, **3:84–91,** 3:86–88
  defined, **3:3–6**
  diapause and, 3:36–37, 3:49–50
  distribution, **3:52–58**
  ecology, **3:41–51**
  as educational tools, 3:82–83
  evolution, **3:7–16,** 3:9–11, 3:44–45
  farming of, 3:80
  feeding ecology, 3:5–6, 3:19–21, 3:41–42, 3:46, 3:62–64
  flight, 3:*19*, 3:22–24, 3:45
  as food, 3:80–81
  function, **3:17–30**
  giant, 3:13–14
  humans and, **3:74–83**
  life histories, **3:31–40**
  medicinal, 3:81
  metamorphosis, 3:*33*–35, 3:44
  migration, 3:36–38, 3:66
  molting, 3:31–33, 3:*32*
  nematodes and, 1:36
  as pests, 3:74–77
  phobias, 3:77
  physical characteristics, 3:4–5, 3:10, 3:12–13, 3:17–30, 3:*23*, 3:43–44
  physical forces and, 3:44
  plants and, 3:14, 3:48–49
  pollination by, 3:49, 3:78–79

reproduction, 3:4, 3:28–*30*, **3:31–40**, 3:35, 3:44, 3:46–47, 3:59–62
social, 1:37
structure, **3:17–30**
succession, 3:47
taxonomy, 3:4–5, **3:7–16**
temperatures and, 3:36–38, 3:44
terrestrial, 3:11–12, 3:22, 3:82
See also Social insects; specific insects; specific types of insects
Insemination, traumatic, 3:263
See also Copulation; Reproduction
Inserted sequences. *See* Retroposons
Inshore sand lances, 5:*336*, 5:*337*
Inshore squids, 2:475, 2:477, 2:478, 2:*481*, 2:*482*–483
Insight, problem solving by, 12:140
See also Cognition
Instars, larval, 3:31
Instrumental conditioning, 1:41
See also Behavior
Insuetophrynus spp., 6:157
Insular fruit bats. *See* Tongan flying foxes
Insulation, hair for, 12:38
See also Physical characteristics
Integrated pest management (IPM), 3:50, 3:76, 3:80, 12:186–188, 12:193
Integrated rabbit control, 12:186–188
Integripalpia. *See* Tube case makers
Integumental derivatives, 12:7–8
Integumentary sense organs (ISOs), 7:160
Integumentary system, 4:15–16, 7:29–30, 12:36–39
chameleons, 7:224–225, 7:*226*
impermeability of, 7:3
lissamphibians, 6:15, 6:*26*
skin shedding, in Squamata, 7:198
toxic, 6:198
turtles, 7:67
See also Physical characteristics
Intelligence, **12:149–164**
avian, 8:18
of Varanidae, 7:360
Inter-sexual selection, 12:101–102
See also Reproduction
Interdependence, in ecosystems, 12:216–217
Interindividual discrimination, 12:9–10
Intermediate horseshoe bats, 13:390, 13:392
Intermediate slit-faced bats, 13:373, 13:376*t*
Internal fertilization, 2:17–18, 2:23–24
See also Reproduction
Internal organs, 12:174
crocodilian, 7:*160–161*
lizards, 7:*199*
snakes, 7:*200*
Testudines, 7:*68*
tuatara, 7:*191*
See also Hearts; Lungs; Physical characteristics
International Code of Zoological Nomenclature, 2:4, 2:8
International Committee on Zoological Nomenclature
bushbabies, 14:23
primates
Lorisidae, 14:13
International Coral Reef Initiative, 1:107
International Crane Foundation, 9:29
International Shrike Working Group, 10:430
International Species Information System (ISIS), 12:205, 12:206
International Symposium on Phylogeny and Classification of Neotropical Fishes, 4:351
International Union for Conservation of Nature. *See* IUCN Red List of Threatened Species
International Whaling Commission, 15:9
Balaenidae, 15:*111*
gray whales, 15:99
northern bottlenosed whales, 15:62, 15:65
pygmy right whales, 15:105
rorquals, 15:124–125
sperm whales, 15:78
Interneurons, 3:25–26
Interspecific behavior, 1:26–27, 2:28, 3:45, 3:46
See also Behavior; Symbiosis
Intestines, 12:47–48, 12:123–124
See also Physical characteristics
Intoshia spp., 1:99
Intoshia linei, 1:*100*, 1:101–*102*
Intra-sexual selection, 12:101–102
See also Reproduction
Intracellular symbiosis, 1:31–32
Intraspecific behavior, 1:26–27, 2:28, 3:45, 3:46
See also Behavior
Intrauterine cannibalism, 4:133
See also Reproduction
Introduced species, 1:51, 2:43, 4:75, 12:139, 12:215
amphibians and, 6:54, 6:58, 6:191, 6:251, 6:262
Australian honeyeaters and, 11:238
Australian region, 12:137–138, 12:185–188
BirdLife International on, 11:508
competition and, 12:118
Corvidae and, 11:508
Hawaiian honeycreepers and, 11:343
Hawaiian Islands, 12:188–192
hummingbirds and, 9:449–450
invasives and pests, 12:182–193
piopios and, 11:118
Raphidae and, 9:270–271
as tody predators, 10:28
white-eyes and, 11:229–230
See also Conservation status; Humans
Introverta, 1:9, 1:12–13
Invasive species, **12:182–193**
See also Introduced species
Inverse square law, 8:39
Invertebrates
behavior of, 1:37–38
marine algal symbiosis and, 1:33
Inyo Mountains salamanders, 6:*394*, 6:*397*
Iodopleura spp. *See* Cotingas
Iodopleura pipra, 10:308
Iomys horsfieldii. *See* Javanese flying squirrels
Ioras, **10:415–424**, 10:*418*
IPM (integrated pest management), 3:50, 3:76, 3:80, 12:186–188, 12:193
Ipnopidae, 4:49, 4:432
Ipswich sparrows. *See* Savannah sparrows
Irediparra gallinacea. *See* Comb-crested jacanas
Irena spp. *See* Fairy bluebirds
Irena cyanogaster. *See* Philippine fairy bluebirds
Irena puella. *See* Fairy bluebirds
Irenidae. *See* Fairy bluebirds; Leafbirds
Irenomys spp., 16:264
Irenomys tarsalis. *See* Chilean tree mice
Iridescent shark-catfishes, 4:*355*, 4:*359*, 4:363
Iridictyon spp., 3:55
Iridopterygidae, 3:177
Iriomote cats, 14:369
Irish elk, 12:19, 12:102
Iron, 12:120
Irrawaddy dolphins, 15:6, 15:43, 15:44, 15:50
Irruptive movements, 8:29, 8:31
Irukandjis, 1:149, 1:*150*, 1:*151*
Irwin, M. P. S., 11:207
Isabela orioles, 11:428
Isabelline pratincoles. *See* Australian pratincoles
Isack, H. A., 10:138
Isancistrum spp., 1:214
Ischnocera, 3:249, 3:250, 3:251
Ischnocnema spp., 6:156
Ischnopsyllidae, 3:347, 3:349
Ischnopsyllus octactenus. *See* Bat fleas
Ischyrorhyncus spp., 15:27
Isectolophidae, 15:237
Ishida, Minoru, 5:163
Isidore's babblers. *See* Rufous babblers
ISIS (International Species Information System), 12:205, 12:206
Isistius spp., 4:151
Isistius brasiliensis. *See* Cookie-cutter sharks
Island biogeography, 3:56
Island canaries, 11:323, 11:325, 11:*327*, 11:338–*339*
Island flying foxes, 13:*323*, 13:*325*
Island foxes, 12:192, 14:273, 14:284*t*
Island hornbills. *See* Sulawesi red-knobbed hornbills
Island night lizards, 7:292
Island seals. *See* Harbor seals
Island spotted skunks, 12:192
Islands, 12:131, 12:220–221
Isle of Pines trogons. *See* Cuban trogons
Islets of Langerhans, lissamphibians, 6:20
Iso spp., 5:68
Iso rhothophilus. *See* Flowers of the wave
Isocrinida, 1:355, 1:356
Isogomphodon spp., 4:114
Isolobodon spp., 16:463
Isolobodon portoricensis. *See* Puerto Rican hutias
Isoodon auratus. *See* Golden bandicoots
Isoodon macrourus. *See* Northern brown bandicoots
Isoodon obesulus. *See* Southern brown bandicoots
Isopoda, **2:249–260,** 2:*253*
behavior, 2:250–251
conservation status, 2:252
distribution, 2:250
evolution, 2:249
feeding ecology, 2:29, 2:251–252
habitats, 2:250
humans and, 2:252
physical characteristics, 2:249–250
reproduction, 2:252
species of, 2:*254*–260
taxonomy, 2:249
Isoptera. *See* Termites
ISOs. *See* Integumentary sense organs
Isothrix bistriata. *See* Toros
Isothrix pagurus, 16:450–451
Isotoma viridis. *See* Varied springtails
Istieus spp., 4:249

Istiophoridae, 4:49, 4:61, 4:68, 5:405
Istiophorus platypterus. See Sailfishes
Isurus oxyrinchus. See Shortfin makos
Italian cave salamanders, 6:*394*, 6:400–401
Iteroparous fishes, 4:47
  See also specific fishes
Ithone spp., 3:308
Ithonidae, 3:305, 3:306, 3:307–308
Itoitantulus misophricola, 2:*285*, 2:*286*
Itombwe nightjars, 9:370, 9:405
Itombwe owls, 9:332, 9:336
Itycirrhitus wilhelmi, 5:237–238
IUCN African Elephant Specialist Group, 15:163
IUCN Caprinae Specialist Group (CSG), 16:87
IUCN Chiroptera Specialist Group, 13:358, 13:420
IUCN Hornbill Specialist Group, 10:75
IUCN Red List of Threatened Species, 1:48, 2:43, 3:85
  amphibians
    African treefrogs, 6:283
    Amero-Australian treefrogs, 6:230
    Asian treefrogs, 6:293
    Asiatic salamanders, 6:337, 6:341
    Australian ground frogs, 6:141
    Bombinatoridae, 6:84
    common spadefoot, 6:124
    Cryptobranchidae, 6:346
    Discoglossidae, 6:90
    Eungella torrent frogs, 6:153
    flatwoods salamanders, 6:358
    ghost frogs, 6:132, 6:134
    Gill's plantanna, 6:102
    golden coqui, 6:165
    golden toads, 6:185, 6:192
    goliath frogs, 6:256
    helmeted water toads, 6:170
    Houston toads, 6:190
    leptodactylid frogs, 6:159
    lungless salamanders, 6:392
    Maud Island frogs, 6:74
    micro frogs, 6:255
    mole salamanders, 6:356, 6:359
    narrow-mouthed frogs, 6:305
    New Zealand frogs, 6:70, 6:72, 6:73
    olms, 6:381
    parsley frogs, 6:128
    Philippine barbourulas, 6:88
    Proteidae, 6:379
    Salamandridae, 6:367
    Seychelles frogs, 6:136, 6:138
    Seychelles treefrogs, 6:289
    Southern gastric brooding frogs, 6:153
    tiger salamanders, 6:360
    torrent salamanders, 6:386
    true frogs, 6:251
    vocal sac-brooding frogs, 6:174, 6:176
  birds
    Abbott's boobies, 8:214
    ant thrushes, 10:241–242
    babblers, 10:507, 10:516
    Bell's vireo, 11:269
    biak monarchs, 11:101
    bitterns, 8:244–245
    blue chaffinches, 11:329
    boobies, 8:214
    buttonquails, 9:14
    Caprimulgiformes, 9:370
    Ciconiiformes, 8:236
    cormorants, 8:202–203
    drongos, 11:439
    ducks, 8:372–373
    dull-blue flycatchers, 11:36
    eagles, 8:323
    Estrildidae, 11:356
    Eupetidae, 11:76
    Falconiformes, 8:315
    falcons, 8:351
    fantails, 11:87
    finches, 11:324–325
    frigatebirds, 8:195
    Galliformes, 8:401–402
    geese, 8:372–373
    Gruiformes, 9:3
    guineafowl, 8:429
    Hawaiian honeycreepers, 11:347, 11:350, 11:351
    hawks, 8:323
    herons, 8:244–245
    hornbills, 10:75, 10:179
    ibises, 8:293
    Irenidae, 10:417
    kokakos, 11:449
    larks, 10:346
    manakins, 10:297
    mimids, 10:468
    monarch flycatchers, 11:98
    mynas, 11:420
    New World warblers, 11:290–291, 11:298, 11:299
    New Zealand wrens, 10:204
    nuthatches, 11:169
    Old World flycatchers, 11:27
    Old World warblers, 11:7
    ovenbirds, 10:210–211
    owls, 9:338
    Pelecaniformes, 8:186
    pelicans, 8:227
    penguins, 8:151
    Phasianidae, 8:436–437
    pheasants, 8:436–437
    piculets, 10:150
    pittas, 10:195
    rail-babblers, 11:79
    rails, 9:51
    Sangihe shrike-thrushes, 11:118
    screamers, 8:394
    shrikes, 10:429
    spoonbills, 8:293
    swans, 8:372–373
    swifts, 9:424
    tiekes, 11:450
    tinamous, 8:59
    titmice, 11:158
    trogons, 9:479, 9:484, 9:485
    turacos, 9:301
    tyrant flycatchers, 10:275
    weaverfinches, 11:361, 11:364, 11:365, 11:366, 11:372
    woodcreepers, 10:230
    woodpeckers, 10:150
    Yemen accentors, 10:460
  fishes
    Acipenseriformes, 4:215, 4:217, 4:218, 4:219, 4:220
    Alabama cavefishes, 5:10
    albacore, 5:415
    angelsharks, 4:162, 4:164, 4:165
    anglerfishes, 5:51
    Arno gobies, 5:386
    Atheriniformes, 5:70
    Atlantic bluefin tunas, 5:415
    Atlantic cods, 5:33
    Beloniformes, 5:81
    blennies, 5:343
    blind cave eels, 5:155
    blind cave gudgeons, 5:381
    blue sharks, 4:124
    Boeseman's rainbowfishes, 5:75
    bull sharks, 4:121
    Callionymoidei, 5:366–367
    carps, 4:311
    catfishes, 4:354, 4:357
    cavefishes, 4:328, 4:329, 4:333
    characins, 4:338
    cichlids, 5:280
    coelacanths, 4:193
    Cypriniformes, 4:303, 4:322
    Cyprinodontiformes, 5:94
    Devils Hole pupfishes, 5:98
    dogfish sharks, 4:152, 4:158
    duckbilled buntingis, 5:83
    dwarf pygmy gobies, 5:386
    Esociformes, 4:380
    flatfishes, 5:453
    Gadiformes, 5:29–30
    Gasterosteiformes, 5:136
    gobies, 5:377
    gray reef sharks, 4:116, 4:119
    great hammerhead sharks, 4:127
    haddocks, 5:34
    Haplochromis chilotes, 5:286
    herrings, 4:278
    Hexanchiformes, 4:144, 4:147, 4:148
    hogfishes, 5:302
    humphead wrasses, 5:301
    Labroidei, 5:298
    labyrinth fishes, 5:429
    leafy seadragons, 5:147
    lemon sharks, 4:123
    leopard sharks, 4:128
    lined seahorses, 5:146
    mackerel sharks, 4:133, 4:136, 4:137, 4:138, 4:140, 4:141
    Mary River cods, 5:231
    mountain perches, 5:232
    mullets, 5:60
    Nassau groupers, 5:271
    ngege, 5:287
    northern cavefishes, 5:9
    oceanic whitetip sharks, 4:121
    o'opo alamo'os, 5:385
    Ophidiiformes, 5:18
    Orectolobiformes, 4:105
    Osmeriformes, 4:393
    Osteoglossiformes, 4:234, 4:240
    Ozark cavefishes, 5:8
    pajama catsharks, 4:126
    Percoidei, 5:199, 5:222, 5:242–243, 5:262
    Percopsiformes, 5:6
    Rajiformes, 4:178, 4:184, 4:185, 4:186
    ringed pipefishes, 5:146
    salmons, 4:407
    sawsharks, 4:167, 4:170
    Scombroidei, 5:407
    Scorpaeniformes, 5:166, 5:183
    sculptured seamoths, 5:145
    southern cavefishes, 5:10

splendid coral toadfishes, 5:46
swordfishes, 5:418
Synbranchiformes, 5:152
Tetraodontiformes, 5:471
tiger sharks, 4:122
toadfishes, 5:42
trondo mainties, 5:288
weedy seadragons, 5:147
insects
American burying beetles, 3:*332*–333
caddisflies, 3:377
Cheiromeles torquatus, 3:199
Coleoptera, 3:322
Diptera, 3:360
earwigs, 3:197
Hemiptera, 3:263
Hymenoptera, 3:407
large blues, 3:394–395
Lepidoptera, 3:388
mantids, 3:181
mayflies, 3:127
Odonata, 3:134–135
Orthoptera, 3:207
Phasmida, 3:224
Phthiraptera, 3:252
Queen Alexandra's birdwings, 3:398–399
stoneflies, 3:143
lower metazoans and lesser deuterostomes
Anthozoa, 1:107
nemerteans, 1:247, 1:255
starlet sea anemones, 1:112
mammals
African elephants, 15:175
African wild dogs, 14:278
agoutis, 16:409
Ahaggar hyraxes, 15:183
Alouatta palliata mexicana, 14:162
Alouatta seniculus insulanus, 14:163
annamite striped rabbits, 16:509
aquatic genets, 14:341
Artiodactyla, 15:272–273
Arvicolinae, 16:229
Asian elephants, 15:174
babirusas, 15:288
bactrian camels, 15:317, 15:320
Baikal seals, 14:430
bald uakaris, 14:151
bandicoots, 13:6
eastern barred, 13:4
golden, 13:4
western barred, 13:4–5
bats
Brazilian free-tailed, 13:493
brown long-eared, 13:514
California leaf-nosed, 13:423
common bentwing, 13:525
disk-winged, 13:473, 13:474
false vampire, 13:382
funnel-eared, 13:461
giant mastiff, 13:492
greater house, 13:491
horseshoe, 13:392
lesser-crested mastiff, 13:490
lesser mouse-tailed, 13:*353*
long-snouted, 13:429
mouse-tailed, 13:352
moustached, 13:436
naked, 13:491
Old World fruit, 13:321, 13:337
Old World leaf-nosed, 13:405
Parnell's moustached, 13:440
Schnabeli's thumbless, 13:471
slit-faced, 13:373
smoky, 13:468
southern long-nosed, 13:429
white-striped free-tailed, 13:492
bears, 14:300
black crested gibbons, 14:221
black rhinoceroses, 15:260
black spotted cuscuses, 13:39
black-tailed prairie dogs, 16:155
Blanford's foxes, 14:280
Bovidae, 16:8
Antilopinae, 16:48
Bovinae, 16:14
Caprinae, 16:94–95
duikers, 16:77
Brazilian shrew mice, 16:274
brush-tailed bettongs, 13:39, 13:76
bush pigs, 15:285
capuchins, 14:106
Carnivora, 14:262
cats, 14:374
Cebus apella robustus, 14:111
Cetacea
Baird's beaked whales, 15:65
Balaenidae, 15:112
beaked whales, 15:61–62
belugas, 15:86
Blanville's beaked whales, 15:67
Bryde's whales, 15:129
Burmeister's porpoises, 15:39
common bottlenosed dolphins, 15:54
Cuvier's beaked whales, 15:66
dolphins, 15:48
fin whales, 15:128
franciscana dolphins, 15:25
gray whales, 15:99
harbor porpoises, 15:38
killer whales, 15:53
Longman's beaked whales, 15:68
northern bottlenosed whales, 15:65
northern minke whales, 15:129
porpoises, 15:35–36
rorquals, 15:125
Shepherd's beaked whales, 15:68
sperm whales, 15:78, 15:79
Chacoan peccaries, 15:295, 15:300
cheetahs, 14:382
chevrotains, 15:329
chinchillas, 16:380
Chinese dormice, 16:324
clouded leopards, 14:387
colugos, 13:302
common warthogs, 15:286
dasyurids, 12:290–291
deer
black muntjacs, 15:346, 15:351
Chinese water, 15:376
Fea's muntjacs, 15:346, 15:351
Gongshan muntjacs, 15:346, 15:352
greater Malay mouse, 15:333
lesser Malay mouse, 15:332
musk, 15:338
sika, 15:369
Dendromus vernayi, 16:291
desert dormice, 16:325
Dipodidae, 16:216
Dorcas gazelles, 16:51
dormice, 16:321
dugongs, 15:201, 15:204
eastern tree hyraxes, 15:183, 15:187
edible dormice, 16:326
Ethiopian wolves, 14:277
Eurasian beavers, 16:180, 16:183
Eurasian lynx, 14:388
Eurasian wild pigs, 15:288
falanoucs, 14:343
five-toed pygmy jerboas, 16:220
forest dormice, 16:324
forest hogs, 15:284
garden dormice, 16:325
Genetta genetta isabelae, 14:341
Geoffroy's cats, 14:385
giant anteaters, 13:153
giant pandas, 14:304
Gilbert's potoroos, 13:39, 13:76
giraffes, 15:408
Giraffidae, 15:405
golden-cheeked gibbons, 14:222
Goodfellow's tree kangaroos, 13:39
gray-backed sportive lemurs, 14:79
great apes, 14:235
greater bilbies, 13:4, 13:22
guanacos, 15:317–318, 15:322
hamsters, 16:243
hazel dormice, 16:326
Hemiechinus nudiventris, 13:207
Herpestidae, 14:352
Heteromyidae, 16:203
Hispaniolan hutias, 16:466
Hispaniolan solenodons, 13:241
Hugh's hedgehogs, 13:207
hutias, 16:463
hyenas, 14:362
Hylomys suillus parvus, 13:207
Indian rhinoceroses, 15:259
Indriidae, 14:67
Insectivora, 13:199
jaguars, 14:386
Jamaican hutias, 16:465
Japanese dormice, 16:325
Javan rhinoceroses, 15:258
koalas, 13:48
Lagomorpha, 16:487
linh duongs, 16:14
lions, 14:380
long-beaked echidnas, 12:233, 12:238
long-eared jerboas, 16:219
long-footed potoroos, 13:76
long-nosed potoroos, 13:39
long-tailed chinchillas, 16:383
manatees, 15:209
maned sloths, 13:165, 13:169
maned wolves, 14:282
masked titis, 14:153
Michoacán pocket gophers, 16:193
Milne-Edwards's sportive lemurs, 14:81
Mindanao gymnures, 13:212
moles
golden, 13:217
Grant's desert golden, 13:221
large golden, 13:221
marsupial, 13:27
monitos del monte, 12:274
monkeys
blackish squirrel, 14:110
Colombian woolly, 14:166
gray woolly, 14:166
Old World, 14:175

red-backed squirrel, 14:109
squirrel, 14:105–106
mountain beavers, 16:133
Murinae, 16:253
Mustelidae, 14:324
northern bettongs, 13:76
northern hairy-nosed wombats, 13:39
northern muriquis, 14:165
northern sportive lemurs, 14:83
Novae's bald uakaris, 14:151
numbats, 12:305
octodonts, 16:436
opossums
Chilean shrew, 12:268, 12:270
Old World, 12:254, 12:257–263
Otariidae, 14:399
otter civets, 14:343
pacaranas, 16:388
pangolins, 16:113
Pel's anomalures, 16:303
Perissodactyla, 15:221–222
Philippine colugos, 13:304
Philippine gymnures, 13:207
picas, 16:495, 16:500
pigs, 15:280–281, 15:289*t*–290*t*
Pitheciidae, 14:148
pocket gophers, 16:190
polar bears, 14:305
porcupines
North African, 16:361
Old World, 16:353–354
thick-spined, 16:358
possums
little pygmy, 13:108
mountain pygmy, 13:39
scaly-tailed, 13:66
primates, 14:11
Procyonidae, 14:310
Proserpine rock wallabies, 13:39
pumas, 14:382
pygmy hippopotamuses, 15:308, 15:311
pygmy hogs, 15:287
Querétaro pocket gophers, 16:194
rats
dassie, 16:331
giant kangaroo, 16:207
Hainan moon, 13:207
Heermann's kangaroo, 16:208
Hispid cotton, 16:276
Malagasy giant, 16:291
red viscacha, 16:438
Rio de Janeiro rice, 16:275
red bald uakaris, 14:151
red-tailed sportive lemurs, 14:80
Roach's mouse-tailed dormice, 16:325
rodents, 16:126
Russian desmans, 13:286
sengis, 16:522, 16:525–526, 16:529
short-beaked echidnas, 12:238
shrews
Chinese short-tailed, 13:256
red-toothed, 13:252
white-toothed, 13:268
Siberian zokors, 16:294
Sigmodontinae, 16:270
Sirenia, 15:195
small-toothed sportive lemurs, 14:83
snow leopards, 14:387
solenodons, 13:239
southern African hedgehogs, 13:210
southern bearded sakis, 14:151
southern muriquis, 14:164
spectacled bears, 14:305
spectacled dormice, 16:323
sportive lemurs, 14:76
squirrels
arctic ground, 16:152
European ground, 16:156
flying, 16:137
Idaho ground, 16:158
tree, 16:167
Sumatran rhinoceroses, 15:257
Talpidae, 13:282
tapirs, 15:241–242, 15:243, 15:257–262
tarsiers, 14:95
Tasmanian bettongs, 13:76, 13:78
Tasmanian wolves, 12:309
tenrecs, 13:229
tigers, 14:381
true seals, 14:422
tuco-tucos, 16:427
Ucayali bald uakaris, 14:151
Uta Hick's bearded sakis, 14:151
Vancouver Island marmots, 16:156
Vespertilionidae, 13:502, 13:522
vicuñas, 15:317–318, 15:321
Viverridae, 14:338
volcano rabbits, 16:514
water chevrotains, 15:331
weasel sportive lemurs, 14:82
western barbastelles, 13:507
white bald uakaris, 14:151
white-cheeked gibbons, 14:221
white-footed sportive lemurs, 14:82
white rhinoceroses, 15:262
white-tailed mice, 16:289
wood lemmings, 16:236
yellow-breasted capuchins, 14:112
yellow-footed rock wallabies, 13:39
yellow-spotted hyraxes, 15:183
protostomes
agate snails, 2:418
Amathimysis trigibba, 2:220
Arachnida, 2:336
Bermudamysis speluncola, 2:217
bivalves, 2:454
caenogastropods, 2:447
copepods, 2:300
Decapoda, 2:201
earthworms, 2:67
European medicinal leeches, 2:81
European pearly mussels, 2:463
fairy shrimps, 2:137
giant clams, 2:464
Gippsland giant worms, 2:72
Isopoda, 2:252
leeches, 2:77–78
Neritopsina, 2:441
Onychophora, 2:111
Platyops sterreri, 2:217
Pulmonata, 2:414
Scolopendra abnormis, 2:356
Spelaeomysis bottazzii, 2:*219*
Stygiomysis hydruntina, 2:217
true limpets, 2:425
vernal pool tadpole shrimps, 2:142, 2:145
Vetigastropoda, 2:431
reptiles
African sideneck turtles, 7:130
Afro-American river turtles, 7:138
Agamidae, 7:211
alligator snapping turtles, 7:94, 7:96
Alligatoridae, 7:173
American crocodiles, 7:184
Anguidae, 7:340
armored chameleons, 7:236
Australo-American sideneck turtles, 7:79, 7:83
big-headed turtles, 7:136
blindsnakes, 7:382
boas, 7:411
Central American river turtles, 7:100
chameleons, 7:232
Chinese alligators, 7:173, 7:176
Chinese softshell turtles, 7:155
Chinese stripe-necked turtles, 7:119
Coachella Valley fringe-toed lizards, 7:255
colubrids, 7:470
common chameleons, 7:238
Cordylidae, 7:322
crocodiles, 7:182
desert tortoises, 7:147
diamondback terrapins, 7:109
Eastern box turtles, 7:113
Elapidae, 7:486
European pond turtles, 7:112
false blindsnakes, 7:388
Galápagos tortoises, 7:147
Geoemydidae, 7:116
gharials, 7:169
giant girdled lizards, 7:324
green seaturtles, 7:90
Helodermatidae, 7:357
Hermann's tortoises, 7:148–149
Iguanidae, 7:246–247
jeweled chameleons, 7:239
Kemp's ridley turtles, 7:91
Kinosternidae, 7:122
Komodo dragons, 7:368
KwaZulu-Natal Midlands dwarf chameleons, 7:235
Lacertidae, 7:298–299
leaf-horned agamas, 7:215
leatherback seaturtles, 7:102
loggerhead turtles, 7:89
Madagascan big-headed turtles, 7:142
minor chameleons, 7:240
mugger crocodiles, 7:185
New World pond turtles, 7:106
painted terrapins, 7:118
pancake tortoises, 7:148
pig-nose turtles, 7:76
pond sliders, 7:111
pythons, 7:421
skinks, 7:329
softshell turtles, 7:152
South American river turtles, 7:140
South American yellow-footed tortoises, 7:146
Sphenodon guntheri, 7:192
splitjaw snakes, 7:430
spotted turtles, 7:111
Teiidae, 7:312
tortoises, 7:144
Varanidae, 7:363
Viperidae, 7:448
yellow-margined box turtles, 7:119
See also Conservation status
IUCN/SSC Pigs, Peccaries, and Hippos Specialist Group, 15:275

cebu bearded pigs, 15:281
peccaries, 15:295
pygmy hogs, 15:280, 15:281, 15:287
IUCN Tapir Specialist Group, 15:242
IUCN (World Conservation Union), 12:213–214
Iverson, Volquard, 9:269, 9:272
Ivory-billed woodpeckers, 10:88, 10:150, 10:*152*, 10:166–*167*
Ivory gulls, 9:203
Ixalotriton spp., 6:391
Ixalus pictus. See Painted Indonesian treefrogs
Ixobrychus exilis. See Least bitterns
Ixobrychus minutus. See Little bitterns
Ixobrychus sturmii. See Dwarf bitterns
Ixonotus guttatus. See Spotted greenbuls
Izubrs, 15:358

J

Ja slit-faced bats, 13:373, 13:377*t*
Jabiru mycteria. See Jabirus
Jabiru storks. *See* Jabirus
Jabirus, 8:265, 8:266, 8:*269*, 8:*273–274*
Jacamaralcyon tridactyla. See Three-toed jacamars
Jacamars, 10:85–88, **10:91–99**, 10:*94*
Jacamerops aurea. See Great jacamars
Jacana spinosa. See Northern jacanas
Jacanas, **9:107–114**, 9:*111*
See also Charadriiformes
Jacanidae. *See* Jacanas
Jack crevalles. *See* Crevalle jacks
Jack snipes, 9:177
Jackals, 12:*144*, 12:148, 14:255, 14:260, **14:265–273, 14:283*t***
Jackass kookaburras. *See* Laughing kookaburras
Jackdaws, 11:505, 11:506, 11:*511*, 11:519
Jackrabbits
antelope, 16:487, 16:*506*, 16:515*t*
black-tailed, 16:484, 16:*485*, 16:487, 16:*508*, 16:515*t*
tehuantepec, 16:487
white-sided, 16:516*t*
Jacks, 5:255
amberjacks, 4:68
bar, 5:*257*
crevalle, 5:*264*, 5:*266*, 5:267
deepwater, 5:166
See also Carangidae; Northern pikes
Jackson, Jeremy B. C., 1:50
Jackson, K. L., 5:179
Jackson's bustards. *See* Denham's bustards
Jackson's centipede eaters, 7:462
Jackson's chameleons, 7:228, 7:*229*, 7:*233*, 7:*235*, 7:238–239
Jackson's dancing whydahs. *See* Jackson's widow-birds
Jackson's mongooses, 14:352
Jackson's widow-birds, 11:377, 11:379, 11:*381*, 11:*390*–391
Jackson's widows. *See* Jackson's widow-birds
Jacky winters, 11:*107*, 11:*108*
Jacobson's organ, 7:*31*, 7:195, 12:80, 12:*103*
See also Physical characteristics
Jaculus spp., 16:212
Jaculus blanfordi, 16:216
Jaculus blanfordi margianus, 16:216
Jaculus blanfordi turcmenicus, 16:216
Jaculus jaculus. See Lesser Egyptian jerboas
Jaculus orientalis. See Greater Egyptian jerboas
Jade treefrogs, 6:292
Jaegers, 9:104, 9:203–209
See also Charadriiformes
Jaguars, 12:*123*, 14:371, 14:374, 14:*377*, 14:*386*
Jaguarundis, 12:116, 14:390*t*
Jamaican blackbirds, 11:*308*, 11:*311*–312
Jamaican boas, 7:411
Jamaican doctor birds. *See* Red-billed streamertails
Jamaican fruit-eating bats, 13:*416*, 13:417, 13:419, 13:420, 13:*421*, 13:*428*, 13:432
Jamaican hutias, 16:*462*, 16:*463*, 16:*464*, 16:*465*
Jamaican Iguana Research and Conservation Group, 7:247
Jamaican iguanas, 7:246, 7:247
Jamaican monkeys, 12:129
Jamaican poorwills, 9:370, 9:405
Jamaican streamertails, 9:441
James' flamingos, 8:303, 8:304, 8:305, 8:306, 8:*307*, 8:*307–308*, 8:*310–311*
Jameson, E. W., 16:143
Jameson's antpeckers. *See* Red-fronted flowerpecker weaver-finches
Jameson's firefinches, 11:*358*, 11:*363*
Jamieson, B. G. M., 2:65
Jamoytius spp., 4:9
Janis, Christine, 15:134
Janthinidae, 2:446
Janvier, P., 4:9
Japan. *See* Palaearctic region
Japan sea basses. *See* Japanese perches
Japanese aras, 5:255
Japanese beaked whales. *See* Ginkgo-toothed beaked whales
Japanese boarfishes, 5:*123*
Japanese clawed salamanders, 6:337, 6:*338*, 6:340–341
Japanese click beetles, 3:81
Japanese colored carps, 4:*299*, 4:303, 4:311
Japanese cranes. *See* Red-crowned cranes
Japanese crested ibises. *See* Japanese ibises
Japanese deep-sea girdle wearers, 1:*346*, 1:*348*–349
Japanese deep-sea limpets, 2:*437*
Japanese dormice, 16:317, 16:319, 16:*322*, 16:*324*, 16:325
Japanese fire-bellied newts, 6:*368*, 6:*370*–371
Japanese giant perches, 5:220
Japanese giant salamanders, 6:343, 6:*344*, 6:345, 6:346, 6:347
Japanese gobies, 5:377
Japanese groupers. *See* Japanese aras
Japanese hagfishes, 4:79
Japanese horn-faced bees, 3:78–79
Japanese horseshoe crabs, 2:328, 2:*330*, 2:*331*, 2:332
Japanese ibises, 8:293, 8:*295*, 8:298–299
Japanese ice gobies, 5:376
Japanese lungless salamanders. *See* Japanese clawed salamanders
Japanese macaques, 12:135, 12:*147*, 12:157, 14:*3*, 14:5, 14:*8*, 14:10, 14:*191*, 14:194
Japanese marbled salamanders, 6:*338*, 6:*339*–340
Japanese medakas. *See* Japanese rice fishes
Japanese mountain moles, 13:282
Japanese night herons, 8:236, 8:245
Japanese perches, 5:*223*, 5:*227*, 5:230–231
Japanese pilchards, 4:279
Japanese pineapplefishes. *See* Pineconefishes
Japanese praying mantis, 3:*177*
Japanese rice fishes, 5:*82*, 5:*83*
Japanese rock-crawlers, 3:190, 3:*191*, 3:*192*
Japanese sandperches, 4:66
Japanese seaperches. *See* Japanese perches
Japanese serows, 16:87, 16:*90*, 16:92, 16:93, 16:95, 16:103*t*
Japanese wagtails, 10:372
Japanese waxwings, 10:447
Japanese white-eyes, 11:228, 11:*231*, 11:*233*
Japygidae, 3:107, 3:*108*
Jar seals. *See* Harp seals
Jara, Fernando, 5:99
Jaragua sphaero, 7:259
Jardin des Plantes, Paris, France, 12:203
Jardine's hummingbirds. *See* Velvet-purple coronets
Jarvis, Jennifer U. M., 12:70
Java file snakes, 7:439–*440*, 7:441, 7:*442*, 7:*443*, 7:*485*
Java finches. *See* Java sparrows
Java frogs. *See* Pointed-tongue floating frogs
Java sparrows, 8:*10*, 11:353, 11:*355*, 11:357, 11:*359*, 11:*371–372*
Java temple birds. *See* Java sparrows
Javan frogmouths, 9:377, 9:378, 9:379
Javan gibbons. *See* Moloch gibbons
Javan gold-spotted mongooses. *See* Small Indian mongooses
Javan leaf insects, 3:*226*, 3:*229*, 3:231–232
Javan pigs, 15:280, 15:289*t*
Javan plovers, 9:164
Javan rhinoceroses, 15:*256*
behavior, 15:258
conservation status, 12:218, 15:222, 15:254, 15:258
distribution, 15:*258*
evolution, 15:249
feeding ecology, 15:258
habitats, 15:252, 15:258
humans and, 15:258
physical characteristics, 15:251, 15:258
reproduction, 15:258
taxonomy, 15:249, 15:258
Javan rusa deer, 15:*360*
Javan short-tailed porcupines, 16:*359*
Javan slit-faced bats, 13:373, 13:376*t*
Javan trogons, 9:*480*, 9:*481*–482
Javan wart snakes. *See* Java file snakes
Javan warty pigs. *See* Javan pigs
Javanese flying squirrels, 16:141*t*
Javanese lapwings, 9:163
Javanese tree shrews, 13:297*t*
Javelinas. *See* Collared peccaries
Jaw animals, 1:9, 1:12, **1:327–330**, 1:*328*, 1:*329*
Jaw-closing mechanisms, 6:415–416
Jaw prehension, 7:4, 7:198, 7:200, 7:204
See also Physical characteristics
Jawed fishes (Gnathostomata), 4:10
Jawfishes (Opistognathidae), 4:48, 5:255, 5:256, 5:258–263
Jawless fishes, 4:9–10
Jaws, 12:9, 12:11, 12:*41*
of carnivores, 12:14

hearing and, 12:50
See also Physical characteristics
Jays, **11:503–510, 11:512–515**
Jefferies, R. P. S., 1:13
Jeffrey, Paul, 2:412
Jeholotriton spp., 6:13
Jellyfish, **1:153–168,** *1:158*, *1:159*, 2:36
behavior, 1:24, 1:30, 1:32, 1:38–39, 1:42, 1:155
conservation status, 1:51, 1:156
distribution, 1:154
evolution, 1:3–4, 1:6, *1:9*, 1:153
feeding ecology, 1:29, 1:155–156
habitats, 1:24, 1:154–155
humans and, 1:156–157
physical characteristics, *1:4*, 1:153–154, *1:155*
reproduction, 1:16, 1:21, 1:156
species of, *1:160*–167
Stromateoidei and, 5:422
symbiosis and, 1:33
taxonomy, 1:10, 1:153
See also specific types of jellyfish
Jenkinsia spp. *See* Dwarf herrings
Jenny wrens. *See* Winter wrens
Jentink's dormice, 16:327*t*
Jentink's duikers, 16:76, *16:79*, *16:82*
Jenynsia spp., 5:89, 5:93
Jenynsiidae, 5:89
Jerboa marsupial mice. *See* Kultarrs
Jerboa mice. *See* Australian jumping mice
Jerboas, 16:123, **16:211–224,** *16:218*, 16:223*t*–224*t*
Jerdon's coursers, 9:153
Jerdon's leafbirds. *See* Blue-winged leafbirds
Jerdon's palm civets, 14:338
Jersey Wildlife Preservation Trust, on splitjaw snakes, 7:430
Jerusalem crickets, 3:201, *3:205*
Jerusalem haddocks. *See* Opahs
Jesus birds. *See* Jacanas
Jesus Christ lizards. *See* Common basilisks; Green basilisks
Jewel-babblers, 11:75, 11:76
Jewel beetles, 3:322, 3:323
Jewelbugs, *3:2*
Jeweled anenomes, *1:103*
Jeweled beetles. *See* Ma'kechs
Jeweled blennies, *5:343*
Jeweled chameleons, 7:232, *7:234*, *7:239*
Jeweled ma'kechs. *See* Ma'kechs
Jewelfishes. *See* Yellowtail damselfishes
Jewelry, insect, 3:74, 3:323
See also Humans
Jicotea turtles. *See* Central American river turtles
Jiggers. *See* Chigoes
Jirds. *See* Sundevall's jirds
Joaquin's petrels, 8:132
John dories, 5:123–124, *5:125*, *5:129*–130
Johnny darters, 5:197, 5:198
Johnsgard, Paul A.
on Anatidae, 8:369
on buttonquails, 9:11
on Gruiformes, 9:1
Johnson, David, 5:163
Johnson, G. D., 4:389, 4:390, 4:405, 4:431
on blennies, 5:341
on Scombroidei, 5:405
on Stephanoberyciformes, 5:105
on Synbranchiformes, 5:151
on Trachinoidei, 5:331
on Zeiformes, 5:123
Johnston, Sir Harry, 15:406
Johnstone's crocodiles, 7:163, 7:164, 7:180, *7:183*, 7:184–*185*
Johnston's genets, 14:338
Johnston's hyraxes, 15:177, 15:189*t*
Johnston's mountain turacos. *See* Ruwenzori turacos
Joint-tubed bryozoans, *2:505*, *2:506*
Jointed skeletons, *2:18*
See also Physical characteristics
Jones, C. M., 8:147
Jones, W. J., 5:195
Jordan, D. S., 4:131
on Cyprinodontidae, 5:89
on labyrinth fishes, 5:427
on sailfin sandfishes, 5:339
Jordanella floridae, 5:92
Jordania zonope. *See* Longfin sculpins
Joturus spp., 5:59
Journal für Ornithologie, 8:27
The Journal of Field Ornithology, 8:27
The Journal of Raptor Research, 8:27
Joyful bulbuls. *See* Joyful greenbuls
Joyful greenbuls, *10:401*, 10:406–*407*
Juan de Fuca liparids. *See* Slipskin snailfishes
Juan Fernández firecrowns, 9:418, 9:441, 9:443, 9:449–450
Juan Fernández fur seals, 14:394, 14:395, 14:398, 14:399, 14:407*t*
Judeo-Christian religions, snakes in, 7:57
Juliana's golden moles, 13:216, 13:222*t*
Juliomys spp., 16:264
Julus scandinavius. *See* Snake millipedes
Jumbie bats. *See* Northern ghost bats
Jumping ability, frogs, 6:25–26, *6:53*, 6:66, *6:66*
Jumping characins. *See* Splash tetras
Jumping Frog Jubilee, *6:53*
Jumping mice, 16:123, 16:125, **16:211–224,** 16:223*t*–224*t*
Jumping shrews. *See* Round-eared sengis
Junco hyemalis. *See* Dark-eyed juncos
Junco hyemalis aikeni. *See* White-winged juncos
Junco hyemalis hyemalis. *See* Slate-colored juncos
Junco hyemalis mearnsi. *See* Pink-sided juncos
Junco hyemalis oreganus, 11:271
June beetles, 3:80
Jungle babblers, *10:506*
Jungle cats, 14:290, 14:390*t*
Jungle leeches. *See* Tiger leeches
Jungle nymphs, 3:224, *3:226*, 3:228–*229*
Jungle perches, 5:220, 5:221, *5:224*, 5:228–*229*
Jungle wallabies. *See* Agile wallabies
Junglefowls, 8:434
Junín grebes, 8:170, 8:171
Junk DNA, 12:29
Jurassic period, 3:52
Juscelinomys spp., 16:266
Jutjaws, 5:235–239, 5:241, 5:242
Jynginae. *See* Wrynecks
Jynx torquilla. *See* Northern wrynecks

K

K-strategists, *12:13*, 12:97
K-strategy, 2:27, 3:46–47
Kachuga tentoria. *See* Indian tent turtles
Kadwell, M., 15:313
Kaempfer's tody tyrants, 10:275
Kafka, Franz, 3:74
Kafue lechwes, 16:7, 16:33, 16:34
Kagus, 9:1–4, **9:41–44,** *9:42*, *9:43*
Kahalalide F, 1:45
Kai horseshoe bats, 13:392
Kajika frogs. *See* Buerger's frogs
Kakaduacridini, 3:203
Kakapo, 9:277, 9:278
Kalahari mole-rats, 12:74
Kalahari sand lizards, 7:298
Kali spp., 5:335
Kalinowski's agoutis, 16:409, 16:415*t*
Kalinowski's tinamous, 8:59
Kalligramma haeckelli, 3:305
Kalligrammatidae, 3:305
Kalophrynus spp., 6:304
Kalotermitidae, 3:164, 3:166
Kaloula conjuncta negrosensis, 6:305
Kaloula pulchra. *See* Malaysian painted frogs
Kaloula taprobanica, 6:313
Kalptorhynchia, 1:186
Kamaos, 10:489
Kamptozoa. *See* Entoprocts
Kangaroo Island emus, 8:87
Kangaroo mice, **16:199–210,** *16:204*, 16:208*t*–209*t*
Kangaroo rats, **16:199–210,** *16:204*
behavior, 16:124, 16:200–202
conservation status, 16:203
distribution, 16:123, *16:199*, 16:200
evolution, 16:199
feeding ecology, 12:117, 16:202
habitats, 16:124, 16:200
humans and, 16:203
physical characteristics, 16:123, 16:199
reproduction, 16:203
species of, 16:207–208, 16:208*t*–209*t*
taxonomy, 16:121, 16:199
vibrations, 12:83
water balance, 12:113
Kangaroos, *12:141*, *13:31*, **13:83–103,** *13:84*, *13:85*, *13:91*, *13:92*, 13:101*t*–102*t*
behavior, 13:35, *13:36*, 13:86–87
conservation status, 13:89
digestive systems of, 12:14, 12:123
distribution, 12:137, *13:83*, 13:84–85
evolution, 13:31–32, 13:83
feeding ecology, 13:87–88
habitats, 13:85–86
humans and, 13:89–90
locomotion, 12:43
physical characteristics, 13:32, 13:33, 13:83–84
reproduction, 12:106, 12:108, 13:88–89
scansorial adaptations, 12:14
species of, 13:93–100, 13:101*t*–102*t*
taxonomy, 13:83
Kannabateomys amblyonyx. *See* Southern bamboo rats
Kannan caecilians, *6:426*, *6:427*, *6:428*
Kantharellidae, 1:93
Kaokoveld hyraxes, 15:177
Karabasia evansi, 3:259

Karanisia spp., 14:13
Karaurid salamanders, 6:13
Karaurus spp., 6:13, 6:324
Karkloof blue butterflies, 3:90
Karoo rats. *See* Angoni vlei rats
Karoophasma spp., 3:217
Karoophasma biedouwensis, 3:*217*, 3:*218*
Kassina spp., 6:280, 6:281, 6:283
Kassina senegalensis. *See* Bubbling kassina
Kassininae, 6:280–281
Kassinula spp., 6:280
Katechonemertes nightingaleensis, 1:255
Katharina tunicata. *See* Black katy chitons
Katsuwonus pelamis. *See* Skipjack tunas
Katydids, 2:37, 2:39, 3:81, **3:201–208**, 3:*209*, **3:213–216**
  shield-backed, 3:57, 3:208
  wart biter, 3:85
Kauai creepers. *See* Akikikis
Kauai oo, 8:26
Keas, 9:276, 9:278, 9:*281*, 9:*294–295*
Keast, Allen, 11:229
Keel-bellied watersnakes, 7:466, 7:468
Keel-billed motmots, 10:31, 10:33, 10:*34*, 10:*36*
Keel-billed toucans, 10:*126*
Keel-scaled boas. *See* Splitjaw snakes
Keel-scaled splitjaws. *See* Splitjaw snakes
Keeled-nose toads. *See* Common Sunda toads
Kelaart's long-clawed shrews, 13:*269*, 13:*271*, 13:275
Kellen's dormice, 16:327*t*
Kelloggella spp., 5:376
Kelly's citrus thrips, 3:282
Kelp flies, 3:57
Kelp forests, 1:50–51, 4:48
  See also Conservation status
Kelp greenlings, 5:*184*, 5:*191*
Kelp gulls, 9:171
Kelp perches, 5:277
Kelpfishes, 5:235–240, 5:242–244
Kelpies, 14:292
Kemp, Alan, 10:71, 10:73
Kemp's grass mice, 16:*271*, 16:*272–273*
Kemp's longbills, 11:5
Kemp's ridley turtles, 7:*88*, 7:*89*, 7:91
Kempyninae, 3:307
Kenagy, G. J., 16:202
Keniabitis spp., 7:445
Kennalestids, 13:193
Kenrick's starlings, 11:*413*, 11:*424*
Kentish plovers. *See* Snowy plovers
Kentrogon larvae, 2:33
Kentropyx spp., 7:309, 7:310, 7:312
Kentucky warblers, 11:291
Kenyan pin-tailed sandgrouse. *See* Chestnut-bellied sandgrouse
Kerala caecilians, 6:5, 6:411, 6:*412*, **6:425–429**, 6:*427*
Keratella spp., 1:260
Keratin, 12:7–8
Keratosa. *See* Verongida
Kerguelen Island fur seals. *See* Antarctic fur seals
Kerivoula spp., 13:519, 13:520, 13:521
Kerivoula lanosa, 13:*522*
Kerivoula papillosa. *See* Papillose bats
Kerivoula papuensis. *See* Golden-tipped bats
Kerivoula picta. *See* Painted bats
Kerivoulinae. *See* *Kerivoula* spp.
Kermit the Frog, 6:*52*
Kermode bears. *See* American black bears
Kerodon spp. *See* Rock cavies
Kerodon rupestris. *See* Rock cavies
Keta salmons, 4:*75*
Ketupa spp. *See* Fish-owls
Key brotulas, 5:*19*, 5:*20–21*
Key deer, 15:*382*
Key West quail doves, 9:*254*, 9:*262*
Keystone species, 1:29, 1:49, 1:51, 12:216–217
  See also Conservation status
Khaudum Game Reserves, 12:*221*
Khulans. *See* Kulans
Khurs, 15:222, 15:*229*
  See also Asiatic wild asses
Ki bekko, 4:*299*
Kiangs, 12:134, 15:226, 15:*231*, 15:*234–235*
Kihengo screeching frogs. *See* Common squeakers
Killdeer, 9:162, 9:163, 9:*165*, 9:167–*168*
Killer whales, 15:*52*, 15:*53*
  behavior, 15:7, 15:44, 15:45, 15:46
  conservation status, 15:*49*, 15:50
  distribution, 12:138, 15:43
  echolocation, 12:87
  feeding ecology, 15:8, 15:46, 15:*47*
  habitats, 15:44
  humans and, 15:10–11
  migrations, 12:*165*
  physical characteristics, 15:*4*, 15:42, 15:43
  pygmy, 15:2, 15:56*t*
  reproduction, 15:48
  See also False killer whales
Killies. *See* Northern mummichogs
Killifishes, 4:12, 4:34, 4:35, 4:52, 4:57, 5:89, **5:89–102**, 5:92, 5:94, 5:*95*, 5:*97*
Kilombero weavers, 11:379
Kim, B. J., 5:1
Kimberella quadrata, 1:*9*
Kin recognition, 12:110
  See also Behavior
Kin selection, 3:70, 3:71
Kinesis, 1:38
  See also Behavior
Kinetoplastida, 2:11
King, David, 10:416
King birds of paradise, 11:489, 11:*492*, 11:*494*, 11:499–*500*
King cobras, 7:42, 7:56, 7:484, 7:486, 7:*490*, 7:*493*
King colobus, 14:175, 14:184*t*
King crabs, 2:197, 2:198
  See also American horseshoe crabs
King crows. *See* Black drongos
King eiders, 8:*375*, 8:387–388
King honey suckers. *See* Regent bowerbirds
King hummingbirds. *See* Crimson topaz
King Island emus, 8:87
King Kong finches, 11:342
King Kong grosbeaks, 11:343
King mackerels, 4:*4*, 5:*408*, 5:*413*, 5:414
King Mahendra Trust, 12:224
King mynas. *See* Helmeted mynas
King of Saxony birds of paradise, 11:485, 11:*494*, 11:*498*–499
King-of-six. *See* Pin-tailed whydahs
King of the herrings. *See* Oarfishes
King penguins, 8:147, 8:150, 8:151, 9:198
King quails, 8:399, 8:*440*, 8:446–*447*
King rails, 9:47, 9:51
King salmons. *See* Chinook salmons
King vultures, 8:277, 8:278, 8:*279*, 8:*282–283*
Kingbirds, 9:444, 10:270, 10:*272*
Kingdoms (taxonomy), 2:8
Kingfishers, 10:1–4, **10:5–23**, 10:*9*, 10:*11–12*
  behavior, 10:8, 10:13–23
  conservation status, 10:10, 10:13–23
  distribution, 10:*5*, 10:7–8, 10:13–23
  evolution, 10:5–6, 10:13–23
  feeding ecology, 10:8–9, 10:13–23
  habitats, 10:8, 10:13–23
  humans and, 10:10, 10:13–23
  physical characteristics, 10:6–7, 10:13–23
  reproduction, 10:*9*–10, 10:13–23
  species of, 10:*13–23*
  taxonomy, 10:5–6, 10:13–23
Kinglet calypturas, 10:306, 10:308
Kinglets, 11:1, 11:2
Kingsnakes, 7:42, 7:44, 7:466, 7:470
Kinkajous, 14:260, 14:309, 14:*310*, 14:311, 14:*312*, 14:313–*314*
Kinonchulus spp., 1:12
Kinorhynchs, 1:9, 1:11, 1:13, 1:22, **1:275–281**, 1:*278*
Kinorhynchus yushini, 1:276, 1:*278*, 1:*280*
Kinosternidae, **7:121–127**, 7:*124*
Kinosterninae. *See* American mud turtles
Kinosternon spp. *See* Eastern mud turtles
Kinosternon angustipons, 7:123
Kinosternon baurii. *See* Striped mud turtles
Kinosternon dunni, 7:123
Kinosternon flavescens. *See* Yellow mud turtles
Kinosternon leucostomum. *See* White-lipped mud turtles
Kinosternon sonoriense, 7:123
Kinugasa flying frogs, 6:*294*, 6:299
Kipingere seedeaters, 11:325
Kiricephalus spp., 2:318, 2:320
Kirilchik, S. V., 5:180
Kirk's caecilians, 6:*433*
Kirk's dikdiks, 16:*61*, 16:*62*, 16:*63*, 16:*65*, 16:*68*, 16:69
Kirtland's warblers, 11:288, 11:291, 11:*292*, 11:*298*
Kissing bugs, 2:11, 3:264, 3:*266*, 3:*271*, 3:279
Kissing gouramies, 4:57, 5:*430*, 5:*431*–432
Kit foxes. *See* Swift foxes
Kitchener, Andrew, 9:270
Kitefin sharks, 4:151, 4:152
Kites, 8:317–318, 8:319
  See also specific types of kites
Kitti's hog-nosed bats, 12:12, 12:58, 12:136, 13:307, **13:367–369**, 13:*368*
Kittiwakes, 8:*108*, 9:*204*, 9:205, 9:*210*, 9:*213*–214
Kittlitz's plovers, 9:163, 9:*165*, 9:168–*169*
Kittlitz's sandplovers. *See* Kittlitz's plovers
Kiunga ballochi, 5:70
Kiwis, 8:53–56, **8:89–94**
Klaas's cuckoos, 11:215, 11:219, 11:221
Klais spp. *See* Hummingbirds
Klaver, Charles, 7:223–224
Klegs. *See* Big black horse flies
Kleinschmidt's falcons. *See* Peregrine falcons
Kleptons, 6:263
Klinokinesis, 1:38
  See also Behavior
Klinotaxis, 1:39
  See also Behavior
Klipdassies. *See* Bush hyraxes; Rock hyraxes

Klipspringers, 12:148, 16:5, 16:7, 16:60, 16:*65*, 16:*67–68*
Kloss gibbons, 14:207–212, 14:212, 14:215, 14:*217*, 14:*218*
Kneria spp., 4:290, 4:291
Kneria auriculata, 4:290
Kneria wittei, 4:*292*, 4:*294*
Kneriidae, 4:57, 4:289, 4:290
Knifefishes, 4:11, 4:12, 4:14, 4:68
African, 4:232, 4:233
Asian, 4:232
clown, 4:*231–234*, 4:*235*, 4:239–240, 4:*241*
South American, **4:369–377,** 4:*372*
Knifejaws, 5:235–238, 5:240, 5:242–244
Knifenose chimaeras. *See* Pacific spookfishes
Knight fishes. *See* Pineconefishes
Knipowitschia punctatissima, 5:377
Knipowitschia thessala, 5:377
Knob-billed ducks. *See* Comb ducks
Knob-scaled lizards, **7:347–352,** 7:*350*
Knob-tailed geckos, 7:*262*
Knobbed hornbills. *See* Sulawesi red-knobbed hornbills
Knock-out mice, 16:128
Knocker cockroaches. *See* Madeira cockroaches
Knout gobies, 5:377
Knysna turacos, 9:300
Ko-ayu. *See* Ayu
Koala lemurs, 14:73–74, 14:75
Koalas, 12:*215*, 12:*223*, 13:*37*, **13:43–50,** 13:*44*, 13:*45*, 13:*46*, 13:*47*, 13:*48*, 13:*49*
behavior, 13:35, 13:37, 13:45–46
conservation status, 13:48–49
digestive system, 12:15, 12:*127*
distribution, 12:137, 13:*43*, 13:44
evolution, 13:31, 13:43
feeding ecology, 13:38, 13:46–47
habitats, 13:34, 13:44–45
humans and, 13:40, 13:49
physical characteristics, 13:32, 13:43–44
reproduction, 12:*103*, 13:33, 13:38, 13:39, 13:47–48
taxonomy, 13:31, 13:43
Kobs, 15:272, 16:5, 16:27, 16:29, 16:30, 16:34, 16:42*t*
Ugandan, 16:7
white-eared, 16:7
Kobus spp., 16:27, 16:28
Kobus defassa, 16:29
Kobus ellipsiprymnus. See Waterbucks
Kobus kob. See Kobs
Kobus kob leucotis. See White-eared kobs
Kobus kob thomasi. See Uganda kobs
Kobus leche. See Kafue lechwes
Kobus leche kafuensis. See Kafue lechwes
Kobus leche robertsi. See Roberts' lechwes
Kobus leche smithemani. See Black lechwes
Kobus megaceros. See Nile lechwes
Koch, Robert, 2:33
Kodiak bears, 12:24, 14:*298*
See also Brown bears
Koelliker's glass lizards. *See* Moroccan glass lizards
Koels, 9:315, 9:*316*, 9:322–323, 11:508
Koenig, Otto, 10:521
Koepcke's hermits, 9:438
Kogia spp., 15:73, 15:74–*75*, 15:76, 15:78
Kogia breviceps. See Pygmy sperm whales
Kogia sima. See Dwarf sperm whales
Koh Tao Island caecilians, 6:*420*, 6:*421*, 6:423
Kohaku, 4:*299*
Kohanee salmons. *See* Sockeye salmons
Koi, 4:*299*, 4:303, 4:311
Kokakos, 11:448, 11:*449*
Kokartus spp., 6:13
Koklass, 8:434
Kollokodontidae, 12:228
Koloas. *See* Mallards
Komarekionidae, 2:65
Komba spp., 14:23
Komodo dragons, 7:202, 7:359, 7:360, 7:*361*, 7:363, 7:*364*, 7:*365*, 7:367–368
Komodo monitors. *See* Komodo dragons
Kona crabs. *See* Spanner crabs
Kona grosbeaks, 11:343, 11:349
Kongonis. *See* Red hartebeests
Konishi, Y., 5:2
Kookaburras, 10:6–7, 10:*8*
Ko'okay spurwing long-legged flies, 3:360
Koopman's porcupines, 16:365, 16:374*t*
Kopua spp., 5:355
Korea. *See* Palaearctic region
Kori bustards, 9:1, 9:*2*, 9:*93*
Korrigums, 16:34
Koshima Island, 12:157
Koslov's pikas, 16:487, 16:495
Kosrae crakes, 9:52
Kosrae Mountain starlings, 11:410
Kotlyar, A. N., 5:1
Kott, Patricia, 1:479
Kottelat, M., 5:83–84
Koupreys, 12:136, 12:177, 16:24*t*
Kowalevsky, A., 1:451
Kraemeria samoensis. See Samoan sand darts
Kraemeriidae. *See* Sand gobies
Kraits, **7:483–487,** 7:*490*, **7:495**
Krajewski, Carey, 9:23
Kravetz, F. O., 16:267
Krefft, Johann L. G., 4:197
Kribensis, 5:*276*
Krill, **2:185–193,** 2:*188*
behavior, 2:186
conservation status, 2:187
distribution, 2:185
evolution, 2:185
feeding ecology, 2:186–187
habitats, 2:185
humans and, 2:187
physical characteristics, 2:185, 2:*186*
reproduction, 2:187
species of, 2:*189*–192
taxonomy, 2:185
Kristensen, R. M., 1:12, 1:13, 1:327, 1:351
Krohnittidae, 1:433
Kronborgia amphipodicola, 1:*189*, 1:*192*
Kronosaurus spp., 7:15
Kroodsma, D. E., 8:42
Krøyers deep sea anglerfishes, 4:*4*, 4:*33*
Kryptophanaron alfredi. See Atlantic flashlightfishes
Kryptopterus bicirrhis. See Glass catfishes
Kudaris. *See* Southern cassowaries
Kudus, 15:271, 15:273, 16:11, **16:11–25,** 16:*16*, 16:24*t*–25*t*
Kuehneotherium spp., 12:11
Kughitang blind loaches, 4:*324*, 4:*326*, 4:327
Kuhlia boninensis, 5:220
Kuhlia caudavittata, 5:220
Kuhlia marginata, 5:220
Kuhlia mugil. See Flagtail kuhlias
Kuhlia munda, 5:220
Kuhlia nutabunda, 5:220
Kuhlia rupestris. See Jungle perches
Kuhlia sandivicensis, 5:220
Kuhlias, 4:47, 5:219–222
Kuhliidae. *See* Kuhlias
Kuhl's pipistrelle bats, 13:516*t*
Kukrisnakes, 7:467, 7:469
Kulans, 12:134, 15:222, 15:229
See also Asiatic wild asses
Kultarrs, 12:278, 12:288, 12:*292*, 12:*296*, 12:297
Kunkele, J., 16:391
Kunsia spp., 16:265, 16:266
Kunsia tomentosus, 16:265, 16:266
Kupeornis gilberti. See White-throated mountain babblers
Kuril seals. *See* Harbor seals
Kurlansky, M., 5:30
Kurrichane buttonquails. *See* Barred buttonquails
KwaZulu-Natal Midlands dwarf chameleons, 7:*234*, 7:*235*
KwaZulu-Natal mist-belt grasslands, 3:90
KwaZulu-Natal Nature Conservation Service, 7:*61*
Kyacks. *See* Alewives
Kyarranus spp., 6:34, 6:141
Kynotidae, 2:65
Kyphosidae. *See* Sea chubs
Kyphosinae, 5:241
Kyphosus elegans. See Cortez chubs

L

La Palma glass frogs, 6:*218*, 6:223
La Plata river dolphins. *See* Franciscana dolphins
Labeo spp., 4:298
Labeo victorianus. See Ningu
Labeoninae, 4:297
Labial teeth, 6:40
Labidochromis vellicans, 5:278
Labidura herculeana. See St. Helena earwigs
Labium, 3:20
See also Physical characteristics
Laboratory animals, 12:173
See also specific animals
Laboratory stick insects. *See* Indian stick insects
Labrador herrings. *See* Atlantic herrings
Labridae. *See* Wrasses
Labrisomidae, 4:48, 5:341, 5:343
Labroidei, **5:275–307,** 5:*282*, 5:*283*, 5:*299*, 5:*300*
behavior, 5:277, 5:294–295
conservation status, 5:280–281, 5:298
distribution, 5:276–277, 5:294
evolution, 5:275–276, 5:293
feeding ecology, 5:277–279, 5:295–296
habitats, 5:277, 5:294
humans and, 5:281, 5:298
physical characteristics, 5:276, 5:293–294
reproduction, 5:279–280, 5:296–298
species of, 5:*284*–290, 5:*301*–307
taxonomy, 5:275, 5:293
Labroides spp., 2:34, 4:69, 5:296

Labroides dimidiatus. See Bluestreak cleaner wrasses
Labropsis spp., 4:69
Labrum, 3:19
See also Physical characteristics
Labyrinth catfishes, 4:352
Labyrinth fishes, **5:427–435,** 5:*430*
Lac insects, 3:81
Laccifer lacca. See Lac insects
Lace corals, 2:*509*, 2:*510*
See also Cauliflower corals
Lace gouramies. *See* Pearl gouramies
Lacedo pulchella. See Banded kingfishers
Lacerta spp., 7:297
Lacerta agilis. See Sand lizards
Lacerta caudiverbera. See Helmeted water toads
Lacerta lepida. See Eyed lizards
Lacerta salamandra. See European fire salamanders
Lacerta vulgaris. See Smooth newts
Lacertidae, 7:207, **7:297–302,** 7:*300*
Lacewings, 3:8, 3:34, 3:63–64, **3:305–314,** 3:*310*
Lacher, T. E., 16:389
Lachesis spp., 7:445, 7:446, 7:448
Lachesis melanocephala. See Black-headed bushmasters
Lachesis muta, 7:456
Lachnolaimus maximus. See Hogfishes
Lacistorhynchidae, 1:226
Laconi, M. R., 16:269
Lacotriton spp., 6:13
Lactation, 12:89, 12:125–127
See also Reproduction
Lactophrys polygonius. See Honeycomb cowfishes
Lactophrys triqueter. See Smooth trunkfishes
Lactoria cornuta. See Longhorn cowfishes
Lactose, 12:126–127
Ladakh pikas, 16:502*t*
Ladurner, P., 1:11
Lady Amherst's pheasants, 8:434
Lady Gould finches. *See* Gouldian finches
Lady Ross's violet plantain-eaters. *See* Ross's turacos
Ladybird beetles, 3:63, 3:80, 3:81, 3:90, 3:316, 3:318–319, 3:320, 3:323
Ladybugs. *See* Ladybird beetles
Ladyfishes, 4:11, 4:69, **4:243–248,** 4:*245*, 4:*246*
Laemobothrion spp., 3:249
Laetmogonidae, 1:420
Laevipilina antarctica, 2:*387*, 2:*389*, 2:*390*
Lafresnaya spp. *See* Hummingbirds
Lafresnaye's vangas, 10:439, 10:440, 10:441
Lagenodelphis hosei. See Fraser's dolphins
Lagenorhynchus acutus. See Atlantic white-sided dolphins
Lagenorhynchus obliquidens. See Pacific white-sided dolphins
Lagenorhynchus obscurus. See Dusky dolphins
Lagidium spp., 16:377
Lagidium peruanum. See Mountain viscachas
Lagidium viscacia. See Southern viscachas
Lagidium wolffsohni. See Wolffsohn's viscachas
Lagomorpha, **16:479–489**
digestive systems, 12:14–15
neonatal milk, 12:127
reproduction, 12:12, 12:103, 12:107–108
vibrations, 12:83
See also Rabbits
Lagonimico spp., 14:115
Lagonosticta spp. *See* Firefinches
Lagonosticta rhodopareia. See Jameson's firefinches
Lagonosticta rubricata. See African firefinches
Lagonosticta senegala. See Red-billed firefinches
Lagoons, 4:48
See also Habitats
Lagopus spp. *See* Ptarmigans
Lagopus lagopus. See Willow ptarmigans
Lagopus mutus. See Rock ptarmigans
Lagorchestes spp. *See* Hare-wallabies
Lagostomus maximus. See Plains viscachas
Lagostrophus spp., *13:85*
Lagostrophus fasciatus. See Banded hare-wallabies
Lagostrophus hirsutus. See Rufous hare-wallabies
Lagosuchus spp., 7:19
Lagothrix spp. *See* Woolly monkeys
Lagothrix cana. See Gray woolly monkeys
Lagothrix flavicauda. See Yellow-tailed woolly monkeys
Lagothrix lagotricha. See Humboldt's woolly monkeys
Lagothrix lugens. See Colombian woolly monkeys
Lagothrix poeppigii, 14:160
Lagurus lagurus. See Steppe lemmings
Lagusia micracanthus, 5:220
Lahontan Basin lizards. *See* Long-nosed leopard lizards
Laingia jaumotti, 1:*131*, 1:141
Laingiomedusae, 1:123, 1:125
Lake Ascotan Mountain killifishes. *See* Ascotan Mountain killifishes
Lake charrs. *See* Lake trouts
Lake habitats, 4:38–39, 4:40, 4:52
See also Habitats
Lake Malawi, cichlids, 5:275–276
Lake perches. *See* Yellow perches
Lake sturgeons, 4:*214*, 4:*216*, 4:*217*–218
Lake Tanganyika, cichlids, 5:275–276
Lake Tanganyika lampeyes. *See* Tanganyika pearl lampeyes
Lake trouts, 4:*408*, 4:418–*419*
Lake Victoria, cichlids, 5:275–276, 5:280–281
Lake whitefishes, 4:*408*, 4:*410*
LAL (limulus amebocyte lysate), 2:42
Lalage spp. *See* Trillers
Lalage leucomela. See Varied trillers
Lalage sueurii. See White-winged trillers
Lama spp., 12:132, 15:265, 15:313, 15:314, 15:317
Lama glama. See Llamas
Lama guanicoe. See Guanacos
Lama pacos. See Alpacas
Lamarck, J. B., 1:451
Lambert, Frank, 10:194
Lamellibrachia spp., 2:91
Lamellibrachia luymesi, 2:92, 2:*93*, 2:*94*
Lamellibrachia satsuma, 2:91
Lamini, 15:313
Lamna spp., 4:132
Lamna ditropis. See Salmon sharks
Lamna nasus. See Porbeagles
Lamnidae, 4:4
Lamniformes. *See* Mackerel sharks
Lampanyctinae, 4:441
Lampanyctus spp., 4:442
Lampanyctus güntheri. See Lepidophanes guentheri
Lampea spp., 1:170
Lampetra spp., 4:83
Lampfishes. *See* Cave basses
Lampobattang flycatchers, 11:27
Lampocteis cruentiventer. See Bloody bellies
Lamponius portoricensis. See Puerto Rican walkingsticks
Lampornis spp. *See* Hummingbirds
Lampornis clemenciae. See Blue-throated hummingbirds
Lamprey eels. *See* Sea lampreys
Lampreys, **4:83–90,** 4:*87*
evolution, 4:9–10, 4:83
feeding ecology, 4:69, 4:*84*, 4:85, 4:88–90
physical characteristics, 4:6, 4:23, 4:83, 4:*84*, 4:88–90
reproduction, 4:27, 4:34, 4:85, 4:88–90
Lamprichthys tanganicanus. See Tanganyika pearl lampeyes
Lamprididae, 4:447, 4:448, 4:449
Lampridiformes, **4:447–455,** 4:*451*
Lampris spp., 4:447, 4:448, 4:449
Lampris guttatus. See Opahs
Lampris zatima, 4:448
Lamprolaima spp. *See* Hummingbirds
Lamprologus spp., 4:66, 5:279
Lampropeltis spp. *See* Kingsnakes
Lampropeltis getula. See Common kingsnakes
Lampropeltis getula floridana. See Kingsnakes
Lampropeltis triangulum. See Milksnakes
Lamprophiinae, 7:467, 7:468
Lampropidae, 2:229
Lamproptera meges. See Swallowtails
Lamprotornis cupreocauda. See Copper-tailed glossy-starlings
Lamprotornis purpureiceps, 11:422
Lamprotornis shelleyi. See Shelley's starlings
Lamprotornis spiloptera. See Spot-winged starlings
Lampshells, 2:13, 2:25, 2:29, 2:*524*, 2:*526*–527
articulate, **2:521–527,** 2:*524*
nonarticulate, **2:515–519,** 2:*517*
See also Brachiopoda
Lampyridae, 3:65, 3:320, 3:321
Lanai finches. *See* Lanai hookbills
Lanai hookbills, 11:342, 11:343, 11:*345*, 11:*349*–350
Lanai pomace flies, 3:360
Lancelets, **1:485–497,** 1:*487*, 1:*489*
behavior, 1:487–488
conservation status, 1:488
distribution, 1:487
evolution, 1:485
feeding ecology, 1:488
habitats, 1:487
humans and, 1:488
physical characteristics, 1:485–487, 1:*486*
reproduction, 1:22, 1:488
species of, 1:*490*–497
taxonomy, 1:14, 1:485
Lanceolated honeyeaters. *See* Striped honeyeaters
Lanceolated jays, 11:506
Lanceolated monklets, 10:103, 10:*105*, 10:*108*–109
Lancer dragonets, 5:*368*, 5:*369*–370
Lances. *See* Inshore sand lances

Lancet flukes, 1:*199*, 1:*202*, 1:*208–209*
Lancet liver flukes. *See* Lancet flukes
Lancetfishes, 4:433, 4:434, 4:*435*, 4:*436*
Land carnivores, **14:255–263**
See also specific carnivores
Land crabs, 2:200, 2:201
Land-locked salmons. *See* Sockeye salmons
Land-locked sea lampreys, 4:84
Land planarians, 1:*189*, 1:*190*, 1:194
Land snails. *See* Giant African land snails; Lung-bearing snails
Lander's horseshoe bats, 13:389–390
Landscape attrition, 3:86–87
Landscape with Reptile (Palmer), 7:454
Landscaping, ecological, 3:90–91
Lane, D. F., 10:32
Langaha spp. *See* Madagascan vinesnakes
Language, 12:160–162
bird influence on, 8:20
insects in, 3:74
See also Behavior; Humans
Langurs, 12:123, 14:7, 14:11, 14:172
Laniarius spp. *See* Bush-shrikes
Laniarius amboimensis. *See* Gabela bush-shrikes
Laniarius barbarus. *See* Yellow-crowned gonoleks
Lanice conchilega. *See* Sand masons
Laniidae. *See* Shrikes
Laniinae. *See* Shrikes
Laniisoma elegans, 10:308
Lanioturdus spp. *See* Bush-shrikes
Lanioturdus torquatus. *See* White-tailed shrikes
Lanius spp. *See* Shrikes
Lanius cabanisi. *See* Long-tailed fiscals
Lanius collaris. *See* Common fiscals
Lanius collurio. *See* Red-backed shrikes
Lanius cristatus. *See* Brown shrikes
Lanius excubitor. *See* Northern shrikes
Lanius excubitoroides. *See* Gray-backed fiscals
Lanius frontatus. *See* Eastern shrike-tits
Lanius gubernator. *See* Emin's shrikes
Lanius kirhocephalus. *See* Variable pitohuis
Lanius ludovicianus. *See* Loggerhead shrikes
Lanius meridionalis. *See* Southern gray shrikes
Lanius minor. *See* Lesser gray shrikes
Lanius miocaenus, 10:425
Lanius newtoni. *See* São Tomé fiscals
Lanius schach. *See* Asian long-tailed shrikes; Long-tailed shrikes
Lanius senator. *See* Woodchats
Lanius souzae. *See* Sousa's shrikes
Lanius sphenocercus. *See* Chinese gray shrikes
Lanius torquatus. *See* Gray butcherbirds
Lankanectes spp., 6:249
Lankanectes corrugatus. *See* Corrugated water frogs
Lankanectinae, 6:246
Lankin, K., 15:147
Lantern sharks, 4:151, 4:152
Lanterneye fishes. *See* Flashlightfishes; Splitfin flashlightfishes
Lanternfishes, **4:441–446**, 4:*444–445*
Lanternflies, 3:*6*, 3:65
Lanthanotines, 7:359
Lanthanotus spp., 7:202, 7:359, 7:360
Lanthanotus borneensis. *See* Earless monitors
Lantz, Walter, 10:88
Lanyon, S. M., 10:325
Lanyon, W. E., 10:325
Lanzarana spp., 6:247
Laomedea spp., 1:129
Laomedea flexuosa, 1:26
Laotian shads, 4:278
Laphria spp., 3:360
Lapland buntings. *See* Lapland longspurs
Lapland longspurs, 11:*264*, 11:*268*, 11:*272*, 11:273
Lapparentophis spp., 7:16
Lappet-brown bats. *See* Allen's big-eared bats
Lappet-faced vultures, 8:*320*, 8:321, 8:323, 8:*325*, 8:*335–336*
Laptev walruses, 14:409, 14:412
Lapwings, **9:161–173**, 9:*162*, 9:*165*
behavior, 9:162, 9:166–172
conservation status, 9:163–164, 9:167–173
distribution, 9:*161*, 9:162, 9:*166–172*
evolution, 9:161
feeding ecology, 9:162, 9:166–173
habitats, 9:162, 9:166–172
humans and, 9:164, 9:167–173
physical characteristics, 9:161–162, 9:166–172
reproduction, 9:162–163, 9:166–173
species of, 9:*166–173*
taxonomy, 9:133, 9:161, 9:167–172
Laqueus californianus. *See* California lampshells
Lar gibbons, 14:207–213, 14:*209*, 14:*217*, 14:*219*
Laredo striped whiptails, 7:312
Large-antlered muntjacs. *See* Giant muntjacs
Large bamboo rats, 16:*283*, 16:*287*, 16:*288*, 16:294–295
Large-banded blennies, 5:*342*
Large-billed blue-flycatchers, 11:*28*, 11:*37*
Large-billed catbirds. *See* Green catbirds
Large-billed crows, 11:508
Large-billed scrub-wrens, 11:55, 11:*56*
Large-billed sparrows. *See* Savannah sparrows
Large-billed terns, 9:203
Large-blotched salamanders. *See Ensatina* spp.
Large blues, 3:*390*, 3:394–*395*
Large carnivorous water bears, 2:116, 2:117, 2:118, 2:*119*, 2:120–121
Large carpenter bees, 3:*410*, 3:*413*
Large coucals. *See* Greater coucals
Large-eared horseshoe bats, 13:*390*
Large-eared hutias. *See* Eared hutias
Large-eared pikas, 16:*496*, 16:*498*
Large-eared slit-faced bats, 13:*372*, 13:*373*, 13:376*t*
Large-eared tenrecs, 13:225, 13:226, 13:227, 13:228, 13:*229*, 13:*231*, 13:*232*–233
Large-footed myotis, 13:313
Large-frilled bowerbirds. *See* Spotted bowerbirds
Large frogmouths, 9:378, 9:379, 9:*380*, 9:382–*383*
Large-gilled crows, 11:508
Large golden moles, 13:215, 13:217, 13:*219*, 13:*220*, 13:221
Large green-billed malkohas. *See* Green-billed malkohas
Large hairy armadillos, 13:183, 13:*184*, 13:191*t*
Large Indian civets. *See* Indian civets
Large intestine, 12:47–48, 12:124
See also Physical characteristics
Large Japanese moles, 13:287*t*
Large Malayan leaf-nosed bats. *See* Diadem roundleaf bats
Large naked-soled gerbils. *See* Bushveld gerbils
Large niltavas, 11:*29*, 11:*36*
Large northern bandicoots. *See* Northern brown bandicoots
Large owlet-nightjars. *See* Feline owlet-nightjars
Large pocket gophers, 16:*191*, 16:*193*–194
Large racket-tailed drongos. *See* Greater racket-tailed drongos
Large red-tailed bumblebees, 3:*410*, 3:*412*
Large roundworms. *See* Maw-worms
Large sandgrouse. *See* Black-bellied sandgrouse
Large-scale mullets, 5:*61*, 5:*63–64*
Large-scale spinycheek sleepers, 4:50, 5:*378*, 5:*380*
Large-scaled shieldtails, 7:*392*
Large slit-faced bats, 13:313–314, 13:*374*, 13:*375*
Large-spotted civets. *See* Oriental civets
Large-tailed nightjars, 9:369, 9:*407*, 9:*412*
Large-toothed bandicoots, 13:4, 13:17*t*
Largehead hairtails, 5:*408*, 5:416–*417*
Largemouth basses, 5:195, 5:*196*, 5:197, 5:199, 5:*201*, 5:204–*205*
Largemouth black basses. *See* Largemouth basses
Largenoses, 5:105, 5:106
Larger double-collared sunbirds. *See* Greater double-collared sunbirds
Larger Indian cuckoo doves. *See* Barred cuckoo-doves
Largescale foureyes, 4:*33*, 5:*91*, 5:*95*, 5:*96*
Largescale stonerollers. *See* Stonerollers
Largespot river stingrays, 4:*173*
Lari, 9:101, 9:102, 9:104
Laridae, **9:203–217**, 9:*210*
behavior, 9:206
conservation status, 9:208–209
distribution, 9:*203*, 9:205
evolution, 9:203–204
feeding ecology, 9:206–207
habitats, 9:205–206
humans and, 9:209
physical characteristics, 9:203, 9:204–205
reproduction, 9:207–208
species of, 9:*211–217*
taxonomy, 9:203–204
Lark buntings, 11:*267*, 11:*281*–282
Lark buttonquails, 9:12, 9:14, 9:*15*, 9:20–*21*
Larks, **10:341–355**, 10:*347*
behavior, 10:343–344
conservation status, 10:346
distribution, 10:*341*, 10:342–343
evolution, 10:341
feeding ecology, 10:344–345
habitats, 10:343
humans and, 10:346
physical characteristics, 10:341–342, 10:*345*
reproduction, 10:*344*, 10:345–346
species of, 10:*348–354*
taxonomy, 10:341
Larosterna inca. *See* Inca terns
Larus spp. *See* Gulls
Larus argentatus. *See* Herring gulls
Larus atricilla. *See* Laughing gulls
Larus dominicanus. *See* Kelp gulls
Larus fuliginosus. *See* Lava gulls
Larus marinus. *See* Black-backed gulls

Larus modestus. See Gray gulls
Larus parasiticus. See Arctic skuas
Larus philadelphia. See Bonaparte's gulls
Larus ridibundus. See Common black-headed gulls
Larus saundersi. See Saunder's gulls
Larutia spp., 7:328
Larvaceans, **1:473–478,** 1:*475*
Larvae, 1:19–23, 2:3–4, 2:21, 2:23, 3:18, 3:47, 4:*31*, 4:34–36, 4:43, 4:62, 4:66, 6:28, 6:*36*, **6:39–43**
vs. adults, 3:33–35
Anura, 6:*36*, 6:39–43
Caudata, 6:326
eels, 4:*258*, 4:262
marine fishes, 4:46–47
salamanders, 6:33, 6:39, 6:42
tiger salamanders, 6:360
transportation of, 6:35–37
See also Reproduction; Tadpoles; specific species
Larval instars, 3:31
See also Reproduction
Larvigenesis, 1:19
See also Reproduction
Larynx, 12:8
See also Physical characteristics
Lasioderma serricorne. See Cigarette beetles
Lasionectes entrichoma, 2:*125*, 2:*127*, 2:*128*
Lasionycteris noctivagans. See Silver-haired bats
Lasiorhinus spp. *See* Hairy-nosed wombats
Lasiorhinus krefftii. See Northern hairy-nosed wombats
Lasiorhinus latifrons. See Southern hairy-nosed wombats
Lasiorhyncus barbicornis, 3:74
Lasis spp., 1:*467*
Lasiurus borealis. See Red bats
Lasiurus cinereus. See Hoary bats
Lasiurus cinereus semotus. See Hawaiian hoary bats
Lasius niger. See Garden ants
Latakoo larks, 10:344, 10:346
Lataste's gundis. *See* Mzab gundis
Lateolabrax spp., 5:220, 5:221
Lateolabrax japonicus. See Japanese perches
Lateral undulation, 7:*25*–*26*
Laterallus spp., 9:46, 9:47, 9:50
Laterallus jamaicensis. See American black rails
Laternaria laternaria. See Peanut bugs
Lates spp., 5:220, 5:221, 5:222
Lates calcarifer. See Barramundis
Lates japonica. See Japanese giant perches
Lates niloticus. See Nile perches
Latham's francolins, 8:434
Lathamus discolor. See Swift parrots
Lathicerus spp., 3:203
Latia neritoides. See New Zealand freshwater limpets
Laticauda spp. *See* Sea kraits
Laticauda colubrina. See Sea kraits
Latidens salimalii. See Salim Ali's fruit bats
Latimeria spp., 4:190
Latimeria chalumnae, 4:12, 4:*189*–196, 4:*190*, 4:*191*, 4:*194*, 4:199
Latimeria menadoensis. See Indonesian coelacanths
Latimeriidae, 4:190
Latinozoros spp., 3:239
Latonia spp., 6:89
Latridae. *See* Trumpeters
Latrine flies, 3:361
Latrodectus hesperus. See Black widow spiders
Lau-lau, 4:*355*, 4:*361*, 4:363–364
Laúd turtles. *See* Leatherback seaturtles
Lauder, G. V., 5:427
Laughing falcons, 8:354–355
Laughing gulls, 9:*205*
Laughing kingfishers. *See* Laughing kookaburras
Laughing kookaburras, 10:6, 10:8, 10:10, 10:*12*, 10:*13*
Laughing thrushes, 10:505, 10:506, 10:508, 10:*510*, 10:512, 11:2
Laupala spp., 3:207
Laurasia, 3:52, 3:53, 3:55
Laurasiatheria, 12:26, 12:*30*, 12:33, 15:134
Laurentophryne spp., 6:184
Lava gulls, 9:205, 9:209
Lavenberg, R. J., 5:352, 5:353
Lavia frons. See Yellow-winged bats
Lavoué, S., 4:255
Lawes's parotias, 11:*493*, 11:497–*498*
Lawes's six-plumed birds of paradise. *See* Lawes's parotias
Lawes's six-wired birds of paradise. *See* Lawes's parotias
Lawes's six-wired parotias. *See* Lawes's parotias
Lawrence, D. H., 12:52
Lawrence's warblers, 11:285–286, 11:294
Lawson, D. P., 6:431
Lawyers. *See* Burbots
Layard's beaked whales. *See* Strap-toothed whales
Laysan albatrosses, 8:*114*, 8:*121*–122
Laysan ducks, 8:364
Laysan finches, 11:*342*, 11:343, 11:*345*, 11:*350*
Laysan mollymawks, 8:*114*, 8:*117*, 8:*121*–122
Laysan rails, 9:*54*, 9:*63*
Lazara, Kenneth J., 5:90
Laza's alpine salamanders, 6:367
Le Carre, John, 13:200
Le Conte's sparrows, 11:*265*
Le Gros Clark, Wilfred E., 13:289
"Le Moquer Polyglotte" (Messiaen), 10:332
Leach's giant geckos. *See* New Caledonian giant geckos
Leach's storm-petrels, 8:135, 8:136, 8:137, 8:138, 8:*140*–141
Lead-backed salamanders. *See* Red-backed salamanders
Leadbeater's possums, 13:*126*, 13:*130*, 13:132
behavior, 13:35, 13:37, 13:127
conservation status, 13:39, 13:129
habitats, 13:34
physical characteristics, 13:32, 13:34
reproduction, 13:132
taxonomy, 13:132
Leader prawns. *See* Giant tiger prawns
Leaf beetles, 3:57, 3:316, 3:317–318, 3:319, 3:321, 3:323
Leaf cutter ants, 2:36, 3:70, 3:72, 3:*407*, 3:408, 3:*410*, 3:*412*, 3:415
Leaf cutter bees, 3:405, 3:*408*
See also Alfalfa leaf cutter bees
Leaf deer. *See* Leaf muntjacs
Leaf-eared mice, 16:269, 16:270
Leaf-folding frogs, 6:282–283
Leaf-footed bugs, 3:*4*, 3:*22*
Leaf frogs, 6:41
Leaf-horned agamas, 7:211, 7:*213*, 7:*215*
Leaf insects, **3:221–232,** 3:*225*
Leaf litter frogs, 6:110, 6:111, 6:*112*, 6:113
Leaf-loves, 10:397, 10:*401*, 10:*407*
Leaf miner flies, 3:361
Leaf miners (beetles), 3:321
Leaf miners (moths), 3:383–385, 3:389, 3:*391*, 3:*394*
Leaf-monkeys, **14:171–186,** 14:*176*, 14:*177*
Asian, 14:172
behavior, 12:153, 14:7, 14:174
conservation status, 14:175
distribution, 14:5, 14:*171*, 14:173
evolution, 14:171–172
feeding ecology, 14:8, 14:9, 14:174
habitats, 14:174
humans and, 14:175
physical characteristics, 14:172–173
reproduction, 14:174–175
species of, 14:*178*–183, 14:184*t*–185*t*
taxonomy, 14:4, 14:171–172
Leaf muntjacs, 15:346, 15:*348*, 15:*349*, 15:354
Leaf-nosed bats
Commerson's, 13:401, 13:*402*, 13:*404*
dusky, 13:*403*, 13:*404*
East Asian tailless, 13:409*t*
evolution, 12:11
New World, 13:311, 13:315
northern, 13:*403*, 13:*404*
trident, 13:*406*, 13:*407*
Leaf-rollers, 3:389
Leaf turners. *See* Tooth-billed bowerbirds
Leaf vipers. *See* Green bush vipers
Leafbirds, **10:415–424,** 10:*418*
Leafcreepers, 10:209
Leaffishes, 4:57, 4:69, 5:235–238, 5:240, 5:242–244
Amazon, 5:*246*, 5:251
Asian, 4:57
Gangetic, 5:*246*, 5:*250*, 5:251
Leafhoppers, 3:63, 3:75, **3:259**
See also Hemiptera
Leafloves. *See* Leaf-loves
Leaftires, 3:389
Leafy seadragons, 4:14, 5:133, 5:*134*, 5:*137*, 5:*143*, 5:146–147
Learned behavior, 3:59
See also Behavior
Learning, 1:37, 1:40–41, 2:39–40, 7:39–40, 12:9–10, 12:140–143, 12:151–155
See also Behavior; Cognition
Learning set, 12:152
Least Bell's vireos, 11:269
Least bitterns, 8:*246*, 8:*257*–*258*
Least chipmunks, 16:*149*, 16:*152*
Least dabchicks. *See* Least grebes
Least grebes, 8:*173*, 8:*175*–*176*
Least honeyeaters. *See* Brown honeyeaters
Least pygmy owls, 9:345
Least seedsnipes, 9:190, 9:191, 9:*192*, 9:*195*
Least shrew-moles. *See* American shrew moles
Least shrews, 13:196
Least storm-petrels, 8:107, 8:108, 8:*109*, 8:135
Least weasels, 14:333*t*
distribution, 14:321
habitats, 14:257
physical characteristics, 14:256, 14:321
predation and, 12:115, 12:116

reproduction, 12:107, 14:262, 14:324
Leather carps. *See* Common carps
Leatherback seaturtles, **7:101–103,** 7:*102*
conservation status, 7:*62*
endothermic, 7:28
physical characteristics, 7:67, 7:70
reproduction, 7:71
See also Seaturtles
Leatherback turtles. *See* Spiny softshells
Leatherbacks. *See* Common carps
Leatherheads. *See* Noisy friarbirds
Leatherjackets, 3:56, 3:359–360, 5:467–471
Leatherjackets (larvae). *See* European marsh crane flies
Lebiasinidae, 4:335
Lecane spp., 1:260
Lecanicephalidae, 1:226
Lecanicephalidea, 1:226, 1:227
Lecanogaster spp., 5:355
Lechriodus spp., 6:140, 6:141
Lechwes, 16:27–32, 16:34, 16:42*t*
black, 16:34
Kafue, 16:7, 16:33, 16:34
Nile, 16:4, 16:27, 16:34, 16:*35*, 16:*37*
Roberts', 16:33
Lecithoepitheliata, 1:185
Lecithotrophic larvae, 2:21
See also Reproduction
Leconte's thrashers, 10:467
Leeches, 2:25, **2:75–83,** 2:*79*
behavior, 2:76
conservation status, 2:77–78
distribution, 2:76
evolution, 2:75
feeding ecology, 2:29, 2:76
habitats, 2:76
humans and, 2:78
medical use of, 2:42
oyster, 1:*189*, 1:*191*
physical characteristics, 2:75–76, 2:77
reproduction, 2:16, 2:17, 2:77
species of, 2:*80*–83
taxonomy, 2:75
Lefranc, N., 10:425
Legatus leucophaius. See Piratic flycatchers
Legend of White Snake, 7:56
Legends. *See* Folklore
Legless lizards, 7:339–740
Leguat, François, 9:269, 9:270, 9:273
Legume-*Rhizobium* symbiosis, 1:31
Leiocephalus eremitus, 7:246–247
Leiocephalus herminieri, 7:246–247
Leiodidae, 3:54, 3:317
Leiognathidae. *See* Ponyfishes
Leiolamus spp., 7:246
Leiolepidinae, 7:204, 7:209
Leiolepis belliana. See Butterfly agamas
Leiopelma spp., 6:35, 6:36, 6:70, 6:77
Leiopelma archeyi. See Archey's frogs
Leiopelma auroraensis, 6:69
Leiopelma hamiltoni. See Hamilton's frogs
Leiopelma hochstetteri. See Hochstetter's frogs
Leiopelma markhami, 6:69
Leiopelma pakeka. See Maud Island frogs
Leiopelma waitomoensis, 6:69
Leiopelmatidae. *See* New Zealand frogs
Leiopotherapon spp., 5:220
Leiopython spp., 7:420
Leiopython albertisii. See White-lipped pythons
Leiothrix lutea. See Red-billed leiothrixes
Leiothrix lutea astleyi. See Astley's leiothrix
Leiperia spp., 2:317, 2:319, 2:320
Leipoa ocellata. See Malleefowls
Leis, J. M., 5:365
Leishmania spp., 2:11
Leishmaniasis, 2:11
Leisler's bats, 13:516*t*
Leistes spp. *See* Icteridae
Leitão de Carvalho, Antenor, 6:301
Leithia spp., 16:317
Leithiinae, 16:317
Leks, 6:49, 6:*249*
Lemmings, 12:115, 16:124, **16:225–238,** 16:*231*, 16:*232*
Lemmiscus curtatus. See Sagebrush voles
Lemmus spp., 16:225–226, 16:227
Lemmus lemmus. See Norway lemmings
Lemmus sibiricus. See Brown lemmings
Lemniscomys barbarus. See Zebra mice
Lemon-chested greenlets, 11:*257*, 11:*260*–261
Lemon sharks, 4:*74*, 4:115–116, 4:*117*, 4:*122*, 4:123
Lemon sole. *See* Winter flounders
Lemon-sponges, 1:58, 1:*60*, 1:*61*–*62*
Lemos-Espinal, Julio, 7:348
Lemur catta. See Ringtailed lemurs
Lemur coronatus. See Crowned lemurs
Lemur fulvus. See Brown lemurs
Lemur fulvus rufus. See Red-fronted lemurs
Lemur lice, 3:*250*
Lemur macaco. See Black lemurs
Lemur mongoz. See Mongoose lemurs
Lemur possums. *See* Lemuroid ringtails
Lemur rubriventer. See Red-bellied lemurs
Lemuridae. *See* Lemurs
Lemuroid ringtails, 13:115, 13:*116*, 13:*118*, 13:*119*
Lemurs, **14:47–61,** 14:*49*, 14:*53*, 14:*54*
baboon, 14:63, 14:*68*, 14:*69*, 14:72
behavior, 12:*142*, 14:6, 14:7–8, 14:*49*, 14:50–51
blue-eyed, 14:*2*
cathemeral, 14:8
conservation status, 14:52
distribution, 12:135–136, 14:5, 14:*47*, 14:49
diurnal, 14:7, 14:8, 14:9
evolution, 14:2, 14:3, 14:47
feeding ecology, 14:9, 14:51
fork-marked, 12:38
gentle, 14:3, 14:8, 14:10–11, 14:*50*
greater bamboo, 14:11, 14:*48*
habitats, 14:6, 14:49
humans and, 14:52
nocturnal, 14:7, 14:8
physical characteristics, 12:79, 14:4, 14:48–49
reproduction, 14:10–11, 14:51–52
sexual dimorphism, 12:99
sloth, 14:63, 14:*68*, 14:*70*, 14:71
species of, 14:55–60
taxonomy, 14:1, 14:3–4, 14:47–48
See also Dwarf lemurs; Mouse lemurs; Sportive lemurs; Strepsirrhines
Lentipes concolor. See O'opo alamo'os
Leodia sexiesperforata. See Six keyhole sand dollars
Leonard, Jennifer, 14:287
Leontopithecus spp. *See* Lion tamarins
Leontopithecus caissara. See Black-faced lion tamarins
Leontopithecus chrysomelas. See Golden-headed lion tamarins
Leontopithecus chrysopygus. See Golden-rumped lion tamarins
Leontopithecus rosalia. See Golden lion tamarins
Leontopithecus rosalia rosalia. See Lion tamarins
Leopard frogs, 6:*59*
Leopard geckos, 7:262
Leopard lizards. *See* Long-nosed leopard lizards
Leopard loaches. *See* Coolie loaches
Leopard seals, 12:87, 12:138, 14:256, 14:260, 14:417, 14:420, 14:*421*, 14:*435t*
Leopard sharks, 4:116, 4:*118*, 4:*125*, 4:128
See also Zebra sharks
Leopards, 12:*133*, 14:374, 14:*377*, 14:*385*, 14:391*t*
behavior, 12:*146*, 12:*152*, 14:*371*
conservation status, 14:374
distribution, 12:129, 14:370
habitats, 14:257
humans and, 14:376
Leopardus (Felis) pardalis. See Ocelots
Leopardus pardalis. See Ocelots
Leopardus tigrinus. See Little spotted cats
Leopardus wiedii. See Margays
Leopold, Aldo, 12:223
Leopoldamys sabanus, 12:220
Lepacritis spp., 3:202
Lepadella spp., 1:260
Lepadichthys spp., 5:355
Lepadichthys lineatus, 5:*355*
Lepadogaster spp., 5:355
Lepadogastrini, 5:355, 5:356
Lepas anatifera. See Common goose barnacles
Lepeophtheirus salmonis. See Salmon lice
Lepetella laterocompressa, 2:*436*
Lepetelloidea, 2:429
Lepetidae, 2:423
Lepetodrilus elevatus, 2:*432*, 2:*433*
Lepidiolamprologus kendalli, 5:*283*, 5:*286*–287
Lepidobatrachus spp., 6:39, 6:41, 6:155, 6:157, 6:158–159
Lepidobatrachus laevis. See Budgett's frogs
Lepidochelys kempii. See Kemp's ridley turtles
Lepidocolaptes fuscus. See Lesser woodcreepers
Lepidodactylus lugubris. See Mourning geckos
Lepidodasyidae, 1:269, 1:270
Lepidodasys spp., 1:269
Lepidodermella squamata, 1:270, 1:*271*, 1:*272*
Lepidogalaxias salamandroides. See Salamanderfishes
Lepidogalaxiidae. *See* Salamanderfishes
Lepidomysidae. *See Spelaeomysis* spp.
Lepidophanes guentheri, 4:*423*, 4:*444*, 4:*445*–446
Lepidophyma spp., 7:291, 7:292
Lepidophyma flavimaculatum. See Yellow-spotted night lizards
Lepidoptera, 3:35, **3:383–404,** 3:*386*, 3:*390*, 3:*391*, 3:*392*
behavior, 3:*385*–386
conservation status, 3:388–389
distribution, 3:385
evolution, 3:383
feeding ecology, 3:20, 3:46, 3:48, 3:*384*, 3:386–387
habitats, 3:385
humans and, 3:389

physical characteristics, 3:19, 3:24, 3:384–385
reproduction, 3:29, 3:34, 3:38, 3:387–388
species of, 3:*393–404*
taxonomy, 3:383–384
wing-beat frequencies, 3:23
Lepidopyga spp. *See* Hummingbirds
Lepidopyga liliae. *See* Sapphire-bellied hummingbirds
Lepidosauria, 7:16, 7:23–24
Lepidosiren spp., 4:197, 4:201–205
Lepidosiren annectens. *See* African lungfishes
Lepidosiren paradoxa. *See* South American lungfishes
Lepidosirenidae, 4:57
Lepidosireniformes. *See* Lungfishes
Lepidothrix (Pipra) vilasboasi. *See* Golden-crowned manakins
Lepidotrichidae. *See* Relic silverfish
Lepidurus spp., 2:141, 2:142
Lepidurus arcticus, 2:142
Lepidurus packardi. *See* Vernal pool tadpole shrimps
Lepilemur spp. *See* Sportive lemurs
Lepilemur dorsalis. *See* Gray-backed sportive lemurs
Lepilemur edwardsi. *See* Milne-Edwards's sportive lemurs
Lepilemur leucopus. *See* White-footed sportive lemurs
Lepilemur microdon. *See* Small-toothed sportive lemurs
Lepilemur mustelinus. *See* Weasel sportive lemurs
Lepilemur ruficaudatus. *See* Red-tailed sportive lemurs
Lepilemur septentrionalis. *See* Northern sportive lemurs
Lepilemuridae. *See* Sportive lemurs
Lepisma saccharina. *See* Silverfish
Lepismatidae, 3:119
Lepisosteidae, 4:11, 4:15, 4:221
Lepisosteiformes. *See* Gars
Lepisosteus spp., 4:221
Lepisosteus ferox. *See* Alligator gars
Lepisosteus oculatus. *See* Spotted gars
Lepisosteus osseus. *See* Longnose gars
Lepisosteus platostomus. *See* Shortnose gars
Lepisosteus platyrhincus. *See* Florida gars
Lepisosteus sinensis, 4:221
Lepisosteus tristoechus. *See* Cuban gars
Lepomis spp., 4:15
Lepomis cyanellus. *See* Green sunfishes
Lepomis gibbosus. *See* Pumpkinseed sunfishes
Lepomis macrochirus. *See* Bluegills
Lepomis megalotis. *See* Longear sunfishes
Lepomis microlophus. *See* Red-ear sunfishes
Lepophidium spp., 5:22
Leporidae. *See* Hares; Rabbits
Leporinae, 16:505
Leposoma spp., 7:303, 7:304
Leposoma percarinatum, 7:304
Leposphids, 6:10
Lepospondyls, 6:10
Leptacris spp., 3:203
Leptailurus (Felis) serval. *See* Servals
Leptasterias hexactis, 1:370
Leptasthenura striolata. *See* Striolated tit-spinetails
Leptasthenura xenothorax. *See* White-browed tit-spinetails
Leptestheriidae, 2:147
Leptininae, 3:54
Leptinotarsa decemilineata. *See* Colorado beetles
Leptobos spp., 15:265, 16:1
Leptobrachella spp., 6:110, 6:111
Leptobrachiinae, 6:109–111
Leptobrachium spp., 6:110, 6:111
Leptobrachium banae. *See* Bana leaf litter frogs
Leptochariidae. *See* Barbeled hound sharks
Leptochilichthyes spp., 4:390
Leptochoeridae, 15:264
Leptocoma spp. *See* Sunbirds
Leptocoma minima. *See* Crimson-backed sunbirds
Leptocoma zeylonica. *See* Purple-rumped sunbirds
Leptocottus armatus. *See* Staghorn sculpins
Leptodactylid frogs, **6:155–171**, *6:160–161*
behavior, 6:158, 6:162–171
conservation status, 6:159, 6:162–171
distribution, 6:6, 6:*155*, 6:157–158, 6:*162–171*
evolution, 6:12–13, 6:155–157
feeding ecology, 6:158, 6:162–171
habitats, 6:158, 6:162–171
humans and, 6:159, 6:162–171
metamorphosis, 6:42
physical characteristics, 6:157, 6:162–171
reproduction, 6:32, 6:34, 6:68, 6:158–159, 6:162–171
species of, 6:162–171
tadpoles, 6:157
taxonomy, 6:139, 6:155–157, 6:162–171, 6:197
Leptodactylidae. *See* Leptodactylid frogs
Leptodactylinae, 6:156
Leptodactylus spp., 6:*30*, 6:35, 6:42, 6:156, 6:157, 6:159
Leptodactylus bolivianus, 6:159
Leptodactylus fallax, 6:37, 6:159
Leptodactylus pentadactylus. *See* South American bullfrogs
Leptodeira spp. *See* Cat-eyed snakes
Leptodeira septentrionalis. *See* Northern cat-eyed snakes
Leptoderinae, 3:54
Leptodon forbesi. *See* White-collared kites
Leptodora kindtii, 2:153
Leptogryllus deceptor. *See* Oahu deceptor bush crickets
Leptolalax spp., 6:110, 6:111
Leptolalax pelodytoides. *See* Slender mud frogs
Leptolalax sungi. *See* Sung's slender frogs
Leptomedusae, 1:123, 1:124, 1:125
Leptonychotes weddellii. *See* Weddell seals
Leptonycteris spp., 13:415
Leptonycteris curasoae. *See* Southern long-nosed bats
Leptonycteris nivalis, 13:417, 13:420
Leptopelinae, 6:281
Leptopelis spp., 6:281, 6:283, 6:289
Leptopelis brevirostris, 6:*281*, 6:282, 6:283
Leptopelis bufonides. *See* Toad-like treefrogs
Leptopelis macrotis. *See* Big-eared forest treefrogs
Leptopelis millsoni, 6:289
Leptopelis palmatus, 6:281, 6:289
Leptopelis rufus, 6:289
Leptopelis xenodactylus, 6:283
Leptoperla cacuminis, 3:143
Leptophryne spp., 6:184
Leptophyes punctatissima. *See* Speckled bush-crickets
Leptopius spp. *See* Wattle pigs
Leptopodomorpha, 3:259
Leptopsyllidae, 3:347, 3:348
Leptopterus chabert. *See* Chabert's vangas
Leptopterus viridis. *See* White-headed vangas
Leptopteryx cruenta. *See* Crimson-bellied orioles
Leptoptilos spp. *See* Storks
Leptoptilos crumeniferus. *See* Marabous
Leptoptilos dubius. *See* Greater adjutants
Leptoptilos javanicus. *See* Lesser adjutant storks
Leptoscopidae. *See* Southern sandfishes
Leptosomidae. *See* Courols
Leptosomus discolor. *See* Courols
Leptostomias gladiator, 4:*422*
Leptostracans, **2:161–166**, *2:163*
Leptosynaptinae, 1:417
Leptothorax longispinosus. *See* Acorn-nesting ants
Leptotila plumbeiceps. *See* Gray-headed doves
Leptotyphlopidae. *See* Slender blindsnakes
Leptotyphlops spp., 7:373–375
Leptotyphlops alfredschmidti, 7:373–374
Leptotyphlops blanfordi, 7:374
Leptotyphlops borrichianus, 7:374
Leptotyphlops boulengeri, 7:373
Leptotyphlops broadleyi, 7:373
Leptotyphlops cairi, 7:374
Leptotyphlops conjunctus, 7:375
Leptotyphlops dulcis. *See* Texas blindsnakes
Leptotyphlops filiformis, 7:374
Leptotyphlops goudotii, 7:375
Leptotyphlops humilis, 7:373–375
Leptotyphlops macrops, 7:373
Leptotyphlops macrorhynchus, 7:373, 7:374
Leptotyphlops macrurus, 7:374
Leptotyphlops melanotermus, 7:373
Leptotyphlops natatrix, 7:375
Leptotyphlops nigricans. *See* Black threadsnakes
Leptotyphlops nursii, 7:374
Leptotyphlops occidentalis, 7:373, 7:374
Leptotyphlops parkeri, 7:374
Leptotyphlops rostratus, 7:374
Leptotyphlops scutifrons. *See* Peters' wormsnakes
Leptotyphlops septemstriatus, 7:374
Leptotyphlops teaguei, 7:373–374
Leptotyphlops tricolor, 7:373–374
Leptotyphlops unguirostris, 7:374
Leptotyphlops weyrauchi, 7:373
Leptotyphlops wilsoni, 7:374
Lepus spp. *See* Hares
Lepus alleni. *See* Antelope jackrabbits
Lepus americanus. *See* Snowshoe hares
Lepus arcticus. *See* Arctic hares
Lepus brachyurus. *See* Hares
Lepus californicus. *See* Black-tailed jackrabbits
Lepus callotis. *See* White-sided jackrabbits
Lepus capensis. *See* Cape hares
Lepus europaeus. *See* European hares
Lepus flavigularis. *See* Tehuantepec jackrabbits
Lepus mandshuricus. *See* Manchurian hares
Lepus othus, 16:487
Lepus timidus. *See* Mountain hares
Lerista spp., 7:328, 7:329
Lerista bougainvillii, 7:329

Lerp psyllids, 3:57
Lerwa lerwa. See Snow partridges
Lesbia spp. *See* Trainbearers
Leschenault's forktails. *See* White-crowned forktails
Lesser adjutant storks, 8:236
Lesser African jacanas. *See* Lesser jacanas
Lesser amikihis. *See* Anianiaus
Lesser Andean flamingos. *See* James' flamingos
Lesser anomalures, 16:*301*, 16:*302*, 16:303–304
Lesser anteaters. *See* Northern tamanduas; Southern tamanduas
Lesser apes. *See* Gibbons
Lesser beaked whales. *See* Pygmy beaked whales
Lesser bilbies, 13:2, 13:19, 13:*20*, 13:22
Lesser bulldog bats, 13:443, 13:*444*, 13:*445*, 13:446, 13:*448*, 13:449–*450*
Lesser cane rats, 16:333, 16:334, 16:*336*, 16:337–*338*
Lesser chameleons. *See* Minor chameleons
Lesser-crested mastiff bats, 12:85, 13:315, 13:486, 13:*489*, 13:*490*
Lesser Cuban nesophontes, 13:243
Lesser cuckoos, 9:314
Lesser dog-faced bats, 13:363*t*
Lesser doglike bats. *See* Lesser dog-faced bats
Lesser egrets. *See* Little egrets
Lesser Egyptian jerboas, 16:*212*, 16:*213*, 16:223*t*
Lesser elaenias, 10:272
Lesser electric rays, 4:*181*, 4:*185*
Lesser false vampire bats, 13:*380*
Lesser fat-tailed jerboas, 16:212, 16:216, 16:223*t*
Lesser flame-backed woodpeckers, 10:*152*, 10:*164*
Lesser flamingos, 8:303, 8:*304*, 8:*307*, 8:*309*
Lesser flat-headed bats. *See* Bamboo bats
Lesser floricans, 9:3, 9:92
Lesser forest wallabies, 13:102*t*
Lesser frigatebirds, 8:193, 8:*194*
Lesser gliders, 13:33, 13:37
Lesser golden-backed woodpeckers. *See* Lesser flame-backed woodpeckers
Lesser gray shrikes, 10:428
Lesser green leafbirds, 10:417, 10:*418*, 10:*420*–421
Lesser gymnures, 12:220, 13:204, 13:*208*, 13:*210*, 13:211
Lesser Haitian ground sloths, 13:*159*
Lesser hedgehog tenrecs, 13:*193*, 13:196, 13:199, 13:*228*, 13:234*t*
Lesser honeyguides, 10:138
Lesser horseshoe bats, 13:316, 13:390, 13:393, 13:*394*, 13:*399*–400
Lesser icterine bulbuls. *See* Icterine greenbuls
Lesser jacanas, 9:107, 9:108, 9:*111*, 9:*113*
Lesser kestrels, 8:350, 8:352
Lesser koa finches, 11:343, 11:349
Lesser kudus, 16:4, 16:11
Lesser lily trotters. *See* Lesser jacanas
Lesser Malay mouse deer, 15:325, 15:*327*, 15:*328*, 15:*330*, 15:*331*, 15:332
Lesser marsupial moles. *See* Northern marsupial moles
Lesser masked owls, 9:338, 9:340–341
Lesser melampittas, 11:75
Lesser naked bats, 13:493*t*
Lesser New Zealand short-tailed bats, 13:313, 13:*454*, 13:455, 13:*456*, 13:457
Lesser nighthawks, 9:403
Lesser octopods, 2:27
Lesser pandas. *See* Red pandas
Lesser pied kingfishers. *See* Pied kingfishers
Lesser pin-tailed sandgrouse. *See* Chestnut-bellied sandgrouse
Lesser pink fairy armadillos. *See* Pink fairy armadillos
Lesser racket-tailed drongos, 11:438
Lesser rheas, 8:69–71, 8:*70*, 8:72, 8:*73*–*74*
Lesser rorquals. *See* Northern minke whales
Lesser sac-winged bats, 13:363*t*
Lesser sandhill cranes, 9:26
Lesser scaly-breasted wren-babblers. *See* Pygmy wren-babblers
Lesser seriemas. *See* Black-legged seriemas
Lesser sheath-tailed bats, 13:*359*, 13:*362*–363
Lesser sheathbills. *See* Black-faced sheathbills
Lesser short-nosed fruit bats, 13:311, 13:*336*
Lesser short-tailed bats. *See* Lesser New Zealand short-tailed bats
Lesser shrews. *See* Lesser white-toothed shrews
Lesser sirens, 6:327, 6:*330*, 6:331–332
Lesser sooty owls, 9:336, 9:338
Lesser spectral tarsiers. *See* Pygmy tarsiers
Lesser spotted dogfishes, 4:*34*
Lesser Sundas pythons, 7:420
Lesser thumbless bats. *See* Smoky bats
Lesser tree shrews. *See* Pygmy tree shrews
Lesser tree swifts. *See* Whiskered tree swifts
Lesser weeverfishes. *See* Lesser weevers
Lesser weevers, 5:335, 5:*336*, 5:*338*–339
Lesser whistling-ducks, 8:370
Lesser white-lined bats. *See* Lesser sac-winged bats
Lesser white-nosed monkeys, 14:*192*
Lesser white-tailed shrews, 13:267
Lesser white-toothed shrews, 13:*267*, 13:*268*, 13:276*t*
Lesser woodcreepers, 10:*231*, 10:235–*236*
Lesser woolly horseshoe bats, 13:388, 13:389, 13:*395*, 13:*397*
Lesser yellow bats. *See* House bats
Lesser yellow-billed kingfishers. *See* Yellow-billed kingfishers
Lesser yellow-eared spiderhunters. *See* Yellow-eared spiderhunters
Lesser yellow-headed vultures, 8:276, 8:*279*, 8:280–*281*
Lesson's motmots. *See* Blue-crowned motmots
Lestodelphys spp. *See* Patagonian opossums
Lestodelphys halli. See Patagonian opossums
Lesuerigobius friesii. See Fries' gobies
Lethocerus spp., 3:62
Lethocerus maximus. See Giant water bugs
Lethrinidae. *See* Emperors
Lethrinops furcifer. See Greenface sandsifters
Lethrus apterus, 3:62
Letter-winged kites, 8:319, 8:327, 8:*327*, 8:*332*
Leucandra walfordi, 1:57
Leucetta chagosensis. See Lemon-sponges
Leucetta microraphis, 1:58
Leucippus spp. *See* Hummingbirds
Leuciscinae, 4:297
Leuciscus spp., 4:298
Leuciscus idella. See Grass carps
Leuciscus leuciscus. See Common daces
Leuciscus molitrix. See Silver carps
Leuckatiara octona, 1:*154*
Leucocarbo spp. *See* Shags
Leucocarbo atriceps. See Imperial shags
Leucocarbo bouganvillii. See Guanay shags
Leucochloris spp. *See* Hummingbirds
Leucopeza semperi. See Semper's warblers
Leucophaeus scoresbii. See Dolphin gulls
Leucopsar spp., 11:409
Leucopsar rothschildi. See Bali mynas
Leucopsarion petersi. See Japanese ice gobies
Leucopternis lacernulata. See White-necked hawks
Leucosarcia melanoleuca. See Wonga pigeons
Leucosoleniida, 1:58
Leucosticte arctoa. See Asian rosy-finches
Leucosticte sillemi. See Sillem's mountain-finches
Leucosticte tephrocotis. See Gray-crowned rosy finches
Leuctra fusca, 3:*141*, 3:*142*
Leuctridae, 3:54, 3:141
Leuresthes sardina, 5:69
Leuresthes tenuis. See California grunions
Leuroglossus stilbius. See California smoothtongues
Levantine vipers, 7:*449*, 7:*459*
Levinebalia spp., 2:161
Levinebalia maria, 2:*163*, 2:*164*, 2:165
Lewis' mudpuppies. *See* Neuse River waterdogs
Lexigrams, 12:161–162
l'Hoest's guenons, 14:188, 14:194
Lialis spp., 7:205
Lialis burtonis. See Burton's snake lizards
Liaoxitriton spp., 6:13
Liaoxitriton zhongjiani, 6:335
Liasis spp., 7:420
Liasis fuscus. See Water pythons
Libelloides spp., 3:308
Liberian mongooses, 14:349, 14:352, 14:*353*, 14:*354*, 14:356
Liberian swamp eels, 5:152
Liberiictis spp., 14:347
Liberiictis kuhni. See Liberian mongooses
Libyan wild cats, 12:180, 14:290
Libypithecus spp., 14:172
Lice, 3:31, 3:36, 3:64
  book, **3:243–248,** 3:*244*, 3:*245*
  chewing, **3:249–257**
  in Middle Ages, 3:85
  as parasites, 3:47
  as pests, 3:76, 3:252
  plant, 3:260, 3:262, 3:263
  sucking, **3:249–257**
Lichenostomus spp. *See* Australian honeyeaters
Lichenostomus melanops. See Yellow-tufted honeyeaters
Lichens, 1:31, 2:41
Lichmera spp. *See* Australian honeyeaters
Lichmera indistincta. See Brown honeyeaters
Lichtenstein's hartebeests, 16:27, 16:29, 16:33, 16:*35*, 16:*38*, 16:40
Lichtenstein's jerboas, 16:212, 16:215, 16:223*t*
Lichtenstein's sandgrouse, 9:231, 9:*234*, 9:237–*238*
Lickliter, R., 15:145
Liem, K. F., 5:427
Life, origins of, 2:8–11

Life cycles. *See* Reproduction
Life histories, **1:15–30, 2:15–24, 3:31–40,** 3:68, **12:97–98,** 12:107–108
See also Reproduction; specific species
The Life of Birds (Attenborough), 11:507
The Life of Lyrebird (Smith), 10:330
Life of Mammals, 13:217
Life spans, 12:97–98
Lift, for powered flight, 12:56
Light, 1:24, 1:38–40, 2:26–27
See also Habitats
Light-mantled albatrosses, 8:114, 8:*117*, 8:*119*–120
Light-mantled sooties. *See* Light-mantled albatrosses
Light-shrimps. *See* Krill
Lightfishes, 4:421, 4:422, 4:423, 4:424
Lightiellidae, 2:131
Ligia spp., 2:249, 2:250, 2:251
Ligidae, 2:250
Lignocellulosic matter, 3:166, 3:168
Ligula intestinalis, 1:230
Lilac-cheeked kingfishers, 10:6, 10:*12*, 10:*16*
Lilac-crowned wrens. *See* Purple-crowned fairy-wrens
Lilac kingfishers. *See* Lilac-cheeked kingfishers
Lily trotters. *See* Jacanas
Limacodidae, 3:386, 3:*386*, 3:389
Limanda ferruginea. *See* Yellowtail flounders
Limax spp. *See* Garden slugs
Limax maximus. *See* Great gray slugs
Limbed vertebrates, evolution, 6:7–11
Limbs, 7:24–25, 12:3–4, 12:8, 12:40–41, 12:56–57
crocodilians, 7:160
Lepidosauria, 7:23–24
Squamata, 7:23–24, 7:202
Testudines, 7:*66*, 7:67
See also Physical characteristics
Limidae, 2:452
Limiting factors, ecological, 2:25–27
Limnadia lenticularis, 2:*147*
Limnadiidae, 2:147
Limnephilidae, 3:54
Limnodromus semipalmatus. *See* Asian dowitchers
Limnodynastes spp., 6:139, 6:141
Limnodynastes convexiusculus, 6:140
Limnodynastes dorsalis, 6:140
Limnodynastes lignarius. *See* Woodworker frogs
Limnodynastes ornatus, 6:139, 6:140
Limnodynastes spenceri. *See* Spencer's burrowing frogs
Limnodynastes tasmaniensis, 6:140, 6:141
Limnodynastidae. *See* Australian ground frogs
Limnodytes phyllophila. *See* Nilgiri tropical frogs
Limnofregata azygosternon, 8:186
Limnogale spp., 13:225, 13:226, 13:227, 13:228
Limnogale mergulus. *See* Aquatic tenrecs
Limnognathia maerski. *See* Jaw animals
Limnognathida. *See* Jaw animals
Limnognathiidae. *See* Jaw animals
Limnomedusa spp., 6:156, 6:157
Limnomedusae, 1:123, 1:124, 1:125–126
Limnonectes spp., 6:33, 6:36, 6:246, 6:249, 6:251
Limnonectini, 6:246, 6:251
Limnoria spp., 2:252
Limnoria quadripunctata. *See* Gribbles
Limonia hardyana, 3:358
Limosa spp. *See* Godwits
Limosa lapponica. *See* Bar-tailed godwits
Limosa limosa. *See* Black-tailed godwits
Limpets, 2:439
deep-sea, **2:435–437**
New Zealand freshwater, 2:*416*, 2:417
true, **2:423–427,** 2:*426*, 2:*427*
See also Neritopsina
Limpkins, 9:1–4, 9:23, **9:37–39,** 9:*38*, 9:*39*, 9:69
Limulidae. *See* Horseshoe crabs
Limulus amebocyte lysate (LAL), 2:42
Limulus polyphemus. *See* American horseshoe crabs
Lina's sunbirds, 11:210
Linckia laevigata. *See* Blue starfishes
Lincoln Park Zoo, Chicago, 12:203
Line-tailed pygmy monitors. *See* Stripe-tailed monitors
Lineated woodpeckers, 10:*148*
Lineatriton spp., 6:325, 6:391
Lined chitons, 2:*393*, 2:*396*, 2:*398*–399
Lined seahorses, 5:134, 5:*137*, 5:*139*, 5:146
Lined snakes, 7:468
Lined surgeonfishes, 5:*394*, 5:*396*
Linepithema humile. *See* Argentine ants
LINEs (long interspersed nuclear elements), 12:29–30
Linesiders. *See* Striped sea basses
Lineus spp., 1:245, 1:*247*
Lineus longissimus, 1:245, 1:*248*, 1:249–*250*
Lineus rubber, 1:246
Lineus viridis, 1:246
Lingcods, 5:182, 5:183, 5:*184*, 5:*191*–192
Lings, 5:26
See also Burbots; Lingcods; Lotidae
Linguatuidae. *See Linguatula* spp.
Linguatula spp., 2:318, 2:319, 2:320
Linguatula arctica, 2:319
Linguatula serrata, 2:319, 2:320
Lingula spp., 2:*515*, 2:516
Lingulidae, 2:515–516
Linguliformea, 2:515
Linh duongs, 15:263, 16:1–2, 16:14
Linnaeus, Carolus, 1:11, 1:417–418, 1:451, 2:8
on albatrosses, 8:116
on Archonta, 13:333
on Cetacea, 15:1
on dippers, 10:475
on flickers, 10:155
on herons, 8:239
on Insectivora, 13:193
on killifishes, 5:90
on mimids, 10:465
on owls, 9:345
on palmchats, 10:455
on Piciformes, 10:85
on primates, 14:1
on Scorpaeniformes, 5:157
on todies, 10:25
Linnaeus's pea crabs. *See* Pea crabs
Linnaeus's snapping termites, 3:*169*, 3:*171*, 3:174
Linnets, 11:323
Linsangs, **14:335–341,** 14:*339*, 14:345t
Linstowiidae, 1:226
Linuche unguiculata. *See* Thimble jellys
Liobranchia spp., 5:355
Liocichla spp. *See* Liocichlas
Liocichla omeiensis. *See* Omei Shan liocichlas
Liocichlas, 10:505
Liodesmidae, 4:229
Lioheterodon spp. *See* Madagascan hog-nosed snakes
Liolaeminae, 7:204
Liomys spp., 16:199, 16:200, 16:203
Liomys irroratus. *See* Mexican spiny pocket mice
Liomys salvini. *See* Salvin's spiny pocket mice
Lion beetles, 3:320, 3:*325*, 3:*327*, 3:328
Lion-tailed macaques. *See* Nilgiri langurs
Lion tamarins, 14:7, 14:8, 14:11, 14:115–121, 14:*119*, 14:123, 14:124
Lionfishes, 4:16, 4:48, 4:68, 4:70, 5:164, 5:165, 5:166
Lions, 14:*369*, 14:*370*, 14:*371*, 14:*374*, 14:*377*
behavior, 12:*9*, 12:*144*, 12:146–147, 12:*153*, 14:258, 14:259, 14:260, 14:371
conservation status, 12:*216*
distribution, 12:129–130, 14:370
feeding ecology, 12:*116*, 14:372
habitats, 12:*136*
humans and, 14:262, 14:375, 14:376
physical characteristics, 14:370
reproduction, 12:*91*, 12:103, 12:107, 12:*109*, 12:*136*, 14:261, 14:262, 14:372
Lion's mane jellyfish, 1:25, 1:29, 1:*159*, 1:*164*–165
Liophidium spp., 7:205
Liophis spp., 7:467
Liophryne spp., 6:36, 6:305
Liopropomini. *See* Swissguard basslets
Liosaccus (=Sphoeroides) pachygaster. *See* Blunthead puffers
Liosceles spp. *See* Tapaculos
Liosceles thoracicus. *See* Rusty-belted tapaculos
Liotyphlops spp., 7:369–371
Liotyphlops anops, 7:369–370
Liotyphlops beui, 7:369–371
Liotyphlops schubarti, 7:369–370
Liotyphlops ternetzii, 7:369–*370*, 7:*371*
Liparididae. *See* Snailfishes
Liparis fucensis. *See* Slipskin snailfishes
Lipaugus spp. *See* Cotingas
Lipaugus lanioides, 10:308
Lipaugus uropygialis, 10:308
Lipids, 12:120
Lipogenys spp., 4:249
Liposcelididae, 3:243
Liposcelis spp., 3:244
Liposcelis bostrychophila, 3:*245*, 3:246–*247*
Lipostraca, 2:135
Lipotes spp., 15:19
Lipotes vexillifer. *See* Baijis
Lipotidae. *See* Baijis
Lipotyphla, 13:227
Lirceus fontinalis, 2:*253*, 2:*255*, 2:256
Lirimiris spp., 1:42, 2:38
Liriope tetraphylla, 1:*130*, 1:145
Lirometopum spp. *See* Pitbull katydids
Lironeca spp., 2:252
Lissachatina fulica. *See* Giant African land snails
Lissamphibians
aquatic, 6:17
evolution, 6:9, 6:11–13
physical characteristics, 6:15–27, 6:*16*, 6:*26*
reproduction, 6:17–18
taxonomy, 6:323

Lissemys punctata. See Indian flapshell turtles
Lissodelphinae, 15:41
Lissodelphis spp., 15:42
Lissodendoryx spp., 1:45
Lissonanchus spp., 5:355
Lissotis spp. *See* Bustards
Lissotis melanogaster. See Black-bellied bustards
Listroscelidinae, 3:207
Literature
- albatrosses, 8:116
- amphibians, 6:52, 6:*53*
- birds, 8:20, 8:26
- dodos, 9:270
- fish, 4:72–73
- hedge sparrows, 10:462
- insects, 3:74
- larks, 10:346
- protostomes, 2:41–42
- *See also* Humans; Mythology; specific species

Lithidiidae, 3:203
Lithobates spp., 6:248
Lithobiomorpha, 2:353, 2:354, 2:355, 2:356
Lithobius forficatus. See Stone centipedes
Lithodytes spp., 6:156
Lithodytes lineatus. See Gold-striped frogs
Lithogramma oculatum, 3:305
Lithonida, 1:58
Litobothriidae. *See* Litobothriidea
Litobothriidea, 1:226, 1:229
Litocranius walleri. See Gerenuks
Litopterna, 12:11, 12:132, 15:131, 15:133, 15:135–138
Litoria spp., 6:227–230
Litoria angiana, 6:225
Litoria aurea. See Green and gold bell frogs
Litoria caerulea. See Green treefrogs
Litoria genimaculata, 6:46
Litoria infrafrenata. See White-lipped treefrogs
Litoria microbelos, 6:227
Litoria nasuta. See Rocket frogs
Litoria splendida, 6:46, 6:227
Litter-dwelling bugs, 3:260
Little auks. *See* Dovekies
Little bitterns, 8:*243*
Little blue penguins. *See* Little penguins
Little brown bats, 13:*498,* 13:*499,* 13:*504,* 13:*511*–512
- behavior, 13:499
- conservation status, 13:316
- distribution, 13:310–311
- echolocation, 12:85
- feeding ecology, 13:313
- reproduction, 12:110, 13:314, 13:315, 13:501

Little brown bustards, 9:93
Little brown cranes. *See* Sandhill cranes
Little brown myotis. *See* Little brown bats
Little bustards, 9:92, 9:93, 9:*95,* 9:*99*
Little cassowaries. *See* Bennett's cassowaries
Little chipmunks. *See* Least chipmunks
Little collared fruit bats, 13:349*t*
Little corellas, 9:*276*
Little dorcopsises. *See* Lesser forest wallabies
Little egrets, 8:*247,* 8:253–254
Little file snakes, 7:439–*440,* 7:*441,* 7:*443*
Little fire ants, 3:72
Little five-toed jerboas, 16:215, 16:216, 16:217, 16:*218,* 16:*220,* 16:221–222
Little flying cows, 13:348*t*
Little golden-mantled flying foxes, 13:332*t*
Little grassbirds, 11:*9,* 11:*21–22*
Little gray kiwis. *See* Little spotted kiwis
Little grebes, 8:169, 8:170, 8:171, 8:172, 8:*173,* 8:174–18:*75*
Little green bee-eaters, 10:*40*
Little ground-jays. *See* Hume's ground-jays
Little guitars. *See* Guitarrita
Little hermits, 9:*441*
Little-house flies, 3:361
Little king birds of paradise. *See* King birds of paradise
Little long-fingered bats, 13:*520*
Little marshbirds. *See* Little grassbirds
"Little Miss Muffet," 2:41
Little Nepalese horseshoe bats, 13:389
Little northern cats. *See* Northern quolls
Little owls, 9:*333,* 9:*348*
Little Papuan frogmouths. *See* Marbled frogmouths
Little penguins, 8:148, 8:150, 8:*152,* 8:*157*
Little pikas. *See* Steppe pikas
Little pocket mice, 16:200, 16:*202,* 16:*267*
Little pygmy possums, 13:106, 13:*109,* 13:*110*–111
Little red brocket deer, 15:396*t*
Little red flying foxes, 12:*56,* 13:*321,* 13:*324,* 13:*325,* 13:330
Little red kalutas, 12:282, 12:299*t*
Little reedbirds. *See* Little grassbirds
Little-ringed plovers, 9:163
Little rock wallabies. *See* Narbaleks
Little rush-warblers, 11:4, 11:*9,* 11:*15*–16
Little shearwaters, 8:108
Little shrike-thrushes, 11:*119,* 11:*123*
Little slaty flycatchers, 11:*29,* 11:*33*
Little sparrowhawks. *See* African little sparrowhawks
Little spotted cats, 14:391*t*
Little spotted kiwis, 8:90, 8:*91,* 8:92–93
Little stints, 9:177
Little striped whiptails, 7:310, 7:312
Little treecreepers. *See* White-throated treecreepers
Little turtledoves. *See* Diamond doves
Little water opossums. *See* Thick-tailed opossums
Little weevers. *See* Lesser weevers
Little woodstars, 9:450
Little woodswallows, 11:*461,* 11:*462–463*
Little yellow-shouldered bats, 13:417, 13:418, 13:*421,* 13:*429,* 13:431
Littleneck clams. *See* Northern quahogs
Littorina spp., 2:446
Littorinidae, 2:446
Littorinimorpha, 2:445
Liturgusa charpentieri, 3:*182,* 3:184, 3:*186*
Liturgusidae, 3:177
Liua spp., 6:335, 6:336
Liua shihi, 6:336
Live-bearers, **5:89–94,** 5:*95,* **5:96, 5:99–101**
Live-bird trade. *See* Cagebirds
Live births. *See* Viviparity
Liver flukes, 1:34–35, 1:*202,* 1:*203,* 1:207
Livestock. *See* Domestic cattle
Livezey, Bradley, 8:369, 9:45
Living fossils, 3:*136,* 3:*137*
Livingstone's flying foxes. *See* Livingstone's fruit bats
Livingstone's fruit bats, 13:*323,* 13:*326,* 13:327
Livingstone's turacos, 9:300
Liza spp., 5:59
Liza grandisquamis. See Large-scale mullets
Liza luciae, 5:60
Liza melinoptera, 5:60
Lizardfishes, 4:66–67, **4:431–434,** 4:*432,* 4:*433,* 4:*435,* ***4:438–439***
Lizards, **7:195–208**
- alligator, **7:339–344,** 7:*341*
- Anguidae, **7:339–345,** 7:*341*
- behavior, 7:204–206
- blindskinks, **7:271–272**
- conservation status, 7:59, 7:62, 7:207
- Cordylidae, **7:319–325,** 7:*323*
- distribution, 7:203–204
- dragon, **7:209–211,** 7:*212,* 7:*213,* **7:216–220**
- ears, 7:9
- evolution, 7:16, 7:195–198
- eyes, 7:9, 7:32
- Florida wormlizards, **7:283–285,** 7:*284*
- in folk medicine, 7:51
- as food, 7:48
- Gekkonidae, **7:259–269,** 7:*264*
- girdled, **7:319–325,** 7:*323*
- glass, **7:339–342,** 7:*341*
- habitats, 7:204
- Helodermatidae, **7:353–358,** 7:*356*
- herbivorous, 7:27
- homing, 7:43
- humans and, 7:52–53
- Iguanidae, **7:243–257,** 7:*248,* 7:*249*
- integumentary system, 7:29
- knob-scaled lizards, **7:347–352,** 7:*350*
- Lacertidae, **7:297–302,** 7:*300*
- limbs, 7:25
- locomotion, 7:7–8, 7:25
- Mexican beaded, 7:10, **7:353–358,** 7:*354,* 7:*356*
- mole-limbed wormlizards, **7:279–282**
- in mythology and religions, 7:54, 7:55
- night, 7:207
- night lizards, **7:291–296,** 7:*293*
- physical characteristics, 7:198–203, 7:*199*
- plated, **7:319–325,** 7:*323*
- reproduction, 7:6–7, 7:42, 7:206–207
- rock, **7:297–302**
- salt glands, 7:30
- skinks, **7:327–338,** 7:*330,* 7:*331*
- spade-headed wormlizards, **7:287–290**
- taxonomy, 7:195–198, 7:*196*
- teeth, 7:9
- Teiidae, **7:309–317,** 7:*313*
- territoriality, 7:43–44
- thermoregulation, 7:28
- Varanidae, **7:359–368,** 7:*364*
- venom, 7:10
- wall, **7:297–302**
- whiptail, **7:309–317,** 7:*313*
- wormlizards, **7:273–277,** 7:*274*
- *See also* specific types of lizards

Llamas, 15:137, **15:313–323,** 15:*314,* 15:*317,* 15:*319,* 15:*322,* 15:323
Llanocetus denticrenatus, 15:3
Llanos long-nosed armadillos, 13:183, 13:184, 13:191*t*
Lo spp., 5:391, 5:393
Loach catfishes, 4:352, 4:353
Loach gobies, 5:373–375, 5:*379,* 5:*380,* 5:388

Loaches, 4:321–*322*, 4:*323–324*, 4:*325–327*, 4:331–332
Loango weavers, 11:379
Loas (voodoo gods), 7:57
Lobata, 1:169
Lobate ctenophores, 1:*169*
Lobatocerebrum, 1:11
Lobe-finned fishes, 4:12, 6:7
Lobed ducks. *See* Musk ducks
Lobodon carcinophagus. *See* Crab-eater seals
Loboglomeris pyrenaica, 2:365
Lobopardisea sericea. *See* Yellow-breasted birds of paradise
Lobophasma spp., 3:217, 3:*218*
Lobotes surinamensis. *See* Tripletails
Lobotidae. *See* Tripletails
Lobster cockroaches. *See* Cinereous cockroaches
Lobsters, **2:197–214,** 2:*202*
behavior, 2:36, 2:199–200
conservation status, 2:201
distribution, 2:199
evolution, 2:197–198
feeding ecology, 2:200
habitats, 2:199
humans and, 2:201
physical characteristics, 2:198
reproduction, 2:17, 2:201
species of, 2:*204*–214
taxonomy, 2:197–198
Lobulia spp., 7:329
Loch Ness monster, 7:54
Lockington, W. N., 5:351
Locomotion, 1:*4*, 1:38–40, 2:*19*, 7:7–8, 7:24–26, 7:202, 12:3–4, 12:8, 12:40–41, 12:41–44
evolution of, 12:14
subterranean mammals, 12:71–72
See also Behavior; Movement; Physical characteristics
Locust birds. *See* Collared pratincoles
Locust starlings. *See* Common mynas
Locustella spp., 11:3, 11:4, 11:7
Locustella naevia. *See* Grasshopper warblers
Locusts, 1:42, 3:9, 3:18, 3:201, 3:204
in art, 3:74
education and, 3:83
as food, 3:80, 3:208
humans and, 3:207–208
migratory, 3:66, 3:204
as pests, 3:75
red, 3:35
See also Desert locusts; Seventeen-year cicadas
Loddes. *See* Capelins
Loder's gazelles. *See* Slender-horned gazelles
Loess steppe, 12:17–18
Loggerhead shrikes, 10:425, 10:427, 10:428, 10:*429*, 10:*431*, 10:*437*–438
Loggerhead turtles, 7:*61*, 7:*85*, 7:*87*, 7:*88*, 7:*89*–90
Loggerheads. *See* Loggerhead turtles
Logperches, 5:*200*, 5:*203*, 5:208
Logrunners, **11:69–74,** 11:*72–73*, 11:75
Loligo spp. *See* Squids
Loligo opalescens. *See* California market squids
Loligo pealeii. *See* Longfin inshore squids
Lonchophylla robusta. *See* Orange nectar bats
Lonchophyllinae, 13:413, 13:414, 13:415, 13:420
Lonchoptera spp., 3:360
Lonchorhina aurita. *See* Sword-nosed bats
Lonchothrix emiliae. *See* Tuft-tailed spiny tree rats
Lonchura spp., 11:353
Lonchura cantans. *See* African silverbills
Lonchura maja. *See* White-headed munias
Lonchura punctulata. *See* Spotted munias
Lonchura striata. *See* White-backed munias
Long-armed brittle stars. *See* Dwarf brittle stars
Long-armed chafers, 3:317
Long-beaked echidnas, 12:108, 12:137, 12:227–238, 12:*237*, 12:*238*, 12:*239*, 12:240*t*
Long-billed crombecs, 11:2, 11:4, 11:*8*, 11:*19*
Long-billed curlews, 9:*181*, 9:*183*
Long-billed gnatwrens, 11:*8*, 11:*10*
Long-billed larks. *See* Cape long-billed larks
Long-billed marsh wrens. *See* Marsh wrens
Long-billed murrelets. *See* Marbled murrelets
Long-billed plovers, 9:162
Long-billed prions. *See* Broad-billed prions
Long-billed rhabdornis. *See* Greater rhabdornis
Long-billed starthroats, 9:445
Long-billed sunbirds, 11:207
Long-billed vultures, 8:323
Long-billed woodcreepers, 10:*231*, 10:*233*
Long-bodied cellar spiders, 2:*338*, 2:*341*, 2:346
See also Common harvestmen
Long-clawed ground squirrels, 16:159*t*
Long-clawed mole mice, 16:265, 16:267
Long-clawed mole voles, 16:227, 16:*232*, 16:*233*, 16:236
Long-clawed squirrels, 16:144
Long-crested eagles, 8:321
Long-eared bats, 13:497–498
Long-eared hedgehogs, 13:204, 13:*207*, 13:*208*, 13:*209*, 13:210–211
Long-eared jerboas, 16:211–212, 16:215, 16:216, 16:*218*, 16:*219*
Long-eared lesser gymnures, 13:203
Long-eared mice, 16:265
Long-eared owls, 9:333, 9:348
Long-fingered bats, 13:313, 13:515*t*
Long-fingered slender toads, 6:*187*, 6:188–*189*
Long-fingered stream toads. *See* Long-fingered slender toads
Long-fingered trioks, 13:133*t*
Long-finned pilot whales, 15:9, 15:56*t*
Long-footed potoroos, 13:32, 13:74, 13:76, 13:77, 13:*79*, 13:80
Long-footed tree shrews. *See* Bornean tree shrews
Long-horned beach hoppers. *See* Beach hoppers
Long-horned grasshoppers, 3:201, 3:204, 3:206
Long interspersed nuclear elements (LINEs), 12:29–30
Long-legged bandicoots. *See* Raffray's bandicoots
Long-legged cellar spiders. *See* Long-bodied cellar spiders
Long-legged pratincoles. *See* Australian pratincoles
Long-neck turtles. *See* Common snakeneck turtles
Long-nosed bandicoots, 13:*2*, 13:11, 13:17*t*
Long-nosed bats. *See* Proboscis bats; Southern long-nosed bats
Long-nosed echymiperas. *See* Rufous spiny bandicoots
Long-nosed leopard lizards, 7:244, 7:*249*, 7:250–*251*
Long-nosed mice, 16:267
Long-nosed pitvipers. *See* Hundred-pace pitvipers
Long-nosed potoroos, 13:39, 13:74, 13:*76*, 13:77, 13:78, 13:*80*
Long-snouted bats, 13:415, 13:417, 13:*421*, 13:*428*, 13:429
Long-snouted dolphins. *See* Spinner dolphins
Long-snouted dragons, 7:*212*, 7:*218*
Long-snouted lashtails. *See* Long-snouted dragons
Long-spined hedgehogs. *See* Brandt's hedgehogs
Long-spined sea biscuits. *See* Sea biscuits
Long-spined sea urchins, 1:32, 1:42, 1:402, 1:405, 1:*407*, 1:*408*, 1:409–410
Long-tailed broadbills, 10:179, 10:*180*, 10:182–*183*
Long-tailed buzzards, 8:*327*, 8:331, 8:*331*
Long-tailed chinchillas, 16:377, 16:*378*, 16:*379*, 16:380, 16:*381*, 16:*382*, 16:383
Long-tailed cuckoos doves. *See* Barred cuckoo-doves
Long-tailed ducks. *See* Oldsquaws
Long-tailed dunnarts, 12:*292*, 12:*295*, 12:298
Long-tailed emerald sunbirds. *See* Malachite sunbirds
Long-tailed fantails, 11:87
Long-tailed field mice, 16:*250*, 16:261*t*
Long-tailed fiscals, 10:428
Long-tailed frogmouths. *See* Sunda frogmouths
Long-tailed fruit bats, 13:*335*
Long-tailed gorals, 16:87, 16:91, 16:93, 16:103*t*
Long-tailed ground-rollers, 10:52, 10:*54*, 10:*57*
Long-tailed hermits, 9:441
Long-tailed honey-buzzards. *See* Long-tailed buzzards
Long-tailed koels, 9:*316*, 9:*323*
Long-tailed macaques, 14:10, 14:193
Long-tailed manakins, 10:*297*, 10:*298*, 10:*299*
Long-tailed marsupial mice. *See* Long-tailed dunnarts
Long-tailed meadowlarks, 11:302, 11:*307*, 11:*315*–316
Long-tailed mice, 16:262*t*, 16:270
Long-tailed minivets, 10:386
Long-tailed moles, 13:288*t*
Long-tailed munias. *See* Pin-tailed parrotfinches
Long-tailed pangolins, 16:109, 16:110, 16:112, 16:*114*, 16:*118*–119
Long-tailed paradigallas, 11:492
Long-tailed pipits, 10:375
Long-tailed planigales, 12:288, 12:*292*, 12:*294*, 12:298
Long-tailed porcupines, 16:*355*, 16:*359*, 16:363–364
Long-tailed potoos, 9:397
Long-tailed pygmy possums, 13:*105*, 13:*106*, 13:*107*

Long-tailed shrew tenrecs, 13:*228*
Long-tailed shrews, 13:195, 13:198
Long-tailed shrikes, 10:426, 10:427, 10:*431*, 10:*436*–437
Long-tailed silky flycatchers, 10:447
Long-tailed skuas, 9:204
Long-tailed tailorbirds. *See* Common tailorbirds
Long-tailed tenrecs, 13:*226*
Long-tailed titmice, 10:*169*, 11:*141*, **11:141–146,** 11:*142*, 11:*144*, 11:*145*
Long-tailed utas. *See* Common sagebrush lizards
Long-tailed widows, 11:379
Long-tailed wood-partridges, 8:456
Long-tailed woodcreepers, 10:*231*, 10:232–*233*
Long-term memory, 12:152, 12:153–154
See also Cognition
Long-toed water beetles, 3:319
Long-tongued bats. *See* specific types of long-tongued bats
Long-tongued bumblebees, 3:78
Long-trained nightjars, 9:402
Long-wattled umbrellabirds, 10:308, 10:*311*, 10:*317*–318
Long-whiskered catfishes, 4:352, 4:353
Long-whiskered owlets, 9:332
Long-winged bats. *See* Common bentwing bats
Long-winged coneheads, 3:*209*, 3:*213*, 3:214
Longclaws, 10:371, 10:372–373, 10:373, 10:374, 10:375, 10:*377*, 10:*380*
Longear sunfishes, 5:195, 5:197, 5:198, 5:*201*, 5:*204*
Longevity. *See* Life spans
Longfin angelfishes. *See* Freshwater angelfishes
Longfin inshore squids, 2:*481*, 2:*482*–483
Longfin sculpins, 5:180
Longfin spadefishes, 5:*395*, 5:400, 5:*401*
Longfinned squids. *See* Longfin inshore squids
Longfinned waspfishes, 5:163
Longfins, 5:255
See also Plesiopidae
Longhorn beetles, 1:42, 2:38, 3:13, 3:316, 3:317, 3:319–321, 3:323
See also Titanic longhorn beetles
Longhorn cowfishes, 5:*473*, 5:479, 5:*480*
Longipeditermes longipes, 2:36
Longjaw bigscales, 5:*107*, 5:*108*–109
Longjaw squirrelfishes. *See* Squirrelfishes
Longlure frogfishes, 5:*48*
Longman's beaked whales, 15:*63*, 15:*67*, 15:68
Longnose batfishes, 5:*52*, 5:*55*, 5:56
Longnose butterflyfishes, 5:242
Longnose filefishes, 5:*473*, 5:*478*, 5:479
Longnose gars, 1:51, 4:*4*, 4:68, 4:69, 4:*222*, 4:*223*, 4:*225*, 4:226, 5:406
Longnose hawkfishes, 4:45, 5:239, 5:240, 5:*245*, 5:*248*–249
Longnose lancetfishes, 4:*435*, 4:*436*
Longnose loaches. *See* Horseface loaches
Longnose sawsharks. *See* Common sawsharks
Longnose suckers, 4:*322*, 4:*329*–330
Longsnout blacksmelts, 4:392, 4:393
Longspine snipefishes, 5:*138*, 5:*143*–144
Longspurs, 11:272
Longtail carpet sharks, 4:105
Longtail knifefishes, 4:370, 4:371, 4:*372*, 4:*374*, 4:376–377
Longtail tadpole shrimps, 2:143, 2:*144*, 2:*145*, 2:146
Longtail weasels, 12:*47*
Lontra spp., 14:319
Lontra canadensis. See Northern river otters
Lontra felina. See Marine otters
Lontra provocax. See Southern river otters
Lookups. *See* Upper mouths
Loons, **8:159–167,** 8:*162*
Loosejaws, 4:423
See also Rat-trap fishes
Lophaetus occipitalis. See Long-crested eagles
Lophelia pertusa. See Deep water reef corals
Lophiidae. *See* Monkfishes
Lophiiformes. *See* Anglerfishes
Lophioidei. *See* Monkfishes
Lophiomeryx spp., 15:343
Lophiomys imhausi. See Crested rats
Lophius americanus. See Monkfishes
Lophius budegassa, 5:51
Lophius piscatorius. See Anglerfishes
Lophocebus spp., 14:188, 14:191
Lophocebus albigena. See Gray-cheeked mangabeys
Lophocebus aterrimus. See Black mangabeys
Lophogastridae, 2:225
Lophogastrids, **2:225–228**
Lophognathus longirostris. See Long-snouted dragons
Lopholaimus spp. *See* Fruit pigeons
Lopholatilus spp., 5:256
Lophophanes spp. *See* Crested tits
Lophophorata, 1:9, 1:10, 1:13
Lophophores, 2:25, 2:29
See also Physical characteristics
Lophophorus spp. *See* Monals
Lophophorus impejanus. See Himalayan monals
Lophophorus lhuysii. See Chinese monals
Lophopodidae, 2:497
Lophorina spp. *See* Birds of paradise
Lophornis spp. *See* Hummingbirds
Lophornis brachylophus. See Short-crested coquettes
Lophornis magnificus. See Frilled coquettes
Lophotaspis vallei, 1:200
Lophotibis cristata. See Madagascar crested ibises
Lophotidae, 4:447, 4:448
Lophotis spp. *See* Bustards
Lophotrochozoa. *See* Ciliated animals
Lophotus spp., 4:447, 4:449, 4:453
Lophotus capellei, 4:453
Lophotus cepedianus, 4:453
Lophotus cristatus, 4:453
Lophotus lacepede. See Crestfishes
Lophozosterops spp., 11:228
Lophozosterops dohertyi, 11:228
Lophura bulweri. See Bulwer's pheasants
Lophura edwardsi. See Edwards' pheasants
Lophura hatinhensis, Vietnamese pheasants
Lophura ignita. See Crested firebacks
Lophura imperialis. See Imperial pheasants
Lophura nycthemera berliozi. See Berlioz's silver pheasants
Lord Derby's anomalures, 16:*301*, 16:*302*–303
Lord Derby's mountain pheasants. *See* Horned guans
Lord Derby's scalytails. *See* Lord Derby's anomalures
Lord Howe Island starlings, 11:410
Lord Howe Island stick insects, 3:224
Lord Vinshu, 2:42
Lorenz, Konrad, 4:61, 9:250
Loricariidae, 4:352, 4:353
Loricifera. *See* Girdle wearers
Loriculus spp. *See* Hanging parrots
Loriculus philippensis. See Philippine hanging-parrots
Loridae. *See* Lorises
Lories, 9:276, 9:279
Lorikeets, 9:*277*, 9:*279*, 9:*281*, 9:289, 9:297–298
Loriolella spp., 1:401
Loris lydekkerianus. See Gray slender lorises
Loris tardigradus. See Slender lorises
Lorises, **14:13–22,** 14:*17*
behavior, 12:116, 14:7–8, 14:14–15
conservation status, 14:16
distribution, 14:5, 14:*13*, 14:14
evolution, 14:2, 14:3, 14:13
feeding ecology, 14:9, 14:15
habitats, 14:14
humans and, 14:16
physical characteristics, 12:79, 14:13–14
reproduction, 14:16
species of, 14:18–19, 14:21*t*
taxonomy, 14:1, 14:3–4, 14:13
See also Strepsirrhines
Lorisidae. *See* Lorises; Pottos
Lorisinae, 14:13
Lota spp., 5:26, 5:27
Lota lota. See Burbots
Lotidae, 5:25, 5:26, 5:*27*, 5:29
Lotinae, 5:25
Lottia alveus, 2:425
Lottia pelta. See Shield limpets
Lottiidae, 2:423, 2:424
Louisiana alligators. *See* American alligators
Louisiana crayfish. *See* Red swamp crayfish
Louisiana waterthrush, 11:288, 11:290
Louse-flies, 3:358, 3:359
Louvars, 5:391–393, 5:*395*, 5:*401*
Lovebirds. *See* Budgerigars
Lovebugs, 3:61, 3:359
Lovejoy, N. R., 5:79
Loveridgea spp., 7:274
Lovettia sealii. See Tasmanian whitebaits
Lower Californian fur seals. *See* Guadalupe fur seals
Lower jaw. *See* Mandibles
Lower metazoans and lesser deuterostomes
behavior, 1:27–30, **1:37–43**
conservation status, **1:47–55,** 1:*52*
defined, **1:3–6**
evolution, **1:7–14,** 1:*8*
feeding ecology, **1:27–30**
habitats, 1:24–30
humans and, **1:44–46**
reproduction, **1:15–30,** 1:*16*
taxonomy, **1:7–14**
Lower primates, 14:1
See also specific lower primates
Lowland eupetes. *See* Blue jewel-babblers
Lowland ringtails, 13:113, 13:114, 13:*115*, 13:*118*, 13:*121*
Lowland tapirs, 9:78, 15:223, 15:*238*, 15:*239*, 15:241–*242*, 15:*244*, 15:*245*

Lowland yellow-billed kingfishers. *See* Yellow-billed kingfishers
Loxia americana. *See* Variable seedeaters
Loxia astrild. *See* Common waxbills
Loxia cantans. *See* African silverbills
Loxia colius. *See* White-backed mousebirds
Loxia curvirostra. *See* Red crossbills
Loxia erythrocephala. *See* Red-headed finches
Loxia guttata. *See* Diamond firetails
Loxia madagascariensis. *See* Madagascar fodies
Loxia megaplaga. *See* Hispaniolan crossbills
Loxia oryzivora. *See* Java sparrows
Loxia philippina. *See* Baya weavers
Loxia prasina. *See* Pin-tailed parrotfinches
Loxia punctulata. *See* Spotted munias
Loxia scotica. *See* Scottish crossbills
Loxia socia. *See* Sociable weavers
Loxioides spp., 11:343
Loxioides bailleui. *See* Palilas
Loxocemidae. *See* Neotropical sunbeam snakes
Loxocemus spp. *See* Neotropical sunbeam snakes
Loxocemus bicolor. *See* Neotropical sunbeam snakes
Loxodonta spp., 15:162, 15:165
Loxodonta adaurora, 15:162
Loxodonta africana. *See* African elephants
Loxodonta africana africana. *See* Savanna elephants
Loxodonta africana cyclotis. *See* Forest elephants
Loxodonta atlantica, 15:162
Loxodonta exoptata, 15:162
Loxodonta pumilio. *See* Pygmy elephants
Loxokalypodidae, 1:319
Loxomitra kefersteinii. *See* Solitary entoprocts
Loxommatids. *See* Baphetids
Loxops coccineus. *See* Akepas
Loxosomatidae, 1:319
Loxosomella spp., 1:*319*
Lubber grasshoppers, 3:*79*, 3:201
Lubchenco, Jane, 1:48
Lucania goodei, 5:92
Lucanidae, 3:322
Lucanus cervus. *See* European stag beetles
Lucerne fleas, 3:101, 3:*102*, 3:*104*
Luces. *See* Northern pikes
Lucifer hummingbirds, 9:443
Luciferase, 4:422
Luciferin, 4:422
Lucifugia spp., 5:17
Lucifugia simile, 5:18
Lucifugia spelaeotes, 5:18
Lucifugia (Stygicola) dentata, 5:18
Lucifugia subterranea, 5:18
Lucifugia teresianarum, 5:18
Lucinidae, 2:453
Luciocephalidae, 4:57
Luciocephalinae, 5:427
Luciocephalus spp., 5:427, 5:428, 5:429
Luciocephalus pulcher. *See* Pikeheads
Luciogobius spp., 5:375
Luckhart, G., 15:146
Lucy. *See Australopithecus afarensis*
Lucy's warblers, 11:288
Lugworms, 2:*48*, 2:*51–52*
Lullula spp. *See* Larks
Lullula arborea. *See* Wood larks
Lulus spp. *See* Millipedes
Lumbricidae, 2:65, 2:66
Lumbricus spp. *See* Earthworms
Lumbricus terrestris. *See* European night crawlers
Lumholtz's tree kangaroos, 13:102*t*
Luminescent anchovies. *See* Lanternfishes
Luminous hakes, 5:25, 5:26, 5:*27*, 5:28, 5:*31*, 5:*35*, 5:39–40
Luminous sharks. *See* Cookie-cutter sharks
Lumpenus medius, 5:310
Lumpfishes, 4:14, 4:*22*, 5:179, 5:180, 5:*182*, 5:*184*, 5:*190*–191
Lumpsuckers. *See* Lumpfishes
Luna, L., 16:266
Lunar tail bigeyes. *See* Goggle eyes
Lunda cirrhata. *See* Tufted puffins
Lundberg, J. G., 4:58
Lundin, Kennet, 1:179
Lundomys spp., 16:265, 16:268
Lung-bearing slugs, **2:411–422,** *2:416*
  behavior, 2:414
  conservation status, 2:414
  distribution, 2:413, 2:*418*
  evolution, 2:411
  feeding ecology, 2:414
  habitats, 2:413–414
  humans and, 2:414–415
  physical characteristics, 2:413
  reproduction, 2:414
  species of, 2:419–421
  taxonomy, 2:411–413
Lung-bearing snails, 1:40, 2:25, **2:411–422,** *2:416*
  behavior, 2:*5*, 2:414
  conservation status, 2:414
  distribution, 2:413
  evolution, 2:411
  feeding ecology, 2:414
  habitats, 2:413–414
  humans and, 2:414–415
  physical characteristics, 2:*412*, 2:413
  reproduction, 2:414
  species of, 2:417–421
  taxonomy, 2:411–413
Lung flukes, 1:35, 1:200
Lunges. *See* Muskellunges
Lungfishes, 4:3, 4:12, 4:*55*, **4:201–207,** 4:*204*
  behavior, 4:*202*, 4:203, 4:205–206
  conservation status, 4:203, 4:205–206
  distribution, 4:202, 4:*205–206*
  evolution, 4:201, 6:7
  feeding ecology, 4:203, 4:205–206
  habitats, 4:202–203, 4:205–206
  humans and, 4:203, 4:205–206
  physical characteristics, 4:*18*, 4:201–202, 4:205
  reproduction, 4:203, 4:205–206
  taxonomy, 4:201, 4:205
  See also Australian lungfishes
Lungless salamanders, **6:389–404,** 6:*393–394*
  behavior, 6:44, 6:392, 6:395–403
  biogeography, 6:4
  conservation status, 6:392, 6:395–403
  distribution, 6:5, 6:*389*, 6:392, 6:*395–403*
  evolution, 6:13, 6:389–392
  feeding ecology, 6:326, 6:392, 6:395–403
  habitats, 6:392, 6:395–403
  humans and, 6:392, 6:395–403
  physical characteristics, 6:17, 6:24, 6:323, 6:325, 6:392, 6:395–403
  reproduction, 6:31, 6:34, 6:35, 6:392, 6:395–403
  species of, 6:395–403
  taxonomy, 6:323, 6:389–392, 6:395–403
Lungs, 7:28–29, 12:8, 12:*40*, 12:45
  bats, 12:60
  lissamphibians, 6:17
  marine mammals, 12:68
  snakes, 7:10
  subterranean mammals, 12:73
  tadpoles, 6:41
  See also Physical characteristics; Respiration
Lungworms, rat, 1:*296*, 1:301–302
Luolishania spp., 2:109
Lurocalis semitorquatus. *See* Short-tailed nighthawks
Luscinia megarhynchos. *See* Nightingales
Luscinia svecica. *See* Bluethroats
Lushes. *See* Burbots
Lushilagus, 16:480
Luteinizing hormone, 12:103
Luth turtles. *See* Leatherback seaturtles
Lutjanidae. *See* Snappers
Lutjanus campechanus. *See* Northern red snappers
Lutjanus kasmira. *See* Blue banded sea perches
Lutodrilidae, 2:65
Lutra spp., 14:319
Lutra lutra. *See* European otters
Lutra maculicollis. *See* Spotted-necked otters
Lutra sumatrana. *See* Hairy-nosed otters
Lutreolina spp. *See* Lutrine opossums
Lutreolina crassicaudata. *See* Thick-tailed opossums
Lutrinae. *See* Otters
Lutrine opossums, 12:250–253
Lutrogale perspiciallata. *See* Smooth-coated otters
Luvaridae. *See* Louvars
Luvarus imperialis. *See* Louvars
Luxilus spp., 4:301
Luxury organs, 12:22–24
  See also Physical characteristics
Luzara spp., 3:203
Luzon bleeding hearts, 9:*253*, 9:*262–263*
Luzon bubble-nest frogs, 6:*294*, 6:298–299
Luzon tarictic hornbills, 10:72
Lybius bidentatus. *See* Double-toothed barbets
Lybius dubius. *See* Bearded barbets; Groove-billed barbets
Lybius rolleti. *See* Black-breasted barbets
Lybius torquatus. *See* Black-collared barbets
Lycaenids, 3:72, 3:386, 3:387, 3:389
Lycaon pictus. *See* African wild dogs
Lycenchelys spp., 5:310
Lycenchelys antarctica, 5:311
Lycidae, 3:65, 3:320
Lycocorax spp. *See* Birds of paradise
Lycocorax pyrrhopterus. *See* Paradise crows
Lycodapus spp., 5:310, 5:311, 5:312
Lycodapus mandibularis, 5:311
Lycodes spp., 5:310
Lycodes frigidus, 5:310
Lycodes pacificus, 5:309
Lycodonomorphus spp. *See* African watersnakes
Lycodontis funebris. *See* Green morays
Lyconus spp., 5:26–28
Lycosidae. *See* Wolf spiders
Lyell, George, 10:207
Lygosoma spp., 7:327–328

Lygosominae, 7:327–328
Lyle's flying foxes, 13:331*t*
Lymantria dispar. See Gypsy moths
Lymantriidae, 3:386
Lyme disease, 16:127
Lymnaeidae, 2:412
Lymnocryptes minimus. See Jack snipes
Lymphatic system, lissamphibians, 6:17
Lynceidae, 2:147, 2:148
Lynceus gracilicornis. See Graceful clam shrimps
Lynch's Cochran frogs, 6:215, 6:*216*, 6:*218*, 6:221
Lyncodon patagonicus. See Patagonian weasels
Lynx, 12:116, 12:139, 14:370–371, 14:374
Lynx canadensis. See Canada lynx
Lynx (Felis) canadensis. See Canada lynx
Lynx (Felis) lynx. See Eurasian lynx
Lynx (Felis) pardinus. See Iberian lynx
Lynx (Felis) rufus. See Bobcats
Lynx lynx. See Lynx
Lynx pardinus. See Iberian lynx
Lynx rufus. See Bobcats
Lyons, D. M., 15:145
Lyre-tailed honeyguides, 10:137, 10:*140*, 10:*142*
Lyre-tailed nightjars, 9:402
Lyrebirds, 8:41, 10:169, 10:170, 10:173, **10:329–335,** 10:*333*, 11:477
Lyretail panchaxes. *See* Chocolate lyretails
Lysiosquilldea, 2:168, 2:170
Lysiosquillina maculata, 2:*172*, 2:173–*174*
Lysiosquilloidea, 2:168, 2:171
Lysorophids, 6:10, 6:11
Lystrophis spp. *See* Neotropical hog-nosed snakes
Lytechinus variegatus. See West Indian sea eggs
Lytocarpia myriophyllum, 1:25
Lytocarpus philippinus. See Stinging hydroids
Lytocestidae, 1:225
Lytta vesicatoria. See Spanish flies

M

Maamingidae, 3:55
Mabuya spp., 7:327–329
Mabuya heathi. See South American skinks
Mabuya striata. See Striped skinks
Macaca spp. *See* Macaques
Macaca arctoides. See Stump-tailed macaques
Macaca assamensis, 14:194
Macaca cyclopis, 14:194
Macaca fascicularis, 14:194
Macaca fuscata. See Japanese macaques
Macaca hecki, 14:194
Macaca leonina, 14:194
Macaca maurus, 14:193–194
Macaca mulatta. See Rhesus macaques
Macaca nemestrina, 14:194
Macaca nigra. See Celebes macaques
Macaca nigrescens, 14:194
Macaca ochreata, 14:194
Macaca pagensis, 14:11, 14:193
Macaca radiata. See Bonnet macaques
Macaca silenus, 14:193–194
Macaca sinica. See Toque macaques
Macaca sylvanus. See Barbary macaques
Macaca thibetana, 14:194
Macaca tonkeana, 14:194
Macaques, 14:*195*
  conservation status, 14:11, 14:193–194
  distribution, 12:137, 14:5, 14:191
  evolution, 14:189
  habitats, 14:191
  humans and, 12:173, 14:194
  learning and, 12:143, 12:151–153
  reproduction, 14:10, 14:193
  species of, 14:200–201, 14:205*t*
  taxonomy, 14:188
Macaroni penguins, 8:148, 8:151, 8:*152*, 8:*154*–155, 9:*198*
MacArthur Foundation, 12:218
Macaws, 9:275, 9:279, 9:280
Maccullochella spp., 5:220
Maccullochella ikei. See Eastern freshwater cods
Maccullochella macquariensis. See Trout cods
Maccullochella peelii. See Mary River cods; Murray cods
Maccullochella peelii mariensis. See Mary River cods
Maccullochella peelii peelii. See Murray cods
Macdonald, D. W., 16:392
Macey, J. Robert, 7:340, 7:347
Macgregoria spp. *See* Birds of paradise
Macgregoria pulchra. See Macgregor's birds of paradise
Macgregor's birds of paradise, 11:235, 11:492
Macgregor's bowerbirds, 11:*478*, 11:*482*, 11:*484*–485
Macgregor's gardenerbirds. *See* Macgregor's bowerbirds
Machaeropterus regulus. See Striped manakins
Machilidae, 3:113
MacInnes' mouse-tailed bats, 13:352
Mackerel icefishes, 5:323, 5:*324*, 5:*326*–327
Mackerel sharks, **4:131–142,** 4:*134*
  behavior, 4:132–133
  conservation, 4:133
  distribution, 4:132
  evolution, 4:131–132
  feeding ecology, 4:133
  habitats, 4:132
  humans and, 4:133
  physical characteristics, 4:132
  reproduction, 4:133
  species of, 4:135–141
  taxonomy, 4:11, 4:131–132
Mackerels, 4:4, 4:68, 5:406, 5:407, 5:*408*, 5:*409*, 5:413–414
Mackinaw trouts. *See* Lake trouts
Macklot's trogons. *See* Javan trogons
Macleay's dorcopsises. *See* Papuan forest wallabies
Macleay's marsupial mice. *See* Brown antechinuses
Macleay's moustached bats, 13:435, 13:436, 13:442*t*
Macleay's spectres, 3:*225*, 3:*230*, 3:231
Macquaria spp., 5:220
Macquaria ambigua. See Golden perches
Macquaria australasica. See Mountain perches
Macquaria novemaculeata. See Australian basses
Macquarie perches. *See* Mountain perches
Macqueen's bustards, 9:93, 9:94
Macracanthorhynchus hirudinaceus. See Giant thorny headed worms
Macrauchenids, 15:133
Macrelaps microlepidotus. See Natal black snakes
Macro-moths, 3:384
Macrobdella decora. See North American medicinal leeches
Macrobiotus spp., 2:116
Macrobiotus hufelandi, 2:116, 2:118
Macrobiotus richtersi, 2:117
Macrobothriidae, 1:225
Macrocephalon maleo. See Maleos
Macrocerastes spp., 7:445
Macrocheira kaempferi. See Giant Japanese spider crabs
Macrochelys spp., 7:93
Macrochelys temminckii. See Alligator snapping turtles
Macrodasyidae, 1:269
Macroderma gigas. See Australian false vampire bats
Macrodipteryx spp., 9:402
Macrodipteryx longipennis. See Standard-winged nightjars
Macrodipteryx vexillarius. See Pennant-winged nightjars
Macrogalidia musschenbroekii. See Sulawesi palm civets
Macrogenioglottus spp., 6:156
Macroglossus sobrinus. See Greater long-tongued fruit bats
Macrognathus aral. See One-stripe spiny eels
Macroherbivory feeding, 2:36
  See also Feeding ecology
Macrolyristes spp., 3:202
Macromeres, 1:18, 2:3, 2:20
Macronectes giganteus. See Southern giant petrels
Macronectes halli. See Northern giant petrels
Macronyx ameliae. See Rosy-breasted longclaws
Macronyx aurantiigula. See Pangani longclaws
Macronyx croceus. See Yellow-throated longclaws
Macronyx flavicollis. See Abyssinian longclaws
Macronyx fuellebornii. See Fuelleborn's longclaws
Macronyx grimwoodi. See Grimwood's longclaws
Macronyx sharpei. See Sharpe's longclaws
Macropanesthia spp., 3:149, 3:150
Macropanesthia rhinoceros, 3:147, 3:150
Macroperipatus torquatus, 2:37
Macropharyngodon meleagris. See Blackspotted wrasses
Macropinna microstoma. See Barreleyes
Macropodidae, **13:83–103,** 13:*91*, 13:*92*, 13:101*t*–102*t*
Macropodinae, 5:427, 13:83
Macropodoidea, 13:31–33, 13:38, 13:39, 13:69, 13:70, 13:71, 13:73
Macropodus spp., 5:427, 5:428
Macropodus ocellatus, 5:428
Macropsalis spp., 9:369
Macropsalis forcipata. See Long-trained nightjars
Macropsalis longipennis. See Standard-winged nightjars
Macropsalis lyra. See Lyre-tailed nightjars
Macropsalis segmentata. See Swallow-tailed nightjars
Macropsalis vexillarius. See Pennant-winged nightjars
Macropteryx spodiopygius. See White-rumped swiftlets
Macropus spp., 13:33, 13:83, 13:85, 13:86–87

Macropus agilis. See Agile wallabies
Macropus antilopinus. See Antilopine wallaroos
Macropus bernardus. See Black wallaroos
Macropus dorsalis. See Black-striped wallabies
Macropus eugenii. See Tammar wallabies
Macropus fuliginosus. See Western gray kangaroos
Macropus giganteus. See Eastern gray kangaroos
Macropus greyi. See Toolache wallabies
Macropus parma. See Parma wallabies
Macropus parryi. See Whiptail wallabies
Macropus phasianellus. See Pheasant cuckoos
Macropus robustus. See Common wallaroos
Macropus rufogriseus. See Red-necked wallabies
Macropus rufogriseus banksianus. See Red-necked wallabies
Macropus rufus. See Red kangaroos
Macropygia unchall. See Barred cuckoo-doves
Macroramphosidae. *See* Snipefishes
Macroramphosus scolopax. See Longspine snipefishes
Macrorhamphosodes uradoi, 5:470
Macroscelidae. *See* Sengis
Macroscelides spp., 16:518
Macroscelides proboscideus. See Round-eared sengis
Macroscelididae. *See* Sengis
Macroscelidinae. *See* Soft-furred sengis
Macrosphenus kempi. See Kemp's longbills
Macrosphenus kretschmeri, 11:5
Macrostomata, 7:198
Macrostomida, 1:185–186, 1:188
Macroteiids, 7:309, 7:311, 7:312
Macrotermes spp., 3:69, 3:165, 3:*165*
Macrotermes carbonarius. See Black macrotermes
Macrotermitidae, 3:165
Macrotermitinae, 3:165, 3:166–167, 3:168
Macrotis spp. *See* Bilbies
Macrotis lagotis. See Greater bilbies
Macrotis leucura. See Lesser bilbies
Macrotus californicus. See California leaf-nosed bats
Macrouridae, 5:25–30, 5:27
Macrourogaleus spp., 4:113
Macrovipera spp., 7:445, 7:447
Macrovipera lebetina. See Levantine vipers
Macrurocyttus spp., 5:123
Macrurocyttus acanthopodus, 5:123
Macruronus spp., 5:26, 5:27, 5:30
Maculinea arion. See Large blues
Macynia labiata. See Thunberg's stick insects
Madagascan big-headed turtles, 7:*139*, 7:*141*–142
Madagascan fruit bats, 13:*334*
Madagascan hissing cockroaches, 3:83, 3:147, 3:150, 3:151
Madagascan hog-nosed snakes, 7:467
Madagascan mantellas, 6:55
Madagascan plated lizards, 7:320, 7:*323*, 7:*324*, 7:325
Madagascan toadlets, 6:6, 6:301, **6:317–321,** 6:*319–320*
Madagascan vinesnakes, 7:467
Madagascan white-eyes. *See* Madagascar pochards
Madagascar. *See* Ethiopian region
Madagascar bazas. *See* Madagascar cuckoo-hawks
Madagascar black bulbuls. *See* Black bulbuls
Madagascar boas, 7:411
Madagascar bulbuls. *See* Black bulbuls
Madagascar buttonquails, 9:17
Madagascar crested ibises, 8:293
Madagascar cuckoo-falcons. *See* Madagascar cuckoo-hawks
Madagascar cuckoo-hawks, 8:327, 8:329–330
Madagascar day geckos, 7:*264*, 7:*267*, 7:268
Madagascar fish-eagles, 8:318, 8:324
Madagascar flying foxes, 13:*322*, 13:*323*, 13:325–*326*
Madagascar fodies, 11:*381*, 11:*388*–389
Madagascar grebes, 8:171
Madagascar harriers, 8:323
Madagascar hedgehogs, 12:*145*, 13:196, 13:226
Madagascar hoopoes, 10:61–62
Madagascar iguanas, 7:243
Madagascar larks, 10:343
Madagascar nightjars, 9:403
Madagascar plovers, 9:164
Madagascar pochards, 8:367, 8:373, 8:*375*, 8:*387*
Madagascar red owls, 9:336, 9:338
Madagascar scops-owls, 9:346
Madagascar slit-faced bats, 13:373
Madagascar snipes, 9:176–177
Madagascar sparrowhawks, 8:323
Madagascar teals, 8:364, 8:367
Madagascar tomato frogs. *See* Tomato frogs
Madagascar tree boas, 7:411
Madagascar weavers. *See* Madagascar fodies
Madagascar white-eyes, 11:227
Madeira cockroaches, 3:150, 3:*152*, 3:*153*, 3:154–155
Madoqua spp. *See* Dikdiks
Madoqua guentheri. See Guenther's dikdiks
Madoqua kirkii. See Kirk's dikdiks
Madoqua piacentinii. See Silver dikdiks
Madoqua saltiana. See Salt's dikdiks
Madras Crocodile Bank, 7:185
Madreporaria. *See* True corals
Madrid Zoo, Spain, 12:203
Magdalena tinamous, 8:59
Magellan diving-petrels, 8:143, 8:*145–146*
Magellanic diving-petrels. *See* Magellan diving-petrels
Magellanic flightless steamerducks. *See* Magellanic steamerducks
Magellanic oystercatchers, 9:126, 9:127
Magellanic penguins, 8:148, 8:149, 8:151, 8:*152*, 8:*155–156*
Magellanic plovers, 9:161–164, 9:*165*, 9:171–*172*
Magellanic steamerducks, 8:*374*, 8:*380*
Magellanic tuco-tucos, 16:427, 16:430*t*
Maggenti, A. R., 1:283, 1:293
Maggots, 3:64, 3:66, 3:76, 3:81, 3:359
Magic eyes, 9:396
Magicicada septendecim. See Seventeen-year cicadas
Magnetic compasses, 8:34
Magnetoreception, in subterranean mammals, 12:75
Magnificent Calypto clams. *See* Giant vent clams
Magnificent frigatebirds, 8:193, 8:*196*, 8:*197*
Magnificent fruit doves. *See* Wompoo fruit doves
Magnificent green broadbills. *See* Hose's broadbills
Magnificent ground pigeons. *See* Pheasant pigeons
Magnificent hummingbirds, 9:443
Magnificent night herons. *See* White-eared night herons
Magnificent quetzals. *See* Resplendent quetzals
Magnificent riflebirds, 11:490
Magnificent sea anemones, 1:*110*, 1:*113*–114
Magnificent treefrogs, 6:46
Magnificent urchins, 1:*407*, 1:*409*
Magosternarchus duccis, 4:371
Magpie geese, 8:363, 8:367, 8:369, 8:370, 8:371, 8:*374*, 8:*376*, 8:393
Magpie-larks. *See* Australian magpie-larks
Magpie-robins, 10:485, 10:*490*, 10:*492*–493
Magpie-shrikes, 10:426, **11:467–475,** 11:*470*
Magpie starlings, 11:*413*, 11:*422–423*
Magpies, 8:20, 9:313, 11:503–505, 11:506, 11:507, 11:509–*510*, 11:515–517
See also Australian magpies
Maguari storks, 8:266
Maharashtra stream toads. *See* Common Sunda toads
Maher, C. R., 15:413
Mahi-mahi. *See* Common dolphinfishes
Mahogany flats. *See* Bed bugs
Mahogany gliders, 13:40, 13:129, 13:133*t*
Maias. *See* Bornean orangutans
Mail-cheeked fishes. *See* Scorpaeniformes
Maindroniidae, 3:119
Mainland Holdings (Papua, New Guinea), 7:182
Maintenance behavior, 1:37
See also Behavior
Major, Richard, 11:86
Majorca midwife toads, 6:59, 6:89, 6:90
Makaira nigricans. See Blue marlins
Makalata armata. See Armored spiny rats
Makalata occasius, 16:450
Ma'kechs, 3:323, 3:*328*, 3:333
Malabar civets, 14:262, 14:338
Malabar gray hornbills, 10:72
Malabar night frogs, 6:*253*, 6:*259*
Malabar spiny dormice, 16:*287*, 16:*290*, 16:293
Malabar spiny mice. *See* Malabar spiny dormice
Malabar squirrels. *See* Indian giant squirrels
Malabar whistling-thrushes, 10:487
Malacanthidae. *See* Tilefishes
Malacanthus spp., 5:256
Malachite sunbirds, 11:*208*, 11:*212*, 11:*220*–221
Malaclemys terrapin. See Diamondback terrapins
Malacochersus tornieri. See Pancake tortoises
Malaconotinae. *See* Bush-shrikes
Malaconotus spp. *See* Bush-shrikes
Malaconotus alius. See Uluguru bush-shrikes
Malaconotus blanchoti. See Gray-headed bush-shrikes
Malaconotus cruentus. See Fiery-breasted bush-shrikes
Malaconotus gladiator. See Green-breasted bush-shrikes
Malacopsylla spp., 3:350
Malacopsyllidae, 3:347, 3:348–349

Malacopsylloidea, 3:347
Malacoptila fulvogularis. *See* Black-streaked puffbirds
Malacoptila panamensis. *See* White-whiskered puffbirds
Malacorhynchus membranaceus. *See* Pink-eared ducks
Malacosteus niger. *See* Rat-trap fishes
Malacostraca, 2:185, 2:197
Malagasy brown mongooses, 14:357*t*
Malagasy bulbuls, 10:395
Malagasy civets, 14:335
Malagasy crickets, 3:203
Malagasy day geckos, 7:260, 7:263
Malagasy flying foxes. *See* Madagascar flying foxes
Malagasy giant rats, 16:*121*, 16:*286*, 16:*289*, 16:291
Malagasy helmet katydids, 3:203
Malagasy jacanas, 9:108
Malagasy mice. *See* Malagasy giant rats
Malagasy mongooses. *See* Falanoucs
Malagasy rainbowfishes, 4:57, 5:67, 5:68
Malagasy rats. *See* Malagasy giant rats
Malagasy reed rats, 16:*284*
Malaita fantails, 11:87
Malania anjouanae, 4:189
Malapteruridae, 4:69, 4:352
Malapterurus electricus. *See* Electric catfishes
Malaria, 2:11–12, 2:31, 11:237–238, 11:343, 11:348
See also Humans
Malas. *See* Rufous hare-wallabies
Malate dehydrogenase, 9:416
Malawi bushbabies, 14:32*t*
Malay eagle-owls. *See* Barred eagle-owls
Malayan box turtles, 7:*12*, 7:*115*
Malayan colugos, 13:*299*, 13:*300*, 13:*301*, 13:*302*, 13:*303*, 13:*304*
Malayan flying lemurs. *See* Malayan colugos
Malayan gharials, 7:179, 7:180
Malayan gymnures. *See* Malayan moonrats
Malayan horned frogs, 6:*111*
Malayan moonrats, 13:194, 13:198, 13:*204*, 13:*208*, 13:212
Malayan pangolins, 16:110, 16:113, 16:*114*, 16:*115*, 16:116
Malayan porcupines. *See* Common porcupines
Malayan softshell turtles, 7:152
Malayan sun bears, 14:295, 14:296, 14:*297*, 14:298, 14:299, 14:300, 14:306*t*
Malayan tapirs, 15:222, 15:237, 15:238, 15:*239*, 15:*240*, 15:241, 15:242, 15:*244*, 15:*247*
Malaysian eagle-owls. *See* Barred eagle-owls
Malaysian honeyguides, 10:139, 10:*140*, 10:141–*142*
Malaysian long-tailed porcupines. *See* Long-tailed porcupines
Malaysian painted frogs, 6:*67*, 6:*303*, 6:*307*, 6:313–314
Malaysian peacock-pheasants, 8:436
Malaysian pitvipers, 7:52, 7:447, 7:448
Malaysian plovers, 9:164
Malaysian rail-babblers, 11:75, 11:76, 11:77, 11:*79*
Malaysian short-tailed porcupines. *See* Javan short-tailed porcupines
Malaysian stink badgers. *See* Indonesian stink badgers
Malaysian tapirs. *See* Malayan tapirs
Malaysian wood-owls. *See* Asian brown wood-owls
Malcolm's Ethiopian toads, 6:35, 6:*186*, 6:*188*
Male reproductive system, 12:91
See also Physical characteristics; Reproduction
Maleofowls. *See* Maleos
Maleos, 8:404, 8:*408*, 8:*410*
Malgasiidae. *See* Malagasy crickets
Malia grata. *See* Malias
Malias, 10:506
Malimbus spp., 11:376, 11:377
Malimbus ballmanni. *See* Gola malimbes
Malimbus ibadanensis. *See* Ibadan malimbes
Malimbus nitens. *See* Blue-billed malimbes
Malindi pipits, 10:375
Malkohas, 9:311, 9:314
Mallada spp., 3:307, 3:309
Mallada albofascialis. *See* Green lacewings
Mallards, 8:21, 8:*50*, 8:*367*, 8:368, 8:*371*, 8:373, 8:*375*, 8:384–385
Mallee kangaroos. *See* Western gray kangaroos
Malleefowls, 8:404, 8:*405*, 8:*408*, 8:*409*–410
Malleus, 12:36
See also Physical characteristics
Mallotus villosus. *See* Capelins
Malpolon monspessulanus. *See* Montpellier snakes
Malpulutta spp. *See Malpulutta kretseri*
Malpulutta kretseri, 5:427, 5:429
Maluridae. *See* Australian fairy-wrens
Malurinae. *See* Australian fairy-wrens
Malurus coronatus. *See* Purple-crowned fairy-wrens
Malurus cyaneus. *See* Superb fairy-wrens
Malurus lamberti. *See* Variegated fairy-wrens
Malurus splendens. *See* Splendid fairy-wrens
Malzahn, E., 2:161
Mambas, 7:483–486, 7:*490*, 7:*491*
Mammae. *See* Mammary glands
Mammal fleas, 3:348
Mammals
aquatic, 12:3, 12:14, 12:44, **12:62–68,** 12:82, 12:91, 12:127, 12:201
arboreal, 12:14, 12:43, 12:79, 12:202
behavior, 12:9–10, **12:140–148**
vs. birds, 12:6
conservation status, 12:15, **12:213–225**
defined, **12:3–15**
dinosaurs and, 12:33
distribution, **12:129–139**
diving, 12:67–68
draft, 12:15
evolution, 12:10–11
feeding ecology, 12:12–15
fossorial, 12:13, 12:69, 12:82
lice and, 3:249, 3:250, 3:251
marine, 12:14, 12:53, 12:62–68, 12:82, 12:126–127, 12:138, 12:164
nocturnal, 12:79–80, 12:115
physical characteristics, **12:36–51**
reproduction, 12:12, 12:51, **12:89–112**
reptiles and, 12:36
saltatory, 12:42–43
subterranean, **12:69–78,** 12:83, 12:113–114, 12:201
terrestrial, 12:13–14, 12:79, 12:200–202
volant, 12:200–201
See also specific mammals and classes of mammals
Mammary glands, 12:4–5, 12:6, 12:14, 12:36, 12:38, 12:*44*, 12:51, 12:89
See also Physical characteristics
Mammary nipples. *See* Nipples
Mammoths, 15:138, 15:139
See also Woolly mammoths
Mammuthus spp., 15:162, 15:165
Mammuthus columbi. *See* Columbian mammoths
Mammuthus primigenius. *See* Woolly mammoths
Man-of-war fishes, 5:421, 5:*423*, 5:424–*425*
Manacled sculpins, 5:180
Management
conservation, 3:90–91
zoo, 12:205–212
Manakins, 10:170–171, **10:295–304,** 10:*296*, 10:*298*
Manatees, 12:*13*, 12:*67*, 15:*193*, **15:205–213,** 15:*210*
behavior, 15:193
conservation status, 15:194–195, 15:208–209
distribution, 15:192, 15:*205*
evolution, 15:191, 15:205
feeding ecology, 15:193, 15:206, 15:208
field studies of, 12:201
habitats, 15:192, 15:206
humans and, 15:195–196, 15:209
physical characteristics, 15:*192*, 15:205
reproduction, 12:103, 12:*106*, 15:193–194, 15:208
taxonomy, 15:191
Manaus long-legged treefrogs, 6:230, 6:*231*, 6:237
Manchurian cranes. *See* Red-crowned cranes
Manchurian eared-pheasants. *See* Brown eared-pheasants
Manchurian falcons. *See* Amur falcons
Manchurian hares, 16:481, 16:506
Mandarin ducks, 8:365, 8:369, 8:370, 8:*374*, 8:381–382
Mandarin salamanders, 6:*368*, 6:375
Mandarinfishes, 5:367, 5:*368*, 5:*370*–371
Mandibles, 3:19–20, 12:9, 12:36
See also Physical characteristics
Mandrills, 14:*189*, 14:*195*, 14:*202*, 14:203–204
conservation status, 14:194
habitats, 14:191
physical characteristics, 14:190, 14:191
reproduction, 14:193
sexual dimorphism, 12:99
taxonomy, 14:188
Mandrillus leucophaeus. *See* Drills
Mandrillus sphinx. *See* Mandrills
Manduca sexta. *See* Tobacco hornworms
Maned ducks, 8:365, 8:370
Maned rats. *See* Crested rats
Maned sloths, 13:165, 13:*166*, 13:*167*, 13:169
Maned wolves, 14:265–266, 14:271, 14:272, 14:*275*, 14:*276*, 14:282
Mangabeys, 14:188, 14:191, 14:193
Mango spp., 9:443
Mangrove crabs, 2:*203*, 2:*205*, 2:212
Mangrove dragonets, 5:366
Mangrove fantails, 11:83, 11:85
Mangrove gerygones, 11:*58*, 11:*60*
Mangrove hummingbirds, 9:418, 9:443
Mangrove pittas, 10:195

Mangrove rivulus, 4:29, 5:92–94, 5:*95*, 5:*96*, 5:102
Mangrove warblers. *See* Mangrove gerygones
Mangrove whistlers, 11:116
Mangroves, 1:*52*, 12:219
See also Conservation status; Habitats
Manidae. *See* Pangolins
Manipur bush-quails, 8:434
Manis spp., 12:135
Manis crassicaudata. *See* Indian pangolins
Manis gigantea. *See* Giant pangolins
Manis javanica. *See* Malayan pangolins
Manis pentadactyla. *See* Chinese pangolins
Manis temminckii. *See* Ground pangolins
Manis tetradactyla. *See* Long-tailed pangolins
Manis tricuspis. *See* Tree pangolins
Manitou darters. *See* Logperches
Mann, G., 16:270
Mannikins, 11:353, 11:354, 11:357
Mannophryne spp., 6:197, 6:198, 6:200
Mannophryne trinitatis. *See* Trinidad poison frogs
Manorina flavigula. *See* Yellow-throated miners
Manorina melanocephala. *See* Noisy miners
Manorina melanophrys. *See* Bell miners
Manorina melanotis. *See* Black-eared miners
Mansadevi (Indian goddess), 7:56
Manta spp. *See* Manta rays
Manta birostris. *See* Atlantic manta rays
Manta rays, 4:*20*, 4:27, 4:67, 4:175–177, 4:*181*
See also Atlantic manta rays
Mantella spp., 6:291, 6:292, 6:293
Mantella aurantiaca. *See* Golden mantellas
Mantella betsileo. *See* Betsileo golden frogs
Mantella cowanii. *See* Golden mantellas
Mantellidae, 6:4
Mantellinae, 6:291
Mantid lacewings, 3:*310*, 3:*311*, 3:312
Mantidactylus spp., 6:68, 6:291, 6:293
Mantidactylus depressiceps, 6:293
Mantidactylus liber. *See* Free Madagascar frogs
Mantidae, 3:177
Mantids, 3:47, **3:177–187,** 3:*180*, 3:*182*
behavior, 3:41, 3:65, 3:*179*, 3:180
conservation status, 3:181
distribution, 3:179
education and, 3:83
evolution, 3:9, 3:177
feeding ecology, 3:*2*, 3:63, 3:180
habitats, 3:180
humans and, 3:181
physical characteristics, 3:19, 3:22, 3:178–179
reproduction, 3:30, 3:31, 3:33–34, 3:38, 3:39, 3:62, 3:*178*, 3:180–*181*
species of, 3:*183*–187
taxonomy, 3:177–178
See also specific types of mantids
Mantis religiosa. *See* European mantids
Mantis shrimps, **2:167–175,** 2:*172*
behavior, 2:169–170
conservation status, 2:171
distribution, 2:168
evolution, 2:167
feeding ecology, 2:170
habitats, 2:168
humans and, 2:171
physical characteristics, 2:167–168
reproduction, 2:170–171
species of, 2:*173*–175
taxonomy, 2:167
Mantisflies, 3:*305*
Mantispidae, 3:305, 3:306, 3:307, 3:308
Mantled baboons. *See* Hamadryas baboons
Mantled black-and-white colobus. *See* Mantled guerezas
Mantled guerezas, 14:*175*, 14:*176*, 14:*178*
Mantled howler monkeys, 14:*161*, 14:*162*
Mantled mangabeys. *See* Gray-cheeked mangabeys
Mantodea. *See* Mantids
Mantoididae, 3:177
Mantophasma zephyra, 3:217
Mantophasmatidae. *See* Mantophasmatodea
Mantophasmatodea, **3:217–220,** 3:*218*
Mantophryne robusta. *See* Boulenger's callulops frogs
Manucodia spp. *See* Birds of paradise
Manucodia chalybata. *See* Crinkle-collared manucodes
Manucodia comrii. *See* Curl-crested manucodes
Manucodia keraudrenii. *See* Trumpet manucodes
Manus fantails, 11:87
Manus masked owls, 9:340–341
Manx shearwaters, 8:108, 8:110, 8:*127*, 8:128–129
Many-colored bush-shrikes, 10:426
Maori chiefs, 5:*324*, 5:*326*, 5:327
Maori wrasses. *See* Humphead wrasses
Map turtles, 7:106
Maps, navigational, 8:35
Maqueen's bustards, 9:92
Marabou storks. *See* Marabous
Marabous, 8:233, 8:235, 8:265, 8:*269*, 8:*274*
Maras, 16:123, 16:125, **16:389–400,** 16:*390*, 16:*391*, 16:*392*, 16:*394*, 16:*397*
Marble eyes. *See* Walleyes
Marble gobies. *See* Marble sleepers
Marble sleepers, 5:374, 5:*379*, 5:*381*–382
Marbled caecilians, 6:416, 6:*417*
Marbled ducks. *See* Marbled teals
Marbled frogmouths, 9:368, 9:377, 9:*380*, 9:*381*
Marbled galaxias, 4:392
Marbled geckos, 7:261
Marbled murrelets, 9:222, 9:*223*, 9:*227–228*
Marbled newts, 6:45, 6:*63*, 6:374
Marbled polecats, 14:323, 14:333*t*
Marbled river cods. *See* River blackfishes
Marbled rockcods, 5:323
Marbled sand gobies. *See* Marble sleepers
Marbled shovel-nosed frogs. *See* Marbled snout-burrowers
Marbled snout-burrowers, 6:*274*, 6:*276*, 6:*277*
Marbled stargazers, 5:*333*
Marbled swamp eels, 5:*153*, 5:155–156
Marbled teals, 8:364, 8:365, 8:370, 8:*375*, 8:386–387
Marblefishes. *See* Seacarps
Marcano's solenodons, 13:237
March crane flies. *See* European marsh crane flies
Marcusenius spp., 4:233
Mares, M. A., 16:269, 16:433–434, 16:436
Margaritifera margaritifera. *See* European pearly mussels
Margaritiferidae, 2:451
Margarops fuscatus. *See* Pearly-eyed thrashers
Margays, 12:116, 14:391*t*
Margelopsis spp., 1:124
Mariana crows, 11:508
Marianas flying foxes. *See* Marianas fruit bats
Marianas fruit bats, 13:*323*, 13:*326*, 13:327
Marianna flying foxes. *See* Marianas fruit bats
Marine algal-invertebrate symbiosis, 1:33
Marine bryozoans, **2:503–508,** 2:*505*, **2:509–514,** 2:*512*
behavior, 2:504, 2:509
conservation status, 2:504, 2:511
distribution, 2:503–504, 2:509
evolution, 2:503, 2:509
feeding ecology, 2:504, 2:510
habitats, 2:504, 2:509
humans and, 2:504, 2:511
physical characteristics, 2:503, 2:509
reproduction, 2:504, 2:510–511
species of, 2:*506*–508, 2:*513*–514
taxonomy, 2:503, 2:509
Marine carnivores, **14:255–263**
See also specific animals
Marine colonial entoprocts, 1:*320*, 1:*322*, 1:*323*
Marine conservation, 1:50–52
See also Conservation status
Marine crocodiles, 7:157
Marine damselfishes, 5:297
Marine decapods, 2:199
See also Decapoda
Marine environment, 1:24–30, 1:*26*, 1:47–53
See also specific marine animals
Marine fishes, **4:42–51,** 4:*44*, 4:50, 4:52, 4:53–56
See also specific fishes
Marine habitats, 4:47–50
See also Habitats
Marine hatchetfishes, 4:421, 4:422
See also Silver hatchetfishes
Marine iguanas, 7:27–28, 7:243, 7:244
Marine insects, 3:48
See also specific marine insects
Marine leeches, 2:76
See also Leeches
Marine Mammal Protection Act (U.S.), 15:35, 15:38, 15:125
Marine mammals, 12:14, 12:138
echolocation, 12:53
evolution, 12:62–68
hearing, 12:82
migrations, 12:164
neonatal milk, 12:126–127
physical characteristics, 12:63, 12:*63*, 12:*64*, 12:*65*, 12:65–67, 12:*66*, 12:*67*
rehabilitation pools for, 12:*204*
water adaptations, 12:62–68
See also specific marine mammals
Marine otters, 14:324
Marine prosobranchs, 2:17
Marine reptiles, 7:14–15, 7:30
See also specific species
Marine reserves, 1:52
Marine sculpins, 4:47
Marine snails, 2:24s
sea slugs, **2:403–410,** 2:*406*
Vetigastropoda, **2:429–434,** 2:*432*
See also Lung-bearing snails; True limpets
Marine spiny eels, 4:249, 4:250, 4:*251*, 4:*252*, 4:253
Marine Stewardship Council, 4:412, 4:415

Marine tardigrades, 2:115, 2:*116*
Marine toads, 6:*186*, 6:190–*191*
behavior, 6:*46*, 6:191
feeding ecology, 6:67, 6:185, 6:191
humans and, 6:51, 6:191
introduction of, 6:54, 6:58, 6:191
metamorphosis, 6:42
Marion Island, 12:185
Maritigrella fuscopunctata. See Tiger flatworms
Maritrema arenaria, 1:199
Mark/recapture, 12:196–197
Markham's storm-petrels, 8:138
Markhors, 16:87, 16:90–94, 16:104*t*
Markle, D. F., 5:25, 5:29
Marlins, 4:49, 4:61, 5:*405*, **5:405–407**, 5:*408*, **5:411**
Marmaronetta angustirostris. See Marbled teals
Marmosa spp. *See* Mouse opossums
Marmosa andersoni, 12:254
Marmosa canescens. See Grayish mouse opossums
Marmosa mexicana. See Mexican mouse opossums
Marmosa murina. See Murine mouse opossums
Marmosa robinsoni, 12:252
Marmosa rubra. See Red mouse opossums
Marmosa xerophila. See Orange mouse opossums
Marmosets, **14:115–133**, 14:*125*, 14:*126*
behavior, 14:6–7, 14:8, 14:116–120, 14:*123*
conservation status, 14:124
distribution, 14:116
evolution, 14:115
feeding ecology, 14:8–9, 14:120, 14:121
habitats, 14:116
humans and, 14:11, 14:124
physical characteristics, 14:116, 14:*122*
reproduction, 14:10, 14:121–124
species of, 14:129–131, 14:132*t*
taxonomy, 14:4, 14:115
Marmosops spp. *See* Slender mouse opossums
Marmosops cracens, 12:254
Marmosops dorothea, 12:251
Marmosops handleyi, 12:254
Marmosops incanus. See Gray slender mouse opossums
Marmosops invictus. See Slaty slender mouse opossums
Marmota spp. *See* Marmots
Marmota caligata. See Hoary marmots
Marmota camtschatica. See Black-capped marmots
Marmota caudata. See Himalayan marmots
Marmota flaviventris. See Yellow-bellied marmots
Marmota marmota. See Alpine marmots
Marmota menzbieri, 16:147
Marmota monax. See Woodchucks
Marmota olympus. See Olympic marmots
Marmota vancouverensis. See Vancouver Island marmots
Marmotini, 16:143
Marmots
alpine, 12:*76*, 12:147, 12:148, 16:*144*, 16:*150*, 16:*155*, 16:157
behavior, 12:147, 16:124, 16:145
black-capped, 16:*146*, 16:159*t*
conservation status, 16:147
evolution, 16:143
feeding ecology, 16:145, 16:147
habitats, 16:124
Himalayan, 16:*146*
hoary, 16:*144*, 16:145, 16:158*t*
humans and, 16:148
Olympic, 16:*148*
reproduction, 16:147
taxonomy, 16:143
Vancouver Island, 16:147, 16:*150*, 16:*152*, 16:156
yellow-bellied, 16:*147*, 16:158*t*
Maroon anemones. *See* Magnificent sea anemones
Maroon-backed whistlers, 11:115
Marosatherina ladigesi. See Celebes rainbowfishes
Marquesas kingfishers, 10:10
Marsh crocodiles. *See* Mugger crocodiles
Marsh deer, 15:379, 15:383, 15:*384*, 15:*386*, 15:*391*–392
Marsh harriers. *See* Hen harriers
Marsh mongooses, 14:349, 14:*352*
Marsh owls, 9:333, 9:347
Marsh pheasants. *See* Bearded reedlings
Marsh rabbits, 16:481, 16:*481*, 16:*506*
Marsh rats, 16:265, 16:266, 16:268
Marsh rice rats, 16:268, 16:*271*, 16:*273*–274
Marsh sandpipers, 9:177
Marsh sparrows. *See* Savannah sparrows
Marsh tchagras, 10:426, 10:427
Marsh terns, 9:203
Marsh treaders. *See* Water measurers
Marsh warblers, 11:2, 11:4, 11:*9*, 11:*16*–17
Marsh wrens, 8:38, 10:*526*, 10:527, 10:528, 10:*530*, 10:*533*–534
Marshall, N. B., 4:442, 5:25, 5:26
Marshall's ioras, 10:416, 10:417
Marshbirds, 11:293
Marshfishes. *See* Bowfins
Marsupial cats. *See* Dasyurids
Marsupial frogs, 6:28–29, 6:37, 6:225–230
Marsupial mice. *See* Dasyurids
Marsupial moles, 12:137, **13:25–29**, 13:*26*, 13:*27*, 13:*28*
behavior, 12:71–72, 12:113, 13:26–27
distribution, 12:137, 13:*25*, 13:26
evolution, 12:13, 13:25
physical characteristics, 12:13, 12:71, 13:25–26
Marsupials
brain, 12:49
digestive system, 12:123
distribution, 12:129, 12:136–137
evolution, 12:11, 12:33
locomotion, 12:14
neonatal requirements, 12:126–127
physical characteristics, 13:1
reproduction, 12:12, 12:15, 12:51, 12:89–93, 12:*102*, 12:106, 12:108–109, 12:110
See also specific types of marsupials
Marsupionatia, 12:33
Martens, 14:256, 14:257, 14:319, 14:323, 14:324
American, 14:*320*, 14:*324*
yellow-throated, 14:*320*
Martes spp. *See* Martens
Martes americana. See American martens
Martes flavigula. See Yellow-throated martens
Martes foina. See Stone martens
Martes martes. See European pine martens
Martes pennanti. See Fishers
Martes zibellina. See Sables
Martinique orioles, 11:305
Martins, 10:174, 10:357–361
See also specific types of martins
Mary River cods, 5:222, 5:*226*, 5:231
Masai giraffes, 15:*402*
Mascalonges. *See* Muskellunges
Mascarene martins, 10:*362*, 10:*366*
Mascarene swallows. *See* Mascarene martins
Mascarene swiftlets, 9:418
Mashona mole-rats, 16:344–345, 16:349*t*
Mashonaland toads. *See* Chirinda toads
Masked bobwhites, 8:455
Masked boobies, 8:*216*, 8:*221*
Masked finfoot, 9:69, 9:70
Masked mouse-tailed dormice, 16:327*t*
Masked palm civets, 14:336, 14:344*t*
Masked shrews, 13:*251*
Masked titis, 14:144, 14:145, 14:148, 14:*149*, 14:*150*, 14:152–153
Masked weavers, 11:*377*, 11:378
Masoala fork-crowned lemurs, 12:38, 14:35, 14:39, 14:*40*, 14:*41*, 14:44
Mason, Ian
on fantails, 11:83
on lyrebirds and scrub-birds, 10:329
on woodswallows, 11:459
Mason wasps. *See* Potter wasps
Massasauga snakes, 7:39, 7:447
Massoutiera spp., 16:311
Massoutiera mzabi. See Mzab gundis
Mastacembelidae, 5:151
Mastacembelinae, 5:151
Mastacembeloidei, 5:151
Mastacembelus spp., 5:*151*, 5:*152*
Mastacembelus armatus. See Zig-zag eels
Mastacembelus erythrotaenia. See Fire eels
Masted crabs, 2:26
Masticophis spp. *See* Whipsnakes
Masticophis taeniatus. See Striped whipsnakes
Mastiff bats, **13:483–496**, 13:*489*, 14:288
behavior, 13:485
conservation status, 13:487
distribution, 13:*483*
evolution, 13:483
feeding ecology, 13:485
physical characteristics, 13:483
reproduction, 13:486–487
species of, 13:*490*–*492*, 13:493*t*–495*t*
taxonomy, 13:483
Mastigias papua. See Golden jellyfish
Mastigoproctus giganteus. See Giant whip scorpions
Mastigoteuthis magna, 2:*481*, 2:*486*
Mastodons, 12:*18*, 15:138
Mastomys natalensis. See Natal multimammate mice
Mastotermes darwiniensis. See Giant Australian termites
Mastotermitidae. *See* Giant Australian termites
Matamatas, 7:78, 7:*80*, 7:*81*–82
Mate selection, 12:101–102, 12:143–144
See also Reproduction
Maternal behavior, 7:42, 12:96–97, 12:107, 12:143–144
See also Reproduction
Maternal recognition, 12:84–85
See also Behavior
Mathevolepis petrotschenkoi, 1:227, 1:228

Mating. *See* Reproduction
Matschie's bushbabies. *See* Dusky bushbabies
Matschie's tree kangaroos, 13:*37*, 13:*86*, 13:*91*, 13:*98*, 13:100
Matsudaira's storm-petrels, 8:136, 8:138
Matsuzawa, Tetsuro, 12:161–162
Matthias fantails, 11:85
Maturation, 1:20, 2:21–24, 3:35
See also Reproduction
Maud Island frogs, 6:69, 6:70, 6:*71*, 6:72, 6:*74*
Maui nukupuus, 11:343
Maui parrotbills, 11:343
Mauremys annamensis. *See* Annam leaf turtles
Mauritian tomb bats, 13:*359*, 13:*362*, 13:363
Mauritius cuckoo-shrikes, 10:387, 10:*388*, 10:*389*–390
Mauritius dodos, 9:269
Mauritius kestrels, 8:27, 8:348
Mauritius parakeets, 9:280
Maurolicinae, 4:422
Mauve baublers. *See* Nightlight jellyfish
Maw-worms, 1:35, 1:*296*, 1:297–298
Mawas. *See* Sumatran orangutans
Mawsonites spriggi, 1:*9*
Maxilla, 12:9
See also Physical characteristics
Maxillae, 3:20
Maxillopoda, 2:273, 2:283, 2:295, 2:299
Maxwell's duikers, 16:*74*, 16:*79*, 16:*80*
Mayan culture, 6:51, 7:55
Mayflies, 3:*29*, 3:52, **3:*125*–131**, 3:*126*, 3:*127*, 3:*128*
Austral kingdom and, 3:55
behavior, 3:60, 3:126
as bioindicators, 3:82
brown, 3:*128*, 3:*129*–130
distribution, 3:54, 3:125
fossil, 3:7, 3:9, 3:10, 3:12, 3:13
giant, 3:13
physical characteristics, 3:12, 3:22
pollution and, 3:88
reproduction, 3:31, 3:33, 3:34, 3:62, 3:127
Mayotte drongos, 11:439
Maypole bowers, 11:*478*, 11:479, 11:*480*, 11:*481*, 11:486
Mayr, G., 9:275
Mazama spp., 15:382, 15:384
Mazama americana. *See* Red brocket deer
Mazama americana temana. *See* Mexican red brocket deer
Mazama bororo. *See* Brocket deer
Mazama bricenii. *See* Merioa brocket deer
Mazama chunyi. *See* Dwarf brocket deer
Mazama gouazoubira. *See* Gray brocket deer
Mazama pandora. *See* Yucatán brown brocket deer
Mazama pocket gophers, 16:195*t*
Mazama rufina. *See* Little red brocket deer
Mazes, in laboratory experiments, 12:153, 12:*154*
Mazon Creek fossils, 3:11
McDowall, R. M., 5:331
McGregor, P. K., 8:41
McGuire, Mike, 11:70
McGuire, Tamara L., 15:29
Meadow jumping mice, 16:*212*, 16:216
Meadow katydids, 3:203
Meadow pipits, 10:372, 10:373
Meadow voles, 16:125–126, 16:*226*, 16:227, 16:238*t*
Meadowlarks, 11:293, 11:301–306, 11:315–316, 11:342
Mealmoths, Indian, 3:76, 3:*392*, 3:401–*402*
Mealworms, 3:80, 3:83, 3:323
Mealybugs, 3:62, 3:66, 3:75, **3:261, 3:274, 3:278**
cactus, 3:81
cassava, 3:415
rhodesgrass, 3:*265*, 3:*274*, 3:278
soil, 3:63
Mearns's flying foxes, 13:332*t*
Mechanoreceptors, 3:26, 4:22, 6:15, 7:9, 7:32
See also Physical characteristics; specific species
Meconema spp. *See* Oak katydids
Meconema thalassinum. *See* Oak brush crickets
Mecopodinae, 3:205
Mecoptera, **3:341–346,** 3:*342*, 3:*344*
Medem's collared titis, 14:148
Medical entomologists, 3:78
Medicinal insects, 3:81
See also specific species
Medicinal leeches, 2:75, 2:76, 2:77
European, 2:75, 2:77–*78*, 2:*79*, 2:80–81
North American, 2:*79*, 2:81–82
Medicine
amphibians and, 6:53
conservation, 12:222–223
reptiles in, 7:50–52
zoological, 12:210–211
See also Diseases; Humans
Medicines. *See* Pharmaceuticals
Medina, M., 1:57
Mediterranean blind mole-rats. *See* Blind mole-rats
Mediterranean chameleons. *See* Common chameleons
Mediterranean fruit flies, 3:59, 3:60, 3:*363*, 3:*372*–*373*
Mediterranean geckos, 7:260
Mediterranean horseshoe bats, 13:391, 13:393, 13:*394*, 13:*399*
Mediterranean monk seals, 14:262, 14:418, 14:422, 14:435*t*
Mediterranean/(North) Atlantic shearwaters. *See* Cory's shearwaters
Mediterranean pen shells. *See* Noble pen shells
Mediterranean stick insects, 3:*227*, 3:*228*
Medusa theory, 1:10
Medusafishes, 5:421
Medusiform sea daisies, 1:*381*, 1:382, 1:*383*, 1:*384*, 1:*385*
Meerkats, 14:259, 14:*347*, 14:357*t*
Meester, J., 16:329
Megabyas flammulatus. *See* Shrike-flycatchers
Megaceryle alcyon. *See* Belted kingfishers
Megachasma pelagios. *See* Megamouth sharks
Megachile spp., 3:406
Megachile centuncularis, 3:*408*
Megachile rotundata. *See* Alfalfa leaf cutter bees
Megachiroptera. *See* Pteropodidae
Megacrania batesii. *See* Peppermint stick insects
Megaderma lyra. *See* Greater false vampire bats
Megaderma spasma. *See* Lesser false vampire bats
Megadermatidae. *See* False vampire bats
Megadyptes antipodes. *See* Yellow-eyed penguins
Megaelosia spp., 6:156
Megaerops niphanae. *See* Ratanaworabhan's fruit bats
Megaladapis spp. *See* Koala lemurs
Megaladapis edwardsi, 14:73–74
Megaladapis grandidieri, 14:73
Megaladapis madagascariensis, 14:73
Megalaima australis. *See* Blue-eared barbets
Megalaima haemacephala. *See* Coppersmith barbets
Megalaima rafflesii. *See* Red-crowned barbets
Megalaima rubricapilla. *See* Crimson-throated barbets
Megalania spp., 7:16, 7:17
Megalania prisca, 7:359
Megalapteryx didinus. *See* Moas
Megaledone spp., 2:478
Megalithone tillyardi. *See* Moth lacewings
Megalodicopia spp., 1:454
Megalodon sharks, 4:*10*, 4:131
Megaloglossus woermanni. *See* African long-tongued fruit bats
Megalomycteridae. *See* Largenoses
Megalomys spp., 16:266
Megalonychidae, 12:11, 13:148, **13:155–159**
Megalophrys intermedius. *See* Annam broad-headed toads
Megalopidae, 4:69, 4:243
Megaloprepus caerulatus. *See* Forest giants
Megalops spp., 4:243
Megalops atlanticus. *See* Atlantic tarpons
Megaloptera, **3:289–295,** 3:*292*
Megalotheca spp., 3:203
Megalurus spp., 11:2, 11:3, 11:7
Megalurus gramineus. *See* Little grassbirds
Megamouth sharks, 4:133, 4:*134*, 4:*136*, 4:139–140
Megamuntiacus spp., 12:129, 15:263
Megamuntiacus vuquangensis. *See* Giant muntjacs
Meganeuridae, 3:13
Meganeuropsis spp., 3:13
Megantereon spp. *See* Saber-toothed cats
Meganyctiphanes spp., 2:185
Meganyctiphanes norvegica. *See* Nordic krill
Megaoryzomys spp., 16:266
Megapedetes spp., 16:307
Megapodiidae. *See* Moundbuilders
Megapodius cumingii. *See* Philippine megapodes
Megapodius eremita. *See* Melanesian scrubfowl
Megapodius laperouse. *See* Micronesian megapodes
Megapodius nicobariensis. *See* Nicobar megapodes; Nicobar scrubfowl
Megapodius pritchardii. *See* Polynesian megapodes
Megapodius reinwardt. *See* Orange-footed megapodes; Orange-footed scrubfowl
Megaptera novaeangliae. *See* Humpback whales
Megarhyssa nortoni, 3:*410*, 3:416–*417*
Megaroyzomys spp., 16:270
Megascolecidae, 2:65, 2:66
Megascolides australis. *See* Gippsland giant worms
Megasecoptera, 3:52
Megasores gigas. *See* Mexican giant shrews
Megatapirus spp., 15:237
Megatheriidae, 13:148
Megazosterops palauensis. *See* Giant white-eyes
Megistolotis lignarius. *See* Woodworker frogs

Megophryidae. *See* Asian toadfrogs
Megophryinae, 6:109, 6:110
Megophrys spp., 6:66, 6:111
Megophrys nasuta. See Malayan horned frogs
Mehelya capensis. See Cape filesnakes
Mehely's horseshoe bats, 13:393
Meiacanthus grammistes. See Striped poison-fang blennies
Meiacanthus migrolineatus. See Saber-toothed blennies
Meinertellidae, 3:113
Meiosis, 1:15–16
See also Reproduction
Meise, Wilhelm, 9:1, 10:317
Mekong wagtails, 10:372, 10:375, 10:*377*, 10:*379*
Melaenornis annamarulae. See Nimba flycatchers
Melaenornis chocolatinus. See Abyssinian slaty flycatchers
Melamphaes spp., 5:105, 5:106
Melamphaidae. *See* Bigscales
Melampitta spp., 11:75, 11:76
Melampitta lugubris. See Lesser melampittas
Melamprosops phaesoma. See Po'oulis
Melanerpes carolinus. See Red-bellied woodpeckers
Melanerpes erythrocephalus. See Red-headed woodpeckers
Melanerpes formicivorus. See Acorn woodpeckers; California woodpeckers
Melanerpes herminieri. See Guadeloupe woodpeckers
Melanesian scrubfowl, 8:401
Melanitta spp. *See* Scoters
Melanitta nigra. See Blackscoters
Melanobatrachinae, 6:302
Melanobatrachus indicus. See Black microhylids
Melanocharis spp. *See* Berrypeckers
Melanocharis arfakiana. See Obscure berrypeckers
Melanocharis versteri. See Fan-tailed berrypeckers
Melanochlora spp. *See* Sultan tits
Melanochlora sultanea. See Sultan tits
Melanocorypha spp. *See* Larks
Melanocorypha calandra. See Calandra larks
Melanocorypha maxima. See Tibetan larks
Melanocorypha mongolica. See Mongolian larks
Melanocorypha yeltoniensis. See Black larks
Melanogrammus aeglefinus. See Haddocks
Melanonidae, 5:25–29, 5:*27*
Melanonus spp. *See* Melanonidae
Melanopareia spp. *See* Tapaculos
Melanopareia elegans. See Elegant crescentchests
Melanopareia maranonica, 10:259
Melanopareia torquata. See Collared crescentchests
Melanophidium spp., 7:391
Melanophryniscus spp., 6:67, 6:184, 6:189, 6:193
Melanophryniscus rubriventris. See Yungus redbelly toads
Melanoplus femurrubrum. See Red-legged locusts
Melanoptila glabrirostris. See Black catbirds
Melanoseps spp., 7:328
Melanostigma spp., 5:310, 5:311
Melanostigma atlanticum. See Atlantic midwater eelpouts
Melanostigma pammelas, 5:311
Melanostomias spp., 4:*423*
Melanosuchus niger. See Black caimans
Melanotaenia boesemani. See Boeseman's rainbowfishes
Melanotaeniidae. *See* Rainbowfishes
Melanotis caerulescens. See Blue mockingbirds
Melanotis hypoleucus. See Blue-and-white mockingbirds
Melanthripidae, 3:281
Melatonin, subterranean mammals and, 12:74–75
Melba finches. *See* Green-winged pytilias
Melba waxbills. *See* Green-winged pytilias
Meleagridinae. *See* Turkeys
Meleagris gallopavo. See Wild turkeys
Meleagris ocellata. See Ocellated turkeys
Meles meles. See European badgers
Melgrims. *See* Pacific sanddabs
Melias tristis. See Green-billed malkohas
Melichneutes spp., 10:137–144
Melichneutes robustus. See Lyre-tailed honeyguides
Melidectes belfordi. See Belford's melidectes
Melidectes rufocrissalis. See Yellow-browed melidectes
Melidora macrorrhina. See Hook-billed kingfishers
Melignomon spp., 10:137–144
Melignomon eisentrauti. See Yellow-footed honeyguides
Melignomon zenkeri. See Zenker's honeyguides
Melinae. *See* Badgers
Meliphaga spp. *See* Australian honeyeaters
Meliphaga aruensis. See Puff-backed meliphagas
Meliphaga auriculata. See Puff-backed meliphagas
Meliphaga cincta. See Stitchbirds
Meliphaga indistincta. See Brown honeyeaters
Meliphagidae. *See* Australian honeyeaters
Melipona beechei, 3:408
Melipotes fumigatus. See Common smoky honeyeaters
Melirrhophetes belfordi. See Belford's melidectes
Melithreptus validirostris. See Strong-billed honeyeaters
Melitta spp., 3:407–408
Melittophagus bullockoides. See White-fronted bee-eaters
Meller's duck, 8:367
Mellisuga spp. *See* Hummingbirds
Mellisuga helenae. See Bee hummingbirds
Mellivora capensis. See Honey badgers
Mellivorinae, 14:319
Melodious blackbirds, 11:302–303, 11:*307*, 11:*318*
Melodious jay thrushes. *See* Hwameis
Melodious larks. *See* Latakoo larks
Melodious warblers, 11:4, 11:6
Melogale moschata. See Chinese ferret badgers
Meloidae, 3:64, 3:65, 3:320
Meloidogyne spp. *See* Root-knot nematodes
Melolontinae, 3:55
Melon-headed whales, 15:1
Melophorus spp., 3:408
Melophus lathami. See Crested buntings
Melopsittacus undulatus. See Budgerigars
Melospiza melodia. See Song sparrows
Melursus ursinus. See Sloth bears
Melville, Herman, 15:78
Membranes, flight. *See* Flight membranes
Memory, 12:152–155
See also Cognition
Men. *See* Humans
Mendosoma spp., 5:239, 5:242
Menetia spp. *See* Australian skinks
Menetia greyii, 7:*330*, 7:*334*, 7:335–336
Mengeidae, 3:335
Mengenillidia, 3:335, 3:336
Menhadens. *See* Atlantic menhadens
Menidae. *See* Moonfishes
Menidia spp., 5:70
Menidia beryllina. See Inland silversides
Menidia clarkhubbsi, 5:70
Menidia extensa. See Waccamaw silversides
Menidiinae, 5:68
Menobranchus punctatus. See Dwarf waterdogs
Menstrual cycles, 12:109
Menstruation, 12:109
Mentawai gibbons. *See* Kloss gibbons
Mentawai Island langurs, 14:172, 14:173, 14:175, 14:*177*, 14:*181*, 14:183
Mentawai Island leaf-monkeys, 14:175, 14:184*t*
Mentawai Islands snub-nosed leaf-monkeys. *See* Mentawai Island langurs
Mentawai langurs, 14:6–7
Menura spp. *See* Lyrebirds
Menura alberti. See Albert's lyrebirds
Menura novaehollandiae. See Lyrebirds; Superb lyrebirds
Menura tyawanoides, 10:329
Menuridae. *See* Lyrebirds
Menzies' echymiperas, 13:10, 13:12
Mephitinae. *See* Skunks
Mephitis spp., 14:319, 14:321
Mephitis macroura. See Hooded skunks
Mephitis mephitis. See Striped skunks
Mercenaria mercenaria. See Northern quahogs
Mercer, J. M., 16:143
Merck's rhinoceroses, 15:249
Merganetta armata. See Torrent ducks
Merganser ducks, 8:365, 8:370
Merginae. *See* Sea ducks
Mergus octosetaceus. See Brazilian mergansers
Mérida small-eared shrews, 13:*253*
Meridozoros spp., 3:239
Merioa brocket deer, 15:396*t*
Meriones spp., 16:127
Meriones crassus. See Sundevall's jirds
Meriones unguiculatus. See Gerbils
Meristogenys spp., 6:247
Merker, St., 14:95
Merlet's scorpionfishes, 5:*168*, 5:173
Merlin falcons, 8:352
Merlucciidae, 5:25–30, 5:*27*
Merluccius spp., 5:27–30, 5:37
Merluccius bilinearis. See Silver hakes
Merluccius productus, 5:*27*
Mermaid's purse skates, 4:*178*
Mermis nigrescens, 1:*286*, 1:288–289
Meroles spp., 7:297
Meroles anchietae. See Shovel-snouted lizards
Meropeidae, 3:341, 3:342
Meropes. *See* Bee-eaters
Meropidae. *See* Bee-eaters

Meropogon forsteni. *See* Purple-bearded bee-eaters
Merops albicollis. *See* White-throated bee-eaters
Merops apiaster. *See* European bee-eaters
Merops bullockoides. *See* White-fronted bee-eaters
Merops cafer. *See* Cape sugarbirds
Merops carunculata. *See* Red wattlebirds
Merops corniculatus. *See* Noisy friarbirds
Merops gularis. *See* Black bee-eaters
Merops malimbicus. *See* Rosy bee-eaters
Merops novaeseelandiae. *See* Tuis
Merops nubicoides. *See* Carmine bee-eaters
Merops orientalis. *See* Little green bee-eaters
Merops ornatus. *See* Rainbow bee-eaters
Merops phrygius. *See* Regent honeyeaters
Merostomata. *See* Horseshoe crabs
Merothripidae, 3:281
Merrett, N. R., 5:37
Merriam's desert shrews. *See* Mexican giant shrews
Merriam's kangaroo rats, 16:202
Merriam's pocket gophers, 16:197*t*
Merrins. *See* Bridled nail-tailed wallabies
Mertensiella caucasica, 6:38
Mertensiella luschani. *See* Caucasian salamanders
Mertensophryne micranotis, 6:33
Mertila spp., 3:55
Merulaxis spp. *See* Tapaculos
Merulaxis ater. *See* Slaty bristlefronts
Merulaxis stresemanni, 10:259
Merychippus spp., 15:137
Merycodontinae, 15:265–266, 15:411
Merycoidodontidae, 15:264
Mesalina spp., 7:297
Mesaspis spp. *See* Alligator lizards
Mesaspis monticola. *See* Montane alligator lizards
Mesaxonia, 15:131
Mesechinus dauuricus. *See* Daurian hedgehogs
Mesechinus hughi. *See* Hugh's hedgehogs
Mesentoblasts, 2:3–4
Mesh herrings. *See* Atlantic herrings
Mesites, 9:1–4, **9:5–10**, 9:7
Mesitornis spp. *See* Roatelos
Mesitornis unicolor. *See* Brown mesites
Mesitornis variegata. *See* White-breasted mesites
Mesitornithidae. *See* Mesites
Meso-eusocial termites, 3:164, 3:168
Mesoamerican burrowing toads, **6:95–97**, 6:*96*
  distribution, 6:6, 6:96
  feeding ecology, 6:67, 6:97
  humans and, 6:51, 6:97
  larvae, 6:39, 6:41, 6:*96*, 6:97
  taxonomy, 6:95
Mesoamerican pythons. *See* Neotropical sunbeam snakes
Mesobdella gemmata, 2:78
Mesoborus spp., 4:69
Mesocapromys spp., 16:462
Mesocapromys angelcabrerai, 16:461
Mesocapromys auritus. *See* Eared hutias
Mesocapromys nanus. *See* Dwarf hutias
Mesocestoididae, 1:226, 1:231
Mesoclemmys spp., 7:77
Mesoclemmys gibba. *See* Gibba turtles
Mesocricetus spp., 16:239
Mesocricetus auratus. *See* Golden hamsters
Mesocricetus brandti. *See* Brandt's hamsters
Mesocricetus newtoni. *See* Romanian hamsters
Mesocyclops leuckarti, 2:*303*, 2:306–307
Mesoderm, 2:3, 2:13, 2:15, 2:20, 2:35
  See also Physical characteristics
Mesogobius batrachocephalus. *See* Knout gobies
Mesomeres, 1:18, 2:20
Mesomys didelphoides. *See* Brazilian spiny tree rats
Mesomys obscurus, 16:451
Mesomyzostomatidae, 2:59
Mesonychidae, 12:30, 15:2, 15:41, 15:131–132, 15:133
Mesophryne spp., 6:12
Mesopithecus spp., 14:172
Mesoplodon spp., 15:2, 15:59
Mesoplodon bidens. *See* Sowerby's beaked whales
Mesoplodon bowdoini. *See* Andrew's beaked whales
Mesoplodon carlhubbsi. *See* Hubb's beaked whales
Mesoplodon densirostris. *See* Blainville's beaked whales
Mesoplodon europaeus. *See* Gervais' beaked whales
Mesoplodon ginkgodens. *See* Ginkgo-toothed beaked whales
Mesoplodon grayi. *See* Gray's beaked whales
Mesoplodon hectori. *See* Hector's beaked whales
Mesoplodon layardii. *See* Strap-toothed whales
Mesoplodon mirus. *See* True's beaked whales
Mesoplodon perrini. *See* Perrin's beaked whales
Mesoplodon peruvianus. *See* Pygmy beaked whales
Mesoplodon stejnergeri. *See* Stejneger's beaked whales
Mesoplodon traversii. *See* Spade-toothed whales
Mesopolystoechotidae, 3:305
Mesopotamians, snakes and, 7:55
Mesopristes spp., 5:220
Mesopropithecus spp., 14:63
Mesopsocus unipunctatus, 3:*245*, 3:*247*
Mesosauria, 7:5, 7:14
Mesotardigrada. *See Thermozodium esakii*
Mesozoa, 2:33
Mespilia globulus. *See* Tuxedo pincushion urchins
Messel fossils, 3:9
Messenger, J. B., 2:477–478
Messiaen, Olivier, 10:332
Messor pergandei. *See* Desert ants
Mesurethra, 2:411, 2:412
Meta-eusocial termites, 3:164–165
Metabolism, body size and, 12:21
  See also Energy
Metabolites, 1:44–45
Metachirus spp. *See* Brown four-eyed opossums
Metachirus nudicaudatus. *See* Brown four-eyed opossums
Metadilepididae, 1:226
Metagonimus yokogawai, 1:200
Metallic hunting wasps, 2:37
Metallic jewel scarabs, 3:320
Metallic wood-boring beetles, 3:316, 3:320–321, 3:323
Metallura spp. *See* Metaltails
Metallura baroni. *See* Violet-throated metaltails
Metallura iracunda. *See* Perijas
Metallura odomae. *See* Neblina metaltails
Metallyticidae, 3:177
Metalstars, 9:450
Metaltails, 9:418, 9:438, 9:442, 9:447
Metamorphosis, 1:20, 2:21
  amphibians, 6:28, 6:39, 6:42–43
  insects, 3:*33*–34, 3:44
  See also Life histories; Reproduction; specific species
Metaphryniscus spp., 6:184
Metasyrphus americanus. *See* American hover flies
Metavelifer spp., 4:447
Metorchis conjunctus, 1:200
Metridium senile. *See* Frilled anemones
Metrocephala anderseni, 3:259
Metylophorus spp., 3:243
Meunier, F. J., 4:369
Mexican axolotl, 6:356, 6:*357*, 6:*359*
Mexican beaded lizards, 7:10, **7:353–358**, 7:*354*, 7:*356*
Mexican big-eared bats. *See* Allen's big-eared bats
Mexican black agoutis, 16:409, 16:*410*, 16:*412*, 16:413
Mexican black howler monkeys, 14:*158*, 14:166*t*
Mexican burrowing pythons. *See* Neotropical sunbeam snakes
Mexican burrowing toads. *See* Mesoamerican burrowing toads
Mexican caecilians, 6:*16*, 6:*21*, 6:*438*, 6:*439*
Mexican chickadees, 11:156
Mexican cloud forest anguid lizards, 7:207
Mexican dippers. *See* American dippers
Mexican fishing bats, 13:313, 13:500
Mexican fox squirrels, 16:175*t*
Mexican free-tailed bats. *See* Brazilian free-tailed bats
Mexican giant musk turtles, 7:121, 7:122
Mexican giant shrews, 13:263*t*
Mexican ground squirrels, 12:*188*
Mexican hairless, 14:287
Mexican hairy porcupines, 16:*366*, 16:367, 16:*370*, 16:*372*, 16:373
Mexican hog-nosed bats, 13:*414*, 13:417, 13:420, 13:435*t*
Mexican mouse opossums, 12:*256*, 12:*260*–261
Mexican red brocket deer, 15:*384*
Mexican ridleys. *See* Kemp's ridley turtles
Mexican spiny pocket mice, 16:208*t*
Mexico. *See* Nearctic region; Neotropical region
Meylan, Peter A., 7:151
Mhorr gazelles, 16:*47*
Miacids, 14:256
Mice
  behavior, 12:142–143, 16:124–125
  birch, **16:211–224**
  Delany's swamp, 16:297*t*
  distribution, 12:137, 12:138, 16:123
  dormice, 12:134, 16:122, 16:125, **16:317–328**
  evolution, 16:121
  fat, 16:296*t*
  grasshopper, 16:125
  gray climbing, 16:*286*, 16:*290*–291
  humans and, 12:139, 16:128

jumping, 16:123, 16:125, **16:211–224**
kangaroo, **16:199–201,** 16:*204*, 16:208*t*–209*t*
knock-out, 16:128
learning and, 12:154
meadow jumping, 16:*212*, 16:216
metabolism, 12:21
old-field, 16:125
Old World, **16:249–262**
pocket, 12:117, 16:121, 16:123, 16:124, **16:199–210**
pouched, 16:297*t*
pygmy, 16:123
pygmy rock, 16:*286*, 16:*289*, 16:290
reproduction, 12:103, 16:125–126
South American, **16:263–279**
taxonomy, 16:121
white-tailed, 16:*286*, 16:288–*289*
See also Flying mice; Muridae
Michoacán pocket gophers, 16:190, 16:*191*, 16:*192*, 16:193
Mico spp. *See* Amazonian marmosets
Mico intermedius. See Aripuanã marmosets
Micoureus spp. *See* Woolly mouse opossums
Micoureus alstoni. See Alston's woolly mouse opossums
Micoureus constantiae, 12:251
Micrastur plumbeus. See Plumbeous forest falcons
Micrathene whitneyi. See Elf owls
Micrelaps muelleri, 7:462
Micrixalinae, 6:246, 6:249
Micrixalus spp., 6:246
Micrixalus phyllophilus. See Nilgiri tropical frogs
Micro frogs, 6:*253*, 6:*255*
Micro-moths, 3:384, 3:385, 3:388–389
Microbatrachella spp., 6:245
Microbatrachella capensis. See Micro frogs
Microbats. *See* Microchiroptera
Microbial fermentation, 12:121–124
Microbiotheridae. *See* Monitos del monte
Microbiotherium spp., 12:273
Microbrotula spp., 5:16
Microcanthinae, 5:241
Microcavia spp. *See* Mountain cavies
Microcavia australis. See Mountain cavies
Microcavia niata. See Andean mountain cavies
Microcebus spp., 14:36, 14:37–38
Microcebus berthae. See Pygmy mouse lemurs
Microcebus griseorufus. See Mouse lemurs
Microcebus murinus. See Gray mouse lemurs
Microcebus ravlobensis, 14:39
Microcebus rufus. See Red mouse lemurs
Microcerculus spp., 10:527, 10:528
Microcerculus marginatus. See Nightingale wrens
Microchaetidae, 2:65
Microchera spp. *See* Hummingbirds
Microchiroptera, 12:58, 12:79, 12:85–87, 13:309
Microcoryphia. *See* Bristletails
Microctenopoma spp., 5:427, 5:428, 5:429
Microcyema vespa, 1:*95*, 1:*96*, 1:97–98
Microdajidae, 2:284
Microdajus langi, 2:*283*, 2:*285*, 2:*286*
Microdesmidae, 5:373, 5:374
Microdictyon spp., 2:109, 12:64
Microdipodops spp., 16:199, 16:200
Microdipodops megacephalus. See Dark kangaroo mice
Microdipodops pallidus. See Pale kangaroo mice
Microdon mutabilis, 3:360
Microeca spp., 11:105, 11:106
Microeca fascinans. See Jacky winters
Microgadus spp., 5:27
Microgadus tomcod. See Atlantic tomcods
Microgale spp., 13:196, 13:225, 13:226–227, 13:228, 13:229
Microgale brevicaudata, 13:227
Microgale dobsoni. See Dobson's shrew tenrecs
Microgale gracilis, 13:226
Microgale gymnorhyncha, 13:226
Microgale longicaudata. See Madagascar hedgehogs
Microgale nasoloi. See Nasolo's shrew tenrecs
Microgale principula. See Long-tailed shrew tenrecs
Microgale pusilla, 13:227
Micrognathozoa. *See* Jaw animals
Microhabitats, 4:48–49, 4:64
See also Habitats
Microhexura montivaga. See Spruce-fir moss spiders
Microhierax caerulescens. See Collared falconets
Microhierax fringillarius. See Black-thighed falconet
Microhierax latifrons. See White-fronted falconets
Microhydrula spp., 1:127
Microhydrulidae, 1:127
Microhyla ornata. See Ornate narrow-mouthed frogs
Microhyla palmata. See Web-foot frogs
Microhylidae. *See* Narrow-mouthed frogs
Microhylinae, 6:302
Microlepidopterans, 3:387
Micromacronus leytensis. See Miniature titbabblers
Micromalthidae, 3:321
Micromeres, 1:18, 2:20
Micromonacha lanceolata. See Lanceolated monklets
Micromus spp., 3:309
Micromys minutus. See Harvest mice
Micronesian flying foxes. *See* Marianas fruit bats
Micronesian honeyeaters, 11:*238*
Micronesian megapodes, 8:407
Micronycteris microtis. See Common big-eared bats
Microorganisms, Ice Age mammals and, 12:25
Micropanyptila furcata. See Pygmy palm-swifts
Microparra capensis. See Lesser jacanas
Microperoryctes spp., 13:2
Microperoryctes longicauda, 13:10
Microperoryctes murina. See Mouse bandicoots
Microperoryctes papuensis. See Papuan bandicoots
Microphiopholis gracillima, 1:389
Microphis spp., 5:134
Microphotolepis schmidti, 4:392
Microphotus angustus. See Pink glowworms
Micropilina arntzi, 2:*389*, 2:*390*–391
Microplana termitophaga, 2:37
Micropotamogale spp., 13:225
Micropotamogale lamottei. See Nimba otter shrews
Micropotamogale ruwenzorii. See Ruwenzori otter shrews
Micropsitta bruijnii. See Red-breasted pygmy parrots
Micropsyche ariana, 3:384
Micropterigids, 3:387
Micropterix calthella, 3:*391*, 3:*394*, 3:396–397
Micropteropus pusillus. See Dwarf epauletted fruit bats
Micropterus salmoides. See Largemouth basses
Microsaurs, 6:10, 6:11
Microsciurus spp., 16:163
Microscopy, 2:4
Microspathodon chrysurus. See Yellowtail damselfishes
Microspathodon dorsalis. See Giant damselfishes
Microstilbon spp. *See* Hummingbirds
Microstoma microstoma. See Slender argentines
Microstomatidae. *See* Deep-sea smelts
Microstomum spp., 1:188
Microteiids, 7:204, **7:303–308,** 7:*305*
Microtinae. *See* Arvicolinae
Microtus spp. *See* Voles
Microtus agrestis. See Field voles
Microtus cabrearae, 16:270
Microtus californicus. See California meadow voles
Microtus chrotorrhinus. See Rock voles
Microtus evoronensis, 16:229
Microtus mujanensis, 16:229
Microtus ochrogaster. See Prairie voles; Voles
Microtus oeconomus. See Root voles
Microtus pennsylvanicus. See Meadow voles
Microtus pinetorum. See Woodland voles
Micrurus spp. *See* Coral snakes
Micrurus fulvius. See North American coral snakes
Mictaceans, **2:241–242**
Mictocaris halope, 2:*241*, 2:*242*
Mid-mountain mouse-warblers. *See* Bicolored mouse-warblers
Midas free-tailed bats, 13:494*t*
Midas tamarins, 14:124, 14:132*t*
Middle Devonian Rhynie chert, 3:10
Middle ear, 12:8–9, 12:36
See also Physical characteristics
Midges, 3:37, 3:56, 3:57, **3:357–361,** 3:*359*
Midget flowerpeckers, 11:*192*, 11:*195*–196
Midgut, 3:20, 3:21
See also Physical characteristics
Midwater scorpionfishes, 5:163, 5:165
Midwife toads, 6:*32*, 6:36, 6:47, 6:*63*, **6:89–94,** 6:*90–91*
Migadopini, 3:55
Migrant shrikes. *See* Loggerhead shrikes
Migration, 3:386
amphibian, 6:59, 6:356
antelopes, 12:87
birds, 8:23, **8:29–35,** 8:*32*, 8:*33*, 8:42, 9:450, 11:4–5
diel, 1:24
fishes, 4:50, 4:61, 4:*73*
insects, 3:36–38, 3:66
lower metazoans and lesser deuterostomes, 1:42–43
mammals, **12:164–170**
protostomes, 2:25
reindeer, 12:116
reptiles, 7:43, 7:*86*, 7:89
vertical, 4:49, 4:60, 4:61, 4:442, 4:443

See also Behavior; specific species
Migratory Bird Treaty Act, 11:509
Migratory squirrels. *See* Gray squirrels
Miles, Lynn, 12:161
Miliaria calandra. See Corn buntings
Military dragons, 7:*212*, 7:*216*–217
Milk, 12:96, 12:106, 12:126–127
See also Lactation
Milkfishes, 4:*289*, **4:289–296,** 4:*292*, 4:*293*
Milksnakes, 7:*466*, 7:*472*, 7:*476*
Milky storks, 8:236
Millepora spp., 1:127, 1:129
Millepora alcicornis. See Fire corals
Milleporidae, 1:129
Millerbirds, 11:7
Millericrinida, 1:355
Miller's mastiff bats, 13:494*t*
Miller's monk sakis, 14:148
Millets, 5:*282*, 5:*284*, 5:286
Millipedes, 2:25, **2:363–373,** 2:*366*
behavior, 2:364–365
conservation status, 2:365
distribution, 2:364
evolution, 2:363
feeding ecology, 2:365
habitats, 2:364
humans and, 2:365
locomotion, 2:*19*
physical characteristics, 2:*18*, 2:363–364
reproduction, 2:23, 2:365
species of, 2:367–370
taxonomy, 2:363
Millot, J., 4:192
Millotauropodidae, 2:375
Milne-Edwards's sifakas, 14:63, 14:*68*, 14:*70*
Milne-Edwards's sportive lemurs, 14:76, 14:*78*, 14:80–81, 14:*83*
Milnesium tardigradum. See Large carnivorous water bears
Milton, K., 14:164
Milus. *See* Pere David's deer
Milvago chimango. See Chimango caracaras
Milvus migrans. See Black kites
Milvus milvus. See Red kites
Milyeringa veritas. See Blind cave gudgeons
Mimetic radiation, 6:206
Mimetica mortuifolia. See Dead leaf mimeticas
Mimic blennies, 5:*341*
Mimic filefishes, 4:70, 5:469
Mimicry, 1:42, 2:37–38, 3:65
aggressive, 4:68, 5:342
amphibian, 6:66, 6:206
in beetles, 3:320
bird songs and, 8:40
blackcaps, 11:22
Corvidae, 11:506
in flies, 3:360
katydid, 3:204
in Lepidoptera, 3:386
in lyrebirds, 10:330–331
Old World warblers, 11:4
in passerines, 10:172–173
See also Behavior; Songs, bird
Mimidae. *See* Mimids
Mimids, **10:465–474,** 10:*469*
Miminis, 11:407
Mimodes graysoni. See Socorro mockingbirds
Mimolagus, 16:480
Mimotoma, 16:480
Mimotonids. *See* Lagomorpha
Mimus spp. *See* Mimids
Mimus magnirostris. See San Andres mockingbirds
Mimus polyglottos. See Northern mockingbirds
Minahasa masked-owls, 9:338, 9:340–341
Minas Gerais tyrannulets, 10:275
Mindanao broadbills, 10:178, 10:179
Mindanao gymnures, 13:207, 13:*208*, 13:212
Mindanao moonrats. *See* Mindanao gymnures
Mindanao tarictic hornbills, 10:72
Mindanao wood-shrews. *See* Mindanao gymnures
Mindoro tarictic hornbills, 10:72, 10:75
Mine urchins. *See* Slate-pencil urchins
Minerals, insect preservation and, 3:8
Miners, 10:210
Miner's cats. *See* Ringtails (Procyonidae)
Minervarya spp., 6:246
Miniature donkeys, 12:*172*
Miniature horses, 15:*216*
Miniature kingfishers. *See* African pygmy-kingfishers
Miniature titbabblers, 10:507
Miniopterinae. *See Miniopterus* spp.
Miniopterus spp., 13:519, 13:520, 13:521, 13:522
Miniopterus australis. See Little long-fingered bats
Miniopterus medius. See Southeast Asian bent-winged bats
Miniopterus schreibersi. See Common bentwing bats
Minipera pedunculata, 1:454
Minister of the Environment and Heritage of Australia, 3:214
Ministry of Environment and Forests (Government of India), on pygmy hogs, 15:287
Minivets, 10:385–387
Mink, 14:319, 14:321, 14:323, 14:324
learning set, 12:152
neonatal requirements, 12:126
reproduction, 12:107
See also American mink
Mink (Keeping) Regulations of 1975 (United Kingdom), 12:183
Minke whales, 15:5, 15:8, 15:9, 15:120–121, 15:123–124
Antarctic, 15:1, 15:120, 15:125, 15:130*t*
dwarf, 12:*200*
northern, 15:120, 15:*122*, 15:125, 15:*126*, 15:128–129
Minnesota Zoo, 12:210
Minnows, **4:297–306,** 4:*304*, 4:*306*, **4:314–315**
behavior, 4:298–299
conservation, 4:303
distribution, 4:52, 4:58, 4:298
evolution, 4:297
feeding ecology, 4:67, 4:69, 4:299–300, 4:322
humans and, 4:303
physical characteristics, 4:*4*, 4:62, 4:297–298
reproduction, 4:35, 4:*300*–303, 4:*302*
species of, 4:307–319
taxonomy, 4:12, 4:297
See also specific types of minnows
Mino spp., 11:408
Minor chameleons, 7:*234*, 7:*239*–240
Minotilta varia. See Black-and-white warblers
Mioeuoticus spp., 14:13
Miomba tits, 11:155
Miopelodytes spp., 6:127
Miopithecus spp. *See* Talapoins
Miopithecus talapoin. See Angolan talapoins
Miracle triplefins, 5:*344*, 5:*345*, 5:348
Mirafra spp. *See* Larks
Mirafra africana. See Rufous-naped bush larks
Mirafra africanoides. See Fawn-colored larks
Mirafra albicauda. See White-tailed larks
Mirafra apiata. See Clapper larks; Rufous-naped larks
Mirafra ashi. See Ash's larks
Mirafra cheniana. See Latakoo larks
Mirafra degodiensis. See Degodi larks
Mirafra hova. See Madagascar larks
Mirafra hypermetra. See Red-winged larks
Mirafra javanica. See Australasian bushlarks; Australasian larks
Mirafra pulpa. See Friedmann's larks
Mirafra rufocinnamomea. See Flapped larks
Mirafra williamsi. See Williams's larks
Mirapinna spp., 5:106
Mirapinna esau. See Hairyfishes
Mirapinnidae, 5:105, 5:106
Miridae, 3:54, 3:55, 3:264
Miro traversi. See Chatham Islands black robins
Mirounga spp. *See* Elephant seals
Mirounga angustirostris. See Northern elephant seals
Mirounga leonina. See Southern elephant seals
Mirror carps. *See* Common carps
Mirror dories. *See* Buckler dories
Mirror self-recognition (MSR), 12:159
Mirrorbellies, 4:392–393
Mirza spp., 14:37, 14:39
Mirza coquereli. See Coquerel's mouse lemurs
Misgurnus fossilis. See Weatherfishes
Misophrioida, 2:299
Miss Waldron's red colobus, 12:214
Mississippi alligators. *See* American alligators
Mississippi cats. *See* Channel catfishes
Mississippi paddlefishes. *See* American paddlefishes
Mistichthys luzonensis. See Sinarapans
Mistle thrushes, 10:486, 10:487–488
Mistletoe flowerpeckers. *See* Mistletoebirds
Mistletoebirds, 11:190, 11:*192*, 11:*197*
Mitchell's plovers, 9:161, 9:162, 9:164
Mitered leaf-monkeys. *See* Banded leaf-monkeys
Mites, **2:333–352**
behavior, 2:335
conservation status, 2:336
distribution, 2:335
evolution, 2:333
feeding ecology, 2:335
habitats, 2:335
humans and, 2:336
physical characteristics, 2:333–334
reproduction, 2:335
taxonomy, 2:333
Mitochondria, 2:8
Mitochondrial DNA (mtDNA), 12:26–27, 12:28–29, 12:31, 12:33
Mitosis, 1:15
See also Reproduction
Mitred grebes. *See* Hooded grebes
Mitred horseshoe bats, 13:389

Mitrella carinata, 2:*264*
Mitsukurina owstoni. See Goblin sharks
Mitsukurinidae, 4:131
Mitu spp. *See* Razor-billed curassows
Mitu mitu. See Alagoas curassows
Mitu salvini. See Salvin's curassows
Mixed-species troops, Callitrichidae and, 14:119–120
Mixodigmatidae, 1:226
Mixophyes spp., 6:139–140, 6:147
Mixophyes balbus, 6:141
Mixophyes fasciolatus. See Giant barred frogs
Mixophyes fleayi, 6:141
Mixophyes hihihorlo, 6:140
Mixophyes intermandibularus, 6:140
Mixophyes iteratus, 6:141
Mixophyes submentalis, 6:140
Mnemiopsis spp., 1:171
Mnemiopsis leidyi. See Sea walnuts
Moas, **8:95–98,** 8:*96*, 8:*97*, 12:*183*, 12:184
Mobbing behavior, 4:70
See also Behavior
Moberg, G. P., 15:145
Mobulidae. *See* Manta rays
Moby Dick, 15:78
Mochokid catfishes, 4:354, 5:279
Mochokidae, 4:62, 4:352
Mocking chats, 10:486
Mocking thrushes. *See* Northern mockingbirds
Mockingbirds. *See* Mimids
Mocquard's rain frogs, 6:317, 6:*318*, 6:*319–320*
Moctezuma, 9:451
Mode, C. J., 3:48
Modicus spp., 5:355
Mogas, 5:277
Mogera spp., 13:279
Mogera etigo. See Echigo Plain moles
Mogera robusta. See Large Japanese moles
Mogera tokudae. See Sado moles
Mogoplistidae, 3:203
Mogoplistoidea. *See* Scale crickets
Mogurnda spp., 5:377
Moho braccatus. See Kauai oo
Mohobisopi spp. *See* Bishop's oo
Moholi bushbabies, 14:10, 14:25, 14:27, 14:*29*
Mohoua spp., 11:115, 11:116, 11:118
Mohoua albicilla. See Whiteheads
Mohoua novaeseelandiae. See Brown creepers
Mohoua ochrocephala. See Yellowheads
Mohua. *See* Yellowheads
Mohua ochrocephala. See Yellowheads
Mohuinae, 11:55
Mojarras, 5:255, 5:257–263
Mojave zebratails. *See* Zebra-tailed lizards
Mola mola. See Molas
Molar teeth, 12:9, 12:14, 12:46
See also Teeth
Molas, 4:*26*, 4:32, 4:50, 4:67, 5:467–471, 5:*472*, 5:477–*478*
Mole crickets, 3:80, 3:201, 3:202, 3:203, 3:207, 3:208
Mole-like rice tenrecs, 13:234*t*
Mole-limbed wormlizards, **7:279–282**
Mole lizards. *See* Two-legged wormlizards
Mole mice, 16:265
Mole-rats, 12:70, 12:72–77, 12:115, 12:116
See also African mole-rats
Mole salamanders, **6:355–361,** 6:*357*
behavior, 6:44, 6:356, 6:358–360
distribution, 6:5, 6:*355*, 6:356, 6:*358–360*
evolution, 6:13, 6:355
reproduction, 6:28
taxonomy, 6:3, 6:323, 6:355, 6:358–359
Mole-shrews, 13:*253*, 13:*255*
Mole voles, 12:71, 16:225, 16:228, 16:230
Molecular clocks, 2:10–11, 12:31, 12:33
Molecular genetics, **12:26–35**
Molecular systematics, 1:4–6, 2:4–5, 2:13–14
See also Taxonomy
Moles, **13:279–288,** 13:*283*
behavior, 12:113, 13:197–198, 13:280–281
claws, 12:39
conservation status, 13:282
distribution, 12:131, 13:197, 13:*279*, 13:280
evolution, 12:13, 13:279
feeding ecology, 12:75–76, 13:198, 13:281–282
habitats, 13:280
humans and, 13:200, 13:282
physical characteristics, 13:195–196, 13:279–280
reproduction, 13:199, 13:282
species of, 13:284–287, 13:287*t*–288*t*
taxonomy, 13:193, 13:194, 13:279
See also Marsupial moles; specific types of moles
Molge pyrrhogaster. See Japanese fire-bellied newts
Molgula spp., 1:452
Molgula manhattensis, 1:454
Molgulidae, 1:453
Molicolidae, 1:226
Molidae. *See* Molas
Mollusks, 2:3, 2:25
evolution, 2:9–10, 2:13–14
feeding ecology, 2:29
humans and, 2:41
opisthobranch, 1:45
parasitism and, 2:33
reproduction, 2:17, 2:21, 2:23–24
taxonomy, 2:35
See also specific mollusks
Mollymawks, 8:114–116, 8:*117*, 8:*120*–121
Moloch gibbons, 14:*215*, 14:*217*, 14:*219*, 14:220
behavior, 14:210, 14:211
conservation status, 14:11, 14:215
distribution, 14:208
evolution, 14:207
feeding ecology, 14:213
Moloch horridus. See Thorny devils
Moloki oo. *See* Bishop's oo
Molossidae. *See* Free-tailed bats; Mastiff bats
Molossops spp., 13:483
Molossops greenhalli. See Greenhalli's dog-faced bats
Molossus ater. See Greater house bats
Molossus molossus. See Pallas's mastiff bats
Molossus pretiosus. See Miller's mastiff bats
Molothrus spp. *See* Icteridae
Molothrus ater. See Brown-headed cowbirds
Molothrus bonariensis. See Shiny cowbirds
Molothrus rufoaxillaris. See Screaming cowbirds
Molpadia oolitica. See Rat-tailed sea cucumbers
Molpadiida, 1:417–421
Molpadiidae, 1:417, 1:420
Molsher, Robyn, 12:185
Molting, 3:31–33, 3:*32*, 8:*9*, 12:38
Molting animals. *See* Ecdysozoa
Moluccan cuscuses, 13:*66t*
Moluccan owlet-nightjars, 9:388
Moluccan woodcocks, 9:176, 9:180
Molva spp., 5:26
Mombasa killifishes, 5:*92*
Momotidae. *See* Motmots
Momotus aequatorialis. See Highland motmots
Momotus mexicanus. See Russet-crowned motmots
Momotus momota. See Blue-crowned motmots
Mona boas, 7:411
Mona Island blindsnakes, 7:382
Monacanthidae, 4:48, 4:67, 5:467
Monachinae, 14:417
Monachus monachus. See Mediterranean monk seals
Monachus schauinslandi. See Hawaiian monk seals
Monachus tropicalis. See West Indian monk seals
Monals, 8:434, 8:*438*, 8:*449–450*
Monarch butterflies, 1:42, 2:27, 3:37, 3:*63*, 3:66, 3:*87*, 3:386, 3:*389*, 11:304
Monarch flycatchers, **11:97–103,** 11:*99*
Monarcha brehmii. See Biak monarchs
Monarcha everetti. See White-tipped monarchs
Monarcha sacerdotum. See Flores monarchs
Monarcha trivigatus. See Spectacled monarch flycatchers
Monarchidae. *See* Monarchs
Monarchs, 11:437
Monasa spp. *See* Nunbirds
Monasa atra. See Black nunbirds
Monasa nigrifrons. See Black-fronted nunbirds
Mongolian gazelles, 12:134, 16:48, 16:*49*, 16:*53–54*
Mongolian hamsters, 16:239, 16:247*t*
Mongolian larks, 10:344
Mongolian pikas. *See* Pallas's pikas
Mongolian wild horses. *See* Przewalski's horses
Mongoloraphidia (Usbekoraphidia) josifovi. See Wart-headed Uzbekian snakeflies
Mongoose lemurs, 14:6, 14:8, 14:*48*, 14:50, 14:*53*, 14:*56–57*
Mongooses, **14:347–358,** 14:*353*
behavior, 12:114, 12:115, 12:146–147, 14:259, 14:349–351
in biological control, 12:190–191
conservation status, 14:262, 14:352
distribution, 12:136, 14:348–349
evolution, 14:347
feeding ecology, 14:260, 14:351–352
habitats, 14:349
humans and, 14:352
physical characteristics, 14:347–348
reproduction, 14:261, 14:352
species of, 14:*354*–356, 14:357*t*
taxonomy, 14:256, 14:347
Monias. *See* Subdesert mesites
Monias benschi. See Subdesert mesites
Moniezia benedeni, 1:*233*, 1:*234*, 1:236–237
Monilifera, 2:85, 2:86
Moniliformis moniliformis, 1:312, 1:*314*–315
Moniligastridae, 2:65
Monitors, 7:206, **7:359–368,** 7:*364*
in folk medicine, 7:51
as food, 7:48
humans and, 7:52, 7:53

Monitos del monte, 12:132, **12:273–275,** 12:*274*
Monk parakeets, 8:24, 8:25, 9:279, 9:*281*, 9:*292*
Monk sakis, 14:143, 14:144–145, 14:147, 14:153*t*
Monk seals, 12:138, 14:256, 14:418
Hawaiian, 12:138, 14:*420*, 14:422, 14:*425*, 14:*429*, 14:431–432
Mediterranean, 14:262, 14:418, 14:422, 14:435*t*
West Indian, 14:422, 14:435*t*
Monk sloths, 13:*166*, 13:*168*
Monkey grasshoppers, 3:201
Monkey spring pupfishes, 5:94
Monkeys, 14:3, 14:8
Allen's swamp, 14:194, 14:*196*, 14:*197*
behavior, 12:148
coloration in, 12:38
encephalization quotient, 12:149
learning set, 12:152
lesser white-nosed, 14:*192*
memory, 12:152–154
numbers and, 12:155
theory of mind, 12:159–160
tool use, 12:157
in zoos, 12:*207*
See also specific types of monkeys
Monkfishes, 5:47–51, 5:*52*, 5:*55–56*
Monniot, Claude, 1:479
Monniot, Françoise, 1:479
Monoblastozoa. *See Salinella salve*
Monocentridae, 5:113, 5:114, 5:116
Monocentris japonica. See Pineconefishes
Monocirrhus polyacanthus. See Amazon leaffishes
Monocle breams, 5:255
See also Nemipteridae
Monocled cobras, 7:*486*
Monocoryne gigantea, 1:127
Monodactylidae. *See* Monos
Monodactylus argenteus. See Monos
Monodactylus falciformis, 5:237
Monodactylus kottelati, 5:237
Monodactylus sebae, 5:237
Monodelphis spp. *See* Short-tailed opossums
Monodelphis americana, 12:251
Monodelphis brevicaudata. See Red-legged short-tailed opossums
Monodelphis dimidiata. See Southern short-tailed opossums
Monodelphis domestica, 12:251
Monodelphis iheringi, 12:251
Monodelphis kunsi. See Pygmy short-tailed opossums
Monodelphis osgoodi, 12:250
Monodelphis rubida, 12:250
Monodelphis scallops, 12:250
Monodelphis sorex, 12:250
Monodelphis unistriata, 12:250
Monodon monoceros. See Narwhals
Monodontidae, **15:81–91,** 15:*89*
Monogamy, 3:59, 3:167, 12:98, 12:107
Canidae, 14:271
Carnivora, 14:261
gibbons, 14:213, 14:214
monkeys, 14:6–7
salmons, 4:63
See also Behavior; Reproduction
Monogeneans, **1:213–224,** 1:*216*, 1:*217*
behavior, 1:38–40, 1:214
conservation status, 1:215
distribution, 1:214
evolution, 1:213
feeding ecology, 1:214
habitats, 1:214
humans and, 1:215
physical characteristics, 1:213–214
reproduction, 1:214–215
species of, 1:*218–223*
taxonomy, 1:11, 1:213
Monognathidae, 4:271, 4:272
Monognathus spp., 4:271
Monognathus rosenblatti, 4:*273*, 4:*275*–276
Monogononta, 1:259, 1:260, 1:261, 1:262
Monomachiidae, 3:55
Monomorphism, sexual, 1:18
See also Reproduction
Monopeltis spp., 7:274
Monophyllus redmani, 13:415
Monophyly, 2:7
Monopis spp., 3:387
Monopisthocotylea, 1:213, 1:214
Monoplacophorans, **2:387–391,** 2:*389*
Monopterus albus, 5:*153*, 5:154–155
Monopterus boueti. See Liberian swamp eels
Monopterus eapeni, 5:151
Monopterus indicus. See Bombay swamp eels
Monopterus roseni, 5:151
Monorhaphis chuni, 1:*70*, 1:*71–72*
Monos, 5:235–239, 5:241, 5:242, 5:244, 5:*246*, 5:249–*250*
Monostilifera, 1:253
Monotaxis grandoculis. See Humpnose big-eye breams
Monotremata. *See* Monotremes
Monotrematum sudamericanum. See South American monotremes
Monotremes, **12:227–234,** 15:133
behavior, 12:230–231
conservation status, 12:233–234
development, 12:106
distribution, 12:136–137, 12:230
evolution, 12:11, 12:33, 12:227–228
feeding ecology, 12:231
habitats, 12:230
humans and, 12:234
physical characteristics, 12:38, 12:228–230
reproduction, 12:12, 12:51, 12:89–93, 12:108, 12:231–233
skeletons, 12:41
taxonomy, 12:11, 12:33, 12:227–228
Monotrysian lepidopterans, 3:384
Monroe, Burt Jr.
on Australian creepers, 11:133
on bee-eaters, 10:39
on cuckoo-shrikes, 10:385
on Irenidae, 10:415
on Passeriformes, 10:169
on plantcutters, 10:325
on Sturnidae, 11:407
on sunbirds, 11:207
on thick-knees, 9:143
on tyrant flycatchers, 10:275
on weaverfinches, 11:353
on woodswallows, 11:459
Monstrilla grandis, 2:*302*, 2:309
Monstrilloida, 2:299
Montagu's harriers, 8:322
Montane alligator lizards, 7:*341*, 7:*343*, 7:344
Montane bamboo rats, 16:450–451, 16:*452*, 16:*455*, 16:457
Montane guinea pigs, 12:181, 16:399*t*
Montane tree shrews, 13:297*t*
Montane woolly flying squirrels, 16:136
Montastrea annularis, 1:30
Montatheris spp., 7:445
Monteiro's hornbills, 10:73, 10:77, 10:*79*
Monterey Bay Aquarium, 1:52
Monterey ensatina. *See Ensatina* spp.
Monterrey Spanish mackerels, 5:407
Monteverde toads. *See* Golden toads
Montezuma (Aztec chief), 12:203
Montezuma's oropendolas, 11:301, 11:*308*, 11:*309*
Monticelliidae, 1:226
Monticola saxatilis. See Rock thrushes
Montifringilla nivalis. See Snow finches
Montivipera spp., 7:445, 7:447
Montpellier snakes, 7:466
Mooi, Randall D., 5:163, 5:331
Moon jellies, 1:24, 1:26, 1:28, 1:*156*, 1:*158*, 1:*166*–167
Moon-toothed degus, 16:433, 16:440*t*
Mooneyes, 4:7, 4:11, 4:57, 4:231–233, 4:*236*, 4:*237*–238
Moonfishes, 5:255, 5:257–263
See also Black crappies; Monos; Opahs
Moonies. *See* Monos
Moonrats, 13:194, 13:197, 13:198, 13:199, 13:203, 13:*204*, 13:*208*, 13:212
Moontail bullseyes. *See* Goggle eyes
Moontails. *See* Stoplight parrotfishes
Moor frogs, 6:63
Moorhens, **9:45–67**
Mooriis. *See* Blunthead cichlids
Moorish idols, 5:391–393, 5:*395*, 5:*402*, 5:403
Moose, 15:*133*, 15:145–146, 15:*263*, 15:*269*, 15:*387*
behavior, 15:383, 15:394
conservation status, 15:384, 15:394
distribution, 12:132, 15:*391*, 15:394
evolution, 15:379, 15:380, 15:381
feeding ecology, 15:394
habitats, 15:383, 15:394
humans and, 15:384–385
migrations, 12:164–165
physical characteristics, 15:140, 15:381, 15:394
reproduction, 15:384, 15:394
taxonomy, 15:394
Mopalia muscosa. See Mossy chitons
Mopane worms, 3:80
Mops midas. See Midas free-tailed bats
Mops niangarae, 13:487
Mops spurelli. See Spurelli's free-tailed bats
Moraes, P. L. R., 14:164
Moras, 5:26
See also Moridae
Moray cods, 5:26
See also Muraenolepididae
Moray eels. *See* Morays
Morays, 2:37, **4:255–270,** 4:*260*
behavior, 4:*256*
conservation, 4:258
distribution, 4:255
evolution, 4:255
feeding ecology, 4:68, 4:256
habitats, 4:47, 4:48, 4:255–256
humans and, 4:258

physical characteristics, 4:44, 4:255
reproduction, 4:256–*258*
species of, 4:265–267
taxonomy, 4:255
Mordacia spp., 4:83
Mordacia mordax. See Short-headed lampreys
Mordaciidae, 4:83
Mordan, Peter B., 2:411
Moreau's sunbirds, 11:210
Morelia spp., 7:420
Morelia amethistina, 7:426
Morelia boeleni. See Black pythons
Morelia carinata. See Rough-scaled pythons
Morelia oenpelliensis, 7:420
Morelia tracyae. See Halmahera pythons
Morelia viridis. See Green pythons
Moreno glacier, 12:*20*
Morenocetus parvus, 15:107
Moreporks. *See* Tawny frogmouths
Moreton Bay bugs. *See* Flathead locust lobsters
Morganucodon spp., 12:11
Moridae, 5:25–30, 5:*27*
Moringua ferruginea. See Rusty spaghetti eels
Moringuidae, 4:255
Morionini, 3:55
Mormon crickets, 3:204, 3:208
Mormonilloida, 2:299
Mormoopidae. *See* Moustached bats
Mormoops spp. *See* Ghost-faced bats
Mormoops blainvillii. See Antillean ghost-faced bats
Mormoops megalophylla. See Ghost-faced bats
Mormopterus spp., 13:483
Mormopterus acetabulosus. See Natal free-tailed bats
Mormopterus beccarii. See Beccari's mastiff bats
Mormopterus phrudus, 13:487
Mormotomyia hirsuta, 3:359
Mormyridae. *See* Elephantfishes
Mormyroidea, 4:231
Mormyrops spp., 4:233
Mormyrus spp., 4:233
Mormyrus rume proboscirostris, 4:233, 4:*236*, 4:*238*–239
Morning warblers. *See* Spotted palm-thrushes
Moroccan glass lizards, 7:*341*, 7:*342*
Morone spp., 5:197
Morone chrysops. See White basses
Morone saxatilis. See Striped sea basses
Moronidae. *See* Striped basses; Temperate basses (Moronidae)
Morpho butterflies, 3:88
Morpho menelaus. See Blue morphos
Morphogenesis, 1:19
See also Reproduction
Morphos, blue, 3:*392*, 3:397, 3:*398*
Morrocoy amarillos. *See* South American yellow-footed tortoises
Morulae, spermatogenic, 2:17
See also Reproduction
Morus spp. *See* Gannets
Morus bassanus. See Northern gannets
Morus capensis. See Cape gannets
Morus serrator. See Australasian gannets
Morwongs, 5:235, 5:237–240, 5:242–244
Mosaic development, 2:35
See also Determinate cleavage
Mosaic gouramies. *See* Pearl gouramies
Mosasaurs, 7:*16*
Moschidae. *See* Musk deer
Moschinae. *See* Musk deer
Moschus berezovskii, 15:335
Moschus chrysogaster. See Himalayan musk deer
Moschus fuscus, 15:335
Moschus moschiferus. See Siberian musk deer
Mosquitofishes, 4:*22*, 5:*94*
Mosquitos, 3:*78*, **3:357–361,** 3:*358*, **3:365–366**
behavior, 3:59, 3:359
biomes, 3:56
feeding ecology, 3:64, 3:360
flight, 3:45
as pests, 3:75, 3:76
physical characteristics, 3:*24*, 3:358–359
reproduction, 3:45, 3:360
yellow fever, 3:76, 3:*362*, 3:365–*366*
Moss, Cynthia, 15:167
Moss bugs, 3:259, 3:*265*, 3:*270*, 3:277
Moss-forest ringtails, 13:114, 13:115, 13:122*t*
Moss frogs, 6:34, 6:147, 6:149, 6:*150*, 6:*152*
Mossbunkers. *See* Atlantic menhadens
Mosshead sculpins, 5:182
Mosshead warbonnets, 5:*311*
Mossy chitons, 2:395, 2:*396*, 2:*399*, 2:400
Motacilla aguimp. See African pied wagtails
Motacilla alba. See White wagtails
Motacilla australis. See Eastern yellow robins
Motacilla capensis. See Cape wagtails
Motacilla cinerea. See Gray wagtails
Motacilla citreola. See Yellow-hooded wagtails
Motacilla clara. See Mountain wagtails
Motacilla flava. See Yellow wagtails
Motacilla flaviventris, 10:372
Motacilla grandis. See Japanese wagtails
Motacilla hirundinaceum. See Mistletoebirds
Motacilla samveasnae. See Mekong wagtails
Motacilla singalensis. See Ruby-cheeked sunbirds
Motacilla subflava. See Tawny-flanked prinias
Motacillidae. *See* Pipits; Wagtails
Moth caterpillars, 3:80
Moth flies, sugarfoot, 3:360
Moth lacewings, 3:307, 3:*310*, 3:*311*, 3:312
Mother Carey's chickens. *See* Wilson's storm-petrels
Moths, 3:35, **3:383–404,** 3:*390*, 3:*391*, 3:*392*
behavior, 3:60, 3:65, 3:385–386
biological control and, 3:80
conservation status, 3:90, 3:388–389
feeding ecology, 3:46, 3:48, 3:63, 3:386–387
habitats, 3:385
humans and, 3:389
logging and, 3:87
olfaction, 3:27
as pests, 3:75
physical characteristics, 3:384–385
pollination by, 3:14
reproduction, 3:30, 3:34, 3:38, 3:59, 3:387–388
species of, 3:*393*–404
symbolism, 3:74, 3:389
taxonomy, 3:383–384
See also specific types of moths
Motility, sperm, 2:16–17
See also Reproduction
Motmots, 10:1–4, **10:31–38,** 10:*34*
Motor neurons, 3:25–26
Mottle-faced tamarins, 14:132*t*
Mottled burrowing frogs. *See* Marbled snout-burrowers
Mottled piculets, 10:150
Mottled sanddabs. *See* Pacific sanddabs
Mottled shovel-nosed frogs. *See* Marbled snout-burrowers
Mottled tuco-tucos, 16:427, 16:430*t*
Mottled whistlers, 11:115, 11:117
Mottlefin parrotfishes. *See* Striped parrotfishes
Motyxia spp., 2:365
Mouflons, 12:178, 15:147, 16:87, 16:91
Moundbuilders, 8:401, **8:403–411**
behavior, 8:404–405
conservation status, 8:407
distribution, 8:*403*, 8:404
evolution, 8:403–404
feeding ecology, 8:405
habitats, 8:404
humans and, 8:407
physical characteristics, 8:404
reproduction, 8:405–407
species of, 8:*409*–410
taxonomy, 8:403
Mounds, termite, 3:165, 3:*166*
Mt. Graham red squirrels, 16:167
Mount Cameroon francolins, 8:437
Mount Kupé bush-shrikes, 10:426, 10:427, 10:429
Mount Omei babblers. *See* Omei Shan liocichlas
Mountain accentors. *See* Siberian accentors
Mountain beavers, 12:131, 16:122, 16:123, 16:126–127, **16:131–134,** 16:*132*, 16:*133*
Mountain blackeyes, 11:228
Mountain bluebirds, 10:*486*
Mountain boomers. *See* Common collared lizards
Mountain brushtail possums, 13:*59*, 13:60, 13:66*t*
Mountain bulbuls, 10:397
Mountain cassowaries. *See* Bennett's cassowaries
Mountain cavies, 16:389, 16:390, 16:391, 16:392, 16:393, 16:*394*, 16:*395*, 16:399*t*
Mountain chickadees, 11:*157*
Mountain cottontails, 16:481, 16:516*t*
Mountain cows. *See* Central American tapirs
Mountain deer, 15:384
Mountain devils. *See* Thorny devils
Mountain earless lizards. *See* Common lesser earless lizards
Mountain gazelles, 12:139, 16:48, 16:*49*, 16:54–*55*
Mountain goats, 15:*134*, 16:*96*
behavior, 16:5, 16:*92*, 16:99
conservation status, 16:99
distribution, 16:*99*
evolution, 15:138, 16:87
feeding ecology, 12:*21*, 16:99
habitats, 12:131, 16:93–94, 16:99
humans and, 16:99
migrations, 12:165
physical characteristics, 16:99
reproduction, 16:94, 16:99
taxonomy, 15:138, 16:87, 16:99
Mountain gorillas, 12:*218*, 14:225, 14:227
See also Eastern gorillas
Mountain hares, 12:129, 16:481, 16:506, 16:*510*, 16:*512*–513

Mountain hens. *See* Great tinamous
Mountain lions. *See* Pumas
Mountain marmots. *See* Hoary marmots
Mountain mullets, 5:*61*, 5:*62*
Mountain nyalas, 16:11, 16:25*t*
Mountain owlet-nightjars, 9:389, 9:*390*, 9:*391*–392
Mountain pacas, 16:*422*, 16:*423*–424
Mountain perches, 5:222, 5:*223*, 5:*226*, 5:232
Mountain plovers, 9:104, 9:162, 9:163, 9:164
Mountain pocket gophers, 16:*185*, 16:*186*, 16:195*t*
Mountain pygmy parrots. *See* Red-breasted pygmy parrots
Mountain pygmy possums, 13:35, 13:*109*, 13:*110*, 13:111
  behavior, 13:37, 13:106–107
  conservation status, 13:39, 13:108
  evolution, 13:105
  feeding ecology, 13:38, 13:107
  habitats, 13:34, 13:106
  reproduction, 13:38
Mountain reedbucks, 16:27, 16:29, 16:30, 16:32
Mountain sheep, 12:165–167, 15:138, 16:4, 16:95
Mountain short-legged toads. *See* Slender mud frogs
Mountain spiny lizards, 7:*249*, 7:*251*, 7:254–255
Mountain squirrels, 16:167, 16:175*t*
Mountain starlings, 11:410
Mountain tapirs, 15:*244*, 15:*245*, 15:246–247
  behavior, 15:239, 15:240
  conservation status, 15:222, 15:*242*
  feeding ecology, 15:*241*
  habitats, 15:218, 15:238
  humans and, 15:*242*
  physical characteristics, 15:216, 15:217, 15:*238*
Mountain tarsiers. *See* Pygmy tarsiers
Mountain thornbills, 11:*56*
Mountain toads, 6:110
Mountain toucans, 10:125, 10:126, 10:128, 10:132
Mountain viscachas, 16:377, 16:378, 16:*379*, 16:380, 16:*381*, 16:*382*–383
Mountain wagtails, 10:372, 10:374
Mountain zebras, 15:215, 15:218, 15:220, 15:222, 15:226–227, 15:*231*, 15:*233*–234
  behavior, 12:146
Mt. Meru chameleons. *See* Jackson's chameleons
Moupin pikas, 16:502*t*
Mourning cloaks, 3:388
Mourning geckos, 7:260, 7:262
Mourning gnats, 3:66
Mouse bandicoots, 13:2, 13:12, 13:17*t*
Mouse-colored thistletails, 10:*213*, 10:*223*–224
Mouse deer, 15:138, 15:266, 15:267, 15:325–328, 15:*330*, 15:*332*–333
Mouse-eared bats, 13:497, 13:498, 13:499
Mouse lemurs, 12:*117*, 14:3, 14:4, 14:6, 14:7, 14:*10*, **14:35–45**, 14:*40*
  behavior, 14:36–37
  conservation status, 14:39
  distribution, 14:*35*, 14:36
  evolution, 14:35
  feeding ecology, 14:37–38
  habitats, 14:36
  humans and, 14:39
  physical characteristics, 14:35
  reproduction, 14:38
  species of, 14:*41*–44
  taxonomy, 14:35
Mouse-like hamsters, 16:241, 16:*286*, 16:*288*
  Afghan, 16:243
  Tsolov's, 16:243
  Urartsk, 16:243
Mouse opossums, 12:250–254, 12:*252*, 12:*256*, 12:*261*–262, 12:264*t*–265*t*
Mouse shrews. *See* Forest shrews
Mouse-tailed bats, 13:308, **13:351–354**
Mouse-tailed dormice, 16:317, 16:318
  masked, 16:327*t*
  Setzer's, 16:328*t*
  See also Roach's mouse-tailed dormice
Mouse-warblers, 11:*58*, 11:62–63
Mousebirds, **9:469–476**, 9:*472*
Mousefishes. *See* Sandfishes; Sargassumfishes
Moustache toads, 6:35, 6:110, 6:111
Moustached bats, 13:308, **13:435–442**, 13:*437*, 13:*439*, 13:442*t*
Moustached guenons, 14:*195*, 14:*198*, 14:199–200
Moustached monkeys. *See* Moustached guenons
Moustached tamarins, 14:117, 14:118, 14:119, 14:120, 14:121, 14:122, 14:*123*, 14:124
Moustached tree swifts, 9:433
Moustached turcas, 10:*261*, 10:*262*–263
Moustached woodcreepers, 10:230
Mouth-brooding cichlids, 4:*34*, 5:*279*
Mouthparts, 3:19–20, 3:*24*
  See also Physical characteristics
Mouths, 2:3, 2:21, 2:35
  See also Physical characteristics
Movement, 4:14, 4:17–20, 4:*26*
Moxostoma lacerum, 4:322
Mozambique spitting cobras, 7:*195*
Mrs. Gould's sunbirds. *See* Gould's sunbirds
Mrs. Gray's lechwes. *See* Nile lechwes
MSR (Mirror self-recognition), 12:159
Mt. Graham red squirrels, 16:167
Mt. Kenya shrews, 13:277*t*
Mt. Kupé bush-shrikes. *See* Mount Kupé bush-shrikes
Mt. Lyell salamanders, 6:325, 6:*393*, 6:401
mtDNA. *See* Mitochondrial DNA
Mubondos. *See* Sharptooth catfishes
Mud crabs. *See* Mangrove crabs
Mud dauber wasps, 3:83, 3:*410*, 3:*420*, 3:422
Mud digger whales. *See* Gray whales
Mud hakes. *See* White hakes
Mud mullets. *See* Shark mullets
Mud puppies. *See* Mudpuppies
Mud salamanders, 6:390
Mud shads. *See* Gizzard shads
Mud shrimps, 2:197, 2:198
Mud sirens, 6:323
Mud turtles
  American, **7:121–127**, 7:*124*
  East African black, 7:*131*, 7:*133*
  East African serrated, 7:*131*, 7:132–*133*
  eastern, 7:121, 7:*122*
  striped, 7:*122*
  white-lipped, 7:*124*, 7:*125*, 7:126
  yellow, 7:72, 7:*124*, 7:*125*
Mudfishes, 4:*292*, 4:294–295, 4:391
  See also Bowfins; Central mudminnows
Mudlarks. *See* Australian magpie-larks
Mudminnows, **4:379–387**, 4:*382*, 4:393
Mudnest builders, **11:453–458**, 11:*455*
Mudpuppies, 6:17, 6:34, 6:35, **6:377–383**, 6:*379*, 6:*380*
  See also Central mudminnows
Mudskippers, 4:50, 5:375, 5:*376*
  See also Atlantic mudskippers
Mudsnakes, 7:467–470
Mueller's clawed frogs. *See* Müller's plantanna
Mueller's gibbons, 14:*214*, 14:*217*, 14:220–*221*
  behavior, 14:210, 14:211
  distribution, 14:208, 14:*212*
  evolution, 14:207, 14:208–209
  feeding ecology, 14:213
  taxonomy, 14:208
Mugger crocodiles, 7:163, 7:179, 7:182, 7:*183*, 7:*185*–186
Mugil spp. *See* Mullets
Mugil cephalus. *See* Flathead mullets
Mugil cerema. *See* White mullets
Mugil princeps, 5:59
Mugilidae. *See* Mullets
Mugiliformes. *See* Mullets
Mugilomorpha, 4:12
Mukenes. *See* Dagaa
Mulberry silkworms. *See* Silkworms
Mule deer. *See* Black-tailed deer
Mule-killers. *See* Velvet ants
Mules, 12:177, 15:223
Mulgaras, 12:*278*, 12:*293*, 12:*294*
Mulhadens. *See* Alewives
Mulicalyx cristata, 1:200
Müller, Johannes, 4:197, 5:275
Müllerian mimicry, 3:65, 3:320, 5:342
Mulleripicus pulverulentus. *See* Great slaty woodpeckers
Mullerornis agilis. *See* Elephant birds
Mullerornis betsilei. *See* Elephant birds
Mullerornis rudis. *See* Elephant birds
Müller's plantanna, 6:*101*, 6:*104*–105
Mullets, 4:12, 4:17, 4:47, 4:52, 4:69, **5:59–66**, 5:*60*, 5:*61*
Mullidae. *See* Goatfishes
Mulloidichthys spp., 5:239
Mulloidichthys martinicus. *See* Yellow goatfishes
Mulloidichthys vanicolensis. *See* Yellowfin goatfishes
Mullus surmuletus. *See* Red goatfishes
Multi-crested birds of paradise. *See* Crested birds of paradise
Multicalycidae, 1:197
Multicellularity, 1:18, 1:37, 2:19
Multituberculata, 12:11, 16:122
Multitubulatina, 1:269, 1:270
Munehara, H., 5:182
Mungos spp., 14:347, 14:348
Mungos mungo. *See* Banded mongooses
Mungotictis spp., 14:347, 14:348
Mungotictis decemlineata. *See* Narrow-striped mongooses
Mungotinae, 14:347, 14:349
Munias, 11:353, 11:*354*, 11:357, 11:*359*, 11:370–371
Munna armoricana, 2:249
Munnings. *See* Banded hare-wallabies
Munro, George C., 11:350
Muntiacinae. *See* Muntjacs
Muntiacus spp. *See* Feral muntjacs; Muntjacs

Muntiacus atherodes. See Bornean yellow muntjacs
Muntiacus crinifrons. See Black muntjacs
Muntiacus feae. See Fea's muntjacs
Muntiacus gongshanensis. See Gongshan muntjacs
Muntiacus muntjak. See Indian muntjacs
Muntiacus muntjak vaginalis. See North Indian muntjacs
Muntiacus putaoensis. See Leaf muntjacs
Muntiacus reevesi. See Reeve's muntjacs
Muntiacus rooseveltorum. See Roosevelt's muntjacs
Muntiacus truongsonensis. See Truong Son muntjacs
Muntjacs, **15:343–355,** 15:*347–348*
  behavior, 12:118, 15:344–345
  conservation status, 15:346
  distribution, 15:*343*, 15:344
  evolution, 12:19, 15:343
  feeding ecology, 15:345
  habitats, 15:344
  humans and, 15:346
  physical characteristics, 15:267, 15:268, 15:269, 15:343–344
  reproduction, 15:346
  species, 15:349–354
  taxonomy, 12:129, 15:343
Muraena anguilla. See European eels
Muraena rostrata. See American eels
Muraenesocidae, 4:255
Muraenidae, 4:44, 4:47, 4:68, 4:255
Muraenolepididae, 5:25–29, 5:*27*
Muraenolepis spp. *See* Muraenolepididae
Muraenophis sathete. See Slender giant morays
Murex, 2:41, 2:447
Murex trunculus, 1:44
Murexia spp., 12:279, 12:288, 12:290
Murexia longicaudata. See Short-furred dasyures
Murexia rothschildi. See Broad-striped dasyures
Murgantia histrionica. See Harlequin stink bugs
Muricidae, 2:447
Muridae, **16:225–298,** 16:*286–287*
  Arvicolinae, 16:124, **16:225–238,** 16:*231*, 16:*232*, 16:270
  behavior, 16:284
  color vision, 12:79
  distribution, 12:134, 16:123, 16:*281*, 16:283
  evolution, 16:122, 16:281–282
  feeding ecology, 16:284
  habitats, 16:283–284
  hamsters, **16:239–248**
  humans and, 16:127–128
  Murinae, **16:249–262,** 16:*254*, 16:261*t*–262*t*
  physical characteristics, 16:123, 16:283
  reproduction, 16:125
  Sigmodontinae, 16:128, **16:263–279,** 16:*271*, 16:277*t*–278*t*
  species of, 16:288–295, 16:296*t*–297*t*
  taxonomy, 16:121, 16:282–283
Murina spp., 13:519, 13:521
Murina huttoni. See Hutton's tube-nosed bats
Murina leucogaster. See Tube-nosed bats, insectivorous
Murina suilla. See Brown tube-nosed bats
Murinae, **16:249–262,** 16:*254*, 16:261*t*–262*t*, 16:283
Murine mouse opossums, 12:*252*
Murininae, 13:519, 13:520–521
Muriquis, 14:155, 14:156, 14:157, 14:158, 14:159, 14:160
Murman herrings. *See* Atlantic herrings
Murofushi, Kiyoko, 12:161–162
Muroidea, 16:121
Murphy, John C., 7:410
Murphy, Robert Cushman, 8:154
Murray, A. M., 5:5
Murray breams. *See* Golden perches
Murray cods, 4:68, 5:*223*, 5:*226*, 5:231
Murray lampreys. *See* Short-headed lampreys
Murray magpies. *See* Australian magpie-larks
Murrayonida, 1:58
Murrelets, 9:102, 9:219–220, 9:222, 9:*223*, 9:227–228
Murres, 8:*108*, **9:219–229,** 9:*220*, 9:*223*
  behavior, 9:221
  conservation status, 9:222
  distribution, 9:*219*, 9:221
  evolution, 9:219
  feeding ecology, 9:221–222
  habitats, 9:221
  humans and, 9:222
  physical characteristics, 9:219, 9:220
  reproduction, 8:*108*, 9:222
  species of, 9:*224–228*
  taxonomy, 9:219–220
Mus musculus. See House mice
Mus spicilegus. See Steppe harvesting mice
Musca autumnalis. See Face flies
Musca domestica. See House flies
Muscardinus avellanarius. See Hazel dormice
Muscicapa caerulescens. See Ashy flycatchers
Muscicapa cassini. See Cassin's flycatchers
Muscicapa coeruleocephala. See Black-naped monarchs
Muscicapa cristata. See African paradise-flycatchers
Muscicapa megarhyncha. See Little shrike-thrushes
Muscicapa multicolor. See Scarlet robins
Muscicapa rufifrons. See Rufous fantails
Muscicapa striata. See Spotted flycatchers
Muscicapidae. *See* Old World warblers
Muscicapinae, 11:25
Muscidae, 3:361
Muscipeta cyaniceps. See Blue-headed fantails
Muscles, 7:23–26
  avian, 8:*6*
  in bat wings, 12:56–57
  See also Physical characteristics
Muscovy ducks, 8:21, 8:368, 8:373
Muscular diaphragm, 12:8
Musculoskeletal system, lissamphibians, 6:15–17, 6:*21*
Musellifer delamarei, 1:270
Mushroom anemones. *See* Mushroom corals
Mushroom corals, 1:106, 1:*109*, 1:*119*–120
Music
  birds and, 8:19–20
  insects in, 3:74
  See also Humans
Musk deer, 12:136, 15:265, 15:266, **15:335–341,** 15:*339*
Musk ducks, 8:364, 8:*375*, 8:*391–392*
Musk oxen. *See* Muskoxen
Musk shrews, 12:107, 13:198, 13:199, 13:266
Musk turtles, 7:72, 7:121–127, 7:*124*
Muskellunges, 4:379, 4:380, 4:*382*, 4:*384*–385
Muskies. *See* Muskellunges
Muskoxen, 16:*97*, 16:*98*, 16:102–103
  behavior, 16:92, 16:93
  distribution, 12:132, 16:*98*
  domestication and, 15:145–146
  evolution, 15:138, 16:87, 16:88
  feeding ecology, 16:93, 16:94
  habitats, 16:91
  physical characteristics, 15:139, 16:*89*, 16:90
  reproduction, 12:102, 16:94
  taxonomy, 16:87, 16:88
Muskrats, 16:*231*, 16:*233*, 16:238*t*
  behavior, 16:227
  distribution, 16:226
  feeding ecology, 16:228
  habitats, 16:124
  humans and, 16:126, 16:230
  physical characteristics, 16:123, 16:226
Musky rat-kangaroos, 13:35, 13:*69*, **13:69–72,** 13:*70*, 13:*71*, 13:*80*
Musophaga spp., 9:300–301
Musophaga cristata. See Great blue turacos
Musophaga rossae. See Ross's turacos
Musophaga violacea. See Violet turacos
Musophagidae. *See* Turacos
Musophagiformes. *See* Turacos
Mussel shrimps, 2:36, **2:311–315,** 2:*312*, 2:*313*
Mussels, 2:25, 2:43, 2:454
  common blue, 2:*455*, 2:*458*–459
  European pearly, 2:*455*, 2:*462*–463
  freshwater, 2:451, 2:453–454
  freshwater pearl, 2:454
  See also Bivalves; specific types of mussels
Musser, G. G., 16:282
Mussuranas, 7:205, 7:467, 7:*471*, 7:*479*–480
Mustangs. *See* Domestic horses
Mustela erminea. See Ermines
Mustela eversmanni. See Steppe polecats
Mustela felipei. See Colombian weasels
Mustela frenata. See Longtail weasels
Mustela lutreola. See European mink
Mustela lutreolina. See Indonesian mountain weasels
Mustela macrodon. See Sea minks
Mustela nigripes. See Black-footed ferrets
Mustela nivalis. See Least weasels
Mustela putorius. See Ferrets
Mustela putorius furo, 12:222
Mustela vison. See American mink
Mustelicolidae, 1:226
Mustelidae, **14:319–334,** 14:*326*
  behavior, 12:145, 14:258–259, 14:321–323
  conservation status, 14:262, 14:324
  distribution, 12:131, 14:*319*, 14:321
  evolution, 14:319
  feeding ecology, 14:260, 14:323
  habitats, 14:321
  humans and, 14:324–325
  physical characteristics, 12:37, 14:319–321
  reproduction, 12:109, 12:110, 14:324
  species of, 14:*327*–331, 14:332*t*–333*t*
  taxonomy, 14:256, 14:319
Mustelinae, 14:319
Mustelus antarcticus, 4:116
Mustelus henlei. See Brown smoothhound sharks
Musth, 15:167, 15:168, 15:170
Mute swans, 8:*363*, 8:*366*, 8:368, 8:369, 8:*372*, 8:373, 8:*374*, 8:*377*–378, 9:1
Mutilla europaea. See Velvet ants

Mutillids, 3:406
Muttonfishes. *See* Ocean pouts
Mutualism, 1:31–32, 2:32, 2:33–34, 3:49, 9:445, 9:446
See also Behavior
Myadestes spp. *See* Solitaires
Myadestes lanaiensis. See Olomaos
Myadestes myadestinus. See Kamaos
Myadestes obscurus. See Omaos
Myadestes palmeri. See Puaiohis
Myadestes townsendi. See Townsend's solitaires
Mycetobiidae, 3:55
Mycetophilidae, 3:359
Mycteria spp. *See* Storks
Mycteria americana. See Wood storks
Mycteria cinerea. See Milky storks
Mycteria leucocephala. See Painted storks
Mycteriini spp. *See* Storks
Myctophidae. *See* Lanternfishes
Myctophiformes. *See* Lanternfishes
Myctophinae, 4:441
Myctophum spp., 4:442, 4:443
Mydaus spp. *See* Stink badgers
Mydaus javensis. See Indonesian stink badgers
Mydaus marchei. See Philippine stink badgers
Myers, George S., 4:351, 5:421
Myers, Norman, 12:218
Myers, Ransom A., 1:50
Myersiella microps, 6:35
Myerslopiidae, 3:55
Mygalomorph spiders, 2:37
Myiagra freycineti. See Guam flycatchers
Myiagra longicauda. See African blue-flycatchers
Myiarchus spp. *See* Tyrant flycatchers
Myiarchus cinerascens. See Ash-throated flycatchers
Myiarchus crinitus. See Great crested flycatchers
Myiarchus nuttingi. See Nutting's flycatchers
Myioborus pariae. See Paria redstarts
Myioborus pictus. See Painted redstarts
Myioceyx lecontei. See African dwarf kingfishers
Myiodynastes luteiventris. See Sulphur-bellied flycatchers
Myiotheretes pernix. See Santa Marta bush tyrants
Myleus pacu. See Silver dollars
Myliobatidae. *See* Eagle rays
Myliobatis californica. See Bat rays
Myliobatoidei, 4:11
Myllokunmingia, 4:9
Mylodontidae, 12:11, 13:148
Mymaridae, 3:43
Mynahs. *See* Mynas
Mynas, **11:407–412, 11:415–420**
Myobatrachidae, **6:147–154,** 6:*150*
distribution, 6:6, 6:*147*, 6:148, 6:*151–153*
reproduction, 6:35, 6:36, 6:68, 6:148–149, 6:151–153
taxonomy, 6:139, 6:147, 6:151–153, 6:155
Myobatrachus spp., 6:35, 6:148, 6:149
Myobatrachus gouldii. See Turtle frogs
Myocapsus spp., 3:55
Myocastor coypus. See Coypus
Myocastoridae. *See* Coypus
Myodocopida, 2:311, 2:312
Myoictis spp., 12:288
Myoictis melas. See Three-striped dasyures
Myoisophagus spp., 1:247
Myomimus spp. *See* Mouse-tailed dormice
Myomimus personatus. See Masked mouse-tailed dormice
Myomimus roachi. See Roach's mouse-tailed dormice
Myomimus setzeri. See Setzer's mouse-tailed dormice
Myoncyteris torquata. See Little collared fruit bats
Myophonus spp. *See* Whistling-thrushes
Myophonus horsfieldii. See Malabar whistling-thrushes
Myoprocta spp., 16:407
Myoprocta acouchy. See Green acouchis
Myoprocta exilis. See Red acouchis
Myopsida. *See* Inshore squids
Myopterus whitleyi. See Bini free-tailed bats
Myopus schisticolor. See Wood lemmings
Myosciurus spp., 16:163
Myosciurus pumilio. See West African pygmy squirrels
Myosorex spp., 13:265, 13:266
Myosorex schalleri. See Schaller's mouse shrews
Myosorex varius. See Forest shrews
Myospalacinae, 16:282, 16:284
Myospalax myospalax. See Siberian zokors
Myotis
gray, 13:310–311, 13:316
large-footed, 13:313
Myotis spp. *See* Mouse-eared bats
Myotis adversus. See Large-footed myotis
Myotis bechsteinii. See Bechstein's bats
Myotis bocagei. See Rufous mouse-eared bats
Myotis capaccinii. See Long-fingered bats
Myotis dasycneme. See Pond bats
Myotis daubentonii. See Daubenton's bats
Myotis formosus, 13:497
Myotis grisescens. See Gray myotis
Myotis lucifugus. See Little brown bats
Myotis myotis. See Greater mouse-eared bats
Myotis mystacinus. See Whiskered bats
Myotis nattereri. See Natterer's bats
Myotis rickettii. See Rickett's big-footed bats
Myotis sodalis. See Indiana bats
Myotis vivesi. See Mexican fishing bats
Myotis welwitschii. See Welwitch's hairy bats
Myoxidae. *See* Dormice
Myoxus glis. See Edible dormice
Myoxus japonicus. See Japanese dormice
Myriapoda, 2:25, 2:353, 2:371, 3:9–10, 3:11
Myrichthys maculosus. See Tiger snake eels
Myrichthys magnificus, 4:267
Myriotrochidae, 1:420
Myripristis jacobus. See Blackbar soldierfishes
Myripristis murdjan. See Red soldierfishes
Myripristus berndti. See Blotcheye soldierfishes
Myrmeciza atrothorax. See Black-throated antbirds
Myrmecobiidae. *See* Numbats
Myrmecobius fasciatus. See Numbats
Myrmecocichla aethiops. See Anteater chats
Myrmecocystus spp., 3:408
Myrmecophaga spp., 13:149, 13:150
Myrmecophaga tridactyla. See Giant anteaters
Myrmecophagidae. *See* Anteaters
Myrmecophilus spp., 3:202, 3:203
Myrmeleon spp., 3:*306*, 3:307
Myrmeleon formicarius. See Antlions
Myrmeleontidae, 3:305, 3:306, 3:307
Myrmeleontiformia, 3:305, 3:306
Myrmica rubra. See Red ants
Myrmoborus myotherinus. See Black-faced antbirds
Myrmothera campanisona. See Thrush-like ant-pittas
Myrmotherula menetriesii. See Gray antwrens
Myrmotherula schisticolor. See Slaty antwrens
Myrmotherula snowi. See Alagoas antwrens
Myrocongridae, 4:255
Myrtle warblers. *See* Yellow-rumped warblers
Mysateles spp., 16:461
Mysateles garridoi, 16:461
Mysateles melanurus. See Black-tailed hutias
Mysateles prehensilis. See Prehensile-tailed hutias
Mysidae, 2:215
Mysids, **2:215–223,** 2:*218*
behavior, 2:216
conservation status, 2:217
distribution, 2:216
evolution, 2:215
feeding ecology, 2:216
habitats, 2:216
humans and, 2:217
physical characteristics, 2:215
reproduction, 2:217
species of, 2:219–222
taxonomy, 2:215
Mysinae, 2:215
Mysis relicta, 2:*215*, 2:*218*, 2:*219*, 2:221
Mystacina robusta. See Greater New Zealand short-tailed bats
Mystacina tuberculata. See Lesser New Zealand short-tailed bats
Mystacinidae. *See* New Zealand short-tailed bats
Mystacocarids, **2:295–297**
Mystacornis crossleyi. See Crossley's babblers
Mysticeti. *See* Baleen whales
Mystromyinae, 16:282
Mystromys albicaudatus. See White-tailed mice
Mythology, 3:74
amphibians in, 6:51–52
Corvidae in, 11:509
protostomes in, 2:41–42
reptiles in, 7:54–57
wagtails in, 10:376
wrens in, 10:529
wrynecks in, 10:154
See also Humans; specific species
Mytilidae, 2:451, 2:452
Mytilus spp. *See* Mussels
Mytilus edulis. See Common blue mussels
Mytonolagus, 16:480
Myxine spp., 4:77
Myxine glutinosa. See Atlantic hagfishes
Myxinidae, 4:4, 4:77
Myxiniformes. *See* Hagfishes
Myxinikela siroka, 4:77
Myxocyprinus asiaticus. See Chinese suckers
Myxoma virus, 12:187
Myxophaga, 3:315, 3:316
See also Beetles
Myxus spp., 5:59
Myza sarasinorum. See Greater Sulawesi honeyeaters
Myzante ignipectus. See Fire-breasted flowerpeckers
Myzomela spp. *See* Australian honeyeaters

Myzomela cardinalis. See Micronesian honeyeaters
Myzomela cruentata. See Red myzomelas
Myzopoda spp. *See* Old World sucker-footed bats
Myzopoda aurita. See Old World sucker-footed bats
Myzopodidae. *See* Old World sucker-footed bats
Myzornis pyrrhoura. See Fire-tailed myzornis
Myzostoma cirriferum, 2:*61,* 2:*62*
Myzostoma parasiticum, 2:59
Myzostoma polycyclus, 2:*59*
Myzostomatidae. *See* Elephants
Myzostomids, 2:31–32, **2:59–63,** 2:*61,* 2:*62*
Mzab gundis, 16:311, 16:312, 16:313, 16:*314*–315

N

Nacellidae, 2:423, 2:424
Nacunda nighthawks, 9:401, 9:*405*
Naemorhedus spp., 16:87
Naemorhedus baileyi. See Red gorals
Naemorhedus caudatus. See Long-tailed gorals
Naemorhedus goral. See Himalayan gorals
Naga Panchami, 7:56
Nagmia spp., 1:197
Nahan's francolins, 8:437
Nail-tailed wallabies, 13:83, 13:85, 13:86, 13:88
Nails, 12:39
  See also Physical characteristics
Naja melanoleuca. See Forest cobras
Naja mossambica. See Mozambique spitting cobras
Naja naja. See Indian cobras
Naja naja annulifera. See Egyptian banded cobras
Naja naja kaouthia. See Monocled cobras
Naja nigricollis. See Black-necked spitting cobras
Nakabo, T., 5:365
Nakamura, I., 5:417
Naked-backed bats
  big, 13:435, 13:442*t*
  Davy's, 13:435, 13:*437,* 13:442*t*
Naked-backed fruit bats, 13:307, 13:308, 13:*335,* 13:348*t*
Naked-backed moustached bats, 13:308
Naked bats, 3:199, 13:*485,* 13:*489,* 13:*490,* 13:491
Naked bulldog bats. *See* Naked bats
Naked characins, 4:338
Naked dragonfishes, 5:*324,* 5:325–*326*
Naked mole-rats, 16:*343,* 16:*346,* 16:*348*–349
  behavior, 12:74, 12:147, 16:124, 16:343
  communication, 12:76, 12:83
  distribution, 16:342, 16:*342*
  feeding ecology, 16:344
  heterothermia, 12:73
  physical characteristics, 12:71–74, 12:*72,* 12:78, 12:*86,* 16:340, 16:341
  reproduction, 12:107, 16:126, 16:344–345
  studies of, 12:72
  taxonomy, 16:339
Naked-tailed armadillos, 13:182, 13:183
Naked-throated birds. *See* Bare-throated bellbirds
Nalacetus spp., 15:2
Namaqua doves, 9:*254,* 9:*258*
Namaqua dune mole-rats, 16:*344,* 16:345, 16:349*t*
Namaqua sandgrouse, 9:*232,* 9:*234*
Namaquophasma spp., 3:217
Namdapha flying squirrels, 16:141*t*
Namib day geckos, 7:261
Namib golden moles. *See* Grant's desert golden moles
Namkungia biryongensis. See Biryong rock-crawlers
Nanaloricida. *See* Girdle wearers
Nanaloricidae, 1:343, 1:344, 1:345
Nanaloricus mysticus. See French girdle wearers
Nancy Ma's douroucoulis. *See* Nancy Ma's night monkeys
Nancy Ma's night monkeys, 14:135, 14:138, 14:*139,* 14:*140*–141
Nancy Ma's owl monkeys. *See* Nancy Ma's night monkeys
Nandidae. *See* Leaffishes
Nandinia binotata. See African palm civets
Nandiniinae. *See* African palm civets
Nandus nandus. See Gangetic leaffishes
Nanjaats. *See* Gerenuks
Nannastacidae, 2:229, 2:230
Nannatherina spp. *See Nannatherina balstoni*
Nannatherina balstoni, 5:222
Nannocharax fasciatus. See Striped African darters
Nannochoristidae, 3:55, 3:341, 3:342, 3:343
Nannoperca spp., 5:220
Nannoperca australis. See Southern pygmy perches
Nannoperca obscura. See Yarra pygmy perches
Nannoperca oxleyana. See Oxleyan pygmy perches
Nannoperca variegata. See Variegated pygmy perches
Nannopercidae. *See* Pygmy perches
Nannophrys spp., 6:246, 6:249
Nannopterum harrisi. See Galápagos cormorants
Nannosciurus spp., 16:163
Nannospalax spp., 16:124
Nannospalax ehrenbergi. See Palestine mole rats
Nannosquilla decemspinosa, 2:*172,* 2:*173,* 2:174
Nannosquillidae, 2:170
Nanny shads. *See* Gizzard shads
Nanonycteris veldkampi. See Little flying cows
Nanophyetus salmincola. See Salmon-poisoning flukes
Nanophyllium pygmaeum, 3:221
Nanorana spp., 6:247–248
Nansenia pelagica, 4:392
Napaeozapus spp., 16:212, 16:213, 16:216
Napaeozapus insignis. See Woodland jumping mice
Napo monk sakis, 14:148
Napoleon wrasses. *See* Humphead wrasses
Napoleon's peacock-pheasants. *See* Palawan peacock-pheasants
Narbaleks, 13:*92,* 13:*96*
Narceus spp., 2:365
Narceus americanus, 2:365
Narcine bancrofti. See Lesser electric rays
Narcinidae. *See* Electric rays
Narcomedusae, 1:123–124, 1:129
Narcondam hornbills, 10:73, 10:75
Nardovelifer altipinnis, 4:448
Nares, 12:67
Narina trogons, 9:*480,* 9:*481*
Narrow-billed bronze cuckoos. *See* Horsfield's bronze-cuckoos
Narrow-billed todies, 10:26, 10:27
Narrow-headed golden moles. *See* Hottentot golden moles
Narrow-mouthed frogs, **6:301–316,** 6:*306*–*307*
  behavior, 6:303, 6:*304,* 6:308–316
  conservation status, 6:305, 6:308–316
  distribution, 6:4, 6:5, 6:6, 6:*301,* 6:303, 6:*308*–*315*
  evolution, 6:301–302
  feeding ecology, 6:303–304, 6:308–316
  habitats, 6:303, 6:308–315
  humans and, 6:305, 6:308–316
  larvae, 6:39, 6:41
  physical characteristics, 6:302–*303,* 6:308–315
  reproduction, 6:32, 6:35, 6:68, 6:304–305, 6:308–316
  species of, 6:308–316
  taxonomy, 6:301–302, 6:308–315
Narrow-nosed nasute termites, 3:163–164
Narrow-snouted scalytails. *See* Beecroft's anomalures
Narrow-striped marsupial shrews, 12:300*t*
Narrow-striped mongooses, 14:*352,* 14:357*t*
Narrow-toed frogs, 6:65
Narwhal-beluga hybrids, 15:81
Narwhals, 15:7, 15:8, 15:9, **15:81–91,** 15:*83,* 15:*86,* 15:*88,* 15:*89*
Nasal cavity, 12:8
  See also Physical characteristics
Nasalis spp., 14:172, 14:173
Nasalis concolor. See Pig-tailed langurs
Nasalis larvatus. See Proboscis monkeys
Nasica longirostris. See Long-billed woodcreepers
Nasirana spp., 6:248, 6:249, 6:250
Nasiternsa bruijnii. See Red-breasted pygmy parrots
Naso lituratus. See Orangespine unicornfishes
Nasolo's shrew tenrecs, 13:*231,* 13:*233*
Nassariidae, 2:446
Nassau groupers, 5:*265,* 5:*269,* 5:270–271
Nasua spp., 14:309, 14:310
Nasua narica. See White-nosed coatis
Nasua nasua. See Ring-tailed coatis
Nasua nelsoni. See Cozumel Island coatis
Nasuella olivacea, 14:310
Nasute termites, 3:*161*
  arboreal, 3:165
  black-headed, 3:*169,* 3:173–*174*
  narrow-nosed, 3:163–164
Nasutitermes spp., 3:165
Nasutitermes nigriceps. See Black-headed nasute termites
Nasutitermes rippertii, 3:165
Nasutitermes triodiae. See Cathedral-mound building termites
Nasutitermitinae, 3:165
Natal black snakes, 7:462
Natal buttonquails. *See* Black-rumped buttonquails
Natal duikers, 16:84*t*

Natal forest frogs, 6:*36*
Natal free-tailed bats, 13:494*t*
Natal ghost frogs, 6:*133*
Natal moonies. *See* Monos
Natal multimammate mice, 16:*249*
Natalidae. *See* Funnel-eared bats
Natalobatrachus spp., 6:247
Natalobatrachus bonebergi. *See* Natal forest frogs
Natalus spp. *See* Funnel-eared bats
Natalus lepidus. *See* Gervais' funnel-eared bats
Natalus major, 13:*460*
Natalus micropus. *See* Small-footed funnel-eared bats
Natalus stramineus. *See* Funnel-eared bats
Natalus tumidifrons. *See* Bahamian funnel-eared bats
Natalus tumidirostris. *See* White-bellied funnel-eared bats
Nathusius's pipistrelle bats, 13:516*t*
Nation-states, 14:252
National Association of Audubon Societies, 8:22
National Audubon Society
on Eastern phoebes, 10:285
on Peruvian plantcutters, 10:327
National Invasive Species Council, 12:184
National Marine Fisheries Service (U.S.). *See* United States National Marine Fisheries Service
National Park Service, on feral pigs, 12:192
National Strategy of Japan on Biological Diversity, 6:342
National Trust for the Turks and Caicos Islands, 7:247
National Zoo (U.S.), 12:205
Natricinae, 7:466, 7:468
Natrix spp. *See* Eurasian watersnakes
Natrix natrix. *See* Grass snakes
Natterer, Johann, 4:201
Natterer's bats, 13:516*t*
Natterer's tuco-tucos, 16:431*t*
Natterjack toads, 6:59, 6:*63*
Natural control programs, 3:77
Natural history field studies, 12:198–200
Natural products, marine, 1:44–46
See also Humans
Natural Resources Conservation Authority, 7:247
Natural selection, 12:97, 12:144–145
See also Evolution; Selection
Nature Biotechnology, 15:151
Naucrates ductor. *See* Pilotfishes
Nauphoeta cinerea. *See* Cinereous cockroaches
Nausithoe rubra, 1:25
Nautichthys oculofasciatus. *See* Sailfin sculpins
Nautilids, **2:475–483**, 2:*481*
behavior, 2:477–478
conservation status, 2:479
distribution, 2:477, 2:*482*
evolution, 2:475
feeding ecology, 2:478
habitats, 2:477
humans and, 2:479–480
physical characteristics, 2:475–477
reproduction, 2:478–479
species of, 2:483
taxonomy, 2:475
Nautiloidea. *See* Nautilids
Nautilus spp., 2:477
Nautilus pompilius. *See* Pearly nautilus
Navahoceros spp., 15:379, 15:380
Navanax inermis, 2:404–405
Nava's wrens, 10:529
Navassa woodsnakes, 7:435
Navigation, **8:32–35**, 12:53–55, 12:87
See also Behavior
Nayarit squirrels. *See* Mexican fox squirrels
Nazarkin, M. V., 5:331
nDNA. *See* Nuclear DNA
Neacomys spp., 16:265
Neafrapus spp., 9:424
Neamblysomus spp., 13:216
Neanderthals, 14:243–*244*, 14:249, 14:251
body size of, 12:19, 12:24–25
brains, 12:24
evolution of, 12:21
See also Homo sapiens
Neanthes succinea. *See* Common shipworms
Nearctic region, 12:131–132, 12:134
Nebalia spp., 2:161
Nebalia bipes, 2:162
Nebalia hessleri, 2:*163*, 2:*164*, 2:165
Nebaliella spp., 2:161
Nebaliidae, 2:161
Nebaliopsidae. *See Nebaliopsis typica*
Nebaliopsis spp. *See Nebaliopsis typica*
Nebaliopsis typica, 2:161, 2:162
Neblina metaltails, 9:438
Nebriini, 3:54
Necator americanus, 1:35
Nechisar nightjars, 9:405
Neck-banded snakes, 7:205, 7:468, 7:469
Necora puber. *See* Shore crabs
Necrolemur spp., 14:1
Necromys benefactus, 16:270
Necromys lasiurus, 16:267
Necrophorus spp., 3:62
Necrosyrtes monachus. *See* Hooded vultures
Necrotaulidae, 3:375
Nectar
hummingbirds and, 9:438–439, 9:441, 9:443, 9:445–446
sunbirds and, 11:209
Nectar bats, 12:13
See also specific types of nectar bats
Nectarinia spp. *See* Sunbirds
Nectarinia chlorophygia. *See* Olive-bellied sunbirds
Nectarinia chrysogenys. *See* Yellow-eared spiderhunters
Nectarinia famosa. *See* Malachite sunbirds
Nectarinia johnstoni. *See* Scarlet-tufted sunbirds
Nectarinia kilimensis. *See* Bronze sunbirds
Nectarinia reichenbachii. *See* Reichenbach's sunbirds
Nectarinia thomensis. *See* São Tomé sunbirds
Nectarinia venusta. *See* Variable sunbirds
Nectariniidae. *See* Sunbirds
Nectogale spp., 13:248, 13:249
Nectogale elegans. *See* Elegant water shrews
Nectomys spp., 16:268
Nectomys parvipes, 16:270
Nectomys squamipes, 16:266, 16:267–268, 16:269
Nectonema agile, 1:*307*, 1:*308*, 1:309
Nectonematoida, 1:305, 1:306
Nectophryne spp., 6:35
Nectophryne afra, 6:35
Nectophryne gardineri. *See* Gardiner's frogs
Nectophryne hosii. *See* Brown tree toads
Nectophrynoides spp., 6:33, 6:68
Nectophrynoides malcolmi. *See* Malcolm's Ethiopian toads
Nectophrynoides occidentalis, 6:38
Nectophrynoides tornieri, 6:38
Nectophrynoides viviparus, 6:38
Nectrideans, 6:10, 6:11
Necturus spp. *See* Mudpuppies
Necturus alabamensis, 6:377, 6:378
Necturus beyeri, 6:377, 6:378
Necturus lewisi. *See* Neuse River waterdogs
Necturus maculosus, 6:377–379, 6:*381*, 6:382
Necturus maculosus lewisis. *See* Neuse River waterdogs
Necturus punctatus. *See* Dwarf waterdogs
Neduba extincta. *See* Antioch dunes shieldbacks
Needle-clawed bushbabies, 14:8–9, 14:26, 14:*27*, 14:*28*
Needle-spined urchins. *See* Rock boring urchins
Needle-tailed swifts. *See* White-throated needletails
Needlefishes, 4:12, 4:49, 4:70, **5:79–81**, 5:*82*, **5:84–85**
Needleflies, common, 3:*144*, 3:*145*
Needlenose gars. *See* Longnose gars
Needletail swifts *(Hirundpus)*, 9:424
Neelipleona, 3:99
Neergaard's sunbirds, 11:210
Neetroplus nematopus. *See* Poor man's tropheuses
Negaprion brevirostris. *See* Lemon sharks
Negro-finches, 11:354, 11:*358*, 11:360–361
Negros bleeding hearts, 9:250
Negros shrews, 13:277*t*
Negros striped-babblers, 10:507
Negros truncate-toed chorus frogs, 6:305
Nelson, J. B., 8:212
Nelson, Joseph S.
on Atheriniformes, 5:67
on Cypriniformes, 4:321
on fathead sculpins, 5:179
on Osmeriformes, 4:389
on Salmoniformes, 4:405
on Scorpaeniformes, 5:157
on Scorpaenoidei, 5:163
Nelson's antelope squirrels, 16:124, 16:145, 16:147, 16:160*t*
Nelson's kangaroo rats, 16:209*t*
Nelson's sharp-tailed sparrows, 11:*268*, 11:*269*, 11:304
Nelson's sparrows. *See* Nelson's sharp-tailed sparrows
Nelson's spiny pocket mice, 16:208*t*
Nemacheilinae, 4:321
Nemacheilus evezardi. *See* Hillstream loaches
Nemacheilus masyae. *See* Arrow loaches
Nemacheilus multifasciatus. *See* Siju blind cavefishes
Nemacheilus oedipus. *See* Cavefishes of Nam Lang
Nemacheilus starostini. *See* Kughitang blind loaches
Nemacheilus toni. *See* Siberian stone loaches
Nemadactylus spp., 5:239, 5:240
Nemapalpus nearcticus. *See* Sugarfoot moth flies
Nemaster rubiginosa. *See* Orange sea lilies
Nemateleotris magnifica. *See* Fire gobies
Nemathelminthes, 1:9

Nematistiidae. *See* Roosterfishes
Nematistius pectoralis. See Roosterfishes
Nematobothrium texomensis, 1:*202,* 1:*205*
Nematobrachion spp., 2:185
Nematocera, 3:357, 3:358, 3:359, 3:360
Nematodes
  behavior, 1:38, 1:43
  evolution, 1:5
  as parasites, 1:34, 1:35–36
  reproduction, 1:19, 1:21
  roundworms, 1:*4,* 1:12, 1:18, 1:20, 1:21, **1:283–292,** 1:*284*
  secernenteans, 1:12, **1:293–304,** 1:*296*
  taxonomy, 1:12
Nematogenyidae, 4:352
Nematomorpha. *See* Hair worms
Nematoscelis spp., 2:185
Nematostella vectensis. See Starlet sea anemones
Nematotaeniidae, 1:226
Nemeritis spp., 3:298
Nemerteans, 1:11, 1:18, 1:21, 1:*253,* 1:*254,* 2:13, 2:21
  See also Anoplans; Enoplans
Nemertesia antenna, 1:25
Nemertodermatida, 1:11
Nemichthydae, 4:255
Nemichthys scolopaceus. See Slender snipe eels
Nemipteridae, 5:255, 5:258–263
Nemobiinae, 3:207
Nemoptera spp., 3:308
Nemoptera sinuata. See Spoonwing lacewings
Nemopteridae, 3:9, 3:305, 3:306, 3:307
Nemouridae, 3:54, 3:141
Neoancistrocrania spp., 2:515
Neoaulocystis grayi, 1:*70,* 1:*71,* 1:74
Neobalaenidae. *See* Pygmy right whales
Neobarrettia vannifera. See Katydids
Neobatrachus spp., 6:140, 6:141
Neobatrachus pictus. See Painted frogs (Discoglossidae)
Neobola argentea. See Dagaa
Neobythitinae, 5:16, 5:18
Neocentrophyidae, 1:275
Neoceratodus spp., 4:197–199, 4:201, 4:202, 4:203
Neoceratodus forsteri. See Australian lungfishes
Neochmia spp., 11:354
Neochmia ruficauda. See Star finches
Neocichla gutteralis. See Babbling starlings
Neocirrhites armatus, 4:42, 5:239, 5:240
Neocoleoids, 2:475, 2:476, 2:477, 2:478, 2:479
Neocondeellum japonicum, 3:*95,* 3:96–97
Neoconocephalus spp., 3:206
Neocortex, 12:6, 12:36, 12:49
  See also Physical characteristics
Neocossyphus spp., 10:487
Neocrania spp., 2:515, 2:516
Neocrania anomala, 2:*517,* 2:*518*
Neocyamus physeteris. See Sperm whale lice
Neocyema erythrostoma, 4:271
Neocynchiropus spp., 5:365
Neodasyidae. *See* Multitubulatina
Neodasys spp., 1:269, 1:270
Neodermata, 1:11
Neodrepanis spp. *See* Sunbird-asities
Neodrepanis coruscans. See Common sunbird-asities
Neodrepanis hypoxantha. See Yellow-bellied sunbird-asities
Neofelis nebulosa. See Clouded leopards
Neofiber alleni. See Round-tailed muskrats
Neogastropoda, 2:445, 2:446
Neogeoscapheus spp., 3:150
Neogobius melanostomus. See Round gobies
Neogosseidae, 1:269
Neohermes spp., 3:*289*
Neohylomys spp., 13:203
Neolamprologus callipterus, 5:279–280
Neolestes spp., 10:395
Neolestes torquatus. See Black-collared bulbuls
Neomachilellus scandens, 3:*115,* 3:*116*
Neomenia spp., 2:380
Neomeniomorpha, 2:379, 2:380
Neomorphinae. *See* Cuckoos
Neomphalida, 2:429
Neomyini, 13:247
Neomys spp., 13:249
Neomys fodiens. See Eurasian water shrews
Neomysis spp., 2:216
Neomysis intermedia, 2:217
Neomysis japonica, 2:217
Neon gobies, 5:*374,* 5:375, 5:376, 5:*378,* 5:384, 5:*385*
Neon tetras. *See* Cardinal tetras
Neonates
  altricial, 12:95–96, 12:97
  body fat, 12:125
  marsupial, 12:108
  nutritional requirements, 12:126–127
  precocial, 12:12, 12:95–96, 12:97
  See also Reproduction
Neoniphon spp., 5:113
Neopallium. *See* Neocortex
Neopelma aurifrons, 10:297
Neoperla spp., 3:142
Neophascogale loreatzi. See Speckled dasyures
Neophema spp. *See* Budgerigars
Neophema chrysogaster. See Orange-bellied parrots
Neophema chrysostoma. See Blue-winged parrots
Neophema petrophila. See Rock parrots
Neophoca cinerea. See Australian sea lions
Neophocaena phocaenoides. See Finless porpoises
Neophron percnopterus. See Egyptian vultures
Neopilina spp., 2:387
Neopilina ewingi, 2:387
Neopilina galathea, 2:387
Neopilininae, 2:387
Neoplatymops spp., 13:311
Neopterygii, 4:11
Neopteryx frosti. See Small-toothed fruit bats
Neorenalia, 1:10
Neosciara spp. *See* Mourning gnats
Neoscopelidae. *See* Blackchins
Neoscopelus microchir, 4:442
Neosebastidae. *See* Gurnard perches
Neoselachii, 4:10, 4:11
Neoseps spp., 7:328
Neosomy, 3:350
Neospiza concolor. See São Tomé grosbeaks
Neostethus bicornis, 5:72, 5:*75,* 5:76
Neotamias spp., 16:143
Neotanaidomorpha, 2:235
Neoteny, 6:377, 9:245
Neotetracus spp., 13:203
Neotis spp. *See* Bustards
Neotis denhami. See Denham's bustards
Neotis nuba. See Nubian bustards
Neotoma spp. *See* Woodrats
Neotoma albigula. See White-throated woodrats
Neotoma cinerea. See Bushy-tailed woodrats
Neotoma floridana. See Eastern woodrats
Neotoma fuscipes. See Dusky-footed woodrats
Neotraginae, 16:1, **16:59–72,** 16:*65,* 16:71*t*
Neotragus batesi. See Dwarf antelopes
Neotragus moshcatus. See Sunis
Neotragus pygmeus. See Royal antelopes
Neotropic cormorants. *See* Olivaceous cormorants
Neotropical burrowing snakes, 7:468
Neotropical hog-nosed snakes, 7:467
Neotropical region, 12:131, 12:132–133
Neotropical sunbeam snakes, 7:*405,* **7:405–407,** 7:*406*
Neotropical swampsnakes, 7:468
Neotropical vultures. *See* New World vultures
Neotropical watersnakes, 7:467, 7:468
Neotropical wood turtles, **7:115–120**
Neotunga spp., 3:350
Nephasoma minutum, 2:98
Nephelobates spp., 6:200
Nephrurus stellatus. See Knob-tailed geckos
Nephtys hombergii. See Catworms
Nepomorpha, 3:259
Nepticulidae, 3:384
Nereids. *See* Clamworms
Nereis spp. *See* Clamworms
Neri-Arboleda, Irene, 14:95
Nerita plicata. See Plicate nerites
Nerites, **2:439–443**
Neritidae, 2:439, 2:440, 2:441, 2:447
Neritopsidae, 2:439
Neritopsina, **2:439–443**
Nerodia spp. *See* North American watersnakes
Nerodia rhombifera. See Diamond-backed watersnakes
Nervous system
  fishes, 4:20–23
  insects, 3:25–*26*
  lissamphibians, 6:21–23
  mammals, 12:49–50
  reptiles, 7:31
  See also Physical characteristics
Nesameletidae, 3:55
Nesbit, P. H., 16:333
Nesionixalus spp., 6:279
Nesocharis spp. *See* Olive-backs
Nesoclopeus woodfordi. See Woodford's rails
Nesofregetta spp. *See* Storm-petrels
Nesokia indica. See Short-tailed bandicoot rats
Nesolagus spp. *See* Striped rabbits
Nesolagus limminsi. See Annamite striped rabbits
Nesolagus timminsi. See Annamite striped rabbits
Nesomantis thomasseti. See Thomasset's frogs
Nesomimus macdonaldi. See Hood mockingbirds
Nesomimus trifasciatus. See Charles mockingbirds; Galápagos mockingbirds
Nesomyinae, 16:282, 16:285
Nesomys rufus. See Red forest rats
Nesophontes spp. *See* West Indian shrews
Nesophontes edithae. See Puerto Rican nesophontes
Nesophontes hypomicrus. See Atalaye nesophontes

Nesophontes longirostris. *See* Slender Cuban nesophontes
Nesophontes major. *See* Greater Cuban nesophontes
Nesophontes micrus. *See* Western Cuban nesophontes
Nesophontes paramicrus. *See* St. Michel nesophontes
Nesophontes submicrus. *See* Lesser Cuban nesophontes
Nesophontes zamicrus. *See* Haitian nesophontes
Nesophontidae. *See* West Indian shrews
Nesopsar nigerrimus. *See* Jamaican blackbirds
Nesoryzomys spp., 16:266
Nesoryzomys darwini, 16:270
Nesoryzomys indefessus, 16:270
Nesoscaptor uchidai. *See* Ryukyu moles
Ness, J. W., 15:145
Nest construction, frogs, 6:50
Nest cooperation. *See* Cooperative breeding
Nesting, crocodilians, 7:163–164, 7:181
  See also Reproduction; specific species
Nestor notabilis. *See* Keas
Nests, 12:96
  See also Reproduction
Net-makers, 3:376
Net spinners, 3:375
Net-winged beetles, 3:317, 3:320, 3:321–322
Netta erythrophthalma. *See* Southern pochards
Netta peposaca. *See* Rosybills
Nettapus auritus. *See* African pygmy geese
Nettastoma brevirostre. *See* Pignosed arrowtooth eels
Nettastomatidae, 4:255
Neurons, 3:25–26
  See also Physical characteristics
Neuroptera. *See* Lacewings
Neurotoxins, 4:15, 4:179, 7:52
Neurotrichus gibbsii. *See* American shrew moles
Neuse River waterdogs, 6:377, 6:378, 6:*380*, 6:*381*–382
Neusticomys oyapocki, 16:265
Neusticurus spp., 7:303, 7:304
Neusticurus ecpleopus, 7:*305*, 7:*306*, 7:307
Neuters. *See* Workers
Nevada side-blotched lizards. *See* Common side-blotched lizards
Nevada zebratails. *See* Zebra-tailed lizards
Nevo, Eviatar, 12:70
Nevrorthidae, 3:305, 3:306, 3:307, 3:308
Nevrorthiformia. *See* Nevrorthidae
New Caledonian crows, 11:507
New Caledonian geckos, 7:262
New Caledonian giant geckos, 7:259, 7:*264*, 7:*265*–266
New Caledonian owlet-nightjars, 9:370, 9:387, 9:388, 9:389
New Granada sea catfishes, 4:*355*, 4:*357*
New Guinea. *See* Australian region
New Guinea babblers. *See* Rufous babblers
New Guinea bush frogs, 6:303, 6:*306*, 6:*308*
New Guinea crocodiles, 7:181
New Guinea forest-rails, 9:46
New Guinea ground boas. *See* Viper boas
New Guinea logrunners, 11:69, 11:71, 11:*72*–*73*
New Guinea olive pythons. *See* Papuan pythons
New Guinea singing dogs, 14:266
New Guinea snakeneck turtles, 7:77
New Guinea snapping turtles, 7:*68*
New Guinea white-eyes, 11:227
New Guinean quolls, 12:299*t*
New Holland honeyeaters, 11:*237*, 11:*239*, 11:*247*–248
New Mexico National Heritage Program, 16:156
New World blackbirds, **11:301–322**, 11:*307*–*308*
  behavior, 11:293
  conservation status, 11:305–306
  distribution, 11:*301*, 11:302
  evolution, 11:301
  feeding ecology, 11:293–294
  habitats, 11:302–303
  humans and, 11:306
  physical characteristics, 11:301–302
  reproduction, 11:*303*, 11:304–305
  species of, 11:*309*–*322*
  taxonomy, 11:301
New World deer, **15:379–397**, 15:*386*, 15:*387*
  conservation status, 15:384
  distribution, 15:382
  evolution, 12:20, 15:379–382
  feeding ecology, 15:384
  habitats, 15:382–384
  humans and, 15:384–385
  physical characteristics, 15:382
  reproduction, 15:384
  species of, 15:388–395, 15:396*t*
  taxonomy, 12:20, 15:379–382
New World dogs, 14:287
  See also Domestic dogs
New World finches, **11:263–283**, 11:*267*, 11:*268*
  behavior, 11:263–264
  conservation status, 11:266
  distribution, 11:*263*
  evolution, 11:263
  feeding ecology, 11:264
  habitats, 11:263
  humans and, 11:266
  physical characteristics, 11:263
  reproduction, 11:264–265
  species of, 11:*269*–*283*
  taxonomy, 11:263, 11:286
New World ground cuckoos, 9:311
New World leaf-nosed bats, 13:311, 13:315
New World monkeys, **14:101–133**, 14:*107*, 14:*125*, 14:*126*
  behavior, 14:6–7, 14:103–104, 14:116–120
  conservation status, 14:105–106, 14:124
  distribution, 14:5, 14:*101*, 14:102–103, 14:*115*, 14:116
  evolution, 14:2, 14:101, 14:115
  feeding ecology, 14:9, 14:104–105, 14:120–121
  habitats, 14:6, 14:103, 14:116
  humans and, 14:106, 14:124
  physical characteristics, 14:101–102, 14:115–116
  reproduction, 14:10, 14:105, 14:121–124
  species of, 14:*108*–112, 14:*127*–131, 14:132*t*
  taxonomy, 14:1, 14:4, 14:101, 14:115
New World natricine snakes, 7:206
New World opossums, 12:132, **12:249–265**, 12:*255*, 12:*256*
  behavior, 12:251–252, 12:257–263, 12:264*t*–265*t*
  conservation status, 12:254, 12:257–263, 12:264*t*–265*t*
  distribution, 12:*249*, 12:250–251, 12:*257*–263, 12:264*t*–265*t*
  evolution, 12:249–250
  feeding ecology, 12:252–253, 12:257–263, 12:264*t*–265*t*
  habitats, 12:251, 12:257–263, 12:264*t*–265*t*
  humans and, 12:254, 12:257–263
  physical characteristics, 12:250, 12:257–263, 12:264*t*–265*t*
  reproduction, 12:253–254, 12:257–263
  taxonomy, 12:249–250, 12:257–263, 12:264*t*–265*t*
New World peccaries. *See* Peccaries
New World pond turtles, **7:105–113**, 7:*108*
New World porcupines, 16:125, **16:365–375**, 16:*370*, 16:374*t*
New World primary screwworms, 3:57, 3:*362*, 3:*365*
New World pythons. *See* Neotropical sunbeam snakes
New World quails, **8:455–463**
  behavior, 8:457
  conservation status, 8:401, 8:458
  distribution, 8:400, 8:*455*, 8:456
  evolution, 8:455
  feeding ecology, 8:457
  habitats, 8:456–457
  humans and, 8:458
  physical characteristics, 8:455–456
  reproduction, 8:457
  species of, 8:*460*–*462*
  taxonomy, 8:399, 8:455
New World soldierless termites, 3:161
New World sparrows. *See* New World finches
New World thrashers, 10:512
New World vultures, 8:233–236, **8:275–285**
  behavior, 8:234–235, 8:277–278
  conservation status, 8:278
  distribution, 8:234, 8:*275*, 8:276–277
  evolution, 8:233, 8:275
  feeding ecology, 8:235, 8:278
  habitats, 8:277
  humans and, 8:278
  physical characteristics, 8:233, 8:276
  reproduction, 8:236, 8:278
  species, 8:280–284
  taxonomy, 8:233, 8:275
New World warblers, **11:285–300**, 11:*292*
  behavior, 11:289
  conservation status, 11:290–291
  distribution, 11:*285*, 11:288
  evolution, 11:285–287
  feeding ecology, 11:*287*, 11:289–290
  habitats, 11:288–289
  humans and, 11:291
  physical characteristics, 11:287–288
  reproduction, 11:290
  species of, 11:*293*–*299*
  taxonomy, 11:285–287
New Zealand. *See* Australian region
New Zealand black oystercatchers. *See* Variable oystercatchers
New Zealand dotterels, 9:164
New Zealand falcons, 8:347
New Zealand fantails, 11:83
New Zealand freshwater limpets, 2:*416*, 2:417
New Zealand frogs, 6:4–5, 6:41, 6:64, **6:69–75**, 6:*71*

New Zealand fungus gnats. *See* New Zealand glowworms
New Zealand fur seals, 14:394, 14:*397*, 14:398, 14:407*t*
New Zealand giant wetas, 3:202, 3:207
New Zealand glowworms, 3:*362*, 3:*366*, 3:368–369
New Zealand graylings, 4:391
New Zealand grebes, 8:171
New Zealand king shags, 8:*204*, 8:208
New Zealand lesser short-tailed bats. *See* Lesser New Zealand short-tailed bats
New Zealand long-eared bats. *See* Lesser New Zealand short-tailed bats
New Zealand pigeons, 9:248
New Zealand quails, 8:437
New Zealand Red Data Book, on lesser New Zealand short-tailed bats, 13:457
New Zealand sea lions. *See* Hooker's sea lions
New Zealand short-tailed bats, **13:453–458,** 13:*456*
New Zealand smelts, 4:390
New Zealand sooty oystercatchers. *See* Variable oystercatchers
New Zealand takahes, 10:326
New Zealand teals. *See* Brown teals
New Zealand wattle birds, **11:447–451,** 11:*449–450*
New Zealand Wildlife Act
  Archey's frogs, 6:72
  Hamilton's frogs, 6:73
  Hochstetter's frogs, 6:74
New Zealand wrens, 10:170, **10:203–208,** 10:*205*
New Zealand's Department of Conservation, 3:322
Newborns. *See* Neonates
Newits. *See* Northern lapwings
Newman's knob-scaled lizards, 7:347–349, 7:*350*, 7:*351*
Newtonia fanovanae. See Red-tailed newtonias
Newton's bowerbirds. *See* Golden bowerbirds
Newts, **6:323–326, 6:363–375,** 6:*368–369*
  behavior, 6:45, 6:*365–366*, 6:370–375
  conservation status, 6:367, 6:370–375
  courtship behavior, 6:49
  definition, 6:363
  distribution, 6:325, 6:*363*, 6:364, 6:*370–375*
  evolution, 6:323–324, 6:363
  feeding ecology, 6:326, 6:366–367, 6:370–375
  habitats, 6:*63*, 6:364–365, 6:370–375
  humans and, 6:367, 6:370–375
  as pets, 6:54–55
  physical characteristics, 6:323–324, 6:363, 6:370–375
  reproduction, 6:30–31, 6:33, 6:326, 6:*363*, 6:*364*, 6:367, 6:370–375
  taxonomy, 6:363, 6:370–375
Nezara viridula. See Southern green stink bugs
Ngege, 5:280, 5:*283*, 5:*286*, 5:287
Nicaragua glass frogs, 6:*218*, 6:220
Nicaraguan pocket gophers, 16:196*t*
Nicator spp., 10:395
Nicator vireo. See Yellow-throated nicators
Niceforo's wrens, 10:529
Niches, ecological. *See* Ecological niches
Nicobar bulbuls, 10:396, 10:398
Nicobar megapodes, 8:404
Nicobar scrubfowl, 8:401
Nicobar tree shrews, 13:292, 13:297*t*
Nicolai, J., 11:375
Nicoletiidae, 3:119
Nicrophorus spp., 3:322
Nicrophorus americanus. See American burying beetles
Nidhogger, 7:54
Nidicolous species, 8:15
Nidirana spp., 6:248, 6:251
Nielsen, Claus, 1:13
Niethammer, Günther, 9:255
Nigerian mole-rats, 16:350*t*
Night-adapted vision, owls, 9:332
Night adders, 7:445–448, 7:*449*, 7:*451*–452, 7:*486*
Night herons, 8:233, 8:234, 8:242
  See also specific types of night herons
Night lizards, 7:207, **7:291–296,** 7:*293*
Night monkeys, 14:6, 14:8, 14:10, **14:135–142,** 14:*139*, 14:141*t*
Night parrots, 9:277
Night skinks, 7:*331*, 7:*334*
Nightfishes, 5:221
Nighthawks, 9:367–370, 9:401–*402*, 9:*404–405*, 9:409
Nightingale wrens, 10:527
Nightingales, 8:19, 10:484, 10:485–486, 10:488, 10:489, 10:*490*, 10:*492*
Nightjars, 9:367–371, **9:401–414,** 9:*407*
  behavior, 9:403
  conservation status, 9:405
  distribution, 9:*401*, 9:402
  evolution, 9:401
  feeding ecology, 9:403–404
  habitats, 9:402–403
  humans and, 9:405–406
  physical characteristics, 9:401–402
  reproduction, 9:404–405
  species of, 9:*408–414*
  taxonomy, 9:335, 9:401
Nightlight jellyfish, 1:26, 1:*159*, 1:165–*166*
Nigripinnis. *See* Blackfin pearl killifishes
Nigrita fusconota. See White-breasted negro-finches
Nihoa finches, 11:342, 11:343
Nilaus spp. *See* Bush-shrikes
Nilaus afer. See Brubrus
Nile crocodiles, 7:48, 7:161–164, 7:*183*, 7:*186*
  conservation status, 7:181
  feeding ecology, 7:*180*
  as food, 7:48
  habitats, 7:179
  reproduction, 7:*182*
Nile lechwes, 15:271, 16:4, 16:27, 16:34, 16:*35*, 16:*38*
Nile monitors, 7:48, 7:360, 7:*362–363*
Nile perches, 4:75, 4:313, 5:220, 5:*224*, 5:*226–227*, 5:280–281
Nile Valley sunbirds, 11:208
Nilgais, 12:136, 15:265, 16:4, 16:11, 16:24*t*
Nilgiri burrowing snakes, 7:*393*
Nilgiri ibex. *See* Nilgiri tahrs
Nilgiri langurs, 14:184*t*
Nilgiri laughing thrushes, 10:507
Nilgiri pipits, 10:375
Nilgiri tahrs, 16:90, 16:92, 16:*93*, 16:94, 16:95, 16:103*t*
Nilgiri tropical frogs, 6:*253*, 6:259
Niltava grandis. See Large niltavas
Nimba flycatchers, 11:27
Nimba otter shrews, 13:234*t*
Nine-banded armadillos, 12:111, 12:131, 13:181, 13:*182*, 13:183–184, 13:*185*, 13:*186*, 13:*187*
Nineline gobies, 5:376
1979 Convention on the Conservation of European Wildlife and Natural Habitats, 2:463
1982 Convention on the Law of the Sea
  common dolphinfishes, 5:215
  pompano dolphinfishes, 5:214
Ningaui ridei. See Wongai ningauis
Ningaui timealeyi. See Pilbara ningauis
Ningu, 4:*303*, 4:*313*
Ninia spp. *See* Coffeesnakes
Ninox spp., 9:332, 9:346, 9:347, 9:349
Ninox boobook. See Southern boobook owls
Ninox novaeseelandiae. See Owl moreporks
Ninox scutulata. See Brown hawk-owls
Ninox superciliaris. See White-browed hawk-owls
Ninoxini, 9:347
Niphates digitalis, 1:*80*
Niphonini. *See* Japanese aras
Nipples, 12:5, 12:89, 12:96–97, 12:106
  See also Physical characteristics
Nipponia nippon. See Japanese ibises
Nipponnemertes pulcher, 1:*256*, 1:*257*
Nippotaeniidae. *See* Nippotaeniidea
Nippotaeniidea, 1:226, 1:229, 1:231
Nit flies. *See* Horse bot flies
Nitlins. *See* Least bitterns
Nitrates, amphibians and, 6:56
Nitrogen
  diving mammals and, 12:68
  from fossil fuels and fertilizers, 12:213
Nitrous oxides, 1:52
Nkulengu rails, 9:45
No-see-ums, 3:64, 3:357, 3:360
Noack's African leaf-nosed bats. *See* Noack's roundleaf bats
Noack's roundleaf bats, 13:*406*, 13:*407*, 13:408
Noble, Gladwyn, 6:135
Noble pen shells, 2:*455*, 2:*458*, 2:460–461
Noble snipes, 9:177
Nocticola spp., 3:147
Noctilio spp. *See* Bulldog bats
Noctilio albiventris. See Lesser bulldog bats
Noctilio leporinus. See Greater bulldog bats
Noctilionidae. *See* Bulldog bats
Noctuids, 3:5, 3:65, 3:383, 3:386, 3:387–388, 3:389
Noctules, 13:310, 13:498, 13:500, 13:*501*, 13:*504*, 13:*510*, 13:512
  giant, 13:313
  greater, 13:516*t*
  reproduction, 12:110
Nocturnal behavior. *See* Behavior
Nocturnal curassows, 8:413
Nocturnal desert skinks. *See* Night skinks
Nocturnal flight, bats, 12:52–55, 12:59
Nocturnal mammals
  body size and, 12:115
  color vision, 12:79
  vision, 12:79–80
  See also specific nocturnal mammals
Noddies, 9:203
Noemacheilus mesonoemachilus sijuensis. See Siju blind cavefishes
Noemacheilus sijuensis. See Siju blind cavefishes

Noemacheilus smithi. See Blind loaches
Noisy friarbirds, 11:*240*, 11:*242*
Noisy miners, 11:238
Noisy night monkeys, 14:135, 14:138, 14:141*t*
Noisy scrub-birds, 10:337–338, 10:*339–340*
Nolf, D., 5:341
Nollman, Jim, 15:93
Nomadism, avian, 8:31
Nomascus spp. *See* Crested gibbons
Nomascus concolor. See Black crested gibbons
Nomascus gabriellae. See Golden-cheeked gibbons
Nomascus leucogenys. See White-cheeked gibbons
Nomeidae. *See* Driftfishes
Nomenclature. *See* Taxonomy
Nomeus gronovii. See Man-of-war fishes
Nomia melanderi. See Alkali bees
Nomorhamphus spp., 5:81
Non-associative learning, 1:40–41, 2:39
See also Behavior
Non-declarative memory, 12:154
Non-native species introduction. *See* Introduced species
Non-shivering thermogenesis (NST), 12:113
Non-spiny bandicoots, 13:9
Nonarticulate lampshells, **2:515–519,** 2:*517*
Nonlinearity, of ecosystems, 12:217
Nonnula ruficapilla. See Rufous-capped nunlets
Nonsporting dogs, 14:288, 14:289
Noodlefishes, 4:391
Nordic krill, 2:*187*, 2:*188*, 2:*189*, 2:191
Nordmann's greenshanks, 9:177, 9:179, 9:180
Norfolius, 3:*310*, 3:*313*, 3:314
Norfolius howensis. See Norfolius
Norfolk Island parakeets, 9:280
Norfolk Island starlings, 11:410
Normanichthyidae. *See Normanichthys crockery*
Normanichthys crockery, 5:179
Noronha vireos. *See* Red-eyed vireos
Noronha wrasses, 5:296
Noronhomys spp., 16:266
North Africa. *See* Palaearctic region
North African catfishes. *See* Sharptooth catfishes
North African crested porcupines. *See* North African porcupines
North African gundis, 16:312, 16:313
North African porcupines, 16:*353*, 16:354, 16:*355*, 16:*360*–361
North African sengis, 16:*519*, 16:*523*, 16:*527*, 16:529–530
North African striped weasels, 14:333*t*
North America. *See* Nearctic region
North American badgers, 14:321, 14:*322*, 14:333*t*
North American banana slugs, 2:413
North American basses, **5:195–206,** 5:*200*, 5:*201*
North American beavers, 16:*178*, 16:*181*, 16:*182*, 16:*183*–184
behavior, 16:*177*, 16:*180*
feeding ecology, 12:*126*, 16:*179*
habitats, 12:*138*
physical characteristics, 12:111
North American black bears. *See* American black bears
North American bullfrogs, 6:*254*, 6:261–*262*
behavior, 6:49, 6:313
introduction of, 6:54, 6:58, 6:251, 6:262
larvae, 6:40, 6:42, 6:262
reproduction, 6:251, 6:262
North American coral snakes, 7:*490*, 7:*491–492*
North American earthsnakes, 7:470
North American freshwater catfishes, 4:352
North American garter snakes, 7:466
North American kangaroo rats, 16:124
North American medicinal leeches, 2:*79*, 2:*81*–82
North American mountain goats, 12:131
North American mudsucker gobies, 4:6
North American paddlefishes. *See* American paddlefishes
North American plains bison. *See* American bison
North American pocket gophers, 16:126
North American porcupines, 16:*365*, 16:*370*, 16:*371–372*
behavior, 16:367–368
distribution, 16:123, 16:367
feeding ecology, 12:*123*
humans and, 16:127
physical characteristics, 12:38
reproduction, 16:*366*
North American pronghorns, 12:40
North American red squirrels, 16:*168*, 16:*173*
See also Red squirrels
North American sheep, 16:92
See also Sheep
North American tube anemones. *See* American tube dwelling anemones
North American watersnakes, 7:466, 7:468
North Atlantic gannets. *See* Northern gannets
North Atlantic right whales, 15:10, 15:107, 15:*108*, 15:*109*, 15:*111*, 15:*114*, 15:*115*, 15:116
North Chinese flying squirrels, 16:141*t*
See also Flying squirrels
North Indian muntjacs, 15:*345*
See also Muntjacs
North Island rifleman. *See* Riflemen
North Moluccan flying foxes, 13:331*t*
North Pacific Anadromous Fish Commission, 4:413
North Pacific anchovies. *See* Northern anchovies
North Pacific herrings. *See* Pacific herrings
North Pacific krill, 2:*188*, 2:*189*
North Pacific right whales, 15:107, 15:109, 15:111–112, 15:*114*, 15:116–*117*
See also Right whales
North River shads. *See* American shads
North Sea beaked whales. *See* Sowerby's beaked whales
Northern anchovies, 4:*4*, 4:*278*, 4:*280*, 4:*285*, 4:287
Northern anteater chats. *See* Anteater chats
Northern Australian snapping turtles. *See* Victoria river snappers
Northern Bahian blond titis, 14:11, 14:148
Northern bald ibises. *See* Hermit ibises
Northern basket stars, 1:*391*, 1:*393*, 1:394
Northern bats, 13:498
Northern beardless flycatchers. *See* Northern beardless-tyrannulets
Northern beardless-tyrannulets, 10:269, 10:270, 10:*276*, 10:*281*–282
Northern bettongs, 13:74, 13:76, 13:77, 13:*79*
Northern bluefin tunas. *See* Atlantic bluefin tunas
Northern bobwhite quails, 8:455, 8:*456*, 8:*457*, 8:*458*, 8:*459*, 8:*461*–462
Northern bobwhites. *See* Northern bobwhite quails
Northern bog lemmings, 16:*227*, 16:238*t*
Northern boobooks. *See* Southern boobook owls
Northern bottlenosed whales, 15:*60*, 15:*61*, 15:62, 15:*63*, 15:*64*–65
Northern brown bandicoots, 13:2, 13:4, 13:*6*, 13:*10*, 13:*11*, 13:*14*–15
Northern brown-shouldered lizards. *See* Common side-blotched lizards
Northern brownbuls, 10:397
Northern brushtail possums, 13:*66t*
Northern casemaker caddisflies. *See* October caddisflies
Northern cassowaries. *See* One-wattled cassowaries
Northern cat-eyed snakes, 7:*471*, 7:*479*, 7:481
Northern cave-lions, 12:24
Northern cavefishes, 5:6, 5:7, 5:*8*–9
Northern chamois, 16:6, 16:87, 16:*96*, 16:*100*
Northern clingfishes, 5:*358*, 5:*360*–361
Northern collared lemmings, 16:*226*, 16:*237*, 16:237*t*
Northern common cuscuses. *See* Common cuscuses
Northern coral snakes. *See* North American coral snakes
Northern crested lizards. *See* Desert iguanas
Northern cricket frogs, 6:*228*, 6:229, 6:*232*, 6:*234*
Northern dusky salamanders. *See* Dusky salamanders
Northern dwarf sirens, 6:327, 6:*330*, 6:*331*
Northern eagle-owls. *See* Eurasian eagle-owls
Northern earless lizards. *See* Common lesser earless lizards
Northern elephant seals, 14:*425*, 14:431
behavior, 14:*421*, 14:*423*
distribution, 14:*429*
feeding ecology, 14:419
humans and, 14:424
lactation, 12:127
physical characteristics, 12:67
reproduction, 14:421–422
Northern fairy flycatchers. *See* African blue-flycatchers
Northern false iguanas. *See* Cape spinytail iguanas
Northern fantails, 11:83, 11:*88*, 11:*93*
Northern flickers, 10:88, 10:149, 10:150, 10:*152*, 10:155–*156*
Northern flying squirrels, 16:135
See also Flying squirrels
Northern four-toothed whales. *See* Baird's beaked whales
Northern fulmars, 8:108, 8:*124*, 8:*125*, 8:*127*, 8:129–130
Northern fur seals, 14:*401*, 14:*402*
conservation status, 12:*219*
distribution, 14:394, 14:395
evolution, 14:393
reproduction, 12:110, 14:396, 14:398, 14:399
Northern gannets, 8:*214*, 8:*216*, 8:217–218
Northern ghost bats, 13:*359*, 13:*360*, 13:362

Northern giant petrels, 8:109
Northern gliders, 13:133*t*
Northern goshawks, 8:323, 8:*326*, 8:*338*
Northern gracile mouse opossums, 12:264*t*
Northern grasshopper mice, 16:*267*
Northern greater bushbabies, 14:25, 14:*25*, 14:*27*, 14:*29*, 14:31–32
Northern ground-hornbills, 10:72
Northern ground utas. *See* Common side-blotched lizards
Northern hairy-nosed wombats, 13:39, 13:40, 13:51, 13:52, 13:53, 13:*54*, 13:*55*, 13:56
Northern halibuts. *See* Pacific halibuts
Northern harriers. *See* Hen harriers
Northern hawk-owls, 9:335, 9:346, 9:349–350, 9:*353*, 9:*361–362*
Northern helmeted curassows, 8:413, 8:*417*, 8:421–422
Northern horned screamers, 8:393
Northern house-martins. *See* House martins
Northern house wrens. *See* House wrens
Northern jacanas, 9:108, 9:*111*, 9:*112*
Northern krill. *See* Nordic krill
Northern lampshells, 2:*522*
Northern lapwings, 9:*165*, 9:*172*–173
Northern largemouth basses. *See* Largemouth basses
Northern leaf-nosed bats, 13:*403*, 13:*404*
Northern leopard frogs, 6:*65*, 6:67
Northern lesser bushbabies. *See* Senegal bushbabies
Northern logrunners. *See* Chowchillas
Northern long-nosed armadillos. *See* Llanos long-nosed armadillos
Northern longears. *See* Longear sunfishes
Northern lyrebirds. *See* Albert's lyrebirds
Northern marsupial moles, 13:26, 13:*28*
Northern mastiff bats, 13:493*t*
Northern minke whales, 12:*200*, 15:120, 15:*122*, 15:125, 15:*126*, 15:128–129
Northern mockingbirds, 10:465, 10:*466*, 10:*468*, 10:*471*
Northern mummichogs, 5:*95*, 5:*97*, 5:99
Northern muriquis, 14:11, 14:156, 14:159, 14:*161*, 14:*164*, 14:165
Northern naked-tailed armadillos, 13:183, 13:190*t*
Northern needle-clawed bushbabies, 14:26, 14:32*t*
Northern needletails. *See* White-throated needletails
Northern orioles. *See* Baltimore orioles
Northern oystercatchers. *See* Variable oystercatchers
Northern Pacific albatrosses, 8:113, 8:114
Northern Pacific sea stars, 1:368, 1:370, 1:*371*, 1:*374*
Northern paper wasps. *See* Golden paper wasps
Northern parulas, 11:*287*
Northern pikas, 16:*493*, 16:*496*, 16:*498*–499
Northern pikeminnows, 4:7, 4:*298*
Northern pikes, 4:14, 4:379, 4:*382*, 4:383–*384*, 5:198
Northern pintails, 8:364
Northern plains gray langurs, 14:10, 14:*173*, 14:175, 14:*176*, 14:*180*
Northern planigales. *See* Long-tailed planigales
Northern plantanna. *See* Müller's plantanna
Northern pocket gophers, 16:*187*, 16:*189*, 16:195*t*
Northern potoos, 9:395
Northern pudus, 15:379, 15:396*t*
Northern puffbacks, 10:*431*, 10:432–*433*
Northern pygmy owls, 9:349, 9:*350*
Northern quahogs, 2:*456*, 2:*461*, 2:466
Northern quetzals. *See* Resplendent quetzals
Northern quolls, 12:281, 12:282, 12:299*t*
Northern raccoons, 12:*189*, 14:*309*, 14:310, 14:*311*, 14:*312*, 14:*313*
Northern ravens, 11:505–507, 11:*511*, 11:522–*523*
Northern red-legged pademelons. *See* Red-legged pademelons
Northern red snappers, 5:*264*, 5:268, 5:*269*, 5:270
Northern right whale dolphins, 15:42
Northern river otters, 12:109, 14:*322*, 14:*323*
Northern rock barnacles, 2:5, 2:29, 2:*275*
Northern rock basses. *See* Rock basses
Northern rock-crawlers, 3:189, 3:*191*, 3:*192*–193
Northern ronquils, 5:311, 5:312
Northern royal albatrosses, 8:113, 8:114, 8:115, 8:*117*
Northern sagebrush lizards. *See* Common sagebrush lizards
Northern saw-whet owls, 9:*353*, 9:*363*–364
Northern screamers, 8:*395*, 8:*396*–397
Northern sea elephants. *See* Northern elephant seals
Northern sea lions. *See* Steller sea lions
Northern sea robins, 5:*164*
Northern sea stars, 1:*6*
Northern seadevils. *See* Krøyers deep sea anglerfishes
Northern short-tailed bats. *See* Lesser New Zealand short-tailed bats
Northern short-tailed shrews, 13:*195*, 13:*250*, 13:*253*, 13:*256*–257
Northern shovelers, 8:*375*, 8:385–386
Northern shovelnose lobsters. *See* Flathead locust lobsters
Northern shrikes, 10:425, 10:427, 10:428, 10:429
Northern side-blotched lizards. *See* Common side-blotched lizards
Northern smooth-tailed tree shrews, 13:297*t*
Northern snakeheads, 5:438, 5:*441*, 5:*442*
Northern snakenecks, 7:78
Northern spadefoot toads, 6:*142*, 6:145
Northern sportive lemurs, 14:*78*, 14:*83*
Northern stargazers, 5:*336*, 5:*337*, 5:339–340
Northern starlings. *See* European starlings
Northern tamanduas, 13:171, 13:*176*, 13:*178*
See also Anteaters
Northern three-toed jerboas. *See* Hairy-footed jerboas
Northern three-toed woodpeckers. *See* Three-toed woodpeckers
Northern treecreepers. *See* Eurasian treecreepers
Northern tropical plantanna. *See* Müller's plantanna
Northern water shrews. *See* Eurasian water shrews
Northern water voles, 12:70–71, 12:182–183, 16:229, 16:*231*, 16:*233*–234
Northern waterthrush, 11:*287*, 11:288
Northern wheatears. *See* Wheatears
Northern wrens. *See* Winter wrens
Northern wrynecks, 10:*149*, 10:*153*, 10:*154*
Northern zebra-tailed lizards. *See* Zebra-tailed lizards
Northwest snapping turtles. *See* Victoria river snappers
Northwestern crows, 11:506
Northwestern salamanders, 6:*356*, 6:*357*, 6:358–359
Norway lemmings, 16:*227*, 16:*231*, 16:*233*, 16:235
Norway rats. *See* Brown rats
Norwegian herrings. *See* Atlantic herrings
Norwegian tubeworms, 2:*86*, 2:*87*, 2:*88*
Noses, 12:8, 12:80, 12:*81*
See also Physical characteristics
Nosopsyllus spp., 3:349
Nostrils, 12:8
See also Physical characteristics
Notacanthidae. *See* Marine spiny eels
Notacanthus spp., 4:249, 4:250
Notaden spp., 6:140, 6:141
Notaden bennettii, 6:140, 6:141
Notaden melanoscaphus. *See* Northern spadefoot toads
Notaden weigeli, 6:141
Notechis scutatus. *See* Tiger snakes
Notemigonus spp., 4:301
Notesthes robusta. *See* Bullrouts
Notharchus macrorhynchos. *See* White-necked puffbirds
Notharctus spp., 14:1
Nothobranchius spp., 5:94
Nothobranchius guentheri. *See* Mombasa killifishes
Nothocercus bonapartei. *See* Highland tinamous
Nothochrysa californica, 3:308
Nothocrax urumutum. *See* Nocturnal curassows
Notholca spp., 1:260
Notholca ikaitophila. *See* Ikaite rotifers
Nothophryne spp., 6:245
Nothoprocta cinerascens. *See* Brushland tinamous
Nothoprocta kalinowskii. *See* Kalinowski's tinamous
Nothoprocta ornata. *See* Ornate tinamous
Nothoprocta taczanowskii. *See* Taczanowski's tinamous
Nothosaurs, 7:14–15
Nothrotheriops spp., 13:151
Nothura maculosa. *See* Spotted nothuras
Notidanoides muensteri, 4:143
Notiomys edwardsii, 16:266
Notiomystis spp. *See* Stitchbirds
Notiosorex spp., 13:248, 13:250
Notiosorex crawfordi. *See* Desert shrews
Notiosoricini, 13:247
Notobathynella williamsi, 2:177
Notobatrachus spp., 6:12, 6:62, 6:64
Notocheiridae. *See* Surf silversides
Notocheirus hubbsi, 5:68
Notocirrhitus splendens. *See* Splendid hawfishes
Notodontids, 3:387
Notogoneus spp., 4:290
Notogoneus osculus, 4:*293*
Notolychnus valdiviae, 4:443
Notomys alexis. *See* Australian jumping mice
Notomys fuscus. *See* Dusky hopping mice
Notonecta sellata. *See* Backswimmers
Notonemouridae, 3:141, 3:142

Notoneuralia, 2:4
Notonycteris spp., 13:413
Notophthalmus spp., 6:49, 6:363, 6:367
Notophthalmus meridionalis. See Black-spotted salamanders
Notophthalmus perstriatus. See Striped newts
Notophthalmus viridescens. See Eastern newts
Notoplana acticola, 1:*189*, 1:*190*
Notopteridae. *See* Knifefishes
Notopteris macdonaldi. See Long-tailed fruit bats
Notopteroidea, 4:231
Notopterus notopterus, 4:232, 4:233, 4:234
Notornis mantelli. See Takahe rails
Notoryctemorphia. *See* Marsupial moles
Notoryctes spp., 13:25, 13:26
Notoryctes caurinus. See Northern marsupial moles
Notoryctes typhlops. See Marsupial moles
Notoryctidae. *See* Marsupial moles
Notorynchus spp., 4:144
Notorynchus cepedianus. See Broadnose sevengill sharks
Notostigmophora. *See* Scutigeromorpha
Notostraca. *See* Tadpole shrimps
Notosudidae. *See* Waryfishes
Notothenia angustata. See Maori chiefs
Notothenia [Parenotothenia] angustata. See Maori chiefs
Notothenia rossii. See Marbled rockcods
Nototheniidae. *See* Notothens
Notothenioidei. *See* Southern cod-icefishes
Notothens, 4:49, 5:321, 5:322
Nototriton spp., 6:392
Notoungulata, 12:11, 12:132, 15:131, 15:133, 15:134, 15:135, 15:136, 15:137
Notropis spp., 4:301
Nourishment. *See* Feeding ecology
Novae's bald uakaris, 14:148, 14:151
Novodinia antillensis. See Velcro sea stars
Novumbra spp. *See* Olympic mudminnows
Novumbra hubbsi. See Olympic mudminnows
Nowicki, S., 8:39
NST (non-shivering thermogenesis), 12:113
Nubian bustards, 9:93
Nubian ibex, 15:146, 16:87, 16:91, 16:92, 16:94, 16:95
Nubian vultures. *See* Lappet-faced vultures
Nubian wild asses, 12:176–177
  See also African wild asses
Nubra pikas, 16:502*t*
Nucifraga spp. *See* Corvidae
Nucifraga caryocatactes. See Spotted nutcrackers
Nucifraga columbiana. See Clark's nutcrackers
Nuciruptor spp., 14:143
Nuclear DNA (nDNA), 12:26–27, 12:29–30, 12:31, 12:33
Nucleosides, 1:44
Nucleotide bases, 12:26–27
Nucras tessellata. See Western sandveld lizards
Nuculidae, 2:451
Nuculoida, 2:453
Nuda, 1:169
Nudibranchs, 2:*38*, 2:*403*, 2:404, 2:*405*
Nukupu'us, 11:343
Numbats, 12:137, 12:277–284, **12:303–306,** 12:*304*, 12:*305*, 12:*306*
Numbers, understanding, 12:155–156, 12:162
  See also Cognition
Numeniini. *See* Curlews; Godwits
Numenius spp. *See* Curlews
Numenius americanus. See Long-billed curlews
Numenius borealis. See Eskimo curlews
Numenius phaeopus. See Whimbrels
Numenius tahitiensis. See Bristle-thighed curlews
Numenius tenuirostris. See Slender-billed curlews
Numida meleagris. See Helmeted guineafowl
Numididae. *See* Guineafowl
Nunbirds, 10:101–103
Nuptial pads, 6:31
Nurse sharks. *See* Orectolobiformes
Nussbaum, R. A., 6:415, 6:419, 6:425, 6:431, 6:432
Nut weevils, 3:*2*
Nutcrackers. *See* Spotted nutcrackers
Nuthatch vangas, 10:439, 10:440
Nuthatches, **11:167–175**, 11:*170*
  behavior, 11:168, 11:171–175
  conservation status, 11:169, 11:171–175
  distribution, 11:*167*, 11:168, 11:*171–174*
  evolution, 11:167
  feeding ecology, 11:168, 11:171–175
  habitats, 11:168, 11:171–174
  humans and, 11:169, 11:171–175
  physical characteristics, 11:167–168, 11:171–174
  reproduction, 11:168–169, 11:171–175
  species of, 11:*171–175*
  taxonomy, 11:167, 11:171–174
Nutmeg finches. *See* Spotted munias
Nutmeg mannikins. *See* Spotted munias
Nutmeg pigeons. *See* White imperial pigeons
Nutrias. *See* Coypus
Nutritional adaptations, **12:120–128**
Nutritional ecology. *See* Feeding ecology
Nutting's flycatchers, 10:*276*, 10:*283*–284
Nyalas, 16:6, 16:11, 16:25*t*
Nyctalus spp. *See* Noctules
Nyctalus lasiopterus. See Greater noctules
Nyctalus leisleri. See Leisler's bats
Nyctalus noctula. See Noctules
Nyctanassa violacea. See Yellow-crowned night herons
Nyctanolis spp., 6:392
Nyctea scandiaca. See Snowy owls
Nyctereutes procyonoides. See Raccoon dogs
Nycteribiidae, 3:358
Nycteridae. *See* Slit-faced bats
Nycteris spp. *See* Slit-faced bats
Nycteris arge. See Bate's slit-faced bats
Nycteris avrita, 13:373
Nycteris gambiensis. See Gambian slit-faced bats
Nycteris grandis. See Large slit-faced bats
Nycteris hispida. See Hairy slit-faced bats
Nycteris intermedia. See Intermediate slit-faced bats
Nycteris javanica. See Javan slit-faced bats
Nycteris macrotis. See Large-eared slit-faced bats
Nycteris madagascarensis. See Madagascar slit-faced bats
Nycteris major. See Ja slit-faced bats
Nycteris thebaica. See Egyptian slit-faced bats
Nycteris woodi. See Wood's slit-faced bats
Nyctibatrachinae, 6:247, 6:249
Nyctibatrachus spp., 6:68, 6:247, 6:249
Nyctibatrachus major. See Malabar night frogs
Nyctibiidae. *See* Potoos
Nyctibius aethereus. See Long-tailed potoos
Nyctibius bracteatus. See Rufous potoos
Nyctibius grandis. See Great potoos
Nyctibius griseus. See Gray potoos
Nyctibius jamaicensis. See Northern potoos
Nyctibius leucopterus. See White-winged potoos
Nyctiborinae, 3:149
Nycticebus bengalensis. See Bengal slow lorises
Nycticebus coucang. See Sunda slow lorises
Nycticebus pygmaeus. See Pygmy slow lorises
Nycticeius humeralis. See Evening bats
Nycticorax magnificus. See White-eared night herons
Nycticorax nycticorax. See Black-crowned night herons
Nycticryphes semicollaris. See South American painted-snipes
Nyctimene spp., 13:334
Nyctimene rabori. See Philippine tube-nosed fruit bats
Nyctimene robinsoni. See Queensland tube-nosed bats
Nyctimystes spp., 6:225, 6:230
Nyctimystes dayi, 6:230
Nyctinomops femorosaccus. See Pocketed free-tailed bats
Nyctiperdix spp., 9:231
Nyctiphanes spp., 2:185
Nyctiphrynus rosenbergi. See Chocó poorwills
Nyctiphrynus vielliardi. See Plain-tailed nighthawks
Nyctixalus spp., 6:291, 6:292
Nyctixalus pictus. See Painted Indonesian treefrogs
Nyctixalus spinosus, 6:293
Nyctyornis athertoni. See Blue-bearded bee-eaters
Nygolaimus parvus, 1:*286*, 1:288
Nyingchi lazy toads, 6:110
Nymphalids, 3:386
Nymphalis antiopa. See Mourning cloaks
Nymphes spp., 3:307
Nymphicus hollandicus. See Cockatiels
Nymphidae, 3:55, 3:305, 3:306, 3:307, 3:308
Nymphitidae, 3:305
Nystactes noanamae. See Sooty-caped puffbirds
Nystalus chacuru. See White-eared puffbirds

O

Oahu alauahios, 11:343
Oahu deceptor bush crickets, 3:207
Oahu tree snails. *See* Agate snails
Oak brush crickets, 3:*201*
Oak katydids, 3:206
Oak toads, 6:42
Oarfishes, 4:*447*–450, 4:*448*, 4:*449*, 4:*451*, 4:*452*, 4:454
Oasis hummingbirds, 9:443
Oasisia spp., 2:91
Oaxaca hummingbirds, 9:450
Obbia larks, 10:346
Obdurodon spp., 12:228
Obelia spp., 1:125
Obelia dichotoma, 1:*131*, 1:143–144
Obelia geniculata, 1:24
Obligatory migration, 8:30
Obligatory symbiosis, 1:32, 2:31
  See also Behavior

Obligatory vivipary, 12:6
See also Reproduction
Obscure berrypeckers, 11:191
Observational learning, 12:158
See also Cognition
Ocadia sinensis. *See* Chinese stripe-necked turtles
Occidozyga spp., 6:247, 6:249
Occidozyga lima. *See* Pointed-tongue floating frogs
Occidozyginae, 6:247, 6:249
Ocean currents, 1:25
Ocean pouts, 5:*309*, 5:310, 5:312, 5:313, 5:*314*, 5:*317*–318
Ocean sunfishes. *See* Molas
Oceanic bugs, 3:262
Oceanic leeches. *See* Marine leeches
Oceanic squids, 2:475, 2:477, 2:478
Oceanic whitetip sharks, 4:116, 4:*117*, 4:*120*, 4:121
Oceanites gracilis. *See* White-vented storm-petrels
Oceanites oceanicus. *See* Wilson's storm-petrels
Oceanites zalascarthmus, 8:135
Oceanitinae. *See* Storm-petrels
Oceanodroma castro. *See* Band-rumped storm-petrels; Wedge-rumped storm-petrels
Oceanodroma furcata. *See* Fork-tailed storm-petrels
Oceanodroma homochroa. *See* Ashy storm-petrels
Oceanodroma hornbyi. *See* Hornby's storm-petrels
Oceanodroma hubbsi, 8:135
Oceanodroma leucorhoa. *See* Leach's storm-petrels
Oceanodroma macrodactyla. *See* Guadalupe storm-petrels
Oceanodroma markhami. *See* Markham's storm-petrels
Oceanodroma matsudairae. *See* Matsudaira's storm-petrels
Oceanodroma melania. *See* Black storm-petrels
Oceanodroma monorhis. *See* Swinhoe's storm-petrels
Oceanodroma tristrami. *See* Tristram's storm-petrels
Ocellated piculets, 10:*150*
Ocellated snake eels. *See* Tiger snake eels
Ocellated snakeheads, 5:*441*, 5:*442*, 5:445
Ocellated tapaculos, 10:*259*, 10:260, 10:*261*, 10:*267*
Ocellated turkeys, 8:434
Ocellated waspfishes, 5:*167*, 5:*169*
Ocelli, 3:28
Ocelots, 14:374, 14:*378*, 14:382–*383*
Ocher stars, 1:369, 1:*372*, 1:*375*
Ochetorhynchus harterti. *See* Bolivian earthcreepers
Ochotona spp. *See* Pikas
Ochotona alpina. *See* Alpine pikas
Ochotona argentata. *See* Silver pikas
Ochotona cansus. *See* Gansu pikas
Ochotona collaris. *See* Collared pikas
Ochotona curzoniae. *See* Plateau pikas
Ochotona dauurica. *See* Daurian pikas
Ochotona erythrotis. *See* Chinese red pikas
Ochotona gaoligongensis. *See* Gaoligong pikas
Ochotona gloveri. *See* Glover's pikas
Ochotona himalayana. *See* Himalayan pikas
Ochotona hoffmanni, 16:495
Ochotona huangensis, 16:495
Ochotona hyperborea. *See* Northern pikas
Ochotona iliensis. *See* Ili pikas
Ochotona koslowi. *See* Koslov's pikas
Ochotona ladacensis. *See* Ladakh pikas
Ochotona macrotis. *See* Large-eared pikas
Ochotona nubrica. *See* Nubra pikas
Ochotona pallasi. *See* Pallas's pikas
Ochotona pallasi hamica, 16:500
Ochotona pallasi sunidica, 16:500
Ochotona princeps. *See* American pikas
Ochotona pusilla. *See* Steppe pikas
Ochotona rufescens. *See* Afghan pikas
Ochotona rutila. *See* Turkestan red pikas
Ochotona thibetana. *See* Moupin pikas
Ochotonidae. *See* Pikas
Ochraceous piculets, 10:150
Ochre-breasted pipits, 10:375
Ochre-flanked tapaculos, 10:*261*, 10:*266*
Ochre mole-rats, 16:350*t*
Ochrotomys nuttalli. *See* Golden mice
Ochrotomys torridus. *See* Southern grasshopper mice
Ocnerodrilidae, 2:65
Ocreatus spp. *See* Hummingbirds
Octacnemidae, 1:454
Octacnemus kottae, 1:*457*, 1:*462*, 1:464
October caddisflies, 3:*378*, 3:*379*, 3:380–381
Octochaetidae, 2:65
Octocorallia, 1:103, 1:104–105, 1:106, 1:108
Octocorallian blue corals, 1:107
Octodon spp., 16:433
Octodon bridgesi. *See* Bridges's degus
Octodon degus. *See* Degus
Octodon lunatus. *See* Moon-toothed degus
Octodon pacificus, 16:433, 16:436
Octodontidae. *See* Octodonts; Pikas
Octodontomys spp. *See* Chozchozes
Octodontomys gliroides. *See Chozchozes*
Octodonts, 12:72, **16:433–441,** 16:*437*, 16:440*t*
Octomys spp., 16:433, 16:436
Octomys mimax. *See* Viscacha rats
Octopodiformes, 2:475
Octopods, 1:42, 2:25, **2:475–485,** 2:*481*
behavior, 2:37, 2:477–478
conservation status, 2:479
distribution, 2:477
evolution, 2:475
feeding ecology, 2:29, 2:478
habitats, 2:477
humans and, 2:479–480
physical characteristics, 2:475–477
reproduction, 2:23, 2:478–479
species of, 2:483–*485*
taxonomy, 2:475
Octopus vulgaris. *See* Common octopods
Octopuses. *See* Octopods
Oculotrema hippopotami, 1:214
Ocyceros birostris. *See* Indian gray hornbills
Ocyceros gingalensis. *See* Sri Lankan gray hornbills
Ocyceros griseus. *See* Malabar gray hornbills
Ocypus olens. *See* Devil's coach-horses
Ocyropsis spp., 1:171
Odacidae. *See* Rock whitings
Odatria spp., 7:360
Odax pullus. *See* Butterfishes
Odd-nosed leaf-monkeys, 14:172
Odd-toed ungulates. *See* Perissodactyla
Odobenidae. *See* Walruses
Odobenus rosmarus. *See* Walruses
Odobenus rosmarus rosmarus. *See* Atlantic walruses
Odocoileinae, 15:380–381, 15:382
Odocoileus spp., 12:102, 15:269, 15:380, 15:382, 15:383
Odocoileus bezoarticus. *See* Pampas deer
Odocoileus hemionus. *See* Black-tailed deer
Odocoileus virginianus. *See* White-tailed deer
Odocoileus virginianus clavium. *See* Key deer
Odocoilid deer. *See Odocoileus* spp.
Odonata, 3:23, **3:133–139,** 3:*136*
Odontaspididae, 4:131
Odontaspis noronhai, 4:133
Odontaspis taurus. *See* Sand sharks
Odontaster validus. *See* Cushion stars
Odontesthes spp., 5:71
Odontobutidae, 5:373, 5:374
Odontoceti. *See* Toothed whales
Odontodactylus scyllarus. *See* Peacock mantis shrimps
Odontophoridae. *See* New World quails
Odontophorus columbianus. *See* Venezuelan wood-quails
Odontophorus strophium. *See* Gorgeted wood-quails
Odontophryinae, 6:156, 6:158
Odontophrynus spp., 6:156, 6:157, 6:159
Odontophrynus occidentalis. *See* Cururu lesser escuerzo
Odontorchilus spp. *See* Tooth-billed wrens
Odontotermes spp., 3:165
Odorrana spp., 6:247, 6:248, 6:249
Oecanthus spp. *See* Tree crickets
Oecomys spp., 16:266
Oedemognathus spp., 4:369
Oedemognathus exodon, 4:371
Oedicnemus capensis. *See* Spotted dikkops
Oedipina spp., 6:325, 6:392
Oedipina gracilis, 6:401
Oedipina pacificensis, 6:401
Oedipina uniformis. *See* Costa Rican worm salamanders
Oedipodinae. *See* Band-winged grasshoppers
Oegophiuridea. *See* Ophiocanopidae
Oegopsida. *See* Oceanic squids
Oena capensis. *See* Namaqua doves
Oenanthe oenanthe. *See* Wheatears
Oenanthe pileata. *See* Capped wheatears
Oenonidae, 2:46
Oerstedia dorsalis, 1:*256*–*257*
Oestridae. *See* Bot flies
Oestrus ovis, 3:64
Ogcocephaloidei. *See* Batfishes (*Ogcocephalus* spp.)
Ogcocephalus spp. *See* Batfishes (*Ogcocephalus* spp.)
Ogcocephalus corniger. *See* Longnose batfishes
Ogilbia spp., 5:17, 5:20
Ogilbia cayorum. *See* Key brotulas
Ogilbia galapagosensis, 5:18
Ogilbia pearsei, 5:18
Ogilby's duikers, 16:*78*, 16:*81*–82
Ogilby's woobegongs. *See* Tasseled wobbegongs
Ogmodon vitianus, 7:484, 7:486
Ognorhynchus icterotis. *See* Yellow-eared parrots
Ogrefishes. *See* Common fangtooths

Ogygoptynx wetmorei, 9:331, 9:345
Oikopleura spp., 1:473, 1:*474*
Oikopleura dioica, 1:473, 1:474, 1:*475*, 1:476–477
Oikopleura labradoriensis, 1:*473*, 1:*475*, 1:477
Oil flies, 3:357
  See also Petroleum flies
Oil palms, 3:79
Oil spills, 8:16, 8:161, 8:368
Oilbirds, 9:367–370, **9:373–376,** 9:*374*, 9:*375*, 12:53
Oilfishes. *See* Eulachons
Oioceros spp., 15:265, 16:1
Oithona plumifera, 2:*302*, 2:307–308
Okapia spp., 15:265
Okapia johnstoni. *See* Okapis
Okapis, 15:267, 15:269, **15:399–409,** 15:*404*, 15:*405*, 15:*406*, 15:*407*, 15:*408*
Okinawa woodpeckers, 10:88, 10:150, 10:*152*, 10:*165*–166
Okiyama, M., 5:2
Olalla, Alfonso, 16:457
Olalla, Carlos, 16:457
Olalla, Manuel, 16:457
Olalla, Ramón, 16:457
Olalla, Rosalino, 16:457
Olallamys albicauda. *See* Montane bamboo rats
Olallamys edax, 16:450–451
Olbiogasteridae, 3:55
Old Dutch Capuchine pigeons, 8:*24*
Old English sheepdogs, 14:292
Old-field mice, 16:125
Old World bats, 12:13, 12:44
Old World cuckoos, 9:311
Old World deer, 12:19, **15:357–372,** 15:*362*, 15:*363*, 15:371*t*
Old World flycatchers, **11:25–43,** 11:*28–29*
  behavior, 11:26
  conservation status, 11:27
  distribution, 11:*25*, 11:26
  evolution, 11:25
  feeding ecology, 11:26–27
  habitats, 11:26
  humans and, 11:27
  physical characteristics, 11:25–26
  reproduction, 11:27
  species of, 11:*30–43*
  taxonomy, 11:25
Old World fruit bats, 12:13, 13:*310*, 13:311, **13:319–350,** 13:*323–324*, 13:*338*–339
  behavior, 13:312, 13:319–320, 13:335
  conservation status, 13:321–322, 13:337
  distribution, 13:*319*, 13:*333*, 13:334
  evolution, 13:319, 13:333
  feeding ecology, 13:320, 13:335–336
  habitats, 13:319, 13:334
  humans and, 13:322, 13:337
  physical characteristics, 12:54, 12:*54*, 12:58, 13:319, 13:333–334
  reproduction, 13:320–321, 13:336–337
  species of, 13:*325*–330, 13:331*t*–332*t*, 13:*340*–347, 13:348*t*–349*t*
  taxonomy, 13:319, 13:333
Old World harvest mice. *See* Harvest mice
Old World leaf-nosed bats, 13:310, **13:401–411,** 13:*406*
  behavior, 13:402, 13:405
  conservation status, 13:405
  distribution, 13:*401*, 13:402
  echolocation, 12:86
  evolution, 13:401
  feeding ecology, 13:405
  habitats, 13:402
  humans and, 13:405
  physical characteristics, 13:401–402
  reproduction, 13:405
  species of, 13:*407*–409, 13:409*t*–410*t*
  taxonomy, 13:401
Old World mice, **16:249–262,** 16:*254*, 16:261*t*–262*t*
Old World monkeys, **14:171–186,** 14:*176*, 14:*177*, **14:187–206,** 14:*195*, 14:*196*
  behavior, 14:6–7, 14:174, 14:191–192
  conservation status, 14:175, 14:194
  distribution, 12:135, 14:5, 14:*171*, 14:173, 14:*187*, 14:191
  evolution, 14:2, 14:171–172, 14:187–189
  feeding ecology, 14:9, 14:174, 14:192–193
  habitats, 14:6, 14:174, 14:191
  humans and, 14:175, 14:194
  physical characteristics, 14:172–173, 14:189–191
  reproduction, 14:10, 14:174–175, 14:193
  species of, 14:*178*–183, 14:184*t*–185*t*, 14:*197*–204, 14:204*t*–205*t*
  taxonomy, 14:1, 14:4, 14:171–172, 14:187–188
Old World newts, 6:45
Old World orioles, **11:427–430, 11:432–434**
Old World palm swifts. *See* African palm swifts
Old World partridges, 8:433–435
Old World pigs. *See* Pigs
Old World porcupines, 16:121, **16:351–364,** 16:*355*, 16:*356*
Old World rats, **16:249–262,** 16:*254*, 16:261*t*–262*t*
Old World sucker-footed bats, 12:136, 13:311, **13:479–481,** 13:*480*
Old World vultures, 8:318
Old World warblers, 8:31, 10:172, **11:1–23,** 11:*8–9*
  behavior, 11:4–5
  conservation status, 11:6–7
  distribution, 11:*1*, 11:3
  evolution, 11:1–2
  feeding ecology, 11:5
  habitats, 11:3–4
  humans and, 11:7
  nest construction, 11:*2*, 11:6
  physical characteristics, 11:1, 11:2–3
  reproduction, 11:5–6
  species of, 11:*10–23*
  taxonomy, 11:1–2, 11:286–287
Old World water shrews, 12:14
Olds, N., 15:177
Oldsquaws, 8:364, 8:365, 8:366, 8:370, 8:*375*, 8:*389*
Oldwives, 5:235–238, 5:240, 5:242–244
Olfaction, 3:27, 3:60
  See also Physical characteristics
Olfactory maps, 8:35
Olfactory marking. *See* Scent-marking
Olfactory senses, 7:31, 7:34–36
  See also Physical characteristics
Olfactory system, amphibian, 6:22–23, 6:24
  See also Smell
Oliarces spp., 3:308
Oliarces clara, 3:307, 3:308
Oligobrachia ivanovi. *See* Black oligobrachias
Oligochaeta. *See* Earthworms
Oligocottus spp., 5:180
Oligocottus maculosus. *See* Tidepool sculpins
Oligodon spp. *See* Asian kukrisnakes
Oligometra serripinna, 1:*360*, 1:362
Oligoneuriidae, 3:54, 3:126
Oligoryzomys spp., 16:128, 16:265–266, 16:270
Oligoryzomys longicaudatus, 16:269
Oligoryzomys nigripes, 16:270
Oligotoma saundersii. *See* Saunders embiids
Oligotrema spp., 1:479
Oligotrema psammites, 1:480
Oligotrema sandersi, 1:*481*, 1:*482*, 1:483
Olindias phosphorica, 1:*133*, 1:142
Olingos, 14:309, 14:310, 14:*311*, 14:316*t*
Olivaceous cormorants, 8:*204*, 8:207–208
Olivaceous piculets, 10:*153*, 10:154–*155*
Olivaceous puyas, 9:442
Olivaceous thornbills, 9:442
Olive baboons, 14:*6*, 14:*190*
Olive-backed jungle-flycatchers. *See* Fulvous-chested jungle-flycatchers
Olive-backed orioles, 11:*428*, 11:*430*, 11:*432*
Olive-backed pipits, 10:372, 10:373, 10:374
Olive-backed pocket mice, 16:208*t*
Olive-backed sunbirds (*Cinnyris* spp.), 11:208, 11:*212*, 11:*224*
Olive-backed sunbirds (*Cyanomitra* spp.). *See* Green-headed sunbirds
Olive-backs, 11:353
Olive-bellied sunbirds, 11:*212*, 11:*222*
Olive bush-shrikes, 10:428
Olive colobus, 14:173, 14:175, 14:*176*, 14:*178*, 14:179
Olive ridley seaturtles, 7:71
Olive seasnakes, 7:*489*, 7:*494*
Olive-sided flycatchers, 10:275, 10:*276*, 10:*282*
Olive thrushes, 10:485, 10:487, 10:*491*, 10:501–*502*
Olive warblers, 11:286
Olive whistlers, 11:117
Olms, 6:35, 6:323, 6:325, **6:377–383,** 6:*380*, 6:*381*
Olomaos, 10:489
Olson, Storrs, 9:45, 10:385
Olympic marmots, 16:*148*
Olympic mudminnows, 4:379, 4:380, 4:*382*, 4:*383*, 4:385
Olympic torrent salamanders, 6:31, 6:*386*
Omanosaura spp., 7:297
Omaos, 10:489
Ombrana spp., 6:251
Omei Shan liocichlas, 10:*510*, 10:516–*517*
Omiltimi rabbits, 16:481
Ommastrephids, 2:476
Ommatidia, 3:27
Ommatoiulus spp., 2:365
Ommatolampinae, 3:206
Ommatolampis spp., 3:204
Ommatophoca rossii. *See* Ross seals
Ommexechidae, 3:203
Omnivores, 3:46
  digestive system, 12:124
  periglacial environments and, 12:25
  See also Feeding ecology; specific omnivores
Omobranchini, 5:342
Omomyiformes, 14:1, 14:3

Omosudidae. *See* Hammerjaws
On the Origin of Species, 2:7
Onagers. *See* Asiatic wild asses
Onchidiidae, 2:411, 2:413
Onchidium verruculatum, 2:*416*, 2:421
Onchnesoma spp., 2:98
Onchobothriidae, 1:226
Onchocerca volvulus. *See* African river blindness nematodes
Oncifelis colocolo. *See* Pampas cats
Oncifelis (Felis) geoffroyi. *See* Geoffroy's cats
Oncifelis geoffroyi. *See* Geoffroy's cats
Oncorhynchus aguabonita. *See* Golden trouts
Oncorhynchus gorbuscha. *See* Pink salmons
Oncorhynchus kisutch. *See* Coho salmons
Oncorhynchus masou. *See* Cherry salmons
Oncorhynchus masou ishikawae. *See* Yamames
Oncorhynchus mykiss. *See* Rainbow trouts
Oncorhynchus nerka. *See* Sockeye salmons
Oncorhynchus tshawytscha. *See* Chinook salmons
Oncothrips waterhousei. *See* Australian acacia gall thrips
Ondatra zibethicus. *See* Muskrats
One-humped camels. *See* Dromedary camels
One-male groups, 14:7
See also Behavior
One-stripe spiny eels, 5:152
One-toed amphiumas, 6:405, 6:407, 6:*408*, 6:409–410
One-wattled cassowaries, 8:*78*, 8:*80–81*
Oniscidae, 2:250
Oniscidea, 2:249
Oniscigastridae, 3:55
Oniscus asellus. *See* Common shiny woodlice
Onnoco agoutis, 16:415*t*
Ontogeny, 2:15, 12:106, 12:141–143, 12:148
Ontogeny recapitulates phylogeny. *See* Biogenetic law
Onuxodon spp., 5:17
Onychodactylus spp., 6:24, 6:335, 6:336
Onychodactylus japonicus. *See* Japanese clawed salamanders
Onychodiaptomus sanguineus, 2:*302*, 2:305
Onychogalea spp. *See* Nail-tailed wallabies
Onychogalea fraenata. *See* Bridled nail-tailed wallabies
Onychognathus morio. *See* Red-winged starlings
Onychognathus tristramii. *See* Tristram's red-winged starlings
Onychomys spp., 16:125
Onychomys leucogaster. *See* Northern grasshopper mice
Onychophorans, 2:14, 2:*109*, **2:109–114**
behavior, 2:37, 2:110
conservation status, 2:111
distribution, 2:110
evolution, 2:9, 2:109
feeding ecology, 2:110
habitats, 2:110
humans and, 2:111
physical characteristics, 2:109–110
reproduction, 2:17, 2:23, 2:110
species of, 2:*113*–114
taxonomy, 2:109
Onychorhynchus coronatus. *See* Royal flycatchers
Onychorhynchus coronatus swainsoni. *See* Atlantic royal flycatchers
Onychoteuthis spp., 1:42, 2:37
Oocytes, 2:16–17, 2:23
See also Reproduction
Oogenesis, 1:17–18, 1:21–23, 2:17, 4:32
See also Reproduction
O'opo alamo'os, 5:375, 5:376, 5:*378*, 5:384–*385*
Oopsacas minuta, 1:69
Opahs, **4:447–450,** 4:*450*, 4:*451*, **4:*452*–453**
Opalina spp., 1:180
Opeatogenys spp., 5:355
Operant conditioning, 1:41
See also Behavior
Opheodrys aestivus. *See* Rough greensnakes
Ophiactidae, 1:387
Ophiactis savignyi. *See* Tropical brittle stars
Ophichitonidae, 1:387
Ophichthidae, 4:48, 4:255
Ophidiidae. *See* Cusk-eels
Ophidiiformes, **5:15–23**, 5:*19*
Ophidiinae. *See* Cusk-eels
Ophidioidei, 5:16
Ophidion spp., 5:22
Ophidion holbrooki. *See* Band cusk-eels
Ophidion scrippsae, 5:*16*
Ophioblennius steindachneri. *See* Large-banded blennies
Ophiocanopidae, 1:387
Ophiocoma paucigranulata. *See* Spiny brittle stars
Ophiocoma wendtii, 1:388
Ophiocomidae, 1:387
Ophiodermatidae, 1:387
Ophiodes spp., 7:339
Ophiodon elongatus. *See* Lingcods
Ophiognathus ampullaceus. *See* Gulper eels
Ophiolepididae, 1:387
Ophiomorus spp., 7:328
Ophiomyxidae, 1:387
Ophionereididae, 1:387
Ophiophagus hannah. *See* King cobras
Ophiophragmus filegranus, 1:388
Ophiopsila californica, 1:389
Ophiopsila riisei, 1:389
Ophiothricidae, 1:387
Ophiothrix spp., 1:*389*
Ophiothrix fragilis. *See* Common brittle stars
Ophisaurus spp. *See* Glass lizards
Ophisaurus apodus, 7:339
Ophisaurus attenuatus, 7:340
Ophisaurus attenuatus attenuatus. *See* Western slender glass lizards
Ophisaurus koellikeri. *See* Moroccan glass lizards
Ophisops spp., 7:297
Ophisternon bengalense, 5:151
Ophisternon candidum. *See* Blind cave eels
Ophisternon infernale. *See* Blind swamp cave eels
Ophiura albida, 1:388
Ophiura ophiura, 1:*392*, 1:*397*, 1:398
Ophiurida. *See* Brittle stars
Ophiuridea, 1:387
Ophiuroidea, **1:387–399,** 1:*391*
behavior, 1:388–389
conservation status, 1:48, 1:390
distribution, 1:387
evolution, 1:387
feeding ecology, 1:389
habitats, 1:388
humans and, 1:390
physical characteristics, 1:387, 1:*388*
reproduction, 1:389–390
species of, 1:*393*–398
taxonomy, 1:387
Ophryacus spp., 7:445, 7:446
Ophryophryne spp., 6:110, 6:111
Ophryophryne microstoma. *See* Asian mountain toads
Ophrysia superciliosa. *See* Himalayan quails
Opiliones, 2:333, 2:*335*
Opisthiini, 3:54
Opisthobranch mollusks, 1:45
Opisthobranchia. *See* Sea slugs
Opisthocentrus ocellatus, 5:310
Opisthocomiformes. *See* Hoatzins
Opisthocomus hoazin. *See* Hoatzins
Opisthoplatia orientalis, 3:150
Opisthoproctidae. *See* Barreleyes
Opisthoproctus grimaldii. *See* Mirrorbellies
Opisthoproctus soleatus, 4:392
Opisthopthalmus spp. *See* Smooth-headed scorpions
Opisthorchis spp., 1:201
Opisthorchis felineus, 1:200
Opisthothylax spp., 6:281, 6:283
Opisthothylax immaculatus. *See* Gray-eyed frogs
Opistognathidae. *See* Jawfishes (Opistognathidae)
Opistoprora spp. *See* Hummingbirds
Oplegnathidae. *See* Knifejaws
Oplurinae, 7:243, 7:244
Opomyzidae, 3:54
Oporornis formosus. *See* Kentucky warblers
Opossum shrimps. *See* Mysids
Opossums, 12:*96*, 12:*143*, 12:*191*
See also New World opossums
Opportunistic symbiosis, 2:31
See also Behavior
Opsanus spp., 5:41, 5:42
Opsanus tau. *See* Oyster toadfishes
Opus 96, 8:19
Orachrysops ariadne. *See* Karkloof blue butterflies
Orange-barred garden eels. *See* Splendid garden eels
Orange-bellied leafbirds, 10:*418*, 10:*422*
Orange-bellied parrots, 9:277
Orange-breasted blue flycatchers, 11:*28*, 11:36–*37*
Orange-breasted bush-shrikes, 10:427, 10:429
Orange-breasted sunbirds, 11:207, 11:210, 11:*211*, 11:*218*
Orange-breasted trogons, 9:*480*, 9:*482*
Orange chats, 11:65, 11:66
Orange-chinned parakeets, 9:277–278
Orange-crowned fairy-wrens. *See* Orange-crowned wrens
Orange-crowned wrens, 11:45, 11:*47*, 11:*52*–53
Orange-footed megapodes, 8:407
Orange-footed scrubfowl, 8:399
Orange leaf-nosed bats. *See* Golden horseshoe bats
Orange-lined triggerfishes. *See* Orange-striped triggerfishes
Orange mantis shrimps, 2:*167*
Orange mouse opossums, 12:254, 12:265*t*
Orange-necked hill-partridges, 8:437
Orange nectar bats, 13:433*t*

Orange roughies, 5:106, 5:114, 5:116, 5:*117*, 5:*119*, 5:121
Orange-rumped agoutis. *See* Red-rumped agoutis
Orange-rumped honeyguides. *See* Yellow-rumped honeyguides
Orange sea lilies, 1:*360*, 1:*363*–364
Orange-spined hairy dwarf porcupines, 16:366, 16:374*t*
Orange-spotted snakeheads, 5:438, 5:*440*, 5:442–*443*
Orange-striped triggerfishes, 5:*473*, 5:*474*–475
Orangespine unicornfishes, 5:*394*, 5:*396*, 5:397
Orangutans, 14:7, 14:*227*, 14:*232*, 14:*234*
behavior, 12:*156*, 12:157–158, 14:6, 14:228–229, 14:232
Bornean, 14:225, 14:*236*, 14:*237*
conservation status, 14:235
distribution, 12:136, 14:5, 14:227
evolution, 12:33, 14:225
feeding ecology, 14:233
habitats, 14:227–228
language, 12:161
learning, 12:151
memory, 12:153
numbers and, 12:155
physical characteristics, 14:226–227, 14:*233*
reproduction, 14:234, 14:235
social cognition, 12:160
taxonomy, 14:225
tool use, 12:157
See also Sumatran orangutans
Oras. *See* Komodo dragons
Orb snails, 2:412
Orbicular velvetfishes, 5:163, 5:166
Orcaella brevirostris. *See* Irrawaddy dolphins
Orcas. *See* Killer whales
Orchard dormice. *See* Garden dormice
Orchard orioles, 11:301, 11:304
Orchestoidea californiana. *See* Beach hoppers
Orchid mantids, 3:181, 3:*182*, 3:*183*–184
Orchopeas howardi. *See* Gray squirrel fleas
Orcininae, 15:41
Orcinus orca. *See* Killer whales
Ordinal relationships, 12:155
Ord's kangaroo rats, 16:*200*, 16:*202*, 16:209*t*
Oreailurus jacobita. *See* Andean cats
Oreamnos spp., 16:87, 16:88
Oreamnos americanus. *See* Mountain goats
Oreaster reticulatus. *See* Cushion stars
Orectolobidae. *See* Wobbegons
Orectolobiformes, 4:11, **4:105–112,** 4:*107*
Orectolobus maculatus. *See* Carpet sharks
Oregon ensatina. *See* *Ensatina* spp.
Oregon sturgeons. *See* White sturgeons
Oreinus prenanti. *See* *Schizothorax prenanti*
Orenji ogons, 4:*299*
Oreocharis spp. *See* Berrypeckers
Oreocharis arfaki. *See* Arfak berrypeckers
Oreochromis spp., 5:277, 5:*279*
Oreochromis esculentus. *See* Ngege
Oreochromis leucostictus, 5:280–281
Oreochromis niloticus, 5:280–281
Oreochromis variabilis, 5:280
Oreodontae, 15:136, 15:264
Oreoica gutturalis. *See* Crested bellbirds
Oreolalax spp., 6:110
Oreolalax schmidti, 6:*112*
Oreomystis bairdi. *See* Akikikis
Oreonax spp. *See* Yellow-tailed woolly monkeys
Oreonax flavicauda. *See* Yellow-tailed woolly monkeys
Oreonectes evezardi. *See* Hillstream loaches
Oreophasis derbianus. *See* Horned guans
Oreopholus spp., 9:161
Oreopholus ruficollis. *See* Tawny-throated dotterels
Oreophryne spp., 6:303
Oreophryne nana. *See* Camigiun narrow-mouthed frogs
Oreophrynella spp., 6:184, 6:193
Oreophrynella huberi, 6:193
Oreophrynella quelchii. *See* Roraima bush toads
Oreopsar bolivianus. *See* Bolivian blackbirds
Oreosomatidae, 5:123
Oreotragus oreotragus. *See* Klipspringers
Oreotrochilus spp. *See* Hillstars
Oreotrochilus chimborazo. *See* Ecuadorian hillstars
Oreotrochilus estella. *See* Andean hillstars
Orestias spp., 5:90, 5:91, 5:94
Orestias ascotanensis. *See* Ascotan Mountain killifishes
Organ-pipe corals, 1:107
Organbirds. *See* Kokakos
Organelles, 2:8
Organochlorine pesticides. *See* Pesticides
Oribis, 16:*60*, 16:62, 16:*65*, 16:*66*
Oriental bay owls, 9:331–332, 9:335–336, 9:*339*, 9:*343*
Oriental bell toads. *See* Oriental fire-bellied toads
Oriental civets, 14:345*t*
Oriental cockroaches, 3:57, 3:76, 3:*82*, 3:*148*, 3:151, 3:*152*, 3:*156*, 3:157
Oriental fire-bellied toads, 6:*84*, 6:*85*, 6:86–*87*
Oriental grass lizards, 7:297
Oriental helmet gurnards, 5:*157*, 5:159, 5:*160*, 5:*161*
Oriental herrings. *See* Pacific herrings
Oriental liver flukes, 1:35, 1:*202*, 1:207–*208*
Oriental magpie-robins. *See* Magpie-robins
Oriental pied hornbills, 10:75
Oriental rat fleas, 3:76, 3:*351*, 3:*352*–353
Oriental region, 12:134, 12:135, 12:136
Oriental sandgrouse. *See* Black-bellied sandgrouse
Oriental silkworms. *See* Silkworms
Oriental six-lined runners. *See* Six-lined grass lizards
Oriental skylarks, 10:345
Oriental white storks, 8:236, 8:267
Orientation behavior, 1:38–40, 8:32–35
See also Behavior
The Origin and Evolution of Birds (Feduccia), 10:51
Origins of life, 2:8–11
Origma solitaria. *See* Rockwarblers
Orinoco crocodiles, 7:164
Oriole blackbirds, 11:*308*, 11:*312*
Orioles, **11:301–306,** 11:*302*, 11:*308*, 11:310–311
See also Old World orioles
Oriolia bernieri. *See* Bernier's vangas
Oriolidae, **11:427–435,** 11:*430*
Oriolus spp. *See* Orioles
Oriolus ater. *See* Brown-headed cowbirds
Oriolus cruentus. *See* Crimson-bellied orioles
Oriolus forsteni. *See* Seram orioles
Oriolus hosii. *See* Black orioles
Oriolus icterocephalus. *See* Yellow-hooded blackbirds
Oriolus isabellae. *See* Isabela orioles
Oriolus larvatus. *See* Eastern black-headed orioles
Oriolus mellianus. *See* Silver orioles
Oriolus mexicanus. *See* Oriole blackbirds
Oriolus oriolus. *See* Eurasian golden orioles
Oriolus phaeochromus. *See* Dusky brown orioles
Oriolus phoeniceus. *See* Red-winged blackbirds
Oriolus sagittatus. *See* Olive-backed orioles
Oriolus xanthonotus. *See* Dark-throated orioles
Oriolus xanthornus. *See* Black-hooded orioles
Orizaba squirrels. *See* Deppe's squirrels
Ormalas. *See* Rufous hare-wallabies
Ornate box turtles, 7:72
Ornate ghost pipefishes, 5:*131*, 5:*137*, 5:*144*, 5:145
Ornate narrow-mouthed frogs, 6:*307*, 6:314
Ornate shrews, 13:*251*
Ornate tinamous, 8:58
Ornate titis, 14:148
Ornate tree shrews. *See* Painted tree shrews
Ornismya anna. *See* Anna's hummingbirds
Ornismya ensifera. *See* Sword-billed hummingbirds
Ornithischia, 7:5
Ornithodira, 7:19
Ornithophily, 9:445
Ornithoptera spp., 3:389
Ornithoptera alexandrae. *See* Queen Alexandra's birdwings
Ornithorhynchidae. *See* Duck-billed platypuses
Ornithorhynchus anatinus. *See* Duck-billed platypuses
Oropendolas, 11:301, 11:302
Orrorin tugenensis, 14:241–242
Ortalis ruficauda. *See* Rufous-vented chachalacas
Ortalis vetula. *See* Plain chachalacas
Orthogeomys spp., 16:185
Orthogeomys cavator. *See* Chiriquá pocket gophers
Orthogeomys cherriei. *See* Cherrie's pocket gophers
Orthogeomys cuniculus, 16:190
Orthogeomys dariensis. *See* Darién pocket gophers
Orthogeomys grandis. *See* Large pocket gophers
Orthogeomys heterodus. *See* Variable pocket gophers
Orthogeomys hispidus. *See* Hispid pocket gophers
Orthogeomys lanius. *See* Big pocket gophers
Orthogeomys matagalpae. *See* Nicaraguan pocket gophers
Orthogeomys underwoodi. *See* Underwood's pocket gophers
Orthokinesis, 1:38
See also Behavior
Orthonectida, 1:19, 1:21, **1:99–102,** 1:*100*
Orthonychidae. *See* Chowchillas; Logrunners
Orthonyx novaeguineae. *See* New Guinea logrunners
Orthonyx spaldingii. *See* Darker chowchillas
Orthonyx temminckii. *See* Southern logrunners

Orthoporus ornatus, 2:364, 2:365
Orthoptera, 2:36, 3:31, 3:33–34, 3:59, **3:201–216,** 3:*209*, 3:*210*
behavior, 3:204–206
conservation status, 3:207
distribution, 3:203
evolution, 3:201
feeding ecology, 3:206–207
fossil, 3:10, 3:13
habitats, 3:203–204
humans and, 3:207–208
physical characteristics, 3:19, 3:202–203
reproduction, 3:44, 3:207
species of, 3:*211*–216
taxonomy, 3:201
wing-beat frequencies, 3:23–24
wings of, 3:22
Orthoramphus spp. *See* Thick-knees
Orthorhyncus spp. *See* Hummingbirds
Orthorrhapha, 3:357, 3:358, 3:359
Orthotomus, 11:3, 11:6, 11:7
Orthotomus sutorius. See Common tailorbirds
Orthurethra, 2:411
Ortygospiza spp. *See* Quailfinches
Ortygospiza atricollis. See African quailfinches
Ortyxelos meiffrenii. See Lark buttonquails
Orycteropodidae. *See* Aardvarks
Orycteropus afer. See Aardvarks
Oryctes rhinoceros. See Rhinoceros scarabs
Oryctolagus spp. *See* European rabbits
Oryctolagus cuniculus. See European rabbits
Oryctolagus domesticus. See Domestic rabbits
Oryctoperus afer. See Aardvarks
Oryx, 12:204, 15:*137*, 16:5, **16:27–43,** 16:41*t*–42*t*
Oryx dammah. See Scimitar-horned oryx
Oryx gazella. See Gemsboks; Oryx
Oryx leucoryx. See Arabian oryx
Oryzaephilus surinamensis. See Sawtoothed grain beetles
Oryzias spp., 5:80, 5:81
Oryzias latipes. See Japanese rice fishes
Oryzias orthognathus, 5:81
Oryzomyines, 16:264
Oryzomys spp. *See* Rice rats
Oryzomys bauri, 16:270
Oryzomys couesi, 16:268, 16:270
Oryzomys galapagoensis, 16:270
Oryzomys gorgasi, 16:270
Oryzomys intermedius, 16:267–268, 16:270
Oryzomys nelsoni, 16:270
Oryzomys palustris. See Marsh rice rats
Oryzomys russatus, 16:270
Oryzomys xantheolus, 16:269
Oryzorictes spp. *See* Rice tenrecs
Oryzorictes hova. See Rice tenrecs
Oryzorictes talpoides. See Mole-like rice tenrecs
Oryzorictinae. *See* Shrew tenrecs
Orzeliscus spp., 2:118
os penis. *See* Baculum
Osbornictis spp., 14:335
Osbornictis piscivora. See Aquatic genets
Oscaecilia spp., 6:436
Oscines. *See* Songbirds
Osmeridae. *See* Smelts
Osmeriformes, **4:389–403,** 4:*394*
behavior, 4:392
conservation, 4:393
distribution, 4:390–391
evolution, 4:389–390
feeding ecology, 4:392–393
Habitats, 4:391–392
humans and, 4:393
physical characteristics, 4:390
reproduction, 4:393
species of, 4:395–401
taxonomy, 4:389–390
Osmerini, 4:391
Osmeroidea, 4:390–391
Osmeroidei, 4:389, 4:390, 4:393
Osmerus eperlanus. See European smelts
Osmia cornifrons. See Japanese horn-faced bees
Osmoregulation, 4:24–25, 4:50
Osmylidae, 3:305, 3:306, 3:307, 3:308
Osmylitidae, 3:305
Osmylus fulvicephalus, 3:307
Osornophryne spp., 6:64, 6:184
Osphronemidae, 4:24, 4:57, 5:427, 5:428
Osphroneminae. *See Osphronemus* spp.
Osphronemus spp., 5:427, 5:428
Osphronemus exodon, 5:428
Osphronemus goramy. See Giant gouramies
Ospreys, 8:313, 8:317, 8:320, 8:*328*, 8:*329*
Ostariophysi, 4:12, 4:22
Osteichthyans, 6:7
Osteichthyes, 4:10, 4:15
Osteocephalus spp., 6:226, 6:227
Osteocephalus oophagus, 6:37, 6:230
Osteocephalus taurinus. See Manaus long-legged treefrogs
Osteoglossidae, 4:11, 4:231, 4:232
Osteoglossiformes, **4:231–241,** 4:*235–236*
Osteoglossomorpha, 4:11
Osteoglossum spp., 4:4, 4:233
Osteoglossum bicirrhosum. See Arawanas
Osteoglossum ferreira, 4:232
Osteolaemus tetraspis. See Dwarf crocodiles
Osteolepimorpha, 4:12
Osteolepis macrolepidotys, 4:*11*
Osteopilus brunneus, 6:37, 6:230
Osteopilus septentrionalis. See Cuban treefrogs
Osteostraci, 4:9
Ostraciidae. *See* Boxfishes
Ostracion cubicus. See Yellow boxfishes
Ostracoda. *See* Mussel shrimps
Ostracoderms, 4:9
Ostreidae, 2:452
Ostriches, 8:*9*, 8:21, 8:22, 8:53–56, **8:99–102,** 8:*100*, 8:*101*, 9:440
Oswego basses. *See* Black crappies
Otaria byronia. See South American sea lions
Otariidae, **14:393–408,** 14:*401*
behavior, 14:396–397
conservation status, 14:399–400
distribution, 14:*393*, 14:394–395
evolution, 14:393
feeding ecology, 14:397–398
habitats, 14:395–396
humans and, 14:400
physical characteristics, 14:393–394, 14:419
reproduction, 14:398–399
species of, 14:*402*–406, 14:407*t*–408*t*
taxonomy, 14:256, 14:393
Otariinae. *See* Sea lions
Otididae. *See* Bustards
Otidides, 9:91
Otidiphabinae, 9:247
Otidiphaps spp. *See* Columbiformes
Otidiphaps nobilis. See Pheasant pigeons
Otiorhynchus sulcatus. See Vine weevils
Otis spp. *See* Bustards
Otis afraides. See White-quilled bustards
Otis bengalensis. See Bengal floricans
Otis caerulescens. See Blue bustards
Otis tarda. See Great bustards
Otis tetrax. See Little bustards
Otobothriidae, 1:226
Otocolobus manul. See Pallas's cats
Otocryptis spp., 7:210
Otocyon megalotis. See Bat-eared foxes
Otolemur spp., 14:23
Otolemur crassicaudatus. See Brown greater bushbabies
Otolemur garnettii. See Northern greater bushbabies
Otolemur monteiri. See Silvery greater bushbabies
Otoliths, Stromateoidei and, 5:421
Otomops martiensseni. See Giant mastiff bats
Otomops papuensis. See Big-eared mastiff bats
Otomops wroughtoni. See Wroughton's free-tailed bats
Otomyinae, 16:283
Otomys angoniensis. See Angoni vlei rats
Otomys sloggetti. See Ice rats
Otophidion spp., 5:22
Otophryne spp., 6:41
Otophryne pyburni. See Pyburn's pancake frogs
Otophryninae, 6:302
Otothrix hodgsoni. See Hodgson's frogmouths
Ototylomys phyllotis. See Big-eared climbing rats
Otter civets, 14:335, 14:337, 14:338, 14:*339*, 14:*342*, 14:343
Otter shrews, 13:194, 13:199, 13:225
Otters, 12:14, **14:319–325,** 14:*326*, **14:330, 14:332–334**
behavior, 14:321–323
conservation status, 14:324
distribution, 14:*319*, 14:321
evolution, 14:319
feeding ecology, 14:260, 14:323
field studies of, 12:201
habitats, 14:255, 14:257, 14:321
humans and, 14:324–325
physical characteristics, 14:319–321
reproduction, 14:324
species of, 14:*330*, 14:332*t*
taxonomy, 14:256, 14:319
Ottoia spp., 12:64
Ottoia prolifica, 1:13, 2:*97*
Otus spp. *See* Scops-owls
Otus asio. See Eastern screech owls
Otus insularis. See Seychelles scops-owls
Otus nudipes. See Puerto Rican screech owls
Otus rutilus. See Madagascar scops-owls
Otus scops. See Eurasian scops-owls
O'u. *See* Lanai hookbills
Ounce. *See* Snow leopards
Ourebia ourebi. See Oribis
Ous. *See* Lanai hookbills
Ovarian cycles, 12:91–92
Ovenbirds, 10:170–171, 10:173–174, **10:209–228,** 10:*212–213*, 11:288, 11:290, 11:*292*, 11:*296–297*, 11:304
behavior, 10:210
conservation status, 10:210–211
distribution, 10:*209*–210
evolution, 10:209

feeding ecology, 10:210
habitats, 10:210
humans and, 10:211
physical characteristics, 10:209
reproduction, 10:210
species of, 10:*214–228*
taxonomy, 10:209
Overfishing, 1:50–51, 4:75
See also Conservation status
Overpopulation, 1:47–48
See also Conservation status; Humans
Ovibos moschatus. See Muskoxen
Ovibovini, 16:87, 16:89
Oviparous reproduction, 4:30–31, 4:65, 4:115–116, 6:28, 6:38, 6:39, 7:6–7, 12:106
See also Reproduction; specific species
Oviposition, 3:39, 3:61–62
Ovis spp., 16:3, 16:4, 16:87, 16:88, 16:90, 16:91, 16:95
Ovis ammon. See Argalis
Ovis ammon hodgsoni. See Tibetan argalis
Ovis ammon musimon. See European mouflons
Ovis aries. See Domestic sheep
Ovis canadensis. See Bighorn sheep
Ovis dalli. See Dall's sheep
Ovis musimon. See Mouflons
Ovis nivicola. See Snow sheep
Ovis orientalis. See Asiatic mouflons; Urials
Ovis vignei. See Urials
Ovophis spp., 7:445
Ovoviviparity, 3:39, 4:31–32, 4:65
See also Reproduction
Ovulation, 12:92, 12:109
Ovum. *See* Eggs, mammalian
Owen, R., 4:201, 13:387
Owenetta spp., 7:13
Oweniidae, 2:47
Owl butterflies, 3:*65*
Owl-faced finches. *See* Double-barred finches
Owl monkeys. *See* Night monkeys
Owl moreporks, 9:163
Owlet moths, 3:13, 3:384
Owlet-nightjars, 9:331, 9:367–369, **9:387–393,** 9:*390*
Owlflies, 3:307, 3:*308*
Owls, 8:20, **9:331–334,** 9:*348*
See also Barn owls; Eagle-owls; Typical owls
Owls of the World (Hume), 9:331
Owston's palm civets, 14:338, 14:344*t*
Owston's woodpeckers. *See* White-backed woodpeckers
Oxen, 15:*149*
Oxeye tarpons, 4:243
Oxleyan pygmy perches, 5:222
Oxpeckers, 11:407–411, 11:*409*, 11:*412*, 11:425
Oxudercinae, 5:373
Oxybelis spp., 7:466
Oxycirrhites typus. See Longnose hawkfishes
Oxydactyla spp., 6:303
Oxyeleotris caeca, 5:375
Oxyeleotris marmorata. See Marble sleepers
Oxyethira spp., 3:376
Oxygen
levels, 4:37, 4:38–39
origins of life and, 2:8–9
subterranean mammals and, 12:72–73, 12:114
Oxyhaloinae, 3:149
Oxylebias pictus. See Painted greenlings
Oxymonacanthus longirostris. See Longnose filefishes
Oxymycterus josei, 16:263
Oxymycterus nasutus, 16:267, 16:268
Oxynotidae. *See* Rough sharks
Oxynotus spp., 4:151, 4:152
Oxyopidae, 2:37
Oxypogon spp. *See* Helmetcrests
Oxyporhamphus spp., 5:79–80
Oxyruncidae. *See* Sharpbills
Oxyruncus cristatus. See Sharpbills
Oxytocin, 12:141
Oxyura jamaicensis. See Ruddy ducks
Oxyuranus spp. *See* Taipans
Oxyuranus microlepidotus. See Taipans
Oxyuranus scutellatus. See Taipans
Oxyurinae. *See* Stiff-tailed ducks
Oxyzygonectes spp., 5:89
Oyans. *See* African linsangs
Oyster crushers. *See* Port Jackson sharks
Oyster leeches, 1:*189*, 1:*191*
Oyster toadfishes, 5:*41*, 5:*42*, 5:*43*, 5:*44*
Oystercatchers, 9:104, **9:125–132,** 9:*129*, 9:133
Oysters, 2:43, 2:454
black-lipped pearl, 2:*455*, 2:*461*–462
eastern American, 2:*455*, 2:*457*, 2:459
pearl, 2:454
See also Bivalves
Ozark cavefishes, 5:6, 5:7, 5:*8*
Ozobranchus spp., 2:76
Ozotocerus spp., 15:382
Ozotocerus bezoarticus. See Pampas deer

P

Paa spp., 6:246, 6:249, 6:252
Paa liebigii. See Spiny-armed frogs
Pacaranas, 16:123, **16:385–388,** 16:*386*, 16:*387*
Pacas, 16:124, **16:417–424,** 16:*418*, 16:*419*, 16:*420*, 16:*422*, 16:*423*
Pachybatrachus spp., 6:12
Pachybatrachus taqueti, 6:99
Pachycara spp., 5:310
Pachycare flavogrisea. See Goldenfaces
Pachycephala spp. *See* Whistlers
Pachycephala albispecularis. See Gray-headed robins
Pachycephala grisola. See Mangrove whistlers
Pachycephala jacquinoti. See Tongan whistlers
Pachycephala lanioides. See White-breasted whistlers
Pachycephala nudigula. See Bare-throated whistlers
Pachycephala olivacea. See Olive whistlers
Pachycephala pectoralis. See Golden whistlers
Pachycephala rufinucha. See Rufous-naped whistlers
Pachycephala rufiventris. See Rufous whistlers
Pachycephala rufogularis. See Red-lored whistlers
Pachycephala schlegelii. See Regent whistlers
Pachycephalidae. *See* Whistlers
Pachycoccyx sonneratii. See Thick-billed cuckoos
Pachycrocuta spp., 14:359
Pachydactylus bibronii. See Bibron's geckos
Pachyderms, 12:91
See also Elephants; Rhinoceroses
Pachygazella, 16:88
Pachylebias spp., 5:90
Pachymerium ferrugineum. See Earth centipedes
Pachyornis elephantopus. See Moas
Pachyornis mappini. See Moas
Pachyplichas jagmi, 10:203
Pachyplichas yaldwyni, 10:203
Pachyporlax spp., 15:265
Pachyptila vittata. See Broad-billed prions
Pachyramphus spp. *See* Tyrant flycatchers
Pachyramphus aglaiae. See Rose-throated becards
Pachyrhachis spp., 7:16
Pachyuromys dupras. See Fat-tailed gerbils
Pacific American sardines. *See* South American pilchards
Pacific angelsharks, 4:*161*, 4:162, 4:*163*, 4:*164*
Pacific barreleyes. *See* Barreleyes
Pacific divers. *See* Pacific loons
Pacific endodontoid snails, 2:414
Pacific flying foxes. *See* Tongan flying foxes
Pacific foureyed fishes, 5:*90*, 5:93
Pacific giant glass frogs, 6:215, 6:216, 6:217, 6:*218*, 6:*219*
Pacific giant salamanders, **6:349–353**
distribution, 6:5, 6:*349*, 6:350, 6:*352*
evolution, 6:13, 6:349
reproduction, 6:34, 6:350, 6:352
taxonomy, 6:3, 6:323
Pacific golden plovers, 8:29, 9:102, 9:162
Pacific green turtles, 7:90
Pacific gulls, 9:203
Pacific hagfishes, 4:77, 4:78, 4:*81*
Pacific halibuts, 5:451, 5:*454*, 5:*457*, 5:458–459
Pacific herrings, 4:*280*, 4:*281*, 4:284
Pacific leopard flounders, 5:*449*
Pacific loons, 8:159, 8:*162*, 8:*164*
Pacific marsh harriers, 8:320
Pacific mole crabs. *See* Pacific sand crabs
Pacific mollymawks, 8:114
Pacific monarchs. *See* Black-naped monarchs
Pacific octopods, giant, 2:476
Pacific rats. *See* Polynesian rats
Pacific salmons. *See* Chinook salmons
Pacific sand crabs, 2:*202*, 2:*206*–207
Pacific sanddabs, 5:*454*, 5:*457*–458
Pacific sea swallows. *See* House swallows
Pacific sheath-tailed bats, 13:358, 13:364*t*
Pacific sleeper sharks, 4:151, 4:152
Pacific spookfishes, 4:*93*, 4:*94*, 4:95
Pacific spotted scorpionfishes, 5:*164*
Pacific sturgeons. *See* White sturgeons
Pacific swallows. *See* House swallows
Pacific swifts, 9:422
Pacific walruses, 14:*409*–412, 14:*411*, 14:*414*, 14:415
Pacific white-rumped swiftlets. *See* White-rumped swiftlets
Pacific white-sided dolphins, 12:*63*, 15:*10*, 15:49
Pacific wolfeels. *See* Wolf-eels
Packman frogs. *See* Suriname horned frogs
Packrats. *See* Woodrats
Pacupeba. *See* Silver dollars
Pacus. *See* Pirapitingas
Padda fuscata. See Timor sparrows
Padda oryzivora. See Java sparrows
Padded sculpins, 5:181

Paddlefishes, 4:*4*, 4:11, 4:18, **4:213–216,** 4:*215*, 4:*216*, **4:220**
Paddy. *See* Black-faced sheathbills
Paddy birds. *See* Java sparrows
Pademelons, 13:34, 13:35, 13:83, 13:85, 13:86, 13:92, 13:97–*98*
Padogobius nigricans. See Arno gobies
Paedogenesis, 3:47
Paedomorphosis, 6:356, 6:363, 9:245
Paenungulata, 15:131, 15:133
Pagodroma nivea. See Snow petrels
Pagophila eburnea. See Ivory gulls
Pagophilus groenlandicus. See Harp seals
Pagothenia borchgrevinki, 4:50
Paguma larvata. See Masked palm civets
Pagurapseudes spp., 2:236
Pagurus bernhardus. See Common hermit crabs
Paini, 6:246
Painkillers, 1:45
Painted batagurs. *See* Painted terrapins
Painted bats, 13:*523*, 13:*525*
Painted buntings, 11:*264*
Painted buttonquails, 9:*15*, 9:*19*–20
Painted conures. *See* Painted parrots
Painted finches. *See* Gouldian finches
Painted francolins, 8:434
Painted frogfishes, 5:*51*
Painted frogs (Discoglossidae), **6:89–94,** 6:*91*, 6:*93*
Painted frogs (Limnodynastidae), 6:*141*, 6:144–145
Painted greenlings, 5:*181*
Painted Indonesian treefrogs, 6:*294*, 6:*298*
Painted lizardfishes, 4:*435*, 4:*437*, 4:439
Painted parrots, 9:*281*, 9:*291*
Painted quaggas. *See* Plains zebras
Painted quail-thrushes, 11:76
Painted quails. *See* King quails
Painted redstarts, 11:289
Painted reed frogs, 6:*280*, 6:*281*, 6:*284*, 6:*287*
Painted salamanders. *See Ensatina* spp.
Painted sandgrouse, 9:231
Painted shrimps. *See* Harlequin shrimps
Painted snipes, 9:104, **9:115–119,** 9:*116*, 9:*118*
Painted storks, 8:*269*, 8:*270–271*
Painted terrapins, 7:*117*, 7:*118*
Painted tree shrews, 13:298*t*
Painted tunicates, 1:*453*
Painted turtles, 7:*65*, 7:72, 7:106, 7:*108*, 7:109, 7:*110*
Painted wolves. *See* African wild dogs
Painted wood storks. *See* Painted storks
Pairs, 12:145–146
See also Behavior
Pajama cardinalfishes, 5:*264*, 5:*266–267*
Pajama catsharks, 4:116, 4:*118*, 4:*122*, 4:126
Pakicetidae, 15:2
Pakicetus spp., 15:2–3, 15:119
Pakistani toads, 6:63
Palaeacanthocephala, 1:311
Palaeanodonts, 15:135
Palaearctic region, 12:131–132, 12:134–135, 12:136
Palaeocopida, 2:311
Palaeocucumaria hunsrueckiana, 1:417
Palaeohirudo eichstaettensis, 2:75
Palaeomastodon, 15:161
Palaeomerycids, 15:137, 15:265, 15:343
Palaeonemertea, 1:245, 1:246
Palaeopalaemon newberryi, 2:198
Palaeopropithecus ingens. See Greater sloth lemurs
Palaeopsylla baltica, 3:347
Palaeopsylla dissimilis, 3:347
Palaeopsylla klebsiana, 3:347
Palaeotodus emryi, 10:25
Palaeotragus spp., 15:265
Palaeotupaia spp., 13:291
Palamedea cornuta. See Horned screamers
Palamedea cristata. See Red-legged seriemas
Palawan hornbills, 10:75
Palawan peacock-pheasants, 8:*438*, 8:*452*
Palawan porcupines. *See* Indonesian porcupines
Palawan stink badgers. *See* Philippine stink badgers
Palawan tits, 11:158
Palawan tree shrews, 13:292, 13:298*t*
Pale-billed aracaris, 10:127
Pale-billed sicklebills, 11:492
Pale-breasted spinetails, 10:*213*, 10:*221–222*
Pale-browed tinamous, 8:59
Pale-faced sheathbills, 9:197, 9:198, 9:200–*201*
Pale field rats. *See* Tunney's rats
Pale foxes, 14:265
Pale-headed brush finches, 11:266
Pale-headed jacamars, 10:93
Pale hedgehogs. *See* Indian desert hedgehogs
Pale kangaroo mice, 16:200, 16:*204*, 16:*206*–207
Pale leopard lizards. *See* Long-nosed leopard lizards
Pale rock sparrows, 11:397, 11:398, 11:*400*, 11:403–*404*
Pale spear-nosed bats, 13:418, 13:433*t*
Pale-throated flycatchers. *See* Nutting's flycatchers
Pale-throated three-toed sloths, 13:*162*, 13:*164*, 13:*165*, 13:*166*
Pale white-eyes. *See* Cape white-eyes
Pale-winged trumpeters, 9:*78*, 9:*80*, 9:*81*–82
Pale-yellow robins, 11:*106*
Palea steindachneri. See Chinese softshell turtles
Paleaopropithecus spp., 14:63
Paleobatrachidae, 6:4
Paleochoerus spp., 15:265
Paleodictyoptera, 3:12, 3:13, 3:52
Paleodonta, 15:264
Paleoentomology, 3:9
Paleoheterodonta, 2:451
Paleolaginae, 16:505
Paleomeryx spp., 15:265
Paleopropithecinae. *See* Sloth lemurs
Paleopteran insects, 3:52
Paleosuchus palpebrosus. See Cuvier's dwarf caimans
Paleosuchus trigonatus. See Smooth-fronted caimans
Paleotheres, 15:136
Paleotraginae, 15:399
Palestine mole rats, 16:*287*, 16:*289*, 16:294
See also Blind mole-rats
Palestine sunbirds, 11:209, 11:210
Palette tangs, 4:*57*, 5:*394*, 5:397–*398*
Paliguana spp., 7:16
Palilas, 11:343, 11:*345*, 11:350–*351*
Palinura, 2:197, 2:198, 2:200
Pallas's cats, 14:391*t*
Pallas's dippers. *See* Brown dippers
Pallas's long-tongued bats, 13:*415*, 13:417, 13:*422*, 13:*427*
Pallas's mastiff bats, 13:486, 13:494*t*
Pallas's pikas, 16:488, 16:*492*, 16:*496*, 16:*499*–500
Pallas's sandgrouse, 9:*234*, 9:*238–239*
Pallas's sea-eagles, 8:323
Pallas's squirrels, 16:*168*, 16:*173*–174
Pallid bats, 13:*308*, 13:313, 13:497, 13:*500*, 13:*505*, 13:*506*
Pallid cuckoos, 11:86, 11:241, 11:242
Pallid finch-larks. *See* Black-crowned sparrow-larks
Pallid harriers, 8:320, 8:321
Pallid swifts, 9:422
Palm bugs, 3:9
Palm civets, 14:258, 14:260, 14:335, 14:337, 14:344*t*
Palm cockatoos, 9:278, 9:*282*, 9:*283*
Palm nut vultures, 8:321
Palm swifts. *See* African palm swifts
Palm weevils, 3:323
Palma, R., 5:99
Palmatogecko rangei. See Web-footed geckos
Palmatorappia spp., 6:246
Palmchats, **10:455–457,** 10:*456*
Palmer, Thomas, 7:454
Palmer, W. M., 7:315
Palmeria dolei. See Akohekohes
Palms, oil, 3:79
Palolo worms, 2:47
Palophus. See Australian phasmid
Palpigradi, 2:333
Paltoperlidae, 3:141
Palythoa caesia. See Rubber corals
Palythoa sea mats. See Rubber corals
Pamexis bifasciatus, 3:309
Pamexis contamminatus, 3:309
Pampas cats, 14:391*t*
Pampas deer, 15:379, 15:*381*, 15:383, 15:384, 15:396*t*
Pampas foxes, 14:284*t*
Pampas meadowlarks, 11:305
Pampatheres. *See* Giant herbivorous armadillos
Pamphagidae, 3:205
Pamphiloidea, 3:54
Pampus argenteus. See Silver pomfrets
Pan hinged terrapins. *See* East African black mud turtles
Pan paniscus. See Bonobos
Pan terrapins. *See* East African black mud turtles
Pan troglodytes. See Chimpanzees
Pan troglodytes schweinfurthii, 14:225
Pan troglodytes troglodytes, 14:225
Pan troglodytes vellerosus, 14:225
Pan troglodytes verus, 14:225
Panama Canal, 12:220
Panamanian squirrel monkeys. *See* Red-backed squirrel monkeys
Panarthropoda, 2:115
Panbiogeography, 3:53, 3:54
Pancake tortoises, 7:66, 7:*145*, 7:147–*148*
Panda prawns. *See* Giant tiger prawns
Pandaka pygmaea. See Dwarf pygmy gobies
Pandas, red, 12:*28*, 14:309–*311*, 14:*312*, 14:*315*

See also Giant pandas
Panderichthys spp., 6:7, 6:*10*
Pandinus imperator. See Emperor scorpions
Pandion haliaetus. See Ospreys
Pandioninae. *See* Ospreys
Pandoridae, 2:451
Panesthiinae, 3:147–148, 3:150
Panfishes. *See* Sunfishes
Pangani longclaws, 10:373
Pangasiidae. *See* Shark catfishes
Pangasius hypophthalmus. See Iridescent shark-catfishes
Pangea, 3:52
Pangio kuhlii. See Coolie loaches
Pangio semicincta. See Coolie loaches
Pangolins, 13:196, **16:107–120**, 16:*109*, 16:*111*, 16:*114*
  behavior, 16:*108*, 16:110–112
  conservation status, 16:113
  distribution, 16:110
  evolution, 16:107
  feeding ecology, 16:112–113
  habitats, 12:135, 16:110
  humans and, 16:113
  physical characteristics, 16:107–110
  reproduction, 16:113
  species of, 16:115–120
  taxonomy, 16:107
Panorpa spp., 3:61
Panorpa communis, 3:*341*, 3:*343*
Panorpa nuptialis, 3:*342*, 3:*344*, 3:*345*, 3:346
Panorpidae, 3:341, 3:342, 3:343
Panorpodidae, 3:341, 3:342
Pantala flavescens. See Wandering gliders
Panterpe spp. *See* Hummingbirds
Panterpe insignis. See Fiery-throated hummingbirds
Panther chameleons, 7:*37*, 7:224, 7:225, 7:*227*, 7:*230*, 7:*233*, 7:*239*, 7:240
Panthera spp., 14:369
Panthera leo. See Lions
Panthera onca. See Jaguars
Panthera pardus. See Leopards
Panthera tigris. See Tigers
Panthera tigris altaica. See Amur tigers
Pantherana spp., 6:248
Pantherinae, 14:369
Panthers. *See* Pumas
Pantholops hodgsonii. See Tibetan antelopes
Panthophthalmidae, 3:359
Pantodon buchholzi. See Freshwater butterflyfishes
Pantodonta, 12:11, 15:131, 15:134, 15:135, 15:136
Pantodontidae. *See* Butterflyfishes
Pantry beetles, 3:57
Pantylus spp., 6:11
Panulirus argus. See Caribbean spiny lobster
Panurus biarmicus. See Bearded reedlings
Panurus biarmicus kosswigi. See Turkish bearded reedlings
Panyptila spp. *See* Swallow-tailed swifts
Panyptila sanctihieronymi. See Greater swallow-tailed swifts
Papasula spp. *See* Abbott's boobies
Papasula abbotti. See Abbott's boobies
Paper moths. *See* Silverfish
Paper nautilus. *See* Greater argonauts
Paper wasps, 3:68, 3:69, 3:*71*, 3:406
  See also Golden paper wasps
Papilio aristodemus ponceanus. See Schaus's swallowtail butterflies
Papilio glaucus. See Swallowtail tigers
Papilio victorinus, 3:*386*
Papillose bats, 13:526*t*
Papio spp. *See* Baboons
Papio anubis. See Olive baboons
Papio cynocephalus. See Yellow baboons
Papio hamadryas. See Hamadryas baboons
Papio hamadryas anubis. See Olive baboons
Papio papio. See Guinea baboons
Papio ursinus. See Chacma baboons
Papionini, 14:188
Pappogeomys spp., 16:185
Pappogeomys alcorni, 16:190
Pappogeomys bulleri. See Buller's pocket gophers
Papua mountain pigeons, 9:250, 9:*380*
Papuan bandicoots, 13:10, 13:12, 13:17*t*
Papuan drongos. *See* Pygmy drongos
Papuan forest wallabies, 13:*91*, 13:*99*, 13:100–101
Papuan frogmouths, 8:*10*, 9:377, 9:*381*–382
Papuan ground boas. *See* Viper boas
Papuan lories, 9:276
Papuan pythons, 7:420, 7:*423*, 7:*424*–425
Papuan treecreepers. *See* White-throated treecreepers
Papuan whipbirds, 11:75, 11:76
Papyrocranus spp., 4:232
Papyrocranus afer, 4:232
Parabathynellidae, 2:177
Parabos spp., 15:265, 16:1
Parabrotulidae, 5:15
Parabuteo unicinctus. See Harris' hawks
Paracallionymus spp., 5:365
Paracanthopterygii. *See* Codlike fishes
Paracanthurus hepatus. See Palette tangs
Paracentrotus lividus, 1:405
Paraceratherium [Indricotherium] ransouralicum, 15:138
Parachanna spp., 5:437
Parachanna obscura. See African snakeheads
Paracheirodon axelrodi. See Cardinal tetras
Parachela, 2:115
Parachute geckos, 7:*261*
Paracirrhites spp., 5:236, 5:239
Paracirrhites arcatus, 4:42, 5:236
Paracirrhites forsteri, 5:236
Paracirrhites hemistictus, 5:236, 5:238
Paracobitis smithi. See Blind loaches
Paracobitis starostini. See Kughitang blind loaches
Paracolobus spp., 14:172
Paracoprids, 3:322
Paracorpididae. *See* Jutjaws
Paracoryne huvei, 1:*132*, 1:135
Paracosoryx prodromus, 15:265
Paracrinia haswelli, 6:149
Paracrocidura spp., 13:265
Paracrocidura graueri. See Grauer's shrews
Paracrocidura maxima. See Greater shrews
Paracrocidura schoutedeni. See Schouteden's shrews
Paractiornis spp., 9:151
Paracynictis spp., 14:347
Paradiceros mukiri, 15:249
Paradigalla spp. *See* Birds of paradise
Paradigalla brevicauda. See Short-tailed paradigallas
Paradiplogrammus spp., 5:365
Paradiplogrammus bairdi. See Lancer dragonets
Paradiplogrammus enneactis. See Mangrove dragonets
Paradipodinae, 16:211, 16:212
Paradipus spp., 16:211
Paradipus ctenodactylus. See Comb-toed jerboas
Paradisaea spp. *See* Birds of paradise
Paradisaea apoda. See Greater birds of paradise
Paradisaea decora. See Goldie's birds of paradise
Paradisaea rubra. See Red birds of paradise
Paradisaeidae. *See* Birds of paradise
Paradisaeinae. *See* Birds of paradise
Paradiscoglossus, 6:*25*, 6:89
Paradise crows, 11:490
Paradise-flycatchers, 11:97
Paradise jacamars, 10:92, 10:*94*, 10:*98*
Paradise kingfishers, 10:6, 10:7, 10:14–15
Paradise riflebirds, 11:490
Paradise whydahs, 11:*361*
Paradisea tristis. See Common mynas
Paradox frogs, 6:42, 6:228, 6:230, 6:*231*, 6:242
Paradoxophyla spp. *See* Web-foot frogs
Paradoxophyla palmata. See Web-foot frogs
Paradoxornis spp. *See* Parrotbills
Paradoxornis paradoxus, 10:506
Paradoxornis webbianus. See Vinous-throated parrotbills
Paradoxurinae, 14:335
Paradoxurus spp., 14:336
Paradoxurus hermaphroditus. See Palm civets
Paradoxurus jerdoni. See Jerdon's palm civets
Paraechinus aethiopicus. See Ethiopian hedgehogs
Paraechinus hypomelas. See Brandt's hedgehogs
Paraechinus micropus. See Indian desert hedgehogs
Paraescarpia spp., 2:91
Parafontaria laminata, 2:365
Paragobiodon spp., 4:42, 4:45
Paragonimus spp., 1:201
Paragonimus westermani. See Lung flukes
Paragordius varius, 1:*305*, 1:*307*, 1:*308*, 1:309
Paraguay hairy dwarf porcupines, 16:366, 16:367, 16:368, 16:374*t*
Paraguayan caiman lizards, 7:310, 7:*313*, 7:*316*
Parahyaena brunnea. See Brown hyenas
Parainocellia braueri. See Brauer's inocelliid snakeflies
Parakeet auklets, 9:220
Parakeets. *See* Budgerigars
Parakneria spp., 4:290, 4:291
Parakuhlia spp., 5:220
Paralabidochromis chilotes, 5:*282*, 5:*285*, 5:286
Paraleiopus macrochelis, 2:*237*, 2:*238*–239
Paralepididae, 4:432, 4:433
Paralichthodidae, 5:450
Paralichthyidae, 5:450, 5:453
Paralichthys dentatus. See Summer flounders
Paralithodes camtschaticus. See Red king crabs
Paralouatta spp., 14:143
Paraluteres prionurus. See Mimic filefishes
Paralysis ticks. *See* Rocky Mountain wood ticks
Paramecium spp., 1:180
Paramyidae, 16:122
Paramysis spp., 2:216
Paramythia spp. *See* Berrypeckers
Paramythia montium. See Crested berrypeckers

Paranebalia spp., 2:161
Paranebaliidae, 2:161
Parantechinus spp. *See* Dibblers
Parantechinus apicalis. See Southern dibblers
Parantechinus bilarni. See Sandstone dibblers
Paranybeliniidae, 1:226
Paranyctimene spp., 13:334
Paranyctoides spp., 13:193
Parapanesthia spp., 3:150
Parapedetes gracilis, 16:307
Parapedetes namaquensis, 16:307
Parapercis spp., 5:334
Parapercis colias. See Blue cods
Parapercis cylindrica, 5:333
Parapercis snyderi. See Japanese sandperches
Parapercis tetracantha. See Four-spined sandperches
Paraphlebia zoe, 3:65
Paraplesiops spp., 5:256
Parapontoporia spp., 15:19, 15:23
Parapsid reptiles, 7:*5*
Parargyractis confusalis, 3:*392*, 3:*398*, 3:401
Parartemia spp., 2:135, 2:136
Parartemiidae, 2:135
Parasambonia spp., 2:318
Parascalops breweri. See Hairy-tailed moles
Parascaptor spp., 13:279
Parascorpididae, 5:243
Parascyllidae. *See* Collared carpet sharks
Parasitic catfishes. *See* Pencil catfishes
Parasitic flatworms, 1:*187*, **1:197–205**, 1:*202*, **1:210**, 2:13, 2:15, 2:33
behavior, 1:37, 1:41, 1:199–200
conservation status, 1:200
distribution, 1:198
evolution, 1:197
feeding ecology, 1:200
habitats, 1:198–199
humans and, 1:201
physical characteristics, 1:197–198
reproduction, 1:16, 1:17, 1:18, 1:20, 1:21, 1:200
species of, 1:*203–205*, 1:210
taxonomy, 1:11, 1:197
See also Monogeneans; Tapeworms
Parasitic jaegers. *See* Arctic skuas
Parasitic skuas. *See* Arctic skuas
Parasitic weavers. *See* Cuckoo finches
Parasitism, 1:32, 1:34–36, 2:28, 2:31–32, 2:*32*, 2:33
bat, 3:263
candiru, 4:367
cockroaches and, 3:148
fishes, 4:83, 4:84
fleas, 3:349
flies, 3:39, 3:47, 3:76, 3:360
Ice Age mammals and, 12:25
infections and, 1:34–36, 2:28, 2:33
insects, 3:47, 3:63, 3:64
lampreys, 4:50, 4:69
social, 3:69
subterranean mammals and, 12:77–78
wasps, 3:5, 3:39, 3:80
See also Behavior; Brood parasitism; specific parasitic species
Parasitism, brood. *See* Brood parasitism
Parasitoids, 3:47, 3:64
Parasitosis, delusory, 3:77
Parasphaerichthys spp., 5:427, 5:433
Parastygarctus sterreri, 2:118
Parataeniophorus spp., 5:106
Paratelmatobius spp., 6:155–156, 6:157
Paratenodera aridifolia. See Japanese praying mantis
Paratopithecus brasiliensis, 14:155
Paratrachichthys spp., 5:115
Paratriakis spp., 4:113
Paratrygon aiereba. See Freshwater stingrays
Paraturbanella spp., 1:*269*
Paravinciguerria spp., 4:421
Paravortex scrobiculariae, 1:*189*, 1:*192*–193
Paraxerus spp., 16:163
Paraxerus alexandri, 16:167
Paraxerus cepapi. See Smith's tree squirrels
Paraxerus cooperi, 16:167
Paraxerus palliatus, 16:167
Paraxerus vexillarius, 16:167
Paraxerus vincenti, 16:167
Paraxonia, 15:131
Parazenidae, 5:123
Parazoa, 1:9, 1:10
Parborlasia spp., 1:*245*
Parborlasia corrugatus, 1:245, 1:*246*
Parchment worms, 2:*49*, 2:*55*–56
Pardalotes, 11:55, 11:189, **11:201–206**, 11:*203*
Pardalotidae. *See* Pardalotes
Pardalotus spp. *See* Pardalotes
Pardalotus punctatus. See Spotted pardalotes
Pardalotus quadragintus. See Forty-spotted pardalotes
Pardalotus rubricatus. See Red-browed pardalotes
Pardalotus striatus. See Striated pardalotes
Pardalotus substriatus. See Striated pardalotes
Pardel lynx. *See* Iberian lynx
Pardinas, U. F. J., 16:264, 16:266
Pardirallus maculatus. See Spotted rails
Pareatinae, 7:465, 7:468
Pareiasaurs, 7:14
Parental behavior, 6:34–35
See also Reproduction
Parental care, 3:44, 3:62, 4:46, 4:66
See also Reproduction
Parenti, Lynne R.
on Adrianichthyids, 5:79
on Atheriniformes, 5:67
on Cyprinodontiformes, 5:89
Parexocoetus spp., 5:79–80
Parhomaloptera normani. See Anamia
Paria redstarts, 11:291
Pariah dogs, 12:173
Paridae. *See* Chickadees; Titmice
Parietal organs, 7:32
Parin, N., 4:421, 5:417
Parioglossus spp., 5:375
Parka squirrels. *See* Arctic ground squirrels
Parker, H. W.
on Australian ground frogs, 6:139
on narrow-mouthed frogs, 6:301
Parma wallabies, 13:40, 13:*87*, 13:*91*, 13:*94*–95
Parmoptila rubrifrons. See Red-fronted flowerpecker weaver-finches
Parmoptila woodhousei. See Flowerpecker weaver-finches
Parnell's moustached bats, 12:86, 13:435, 13:*436*, 13:*437*, 13:*439*, 13:*440*–441
Parophidion spp., 5:22
Paroreomyza spp., 11:343
Paroreomyza maculata. See Oahu alauahios
Parosphromenus spp., 5:427, 5:428
Parosphromenus deissneri, 5:427
Parosphromenus harveyi, 5:429
Parotia spp. *See* Birds of paradise
Parotia lawesii. See Lawes's parotias
Parotia wahnesi. See Wahnes's parotias
Parotias, 11:489, 11:490, 11:492, 11:*493*, 11:497–398
Parotid glands, 6:53, 6:184
Parra africana. See African jacanas
Parra capensis. See Lesser jacanas
Parra chavaria. See Northern screamers
Parrea occa, 1:*70*, 1:*73*
Parrini, Ricardo, 10:308
Parrot chubs. *See* Stoplight parrotfishes
Parrot fever, 8:25
Parrotbills, 10:505, 10:506, 10:508, 10:*509*
Parrotfinches, 11:353–354, 11:356–357, 11:*359*, 11:368–369
Parrotfishes, 4:17, 4:45, **5:293–298**, 5:*295*, 5:*297*, 5:298, 5:*300*, **5:306–307**
feeding ecology, 4:17, 4:50, 4:67, 5:295–296, 5:306–307
habitats, 4:48, 5:294, 5:306
physical characteristics, 4:*4*, 4:7, 4:18, 5:293–294, 5:306
reproduction, 4:45, 4:47, 5:296–298, 5:306–307
Parrots, **9:275–298**, 9:*281–282*
behavior, 9:276–277
conservation status, 9:279–280
distribution, 9:*275*, 9:276
evolution, 9:275
feeding ecology, 9:277–278
habitats, 9:276
humans and, 9:280
physical characteristics, 9:275–276
reproduction, 9:278–279
sounds of, 8:37
species of, 9:*283–298*
taxonomy, 9:275
See also specific types of parrots
Parrots of the World (Forshaw), 9:272
Parsley frogs, 6:5, 6:64, **6:127–130**, 6:*128*, 6:*129*
Parson birds. *See* Tuis
Parson's chameleons, 7:*8*, 7:227, 7:*233*, 7:*236*–237
Partbelt barbs. *See* Tiger barbs
Parthenogenesis, 2:17, 2:22–24, 3:39, 3:59, 4:30, 4:66, 7:6, 7:207, 7:312
See also Reproduction
Parthenos sylvia. See Blue butterflies
Parti-colored bats, 13:516*t*
Partridge finches. *See* African quailfinches
Partridges, 8:399, 8:400, 8:433–437
Parula americana. See Northern parulas
Parulidae. *See* New World warblers
Parulinae. *See* New World warblers
Parupeneus cyclostomus, 5:239
Parus spp., 11:155
Parus afer. See Southern gray tits
Parus amabilis. See Palawan tits
Parus bicolor. See Tufted titmice
Parus caeruleus. See Blue titmice
Parus cela. See Yellow-rumped caciques
Parus cinctus. See Siberian tits
Parus cyanus. See Azure tits
Parus davidi. See Père David's tits

Parus gambeli. *See* Mountain chickadees
Parus griseiventris. *See* Miomba tits
Parus guineenis. *See* White-shouldered tits
Parus lugubris. *See* Somber tits
Parus major. *See* Great tits
Parus montanus. *See* Willow tits
Parus rufiventris. *See* Rufous-bellied tits
Parus venustulus. *See* Yellow-bellied tits
Parus wollweberi. *See* Bridled titmice
Paruterinidae, 1:226
Parvalacerta spp., 7:297
Parvicrepis spp., 5:355
Parvimolge spp., 6:391
Passenger pigeons, 9:243, 9:245, 9:251–252
Passer spp. *See* Sparrows
Passer domesticus. *See* House sparrows
Passer luteus. *See* Golden sparrows
Passer montanus. *See* Tree sparrows
Passer motitensis. *See* Southern rufous sparrows
Passerculus sandwichensis. *See* Savannah sparrows
Passeri. *See* Songbirds
Passerida, 10:171
Passeridae. *See* Sparrows
Passeriformes, **10:169–175**
Passerina ciris. *See* Painted buntings
Pasteur, Louis, 2:33, 3:80
Pastor spp., 11:409
Pastoral Symphony, 8:19
Pataecidae. *See* Australian prowfishes
Pataecus fronto. *See* Red indianfishes
Patagium. *See* Flight membranes
Patagona spp. *See* Hummingbirds
Patagona gigas. *See* Giant hummingbirds
Patagonia frogs, 6:*161*, 6:*169*–170
Patagonia seedsnipes. *See* Least seedsnipes
Patagonian catfishes, 4:352
Patagonian cavies. *See* Maras
Patagonian hares. *See* Maras
Patagonian lanceheads, 7:*450*, 7:*453*–454
Patagonian opossums, 12:249–253, 12:*255*, 12:*258*, 12:259
Patagonian toothfishes, 5:323
Patagonian weasels, 14:332*t*
Patas monkeys, 12:146, 14:7, 14:10, 14:188, 14:191, 14:192, 14:*196*, 14:197–*198*
Patekes. *See* Brown teals
Patella spp., 2:423, 2:*424*, 2:425
Patella vulgata. *See* Common limpets
Patellidae, 2:423, 2:424, 2:425
Patellina, 2:423
Patellogastropoda. *See* True limpets
Paternal care, 12:96, 12:107
  See also Reproduction
Paternity, sperm competition and, 12:105–106
  See also Reproduction
Patiria miniata. *See* Bat stars
Patiriella parvivipara. *See* Cushion stars
Patterns, hair, 12:38
Patterson, B. D., 16:266
Patterson, C.
  on beardfishes, 5:1
  on Gonorynchiformes, 4:289
  on Osmeriformes, 4:389, 4:390
  on Polypteriformes, 4:209
  on Salmoniformes, 4:405
  on Stephanoberyciformes, 5:105
  on Synbranchiformes, 5:151
  on Zeiformes, 5:123
Patterson, Francine, 12:162
Pattipola caecilians, 6:*421*, 6:423
Patton, J. L., 16:263
Patuki blennies, 5:343
Patuxent Wildlife Research Center, 9:28
Paucituberculata. *See* Shrew opossums
Paucitubulatina, 1:269, 1:270
Paulinia acuminata, 3:203, 3:206
Paulissen, Mark, 7:315
Pauropodidae, 2:375
Pauropods, **2:375–377**
Pauxi pauxi. *See* Northern helmeted curassows
Pavo spp. *See* Peafowls
Pavo cristatus. *See* Indian peafowls; Peacocks
Pavo muticus. *See* Green peafowls
PCBs. *See* Polychlorinated biphenyls
PCR. *See* Polymerase chain reaction
Pea aphids, 3:*265*, 3:267–*268*
Pea crabs, 2:*202*, 2:*208*, 2:211–212
Pea urchins, 1:*407*, 1:*411*–412
Peach-faced lovebirds. *See* Rosy-faced lovebirds
Peacock basses. *See* Speckled pavons
Peacock flounders, 4:*4*, 5:*454*, 5:455–*456*
Peacock mantis shrimps, 2:*169*, 2:170, 2:*172*, 2:*174*–175
Peacock-pheasants, 8:434
Peacock soles, 4:48
Peacocks, 8:20
Peafowls, 8:433–436
Peak-saddle horseshoe bats. *See* Blasius's horseshoe bats
Peale's falcons. *See* Peregrine falcons
Peanut bugs, 2:39
Peanut worms, 2:13–14, 2:21, 2:24, 2:97, **2:97–101**, 2:*99*
Pear and cherry slugs, 3:*410*, 3:*414*, 3:422–423
Pear sawflies. *See* Pear and cherry slugs
Pear slug sawflies. *See* Pear and cherry slugs
Pear slugworms. *See* Pear and cherry slugs
Pearl gouramies, 5:*430*, 5:*433*, 5:434
Pearl mussels, freshwater, 2:454
Pearl oysters, 2:454, 2:*455*, 2:*461*–462
Pearl perches, 5:255
  See also Glaucosomatidae
Pearl-spotted owlets, 9:347, 9:*353*, 9:*362*
Pearl-spotted owls. *See* Pearl-spotted owlets
Pearleyes, 4:432
Pearlfishes, 2:32, 4:69, 5:*15*–18, 5:*17*, 5:*19*, 5:*20*, 5:21
Pearly-eyed thrashers, 10:467
Pearly mussels, European, 2:*455*, 2:*462*–463
Pearly nautilus, 2:*481*, 2:*482*, 2:483
Pearson, O. P., 16:267, 16:269
Pearsonomys annectens, 16:266
Pearsonothuria graeffei, 1:421
Pearson's long-clawed shrews, 13:*270*, 13:*272*, 13:274
Pearson's tuco-tucos, 16:*428*, 16:*429*
Pebble eating assistance. *See* Feeding ecology
Pebbled butterflyfishes, 4:*4*, 5:*245*, 5:*247*
Pecari spp., 15:267
Peccaries, **15:291–300**, 15:*297*
  behavior, 15:293–294
  conservation status, 15:295–296
  distribution, 15:*291*, 15:292
  evolution, 15:137, 15:264, 15:265, 15:291
  feeding ecology, 15:141, 15:271, 15:294–295
  habitats, 15:292–293
  humans and, 15:296
  physical characteristics, 15:140, 15:291–292
  reproduction, 15:295
  species of, 15:298–300
  taxonomy, 15:291
Pechora pipits, 10:372, 10:373
Pecoran ruminants, 15:136
Pectenocypris balaena, 4:299–300
Pectinatella magnifica, 2:*499*, 2:*500*, 2:501
Pectinatellidae, 2:497
Pectinator spp., 16:311
Pectinator spekei. *See* Speke's pectinators
Pectinidae, 2:452
Pectinura maculata. *See* Snake stars
Pectoral girdle, 12:8, 12:41
  See also Physical characteristics
Pectoral nightjars. *See* Fiery-necked nightjars
Pedal sea cucumbers. *See* Slipper sea cucumbers
Pedder galaxias, 4:391
Pederson cleaner shrimps, 2:34
Pedetes capensis. *See* Springhares
Pedetidae. *See* Springhares
Pedicellinidae, 1:319
Pediculus humanus. *See* Human head/body lice
Pediolagus spp., 16:389, 16:390
Pediolagus salinicola. *See* Salt-desert cavies
Pedionomus torquatus. *See* Plains-wanderers
Pedioplanis spp., 7:297
Pedostibes spp., 6:184, 6:194
Pedostibes hosii. *See* Brown tree toads
Pegasidae. *See* Seamoths
Pegasus lancifer. *See* Sculptured seamoths
Pekin robins. *See* Red-billed leiothrixes
Pekinese, 14:289
Pelage. *See* Hairs
Pelagia noctiluca. *See* Nightlight jellyfish
Pelagic cormorants, 8:*204*, 8:*207*
Pelagic fishes, 4:39, 4:47, 4:48, 4:49, 4:62
  See also specific fishes
Pelagic sea cucumbers, 1:*424*, 1:*426*, 1:430
Pelagic shags. *See* Pelagic cormorants
Pelagic stingrays, 4:176
Pelagodiscus spp., 2:515
Pelagohydra spp., 1:124
Pelagothuria spp., 1:421
Pelagothuria natatrix. *See* Pelagic sea cucumbers
Pelamis platurus. *See* Yellow-bellied seasnakes
Pelargopsis capensis. *See* Stork-billed kingfishers
Pelates spp., 5:220
Pelates quadrilineatus, 5:220
Pelea spp., 16:28
Pelea capreolus. *See* Gray rheboks
Pelecanidae. *See* Pelicans
Pelecaniformes, **8:183–186**, 8:*184*, 8:*185*
Pelecanoides magellani. *See* Magellan diving-petrels
Pelecanoides urinatrix. *See* Common diving-petrels
Pelecanoididae. *See* Diving-petrels
Pelecanus conspicillatus. *See* Australian pelicans
Pelecanus crispus. *See* Dalmatian pelicans
Pelecanus erythrorhynchos. *See* American white pelicans
Pelecanus grandis, 8:225
Pelecanus occidentalis. *See* Brown pelicans
Pelecanus olivaceus. *See* Olivaceous cormorants
Pelecanus onocrotalus. *See* Great white pelicans
Pelecanus philippensis. *See* Spot-billed pelicans
Pelecanus rufescens. *See* Pink-backed pelicans
Pelecinus polyturator, 3:*409*, 3:418–*419*

Peleinae, 16:1
Pelias spp., 7:445, 7:447
Pelican eels, 4:*4*, 4:271, 4:*273*, 4:*274–275*
Pelican fishes. *See* Gulper eels; Pelican eels
Pelican gulper fishes. *See* Pelican eels
Pelican gulpers. *See* Pelican eels
Pelicans, 8:183–186, **8:225–232**
Pelobates spp., 6:119
Pelobates fuscus. *See* Common spadefoot
Pelobatidae. *See* Spadefoot toads
Pelochelys cantorii. *See* Asian giant softshell turtles
Pelodiscus sinensis. *See* Chinese softshell turtles
Pelodryadinae, 6:225, 6:230
Pelodytes caucasicus, 6:128
Pelodytes nasutus. *See* Rocket frogs
Pelodytes punctatus. *See* Parsley frogs
Pelodytidae. *See* Parsley frogs
Pelomedusa spp., 7:129
Pelomedusa subrufa. *See* Helmeted turtles
Pelomedusidae. *See* African sideneck turtles
Pelomys fallax. *See* Creek rats
Pelonaia corrugata, 1:*457*, 1:*460*, 1:466
Pelonia spp., 2:318
Pelophilini, 3:54
Pelophilus spp., 6:89
Pelophryne spp., 6:34, 6:184
Pelophylax spp., 6:248, 6:250, 6:252, 6:263
Peloridiidae, 3:55
Peloridium hammoniorum. *See* Moss bugs
Pel's anomalures, 16:300, 16:*301*, 16:303, 16:*305*
Pel's scaletails. *See* Pel's anomalures
Pelsartia spp., 5:220
Peltohyas australis. *See* Inland dotterels
Peltops blainvillii. *See* Clicking peltops
Peltropses, 11:468
Pelusios spp., 7:129
Pelusios broadleyi, 7:130
Pelusios seychellensis, 7:130
Pelusios sinuatus. *See* East African serrated mud turtles
Pelusios subniger. *See* East African black mud turtles
Pelvic bones, 12:8
  See also Physical characteristics
Pelvicachromis pulcher. *See* Kribensis
Pelycosauria, 7:5
Pempheridae, 4:48, 4:68
Pemphiginae, 3:68
Pen shells. *See* Noble pen shells
Pen-tailed tree shrews, 13:291, 13:*293*, 13:*294*, 13:296
Penaeoidea, 2:197, 2:198, 2:201
Penaeus monodon. *See* Giant tiger prawns
Penang Taylor's frogs, 6:*253*, 6:*257*
Penant's red colobus, 14:185*t*
Pencil catfishes, 4:69, 4:352, 4:353, 4:354
Penduline titmice, **11:147–153,** 11:*150*
Penelope spp. *See* Guans
Penelope jacquacu. *See* Spix's guans
Penelope purpurascens. *See* Crested guans
Penelope superciliaris. *See* Rusty-margined guans
Penelopides affinis. *See* Mindanao tarictic hornbills
Penelopides exarhatus. *See* Sulawesi tarictic hornbills
Penelopides manillae. *See* Luzon tarictic hornbills
Penelopides mindorensis. *See* Mindoro tarictic hornbills
Penelopides panini. *See* Visayan tarictic hornbills
Penelopina nigra. *See* Highland guans
Peneothello sigillatus. *See* White-winged robins
Penguin Encounter exhibit (Sea World), 8:153
Penguins, **8:147–158,** 8:*152*, 9:198–199
  behavior, 8:149–150
  conservation status, 8:151
  distribution, 8:*147*, 8:149
  evolution, 8:147–148
  feeding ecology, 8:150
  habitats, 8:149
  humans and, 8:151
  physical characteristics, 8:148–149
  reproduction, 8:*150*–151
  species of, 8:*153–158*
  taxonomy, 8:147
Peniagone spp., 1:420
Penicillata, 2:363
Penicillin, 1:49
  See also Humans
Peninsula cooter turtles, 7:*36*
Peninsular pronghorns, 15:415
Penises, 12:91, 12:104, 12:105, 12:111
  See also Physical characteristics
Pennant-winged nightjars, 9:369, 9:370, 9:402, 9:404
Pennatulacea, 1:103
Pentaceros japonicus. *See* Japanese boarfishes
Pentacerotidae, 5:235, 5:237, 5:243–244
Pentacrinids, 1:356
Pentadactyly, 12:42
  See also Physical characteristics
Pentalagus spp. *See* Amami rabbits
Pentalagus furnessi. *See* Amami rabbits
Pentanchus profundicolus, 4:114
Pentastomida. *See* Tongue worms
Pentatomidae, 3:264
Pentatomomorpha, 3:259
Penthoceryx sonneratii. *See* Banded bay cuckoos
Penultimate Glaciation, 12:20–21
People for the Ethical Treatment of Animals (PETA), 14:294
Peponocephala electra. *See* Melon-headed whales
Peppermint stick insects, 3:*223*
Peppershrikes, **11:255–262,** 11:*257*
Peprilus triacanthus, 5:*421*, 5:422, 5:*423*, 5:*424*, 5:425–426
Pepsis grossa. *See* Tarantula hawks
Peptides, 1:45
Peracarida, 2:229, 2:241, 2:243, 2:245
Peradeniidae, 3:55
Peramelemorphia. *See* Bandicoots
Perameles spp. *See* Bandicoots
Perameles bougainville. *See* Western barred bandicoots
Perameles eremiana. *See* Desert bandicoots
Perameles gunnii. *See* Eastern barred bandicoots
Perameles nasuta. *See* Long-nosed bandicoots
Peramelidae. *See* Dry-country bandicoots
Peramelinae, 13:9
Perameloidea, 13:9
Perca flavescens. *See* Yellow perches
Perca fluviatilis. *See* European perches
Percheron horses, 12:176
Perches, 4:52, 4:57, 4:58, 4:75, **5:195–199,** 5:*200*, 5:*201*, 5:*203*, 5:*204*, **5:207–208**
  blue banded sea, 5:*255*
  climbing, 5:428, 5:*430*, 5:*431*
  European, 5:49, 5:198
  gurnard, 5:163
  kelp, 5:277
  log, 5:*200*, 5:*203*, 5:208
  pirate, 4:57, 5:5, 5:*7*, 5:10–*11*
  splendid, 5:255, 5:*256*, 5:258–263
  tule, 5:276
  See also specific types of perches
Perching birds. *See* Passeriformes
Perchlets, 5:255
  See also Anthiinae
Perchlike fishes, 4:36, 4:448
Perchoerus spp., 15:265
Percichthyidae, 4:68, 5:219–222
Percichthys spp., 5:220
Percidae. *See* Perches
Perciformes. *See* Perchlike fishes
Percilia spp., Chilean perches
Percilidae. *See* Chilean perches
Percina caprodes. *See* Logperches
Percival's trident bats, 13:*402*, 13:410*t*
Percoidei, **5:195–274,** 5:*200*, 5:*201*, 5:*213*, 5:*223*, 5:*224*, 5:*245*, 5:*246*, 5:*264*, 5:*265*
  behavior, 5:197, 5:212, 5:221, 5:239–240, 5:261
  conservation status, 5:199, 5:212, 5:222, 5:243–244, 5:262
  distribution, 5:196, 5:211, 5:220, 5:237–238, 5:258–259
  evolution, 5:195, 5:211, 5:219, 5:235, 5:255
  feeding ecology, 5:197–198, 5:212, 5:221, 5:240–242, 5:260–261
  habitats, 5:196–197, 5:211–212, 5:220–221, 5:238–239, 5:259–260
  humans and, 5:199, 5:212, 5:222, 5:244, 5:263
  physical characteristics, 5:195–196, 5:211, 5:219–220, 5:235–237, 5:255–258
  reproduction, 5:198, 5:212, 5:221–222, 5:242–243, 5:261–262
  species of, 5:*202*–208, 5:*214–217*, 5:*225*–233, 5:*247–253*, 5:*266–273*
  taxonomy, 5:195, 5:211, 5:219, 5:235, 5:255
Percomorpha, 4:12
Percophidae. *See* Duckbills
Percopsidae. *See* Troutperches
Percopsiformes, **5:5–13,** 5:*7*
Percopsis omiscomaycus. *See* Troutperches
Percopsis transmontana. *See* Sand rollers
Percopsoidei, 5:5
Perdicini. *See* Partridges
Perdicula spp. *See* Bush-quails
Perdicula manipurensis. *See* Manipur bush-quails
Perdix hodgsoniae. *See* Tibetan partridges
Perdix perdix. *See* Gray partridges
Perdix varia. *See* Painted buttonquails
Pere David's deer, 15:*267*, 15:358, 15:360, 15:*362*, 15:366–367
Père David's ground sparrows, 11:*400*, 11:*405*–406
Père David's snow finches. *See* Père David's ground sparrows
Père David's tits, 11:156
Peregrine falcons, 8:*10*, 8:21, 8:27, 8:314, 8:*348*, 8:349, 8:351, 8:352, 8:353, 8:360–361
Perenties, 7:361
Peres, C. A., 14:165
Perez's snouted frogs, 6:*161*, 6:166

Pericharax spp., 1:*59*
Pericharax heteroraphis, 1:58, 1:*60*, 1:*61*, 1:62
Periclimenes spp., 2:*200*
Periclimenes colemani. See Coleman's shrimps
Periclimenes pedersoni. See Pederson cleaner shrimps
Pericrocotus spp. *See* Minivets
Pericrocotus divaricatus. See Ashy minivets
Pericrocotus ethologus. See Long-tailed minivets
Pericrocotus igneus. See Fiery minivets
Pericrocotus roseus. See Rosy minivets
Perigoninae, 3:55
Perijas, 9:438
Perikoala palankarinnica, 13:43
Peringueyella spp., 3:203
Periodical cicadas. *See* Seventeen-year cicadas
Periodicity, reproductive, 2:18
See also Reproduction
Perionyx excavatus, 2:68
Periophthalmus spp., 5:375
Periophthalmus barbarus. See Atlantic mudskippers
Peripatidae, 2:109, 2:110
Peripatopsidae, 2:109, 2:110
Peripatuses, **2:109–114**
behavior, 2:110
conservation status, 2:111
distribution, 2:110
evolution, 2:109
feeding ecology, 2:110
habitats, 2:110
humans and, 2:111
physical characteristics, 2:109–110
reproduction, 2:110
species of, 2:*113*–114
taxonomy, 2:109
Periphylla periphylla. See Crown jellyfish
Periplaneta spp. *See* Cockroaches
Periplaneta americana. See American cockroaches
Peripodida. *See* Sea daisies
Periptychoidea, 15:134
Perischoechinoidea, 1:401
Perisoreus spp. *See* Corvidae
Perisoreus canadensis. See Gray jays
Perissocephalus spp. *See* Cotingas
Perissocephalus tricolor. See Capuchin birds
Perissodactyla, 15:131–132, 15:133, 15:135–141, **15:215–224**
digestive systems of, 12:14–15, 12:121
distribution, 12:132
neonatal milk, 12:127
vision, 12:79
Perissodus spp., 5:279
Perissodus microlepis, 5:279
Peristediidae. *See* Armored sea robins
Perlidae, 3:141, 3:142
Permithonidae, 3:305
Permithonopsis obscura, 3:305
Permotemisthida, 3:52
Permotipula patricia, 3:357
Perodicticus potto. See Pottos
Perodictinae, 14:13
Perognathinae, 16:199
Perognathus spp., 16:199, 16:200, 16:203
Perognathus fasciatus. See Olive-backed pocket mice
Perognathus flavus. See Silky pocket mice
Perognathus inornatus. See San Joaquin pocket mice
Perognathus longimembris. See Little pocket mice
Perognathus longimembris pacificus, 16:203
Perognathus parvus. See Great Basin pocket mice
Peromyscus attwateri. See Texas mice
Peromyscus boylii. See Brush mice
Peromyscus californicus. See California mice
Peromyscus eremicus. See Cactus mice
Peromyscus leucopus. See White-footed mice
Peromyscus maniculatus. See Deer mice
Peromyscus polinotus. See Old-field mice
Peropteryx kappleri. See Greater dog-faced bats
Peropteryx macrotis. See Lesser dog-faced bats
Peroryctes spp., 13:1–2
Peroryctes broadbenti. See Giant bandicoots
Peroryctes longicauda. See Striped bandicoots
Peroryctes papuensis. See Papuan bandicoots
Peroryctes raffrayana. See Raffray's bandicoots
Peroryctidae. *See* Rainforest bandicoots
Perridodus spp., 4:69
Perrin's beaked whales, 15:1, 15:61, 15:70*t*
Persian fallow deer. *See* Fallow deer
Persian gazelles, 16:48, 16:57*t*
Persian moles, 13:282
Persian onagers, 15:*219*, 15:222
Persian squirrels, 16:167, 16:175*t*
Persian trident bats, 13:409*t*
Persian wild asses. *See* Persian onagers
Perulibatrachus spp., 5:42
Peruvian anchoveta. *See* Anchoveta
Peruvian beaked whales. *See* Stejneger's beaked whales
Peruvian boobies, 8:213–214, 8:215, 8:*216*, 8:220–221
Peruvian crevice-dwelling bats, 13:519, 13:520, 13:521, 13:526*t*
Peruvian diving-petrels, 8:143, 8:144
Peruvian guemals. *See* Peruvian huemuls
Peruvian huemuls, 15:379, 15:384, 15:396*t*
Peruvian mountain viscachas. *See* Mountain viscachas
Peruvian piedtails, 9:418
Peruvian plantcutters, 10:325, 10:326, 10:*327*
Peruvian sheartails, 9:443
Peruvian spider monkeys, 14:*161*, 14:*163*–164
Peruvian thick-knees, 9:145–146
Pesquet's parrots, 9:279, 9:*281*, 9:*295*
Pest control, 3:50, 3:76–77, 3:168, 12:186–188
See also Biological control; specific animals
Pesticides, 3:*42*, 3:77, 6:56, 6:57, 8:16, 8:21, 8:26, 8:27
See also Conservation status
Pests, 3:74–77
beetle, 3:50, 3:75, 3:76, 3:323
caterpillar, 3:389
cockroach, 3:76, 3:151
Diptera, 3:361
fly, 3:75, 3:76, 3:361
Hemiptera, 3:264
humans and, 3:57
mammalian, **12:182–193**
thrips, 3:282, 3:283
See also Agricultural pests; specific species
PETA (People for the Ethical Treatment of Animals), 14:294
Petalophthalmidae, 2:215
Petalura ingentissima. See Australian dragonflies
Petauridae, 13:35, **13:125–133**, 13:*130*, 13:133*t*
Petaurillus hosei. See Hose's pygmy flying squirrels
Petaurinae, 13:125, 13:126, 13:127–128
Petaurirodea, 13:113
Petaurista spp., 16:135
Petaurista petaurista. See Red giant flying squirrels
Petauroidea, 13:32, 13:34, 13:38
Petauroides spp., 13:113, 13:114, 13:116
Petauroides volans. See Greater gliders
Petaurus spp., 13:113, 13:125, 13:126–127, 13:128
Petaurus abidi. See Northern gliders
Petaurus australis. See Yellow-bellied gliders
Petaurus breviceps. See Sugar gliders
Petaurus gracilis. See Mahogany gliders
Petaurus norfolcensis. See Squirrel gliders
Peters Checklist
on Charadriidae, 9:161
on cotingas, 10:305
on Cracidae, 8:413
on kingfishers, 10:6
on larks, 10:341
on manakins, 10:295
on mimids, 10:465
on penduline titmice, 11:147
on sharpbills, 10:291
on toucans, 10:125, 10:131
on Tyrannidae, 10:269
on wrens, 10:525
Peters' earthsnakes. *See* Peters' wormsnakes
Peter's elephantnoses. *See* Elephantnose fishes
Peter's finfoot. *See* African finfoot
Peter's spotted firefinches. *See* Peter's twinspots
Peters' threadsnakes. *See* Peters' wormsnakes
Peter's treefrogs. *See* Painted Indonesian treefrogs
Peter's twinspots, 11:*358*, 11:*362*–363
Peters' wormsnakes, 7:375, 7:*376*–*377*
Peters's disk-winged bats, 13:*475*, 13:*476*–477
Peters's duikers, 16:84*t*
Peters's dwarf epauleted fruit bats. *See* Dwarf epauletted fruit bats
Peters's sac-winged bats. *See* Lesser dog-faced bats
Peters's squirrels, 16:*168*, 16:*171*, 16:172
Petinomys spp., 16:136
Petinomys genibarbis. See Whiskered flying squirrels
Petitella georgiae. See False rummynose tetras
Petrels, 8:107–110, **8:123–133**, 8:*125*, 8:*127*
behavior, 8:109, 8:124, 8:131–132
conservation status, 8:110, 8:123
distribution, 8:123, 8:*123*
evolution, 8:123
feeding ecology, 8:109, 8:124, 8:*125*
habitats, 8:108, 8:124
humans and, 8:110, 8:123
physical characteristics, 8:107, 8:123
reproduction, 8:109–110, 8:125–126
species of, 8:*131*–133
taxonomy, 8:123, 8:131–133
The Petrels (Warham), 8:107
Petrobiona masselina, 1:*60*, 1:*61*, 1:64–65
Petrobius brevistylis, 3:*115*, 3:*116*, 3:117
Petrobius maritimus, 3:*113*
Petrocephalus soudanensis, 4:233

Petrodromus spp., 16:518
Petrodromus tetradactylus. See Four-toed sengis
Petrogale spp. *See* Rock wallabies
Petrogale assimilis. See Allied rock-wallabies
Petrogale concinna. See Narbaleks
Petrogale penicillata. See Brush-tailed rock wallabies
Petrogale persephone. See Proserpine rock wallabies
Petrogale xanthopus. See Yellow-footed rock wallabies
Petroica spp., 11:106
Petroica goodenovii. See Red-capped robins
Petroica multicolor. See Scarlet robins
Petroica phoenicea. See Flame robins
Petroica traversi. See Chatham Islands black robins
Petroicidae. *See* Australian robins
Petrolacosaurus spp., 7:15
Petroleum flies, 3:359, 3:*362*, 3:*365*, 3:367
Petromuridae. *See* Dassie rats
Petromus typicus. See Dassie rats
Petromyscinae, 16:282, 16:285
Petromyscus collinus. See Pygmy rock mice
Petromyzon spp., 4:83
Petromyzon marinus. See Sea lampreys
Petromyzonidae, 4:6, 4:83
Petromyzoniformes. *See* Lampreys
Petronia brachydactyla. See Pale rock sparrows
Petronia petronia. See Rock sparrows
Petropedetes spp., 6:35, 6:247, 6:249
Petropedetinae, 6:247, 6:251
Petropseudes spp., 13:113
Petropseudes dahli. See Rock ringtails
Petroscirtes breviceps, 5:345–346
Petrosina, 1:78
Petrosum, 12:9
Pets, 12:15, 12:173
  amphibians as, 6:54–55, 6:58
  cats and dogs as, 14:291
  humans and, 14:*291*
  reptilian, 7:53, 7:*53t* (*See also* Humans; specific species)
  See also Cagebirds; specific animals
Pettigrew, Jack, 13:333
Petzold, H.-G., 7:448
Peucedramidae, 11:286
Peucedramus taeniatus. See Olive warblers
Pewees, 10:270, 10:274
Pezophaps solitaria. See Rodrigues solitaires
Pezoporus occidentalis. See Night parrots
Pezoporus wallicus. See Ground parrots
Pfrender, M., 6:432
Pfungst, Oskar, 12:155
pH levels, 4:37
Phacellodomus spp. *See* Thornbirds
Phacellodomus ruber. See Greater thornbirds
Phacochoerinae, 15:275
Phacochoerus spp., 15:276
Phacochoerus aethiopicus. See Warthogs
Phacochoerus aethiopicus delamerei. See Warthogs
Phacochoerus africanus. See Common warthogs
Phaenicophaeinae. *See* Cuckoos
Phaenicophaeus pyrrhocephalus, 9:315
Phaenomys spp., 16:264
Phaenomys ferrugineus. See Rio de Janeiro rice rats
Phaeognathus spp. *See* Red Hills salamanders
Phaeognathus hubrichti. See Red Hills salamanders
Phaeophilacris spectrum, 3:206
Phaeostigma (Phaeostigma) notata. See Common European snakeflies
Phaethon spp. *See* Tropicbirds
Phaethon aethereus. See Red-billed tropicbirds
Phaethon lepturus. See White-tailed tropicbirds
Phaethon rubricauda. See Red-tailed tropicbirds
Phaethontidae. *See* Tropicbirds
Phaethornis spp., 9:443, 9:447–448
Phaethornis guy. See Green hermits
Phaethornis koepckeae. See Koepcke's hermits
Phaethornis longuemareus. See Little hermits
Phaethornis ruber. See Reddish hermits
Phaethornis yaruqui. See Hermit hummingbirds
Phaethornithinae. *See* Hermits
Phaetusa simplex. See Large-billed terns
Phago spp., 4:69
Phagocata spp., 1:188
Phainopepla nitens. See Phainopeplas
Phainopeplas, 10:447, 10:*448*, 10:*450*, 10:452–*453*
Phainoptila melanoxantha. See Black-and-yellow silky flycatchers
Phalacrocoracidae. *See* Cormorants
Phalacrocorax spp. *See* Cormorants
Phalacrocorax atriceps. See Blue-eyed imperial shags
Phalacrocorax auritus. See Double-crested cormorants
Phalacrocorax carbo. See Great cormorants
Phalacrocorax carunculatus. See New Zealand king shags
Phalacrocorax olivaceus. See Olivaceous cormorants
Phalacrocorax pelagicus. See Pelagic cormorants
Phalacrocorax penicillatus. See Brandt's cormorants
Phalacrocorax perspicillatus. See Spectacled cormorants
Phalacrognathus muelleri, 3:322
Phalaenoptilus nuttallii. See Common poorwills
Phalanger spp., 13:57–58
Phalanger alexandrae. See Gebe cuscuses
Phalanger gymnotis. See Ground cuscuses
Phalanger lullulae. See Woodlark Island cuscuses
Phalanger maculatus. See Spotted cuscuses
Phalanger matanim. See Telefomin cuscuses
Phalanger orientalis. See Common cuscuses
Phalanger ornatus. See Moluccan cuscuses
Phalanger sericeus. See Silky cuscuses
Phalangeridae, 13:31, 13:32, **13:57–67**, 13:*62*, 13:*66t*
Phalangeroidea, 13:31
Phalangium opilio. See Common harvestmen
Phalangopsidae, 3:203, 3:204
Phalanx growth strategy, 1:26–27
Phalaropes, 9:102, 9:104, 9:175–180, 9:*178*, 9:*181*, 9:187–188
Phalaropodinae. *See* Phalaropes
Phalaropus fulicaria. See Gray phalaropes
Phalcoboenus australis. See Striated caracaras
Phallocaecilius spp., 3:244
Phallodeum, 6:32–33
Phallodrilus macmasterae, 2:67
Phallostethidae. *See* Priapium fishes
Phallusia julinea, 1:*451*
Phaner spp. *See* Masoala fork-crowned lemurs
Phaner furcifer. See Masoala fork-crowned lemurs
Phaneropterinae, 3:205, 3:206
Phanidae, 3:361
Phantasmal poison frogs, 6:53, 6:*202*, 6:208
Phantom crane flies, 3:*360*
Phantom glass catfishes. *See* Glass catfishes
Phaon spp., 3:55
Phaps chalcoptera. See Common bronzewings
Pharetronida, 1:57
Pharmaceuticals, 1:44–45, 1:49, 2:42–43
  See also Humans
Pharomachrus auriceps. See Golden-headed quetzels
Pharomachrus mocinno. See Resplendent quetzals
Pharyngidea, 2:59
Pharyngodon petasatus. See Yucatecan shovel-headed treefrogs
Phascalonus gigas, 13:51
Phascogale calura. See Red-tailed phascogales
Phascogale tapoatafa. See Brush-tailed phascogales
Phascogales, 12:281–282, 12:284, 12:*289*, 12:290
Phascolarctidae. *See* Koalas
Phascolarctos cinereus. See Koalas
Phascolarctos cinereus adustus. See Queensland koalas
Phascolarctos cinereus cinereus, 13:43
Phascolarctos cinereus victor, 13:43
Phascolarctos stirtoni, 13:43
Phascolion spp., 2:98
Phascolion strombus, 2:98
Phascolosomatidea, 2:97
Phascolosorex spp., 12:288, 12:290
Phascolosorex dorsalis. See Narrow-striped marsupial shrews
Phasianidae, **8:433–453**, 8:*438–440*
  behavior, 8:435
  conservation status, 8:436–437
  distribution, 8:*433*, 8:434
  evolution, 8:433
  feeding ecology, 8:435–436
  habitats, 8:434–435
  humans and, 8:437
  physical characteristics, 8:433–434
  reproduction, 8:436
  species of, 8:*441–452*
  taxonomy, 8:433
Phasianini. *See* Pheasants
Phasianus colchicus. See Ring-necked pheasants
Phasmahyla spp., 6:227
Phasmatidae, 3:221, 3:222
Phasmida, **3:221–232**, 3:*225*, 3:*226*, 3:*227*
  behavior, 3:222–223
  conservation status, 3:223–224
  distribution, 3:222
  evolution, 3:221
  feeding ecology, 3:223
  habitats, 3:222
  humans and, 3:224
  physical characteristics, 3:221–222
  reproduction, 3:223
  species of, 3:*228–232*
  taxonomy, 3:221
Phasmodinae, 3:203
Pheasant cuckoos, 9:*317*, 9:*328*
Pheasant finches. *See* Common waxbills

Pheasant pigeons, 9:241, 9:247, 9:*254*, 9:*263*
Pheasant-tailed jacanas, 9:107, 9:108, 9:*111*, 9:*112*–113
Pheasants, **8:433–453**, 8:*436*, 8:*438*
behavior, 8:401, 8:435
conservation status, 8:436–437
distribution, 8:400, 8:*433*, 8:434
evolution, 8:433
feeding ecology, 8:435–436
habitats, 8:434–435
humans and, 8:437
physical characteristics, 8:433–434
reproduction, 8:436
species of, 8:*441–452*
taxonomy, 8:399, 8:433
Phedina borbonica. *See* Mascarene martins
Phedina brazzae. *See* Brazza's swallows
Phegornis mitchellii. *See* Mitchell's plovers
Pheidole megacephala. *See* Big-headed ants
Phelsuma spp. *See* Malagasy day geckos
Phelsuma edwardnewtoni. *See* Giant geckos
Phelsuma madagascariensis. *See* Madagascar day geckos
Phenablennius spp., 5:342
Phenacoccus manihoti. *See* Cassava mealybugs
Phenacodontidae, 15:132
Phenacolepadidae, 2:439
Phenacomys intermedius. *See* Western heather voles
Pherallodichthys spp., 5:355
Pherallodiscus spp., 5:355
Pherallodus spp., 5:355
Pheromones, 3:60, 4:62, 4:335, 6:44–45, 6:45–46, 12:37, 12:80
See also Physical characteristics
Pheronema carpenteri. *See* Bird's nest sponges
Phialella zappai. *See* Zappa's jellyfish
Phibalura spp. *See* Cotingas
Phibalura flavirostris, 10:307
Philadelphia Zoo, 12:203
Philander spp. *See* Gray four-eyed opossums
Philander andersoni. *See* Black four-eyed opossums
Philander opossum. *See* Gray four-eyed opossums
Philansius plebeius, 3:376
Philantomba spp., 16:73
Philautus spp., 6:37, 6:291, 6:292, 6:293
Philautus alticola, 6:293
Philautus carinensis, 6:292
Philautus poecilus, 6:293
Philautus schmackeri, 6:293
Philautus surdus. *See* Luzon bubble-nest frogs
Philby's rock partridges, 8:434
Philemon spp., 11:235
Philemon corniculatus. *See* Noisy friarbirds
Philemon fuscicapillus. *See* Dusky friarbirds
Philemon subcorniculatus. *See* Seram friarbirds
Philepitta spp. *See* Asities
Philepitta castanea. *See* Velvet asities
Philepitta schlegeli. *See* Schelgel's asites
Philepittidae. *See* Asities; Sunbird-asities
Philesturnus carunculatus. *See* Tiekes
Philetairus socius. *See* Sociable weavers
Philippine badgers. *See* Philippine stink badgers
Philippine barbourulas, 6:84, 6:88
Philippine brown deer, 15:371*t*
Philippine buttonquails. *See* Barred buttonquails
Philippine colugos, 13:302, 13:*303*, 13:*304*
Philippine creepers, 10:505, **11:183–188**, 11:*185*
Philippine crocodiles, 7:59
Philippine fairy bluebirds, 10:416, 10:*418*, 10:*423*–424
Philippine flying lemurs. *See* Philippine colugos
Philippine Gray's monitors, 7:363
Philippine gymnures, 13:207
Philippine hanging-parrots, 9:*281*, 9:295–*296*
Philippine leafbirds, 10:415, 10:417, 10:*418*, 10:*420*
Philippine megapodes, 8:406
Philippine pelicans. *See* Spot-billed pelicans
Philippine sambars. *See* Philippine brown deer
Philippine spotted deer. *See* Visayan spotted deer
Philippine stink badgers, 14:333*t*
Philippine tarsiers, 14:*92*, 14:*93*, 14:94, 14:*95*, 14:*96*, 14:*97*
Philippine tree shrews, 13:292, 13:*293*, 13:295–*296*
Philippine tube-nosed fruit bats, 13:*334*
Philippine warty pigs, 12:179, 15:280, 15:290*t*
Philippine wood shrews. *See* Mindanao gymnures
Philippines. *See* Oriental region
Phillips, John, 7:360
Philobythiidae, 1:225
Philodina spp., 1:*259*, 1:260
Philodina gregaria, 1:*261*
Philodryas spp., 7:467
Philomachus pugnax. *See* Ruffs
Philopatry, 11:5
Philoria spp., 6:34, 6:141
Philoria frosti. *See* Baw Baw frogs
Philortyx fasciatus. *See* Barred quails
Philydor novaesi. *See* Alagoas foliage-gleaners
Philydor pyrrhodes. *See* Cinnamon-rumped foliage-gleaners
Philydorinae. *See* Leafcreepers
Phiomia, 15:161
Phipson's shieldtail snakes, 7:392, 7:*393*–394
Phlaeothripidae, 3:68, 3:281, 3:282
Phlebobranchia, 1:452
Phlebonotus spp., 3:150
Phleocryptes melanops. *See* Wren-like rushbirds
Phlitrum, 12:8
Phlogophilus harterti. *See* Peruvian piedtails
Phlyctimantis spp., 6:281
Phobaeticus kirbyi, 3:13, 3:221
Phobias, insect, 3:77
Phoca caspica. *See* Caspian seals
Phoca hispida. *See* Ring seals
Phoca sibirica. *See* Baikal seals
Phoca vitulina. *See* Harbor seals
Phocanema decipiens. *See* Cod worms
Phocarctos hookeri. *See* Hooker's sea lions
Phocidae. *See* True seals
Phocinae, 14:417
Phocoena spp. *See* Porpoises
Phocoena dalli. *See* Dall's porpoises
Phocoena dioptrica. *See* Spectacled porpoises
Phocoena phocoena. *See* Harbor porpoises
Phocoena sinus. *See* Vaquitas
Phocoena spinipinnis. *See* Burmeister's porpoises
Phocoenidae. *See* Porpoises
Phocoenoides spp., 15:33
Phodilus spp., 9:335
Phodilus badius. *See* Oriental bay owls
Phodopus spp. *See* Dwarf hamsters
Phodopus sungorus. *See* Dzhungarian hamsters
Phoebastria spp. *See* Albatrosses
Phoebes, 10:270, 10:273, 10:274
Phoebetria spp. *See* Albatrosses
Phoebetria fusca. *See* Sooty albatrosses
Phoebetria palpebrata. *See* Light-mantled albatrosses
Phoenicircus spp. *See* Cotingas
Phoenicircus nigricollis. *See* Black-necked red cotingas
Phoeniconaias minor. *See* Lesser flamingos
Phoenicoparrus andinus. *See* Andean flamingos
Phoenicoparrus chilensis. *See* Chilean flamingos
Phoenicoparrus jamesi. *See* James' flamingos
Phoenicopteridae. *See* Flamingos
Phoenicopteriformes. *See* Flamingos
Phoenicopterus chilensis. *See* Chilean flamingos
Phoenicopterus ruber. *See* Greater flamingos
Phoeniculidae. *See* Woodhoopoes
Phoeniculinae. *See* Woodhoopoes
Phoeniculus bollei. *See* White-headed hoopoes
Phoeniculus castaneiceps. *See* Forest woodhoopoes
Phoeniculus damarensis. *See* Violet woodhoopoes
Phoeniculus purpureus. *See* Green woodhoopoes
Phoeniculus somaliensis. *See* Black-billed woodhoopoes
Phoenicurus ochruros. *See* Black redstarts
Phoenicurus phoenicurus. *See* Redstarts
Phoenix Zoo, 12:*209*
Pholadidae, 2:452
Pholcus phalangioides. *See* Long-bodied cellar spiders
Pholidae. *See* Gunnels
Pholidichthyidae. *See* Convict blennies
Pholidornis rushiae. *See* Tit-hylias
Pholidoskepia, 2:379
Pholidota. *See* Pangolins
Pholis gunnellus. *See* Atlantic butterfishes
Pholis laeta, 5:312
Pholis nebulosa. *See* Tidepool gunnels
Pholis ornata. *See* Saddleback gunnels
Phoraspis spp., 3:150
Phoresy, 2:32, 3:251
See also Behavior
Phoridae, 3:358, 3:361
Phoronids, 1:9, 1:13, 1:23, 2:13, **2:491–495**, 2:*493*
Phoronis spp., 2:491
Phoronis australis, 2:491
Phoronis hippocrepia, 2:491
Phoronis ijimai, 2:491, 2:*493*, 2:494, 2:*495*
Phoronis muelleri, 2:491
Phoronis ovalis, 2:491
Phoronis pallida, 2:491
Phoronis psammophila, 2:491
Phoronopsis spp., 2:491
Phoronopsis albomaculata, 2:491
Phoronopsis californica. *See* Golden phoronoids
Phoronopsis harmeri, 2:491, 2:*493*, 2:494, 2:*495*
Phorusrhacidae, 9:1
Phosichthyidae. *See* Lightfishes
Photoblepharon palpebratus. *See* Eyelight fishes
Photokinesis, 1:38
See also Behavior
Photoperiodicity, 12:74–75
Photophores, 4:392, 4:422, 4:441

Photosynthesis, 1:24, 1:32–34
Phototaxis, 1:39
See also Behavior
Phototelotaxis, 1:39
See also Behavior
Phototropotaxis, 1:39
See also Behavior
Phoxinus spp., 4:298
Phoxinus phoxinus. See Eurasian minnows
Phractocephalus hemioliopterus. See Redtail catfishes
Phractolaemidae, 4:57
Phractolaemus spp., 4:289, 4:290, 4:291
Phractolaemus ansorgii. See African mudfishes
Phragmophora, 1:433
Phronima spp., 2:261
Phronima sedentaria. See Coopers of the sea
Phrygilus unicolor. See Plumbeous sierra-finches
Phrynarachne spp. *See* Crab spiders
Phrynarachne spp. *See* Crab spiders
Phrynidium varium. See Harlequin frogs
Phrynixini, 3:55
Phrynobatrachus spp., 6:247, 6:248, 6:250
Phrynobatrachus capensis. See Micro frogs
Phrynocephalus spp., 7:211
Phrynocephalus mystaceus. See Toad-headed agamas
Phrynodon spp., 6:35, 6:247, 6:251
Phrynodon sandersoni. See Sanderson's hook frogs
Phrynoglossus spp., 6:247, 6:251
Phrynohyas spp., 6:227
Phrynohyas resinifictrix, 6:37, 6:230
Phrynohyas venulosa, 6:230
Phrynomantis spp., 6:302, 6:303
Phrynomantis bifasciatus. See Banded rubber frogs
Phrynomerinae, 6:302
Phrynops spp., 7:77
Phrynops hilarii. See Toad head turtles
Phrynopus spp., 6:156, 6:158
Phrynosoma spp. *See* Horned lizards
Phrynosoma cornutum. See Texas horned lizards
Phrynosoma mcalli. See Flat-tailed horned lizards
Phrynosomatinae, 7:243, 7:244
Phrynus parvulus, 2:*337*, 2:*341–342*
Phtheirichthys spp., 5:211
Phthia picta. See Tomato bugs
Phthiraptera, **3:249–257**, 3:*253*
Phycidae, 5:25, 5:26, 5:27, 5:28, 5:29
Phycinae, 5:25
Phycis spp., 5:28
Phycodurus eques. See Leafy seadragons
Phylactolaemata. *See* Freshwater bryozoans
Phylidonyris novaehollandiae. See New Holland honeyeaters
Phyllastrephus spp. *See* African greenbuls
Phyllastrephus albigularis. See White-throated greenbuls
Phyllastrephus cabanisi. See Cabanis's greenbuls
Phyllastrephus cerviniventris. See Gray-olive bulbuls
Phyllastrephus flavostriatus. See Yellow-streaked greenbuls
Phyllastrephus icterinus. See Icterine greenbuls
Phyllastrephus poliocephalus. See African gray-headed greenbuls
Phyllastrephus scandens. See Leaf-loves
Phyllastrephus strepitans. See Northern brownbuls
Phyllidia coelestis. See Nudibranchs
Phylliidae, 3:221, 3:222, 3:231
Phyllium bioculatum, 3:*221*
See also Javan leaf insects
Phyllium (Pulchriphyllium) bioculatum. See Javan leaf insects
Phyllium (Pulchriphyllium) giganteum, 3:221
Phyllobates spp., 6:66, 6:197, 6:198, 6:200
Phyllobates terribilis. See Golden dart-poison frogs
Phyllobothriidae, 1:226
Phyllobothrium squali, 1:*232*, 1:*240*, 1:241–242
Phyllocarida. *See* Leptostracans
Phyllocnistis citrella. See Citrus leaf miners
Phyllodocids, 2:47
Phyllodytes spp., 6:37, 6:227
Phyllomedusa spp., 6:49, 6:193, 6:227, 6:228–229
Phyllomedusa bicolor, 6:227, 6:230
Phyllomedusa hypochondrialis. See Tiger-leg monkey frogs
Phyllomedusa vaillanti. See White-lined treefrogs
Phyllomedusinae, 6:41, 6:226, 6:230
Phyllomyias virescens urichi. See Urich's tyrannulets
Phyllonastes spp., 6:156
Phyllonycterinae, 13:413, 13:415
Phyllonycteris aphylla, 13:420
Phyllonycteris poeyi. See Cuban flower bats
Phyllophora spp., 3:202
Phyllophorinae. *See* Giant helmet katydids
Phyllopteryx teaniolatus. See Weedy seadragons
Phylloriza peronlesueri, 1:25
Phylloscartes beckeri. See Bahia tyrannulets
Phylloscartes ceciliae. See Alagoas tyrannulets
Phylloscartes lanyoni. See Antioquia bristle tyrants
Phylloscartes roquettei. See Minas Gerais tyrannulets
Phylloscopus spp., 11:3, 11:4–5, 11:7
Phylloscopus borealis. See Arctic warblers
Phylloscopus collybita. See Chiffchaffs
Phylloscopus sibilatrix. See Wood warblers
Phyllostomidae. *See* American leaf-nosed bats
Phyllostominae, 13:413, 13:414, 13:416, 13:417, 13:420
Phyllostomus discolor. See Pale spear-nosed bats
Phyllostomus hastatus. See Greater spear-nosed bats
Phyllotines, 16:264
Phyllotis spp., 16:265
Phyllotis darwini. See Leaf-eared mice
Phyllotis sublimis, 16:267
Phylloxera, 3:264
Phylloxoridae, 3:54
Phylobythoides stunkardi, 1:229
Phylogenetics, 2:7–14, 2:15, 3:53
cognition and, 12:149–150
molecular genetics and, **12:26–35**
social behavior and, 12:147–148
Phylogeny and Classification of Birds (Sibley and Ahlquist), 10:505
Phymateus spp., 3:208
Physalaemus spp., 6:66, 6:67, 6:156, 6:158
Physalaemus nattereri, 6:*45*, 6:*64*
Physalaemus pustulosus. See Túngara frogs
Physalia spp., 1:29
Physalia physalis. See Portuguese men of war
Physeter spp., 15:73, 15:75
Physeter macrocephalus. See Sperm whales
Physeteridae. *See* Sperm whales
Physical characteristics
amphibians, 6:3, **6:15–27**, 6:*16*
anurans, 6:*19*, 6:*22*, 6:25–26, 6:62–*63*, 6:66
African treefrogs, 6:281, 6:285–289
Amero-Australian treefrogs, 6:215, 6:226–228, 6:233–242, 6:281
Arthroleptidae, 6:265, 6:268–270
Asian toadfrogs, 6:110, 6:113–116
Asian treefrogs, 6:248, 6:281, 6:291–292, 6:295–299
Australian ground frogs, 6:140, 6:143–145
Bombinatoridae, 6:26, 6:83, 6:86–88
Bufonidae, 6:183–184, 6:188–194
Discoglossidae, 6:89, 6:92–94
Eleutherodactylus spp., 6:*157*
frogs, 6:*20*, 6:*23*, 6:25–26, 6:62–63
ghost frogs, 6:131, 6:133
glass frogs, 6:215–216, 6:219–223
leptodactylid frogs, 6:157, 6:162–171
Madagascan toadlets, 6:317, 6:319–320
Mesoamerican burrowing toads, 6:95–96
Myobatrachidae, 6:148, 6:151–153
narrow-mouthed frogs, 6:302–*303*, 6:308–315
New Zealand frogs, 6:69, 6:72–74
parsley frogs, 6:127, 6:129
Pipidae, 6:99–100, 6:103–106
poison frogs, 6:197–198, 6:203–209
Ruthven's frogs, 6:211
Seychelles frogs, 6:136, 6:137
shovel-nosed frogs, 6:273
spadefoot toads, 6:119, 6:124–125
tadpoles, 6:39–41, 6:62–63, 6:*66*
tailed frogs, 6:26, 6:77, 6:80–81
three-toed toadlets, 6:179, 6:181–182
toads, 6:25–26, 6:62–63
true frogs, 6:248–250, 6:255–263
vocal sac-brooding frogs, 6:175–176
Xenopus spp., 6:100
caecilians, 6:15, 6:26–27, 6:411–412, 6:415–416
American tailed, 6:416, 6:417
Asian tailed, 6:416, 6:419, 6:422–423
buried-eye, 6:431–432, 6:433
Kerala, 6:426, 6:428
tailless, 6:435, 6:439–440
jaw-closing mechanisms, 6:415–416
lissamphibians, 6:15–27, 6:*16*, 6:26
salamanders, 6:15, 6:*18*, 6:23–*25*, 6:*24*, 6:323–324
amphiumas, 6:405–406, 6:409–410
Asiatic, 6:24, 6:336, 6:339–341
Asiatic giant, 6:343
Caudata, 6:323–324
lungless, 6:17, 6:24, 6:323, 6:325, 6:392, 6:395–403
mole, 6:355–356, 6:358–360
Pacific giant, 6:349–350, 6:352
Proteidae, 6:378, 6:381–382
Salamandridae, 6:17, 6:24, 6:323–324, 6:325, 6:363, 6:370–375
sirens, 6:327–328, 6:331–332
torrent, 6:385, 6:387
birds, 8:1–9, 8:*5*, 8:*6*, 8:*15*
albatrosses, 8:107, 8:113–114, 8:118–121

Alcidae, 9:220, 9:224–228
anhingas, 8:183, 8:201
Anseriformes, 8:363–364
ant thrushes, 10:239–240, 10:245–256
Apodiformes, 9:415–417
asities, 10:187–188, 10:190
Australian chats, 11:65, 11:67
Australian creepers, 11:133, 11:137–139
Australian fairy-wrens, 11:45, 11:48–53
Australian honeyeaters, 11:235–236, 11:241–253
Australian robins, 11:105, 11:108–112
Australian warblers, 11:55, 11:59–63
babblers, 10:506, 10:511–523
barbets, 10:113–114, 10:119–122
barn owls, 9:335, 9:340–343
bee-eaters, 10:39
birds of paradise, 11:489–490, 11:495–501
bitterns, 8:233–234, 8:239–241
body temperature, 8:9–10
Bombycillidae, 10:447, 10:451–454
boobies, 8:184, 8:211
bowerbirds, 11:477–478, 11:483–488
broadbills, 10:177, 10:181–185
bulbuls, 10:395–396, 10:402–412
bustards, 9:91, 9:96–99
buttonquails, 9:12, 9:16–21
Caprimulgiformes, 9:367–368
cassowaries, 8:75
Charadriidae, 9:161–162, 9:166–172
chats, 10:483–485
chickadees, 11:155–156, 11:160–165
chowchillas, 11:69–70, 11:72–73
Ciconiiformes, 8:233–234
circulation, 8:9–10
Columbidae, 9:247, 9:255–266
Columbiformes, 9:241–243
condors, 8:276
Coraciiformes, 10:2
cormorants, 8:184, 8:201, 8:205–206
Corvidae, 11:503–504, 11:512–522
cotingas, 10:305–306, 10:312–321
crab plovers, 9:121
Cracidae, 8:413–414
cranes, 9:24, 9:31–35
cuckoo-shrikes, 10:385, 10:389–393
cuckoos, 9:311–312, 9:318–329
dippers, 10:475, 10:479–481
diving-petrels, 8:143, 8:145–146
doves, 9:247, 9:255–266
drongos, 11:437–438, 11:441–444
ducks, 8:369–370
eagles, 8:318–319
ears, 8:6
elephant birds, 8:103–104
emus, 8:83–84
Eupetidae, 11:75–76, 11:78–80
eyes, 8:5–6
fairy bluebirds, 10:415–416, 10:423–424
Falconiformes, 8:313–314
false sunbirds, 10:187–188, 10:191
fantails, 11:84, 11:89–94
feathers, 8:7, 8:8–9, 8:*11*, 8:*12*, 8:*14*, 8:*16*
feet, 8:*5*, 8:*9*
finches, 11:323, 11:328–339
flamingos, 8:303–304
flowerpeckers, 11:189, 11:193–198
frigatebirds, 8:193, 8:197–198
frogmouths, 9:377–378, 9:381–385
Galliformes, 8:399
gannets, 8:211
geese, 8:369–370
Glareolidae, 9:151, 9:155–160
grebes, 8:169–170, 8:174–180
Gruiformes, 9:1
guineafowl, 8:425–426
gulls, 9:211–217
hammerheads, 8:261
Hawaiian honeycreepers, 11:341, 11:346–351
hawks, 8:318–319
hedge sparrows, 10:459, 10:462–464
herons, 8:233–234, 8:239–241
hoatzins, 8:466
honeyguides, 10:137, 10:141–144
hoopoes, 10:61
hornbills, 10:71–72, 10:78–83
hummingbirds, 9:437, 9:438–442, 9:455–467
ibises, 8:233–234, 8:291
Icteridae, 11:301–302, 11:309–321
ioras, 10:415–416, 10:419–420
jacamars, 10:91–92
jacanas, 9:107–108, 9:112–114
kagus, 9:41
kingfishers, 10:6–7, 10:12–13
kiwis, 8:89
lapwings, 9:167–172
Laridae, 9:204–205, 9:211–217
larks, 10:341–342, 10:*345*, 10:348–354
leafbirds, 10:415–416, 10:420–423
limpkins, 9:37
logrunners, 11:69–70, 11:72–73
long-tailed titmice, 11:141, 11:145–146
lyrebirds, 10:329–330, 10:334
magpie-shrikes, 11:467–468, 11:471–475
manakins, 10:295, 10:299–303
mesites, 9:5, 9:8–9
moas, 8:95–96
monarch flycatchers, 11:97, 11:100–102
motmots, 10:32, 10:35–37
moundbuilders, 8:404
mousebirds, 9:469, 9:473–476
mudnest builders, 11:453, 11:456–457
muscles, 8:*6*
New World blackbirds, 11:309–321
New World finches, 11:263, 11:269–282
New World quails, 8:455–456
New World vultures, 8:233–234, 8:276
New World warblers, 11:287–288, 11:293–299
New Zealand wattle birds, 11:447, 11:449–450
New Zealand wrens, 10:204, 10:206–207
nightjars, 9:401–202, 9:408–413
nuthatches, 11:167–168, 11:171–174
oilbirds, 9:373, 9:374
Old World flycatchers, 11:25–26, 11:30–43
Old World warblers, 11:2–3, 11:10–23
Oriolidae, 11:427–428, 11:431–434
ostriches, 8:99–100
ovenbirds, 10:209, 10:214–228
owlet-nightjars, 9:387–388, 9:391–392
owls, 9:331–332
oystercatchers, 9:125, 9:130–131
painted-snipes, 9:118–119
palmchats, 10:455
pardalotes, 11:201, 11:204–206
parrots, 9:275–276, 9:283–297
Passeriformes, 10:171–172
Pelecaniformes, 8:183–184
pelicans, 8:183–184, 8:225
penduline titmice, 11:147, 11:151–152
penguins, 8:148–149, 8:153–157
Phasianidae, 8:433–434
pheasants, 8:433–434
Philippine creepers, 11:183, 11:186–187
Picidae, 10:147–148, 10:154–167
Piciformes, 10:86
pigeons, 9:247, 9:255–266
pipits, 10:371–372, 10:378–383
pittas, 10:193, 10:197–200
plantcutters, 10:325–326, 10:327–328
plovers, 9:166–172
potoos, 9:395, 9:399–400
pratincoles, 9:155–160
Procellariidae, 8:123, 8:129–130
Procellariiformes, 8:107–108
pseudo babblers, 11:127
puffbirds, 10:101–102, 10:106–110
rails, 9:46, 9:57–67
Raphidae, 9:270, 9:272–273
Recurvirostridae, 9:133–134, 9:137–140
respiration, 8:9–10
rheas, 8:69
rollers, 10:51–52, 10:55–58
sandgrouse, 9:231, 9:235–238
sandpipers, 9:175–176, 9:182–188
screamers, 8:393–394
scrub-birds, 10:337, 10:339
secretary birds, 8:343
seedsnipes, 9:189, 9:193–195
seriemas, 9:85, 9:88–89
sharpbills, 10:291–292
sheathbills, 9:197–198, 9:200
shoebills, 8:287
shrikes, 10:425–426, 10:432–438
skeletal system, 8:*6*, 8:9, 8:*15*
smell, 8:6
sparrows, 11:397, 11:401–405
spoonbills, 8:233, 8:291
storks, 8:233–234, 8:265
storm-petrels, 8:135, 8:139–141
Struthioniformes, 8:54
Sturnidae, 11:407–408, 11:414–425
sunbirds, 11:207–208, 11:213–225
sunbitterns, 9:73
sungrebes, 9:69, 9:71–72
swallows, 10:357–358, 10:363–369
swans, 8:369–370
swifts, 9:*417*, 9:421–422, 9:426–431
tapaculos, 10:257–258, 10:262–267
terns, 9:211–217
thick-knees, 9:143, 9:147–148
thrushes, 10:483–485
tinamous, 8:57
titmice, 11:155–156, 11:160–165
todies, 10:25–26, 10:29
toucans, 10:125–126, 10:130–134
touch, 8:6, 8:8
tree swifts, 9:433, 9:435
treecreepers, 11:177, 11:180–181
trogons, 9:477, 9:481–485
tropicbirds, 8:184, 8:187, 8:190–191
trumpeters, 9:77–78, 9:81–82
turacos, 9:299–300, 9:304–310
typical owls, 9:345–346, 9:354–365
tyrant flycatchers, 10:270, 10:278–288
vanga shrikes, 10:439–440, 10:443–444

Vireonidae, 11:255, 11:258–262
wagtails, 10:371–372, 10:378–383
weaverfinches, 11:353–354, 11:*354*, 11:*355*, 11:360–372
weavers, 11:376, 11:382–394
whistlers, 11:116, 11:120–125
white-eyes, 11:227–228, 11:232–233
wings, 8:*4*, 8:5
woodcreepers, 10:229, 10:232–236
woodhoopoes, 10:65, 10:68
woodswallows, 11:459, 11:462–464
wrens, 10:525–526, 10:531–538
fishes, 4:3–6, 4:*4*–7, **4:14–28,** 4:*15–19*, 4:*21*, 4:*22*, 4:*24–27*
Acanthuroidei, 5:391, 5:396–403
Acipenseriformes, 4:213, 4:217–220
Albuliformes, 4:249, 4:252–253
angelsharks, 4:161–162, 4:164
anglerfishes, 5:47, 5:53–56
Anguilliformes, 4:255, 4:260–268
Atheriniformes, 5:68, 5:73–76
Aulopiformes, 4:431–432, 4:436–439
Australian lungfishes, 4:197–198
beardfishes, 5:1, 5:3
Beloniformes, 5:79–80, 5:83–85
Beryciformes, 5:113–114, 5:118–121
bichirs, 4:15, 4:24, 4:209–210, 4:211
blennies, 5:341–342, 5:345–348
bowfins, 4:6, 4:7, 4:14, 4:23, 4:229–230
bullhead sharks, 4:97–98, 4:101–102
Callionymoidei, 5:365–366, 5:369–371
carps, 4:297–298, 4:*299*, 4:*301*, 4:308–313
catfishes, 4:*4*, 4:6, 4:*22*, 4:24, 4:60, 4:62, 4:351–*352*, 4:357–366
characins, 4:336, 4:*337*, 4:342–349
chimaeras, 4:91, 4:94–95
cichlids, 4:62
coelacanths, 4:190–191
Cypriniformes, 4:297–298, 4:*301*, 4:*302*, 4:307–319, 4:321, 4:325–333
Cyprinodontiformes, 5:90–91, 5:96–102
dogfish sharks, 4:151, 4:155–157
dories, 5:123–124, 5:126–129
eels, 4:255, 4:260–268
electric eels, 4:369–370, 4:373–376
elephantfishes, 4:231, 4:238
Elopiformes, 4:243, 4:246–247
Esociformes, 4:379–380, 4:383–386
flatfishes, 5:*450*–452, 5:455–463
Gadiformes, 5:25–26, 5:32–34, 5:36–39
gars, 4:23, 4:221, 4:224–227
Gasterosteiformes, 5:132–133, 5:139–147
gobies, 4:44, 5:374, 5:380–388
Gobiesocoidei, 5:355–356, 5:359–363
Gonorynchiformes, 4:290, 4:293–295
ground sharks, 4:113–114, 4:119–128
Gymnotiformes, 4:369–370, 4:373–376
hagfishes, 4:4, 4:6, 4:25, 4:77, 4:*78*, 4:80–81
herrings, 4:277, 4:*278*, 4:281–287
Hexanchiformes, 4:143–144, 4:146–148
Labroidei, 5:276, 5:284–290, 5:293–294, 5:301–306
labyrinth fishes, 5:427, 5:431–434
lampreys, 4:6, 4:23, 4:83, 4:*84*, 4:88–89
Lampridiformes, 4:*449*, 4:452–454
lanternfishes, 4:441–442, 4:444–446
lungfishes, 4:*18*, 4:201–202, 4:205
mackerel sharks, 4:132, 4:135–141
marine fishes, 4:44–45
minnows, 4:*4*, 4:62, 4:297–298, 4:314–315
morays, 4:44, 4:255, 4:264–266
mudminnows, 4:379–380, 4:383–386
mullets, 5:59, 5:62–65
Ophidiiformes, 5:16, 5:20–22
Orectolobiformes, 4:105, 4:108–110
Osmeriformes, 4:390, 4:395–401
Osteoglossiformes, 4:231–232, 4:237–240
paddlefishes, 4:7, 4:18
Percoidei, 5:195–196, 5:202–208, 5:211, 5:214–217, 5:219–220, 5:225–233, 5:235–237, 5:247–253, 5:255–258, 5:266–268, 5:270–272
Percopsiformes, 5:5, 5:8–12
pikes, 4:379–380, 4:383–386
ragfishes, 5:351–352
Rajiformes, 4:27, 4:*174*–176, 4:*175*, 4:182–188
rays, 4:60, 4:*174*–176, 4:*175*, 4:182–188
Saccopharyngiformes, 4:271, 4:274–276
salmons, 4:60, 4:405, 4:410–419
sawsharks, 4:167–168, 4:170
Scombroidei, 5:405–406, 5:409–418
Scorpaeniformes, 5:157–158, 5:161–162, 5:164, 5:169–178, 5:180, 5:186–193
Scorpaenoidei, 5:164
seahorses, 4:14, 4:26
sharks, 4:18, 4:60
skates, 4:60, 4:*174*–176, 4:*175*, 4:187
snakeheads, 5:437, 5:*438*, 5:442–446
South American knifefishes, 4:369–370, 4:373–376
southern cod-icefishes, 5:321–322, 5:325–328
Stephanoberyciformes, 5:105–106, 5:108–110
stingrays, 4:174–176, 4:190
Stomiiformes, 4:421–*422*, 4:426–429
Stromateoidei, 5:421–422
sturgeons, 4:18, 4:213, 4:217–220
Synbranchiformes, 5:151, 5:154–156
Tetraodontiformes, 5:467–468, 5:474–484
toadfishes, 5:41, 5:44–45
Trachinoidei, 5:332–333, 5:337–339
tunas, 4:*4*, 4:14, 4:19
whale sharks, 4:*3*, 4:14, 4:44, 4:110
wrasses, 4:*4*, 4:*17*, 4:26
Zoarcoidei, 5:310, 5:315–317
insects, 3:4–5, 3:10, **3:17–30,** 3:*23*, 3:68–69
beetles, 3:10, 3:18, 3:*24*, 3:317–319
book lice, 3:243, 3:246–247
bristletails, 3:113, 3:116–117
butterflies, 3:10, 3:*24*, 3:384–385
caddisflies, 3:375, 3:379–381
caterpillars, 3:18
cockroaches, 3:19, 3:21–23, 3:27, 3:147–148, 3:153–158
Coleoptera, 3:22, 3:317–319, 3:327–333
diplurans, 3:4, 3:107, 3:110–111
Diptera, 3:358–359, 3:364–374
dragonflies, 3:5, 3:10, 3:21–22, 3:27, 3:133
earwigs, 3:195–196, 3:199–200
fleas, 3:347–348, 3:352–354
flies, 3:18, 3:358–359
grasshoppers, 3:21–22, 3:27, 3:202–203
Hemiptera, 3:260–261, 3:267–268, 3:270–279
Hymenoptera, 3:19, 3:22, 3:405, 3:411–424
lacewings, 3:306, 3:311–314
Lepidoptera, 3:19, 3:24, 3:384–385, 3:393–404
mantids, 3:19, 3:22, 3:178–179, 3:183–186
Mantophasmatodea, 3:217–218
mayflies, 3:125, 3:129–130
Mecoptera, 3:341–342, 3:345–346
Megaloptera, 3:289–290, 3:293–294
mosquitos, 3:*24*, 3:358–359
noctuid moths, 3:5
Odonata, 3:133, 3:137–138
Orthoptera, 3:202–203, 3:211–215
Phasmida, 3:221–222, 3:228–232
Phthiraptera, 3:249–250, 3:254–256
proturans, 3:4, 3:93, 3:96–97
rock-crawlers, 3:189, 3:192–193
snakeflies, 3:297, 3:300–302
springtails, 3:99, 3:103–104
stoneflies, 3:141–142, 3:145–146
strepsipterans, 3:335, 3:338–339
termites, 3:163, 3:170–174
thrips, 3:281, 3:285–287
Thysanura, 3:119, 3:122–123
webspinners, 3:233, 3:236–237
zorapterans, 3:239–240, 3:241
lower metazoans and lesser deuterostomes
acoels, 1:179–180, 1:181–182
anoplans, 1:245, 1:249–250
arrow worms, 1:433–434, 1:437–441
box jellies, 1:147, 1:*148*, 1:151–152
calcareous sponges, 1:*58*, 1:61–64
cnidarians
Anthozoa, 1:103–*105*, 1:111–120
Hydrozoa, 1:123–127, 1:*124*, 1:134–145
comb jellies, 1:169, 1:173–177
demosponges, 1:*78*–79, 1:82–86
echinoderms
Crinoidea, 1:355–*356*, 1:361–364
Echinoidea, 1:401–402, 1:*403*, 1:408–414
Ophiuroidea, 1:387, 1:*388*, 1:393–398
sea cucumbers, 1:*419–420*, 1:425–431
sea daisies, 1:381–*382*, 1:385–386
sea stars, 1:367–*368*, 1:373–379
enoplans, 1:253, 1:*254*, 1:256–257
entoprocts, 1:319, 1:323–325
flatworms
free-living flatworms, 1:*186–187*, 1:190–195
monogeneans, 1:213–214, 1:218–223
tapeworms, 1:227–229, 1:234–241
gastrotrichs, 1:269–270, 1:272–273
girdle wearers, 1:343–344, 1:347–349
glass sponges, 1:67–68, 1:*69*, 1:71–75
gnathostomulids, 1:331, 1:334–335
hair worms, 1:305, 1:308–309
hemichordates, 1:443–444, 1:447–449
jaw animals, 1:327–*328*
jellyfish, 1:*4*, 1:153–154, 1:*155*, 1:160–167
kinorhynchs, 1:275–276, 1:279–280
lancelets, 1:485–487, 1:*486*, 1:490–496
larvaceans, 1:473, 1:476–478
nematodes
roundworms, 1:*4*, 1:283–*284*, 1:287–291
secernenteans, 1:293–294, 1:297–304
Orthonectida, 1:99, 1:101–102
placozoans, 1:87–88
priapulans, 1:337, 1:340–341
rhombozoans, 1:93, 1:96–98
rotifers, 1:259–260, 1:264–267
Salinella salve, 1:91

salps, 1:467, 1:471–472
sea squirts, 1:*452*–454, 1:458–466
sorberaceans, 1:479–480, 1:482–483
sponges, 1:*5*
thorny headed worms, 1:311–312, 1:314–316
Trematoda, 1:197–198, 1:203–205, 1:207, 1:209–211
wheel wearers, 1:351–352
mammals, **12:36–51**
aardvarks, 15:155–156
agoutis, 16:407, 16:411–414, 16:415*t*
anteaters, 12:39, 12:46, 13:171–172, 13:177–179
armadillos, 13:182, 13:187–190, 13:190*t*–192*t*
Artiodactyla, 12:40, 12:79, 15:138–140, 15:267–269
aye-ayes, 14:85, 14:*86*
baijis, 15:19–20
bandicoots, 13:1–2
dry-country, 13:9–10, 13:14–16, 13:16*t*–17*t*
rainforest, 13:9–10, 13:14–17*t*
bats, 12:56–60, 13:307–308, 13:310
American leaf-nosed, 13:413–415, 13:*414*, 13:423–432, 13:433*t*–434*t*
bulldog, 13:443, 13:445, 13:449–450
disk-winged, 13:473, 13:476–477
Emballonuridae, 13:355, 13:360–363, 13:363*t*–364*t*
false vampire, 13:310, 13:379, 13:384–385
funnel-eared, 13:459–*460*, 13:*461*, 13:463–465
horseshoe, 13:387–388, 13:396–399
Kitti's hog-nosed, 13:367–368
Molossidae, 13:483, 13:490–493, 13:493*t*–495*t*
mouse-tailed, 13:351, 13:353
moustached, 13:435, 13:440–441, 13:442*t*
New Zealand short-tailed, 13:453, 13:457
Old World fruit, 13:319, 13:325–330, 13:331*t*–332*t*, 13:333–334, 13:340–347, 13:348*t*–349*t*
Old World sucker-footed, 13:479
slit-faced, 13:371, 13:*372*, 13:375, 13:376*t*–377*t*
smoky, 13:467, 13:470
Vespertilionidae, 13:497–498, 13:506–514, 13:515*t*–516*t*, 13:519–520, 13:524–525, 13:526*t*
bats Old World leaf-nosed, 13:401–402, 13:407–409, 13:409*t*–410*t*
bears, 14:296, 14:302–305, 14:306*t*
beavers, 12:111, 16:177–179, 16:183
bilbies, 13:19
botos, 15:27–28
Bovidae, 16:2–4
Antilopinae, 16:45, 16:50–55, 16:56*t*–57*t*
Bovinae, 16:18–23, 16:24*t*–25*t*
Caprinae, 16:88–89, 16:98–103, 16:103*t*–104*t*
duikers, 16:73, 16:80–84, 16:84*t*–85*t*
Hippotraginae, 16:28, 16:37–42, 16:42*t*–43*t*
Neotraginae, 16:59, 16:66–71, 16:71*t*
bushbabies, 14:23–24, 14:28–32, 14:32*t*–33*t*
Camelidae, 15:314–315, 15:320–323
Canidae, 14:266–267, 14:276–282, 14:283*t*–284*t*
capybaras, 16:401–402
Carnivora, 12:14, 12:79, 12:82, 12:120, 12:123–124, 14:256–257
cats, 12:84, 12:120, 14:370, 14:379–389, 14:390*t*–391*t*
Caviidae, 16:390, 16:395–398, 16:399*t*
Cetacea, 12:44, 12:67, 12:68, 15:4–5
Balaenidae, 15:4, 15:107, 15:109, 15:115–117
beaked whales, 15:59, 15:61, 15:64–68, 15:69*t*–70*t*
bottlenosed dolphins, 12:67–68
dolphins, 12:66, 15:42–43, 15:53–55, 15:56*t*–57*t*
franciscana dolphins, 15:23–24
Ganges and Indus dolphins, 15:13–14
gray whales, 15:93–94
Monodontidae, 15:81–82, 15:90
porpoises, 15:33–34, 15:38–39, 15:39*t*
pygmy right whales, 15:103–104
rorquals, 12:12, 12:63, 12:79, 12:82, 12:213, 15:4, 15:119–120, 15:127–130, 15:130*t*
sperm whales, 15:73, 15:79–80
chevrotains, 15:326–327, 15:331–333
Chinchillidae, 16:377, 16:382–383, 16:383*t*
colugos, 13:*300*–301, 13:304
coypus, 16:123, 16:473–474
dasyurids, 12:287–288, 12:294–298, 12:299*t*–301*t*
Dasyuromorphia, 12:277–279
deer
Chinese water, 15:373–374
muntjacs, 15:343–344, 15:349–354
musk, 15:335, 15:340–341
New World, 15:382, 15:388–396, 15:396*t*
Old World, 15:359, 15:364–371, 15:371*t*
Dipodidae, 16:211–212, 16:219–222, 16:223*t*–224*t*
Diprotodontia, 13:32–33
dormice, 16:317, 16:323–326, 16:327*t*–328*t*
duck-billed platypuses, 12:228–230, 12:243–244
Dugongidae, 15:199, 15:203
echidnas, 12:235–236, 12:240*t*
elephants, 12:11–12, 12:79, 12:82, 12:83, 15:165–166, 15:174–175
Equidae, 15:225–226, 15:232–235, 15:236*t*
Erinaceidae, 13:203, 13:209–213, 13:212*t*–213*t*
giant hutias, 16:469, 16:471*t*
gibbons, 14:209–210, 14:218–222
Giraffidae, 12:40, 12:102, 15:399–401, 15:408
great apes, 14:225–227, 14:237–239
gundis, 16:311–*312*, 16:314–315
Herpestidae, 14:347–348, 14:354–356, 14:357*t*
Heteromyidae, 16:199, 16:205–209, 16:208*t*–209*t*
hippopotamuses, 12:68, 15:304–306, 15:310–311
humans, 12:23–24, 14:244–247
hutias, 16:461, 16:467*t*
Hyaenidae, 14:360, 14:364–366
hyraxes, 15:178–179, 15:186–188, 15:189*t*
Indriidae, 14:63–65, 14:69–72
Insectivora, 13:194–197
koalas, 13:43–44
Lagomorpha, 16:480–481
lemurs, 14:48–49, 14:55–60
Cheirogaleidae, 14:35, 14:41–44
sportive, 14:75, 14:79–83
Leporidae, 12:46, 12:121, 12:122, 16:505, 16:511–514, 16:515*t*–516*t*
Lorisidae, 14:13–14, 14:18–20, 14:21*t*
Macropodidae, 13:32, 13:33, 13:83–84, 13:93–100, 13:101*t*–102*t*
manatees, 15:205, 15:211–212
marine mammals, 12:63, 12:*63*, 12:*64*, 12:*65*, 12:65–67, 12:*66*, 12:*67*
Megalonychidae, 13:155, 13:159
moles
golden, 13:195–197, 13:216, 13:220–222, 13:222*t*
marsupial, 13:25–26, 13:28
monitos del monte, 12:273–274
monkeys
Atelidae, 14:155–156, 14:162–166, 14:166*t*–167*t*
Callitrichidae, 12:39, 14:115–116, 14:127–131, 14:132*t*
Cebidae, 14:101–102, 14:108–112
cheek-pouched, 14:189–191, 14:197–204, 14:204*t*–205*t*
leaf-monkeys, 14:172–173, 14:178–183, 14:184*t*–185*t*
night, 14:135, 14:137, 14:141*t*
Pitheciidae, 14:143–144, 14:150–152, 14:153*t*
monotremes, 12:38, 12:228–230
mountain beavers, 16:131
Muridae, 12:79, 16:123, 16:283, 16:288–295, 16:296*t*–297*t*
Arvicolinae, 16:225–226, 16:233–238, 16:237*t*–238*t*
hamsters, 16:239, 16:245–246, 16:247*t*
Murinae, 16:249–250, 16:255–260, 16:261*t*–262*t*
Sigmodontinae, 16:264–265, 16:272–275, 16:277*t*–278*t*
musky rat-kangaroos, 13:70
Mustelidae, 12:37, 12:38, 14:319–321, 14:327–331, 14:332*t*–333*t*
numbats, 12:303
octodonts, 16:434, 16:438–440, 16:440*t*
opossums
New World, 12:250, 12:257–263, 12:264*t*–265*t*
shrew, 12:267, 12:270
Otariidae, 14:393–394, 14:402–406
pacaranas, 16:386
pacas, 16:417–418, 16:423–424
pangolins, 16:107–110, 16:115–119
peccaries, 15:291–292, 15:298–299
Perissodactyla, 15:216–217
Petauridae, 13:125, 13:131–133, 13:133*t*
Phalangeridae, 13:57–58, 13:*63–66t*
pigs, 15:275–276, 15:284–288, 15:289*t*–290*t*
pikas, 16:491, 16:497–501, 16:501*t*–502*t*

pocket gophers, 16:186–187, 16:192–194, 16:195*t*–197*t*
porcupines
New World, 16:366–*367*, 16:371–373, 16:374*t*
Old World, 16:351–352, 16:357–363
possums
feather-tailed, 13:140, 13:144
honey, 13:135–136
pygmy, 13:106, 13:110–111
primates, 14:4–5
Procyonidae, 14:309, 14:313–315, 14:315*t*–316*t*
pronghorns, 15:411–412
Pseudocheiridae, 13:113–114, 13:113–123, 13:119–123, 13:122*t*–123*t*
rat-kangaroos, 13:73–74, 13:78–80
rats
African mole-rats, 12:71, 12:*72*, 12:72–75, 12:73, 12:76–77, 12:78, 12:*86*, 16:339–342, 16:*341*, 16:347–349, 16:349*t*–350*t*
cane, 16:333–334, 16:337
chinchilla, 16:443–444, 16:447–448
dassie, 16:329–330
spiny, 16:*449*–450, 16:453–457, 16:458*t*
rhinoceroses, 15:251–252, 15:257–262
rodents, 16:122–123
sengis, 16:518–519, 16:524–529, 16:531*t*
shrews
red-toothed, 13:247–248, 13:255–263, 13:263*t*
West Indian, 13:243–244
white-toothed, 13:265–266, 13:271–275, 13:276*t*–277*t*
Sirenia, 15:192
size, 12:11–12
solenodons, 13:237–238, 13:240
springhares, 16:307–308
squirrels
flying, 16:135, 16:139–140, 16:141*t*–142*t*
ground, 12:113, 16:143–144, 16:151–158, 16:158*t*–160*t*
scaly-tailed, 16:299–300, 16:302–305
tree, 16:163, 16:169–174*t*, 16:175*t*
subterranean mammals, 12:71–78
Talpidae, 13:279–280, 13:284–287, 13:287*t*–288*t*
tapirs, 15:238, 15:245–247
tarsiers, 14:92–93, 14:97–99
tenrecs, 13:225–226, 13:232–234, 13:234*t*
three-toed tree sloths, 13:161–162, 13:167–169
tree shrews, 13:291, 13:294–296, 13:297*t*–298*t*
true seals, 14:256, 14:417, 14:*419*, 14:427–432, 14:434, 14:435*t*
tuco-tucos, 16:425, 16:429–430, 16:430*t*–431*t*
ungulates, 15:138–141, 15:141–142
Viverridae, 14:335–336, 14:340–343, 14:344*t*–345*t*
walruses, 12:46, 12:68, 14:409–*410*, 14:*413*
whales, 12:67, 12:68
wombats, 13:51, 13:55–56
Xenartha, 13:149
protostomes
amphionids, 2:195–196
amphipods, 2:261, 2:265–271
anaspidaceans, 2:181, 2:183
aplacophorans, 2:379–380, 2:382–385
Arachnida, 2:333–*334*, 2:*336*, 2:339–352
articulate lampshells, 2:521–522, 2:525–527
bathynellaceans, 2:177–178, 2:179
beard worms, 2:85, 2:88
bivalves, 2:451–452, 2:*453*, 2:457–461, 2:463–466
caenogastropods, 2:445–446, 2:449
centipedes, 2:*19*, 2:353–*354*, 2:358–362
cephalocarids, 2:131, 2:133
Cephalopoda, 2:*19*, 2:475–477, 2:*476*, 2:482–488
chitons, 2:393–395, 2:397–400
clam shrimps, 2:147, 2:150–151
copepods, 2:299–300, 2:304–309
cumaceans, 2:229, 2:232
Decapoda, 2:198, 2:*199*, 2:204–213
deep-sea limpets, 2:435, 2:437
earthworms, 2:*18*, 2:65–*66*, 2:*67*, 2:70–73
echiurans, 2:103, 2:106–107
fairy shrimps, 2:135–136, 2:139
fish lice, 2:289, 2:292–293
freshwater bryozoans, 2:497, 2:500–502
horseshoe crabs, 2:327–328, 2:*329*, 2:331–332
Isopoda, 2:249–250, 2:254–259
krill, 2:185, 2:*186*, 2:189–192
leeches, 2:75–76, 2:77, 2:80–82
leptostracans, 2:161, 2:164–165
lophogastrids, 2:225, 2:227
mantis shrimps, 2:167–168, 2:173–175
marine bryozoans, 2:503, 2:506–508, 2:509, 2:513–514
mictaceans, 2:241, 2:242
millipedes, 2:*18*, 2:*19*, 2:363–364, 2:367–370
monoplacophorans, 2:387–388, 2:390–391
mussel shrimps, 2:311–312, 2:314
mysids, 2:215, 2:219–221
mystacocarids, 2:295, 2:297
myzostomids, 2:59, 2:62
Neritopsina, 2:439–440, 2:442–443
nonarticulate lampshells, 2:515, 2:518–519
Onychophora, 2:109–110, 2:113–114
pauropods, 2:375, 2:377
peanut worms, 2:97–98, 2:100–101
phoronids, 2:491, 2:494
Polychaeta, 2:45–46, 2:50–56
Pulmonata, 2:413, 2:417–421
remipedes, 2:125, 2:128
sea slugs, 2:403, 2:*404*, 2:407–409
sea spiders, 2:321, 2:324–325
spelaeogriphaceans, 2:243, 2:244
symphylans, 2:371, 2:373
tadpole shrimps, 2:141, 2:145–146
tanaids, 2:235, 2:238–239
tantulocaridans, 2:283, 2:286
Thecostraca, 2:273–275, 2:278–280
thermosbaenaceans, 2:245, 2:247
tongue worms, 2:318–319
true limpets, 2:423, 2:426–427
tusk shells, 2:469, 2:473–474
Vestimentifera, 2:91, 2:94
Vetigastropoda, 2:429–430, 2:433–434
water bears, 2:115–116, 2:120–123
water fleas, 2:153, 2:157–158
reptiles, 7:4–*5*, 7:8–9, **7:23–33**
African burrowing snakes, 7:461, 7:463
African sideneck turtles, 7:129, 7:132–133
Afro-American river turtles, 7:137, 7:140–141
Agamidae, 7:209–210, 7:214–221
Alligatoridae, 7:171–172, 7:175–177
Anguidae, 7:339–340, 7:342–344
Australo-American sideneck turtles, 7:77, 7:81–83
big-headed turtles, 7:135
blindskinks, 7:271
blindsnakes, 7:379–380, 7:383–385
boas, 7:409–410, 7:413–417
Central American river turtles, 7:99
chameleons, 7:224–227, 7:*225*, 7:*226*, 7:*231*, 7:235–241
colubrids, 7:467, 7:473–481
Cordylidae, 7:319–320, 7:324–325
crocodilians, 7:157–*161*, 7:*159*
Crocodylidae, 7:179, 7:184–187
early blindsnakes, 7:369–370
Elapidae, 7:484, 7:491–498
false blindsnakes, 7:387
false coral snakes, 7:400
file snakes, 7:439–440
Florida wormlizards, 7:284
Gekkonidae, 7:259–*260*, 7:265–269
Geoemydidae, 7:115, 7:118–119
gharials, 7:167
Helodermatidae, 7:353–354
Iguanidae, 7:243–244, 7:250–256
Kinosternidae, 7:121, 7:125–126
knob-scaled lizards, 7:347–348, 7:351
Lacertidae, 7:297, 7:301–302
leatherback seaturtles, 7:101
Microteiids, 7:303, 7:306–307
mole-limbed wormlizards, 7:279–280, 7:281
Neotropical sunbeam snakes, 7:405, 7:*406*
New World pond turtles, 7:105, 7:109–112
night lizards, 7:291, 7:294–295
Parson's chameleons, 7:*8*
pig-nose turtles, 7:75
pipe snakes, 7:395–396
pythons, 7:420, 7:424–428
seaturtles, 7:85, 7:89–91
shieldtail snakes, 7:391–393
skinks, 7:328, 7:332–337
slender blindsnakes, 7:373–374, 7:376–377
snapping turtles, 7:93, 7:95–96
softshell turtles, 7:151, 7:154–155
spade-headed wormlizards, 7:287–288, 7:289
splitjaw snakes, 7:429–430
Squamata, 7:*197*, 7:198–203, 7:*199–201*
sunbeam snakes, 7:401–402
Teiidae, 7:310, 7:314–316
tortoises, 7:143, 7:146–148
Tropidophiidae, 7:433–434, 7:436–437
tuatara, 7:189–*190*, 7:*191*
turtles, 7:*65–70*
Varanidae, 7:359–360, 7:365–368
Viperidae, 7:446–447, 7:451–459
wormlizards, 7:273–274
Physics, of powered flight, 12:55–56
Physiculus helenaensis, 5:29
Physidae. *See* Tadpole snails
Phytosaurs, 7:17
Phytotoma raimondii. *See* Peruvian plantcutters
Phytotoma rara. *See* Rufous-tailed plantcutters
Phytotoma rutila. *See* Red-breasted plantcutters
Phytotomidae. *See* Plantcutters

Phyzelaphryne spp., 6:156
Piagetiella spp., 3:250
Piaggo, A. J., 16:143
Piano Concerto no. 3, 8:19
Piapiacs, 11:503
Piaractus brachypomus. *See* Pirapitingas
Pica spp. *See* Corvidae
Pica nuttalli. *See* Yellow-billed magpies
Pica pica. *See* Eurasian magpies; Magpies
Picassofishes. *See* Blackbar triggerfishes
Picathartes gymnocephalus. *See* White-necked picathartes
Picathartes oreas. *See* Gray-necked picathartes
Pichi armadillos, 13:*149*, 13:153, 13:184, 13:192*t*
Pichincha glass frogs, 6:*218*, 6:219–220
Picidae, **10:147–168,** 10:*152–153*
behavior, 10:149
conservation status, 10:150
distribution, 10:*147*, 10:148
evolution, 10:147
feeding ecology, 10:149
habitats, 10:148–149
humans and, 10:150–151
physical characteristics, 10:147–148
reproduction, 10:150
species of, 10:*154–167*
taxonomy, 10:147
Piciformes, **10:85–89**
Picinae. *See* Woodpeckers
Pickerels. *See* Walleyes
Picket pins. *See* Richardson's ground squirrels
Picoides borealis. *See* Red-cockaded woodpeckers
Picoides ramsayi. *See* Sulu woodpeckers
Picoides tridactylus. *See* Three-toed woodpeckers
Piculets, **10:147–155,** 10:*153*
Picumninae. *See* Piculets
Picumnus cirratus. *See* Ocellated piculets
Picumnus fulvescens. *See* Tawny piculets
Picumnus fuscus. *See* Rusty-necked piculets
Picumnus limae. *See* Ochraceous piculets
Picumnus nebulosus. *See* Mottled piculets
Picumnus olivaceus. *See* Olivaceous piculets
Picumnus steindachneri. *See* Speckle-chested piculets
Picus canus. *See* Gray-faced woodpeckers
Picus viridis. *See* Eurasian green woodpeckers
Piebald shrews, 13:*269*, 13:*271*, 13:272–273
Pied avocets, 9:134, 9:*136*, 9:*140*–141
Pied-billed grebes, 8:169, 8:*173*, 8:*176*
Pied crows, 11:505
Pied currawongs, 11:86, 11:468, 11:469, 11:*470*, 11:*474*–475
Pied flycatchers, 11:*29*, 11:31–*32*
Pied geese. *See* Magpie geese
Pied honeyeaters, 11:237
Pied kingfishers, 10:6, 10:7, 10:8, 10:*11*, 10:*23*
Pied magpies, 11:504
Pied starlings. *See* Magpie starlings
Pied tamarins, 14:124, 14:132*t*
Pied trillers. *See* Varied trillers
Pieridae, 3:386, 3:389
Pieris spp., 3:63
Pieris brassicae. *See* White butterflies
Pieris rapae. *See* European cabbage whites
Pietsch, T. W.
on anglerfishes, 5:47
on frogfishes, 5:49
on Trachinoidei, 5:331
Pig-footed bandicoots, 13:2, 13:*5*, 13:9–10, 13:11, 13:*11*, 13:*15*
Pig-nose turtles, **7:75–76**
Pig-nosed frogs. *See* Marbled snout-burrowers
Pig-tailed langurs, 14:184*t*
Pig-tailed snub-nosed langurs. *See* Mentawai Island langurs
Pigeon guillemots, 9:*219*
Pigeon lice, slender, 3:*255*, 3:256
Pigeons, 8:*24*, 9:241–246, 9:*242*, **9:247–267,** 9:*253–254*
behavior, 9:243, 9:249
conservation status, 9:251–252
distribution, 9:243, 9:*247*–248
evolution, 9:241, 9:245, 9:247
feeding ecology, 9:243–244, 9:249–250
habitats, 9:248–249
homing, 8:22, 8:33–35
humans and, 9:252
physical characteristics, 9:241–243, 9:247
reproduction, 9:244, 9:250–251
species of, 9:*255–266*
taxonomy, 9:241, 9:247
Pigfishes, 5:163, 5:164, 5:166
See also Grunt sculpins
Pignosed arrowtooth eels, 4:*260*, 4:*266*, 4:268–269
Pigs, 15:263, **15:275–290,** 15:*283*
Asian, 15:149
behavior, 15:278
Chinese Meishan, 15:149
conservation status, 15:280–281
distribution, 12:136, 15:*275*, 15:276–277
European, 12:191–192
evolution, 15:137, 15:265, 15:267, 15:275
feeding ecology, 15:141, 15:271, 15:278–279
feral, 12:*190*, 12:191–192
habitats, 15:277
humans and, 15:273, 15:281–282
physical characteristics, 15:140, 15:268, 15:275–276
reproduction, 15:143, 15:272, 15:279–280
species of, 15:284–288, 15:289*t*–290*t*
taxonomy, 15:275
Vietnam warty, 15:263, 15:280, 15:289*t*
vision, 12:79
See also Domestic pigs; Eurasian wild pigs
Pigs, Peccaries, and Hippos Specialist Group (IUCN/SSC), 15:275
cebu bearded pigs, 15:281
peccaries, 15:295
pygmy hogs, 15:280, 15:281, 15:287
Pikas, **16:491–503,** 16:*496*
behavior, 16:482–483, 16:491–493
conservation status, 16:487, 16:495
distribution, 12:134, 16:481, 16:*491*
evolution, 16:479–480, 16:491
feeding ecology, 16:484, 16:485, 16:493–494
habitats, 16:124, 16:481, 16:491
humans and, 16:487–488, 16:495
physical characteristics, 16:123, 16:480–481, 16:491
reproduction, 16:486, 16:494–495
species of, 16:497–501, 16:501*t*–502*t*
taxonomy, 16:479–480, 16:491
Pike glassfishes. *See* Mackerel icefishes
Pike-headed alligators. *See* American alligators
Piked dogfishes, 4:15, 4:*22*, 4:35, 4:124, 4:*151–155*, 4:157–158
Piked whales. *See* Northern minke whales
Pikeheads, 4:57, 5:*430*, 5:*432*, 5:433
Pikeperches. *See* Walleyes
Pikes, 4:*5*, 4:14, 4:50, 4:58, 4:68, **4:379–387,** 4:*382*, 5:198
See also African pikes
Pilbara ningauis, 12:*292*, 12:*294*, 12:297–298
Pilchards, 4:277, 4:279, 4:*280*, 4:285–*286*
Pile-worms. *See* Common shipworms
Pileated antwrens. *See* Black-capped antwrens
Pileated gibbons, 14:208, 14:*211*, 14:*212*, 14:*213*, 14:*217*, 14:*218*–219
behavior, 14:210, 14:211
conservation status, 14:215
evolution, 14:207
feeding ecology, 14:213
Pileated woodpeckers, 10:*88*, 10:149, 10:150, 10:151
Pileworms. *See* Common shipworms
Pilherodius pileatus. *See* Capped herons
Pilina spp., 2:387
Piliocolobus spp. *See* Red colobus
Piliocolobus badius waldronae, 14:179
Piliocolobus gordonorum, 14:175
Piliocolobus kirkii, 14:175
Piliocolobus pennantii, 14:175
Piliocolobus rufomitratus, 14:175
Pill millipedes, 2:364, 2:*366*, 2:*367*–368
Pillbugs, **2:249–260,** 2:*251*
behavior, 2:250–*251*
conservation status, 2:252
distribution, 2:250
evolution, 2:249
feeding ecology, 2:251–252
habitats, 2:250
humans and, 2:252
physical characteristics, 2:249–250
reproduction, 2:252
taxonomy, 2:249
Pilot whales, 15:2, 15:*3*, 15:8, 15:43–46, 15:48, 15:50
Pilotbirds, 10:331, 11:55
Pilotfishes, 5:256–257, 5:260
Pimelodidae, 4:352
Pimelodus filamentosus. *See* Lau-lau
Pimephales spp., 4:35
Pimephales promelas. *See* Fathead minnows
Pimm, Stuart, 1:48, 1:50
Pin-tailed nonpareils. *See* Pin-tailed parrotfinches
Pin-tailed parrotfinches, 11:*359*, 11:*368*
Pin-tailed whydahs, 11:*381*, 11:*392*–393
Pinablattella spp., 3:147
Pinarocorys spp. *See* Larks
Pinctada fucata, 2:454
Pinctada margaritifera. *See* Black-lipped pearl oysters
Pine grosbeaks, 11:*327*, 11:*338*
Pine rosefinches. *See* Pine grosbeaks
Pine sculpins. *See* Pineconefishes
Pine squirrels. *See* North American red squirrels
Pine-wood sparrows. *See* Bachman's sparrows
Pineapplefishes, 5:113–116, 5:*116*
Pinebirds. *See* Gray-crowned babblers
Pineconefishes, 5:113–116, 5:*117*, 5:*120*, 5:121
Pingalla spp., 5:220
Pinguinus impennis. *See* Great auks
Pinguipedidae. *See* Sandperches

Pinheads. *See* Northern anchovies
Pinicola enucleator. *See* Pine grosbeaks
Pink-backed firefinches. *See* Jameson's firefinches
Pink-backed pelicans, 8:227
Pink-billed parrotfinches, 11:357, 11:*359*, 11:*369*
Pink boarfishes. *See* Red boarfishes
Pink curlews. *See* Roseate spoonbills
Pink-eared ducks, 8:365
Pink-faced sittellas. *See* Black sittellas
Pink fairy armadillos, 13:*185*, 13:*186*, 13:*189*, 13:190
Pink glowworms, 3:*324*, 3:*330*
Pink-headed ducks, 8:365, 8:367, 8:372–373
Pink-headed warblers, 11:291
Pink paddles, 1:*9*
Pink river dolphins. *See* Botos
Pink salmons, 4:406, 4:*409*, 4:*411*–412
Pink-sided juncos, 11:271
Pink-throated longclaws. *See* Rosy-breasted longclaws
Pinna nobilis. *See* Noble pen shells
Pinnae, 12:54, 12:82
Pinnate batfishes, 5:*393*
Pinnated grouse. *See* Greater prairie chickens
Pinnipeds, 12:14, 12:63, 12:67, 12:68
  distribution, 12:129, 12:137, 12:138
  humans and, 12:173
  reproduction, 12:95, 12:98, 12:107, 12:110
  vision, 12:79–80
Pinnotheres pisum. *See* Pea crabs
Pinworms, 1:*285*
Pinyon jays, 11:506, 11:*510*, 11:*512*
Pinzon, Martín, 15:242
Pionopsitta vulturina. *See* Pesquet's parrots
Piophilidae, 3:54
Piopios, 11:115, 11:118
Pipa spp., 6:36–37
Pipa aspera, 6:36–37
Pipa carvalhoi, 6:*32*, 6:36–37
Pipa myersi, 6:36–37, 6:*101*, 6:102
Pipa parva, 6:102
Pipa pipa. *See* Suriname toads
Pipa snethlageae, 6:100
Pipanacoctomys spp., 16:434, 16:436
Pipanacoctomys aureus, 16:433–434
Pipe snakes, **7:395–397**
Pipefishes, 4:26, 4:46, 4:48, 4:69, 5:131, 5:132, 5:134–136
Pipidae, **6:99–107**, 6:*103–106*
  biogeography, 6:4
  distribution, 6:6, 6:*99*, 6:100, 6:*103–106*
  feeding ecology, 6:67, 6:100–101, 6:103–106
  tadpoles, 6:41, 6:43, 6:*101*, 6:102
  taxonomy, 6:99, 6:103–106
Pipile spp. *See* Piping guans
Pipile pipile. *See* Trinidad piping-guan
Pipilo spp. *See* Towhees
Pipilo erythrophthalmus. *See* Eastern towhees
Pipinae. *See* Suriname toads
Piping guans, 8:413
Piping plovers, 9:104, 9:162, 9:*164*
Pipistrelles, 12:*58*, 13:312, 13:497, 13:498, 13:499
  banana, 13:311
  Bodenheimer's, 13:313
  smell, 12:85
Pipistrellus spp. *See* Pipistrelles
Pipistrellus babu. *See* Himalayan pipistrelles
Pipistrellus bodenheimeri. *See* Bodenheimer's pipistrelles
Pipistrellus kuhlii. *See* Kuhl's pipistrelle bats
Pipistrellus nanus. *See* African banana bats
Pipistrellus nathusii. *See* Nathusius's pipistrelle bats
Pipistrellus pipistrellus. *See* Pipistrelles
Pipistrellus subflavus. *See* Eastern pipistrelles
Pipits, 9:314, **10:371–384,** 10:*377*, 11:272
  behavior, 10:373–374
  conservation status, 10:375–376
  distribution, 10:*371*, 10:372–373
  evolution, 10:371
  feeding ecology, 10:374
  habitats, 10:373
  humans and, 10:376
  physical characteristics, 10:371–372
  reproduction, 10:374–375
  taxonomy, 10:371
Pipra aureola. *See* Crimson-hooded manakins
Pipra cornuta. *See* Scarlet-horned manakins
Pipra erythrocephala. *See* Golden-headed manakins
Pipra filicauda. *See* Wire-tailed manakins
Pipra mentalis. *See* Red-capped manakins
Pipreola spp. *See* Cotingas
Pipreola arcuata. *See* Barred fruiteaters
Pipridae. *See* Manakins
Piprites pileatus. *See* Black-capped manakins
Pipunculidae, 3:360
Piranhas, 4:69, 4:*335*, 4:*336*, 4:337, 4:*341*
Pirapitingas, 4:*341*, 4:*343*, 4:346
Pirarara. *See* Redtail catfishes
Pirarucu. *See* Arapaima
Pirate perches, 4:57, 5:5, 5:7, 5:10–*11*
Piratic flycatchers, 10:274
Piratinga. *See* Lau-lau
Pirre warblers, 11:291
Pisaster ochraceus. *See* Ocher stars
Pisces, 4:6–7
Piscicolidae. *See* Fish leeches
Piscivores, 4:50
Pistol shrimps, 2:32, 2:200, 2:201
Pit-headed poachers. *See* Rockheads
Pitangus sulphuratus. *See* Great kiskadees
Pitbull katydids, 3:206
Pithead poachers. *See* Rockheads
Pithecia spp. *See* Sakis
Pithecia aequatorialis, 14:143
Pithecia albicans, 14:143, 14:146, 14:147
Pithecia irrorata, 14:143
Pithecia monachus. *See* Monk sakis
Pithecia monachus milleri. *See* Miller's monk sakis
Pithecia monachus napiensis. *See* Napo monk sakis
Pithecia pithecia. *See* White-faced sakis
Pithecia pithecia chrysocephala. *See* Golden-faced sakis
Pitheciidae, **14:143–154,** 14:*149*
  behavior, 14:145–146
  conservation status, 14:148
  distribution, 14:*143*, 14:144–145
  evolution, 14:143
  feeding ecology, 14:146–147
  habitats, 14:145
  humans and, 14:148
  physical characteristics, 14:143–144
  reproduction, 14:147
  species of, 14:*150*–153, 14:153*t*
  taxonomy, 14:143
Pitheciinae, 14:143, 14:146–147
Pithecophaga jefferyi. *See* Great Philippine eagles
Pitohui cristatus. *See* Crested pitohuis
Pitohui ferrugineus. *See* Rusty pitohuis
Pitohui incertus. *See* White-bellied pitohuis
Pitohui kirhocephalus. *See* Variable pitohuis
Pitohuis, 11:76, 11:116–118, 11:*119*, 11:125
Pitta angolensis. *See* African pittas
Pitta baudii. *See* Blue-headed pittas
Pitta brachyura. *See* Indian pittas
Pitta caerulea. *See* Giant pittas
Pitta dohertyi. *See* Sula pittas
Pitta elegans. *See* Elegant pittas
Pitta erythrogaster. *See* Red-bellied pittas
Pitta granatina. *See* Garnet pittas
Pitta guajana. *See* Banded pittas
Pitta gurneyi. *See* Gurney's pittas
Pitta iris. *See* Rainbow pittas
Pitta kochi. *See* Whiskered pittas
Pitta megarhyncha. *See* Mangrove pittas
Pitta moluccensis. *See* Blue-winged pittas
Pitta nympha. *See* Fairy pittas
Pitta schneideri. *See* Schneider's pittas
Pitta sordida. *See* Hooded pittas
Pitta steerii. *See* Azure-breasted pittas
Pitta superba. *See* Superb pittas
Pitta venusta. *See* Graceful pittas
Pittas, 10:170, **10:193–201,** 10:*194*, 10:*196*
Pittas, Broadbills, and Asities (Lambert and Woodcock), 10:194
Pittas of the World (Erritzoe and Erritzoe), 10:194
Pitted poachers. *See* Rockheads
Pitted-shelled turtles. *See* Pig-nose turtles
Pittidae. *See* Pittas
Pituitary gland, lissamphibians, 6:19
Pituophis spp., 7:42
Pituophis catenifer. *See* Bullsnakes
Pituriaspida spp., 4:9
Pitvipers, 7:9, 7:32–33, 7:41, 7:42, 7:206, **7:445–456,** 7:*450*
Pityriasinae. *See* Bornean bristleheads
Pityriasis. *See* Bristleheads
Pityriasis spp. *See* Bornean bristleheads
Pityriasis gymnocephala. *See* Bornean bristleheads
Pizonyx visesi. *See* Mexican fishing bats
Placentals, 14:255–256
  evolution, 12:11, 12:*31*, 12:33
  neonatal milk, 12:127
  phylogenetic tree of, 12:11, 12:*32*, 12:33
  reproduction, 12:51, 12:89–93, 12:*102*, 12:106, 12:109
  See also specific placentals
Placentas, 12:93–94, 12:101, 12:106, 12:109
  See also Reproduction
Placentation, 7:7
  See also Reproduction
Placiphorella spp., 2:395
Placiphorella velata. *See* Veiled chitons
Placobdelloides jaegerskioeldi. *See* Hippopotamus leeches
Placodermi, 4:10
Placodontia, 7:5, 7:15
Placodus spp., 7:15
Placozoans, 1:9, 1:10, 1:18–19, 1:20, **1:87–89,** 1:*88*
Plagiobrissus grandis. *See* Sea biscuits

Plagiodontia spp., 16:462
Plagiodontia aedium. *See* Hispaniolan hutias
Plagiodontia ipnaeum. *See* Samana hutias
Plagioglypta iowensis, 2:469
Plagioporus sinitsini, 1:199
Plagiorhynchus cylindraceus, 1:316
Plague, 3:76, 3:353
Plaice, 5:*165*, 5:452, 5:*454*, 5:459–*460*
Plain antvireos, 10:*240*
Plain-backed sunbirds, 11:210
Plain-brown woodcreepers, 10:*231*, 10:*232*
Plain chachalacas, 8:*399*, 8:*414*, 8:*417*, 8:418
Plain-flanked rails, 9:51
Plain flowerpeckers, 11:*192*, 11:*194*–195
Plain-headed creepers. *See* Stripe-breasted rhabdornis
Plain-headed rhabdornis. *See* Stripe-breasted rhabdornis
Plain-nosed bats, 13:310, 13:311, 13:313, 13:315
See also Vespertilionidae
Plain-pouched hornbills, 10:73, 10:77, 10:*82*–83
Plain-pouched wreathed hornbills. *See* Plain-pouched hornbills
Plain prinias. *See* Tawny-flanked prinias
Plain spinetails, 10:211
Plain-tailed motmots. *See* Broad-billed motmots
Plain-tailed nighthawks, 9:405
Plain-throated sunbirds, 11:*211*, 11:*214*
Plainfin midshipmen, 5:*43*, 5:44–*45*
Plains baboons, 14:7, 14:10
Plains bison. *See* American bison
Plains garter snakes, 7:44, 7:470
See also Garter snakes
Plains harvest mice, 16:*264*
Plains kangaroos. *See* Red kangaroos
Plains mice, 16:*252*
Plains pocket gophers, 12:74, 16:*186*, 16:*191*, 16:*192*–193
Plains spadefoot toads, 6:*123*, 6:125
Plains viscacha rats. *See* Red viscacha rats
Plains viscachas, 16:124, 16:377, 16:378, 16:379–380, 16:*381*, 16:*382*
Plains-wanderers, 9:11, 9:14–*15*, 9:*21*, 9:104
Plains zebras, 12:*94*, 15:*220*, 15:*221*, 15:*228*, 15:*231*, 15:234
behavior, 12:146, 15:219, 15:*226*, 15:227
coevolution and, 12:216
conservation status, 15:229
distribution, 15:*233*
feeding ecology, 15:234
habitats, 15:218
true, 15:222
Planarians, 1:38–41, 1:*189*, 1:*190*, 1:*191*, 1:194–195, 2:35, 2:36, 2:37
Planctosphaera pelagica, 1:443, 1:444, 1:445
Planctosphaeroidea. *See Planctosphaera pelagica*
Planigale gilesi. *See* Gile's planigales
Planigale ingrami. *See* Long-tailed planigales
Planigale maculata. *See* Pygmy planigales
Planigales, 12:280, 12:281
Planktivore fishes, 4:67
Planktonic dispersal, 1:19, 2:21
Planktotrophic larvae, 2:21
See also Reproduction
Planodasyidae, 1:269
Planorbidae, 2:412
Plant bugs, 3:75
Plant-feeding insects, 3:75
See also Herbivores
Plant galls, 3:63
Plant-hoppers, 3:56, 3:259
Plant lice, 3:260, 3:262, 3:263
Plantain eaters, 9:299, 9:301, 9:*303*
Plantanna. *See* Common plantanna
Plantcutters, 10:170–171, **10:325–328**
Plante, Rafael, 4:193
Plantigrades, 12:*39*
Plants, 3:48–49
ants and, 3:49
beetles and, 3:315
toxins in, 3:46, 3:48
Planula, 1:7
Plasmids, 1:31
Plasmodium spp., 2:11
Plasmodium falciparum, 2:12, 2:31
Plasticity
behavioral, 12:140–143
developmental, 2:21–22
See also Behavior
Platacanthomys lasiurus. *See* Malabar spiny dormice
Platalea ajaja. *See* Roseate spoonbills
Platalea leucorodia. *See* Spoonbills
Platalea minor. *See* Black-faced spoonbills
Platalea pygmea. *See* Spoon-billed sandpipers
Platalina genovensium. *See* Long-snouted bats
Platanista spp., 15:13
Platanista gangetica. *See* Ganges and Indus dolphins
Platanista gangetica gangetica, 15:13
Platanista gangetica minor, 15:13
Platanistidae. *See* Ganges and Indus dolphins; River dolphins
Platanistoidea, 15:2, 15:13
Platax pinnatus. *See* Pinnate batfishes
Platax teira. *See* Longfin spadefishes
Plate-billed mountain toucans, 10:128, 10:*129*, 10:*132*
Plate corals. *See* Mushroom corals
Plateau pikas, 16:482, 16:488, 16:492, 16:494, 16:*496*, 16:*497*–498
Plateau side-blotched lizards. *See* Common side-blotched lizards
Plated catfishes, 4:352
Plated lizards, **7:319–325**, 7:*323*
Platyarthrus spp., 2:251
Platycephalidae. *See* Flatheads
Platycephaloidei. *See* Flatheads
Platycephalus spp., 5:157
Platycephalus indicus. *See* Indian flatheads
Platycephalus janeti, 5:157
Platycercus elegans. *See* Crimson rosellas
Platycercus eximius. *See* Eastern rosellas
Platycopioida, 2:299, 2:300
Platyctenida, 1:169, 1:170, 1:171
Platydesmids, 2:365
Platygonus spp., 15:291
Platyhelminthes. *See* Parasitic flatworms
Platylophus galericulatus. *See* Crested jays
Platymantis spp., 6:246, 6:248, 6:251
Platymops spp., 13:311
Platyops sterreri, 2:217
Platypelis spp., 6:35, 6:301, 6:304
Platyplectrurus spp., 7:391
Platyplectrurus madurensis, 7:392
Platypuses. *See* Duck-billed platypuses
Platyrhinidae, 4:11
Platyrhynchos albicollis. *See* White-throated fantails
Platyrhynchos rufiventris. *See* Northern fantails
Platyrhynchus ceylonensis. *See* Gray-headed flycatchers
Platyrinchinae. *See* Tyrant flycatchers
Platyrrhinus helleri, 13:*414*
Platysaurus spp. *See* Flat lizards
Platysaurus capensis. *See* Cape flat lizards
Platysmurus leucopterus. *See* White-winged choughs
Platystacus chaca. *See* Squarehead catfishes
Platysteirinae. *See* African flycatchers
Platysteria laticincta. *See* Banded wattle-eyes
Platysteria peltata. *See* Black-throated wattle-eyes
Platysternidae. *See* Big-headed turtles
Platysternon megacephalum. *See* Big-headed turtles
Platystictidae, 3:55
Platytroctidae. *See* Tubeshoulders
Plautus alle. *See* Dovekies
Play, 7:39, 12:142
See also Behavior
Plecodus spp., 4:69
Plecoglossinae, 4:390, 4:391
Plecoglossus altivelis. *See* Ayu
Plecoptera. *See* Stoneflies
Plecostomus spp., 4:6
Plecotus auritus. *See* Brown long-eared bats; Long-eared bats
Plecotus austriacus. *See* Gray long-eared bats
Plectorhyncha lanceolata. *See* Striped honeyeaters
Plectrohyla spp., 6:227, 6:228–229, 6:230
Plectrohyla glandulosa, 6:228–229
Plectrohyla hartwegi. *See* Hartweg's spike-thumb frogs
Plectrophenax nivalis. *See* Snow buntings
Plectropomus spp., 5:262
Plectropomus laevis. *See* Blacksaddled coral groupers
Plectrurus perrotetii. *See* Nilgiri burrowing snakes
Plegadis chihi. *See* White-faced glossy ibises
Pleistocene period, 3:53
Pleocyemata, 2:197
Plesiobatidae, 4:176
Plesiochrysa spp., 3:307
Plesiopidae, 5:255, 5:256, 5:258–263
Plesiorobius spp., 3:305
Plesiosaurs, 7:15
Plethobasus cicatricosus. *See* White wartyback mussels
Plethodon spp., 6:35, 6:44, 6:391
Plethodon cinereus. *See* Red-backed salamanders
Plethodon jordani. *See* Appalachian woodland salamanders
Plethodon shermani. *See* Red-legged salamanders
Plethodonini, 6:390
Plethodontidae. *See* Lungless salamanders
Plethodontinae, 6:390
Plethodontohyla spp., 6:35, 6:304
Pleuragramma antarcticum, 5:322, 5:323
Pleurobrachia spp., 1:*170*, 1:171
Pleurobrachia bachei. *See* Sea gooseberries
Pleurobrachia pileus, 1:28
Pleurodeles waltl. *See* Spanish sharp-ribbed newts

Pleurodema spp., 6:66, 6:156, 6:158
Pleurodema bufonina. See Gray four-eyed frogs
Pleurodema marmorata. See Puna frogs
Pleurodira, 7:13, 7:65
Pleurogona. *See* Stolidobranchia
Pleurogyra sinuosa. See Bubble corals
Pleuronectes platessa. See Plaice
Pleuronectidae, 5:450, 5:453
Pleuronectiformes. *See* Flatfishes
Pleuronectoidei, 5:449, 5:450
Pleurotomarioidea, 2:429, 2:430, 2:431
Pleustes platypa, 2:*264*, 2:*268–269*
Plexaura homomala, 1:44
Pliauchenia spp., 15:265
Plicate nerites, 2:*442*
Pliciloricidae, 1:343, 1:344, 1:345
Pliciloricus hadalis. See Japanese deep-sea girdle wearers
Pliny, 4:73, 11:509, 13:207
Pliopontes spp., 15:23
Pliotrema warreni. See Sixgill sawsharks
Ploceidae. *See* Weavers
Ploceinae. *See Ploceus* spp.
Plocepasser mahali. See White-browed sparrow weavers
Plocepasserinae. *See* Sparrow-weavers
Ploceus spp., 11:375–376, 11:377
Ploceus aureonucha. See Yellow-footed weavers
Ploceus bannermani. See Bannerman's weavers
Ploceus batesi. See Bates's weavers
Ploceus bicolor. See Dark-backed weavers
Ploceus burnieri. See Kilombero weavers
Ploceus capensis. See Cape weavers
Ploceus cucullatus. See Village weavers
Ploceus galbula. See Rüppells' weavers
Ploceus golandi. See Clarke's weavers
Ploceus hypoxanthus. See Asian golden weavers
Ploceus manyar. See Streaked weavers
Ploceus nicolli. See Tanzanian mountain weavers
Ploceus nigerrimus. See Vieillot's black weavers
Ploceus nigrimentum. See Black-chinned weavers
Ploceus nitens. See Blue-billed malimbes
Ploceus ocularis. See Spectacled weavers
Ploceus philippinus. See Baya weavers
Ploceus rubriceps. See Red-headed weavers
Ploceus sakalava. See Sakalava weavers
Ploceus spekeoides. See Fox's weavers
Ploceus subpersonatus. See Loango weavers
Plodia interpunctella. See Indian mealmoths
Plotosidae, 4:352
Plotosus lineatus. See Coral catfishes
Ploughbills, 11:115, 11:117
Plovers, **9:161–173,** 9:*165*
  behavior, 9:162, 9:166–172
  conservation status, 9:163–164, 9:167–173
  distribution, 9:*161*, 9:162, 9:*166–172*
  evolution, 9:161
  feeding ecology, 9:162, 9:166–173
  habitats, 9:102, 9:162, 9:166–172
  humans and, 9:164, 9:167–173
  physical characteristics, 9:161–162, 9:166–172
  reproduction, 9:162–163, 9:166–173
  species of, 9:*166–173*
  taxonomy, 9:104, 9:133, 9:161
Plum-headed parakeets, 9:279
Plum-throated cotingas, 10:306, 10:*310*, 10:312, 10:*314*–315
Plumage. *See* Physical characteristics
Plumatella fungosa, 2:*499*, 2:*501–502*
Plumatellidae, 2:497
Plumbeous forest falcons, 8:353, 8:*355–356*
Plumbeous kites, 8:322
Plumbeous sierra-finches, 11:*267*, 11:*279–280*
Plumed frogmouths. *See* Marbled frogmouths
Plumed guineafowl, 8:426–428
Plumed whistling ducks, 8:370
Plumose anemones. *See* Frilled anemones
Plumularia setacea, 1:25
Plunderfishes, 5:321, 5:322
Plush-crested jays, 11:504
Plutellidae, 3:389
Pluvialis spp., 9:161
Pluvialis dominica. See American golden-plovers
Pluvialis fulva. See Pacific golden plovers
Pluvialis squatarola. See Black-bellied plovers
Pluvianellidae. *See* Magellanic plovers
Pluvianellus socialis. See Magellanic plovers
Pluvianidae. *See* Egyptian plovers
PM2000 (software), 12:207
Pneumoridae, 3:203
Pneumoroidea. *See* Bladder grasshoppers
Pnoepyga pusilla. See Pygmy wren-babblers
Poachers, 5:179, 5:180, 5:182, 5:183
Poblana alchichica, 5:70
Poblana letholepis, 5:70
Poblana squamata, 5:70
Pocillopora damicornis. See Cauliflower corals
Pocillopora eydouxi, 4:42
Pocket gophers, 12:71, **16:185–197,** 16:*186*, 16:*187*, 16:*191*, 16:283
  behavior, 16:124, 16:188–189
  conservation status, 16:190
  distribution, 12:131, 16:123, 16:*185*, 16:187
  evolution, 16:121, 16:185
  feeding ecology, 16:189–190
  habitats, 16:187
  humans and, 16:126, 16:127, 16:190
  photoperiodicity, 12:74
  physical characteristics, 16:123, 16:186–187
  reproduction, 16:125, 16:190
  species of, 16:192–195, 16:195*t*–197*t*
  taxonomy, 16:121, 16:185
Pocket mice, 12:117, 16:121, 16:123, 16:124, **16:199–210,** 16:*204*, 16:208*t*–209*t*
Pocketed free-tailed bats, 13:495*t*
Podager nacunda. See Nacunda nighthawks
Podarcis lilfordi, 7:*298*
Podargidae. *See* Oilbirds
Podarginae. *See* Frogmouths
Podargus spp. *See* Frogmouths
Podargus auritus. See Large frogmouths
Podargus cornutus. See Sunda frogmouths
Podargus javensis. See Javan frogmouths
Podargus ocellatus. See Marbled frogmouths
Podargus papuensis. See Papuan frogmouths
Podargus poliolophus. See Short-tailed frogmouths
Podargus stellatus. See Gould's frogmouths
Podargus strigoides. See Tawny frogmouths
Podica senegalensis. See African finfoot
Podiceps andinus. See Colombian grebes
Podiceps auritus. See Horned grebes
Podiceps cristatus. See Great crested grebes
Podiceps gallardoi. See Hooded grebes
Podiceps grisegena. See Red-necked grebes
Podiceps major. See Great grebes
Podiceps nigricollis. See Eared grebes
Podiceps occipitalis. See Silvery grebes
Podicipediformes. *See* Grebes
Podilymbus gigas. See Atitlán grebes
Podilymbus podiceps. See Pied-billed grebes
Podisus maculiventris. See Spiny soldier bugs
Podoces spp., 11:503, 11:504
Podocnemididae. *See* Afro-American river turtles
Podocnemis expansa. See South American river turtles
Podocnemis unifilis. See Yellow-spotted river turtles
Podocnemis vogli, 7:137
Podocopida, 2:311, 2:312
Podocoryne spp., 1:27
Podocoryne carnea, 1:26–27
Podogymnura aureospinula. See Dinagat gymnures
Podogymnura truei. See Mindanao gymnures
Podophis spp., 7:16
Podura aquatica. See Water springtails
Poebrodon spp., 15:265
Poebrotherium spp., 15:265
Poecile spp., 11:155, 11:156
Poecile atricapilla. See Black-capped chickadees
Poecile hudsonica. See Boreal chickadees
Poecile sclateri. See Mexican chickadees
Poecilia formosa, 5:93
Poecilictis libyca. See North African striped weasels
Poeciliidae, 4:12, 4:15, 4:32, 5:89–90, 5:92
Poeciliinae, 5:92–93
Poeciliopsis spp., 4:30
Poeciliopsis monacha-lucida, 5:93
Poecilogale albinucha. See White-naped weasels
Poecilopsettidae, 5:450
Poecilopsorus iridescens, 3:*243*
Poecilosclerida, 1:77, 1:78, 1:79
Poelagus spp. *See* Bunyoro rabbits
Poelagus marjorita. See Bunyoro rabbits
Poeoptera kenricki. See Kenrick's starlings
Poeoptera stuhlmanni, 11:424
Poephila bichenovii. See Double-barred finches
Poephila cincta. See Black-throated finches
Pogona barbata. See Bearded dragons
Pogona minor. See Bearded dragons
Pogona vitticeps. See Bearded dragons
Pogoniulus bilineatus. See Yellow-rumped tinkerbirds
Pogoniulus chrysoconus. See Yellow-fronted tinkerbirds
Pogoniulus pusillus. See Red-fronted tinkerbirds
Pogonophora. *See* Beard worms
Pogy. *See* Atlantic menhadens
Pohnpei flying foxes, 13:322
Pohnpei Mountain starlings, 11:410
Poiana spp., 14:335
Poiana richardsonii. See African linsangs
Poicephalus rueppellii. See Rüppell's parrots
Poikilothermy, 2:26
Point mutations, 12:27, 12:30
Pointed-tongue floating frogs, 6:*254*, 6:*259–260*
Pointers, 14:288
Poison-fangs, 4:69
Poison frogs, **6:197–210,** 6:*198*, 6:*201–202*
  behavior, 6:66–67, 6:199, 6:203–209
  conservation status, 6:200, 6:203–209
  for dart poison, 6:200, 6:*201*, 6:209
  distribution, 6:6, 6:*197*, 6:198, 6:*203–209*

evolution, 6:197
feeding ecology, 6:199, 6:203–209
habitats, 6:198–199, 6:203–209
humans and, 6:200, 6:203–209
as pets, 6:55
physical characteristics, 6:197–198, 6:203–209
reproduction, 6:34, 6:36, 6:68, 6:199–200, 6:203–209
species of, 6:203–209
tadpoles, 6:*198*, 6:*199*, 6:200
taxonomy, 6:197, 6:203–209
toxins, 6:*45*, 6:54, 6:189, 6:198–199, 6:200, 6:208–209
Poison glands, 6:66–67
Poisoning
common dolphinfishes, 5:215–216
shellfish, 2:12
striped bristletooths, 5:399
See also specific species
Poisonous fishes
boxfishes, 5:469
Tetraodontiformes, 5:471
Poisons
in pest control, 12:186 (*See also* Toxins)
sponges and, 1:29
Polar bears, 12:*132*, 14:*301*, 14:*304*–305
behavior, 12:*143*, 14:258, 14:298
conservation status, 14:300
distribution, 12:129, 12:138, 14:296
evolution, 14:295–296
feeding ecology, 14:260, 14:298–299
habitats, 14:257, 14:296
Ice Age, 12:24
migrations, 12:*166*
physical characteristics, 14:296
reproduction, 12:106, 14:*296*, 14:299
taxonomy, 14:295
Polarized-light compasses, 8:34
Polecats, 14:256, 14:319
marbled, 14:323
steppe, 14:321
Polemon spp. *See* Snake eaters
Polihierax insignis. *See* White-rumped falcons
Polihierax semitorquatus. *See* African pygmy falcons
Poliocephalus poliocephalus. *See* Hoary-headed grebes
Poliocephalus rufopectus. *See* New Zealand grebes
Polioptila caerulea. *See* Blue-gray gnatcatchers
Polioptila lactea. *See* Creamy-bellied gnatcatchers
Polistes spp., 3:69
Polistes fuscatus. *See* Golden paper wasps
Polistes hebraeus, 3:*71*
Politics, conservation and, 12:224–225
Polka-dot cardinalfishes. *See* Pajama cardinalfishes
Pollachius spp., 5:27, 5:28
Pollachius virens. *See* Pollocks
Pollen's vangas, 10:439, 10:*440*, 10:441
Pollimyrus isidori. *See* Elephantfishes
Pollination, 3:49, 3:78–79
bee, 3:14, 3:72
beetle, 3:14
See also Reproduction
Pollocks, 5:26, 5:29, 5:*31*, 5:*33*, 5:34–35
See also Alaska pollocks; Gadidae
Poll's shrews, 13:276*t*
Pollution, 1:25–26, 1:47–48, 1:52, 7:60–62, 12:213
See also Conservation status; Humans
Polo, Marco, 8:26, 8:104, 15:254–255
Polophilus sinensis. *See* Greater coucals
Polyacanthonotus spp., 4:249
Polyacanthonotus merretti. *See* Marine spiny eels
Polyancistrini, 3:203
Polyancistrus spp., 3:204
Polyancistrus serrulatus. *See* Hispaniola hooded katydids
Polyandry, 3:59, 12:107, 12:146
Polyartemia spp., 2:135
Polyartemiella spp., 2:135
Polyarthra spp., 1:261
Polyboroides spp. *See* Harrier hawks
Polyborus plancus. *See* Crested caracaras
Polybrachiorhynchus dayi, 1:247
Polycarpa spp., 1:452
Polycentrus schomburgkii, 5:237
Polychaeta, 1:38, 1:40, **2:45–57**, 2:*48*, 2:*49*
behavior, 2:32, 2:43, 2:46
conservation status, 2:47
distribution, 2:46
evolution, 2:45
feeding ecology, 2:36, 2:46–47
habitats, 2:46
humans and, 2:47
physical characteristics, 2:45–46
reproduction, 2:21–22, 2:21–23, 2:47
species of, 2:*50*–*56*
taxonomy, 2:45
Polychlorinated biphenyls, 8:27
Polychrotinae. *See* Anoles
Polycladida, 1:19, 1:186, 1:188
Polydesmida, 2:365
Polydesmus spp., 2:*364*
Polydesmus angustus. *See* Flat-backed millipedes
Polyembryony, 2:22
Polyerata spp. *See* Hummingbirds
Polyerata boucardi. *See* Mangrove hummingbirds
Polyerata fimbriata. *See* Glittering-throated emeralds
Polyglyphanodontines, 7:309
Polygonia spp., 3:388
Polygyny, 3:59, 12:98, 12:107, 12:146, 12:147–148
Polymerase chain reaction (PCR), 1:49, 2:7–8, 2:49, 12:26
Polymitarcyidae, 3:125, 3:126
Polymixia spp. *See* Beardfishes
Polymixia berndti, 5:*1*
Polymixia lowei, 5:2
Polymixia nobilis. *See* Stout beardfishes
Polymixiidae. *See* Beardfishes
Polymixiiformes. *See* Beardfishes
Polynemidae. *See* Threadfins
Polynesian megapodes, 8:407
Polynesian rats, 12:189
Polynesian sandpipers, 9:175
Polyodon spathula. *See* American paddlefishes
Polyodontidae, 4:11, 4:213
Polyonymus spp. *See* Hummingbirds
Polyopisthocotylea, 1:213, 1:214
Polyorchis penicillatus, 1:*132*, 1:136
Polyp theory, 1:10
Polypedates spp., 6:291, 6:292
Polypedates arboreus. *See* Kinugasa flying frogs
Polypedates beddomii. *See* Beddome's Indian frogs
Polypedates formosus. *See* Beautiful torrent frogs
Polypedates hascheanus. *See* Penang Taylor's frogs
Polypedates surdus. *See* Luzon bubble-nest frogs
Polypedilum spp., 3:37
Polyphaga, 3:315, 3:316, 3:318
See also Beetles
Polyplacophora. *See* Chitons
Polyplectron spp. *See* Peacock-pheasants
Polyplectron emphanum. *See* Palawan peacock-pheasants
Polyplectron malacense. *See* Malaysian peacock-pheasants
Polyplectron schleiermacheri. *See* Bornean peacock-pheasants
Polypocephalidae, 1:226
Polypodiozoa. *See Polypodium hydriforme*
Polypodium hydriforme, 1:123, 1:144–145
feeding ecology, 1:129
humans and, 1:129
physical characteristics, 1:126–127
Polyprionidae. *See* Wreckfishes
Polyps, elephant ear, 1:*110*, 1:115–*116*
Polypteridae. *See* Bichirs
Polypteriformes. *See* Bichirs
Polypterus spp., 4:209, 4:210
Polypterus ornatipinnis. *See* Bichirs
Polypterus senegalus, 4:210
Polypterus weeksii. *See* Week's bichirs
Polystilifera, 1:253
Polystoechotidae, 3:305, 3:307–308
Polystoma integerrimum, 1:*217*, 1:222–*223*
Polystomatids, 1:214
Polytmus spp. *See* Hummingbirds
Polyxenus lagurus. *See* Bristly millipedes
Polyzonium spp., 2:364
Polyzosteriinae, 3:149
Pomacanthidae. *See* Angelfishes
Pomacanthus imperator. *See* Emperor angelfishes
Pomacea canaliculata. *See* Golden apple snails
Pomacentridae. *See* Damselfishes
Pomacentrus albifasciatus, 4:70
Pomarea nigra. *See* Tahiti monarchs
Pomarine skuas, 9:204
Pomatodelphis spp., 15:13
Pomatomidae. *See* Bluefishes
Pomatomus saltatrix. *See* Bluefishes
Pomatorhinus isidorei. *See* Rufous babblers
Pomatorhinus montanus. *See* Chestnut-backed scimitar-babblers
Pomatoschistus spp., 5:375
Pomatoschistus (?) cf. *bleicheri*, 5:374
Pomatoschistus microps. *See* Common gobies
Pomatostomidae. *See* Pseudo babblers
Pomatostomus spp. *See* Pseudo babblers
Pomatostomus halli. *See* Hall's babblers
Pomatostomus ruficeps. *See* Crowned babblers
Pomatostomus superciliosus. *See* White-browed babblers
Pomatostomus temporalis. *See* Gray-crowned babblers
Pomfrets, 5:255, 5:257–263
Pomoxis nigromaculatus. *See* Black crappies
Pompano dolphinfishes, 5:*213*, 5:*214*
Pompano dolphins. *See* Pompano dolphinfishes
Pompanos, 5:255
See also Carangidae

Pomphorhynchus laevis, 1:*315,* 1:316–317
Pompilidae, 3:406
Ponapé fantails, 11:85
Pond bats, 13:313, 13:515*t*
Pond sliders, 7:*108,* 7:*110*–111
Pondaungia spp., 14:2
Ponde. *See* Murray cods
Ponds, for insect conservation, 3:90–91
Ponerine ants, 3:71
Pongidae. *See* Great apes
Ponginae, 14:225
Pongo spp. *See* Orangutans
Pongo abelii. See Sumatran orangutans
Pongo pygmaeus. See Bornean orangutans
Pontinus nigropunctatus. See Deepwater jacks
Pontomyia spp., 3:358
Pontoporia blainvillei. See Franciscana dolphins
Pontoporiidae. *See* Franciscana dolphins; River dolphins
Pontoscolex corethrurus, 2:*69,* 2:72
Ponyfishes, 5:255, 5:257–263
Poodles, 14:289
Poor man's tropheuses, 5:277
Poorwills. *See* Common poorwills
Po'oulis, 11:342, 11:*345,* 11:*351*
Popelairia conversii. See White-rumped green thorntails
Popes. *See* Ruffes
Pope's pitvipers, 7:*446*
Popeyed sea goblins. *See* Bearded ghouls
Poptella orbicularis. See Silver tetras
Population control, 1:47–48
  See also Conservation status; Humans
Population ecology, 2:25, 2:26–27, 4:42–44
  See also Reproduction
Populations
  growth, 12:97
  human, 12:213
  indexes, 12:195–196, 12:200–202
  size of, genetic diversity and, 12:221–222
  in zoos, 12:207–209
Porbeagles, 4:133, 4:*134,* 4:138–*139*
Porcelain crabs, 2:197, 2:198
Porcellio scaber. See Common rough woodlice
Porcellionidae, 2:250
Porcupine quills, 16:*367*
Porcupinefishes, 4:70, 5:467, 5:469–471, 5:*472,* 5:*475,* 5:477
Porcupines, 12:38, 12:*123,* 16:121, 16:123, 16:127
  New World, 16:125, **16:365–375**
  Old World, 16:121, **16:351–364**
Porgies, 4:67, 5:255, 5:258–260, 5:262, 5:263
Porichthyinae, 5:41
Porichthys spp., 5:41, 5:42
Porichthys notatus. See Plainfin midshipmen
Porichthys plectrodon. See Atlantic midshipmen
Porifera. *See* Sponges
Pork tapeworms, 1:35, 1:*225,* 1:*227,* 1:231
Pork worms. *See* Trichina worms
Porocephalida, 2:317–318, 2:319, 2:320
Porocephalidae, 2:318
Porocephalus spp., 2:318, 2:320
Porocephalus crotali, 2:*317,* 2:*318,* 2:319, 2:320
Poroderma africanum. See Pajama catsharks
Porolepimorpha, 4:12
Poromera spp., 7:297
Poromitra spp., 5:106
Porphyrio spp. *See* Gallinules
Porphyrio mantelli. See Takahe rails
Porphyrio martinica. See Purple gallinules
Porphyrio porphyrio. See Purple swamphens
Porphyrolaema spp. *See* Cotingas
Porpita spp., 1:29
Porpitidae, 1:124
Porpoises, 15:1–2, **15:33–40,** 15:*37*
  behavior, 15:6, 15:35
  conservation status, 15:35–36
  distribution, 15:*33,* 15:34
  echolocation, 12:87
  evolution, 15:3, 15:33
  feeding ecology, 15:35
  habitats, 15:6, 15:34–35
  humans and, 15:36
  physical characteristics, 15:4, 15:33–34
  reproduction, 15:9, 15:35
  species of, 15:38–39, 15:39*t*
  taxonomy, 15:33
Port Jackson sharks, 4:*17,* 4:97, 4:*98,* 4:*100,* 4:101–*102*
Porthidium spp., 7:445
Porthidium nasutum, 7:448
Porthidium ophryomegas, 7:447, 7:448
Portuguese men of war, 1:33, 1:*133,* 1:138–139
Portuguese watchdogs, 14:292
Porzana spp., 9:46, 9:47
Porzana carolina. See Soras rails
Porzana cinerea. See White-browed crakes
Porzana marginalis. See Striped crakes
Porzana monasa. See Kosrae crakes
Porzana palmeri. See Laysan rails
Porzana porzana. See Spotted crakes
Porzana pusilla. See Baillon's crakes
Porzana tabuensis. See Spotless crakes
Posidonichthys spp., 5:355
Positive reinforcement, in zoo training, 12:211
Poso bungus, 5:377
Possums, 12:137, 12:145, 13:31–35, 13:37–40, **13:57–67,** 13:*59,* 13:*62,* 13:66*t*
  See also specific types of possums
Postcanines. *See* Molar teeth; Premolars
Postembryonic development, 1:19–23, 2:21–22, 2:23
  See also Reproduction
Postnatal care, 12:5
  See also Reproduction
Postnatal development, 12:96–97
Pot-bellied pigs, 12:191, 15:*278*
Potamobatrachus spp., 5:42
Potamochoerus spp., 15:275, 15:276
Potamochoerus larvatus. See Bush pigs
Potamochoerus porcus. See Red river hogs
Potamochoerus porcus nyassae. See Red river hogs
Potamogale spp. *See* Otter shrews
Potamogale velox. See Giant otter shrews
Potamorrhaphis spp., 5:80
Potamotrygonidae. *See* Freshwater stingrays
Potato bugs. *See* Tomato bugs
Potomac shads. *See* American shads
Potomaglinae. *See* Otter shrews
Potoos, 9:367–370, **9:395–400,** 9:*398*
Potoroidae. *See* Rat-kangaroos
Potoroinae, 13:73
Potoroos, 13:32, 13:73
  See also specific types of potoroos
Potorous spp. *See* Potoroos
Potorous gilberti. See Gilbert's potoroos
Potorous longipes. See Long-footed potoroos
Potorous tridactlus. See Long-nosed potoroos
Potos spp., 14:309
Potos flavus. See Kinkajous
Potter, Beatrix, 13:200, 13:207
Potter wasps, 3:*409,* 3:*418,* 3:423
Pottos, 12:116, **14:13–21,** 14:*14,* 14:*15,* 14:*17,* 14:21*t*
Pouch snails. *See* Tadpole snails
Pouched lampreys, 4:83, 4:*87,* 4:*88*
Pouched mice, 16:297*t*
Pouched talegallus. *See* Australian brush-turkeys
Poverty, effect on conservation, 12:220–221
Powder blue tangs, 5:*392*
Powderpost beetles, 3:75–76
Powered flight, 12:43–44, 12:55–56
Pox, avian, 11:343, 11:348
Poyato-Ariza, F. J., 4:289
Poyntonia spp., 6:245
Praedatophasma spp. *See* Gladiators
Praedatophasma maraisi. See Gladiators
Prairie bobolinks. *See* Lark buntings
Prairie dogs, 12:*115,* 16:124, 16:126, 16:127, 16:*128,* 16:143, 16:144
  See also Black-tailed prairie dogs; Gunnison's prairie dogs
Prairie grouse. *See* Greater prairie chickens
Prairie rattlesnakes, 7:34, 7:38, 7:44, 7:50
Prairie voles, 16:125, 16:*231,* 16:*234*
Prairie warblers, 11:*286,* 11:*291*
Prasinohaema spp., 7:329
Pratincoles, **9:151–157,** 9:*154*
Pravdin's rock-crawlers, 3:*191,* 3:*192,* 3:193
Prawn killers. *See* Mantis shrimps
Prawns, 2:198
Praxibulini, 3:203
Praying mantids. *See* Mantids
Precambrian evolution, 1:8–*9*
  See also Evolution
Precautionary principle, 3:90
Precious corals. *See* Red corals
Precocial offspring, 12:12, 12:95–96, 12:97, 12:106
Predation, 3:47, 3:63–64, 3:148
  amphibians and, 6:6, 6:58
  Anura and, 6:65–*67*
  bird songs and, 8:39–40
  cnidarians, 1:28–30
  fishes, 4:43–44, 4:49, 4:68–70
  mammals, 12:21, 12:22–23, 12:64, 12:114–116
  protostomes, 2:26, 2:29, 2:36–37
  reptiles, 7:39, 7:162–163
  salamanders and, 6:325
  of tadpoles, 6:65
  true frogs and, 6:250–251
  See also Behavior
Prefica nivea, 9:373
Prehensile-tailed hutias, 16:461, 16:*464,* 16:*466*
Prehensile-tailed porcupines, 16:365, 16:366, 16:*367,* 16:*368,* 16:*369,* 16:*370,* 16:*372–373*
Prehensile-tailed skinks, 7:329, 7:*331,* 7:*332*–333
Premack, David, 12:161
Premaxilla, 12:9
Premnas spp., 2:34, 5:294–295
Premolars, 12:9, 12:14, 12:46
  See also Teeth
Presbytis spp., 14:172, 14:173

Presbytis comata, 14:175
Presbytis femoralis, 14:175
Presbytis fredericae, 14:175
Presbytis frontata, 14:175
Presbytis hosei. See Hose's leaf-monkeys
Presbytis johni. See Nilgiri langurs
Presbytis melalophos. See Banded leaf-monkeys
Presbytis potenziani. See Mentawai Island leaf-monkeys
Presbytis sivalensis, 14:172
Presbytis thomasi, 14:175
Preservation, fossil, 3:7–9
Prettyface wallabies. *See* Whiptail wallabies
Prevolitans faedoensis, 5:157
Prévost, Bénédicte, 2:33
Priacanthidae. *See* Bigeyes
Priacanthus hamrur. See Goggle eyes
Priacma serrata. See Cupedid beetles
Priapium fishes, 5:67, 5:68, 5:*70*
Priapulans, 1:5, 1:12–13, 1:48, **1:337–341**, 1:*339*
Priapulida, 1:9, 1:22, 2:5
Priapulus caudatus, 1:337, 1:338, 1:*339*, 1:*340*
Price, E. O., 15:145
Price, Trevor, 11:4
Pricklebacks, 4:67, 5:309–313, 5:*314*, 5:*317*
Pricklefishes, 5:105, 5:106, 5:*107*, 5:*109*–110
Pricklepigs. *See* North American porcupines
Prickly forest skinks, 7:*330*, 7:*333*, 7:335
Prides, lion, 14:258
Primates, **14:1–12**
  behavior, 14:6–8
  body size of, 12:21
  color vision, 12:79
  conservation status, 14:11
  distribution, 12:132, 14:5–6
  evolution, 12:33, 14:1–3
  feeding ecology, 14:8–9
  habitats, 14:6
  haplorhine, 12:11
  humans and, 14:11
  as keystone species, 12:216–217
  locomotion, 12:14
  memory, 12:154
  neonatal requirements, 12:126–127
  physical characteristics, 14:4–5
  reproduction, 12:90–91, 12:92, 12:95, 12:97, 12:98, 12:104, 12:105–106, 12:109, 14:10–11
  sexual dimorphism, 12:99
  taxonomy, 14:1–4
  touch, 12:83
  trumpeters and, 9:77, 9:78, 9:79
  See also specific types of primates
Primelephas spp., 15:162
Primitive caddisflies, 3:375, 3:376
Primitive ghost moths, 3:385
Primitive similarities, 12:28
Primodroma bournei, 8:135
Prince Albert's lyrebirds. *See* Albert's lyrebirds
Prince Edward lyrebirds. *See* Superb lyrebirds
Prince Ruspoli's turacos, 9:300, 9:302
Prinia spp., 11:3, 11:4, 11:394
Prinia bairdii, 11:4, 11:6
Prinia hodgsonii, 11:5
Prinia subflava. See Tawny-flanked prinias
Prinias, 11:*2*–3, 11:*8*, 11:13
Priodontes maximus. See Giant armadillos
Prionace glauca. See Blue sharks
Prionailurus bengalensis. See Leopards
Prionailurus bengalensis iriomotensis. See Iriomote cats
Prioniturus spp. *See* Racket-tailed parrots
Prionochilus spp. *See* Flowerpeckers
Prionochilus thoracicus. See Scarlet-breasted flowerpeckers
Prionodactylus argulus. See Elegant-eyed lizards
Prionodon spp., 14:335
Prionodon linsang. See Banded linsangs
Prionodura spp., 11:477, 11:478, 11:480
Prionodura newtoniana. See Golden bowerbirds
Prionoglaris stygia, 3:243
Prionopinae. *See* Helmet-shrikes
Prionops spp. *See* Helmet-shrikes
Prionops alberti. See Yellow-crested helmet-shrikes
Prionops gabela. See Gabela helmet-shrikes
Prionops plumatus. See White helmet-shrikes
Prionotus spp. *See* Sea robins
Prionotus carolinus. See Northern sea robins
Prionotus evolans. See Striped sea robins
Prions, 8:123
Prionurus laticlavius. See Yellowtailed surgeonfishes
Prionurus maculatus. See Yellowspotted sawtails
Priotelus temnurus. See Cuban trogons
Pristigasteridae. *See* Sawbelly herrings
Pristiophoriformes. *See* Sawsharks
Pristiophorus spp., 4:167
Pristiophorus cirratus. See Common sawsharks
Pristiophorus japonicus, 4:167
Pristiophorus nudipinnis, 4:167
Pristiophorus schroederi, 4:167
Pristis cirratus. See Common sawsharks
Pristis pectinata. See Smalltooth sawfishes
Pristoidei, 4:11
Pristolepidae, 4:57
Pristolepis fasciata. See Catopras
Pristorhamphus versteri. See Fan-tailed berrypeckers
Pro-eusocial termites, 3:164, 3:167–168
Proatheris spp., 7:445
ProAves Peru, 10:327
Problem solving, by insight, 12:140
Probolitrema spp., 1:197
Proboscidactyla flavicirrata, 1:28
Proboscidea. *See* Elephants
Probosciger aterrimus. See Palm cockatoos
Proboscis bats, 13:356, 13:357, 13:*359*, 13:*361*
Proboscis monkeys, 12:219, 14:171–*172*, 14:173, 14:175, 14:*177*, 14:*180*, 14:181–182
Probothriocephalus alaini, 1:229
Probothriocephalus muelleri, 1:229
Procambarus clarkii. See Red swamp crayfish
Procamelus spp., 15:265
Procapra spp. *See* Central Asian gazelles
Procapra gutturosa. See Mongolian gazelles
Procapra picticaudata. See Tibetan gazelles
Procapra przewalskii, 16:48
Procatopus spp., 5:93
Procavia spp., 15:177
Procavia capensis. See Rock hyraxes
Procavia habessinica. See Abyssinian hyraxes
Procavia johnstoni. See Johnston's hyraxes
Procavia ruficeps. See Red-headed rock hyraxes
Procavia syriacus. See Syrian rock hyraxes
Procavia welwitschii. See Kaokoveld hyraxes
Procaviidae. *See* Hyraxes
Procellaria cinerea. See Gray petrels
Procellariidae, **8:123–133**
Procellariiformes, **8:107–111**, 8:*108*, 8:*109*, 8:110
Procelsterna spp. *See* Noddies
Proceratophrys spp., 6:156, 6:157
Proceratophrys appendiculata, 6:158
Prochaetoderma yongei, 2:*381*, 2:*384*, 2:385
Prochloron, 1:33–34
Procnias spp. *See* Cotingas
Procnias alba. See White bellbirds
Procnias averano. See Bearded bellbirds
Procnias nudicollis. See Bare-throated bellbirds
Procnias tricarunculata. See Three-wattled bellbirds
Procolobus spp., 14:172, 14:173
Procolobus badius. See Western red colobus
Procolobus badius waldroni. See Miss Waldron's red colobus
Procolobus pannantii. See Penant's red colobus
Procolobus preussi, 14:175
Procolobus rufomitratus, 14:11
Procolobus verus. See Olive colobus
Proctoporus spp., 7:303
Proctrotrupoid wasps, 3:64
Procuticle, 3:31, 3:32–33
Procyon spp. *See* Raccoons
Procyon cancrivorus. See Crab-eating raccoons
Procyon gloveralleni. See Barbados raccoons
Procyon insularis, 14:310
Procyon lotor. See Northern raccoons
Procyon maynardi. See Bahaman raccoons
Procyon minor. See Guadeloupe raccoons
Procyon pygmaeus. See Cozumel Island raccoons
Procyonidae, 14:256, 14:259, 14:262, **14:309–317**, 14:*312*, 14:315*t*–316*t*
Prodendrogale spp., 13:291
Prodotiscus spp., 10:137–144
Prodotiscus regulus. See Cassin's honeyguides
Prodotiscus zambesiae. See Green-backed honeyguides
Prodromus spp., 3:55
Prodryas persephone, 3:383
Proechimys spp., 16:449, 16:*449*, 16:*451*
Proechimys albispinus, 16:450–451
Proechimys amphichoricus. See Venezuelan spiny rats
Proechimys gorgonae, 16:450–451
Proechimys semispinosus. See Spiny rats
Profelis aurata. See African golden cats
Profundulidae, 4:57
Progalago spp., 14:23
Proganochelys spp., 7:13
Progne sinaloae. See Sinaloa martins
Progne subis. See Purple martins
Progogrammus spp., 5:365
Progynotaenia odhneri, 1:*233*, 1:237–*238*
Progynotaeniidae, 1:226, 1:230–231
Prohylobates spp., 14:172, 14:188–189
Proichthydidae, 1:269
Project Feeder Watch, 8:23
Prokaryotes, 2:8
  See also specific prokaryotes
Prolactin, 12:148
Prolagus spp., 16:480, 16:491
Prolagus sardus. See Sardinian pikas
Prolebias spp., 5:90
Prolecithophora, 1:186, 1:188
Promerops cafer. See Cape sugarbirds
Prometheomys spp., 16:225

Prometheomys schaposchnikowi. See Long-clawed mole voles
Promiscuity, 12:98, 12:103, 12:107
See also Behavior
Promops centralis. See Big crested mastiff bats
Prong-billed barbets, 10:115
Pronghorns, 12:40, **15:411–417,** 15:*412,* 15:*413,* 15:*414,* 15:*415,* 15:*416*
behavior, 15:271, 15:412–414
conservation status, 15:415–416
distribution, 12:131, 12:*132,* 15:*411,* 15:412
evolution, 15:137, 15:138, 15:265, 15:266, 15:267, 15:411
feeding ecology, 15:414
habitats, 15:269, 15:412
humans and, 15:416–417
physical characteristics, 15:268, 15:411–412
reproduction, 15:143, 15:272, 15:414–415
scent marking, 12:80
taxonomy, 15:411
Pronolagus spp. *See* Rockhares
Pronuclear fusion, 1:18
Pronycticeboides spp., 14:13
Propagation programs, 6:347
Propalaeochoerus spp., 15:265
Propelodytes spp., 6:127
Propelodytes arevacus, 6:127
Prophalangopsidae, 3:201
Propherallodus spp., 5:355
Propithecus spp., 14:8
Propithecus candidus, 14:63, 14:67
Propithecus diadema, 14:63, 14:67
Propithecus diadema diadema. See Diademed sifakas
Propithecus edwardsi. See Milne-Edwards's sifakas
Propithecus perrieri, 14:63, 14:67
Propithecus tattersalli. See Golden-crowned sifakas
Propithecus verreauxi, 14:7, 14:63
Propithecus verreauxi coquereli. See Coquerel's sifakas
Propleopinae. *See* Giant rat-kangaroos
Propleopus oscillans, 13:70
Proplicastomata, 1:186
Proplina spp., 2:387
Propolis, 3:79
Proprioceptors, 7:9
Prosalirus spp., 6:11
Prosalirus bitis, 6:*12*
Prosciurillus spp., 16:163
Prosciurillus abstrusus, 16:167
Prosciurillus weberi, 16:167
Proscylliidae. *See* Finback catsharks
Proscyllium habereri, 4:115
Proseriata, 1:186
Proserpine rock wallabies, 13:32, 13:39
Prosigmodon spp., 16:264
Prosimian primates
behavior, 12:145–146, 14:7–8
distribution, 14:5
evolution, 14:1
feeding ecology, 14:9
habitats, 14:6
reproduction, 12:90, 14:10
taxonomy, 14:1, 14:3
See also Lemurs; Lorises; Primates; Tarsiers
Prosobonia cancellata. See Tuamotu sandpipers
Prosoboniini. *See* Polynesian sandpipers
Prosobothriidae, 1:226
Prosobranchs, 2:17
Prosorrhyncha, 3:259
Prosqualondotidae, 15:13
Prostaglandins, 1:44
Prostate gland, 12:104
Prosthemadera novaeseelandiae. See Tuis
Prostoma spp., 1:254
Prostoma eilhardi, 1:254
Prostoma graecense, 1:254
Prosymna spp. *See* Shovel-snouted snakes
Protapirus spp., 15:237
Proteidae, **6:377–383,** 6:*380*
distribution, 6:5, 6:*377,* 6:378, 6:*381–382*
evolution, 6:13
physical characteristics, 6:21, 6:323, 6:378, 6:381–382
reproduction, 6:29, 6:34–35, 6:378–379, 6:381–383
taxonomy, 6:323, 6:377, 6:381–382
Protein synthesis, 12:26–27, 12:29–31, 12:122–123
Proteles cristatus. See Aardwolves
Proteles cristatus cristatus, 14:359
Proteles cristatus septentrionalis, 14:359
Protemblemaria punctata, 5:343
Protentomidae, 3:93
Proteocephalidae, 1:226
Proteocephalidea, 1:226, 1:227, 1:231
Proteocephalus longicollis, 1:*232,* 1:239–*240*
Proterometra dickermani, 1:200
Proteropithecia spp., 14:143
Proterospongia, 1:8, 1:10
Proterospongia spp., 2:12
Proteus spp. *See* Olms
Proteus anguinus. See Olms
Prothonotary warblers, 11:290, 11:*292,* 11:294–*295*
Prothoracic glands, 3:28
See also Physical characteristics
Protists, 2:11–12
Protobothrops spp., 7:445
Protobranchia, 2:451
Protocetidae, 15:2
Protocobitis typhlops. See Blind cave loach of Xiaao
Protocoleopterans, 3:315
Protogobius spp., 5:373
Protohynobiinae, 6:336
Protohynobius puxiongensis, 6:335, 6:336
Protolophotus spp., 4:448
Protomecicthys spp., 4:448
Protomeryx spp., 15:265
Protomyzostomatidae, 2:59
Protonotaria citrea. See Prothonotary warblers
Protopteridae, 4:57
Protopterus. See African lungfishes
Protopterus spp., 4:201–205
Protopterus aethiopicus, 4:201, 4:202, 4:203
Protopterus amphibius. See Southeastern African lungfishes
Protopterus annectens. See African lungfishes
Protopterus dolloi, 4:201, 4:202, 4:203
Protoreaster nodosus. See Chocolate-chip sea stars
Protorthopterans, 3:12
Protosciurus spp., 16:135, 16:163
Protostomes, 1:*4,* 1:9, 1:19, 1:40, 1:41
behavior, **2:35–40**
defined, **2:3–6**
diversity of, 2:25
evolution, **2:7–14,** 2:*9,* 2:*10*
feeding ecology, 2:29, 2:36–37
humans and, **2:41–44**
reproduction, **2:14–24,** 2:31–32
taxonomy, 2:4–5, 2:11–14
See also specific species and topics
Prototheria. *See* Monotremes
Prototroctes oxyrhynchus. See New Zealand graylings
Prototroctes semoni. See Australian smelts
Prototroctidae, 4:391
Protoungulata, 15:131, 15:136
Protoxerus spp., 16:163
Protungulatum spp., 15:132
Proturans, 3:4, 3:10, **3:93–98,** 3:*95*
Prowfishes, 5:309–313
See also Australian prowfishes
Prum, Richard
on broadbills, 10:177
on philepittids, 10:187–188
on plantcutters, 10:325
Prunella collaris. See Alpine accentors
Prunella fagani. See Yemen accentors
Prunella modularis. See Dunnocks
Prunella montanella. See Siberian accentors
Prunella rubeculoides. See Robin accentors
Prunellidae. *See* Hedge sparrows
Prunellinae. *See* Hedge sparrows
Pryer's flowerpeckers. *See* Scarlet-backed flowerpeckers
Pryer's woodpeckers. *See* Okinawa woodpeckers
Przewalski's horses, 15:*231,* 15:*234*–235
behavior, 15:226
conservation status, 12:175, 12:222, 15:149, 15:222
distribution, 12:134
domestication and, 12:175, 15:149
feeding ecology, 15:*217*
genetic diversity and, 12:222
physical characteristics, 15:226
Przewalski's wild horses. *See* Przewalski's horses
Psalidoprocne nitens. See Square-tailed saw-wings
Psaltria exilis. See Pygmy tits
Psaltriparus spp. *See* Bushtits
Psaltriparus minimus. See Bushtits
Psammodes spp. *See* Tok-tokkies
Psammodromas spp., 7:297
Psammoperca spp. *See Psammoperca waigiensis*
Psammoperca waigiensis, 5:220, 5:221, 5:222
Psammophis spp. *See* Sandsnakes
Psarisomus dalhousiae. See Long-tailed broadbills
Psarocolius spp. *See* Oropendolas
Psarocolius cassini. See Baudó oropendolas
Psarocolius decumanus. See Crested oropendolas
Psarocolius montezuma. See Montezuma's oropendolas
Psenicubiceps alatus, 5:421
Psettodes spp. *See* Spiny flounders
Psettodidae. *See* Spiny flounders
Psettoidei, 5:449
Pseudacris crucifer. See Spring peepers
Pseudacris ocularis, 6:230
Pseudacris triseriata. See Chorus frogs
Pseudacteon curvatus. See Fire ant decapitating flies
Pseudailurus spp., 14:369

Pseudaletia unipuncta, 3:66
Pseudalopex culpaeus. See Culpeos
Pseudalopex fulvipes. See Darwin's foxes
Pseudalopex gymnocercus. See Pampas foxes
Pseudantechinus macdonnellensis. See Fat-tailed pseudantechinuses
Pseudanthias bicolor. See Bicolor anthias
Pseudanthias ventralis. See Hawaiian anthias
Pseudaphritidae. *See* Australian congollis
Pseudaphritis urvillii. See Australian congollis
Pseudechis spp. *See* Black snakes
Pseudechis porphyriacus. See Red-bellied black snakes
Pseudemydura spp., 7:78
Pseudemydura umbrina. See Western swamp turtles
Pseudemys concinna. See River cooters
Pseudemys nelsoni. See Red-bellied turtles
Pseudeuphausia spp., 2:185
Pseudhymenochirus spp., 6:100, 6:102
Pseudibis davisoni. See White-shouldered ibis
Pseudibis gigantea. See Giant ibises
Pseudidae, 6:4
Pseudinae, 6:226, 6:228–229, 6:230
Pseudione spp., 2:252
Pseudis paradoxa. See Paradox frogs
Pseudo babblers, **11:127–131,** 11:*129*
Pseudoamolops spp., 6:248, 6:249
Pseudobalistes spp., 4:68
Pseudobathylagus milleri. See Stout blacksmelts
Pseudoboines, 7:467
Pseudobranchus spp. *See* Dwarf sirens
Pseudobranchus axanthus, 6:327, 6:331
Pseudobranchus striatus. See Northern dwarf sirens
Pseudobufo spp., 6:184
Pseudobufo subasper. See Aquatic swamp toads
Pseudocalliurichthys spp., 5:365
Pseudocalyptomena spp. *See* Broadbills
Pseudocalyptomena graueri. See African green broadbills
Pseudocarcharias spp., 4:131
Pseudocarcharias kamoharai. See Crocodile sharks
Pseudocerastes spp., 7:445
Pseudoceratinidae, 1:79
Pseudoceros bifurcus. See Bifurcating flatworms
Pseudoceros bimarginatus, 1:*185*
Pseudoceros ferrugineus, 1:*189*, 1:190–*191*
Pseudoceros tristriatus. See Three striped flatworms
Pseudocheiridae, 13:34, 13:35, **13:113–123,** 13:*118*, 13:122*t*–123*t*
Pseudocheirus spp., 13:115
Pseudocheirus caroli. See Weyland ringtails
Pseudocheirus forbesi. See Moss-forest ringtails
Pseudocheirus peregrinus. See Common ringtails
Pseudocheirus schlegeli. See Arfak ringtails
Pseudochelidon eurystomina. See African river-martins
Pseudochelidoninae. *See* Swallows
Pseudochirops spp., 13:113
Pseudochirops albertisi. See d'Albertis's ringtail possums
Pseudochirops archeri. See Green ringtails
Pseudochirops corinnae. See Golden ringtail possums
Pseudochirops cupreus. See Coppery ringtail possums
Pseudochirulus spp., 13:113
Pseudochirulus canescens. See Lowland ringtails
Pseudochirulus herbertensis. See Herbert River ringtails
Pseudochromidae, 4:48, 5:255, 5:256, 5:258–263
Pseudochrominae, 5:256
Pseudococculina, 2:429
Pseudocoelomates, 1:4–5, 2:35
Pseudocolochirus violaceus. See Sea apples
Pseudocordylus spp. *See* Crag lizards
Pseudocumatidae, 2:229
Pseudodiplorchis americanus, 1:*217*, 1:*223*
Pseudoeryx spp., 7:468
Pseudoeurycea spp., 6:391
Pseudoeurycea bellii. See Bell's salamanders
Pseudoeurycea gadovii, 6:325
Pseudogastromyzon cheni. See Chinese hillstream loaches
Pseudohemisus calcaratus. See Mocquard's rain frogs
Pseudohynobius spp., 6:335
Pseudohynobius flavomaculatus, 6:336
Pseudois spp., 16:87, 16:90, 16:91
Pseudois nayaur. See Blue sheep
Pseudois schaeferi. See Dwarf blue sheep
Pseudoleistes spp. *See* Marshbirds
Pseudoleistes virescens. See Brown-and-yellow marshbirds
Pseudomugil mellis, 5:70
Pseudomugilidae. *See* Blue-eyes
Pseudomyrmex spp., 3:49
Pseudomys australis. See Plains mice
Pseudomys higginsi. See Long-tailed mice
Pseudomys nanus. See Western chestnut mice
Pseudonaja spp. *See* Brown snakes
Pseudonaja textilis. See Brown snakes
Pseudonestor spp., 11:343
Pseudonestor xanthophrys. See Maui parrotbills
Pseudonovibos spiralis. See Linh duongs
Pseudopaludicola spp., 6:156, 6:158
Pseudoperilampus ocellatus. See Rosy bitterlings
Pseudophasmatidae, 3:221, 3:222
Pseudophryne spp., 6:149
Pseudophryne corroborree. See Corroboree toadlets
Pseudophryne douglasi, 6:149
Pseudophyllidea, 1:225, 1:227, 1:229, 1:231
Pseudophyllinae, 3:205, 3:206
Pseudophyllodromiinae, 3:149
Pseudopimelodidae, 4:352
Pseudoplatystoma fasciatum. See Tiger shovelnose catfishes
Pseudopleuronectes americanus. See Winter flounders
Pseudopodia, 1:17, 2:12
Pseudopodoces spp., 11:504
Pseudopotto martini. See False pottos
Pseudopsectra usingeri, 3:309
Pseudopterogorgia elisabethae. See Sea whips
Pseudopterosins, 1:45
Pseudorana spp., 6:248, 6:249
Pseudorca crassidens. See False killer whales
Pseudorhina spp., 4:161
Pseudoryx spp., 15:263
Pseudoryx nghetinhensis. See Saolas
Pseudoryzomys spp., 16:266
Pseudoscaphirhynchus spp., 4:213
Pseudoscorpiones, 2:333
Pseudosphromenus spp., 5:427
Pseudosuchia, 7:17
Pseudotriakidae. *See* False catsharks
Pseudotriakis spp., 4:114
Pseudotriakis microdon. See False catsharks
Pseudotrichonitidae, 4:431, 4:432
Pseudotriton spp., 6:34, 6:390
Pseudotriton ruber, 6:*45*
Pseustes spp. *See* Bird snakes
Psilopsocus spp., 3:244
Psilopsocus mimulus, 3:243
Psilorhamphus spp. *See* Tapaculos
Psilorhamphus guttatus. See Spotted bamboowrens
Psilorhynchus aymonieri. See Algae eaters
Psittacidae. *See* Parrots
Psittaciformes. *See* Parrots
Psittacosis, 8:25, 9:256
Psittacula cyanocephala. See Plum-headed parakeets
Psittacula echo. See Mauritius parakeets
Psittacula himalayana. See Slaty-headed parakeets
Psittacula krameri. See Rose-ringed parakeets
Psittacus aterrimus. See Palm cockatoos
Psittacus erithacus. See African gray parrots; Gray parrots
Psittacus galeritus. See Sulphur-crested cockatoos
Psittacus haematod. See Rainbow lorikeets
Psittacus hollandicus. See Cockatiels
Psittacus hyacinthinus. See Hyacinth macaws
Psittacus macao. See Scarlet macaws
Psittacus ochrocephalus. See Yellow-crowned Amazons
Psittacus pertinax. See Brown-throated parakeets
Psittacus philippensis. See Philippine hanging-parrots
Psittacus roratus. See Eclectus parrots
Psittacus roseicollis. See Rosy-faced lovebirds
Psittacus undulatus. See Budgerigars
Psittirostra spp., 11:343
Psittirostra cantans. See Laysan finches
Psittirostra palmeri. See Greater koa finches
Psittrichas fulgidus. See Pesquet's parrots
Psocids, 3:243–244
Psococerastis spp., 3:243
Psocomorpha, 3:243
Psocoptera. *See* Book lice
Psocopterans. *See* Book lice
Psolidae, 1:420
Psolus chitinoides. See Slipper sea cucumbers
Psophia crepitans. See Common trumpeters
Psophia leucoptera. See Pale-winged trumpeters
Psophia undulata. See Houbara bustards
Psophia viridis. See Dark-winged trumpeters
Psophiidae. *See* Trumpeters
Psophodes olivaceus. See Eastern whipbirds
Psoquilla marginepunctata, 3:*245*, 3:*246*
Psychids, 3:387
Psychoactive drugs, 6:53
Psychopsidae, 3:305, 3:306, 3:307, 3:308
Psychrolutes paradoxus. See Tadpole sculpins
Psychrolutes sigalutes. See Soft sculpins
Psychrolutidae. *See* Fathead sculpins
Psychropotidae, 1:420
Psyllids, 3:63, 3:64, 3:66, 3:75
Psyllophryne didactyla. See Two-toed toadlets
Ptarmigans, 8:434, 8:*440*, 8:441–442
Ptarmus jubatus. See Crested scorpionfishes
Ptenoglossa, 2:445

Ptenopus spp. *See* Bell geckos
Pteraclis spp., 5:257
Pteranodon spp., 7:19
Pterapogon kauderni. *See* Banggai cardinalfishes
Pteraspidomorphi, 4:77
Ptereleotridae, 5:373, 5:374
Pteridophora spp. *See* Birds of paradise
Pteridophora alberti. *See* King of Saxony birds of paradise
Pteriidae, 2:451, 2:454
Pteriomorphia, 2:451
Pternohyla spp., 6:226, 6:227, 6:228–229
Pterobothriidae, 1:226
Pterobranchs, 1:13, 1:443, 1:444, 1:445
Pterocles spp. *See* Sandgrouse
Pterocles burchelli. *See* Burchell's sandgrouse
Pterocles coronatus. *See* Crowned sandgrouse
Pterocles exustus. *See* Chestnut-bellied sandgrouse
Pterocles lichtensteinii. *See* Lichtenstein's sandgrouse
Pterocles namaqua. *See* Namaqua sandgrouse
Pterocles orientalis. *See* Black-bellied sandgrouse
Pterocles senegallus. *See* Spotted sandgrouse
Pteroclididae. *See* Sandgrouse
Pterocliformes. *See* Sandgrouse
Pterocnemia pennata. *See* Lesser rheas
Pterodactyloids, 7:19
Pterodaustro spp., 7:19
Pterodroma cahow. *See* Bermuda petrels
Pterodroma nigripennis. *See* Black-winged petrels
Pteroglossus spp. *See* Aracaris
Pteroglossus castanotis. *See* Chestnut-eared aracaris
Pteroglossus flavirostris. *See* Pale-billed aracaris
Pteroglossus torquatus. *See* Collared aracaris
Pteroglossus viridis. *See* Green aracaris
Pterognathiidae, 1:331
Pterois spp. *See* Lionfishes
Pterois volitans. *See* Red lionfishes
Pteromyinae. *See* Flying squirrels
Pteromys volans. *See* Siberian flying squirrels
Pteronarcyidae, 3:141
Pteronarcys spp. *See* Giant stoneflies
Pteronarcys californica. *See* Giant salmonflies
Pteronotus spp., 13:308, 13:*437*
Pteronotus davyi. *See* Davy's naked-backed bats
Pteronotus gymnonotus. *See* Big naked-backed bats
Pteronotus macleayii. *See* Macleay's moustached bats
Pteronotus parnellii. *See* Parnell's moustached bats
Pteronotus personatus. *See* Wagner's moustached bats
Pteronotus quadridens. *See* Sooty moustached bats
Pteronura spp., 14:319
Pteronura brasiliensis. *See* Giant otters
Pterophanes spp. *See* Hummingbirds
Pterophanes cyanopterus. *See* Great sapphirewings
Pterophyllum spp., 5:279
Pterophyllum scalare. *See* Freshwater angelfishes
Pteroplatytrygon violacea. *See* Pelagic stingrays
Pteropodidae, 12:86, 13:307, 13:308, 13:309, 13:314
See also Flying foxes
Pteropodocys spp. *See* Cuckoo-shrikes
Pteropodocys maxima. *See* Ground cuckoo-shrikes
Pteroptochos spp. *See* Tapaculos
Pteroptochos megapodius. *See* Moustached turcas
Pteroptochos tarnii. *See* Black-throated huet-huets
Pteropus spp. *See* Flying foxes
Pteropus admiralitatum. *See* Admiralty flying foxes
Pteropus alecto. *See* Black flying foxes
Pteropus argentatus. *See* Ambon flying foxes
Pteropus brunneus. *See* Dusky flying foxes
Pteropus caniceps. *See* North Moluccan flying foxes
Pteropus conspicillatus. *See* Spectacled flying foxes
Pteropus dasymallus. *See* Ryukyu flying foxes
Pteropus fundatus. *See* Banks flying foxes
Pteropus giganteus. *See* Indian flying foxes
Pteropus gilliardi. *See* Gilliard's flying foxes
Pteropus hypomelanus. *See* Island flying foxes
Pteropus livingstonii. *See* Livingstone's fruit bats
Pteropus lylei. *See* Lyle's flying foxes
Pteropus macrotis. *See* Big-eared flying foxes
Pteropus mahaganus. *See* Sanborn's flying foxes
Pteropus mariannus. *See* Marianas fruit bats
Pteropus mearnsi. *See* Mearns's flying foxes
Pteropus melanotus. *See* Blyth's flying foxes
Pteropus molossinus. *See* Pohnpei flying foxes
Pteropus niger. *See* Greater Mascarene flying foxes
Pteropus poliocephalus. *See* Gray-headed flying foxes
Pteropus pumilus. *See* Little golden-mantled flying foxes
Pteropus rodricensis. *See* Rodricensis flying foxes
Pteropus rufus. *See* Madagascar flying foxes
Pteropus samoensis. *See* Samoan flying foxes
Pteropus scapulatus. *See* Little red flying foxes
Pteropus tokudae. *See* Guam flying foxes
Pteropus tonganus. *See* Tongan flying foxes
Pterorana spp., 6:248
Pterosagitta draco, 1:434, 1:435, 1:*436*, 1:*437*
Pterosagittidae, 1:433
Pterosaurs, 3:13–14, 7:5, 7:19, 8:*47*, 8:51, 12:52
Pteroscyllium spp., 4:113
Pterotermes occidentis. *See* Giant Sonoran drywood termites
Pterygocephalus paradoxus, 5:157
Pthirus pubis. *See* Crab lice; Pubic lice
Ptilichthyidae. *See* Quillfishes
Ptiliidae, 3:43
See also Beetles
Ptilinopus spp. *See* Fruit pigeons
Ptilinopus coralensis. *See* Atoll fruit doves
Ptilinopus magnificus. *See* Wompoo fruit doves
Ptilocercus spp., 13:292
Ptilocercus lowii. *See* Pen-tailed tree shrews
Ptilogonatinae. *See* Silky flycatchers
Ptilogonys caudatus. *See* Long-tailed silky flycatchers
Ptilogonys cinereus. *See* Gray silky flycatchers
Ptilonorhynchidae. *See* Bowerbirds
Ptilonorhynchus spp., 11:477, 11:478, 11:480
Ptilonorhynchus violaceus. *See* Satin bowerbirds
Ptilopsis spp. *See* Southern white-faced owls
Ptilopsis granti. *See* Southern white-faced owls
Ptiloris spp. *See* Birds of paradise
Ptiloris magnificus. *See* Magnificent riflebirds
Ptiloris paradiseus. *See* Paradise riflebirds
Ptiloris victoriae. *See* Victoria's riflebirds
Ptilorrhoa spp. *See* Jewel-babblers; Rail babblers
Ptilorrhoa caerulescens. *See* Blue jewel-babblers
Ptilostomus afer. *See* Piapiacs
Ptyas spp. *See* Banded ratsnakes
Ptyas korros, 7:51
Ptychadena spp., 6:250
Ptychadena oxyrhynchus. *See* Sharp-nosed grass frogs
Ptychadeninae, 6:247
Ptychocheilus lucius. *See* Colorado pikeminnows
Ptychocheilus oregonensis. *See* Northern pikeminnows
Ptychochromoides betsileanus. *See* Trondo mainties
Ptychodactarian anemones, 1:106
Ptychodera spp., 1:445
Ptychodera flava. *See* Hawaiian acorn worms
Ptychoglossus spp., 7:303
Ptychohyla spp., 6:227, 6:230
Ptychopteriidae, 3:54
Ptychoramphus spp. *See* Auks
Ptychozoon spp., 7:*261*
Ptyctodontiformes, 4:10
Ptyonoprogne rupestris. *See* Crag martins
Puaiohis, 10:489
Pubic lice, 3:*249*, 3:252
Pucrasia macrolopha. *See* Koklass
Puddling, 3:38
Pudu spp., 15:382, 15:384
Pudu mephistophiles. *See* Northern pudus
Pudu pudu. *See* Southern pudus
Puerto Rican Amazon parrots, 8:27, 9:280
Puerto Rican boas, 7:411
Puerto Rican coqui, 6:159, 6:*161*, 6:*164*
Puerto Rican emeralds, 9:*453*, 9:*464*
Puerto Rican giant hutias, 16:*470*, 16:471*t*
Puerto Rican hutias, 16:467*t*
Puerto Rican nesophontes, 13:243–244
Puerto Rican nightjars, 9:370, 9:405
Puerto Rican screech owls, 9:346
Puerto Rican todies, 10:26–27, 10:*29–30*
Puerto Rican walkingsticks, 3:*35*
Puff adders. *See* Eastern hog-nosed snakes
Puff-backed honeyeaters. *See* Puff-backed meliphagas
Puff-backed meliphagas, 11:*239*, 11:*251*
Puff-faced watersnakes, 7:468
Puff-throated bulbuls, 10:397
Puffbacks, 10:425–430, 10:*431*, 10:432–*433*
Puffbirds, 10:85–88, **10:101–111,** 10:*102*, 10:*105*
  behavior, 10:102
  conservation status, 10:103–104
  distribution, 10:*101*, 10:102
  evolution, 10:101
  feeding ecology, 10:103
  habitats, 10:102
  humans and, 10:104
  physical characteristics, 10:101–102
  reproduction, 10:103
  species of, 10:*106–111*
  taxonomy, 10:101
Pufferfishes, 4:14, 4:70, **5:467–471,** 5:*472*, 5:*474*, 5:*476*, **5:481–482,** 5:*483*, 5:484
Puffins, **9:219–229,** 9:*223*
  behavior, 9:221, 9:226–227

conservation status, 9:222, 9:226–227
distribution, 9:*219*, 9:221, 9:226
evolution, 9:219
feeding ecology, 9:221–222, 9:227
habitats, 9:221, 9:226
humans and, 9:222, 9:227
physical characteristics, 9:220, 9:226
reproduction, 9:222, 9:227
species of, 9:*226–227*
taxonomy, 9:219–220, 9:226
Puffinus assimilis. See Little shearwaters
Puffinus carneipes. See Flesh-footed shearwaters
Puffinus griseus. See Sooty shearwaters
Puffinus puffinus. See Manx shearwaters
Puffinus tenuirostris. See Short-tailed shearwaters
Pufflegs, 9:438, 9:442, 9:443, 9:447, 9:450
Pukus, 16:27, 16:29, 16:30, 16:32
Pulchrana spp., 6:249
Pulex spp., 3:349, 3:350
Pulex larimerius, 3:347
Pulicidae, 3:347, 3:349
Pulicoidea, 3:347
Pulmonata, **2:411–422**, 2:*416*
behavior, 2:414
conservation status, 2:414
distribution, 2:413
evolution, 2:411
feeding ecology, 2:414
habitats, 2:413–414
humans and, 2:414–415
physical characteristics, 2:*412*, 2:413
reproduction, 2:16, 2:414
species of, 2:417–421
taxonomy, 2:411–413
Pulsatrix perspicillata. See Spectacled owls
Pulvinomyzostomatidae, 2:59
Pulvinomyzostomum pulvinar, 2:*61*, 2:*62*
Puma concolor. See Pumas
Puma (Felis) concolor. See Pumas
Pumas, 14:*378*, 14:382
behavior, 12:*121*
distribution, 12:129, 14:371, 14:*386*
feeding ecology, 14:372
as keystone species, 12:217
Pumpkin toadlets, 6:179, 6:*180*, 6:*181*
Pumpkinseed sunfishes, 5:195, 5:*195*, 5:198
Puna flamingos. *See* James' flamingos
Puna frogs, 6:64, 6:157
Puna tinamous, 8:*58*
Punares, 16:*450*, 16:458*t*
Punciidae, 2:311
Punctum pygmaeum, 2:413
Pungitius hellenicus, 5:136
Punkies, 3:357
Punomys spp., 16:264
Punomys lemminus, 16:266
Puntius tetrazona tetrazona. See Tiger barbs
Pupa solidula, 2:*406*, 2:*407*
Pupal stage, 3:34, 3:35
See also Reproduction
Pupfishes, 4:12, 5:94, 5:*95*, 5:97–98
Purcell's ghost frogs, 6:*132*
Purple-backed sunbeams, 9:450
Purple-bearded bee-eaters, 10:*43*, 10:*44*–45
Purple-bellied fruit doves. *See* Wompoo fruit doves
Purple-breasted cotingas, 10:*310*, 10:312–*313*
Purple-breasted finches. *See* Gouldian finches
Purple-crested louries. *See* Purple-crested turacos
Purple-crested turacos, 9:300, 9:*303*, 9:*306*–307
Purple-crowned fairies, 9:*439*
Purple-crowned fairy-wrens, 11:46, 11:*47*, 11:*48*–49
Purple gallinules, 9:47, 9:52, 9:110
Purple-glossed snakes, 7:461, 7:462
Purple guans. *See* Crested guans
Purple-heart urchins, 1:403
Purple herons, 8:235
Purple indigobirds, 11:363
Purple martins, 8:24, 10:*362*, 10:*367*–368
Purple-naped sunbirds, 11:*211*, 11:*216*
Purple needletails, 9:421
Purple-rumped sunbirds, 11:210
Purple sandpipers, 9:177, 9:178
Purple satins. *See* Satin bowerbirds
Purple sea stars. *See* Ocher stars
Purple sea urchins, 1:405
See also Short-spined sea urchins
Purple sunbirds, 11:210, 11:*212*, 11:*223*
Purple swamphens, 9:47, 9:50, 9:52
Purple-throated caribs, 9:443–444, 9:447
Purple-throated fruitcrows, 10:306, 10:307, 10:312
Purple-winged starlings. *See* European starlings
Purrhula albifrons. See Thick-billed weavers
Purring catfishes. *See* Blue-eye catfishes
Purse case makers, 3:375, 3:376
Purse web spiders, 2:*338*, 2:342–343
Puya raimondii. See Arborescent puyas
Puzanovia rubra, 5:310
Pyburn's pancake frogs, 6:*307*, 6:315
Pycnogonida. *See* Sea spiders
Pycnogonum littorale, 2:*322*
Pycnonotidae. *See* Bulbuls
Pycnonotus spp., 10:396–397
Pycnonotus barbatus. See Common bulbuls
Pycnonotus cafer. See Red-vented bulbuls
Pycnonotus capensis. See Cape bulbuls
Pycnonotus goiaver. See Yellow-vented bulbuls
Pycnonotus importunus. See Somber greenbuls
Pycnonotus jocosus. See Andaman red-whiskered bulbuls; Red-whiskered bulbuls
Pycnonotus latirostris. See Yellow-whiskered greenbuls
Pycnonotus masukuensis. See Shelley's greenbuls
Pycnonotus nigricans. See African red-eyed bulbuls
Pycnonotus pencillatus. See Yellow-eared bulbuls
Pycnonotus sinensis. See Chinese bulbuls
Pycnonotus striatus. See Striated bulbuls
Pycnonotus taivanus. See Styan's bulbuls
Pycnonotus xantholaemus. See Yellow-throated bulbuls
Pycnonotus zeylanicus. See Straw-headed bulbuls
Pycnophyes spp., 1:275
Pycnophyes kielensis, 1:276
Pycnophyidae, 1:275, 1:277
Pycnopodia helianthoides. See Sunflower stars
Pycnoptilus floccosus. See Pilotbirds
Pygarrhichas albogularis. See White-throated treerunners
Pygathrix spp., 14:172, 14:173
Pygathrix nemaeus. See Red-shanked douc langurs
Pygathrix nigripes, 14:175
Pygathrix roxellana. See Golden snub-nosed monkeys
Pygeretmus spp., 16:215
Pygeretmus platyurus. See Lesser fat-tailed jerboas
Pygeretmus pumilio, 16:213, 16:214, 16:216, 16:217
Pygeretmus shitkovi. See Greater fat-tailed jerboas
Pygiopsyllidae, 3:347, 3:348–349
Pygmy angelfishes, 5:238, 5:242
Pygmy anteaters. *See* Silky anteaters
Pygmy antelopes. *See* Dwarf antelopes
Pygmy backswimmers, aquatic, 3:260
Pygmy beaked whales, 15:1, 15:59, 15:70*t*
Pygmy blues, Western, 3:384
Pygmy Bryde's whales, 15:1
Pygmy chimpanzees. *See* Bonobos
Pygmy corydoras. *See* Dwarf corydoras
Pygmy dormice, Chinese, 16:297*t*
Pygmy drongos, 11:437, 11:438, 11:*440*, 11:*441*
Pygmy elephants, 15:163
Pygmy falcons. *See* African pygmy falcons
Pygmy fruit bats. *See* Black flying foxes
Pygmy geese. *See* African pygmy geese
Pygmy gliders, 13:33, 13:35, 13:38, 13:*139*, 13:*140*, 13:*141*, 13:142, 13:*143*, 13:*144*
Pygmy gliding possums. *See* Pygmy gliders
Pygmy goannas, 7:361
Pygmy goats, 16:*95*
Pygmy grasshoppers, 3:201, 3:202, 3:203, 3:206
Pygmy hippopotamuses, 15:*307*, 15:*309*
behavior, 15:306, 15:311
conservation status, 15:308, 15:311
distribution, 15:306, 15:*310*
evolution, 15:301
feeding ecology, 15:308, 15:311
habitats, 15:306, 15:311
humans and, 15:308, 15:311
physical characteristics, 15:311
reproduction, 15:*305*, 15:308, 15:311
taxonomy, 15:301, 15:311
Pygmy hogs, 15:275, 15:277, 15:280, 15:281, 15:*283*, 15:286–*287*
Pygmy jerboas, 16:211, 16:212, 16:213, 16:215
Pygmy killer whales, 15:2, 15:56*t*
Pygmy kingfishers. *See* African pygmy-kingfishers
Pygmy lorises. *See* Pygmy slow lorises
Pygmy marmosets, 14:*120*, 14:*125*, 14:*130*, 14:131
behavior, 14:117, 14:118
distribution, 14:116
feeding ecology, 14:120
habitats, 14:116
physical characteristics, 14:116
reproduction, 14:121, 14:122, 14:123
taxonomy, 14:115
Pygmy mice, 16:123
Pygmy mole crickets, 3:201, 3:202, 3:203, 3:205
Pygmy mouse lemurs, 14:4, 14:35, 14:*36*, 14:39
Pygmy nuthatches, 11:*167*, 11:*168*
Pygmy owls, 9:345, 9:347, 9:349, 9:*350*
Pygmy palm-swifts, 9:421
Pygmy parrots, 9:275, 9:277
Pygmy perches, 5:219–222
southern, 5:*224*, 5:*229*–230

western, 5:*224*, 5:*229*, 5:230
Pygmy phalangers. *See* Pygmy gliders
Pygmy planigales, 12:300*t*
Pygmy possums, 13:31, 13:33, 13:38, 13:39, **13:105–111,** 13:*109*
Pygmy pythons, 7:421, 7:*423*, 7:*424*
Pygmy rabbits, 16:481, 16:487, 16:505, 16:515*t*
Pygmy right whales, **15:103–106,** 15:*104*, 15:*105*
Pygmy ringtails, 13:*118*, 13:*121*, 13:122
Pygmy rock mice, 16:*286*, 16:*289*, 16:290
Pygmy seahorses, 5:133
Pygmy seedsnipes. *See* Least seedsnipes
Pygmy sharks, 4:151
Pygmy shore-eels, 5:*358*, 5:*361*
Pygmy short-tailed opossums, 12:251, 12:254, 12:*256*, 12:*262–263*
Pygmy shrews. *See* American pygmy shrews
Pygmy silversides, 5:67, 5:68
Pygmy slow lorises, 14:*14*, 14:15, 14:*16*, 14:*17*, 14:*18*, 14:19
Pygmy sperm whales, 15:6, 15:73, 15:74, 15:76, 15:77, 15:*78*, 15:*79*, 15:80
Pygmy squirrels, 16:163
Pygmy sunbirds, 11:209
Pygmy sunfishes, 4:57, **5:195–199,** 5:*201*, **5:205–206**
Pygmy swiftlets, 9:421
Pygmy tarsiers, 14:93, 14:95, 14:*97*, 14:99
Pygmy three-toed sloths. *See* Monk sloths
Pygmy tits, 10:169, 11:141, 11:142
Pygmy tree shrews, 13:*290*, 13:*292*, 13:297*t*
Pygmy white-toothed shrews. *See* Savi's pygmy shrews
Pygmy wren-babblers, 10:506, 10:*510*, 10:*512*
Pygocentrus nattereri. *See* Red-bellied piranhas
Pygocephalomorpha, 2:215, 2:225
Pygoplites diacanthus. *See* Regal angelfishes
Pygopodinae. *See* Pygopods
Pygopods, 7:205, **7:259–269,** 7:*264*
  behavior, 7:261
  conservation status, 7:262–263
  distribution, 7:*259*, 7:260
  evolution, 7:259
  feeding ecology, 7:261–262
  habitats, 7:260–261
  humans and, 7:263
  physical characteristics, 7:260
  reproduction, 7:262
  species of, 7:*265–269*
Pygoscelis spp. *See* Penguins
Pygoscelis adeliae. *See* Adelie penguins
Pygoscelis antarctica. *See* Chinstrap penguins
Pygoscelis papua. *See* Gentoo penguins
Pyke, G. H., 9:445
Pylodictis olivaris. *See* Flathead catfishes
Pyralids, 3:386, 3:387, 3:388, 3:389
Pyramid butterflyfishes, 5:*239*
Pyramondontimes, 5:17
Pyrenean brook salamanders, 6:*369*, 6:371
Pyrenean desmans, 13:281, 13:282, 13:287*t*
Pyrenestes sanguineus. *See* Crimson seedcrackers
Pyrgilauda spp. *See* Ground sparrows
Pyrgilauda davidiana. *See* Père David's ground sparrows
Pyrgita motitensis. *See* Southern rufous sparrows
Pyrgocorypha spp., 3:203
Pyrgomorphidae, 3:203, 3:208
Pyrgomorphoidea, 3:201
Pyrocephalus rubinus. *See* Vermilion flycatchers
Pyroderus scutatus. *See* Red-ruffed fruit-crows
Pyrosoma atlanticum. *See* Pyrosomes
Pyrosomatida, 1:467, 1:468
Pyrosomes, 1:*469*, 1:470
Pyrotheria, 12:11, 15:131, 15:136
Pyrrhocorax spp., 11:503, 11:505, 11:507
Pyrrhocorax graculus. *See* Alpine choughs
Pyrrhocorax pyrrhocorax. *See* Red-billed choughs
Pyrrholaemus brunneus. *See* Redthroats
Pyrrhula murina. *See* Azores bullfinches
Pyrrhula pyrrhula. *See* Eurasian bullfinches
Pyrrhura picta. *See* Painted parrots
Python spp., 7:419, 7:420
Python anchietae. *See* Angolan pythons
Python brongersmai. *See* Blood pythons
Python molurus. *See* Indian pythons
Python molurus bivattatus. *See* Burmese pythons
Python regius. *See* Ball pythons
Python reticulatus. *See* Reticulated pythons
Python sebae. *See* African rock pythons
Python timoriensis. *See* Lesser Sundas pythons
Pythonidae. *See* Pythons
Pythons, 7:**419–428,** 7:*423*
  behavior, 7:205, 7:420–421
  boas *vs.*, 7:409, 7:419
  conservation status, 7:421–422
  distribution, 7:*419*, 7:420
  endothermic, 7:28
  evolution, 7:419
  feeding ecology, 7:421
  habitats, 7:420
  humans and, 7:421, 7:422
  muscles, 7:26
  in mythology and religions, 7:55–57
  physical characteristics, 7:202
  reproduction, 7:*36*, 7:42, 7:207, 7:421
  skeleton, 7:24
  snake charming and, 7:56
  species of, 7:*424*–428
  taxonomy, 7:419
  See also specific types of pythons
Pytilia melba. *See* Green-winged pytilias
Pytilias, 11:353, 11:354, 11:*355*, 11:*358*, 11:361–362
Pytilla melba. *See* Green-winged pytilias
Pyura spp., 1:452
Pyxicephalinae, 6:247
Pyxicephalus spp., 6:67, 6:247–248, 6:250, 6:251, 6:252
Pyxicephalus adspersus. *See* African bullfrogs

Q

Quaggas, 15:146, 15:221, 15:236*t*
Quahogs. *See* Northern quahogs
Quail-plovers. *See* Lark buttonquails
Quail thrushes, **11:75–81,** 11:77
Quailfinches, 11:353, 11:354, 11:*358*, 11:365–366
Quails, 8:19, 8:433, 8:*440*, 9:11
Quails, New World. *See* New World quails
Quanacaste squirrels. *See* Deppe's squirrels
Quarrions. *See* Cockatiels
Quattro, J. M., 5:195
Queen Alexandra's birdwings, 3:384, 3:389, 3:*390*, 3:*396*, 3:398–399
Queen angelfishes, 5:*235*
Queen Charlotte owls. *See* Northern saw-whet owls
Queen conches, 2:447
Queen scallops, 2:*455*, 2:459–460, 2:*462*
Queen triggerfishes, 5:471
Queen Victoria lyrebirds. *See* Superb lyrebirds
Queen Victoria's rifle-birds. *See* Victoria's riflebirds
Queens, 3:68
  honeybee, 3:*64*
  termite, 3:*165*
  vs. workers, 3:69, 3:70, 3:71
Queensland blossom bats. *See* Southern blossom bats
Queensland gardeners. *See* Golden bowerbirds
Queensland koalas, 13:43, 13:*44*, 13:*45*, 13:*48*
Queensland lungfishes. *See* Australian lungfishes
Queensland tube-nosed bats, 13:*335*, 13:*337*, 13:*339*, 13:*346*, 13:347, 13:349*t*
Quelea spp., 11:375
Quelea quelea. *See* Red-billed queleas
Quemisia spp., 16:469
Quemisia gravis. *See* Hispaniolan giant hutias
Querétaro pocket gophers, 16:190, 16:*191*, 16:*192*, 16:194
Querula spp. *See* Cotingas
Querula purpurata. *See* Purple-throated fruitcrows
Quetzalcoatl (Mayan deity), 7:55, 9:451, 9:479
Quetzalcoatlas, 7:19
Quill-snouted snakes, 7:461, 7:462, 7:*463*
Quillers. *See* North American porcupines
Quillfishes, 5:309–313
Quillpigs. *See* North American porcupines
Quills. *See* Spines
Quiscalus spp. *See* Grackles
Quiscalus mexicanus. *See* Great-tailed grackles
Quiscalus niger. *See* Antillean grackles
Quiscalus quiscula. *See* Common grackles
Quokkas, 12:123, 13:34, 13:*92*, 13:*93*, 13:95–96
Quolls, 12:279–284, 12:288–291, 14:256

R

R-complex. *See* Reptilian brain
r-strategists, *12:13*, 12:97
r-strategy, 2:27, 3:46, 3:47
Raasch, Maynard S., 5:99
Rabbit calicivirus disease (RCD), 12:187–188, 12:223
Rabbit-eared bandicoots. *See* Bilbies
Rabbit Nuisance Bill of 1883 (Australia), 12:185
Rabbit rats, 16:265, 16:266, 16:269, 16:*271*, 16:*272*
Rabbitfishes, 4:67, 5:391, 5:392, 5:393
  See also Foxface rabbitfishes
Rabbits, 12:180, 16:479–488, **16:505–516,** 16:*510*, 16:515*t*–516*t*
  behavior, 16:506–507
  conservation status, 16:509
  distribution, 12:137, 16:*505*–506

evolution, 16:505
feeding ecology, 16:507
habitats, 16:513–515
humans and, 16:509
as pests, 12:185–186
physical characteristics, 12:46, 12:121, 12:122, 16:505
reproduction, 12:103, 12:126, 16:507–509
species of, 16:513–515, 16:515*t*–516*t*
taxonomy, 16:505
Rabies, 13:316
Raccoon dogs, 14:265, 14:266, 14:267, 14:*275*, 14:*278*, 14:281–282
Raccoons, 12:79, 12:224, 14:256, 14:260, **14:309–317,** 14:*312*, 14:315*t*–316*t*
Racers, 7:26, 7:205–206, 7:466, 7:469, 7:470
Racey, Paul A., 13:502
Rachycentridae. *See* Cobias
Rachycentron canadum. See Cobias
Racket-tailed parrots, 9:279
Racket-tailed rollers, 10:52
Radial cleavage, 1:19, 1:22, 1:23, 2:3, 2:19–20
See also Reproduction
Radial symmetry, 2:35
See also Physical characteristics
Radiata, 1:8–9
Radiated tortoises, 7:*30*
Radiation
adaptive, 8:3, 11:342, 11:350
mammalian, 12:11, 12:12
mimetic, 6:206
Radiicephalidae, 4:447, 4:448
Radiicephalus spp., 4:447, 4:449
Radiicephalus elongatus, 4:453
Radinsky, Leonard, 15:132
Radio telemetry, 12:199–202
Radio tracking devices, 6:*58*
Raffray's bandicoots, 13:*4*, 13:*11*, 13:*14*, 13:15–16
Ragfishes, 5:*351*, **5:351–353,** 5:*352*, 5:*353*
Ragworms, 2:*49*, 2:*51*, 2:53–54
Raia jamaicensis. See Yellow stingrays
Raikova, Olga, 1:179
Raikow, Robert J., 10:169
Rail babblers, 10:505
See also Malaysian rail-babblers
Raillietiella spp., 2:317
Rails, 9:1–4, **9:45–67,** 9:*53–56*
behavior, 9:48–49
conservation status, 9:51–52
distribution, 9:*45*, 9:47
evolution, 9:45
feeding ecology, 9:48, 9:49–50
habitats, 9:47–48, 9:51–52
humans and, 9:52
physical characteristics, 9:45, 9:46
reproduction, 9:48, 9:50–51
species of, 9:57–67
taxonomy, 9:5, 9:45–46
Rails: A Guide to the Rails, Crakes, Gallinules and Coots of the World (Taylor), 9:45
Rainbow bee-eaters, 10:42, 10:*43*, 10:*46–47*
Rainbow birds. *See* Rainbow bee-eaters
Rainbow boas, 7:410
Rainbow darters, 5:*198*, 5:*200*, 5:*203*, 5:206
Rainbow finches. *See* Gouldian finches
Rainbow lories. *See* Rainbow lorikeets
Rainbow lorikeets, 8:*10*, 9:277, 9:*279*, 9:*281*, 9:297–298
Rainbow pittas, 10:194, 10:195, 10:*196*, 10:200–*201*
Rainbow seaperches, 5:*283*, 5:*287*, 5:290
Rainbow smelts, 4:391
Rainbow snakeheads, 5:438, 5:439, 5:*440*, 5:*443*
Rainbow snakes, 7:469
Rainbow surfperches. *See* Rainbow seaperches
Rainbow trouts, 4:*26*, 4:*39*, 4:*409*, 4:*410*, 4:411, 4:413–414
Rainbowfishes, 4:12, 4:57, **5:67–76,** 5:*72*
Rainforest bandicoots, 13:1, **13:9–18,** 13:*11*, 13:16*t*–17*t*
Rainforests, 12:218
See also Habitats
Rainpool gliders. *See* Wandering gliders
Raja altavela. See Spiny butterfly rays
Raja binoculata. See Mermaid's purse skates
Raja birostris. See Atlantic manta rays
Raja centroura. See Roughtail stingrays
Raja clavata, 4:*174*
Raja eglanteria. See Clearnose skates
Raja narinari, 4:*184*–185
Raja ocellata. See Winter skates
Rajidae. *See* Skates
Rajiformes, **4:173–188,** 4:*180–181*
behavior, 4:177
conservation, 4:178
distribution, 4:176
evolution, 4:173–174
feeding ecology, 4:177–178
habitats, 4:176–177
humans and, 4:178–179
physical characteristics, 4:27, 4:*174*–176, 4:*175*
reproduction, 4:*178*
species of, 4:182–188
taxonomy, 4:11, 4:173–174
Rallicula forbesi. See Forbes's forest-rails
Rallidae. *See* Rails
Rallidontidae, 3:55
Ralliformes, 9:45
Rallina forbesi. See Forbes's forest-rails
Rallinae, 9:45
Rallus spp., 9:47
Rallus aquaticus. See Water rails
Rallus crex. See Corncrakes
Rallus cuvieri. See White-throated rails
Rallus elegans. See King rails
Rallus jamaicensis. See Black rails
Rallus limicola. See Virginia rails
Rallus longirostris. See Clapper rails
Rallus maculatus. See Spotted rails
Rallus wetmorei. See Plain-flanked rails
Rallus ypecha. See Giant wood-rails
Ramonellus spp., 6:13
Ramphastidae. *See* Toucans
Ramphastos spp. *See* Toucans
Ramphastos dicolorus. See Red-breasted toucans
Ramphastos sulfuratus. See Keel-billed toucans
Ramphastos swainsonii. See Chestnut-mandibled toucans
Ramphastos toco. See Toco toucans
Ramphastos tucanus. See White-throated toucans
Ramphastos vitellinus. See Channel-billed toucans
Ramphastos vitellinus culminatus. See Yellow-ridged toucans
Ramphocaenus melanurus. See Long-billed gnatwrens
Ramphocinclus brachyurus. See White-breasted thrashers
Ramphocoris spp. *See* Larks
Ramphocoris clotbey. See Thick-billed larks
Ramphodon naevius. See Saw-billed hermits
Ramphomicron spp. *See* Hummingbirds
Ramphotyphlops spp., 7:380–382
Ramphotyphlops bituberculatus, 7:380
Ramphotyphlops braminus, 7:379, 7:381, 7:382
Ramphotyphlops cumingii, 7:380, 7:381
Ramphotyphlops exocoeti. See Christmas Island blindsnakes
Ramphotyphlops grypus, 7:380
Ramphotyphlops nigrescens. See Blackish blindsnakes
Ramphotyphlops proximus, 7:379
Ramphotyphlops unguirostris, 7:379
Ramphotyphlops wiedii, 7:382
Ram's horn squids, 2:475, 2:*481*, 2:*487*, 2:488
Ramsar Convention, 1:330, 8:368
Ramsayornis modestus. See Brown-backed honeyeaters
Ramsdelepidion schusteri, 3:119
Ramshorn snails, 2:412
Ramus mandibulae, 12:9, 12:11
Rana spp.
behavior, 6:66
conservation status, 6:251
humans and, 6:252
physical characteristics, 6:183
reproduction, 6:29, 6:31–32
taxonomy, 6:248
Rana arborea. See European treefrogs
Rana areolata. See Crawfish frogs
Rana arvalis. See Moor frogs
Rana bombina. See Fire-bellied toads
Rana boylii. See Foothill yellow-legged frogs
Rana caerulea. See Green treefrogs
Rana catesbeiana. See North American bullfrogs
Rana clamitans. See Green frogs
Rana cornuta. See Suriname horned frogs
Rana corrugata. See Corrugated water frogs
Rana eiffingeri. See Eiffinger's Asian treefrogs
Rana erythraea, 6:250
Rana esculenta. See Roesel's green frogs
Rana fisheri, 6:251
Rana goliath. See Goliath frogs
Rana lessonae, 6:262, 6:263
Rana leucophyllata. See Hourglass treefrogs
Rana liebigii. See Spiny-armed frogs
Rana lima. See Pointed-tongue floating frogs
Rana lineata. See Gold-striped frogs
Rana marina. See Marine toads
Rana miliaris. See Rock River frogs
Rana opisthodon. See Faro webbed frogs
Rana oxyrhynchus. See Sharp-nosed grass frogs
Rana paradoxa. See Paradox frogs
Rana pentadactyla. See South American bullfrogs
Rana pipa. See Suriname toads
Rana pipiens. See Northern leopard frogs
Rana punctata. See Parsley frogs
Rana ridibunda, 6:262, 6:263
Rana sensu stricto, 6:250
Rana sylvatica. See Wood frogs
Rana temporaria. See Brown frogs
Rana tigerina. See Indian tiger frogs

Rana tlaloci, 6:251
Rana variegata. See Yellow-bellied toads
Ranacephala spp., 7:77
Ranacephala hogei. See Hoge's sideneck turtles
Ranching, reptiles, 7:48–*49*
Randall, Jan A., 16:200–201, 16:203
Random genetic drift, 12:222
Rangifer spp., 15:382, 15:383
Rangifer tarandus. See Reindeer
Rangifer tarandus caribou. See Canadian caribou
Ranidae. *See* True frogs
Ranina ranina. See Spanner crabs
Raninae, 6:247–248
Ranini, 6:247–248
Ranixalinae, 6:248
Ranodon spp., 6:335, 6:336
Ranodon sibiricus. See Semirechensk salamanders
Rapa Island tobies, 5:471
Raphicerus campestris. See Steenboks
Raphicerus melanotis. See Cape grysboks
Raphicerus sharpei. See Sharpe's grysboks
Raphidae, **9:269–274,** 9:*272–273*
Raphidia notata, 3:*298*
Raphidiidae, 3:297, 3:298
Raphidioptera. *See* Snakeflies
Raphus cucullatus. See Dodos
Raphus solitarius. See Solitaires
Rapismatidae, 3:306
Rappia tuberculata, 6:287
Rappole, John, 10:416
Raptophasma spp., 3:217, 3:218
Raptors. *See* Falconiformes
Raptors of the World (Ferguson-Lees and Christie), 8:233
Rare Breeds Survival Trust, on pigs, 15:282
Rarotonga starlings, 11:410, 11:*413*, 11:*414*–415
Rasbora heteromorpha. See Harlequins
Rasborinae, 4:297
Raso larks, 10:346
Raspy crickets, 3:207
Rastrineobola argentea. See Dagaa
Rat-kangaroos, 12:145, 13:31–32, 13:34, 13:35, 13:38, **13:73–81,** 13:77
Rat-like hamsters, 16:239, 16:240
Rat lungworms, 1:*296*, 1:300
Rat-tailed maggots, 3:359
Rat-tailed sea cucumbers, 1:*424*, 1:*427*, 1:430–431
Rat tapeworms, 1:35, 1:230
Rat-trap fishes, 4:*425*, 4:*427*, 4:429–430
Ratabulus megacephalus, 5:158
Ratanaworabhan's fruit bats, 13:349*t*
Ratchet-tailed treepies, 11:503
Ratels. *See* Honey badgers
Ratfishes, 4:91
Rathbunella spp., 5:310
Rathbunella hypoplecta. See Stripedfin ronquils
Rathousiidae, 2:411, 2:413
Ratites, **8:53–56,** 8:99, 8:103
See also specific families
Rats, 12:*94*
behavior, 12:142–143, 12:*186*, 12:*187*, 16:124–125
cane, 16:121, **16:333–338**
chinchilla, **16:443–448**
dassie, 16:121, **16:329–332,** 16:*330*
distribution, 12:137, 16:123
evolution, 16:121
Hawaiian Islands, 12:189–191
humans and, 12:139, 12:173, 16:128
kangaroo, 16:121, 16:123, 16:124, **16:199–210**
learning set, 12:152
memory and, 12:153
mole, 12:70, 12:72–77, 12:115, 12:116
neonatal requirements, 12:126
numbers and, 12:155
Old World, **16:249–262**
Polynesian, 12:189
reproduction, 12:106, 16:125
South American, **16:263–279**
spiny, 16:253, **16:449–459**
taxonomy, 16:121
See also Black rats; Moonrats; Muridae
Ratsnakes, 7:41, 7:42, 7:43, 7:466, 7:468, 7:470
Rattails, 5:26, 5:27
See also Macrouridae
Rattlesnakes
behavior, 7:44
chemosensory system, 7:36–37
eastern diamondback, 7:40, 7:50, 7:475
as food, 7:48
homing, 7:43
prairie, 7:34, 7:38, 7:44, 7:50
in religions, 7:56–57
reproduction, 7:42
roundups, 7:49–50
timber, 7:*50*, 7:448, 7:*450*, 7:*452*, 7:454
venom, 7:10
western diamondback, 7:38, 7:40, 7:42, 7:44, 7:50, 7:*197*
See also Snakes
Rattus exulans. See Polynesian rats
Rattus norvegicus. See Brown rats
Rattus rattus. See Black rats
Rattus tunneyi. See Tunney's rats
Ratufa spp., 16:163
Ratufa affinis. See Giant squirrels
Ratufa indica. See Indian giant squirrels
Ratufa macroura, 16:167
Rauisuchids, 7:17
Raven, Peter, 1:50, 3:48
Ravens, 8:20, 9:313, 10:169, 11:505–507, 11:509, 11:*511*, 11:522
Ray, J., 8:369
Ray-finned fishes, 4:11, 6:7
Ray-tailed trogons. *See* Collared trogons
Rayner, J., 12:57
Rays, **4:173–188,** 4:*180–181*
behavior, 4:177
conservation, 4:178
distribution, 4:176
evolution, 4:10, 4:173–174
feeding ecology, 4:177–178
habitats, 4:176–177
humans and, 4:178–179
physical characteristics, 4:60, 4:*174*–176, 4:*175*
reproduction, 4:*178*
species of, 4:182–188
taxonomy, 4:10–11, 4:173–174
See also specific types of rays
Razor-billed auks. *See* Razorbills
Razor-billed curassows, 8:413
Razorbacks. *See* Fin whales
Razorbills, 9:219, 9:*223*, 9:*225*
Razorfishes. *See* Common shrimpfishes
RBCs. *See* Erythrocytes
RCD (rabbit calicivirus disease), 12:187–188, 12:223
Reaka, M. J., 2:170
Recall, 12:152, 12:154
Recapitulation, 1:7
See also Evolution
Recognition, 12:152, 12:154
Recombination (genetics), 1:15
Record keeping, zoo, 12:206–207
Recordings, of bird songs, 8:37–38
Records of the Australian Museum, 10:329
Recruitment, 2:27
Rectilinear locomotion, 7:26
Recurve-billed bushbirds, 10:242
Recurvirostra spp. *See* Avocets
Recurvirostra americana. See American avocets
Recurvirostra andina. See Andean avocets
Recurvirostridae, **9:133–141,** 9:*136*
Red abalone, 2:*432*, 2:*433*
Red acouchis, 16:409, 16:*410*, 16:*411*, 16:413–414
Red algae, 1:44, 2:9
Red-and-black giraffe beetles. *See* Giraffe-necked weevils
Red-and-white uakaris. *See* Bald uakaris
Red-and-yellow barbets, 10:116
Red ants, 3:68
Red-armed bats. *See* Natterer's bats
Red avadavats, 11:*359*, 11:*365*
Red-backed buttonquails, 9:16
Red-backed fairy-wrens, 11:*47*, 11:*49*
Red-backed juncos. *See* Dark-eyed juncos
Red-backed kingfishers, 10:19
Red-backed mousebirds. *See* Chestnut-backed mousebirds
Red-backed poison dart frogs, 6:*199*
Red-backed salamanders, 6:44, 6:*393*, 6:402
Red-backed shrikes, 10:425, 10:428, 10:429–430, 10:*431*, 10:435–*436*
Red-backed squirrel monkeys, 14:101, 14:*102*, 14:103, 14:*107*, 14:*108*–109
Red-backed voles, 16:226, 16:227, 16:228
Red-backed wrens. *See* Red-backed fairy-wrens
Red bald uakaris, 14:*145*, 14:148, 14:151
Red-banded frogs. *See* Banded rubber frogs
Red-barred hawkfishes, 5:238
Red bats, 13:310, 13:312, 13:314, 13:315, 13:*498*, 13:499–500, 13:501
Red-bellied black snakes, 7:486, 7:*489*, 7:*497*
Red-bellied grackles, 11:305–306
Red-bellied lemurs, 14:6, 14:*54*, 14:*56*, 14:60
Red-bellied mudsnakes, 7:469
Red-bellied newts, 6:373
Red-bellied pacus. *See* Pirapitingas
Red-bellied piranhas, 4:*4*, 4:*335*, 4:*341*, 4:*345*, 4:346
Red-bellied pittas, 10:194
Red-bellied squirrels. *See* Pallas's squirrels
Red-bellied tamarins, 14:119, 14:124
Red-bellied trogons. *See* Collared trogons
Red-bellied turtles, 7:*106*
Red-bellied woodpeckers, 8:24, 10:150
Red-billed buffalo weavers, 11:*380*, 11:*382*
Red-billed choughs, 11:504, 11:*511*, 11:*518*–519
Red-billed dioches. *See* Red-billed queleas
Red-billed dwarf hornbills, 10:71

Red-billed firefinches, 11:354, 11:*354*
Red-billed hill-partridges, 8:434
Red-billed hill tits. *See* Red-billed leiothrixes
Red-billed hornbills, 10:72, 10:77, 10:*78–79*
Red-billed leiothrixes, 10:*509*, 10:*517*
Red-billed oxpeckers, 11:409, 11:*412*, 11:*425*
Red-billed queleas, 11:377, 11:379, 11:*381*, 11:*388*
Red-billed rollers. *See* Dollarbirds
Red-billed scythebills, 10:*230*, 10:*231*, 10:*236*
Red-billed streamertails, 9:*453*, 9:*463*–464
Red-billed toucans. *See* White-throated toucans
Red-billed tropicbirds, 8:*188*, 8:*191*
Red-billed woodhoopoes. *See* Green woodhoopoes
Red birds of paradise, 11:492
Red bishop-bird. *See* Southern red bishops
Red blood cells. *See* Erythrocytes
Red boarfishes, 5:*125*, 5:*126*
Red boobooks. *See* Southern boobook owls
Red-breasted babblers. *See* Gray-crowned babblers
Red-breasted blackbirds. *See* White-browed blackbirds
Red-breasted buttonquails. *See* Red-chested buttonquails
Red-breasted nuthatches, 11:*168*, 11:*170*, 11:*171*
Red-breasted plantcutters, 10:325, 10:326
Red-breasted pygmy parrots, 9:*281*, 9:*293–294*
Red-breasted sapsuckers, 10:*148*
Red-breasted sunbirds. *See* Greater double-collared sunbirds
Red-breasted toucans, 10:126
Red brocket deer, 15:*384*, 15:*387*, 15:*393*–394
Red-browed pardalotes, 11:*203*, 11:*204*–205
Red-browed treecreepers, 11:*136*, 11:*137*–138
Red caecilians, 6:426, 6:*427*, 6:428
Red canaries, 11:339
See also Crimson chats
Red-capped flowerpeckers, 11:*192*, 11:*195*
Red-capped manakins, 10:*298*, 10:*301*
Red-capped mangabeys. *See* Collared mangabeys
Red-capped robins, 11:105
Red-cheeked cordon-bleu, 11:*358*, 11:363–*364*
Red-cheeked parrots, 9:279
Red-chested buttonquails, 9:*15*, 9:*20*
Red-chested cuckoos, 11:221
Red coats. *See* Bed bugs
Red-cockaded woodpeckers, 10:149, 10:150, 10:*153*, 10:*159*–160
Red-collared widow-birds, 11:*381*, 11:*390*
Red colobus, 14:7, 14:172–175
Red corals, 1:*109*, 1:*112*, 1:114
Red cornetfishes, 5:134
Red-cotingas. *See* Cotingas
Red crayfish. *See* Red swamp crayfish
Red-crested turacos, 9:300
Red crossbills, 11:*325*, 11:*326*, 11:336–*337*
Red-crowned barbets, 10:*114*
Red-crowned cranes, 9:3, 9:*24–25*, 9:*30*, 9:*35*–36
Red-crowned flowerpeckers. *See* Red-capped flowerpeckers
Red Data Book of Threatened Vertebrates (Greek), 7:238
Red Data Books, 16:219, 16:220
aardvarks, 15:158
alpine shrews, 13:260
ghost frogs, 6:134
giant shrews, 13:262
Gill's plantanna, 6:102
lesser New Zealand short-tailed bats, 13:457
parsley frogs, 6:128
wart biter katydids, 3:85
Red deer, 15:*359*, 15:*363*, 15:369–370
behavior, 15:*360*
competition, 12:118
distribution, 15:*364*
evolution, 12:19, 15:357
humans and, 15:273, 15:361
predators and, 12:116
reproduction, 15:*360*
Red drums, 5:*265*, 5:272–273
Red duikers, 16:73
Red-ear sunfishes, 5:197
Red-eared bulbuls. *See* Red-whiskered bulbuls
Red-eared sliders, 7:47, 7:48, 7:*66*, 7:107
See also Pond sliders
Red efts, 6:*368*, 6:372
Red-eyebrowed treecreepers. *See* Red-browed treecreepers
Red-eyed leaf frogs. *See* Red-eyed treefrogs
Red-eyed towhees. *See* Eastern towhees
Red-eyed treefrogs, 6:*232*, 6:*241*
Red-eyed vireos, 11:*256*, 11:*257*, 11:*258*
Red-faced mousebirds, 9:*470*, 9:*471*, 9:*472*, 9:475–*476*
Red-fan parrots, 9:279
Red firefishes. *See* Red lionfishes
Red-flanked duikers, 16:77, 16:*79*, 16:82–*83*
Red fodies. *See* Madagascar fodies
Red-footed boobies, 8:*215*, 8:221–*222*
Red-footed falcons, 8:349
Red forest rats, 16:*281*
Red foxes, 14:*269*, 14:*275*, 14:278–280, 14:*279*, 14:287–288
behavior, 12:145–146, 14:258, 14:259, 14:267–268
conservation status, 14:272
distribution, 12:137
evolution, 14:265
feeding ecology, 14:271
habitats, 14:267
humans and, 14:273
as pests, 12:185
reproduction, 12:107, 14:272
taxonomy, 14:265
Red frog crabs. *See* Spanner crabs
Red-fronted antpeckers. *See* Red-fronted flowerpecker weaver-finches
Red-fronted flowerpecker weaver-finches, 11:*355*, 11:*358*, 11:*360*
Red-fronted gazelles, 16:48, 16:57*t*
Red-fronted lemurs, 14:*51*
Red-fronted parakeets, 9:276
Red-fronted tinkerbirds, 10:115
Red gazelles, 12:129
Red giant flying squirrels, 16:*136*, 16:*138*, 16:*139*
Red goatfishes, 5:*244*
Red gorals, 15:272, 16:87, 16:92
Red grouse. *See* Willow ptarmigans
Red gurnards, 5:*168*, 5:*171*, 5:177
Red hakes, 5:27, 5:28, 5:29, 5:*31*, 5:*38*–39
Red-handed howler monkeys, 14:166*t*
Red haplognathias, 1:*331*, 1:*333*, 1:*334*
Red hartebeests, 12:116, 15:*139*
Red-headed barbets, 10:115, 10:116
Red-headed ducks, 8:367
Red-headed finches, 11:*358*, 11:*372*
Red-headed parrotfinches, 11:356
Red-headed rock hyraxes, 15:177, 15:189*t*
Red-headed rockfowls. *See* Gray-necked picathartes
Red-headed tits. *See* Black-throated tits
Red-headed weaver-finches. *See* Red-headed finches
Red-headed weavers, 11:*378*, 11:*379*, 11:*380*, 11:391–*392*
Red-headed woodpeckers, 10:149, 10:150, 10:*153*, 10:*157*–158
Red Hills salamanders, 6:389, 6:*393*, 6:395–*396*
Red honeyeaters. *See* Red myzomelas
Red howler monkeys. *See* Bolivian red howler monkeys; Venezuelan red howler monkeys
Red imported fire ants, 3:68, 3:70, 3:72
Red indianfishes, 5:*168*, 5:171–*172*
Red Irish lords, 5:*179*
Red jungle fowl, 8:21, 8:399, 8:435, 8:437, 8:*438*, 8:*450*
Red kangaroos, 12:*4*, 13:*91*, 13:95
distribution, 13:34, 13:*93*
feeding ecology, 13:38
habitats, 13:34
humans and, 13:90
physical characteristics, 13:*33*
reproduction, 12:108, 13:*88*, 13:*89*
taxonomy, 13:31
Red king crabs, 2:199–200, 2:*202*, 2:*208*–209
Red kites, 8:316, 8:323
Red-kneed dotterels, 9:161, 9:162
Red-knobbed coots, 9:50
Red-knobbed hornbills. *See* Sulawesi red-knobbed hornbills
Red knots, 9:175, 9:179, 9:180
Red larks. *See* Ferruginous larks
Red-legged locusts, 3:*206*
Red-legged pademelons, 13:*92*, 13:*97–98*
Red-legged salamanders, 6:44
Red-legged seriemas, 9:*1*, 9:*86*, 9:*87*, 9:*88*
Red-legged short-tailed opossums, 12:*256*, 12:*262*
Red-legged thrushes, 10:*491*, 10:*503*
Red lionfishes, 5:*163*, 5:*165*, 5:*168*, 5:*170*, 5:173
Red List. *See* IUCN Red List of Threatened Species
Red locusts, 3:35
Red-lored Amazons, 9:277–278
Red-lored whistlers, 11:117, 11:118
Red meerkats. *See* Yellow mongooses
Red monkeys. *See* Patas monkeys
Red mouse lemurs, 14:*10*, 14:*36*, 14:*38*, 14:39, 14:*40*, 14:*42*, 14:43
Red mouse opossums, 12:251, 12:*252*
Red munias. *See* Red avadavats
Red myzomelas, 11:*240*, 11:*246*
Red-naped trogons, 9:479
Red-necked foliage-gleaners. *See* Rufous-necked foliage-gleaners
Red-necked francolins, 8:*440*, 8:*445–446*
Red-necked grebes, 8:*171*
Red-necked night monkeys, 14:135
Red-necked nightjars, 9:402

Red-necked spurfowls. *See* Red-necked francolins
Red-necked wallabies, 12:*137*, 13:*34*, 13:*87*, 13:102*t*
Red neons. *See* Cardinal tetras
Red-nosed tree rats, 16:268
Red pacus. *See* Pirapitingas
Red pandas, 12:*28*, 14:309–*311*, 14:*312*, 14:*315*
Red piranhas. *See* Red-bellied piranhas
Red Queen hypothesis, 1:31
Red rain frogs, 6:317, 6:318, 6:*320*
Red rasboras. *See* Harlequins
Red river hogs, 15:277, 15:279, 15:*281*, 15:289*t*
Red rock crabs, 2:34
Red rosellas. *See* Eastern rosellas
Red roughies. *See* Orange roughies
Red-ruffed fruit-crows, 10:305, 10:307
Red-rumped agoutis, 16:*409*, 16:*410*, 16:411–*412*
Red salmons. *See* Sockeye salmons
Red Sea cliff-swallows, 10:360
Red sea urchins, 1:*401*, 1:405
Red-shafted flickers. *See* Northern flickers
Red-shanked douc langurs, 12:218, 14:*172*, 14:175, 14:*177*, 14:*182*
Red-shouldered cuckoo-shrikes, 10:*388*, 10:390–*391*
Red-shouldered vangas, 10:439, 10:440, 10:441–442
Red-sided eclectus parrots. *See* Eclectus parrots
Red-sided garter snakes, 7:479
Red-sided parrots. *See* Eclectus parrots
Red-sided suckers. *See* Longnose suckers
Red siskins, 11:324
Red snakeheads. *See* Giant snakeheads
Red snappers. *See* Northern red snappers
Red soft tree corals, 1:*109*, 1:*112*, 1:114–115
Red soldierfishes, 5:116
Red-spotted newts. *See* Eastern newts
Red squirrels, 12:139, 12:183–184, 16:166, 16:*167*, 16:*168*, 16:*169*–170
Red swamp crayfish, 2:*203*, 2:*204*–205
Red-tailed bulbuls. *See* Red-tailed greenbuls
Red-tailed bumblebees. *See* Large red-tailed bumblebees
Red-tailed chipmunks, 16:159*t*
Red-tailed greenbuls, 10:397, 10:*400*, 10:*409*
Red-tailed hawks, 8:22, 9:349
Red-tailed monkeys, 12:38
Red-tailed newtonias, 11:27, 11:*28*, 11:39–*40*
Red-tailed phascogales, 12:*289*, 12:300*t*
Red-tailed pipe snakes, 7:*396*, 7:*397*
Red-tailed sportive lemurs, 14:*74*, 14:*75*, 14:*78*, 14:79–*80*
Red-tailed tropicbirds, 8:*188*
Red-tailed vangas, 10:439, 10:440
Red-tailed wambergers. *See* Red-tailed phascogales
Red-throated caracaras, 8:349
Red-throated divers. *See* Red-throated loons
Red-throated little grebes. *See* Little grebes
Red-throated loons, 8:159, 8:*160*, 8:161, 8:*162*, 8:*163*–164
Red-throated pipits, 10:372, 10:*373*, 10:374, 10:*377*, 10:*381*–382
Red-throated sunbirds, 11:210
Red-throated twinspots. *See* Peter's twinspots
Red tide, 2:12
Red-tipped pardalotes. *See* Striated pardalotes
Red titis, 14:153*t*
Red-toothed shrews, **13:247–264,** 13:*253*, 13:*254*, 13:263*t*
Red tree voles, 16:*231*, 16:*234*–235
Red velvetfishes, 5:163, 5:164
Red-vented bulbuls, 10:*396*, 10:397, 10:*401*, 10:*402*–403
Red viscacha rats, 16:124, 16:433, 16:434, 16:436, 16:*437*, 16:*438*
Red volutes, 2:*445*, 2:*446*
Red wattlebirds, 11:*240*, 11:*241*
Red wattled curassows. *See* Wattled curassows
Red whalefishes, 5:105, 5:*107*, 5:*108*
Red-whiskered bulbuls, 10:396, 10:399, 10:*401*, 10:*403*–404
Red-winged blackbirds, 8:19, 11:293, 11:304–*305*, 11:*308*, 11:313, 11:*314*
Red-winged larks, 10:*345*
Red-winged pratincoles. *See* Collared pratincoles
Red-winged starlings, 11:408, 11:*413*, 11:*423*
Red-winged tinamous, 8:*60*, 8:*64*–*65*
Red-winged warblers, 11:2
Red wolves, 12:*214*, 14:262, 14:273
Redbellies. *See* Stoplight parrotfishes
Reddish hermits, 9:438, 9:448
Reddish owlet-nightjars. *See* Feline owlet-nightjars
Redeyes. *See* Rock basses
Redfishes, 5:113, 5:115, 5:166
See also Red drums
Redford, K. H., 16:434
Redmouth whalefishes, 5:105
Redpolls. *See* Common redpolls
Redside daces, 4:*298*
Redstarts, 9:314, 10:484, 10:486, 10:488, 11:*287*–288, 11:291
Redtail boas. *See* Boa constrictors
Redtail butterflyfishes, 5:*238*
Redtail catfishes, 4:*355*, 4:*357*, 4:364
Redtail splitfins, 5:*95*, 5:99–*100*
Redthroats, 11:56
Reductional division. *See* Meiosis
Redunca spp., 16:5, 16:27, 16:28
Redunca arundinum. *See* Southern reedbucks
Redunca fulvorufula. *See* Mountain reedbucks
Redunca fulvorufula adamauae. *See* Western mountain reedbucks
Redunca redunca. *See* Bohor reedbucks
Reduncinae, 16:1, 16:7
Reduncini, 16:27–31, 16:34
Reduviidae, 3:53
Redwings, 10:485, 10:486
Reed buntings, 11:*264*, 11:*268*, 11:*274*–*275*
Reed frogs. *See* Painted reed frogs
Reed rats, Malagasy, 16:*284*
Reed warblers, 9:313, 11:5, 11:*6*, 11:7
Reedbucks, 16:5, **16:27–43,** 16:*35*, 16:41*t*–42*t*
Reeder, D. M., 16:121, 16:389
Reedfishes, 4:24, 4:209, 4:210, 4:*211*–212
Reedsnakes, 7:466
Reef boring sea-hedgehogs. *See* Rock boring urchins
Reef bugs. *See* Flathead locust lobsters
Reef corals, deep water, 1:*110*, 1:*117*, 1:118
Reef cuskfishes, 4:49
Reef herrings, 4:44, 4:48, 4:67, 4:70
Reef stonefishes, 5:166, 5:*167*, 5:*175*, 5:176–177
Reef terrae, 4:48
Reeve's muntjacs, 15:343, 15:*344*, 15:*345*, 15:346, 15:*347*, 15:*349*
Reference memory. *See* Long-term memory
Referential pointing/gazing, 12:160
Regal angelfishes, 5:240
Regal pythons. *See* Reticulated pythons
Regalecidae, 4:447, 4:448
Regalecus spp., 4:447, 4:449
Regalecus glesne. *See* Oarfishes
Regan, C. T., 5:405
Regeneration, of tails, 7:*24*
Regent birds. *See* Regent bowerbirds
Regent bowerbirds, 11:477, 11:480, 11:*482*, 11:*486*–487
Regent honeyeaters, 11:238, 11:*240*, 11:242–*243*
Regent whistlers, 11:*119*, 11:*121*
Regina alleni. *See* Striped crayfish snakes
Regulative development. *See* Indeterminate cleavage
Regulus satrapa. *See* Golden-crowned kinglets
Rehabilitation pools, 12:*204*
Rehbachiella spp., 2:135
Rehbachiella kinnekullensis, 2:135
Reichard, U., 14:214
Reichenbach's sunbirds, 11:*211*, 11:216–*217*
Reichenow's woodpeckers. *See* Bennett's woodpeckers
Reig, O. A., 16:265, 16:435
Reighardia spp., 2:317, 2:319
Reighardia sternae, 2:319
Reighardiidae. *See Reighardia* spp.
Reindeer, 15:*270*, 15:*387*, 15:394–395
behavior, 15:383
conservation status, 15:384
distribution, 12:131, 12:138, 15:*393*
domestication of, 12:172–173, 12:180, 15:145
evolution, 15:379, 15:380, 15:381–382
feeding ecology, 12:*125*
field studies of, 12:*197*
habitats, 15:383
humans and, 15:361, 15:384
migrations, 12:87, 12:116, 12:164, 12:*165*, 12:*168*, 12:170
physical characteristics, 15:268, 15:*383*
Reintroductions, 12:139, 12:223–224
red wolves, 12:*214*
zoos and, 12:204, 12:208
See also Conservation status
Reinwardtoena crassirostris. *See* Crested cuckoo-doves
Reinwardt's trogons. *See* Javan trogons
Reithrodon spp., 16:265
Reithrodon auritus. *See* Rabbit rats
Reithrodon typicus, 16:268
Reithrodontines, 16:263–264
Reithrodontomys montanus. *See* Plains harvest mice
Reithrodontomys raviventris. *See* Saltmarsh harvest mice
Reithrodontomys sumichrasti. *See* Sumichrast's harvest mice
Reithrosciurus spp., 16:163
Reitner, J., 1:57
Relic silverfish, 3:*121*, 3:122, 3:*123*
Religion, 3:74
birds and, 8:26 (*See also* Mythology)

fishes and, 4:72
protostomes in, 2:42
reptiles and, 7:*50*, 7:55–57 (*See also* specific species)
See also Humans
Rembrandt, 2:447
Remingtonocetidae, 15:2
Remipedes, **2:125–129**
Remiz consobrinus. See Chinese penduline tits
Remiz coronatus. See White-crowned tits
Remiz macronyx. See Black-headed penduline titmice
Remiz pendulinus. See European penduline tits
Remora spp., 5:211
Remora remora. See Common remoras
Remoras, 4:49, 4:69, 4:177, **5:211–216,** 5:*212*, 5:*213*, 5:*214*
Remorina spp., 5:211
Removal sampling, 12:197, 12:200
Remsen, James V., 10:127, 10:132, 10:133
Renal system, 12:89–90
See also Physical characteristics
Renenutet (Egyptian goddess), 7:55
Renewable resources, 12:215
Renilla reniformis. See Sea pansies
Rensch's Rule, 12:99
Renyxidae, 3:54
Rephicerus campestris. See Steenboks
Repletes, 3:81
Repomucenus spp., 5:365, 5:366, 5:367
Repomucenus richardsonii. See Richardson's dragonets
Reproduction
amphibians, 6:3, 6:18, **6:28–38**
amplexus in, 6:*65*, 6:68, 6:304, 6:365–366
anurans, 6:28–*29*, 6:*30*–34, 6:*32*, 6:38, 6:65, 6:68
African treefrogs, 6:282–283, 6:285–289
Amero-Australian treefrogs, 6:229–230, 6:233–242
Arthroleptidae, 6:266, 6:268–271
Asian toadfrogs, 6:111, 6:113–117
Asian treefrogs, 6:*30*, 6:32, 6:292–293, 6:295–299
Australian ground frogs, 6:32, 6:34, 6:141, 6:143–145
Bombinatoridae, 6:41, 6:84, 6:86–88
brown frogs, 6:*31*, 6:49, 6:263
Bufonidae, 6:25, 6:35, 6:*185*, 6:188–194
Discoglossidae, 6:90, 6:92–94
Eleutherodactylus spp., 6:33, 6:35, 6:68, 6:158
frogs, 6:28–38, 6:*31*, 6:*34*, 6:*35*, 6:*36*, 6:68
gastric brooding frogs, 6:37, 6:149, 6:*150*, 6:153
ghost frogs, 6:132, 6:133–134
glass frogs, 6:*30*, 6:32, 6:35, 6:*36*, 6:*216*–217, 6:219–223
leptodactylid frogs, 6:32, 6:34, 6:68, 6:158–159, 6:162–171
Madagascan toadlets, 6:318, 6:319–320
Mesoamerican burrowing toads, 6:97
Myobatrachidae, 6:35, 6:36, 6:68, 6:148–149, 6:151–153
narrow-mouthed frogs, 6:32, 6:35, 6:68, 6:304–305, 6:308–316
New Zealand frogs, 6:70, 6:72–74
North American bullfrogs, 6:251, 6:262
parsley frogs, 6:128, 6:129
Pipidae, 6:101–102, 6:104–106
poison frogs, 6:34, 6:36, 6:68, 6:199–200, 6:203–209
Ruthven's frogs, 6:212
Seychelles frogs, 6:136, 6:137–138
shovel-nosed frogs, 6:275
spadefoot toads, 6:121–122, 6:124–125
tailed frogs, 6:78, 6:80–81
three-toed toadlets, 6:180, 6:181–182
toads, 6:33, 6:35, 6:36, 6:*37*, 6:68
treefrogs, 6:28, 6:32, 6:35, 6:50
true frogs, 6:28, 6:35, 6:68, 6:251, 6:255–263
vocal sac-brooding frogs, 6:*174*, 6:175–176
Xenopus spp., 6:41, 6:102
caecilians, 6:29–30, 6:32, 6:35, 6:38, 6:39, 6:412
American tailed, 6:416, 6:417–418
Asian tailed, 6:420, 6:422–423
buried-eye, 6:432, 6:433–434
Kerala, 6:426, 6:428
tailless, 6:436–437, 6:439–441
lissamphibians, 6:17–18
salamanders, 6:28–*29*, 6:*31*–33, 6:34–35, 6:*37*–38, 6:325, 6:326
amphiumas, 6:*406*, 6:407, 6:409–410
Asiatic, 6:32, 6:34, 6:337, 6:339–342
Caudata, 6:326
Cryptobranchidae, 6:29, 6:32, 6:34, 6:345–346
lungless, 6:32, 6:34, 6:35, 6:392, 6:395–403
mole, 6:356, 6:358–360
Pacific giant, 6:34, 6:350, 6:352
Proteidae, 6:378–379, 6:381–383
Salamandridae, 6:30–31, 6:33, 6:326, 6:*363–364*, 6:367, 6:370–375
sirens, 6:32, 6:329, 6:331–332
torrent, 6:386, 6:387
vocalizations and, 6:35, 6:47, 6:49
birds
albatrosses, 8:109, 8:*114*, 8:115–116, 8:118–122
Alcidae, 9:222, 9:224–228
anhingas, 8:202, 8:207–209
Anseriformes, 8:366–367
ant thrushes, 10:241, 10:245–256
Apodiformes, 9:417–418
asities, 10:188, 10:190
Australian chats, 11:66, 11:67–68
Australian creepers, 11:135, 11:137–139
Australian fairy-wrens, 11:46, 11:48–53
Australian honeyeaters, 11:237, 11:241–253
Australian robins, 11:106, 11:108–112
Australian warblers, 11:56–57, 11:59–63
babblers, 10:507, 10:511–523
barbets, 10:116–117, 10:119–122
barn owls, 9:337–338, 9:340–343
bee-eaters, 10:*41*–42
bird songs and, 8:37, 8:39, 8:41–42
birds, 8:*4*, 8:11–14
birds of paradise, 11:491–492, 11:495–502
bitterns, 8:244
Bombycillidae, 10:*448*, 10:451–454
boobies, 8:213–214
bowerbirds, 11:479–480, 11:483–488
breeding seasons, 8:12
broadbills, 10:178–179, 10:181–186
bulbuls, 10:398, 10:402–413
bustards, 9:93, 9:96–99
buttonquails, 9:13–14, 9:16–21
Caprimulgiformes, 9:370
caracaras, 8:350–351
cassowaries, 8:76
Charadriidae, 9:162–163, 9:166–173
Charadriiformes, 9:102
chats, 10:487–488
chickadees, 11:157–158, 11:160–166
chowchillas, 11:70–71, 11:72–74
Ciconiiformes, 8:235–236
Columbidae, 9:250–251, 9:255–266
Columbiformes, 9:244
communcation systems, 8:18
condors, 8:278
Coraciiformes, 10:3
cormorants, 8:202, 8:205–206, 8:207–209
Corvidae, 11:507, 11:512–522
cotingas, 10:307, 10:312–322
courtships displays, 8:9, 8:18, 8:20
crab plovers, 9:122–123
Cracidae, 8:416
cranes, 9:24, 9:27, 9:28, 9:31–36
cuckoo-shrikes, 10:387
cuckoos, 9:313–315, 9:318–329
dippers, 10:476–477, 10:479–482
diving-petrels, 8:144, 8:145–146
doves, 9:244, 9:250–251, 9:255–266
drongos, 11:*438*, 11:439, 11:441–445
ducks, 8:372
eagles, 8:321, 8:322–323
elephant birds, 8:104
embryos, 8:*4*
emus, 8:85
Eupetidae, 11:76, 11:78–81
fairy bluebirds, 10:417, 10:420–424
Falconiformes, 8:315
falcons, 8:350–351
false sunbirds, 10:188, 10:191
fantails, 11:86, 11:89–95
finches, 11:*324*, 11:328–339
flamingos, 8:303, 8:305–306
flowerpeckers, 11:190–191, 11:193–198
frigatebirds, 8:185, 8:*194*, 8:*195*, 8:197–198
frogmouths, 9:379, 9:381–385
Galliformes, 8:400
gannets, 8:213–214
geese, 8:372
Glareolidae, 9:153, 9:155–160
grebes, 8:*171*, 8:174–181
Gruiformes, 9:3
guineafowl, 8:428–429
gulls, 9:211–217
hammerheads, 8:262–263
Hawaiian honeycreepers, 11:*343*, 11:346–351
hawks, 8:321, 8:322–323
hedge sparrows, 10:459–460, 10:462–464
herons, 8:234, 8:236, 8:244
hoatzins, 8:467
honeyguides, 10:139, 10:141–144
hoopoes, 10:63
hornbills, 10:74–75, 10:78–84
hummingbirds, 9:447–449, 9:455–467
ibises, 8:236, 8:292–293
Icteridae, 11:304–305, 11:309–322
ioras, 10:417, 10:419–420
jacamars, 10:92–93

jacanas, 9:*108*, 9:*109*–110, 9:112–114
kagus, 9:42–43
kingfishers, 10:*9*–10, 10:13–23
kiwis, 8:90
lapwings, 9:167–173
Laridae, 9:207–208, 9:211–217
larks, 10:*344*, 10:345–346, 10:348–354
leafbirds, 10:417, 10:420–423
limpkins, 9:38
logrunners, 11:70–71, 11:72–74
long-tailed titmice, 11:142–143, 11:145–146
lyrebirds, 10:331–332, 10:334–335
magpie-shrikes, 11:469, 11:471–475
manakins, 10:296, 10:299–303
mating systems, 8:12
mesites, 9:6, 9:8–9
moas, 8:98
monarch flycatchers, 11:98, 11:100–103
motmots, 10:33, 10:35–37
moundbuilders, 8:405–407
mousebirds, 9:470, 9:473–476
mudnest builders, 11:453–*454*, 11:456–458
New World finches, 11:264–265, 11:269–283
New World quails, 8:457
New World vultures, 8:236, 8:278
New World warblers, 11:290, 11:293–299
New Zealand wattle birds, 11:447–448, 11:449–451
New Zealand wrens, 10:204, 10:206–207
night herons, 8:234
nightjars, 9:404–*405*, 9:408–414
nuthatches, 11:168–169, 11:171–175
oilbirds, 9:375
Old World flycatchers, 11:27, 11:30–43
Old World warblers, 11:5–6, 11:10–23
Oriolidae, 11:*429*, 11:431–434
ostriches, 8:*100*, 8:101, 8:*102*
ovenbirds, 10:210, 10:214–228
owlet-nightjars, 9:389, 9:391–393
owls, 9:333, 9:337–338, 9:340–343, 9:*348*, 9:350–351, 9:354–365
oystercatchers, 9:127, 9:130–131
painted-snipes, 9:116–117, 9:118–119
palmchats, 10:456
pardalotes, 11:202, 11:204–206
parrots, 9:278–279, 9:283–298
Passeriformes, 10:173–174
Pelecaniformes, 8:*184*, 8:*185*–186
pelicans, 8:185–186, 8:226–227
penduline titmice, 11:148, 11:151–153
penguins, 8:*150*–151, 8:153–157
Phalaropes, 9:179
Phasianidae, 8:*436*
pheasants, 8:436
Philippine creepers, 11:184, 11:186–187
Picidae, 10:150, 10:154–167
Piciformes, 10:87–88
pigeons, 9:244, 9:250–251, 9:255–266
pipits, 10:374–375, 10:378–383
pittas, 10:194–195, 10:197–201
plantcutters, 10:326, 10:327–328
plovers, 9:166–173
potoos, 9:396–397, 9:399–400
pratincoles, 9:155–160
Procellariidae, 8:125–126, 8:129–130
Procellariiformes, 8:109–110
pseudo babblers, 11:128, 11:130–131
puffbirds, 10:103, 10:106–111
rails, 9:48, 9:50–51, 9:57–67, 9:59, 9:66
Raphidae, 9:270, 9:272–273
Recurvirostridae, 9:135, 9:137–140
rheas, 8:70–71
rollers, 10:53, 10:55–58
sandgrouse, 9:232–233, 9:235–238
sandpipers, 9:179, 9:182–188
screamers, 8:394
scrub-birds, 10:338, 10:339–340
secretary birds, 8:344–345
seedsnipes, 9:191, 9:193–195
seriemas, 9:86, 9:88–89
sharpbills, 10:292–293
sheathbills, 9:199, 9:200–201
shoebills, 8:288
shrikes, 10:428–429, 10:432–438
sparrows, 11:398, 11:401–406
spoonbills, 8:236, 8:292–293
storks, 8:266–267
storm-petrels, 8:137–138, 8:140–141
Struthioniformes, 8:55
Sturnidae, 11:409–410, 11:414–425
sunbirds, 11:210, 11:213–225
sunbitterns, 9:74–75
sungrebes, 9:70, 9:71–72
swallows, 10:359–360, 10:363–369
swans, 8:372
swifts, 9:423–424, 9:426–431
tapaculos, 10:259, 10:262–267
terns, 9:211–217
thick-knees, 9:145, 9:147–148
thrushes, 10:487–488
tinamous, 8:57–59
titmice, 11:157–158, 11:160–166
todies, 10:27, 10:29–30
toucans, 10:127–128, 10:130–135
tree swifts, 9:434
treecreepers, 11:178, 11:180–181
trogons, 9:479, 9:481–485
tropicbirds, 8:*188*–189, 8:190–191
trumpeters, 9:79, 9:81–82
turacos, 9:301, 9:304–310
typical owls, 9:350–351, 9:354–365
tyrant flycatchers, 10:273–274, 10:278–288
vanga shrikes, 10:441, 10:443–444
Vireonidae, 11:256, 11:258–262
wagtails, 10:374–375, 10:378–383
weaverfinches, 11:355–356, 11:360–372
weavers, 11:377–378, 11:382–394
whistlers, 11:117–118, 11:120–126
white-eyes, 11:229, 11:232–233
woodcreepers, 10:230, 10:232–236
woodhoopoes, 10:66, 10:68–69
woodswallows, 11:460, 11:462–465
wrens, 10:528–529, 10:531–538
young, 8:14–15
fishes, 4:25–27, **4:29–35**, 4:*31–33*, 4:64–67
Acanthuroidei, 5:393, 5:397–403
Acipenseriformes, 4:214, 4:217–220
Albuliformes, 4:250, 4:252–253
angelsharks, 4:162, 4:164–165
anglerfishes, 5:50–51, 5:53–56
Anguilliformes, 4:256–*258*, 4:262–269
Atheriniformes, 5:69–70, 5:73–76
Aulopiformes, 4:45, 4:433, 4:436–439
Australian lungfishes, 4:199
beardfishes, 5:2, 5:3
Beloniformes, 5:81, 5:83–86
Beryciformes, 5:116, 5:118–121
bichirs, 4:210, 4:211–212
bisexual, 4:29
blennies, 5:343, 5:345–348
bowfins, 4:230
bullhead sharks, 4:98, 4:101–102
Callionymoidei, 5:366, 5:369–371
carps, 4:67, 4:69, 4:*300*–303, 4:308–313
catfishes, 4:30, 4:66, 4:354, 4:357–366
characins, 4:66, 4:67, 4:*336*, 4:*337*, 4:*338*, 4:342–350
chimaeras, 4:92, 4:94–95
cichlids, 4:66
coelacanths, 4:192
Cypriniformes, 4:297, 4:*300*–303, 4:*302*, 4:307–319, 4:322, 4:325–333
Cyprinodontiformes, 5:92–93, 5:96–102
damselfishes, 4:35
dogfish sharks, 4:*152*, 4:156–158
dories, 5:124, 5:126–127, 5:129–130
eels, 4:47, 4:256–*258*, 4:262–269
eggs, 4:*29*–33, 4:*31*, 4:*34*, 4:46, 4:65, 4:66
electric eels, 4:66, 4:371, 4:373–377
Elopiformes, 4:244, 4:247
Esociformes, 4:380, 4:383–386
fertilization, 4:33–34
flatfishes, 5:*451*, 5:453, 5:455–463
Gadiformes, 5:29, 5:33–40
gars, 4:222, 4:224–227
Gasterosteiformes, 5:135–136, 5:139–143, 5:145–147
gobies, 5:376–377, 5:380–388
Gobiesocoidei, 5:357, 5:359–363
Gonorynchiformes, 4:291, 4:293–295
ground sharks, 4:115–116, 4:119–128
groupers, 4:46, 4:47, 4:61
Gymnotiformes, 4:371, 4:373–377
hagfishes, 4:26, 4:*30*, 4:34, 4:79, 4:80–81
hawkfishes, 4:46, 4:47
hermaphroditic, 4:29–30, 4:45
herrings, 4:278, 4:281–288
Hexanchiformes, 4:144, 4:146–148
Labroidei, 5:279–280, 5:284–290, 5:296–298, 5:301–307
labyrinth fishes, 5:428–429, 5:431–434
lampreys, 4:26, 4:34, 4:85, 4:88–90
Lampridiformes, 4:450, 4:452–454
lanternfishes, 4:443, 4:445–446
lungfishes, 4:203, 4:205–206
mackerel sharks, 4:133, 4:135–141
marine fishes, 4:44–47
minnows, 4:35, 4:*300*–303, 4:*302*, 4:314–315
morays, 4:256–*258*, 4:265–267
mudminnows, 4:380, 4:383–386
mullets, 5:60, 5:62–65
Ophidiiformes, 5:18, 5:20–22
Orectolobiformes, 4:105, 4:109–111
Osmeriformes, 4:393, 4:395–401
Osteoglossiformes, 4:233–234, 4:237–241
oviparous, 4:30–31, 4:65, 4:115–116
ovoviviparous, 4:31–32, 4:65
parrotfishes, 4:45, 4:47
Percoidei, 5:198, 5:202–208, 5:212, 5:214–217, 5:221–222, 5:225–226, 5:228–233, 5:242, 5:247–253, 5:261–262, 5:266–268, 5:270–273
Percopsiformes, 5:5, 5:8–12
pikes, 4:380, 4:383–386
pipefishes, 4:46
ragfishes, 5:353
Rajiformes, 4:*178*, 4:182–188

rays, 4:*178*, 4:182–188
Saccopharyngiformes, 4:272, 4:274–276
salmons, 4:35, 4:*47*, 4:*407*, 4:410–419
sawsharks, 4:168, 4:170–171
Scombroidei, 5:406, 5:409–418
Scorpaeniformes, 5:158, 5:161–162, 5:165–166, 5:169–178, 5:182–183, 5:186–193
Scorpaenoidei, 5:165–166
sea basses, 4:25, 4:29, 4:*31*, 4:46, 4:61
seahorses, 4:46
sharks, 4:26, 4:34, 4:35
skates, 4:27, 4:46, 4:*178*, 4:187
snakeheads, 5:*438*, 5:439, 5:442–446
South American knifefishes, 4:371, 4:373–377
southern cod-icefishes, 5:322, 5:325–328
spawning, 4:46, 4:47, 4:51, 4:65, 4:66, 4:*336*
Stephanoberyciformes, 5:106, 5:108–110
stingrays, 4:46, 4:178
Stomiiformes, 4:*422*, 4:424, 4:426–429
Stromateoidei, 5:422
Synbranchiformes, 5:152, 5:154–156
Tetraodontiformes, 5:470–471, 5:474–475, 5:477–484
toadfishes, 5:42, 5:44–45
Trachinoidei, 5:334–335, 5:337–339
trouts, 4:*32*, 4:*405*, 4:*407*
viviparous, 4:32, 4:65–66, 4:109, 4:110, 4:115–116
wrasses, 4:29, 4:*33*, 4:45
Zoarcoidei, 5:312–313, 5:315–318
insects, 3:4, 3:28–*30*, 3:*31*, **3:31–40,** 3:44, 3:46–47, 3:59–62
ants, 3:36, 3:59, 3:*69*, 3:407
aphids, 3:*34*, 3:37–38, 3:39, 3:44, 3:59
aquatic beetles, 3:62
bacterial infections and, 3:39
bark beetles, 3:62, 3:322
bees, 3:36, 3:38, 3:39, 3:59, 3:407
beetles, 3:29, 3:38, 3:45, 3:59, 3:62, 3:321–322
book lice, 3:244, 3:246–247
bristletails, 3:114, 3:116–117
butterflies, 3:34, 3:36, 3:38, 3:387–388
caddisflies, 3:34, 3:38, 3:376–377, 3:379–381
cockroaches, 3:30, 3:31, 3:33–34, 3:39, 3:62, 3:149–150, 3:153–158
Coleoptera, 3:44, 3:321–322, 3:327–333
crickets, 3:31, 3:38, 3:207
damselflies, 3:*60*, 3:61, 3:62, 3:134
diplurans, 3:108, 3:110–111
Diptera, 3:360, 3:364–374
dragonflies, 3:34, 3:61, 3:62, 3:134
earwigs, 3:*196*–197, 3:199–200
fleas, 3:349–350, 3:352–355
flies, 3:38, 3:39, 3:360
fruit flies, 3:4, 3:29
grasshoppers, 3:29, 3:30, 3:31, 3:39, 3:207
Hemiptera, 3:262–263, 3:267–280
honeybees, 3:61
Hymenoptera, 3:34, 3:38, 3:39, 3:44, 3:71, 3:407, 3:412–424
lacewings, 3:308, 3:311–314
Lepidoptera, 3:34, 3:387–388, 3:393–404
mantids, 3:30, 3:31, 3:33–34, 3:38, 3:39, 3:62, 3:*178*, 3:180–*181*, 3:183–186
Mantophasmatodea, 3:219
mayflies, 3:31, 3:33, 3:34, 3:62, 3:127, 3:129–130
Mecoptera, 3:343, 3:345–346
Megaloptera, 3:291, 3:293–294
mosquitos, 3:360
moths, 3:29, 3:34, 3:59, 3:60, 3:387–388
Odonata, 3:*134*, 3:137–138
Orthoptera, 3:44, 3:207, 3:211–216
Phasmida, 3:223, 3:228–232
Phthiraptera, 3:251–252, 3:254–256
proturans, 3:94, 3:96–97
rock-crawlers, 3:190, 3:192–193
scale insects, 3:59, 3:62, 3:263
snakeflies, 3:298, 3:300–302
springtails, 3:61, 3:100, 3:103–104
stoneflies, 3:31, 3:34, 3:143, 3:145–146
strepsipterans, 3:336, 3:338–339
termites, 3:35–36, 3:167–168, 3:170–175
thrips, 3:282, 3:285–287
Thysanura, 3:119–120, 3:122–123
wasps, 3:36, 3:39, 3:59, 3:62, 3:407
water bugs, 3:62, 3:263
webspinners, 3:234, 3:236–237
winged insects, 3:33–34
zorapterans, 3:240, 3:241
lower metazoans and lesser deuterostomes, **1:15–30,** 1:*16*
acoels, 1:180, 1:181–182
anoplans, 1:246–247, 1:249–251
arrow worms, 1:22, 1:435, 1:437–442
box jellies, 1:148, 1:151–152
calcareous sponges, 1:20, 1:21, 1:59, 1:61–65
cnidarians, 1:19–20, 1:21
Anthozoa, 1:107, 1:111–121
Hydrozoa, 1:16, 1:20, 1:21, 1:*126*, 1:129, 1:134–145
comb jellies, 1:19, 1:21, 1:171, 1:173–177
demosponges, 1:80, 1:82–86
echinoderms
Crinoidea, 1:358–359, 1:362–365
Echinoidea, 1:22, 1:405, 1:408–415
Ophiuroidea, 1:389–390, 1:394–398
sea cucumbers, 1:421–422, 1:425–431
sea daisies, 1:382–383, 1:385–386
sea stars, 1:369–370, 1:373–379
enoplans, 1:255, 1:256–257
entoprocts, 1:321, 1:323–325
flatworms
free-living flatworms, 1:18, 1:20, 1:21, 1:188, 1:190–195
monogeneans, 1:214–215, 1:218–223
tapeworms, 1:20, 1:21, 1:230–231, 1:234–242
gastrotrichs, 1:19, 1:22, 1:270, 1:272–273
genomic conflict and, 1:31
girdle wearers, 1:22, 1:345, 1:348–349
glass sponges, 1:69, 1:71–75
gnathostomulids, 1:19, 1:22, 1:332, 1:334–335
hair worms, 1:21–22, 1:306, 1:308–309
hemichordates, 1:22, 1:445, 1:447–450
jaw animals, 1:329
jellyfish, 1:16, 1:21, 1:156, 1:160, 1:162–167
kinorhynchs, 1:22, 1:276–277, 1:279–280
lancelets, 1:488, 1:490–497
larvaceans, 1:474, 1:476–478
nematodes
roundworms, 1:18, 1:20, 1:285, 1:287–292
secernenteans, 1:294–295, 1:298–304
Orthonectida, 1:19, 1:21, 1:99, 1:101–102
placozoans, 1:18–19, 1:20, 1:89
priapulans, 1:338, 1:340–341
rhombozoans, 1:19, 1:21, 1:93–94, 1:96–98
rotifers, 1:16, 1:19, 1:22, 1:261–262, 1:264–267
Salinella salve, 1:19, 1:20–21, 1:92
salps, 1:468, 1:471–472
sea squirts, 1:455, 1:458–466
sorberaceans, 1:480, 1:482–483
sponges, 1:16–18, 1:19–20, 1:21
thorny headed worms, 1:22, 1:312–313, 1:315–316
Trematoda, 1:16, 1:17, 1:18, 1:20, 1:21, 1:200, 1:203–211
wheel wearers, 1:23, 1:352–354
mammals, 12:12, 12:51, **12:89–112**
aardvarks, 15:158
agoutis, 16:409, 16:411–414
anteaters, 12:94, 13:174–175, 13:177–179
armadillos, 12:94, 12:103, 12:110–111, 13:185, 13:187–190
Artiodactyla, 12:127, 15:272
aye-ayes, 14:88
baijis, 15:21
bandicoots, 13:4
dry-country, 13:11, 13:14–16
rainforest, 13:11, 13:14–17*t*
bats, 13:314–315
American leaf-nosed, 13:419–420, 13:423–432
bulldog, 13:447, 13:449–450
disk-winged, 13:474, 13:476–477
Emballonuridae, 13:358, 13:360–363
false vampire, 13:382, 13:384–385
funnel-eared, 13:461, 13:463–465
horseshoe, 13:392, 13:396–400
Kitti's hog-nosed, 13:368
Molossidae, 13:486–487, 13:490–493
mouse-tailed, 13:352, 13:353
moustached, 13:436, 13:440–441
New Zealand short-tailed, 13:454, 13:457–458
Old World fruit, 13:320–321, 13:325–330, 13:336–337, 13:340–347
Old World leaf-nosed, 13:405, 13:407–409
Old World sucker-footed, 13:480
slit-faced, 13:373, 13:375–376
smoky, 13:468, 13:470–471
Vespertilionidae, 13:500–501, 13:506–510, 13:512–514, 13:522, 13:524–525
bears, 14:299, 14:302–305
beavers, 16:180, 16:183–184
bilbies, 13:22
body size and, 12:15, 12:22–24
botos, 15:29
bottleneck species and, 12:221–222
Bovidae, 16:7–8
Antilopinae, 16:48, 16:50–56
Bovinae, 16:14, 16:18–23
Caprinae, 16:93–94, 16:98–103
duikers, 16:76–77, 16:80–84
Hippotraginae, 16:32–33, 16:37–42
Neotraginae, 16:62–63, 16:66–70
bushbabies, 14:26, 14:28–32

Camelidae, 15:317, 15:320–323
Canidae, 14:271–272, 14:277–283
capybaras, 16:404–405
Carnivora, 12:90, 12:92, 12:94, 12:95, 12:98, 12:109, 12:110, 12:126–127, 14:261–262
cats, 12:103, 12:106, 12:109, 12:111, 12:126, 14:372–373, 14:379–389
Caviidae, 16:393, 16:395–398
Cetacea, 12:92, 12:94, 12:103–104, 15:9
  Balaenidae, 15:110–111, 15:115–117
  beaked whales, 15:61
  dolphins, 15:47–48, 15:53–55
  franciscana dolphins, 15:24–25
  Ganges and Indus dolphins, 15:15
  gray whales, 15:98–99
  Monodontidae, 15:85–86, 15:90–91
  porpoises, 15:35, 15:38–39
  pygmy right whales, 15:105
  rorquals, 15:124, 15:127–130
  sperm whales, 15:76–77, 15:79–80
chevrotains, 15:328–329, 15:331–333
Chinchillidae, 16:380, 16:382–383
colugos, 13:302, 13:304
coypus, 16:477
dasyurids, 12:289–290, 12:294–298
Dasyuromorphia, 12:281–283
deer
  Chinese water, 15:375–376
  muntjacs, 15:346, 15:349–354
  musk, 15:337–338, 15:340–341
  New World, 15:384, 15:388–395
  Old World, 15:360, 15:364–371
Dipodidae, 16:215–216, 16:219–222
Diprotodontia, 13:38–39
domestication and, 12:174
dormice, 16:320–321, 16:323–326
duck-billed platypuses, 12:101, 12:106, 12:108, 12:*228*, 12:231–233, 12:*232*, 12:245–246
Dugongidae, 15:201, 15:203
echidnas, 12:237–238
elephants, 12:90, 12:91, 12:94, 12:95, 12:98, 12:103, 12:106–108, 12:*108*, 12:209, 15:170–171, 15:174–175
Equidae, 15:228–229, 15:232–235
Erinaceidae, 13:206–207, 13:209–212
giant hutias, 16:470
gibbons, 14:213–214, 14:218–222
Giraffidae, 15:404–405, 15:408–409
great apes, 14:234–235, 14:237–240
gundis, 16:313, 16:314–315
Herpestidae, 14:352, 14:354–356
Heteromyidae, 16:203, 16:205–208
hippopotamuses, 15:308, 15:310–311
humans, 14:248–249
hutias, 16:463
Hyaenidae, 14:261, 14:362, 14:365–367
hyraxes, 15:182–183, 15:186–188
Indriidae, 14:66–67, 14:69–72
Insectivora, 13:199
koalas, 13:47–48
Lagomorpha, 16:485–487
lemurs, 14:10–11, 14:51–52, 14:56–60
  Cheirogaleidae, 14:38, 14:41–44
  sportive, 14:76, 14:79–83
Leporidae, 12:126, 16:507–509, 16:511–515
Lorisidae, 14:16, 14:18–20
luxury organs and, 12:23–24
Macropodidae, 13:88–89
manatees, 15:208, 15:211–212
Megalonychidae, 13:157, 13:159
migrations and, 12:170
moles
  golden, 13:217, 13:220–221
  marsupial, 13:27, 13:28
monitos del monte, 12:274
monkeys
  Atelidae, 14:158–159, 14:162–166
  Callitrichidae, 14:121–124, 14:127–131
  Cebidae, 14:105, 14:108–112
  cheek-pouched, 14:193, 14:197–204
  leaf-monkeys, 14:174–175, 14:178–183
  night, 14:138
  Pitheciidae, 14:147, 14:150–153
monotremes, 12:12, 12:51, 12:89–93, 12:108, 12:231–233
mountain beavers, 16:133
Muridae, 16:125, 16:284–285, 16:288–295
  Arvicolinae, 16:228–229, 16:233–237
  hamsters, 16:242–243, 16:245–246
  Murinae, 16:252, 16:256–260
  Sigmodontinae, 16:269–270, 16:272–276
musky rat-kangaroos, 13:71
Mustelidae, 14:324, 14:327–331
numbats, 12:305
nutritional requirements and, 12:125–127
octodonts, 16:436, 16:438–440
opossums
  New World opossums, 12:253–254, 12:257–263
  pygmy, 13:107–108, 13:110–111
  shrew, 12:268, 12:270
Otariidae, 14:398–399, 14:402–406
pacaranas, 16:387–388
pacas, 16:419–420, 16:423–424
pangolins, 16:113, 16:115–120
peccaries, 15:295, 15:298–300
Perissodactyla, 15:220–221
Petauridae, 13:128–129, 13:131–132
Phalangeridae, 13:60, 13:*63–66t*
photoperiodicity and, 12:75
pigs, 15:279–280, 15:284–288, 15:*289t–290t*
pikas, 16:494–495, 16:497–501
pocket gophers, 16:190, 16:192–195
porcupines
  New World, 16:369, 16:372–373
  Old World, 16:353, 16:357–364
possums
  feather-tailed, 13:142, 13:144
  honey, 13:137–138
primates, 14:10–11
Procyonidae, 14:310, 14:313–315
pronghorns, 15:414–415
Pseudocheiridae, 13:116–117, 13:119–123
rat-kangaroos, 13:76, 13:78–81
rats
  African mole-rats, 12:107, 16:344–345, 16:347–349
  cane, 16:334–335, 16:337–338
  chinchilla, 16:445, 16:447–448
  dassie, 16:331
  spiny, 16:450, 16:453–457
rhinoceroses, 15:254, 15:257–262
rodents, 16:125–126
sengis, 16:521–522, 16:524–530
shrews
  red-toothed, 13:251–252, 13:255–263
  West Indian, 13:244
  white-toothed, 13:268, 13:271–275
Sirenia, 15:193–194
solenodons, 13:239, 13:241
springhares, 16:*309*
squirrels
  flying, 16:136–137, 16:139–140
  ground, 16:147, 16:151–158
  scaly-tailed, 16:300, 16:302–305
  tree, 16:166–167, 16:169–174
subterranean mammals, 12:76
Talpidae, 13:282, 13:284–287
tapirs, 15:241, 15:245–247
tarsiers, 14:95, 14:97–99
tenrecs, 13:229, 13:232–233
three-toed tree sloths, 13:165, 13:167–169
tree shrews, 13:292, 13:294–296
true seals, 12:126, 12:127, 14:261, 14:262, 14:420–422, 14:427, 14:429–434
tuco-tucos, 16:427, 16:429–430
ungulates, 15:142–143
Viverridae, 14:338, 14:340–343
viviparous, 12:5
walruses, 12:111, 14:414–415
wombats, 13:53, 13:55–56
Xenartha, 13:152
in zoo populations, 12:206–209
protostomes, **2:14–24**, 2:27
  amphionids, 2:196
  amphipods, 2:261–262, 2:265–271
  anaspidaceans, 2:182, 2:183
  aplacophorans, 2:380, 2:382–385
  Arachnida, 2:*20*, 2:23, 2:335, 2:339–352
  articulate lampshells, 2:523, 2:525–527
  bathynellaceans, 2:178, 2:179
  beard worms, 2:86, 2:88
  bivalves, 2:24, 2:453–454, 2:457–466
  caenogastropods, 2:447, 2:449
  centipedes, 2:355–356, 2:358–362
  cephalocarids, 2:132, 2:133
  Cephalopoda, 2:17, 2:478–479, 2:483–489
  chitons, 2:395, 2:398–401
  clam shrimps, 2:148, 2:150–151
  copepods, 2:300, 2:304–310
  cumaceans, 2:230, 2:232
  Decapoda, 2:*16*, 2:17, 2:18, 2:201, 2:204–214
  deep-sea limpets, 2:436, 2:437
  earthworms, 2:22, 2:67, 2:70–73
  echiurans, 2:16, 2:104, 2:106–107
  fairy shrimps, 2:137, 2:139–140
  fish lice, 2:290, 2:292–293
  freshwater bryozoans, 2:497–498, 2:500–502
  horseshoe crabs, 2:328, 2:331–332
  Isopoda, 2:252, 2:254–260
  krill, 2:187, 2:189–192
  leeches, 2:16, 2:17, 2:77, 2:80–83
  leptostracans, 2:162, 2:164–165
  lophogastrids, 2:226, 2:227
  mantis shrimps, 2:170–171, 2:173–175
  marine bryozoans, 2:504, 2:506–508, 2:510–511, 2:513–514
  mictaceans, 2:241, 2:242
  millipedes, 2:23, 2:365, 2:367–370
  monoplacophorans, 2:388, 2:390–391
  mussel shrimps, 2:312, 2:314–315
  mysids, 2:217, 2:219–222
  mystacocarids, 2:295–296, 2:297
  myzostomids, 2:60, 2:62

Neritopsina, 2:441, 2:442–443
nonarticulate lampshells, 2:516, 2:518–519
Onychophora, 2:17, 2:23, 2:110, 2:113–114
pauropods, 2:376, 2:377
peanut worms, 2:98, 2:100–101
phoronids, 2:492, 2:494
Polychaeta, 2:21–23, 2:47, 2:50–56
Pulmonata, 2:16, 2:414, 2:417–421
remipedes, 2:126, 2:128
sea slugs, 2:405, 2:407–410
sea spiders, 2:322, 2:324–325
spelaeogriphaceans, 2:243, 2:244
symphylans, 2:372, 2:373
tadpole shrimps, 2:142, 2:145–146
tanaids, 2:236, 2:238–239
tantulocaridans, 2:284, 2:*286*
Thecostraca, 2:275–276, 2:278–281
thermosbaenaceans, 2:246, 2:247
tongue worms, 2:320
true limpets, 2:424, 2:426–427
tusk shells, 2:470, 2:473–474
Vestimentifera, 2:92, 2:94–95
Vetigastropoda, 2:430, 2:433–434
water bears, 2:23, 2:117–118, 2:121–123
water fleas, 2:154, 2:157–158
reptiles, **7:6–7**, 7:*36*
activity patterns and, 7:44
African burrowing snakes, 7:462–464
African sideneck turtles, 7:130, 7:132–133
Afro-American river turtles, 7:137–138, 7:140–142
Agamidae, 7:211, 7:214–221
Alligatoridae, 7:*173*, 7:175–177
American alligators, 7:*4*, 7:*36*
Anguidae, 7:340, 7:342–344
Australo-American sideneck turtles, 7:78, 7:81–83
big-headed turtles, 7:136
blindskinks, 7:272
blindsnakes, 7:382–385
boas, 7:411, 7:413–417
Central American river turtles, 7:100
chameleons, 7:*230*–232, 7:235–241
colubrids, 7:469–470, 7:473–481
Cordylidae, 7:321–322, 7:324–325
crocodilians, 7:6–7, 7:162–163, 7:163–164
Crocodylidae, 7:180–*181*, 7:184–187
early blindsnakes, 7:371
Elapidae, 7:486, 7:491–498
false blindsnakes, 7:388
false coral snakes, 7:400
file snakes, 7:442
Florida wormlizards, 7:284–285
Gekkonidae, 7:262, 7:265–269
Geoemydidae, 7:116, 7:118–119
gharials, 7:168–169
Helodermatidae, 7:357
Iguanidae, 7:246, 7:250–257
Kinosternidae, 7:122, 7:125–126
knob-scaled lizards, 7:348, 7:351
Lacertidae, 7:298, 7:301–302
leatherback seaturtles, 7:*102*
lizards, 7:6–7
Microteiids, 7:304, 7:306–307
mole-limbed wormlizards, 7:280, 7:281
Neotropical sunbeam snakes, 7:406
New World pond turtles, 7:*106*, 7:109–113
night lizards, 7:292, 7:294–295
oviparous, 7:6–7
pig-nose turtles, 7:76
pipe snakes, 7:397
pythons, 7:421, 7:424–428
seaturtles, 7:86–87, 7:89–91
shieldtail snakes, 7:392–394
skinks, 7:329, 7:332–337
slender blindsnakes, 7:375–377
snakes, 7:6–7
snapping turtles, 7:*6*, 7:*94*, 7:95–96
softshell turtles, 7:152, 7:154–155
spade-headed wormlizards, 7:288, 7:289
splitjaw snakes, 7:430
Squamata, 7:*206*–207
sunbeam snakes, 7:403
tactile cues, 7:40–42
Teiidae, 7:312, 7:314–316
tortoises, 7:7, 7:70–72, 7:*71*, 7:143–144, 7:146–148
Tropidophiidae, 7:435–437
tuatara, 7:6–7, 7:190–191
turtles, 7:6–7, 7:70–72, 7:*71*
Varanidae, 7:*360*, 7:362–363, 7:365–368
Viperidae, 7:448, 7:451–459
viviparous, 7:6–7
vomeronasal system and, 7:35–36
wormlizards, 7:274, 7:276
See also Cooperative breeding; Physical characteristics
Reproductive duality, 1:16
See also Reproduction
Reproductive technology, in zoos, 12:208
Reptiles
vs. amphibians, 7:3
behavior, **7:34–46**
brains, 7:8
conservation status, 7:10–11, **7:59–63**
defined, **7:3–11**
definition and description, **7:3–11**
diseases, 7:61, 7:62
diversity, 7:7–8
ears, 7:9
evolution, **7:12–22**, 7:*13*
eyes, 7:8–9
feeding ecology, 7:3–4, 7:26–28
in folk medicine, 7:50–52, 7:*54*
humans and, **7:47–58**
locomotion, 7:8, 7:24–26
mammals and, 12:36
as pets, 7:53
physical characteristics, 7:4–*5*, 7:8–9, **7:23–33**
reproduction, **7:6–7**, 7:*36*
See also specific reptiles
Reptilian brain, 7:8
Reptiliomorphs, 6:8
Requiem sharks, 4:11, 4:48, 4:113, 4:115
Resilin, 3:17–18
Resistance, viral, 12:187
Resource Dispersion Hypothesis, 14:259–260
Resource-holding power (RHP), 12:144
Respiration, 4:6, 4:*19*, 4:23–24, 4:*25*, 7:28–29, 7:*67*, 7:160–161
avian, 8:9–10
bats, 12:60
marine mammals, 12:67–68
See also Physical characteristics
Respiratory system, 3:*21*–22, 12:45–46, 12:73, 12:114
See also Physical characteristics
Resplendent quetzals, 9:477, 9:*478*, 9:479, 9:*480*, 9:484–*485*
Reticulated giraffes, 15:*401*, 15:*402*, 15:*405*
Reticulated pythons, 7:56, 7:202, 7:420–422, 7:*423*, 7:*426*, 7:427–428
Reticulated salamanders. *See* Flatwoods salamanders
Reticulated toadfishes, 5:42
Reticulitermes flavipes. *See* Eastern subterranean termites
Retinas, 12:79
See also Eyes
Retreat makers, 3:375
Retrievers, 14:*288*
Retrieving information, 12:152
Retropinna retropinna. *See* Cucumberfishes
Retropinna semoni. *See* Australian smelts
Retropinna tasmanica. *See* Tasmanian smelts
Retropinnidae, 4:57, 4:390, 4:391
Retroposons, 12:29–30
Reunion cuckoo-shrikes, 10:387
Reunion flightless ibises, 8:291
Reunion harriers, 8:323
Reunion solitaires, 9:269
Reunion starlings, 11:410
Reverse sexual dimorphism, 12:99
Reyer, H. U., 10:138
Rhabditophora, 1:11
Rhabdocalyptus dawsoni. *See* Sharp-lipped boot sponges
Rhabdocoela, 1:186, 1:187, 1:188
Rhabdoderma exiguum spp., 4:192
Rhabdolichops spp., 4:369, 4:370, 4:371
Rhabdomys pumilio. *See* Four-striped grass mice
Rhabdophis tigrinus. *See* Yamakagashis
Rhabdopleura normani, 1:445, 1:*446*, 1:*448*, 1:449
Rhabdornis spp. *See* Philippine creepers
Rhabdornis grandis. *See* Greater rhabdornis
Rhabdornis inornatus. *See* Stripe-breasted rhabdornis
Rhabdornis mysticalis. *See* Stripe-headed rhabdornis
Rhabdornithidae. *See* Philippine creepers
Rhabdosoma brevicaudatum, 2:264, 2:*265*, 2:271
Rhabdotogryllus caraboides. *See* Beetle crickets
Rhabdouraea bentzi, 2:161
Rhabdus rectius, 2:470
Rhachiberothidae, 3:305, 3:306, 3:307
Rhacodactylus auriculatus. *See* New Caledonian geckos
Rhacodactylus leachianus. *See* New Caledonian giant geckos
Rhacophoridae. *See* Asian treefrogs
Rhacophorinae, 6:291
Rhacophorus spp., 6:291, 6:292
Rhacophorus aboreus. *See* Kinugasa flying frogs
Rhacophorus dulitensis. *See* Jade treefrogs
Rhacophorus liber. *See* Free Madagascar frogs
Rhacophorus nigropalmatus. *See* Wallace's flying frogs
Rhacostoma atlanticum, 1:127
Rhadinosteus spp., 6:12
Rhagoletis zephyria. *See* Tephritid flies
Rhagologus leucostigma. *See* Mottled whistlers
Rhagomys spp., 16:264, 16:265, 16:266
Rhagomys rufescens, 16:270
Rhamphichthyidae, 4:369
Rhamphichthys marmoratus, 4:371
Rhamphichthys rostratus. *See* Bandfishes

Rhamphocharis spp. *See* Berrypeckers
Rhamphocottus richardsoni. See Grunt sculpins
Rhampholeon spp., 7:224–225, 7:227
Rhampholeon brevicaudatus. See Short-tailed chameleons
Rhamphophryne spp., 6:184
Rhamphorhyncoids, 7:19
Rhaphidophoridae, 3:203, 3:204
Rhaphiomidas terminatus. See Flower-loving flies
Rhaphiomidas terminatus abdominalis. See Delhi sands flower-loving flies
Rhaphiomidas terminatus terminatus. See El Segundo flower-loving flies
Rhea americana. See Greater rheas
Rhea lice, 3:*251*
Rheas, 8:21–22, 8:53–56, **8:69–74**
Rheboks, 16:31
Rheidae. *See* Rheas
Rhenaniformes, 4:10
Rheobatrachus spp., 6:37, 6:139, 6:147, 6:148, 6:149
Rheobatrachus silus. See Southern gastric brooding frogs
Rheobatrachus vitellinus, 6:149
Rheocles derhami, 5:72, 5:74–*75*
Rheocles sikorae, 5:70
Rheocles wrightae, 5:70
Rheodytes leukops. See Fitzroy River turtles
Rheokinesis, 1:39
See also Behavior
Rheomys mexicanus, 16:268
Rheotaxis, 1:40
See also Behavior
Rhesus macaques, 12:151–153, 14:5, 14:194, 14:*195*, 14:*200*
Rhesus monkeys. *See* Rhesus macaques
Rhims. *See* Slender-horned gazelles
Rhinarium, 12:8
Rhinatrema bivittatum. See Two-lined caecilians
Rhinatrematidae. *See* American tailed caecilians
Rhincodon typus. See Whale sharks
Rhincodontidae, 4:105
Rhinecanthus aculeatus. See Blackbar triggerfishes
Rhinemys spp., 7:77
Rhineura floridana. See Florida wormlizards
Rhineuridae. *See* Florida wormlizards
Rhinidae, 4:11
Rhinobatidae. *See* Guitarfishes
Rhinobatos lentiginosus. See Atlantic guitarfishes
Rhinobatos productus. See Guitarfishes
Rhinoceros auklets, 9:219
Rhinoceros beetles, 3:81
Rhinoceros hornbills, 10:72, 10:75–76
Rhinoceros iguanas, 7:246
Rhinoceros katydids, 3:*202*, 3:206–207
Rhinoceros scarabs, 3:323
Rhinoceros spp., 15:216, 15:217, 15:252
Rhinoceros sondiacus. See Javan rhinoceroses
Rhinoceros sondiacus annamiticus, 12:220
Rhinoceros unicornis. See Indian rhinoceroses
Rhinoceroses, 15:146, 15:237, **15:249–262**, 15:*256*
behavior, 15:218, 15:219, 15:253
conservation status, 15:221, 15:222, 15:254
distribution, 12:136, 15:217, 15:*249*, 15:252
evolution, 15:136, 15:137, 15:215, 15:216, 15:249, 15:251
feeding ecology, 15:220, 15:253
habitats, 15:218, 15:252–253
horns, 12:40
humans and, 12:184, 15:223, 15:254–255
migrations, 12:167
neonatal milk, 12:126
physical characteristics, 15:138, 15:139, 15:216, 15:217, 15:251–252
population indexes, 12:195
reproduction, 12:91, 15:143, 15:221, 15:254
species of, 15:257–262
taxonomy, 15:249, 15:251
woolly, 15:138, 15:249
in zoos, 12:209
Rhinocerotidae. *See* Rhinoceroses
Rhinocerotinae, 15:249
Rhinochimaera pacifica. See Pacific spookfishes
Rhinocrypta spp. *See* Tapaculos
Rhinocrypta lanceolata. See Crested gallitos
Rhinocryptidae. *See* Tapaculos
Rhinoderma darwinii. See Darwin's frogs
Rhinoderma rufum. See Chile Darwin's frogs
Rhinodermatidae. *See* Vocal sac-brooding frogs
Rhinoleptus spp., 7:373, 7:374
Rhinoleptus koniagui, 7:373, 7:374
Rhinolophidae. *See* Horseshoe bats
Rhinolophus spp. *See* Horseshoe bats
Rhinolophus affinis. See Intermediate horseshoe bats
Rhinolophus alcyone, 13:388
Rhinolophus beddomei. See Lesser woolly horseshoe bats
Rhinolophus blasii. See Blasius's horseshoe bats
Rhinolophus capensis. See Cape horseshoe bats
Rhinolophus clivosus. See Geoffroy's horseshoe bats
Rhinolophus cognatus. See Andaman horseshoe bats
Rhinolophus convexus, 13:392
Rhinolophus denti. See Dent's horseshoe bats
Rhinolophus euryale. See Mediterranean horseshoe bats
Rhinolophus ferrumequinum. See Greater horseshoe bats
Rhinolophus fumigatus. See Ruppell's horseshoe bats
Rhinolophus hildebrandti. See Hildebrandt's horseshoe bats
Rhinolophus hipposideros. See Lesser horseshoe bats
Rhinolophus imaizumii, 13:389, 13:392
Rhinolophus keyensis. See Kai horseshoe bats
Rhinolophus landeri. See Lander's horseshoe bats
Rhinolophus lepidus. See Blyth's horseshoe bats
Rhinolophus luctus. See Woolly horseshoe bats
Rhinolophus maclaudi, 13:389
Rhinolophus macrotis. See Rufous big-eared horseshoe bats
Rhinolophus megaphyllus. See Eastern horseshoe bats
Rhinolophus mehelyi. See Mehely's horseshoe bats
Rhinolophus mitratus. See Mitred horseshoe bats
Rhinolophus monoceros, 13:389
Rhinolophus paradoxolophus, 13:389
Rhinolophus philippinensis. See Large-eared horseshoe bats
Rhinolophus rouxii. See Rufous horseshoe bats
Rhinolophus simulator. See Bushveld horseshoe bats
Rhinolophus subbadius. See Little Nepalese horseshoe bats
Rhinolophus trifoliatus. See Trefoil horseshoe bats
Rhinolophus yunanensis. See Dobson's horseshoe bats
Rhinomugil spp., 5:65
Rhinomugil corsula, 5:65
Rhinomugil nasutus. See Shark mullets
Rhinomuraena quaesita. See Ribbon morays
Rhinomyias addita. See Streaky-breasted jungle-flycatchers
Rhinomyias brunneata. See Brown-chested jungle-flycatchers
Rhinomyias olivacea. See Fulvous-chested jungle-flycatchers
Rhinonicteris aurantia. See Golden horseshoe bats
Rhinophis spp., 7:391, 7:392
Rhinophis oxyrhynchus, 7:391
Rhinophrynidae. *See* Mesoamerican burrowing toads
Rhinophrynus spp., 6:67
Rhinophrynus canadensis, 6:95
Rhinophrynus dorsalis. See Mesoamerican burrowing toads
Rhinophylla spp., 13:416
Rhinophylla pumilio. See Dwarf little fruit bats
Rhinopias spp. *See* Weedy scorpionfishes
Rhinopias aphanes. See Merlet's scorpionfishes
Rhinopithecus spp., 14:172, 14:173
Rhinopithecus avunculus, 14:11, 14:175
Rhinopithecus bieti, 14:175
Rhinopithecus brelichi. See Gray snub-nosed monkeys
Rhinopithecus roxellana. See Golden snub-nosed monkeys
Rhinoplax vigil. See Helmeted hornbills
Rhinopoma hardwickei. See Hardwicke's lesser mouse-tailed bats
Rhinopoma macinnesi. See MacInnes' mouse-tailed bats
Rhinopoma microphyllum. See Greater mouse-tailed bats
Rhinopoma muscatellum. See Small mouse-tailed bats
Rhinopomastinae. *See* Common scimitarbills
Rhinopomastus aterrimus. See Black woodhoopoes
Rhinopomastus cyanomelas. See Common scimitarbills
Rhinopomastus minor. See Abyssinian scimitarbills
Rhinopomatidae. *See* Mouse-tailed bats
Rhinopsar brunneicapilla. See White-eyed starlings
Rhinoptericolidae, 1:226
Rhinopteridae. *See* Cownose rays
Rhinoptilus cinctus. See Three-banded coursers
Rhinos. *See* Rhinoceroses
Rhinotermitidae, 3:164, 3:166
Rhinotyphlops spp., 7:380–382
Rhinotyphlops acutus, 7:379, 7:381
Rhinotyphlops feae, 7:381
Rhinotyphlops newtoni, 7:381

Rhinotyphlops schlegelii. See Schlegel's blindsnakes
Rhinotyphlops simoni, 7:381
Rhinotyphlops unitaeniatus, 7:380
Rhionaesetina variegata, 3:*134*
Rhipidomys spp., 16:265, 16:266
Rhipidura spp. *See* Fantails
Rhipidura albiscapa. See Gray fantails
Rhipidura atra. See Black fantails
Rhipidura brachyrhyncha. See Dimorphic fantails
Rhipidura cockerelli. See Cockerell's fantails
Rhipidura cyaniceps. See Blue-headed fantails
Rhipidura fuliginosa. See New Zealand fantails
Rhipidura fuscorufa. See Cinnamon-tailed fantails
Rhipidura hypoxantha. See Yellow-breasted fantails
Rhipidura leucophrys. See Willie wagtails
Rhipidura malaitae. See Malaita fantails
Rhipidura nigrocinnamomea. See Black-and-cinnamon fantails
Rhipidura opistherythra. See Long-tailed fantails
Rhipidura phasiana. See Mangrove fantails
Rhipidura rufifrons. See Rufous fantails
Rhipidura rufiventris. See Northern fantails
Rhipidura semirubra. See Manus fantails
Rhipidura spilodera. See Streaked fantails
Rhipidura superciliosa. See Blue fantails
Rhipidura tenebrosa. See Dusky fantails
Rhipidura threnothorax. See Sooty thicket-fantails
Rhipiduridae. *See* Fantails
Rhithrogena semicolorata, 3:*126*
Rhizobium-legume symbiosis, 1:31
Rhizocephala, 2:273, 2:274, 2:276
Rhizomyinae, 16:123, 16:282, 16:284
Rhizomys sumatrensis. See Large bamboo rats
Rhizopoda, 2:12
Rhizoprionodon spp., 4:114
Rhizosomichthys totae. See Bagre graso
Rhizostoma spp., 1:26
Rhizostoma pulmo, 1:25
Rhizostomeae, 1:153, 1:154, 1:155, 1:156, 1:157
Rhodacanthis spp., 11:343
Rhodacanthis flaviceps. See Lesser koa finches
Rhodacanthis palmeri. See Greater koa finches
Rhodesgrass mealybugs, 3:*265*, 3:*274*, 3:278
Rhodesgrass scales. *See* Rhodesgrass mealybugs
Rhodeus spp., 4:298
Rhodeus ocellatus. See Rosy bitterlings
Rhodonessa caryophyllacea. See Pink-headed ducks
Rhodopis spp. *See* Hummingbirds
Rhodopis vesper. See Oasis hummingbirds
Rhodostethia rosea. See Ross's gulls
Rhogeessa anaeus. See Yellow bats
Rhombic egg-eaters. *See* Common egg-eaters
Rhombic night adders, 7:*449*, 7:*451*–452
Rhombomys spp., 16:127
Rhombosoleidae, 5:450
Rhombozoans, 1:19, 1:21, **1:93–98**, 1:*95*
Rhopalodina lageniformis. See Flask-shaped sea cucumbers
Rhopalodinidae, 1:419
Rhopalopsyllidae, 3:347, 3:348–349
Rhopalopsyllus spp., 3:347
Rhopalura spp., 1:99
Rhopalura ophiocomae, 1:*99*, 1:*100*, 1:101, 1:*102*
Rhopaluridae. *See* Orthonectida
Rhopiena esculenta, 1:25
Rhopodytes tristis. See Green-billed malkohas
Rhoptropus afer. See Namib day geckos
RHP (resource-holding power), 12:144
Rhyacichthyidae. *See* Loach gobies
Rhyacichthys spp., 5:373
Rhyacichthys aspro. See Loach gobies
Rhyacichthys guilberti, 5:375
Rhyacontritonidae, 6:13
Rhyacophila amabilis, 3:377
Rhyacophiloidea. *See* Primitive caddisflies
Rhyacotriton cascadae. See Cascade torrent salamanders
Rhyacotriton olympicus. See Olympic torrent salamanders
Rhyacotriton variegatus, 6:386
Rhyacotritonidae. *See* Torrent salamanders
Rhynchobdellida, 2:75
Rhynchocephalia. *See* Tuatara
Rhynchocyon spp. *See* Giant sengis
Rhynchocyon chrysopygus. See Golden-rumped sengis
Rhynchocyon cirnei. See Checkered sengis
Rhynchocyon petersi. See Black-and-rufous sengis
Rhynchocyoninae. *See* Giant sengis
Rhynchogale spp., 14:347
Rhyncholestes spp., 12:267–268
Rhyncholestes raphanurus. See Chilean shrew opossums
Rhynchomeles spp., 13:10
Rhynchomeles prattorum. See Seram Island bandicoots
Rhynchonycteris naso. See Proboscis bats
Rhynchopelates oxyrhynchus, 5:220
Rhynchophorus palmarum. See Palm weevils
Rhynchophthirina, 3:249, 3:250, 3:251
Rhynchopsitta pachyrhyncha. See Thick-billed parrots
Rhynchortyx cinctus. See Tawny-faced quails
Rhynchotus rufescens. See Red-winged tinamous
Rhyniella praecursor, 3:99
Rhynochetos jubatus. See Kagus
Rhyparobia maderae. See Madeira cockroaches
Rhyticeros everetti. See Sumba hornbills
Rhyticeros narcondami. See Narcondam hornbills
Rhyticeros subruficollis. See Plain-pouched hornbills
Rhyticeros undulatus. See Wreathed hornbills
Rhytidochrotinae, 3:206
Rhytinas. *See* Steller's sea cows
Rhytiodentalium kentuckyensis, 2:469
Ribbon eels. *See* Ribbon morays
Ribbon morays, 4:*260*, 4:*266*–267
Ribbon-tailed astrapias, 11:*490*, 11:492, 11:*494*, 11:*497*
Ribbon-tailed birds of paradise. *See* Ribbon-tailed astrapias
Ribbon-tailed drongos, 11:*440*, 11:*442*–443
Ribbon tails. *See* Ribbon-tailed astrapias
Ribbon worms, 1:19, 1:*245*, 1:*247*, 1:254
See also specific ribbon worms
Ribbonfishes, 4:447
Ribeiroia spp., 1:199–200
Ribs, 12:41
See also Physical characteristics
Rice munias. *See* Java sparrows
Rice (paddy field) eels. *See* Swamp eels
Rice rats
behavior, 16:267, 16:268
distribution, 16:265–266
humans and, 16:128, 16:270
marsh, 16:268, 16:*271*, 16:*273*–*274*
Rio de Janeiro, 16:*271*, 16:*273*, 16:*275*
taxonomy, 16:263
Rice tadpole shrimps. *See* Longtail tadpole shrimps
Rice tenrecs, 13:225, 13:*226*, 13:227
Rice weevils, 3:76
Ricebirds. *See* Java sparrows; Spotted munias
Ricefishes, 5:79, 5:80, 5:81
Richardiid flies, 3:360
Richard's pipits, 10:*375*
Richardson's dragonets, 5:*368*, 5:*369*, 5:370
Richardson's ground squirrels, 16:159*t*
Richdale, L. E., 8:118
Richman, Adam, 11:4
Richtersius coronifer. See Giant yellow water bears
Ricinoides afzelii, 2:*337*, 2:*344*, 2:348
Ricinulei, 2:333
Rickets, 12:74
Rickett's big-footed bats, 13:313
Ricochet saltation, 12:43
Ride, David, 13:105–106
Ridgeheads. *See* Bigscales; Longjaw bigscales
Ridgeia spp., 2:91
Ridgeia piscesae, 2:*85*, 2:91
Ridgley, Robert S., 10:303
Ridgway's cotingas. *See* Turquoise cotingas
Ridley's leaf-nosed bats. *See* Ridley's roundleaf bats
Ridley's roundleaf bats, 13:410*t*
Rieffer's hummingbirds. *See* Rufous-tailed hummingbirds
Rieger, R. M., 1:11
Riekoperla darlingtoni, 3:143
Riffle beetles, 3:319
Riflemen, 10:203, 10:204, 10:*205*, 10:*206*
Rift lake cichlids, 5:*280*
Riftia spp., 2:91
Riftia pachyptila. See Hydrothermal vent worms
Right halibuts. *See* Pacific halibuts
Right whale dolphins, 15:4, 15:46
Right whales, **15:107–118**, 15:*114*
behavior, 15:109–110
conservation status, 15:9, 15:10, 15:111–112
distribution, 15:5, 15:*107*, 15:109
evolution, 15:3, 15:107
feeding ecology, 15:110
habitats, 15:109
humans and, 15:112
migrations, 12:87
physical characteristics, 15:4, 15:107, 15:109
pygmy, **15:103–106**, 15:*104*, 15:*105*
reproduction, 15:110–111
species of, 15:*115*–118
taxonomy, 15:107
Rileyiella spp., 2:317, 2:318
The Rime of the Ancient Mariner (Coleridge), 8:116
Rimicola spp., 5:355
Rimski-Korsakov, Nikolai, 3:408
Ring-necked parakeets. *See* Rose-ringed parakeets

Ring-necked pheasants, 8:24, 8:402, 8:433–434, 8:*435*, 8:436, 8:*438*, 8:451–*452*
Ring-necked snakes, 7:*35*, 7:467–469
Ring ouzels, 10:487
Ring seals, 12:129, 14:*418*, 14:435*t*
Ring-tailed cats. *See* Ringtails (Procyonidae)
Ring-tailed coatis, 14:316*t*
Ring-tailed mongooses, 14:349, 14:350, 14:*353*, 14:*354*
Ring-tailed rock wallabies. *See* Yellow-footed rock wallabies
Ringed finches. *See* Double-barred finches
Ringed fruit bats. *See* Little collared fruit bats
Ringed kingfishers, 10:*6*
Ringed pipefishes, 5:134, 5:*135*, 5:*137*, 5:*142*, 5:145–146
Ringed plovers, 9:161, 9:*165*, 9:*167*
Ringed pythons, 7:420
Ringed turtledoves, 8:24
Ringlet butterflies, 3:57
Ringtail possums, 13:31, 13:34, 13:35, 13:39, **13:113–123**, 13:122*t*–123*t*
Ringtailed lemurs, 12:105, 14:6, 14:7, 14:*48*, 14:*49*, 14:50, 14:*51*, 14:*53*, 14:*55*–56
Ringtails (Procyonidae), 14:7, 14:259, 14:*310*, 14:*312*, 14:*313*, 14:314–315
Rinkhal's cobras, 7:485
Rio Cauca caecilians, 6:*411*, 6:440, 6:441
Rio de Janeiro greenlets. *See* Lemon-chested greenlets
Rio de Janeiro rice rats, 16:*271*, 16:*273*, 16:275
Rio Negro tuco-tucos, 16:*428*, 16:*429*
Riobamba marsupial frogs, 6:*6*, 6:*233*
Riopa spp., 7:328
Riparia riparia. *See* Sand martins
Risk assessment, of extinct species, 1:49
See also Conservation status
Risor spp., 5:374
Rissa tridactyla. *See* Black-legged kittiwakes
Rissoam dwarf. *See* Dzhungarian hamsters
Risso's dolphins, 15:1, 15:8, 15:57*t*
Ritterazines, 1:46
Ritterella tokioka, 1:46
Ritteri anemones. *See* Magnificent sea anemones
River blackfishes, 5:*223*, 5:*227*–228
River blindness nematodes. *See* African river blindness nematodes
River cats. *See* Channel catfishes
River cooters, 7:*107*
River dolphins, 12:79, 15:2, 15:5–6, 15:13, 15:15, 15:19
See also Baijis; Botos; Franciscana dolphins; Ganges and Indus dolphins
River habitats, 4:39–40, 4:52
See also Habitats
River hatchetfishes, 4:*339*, 4:*340*, 4:*349*
River hippopotamuses. *See* Common hippopotamuses
River-martins. *See* Swallows
River nerites, 2:*442*, 2:*443*
River otters, 12:80, 12:132, 14:*322*, 14:*323*
River pearl mussels. *See* European pearly mussels
River pipefishes, 5:136
River stingrays, 4:57
River worms, 2:68, 2:*69*, 2:*70*, 2:71–72
Riverine rabbits, 16:481, 16:487, 16:506, 16:509, 16:515*t*
Riversleigh, Queensland, 12:228
Rivulidae, 5:90, 5:93
Rivulinae, 4:12, 5:90
Rivulus, 5:91–92
See also Mangrove rivulus
Rivulus spp., 5:90–92
Rivulus marmoratus. *See* Mangrove rivulus
Roaches, 4:*297*
Roach's mouse-tailed dormice, 16:318, 16:321, 16:*322*, 16:*324*, 16:325
Roadrunners, 9:311, 9:*312*, 9:*313*, 9:314, 9:327
Roadside hawks, 8:322
Roan antelopes, 16:2, 16:6, 16:27, 16:29–32, 16:41*t*
Roatán Island agoutis, 16:409, 16:*410*, 16:*412*–413
Roatelos, **9:5–10**
Robalo blancos. *See* Common snooks
Robber flies, 3:9, 3:63–64, 3:357
Roberts' lechwes, 16:33
Robin accentors, 10:*461*, 10:*463*
Robins, 10:484–*485*, 10:*491*
See also specific types of robins
Robins, Australian. *See* Australian robins
Robinson, George Augustus, 12:310
Robinson, Michael H., 12:205
Robotic cockroaches, 3:150
Robust chimpanzees. *See* Chimpanzees
Robust white-eyes, 11:229
Roccus. *See* Striped sea basses
Rock barnacles, 2:277, 2:280–*281*
Rock basses, 5:*200*, 5:*202*
Rock bees, 10:144
Rock boring urchins, 1:402, 1:*406*, 1:410–*411*
Rock buntings, 11:*268*, 11:*274*
Rock cavies, 16:*390*, 16:*394*, 16:396–*397*, 16:399*t*
behavior, 16:392
distribution, 16:390
evolution, 16:389
feeding ecology, 16:393–394
habitats, 16:*391*
physical characteristics, 16:390
reproduction, 16:126, 16:393
Rock-crawlers, **3:189–194**, 3:*191*
Rock doves. *See* Rock pigeons
Rock flagtails. *See* Jungle perches
Rock grenadiers. *See* Roundnose grenadiers
Rock-haunting possums. *See* Rock ringtails
Rock hyraxes, 15:177–*180*, 15:*181*, 15:*182*, 15:*183*, 15:*184*, 15:*185*, 15:*187*, 15:188
Rock jumpers, 10:505
Rock lizards, **7:297–302**
Rock mice, pygmy, 16:*286*, 16:*289*, 16:290
Rock parrots, 9:276, 9:278–279
Rock partridges, 8:402, 8:434
Rock pigeons, 8:24, 9:243, 9:*253*, 9:*255*
Rock pipits, 10:373, 10:374
Rock possums. *See* Rock ringtails
Rock pratincoles, 9:*154*, 9:155–*156*
Rock ptarmigans, 8:434, 8:435
Rock-rabbits. *See* American pikas
Rock rats, 16:434, 16:435, 16:*437*, 16:*438*–439, 16:440*t*
Rock ringtails, 13:114, 13:115, 13:117, 13:*118*, 13:*119*, 13:120
Rock River frogs, 6:*156*, 6:*160*, 6:163–164
Rock robins. *See* Rockwarblers
Rock salmon. *See* Bocaccios
Rock sandpipers, 9:178
Rock slaters, 2:250
Rock sparrows, 11:*400*, 11:*404*
Rock thrushes, 10:484, 10:485, 10:487, 10:*491*, 10:*500*
Rock voles, 16:238*t*
Rock wallabies, 13:34, 13:83, 13:84, 13:85, 13:86
Rock whitings, 4:67, 5:275, **5:293–298**, 5:*300*, 5:*301*, **5:304**
Rock wrens, 10:172, 10:174, 10:203, 10:204, 10:526, 10:529
Rockefeller's sunbirds, 11:210
Rocket frogs, 6:66, 6:229, 6:*231*, 6:241
See also Blue-toed rocket frogs
Rockfishes, 4:30, 5:163–166
See also Striped sea basses
Rockhares, 16:481–482, 16:505
Rockheads, 5:*185*, 5:*186*
Rockhopper blennies, 5:342
Rockhopper penguins, 8:148, 8:*150*, 8:151
Rocklings, 5:26
See also Phycidae
Rocks. *See* Striped sea basses
Rocksuckers, 5:355, 5:356, 5:*358*, 5:*361*–362
Rockwarblers, 11:55, 11:56, 11:*58*, 11:*61*
Rocky Mountain goats. *See* Mountain goats
Rocky Mountain grasshoppers, 3:57
Rocky Mountain spotted fever, 16:148
Rocky Mountain spotted fever ticks. *See* Rocky Mountain wood ticks
Rocky Mountain tailed frogs, 6:77, 6:*79*, 6:*80*
Rocky Mountain wood ticks, 2:*337*, 2:*340*–341
Rodentia. *See* Rodents
Rodents, 12:13, **16:121–129**
behavior, 12:141, 12:142–143, 16:124–125
color vision, 12:79
conservation status, 16:126
digestive system, 12:14–15, 12:121
distribution, 12:129, 12:131, 12:132, 12:134, 12:135, 12:136, 12:137, 12:138, 16:123
evolution, 16:121–122
feeding ecology, 16:125
field studies of, 12:*197*
geomagnetic fields, 12:84
habitats, 16:124
humans and, 16:126–128
hystricognathe, 12:11
Lagomorpha and, 16:479–480
locomotion, 12:14
as pests, 12:182
pheromones, 12:80
physical characteristics, 16:122–123
reproduction, 12:90, 12:91, 12:92, 12:94, 12:95, 12:107–108, 12:110, 16:125–126
taxonomy, 16:121–122
teeth, 12:*37*, 12:*46*
See also specific rodents
Rodricensis flying foxes, 13:*324*, 13:*326*, 13:329
Rodrigues flying foxes. *See* Rodricensis flying foxes
Rodrigues solitaires, 9:244, 9:269, 9:270, 9:*273*
Rodrigues starlings, 11:410
Roe deer
behavior, 12:145
competition and, 12:118
European, 15:384, 15:*386*, 15:*388*
evolution, 15:379, 15:380

habitats, 15:*269*, 15:383
humans and, 15:384–385
physical characteristics, 15:382
predation and, 12:116
Siberian, 15:384, 15:*386*, 15:*388*–389
taxonomy, 15:381
Roesel's green frogs, 6:*254*, 6:*262*–263
Rogovin, K. A., 16:216
Rogue fishes. *See* Cockatoo waspfishes
Rollandia spp. *See* Grebes
Rollandia microptera. See Titicaca flightless grebes
Rollandia roland. See White-tufted grebes
Rollers, 10:1–4, **10:51–59,** 10:*54*
Rollulus rouloul. See Crested wood-partridges
Romalea gutatta. See Eastern lubber grasshoppers
Romaleidae, 3:203
Roman Empire, zoos in, 12:203
Roman mythology, reptiles in, 7:54–55
Roman snails, 2:*416*, 2:*419*, 2:420
Romanian hamsters, 16:243
Romanomermis spp., 1:36
Romeo and Juliet (Shakespeare), 10:346
Romer, A. S., 7:13, 16:333
Romerolagus spp. *See* Volcano rabbits
Romerolagus diazi. See Volcano rabbits
Rondeletia spp. *See* Redmouth whalefishes
Rondeletiidae. *See* Redmouth whalefishes
Rondo bushbabies, 14:26, 14:33*t*
Ronquils, 5:309–313
Ronquilus jordani. See Northern ronquils
Rood, J. P., 16:391
Rooks, 8:18, 11:505, 11:506, 11:*511*, 11:*520*
Roosevelt's muntjacs, 15:343, 15:*347*, 15:*350*
Roosterfishes, **5:211–212,** 5:*213*, 5:*215*, **5:216**
Roosting bats, 13:310, 13:311, 13:312
See also specific roosting bats
Root-feeding wireworms, 3:56
Root-knot nematodes, 1:36
Root-like barnacles, 2:33, 2:*277*, 2:*278*, 2:279–280
Root miner flies, 3:361
Root rats. *See* Large bamboo rats
Root voles, 16:229
Ropalidia spp. *See* Wasps
Ropefishes. *See* Reedfishes
Roproniidae, 3:54
Roraima bush toads, 6:*186*, 6:193
Rorquals, **15:119–130,** 15:*126*
behavior, 15:6, 15:7, 15:122–123
conservation status, 15:9, 15:10, 15:124–125
distribution, 15:5, 15:*119*, 15:120–121
evolution, 12:*66*, 15:3, 15:119
feeding ecology, 15:8, 15:123–124
habitats, 15:6, 15:121–122
humans and, 15:125
physical characteristics, 12:12, 12:63, 12:79, 12:82, 12:213, 15:4, 15:94, 15:119–120
reproduction, 15:124
species of, 15:127–129, 15:130*t*
taxonomy, 15:1, 15:119
Rose, G. A., 5:32
Rose-breasted pygmy parrots. *See* Red-breasted pygmy parrots
Rose-colored starlings. *See* Rosy starlings
Rose mantids. *See* Wandering violin mantids
Rose-ringed parakeets, 9:251, 9:278, 9:280, 9:*282*, 9:*286*–287
Rose-throated becards, 10:270, 10:*271*, 10:*276*, 10:*278*
Roseate spoonbills, 8:*10*, 8:*235*, 8:291, 8:*293*, 8:*295*, 8:300–*301*
Rosefinches, 11:323
Rosella parrots. *See* Eastern rosellas
Rosellas. *See* Eastern rosellas
Rosen, Donn E.
on Adrianichthyids, 5:79
on Atheriniformes, 5:67
on Ceratodontiformes, 4:197
on Cyprinodontiformes, 5:89
on Gonorynchiformes, 4:289
on Osmeriformes, 4:389
on Stomiiformes, 4:421
on Stromateoidei, 5:421
Rosen knifefishes, 4:371
Rosenberg's gladiator frogs. *See* Rosenberg's treefrogs
Rosenberg's treefrogs, 6:35, 6:230, 6:*232*, 6:237
Rose's ghost frogs, 6:131, 6:132, 6:*133*–134
Ross seals, 12:138, 14:435*t*
Rossetti, Dante Gabriel, 13:53
Rossophyllum maculosum. See Speckled rossophyllums
Rossophyllums, speckled, 3:*209*, 3:*211*, 3:215–216
Ross's gulls, 9:203
Ross's turacos, 9:300, 9:302, 9:*303*, 9:304–*305*
Rostratula australis. See Australian greater painted-snipes
Rostratula bengalensis. See Greater painted-snipes
Rostratulidae. *See* Painted snipes
Rostrhamus sociabilis. See Snail kites
Rostroconchia, 2:469
Rosy bee-eaters, 10:39, 10:42
Rosy bitterlings, 4:299, 4:*305*, 4:*311*, 4:317
Rosy boas, 7:409–411, 7:*412*, 7:*413*, 7:416
Rosy-breasted longclaws, 10:*377*, 10:*380*
Rosy-faced lovebirds, 9:*281*, 9:*296*–297
Rosy feather stars, 1:*360*, 1:*362*
Rosy flamingos. *See* Greater flamingos
Rosy minivets, 10:386
Rosy pastors. *See* Rosy starlings
Rosy-patched shrikes, 10:427
Rosy pipits, 10:372
Rosy-reds. *See* Fathead minnows
Rosy spoonbills. *See* Roseate spoonbills
Rosy starlings, 11:*413*, 11:420–*421*
Rosy weedfishes, 5:*344*, 5:*345*, 5:347
Rosy wolfsnails, 2:*416*, 2:*418*, 2:420–421
Rosybills, 8:367
Rotaria spp., 1:260
Roth, V. L., 16:143
Rothschild, Walter, 10:207
Rothschild's giraffes, 15:*402*
Rothschild's mynas. *See* Bali mynas
Rothschild's porcupines, 16:365, 16:366, 16:374*t*
Rothschild's starlings. *See* Bali mynas
Rotifers, **1:259–268,** 1:*260*, 1:*263*
behavior, 1:38, 1:43, 1:261
conservation status, 1:262
distribution, 1:260
evolution, 1:259
feeding ecology, 1:261
habitats, 1:260–261
humans and, 1:262
physical characteristics, 1:259–260
reproduction, 1:16, 1:19, 1:22, 1:261–262
species of, 1:*264*–*267*
taxonomy, 1:9, 1:11–12, 1:259
Rottenwood termites, 3:162, 3:164, 3:166, 3:167
See also Wide-headed rottenwood termites
Rottweilers, 14:292
Rough-billed pelicans. *See* American white pelicans
Rough earthsnakes, 7:468
Rough-faced cormorants. *See* New Zealand king shags
Rough-faced shags. *See* New Zealand king shags
Rough flounders. *See* Winter flounders
Rough greensnakes, 7:468
Rough-haired golden moles, 13:222*t*
Rough-legged buzzards, 8:323, 8:*339*
Rough-legged hawks. *See* Rough-legged buzzards
Rough-legged jerboas. *See* Hairy-footed jerboas
Rough-scaled lizards, 7:298
Rough-scaled pythons, 7:420
Rough sharks, 4:151, 4:152
Rough-skinned newts, 6:*323*, 6:*364*
Rough-templed tree-babblers. *See* Flame-templed babblers
Rough-toothed dolphins, 15:8, 15:49
Roughies, **5:113–116,** 5:*117*, 5:*119*, **5:121**
Roughtail stingrays, 4:*180*, 4:*182*
Roulin, X., 15:242
Roulin's tapirs. *See* Mountain tapirs
Roulrouls. *See* Crested wood-partridges
Round-eared bats, 13:311
Round-eared sengis, 16:518, 16:519, 16:*520*, 16:522, 16:*523*, 16:525–*526*
Round frogs. *See* Sandhill frogs
Round gobies, 5:377
Round Island boas. *See* Splitjaw snakes
Round rays, 4:174
Round-tailed muskrats, 16:225, 16:238*t*
Roundhead congas. *See* Blackbar soldierfishes
Roundheads, 5:255
See also Plesiopidae
Roundleaf bats
Ashy, 13:410*t*
Noack's, 13:*406*, 13:*407*, 13:408
Ridley's, 13:410*t*
Sundevall's, 13:*403*
See also specific types of roundleaf bats
Roundleaf horseshoe bats, 13:*403*
Roundnose grenadiers, 5:*31*, 5:*36*–*37*
Roundups, rattlesnakes, 7:49–50
Roundworms, **1:283–292,** 1:*284*, 1:*286*
behavior, 1:285
conservation status, 1:285
distribution, 1:284
evolution, 1:283
feeding ecology, 1:285
habitats, 1:284
humans and, 1:285
physical characteristics, 1:*4*, 1:283–*284*
reproduction, 1:18, 1:20, 1:21, 1:285
species of, 1:287–292
taxonomy, 1:12, 1:283
See also specific types of roundworms
Rousette bats, 12:55, 13:333
Rousettus spp., 13:334

Rousettus aegyptiacus. *See* Egyptian rousettes
Route-based navigation, 8:33
Rove beetles, 3:57, 3:316, 3:319
behavior, 3:320
feeding ecology, 3:321
physical characteristics, 3:317, 3:318
reproduction, 3:321, 3:322
See also Devil's coach-horses
Row pore rope sponges, 1:77
Rowe, D. L., 16:389
Rowe, F. W. E., 1:381
Royal albatrosses, 8:110, 8:113, 8:*117*, 8:*118*
Royal antelopes, 16:60, 16:63, 16:71*t*
Royal cinclodes, 10:210–211
Royal cranes. *See* Gray crowned cranes
Royal flycatchers, 10:270, 10:*271*, 10:274, 10:275
Royal jelly, 3:79
Royal parrotfinches, 11:356
Royal penguins. *See* Macaroni penguins
Royal pythons. *See* Ball pythons
Royal sunangels, 9:418, 9:438
Royal urchins. *See* Tuxedo pincushion urchins
Rubber boas, 7:409
Rubber corals, 1:*109*, 1:*120*, 1:121
Rubber frogs. *See* Banded rubber frogs
Rubrisciurus spp., 16:163
Rubus ulmifolius, 9:450
Ruby-cheeked sunbirds, 11:*211*, 11:*213*
Ruby-throated hummingbirds, 9:*439*, 9:446
Ruby topaz, 9:*453*, 9:*462*
Ruby-topaz hummingbirds. *See* Ruby topaz
Rubycheek. *See* Ruby-cheeked sunbirds
Rucker's hermits. *See* Band-tailed barbthroats
Rudarius excelsius. *See* Diamond filefishes
Rudderfishes. *See* Sea chubs
Rudds, 4:*303*, 4:*308*, 4:317–318
Rudd's larks, 10:346
Ruddy ducks, 8:372
Ruddy shelducks, 8:*374*, 8:*379*–380
Ruddy turnstones, 9:*104*, 9:*176*, 9:*181*, 9:184–*185*
Ruddy woodcreepers, 10:*230*
Rufescent bandicoots. *See* Rufous spiny bandicoots
Rufescent tinamous. *See* Thicket tinamous
Ruffed flycatchers, 8:*10*
Ruffed grouse, 8:434, 8:*437*, 8:*439*, 8:443–444
Ruffed lemurs. *See* Variegated lemurs
Ruffes, 5:*200*, 5:*207*
Ruffs, 9:176–180, 9:*181*, 9:*187*
Rufous babblers, 11:127, 11:*129*, 11:*130*
Rufous-backed kingfishers, 10:7
Rufous-backed shrikes. *See* Long-tailed shrikes
Rufous-banded honeyeaters, 11:*239*, 11:*248*
Rufous-bellied seedsnipes, 9:*190*, 9:191, 9:*192*, 9:*193*
Rufous-bellied tits, 11:155, 11:*159*, 11:*163*
Rufous bettongs, 13:*74*, 13:*75*, 13:*76*, 13:*77*, 13:*79*, 13:80–81, 16:*127*
Rufous big-eared horseshoe bats, 13:390
Rufous-breasted buttonquails. *See* Red-chested buttonquails
Rufous-breasted hermits. *See* Hairy hermits
Rufous-breasted honeyeaters. *See* Rufous-banded honeyeaters
Rufous-breasted whistlers. *See* Rufous whistlers
Rufous bristlebirds, 11:57
Rufous-browed peppershrikes, 11:*257*, 11:*261*
Rufous bush chats, 10:*490*, 10:*493*–494
Rufous-capped motmots, 10:31
Rufous-capped nunlets, 10:102, 10:*105*, 10:*109*
Rufous-cheeked nightjars, 9:402
Rufous-collared kingfishers, 10:*12*, 10:*15*–16
Rufous-collared pratincoles. *See* Rock pratincoles
Rufous elephant shrews. *See* Rufous sengis
Rufous fantails, 11:83, 11:85–86, 11:*88*, 11:90–*91*
Rufous-fronted ant thrushes, 10:242
Rufous-fronted dippers, 10:477
Rufous-fronted fantails. *See* Rufous fantails
Rufous hare-wallabies, 13:34, 13:*91*, 13:*98*, 13:99
Rufous-headed crowtits. *See* Vinous-throated parrotbills
Rufous-headed hornbills, 10:75
Rufous hornbills, 10:72
Rufous horneros, 10:173–174, 10:*210*, 10:211, 10:*212*, 10:*218*
Rufous horseshoe bats, 13:389, 13:391, 13:392, 13:*395*, 13:*396*–397
Rufous hummingbirds, 8:24, 9:439, 9:445
Rufous leaf bats. *See* Rufous horseshoe bats
Rufous motmots, 10:31–32, 10:33
Rufous mouse-eared bats, 13:311
Rufous-naped bush larks, 10:*342*
Rufous-naped larks, 10:344
Rufous-naped whistlers, 11:115, 11:*116*, 11:117, 11:*119*, 11:*120*–121
Rufous-necked foliage-gleaners, 10:*213*, 10:*226*
Rufous-necked wood-rails, 9:49
Rufous ovenbirds. *See* Rufous horneros
Rufous piculets, 10:*153*, 10:*155*
Rufous potoos, 9:395, 9:*398*, 9:*399*
Rufous ringtails. *See* Common ringtails
Rufous sabrewings, 9:*454*, 9:*459*
Rufous scrub-birds, 10:337–338, 10:*339*
Rufous sengis, 16:*521*, 16:*523*, 16:*528*, 16:529
Rufous shrike-thrushes. *See* Little shrike-thrushes
Rufous-sided towhees. *See* Eastern towhees
Rufous spiny bandicoots, 13:2, 13:4, 13:*11*, 13:*15*, 13:16
Rufous-tailed bronze cuckoos. *See* Horsfield's bronze-cuckoos
Rufous-tailed chachalacas. *See* Rufous-vented chachalacas
Rufous-tailed hornbills. *See* Visayan tarictic hornbills
Rufous-tailed hummingbirds, 9:444, 9:445, 9:*453*, 9:*465*
Rufous-tailed jacamars, 10:*92*, 10:93, 10:*94*, 10:*95*
Rufous-tailed plantcutters, 10:325, 10:326, 10:*327*–*328*
Rufous-tailed rock thrushes. *See* Rock thrushes
Rufous-tailed scrub-robins. *See* Rufous bush chats
Rufous-tailed tree shrews, 13:298*t*
Rufous-tailed xenops, 10:*213*, 10:*225*–226
Rufous-throated dippers, 10:*478*, 10:*481*–482
Rufous-tipped chachalacas. *See* Rufous-vented chachalacas
Rufous tits. *See* Rufous-bellied tits
Rufous tree rats, 16:458*t*
Rufous treecreepers, 11:*133*, 11:134, 11:*136*, 11:*138*
Rufous treepies, 11:507, 11:*510*, 11:*516*
Rufous vangas, 10:439, 10:440, 10:441, 10:*443*
Rufous-vented chachalacas, 8:*417*, 8:*418*–*419*
Rufous whistlers, 11:117, 11:*119*, 11:*122*–123
Rufous-winged akalats, 10:*510*, 10:*511*
Rufous-winged illadopsis. *See* Rufous-winged akalats
Rufous-winged sunbirds, 11:210
Rufous woodcocks, 9:176
Rufous woodpeckers, 10:*152*, 10:156–*157*
Rugiloricus carolinensis. *See* American diamond girdle wearers
Rugiloricus cauliculus. *See* Bucket-tailed loriciferans
Rugogaster hydrolagi, 1:*202*, 1:*203*
Rugogastridae, 1:197
Rugosa spp., 6:248
Ruiz-Trillo, I., 1:179
Rumbaugh, Duane, 12:161
Rumen, 12:122–124
See also Physical characteristics
Ruminants
evolution, 12:216, 15:136, 15:263–265
feeding ecology, 12:14, 12:118–119, 12:120–124, 15:141, 15:271–272
physical characteristics, 12:10
taxonomy, 15:263, 15:266–267
vitamin requirements, 12:120
See also specific ruminants
Rumination, 12:122–124
See also Feeding ecology
Rumpler, Y., 14:74–75
Running, 12:42
See also Behavior
Running frogs. *See* Bubbling kassina
Runwenzorisorex spp., 13:265
Rupicapra spp. *See* Chamois
Rupicapra pyrenaica. *See* Southern chamois
Rupicapra rupicapra. *See* Northern chamois
Rupicaprini, 16:87, 16:88, 16:89, 16:90, 16:94
Rupicola spp. *See* Cotingas
Rupicola peruviana. *See* Andean cocks-of-the-rock
Rupicola rupicola. *See* Guianan cocks-of-the-rock
Rupirana spp., 6:156
Ruppell's horseshoe bats, 13:390
Rüppell's parrots, 9:276
Rüppell's vultures, 8:322
Rüppell's weavers, 11:376
Rusa, 12:19
Ruspolia spp., 3:203, 3:204, 3:206
Russell, Charlie, 15:416
Russel's vipers, 7:56, 7:448, 7:*449*, 7:*451*, 7:458
Russet bellied spinetails, 10:211
Russet-crowned motmots, 10:31, 10:32
Russet-mantled softtails, 10:211
Russian brown bears. *See* Brown bears
Russian desmans, 13:200, 13:280, 13:281, 13:282, 13:*283*, 13:*285*, 13:286
Russian flying squirrels. *See* Siberian flying squirrels
Russian steppe cockroaches, 3:150
Rusty-backed antwrens, 10:*241*
Rusty-belted tapaculos, 10:*261*, 10:*264*

Rusty blackbirds, 11:302, 11:304, 11:*307*, 11:*319*–320
Rusty-flanked treecreepers, 11:177, 11:178
Rusty-margined guans, 8:414
Rusty-necked piculets, 10:150
Rusty pitohuis, 11:116
Rusty spaghetti eels, 4:*260*, 4:*265*
Rusty-winged starlings, 11:410
Rutelinae, 3:55
Ruthven's frogs, 6:6, **6:211–213,** 6:*212*
Rutilus spp., 4:298
Rutilus anomalous. See Stonerollers
Rutilus rutilus. See Roaches
Ruvettus spp., 5:406
Ruwenzori otter shrews, 13:234*t*
Ruwenzori shrews, 13:*269*, 13:*272*, 13:274–275
Ruwenzori turacos, 9:300, 9:*303*, 9:*305*–306
Ruwenzorisorex suncoides. See Ruwenzori shrews
Ruwenzorornis johnstoni. See Ruwenzori turacos
Rynchopidae. *See* Skimmers
Rynchops albicollis. See Indian skimmers
Rynchops flavirostris. See African skimmers
Rynchops niger. See Black skimmers
Ryukyu flying foxes, 13:331*t*
Ryukyu moles, 13:282
Ryukyu rabbits. *See* Amami rabbits

S

Sabelariids, 2:47
Sabellaria alveolata. See Honeycomb worms
Sabellids, 2:46, 2:47
Saber-toothed blennies, 4:*56*, 5:342
Saber-toothed cats, 14:256
Saber-toothed tigers, 12:*23*, 12:*24*, 14:369, 14:375
See also Saber-toothed cats
Sabertooth fishes, 4:432
Sabertooth viperfishes, 4:*421*
Sabine's gulls, 9:203
Sable antelopes, 16:2, 16:5, 16:27, 16:28, 16:29, 16:31, 16:32, 16:*36*, 16:*39*, 16:41
Sablefishes, 5:179, 5:182, 5:183, 5:*184*, 5:186–*187*
Sables, 14:324
Sac-winged bats, **13:355–365,** 13:*359*, 13:363*t*, 13:364*t*
Saccogaster melanomycter, 5:18
Saccoglossus horsti, 1:445
Saccoglossus kowalevskii, 1:445, 1:*446*, 1:447–*448*
Saccoglossus ruber, 1:444
Saccolaimus peli. See Giant pouched bats
Saccopharyngidae, 4:271, 4:272
Saccopharyngiformes, 4:11, **4:271–276,** 4:*273*
Saccopharyngoidei, 4:271
Saccopharynx spp., 4:271, 4:274, 4:275
Saccopharynx ampullaceus. See Gulper eels
Saccopteryx bilineata. See Greater sac-winged bats
Saccopteryx leptura. See Lesser sac-winged bats
Saccostomus campestris. See Pouched mice
Sacculina carcini. See Root-like barnacles
Sachser, N., 16:392
Sacred baboons. *See* Hamadryas baboons
Sacred ibises, 8:26, 8:236, 8:291, 8:293, 8:*295*, 8:*296*
Sacred kingfishers, 10:6
Sacred langurs. *See* Northern plains gray langurs
Sacred scarabs, 3:74, 3:*316*, 3:322, 3:*324*, 3:*327*, 3:331–332
Sactosoma vitreum, 2:103
Sactosomatidae. *See Sactosoma vitreum*
Sactosomatidea. *See Sactosoma vitreum*
Saddle-back tamarins, 14:116–121, 14:*125*, 14:127–*128*
Saddleback gunnels, 5:312, 5:*314*, 5:*316*
Saddleback tapirs. *See* Malayan tapirs
Saddlebacks. *See* Tiekes
Saddlebilled storks. *See* Saddlebills
Saddlebills, 8:*266*, 8:*269*, 8:272–*273*
Sado moles, 13:282
Saeed, B., 5:67
Saffron-bellied frogs, 6:*306*, 6:311–*312*
Saffron-cowled blackbirds, 11:306
Saffron siskins, 11:324–325
Saffron toucanets, 10:88, 10:125, 10:128, 10:*129*, 10:*131*–132
Safina, C., 5:418
Sage grouse, 8:20, 8:434, 8:437
Sagebrush lizards, 7:*249*, 7:*255*–256
Sagebrush voles, 16:237*t*
Sage's rock rats, 16:440*t*
Saginae, 3:207
Sagitta spp., 1:433
Sagitta bipunctata, 1:*436*, 1:*438*
Sagitta elegans, 1:435
Sagitta enflata, 1:434, 1:435, 1:*436*, 1:*438*–439
Sagitta gazellae, 1:434
Sagitta planctonis, 1:434, 1:*436*, 1:*439*–440
Sagitta regularis, 1:434
Sagitta setosa, 1:434, 1:*436*, 1:*439*, 1:440
Sagitta tasmanica, 1:435
Sagittariidae. *See* Secretary birds
Sagittarius serpentarius. See Secretary birds
Sagittidae. *See Sagitta* spp.
Saguinus spp. *See* Tamarins
Saguinus bicolor. See Pied tamarins
Saguinus fuscicollis. See Saddle-back tamarins
Saguinus geoffroyi. See Geoffroy's tamarins
Saguinus graellsi. See Graell's black-mantled tamarins
Saguinus imperator. See Emperor tamarins
Saguinus inustus. See Mottle-faced tamarins
Saguinus labiatus. See Red-bellied tamarins
Saguinus leucopus. See White-lipped tamarins
Saguinus midas. See Midas tamarins
Saguinus mystax. See Moustached tamarins
Saguinus nigricollis. See Black-mantled tamarins
Saguinus oedipus. See Cotton-top tamarins
Saguinus oedipus oedipus. See Cotton-top tamarins
Sahara gundis. *See* Desert gundis
Sahara sandboas, 7:409, 7:411
Saharagalago spp., 14:23
Saharan horned vipers. *See* Horned vipers
Saharan sandgrouse. *See* Spotted sandgrouse
Sahelanthropus chadensis, 14:241–242
Sahlins, Marshall, 14:248
Saiga antelopes, 16:4, 16:9, **16:45–58,** 16:*49*, 16:*53*
Saiga tatarica. See Saiga antelopes
Saigas. *See* Saiga antelopes
Sailfin dories. *See* Buckler dories
Sailfin lizards, 7:*213*, 7:*215*, 7:218
Sailfin plunderfishes, 5:*324*, 5:325
Sailfin sandfishes, 5:335, 5:*336*, 5:*338*, 5:339
Sailfin sculpins, 5:*185*, 5:187–*188*
Sailfishes, 5:*408*, 5:*410*
Sailor fishes. *See* Sailfin sculpins
Saimiri spp. *See* Squirrel monkeys
Saimiri boliviensis. See Bolivian squirrel monkeys
Saimiri oerstedii. See Red-backed squirrel monkeys
Saimiri oerstedii citrinellus, 14:105
Saimiri oerstedii oerstedii, 14:105
Saimiri sciureus. See Common squirrel monkeys
Saimiri ustus, 14:101, 14:102
Saimiri vanzolinii. See Blackish squirrel monkeys
Saint Andrew vireos, 11:256
Saint Bernards, 14:293
St. Croix ground lizards, 7:312
St. Helena dragonets, 5:366–367
St. Helena earwigs, 3:197, 3:*198*, 3:*199*, 3:200
St. Helena plovers, 9:163–164
St. Helena waxbills. *See* Common waxbills
St. Lucia whiptails, 7:312
St. Michel nesophontes, 13:243
Saint Patrick, 7:54
St. Thomas conures. *See* Brown-throated parakeets
St. Vincent Amazons, 9:276
Saiphos equalis, 7:329
Saithes. *See* Pollocks
Sakalava weavers, 11:*381*, 11:*386*–387
Saker falcons, 8:21, 8:315
Sakis, **14:143–154,** 14:*149*
behavior, 14:145–146
conservation status, 14:148
distribution, 14:144–145
evolution, 14:143
feeding ecology, 14:146–147
habitats, 14:145
humans and, 14:148
physical characteristics, 14:143–144
reproduction, 14:147
species of, 14:*150*–151, 14:153*t*
taxonomy, 14:143
Salamanderfishes, 4:57, 4:390–393, 4:*394*, 4:*396*, 4:397–398
Salamanders, 6:3, **6:323–326,** 6:*344*
aquatic, 6:15, 6:25, 6:325
Asiatic giant, **6:343–347,** 6:*344*
behavior, 6:45, 6:325
black-spotted, 6:372
chemosensory cues, 6:24, 6:44–45
communication, 6:44–45
courtship behavior, 6:44–45, 6:49, 6:365–366
distribution, 6:5, 6:325
European, 6:**363–375,** 6:*368*–*369*
evolution, 6:4, 6:9–10, 6:11, 6:13, 6:15, 6:323–324
feeding ecology, 6:6, 6:24, 6:65, 6:326
as food, 6:54
hybridization, 6:44
larvae, 6:33, 6:39, 6:42
medicinal uses of, 6:53
mythology and, 6:51–52
physical characteristics, 6:15, 6:*18*, 6:23–*25*, 6:*24*, 6:323–324
predators and, 6:325

reproduction, 6:28–*29*, 6:*31*–35, 6:*37*–38, 6:325, 6:326
taxonomy, 6:3, 6:4, 6:323–*324*
torrent, 6:3, 6:5, 6:323, 6:325, 6:349, **6:385–388,** 6:*387*
toxins, 6:325
See also specific types of salamanders
Salamandra atra. See Alpine salamanders
Salamandra japonica. See Japanese clawed salamanders
Salamandra lanzai. See Laza's alpine salamanders
Salamandra salamandra. See European fire salamanders
Salamandra tigrina. See Tiger salamanders
Salamandrella spp., 6:335
Salamandrella keyserlingii. See Siberian salamanders
Salamandridae, **6:363–375,** 6:*368–369*
behavior, 6:*365*–366, 6:370–375
communication, 6:44
conservation status, 6:367, 6:370–375
courtship behavior, 6:49, 6:365–366
distribution, 6:*363*, 6:364, 6:*370–375*
evolution, 6:13, 6:363
feeding ecology, 6:366–367, 6:370–375
habitats, 6:364–365, 6:370–375
humans and, 6:367, 6:370–375
physical characteristics, 6:17, 6:24, 6:325, 6:363, 6:370–375
reproduction, 6:30–31, 6:326, 6:367, 6:370–375
species of, 6:370–375
taxonomy, 6:323, 6:363, 6:370–375
See also Newts
Salamandrina spp., 6:17, 6:24
Salamandrina terdigitata. See Spectacled salamanders
Salamandroidea, 6:13, 6:323
Salanoia spp., 14:347
Salanoia concolor. See Malagasy brown mongooses
Salaria fluviatilis. See Freshwater blennies
Salarias fasciatus. See Jeweled blennies
Saldidae, 3:54
Saldula coxalis. See Shore bugs
Salidae. *See* Boobies; Gannets
Salientia, 6:11
Salim Ali's fruit bats, 13:349*t*
Salinella salve, 1:19, 1:20–21, **1:91–92**
Salinity, 1:25, 2:26, 4:37, 4:50, 4:52
See also Habitats
Salinoctomys spp., 16:434, 16:436
Salinoctomys loschalchalerosorum, 16:433–434
Saliva, 12:122
Salivary glands, 7:9–10
See also Physical characteristics
Salmo alpinus. See Charrs
Salmo altivelis. See Ayu
Salmo arcticus. See Arctic graylings
Salmo clupeaformis. See Lake whitefishes
Salmo fontinalis. See Brook trouts
Salmo gorbuscha. See Pink salmons
Salmo kisatch. See Coho salmons
Salmo masou. See Cherry salmons
Salmo mykiss. See Rainbow trouts
Salmo mykiss aguabonita. See Golden trouts
Salmo namaycush. See Lake trouts
Salmo nerka. See Sockeye salmons
Salmo pacificus. See Eulachons
Salmo salar. See Atlantic salmons
Salmo silus. See Greater argentines
Salmo trutta. See Brown trouts
Salmo tshawytscha. See Chinook salmons
Salmon lice, 2:*303*, 2:309–310
Salmon-poisoning flukes, 1:201, 1:*202*, 1:*206*, 1:209–210
Salmon rainbowfishes, 5:*67*
Salmon sharks, 4:132, 4:133
Salmon trouts. *See* Cherry salmons; Lake trouts
Salmonella, pigeons and, 9:256
Salmonflies, giant, 3:*144*, 3:*145*
Salmonidae, 4:50, 4:52, 4:58, 4:405
Salmoniformes. *See* Salmons
Salmons, **4:405–420,** 4:*406*, 4:*408–409*
behavior, 4:*63*, 4:406–407
conservation, 4:407
distribution, 4:52, 4:58, 4:405, 4:*406*
evolution, 4:405
feeding ecology, 4:68, 4:407
habitats, 4:50, 4:405–406
humans and, 4:407
physical characteristics, 4:60, 4:405
reproduction, 4:35, 4:47, 4:*407*
species of, 4:410–419
taxonomy, 4:405
Salpa fusiformis, 1:*469*, 1:472–473
Salpida, 1:467, 1:468
Salpinctes spp., 10:526
Salpinctes mexicanus. See Canyon wrens
Salpingotus spp., 16:217
Salpingotus crassicauda. See Thick-tailed pygmy jerboas
Salpingotus heptneri, 16:212, 16:216
Salpingotus kozlovi, 16:212, 16:215, 16:216
Salpingotus michaelis, 16:217
Salpingotus pallidus, 16:216
Salpingotus thomasi, 16:217
Salps, **1:*467*–472,** 1:*469*
Salt-and-pepper shrimps. *See* Sevenspine bay shrimps
Salt-desert cavies, 16:389, 16:390–391, 16:393, 16:*394*, 16:*397*, 16:398
Salt glands, 7:30
See also Physical characteristics
Salt licks, 12:17–18
Salta tuco-tucos, 16:431*t*
Saltatory mammals, 12:42–43
Saltenia spp., 6:12
Saltenia ibanezi, 6:99
Salticidae, 2:37
Salticus scenicus. See Zebra spiders
Saltmarsh harvest mice, 16:*265*
Salt's dikdiks, 16:*65*, 16:*67*, 16:69–70
Saltwater crocodiles, 7:157–158, 7:161–163, 7:*183*, 7:*185*, 7:186–187
conservation status, 7:181–182
farming, 7:48
as food, 7:48
humans and, 7:182
muscular limbs of, 7:23
physical characteristics, 7:179
Saltwater fishes. *See* Marine fishes
Salukis, 14:292
Salvadori, T., 9:275
Salvadorina waigiuensis. See Salvadori's teals
Salvadori's ducks. *See* Salvadori's teals
Salvadori's eremomelas. *See* Yellow-bellied eremomelas
Salvadori's nightjars, 9:405
Salvadori's serins, 11:325
Salvadori's teals, 8:365, 8:*374*, 8:*383*
Salvelinus agassizi, 4:407
Salvelinus alpinus. See Charrs
Salvelinus fontinalis. See Brook trouts
Salvelinus namaycush. See Lake trouts
Salvin's curassows, 8:*416*
Salvin's mollymawks, 8:114
Salvin's spiny pocket mice, 16:*204*, 16:*205*–206
Samana hutias, 16:467*t*
Samar broadbills. *See* Visayan wattled broadbills
Samaridae, 5:450
Sambars, 12:19, 15:*358*, 15:*362*, 15:*367*
Sambava tomato frogs, 6:*304*
Sambonia spp., 2:318
Sambonidae, 2:318
Samoan crabs. *See* Mangrove crabs
Samoan flying foxes, 13:332*t*
Samoan moorhens, 9:47
Samoan sand darts, 5:*379*, 5:*385*, 5:387
Samotherium spp., 15:265
Samut Prakan Crocodile Farm (Thailand), 7:182
San Andres mockingbirds, 10:468
San Benito sparrows. *See* Savannah sparrows
San Cristobal moorhens, 9:47
San Diego banded geckos, 7:265
San Diego horned lizards. *See* Texas horned lizards
San Francisco garter snakes, 7:479
San Joaquin antelope squirrels. *See* Nelson's antelope squirrels
San Joaquin pocket mice, 16:*204*, 16:*205*
San Joaquin Valley kangaroo rats, 16:209*t*
San Marcos salamanders. *See* Texas blind salamanders
Sanaga pygmy herrings, 4:277
Sanborn's flying foxes, 13:331*t*
Sanchez-Cordero, V., 16:200
Sand cats, 14:371, 14:390*t*
Sand-colored nighthawks, 9:*404*
A Sand County Almanac (Leopold), 12:223
Sand crabs, 2:198
Sand dabs. *See* Windowpane flounders
Sand darts. *See* Samoan sand darts
Sand dollars, **1:401–*406*,** 1:*403*, 1:*405*, **1:412**
Sand eels, 5:131, 5:133
See also Inshore sand lances; Sandfishes
Sand fiddler crabs, 2:*202*, 2:*210*
Sand fleas. *See* Chigoes
Sand flies, 3:361
Sand gazelles. *See* Slender-horned gazelles
Sand gobies, 5:373–375
Sand hoppers. *See* Beach hoppers
Sand isopods, 2:*253*, 2:*255*, 2:*256*
Sand lances, 5:331–335
Sand lizardfishes. *See* Clearfin lizardfishes
Sand lizards, 7:298, 7:*300*, 7:*301–302*
Sand martins, 10:*362*, 10:*366–367*
Sand masons, 2:*48*, 2:*50*, 2:56
Sand monitors, 7:362
Sand nests, cichlids and, 5:279
See also Reproduction
Sand rats. *See* Sundevall's jirds
Sand rollers, 5:7, 5:*11*, 5:12
Sand sharks, 4:7
Sand shrimps. *See* Sevenspine bay shrimps

Sand stargazers, 5:341–342, 5:*344*, 5:*347*–348
Sand stars, 1:369, 1:*371*, 1:*376*–377
Sand swimmers. *See* Half-girdled snakes
Sand-swimming, 12:71–72
  See also Behavior
Sand tiger sharks, 4:11, 4:131–133, 4:*134*, 4:*136*, 4:140–141
Sand worms, **2:45–57**, 2:*48*, 2:*49*
  behavior, 2:46
  conservation status, 2:47
  distribution, 2:46
  evolution, 2:45
  feeding ecology, 2:46–47
  habitats, 2:46
  humans and, 2:47
  physical characteristics, 2:45–46
  reproduction, 2:47
  species of, 2:*50*–*56*
  taxonomy, 2:45
Sandboas, 7:409–411
Sandburrowers, 5:331–335
Sanddivers, 4:48, 4:70, 5:331–335, 5:*334*
Sandelia spp., 5:427, 5:428
Sandelia bainsii, 5:429
Sandelia capensis, 5:428
Sanderlings, 9:176
Sanderson's hook frogs, 6:247, 6:*254*, 6:260
Sandfishes, 4:*292*, 4:*293*–294, 5:331–335, 7:*330*, 7:*332*, 7:337
Sandfleas. *See* Beach hoppers
Sandgropers, 3:201, 3:202, 3:203
Sandgrouse, **9:231–239**, 9:*234*
Sandhill cranes, 9:23–27, 9:*28*, 9:*30*, 9:*33*
Sandhill dunnarts, 12:301*t*
Sandhill frogs, 6:*148*, 6:*149*, 6:*150*, 6:*151*
Sandperches, 4:48, 4:66, 5:255, 5:*331*, 5:331–335
Sandpipers, 9:102–104, **9:175–188**, 9:*181*
  behavior, 9:178
  conservation status, 9:103, 9:179–180
  distribution, 9:*175*, 9:176–177
  evolution, 9:175
  feeding ecology, 9:178–179
  habitats, 9:175
  humans and, 9:180
  physical characteristics, 9:175–176
  reproduction, 9:179
  species of, 9:*182*–*188*
  taxonomy, 9:104, 9:175
Sandplovers, 9:161
Sandsnakes, 7:466
Sandstone dibblers, 12:300*t*
Sandstone shrike-thrushes, 11:116, 11:117–118
Sandy wallabies. *See* Agile wallabies
Sanford's bowerbirds. *See* Archbold's bowerbirds
Sanford's sea-eagles, 8:323, 8:324
Sangihe shrike-thrushes, 11:118
Sangihe tarsiers, 14:95, 14:*97*, 14:99
Sanguinicolidae, 1:199
Sanguivorous leeches. *See* Blood-feeding leeches
Sanmartín, I., 3:55
Sanopus spp., 5:41, 5:42
Sanopus astrifer. *See* Whitespotted toadfishes
Sanopus greenfieldorum. *See* Whitelined toadfishes
Sanopus reticulatus. *See* Reticulated toadfishes
Sanopus splendidus. *See* Splendid coral toadfishes
Sansom, I. J., 4:9
Santa Marta bush tyrants, 10:275
Santo Domingo trogons. *See* Hispaniolan trogons
São Tomé fiscals, 10:427, 10:429
São Tomé grosbeaks, 11:324
São Tomé orioles, 11:428
São Tomé sunbirds, 11:208, 11:210, 11:*211*, 11:*217*
Saolas, 12:129, 16:4, 16:11
Sap beetles, 3:318
Sapheopipo noguchii. *See* Okinawa woodpeckers
Sapphire-bellied hummingbirds, 9:449
Sapphire-vented pufflegs, 9:447
Sappho spp. *See* Hummingbirds
Saproxylic Invertebrates Project, 3:322
Sarcomastigophora, 2:11
Sarcophagidae. *See* Flesh flies
Sarcophilus laniarius. *See* Tasmanian devils
Sarcopterygians, 6:7
Sarcopterygii, 4:10, 4:12
Sarcoramphus papa. *See* King vultures
Sarcosuchus spp., 7:157, 7:158
Sardina pilchardus. *See* European pilchards
Sardinas. *See* South American pilchards
Sardines, 4:34, 4:75, 4:277
  See also Atlantic herrings; Dagaa; European pilchards; South American pilchards
Sardinian pikas, 16:502*t*
Sardinops melanostictus. *See* Japanese pilchards
Sardinops sagax. *See* South American pilchards
Sargassumfishes, 4:62, 4:69, 5:*50*, 5:51, 5:*52*, 5:*53*–*54*
Sargocentron spp., 5:113
Sargocentron vexillarium. *See* Squirrelfishes
Sargocentron xantherythrum. *See* Hawaiian squirrelfishes
Sarkidiornis melanotos. *See* Comb ducks
Saroglossa spiloptera. *See* Spot-winged starlings
Sarotherodon spp., 5:279
Sarothrura spp. *See* Flufftails
Sarothrura elegans. *See* Buff-spotted flufftails
Sarothruridae, 9:45
Sarus cranes, 9:1, 9:3, 9:25, 9:*30*, 9:33–*34*
Sasia abnormis. *See* Rufous piculets
Sasins. *See* Blackbucks
Sassabies, 16:27, 16:32
Satellite radio-tracking, 8:27
Satin birds. *See* Satin bowerbirds
Satin bowerbirds, 11:477, 11:*479*, 11:*481*, 11:*482*, 11:*487*
Satin grackles. *See* Satin bowerbirds
Satunin's five-toed pgymy jerboas. *See* Five-toed pygmy jerboas
Saturniidae. *See* Atlas moths
Satyr tragopans, 8:*438*, 8:*449*
Saucer water bugs. *See* Creeping water bugs
Saucerottia spp. *See* Hummingbirds
Saugers, 5:198
Saunders embiids, 3:*235*, 3:*236*, 3:237
Saunder's gulls, 9:209, 9:*210*, 9:*212*–*213*
Sauresia spp., 7:339
Saurida gracilis. *See* Slender lizardfishes
Sauries, 4:12, 5:79, 5:80, 5:81, 5:*82*, 5:*84*
Saurischia, 7:5
Sauromalus obesus. *See* Common chuckwallas
Sauromys spp., 13:311
Sauropterygia, 7:5
Saurothera californiana. *See* Greater roadrunners
Savage-Rumbaugh, Sue, 12:162
Savanna baboons, 14:*193*
Savanna dormice, 16:*320*, 16:327*t*
Savanna elephants, 15:162–163, 15:165
Savanna nightjars, 9:403
Savanna squeaking frogs. *See* Common squeakers
Savanna vultures. *See* Lesser yellow-headed vultures
Savannah sparrows, 10:*171*, 10:172, 11:*267*, 11:*282*–283
Savannas
  Ice Age and, 12:18–19
  tropical, 12:219
  See also Habitats
Savigny, J. C., 1:451
Savigny's brittle stars. *See* Tropical brittle stars
Savigny's eagle-owls, 9:347
Savi's pygmy shrews, 12:12, 13:194, 13:266, 13:*267*, 13:*270*, 13:271–*272*
Savolainen, Peter, 14:287
Saw-billed hermits, 9:*454*, 9:*455*
Saw-billed kingfishers. *See* Yellow-billed kingfishers
Saw-jawed turtles. *See* Painted terrapins
Saw-scaled vipers, 7:447, 7:448, 7:*449*, 7:*451*, 7:458
Saw-whet owls. *See* Northern saw-whet owls
Sawbellies. *See* Alewives
Sawbelly herrings, 4:277
Sawbwa barbs, 4:298
Sawbwa resplendens. *See* Sawbwa barbs
Sawfishes, 4:11, 4:167, 4:173–179, 4:186
Sawflies, 3:28, 3:37, 3:63, 3:75, **3:*405*–408, 3:422–423**
Sawsharks, 4:11, **4:167–171**, 4:*169*
Sawtoothed grain beetles, 3:76
Saxicola longirostris. *See* White-browed scrubwrens
Saxicola rubetra. *See* Whinchats
Saxicola torquata. *See* Stonechats
Saxipendium coronatum. *See* Spaghetti worms
Sayornis spp. *See* Phoebes
Sayornis nigricans. *See* Black phoebes
Sayornis phoebe. *See* Eastern phoebes
Sayornis saya. *See* Say's phoebes
Say's phoebes, 10:*276*, 10:*284*
Scads, 4:67
Scala naturae, 12:149
Scale crickets, 3:201
Scale insects, 3:81, **3:*259*–264**
  ants and, 3:49
  biological control and, 3:80
  defense mechanisms, 3:66
  distribution, 3:57
  excretion, 3:21
  reproduction, 3:59, 3:62, 3:263
  sexual dimorphism in, 3:59
Scale-throated earthcreepers, 10:*212*, 10:*215*
Scaled antbirds, 10:*244*, 10:*250*–251
Scaled ground-rollers. *See* Scaly ground-rollers
Scaled lings. *See* Bowfins
Scaleless dragonfishes, 4:*425*, 4:*428*–429
Scales, 4:*6*, 4:15–16
  See also Physical characteristics
Scalloped hammerhead sharks, 4:*115*
Scallops, 2:454
  See also Bivalves; Queen scallops

Scalopinae, 13:279
Scalopus aquaticus. See Eastern moles
Scaly anteaters. *See* Ground pangolins; Pangolins
Scaly-breasted finches. *See* Spotted munias
Scaly-breasted mannikins. *See* Spotted munias
Scaly-breasted munias. *See* Spotted munias
Scaly brittle stars. *See* Dwarf brittle stars
Scaly-fronted wormsnakes. *See* Peters' wormsnakes
Scaly ground-rollers, 10:*54*, 10:56–57
Scaly-tailed possums, 13:57, 13:59, 13:60, 13:*62*, 13:*65*–66
Scaly-tailed squirrels, 16:122, 16:123, **16:299–306,** 16:*301*
Scaly-throated honeyguides, 10:137, 10:*138*, 10:*140*, 10:*143*
Scaly thrushes. *See* White's thrushes
Scalyhead sculpins, 5:183
Scammon, Charles, 15:96
Scandentia. *See* Tree shrews
Scansorial adaptations, 12:14
Scapanorhynchus lewisi, 4:131
Scapanus latimanus. See Broad-footed moles
Scapherpetontids, 6:13
Scaphiodontophis spp. *See* Neck-banded snakes
Scaphiophryne spp., 6:317, 6:318
Scaphiophryne brevis, 6:318
Scaphiophryne calcarata. See Mocquard's rain frogs
Scaphiophryne gottlebei. See Red rain frogs
Scaphiophryne madagascariensis, 6:318
Scaphiophryne marmorata, 6:317, 6:318
Scaphiophrynidae. *See* Madagascan toadlets
Scaphiopus spp., 6:119
Scaphiopus couchii. See Couch's spadefoot toads
Scaphiopus holbrookii. See Eastern spadefoot toads
Scaphirhynchus spp., 4:213
Scaphistostreptus seychellarum, 2:363
Scaphopoda. *See* Tusk shells
Scaponulus oweni. See Gansu moles
Scapteromys spp., 16:268
Scapteromys aquaticus, 16:268
Scapteromys tumidus, 16:266, 16:268
Scaptochirus spp., 13:279
Scaptochirus moschatus. See Short-faced moles
Scaptonyx fuscicaudatus. See Long-tailed moles
Scapula, 12:41
Scarab beetles. *See* Scarabs
Scarabaeid grubs, 3:56
Scarabaeidae, 3:53, 3:322
Scarabaeus sacer. See Sacred scarabs
Scarabs, 3:42, 3:62, 3:80–81, 3:316, 3:320–323
See also Sacred scarabs
Scardefella inca. See Inca doves
Scardinius erythrophthalmus. See Rudds
Scaridae. *See* Parrotfishes
Scaridium spp., 1:261
Scarlet-backed flowerpeckers, 11:190, 11:*192*, 11:197–*198*
Scarlet-breasted flowerpeckers, 11:191, 11:*192*, 11:*193*
Scarlet characins. *See* Cardinal tetras
Scarlet-chested sunbirds, 11:208–210, 11:*212*, 11:*219*–220
Scarlet-collared flowerpeckers, 11:191
Scarlet-crowned barbets, 10:115
Scarlet-fronted parrots, 9:278
Scarlet-headed blackbirds, 11:293
Scarlet-hooded barbets, 10:116
Scarlet-horned manakins, 10:*298*, 10:*301*–302
Scarlet ibises, 8:291, 8:*292*, 8:*295*, 8:*299–300*
Scarlet kingsnakes. *See* Milksnakes
Scarlet macaws, 9:*278*, 9:*282*, 9:*289*–290
Scarlet robins, 11:105, 11:*107*, 11:*108*–109
Scarlet-rumped trogons, 9:479
Scarlet-throated sunbirds. *See* Scarlet-chested sunbirds
Scarlet-tufted sunbirds, 11:209, 11:*211*, 11:*213*–214
Scarletsnakes, 7:469
Scarthyla goinorum. See Amazonian skittering frogs
Scartichthys spp., 5:343
Scarus frenatus. See Bridled parrotfishes
Scarus guacamaia. See Parrotfishes
Scarus iseri. See Striped parrotfishes
Scatophagidae. *See* Scats
Scatophagus argus. See Spotted scats
Scatophagus tetracanthus, 5:393
Scats, 5:391, 5:392–393
Scaturiginichthys vermeilipinnis, 5:70
Scavenger vultures. *See* Egyptian vultures
Scavenging, 2:29
See also Feeding ecology
Sceliphron caementarium. See Mud dauber wasps
Sceloporus spp. *See* Spiny lizards
Sceloporus jarrovii. See Mountain spiny lizards
Scelorchilus spp. *See* Tapaculos
Scelorchilus albicollis, 10:260
Scelorchilus rubecula. See Chucao tapaculos
Scelotes spp., 7:328
Scenella spp., 2:387
Scenopoeetes spp., 11:477, 11:478, 11:480
Scenopoeetes dentirostris. See Tooth-billed bowerbirds
Scent glands, 12:37, 12:81
Scent-marking, 12:80–81, 14:*123*
See also Behavior
Scenthounds, 14:*292*
Schach shrikes. *See* Long-tailed shrikes
Schad roaches. *See* Oriental cockroaches
Schaefer, Carl W., 3:259
Schäfer, Ernst, 9:248
Schaller's mouse shrews, 13:276*t*
Schalow's turacos, 9:300, 9:302
Schaus's swallowtail butterflies, 3:56
Schelgel's asites, 10:188
Schetba rufa. See Rufous vangas
Schilbeidae, 4:352
Schindleria praematura, 5:374
Schindleriidae, 5:373, 5:374
Schismaderma spp., 6:184
Schistocerca gregaria. See Desert locusts; Locusts
Schistolais spp., 11:4, 11:6
Schistometopum thomense, 6:411
Schistosoma spp., 1:22, 1:35, 1:200, 1:201
Schistosoma haematobium, 1:35
Schistosoma japonicum, 1:35
Schistosoma mansoni. See Human blood flukes
Schistosomatidae, 1:18, 1:199, 1:200
Schistosomiasis, 1:34, 1:35
Schistura oedipus. See Cavefishes of Nam Lang
Schistura sijuensis. See Siju blind cavefishes
Schizochoeridae, 1:225
Schizocoely, 2:3, 2:21, 2:35
Schizoeaca griseomurina. See Mouse-colored thistletails
Schizomida, 2:333
Schizophora, 3:357
Schizothoracinae, 4:297, 4:298
Schizothorax prenanti, 4:*306*, 4:*308*, 4:318
Schlegel's beaked snakes. *See* Schlegel's blindsnakes
Schlegel's blindsnakes, 7:379–380, 7:382, 7:*383*
Schlegel's mynas. *See* Bare-eyed mynas
Schlegel's whistlers. *See* Regent whistlers
Schmidly, D. J., 16:200
Schmidt, K., 3:117, 7:475
Schmidt's lazy toads, 6:*112*, 6:114
Schnabeli's thumbless bats, 13:*469*, 13:*470*–471
Schneider, Gustav, 10:72
Schneider's pittas, 10:195
Schneider's smooth-fronted caimans. *See* Smooth-fronted caimans
Schodde, Richard, 10:329, 11:83, 11:459
Schomburgk, Robert, 10:318
Schomburgk's deer, 15:360, 15:371*t*
Schonbrunn Zoo, Vienna, Austria, 12:203
Schools of fish, 4:63–64, 4:277–*278*
See also Behavior
Schoutedenella poecilonotus. See Ugandan squeakers
Schouteden's African squeaker catfishes, 4:*354*
Schouteden's shrews, 13:*269*, 13:*272*, 13:274
Schram, F. R., 2:161
Schreckstoff spp., 4:335
Schreiber's bent-winged bats. *See* Common bentwing bats
Schreiber's long-fingered bats. *See* Common bentwing bats
Schremmer's snakeflies, 3:*299*, 3:*300*, 3:301
Schuchmann, K-L., 9:415, 9:437, 9:444
Schuettea spp., 5:237
Schufeldt's juncos. *See* Dark-eyed juncos
Schultz, R. Jack, 5:93
Schummel's inocelliid snakeflies, 3:*299*, 3:*300*
Sciaenidae, 5:255, 5:258–263
Sciaenops ocellatus. See Red drums
Sciara spp., 3:360
Sciara militaris. See Armyworms
Scientific research, in zoos, 12:204
Scimitar babblers, 10:506, 10:*510*
Scimitar-billed woodcreepers, 10:*231*, 10:*234*
Scimitar-billed woodhoopoes. *See* Common scimitarbills
Scimitar-horned oryx, 16:27–28, 16:29, 16:31, 16:32, 16:33, 16:41*t*
Scimitarbills. *See* Common scimitarbills
Scina borealis, 2:*263*, 2:*268*, 2:271
Scinax spp., 6:227
Scinax rizibilis, 6:230
Scincella spp., 7:328, 7:329
Scincidae. *See* Skinks
Scincinae, 7:327
Scincomorpha, 7:196, 7:198
Scincus scincus. See Sandfishes
Scissor-tailed flycatchers, 10:270, 10:272, 10:273, 10:*276*, 10:*278*–279
Scissurelloidea, 2:430
Sciuridae, 12:79, 16:121–122, 16:123, 16:125, 16:126
Sciurillus spp., 16:163
Sciurognathi, 12:71, 16:121, 16:125

Sciurotamias spp., 16:143, 16:144
Sciurotamias forresti, 16:147
Sciurus spp., 16:163, 16:300
Sciurus aberti. *See* Abert squirrels
Sciurus alleni. *See* Allen's squirrels
Sciurus anomalus. *See* Persian squirrels
Sciurus arizonensis. *See* Arizona gray squirrels
Sciurus carolinensis. *See* Gray squirrels
Sciurus deppei. *See* Deppe's squirrels
Sciurus dubius, 16:163
Sciurus griseus. *See* Western gray squirrels
Sciurus nayaritensis. *See* Mexican fox squirrels
Sciurus niger. *See* Eastern fox squirrels
Sciurus niger avicennia. *See* Big cypress fox squirrels
Sciurus niger cinereus. *See* Delmarva fox squirrels
Sciurus oculatus. *See* Peters's squirrels
Sciurus richmondi, 16:167
Sciurus sanborni, 16:167
Sciurus variegatoides. *See* Variegated squirrels
Sciurus vulgaris. *See* Red squirrels
Sciurus yucatanensis. *See* Yucatán squirrels
Sclater's forest falcons. *See* Plumbeous forest falcons
Sclater's golden moles, 13:198, 13:200
Sclater's larks, 10:346
Sclater's lemurs. *See* Black lemurs
Scleractinia. *See* True corals
Scleroglossa, 7:195–196, 7:197–198, 7:200, 7:202–204, 7:207
Sclerognathus cyprinella. *See* Bigmouth buffaloes
Sclerolinum brattstromi, 2:86
Scleropages spp., 4:233
Scleropages formosus, 4:232, 4:234
Scleropages jardini, 4:232, 4:234
Scleropages leichardti. *See* Australian spotted barramundis
Scleroperalia, 1:331, 1:332
Sclerosponges, 1:79
Sclerotization, 3:17, 3:32–33
Sclerurus rufigularis. *See* Short-billed leaftossers
Scolecomorphidae. *See* Buried-eye caecilians
Scolecomorphus spp., 6:431, 6:432, 6:437
Scolecomorphus kirkii. *See* Kirk's caecilians
Scolecomorphus vittatus. *See* Banded caecilians
Scolecophidia. *See* Blindsnakes
Scolecoseps spp., 7:328
Scolia dubia. *See* Digger wasps
Scoliidae, 3:406
Scoliodon laticaudus, 4:116
Scolomys spp., 16:265
Scolopacidae. *See* Sandpipers
Scolopacids. *See* Sandpipers
Scolopacinae, 9:175–180
Scolopax celebensis. *See* Sulawesi woodcocks
Scolopax erythropus. *See* Spotted redshanks
Scolopax minor. *See* American woodcocks
Scolopax mira. *See* Amami woodcocks
Scolopax rochussenii. *See* Moluccan woodcocks
Scolopax rusticola. *See* Eurasian woodcocks
Scolopax saturata. *See* Rufous woodcocks
Scolopenders, 2:*357*, 2:*360*, 2:361–362
Scolopendra spp., 2:*355*
Scolopendra abnormis, 2:356
Scolopendra morsitans. *See* Scolopenders
Scolopendrellidae, 2:*371*
Scolopendridae, 2:355
Scolopendromorpha, 2:353, 2:354, 2:356
Scoloplacidae, 4:352
Scomber scombrus. *See* Atlantic mackerels
Scomberesocidae. *See* Sauries
Scomberesox saurus saurus. *See* Atlantic sauries
Scomberoides spp., 5:260
Scomberoides lysan. *See* Common carps
Scomberomorus cavalla. *See* King mackerels
Scomberomorus concolor. *See* Monterrey Spanish mackerels
Scombridae, 4:68, 5:405
Scombroidei, **5:405–419,** 5:*408*
behavior, 5:406
conservation status, 5:407
distribution, 5:406
evolution, 5:405
feeding ecology, 5:406
habitats, 5:406
humans and, 5:407
physical characteristics, 5:405–406
reproduction, 5:406
species of, 5:*409*–418
taxonomy, 5:405
Scoop (Waugh), 12:183
Scopelarchidae. *See* Pearleyes
Scopelengys tristis, 4:442
Scopeloberyx spp., 5:106
Scopeloberyx robustus. *See* Longjaw bigscales
Scopelogadus spp., 5:106
Scopelus antarcticus. *See Electrona antarctica*
Scopelus pterotus. *See* Skinnycheek lanternfishes
Scophthalmidae, 5:450, 5:453
Scophthalmus aquosus. *See* Windowpane flounders
Scophthalmus maximus. *See* Turbot fishes
Scopidae. *See* Hammerheads
Scops-owls, 9:332–333, 9:346, 9:349–350, 9:*352*, 9:354–355
Scopuridae, 3:54, 3:141
Scopus umbretta. *See* Hammerheads
Scorpaena guttata. *See* California scorpionfishes
Scorpaena mystes. *See* Pacific spotted scorpionfishes
Scorpaena plumieri. *See* Spotted scorpionfishes
Scorpaenichthys spp., 5:183
Scorpaenichthys marmoratus. *See* Cabezons
Scorpaenidae. *See* Scorpionfishes
Scorpaeniformes, **5:157–194,** 5:*160*, 5:*167*, 5:*168*, 5:*184*, 5:*185*
behavior, 5:158, 5:165, 5:181–182
conservation status, 5:159, 5:166, 5:183
distribution, 5:158, 5:164, 5:180
evolution, 5:157, 5:163–164, 5:179
feeding ecology, 5:158, 5:165, 5:182
habitats, 5:158, 5:164–165, 5:180–181
humans and, 5:159, 5:166, 5:183
physical characteristics, 5:157–158, 5:164, 5:180
reproduction, 5:158, 5:165–166, 5:182–183
species of, 5:161–162, 5:*169*–178, 5:*186*–193
taxonomy, 5:157, 5:163–164, 5:179
Scorpaenoidei, **5:163–178,** 5:*167*, 5:*168*
behavior, 5:165
conservation status, 5:166
distribution, 5:164
evolution, 5:163–164
feeding ecology, 5:165
habitats, 5:164–165
humans and, 5:166
physical characteristics, 5:164
reproduction, 5:165–166
species of, 5:*169*–178
taxonomy, 5:163–164
Scorpaenoideorum prominens, 5:164
Scorpidinae, 5:241
Scorpio spp. *See* Scorpions
Scorpion. *See* Broad-headed skinks
Scorpionfishes, 4:15, 4:30, 4:42, 4:48, 4:68, 4:70, **5:163–177,** 5:*167*, 5:*168*
Scorpionflies, 3:34, 3:60, 3:61, **3:*341–346*,** 3:*343*
Scorpions, 2:25, **2:333–352,** 2:*337*
behavior, 2:335
conservation status, 2:336
distribution, 2:335
evolution, 2:333
feeding ecology, 2:335
habitats, 2:335
humans and, 2:336
physical characteristics, 2:333–334
reproduction, 2:*20*, 2:335
species of, 2:345–352
taxonomy, 2:333
Scortum spp., 5:220
Scoters, 8:365, 8:371, 8:372
Scotiophryne spp., 6:89
Scotomanes ornatus. *See* Harlequin bats
Scotonycteris zenkeri. *See* Zenker's fruit bats
Scotopelia. *See* Fishing-owls
Scotophilus spp., 13:311, 13:497
Scotophilus kuhlii. *See* House bats
Scotoplanes globosa. *See* Sea pigs
Scottish crossbills, 11:325
Scott's orioles, 11:*302*
Scramble competition, 6:48–49
Scrawled filefishes, 5:469, 5:*472*, 5:*476*, 5:478
Screamers, **8:393–397**
Screaming cowbirds, 11:304, 11:320
Screech owls. *See* Eastern screech owls
Screwworms, 3:57, 3:76, 3:80, 3:*362*, 3:*365*
Scrotum, 12:103
See also Physical characteristics
Scrub-birds, 10:170, 10:329, **10:337–340,** 11:477
Scrub blackbirds. *See* Melodious blackbirds
Scrub-jays, 8:25, 11:504, 11:506, 11:507, 11:508
Scrub robins, 11:106
Scrub turkeys. *See* Australian brush-turkeys
Scrub wallabies. *See* Black-striped wallabies
Scrubtits, 11:57
Scrubwrens, 11:56, 11:57, 11:*58*, 11:59–60, 11:70
Scuds, 2:*262*
Sculpins, 4:47, 4:68, **5:179–193,** 5:*185*
See also California scorpionfishes
Sculptured seamoths, 5:*138*, 5:*144*–145
Scurria spp., 2:*423*
Scutellerid bugs, terrestrial, 3:260
Scutiger spp., 6:110, 6:111
Scutiger nyingchiensis. *See* Nyingchi lazy toads
Scutiger schmidti. *See* Schmidt's lazy toads
Scutigera coleoptrata. *See* House centipedes
Scutigerella spp., 2:371
Scutigerella immaculata. *See* Garden symphylans
Scutigerellidae, 2:*371*
Scutigeromorpha, 2:353, 2:355, 2:356
Scutisorex spp., 13:265
Scutisorex somereni. *See* Armored shrews
Scuttles. *See* Common octopods

Scyliorhinidae, 4:113
Scyliorhinus spp., 4:113
Scyliorhinus canicula. See Lesser spotted dogfishes
Scyliorhinus retifer. See Chain catsharks
Scylla serrata. See Mangrove crabs
Scyllium retiferum. See Chain catsharks
Scyllium ventriosum. See Swellsharks
Scymnus brasiliensis. See Cookie-cutter sharks
Scyphophorus yuccae, 3:63
Scyphozoa. *See* Jellyfish
Scytalinidae. *See* Graveldivers
Scytalopus spp. *See* Tapaculos
Scytalopus indigoticus, 10:259
Scytalopus iraiensis, 10:259
Scytalopus novacapitalis, 10:259
Scytalopus panamensis, 10:259
Scytalopus psychopompus, 10:259
Scytalopus robbinsi, 10:259
Scythrodes spp., 6:156
Scythrops novaehollandiae. See Channel-billed cuckoos
Sea anemones, 1:103, 1:*107*, 2:*34*, 2:36
  conservation status, 1:53
  damselfishes and, 5:294–295
  feeding ecology, 1:27–28, 1:30
  magnificent, 1:*110*, 1:113
  starlet, 1:107, 1:*110*, 1:111–*112*
  symbiosis, 1:33
Sea apples, 1:*418*, 1:*424*, 1:*428*–429
Sea Atlantic herrings. *See* Atlantic herrings
Sea basses, 5:255–256, **5:255–263,** 5:258–263
  European, 4:*31*, 5:198
  feeding ecology, 4:44, 4:68, 5:260–261
  reproduction, 4:25, 4:29, 4:*31*, 4:46, 4:61, 5:261–262
Sea bears. *See* Northern fur seals; Polar bears
Sea biscuits, 1:*406*, 1:412–*413*
Sea carps. *See* Seacarps
Sea catfishes, 4:353, 4:354, 4:*355*, 4:*357*
Sea chubs, 4:67, **5:235–244,** 5:*239*, 5:*246*, **5:249**
Sea cows. *See* Dugongidae; Sea pigs; Steller's sea cows
Sea cucumbers, **1:417–431,** 1:*423*, 1:*424*, 2:*38*
  behavior, 1:42, 1:421
  conservation status, 1:48, 1:422
  distribution, 1:420–421
  evolution, 1:417, 1:418–419
  feeding ecology, 1:421
  habitats, 1:421
  humans and, 1:422
  physical characteristics, 1:*419–420*
  reproduction, 1:421–422
  species of, 1:425–431
  taxonomy, 1:417–419
Sea daisies, **1:381–386,** 1:*382*, 1:*384*
Sea ducks, 8:363, 8:369
Sea eels. *See* American congers
Sea eggs, West Indian. *See* West Indian sea eggs
Sea fans, 1:103, 1:107
Sea gingers. *See* Fire corals
Sea gooseberries, 1:4, 1:*172*, 1:*176*
Sea hares, 1:45
Sea kraits, 7:483–488, 7:*490*, 7:*495*
Sea lampreys, 4:*4*, 4:7, 4:*83–87*, 4:89–*90*, 4:214
Sea lilies, **1:355–360,** 1:*359*, **1:363–365**
  behavior, 1:357
  conservation status, 1:358
  distribution, 1:356
  evolution, 1:355
  feeding ecology, 1:357–358
  habitats, 1:356
  humans and, 1:358
  physical characteristics, 1:355–*356*
  reproduction, 1:358
  species of, 1:*362*, 1:*363–364*
  taxonomy, 1:355
Sea lions, 12:14, 12:*145*, 14:256, **14:393–408,** 14:*401*
  behavior, 14:396–397
  conservation status, 14:399–400
  distribution, 12:138, 14:394
  evolution, 14:393
  feeding ecology, 14:397–398
  habitats, 14:395–396
  learning set, 12:152
  locomotion, 12:44
  memory, 12:153
  physical characteristics, 12:63, 14:393–394
  reproduction, 14:398–399
  species of, 14:*403*–406
  taxonomy, 14:393
Sea mats, 2:*512*, 2:*513*, 2:514
Sea mice. *See* Sargassumfishes
Sea-mice, 2:45
Sea minks, 14:262
Sea nettles, 1:25–26, 1:29, 1:*158*, 1:*163*, 1:165
Sea otters, 12:14, 14:258, 14:321, 14:324
  conservation status, 1:50–51, 12:214
  distribution, 12:138
  physical characteristics, 12:63, 12:*64*
  tool use, 12:*158*
Sea pansies, 1:103, 1:*109*, 1:*111*, 1:116–117
Sea pens, 1:*9*, 1:103, 1:105
Sea pigs, 1:*423*, 1:429, 1:*430*, 1:*454*, 1:*457*, 1:*459*, 1:465
Sea robins, 4:14, 4:*22*, 5:159, 5:163–166
Sea salmons. *See* Atlantic salmons
Sea skaters, 3:*266*, 3:*267*, 3:274–275
  behavior, 3:60, 3:261
  biomes, 3:57
  conservation status, 3:264
  distribution, 3:52
  evolution, 3:7
  feeding ecology, 3:262
  fossilized, 3:*9*
Sea slugs, **2:403–410,** 2:*406*
  behavior, 2:403–404
  conservation status, 2:405
  distribution, 2:403
  evolution, 2:403
  feeding ecology, 2:404–405
  habitats, 2:403
  humans and, 2:43, 2:405
  physical characteristics, 2:403, 2:*404*
  reproduction, 2:405
  species of, 2:*407*–410
  taxonomy, 2:403
Sea spiders, 2:*321*, **2:321–325,** 2:*323*
Sea squirts, **1:451–466,** 1:*456*, 1:*457*
  behavior, 1:454
  conservation status, 1:455
  distribution, 1:454
  evolution, 1:5, 1:451–452
  feeding ecology, 1:454
  habitats, 1:454
  humans and, 1:455
  physical characteristics, 1:*452*–454
  reproduction, 1:455
  species of, 1:*458*–466
  taxonomy, 1:451–452
Sea stars, **1:367–380,** 1:*371*, 1:*372*, 2:36
  behavior, 1:369
  conservation status, 1:48, 1:370
  distribution, 1:368
  evolution, 1:367
  feeding ecology, 1:369–370
  habitats, 1:369
  humans and, 1:370
  physical characteristics, 1:367–*368*
  reproduction, 1:369–370
  species of, 1:*373*–379
  taxonomy, 1:367
Sea sticks. *See* Atlantic herrings
Sea sturgeons. *See* Atlantic sturgeons
Sea toads, 5:47
Sea trouts. *See* Brown trouts
Sea urchins, **1:401–416,** 1:*406*, 1:*407*
  behavior, 1:40, 1:403
  conservation status, 1:50–51, 1:405
  distribution, 1:402–403
  evolution, 1:401
  feeding ecology, 1:*402*, 1:404
  habitats, 1:403
  humans and, 1:405
  physical characteristics, 1:401–402, 1:*403*
  reproduction, 1:405
  species of, 1:*408*–415
  taxonomy, 1:401
Sea walnuts, 1:51, 1:171, 1:*172*, 1:*173*, 1:174
Sea wasps, 1:*147*, 1:148, 1:149, 1:*150*, 1:*151*–152
Sea water-striders. *See* Sea skaters
Sea whips, 1:45, 1:103
Sea World (San Diego, California), 8:153
Seacarps, 4:67, 5:235, 5:237–240, 5:242, 5:244
Seadevils, 5:*52*, 5:*53*, 5:54–55
"Seafood Watch Card," 1:52
Seahorses, 4:*4*, 4:14, 4:26, 4:46, 4:48, 4:69, **5:131–136,** 5:*137*, 5:*139*, **5:146**
Seal worms. *See* Cod worms
Sealice, 2:300, 2:301
Seals, 12:14, 14:255
  distribution, 12:129, 12:136, 12:138
  eared, **14:393–408,** 14:*401*
  locomotion, 12:44
  neonatal milk, 12:126–127
  neonates, 12:125
  physical characteristics, 12:63
  reproduction, 12:91
  touch, 12:82–83
  See also True seals
Seamoths, 5:131, 5:133, 5:135, 5:136
Search dogs, 14:293
Seashore sparrows. *See* Savannah sparrows
Seasnakes, 7:30, **7:483–487,** 7:*489*, 7:*490*, **7:494–497**
Seasonality
  of migration, 8:31–32, 12:164–170
  of nutritional requirements, 12:124–125
  of reproduction, 1:20, 12:91–92, 12:103
Seattle Zoo, 12:*210*
Seaturtles, 7:*38*, 7:43, **7:85–92,** 7:*87*, 7:*88*
  behavior, 7:*86*
  conservation status, 1:51, 7:59–61, 7:*60*, 7:87
  evolution, 7:14, 7:85
  feeding ecology, 7:86

as food, 7:47
habitats, 7:86
humans and, 7:87
locomotion, 7:24
physical characteristics, 7:85
reproduction, 7:71, 7:86–87
species of, 7:*89*–91
taxonomy, 7:85
See also Leatherback seaturtles
Sebaceous glands, 12:37
See also Physical characteristics
Seba's short-tailed bats, 13:*312*, 13:417, 13:418, 13:419, 13:*421*, 13:*429*–430
Sebastapistes cyanostigma, 4:42
Sebastes spp., 5:166
Sebastes fasciatus, 5:166
Sebastes paucispinis. *See* Bocaccios
Sebastidae. *See* Rockfishes
Sebastolobus spp., 5:166
Sebastolobus alascanus, 5:166
Sebekia spp., 2:317–318
Sebekidae, 2:317–318
Sebum, 12:37
Secernenteans, **1:293–304**, 1:*296*
behavior, 1:294
conservation status, 1:295
distribution, 1:294
evolution, 1:293
feeding ecology, 1:294
habitats, 1:294
humans and, 1:295
physical characteristics, 1:293–294
reproduction, 1:294–295
species of, 1:297–304
taxonomy, 1:12, 1:293
Secondary bony palate, 12:8, 12:11
Secretary birds, 8:313, **8:343–345**, 8:*344*
Secretary blennies, 4:*4*, 5:*344*, 5:*346*–347
Sedge frogs. *See* Painted reed frogs
Sedge wrens, 10:526, 10:527
Seebohm's coursers. *See* Three-banded coursers
Seed harvesters, 3:408
Seedcrackers, 11:354, 11:*355*, 11:*358*, 11:362
Seedeaters, 9:396, 11:264, 11:266–*267*, 11:277–278, 11:323, 11:325
Seedsnipes, 9:11, 9:104, **9:189–196**, 9:*192*
Seehausen, O., 5:281
Seeing-eye dogs, 14:*293*
Seepiophila spp., 2:91
Segmentation, 2:14
Segmented worms. *See* Annelida
Seguy, E. A., 3:323
Sei whales, 15:5, 15:8, 15:120, 15:121, 15:125, 15:130*t*
Seicercus spp., 11:3, 11:5, 11:6
Seicercus burkii. *See* Golden-spectacled warblers
Seismic communication, 12:76
See also Behavior
Seison nebaliae, 1:*263*, 1:*264*, 1:267
Seisonidea, 1:259, 1:260, 1:261
Seiurus aurocapillus. *See* Ovenbirds
Seiurus motacilla. *See* Louisiana waterthrush
Seiurus noveboracensis. *See* Northern waterthrush
Selasphorus spp. *See* Hummingbirds
Selasphorus platycercus. *See* Broad-tailed hummingbirds
Selasphorus rufus. *See* Rufous hummingbirds
Selasphorus sasin. *See* Allen's hummingbirds
Selection, migration and, 8:30
Selenidera spp. *See* Black toucanets
Seleucidis spp. *See* Birds of paradise
Selevinia spp. *See* Desert dormice
Selevinia betpakdalaensis. *See* Desert dormice
Self-directed behaviors, 12:159
See also Behavior
Self-organization, 3:72
Self recognition, 12:159–160
See also Cognition
Selfia spp., 2:317
Selfish herds, 12:167–168
Selva caciques, 11:306
Semaeostomeae, 1:153–154, 1:155–156
Semantic memory, 12:154
Semaphore crabs, 2:29
Semaprochilodus taeniurus. *See* Flagtail prochilodus
Semelparous fishes, 4:47
Semiaquatic bugs, 3:261
Semiaquatic water measurers, 3:260
Semibalanus balanoides. *See* Rock barnacles
Semicollared flycatchers. *See* Collared flycatchers
Semicossyphus pulcher. *See* California sheepheads
Semifossorial mammals, 12:13
See also Rodents
Semioptera spp. *See* Birds of paradise
Semioptera wallacii. *See* Standardwing birds of paradise
Semipalmated geese. *See* Magpie geese
Semipalmated plovers, 9:162
Semirechensk salamanders, 6:53, 6:335, 6:337, 6:*338*, 6:341
Semites, snakes and, 7:55
Semnodactylus spp., 6:281
Semnoderidae, 1:275
Semnopithecus spp., 14:172, 14:173
Semnopithecus entellus. *See* Northern plains gray langurs
Semnornis frantzii. *See* Prong-billed barbets
Semnornis ramphastinus. *See* Toucan barbets
Semper, Carl, 1:418
Semper's warblers, 11:290, 11:291
Senegal bushbabies, 12:105, 14:*24*, 14:*25*, 14:*26*, 14:27, 14:*28*, 14:30
Senegal galagos. *See* Senegal bushbabies
Senegal thick-knees, 9:144, 9:145
Sengis, 12:135, **16:517–532**, 16:*523*, 16:531*t*
Sense organs, 3:25–28
See also Physical characteristics
Senses, avian, 8:5–6, 8:8
Sensitization, 1:41, 2:39
See also Behavior
Sensory neurons, 3:25–26, 3:27
See also Physical characteristics
Sensory system, 4:21–22, 4:60, 4:68, 7:8–10, 7:159–160
See also Physical characteristics
Sensory system, lissamphibians, 6:15, 6:21–23
Sensory systems, 12:5–6, 12:9, 12:49–50, 12:75–77, **12:79–88**, 12:114
See also Physical characteristics
Sephanoides spp. *See* Hummingbirds
Sephanoides fernandensis. *See* Juan Fernández firecrowns
Sephanoides sephaniodes. *See* Green-backed firecrowns
Sepia officinalis. *See* Common cuttlefishes
Sepiida. *See* Cuttlefishes
Sepioids, 2:479
Sepiolida, 2:475
Sepsisoma spp., 3:360
Sepsophis spp., 7:328
Septaria spp., 2:440
Sequenzoidea, 2:429
Seram friarbirds, 11:*428*
Seram Island bandicoots, 13:10, 13:12
Seram orioles, 11:*428*
Serenoichthys spp., 4:209
Sergeant bakers, 4:434
Sergestoidea, 2:197
Serial position effect, 12:154
Sericolophus hawaiicus, 1:68
Sericornis spp., 11:55
Sericornis citreogularis. *See* Yellow-throated scrubwrens
Sericornis frontalis. *See* White-browed scrubwrens
Sericornis keri. *See* Atherton scrubwrens
Sericornis magnirostris. *See* Large-billed scrubwrens
Sericornis nigro-rufa. *See* Bicolored mouse-warblers
Sericulture, 3:79–80
Sericulus spp., 11:477, 11:478, 11:480
Sericulus aureus. *See* Flame bowerbirds
Sericulus bakeri. *See* Adelbert bowerbirds
Sericulus chrysocephalus. *See* Regent bowerbirds
Seriemas, 9:1–3, **9:85–89**, 9:*87*
Serilophus spp. *See* Broadbills
Serilophus lunatus. *See* Silver-breasted broadbills
Serins. *See* European serins
Serinus ankoberensis. *See* Ankober serins
Serinus canaria. *See* Canaries; Island canaries
Serinus flavigula. *See* Yellow-throated seedeaters
Serinus melanochrous. *See* Kipingere seedeaters
Serinus serinus. *See* European serins
Serinus syriacus. *See* Syrian serins
Serinus xantholaema. *See* Salvadori's serins
Serotine bats, 13:498, 13:515*t*
Serows, 12:135, 16:2, 16:87, 16:89, 16:90, 16:92, 16:95, 16:*96*, 16:*98*
Serpent-eagles, 8:318, 8:323
Serpent stars, 1:388
Serpent Worship in Africa (Hambly), 7:55
Serpentine locomotion, 7:8
Serpula vermicularis. *See* Tubeworms
Serpulids, 2:46, 2:47
Serranidae. *See* Sea basses
Serraninae. *See* Sea basses
Serranus spp., 4:30, 5:262
Serrasalminae, 4:339
Serrasalmus spp., 4:69
Serrated hinged terrapins. *See* East African serrated mud turtles
Serrated swimming crabs. *See* Mangrove crabs
Serrated turtles. *See* East African serrated mud turtles
Serritermitidae, 3:164, 3:166
Serrivomeridae, 4:255
Sertularella miuresis, 1:25
Sertularia spp., 1:129
Servals, 14:*378*, 14:*381*, 14:384
Service dogs, 14:293
Sesiidae, 3:389

Sessile barnacles, 2:273, 2:*274*, 2:*275*
Sessile polyps, 1:24
Setarches guentheri. See Deepwater scorpionfishes
Setarchidae. *See* Midwater scorpionfishes
Setifer spp., 13:225, 13:226, 13:230
Setifer setosus. See Greater hedgehog tenrecs
Setonix spp., 13:83, 13:85
Setonix brachyurus. See Quokkas
Setophaga ruticilla. See American redstarts
Setornis criniger. See Hook-billed bulbuls
Setzer's hairy-footed gerbils, 16:296*t*
Setzer's mouse-tailed dormice, 16:328*t*
Seven-banded armadillos, 13:183, 13:*184*, 13:191*t*
Sevengill sharks, **4:143–149,** 4:*145*
Sevenspine bay shrimps, 2:*202*, 2:*205*
Seventeen-year cicadas, 3:38, 3:*262*, 3:*265*, 3:270, 3:*270*
Severtzov's grouse. *See* Chinese grouse
Sewall Wright effect. *See* Random genetic drift
Sewellels. *See* Mountain beavers
Sex changes
  cichlids, 5:279
  hawkfishes, 5:240, 5:242
  parrotfishes, 5:297
  Percoidei, 5:242
  wrasses, 5:297
Sex chromosomes, 12:30–31
Sex determination, 3:38, 12:10
Sex ratios, 4:45
Sexton beetles, 3:42
Sexual dimorphism, 1:18, 2:23, 3:59, 4:*33*, 4:45–46, 4:64, 12:99
  See also Physical characteristics; Reproduction
Sexual maturity, 3:35, 12:97, 12:107
  See also Reproduction
Sexual monomorphism, 1:18
  See also Reproduction
Sexual parthenogenesis, 1:15
  See also Reproduction
Sexual reproduction, 1:15–16
  See also Reproduction
Seychelles brown white-eyes. *See* Seychelles gray white-eyes
Seychelles frogs, 6:6, 6:36, 6:68, **6:135–138,** 6:*137*
Seychelles gray white-eyes, 11:*231*, 11:*232–233*
Seychelles kestrels, 8:348
Seychelles magpie-robins, 10:485, 10:488
Seychelles scops-owls, 9:346, 9:*352*, 9:*354–355*
Seychelles sheath-tailed bats, 13:358, 13:364*t*
Seychelles sunbirds, 11:208, 11:210
Seychelles swiftlets, 9:418
Seychelles tortoises, 7:59
Seychelles treefrogs, 6:*282*, 6:283, 6:*284*, 6:*289*
Seychelles warblers, 11:6, 11:7
Seychelles white-eyes. *See* Seychelles gray white-eyes
Seymouriamorphs, 6:10
Shadow chipmunks. *See* Allen's chipmunks
Shads, 4:277
  Alabama, 4:278
  American, 4:*280*, 4:*281*, 4:282
  gizzard, 4:*280*, 4:284–*285*
  Laotian, 4:278
  threadfin, 4:*277*
Shaggy-haired bats, 13:364*t*
Shags, 8:183
Shakespeare, William, 6:52, 10:346, 13:200, 13:207
Shallow habitats, 4:53–56
Shamolagus, 16:480
Shanks, 9:175, 9:177, 9:179
Shannon mudminnows. *See* Salamanderfishes
Shanxi hedgehogs. *See* Hugh's hedgehogs
Shark catfishes, 4:353
Shark-fin catfishes. *See* Iridescent shark-catfishes
Shark mullets, 5:*61*, 5:*64–65*
Shark-toothed dolphins, 15:3
Shark-watching, 4:109
Sharks
  angelsharks, 4:11, **4:161–165,** 4:*163*
  attacks by, 4:116, 4:137–138
  banded cat, 4:*105*
  bullhead, 4:11, **4:97–103,** 4:*100*
  carpet, **4:105–112,** 4:*107*
  conservation status, 1:51
  dogfish, 4:11, **4:151–158,** 4:*154*
  evolution, 4:10
  feeding ecology, 4:69
  as food, 4:*75*
  ground, 4:11, **4:113–130,** 4:*117–118*
  mackerel, 4:11, **4:131–142,** 4:*134*
  nurse, 4:11, **4:105–112,** 4:*107*, 4:*108*
  physical characteristics, 4:18, 4:60
  reproduction, 4:26, 4:34, 4:35
  sawsharks, 4:11, **4:167–171,** 4:*169*
  sevengill, **4:143–149,** 4:*145*
  sixgill, **4:143–149,** 4:*145*
  taxonomy, 4:10–11
Sharksuckers, 4:69
Sharp-lipped boot sponges, 1:68–69, 1:*70*, 1:*72*, 1:75
Sharp-nose puffers, 4:70
Sharp-nosed bats. *See* Proboscis bats
Sharp-nosed caecilians. *See* Frigate Island caecilians
Sharp-nosed grass frogs, 6:*254*, 6:260
Sharp-nosed pitvipers. *See* Hundred-pace pitvipers
Sharp-nosed reed frogs, 6:281, 6:*284*, 6:*286*
Sharp-tailed grouse, 8:437
Sharp-tailed sandpipers, 9:176, 9:179
Sharp-tailed sparrows. *See* Nelson's sharp-tailed sparrows
Sharpbills, 10:170–171, **10:291–293,** 10:*292*
Sharpe's grysboks, 16:*61*, 16:71*t*
Sharpe's grysbucks. *See* Sharpe's grysboks
Sharpe's longclaws, 10:371, 10:374, 10:375
Sharpe's steinboks. *See* Sharpe's grysboks
Sharpies. *See* Channel catfishes
Sharpnose sevengill sharks, 4:143, 4:144
Sharptooth catfishes, 4:*355*, 4:359–*360*
Shaw, George, 12:227, 12:243
Shearwaters, **8:123–129,** 8:*127*
Sheatfishes, 4:353
  See also European wels
Sheath-tailed bats, 12:81, **13:355–365,** 13:*359*, 13:364*t*
Sheathbills, 9:104, **9:197–201,** 9:*198*, 9:*200–201*
Sheep, 15:*147*, 15:263, 15:265, 16:1, **16:87–105,** 16:*96*, 16:*97*
  behavior, 15:271, 16:92–93
  conservation status, 16:94–95
  digestive system, 12:122
  distribution, 16:4, 16:*87*, 16:90–91
  evolution, 16:87–89
  feeding ecology, 16:93–94
  feral cats and, 12:185
  habitats, 16:91–92
  humans and, 15:273, 16:8, 16:9, 16:95
  nutritional requirements, 12:124–125
  physical characteristics, 16:3, 16:89–90
  reproduction, 16:94
  smell, 12:85
  species of, 16:98–104*t*
  taxonomy, 16:87–89
  See also specific types of sheep
Sheep and goat fleas, 3:*351*, 3:*354–355*
Sheep keds, 3:76
Sheep liver flukes. *See* Liver flukes
Sheep stomach worms. *See* Barber's pole worms
Sheepdogs, 14:288, 14:292
Sheepsheads, 4:17, 4:53, 5:94
Shelania laurenti, 6:99
Shelania pascuali, 6:99
Shelducks, 8:363, 8:366, 8:367, 8:369
Shelley, Mary, 1:129
Shelley's bulbuls. *See* Shelley's greenbuls
Shelley's greenbuls, 10:*401*, 10:*405*–406
Shelley's starlings, 11:408
Shellfish, as food, 2:41
  See also specific shellfish
Shellfish poisoning, 2:12
Shells
  cone, 2:446, 2:447
  geography cone, 2:447, 2:*448*, 2:449
  slit, **2:429–434**
  top, **2:429–434,** 2:*432*
  tusk, **2:469–474,** 2:*472*
  watering pot, 2:*456*, 2:*458*, 2:460
Shenbrot, G. I., 16:216
Shepherd fishes. *See* Man-of-war fishes
Shepherd spiders. *See* Long-bodied cellar spiders
Shepherd's beaked whales, 15:59, 15:*63*, 15:67–68
Sherbrooke Forest, lyrebirds and, 10:332
Shesh Nag, 7:56
Shetland ponies, 15:*230*
Shield-backed katydids, 3:57, 3:208
Shield bugs, 3:*37*
Shield limpets, 2:*426*–427
Shield shrimps, 2:*142*
Shielded blindsnakes. *See* Peters' wormsnakes
Shieldtail snakes, **7:391–394**
Shiner perches, 5:276
Shiners. *See* Silverfish
Shinglebacks. *See* Bobtails
Shining drongos, 11:438, 11:442
Shiny cowbirds, 11:304
Shiny guinea pigs, 16:399*t*
Ship cockroaches. *See* American cockroaches
Ship rats. *See* Black rats
Ship's worms. *See* Common shipworms
Shipworms, 2:*43*, 2:454
Shirleyrhynchidae, 1:226
Shiro bekko, 4:*299*
Shoals, 4:63–64
Shoebills, 8:233, 8:234, 8:235, 8:239, **8:287–289,** 8:*288*

Shomronella spp., 6:12
Shore bugs, 3:57, 3:263, 3:*266*, 3:*276*, 3:279–280
Shore crabs, 1:38, 2:25, 2:26
Shore eels. *See* Common shore-eels
Shore flies, 3:359
Shore herrings. *See* Atlantic herrings
Shore larks. *See* Horned larks
Shore plovers, 9:161, 9:163–164
Shorebirds, 9:102, 9:104
Short-beaked echidnas, 12:*227*, 12:*239*, 12:*240*, 12:240*t*
behavior, 12:*230*, 12:237, 12:*238*
conservation status, 12:233
distribution, 12:230, 12:236
feeding ecology, 12:*228*, 12:231, 12:*236*, 12:237
habitats, 12:230, 12:237
humans and, 12:238
physical characteristics, 12:235–236
reproduction, 12:108, 12:231–232, 12:233, 12:237
taxonomy, 12:227, 12:235
Short-beaked saddleback dolphins, 15:2, 15:3, 15:42, 15:44, 15:47, 15:56*t*
Short-billed leafscrapers. *See* Short-billed leaftossers
Short-billed leaftossers, 10:*213*, 10:*227*–228
Short-clawed larks, 10:342
Short-crested coquettes, 9:450
Short-eared dogs, 14:283*t*
Short-eared owls, 9:332, 9:333, 9:347, 9:*353*, 9:364–*365*
Short-eared sengis. *See* Round-eared sengis
Short-faced bears, 12:24
Short-faced moles, 13:288*t*
Short-finned pilot whales, 15:56*t*
Short-furred dasyures, 12:299*t*
Short-headed lampreys, 4:83, 4:84, 4:*87*, 4:*88*, 4:89
Short-horned grasshoppers, 3:201, 3:204, 3:206
Short-horned water buffaloes, 16:1
Short interspersed nuclear elements (SINEs), 12:29–30
Short-legged ground-rollers, 10:53
Short-nosed bandicoots. *See* Southern brown bandicoots
Short-nosed fruit bats, 13:311
Short-nosed seahorses, 5:*134*
Short-nosed tripodfishes. *See* Triplespines
Short-palate fruit bats, 13:348*t*
Short pythons. *See* Blood pythons
Short-snouted sengis, 16:*520*, 16:*523*, 16:*527*–528
Short-spined sea urchins, 1:402–403, 1:*407*, 1:*409*, 1:410
Short-spined urchins. *See* West Indian sea eggs
Short-tailed albatrosses, 8:110, 8:113
Short-tailed bandicoot rats, 16:261*t*
Short-tailed bats, chestnut, 13:433*t*
Short-tailed chameleons, 7:*234*, 7:*235*, 7:240–241
Short-tailed chinchillas, 16:377, 16:380, 16:383*t*
Short-tailed frogmouths, 9:377, 9:378, 9:379
Short-tailed gymnures. *See* Lesser gymnures
Short-tailed macaques, 14:193
Short-tailed monitors, 7:359, 7:361, 7:362, 7:*364*, 7:*366*
Short-tailed nighthawks, 9:404
Short-tailed opossums, 12:250–254
Short-tailed paradigallas, 11:*494*, 11:*496*–497
Short-tailed porcupines, 16:365, 16:374*t*
Short-tailed pygmy monitors. *See* Short-tailed monitors
Short-tailed pythons. *See* Blood pythons
Short-tailed shearwaters, 8:108, 8:110, 8:*129*
Short-tailed shrews, 13:195, 13:197, 13:198, 13:200
Short-tailed wallabies. *See* Quokkas
Short-tailed wattled birds of paradise. *See* Short-tailed paradigallas
Short-term memory, 12:152
Short-toed larks. *See* Greater short-toed larks
Short-toed treecreepers, 11:177, 11:*179*, 11:*181*
Short-winged grebes. *See* Titicaca flightless grebes
Shortfin catfishes, 4:353
Shortfin makos, 4:132, 4:133, 4:*134*, 4:*135*, 4:138
Shorthead wormlizards. *See* Spade-headed wormlizards
Shortnose gars, 4:*223*, 4:*226*–227
Shortnose sturgeons, 4:215, 4:*216*, 4:*217*
Shoshani, J., 15:177
Shovel-billed kingfishers. *See* Shovel-billed kookaburras
Shovel-billed kookaburras, 10:6, 10:*12*, 10:*13*–14
Shovel-footed squeakers. *See* Common squeakers
Shovel-nosed frogs, 6:6, **6:273–278,** 6:*274*, 6:*276*
Shovel-nosed snakes, 7:466, 7:468
Shovel-snouted lizards, 7:298
Shovel-snouted snakes, 7:468
Shovelbills, 11:342
Shovelers. *See* Northern shovelers
Shrew gymnures, 13:213*t*
Shrew mice, 13:197, 16:265, 16:*271*, 16:*273*, 16:274
Shrew moles, 13:196, 13:198, **13:279–288,** 13:*283*, 13:288*t*
Shrew opossums, 12:132, **12:267–271,** 12:*269*
Shrew tenrecs, 13:225
Shrews, 13:193–200, 13:247
alpine, 13:*254*, 13:*258*, 13:260
behavior, 12:145
bicolored, 13:*266*
color vision, 12:79
distribution, 12:131
echolocation, 12:53, 12:86
musk, 12:107, 12:115, 13:198, 13:199, 13:266
Nimba otter, 13:234*t*
otter, 13:194, 13:199, 13:225
red-toothed, **13:247–264**
Ruwenzori otter, 13:234*t*
smell, 12:80
West Indian, 12:133, 13:193, 13:197, **13:243–245**
white-toothed, 13:198, 13:199, **13:265–277**
See also Tree shrews
Shrike-flycatchers, 11:*28*, 11:*40*
Shrike-thrushes, 11:115–118, 11:*119*, 11:123–124
Shrike-tits, 11:115, 11:116, 11:117
Shrike-vireos, 11:*257*, 11:261–262
Shrikes, 9:314, 10:173, **10:425–438,** 10:*431*, 11:5
behavior, 10:427–428
conservation status, 10:429
distribution, 10:*425*, 10:426–427
evolution, 10:425
feeding ecology, 10:428
habitats, 10:427
humans and, 10:429–430
physical characteristics, 10:425–426
reproduction, 10:428–429
species of, 10:*432–438*
taxonomy, 10:425
vanga, **10:439–445**
See also Magpie-shrikes; Peppershrikes
Shrimp gobies, 4:64
Shrimpfishes, 5:131–132, 5:133, 5:135, 5:*138*, 5:139–*140*
Shrimps, **2:197–214,** 2:*202*, 2:*203*
behavior, 2:32, 2:33–34, 2:43, 2:199–200
clam, **2:147–151**
Coleman's, 1:*34*
conservation status, 2:43, 2:201
distribution, 2:199
evolution, 2:197–198
fairy, **2:135–140**
feeding ecology, 2:29, 2:200
habitats, 2:199
humans and, 2:201
mantis, **2:167–175**
mussel, **2:311–315**
physical characteristics, 2:198
reproduction, 2:17, 2:201
species of, 2:*204*–214
tadpole, **2:141–146**
taxonomy, 2:197–198
Shu, D. G., 4:9
Shumaker, Robert, 12:161
Shy albatrosses, 8:114, 8:115, 8:*120*
Shy tobies, 5:*468*
Sialia currucoides. *See* Mountain bluebirds
Sialia mexicana. *See* Western bluebirds
Sialia sialis. *See* Eastern bluebirds
Sialidae. *See* Alderflies
Sialis spp., 3:54
Sialis lutaria. *See* Alderflies
Siamangs, 14:*216*, 14:222
behavior, 14:*210*, 14:210–211
distribution, 14:*221*
evolution, 14:207
feeding ecology, 14:212–213
habitats, 14:210
physical characteristics, 14:209
taxonomy, 14:207
Siamese crocodiles, 7:59, 7:164
Siamese fighting fishes, 5:*430*, 5:*432*–433
Siamopithecus spp., 14:2
Siberian accentors, 10:*461*, 10:463–*464*
Siberian-American moose, 15:380
Siberian chipmunks, 16:144, 16:158*t*
Siberian cranes, 9:3, 9:23–29, 9:*30*, 9:32–*33*
Siberian-European moose, 15:380
Siberian flying squirrels, 16:135, 16:*138*, 16:139–*140*
Siberian huskies, 14:289, 14:*292*, 14:293
Siberian ibex, 16:87, 16:91–92, 16:93, 16:97, 16:*101*–102
Siberian jays, 11:507

Siberian marals, 15:357
Siberian musk deer, 15:335, 15:*336*, 15:*337*, 15:*338*, 15:*339*, 15:*340*–341
Siberian pikas. *See* Northern pikas
Siberian roe deer, 15:384, 15:*386*, 15:*388*–389
Siberian rubythroats, 10:*491*, 10:*499*
Siberian salamanders, 6:51, 6:325, 6:336, 6:337, 6:*338*, 6:341–342
Siberian snow sheep, 12:178
Siberian stone loaches, 4:325
Siberian tigers. *See* Amur tigers
Siberian tits, 11:156, 11:157, 11:*159*, 11:*161*–162
Siberian water deer, 15:376
Siberian white cranes. *See* Siberian cranes
Siberian zokors, 16:*287*, 16:*290*, 16:293–294
Sibley, Charles, 10:2
 on Apodiformes, 9:415
 on Australian creepers, 11:133
 on babblers, 10:505, 10:513
 on bee-eaters, 10:39
 on Charadriidae, 9:161
 on Charadriiformes, 9:101
 on Corvidae, 11:503
 on cuckoo-shrikes, 10:385
 on frogmouths, 9:377
 on hummingbirds, 9:437
 on Irenidae, 10:415
 on logrunners, 11:69
 on long-tailed titmice, 11:141
 on magpie geese, 8:363
 on Old World warblers, 11:1
 on Passeriformes, 10:169, 10:171
 on Pelecaniformes, 8:183
 on Piciformes, 10:85–86
 on plantcutters, 10:325
 on Sturnidae, 11:407
 on sunbirds, 11:207
 on thick-knees, 9:143
 on toucans, 10:125
 on treecreepers, 11:177
 on trumpeters, 9:77
 on tyrant flycatchers, 10:275
 on weaverfinches, 11:353
 on weavers, 11:375
 on woodswallows, 11:459
 on wrentits, 10:513
Siboglinum fiordicum. *See* Norwegian tubeworms
Siboglinum poseidoni, 2:86
Sibon spp. *See* Snail-suckers
Sibyllidae, 3:177
Sibynophis spp., 7:205
Sichuan golden snub-nosed monkeys. *See* Golden snub-nosed monkeys
Sichuan hill-partridges, 8:437, 8:*440*, 8:*448*
Sichuan partridges. *See* Sichuan hill-partridges
Sicista spp., 16:216
Sicista armenica, 16:216
Sicista betulina, 16:214, 16:*216*
Sicista caudata, 16:216
Sicista subtilis. *See* Southern birch mice
Sick, Helmut, 10:307
Sickle-crested birds of paradise. *See* Crested birds of paradise
Sickle-winged guans, 8:413
Sickle-winged nightjars, 9:405
Sicklebilled vangas, 10:439, 10:440, 10:441, 10:442
Sicklebills, 11:343
Sicklefishes, 5:235–239, 5:241, 5:242
Sicyases spp., 5:355
Sicydiinae, 5:373, 5:375, 5:376, 5:377
Sicyoptera cuspidata, 3:309
Sicyoptera dilatata, 3:309
Sidamo bushlarks, 10:346
Side-blotched lizards. *See* Common side-blotched lizards
Side-stabbing snakes. *See* Southern burrowing asps
Side-striped jackals, 12:117, 14:283*t*
Sidewinding, 7:*25*–26, 7:39, 7:*40*, 7:447
Sidneyia, 12:64
Sierra Madre sparrows, 11:266
Sierra Nevada ensatina. *See Ensatina* spp.
Sifakas, **14:63–72,** 14:*68*
 behavior, 14:7, 14:8, 14:65–66
 conservation status, 14:67
 distribution, 14:65
 evolution, 14:63
 feeding ecology, 14:66
 habitats, 14:65
 humans and, 14:67
 physical characteristics, 14:63–65
 reproduction, 14:66–67
 species of, 14:*70*
 taxonomy, 14:63
Siganidae. *See* Rabbitfishes
Siganus vulpinus. *See* Foxface rabbitfishes
Sigara platensis. *See* Water boatmen
Sight dogs, 14:288
Sighthounds, 14:292
Sigmatoneura spp., 3:243
Sigmoceros lichtensteinii. *See* Lichtenstein's hartebeests
Sigmodon spp., 16:127, 16:263, 16:264, 16:265–266
Sigmodon alstoni, 16:270
Sigmodon hispidus. *See* Hispid cotton rats
Sigmodontinae, 16:128, **16:263–279,** 16:*271*, 16:277*t*–278*t*, 16:282
Sigmodontini, 16:264
Sigmodontomys aphrastus, 16:270
Sigmoria aberrans, 2:*363*
Sigmurethra, 2:411
Signaling, bird songs for, 8:41
Signigobius biocellatus. *See* Twinspot gobies
Siju blind cavefishes, 4:*324*, 4:*328*, 4:329
Sika deer, 12:19, 15:358, 15:361, 15:*363*, 15:*368*, 15:369
Silence of the Lambs, 3:403
Silent dormice, 16:327*t*
Silicularia bilabiata, 1:24
Silk, commercial, 3:79–80
Silkmoth caterpillars, 3:*388*
Silkmoths, giant, 3:47
 See also Silkworms
Silkworms, 3:79–80, 3:388, 3:389, 3:*390*, 3:*393*
Silky anteaters, 13:*172*, 13:*174*, 13:*175*, 13:*176*, 13:*177*
Silky cuscuses, 13:66*t*
Silky flycatchers, **10:447–450, 10:452–454**
Silky pocket mice, 16:199, 16:208*t*
Silky shrew opossums, 12:*269*, 12:*270*
Sillaginidae. *See* Sillagos
Sillagos, 5:255, 5:256, 5:258–263
Sillem's mountain-finches, 11:325
Silphidae, 3:321
Silt, from glaciers, 12:17
Silurana spp., 6:102
Silurana tropicalis. *See* Tropical clawed frogs
Siluridae. *See* Sheatfishes
Siluriformes. *See* Catfishes
Silurus bicirrhis. *See* Glass catfishes
Silurus electricus. *See* Electric catfishes
Silurus fasciatus. *See* Tiger shovelnose catfishes
Silurus gariepinus. *See* Sharptooth catfishes
Silurus glanis. *See* European wels
Silurus hemioliopterus. *See* Redtail catfishes
Silurus lineatus. *See* Coral catfishes
Silurus punctatus. *See* Channel catfishes
Silva Maia, E. J., 9:452
Silver aruana. *See* Arawanas
Silver barfishes. *See* Dorab wolf herrings
Silver bosunbirds. *See* Red-billed tropicbirds
Silver-breasted broadbills, 10:179, 10:*180*, 10:*183*–184
Silver carps, 4:299–300, 4:303, 4:*305*, 4:*310*, 4:312–313
Silver catfishes. *See* Iridescent shark-catfishes
Silver cats. *See* Channel catfishes
Silver chubs. *See* Troutperches
Silver cyprinids. *See* Dagaa
Silver dikdiks, 16:63, 16:71*t*
Silver dollars, 4:339, 4:*341*, 4:*344*
Silver foxes. *See* Red foxes
Silver-haired bats, 13:310, 13:*503*, 13:*505*, 13:*508*, 13:509
Silver hakes, 5:*31*, 5:*37*–38
Silver hatchetfishes, 4:*425*, 4:426–*427*
 See also River hatchetfishes
Silver lampreys, 4:83, 4:*87*, 4:89, 4:*90*
Silver needlefishes, 5:*79*
Silver orioles, 11:428
Silver perches, 5:*222*
Silver pikas, 16:487, 16:495
Silver pomfrets, 5:422
Silver prochilodus. *See* Flagtail prochilodus
Silver salmons. *See* Coho salmons
Silver tetras, 4:*336*
Silver witches. *See* Silverfish
Silverback jackals, 12:8, 12:117, 12:216, 14:*256*, 14:*257*, 14:*269*, 14:*272*, 14:283*t*
Silverbacks, 12:*141*, 14:228
Silverbills. *See* African silverbills
Silverbirds, 11:*28*, 11:*39*
Silvered langurs. *See* Silvery leaf-monkeys
Silvereyes, 11:227, 11:*228*
Silverfish, 3:10, 3:11–12, 3:13, 3:61, 3:64, **3:*119*–124,** 3:*121*, 3:*123*
Silversides, 4:12, 4:35, **5:67–71,** 5:*68*, 5:*72*, **5:73–76**
Silvery-cheeked hornbills, 10:77, 10:*83*–84
Silvery gibbons. *See* Moloch gibbons
Silvery greater bushbabies, 14:33*t*
Silvery grebes, 8:169
Silvery John dories. *See* Buckler dories
Silvery leaf-monkeys, 14:*177*, 14:*181*
Silvery lutungs. *See* Silvery leaf-monkeys
Silvery marmosets, 14:*124*, 14:132*t*
Silvery mole-rats, 12:72, 12:74, 12:75, 12:76, 16:339, 16:342, 16:344, 16:349*t*
Silvery moonies. *See* Monos
Silvery mountain voles, 16:*232*, 16:*235*, 16:236
Simakobus. *See* Mentawai Island langurs
Simbi (Loa god), 7:57
Simenchelys parasitica. *See* Snubnosed eels
Simenchelys parasiticus. *See* Snubnosed eels

Simian primates, 12:90, 14:8
Simias spp. *See* Mentawai Island langurs
Simias concolor. See Mentawai Island langurs
Simmons, N. B., 13:333
Simony's giant lizards, 7:298–299
Simoselaps spp. *See* Half-girdled snakes
Simoselaps calonotus, 7:486
Simoselaps semifasciatus. See Half-girdled snakes
Simplex uterus, 12:90–91, 12:110–111
Simplicendentata. *See* Rodents
Simpson, George G.
  on deer, 15:380–381
  on Dipodidae, 16:211
  on Lagomorpha, 16:479
  on tree shrews, 13:289
Simulium spp. *See* Black flies
Simultaneous spawning, 1:18
  See also Reproduction
Sinaloa martins, 10:361
Sinamiidae, 4:229
Sinarapans, 5:377
Sinentomidae, 3:93
Sinentomon yoroi, 3:*95*, 3:*97*
Sinerpeton spp., 6:13
SINEs (short interspersed nuclear elements), 12:29–30
Singapore roundleaf horseshoe bats. *See* Ridley's roundleaf bats
Singed sandgrouse. *See* Chestnut-bellied sandgrouse
Singing bushlarks. *See* Australasian larks
Singing crickets, 3:81, 3:208
Singing fishes. *See* Plainfin midshipmen
Singing fruit bats, 13:*338*, 13:*341*, 13:342
Singing katydids, 3:208
Singing rats. *See* Amazon bamboo rats
Single nesting cuckoos, 8:40–41
Singleslits, **5:355–357**, 5:*358*, **5:359**, **5:361**
Singlespot frogfishes. *See* Big-eye frogfishes
Sinoconodon spp., 12:11
Sinoperca spp., 5:220
Siphlonuridae, 3:54
Siphonaptera. *See* Fleas
Siphonaridae, 2:414
Siphonodictyon spp., 1:29
Siphonophora millepeda, 2:363
Siphonophorae, 1:29, 1:51, 1:123, 1:124, 1:125–126, 1:127
Siphonophoridae, 2:365
Siphonops annulatus, 6:411, 6:436
Siphonops mexicanus. See Mexican caecilians
Siphonorhis americana. See Jamaican poorwills
Siphonostomatoida, 2:299
Sipodotus spp. *See* Wallace's wrens
Sipodotus wallacii. See Wallace's wrens
Siptornopsis hypochondriacus. See Great spinetails
Sipuncula. *See* Peanut worms
Sipunculidae, 2:97
Sipunculus spp., 2:98
Sipunculus nudus, 2:*99*, 2:100–101
Siredon gracilis. See Northwestern salamanders
Siredon mexicanum. See Mexican axolotl
Siren intermedia. See Lesser sirens
Siren lacertina. See Greater sirens
Siren striata. See Northern dwarf sirens
Sirena maculosa. See Mudpuppies
Sirenia, **15:191–197**
  behavior, 15:192–193
  conservation status, 15:194–195
  distribution, 12:138, 15:192
  evolution, 15:133, 15:191
  feeding ecology, 15:193
  habitats, 12:14, 15:192
  humans and, 15:195–196
  physical characteristics, 12:63, 12:68, 15:192
  reproduction, 12:103, 15:193–194
  taxonomy, 15:131, 15:134, 15:191–192
  See also Dugongs; Manatees; Steller's sea cows
Sirenidae. *See* Sirens
Sirenoidea. *See* Sirens
Sirens, 6:17, **6:327–333**, 6:*330*
  distribution, 6:5, 6:*327*, 6:328, 6:*331–332*
  dwarf, **6:327–333**, 6:*330*
  evolution, 6:13, 6:327
  physical characteristics, 6:323, 6:325, 6:327–328, 6:331–332
  reproduction, 6:32, 6:329, 6:331–332
  taxonomy, 6:323, 6:327, 6:331–332, 6:335
Sirex. *See* European wood wasps
Sirex noctilio. See European wood wasps
Siricidae, 3:406
Siriella spp., 2:216
Siskins, 11:323
Sisoridae, 4:353
Sistrurus spp., 7:445, 7:446
Sisyridae, 3:305, 3:306, 3:307, 3:308
Sitatungas, 16:2, 16:4, 16:11
Sitophilus granarius. See Grain weevils
Sitophilus oryzae. See Rice weevils
Sitta spp. *See* Nuthatches
Sitta canadensis. See Red-breasted nuthatches
Sitta carolinensis. See White-breasted nuthatches
Sitta europaea. See Nuthatches
Sitta formosa. See Beautiful nuthatches
Sitta ledanti. See Algerian nuthatches
Sitta magna. See Giant nuthatches
Sitta pusilla. See Brown-headed nuthatches
Sitta pygmaea. See Pygmy nuthatches
Sitta solangiae. See Yellow-billed nuthatches
Sitta victoriae. See White-browed nuthatches
Sitta yunnanensis. See Yunnan nuthatches
Sittellas, 11:167
Sittidae. *See* Nuthatches
Situla spp., 1:454
Situla pelliculosa, 1:*457*, 1:*460*, 1:464
Six-banded armadillos. *See* Yellow armadillos
Six keyhole sand dollars, 1:402, 1:*406*, 1:412, 1:*413*
Six-lined grass lizards, 7:*300*, 7:*301*, 7:302
Six-lined racerunners, 7:*313*, 7:*314*–315
Six-stripe soapfishes. *See* Sixline soapfishes
Sixgill sawsharks, 4:167, 4:168, 4:*169*, 4:170–17i
Sixgill sharks, **4:143–149**, 4:*145*
Sixgill stingrays, 4:190
Sixline soapfishes, 5:*265*, 5:*267*, 5:271
Size. *See* Body size
Sjöstedt's honeyguide greenbuls, 10:397, 10:398
Skates, **4:173–181**, 4:*180*, **4:187**
  behavior, 4:177
  conservation, 4:178
  distribution, 4:176
  evolution, 4:10, 4:11, 4:173–174
  feeding ecology, 4:177–178
  habitats, 4:176–177
  humans and, 4:178–179
  physical characteristics, 4:60, 4:*174*–176, 4:*175*
  reproduction, 4:27, 4:46, 4:*178*
  species of, 4:187
  taxonomy, 4:173–174
Skeletal system, 1:*5*
  birds, 8:9, 8:*15*
  clams, 2:*18*
  crocodilians, 7:23
  lepidosauria, 7:23–24
  mammals, 12:40–41, 12:174
  marine mammals, 12:65
  reptiles, 7:23–26
  turtles, 7:23, 7:*70*
  See also Physical characteristics
Skeleton shrimps, 2:*261*, 2:*263*, 2:*265*
Skeneoidea, 2:430, 2:431
Skew-beaked whales. *See* Hector's beaked whales
Skilfishes, 5:179, 5:182, 5:*184*, 5:*187*
Skimmers, 9:102, 9:104, 9:203–209
  See also Charadriiformes
Skin, 12:7, 12:36, 12:*39*
  See also Integumentary system; Physical characteristics
Skin beetles, 3:64, 3:321
Skinks, 7:327, **7:327–338**, 7:328, 7:329, 7:*330*, 7:*331*
  Australian, 7:202
  behavior, 7:204, 7:205, 7:329
  blindskinks, **7:271–272**
  chemosensory systems, 7:36
  conservation status, 7:329
  distribution, 7:*327*, 7:328
  evolution, 7:196, 7:327–328
  feeding ecology, 7:329
  Great Plains, 7:42
  habitats, 7:329
  humans and, 7:329
  physical characteristics, 7:328
  reproduction, 7:206–207, 7:329
  species of, 7:*332–337*
  taxonomy, 7:327–328
Skinner, B. F., 2:35
Skinnycheek lanternfishes, 4:*444*–445
Skipjack tunas, 5:406, 5:*408*, 5:412, 5:*413*
Skipjack winter shads. *See* Gizzard shads
Skippers, 3:38, **3:383**, **3:385**, 3:*389*
  See also Lepidoptera
Skuas, 9:104, 9:199, 9:203–209
  See also Charadriiformes
Skulls, 7:4–*5*, 7:26, 12:8, 12:10, 12:41, 12:174
  crocodilian, 7:26, 7:*159*
  lizards, 7:*199*
  snakes, 7:26, 7:*201*
  Squamata, 7:198, 7:200
  tuatara, 7:*191*
  turtle, 7:26, 7:*65*, 7:67
  See also Physical characteristics
Skunks, **14:319–325**, 14:*320*, 14:*326*, **14:329**, **14:332–334**
  behavior, 12:*186*, 14:321–323
  conservation status, 14:324
  distribution, 14:*319*, 14:321
  evolution, 14:319
  feeding ecology, 12:192, 14:323
  habitats, 14:319–321
  humans and, 14:324–325
  physical characteristics, 12:38, 14:319–321
  reproduction, 14:324

species of, 14:*329*, 14:332*t*, 14:333*t*
taxonomy, 14:256, 14:319
Skutch, A., 10:92, 10:174
Sky larks, 10:343, 10:*344*, 10:345, 10:346, 10:*347*, 10:*353–354*
Sladen's barbets, 10:115
Slant-faced grasshoppers, 3:205
Slate-colored juncos, 11:271
Slate-pencil urchins, 1:401, 1:403, 1:*406*, 1:*408*–409
Slaters, **2:249–260**
behavior, 2:250–251
conservation status, 2:252
distribution, 2:250
evolution, 2:249
feeding ecology, 2:251–252
habitats, 2:250
humans and, 2:252
physical characteristics, 2:249–250
reproduction, 2:252
taxonomy, 2:249
Slaty antwrens, 10:*240*
Slaty-bellied ground warblers. *See* Slaty-bellied tesias
Slaty-bellied tesias, 11:*8*, 11:*14*
Slaty-breasted tinamous, 8:58, 8:59, 8:*60*, 8:*63–64*
Slaty bristlefronts, 10:259, 10:*261*, 10:264–*265*
Slaty-capped shrike-vireos, 11:*257*, 11:261–*262*
Slaty flycatchers. *See* Abyssinian slaty flycatchers
Slaty-headed parakeets, 9:276
Slaty mouse opossums. *See* Slaty slender mouse opossums
Slaty slender mouse opossums, 12:265*t*
Slaty-tailed trogons, 9:478
Sleeper sharks, 4:11, 4:151
See also Greenland sharks
Sleeping sickness, 3:76
Sleepy lizards. *See* Bobtails
Slender argentines, 4:392
Slender-billed curlews, 9:180
Slender-billed parrots, 9:277–278
Slender-billed shearwaters/petrels. *See* Short-tailed shearwaters
Slender-billed wrens, 10:529, 10:*530*, 10:*532–533*
Slender blindsnakes, 7:198, 7:201, 7:202, 7:371, **7:373–377**
Slender Cuban nesophontes, 13:243
Slender falanoucs. *See* Falanoucs
Slender frogs, 6:110
Slender giant morays, 4:*260*, 4:*266*, 4:267
Slender grinners. *See* Slender lizardfishes
Slender-horned gazelles, 16:48, 16:57*t*
Slender lizardfishes, 4:*431*, 4:*435*, 4:*438*–439
Slender lorises, 14:13, 14:14, 14:*15*, 14:16, 14:*17*, 14:*18*, 14:21*t*
Slender mongooses, 14:349, 14:357*t*
Slender mouse opossums, 12:250–253
Slender mud frogs, 6:111, 6:*112*, 6:*113*–114
Slender pigeon lice, 3:*253*, 3:*255*, 3:256
Slender snipe eels, 4:257, 4:*259*, 4:*264*, 4:267–268
Slender-tailed dunnarts, 12:*290*, 12:301*t*
Slender tree shrews, 13:297*t*
Slender worm-eels. *See* Rusty spaghetti eels
Slickers. *See* Silverfish
Slickheads, 4:390, 4:391, 4:392, 4:*394*, 4:*395*
Slime eels. *See* Atlantic hagfishes; Pacific hagfishes
Slime glands, 4:77
See also Physical characteristics
Slimeheads. *See* Orange roughies; Roughies
Slimy loaches. *See* Coolie loaches
Slingjaw wrasses, 4:*17*, 5:*295*
Slipmouths. *See* Ponyfishes
Slipper lobsters, 2:197, 2:198
Slipper sea cucumbers, 1:*423*, 1:*428*, 1:429
Slipskin snailfishes, 5:*184*, 5:*186*, 5:192–193
Slit-faced bats, 13:310, **13:371–377**, 13:*372*, 13:*374*, 13:376*t*–377*t*
See also specific types of slit-faced bats
Slit shells, **2:429–434**
Sloane's viperfishes. *See* Viperfishes
Slobodyanyuk, S. Ya., 5:180
Sloggett's vlei rats. *See* Ice rats
Sloth bears, 14:295, 14:296, 14:297, 14:298, 14:299, 14:300, 14:306*t*
Sloth lemurs, 14:63, 14:*68*, 14:*70*, 14:71
Sloths, 13:147–153, 13:155, 13:156, 13:165
claws, 12:39
digestive systems of, 12:15, 12:123
lesser Haitian ground, 13:*159*
locomotion, 12:43
monk, 13:*166*, 13:*168*
reproduction, 12:94, 12:103
southern two-toed, 13:*148*, 13:*157*
West Indian, **13:155–159**
See also Tree sloths; specific types of sloths
Slow lorises, 14:13, 14:14
See also Pygmy slow lorises; Sunda slow lorises
Slowworms, 7:339, 7:340
Slug-eaters, 7:467, 7:*471*, 7:*477*
Slugs, 2:439
lung-bearing, **2:411–422,** 2:*416*
pear and cherry, 3:*410*, 3:*414*, 3:422–423
sea, **2:403–410,** 2:*406*
See also Neritopsina
Slugworms. *See* Pear and cherry slugs
Small Asian sheath-tailed bats, 13:*357*
Small buttonquails, 9:*13*, 9:*16*
See also Barred buttonquails
Small-eared dogs, 14:266
Small-eared shrews, 13:198
Small fat-tailed opossums, 12:265*t*
Small five-toed jerboas. *See* Little five-toed jerboas
Small flying foxes. *See* Island flying foxes
Small-footed funnel-eared bats, 13:*461*, 13:*462*, 13:*464*–465
Small ground-finches, 11:*266*
Small hairy armadillos, 13:*186*, 13:187–*188*
Small Indian mongooses, 14:349, 14:*353*, 14:*355*
Small intestine, 12:47–48, 12:123
See also Physical characteristics
Small Lander's horseshoe bats, 13:388
Small minivets. *See* Fiery minivets
Small mouse-tailed bats, 13:352
Small pied kingfishers. *See* Pied kingfishers
Small pin-tailed sandgrouse. *See* Chestnut-bellied sandgrouse
Small Population Analysis and Record Keeping System (SPARKS), 12:206–207
Small red brocket deer. *See* Brocket deer
Small-scaled burrowing asps, 7:462
Small-scaled pangolins. *See* Tree pangolins
Small screaming armadillos. *See* Small hairy armadillos
Small-spotted genets. *See* Common genets
Small-spotted lizards. *See* Long-nosed leopard lizards
Small Sulawesi cuscuses, 13:57–58, 13:60, 13:*62*, 13:*63–64*
Small-toothed fruit bats, 13:348*t*
Small-toothed moles, 13:282
Small-toothed mongooses. *See* Falanoucs
Small-toothed palm civets, 14:344*t*
Small-toothed sportive lemurs, 14:*78*, 14:*79*, 14:82–83
Small whites. *See* European cabbage whites
Smaller horseshoe bats. *See* Eastern horseshoe bats
Smallmouth basses, 5:199
Smallscale yellowfins, 4:*305*, 4:*319*
Smalltail lancelets, 1:488, 1:*489*, 1:*490*
Smalltooth sawfishes, 4:*176*, 4:*180*, 4:*186*
Smaragdia viridis, 2:*439*, 2:*440*
Smaragdiinae, 2:440
Smell, 8:6, 12:5–6, 12:49–50, 12:80–81, 12:*81*, 12:*83*
See also Olfactory senses; Physical characteristics
Smelt whitings. *See* Sillagos
Smelts, 4:69, **4:389–393**, 4:*394*, **4:401**
See also Australian smelts
Smicridea spp., 3:376
Smicronrnis brevirostris. *See* Weebills
Smilodon californicus. *See* Saber-toothed cats
Sminthopsis archeri, 12:279
Sminthopsis crassicaudata. *See* Fat-tailed dunnarts
Sminthopsis leucopus. *See* White-footed dunnarts
Sminthopsis longicaudata. *See* Long-tailed dunnarts
Sminthopsis murina. *See* Slender-tailed dunnarts
Sminthopsis psammophila. *See* Sandhill dunnarts
Sminthopsis virginiae, 12:279
Sminthurides aquaticus, 3:100
Sminthurids, 3:100
Sminthurus viridis. *See* Lucerne fleas
Smintinae, 16:211, 16:212
Smith, G. A., 9:275
Smith, J. B. L., 4:189
Smith, J. Maynard, 3:67
Smith, Len, 10:330
Smith, M. F., 16:263
Smithornis spp. *See* Broadbills
Smithornis capensis. *See* African broadbills
Smith's rock sengis. *See* Western rock sengis
Smith's tree squirrels, 16:*165*
Smoke-colored peewees. *See* Greater pewees
Smoky bats, 13:308, **13:467–471,** 13:*468*, 13:*469*, 13:*470*
Smoky-brown woodpeckers, 10:*152*, 10:*161*
Smoky honeyeaters. *See* Common smoky honeyeaters
Smoky pocket gophers, 16:197*t*
Smoky shrews, 13:*194*
Smooth-breasted snakeheads. *See* Walking snakeheads
Smooth-coated otters, 14:325
Smooth-fronted caimans, 7:164, 7:171, 7:*174*, 7:*175*, 7:177
Smooth-headed scorpions, 2:*333*

Smooth lampshells. *See* California lampshells
Smooth newts, 6:*365*, 6:*366*, 6:374–375
Smooth-scaled splitjaws, 7:429, 7:*430*
Smooth snakes, 7:*472*, 7:*473–474*
Smooth-toothed pocket gophers. *See* Valley pocket gophers
Smooth trunkfishes, 5:*470*
Smoothtongues. *See* California smoothtongues
Smutsornis africanus. *See* Double-banded coursers
Smyrna mokarran, 4:116
Smythies, Bertram, 10:507
Snaggletooths, 4:423
Snail kites, 8:322, 8:324–315
Snail-suckers, 7:201, 7:467
Snailfishes, 5:179, 5:180–181, 5:183
Snails
caenogastropods, **2:445–449,** 2:*448*
feeding ecology, 2:36
as food, 2:41
golden apple, 2:*28*
humans and, 2:41
locomotion, 2:*19*
lung-bearing, **2:411–422,** 2:*412*, 2:*416*
reproduction, 2:18, 2:23
sea slugs, **2:403–410,** 2:*406*
skeletons, 2:*18*
true limpets, **2:423–427,** 2:*426*, 2:*427*
Vetigastropoda, **2:429–434,** 2:*432*
See also Neritopsina; specific types of snails
Snake charming, cobras, 7:485
Snake eaters, 7:461
Snake eels, 4:48
Snake lizards, 7:320, 7:321
See also Burton's snake lizards
Snake mackerels, 4:68, 5:*408*, 5:*409*
Snake millipedes, 2:365, 2:*366*, 2:*367*, 2:368
Snake mudheads. *See* African mudfishes
Snake oil, 7:50
Snake stars, 1:*387*, 1:388, 1:*391*, 1:*393–394*
Snakebirds. *See* American anhingas
Snakeflies, 3:9, 3:54, 3:*56*, **3:297–303,** 3:*298*, 3:*299*
Snakehead murrels. *See* Striped snakeheads
Snakeheads, 4:69, 5:*437*, **5:437–447,** 5:*438*, 5:*440*, 5:*441*
behavior, 5:438
conservation status, 5:439
distribution, 5:437–438
evolution, 5:437
feeding ecology, 5:438–439
habitats, 5:438
humans and, 5:439
physical characteristics, 5:437, 5:*438*
reproduction, 5:*438*, 5:439
species of, 5:*442–446*
taxonomy, 5:437
Snakeneck turtles, 7:*78*, 7:*80*, 7:*82*–83
Snakes, **7:195–208**
African burrowing, **7:461–464**
behavior, 7:204–206
blindsnakes, 7:198, 7:201, **7:379–385,** 7:*380*
boas, **7:409–417,** 7:*412*
colubrids, **7:465–482,** 7:*471*, 7:*472*
conservation status, 7:59, 7:207
corn, 7:*39*, 7:470
digestion, 7:27
distribution, 7:203–204
early blindsnakes, 7:198, 7:201, 7:202, **7:369–372**
ears, 7:9, 7:32
Elapidae, **7:483–499,** 7:*489*, 7:*490*
evolution, 7:16–17, 7:195–198
eyes, 7:9, 7:32
false blindsnakes, **7:387–388**
false coral snakes, **7:399–400**
fear of, 7:53–54
file snakes, 7:198, **7:439–444,** 7:*443*
in folk medicine, 7:50–52
as food, 7:48
habitats, 7:204
homing, 7:43
humans and, 7:49–50, 7:52–53
integumentary system, 7:29
locomotion, 7:8, 7:*25*–26
lungs, 7:10
migrations, 7:43
in mythology and religion, 7:*50*, 7:*52*, 7:54–57
Neotropical sunbeam, **7:405–407**
olfactory system, 7:34–36
physical characteristics, 7:*197*, 7:198–203, 7:*200*, 7:*201*
pipe snakes, **7:395–397**
pythons, **7:419–428,** 7:*423*
reproduction, 7:6–7, 7:40–42, 7:206–207
salt glands, 7:30
shieldtail, **7:391–394**
skulls, 7:26, 7:*201*
slender blindsnakes, **7:373–377**
splitjaw, **7:429–431,** 7:*430*
sunbeam, **7:401–403**
taxonomy, 7:195–198, 7:*196*
teeth, 7:9, 7:26
territoriality, 7:44
Tropidophiidae, **7:433–438**
Viperidae, **7:445–460,** 7:*449*, 7:*450*
See also specific types of snakes
Snappers, 4:48, 4:68, **5:255–263,** 5:*264*, **5:268,** 5:*269*, **5:270**
Snapping shrimps. *See* Pistol shrimps
Snapping termites, Linnaeus's, 3:*169*, 3:*171*, 3:174
Snapping turtles, **7:93–97,** 7:*95*
farming, 7:48
as food, 7:47
New Guinea, 7:*68*
reproduction, 7:*6*, 7:72, 7:94, 7:*94*, 7:95–96
See also Alligator snapping turtles
Snares penguins, 8:148, 8:151
Snipefishes, 5:131, 5:133–135
Snipes, 8:37, 9:104, 9:115, 9:175–180, 9:*181*, 9:182–183
See also Painted snipes; Seedsnipes
Snooks, **5:219–225,** 5:*224*
Snow, David, 9:373
Snow buntings, 11:*264*, 11:*268*, 11:*272–273*
Snow finches, 11:397, 11:*400*, 11:*405*
Snow fleas, 3:*100*, 3:101
See also Snow scorpionflies; Varied springtails
Snow geese, 8:365, 8:*371*
Snow leopards, 14:370–371, 14:372, 14:374, 14:*377*, 14:*380*, 14:386–387
Snow partridges, 8:434
Snow petrels, 8:108
Snow pigeons, 9:248, 9:*253*, 9:*256*
Snow scorpionflies, 3:341, 3:342, 3:*343*, 3:*344*, 3:*345*–346
Snow sheep, 16:87, 16:91, 16:92, 16:104*t*
Snow voles, 16:228
Snowbirds. *See* Snow buntings
Snowcocks, 8:434, 8:435
Snowflakes. *See* Snow buntings
Snowshoe hares, 16:*508*, 16:*510*, 16:*511*
feeding ecology, 16:484
habitats, 16:481, 16:506
humans and, 16:488
physical characteristics, 12:38, 16:*479*, 16:*507*
predation and, 12:115
Snowy albatrosses. *See* Wandering albatrosses
Snowy egrets, 8:26, 8:*245*
Snowy owls, 9:332, 9:333, 9:346, 9:*348*, 9:350, 9:*352*, 9:*359*
Snowy plovers, 9:104, 9:162–164, 9:*165*, 9:169–*170*
Snub-nosed monkeys, 14:11
See also Golden snub-nosed monkeys
Snubnosed eels, 4:69, 4:*259*, 4:269
Soa soas. *See* Sailfin lizards
Soapfishes (Diploprionini), 5:255, 5:261, 5:262
Soapfishes (Epinephelini), 5:256
Soapfishes (Grammistini), 5:255, 5:259, 5:261, 5:262
Soapfishes (Serranidae), 4:68
Sobek (Egyptian idol), 7:55
Sociable plovers, 9:161, 9:163, 9:164
Sociable weavers, 11:372, 11:*378*, 11:*380*, 11:*383*–384
Social behavior, 12:4–6, 12:9–10
babblers, 10:514
bee-eaters, 10:*39*–40
boobies, 8:212
boobies and, 8:213
cattle egrets, 8:251
Charadriiformes, 9:102
Ciconiiformes, 8:234
Columbidae, 9:249
cuckoos, 9:312
fishes, 4:62–64
flamingos, 8:306
gannets, 8:212, 8:213
herons, 8:242–243
little egrets, 8:254
Old World warblers, 11:5
penguins, 8:149–150
pseudo babblers, 11:128
trumpeters, 9:78
typical owls, 9:348
See also Behavior; Social insects
Social cognition, 12:160
See also Cognition
Social complexity, human, 14:251–252
Social grooming, 14:7
See also Behavior
Social insects, 1:37, 2:37, 3:35–36, 3:44, 3:49, 3:62, **3:67–73**
See also Eusociality; specific social insects
Social learning, 12:157–159
Social mimicry, 5:342
See also Behavior
Social organization. *See* Behavior
Social parasitism, 3:69
Social systems, 12:144–148
See also Behavior

Social tuco-tucos, 16:124, 16:426, 16:427, 16:*428*, 16:*429*–430
Sociobiology, 12:148
Sociological techniques, species lists and, 12:195
Sockeye salmons, 4:*4*, 4:*405*, 4:*407*, 4:*409*, 4:*411*, 4:414–415
Socorro mockingbirds, 10:468
Socorro wrens, 10:529
Socotra sunbirds, 11:208
Sodium monofluoroacetate (1080), 12:186
Soft corals, 1:103, 1:105, 1:106
Soft flounders. *See* Pacific sanddabs
Soft-furred sengis, 16:518, 16:520–521, 16:525–530
Soft palate, 12:8
See also Physical characteristics
Soft sculpins, 5:*185*, 5:*189*, 5:193
Soft tree corals. *See* Red soft tree corals
Softshell turtles, 7:48, 7:65–72, **7:151–155**, 7:*153*
Soil-dwelling ants, 3:72
Soil mealybugs, 3:63
Sokoke pipits, 10:373, 10:375
Sokolov, V. E., 16:216
Soldier beetles, 3:317, 3:320
Soldier bugs, spiny, 3:*266*, 3:*273*, 3:277–278
Soldier crabs. *See* Common hermit crabs
Soldierfishes, 4:48, 4:67, 5:113, 5:*114*, 5:*116*, 5:*117*, 5:*120*–121
Soldierless termites, New World, 3:161
Soldiers, termite, 3:163, 3:165
Solea solea. *See* Common soles
Soleidae. *See* Soles (Soleidae)
Solemyidae, 2:451, 2:453
Soleneiscus radovani, 1:*60*, 1:62–*63*
Solenidae, 2:452
Solenodon spp. *See* Solenodons
Solenodon arredondoi. *See* Arredondo's solenodons
Solenodon cubanus. *See* Cuban solenodons
Solenodon marconoi. *See* Marcano's solenodons
Solenodon paradoxus. *See* Hispaniolan solenodons
Solenodons, 12:133, 13:193, 13:194, 13:195, 13:197, 13:198, 13:199, **13:237–242**
Solenodontidae. *See* Solenodons
Solenogasters. *See* Aplacophorans
Solenogastres. *See* Neomeniomorpha
Solenopsis invicta. *See* Red imported fire ants
Solenoptilidae, 3:305
Solenostomidae. *See* Ghost pipefishes
Solenostomus spp. *See* Ghost pipefishes
Solenostomus cyanopterus, 5:133
Solenostomus paradoxus. *See* Ornate ghost pipefishes
Soles (Achiridae), 5:453
Soles (fossil), 5:449
Soles (Soleidae), 4:48, 4:70, 5:450, 5:452, 5:453
Solid-billed hornbills. *See* Helmeted hornbills
Solisorex spp., 13:265
Solisorex pearsoni. *See* Pearson's long-clawed shrews
Solitaires, 9:244, **9:269–274**, 10:483, 10:485, 10:*490*, 10:*495*
Solitary behavior, 12:145, 12:148
See also Behavior
Solitary entoprocts, 1:319, 1:*320*, 1:321, 1:*322*, 1:*324*, 1:325
Solitary sandpipers, 9:179
Solitary tinamous, 8:59
Solivomer spp., 4:442
Solmundella bitentaculata, 1:*130*, 1:143
Solnhofenamiinae, 4:229
Solomons flowerpeckers. *See* Midget flowerpeckers
Solos, 14:208
Solpugida, 2:333
Somali bushbabies, 14:26, 14:32*t*
Somali sengis, 16:522, 16:531*t*
Somali wild asses, 12:177
See also African wild asses
Somalian hedgehogs, 13:213*t*
Somaliland pin-tailed sandgrouse. *See* Chestnut-bellied sandgrouse
Somaliland sandgrouse. *See* Lichtenstein's sandgrouse
Somateria fischeri. *See* Spectacled eiders
Somateria mollissima. *See* Common eiders
Somateria spectabilis. *See* King eiders
Somatic cells, 1:18
Somatosensory perception, 12:77
See also Physical characteristics
Somber greenbuls, 10:397
Somber tits, 11:156, 11:*159*, 11:*160*
Sommer, V., 14:214
Somniosidae. *See* Sleeper sharks
Somniosus spp., 4:151
Somniosus microcephalus. *See* Greenland sharks
Somniosus pacificus. *See* Pacific sleeper sharks
Somuncuria spp., 6:157, 6:158
Sonar, in field studies, 12:201
Song sparrows, 8:40, 8:41, 11:*268*, 11:269–*270*
Song thrushes, 10:345, 10:484, 10:486, 10:487, 10:488, 10:489
Song wrens, 10:528
Songbirds, 8:30, 8:32, 8:37–43, 10:169–174, 11:1
Songs, bird, **8:37–43**
anatomy and, 8:39
ant thrushes, 10:240
bird hearing, 8:39–40
counter-singing, 9:300
cranes, 9:24–25
development of, 8:40–41
duets, 8:37, 10:115, 10:120, 10:121, 11:4, 11:49, 11:76, 11:117
functions of, 8:41–42
larks, 10:343–344
loons, 8:160
lyrebirds, 10:330–331, 10:332
method of study, 8:37–39
mimicry, 11:4, 11:22, 11:506
owls, 9:348–349
passerines, 10:172–173
species-specific, 8:40
woodpeckers, 8:37
wrens, 10:527
See also Behavior; specific species
Sonograms, bird songs and, 8:38
Sonora spp. *See* Groundsnakes
Sonora clingfishes, 5:*358*, 5:*360*, 5:362
Sonoran collared lizards. *See* Common collared lizards
Sonoran pronghorns, 15:415, 15:416
Sonoran tiger salamanders, 6:360
Sooglossidae. *See* Seychelles frogs
Sooglossus gardineri. *See* Gardiner's frogs
Sooglossus sechellensis. *See* Seychelles frogs
Sooty albatrosses, 8:110
Sooty-caped puffbirds, 10:103–104
Sooty grunters, 5:*223*, 5:*227*, 5:232
Sooty mangabeys, 14:194, 14:204*t*
Sooty moustached bats, 13:435, 13:436, 13:442*t*
Sooty owls, 9:336, 9:338, 9:*339*, 9:*340*
Sooty oystercatchers, 9:126
Sooty shearwaters, 8:30, 8:*31*, 8:110
Sooty terns, 9:209, 9:*210*, 9:*215*–216
Sooty thicket-fantails, 11:83, 11:85, 11:*88*, 11:*94*
Sooty tits, 11:143
Soras rails, 9:52
Sorbeoconcha, 2:445, 2:446
Sorberacea. *See* Sorberaceans
Sorberaceans, **1:479–483**, 1:*481*
Sorex spp., 13:251, 13:268
Sorex alpinus. *See* Alpine shrews
Sorex araneus. *See* Common shrews
Sorex bendirii, 13:249
Sorex cinereus. *See* Masked shrews
Sorex daphaenodon, 13:248
Sorex dispar. *See* Long-tailed shrews
Sorex fumeus. *See* Smoky shrews
Sorex hoyi. *See* American pygmy shrews
Sorex minutissimus, 13:247
Sorex minutus. *See* Eurasian pygmy shrews
Sorex mirabilis. *See* Giant shrews
Sorex ornatus. *See* Ornate shrews
Sorex palustris. *See* American water shrews
Sorex tundrensis. *See* Tundra shrews
Sorghum headworms. *See* Corn earworms
Soriacebus spp., 14:143
Soricidae. *See* Shrews
Soricini, 13:247
Soriciniae. *See* Red-toothed shrews
Soriculus spp., 13:248
Soriculus caudatus. *See* Hodgson's brown-toothed shrews
Sorosichthys spp., 5:115
Sotalia fluviatilis. *See* Tucuxis
Soto, D., 5:99
Soulé, Michael, 1:48
Sound receptors, 12:9
Sound transmission, bird song and, 8:39
Sounds, 4:61, 12:81–82
See also Physical characteristics; Songs, bird
Sousa spp. *See* Humpback dolphins
Sousa chinensis. *See* Humpback dolphins
Sousa teuszi. *See* Atlantic humpbacked dolphins
Sousa's shrikes, 10:429
South African buttonquails. *See* Black-rumped buttonquails
South African Cape dories, 5:123
South African crested porcupines. *See* South African porcupines
South African ground squirrels, 16:145, 16:*149*, 16:*151*, 16:154
South African pangolins. *See* Ground pangolins
South African porcupines, 16:*352*, 16:*354*, 16:*355*, 16:*361*–362
South African red-billed hornbills. *See* Red-billed hornbills
South America. *See* Neotropical region
South American beavers. *See* Coypus
South American bullfrogs, 6:*157*, 6:*160*, 6:166–167
South American bush rats. *See* Degus

South American butterflies, 3:65
South American deer, 15:380, 15:382, 15:383
South American fur seals, 14:394, 14:399, 14:407*t*
South American hillstream catfishes, 4:352
South American knifefishes, **4:369–377,** 4:*372*
South American leaf-fishes. *See* Amazon leaffishes
South American lungfishes, 4:6, 4:57, 4:*201*, 4:203, 4:*204*, 4:*205*
South American maras, 16:125
South American mice, **16:263–279,** 16:*271*, 16:277*t*–278*t*
South American monotremes, 12:228
South American nightjars, 9:369
South American painted-snipes, 9:115, 9:116–117, 9:*118–119*
South American pearl kites, 8:319
South American pigfishes, 5:*167*, 5:170–*171*
South American pilchards, 4:279, 4:*280*, 4:*286*
South American rats, **16:263–279,** 16:*271*, 16:277*t*–278*t*
South American river turtles, 7:70, 7:*139*, 7:*140*
South American ruddy quail doves, 9:250
South American sea lions, 14:394, 14:396, 14:399, 14:*401*, 14:*404*–405
South American skinks, 7:206, 7:329
South American tapirs. *See* Lowland tapirs
South American tropical lizards, 7:243
South American yellow-footed tortoises, 7:*145*, 7:*146*
South Georgia pipits, 10:*372*, 10:375–376
South Georgian diving-petrels, 8:143
South Island riflemen. *See* Riflemen
South Polar skuas, 9:206, 9:*207*
Southeast Asia. *See* Oriental region
Southeast Asian bent-winged bats, 13:526*t*
Southeast Asian broad-skulled toads. *See* Common Sunda toads
Southeast Asian damselflies, 3:133
Southeast Asian porcupines. *See* Common porcupines
Southeast Asian shrews, 13:276*t*
Southeastern African lungfishes, 4:*18*, 4:201, 4:202, 4:203
Southeastern pocket gophers, 16:187, 16:195*t*
Southern African hedgehogs, 13:204, 13:*206*, 13:207, 13:*208*, 13:*209*–210
Southern Bahian masked titis, 14:148
Southern bamboo rats, 16:450, 16:*452*, 16:*454*, 16:456–457
Southern beaked whales. *See* Gray's beaked whales
Southern bearded sakis, 14:*146*, 14:148, 14:151
Southern birch mice, 16:*214*, 16:*218*, 16:*219*
Southern blossom bats, 13:*336*, 13:*339*, 13:*346*–347
Southern bluefin tunas, 5:407
Southern boobook owls, 9:348, 9:*353*, 9:*364*
Southern bottlenosed whales, 15:69*t*
Southern bromeliad woodsnakes, 7:*436–437*
Southern brown bandicoots, 13:2, 13:5, 13:10–11, 13:*12*, 13:16*t*
Southern burrowing asps, 7:*463*–464
Southern carmine bee-eaters. *See* Carmine bee-eaters
Southern cassowaries, 8:*53*, 8:*76*, 8:*78*, 8:79, 8:*79*
Southern cavefishes, 5:6, 5:7, 5:*9*, 5:10
Southern chamois, 16:87, 16:103*t*
Southern cod-icefishes, 4:49, **5:321–329,** 5:*324*
Southern copperheads, 7:52, 7:447, 7:448
Southern coquinas. *See* Coquina clams
Southern dibblers, 12:281, 12:*293*, 12:*295*–296
Southern elephant seals, 14:256, 14:421–422, 14:*423*, 14:435*t*
Southern emu-wrens, 11:*47*, 11:*48*
Southern figbirds. *See* Australasian figbirds
Southern flying squirrels, 12:*55*, 16:135, 16:*136*, 16:*138*, 16:139, 16:*140*
Southern foam nest treefrogs. *See* Gray treefrogs
Southern four-toothed whales. *See* Arnoux's beaked whales
Southern fulmars, 3:349
Southern fur seals. *See* Galápagos fur seals
Southern gastric brooding frogs, 6:149, 6:*150*, 6:153
Southern giant bottlenosed whales. *See* Arnoux's beaked whales
Southern giant petrels, 8:108, 8:*125*, 8:*127*, 8:132, 8:*133*
Southern grasshopper mice, 16:277*t*
Southern gray shrikes, 10:427
Southern gray tits, 11:155
Southern graylings, 4:390
Southern green bugs. *See* Southern green stink bugs
Southern green stink bugs, 3:*266*, 3:*275*, 3:277
Southern ground-hornbills, 6:*68*, 10:71, 10:73, 10:75, 10:77, 10:*78*
Southern hairy-nosed wombats, 13:40, 13:51, 13:52, 13:53, 13:*54*, 13:*55*
Southern horned screamers, 8:393
Southern house wrens. *See* House wrens
Southern koalas, 13:44
Southern krill. *See* Antarctic krill
Southern lapwings, 9:163
Southern lesser bushbabies. *See* Moholi bushbabies
Southern logrunners, 11:69, 11:*70*, 11:71, 11:*72*
Southern long-nosed armadillos, 13:191*t*
Southern long-nosed bats, 13:415, 13:417, 13:*419*, 13:420, 13:*421*, 13:*428*
Southern long-tongued bats, 13:433*t*
Southern minnows. *See* Galaxiids
Southern muriquis, 14:11, 14:156, 14:158, 14:159, 14:*161*, 14:*164*
Southern naked-tailed armadillos, 13:*185*, 13:190*t*
Southern needle-clawed bushbabies, 14:26, 14:*27*, 14:*28*
Southern opossums, 12:131, 12:*143*, 12:*251*, 12:*253*, 12:264*t*
Southern platypus frogs. *See* Southern gastric brooding frogs
Southern pochards, 8:364
Southern pocket gophers, 16:*188*, 16:195*t*
Southern pudus, 15:379, 15:*382*, 15:384, 15:*387*, 15:*389*, 15:392–393
Southern pygmy perches, 5:*224*, 5:*229*–230
Southern red-backed voles, 16:237*t*
Southern red bishops, 11:*381*, 11:*389*
Southern reedbucks, 16:27–30, 16:*33*, 16:*35*, 16:*37*
Southern right whale dolphins, 15:42
Southern right whales, 15:107, 15:*112*, 15:*114*
behavior, 15:*110*, 15:*113*, 15:117
conservation status, 15:111, 15:112, 15:117–118
distribution, 15:*117*
feeding ecology, 15:*108*, 15:117
habitats, 15:117
humans and, 15:118
physical characteristics, 15:*108*, 15:109, 15:117
reproduction, 15:117
taxonomy, 15:117
See also Right whales
Southern river otters, 14:324
Southern royal albatrosses, 8:113
Southern rufous sparrows, 11:*400*, 11:*403*
Southern sagebrush lizards. *See* Common sagebrush lizards
Southern sandfishes, 5:331–335
Southern screamers, 8:*394*
Southern scrub robins, 11:106, 11:*107*, 11:*112*
Southern sea wasps, 1:*147*, 1:148
Southern short-tailed bats. *See* Greater New Zealand short-tailed bats
Southern short-tailed opossums, 12:250, 12:251, 12:265*t*
Southern short-tailed shrews, 13:198, 13:*252*
Southern smoothtongues. *See* California smoothtongues
Southern stingrays, 4:*26*, 4:*176*, 4:179
Southern tamanduas, 13:*151*, 13:*172*, 13:*176*, 13:177–*178*
Southern three-banded armadillos, 13:*186*, 13:188–*189*
Southern three-toed toadlets, 6:*182*
Southern tree hyraxes, 15:177, 15:*181*, 15:*182*, 15:184, 15:*185*, 15:*186*
Southern tuco-tucos, 16:430*t*
Southern two-toed sloths, 13:*148*, 13:*157*
Southern viscachas, 16:*379*, 16:380, 16:383*t*
Southern waterdogs. *See* Dwarf waterdogs
Southern white-faced owls, 9:331, 9:*352*, 9:355–*356*
Southern whitefaces, 11:*58*, 11:*62*
Southward, Eve C., 2:92
Sowbugs, 2:249, 2:250, 2:251
Sowerby's beaked whales, 15:69*t*
Spaceflight, 8:51
Spade-headed wormlizards, **7:287–290**
Spade-toothed whales, 15:1, 15:70*t*
Spadefishes, 4:69, 5:391, 5:392, 5:393
Spadefoot toads, 6:4–5, 6:64, **6:119–125,** 6:*120–121*, 6:*123*
See also Burmese spadefoot toads
Spadella spp., 1:434
Spadella cephaloptera, 1:434, 1:435, 1:*436*, 1:*441*–442
Spadellidae, 1:433
Spaepen, J. F., 10:346
Spaghetti worms, 1:*446*, 1:*448*, 1:449–450
Spalacinae, 16:123, 16:282, 16:284
Spalacopus spp., 16:433
Spalacopus cyanus. *See* Coruros
Spalacopus cyanus maulinus, 16:439
Spalax spp. *See* Blind mole-rats
Spalax ehrenbergi. *See* East Mediterranean blind mole-rats
Spalax graecus. *See* Bukovin mole rats
Spallanzani, Lazzaro, 2:117, 2:118, 12:54

Spangled cotingas, 10:307, 10:*310*, 10:*312*, 10:314
Spangled drongos, 11:438, 11:442
Spangled owlet-nightjars, 9:387
Spangled pike cichlids. *See* Millets
Spanish dancers, 1:45
Spanish flies, 3:81, 3:*324*, 3:327, 3:331
Spanish ibex, 16:87, 16:91, 16:104*t*
Spanish imperial eagles, 8:323
Spanish mackerels, 5:406, 5:407
Spanish sharp-ribbed newts, 6:*372*
Spanish slugs, 2:*416*, 2:*418*, 2:419–420
Spanish sparrows, 11:398
Spanner crabs, 2:*203*, 2:*209*, 2:212–213
Sparganophilidae, 2:65
Sparidae. *See* Porgies
Sparisoma viride. *See* Stoplight parrotfishes
Sparking violet-ears, 9:442, 9:*454*, 9:*460*–461
SPARKS (Small Population Analysis and Record Keeping System), 12:206–207
Sparrow-larks, 10:342, 10:344
Sparrow-weavers, 11:376, 11:377–378, 11:*380*, 11:382–383
Sparrowhawks, 8:320
Sparrows, **11:397–406,** 11:*398*, 11:*400*
  See also New World finches; specific types of sparrows
Spatangus purpureus. *See* Purple-heart urchins
Spathebothriidae, 1:225
Spathebothriidea, 1:225, 1:228
Spatial memory, 12:140–142, 12:153
Spawning, 1:18, 1:21–23, 2:17–18, 2:23–24, 2:27, 4:46, 4:47, 4:61, 4:65, 4:66, 4:*336*
  See also Reproduction
Spea spp., 6:119
Spea bombifrons. *See* Plains spadefoot toads
Spea intermontana, 6:122
Spear-nosed bats, greater, 13:312
Species
  bottleneck, 12:221–222
  eusocial, 12:146–147
  founder, 12:207–208, 12:221–222
  invasive, **12:182–193**
  keystone, 12:216–217
  lists, 12:194–195
  See also Introduced species; specific species
Species interactions, 2:28
Species Survival Commission (SSC), 1:48, 14:11
Species Survival Plan (SSP), 12:209
Specific symbiosis, 2:31
  See also Behavior
Speckle-chested piculets, 10:150
Speckled antshrikes, 10:242
Speckled basses. *See* Black crappies
Speckled bush-crickets, 3:*209*, 3:*212*, 3:214
Speckled cats. *See* Bowfins; Channel catfishes
Speckled dasyures, 12:281, 12:288, 12:*293*, 12:*295*
Speckled earless lizards. *See* Common lesser earless lizards
Speckled marsupial mice. *See* Southern dibblers
Speckled mousebirds. *See* Bar-breasted mousebirds
Speckled pavons, 5:*283*, 5:*284*, 5:285
Speckled ragfishes. *See* Ragfishes
Speckled rossophyllums, 3:*209*, 3:*211*, 3:215–216
Speckled sea trouts. *See* Kelp greenlings
Speckled trouts. *See* Brook trouts
Speckled warblers, 11:56
Specklethroated woodpeckers. *See* Bennett's woodpeckers
Spectacled bears, 14:*301*, 14:305
  behavior, 14:297, 14:298
  conservation status, 14:300
  distribution, 14:*302*
  evolution, 14:295–296
  habitats, 14:297
  physical characteristics, 14:296, 14:*299*
Spectacled bushbabies. *See* Dusky bushbabies
Spectacled caimans. *See* Common caimans
Spectacled cormorants, 8:186
Spectacled dormice, 16:318, 16:*322*, 16:*323*
Spectacled eiders, 8:*365*
Spectacled flying foxes, 12:*57*, 13:319, 13:*321*, 13:*322*, 13:*324*, 13:*326*, 13:329–330
Spectacled fruit bats. *See* Spectacled flying foxes
Spectacled hare wallabies, 13:34
Spectacled monarch flycatchers, 11:*98*
Spectacled owls, 9:349, 9:*353*, 9:360–*361*
Spectacled porpoises, 15:4, 15:33, 15:34, 15:35, 15:36, 15:39*t*
Spectacled salamanders, 6:*365*
Spectacled squirrels. *See* Peters's squirrels
Spectacled thrushes. *See* Hwameis
Spectacled weavers, 11:*378*, 11:*380*, 11:384–*385*
Spectral bats, 13:*310*, 13:313, 13:418, 13:419, 13:420, 13:*422*, 13:424–*425*
Spectral tarsiers, 14:*92*, 14:*93*, 14:94, 14:95, 14:*96*, 14:97, 14:98
Spectrographs, bird songs and, 8:38
Speculipastor bicolor. *See* Magpie starlings
Speke's pectinators, 16:312, 16:313, 16:*314*
Spelaeogriphaceans, **2:243–244**
Spelaeogriphidae, 2:243
Spelaeogriphus lepidops, 2:*243*, 2:*244*
Spelaeomysis spp., 2:215
Spelaeomysis bottazzii, 2:*218*, 2:*219*–220
Speleomantes spp., 6:391
Speleonectes gironensis, 2:*126*, 2:*127*, 2:*128*
Speleonectes lucayensis, 2:126
Spencer's burrowing frogs, 6:140, 6:*140*, 6:*141*
Speonebalia spp., 2:161
Speoplatyrhinus poulsoni. *See* Alabama cavefishes
Speothos venaticus. *See* Bush dogs
Sperm, 2:16–17, 2:23, 3:38–39, 3:61, 12:91, 12:101, 12:104
  competition, 12:98–99, 12:105–106
  delayed fertilization and, 12:110
  motility, 1:17
  See also Reproduction
Sperm whale lice, 2:*263*, 2:*266*–*267*
Sperm whales, **15:73–80,** 15:*74*, 15:*75*, 15:*76*, 15:*77*
  behavior, 15:6, 15:7, 15:74–75
  conservation status, 15:78, 15:79–80
  distribution, 15:*73*, 15:74
  evolution, 15:3, 15:73
  feeding ecology, 15:75–76
  habitats, 15:74
  humans and, 15:78
  physical characteristics, 15:4, 15:73
  reproduction, 15:9, 15:76–77
  species of, 15:79–80
  taxonomy, 12:14, 15:1, 15:2, 15:73
Spermatogenesis, 1:17–18, 1:21–23, 2:16–17, 4:32, 12:94–95, 12:102–103
  See also Reproduction
Spermatogenic morulae, 2:17
  See also Reproduction
Spermatophores, 1:18, 2:17, 2:23, 6:33, 6:365–366, 6:371
  See also Reproduction
Spermophaga spp. *See* Bluebills
Spermophaga haematina. *See* Western bluebills
Spermophaga niveoguttata. *See* Peter's twinspots
Spermophilopsis spp., 16:143, 16:144, 16:145, 16:147
Spermophilus spp., 16:143
Spermophilus armatus. *See* Uinta ground squirrels
Spermophilus beldingi. *See* Belding's ground squirrels
Spermophilus brunneus. *See* Idaho ground squirrels
Spermophilus citellus. *See* European ground squirrels
Spermophilus columbianus. *See* Columbian ground squirrels
Spermophilus lateralis. *See* Golden-mantled ground squirrels
Spermophilus leptodactylus. *See* Long-clawed ground squirrels
Spermophilus mexicanus. *See* Mexican ground squirrels
Spermophilus mohavensis, 16:147–148
Spermophilus parryii. *See* Arctic ground squirrels
Spermophilus richardsonii. *See* Richardson's ground squirrels
Spermophilus saturatus, 16:143
Spermophilus tridecemlineatus. *See* Thirteen-lined ground squirrels
Spermophilus washingtoni, 16:147–148
Spermophorella maculatissima. *See* Beaded lacewings
Sperosoma giganteum, 1:401
Speyeria spp., 3:54
Speyeria adiaste atossa, 3:388
Sphaenorhynchus lacteus, 6:229
Sphaeramia nematoptera. *See* Pajama cardinalfishes
Sphaerias blanfordi. *See* Blanford's fruit bats
Sphaerichthys spp., 5:427, 5:428, 5:433
Sphaerodactylus ariasae. *See* Jaragua sphaero
Sphaeroeca spp., 2:12
Sphaeroma spp., 2:252
Sphaeroma terebrans, 2:250, 2:*253*, 2:*258*, 2:259
Sphaeromatidae, 2:252
Sphaerotheca spp., 6:246, 6:248
Sphaerotheriida, 2:365
Sphecidae, 3:62, 3:406
Spheciospongia spp., 1:29
Sphecotheres spp. *See* Figbirds
Sphecotheres hypoleucus. *See* Wetar figbirds
Sphecotheres vielloti. *See* Australasian figbirds
Sphecotheres viridis. *See* Timor figbirds
Spheniscidae. *See* Penguins
Sphenisciformes. *See* Penguins
Spheniscus spp. *See* Penguins
Spheniscus demersus. *See* African penguins
Spheniscus humboldti. *See* Humboldt penguins
Spheniscus magellanicus. *See* Magellanic penguins
Spheniscus mendiculus. *See* Galápagos penguins

Sphenodon spp. *See* Tuatara
Sphenodon guntheri, 7:189–*192*
Sphenodon punctatus, 7:189–*192*
 See also Tuatara
Sphenodontidae. *See* Tuatara
Sphenomorphus spp., 7:327–329
Sphenophryne cornuta. *See* Horned land frogs
Sphere urchins. *See* Tuxedo pincushion urchins
Spheritidae, 3:54
Sphiggurus spp., 16:366, 16:367
Sphiggurus mexicanus. *See* Mexican hairy porcupines
Sphiggurus spinosus. *See* Paraguay hairy dwarf porcupines
Sphiggurus vestitus. *See* Brown hairy dwarf porcupines
Sphiggurus villosus. *See* Orange-spined hairy dwarf porcupines
Sphinctozoans, 1:10
Sphingids, 3:90, 3:385, 3:386, 3:387
Sphyraena barracuda. *See* Great barracudas
Sphyraenidae, 4:48, 4:68, 4:115, 5:405
Sphyrapicus ruber. *See* Red-breasted sapsuckers
Sphyrapicus varius. *See* Yellow-bellied sapsuckers
Sphyriocephalidae, 1:226
Sphyrna lewini. *See* Scalloped hammerhead sharks
Sphyrna mokarran. *See* Great hammerhead sharks
Sphyrna tiburo. *See* Bonnethead sharks
Sphyrnidae, 4:27, 4:113
Sphyrometopa spp., 3:203
Spice finches. *See* Spotted munias
Spicer, G. S., 16:143
Spicospina flammocaerulea, 6:148
Spider bat flies, 3:*363*, 3:*369*–370
Spider hazards. *See* Deep water reef corals
Spider-hunting wasps. *See* Tarantula hawks
Spider-Man comics, 2:41
Spider monkeys, **14:155–169,** 14:*161*
 behavior, 14:7, 14:157
 conservation status, 14:159–160
 distribution, 14:156
 evolution, 14:155
 feeding ecology, 14:158
 habitats, 14:156–157
 humans and, 14:160
 reproduction, 14:158–159
 species of, 14:*163*–164, 14:167*t*
 taxonomy, 14:155
Spider wasps. *See* Tarantula hawks
Spiderhunters, 11:208, 11:209, 11:210, 11:*212*, 11:225
Spiders, 2:25, **2:333–352,** 2:*338*
 behavior, 2:36, 2:37, 2:335
 conservation status, 2:336
 distribution, 2:335
 evolution, 2:333
 feeding ecology, 2:335
 habitats, 2:335
 humans and, 2:41, 2:336
 physical characteristics, 2:333–334
 reproduction, 2:23, 2:335
 sea, **2:321–325**
 species of, 2:339–351
 taxonomy, 2:333
Spiders' nests. *See* Deep water reef corals
Spike-heeled larks, 10:346
Spikefishes, 5:467–471, 5:*473*, 5:*482*–483
Spilocuscus spp., 13:57
Spilocuscus maculatus. *See* Common spotted cuscuses
Spilocuscus papuensis. *See* Waigeo cuscuses
Spilocuscus rufoniger. *See* Black-spotted cuscuses
Spilogale spp., 14:319, 14:321
Spilogale gracilis. *See* Western spotted skunks
Spilogale gracilis amphiala. *See* Island spotted skunks
Spilogale putorius. *See* Eastern spotted skunks
Spilopsyllus spp., 3:350
Spiloptila spp., 11:4
Spilornis elgini. *See* Andaman serpent-eagles
Spilotes spp. *See* Tiger ratsnakes
Spinal cord, 7:31
 See also Physical characteristics; Skeleton
Spine-cheek anemonefishes, 5:*294*
Spine-tailed logrunners. *See* Southern logrunners
Spine-tailed swifts. *See* White-throated needletails
Spined pygmy sharks, 4:151
Spinefoots. *See* Rabbitfishes
Spinejaw snakes, **7:433–438**
Spines, 12:38, 12:*43*
Spinetails, 9:421, 10:210
Spinicapitichthys spp., 5:365
Spinner dolphins, 15:*1*, 15:6, 15:44, 15:46, 15:*48*, 15:49, 15:*50*, 15:*52*, 15:*54*–55
Spinophrynoides spp., 6:184
Spiny agamas, 7:*213*, 7:*214*
Spiny-armed frogs, 6:*253*, 6:258
Spiny bandicoots, 13:9, 13:17*t*
 See also Rufous spiny bandicoots
Spiny brittle stars, 1:*390*
Spiny butterfly rays, 4:*181*, 4:*183*
Spiny-cheeked honeyeaters, 11:237
Spiny devilfishes. *See* Bearded ghouls
Spiny dogfishes. *See* Piked dogfishes
Spiny dwarf catfishes, 4:352
Spiny eels, **5:151–156,** 5:*152*
Spiny flounders, 5:449–450, 5:452
Spiny-headed treefrogs, 6:37, 6:226, 6:228–229, 6:230, 6:*232*, 6:234–*235*
Spiny hedgehogs, 13:199
Spiny lizards, 7:41, 7:246, 7:*249*, 7:*251*, 7:254–255
Spiny lobsters, 1:51, 2:197, 2:198, 2:200
Spiny mice. *See* Egyptian spiny mice
Spiny plunderfishes, 5:321, 5:322
Spiny pocket mice, 16:199, 16:208*t*
 Desmarest's, 16:*204*
 forest, 16:199
 Mexican, 16:208*t*
 Nelson's, 16:208*t*
 Salvin's, 16:*204*, 16:*205*–206
Spiny prawns. *See* Banded coral shrimps
Spiny predator stink bugs. *See* Spiny soldier bugs
Spiny rats, 16:253, **16:449–459,** 16:*450*, 16:*451*, 16:*452*, 16:*453*, 16:458*t*
Spiny softshells, 7:*153*, 7:*154*–155
Spiny soldier bugs, 3:*266*, 3:*273*, 3:277–278
Spiny-tailed agamas, 7:*213*, 7:*214*, 7:221
Spiny-tailed iguanas. *See* Cape spinytail iguanas
Spiny tenrecs, 13:225, 13:226, 13:228, 13:230
Spiny tree rats
 Brazilian, 16:451, 16:458*t*
 tuft-tailed, 16:458*t*
Spinyfins, 5:113, 5:115
Spiomenia spiculata, 2:*381*, 2:*382*, 2:383
Spionidae, 2:46, 2:47
Spiracles, 3:21–22, 6:41
Spiral cleavage, 1:19, 1:21, 1:22, 2:3, 2:4, 2:19–20, 2:35
 See also Reproduction
Spiralia. *See* Spiral cleavage
Spirasigmidae, 1:79
Spiraxids, 2:414
Spirobolids, 2:364
Spirobolus spp., 2:365
Spirophorida, 1:77, 1:79
Spirorbids, 2:32, 2:46, 2:47
Spirorchiidae, 1:199
Spirostreptids, 2:364
Spirula spirula. *See* Ram's horn squids
Spirulida. *See* Ram's horn squids
Spitting cobras, 7:*487*, 7:*490*, 7:*492*–493
Spittle bugs, 3:64, 3:*265*, 3:269–*270*
Spittle-insects. *See* Spittle bugs
Spix's disk-winged bats, 13:312, 13:*474*, 13:*475*, 13:*476*
Spix's guans, 8:414
Spix's saddleback toads. *See* Pumpkin toadlets
Spizaetus ornatus. *See* Hawk-eagles
Spizella passerina. *See* Chipping sparrows
Spiziapteryx circumcinctus. *See* Spot-winged falconets
Spizixos canifrons. *See* Crested finchbills
Spizocorys spp. *See* Larks
Spizocorys fringillaris. *See* Botha's larks
Spizocorys obbiensis. *See* Obbia larks
Spizocorys sclateri. *See* Sclater's larks
Splash tetras, 4:338, 4:*340*, 4:*349*–350
Splashing tetras. *See* Splash tetras
Splay-toothed beaked whales. *See* Andrew's beaked whales
Splendid coral toadfishes, 5:*41*, 5:42, 5:*43*, 5:*44*, 5:45–46
Splendid fairy-wrens, 11:*46*, 11:*47*, 11:49–*50*
Splendid garden eels, 4:*259*, 4:*263*, 4:264–*265*
Splendid hawfishes, 5:238
Splendid perches, 5:255, 5:256, 5:258–263
Splendid sunbirds, 11:209
Splendid wrens. *See* Splendid fairy-wrens
Split-toes. *See* Mantis shrimps
Splitfin flashlightfishes, 5:*113*, 5:*117*, 5:*118*
Splitfins, 4:32, 4:57, 5:89, 5:92, 5:93
Splitjaw boas. *See* Splitjaw snakes
Splitjaw snakes, **7:429–431,** 7:*430*
Splits. *See* Atlantic herrings
Spondylidae, 2:452
Spondylophryne, 6:89
Sponges, 2:15
 behavior, 1:43, 2:36
 calcareous, **1:57–65,** 1:*58*, 1:*59*, 1:*60*
 demosponges, **1:77–86,** 1:*78*, 1:*81*
 evolution, 1:3–4, 1:6, 1:8, 2:9, 2:12
 feeding ecology, 1:27
 glass, **1:67–76,** 1:*68*, 1:*69*, 1:*70*
 humans and, 1:*44*–45
 pharmacological products from, 1:29
 physical characteristics, 1:*5*
 reproduction, 1:16–18, 1:19–20, 1:21
 symbiosis, 1:32, 1:*33*, 2:32
 taxonomy, 1:10
Spongia officinalis. *See* Bath sponges
Spongiidae, 1:79

Spongilla flies, 3:308
Spongilla lacustris. See Freshwater sponges
Spongillina. *See* Freshwater sponges
Spongothymidine, 1:44
Spongouridine, 1:44
Spontaneous ovulation, 12:92
Spookfishes. *See* Barreleyes
Spoon-billed sandpipers, 9:176, 9:179, 9:180, 9:*181*, 9:*186*
Spoon worms, 2:24, 2:*103*, 2:*104*
Spoonbill cats. *See* American paddlefishes
Spoonbills, **8:291–301**, 8:300, 9:179
  behavior, 8:235, 8:243, 8:292
  conservation status, 8:293
  distribution, 8:*291–292*
  evolution, 8:291
  feeding ecology, 8:292
  habitats, 8:292
  humans and, 8:293
  physical characteristics, 8:291
  reproduction, 8:236, 8:292–293
  species of, 8:*300–301*
  taxonomy, 8:233, 8:291
Spoonwing lacewings, 3:*310*, 3:*313*
Sporophila spp. *See* Seedeaters
Sporophila americana. See Variable seedeaters
Sporophila torqueola. See White-throated seedeaters
Sporting dogs, 14:288
Sportive lemurs, **14:73–84,** 14:*78*
  behavior, 14:8, 14:75–76
  conservation status, 14:76
  distribution, 14:*73*, 14:75
  evolution, 14:73–75
  feeding ecology, 14:8, 14:76
  habitats, 14:75
  humans and, 14:77
  physical characteristics, 14:75
  reproduction, 14:76
  species of, 14:*79–83*
  taxonomy, 14:3–4, 14:73–75
Spot-backed antbirds, 10:*243*, 10:*252*
Spot-billed pelicans, 8:227, 8:*229*, 8:232, 8:*232*
Spot-crowned antvireos, 10:*244*, 10:*248*
Spot-crowned barbets, 10:116
Spot-fin porcupinefishes, 5:*472*, 5:*475*, 5:477
Spot-fronted swifts, 9:424
Spot-headed honeyeaters. *See* Greater Sulawesi honeyeaters
Spot-winged falconets, 8:*356*
Spot-winged starlings, 11:*413*, 11:*416*
Spot-winged tits, 11:157
Spotless crakes, 9:48
Spots. *See* Channel catfishes
Spotted algae eaters, 4:*322*, 4:*326*, 4:333
Spotted-backed weavers. *See* Village weavers
Spotted bamboowrens, 10:259, 10:*261*, 10:265–*266*
Spotted bats, 13:*314*, 13:500, 13:*505*, 13:*508*–509
Spotted-billed pelicans. *See* Spot-billed pelicans
Spotted blindsnakes, 7:*197*, 7:379
Spotted bowerbirds, 11:*482*, 11:487–*488*
Spotted burrowing frogs. *See* Spotted snout-burrowers
Spotted buttonquails, 9:14, 9:17
Spotted catbirds. *See* Green catbirds
Spotted characins. *See* Splash tetras
Spotted Cochran frogs, 6:*218*, 6:221
Spotted coral crouchers, 4:42, 5:*168*, 5:170, 5:*171*
Spotted crakes, 9:49
Spotted creepers, 11:177
Spotted cuscuses, 13:57, 13:*59*
Spotted deer. *See* Chitals
Spotted dikkops, 9:144, 9:*147*
Spotted dolphins, 12:*164*, 15:49
  See also Atlantic spotted dolphins
Spotted eagle rays, 4:*181*, 4:*184*–185
Spotted eared-nightjars. *See* Spotted nightjars
Spotted flounders. *See* Windowpane flounders
Spotted flycatchers, 11:*29*, 11:*30*
Spotted gars, 4:*221*, 4:222, 4:*223*, 4:*225–226*, 4:227
Spotted greenbuls, 10:397, 10:398
Spotted ground-birds. *See* Spotted quail-thrushes
Spotted handfishes, 5:51
Spotted harlequin snakes, 7:*485*
Spotted honeyguides, 10:*140*, 10:*143*–144
Spotted hyenas, 14:*359*, 14:*361*, 14:*363*
  behavior, 12:145, 14:259, 14:363, 14:364
  conservation status, 14:363, 14:365
  distribution, 14:*364*
  evolution, 14:359–360
  feeding ecology, 12:*4*, 12:*147*, 14:260, 14:*362*, 14:364–365
  habitats, 14:364
  humans and, 14:363, 14:365
  physical characteristics, 14:360, 14:364
  reproduction, 14:365
  taxonomy, 14:359–360, 14:364
Spotted kiwis, 8:90
Spotted lizards. *See* Common lesser earless lizards
Spotted mannikins. *See* Spotted munias
Spotted mouse deer, 15:325, 15:*327*, 15:*330*, 15:*332*, 15:333
Spotted munias, 11:357, 11:*359*, 11:370–*371*
Spotted-necked otters, 14:321
Spotted nightjars, 9:*407*, 9:408–*409*
Spotted nothuras, 8:*60*, 8:*65–66*
Spotted nutcrackers, 11:504, 11:505, 11:507, 11:*511*, 11:*518*
Spotted owls, 9:333
Spotted palm-thrushes, 10:*491*, 10:*499*–500
Spotted pardalotes, 11:*203*, 11:204, 11:*205*
Spotted puffers. *See* Guinea fowl puffers
Spotted quail-thrushes, 11:76, 11:77, 11:*78*
Spotted ragfishes. *See* Ragfishes
Spotted rails, 9:*54*, 9:*64*–65
Spotted ratfishes, 4:*91*, 4:*93*, 4:*94*
Spotted redshanks, 9:177, 9:179, 9:*181*, 9:183–*184*
Spotted salamanders, 6:49
Spotted sanddivers, 5:*332*
Spotted sandgrouse, 9:*234*, 9:235–*236*
Spotted sandpipers, 9:178, 9:179
Spotted scats, 5:393, 5:*395*, 5:401–*402*
Spotted scorpionfishes, 5:*164*
Spotted seals. *See* Harbor seals
Spotted sharpnose puffers. *See* Spotted tobies
Spotted shovel-nosed frogs. *See* Spotted snout-burrowers
Spotted-sided finches. *See* Diamond firetails; Zebra finches
Spotted skunks
  eastern, 14:*321*, 14:324
  island, 12:192
  western, 14:324
Spotted snake eels. *See* Tiger snake eels
Spotted snout-burrowers, 6:*274*, 6:*276*, 6:277
Spotted-tailed quolls, 12:279, 12:280–281, 12:*282*, 12:289, 12:299*t*
Spotted thick-knees. *See* Spotted dikkops
Spotted tinselfishes. *See* Tinselfishes
Spotted tobies, 5:*472*, 5:481, 5:*482*
Spotted turtles, 7:*108*, 7:*110*, 7:111
Sprague's pipits, 10:372, 10:375, 10:*377*, 10:*383*
Sprats, 4:277
Spraying characins. *See* Splash tetras
Spreading adders. *See* Eastern hog-nosed snakes
Sprengel, Christian Konrad, 3:14
Spring, bird songs and, 8:37
Spring cavefishes, 5:7, 5:*8*, 5:9–10
Spring herrings. *See* Alewives; Atlantic herrings
Spring lizards. *See* Salamanders
Spring peepers, 6:47, 6:228–229
Spring salamanders, 6:390, 6:*391*
Springboks, 13:*139*, 16:7, **16:45–58,** 16:*47*, 16:*48*, 16:*49*, 16:*50*, 16:*56t–57t*
Springer, V. G., 4:54, 4:191, 5:341
Springhares, 16:122, 16:123, **16:307–310,** 16:*308*, 16:*309*, 16:*310*
Springtails, **3:99–105,** 3:*100*, 3:*102*
  conservation status, 3:86
  evolution, 3:10, 3:11, 3:12
  habitats, 3:57, 3:99–100
  physical characteristics, 3:4
  reproduction, 3:61, 3:100
Spruce-fir moss spiders, 2:*338*, 2:*340*, 2:343
Spruce grouse, 8:435
Spur-and-groove zones, 4:48
Spur-winged lapwings, 9:162
Spurdogs. *See* Piked dogfishes
Spurelli's free-tailed bats, 13:494*t*
Spurfowl, 8:434
Squacco herons, 8:*247*, 8:*252–253*
Squalea, 4:11, 4:143, 4:161, 4:167
Squalidae. *See* Dogfish sharks
Squaliformes. *See* Dogfish sharks
Squaliolus spp., 4:151
Squalodelphinidae, 15:13
Squalodontidae, 15:3, 15:13
Squalus spp., 4:15, 4:151, 4:152
Squalus acanthias. See Piked dogfishes
Squalus africanus. See Pajama catsharks
Squalus carcharias. See White sharks
Squalus cepedianus. See Broadnose sevengill sharks
Squalus cuvier. See Tiger sharks
Squalus glauca. See Blue sharks
Squalus griseus. See Bluntnose sixgill sharks
Squalus longimanus. See Oceanic whitetip sharks
Squalus maximus. See Basking sharks
Squalus microcephalus. See Greenland sharks
Squalus nasus. See Porbeagles
Squalus portus jacksoni. See Port Jackson sharks
Squalus vulpinus. See Thresher sharks
Squamata, **7:195–208**
  behavior, 7:204–206
  conservation status, 7:59, 7:207
  distribution, 7:203–204
  evolution, 7:195–198

feeding ecology, 7:204–*205*
habitats, 7:204
hearing, 7:32
integumentary system, 7:*30*
physical characteristics, 7:23–24, 7:*197*, 7:*199–201*
reproduction, 7:*205*, 7:*206*–207
skulls of, 7:5
taxonomy, 7:195–198, 7:*196*
teeth, 7:26
See also Lizards; Snakes
Squamous cells, 12:36
See also Physical characteristics
Square-lipped rhinoceroses. *See* White rhinoceroses
Square-tailed drongos, 11:438, 11:*440*, 11:*441*–442
Square-tailed kites, 8:*322*
Square-tailed saw-winged swallows. *See* Square-tailed saw-wings
Square-tailed saw-wings, 10:*362*, 10:*368*–369
Squarehead catfishes, 4:*350*, 4:353, 4:*356*
Squaretails. *See* Brook trouts
Squatina spp., 4:161
Squatina aculeata, 4:162
Squatina africana, 4:162
Squatina argentina, 4:162
Squatina australis, 4:161–162
Squatina californica. *See* Pacific angelsharks
Squatina dumeril, 4:162
Squatina formosa, 4:162
Squatina guggenheim, 4:162
Squatina japonica, 4:162
Squatina nebulosa, 4:162
Squatina occulta, 4:162
Squatina oculata, 4:162
Squatina squatina, 4:162, 4:*163*, 4:164–*165*
Squatina tergocellata, 4:161, 4:162
Squatina tergocellatoides, 4:161, 4:162
Squatinidae, 4:161
Squatiniformes. *See* Angelsharks
Squeakers, 4:352, 4:354, **6:265–271,** 6:*267*
Squids, 2:25, **2:475–483,** 2:*478*, 2:*481*, **2:486–489**
behavior, 2:37, 2:477–478
conservation status, 2:479
distribution, 2:477
evolution, 2:475
feeding ecology, 2:478
habitats, 2:477
humans and, 2:42, 2:479–480
locomotion, 2:*19*
physical characteristics, 2:475–477
reproduction, 2:17, 2:23, 2:478–479
species of, 2:*482*–483, 2:*486*–489
taxonomy, 2:475
Squilla mantis, 2:171
Squilloidea, 2:168, 2:171
Squirrel gliders, 13:37, 13:*127*, 13:*128*, 13:*129*, 13:133*t*
Squirrel hakes. *See* Red hakes
Squirrel monkeys, 14:4, **14:101–110,** 14:*107*
behavior, 14:103, 14:104
conservation status, 14:105–106
distribution, 14:*101*, 14:102–103
evolution, 14:101
feeding ecology, 14:104–105
habitats, 14:103
humans and, 14:106
memory, 12:153
physical characteristics, 14:101–102
reproduction, 14:10, 14:105
species of, 14:*108*–110
taxonomy, 14:101
Squirrelfishes, 4:*31*, 4:48, 4:67, **5:113–120,** 5:*117*
Squirrels
behavior, 16:124
claws, 12:39
complex-footed, 16:136
conservation status, 16:126
distribution, 12:131, 16:123
flying squirrels, **16:135–142**
ground, **16:143–161**
habitats, 16:124
humans and, 16:127
learning set, 12:152
locomotion, 12:43
physical characteristics, 16:123
scaly-tailed, 16:123, **16:299–306**
tree, 12:115, **16:163–176,** 16:*168*, 16:174*t*–175*t*
Sri Lanka frogmouths, 9:368, 9:*380*, 9:383–*384*
Sri Lanka hill mynas. *See* Sri Lanka mynas
Sri Lanka magpies, 11:508
Sri Lanka mynas, 11:410, 11:*412*, 11:*418*
Sri Lankan gray hornbills, 10:72
Sri Lankan hump-nosed pitvipers, 7:*450*, 7:*454*, 7:455–456
SSC (Species Survival Commission), 1:48, 14:11
SSP (Species Survival Plan), 12:209
Stable flies, 3:76
Stachyris negrorum. *See* Negros striped-babblers
Stachyris speciosa. *See* Flame-templed babblers
Stactolaema anchietae. *See* Anchieta's barbets
Stactolaema whytii. *See* Whyte's barbets
Stag beetles, 3:316, 3:317, 3:320
See also European stag beetles
Stagemakers. *See* Tooth-billed bowerbirds
Staghorn sculpins, 5:183
Stagmomantis carolina. *See* Carolina mantids
Stagmomantis limbata, 3:*180*
Stagnolepus spp., 7:17–18
Stagonopleura guttata. *See* Diamond firetails
Staining bugs, 3:*265*, 3:*268*, 3:278–279
Stalk-eyed flies. *See Cyrtodiopsis dalmanni*
Stalk-less barnacles. *See* Rock barnacles
Stalked barnacles, 2:275, 2:276
Stalked crinoids, West Atlantic, 1:*361*, 1:*364*, 1:365
See also Sea lilies
Stalked jellyfish, 1:*38*, 1:153, 1:154–155, 1:156, 1:*159*, 1:*160*, 1:167
Standard-winged nightjars, 9:369, 9:402, 9:404, 9:*407*, 9:*413*
Standardwing birds of paradise, 11:490, 11:*494*, 11:*500*–501
Stansbury's swift. *See* Common side-blotched lizards
Stapes, 12:36
Star finches, 11:356
Star-nosed moles, 13:196, 13:*282*, 13:*283*
behavior, 13:287
conservation status, 13:287
distribution, 13:*285*, 13:287
feeding ecology, 12:76, 13:*281*, 13:287
habitats, 13:*279*, 13:280, 13:287
humans and, 13:287
physical characteristics, 12:*70*, 12:83, 13:195, 13:287
reproduction, 13:287
taxonomy, 13:287
Star-pattern navigation, 8:34
Star Trek (television program), 8:22
Starfishes, 1:40
Stargazers, 4:19, 4:48, 4:69, 5:331–335
Stark's larks, 10:345
Starksia spp., 5:343
Starksia lepicoelia. *See* Blackcheek blennies
Starlet sea anemones, 1:107, 1:*110*, 1:111–*112*
Starlings, 8:25, 10:150, 10:361, **11:407–415,** 11:*412–413*, **11:420–425**
Staroenas, 9:247
Starostin's loaches. *See* Kughitang blind loaches
Starry moray eels, 4:*3*
Stars, migration by, 8:*33*
Startle displays, 1:42, 2:38–39
See also Behavior
States (social organizational structures), 14:252
Statocysts, 1:39
Staurois spp., 6:248
Staurois latopalmatus, 6:46
Stauromedusae. *See* Stalked jellyfish
Stauroteuthis syrtensis, 2:*481*, 2:*485*
Staurotypinae. *See* Musk turtles
Staurotypus spp., 7:121, 7:122
Staurotypus triporcatus. *See* Mexican giant musk turtles
Steambugs. *See* German cockroaches
Steamer-ducks, 8:363, 8:365, 8:370, 8:371, 8:372
Steamflies. *See* German cockroaches
Steatogenys elegans, 4:370
Steatomys pratensis. *See* Fat mice
Steatornis caripensis. *See* Oilbirds
Steatornithi. *See* Oilbirds
Steatornithidae. *See* Oilbirds
Steelheads. *See* Rainbow trouts
Steenboks, 15:*136*, 16:2, **16:59–72,** 16:*61*, 16:*63*, 16:*64*, 16:*65*, 16:*66*–*67*, 16:71*t*
Steenbucks. *See* Steenboks
Stefania spp., 6:37, 6:230
Steganopus tricolor. *See* Wilson's phalaropes
Stegastes planifrons. *See* Threespot damselfishes
Stegastes sanctaehelenae, 5:298
Stegastes sanctipauli, 5:298
Stegias clibanarii, 2:252
Stegobium paniceum. *See* Drugstore beetles
Stegocephalians, 6:7–11
Stegodons, 15:161
Stegostoma fasciatum. *See* Zebra sharks
Stegostomatidae. *See* Zebra sharks
Steinboks. *See* Steenboks
Steinbucks. *See* Steenboks
Steindachner, F.
on *Cynolebias bellottii*, 5:90
on ragfishes, 5:351
on sailfin sandfishes, 5:339
Steindachneria spp. *See* Luminous hakes
Steindachneria argentea. *See* Luminous hakes
Steindachneriidae. *See* Luminous hakes
Steindachner's turtles, 7:78
Steinernema spp., 1:36
Steirodon careovirgulatum, 3:202
Steiromys spp., 16:366

Stejneger's beaked whales, 15:70*t*
Stelechopidae, 2:59
Steller sea lions, 14:*394*, 14:396, 14:400, 14:*401*, 14:*402*, 14:406
Steller's jays, 11:504, 11:506–507
Steller's sea cows, **15:199–204,** 15:*202*, 15:*203*
  distribution, 15:*199*
  evolution, 15:191, 15:199
  feeding ecology, 15:193, 15:200
  habitats, 12:138, 15:192, 15:200
  humans and, 12:215, 15:195, 15:201
  physical characteristics, 15:192, 15:199
  reproduction, 15:194, 15:201
  taxonomy, 15:191–192
Steller's sea-eagles, 8:314, 8:319, 8:320, 8:323, 8:328, 8:*333*–334
Stem-boring flies, 3:56
Stem-caecilians, 6:13
Stem reptiles. *See* Cotylosauria
Stem-salamandroids, 6:13
Stem-tetrapods, 6:8, 6:9, 6:10
Stenella spp., 15:43–44
Stenella attenuata. See Spotted dolphins
Stenella coeruleoalba. See Striped dolphins
Stenella frontalis. See Atlantic spotted dolphins
Stenella longirostris. See Spinner dolphins
Steno bredanensis. See Rough-toothed dolphins
Stenoderma rufum, 13:420
Stenodermatinae, 13:413, 13:414, 13:416–417, 13:418, 13:420
Stenolaemata, **2:503–508,** 2:*505*
Stenonema vicarium, 3:*128*, 3:*129*, 3:130
Stenoninae, 15:41
Stenopelmatoidea, 3:201
Stenopelmatus fuscus. See Jerusalem crickets
Stenopodidea. *See* Boxer shrimps
Stenopodon galmani, 12:228
Stenopsocidae, 3:243
Stenopsyche siamensis, 3:*378*, 3:*380*, 3:381
Stenopterygius spp. *See* Ichthyosaurs
Stenopus hispidus. See Banded coral shrimps
Stenorhynchus seticornis. See Yellowline arrow crabs
Stenorrhina spp., 7:469
Stenostomum spp., 1:188
Stenothecoides spp., 2:387
Stensiö, Erik A., 4:209
Stephanoaetus coronatus. See Crowned eagles
Stephanoberycidae. *See* Pricklefishes
Stephanoberyciformes, **5:105–111,** 5:*107*
Stephanoberycoidea, 5:105, 5:106
Stephanoberyx monae. See Pricklefishes
Stephanocircidae, 3:347, 3:348–349
Stephanocircus dasyuri. See Helmet fleas
Stephanoxis spp. *See* Hummingbirds
Stephens Island wrens, 10:203, 10:204, 10:*205*, 10:206–*207*
Stephen's kangaroo rats, 16:203, 16:209*t*
Stephen's rocket frogs, 6:*201*, 6:204
Stephens wrens. *See* Stephens Island wrens
Stephopaedes spp., 6:184
Stephopaedes anotis. See Chirinda toads
Steppe, 3:56
  human evolution and, 12:20
  loess, 12:17–18
  See also Habitats
Steppe deer, 15:379
Steppe eagles, 8:321
Steppe harvesting mice, 12:15
Steppe lemmings, 16:*232*, 16:*233*, 16:237
Steppe pikas, 16:483, 16:491, 16:495, 16:*496*, 16:501
Steppe polecats, 14:321
Steppe rhinoceroses, 15:249
Steppe sicistas. *See* Southern birch mice
Stercorariidae, 9:203
Stercorarius spp., 9:204, 9:205, 9:206
Stercorarius longicaudus. See Long-tailed skuas
Stercorarius parasiticus. See Arctic skuas
Stercorarius pomarinus. See Pomarine skuas
Sterechinus neumayeri, 1:403
Stereoblastulas, 2:20–21
Stereochilus spp., 6:390
Stereochilus marginatus, 6:390
Stereoscopic vision. *See* Depth perception
Sterna spp., 9:203
Sterna caspia. See Caspian terns
Sterna fuscata. See Sooty terns
Sterna hirundo. See Common terns
Sterna nigra. See Black terns
Sterna paradisaea. See Arctic terns
Sternal bones, 12:41
  See also Skeletal system
Sternal glands, 12:37
  See also Physical characteristics
Sternarchorhynchus spp., 4:369–370
Sternarchorhynchus curvirostris, 4:*372*, 4:373–*374*
Sternarchorhynchus roseni. See Rosen knifefishes
Sternidae. *See* Terns
Sternoptychidae. *See* Marine hatchetfishes
Sternopygidae, 4:369, 4:370
Sternopygus spp., 4:370
Sternopygus astrabes, 4:369
Sternopygus macrurus. See Longtail knifefishes
Sternorrhyncha, 3:259, 3:262, 3:263
  See also Aphids; Scale insects
Sternotherus spp., 7:121
Sternotherus depressus, 7:123
Sternotherus odoratus. See Stinkpots
Steropodontidae, 12:228
Stethacanthid sharks, 4:10
Stethophyma spp., 3:205
Stewart Island short-tailed bats. *See* Greater New Zealand short-tailed bats
Sthenurinae, 13:83
Stiassny, Melanie L. J., 5:83, 5:94
Stichaeidae. *See* Pricklebacks
Stichaeinae, 5:309
Stichocotylidae, 1:197
Stichopodidae, 1:419, 1:420
Stichorchis subtriquetres, 1:200
Stichotrema dallatorreanum, 3:*337*, 3:*338*, 3:339
Stick insects, 3:13, 3:38, **3:221–232,** 3:*226*, 3:*227*
Stick nest rats. *See* Greater stick-nest rats
Sticklebacks, 4:35, **5:131–136,** 5:*137*, 5:*140*, **5:141–143**
Stiff-tailed ducks, 8:363, 8:364, 8:369
Stiles, F. G., 9:446
Stiletto snakes. *See* African burrowing snakes
Stilifer spp., 2:*32*
Stilifer linckiae, 2:33
Stiltia isabella. See Australian pratincoles
Stilts, 9:104, **9:133–139,** 9:*136*
Stinging hydroids, 1:*127*
Stingless bees, 3:68, 3:72
Stingrays, 4:57, 4:173–179, 4:*181*, 4:182–186, 4:188
  distribution, 4:*52*, 4:176
  evolution, 4:11, 4:173
  feeding ecology, 4:68, 4:177–178
  freshwater, 4:57, 4:*173*, 4:174, 4:176–177, 4:*180*, 4:*184*–186
  physical characteristics, 4:174–176, 4:190
  reproduction, 4:46, 4:178
Stings, by reef stonefishes, 5:177
Stink badgers, 14:321
Stink bugs, 3:65, 3:*259–262*, 3:*263*, 3:264, 3:*275*, 3:277
  See also Southern green stink bugs
Stinkers. *See* Southern giant petrels; Swamp wallabies
Stinkpots, 7:*122*, 7:*124*, 7:*125*, 7:126
  See also Southern giant petrels
Stinson, N., 16:215
Stipiturus malachurus. See Southern emu-wrens
Stitchbirds, 11:237, 11:238, 11:*240*, 11:*245*–246
Stizostedion canadense. See Saugers
Stizostedion vitreum. See Walleyes
Stoats. *See* Ermines
Stock doves, 9:243
Stock pigeons, 9:251
Stockwhip birds. *See* Eastern whipbirds
Stoecharthrum spp., 1:99
Stoichactis helianthus, 1:28
Stokellia anisodon. See Stokell's smelts
Stokell's smelts, 4:391
Stoker, Bram, 13:317
Stolidobranchia, 1:452
Stoloteuthis leucoptera. See Butterfly bobtail squids
Stomach oils, of tubenoses, 8:108, 8:*124*, 8:136
Stomachs, 12:47–48, 12:121–122, 12:123
  See also Physical characteristics
Stomatopoda. *See* Mantis shrimps
Stomatorhinus spp., 4:233
Stomiidae, 4:421, 4:423
Stomiiformes, ***4:421–430,*** 4:*425*
Stomolophus meleagris. See Cannonball jellyfish
Stone centipedes, 2:*357*, 2:360–*361*
Stone-curlews, 9:143–146
Stone dormice, 16:327*t*
Stone loaches, 4:*322*, 4:*325*–326
Stone martens, 12:132
Stone tools, 14:242
Stonechats, 10:485, 10:488, 10:*490*, 10:*496*
Stonefishes, 4:15, 4:48, 5:49, 5:163, 5:164, 5:165
  See also Reef stonefishes
Stoneflies, **3:*141*–146,** 3:*142*, 3:*144*
  as bioindicators, 3:82
  conservation status, 3:88, 3:143
  evolution, 3:7, 3:141
  reproduction, 3:31, 3:34, 3:143
Stonemyia volutina. See Volutine stoneyian tabanid flies
Stonerollers, 4:*300*, 4:301, 4:*303*, 4:*307*–308
Stony corals, 1:103
Stoplight parrotfishes, 5:*300*, 5:*306*–307
Storing information, 12:152
Stork-billed kingfishers, 10:*12*, 10:*18*
Storks, 8:233–236, **8:265–274**
Storm-petrels, 8:107–110, **8:135–142,** 14:291
Storm's storks, 8:267
Stout beardfishes, 5:*3*
Stout blacksmelts, 4:393

Stove-pipe sponges, 1:*81*, 1:*83*–84
Strap-toothed whales, 15:70*t*
Stratification, water, 4:38–39
Stratiodrilus spp., 2:46
Stratiomyidae, 3:361
Straw-colored fruit bats, 13:310, 13:*338*, 13:340–*341*
Straw-crowned bulbuls. *See* Straw-headed bulbuls
Straw-headed bulbuls, 10:397, 10:399, 10:*401*, 10:404–*405*
Strawberry basses. *See* Black crappies
Strawberry finches. *See* Red avadavats
Strawberry poison frogs, 6:*45*, 6:198–199, 6:200, 6:*202*, 6:*205*, 6:207
Streak-breasted bulbuls, 10:398
Streak-capped spinetails, 10:*213*, 10:*222*–*223*
Streaked boobooks. *See* Southern boobook owls
Streaked bowerbirds, 11:*478*
Streaked cisticolas. *See* Zitting cisticolas
Streaked fantails, 11:83, 11:*88*, 11:*92*
Streaked weavers, 11:376
Streaky-breasted jungle-flycatchers, 11:27
Stream catfishes, 4:353
Stream habitats, 4:39–40, 4:62
See also Habitats
Strepera graculina. *See* Pied currawongs
Strepera versicolor. *See* Tasmanian gray currawongs
Strepsiptera. *See* Strepsipterans
Strepsipterans, **3:335–339,** 3:*337*
Strepsirrhines, 12:8, 12:94, 14:3, 14:9
See also Lemurs; Lorises
Streptaxidae, 2:411
Streptocephalidae, 2:135, 2:136, 2:137
Streptocephalus spp., 2:137
Streptocephalus proboscideus. *See* Sudanese fairy shrimps
Streptocitta spp., 11:408
Streptocitta albertinae. *See* Bare-eyed mynas
Streptopelia spp., 9:243
Streptopelia capicola. *See* Cape turtledoves
Streptopelia decaocto. *See* Eurasian collared doves
Streptopelia risoria. *See* Ringed turtledoves
Streptopelia turtur. *See* Eurasian turtledoves
Streptoprocne semicollaris. *See* White-naped swifts
Streptoprocne zonaris. *See* White-collared swifts
Streptostyly, 7:195
Stresemann's bush-crows, 11:503
Striated bulbuls, 10:397
Striated caracaras, 8:348, 8:*350*
Striated grassbirds. *See* Little grassbirds
Striated grasswrens, 11:*47*, 11:*52*
Striated pardalotes, 11:*202*, 11:*203*, 11:*205*–206
Striated softtails, 10:211
Striated thornbills, 11:56
Stridulation, 3:59–60
See also Behavior
Strier, K. B., 14:165
Strigidae. *See* Typical owls
Strigiformes. *See* Owls
Strigocuscus spp., 13:57
Strigocuscus celebensis. *See* Small Sulawesi cuscuses
Strigops habroptilus. *See* Bulky kakapos
Striolated tit-spinetails, 10:*212*, 10:*220*
Strip sampling, 12:197–198
Stripe-backed wrens, 10:528
Stripe-breasted rhabdornis, 11:183, 11:184, 11:*185*, 11:*186*–187
Stripe-crowned pardalotes. *See* Striated pardalotes
Stripe-headed creepers. *See* Stripe-headed rhabdornis
Stripe-headed rhabdornis, 11:183, 11:184, 11:*185*, 11:*187*
Stripe-tailed monitors, 7:361, 7:362, 7:*364*, 7:*366*–367
Striped African darters, 4:*340*, 4:*342*, 4:347
Striped anostomus. *See* Striped headstanders
Striped-backed duikers. *See* Zebra duikers
Striped bandicoots, 13:17*t*
See also Eastern barred bandicoots
Striped bark scorpions. *See* Striped scorpions
Striped basses, 4:49, 4:68
See also Striped sea basses
Striped blind legless skinks, 7:*330*, 7:*332*
Striped boarfishes, 5:*245*, 5:251–*252*
Striped bristletooths, 5:*394*, 5:398–*399*
Striped buttonquails. *See* Barred buttonquails
Striped catsharks. *See* Pajama catsharks
Striped crakes, 9:50, 9:*54*, 9:63–*64*
Striped crayfish snakes, 7:469
Striped danios. *See* Zebrafishes
Striped dolphins, 15:48, 15:57*t*
Striped dwarf hamsters, 16:247*t*
Striped eel-catfishes. *See* Coral catfishes
Striped-faced fruit bats, 13:348*t*
Striped field mice, 16:261*t*
Striped flowerpeckers. *See* Thick-billed flowerpeckers
Striped foureyed fishes. *See* Largescale foureyes
Striped hardyheads. *See* Eendracht land silversides
Striped headstanders, 4:*340*, 4:*342*–343
Striped honeyeaters, 11:237, 11:*240*, 11:*241*–242
Striped hyenas, 14:359, 14:360, 14:362, 14:*363*, 14:*365*–366
Striped kingfishers, 10:*11*, 10:*18*–19
Striped manakins, 10:*298*, 10:*300*
Striped mud turtles, 7:*122*
Striped mullets. *See* Flathead mullets
Striped newts, 6:372
Striped parrotfishes, 5:*300*, 5:*306*
Striped phalangers. *See* Striped possums
Striped pipits, 10:373
Striped poison-fang blennies, 5:*344*, 5:*345*–346
Striped polecats. *See* Zorillas
Striped possums, **13:125–133,** 13:*126*, 13:*127*, 13:*130*, 13:*131*, 13:132, 13:133*t*
Striped quill-snouted snakes. *See* Variable quill-snouted snakes
Striped rabbits, 16:482, 16:487, 16:505, 16:506, 16:509
Striped sand lizards. *See* Western sandveld lizards
Striped scorpions, 2:*337*, 2:*349*–350
Striped sea basses, 5:196, 5:198, 5:*201*, 5:*203*, 5:206
Striped sea robins, 5:*167*, 5:177–178
Striped silversides. *See* Eendracht land silversides
Striped skinks, 7:*330*, 7:*332*, 7:335
Striped skunks, 14:*326*, 14:*329*
behavior, 14:321
feeding ecology, 14:*321*
habitats, 14:321
humans and, 12:*186*, 14:325
physical characteristics, 14:321
reproduction, 12:109, 12:110, 14:324
Striped snakeheads, 5:438, 5:*441*, 5:*443*, 5:445–446
Striped tree squirrels, 16:165, 16:*166*
Striped whipsnakes, 7:43
Stripedfin ronquils, 5:311, 5:312
Striper basses. *See* Striped sea basses
Strix spp., 9:349
Strix acadica. *See* Northern saw-whet owls
Strix alba. *See* Common barn owls
Strix aluco. *See* Tawny owls
Strix asio. *See* Eastern screech owls
Strix badia. *See* Oriental bay owls
Strix boobook. *See* Southern boobook owls
Strix brevis, 9:331, 9:345
Strix bubo. *See* Eurasian eagle-owls
Strix butleri. *See* Hume's owls
Strix cunicularia. *See* Burrowing owls
Strix flammea. *See* Short-eared owls
Strix leptogrammica. *See* Asian brown wood-owls
Strix longimembris. *See* Eastern grass owls
Strix occidentalis. *See* Spotted owls
Strix perlatum. *See* Pearl-spotted owlets
Strix perspicillata. *See* Spectacled owls
Strix scandiaca, 9:359
Strix scops. *See* Eurasian scops-owls
Strix tenebricosus. *See* Sooty owls
Strix uralensis. *See* Ural owls
Strix virginiana. *See* Great horned owls
Stromateidae. *See* Butterfishes (Stromateidae)
Stromatoidei, **5:421–426,** 5:*423*
Stromatoporids, 1:10
Strombidae, 2:446
Strombus spp., 2:447
Strong-billed honeyeaters, 11:*240*, 11:244–*245*
Strongylocentrotus droebachiensis. *See* Green sea urchins
Strongylocentrotus franciscanus. *See* Red sea urchins
Strongylocentrotus purpuratus. *See* Purple sea urchins
Strongyloides stercoralis. *See* Threadworms
Strongylopus spp., 6:245
Strongylura exilis. *See* Californian needlefishes
Strophidon sathete. *See* Slender giant morays
Strophocheilus spp., 2:414
Strophurus spp., 7:261
Structural diseases, 2:33
Structural pests, 3:75
Struthidea cinerea. *See* Apostlebirds
Struthideinae. *See* Apostlebirds
Struthio camelus. *See* Ostriches
Struthio cucullatus. *See* Dodos
Struthiolipeurus stresemanni, 3:*251*
Struthionidae. *See* Ostriches
Struthioniformes, **8:53–56**
Stuart, Chris, 8:128
Stuart, Tilde, 8:128
Stuart's antechinuses. *See* Brown antechinuses
Stuhlmann's golden moles, 13:218, 13:*219*, 13:*220*, 13:221
Stump-eared squirrels. *See* Fox squirrels

Stump-tailed macaques, 14:*192*, 14:194
Stump-tailed porcupines. *See* Short-tailed porcupines
Stupendemys spp., 7:14
Stupendemys geographicus, 7:137
Sturgeons, 4:*4*, 4:11, 4:18, 4:35, 4:58, 4:*75*, **4:213–219,** 4:*216*
Sturmbauer, C., 5:276
Sturnella spp. *See* Icteridae
Sturnella defilippii. See Pampas meadowlarks
Sturnella loyca. See Long-tailed meadowlarks
Sturnella magna. See Eastern meadowlarks
Sturnella neglecta. See Western meadowlarks
Sturnella superciliaris. See White-browed blackbirds
Sturnia spp., 11:409
Sturnidae, **11:407–426,** 11:*412–413*
behavior, 11:408
conservation status, 11:410
distribution, 11:*407*, 11:408
evolution, 11:407
feeding ecology, 11:408–409, 11:*410*
habitats, 11:408
humans and, 11:411
physical characteristics, 11:407–408
reproduction, 11:409–410
species of, 11:*414–425*
taxonomy, 11:407
Sturnini, 11:407
Sturnira spp., 13:418
Sturnira lilium. See Little yellow-shouldered bats
Sturnira ludovici. See Highland yellow-shouldered bats
Sturnira thomasi, 13:420
Sturnus spp., 11:408
Sturnus albofrontatus. See White-faced starlings
Sturnus holosericeus. See Yellow-billed caciques
Sturnus rosea. See Rosy starlings
Sturnus vulgaris. See European starlings
Styan's bulbuls, 10:398
Styela spp., 1:452
Stygiomysidae. *See Stygiomysis* spp.
Stygiomysis spp., 2:215
Stygiomysis cokei, 2:*216*, 2:*218*, 2:*220*, 2:221–222
Stygiomysis hydruntina, 2:217
Stygocarididae, 2:181
Stygocaridinea, 2:181
Stylactis spp., 1:27
Stylasteridae, 1:127, 1:129
Stylephoridae, 4:447, 4:448
Stylephorus spp., 4:447, 4:448
Stylephorus chordatus, 4:449
Stylocheiron spp., 2:185
Stylochus spp., 1:188
Stylochus inimicus. See Oyster leeches
Styloctenium wallace. See Striped-faced fruit bats
Stylodipus andrewsi. See Andrews's three-toed jerboas
Stylodipus telum, 16:214, 16:215, 16:216, 16:217
Stylommatophora, 2:411, 2:412, 2:413
Styloperlidae, 3:141
Stylopidia, 3:335, 3:336
Styraconyx spp., 2:117–118
Subantarctic diving-petrels. *See* Common diving-petrels
Subantarctic fur seals, 14:394, 14:395, 14:398, 14:399, 14:407*t*
Subdesert mesites, 9:5, 9:*6*, 9:7, 9:*9*
Suberites ficus, 1:*33*
Subimago, 3:125, 3:*126*
Subitizing, 12:155
Suboscines. *See* Tyranni
Subterranean atmosphere. *See* Atmosphere, subterranean
Subterranean mammals, **12:69–78,** 12:113–114
field studies of, 12:201
vibrations, 12:83
See also specific mammals
Subterranean mice, 16:270
Subterranean termites
eastern, 3:*169*, 3:*171*, 3:172–173
Formosan, 3:57
Subtidal habitats, 4:47
See also Habitats
Subtriquetra spp., 2:318, 2:319
Subtriquetridae. *See Subtriquetra* spp.
Subulinidae, 2:411
Subunguis, 12:39–40
Succession, 3:47
Succinea spp., 2:*413*
Succineidae, 2:411–412
Succineogerris larssoni, 3:259
Sucker catfishes, 4:353
Sucker loaches. *See* Algae eaters
Suckerfishes, 2:31
Suckermouth armored catfishes, 4:352, 4:353
Suckers, 4:321, 4:*322*
Sucking bugs, 3:31, 3:38
See also True bugs
Sucking lice, **3:249–257**
Suckling, 12:89, 12:96–97
See also Mammary glands
Suction feeding, 7:3–4
Sudanese fairy shrimps, 2:*138*, 2:139–*140*
Sudoriferous sweat glands. *See* Apocrine sweat glands
Suffusio predatrix, 7:369
Sugar cane beetles, 3:77
Sugar fish. *See* Silverfish
Sugar gliders, 12:148, 13:33, 13:37, 13:126, 13:*128*–129, 13:*130*, 13:*131*
Sugarbirds, 11:235, 11:237
Sugarfoot moth flies, 3:360
Sugarlice. *See* Silverfish
Suggrundus meerdervoortii, 5:158
Suidae. *See* Pigs
Suiformes, 15:266, 15:275, 15:292, 15:302
Suinae, 15:136, 15:267, 15:275
Suk sandgrouse. *See* Lichtenstein's sandgrouse
Sukumar, Raman, 15:171
Sula spp. *See* Boobies
Sula dactylatra. See Masked boobies
Sula leucogaster. See Brown boobies
Sula magpies. *See* Bare-eyed mynas
Sula mynas. *See* Bare-eyed mynas; Helmeted mynas
Sula nebouxii. See Blue-footed boobies
Sula pittas, 10:195
Sula starlings. *See* Bare-eyed mynas; Helmeted mynas
Sula sula. See Red-footed boobies
Sula variegata. See Peruvian boobies
Sulawesi. *See* Australian region
Sulawesi barn owls, 9:338, 9:341
Sulawesi bear cuscuses, 13:35, 13:58, 13:*58*, 13:60, 13:*62*, 13:*63*
Sulawesi drongos, 11:438
Sulawesi palm civets, 14:338, 14:344*t*
Sulawesi rainbows, 4:57, 5:67, 5:68
Sulawesi red-knobbed hornbills, 10:72–*73*, 10:74, 10:75, 10:76, 10:77, 10:81–*82*
Sulawesi tarictic hornbills, 10:72
Sulawesi tarsiers. *See* Spectral tarsiers
Sulawesi woodcocks, 9:176
Sulawesi wrinkled hornbills. *See* Sulawesi red-knobbed hornbills
Sulci, 3:17
Sulidae. *See* Boobies; Gannets
Sulphur-bellied flycatchers, 10:270, 10:271, 10:*276*, 10:*288*
Sulphur-crested cockatoos, 9:*282*, 9:283–*284*
Sultan tits, 11:155, 11:*159*, 11:*165*
Sulu hornbills, 10:75
Sulu woodpeckers, 10:150
Sumaco horned treefrogs, 6:*232*, 6:233–234
Sumatra barbs. *See* Tiger barbs
Sumatran drongos, 11:439
Sumatran orangutans, 14:11, 14:225, 14:235, 14:*236*, 14:*237*–238
Sumatran porcupines, 16:*356*, 16:*357*–358
Sumatran rabbits, 16:487, 16:516*t*
Sumatran rhinoceroses, 15:*250*, 15:*256*, 15:*257*
behavior, 15:218
conservation status, 12:218, 15:222, 15:254
evolution, 15:249
habitats, 15:252
humans and, 15:255
physical characteristics, 15:251
reproduction, 15:220–221, 15:254
See also Rhinoceroses
Sumatran short-tailed porcupines. *See* Sumatran porcupines
Sumatran surilis. *See* Banded leaf-monkeys
Sumatran thick-spined porcupines. *See* Sumatran porcupines
Sumba buttonquails, 9:14, 9:20
Sumba hornbills, 10:72–73, 10:75, 10:76, 10:77, 10:*83*
Sumba Island hornbills. *See* Sumba hornbills
Sumba wreathed hornbills. *See* Sumba hornbills
Sumichrast's harvest mice, 16:*269*
Sumichrast's wrens. *See* Slender-billed wrens
Summer flounders, 5:452, 5:*454*, 5:*456*, 5:458
Summer herrings. *See* Atlantic herrings
Summer yellow birds. *See* Yellow warblers
Sun, migration by, 8:*32*, 8:34
Sun bears. *See* Malayan sun bears
Sun stars. *See* Sunflower stars
Sunangels, 9:443
Sunbeam snakes, **7:401–403**
Sunbird-asities, **10:187–191,** 10:*189*
Sunbirds, **11:207–226,** 11:*211–212*
behavior, 11:209
conservation status, 11:210
distribution, 11:*207*, 11:208
evolution, 11:207
feeding ecology, 11:*209*
fire-tailed myzornis and, 10:519
habitats, 11:208–209
humans and, 11:210
physical characteristics, 11:207–208
reproduction, 11:210

species of, 11:*213–225*
taxonomy, 11:207
Sunbitterns, 9:1–4, **9:73–76,** 9:*74,* 9:*75*
Suncus spp., 13:265, 13:266, 13:267
Suncus etruscus. See Savi's pygmy shrews
Suncus murinus. See Musk shrews
Sunda frogmouths, 9:378, 9:*380,* 9:*385*
Sunda honeyguides. *See* Malaysian honeyguides
Sunda porcupines. *See* Javan short-tailed porcupines
Sunda sambars. *See* Timor deer
Sunda shrews, 13:277*t*
Sunda slow lorises, 14:*15,* 14:*17,* 14:*18*–19
Sunda teals, 8:364
Sunda water shrews, 13:263*t*
Sundasalangidae, 4:57
Sundasciurus spp., 16:163
Sundasciurus brookei, 16:167
Sundasciurus jentinki, 16:167
Sundasciurus juvencus, 16:167
Sundasciurus moellendorffi, 16:167
Sundasciurus rabori, 16:167
Sundasciurus samarensis, 16:167
Sundasciurus steerii, 16:167
Sundevall's jirds, 16:*286,* 16:291–*292*
See also Gerbils
Sundevall's roundleaf bats, 13:*403*
Sunfishes, 4:15, 4:57, 4:67, 4:68, **5:195–199,** 5:*196,* 5:*200,* 5:*201,* **5:202–205**
See also Molas
Sunflower stars, 1:*371,* 1:375–*376*
Sungazers. *See* Giant girdled lizards
Sungei tuntong. *See* Painted terrapins
Sungrebes, 9:1–4, **9:69–72,** 9:*70*
Sung's slender frogs, 6:111
Sunis, 16:71*t*
Sunlight. *See* Light
Supella longipalpa. See Brownbanded cockroaches
Superb blue wrens. *See* Superb fairy-wrens
Superb fairy-wrens, 11:*47,* 11:*50*
Superb lyrebirds, 10:329–332, 10:*333,* 10:334–*335*
Superb pittas, 10:194, 10:195, 10:*196,* 10:*198*
Superficial cleavage, 2:20, 2:23
See also Reproduction
Superposition eyes, 3:27
Superstitions, reptiles and, 7:57
Supply and demand economies, 12:223
Suprabranchial chamber, snakeheads, 5:437, 5:*438*
Surat Thani Province, Thailand, 12:220
Surdisorex spp., 13:265
Surdisorex norae. See Aberdare shrews
Surdisorex polulus. See Mt. Kenya shrews
Surf sardines. *See* Flowers of the wave
Surf silversides, 5:67, 5:68
Surf smelts, 4:392
Surfbirds, 9:178
Surfperches, 4:32, 4:35, **5:275–281,** 5:*283,* **5:290**
Surgeonfishes, 4:34, 4:67, **5:391–393,** 5:*392,* 5:*394,* **5:396–400**
Suricata spp., 14:347, 14:348
Suricata suricatta. See Meerkats
Suricates. *See* Meerkats
Suriname clicking crickets, 3:*210,* 3:*211*–212
Suriname cockroaches, 3:*152,* 3:*153*
Suriname horned frogs, 6:*158,* 6:*161,* 6:*162*
Suriname toads, 6:36–37, **6:99–107,** 6:*100,* 6:*101,* 6:*106*
Surnia ulula. See Northern hawk-owls
Surniculus lugubris. See Asian drongo-cuckoos
Surveys, biodiversity, 12:194–198, 12:*195*
Survival of the fittest. *See* Evolution; Natural selection
Survivorship, 2:27
See also Reproduction
Sus spp., 15:265, 15:275, 15:276, 15:278
Sus barbatus. See Bearded pigs
Sus bucculentus. See Vietnam warty pigs
Sus cebifrons. See Cebu bearded pigs
Sus celebensis. See Celebes pigs
Sus domestica. See Domestic pigs
Sus philippensis. See Philippine warty pigs
Sus salvanius. See Pygmy hogs
Sus scrofa. See Eurasian wild pigs; Feral pigs
Sus scrofa riukiuanus, 15:280
Sus scrofa vittatus. See Asian banded boars
Sus timoriensis. See Timor wild boars
Sus verrucosus. See Javan pigs
Suspension feeding, 1:27, 2:29, 2:36
See also Behavior; Feeding ecology
Susquehanna shads. *See* American shads
Sustainable utilization, 3:84
Sustained-yield harvests, 12:219
Susus. *See* Ganges and Indus dolphins
Sutchi catfishes. *See* Iridescent shark-catfishes
Suthers, R. A., 8:39
Swai. *See* Iridescent shark-catfishes
Swainson's hawks, 8:320
Swainson's thrushes, 10:484
Swallow plovers. *See* Collared pratincoles
Swallow-shrikes. *See* Woodswallows
Swallow-tailed gulls, 9:203, 9:205, 9:208
Swallow-tailed nightjars, 9:402
Swallow-tailed swifts, 9:424
Swallow-winged puffbirds, 10:101, 10:103, 10:*105,* 10:*110*–111
Swallow wings. *See* Swallow-winged puffbirds
Swallowers, **4:271–276,** 4:*273*
Swallows, 10:174, **10:357–369,** 10:*362*
behavior, 10:358–359
conservation status, 10:360–361
distribution, 10:*357,* 10:358
evolution, 10:357
feeding ecology, 10:359
habitats, 10:358
humans and, 10:361
physical characteristics, 10:357–358
reproduction, 10:359–360
species of, 10:*363–364*
taxonomy, 10:357
See also specific types of swallows
Swallowtail butterflies, Schaus's, 3:56
Swallowtail caterpillars, 3:65
Swallowtail tigers, 3:*387*
Swallowtails, 3:*383*
Swamp deer. *See* Barasinghas
Swamp eels, **5:151–156,** 5:*153*
Swamp francolins, 8:434
Swamp galaxias, 4:391
Swamp greenbuls, 10:397, 10:398, 10:399
Swamp lynx. *See* Jungle cats
Swamp mice, Delany's, 16:297*t*
Swamp monkeys, 14:193
Swamp rabbits, 16:481, 16:516*t*
Swamp rats. *See* Creek rats
Swamp wallabies, 13:*92,* 13:*93,* 13:95
Swampfishes, 5:5, 5:7, 5:*9*
Swan galaxias, 4:391
Swan geese, 8:367, 8:368, 8:373
Swans, **8:369–392**
behavior, 8:371–372
conservation status, 8:372–373
distribution, 8:*369,* 8:370–371
evolution, 8:369
feeding ecology, 8:372
habitats, 8:371
humans and, 8:373
physical characteristics, 8:369–370
reproduction, 8:372
species of, 8:*376*–392
taxonomy, 8:369
Swarm intelligence, 3:72
Swarming behavior
desert locusts, 3:204, 3:208
Diptera, 3:359
Orthoptera, 3:204
See also Behavior; specific species
Sweat bees, 3:68
Sweat glands, 12:7, 12:36–38, 12:89
See also Physical characteristics
Sweeper tentacles, 1:30
Sweepers, 4:48, 4:68
Sweetfishes. *See* Ayu
Sweetlips, 5:255
See also Haemulidae
Swellsharks, 4:*118,* 4:*125*–126
Swift foxes, 12:125, 14:265, 14:267, 14:270, 14:271, 14:284*t*
Swift parrots, 9:277
Swiftlets, 9:421, 9:422–424, 9:*425,* 9:428
Swifts, 9:415–419, 9:*417,* **9:421–432,** 9:*425*
behavior, 9:417, 9:422–423
conservation status, 9:424
distribution, 9:417, 9:*421,* 9:422
evolution, 9:415, 9:421
feeding ecology, 9:417, 9:423
habitats, 9:417, 9:422
humans and, 9:418–419, 9:424
physical characteristics, 9:415–*416,* 9:421–422
reproduction, 9:417–418, 9:423–424
species of, 9:*426–431*
vs. swallows, 10:357
taxonomy, 9:415, 9:421
tree, 9:415, 9:417, **9:433–436,** 9:*435–436*
Swim bladders, 4:6, 4:24, 4:*24*
Swinhoe's storm-petrels, 8:136
Swissguard basslets, 5:255, 5:256, 5:259, 5:261, 5:262
Sword-billed hummingbirds, 9:*439,* 9:443, 9:*453,* 9:*466*–467
Sword-nosed bats, 13:414, 13:*415*
Swordfishes, 4:*4,* 4:69, 5:405, 5:407, 5:*408,* 5:*416,* 5:417–418
Swordtails. *See* Green swordtails
Sycon capricorn, 1:*60,* 1:*63*–64
Sycon gelatinosum, 1:37
Syconycteris australis. See Southern blossom bats
Syllids, 2:47
Sylvia spp 11:3, 11:4, 11:6
Sylvia atricapilla. See Blackcaps
Sylvia borin. See Garden warblers
Sylvia communis. See Whitethroats
Sylvia juncidis. See Zitting cisticolas
Sylvia leucophaea. See Jacky winters

Sylvia nisoria. See Barred warblers
Sylvia rufiventris. See Rufous whistlers
Sylvia solitaria. See Rockwarblers
Sylvicapra spp., 16:73
Sylvicapra grimmia. See Bush duikers
Sylvietta rufescens. See Zitting cisticolas
Sylviidae. *See* Old World warblers
Sylvilagus spp. *See* Cottontails
Sylvilagus aquaticus. See Swamp rabbits
Sylvilagus audubonii. See Desert cottontails
Sylvilagus bachmani, 16:481
Sylvilagus brasiliensis, 16:481
Sylvilagus cunicularius, 16:481
Sylvilagus floridanus. See Eastern cottontails
Sylvilagus graysoni, 16:481
Sylvilagus insonus. See Omiltimi rabbits
Sylvilagus mansuetus, 16:481
Sylvilagus nuttallii. See Mountain cottontails
Sylvilagus palustris. See Marsh rabbits
Sylvilagus transitionalis, 16:481
Sylvioidea, 11:1
Sylviorthorhynchus desmursii. See Des Murs's wiretails
Sylviparus spp. *See* Yellow-browed tits
Sylviparus modestus. See Yellow-browed tits
Sylvisorex spp. *See* Forest musk shrews
Sylvisorex megalura. See Forest musk shrews
Sylvisorex vulcanorum. See Volcano shrews
Sylvoid oscines, 10:171
Syma torotoro. See Yellow-billed kingfishers
Symbiida. *See* Wheel wearers
Symbiidae. *See* Wheel wearers
Symbiodinium spp., 1:33
Symbion spp., 1:351–352
Symbion pandora, 1:351, 1:*352*, 1:*353*
Symbionts, 2:31
Symbiosis, 1:27, **1:31–36**, 1:*34*, **2:31–34**, 3:72, 4:64, 4:69, 9:109
See also Behavior; Feeding ecology
Symbolic communication, 12:161–162
Symbolism, 3:74
See also humans
Symmetrodontomys spp., 16:264
Symmetry, biological success and, 12:23–24
Sympatric occurrence, 12:116
Symphalangus spp., 14:207
Symphalangus syndactylus. See Siamangs
Symphurus spp. *See* Tonguefishes
Symphurus atricaudus. See California tonguefishes
Symphylans, **2:371–373**
Symphypleona, 3:99
Symphypleonids, 3:100
Symphysodon spp., 5:279
Symphysodon aequifasciata. See Blue discus
Symphyta, 3:405, 3:407
See also Sawflies
Symplesiomorphies, 12:7
Symsagittifera corsicae, 1:*181*, 1:*182*
Synagoga n.sp., 2:*273*
Synallactidae, 1:419, 1:420
Synallaxeinae. *See* Bushcreepers
Synallaxis spp. *See* Spinetails
Synallaxis albescens. See Pale-breasted spinetails
Synallaxis infuscata. See Plain spinetails
Synallaxis kollari. See Hoary-throated spinetails
Synallaxis tithys. See Blackish-headed spinetails
Synallaxis whitneyi. See Bahia spinetails
Synallaxis zimmeri. See Russet bellied spinetails
Synalpheus spp. *See* Pistol shrimps
Synamoebium, 1:7
Synanceia spp., 4:15
Synanceia verrucosa. See Reef stonefishes
Synanceiidae. *See* Stonefishes
Synaphobranchidae, 4:255
Synapomorphy, 12:7
See also Homology
Synapsid reptiles, 7:4–*5*
Synapsids, 12:10
Synapta maculata. See Giant medusan worms
Synaptidae, 1:417, 1:420, 1:421
Synaptinae, 1:417, 1:420
Synaptomys borealis. See Northern bog lemmings
Synapturanus salseri. See Timbo disc frogs
Synbranchidae, 5:151, 5:152
Synbranchiformes, **5:151–156**, 5:*153*
Synbranchoidei, 5:151
Synbranchus marmoratus. See Marbled swamp eels
Syncerus caffer. See African buffaloes
Synchiropus spp., 5:365, 5:366
Synchiropus splendidus. See Mandarinfishes
Synchirus gilli. See Manacled sculpins
Syncomistes spp., 5:220, 5:221
Syncope spp., 6:304
Syncytial theory, 2:12
See also Evolution
Syndactyla ruficollis. See Rufous-necked foliage-gleaners
Syndermata, 1:11–12
Syngamy, 1:18
Syngnathidae, 4:26, 4:46, 4:48, 4:69, 5:131, 5:133, 5:135, 5:136
Syngnathoidei, 5:131, 5:135
Syngnathus typhle, 5:133
Syngnathus watermeyeri. See River pipefishes
Synocnus comes. See Lesser Haitian ground sloths
Synodontidae, 4:60, 4:432, 4:433
Synodontis spp., 4:352
Synodontis angelicus. See African polka-dot catfishes
Synodontis multipunctatus. See Mochokid catfishes
Synodontis nigriventris. See Blotched upsidedown catfishes
Synodontis schoutedeni. See Schouteden's African squeaker catfishes
Synodontoidei, 4:431
Synodus spp., 4:*433*
Synodus dermatogenys. See Clearfin lizardfishes
Synodus englemani. See Engleman's lizardfishes
Synodus intermedius. See Lizardfishes
Synoicum spp., 1:452
Syntax, 12:161
Syntermes spp., 3:165
Syntheosciurus spp., 16:163
Syntheosciurus brochus. See Mountain squirrels
Synthliboramphus spp. *See* Murrelets
Synthliboramphus hypoleucus. See Xantus's murrelets
Sypheotides spp. *See* Bustards
Sypheotides indica. See Lesser floricans
Syrbula spp., 3:207
Syrian bears. *See* Brown bears
Syrian golden hamsters. *See* Golden hamsters
Syrian hamsters. *See* Golden hamsters
Syrian onagers, 15:222
Syrian rock hyraxes, 15:189*t*
Syrian serins, 11:325
Syringes, 8:39
bird songs and, 8:39
of passerines, 10:170, 10:171
See also specific species
Syrphid flies, 3:63, 3:360
Syrrhaptes spp. *See* Sandgrouse
Syrrhaptes paradoxus. See Pallas's sandgrouse
Systellognatha, 3:142–143
Systellommatophora, 2:412–413, 2:414
Systema Naturae, 1:11, 2:8, 13:193
Systema Porifera, 1:77
Systematics. *See* Taxonomy
Szathmáry, E., 3:67

T

Taaningichthys spp., 4:442
Tabanid flies, 3:*24*
Belkin's dune, 3:360
Volutine stoneyian, 3:360
Tabanus punctifer. See Big black horse flies
Tabbigaws. *See* Port Jackson sharks
Taber, A. B., 16:392
Tachinidae, 3:38, 3:64, 3:360
Táchira emeralds, 9:449
Tachybaptus spp. *See* Grebes
Tachybaptus dominicus. See Least grebes
Tachybaptus pelzelnii. See Madagascar grebes
Tachybaptus ruficollis. See Little grebes
Tachybaptus rufolavatus. See Alaotra grebes
Tachycines spp., 3:203
Tachycines asynamorus. See Greenhouse camel crickets
Tachycineta bicolor. See Tree swallows
Tachycineta cyaneoviridis. See Bahama swallows
Tachycineta euchrysea. See Golden swallows
Tachycneminae. *See Tachycnemis* spp.
Tachycnemis spp., 6:281
Tachycnemis seychellensis. See Seychelles treefrogs
Tachyeres brachypterus. See Steamer-ducks
Tachyeres pteneres. See Magellanic steamerducks
Tachyglossidae. *See* Echidnas
Tachyglossus spp. *See* Short-beaked echidnas
Tachyglossus aculeatus. See Short-beaked echidnas
Tachyglossus aculeatus acanthion, 12:236
Tachyglossus aculeatus aculeatus, 12:236
Tachyglossus aculeatus lawesi, 12:237
Tachyglossus aculeatus multiaculeatus, 12:236
Tachyglossus aculeatus setosus, 12:236
Tachymarptis melba. See Alpine swifts
Tachyoryctes splendens. See East African mole rats
Tachypleus tridentatus. See Japanese horseshoe crabs
Tachypodoiulus spp., 2:365
Tactile behavioral cues, 7:40–42
See also Behavior
Tactile hairs, 12:38–39
See also Physical characteristics
Tacto-location, 8:235, 8:*266*
Tactostoma macropus, 4:*422*
Taczanowski's tinamous, 8:59

Tadarida aegyptiaca. See Egyptian free-tailed bats
Tadarida australis. See White-striped free-tailed bats
Tadarida brasiliensis. See Brazilian free-tailed bats
Tadarida fulminans, 13:*487*
Tadarida teniotis. See European free-tailed bats
Tadorna cristata. See Crested shelducks
Tadorna ferruginea. See Ruddy shelducks
Tadorninae. *See* Shelducks
Tadpole sculpins, 5:181
Tadpole shrimps, **2:141–146,** 2:*144*
 behavior, 2:142
 conservation status, 2:142
 distribution, 2:141–142
 evolution, 2:141
 feeding ecology, 2:142
 habitats, 2:142
 humans and, 2:143
 physical characteristics, 2:141
 reproduction, 2:142
 species of, 2:*145*–146
 taxonomy, 2:141
Tadpole snails, 2:412, 2:415
Tadpoles, 6:28–29, 6:36–43, 6:*40*, 6:*41*
 Amero-Australian treefrogs, 6:227–228
 behavior, 6:41–42, 6:44
 carnivorous, 6:39
 feeding ecology, 6:6, 6:37–38, 6:39, 6:40–41
 leptodactylid frogs, 6:157
 Mesoamerican burrowing toads, 6:*96*
 metamorphosis, 6:28, 6:39, 6:42–43
 physical characteristics, 6:39–41, 6:62–63, 6:*66*
 Pipidae, 6:41, 6:43, 6:*101*, 6:102
 poison frogs, 6:*198*, 6:*199*, 6:200
 predators of, 6:65
 transportation of, 6:37–38, 6:*199*, 6:200
 true frogs, 6:42, 6:249–251
Taenia spp., 1:*227*, 1:229
Taenia hydatigena, 1:231
Taenia multiceps, 1:231
Taenia ovis, 1:231
Taenia saginata. See Beef tapeworms
Taenia solium. See Pork tapeworms
Taeniidae, 1:226, 1:230, 1:231
Taeniodonta, 12:11
Taeniopterygidae, 3:54, 3:141
Taeniura lymma. See Blue-spotted stingrays
Tagesoidea nigrofascia, 3:*222*
Tagula butcherbirds, 11:469
Tahiti monarchs, 11:98
Tahiti swiftlets, 9:418
Tahitian pearl oysters. *See* Black-lipped pearl oysters
Tahkis. *See* Przewalski's horses
Tahrs, 16:87, 16:88, 16:92
 See also specific types of tahrs
Taiga tits. *See* Siberian tits
Tailed frogs, **6:77–81,** 6:*79*
 distribution, 6:5, 6:77, 6:*80–81*
 evolution, 6:4, 6:64, 6:77
 larvae, 6:41
 physical characteristics, 6:26, 6:77, 6:80–81
 taxonomy, 6:77, 6:80–81
Tailless bats, Geoffroy's, 13:*422*, 13:427–*428*
Tailless caecilians, **6:435–441,** 6:*438*
 distribution, 6:5, 6:*435*, 6:*439–440*
 larvae, 6:39, 6:436
 taxonomy, 6:411, 6:*412*, 6:435
Tailorbirds, 11:*2*, 11:5–6, 11:*9*, 11:18
Tailors. *See* Bluefishes
Tails, 12:*42*
 autotomy, 7:261
 in locomotion, 12:43
 regeneration of, 7:*24*
 subterranean mammals, 12:72
 See also Physical characteristics
Taipans, 7:483, 7:485, 7:487, 7:*490*, 7:*496*
Taisho sanshoku, 4:*299*
Taita falcons, 8:351
Taiwan bulbuls, 10:398–399
Tajik markhors, 16:*8*
Takahe rails, 9:52, 9:*56*, 9:65–*66*, 10:326
Takifugu spp. *See* Fugus
Takifugu rubripes. See Fugus
Takins, 15:*138*, 16:87–90, 16:92, 16:93, 16:94, 16:*97*, 16:*101*, 16:103
Takydromus spp. *See* Oriental grass lizards
Takydromus sexlineatus. See Six-lined grass lizards
Talamancan web-footed salamanders, 6:*393*, 6:397–*398*
Talapoins, 14:188, 14:191, 14:193, 14:*196*, 14:*198*–199
Talas tuco-tucos, 16:*426*, 16:*427*
Talaud rails, 9:47, 9:*54*, 9:*62*
The Tale of Mrs. Tiggy-Winkle, 13:200
Taliabu masked owls, 9:338, 9:340–341
Talking catfishes, 4:352
Talking mynas. *See* Hill mynas
Talpa spp., 13:279
Talpa europaea. See European moles
Talpa streeti. See Persian moles
Talpidae, **13:279–288,** 13:*283*
 behavior, 13:280–281
 conservation status, 13:282
 distribution, 13:*279*, 13:280
 evolution, 13:279
 feeding ecology, 13:281–282
 habitats, 13:280
 physical characteristics, 13:279–280
 reproduction, 13:282
 species of, 13:*284*–287, 13:287*t*–288*t*
 taxonomy, 13:194, 13:279
 vision, 12:79
Talpinae, 13:279, 13:280
Tamandua spp. *See* Tamanduas
Tamandua mexicana. See Northern tamanduas
Tamandua tetradactyla. See Southern tamanduas
Tamanduas, 13:151, 13:172
Tamarins, **14:115–132,** 14:*125*
 behavior, 14:6–7, 14:8, 14:116–120, 14:*123*
 conservation status, 14:11, 14:124
 distribution, 14:116
 evolution, 14:115
 feeding ecology, 14:120–121
 habitats, 14:116
 humans and, 14:124
 physical characteristics, 14:115–116, 14:*122*
 reproduction, 14:10, 14:121–124
 species of, 14:*127*–129, 14:132*t*
 taxonomy, 14:4, 14:115
Tamarugo conebills, 11:291
Tambaqui, 4:339
Tamias spp., 16:143
Tamias minimus. See Least chipmunks
Tamias palmeri, 16:147–148
Tamias ruficaudus. See Red-tailed chipmunks
Tamias senex. See Allen's chipmunks
Tamias sibericus. See Siberian chipmunks
Tamias striatus. See Eastern chipmunks
Tamias townsendii. See Townsend's chipmunks
Tamiasciurus spp., 16:163
Tamiasciurus douglasii. See Douglas's squirrels
Tamiasciurus hudsonicus. See North American red squirrels
Tamiasciurus hudsonicus grahamensis. See Mt. Graham red squirrels
The Taming of the Shrew, 13:200
Tamini, 16:143
Tammar wallabies, 12:185, 13:*89*, 13:101*t*
Tampico cichlids, 5:*275*
Tanagers, 11:286, 11:323
Tanagra erythroryncha. See Red-billed oxpeckers
Tanagra jacarina. See Blue-black grassquits
Tanaidomorpha, 2:235
Tanaids, **2:235–239,** 2:*237*
Tanaoceroidea. *See* Desert grasshoppers
Tanarctus bubulubus. See Balloon water bears
Tandan catfishes, 4:352
Tanganyika killifishes. *See* Tanganyika pearl lampeyes
Tanganyika pearl killifishes. *See* Tanganyika pearl lampeyes
Tanganyika pearl lampeyes, 5:93, 5:*95*, 5:*97*, 5:100–101
Tangs, 5:391, 5:*392*
 blue, 5:*393*
 palette, 4:*57*, 5:*394*, 5:397–*398*
 yellow, 5:*394*, 5:*398*, 5:400
Tanimbar starlings, 11:410
Tanner's litter frogs, 6:265, 6:*267*, 6:268–269
Tanner's squeakers. *See* Tanner's litter frogs
Tantilla spp. *See* Black-headed snakes
Tantulocaridans, **2:283–287,** 2:*285*
Tanymastigidae, 2:135
Tanysiptera spp. *See* Paradise kingfishers
Tanysiptera galatea. See Common paradise kingfishers
Tanysiptera nympha. See Paradise kingfishers
Tanzanian mountain weavers, 11:379
Tanzaniophasma spp., 3:217
Tanzaniophasma subsolana, 3:219
Tapaculos, 10:170–171, **10:257–268,** 10:*261*
 behavior, 10:258
 conservation status, 10:259
 distribution, 10:*257*, 10:258
 evolution, 10:257
 feeding ecology, 10:259
 habitats, 10:258
 humans and, 10:259–260
 physical characteristics, 10:257–258
 reproduction, 10:259
 species of, 10:262–267
 taxonomy, 10:257
Tapera spp. *See* Cuckoos
Tapera naevia. See American striped cuckoos
Tapeta lucida, 12:79–80
Tapeworms, 1:11, 1:35, **1:225–243,** 1:*232*, 1:*233*, 2:28
 behavior, 1:230
 conservation status, 1:231
 distribution, 1:229
 evolution, 1:226–227
 feeding ecology, 1:230
 habitats, 1:229–230

humans and, 1:231, 1:247
physical characteristics, 1:227–229
reproduction, 1:20, 1:21, 1:230–231
species of, 1:*234–242*
taxonomy, 1:225–226
Taphozoinae, 13:355
Taphozous spp. *See* Tomb bats
Taphozous achates. See Brown-bearded sheath-tail bats
Taphozous hildegardeae. See Hildegarde's tomb bats
Taphozous mauritianus. See Mauritian tomb bats
Taphozous melanopogon. See Black-bearded tomb bats
Taphozous perforatus. See Egyptian tomb bats
Taphozous troughtoni, 13:358
Taphrolesbia spp. *See* Hummingbirds
Tapir Specialist Group, IUCN, 15:242
Tapiridae. *See* Tapirs
Tapirs, 12:131, 15:136, 15:137, 15:215–223, **15:237–248,** 15:*244*
Tapirus spp. *See* Tapirs
Tapirus bairdii. See Central American tapirs
Tapirus indicus. See Malayan tapirs
Tapirus pinchaque. See Mountain tapirs
Tapirus terrestris. See Lowland tapirs
Tarachodidae, 3:177
Tarachodula pantherina, 3:*178*
Tarantula hawks, 3:*409*, 3:*419*
Tararira. *See* Trahiras
Tardigrada. *See* Water bears
Target perches, 5:*223*, 5:*229*, 5:232–233
Target training, of zoo animals, 12:211
Taricha spp., 6:325, 6:363, 6:366, 6:367
Taricha granulosa. See Rough-skinned newts
Taricha rivularis. See Red-bellied newts
Taricha torosa. See California newts
Tarichatoxin, 6:325
Tarictic hornbills, 10:72
Tarpans, 12:175, 15:222
Tarpons, 4:11, 4:69, **4:243–248,** 4:*245*
Tarsiers, **14:91–100,** 14:*96*
behavior, 14:8, 14:94–95
conservation status, 14:95
distribution, 12:136, 14:5, 14:*91*, 14:93
evolution, 14:2, 14:91
feeding ecology, 14:8
habitats, 14:6, 14:93–94
humans and, 14:95
locomotion, 12:43
physical characteristics, 14:92–93
reproduction, 14:95
species of, 14:*97–99*
taxonomy, 14:1, 14:3–4, 14:91–92
Tarsiidae. *See* Tarsiers
Tarsipedidae. *See* Honey possums
Tarsipedoidea, 13:32, 13:139
Tarsipes rostratus. See Honey possums
Tarsius spp. *See* Tarsiers
Tarsius bancanus. See Western tarsiers
Tarsius dianae. See Dian's tarsiers
Tarsius eocaenus spp., 14:2
Tarsius pumilus. See Pygmy tarsiers
Tarsius sangirensis. See Sangihe tarsiers
Tarsius spectrum. See Spectral tarsiers
Tarsius syrichta. See Philippine tarsiers
Tarzan (movie), 8:23
Tasmacetus spp., 15:59
Tasmacetus shepherdi. See Shepherd's beaked whales
Tasman beaked whales. *See* Shepherd's beaked whales
Tasmanian barred bandicoots. *See* Eastern barred bandicoots
Tasmanian bettongs, 13:*73*, 13:74, 13:76, 13:77, 13:*78*
Tasmanian devils, 12:*278*, **12:287–291,** 12:*288*, 12:*293*, 12:*295*, **12:296**
behavior, 12:281, 12:*281*
conservation status, 12:290
distribution, 12:137, 12:279
evolution, 14:256
feeding ecology, 12:289
habitats, 12:280, 12:289
humans and, 12:284
physical characteristics, 12:288
reproduction, 12:282
Tasmanian emus, 8:85
Tasmanian gray currawongs, 11:468
Tasmanian mudfishes, 4:391
Tasmanian muttonbirds. *See* Short-tailed shearwaters
Tasmanian native hens, 9:50, 9:52
Tasmanian remarkables, 2:*357*, 2:359–*360*
Tasmanian ringtails. *See* Common ringtails
Tasmanian smelts, 4:391, 4:393
Tasmanian thornbills, 11:57
Tasmanian torrent midges, 3:360
Tasmanian white-eyes, 11:228
Tasmanian whitebaits, 4:390, 4:393
Tasmanian wolves, 12:*31*, **12:307–310,** 12:*308*, 12:*309*
behavior, 12:280, 12:308
conservation status, 12:283, 12:309–310
distribution, 12:137, 12:308
evolution, 12:277, 12:307, 14:256
feeding ecology, 12:281, 12:308–309
habitats, 12:308, 12:*310*
humans and, 12:284, 12:310
physical characteristics, 12:278–279, 12:307–308
reproduction, 12:282, 12:309
taxonomy, 12:277, 12:307, 14:256
Tassel-eared marmosets, 14:*118*, 14:132*t*
Tassel-eared squirrels. *See* Abert squirrels
Tasseled wobbegongs, 4:*107*, 4:*109*–110
Tasseltails. *See* Silverfish
Taste, 3:26–27, 7:30, 12:50, 12:*80*
See also Physical characteristics
Tataupa tinamous, 8:59
Tatera spp., 16:127
Tatera leucogaster. See Bushveld gerbils
Tate's trioks, 13:129, 13:133*t*
Tatria biremis, 1:230, 1:*233*, 1:*236*
Tattersall, Ian
on koala lemurs, 14:73
on *Lepilemur* spp., 14:75
Tattlers, 9:177
Taudactylus spp., 6:148, 6:149
Taudactylus diurnus, 6:149
Taudactylus eungellensis. See Eungella torrent frogs
Tauraco spp., 9:300
Tauraco bannermani. See Bannerman's turacos
Tauraco corythaix. See Knysna turacos
Tauraco erythrolophus. See Red-crested turacos
Tauraco fischeri. See Fischer's turacos
Tauraco hartlaubi. See Hartlaub's turacos
Tauraco leucolophus. See White-crested turacos
Tauraco leucotis. See White-cheeked turacos
Tauraco livingstonii. See Livingstone's turacos
Tauraco macrorhynchus. See Yellow-billed turacos
Tauraco persa. See Green turacos
Tauraco ruspolii. See Prince Ruspoli's turacos
Tauraco schalowi. See Schalow's turacos
Tauraco schuettii. See Black-billed turacos
Taurine, 12:127
Taurotragus spp., 16:11
Taurotragus oryx. See Elands
Tawny-bellied motmots. *See* Blue-crowned motmots
Tawny-faced quails, 8:456
Tawny-flanked prinias, 11:*2*, 11:4, 11:*8*, 11:*13*
Tawny frogmouths, 9:*367*, 9:368–370, 9:377, 9:*378–380*, 9:*382*
Tawny owls, 9:332, 9:350, 9:*353*, 9:*360*
Tawny piculets, 10:150
Tawny pipits, 10:372, 10:373
Tawny-shouldered frogmouths. *See* Tawny frogmouths
Tawny-throated dotterels, 9:161
Tawny-winged woodcreepers, 10:230
Tawny wood-owls. *See* Tawny owls
Taxidea taxus. See North American badgers
Taxis, 1:39–40
See also Behavior
Taxonomy
amphibians, 6:3–*4*, 6:*11*
anurans, 6:4, 6:61–62
African treefrogs, 6:3–4, 6:279–281, 6:285–289
Amero-Australian treefrogs, 6:225–226, 6:233–242
Arthroleptidae, 6:265, 6:268–270
Asian toadfrogs, 6:109–110, 6:113–116
Asian treefrogs, 6:4, 6:245, 6:279, 6:291, 6:295–299
Australian ground frogs, 6:139–140, 6:143–145
Bombinatoridae, 6:83, 6:86–88
Bufonidae, 6:179, 6:188–194, 6:197
Discoglossidae, 6:89, 6:92–94
Eleutherodactylus spp., 6:156
frogs, 6:4–5, 6:11–*13*, 6:15, 6:61–62
ghost frogs, 6:131
glass frogs, 6:215, 6:219–223, 6:225
leptodactylid frogs, 6:139, 6:147, 6:155–157, 6:162–171, 6:173, 6:197
Madagascan toadlets, 6:317, 6:319–320
Mesoamerican burrowing toads, 6:95
Myobatrachidae, 6:139, 6:147, 6:151–153, 6:155
narrow-mouthed frogs, 6:273, 6:301–302, 6:308–315
New Zealand frogs, 6:69, 6:72–74
parsley frogs, 6:127, 6:129
Pipidae, 6:99, 6:103–106
poison frogs, 6:173, 6:197, 6:203–209
Ruthven's frogs, 6:211
Seychelles frogs, 6:135, 6:137
shovel-nosed frogs, 6:273
spadefoot toads, 6:119, 6:124–125
tailed frogs, 6:77, 6:80–81
three-toed toadlets, 6:173, 6:179, 6:181–182
toads, 6:15, 6:61–62
treefrogs, 6:215

true frogs, 6:245–248, 6:255–263, 6:291
vocal sac-brooding frogs, 6:173, 6:175–176
Xenopus spp., 6:99
caecilians, 6:3, 6:411, 6:*412*
American tailed, 6:415–416, 6:417
Asian tailed, 6:411, 6:*412*, 6:415, 6:419, 6:422–423, 6:425–426
buried-eye, 6:431, 6:433
gymnophionans, 6:13
Kerala, 6:425–426, 6:428
tailless, 6:411, 6:*412*, 6:425, 6:431, 6:435, 6:439–440
lissamphibians, 6:9, 6:11–13, 6:323
salamanders, 6:3, 6:4, 6:323–*324*
amphiumas, 6:405, 6:409–410
Asiatic, 6:323, 6:335–336, 6:339–341, 6:343, 6:385
Caudata, 6:323–324, 6:327
Cryptobranchidae, 6:323, 6:335, 6:343
lungless, 6:323, 6:389–392, 6:395–403
mole, 6:13, 6:323, 6:349, 6:355, 6:358–359, 6:385
Pacific giant, 6:3, 6:323, 6:349, 6:352, 6:385, 6:405
Proteidae, 6:377, 6:381–382
Salamandridae, 6:323, 6:363, 6:370–375
sirens, 6:323, 6:327, 6:331–332, 6:335
torrent, 6:3, 6:323, 6:385, 6:387
vocal sac-brooding frogs, 6:175–176
birds
albatrosses, 8:113, 8:118–121
Alcidae, 9:219–220, 9:224–228
anhingas, 8:201, 8:207–209
Anseriformes, 8:363
ant thrushes, 10:239, 10:245–255
Apodiformes, 9:415
asities, 10:187, 10:190
Australian chats, 11:65, 11:67
Australian creepers, 11:133, 11:137–139
Australian fairy-wrens, 11:45, 11:48–53
Australian honeyeaters, 11:235, 11:241–253
Australian robins, 11:105, 11:108–112
Australian warblers, 11:55, 11:59–63
babblers, 10:505, 10:511–523
barbets, 10:113, 10:119–122
barn owls, 9:340–343
bee-eaters, 10:39
birds of paradise, 11:477, 11:489, 11:495–501
bitterns, 8:233, 8:239
Bombycillidae, 10:447, 10:451–454
boobies, 8:211
bowerbirds, 11:477, 11:483–487
bristletails, 3:113, 3:116–117
broadbills, 10:177, 10:181–185
bulbuls, 10:395, 10:402–412
bustards, 9:91, 9:96–99
buttonquails, 9:11–12, 9:16–21
Caprimulgiformes, 9:367
caracaras, 8:347
cassowaries, 8:75
Charadriidae, 9:161, 9:166–172
Charadriiformes, 9:101
chats, 10:483
chickadees, 11:155, 11:160–165
chowchillas, 11:69, 11:72–73
Ciconiiformes, 8:233
Columbidae, 9:247, 9:255–266
Columbiformes, 9:241
condors, 8:275
Coraciiformes, 10:1–2
cormorants, 8:201, 8:205–206, 8:207–209
Corvidae, 11:503, 11:512–522
cotingas, 10:305, 10:312–321
coursers, 9:104, 9:155–159
crab plovers, 9:121
Cracidae, 8:416
cranes, 9:23–24, 9:31–36
cuckoo-shrikes, 10:385, 10:389–393
cuckoos, 9:311, 9:318–329
dippers, 10:475, 10:479–481
diving-petrels, 8:143, 8:145–146
doves, 9:247, 9:255–266
drongos, 11:437, 11:441–444
ducks, 8:369
eagles, 8:317–318
elephant birds, 8:103
emus, 8:53
Eupetidae, 11:75, 11:78–80
fairy bluebirds, 10:415, 10:423
Falconiformes, 8:313
falcons, 8:347
false sunbirds, 10:187, 10:190
fantails, 11:83–84, 11:89–94
finches, 11:286, 11:323, 11:328–338
flamingos, 8:303
flowerpeckers, 11:189, 11:193–198
frigatebirds, 8:183, 8:193, 8:197–198
frogmouths, 9:377, 9:381–385
Galliformes, 8:399
gannets, 8:211
geese, 8:369
Glareolidae, 9:104, 9:151, 9:155–159
grebes, 8:169, 8:174–180
Gruiformes, 9:1
guineafowl, 8:425
gulls, 9:104, 9:211–216
hammerheads, 8:261
Hawaiian honeycreepers, 11:341, 11:346–351
hawks, 8:317–318
hedge sparrows, 10:459, 10:462–464
herons, 8:239
hoatzins, 8:467
honeyguides, 10:141–144
hoopoes, 10:61
hornbills, 10:71, 10:78–83
hummingbirds, 9:437–438, 9:455–467
ibises, 8:233, 8:291
Icteridae, 11:301, 11:309–321
ioras, 10:415, 10:419–420
jacamars, 10:91
jacanas, 9:104, 9:107, 9:112–114
kagus, 9:41
kingfishers, 10:5–6, 10:13–23
kiwis, 8:89
lapwings, 9:133
Laridae, 9:203–204, 9:211–216
larks, 10:341, 10:348–354
leafbirds, 10:415, 10:420–423
logrunners, 11:72–73
long-tailed titmice, 11:141, 11:145–146
lyrebirds, 10:329, 10:334
magpie-shrikes, 11:467, 11:471–475
manakins, 10:295, 10:299–303
mesites, 9:5, 9:8–9
moas, 8:95
monarch flycatchers, 11:97, 11:100–102
motmots, 10:31–32, 10:35–37
moundbuilders, 8:403
mousebirds, 9:469, 9:473–476
mudnest builders, 11:453, 11:456–457
New World blackbirds, 11:309–321
New World finches, 11:263, 11:269–282, 11:286
New World quails, 8:399, 8:455
New World vultures, 8:233, 8:275
New World warblers, 11:285–287, 11:293–298
New Zealand wattle birds, 11:447, 11:449–450
New Zealand wrens, 10:203, 10:206
nightjars, 9:401, 9:408–413
nuthatches, 11:167, 11:171–174
oilbirds, 9:373
Old World flycatchers, 11:25, 11:30–42
Old World warblers, 11:1–2, 11:10–22, 11:286–287
Oriolidae, 11:427, 11:431–434
ostriches, 8:99
ovenbirds, 10:209, 10:214–227
owlet-nightjars, 9:387, 9:391–392
owls, 9:331, 9:335, 9:340–343, 9:354–364
oystercatchers, 9:125, 9:130
painted-snipes, 9:115, 9:118
palmchats, 10:455
pardalotes, 11:201, 11:204–205
parrots, 9:275, 9:283–297
Passeriformes, 10:169–171
Pelecaniformes, 8:183
pelicans, 8:183, 8:225
penduline titmice, 11:147, 11:151–152
penguins, 8:147–148, 8:153–157
Phasianidae, 8:433
pheasants, 8:399, 8:433
Philippine creepers, 11:183, 11:186–187
phylogenetic systematics, 3:53
Picidae, 10:147, 10:154–166
Piciformes, 10:85–86
pigeons, 9:247, 9:255–266
pipits, 10:371, 10:378–383
pittas, 10:193, 10:197–200
plantcutters, 10:325, 10:327
plovers, 9:104, 9:133
potoos, 9:395, 9:399–400
pratincoles, 9:104, 9:155–159
Procellariidae, 8:123, 8:129–130
Procellariiformes, 8:107
pseudo babblers, 11:127, 11:130
puffbirds, 10:101, 10:106–110
rails, 9:5, 9:45–46, 9:57–67
Raphidae, 9:269–270, 9:272–273
Recurvirostridae, 9:104, 9:133, 9:137–140
rheas, 8:69
rollers, 10:51, 10:55–57
sandgrouse, 9:231, 9:235–238
sandpipers, 9:104, 9:175, 9:182–187
screamers, 8:393
scrub-birds, 10:337, 10:339
secretary birds, 8:344
seedsnipes, 9:189, 9:193–195
seriemas, 9:85
sharpbills, 10:291
sheathbills, 9:197, 9:200
shoebills, 8:233, 8:287
shrikes, 10:425, 10:432–438
sparrows, 11:397, 11:401–405
spoonbills, 8:233, 8:291

storks, 8:233, 8:265
storm-petrels, 8:135, 8:139–141
Struthioniformes, 8:53
Sturnidae, 11:407, 11:414–425
sunbirds, 11:207, 11:213–225
sunbitterns, 9:73
sungrebes, 9:69
swallows, 10:357, 10:363–368
swans, 8:369
swifts, 9:421, 9:426–431
tapaculos, 10:257, 10:262–267
terns, 9:104, 9:211–216
thick-knees, 9:143, 9:147
thrushes, 10:483
tinamous, 8:57
titmice, 11:155, 11:160–165
todies, 10:25, 10:29
toucans, 10:125, 10:130–134
tree swifts, 9:433, 9:435
treecreepers, 11:177, 11:180–181
trogons, 9:477, 9:481–484
tropicbirds, 8:183, 8:187, 8:190–191
trumpeters, 9:77, 9:81–82
turacos, 9:299, 9:304–309
typical owls, 9:345, 9:354–364
tyrant flycatchers, 10:269–270, 10:278–288
vanga shrikes, 10:439, 10:443–444
Vireonidae, 11:255, 11:258–262
wagtails, 10:371, 10:378–383
weaverfinches, 11:353, 11:360–372
weavers, 11:375–376, 11:382–393
whistlers, 11:115–116, 11:120–125
white-eyes, 11:227, 11:232
woodcreepers, 10:229, 10:232–236
woodhoopoes, 10:65, 10:68
woodswallows, 11:459, 11:462–464
wrens, 10:525, 10:531–537
fishes, **4:9–13**
Acanthuroidei, 5:391, 5:396–403
Acipenseriformes, 4:213, 4:217–220
Albuliformes, 4:249, 4:252–253
angelsharks, 4:161, 4:164
anglerfishes, 5:47, 5:53–56
Anguilliformes, 4:11, 4:255, 4:260–268
Atheriniformes, 5:67, 5:73–76
Aulopiformes, 4:431, 4:436–439
Australian lungfishes, 4:197
beardfishes, 5:1, 5:3
Beloniformes, 5:79, 5:83–85
Beryciformes, 5:113, 5:118–121
bichirs, 4:11, 4:209, 4:211
blennies, 5:341, 5:345–348
bowfins, 4:11, 4:229
bullhead sharks, 4:97, 4:101
Callionymoidei, 5:365, 5:369–371
carps, 4:297, 4:308–313
catfishes, 4:12, 4:351, 4:357–366
characins, 4:335, 4:342–349
chimaeras, 4:91, 4:94–95
coelacanths, 4:189–190
Cypriniformes, 4:36, 4:297, 4:307–319, 4:321, 4:325–333
Cyprinodontiformes, 5:89–90, 5:96–102
dogfish sharks, 4:151, 4:155–157
dories, 5:123, 5:126–129
eels, 4:11, 4:255, 4:260–268
electric eels, 4:369, 4:373–376
elephantfishes, 4:11, 4:238
Elopiformes, 4:243, 4:246–247
Esociformes, 4:379, 4:383–386
flatfishes, 5:449–450, 5:455–460, 5:462
Gadiformes, 5:25, 5:32–39
gars, 4:11, 4:221, 4:224–227
Gasterosteiformes, 5:131, 5:139–147
gobies, 5:373, 5:380–388
Gobiesocoidei, 5:355, 5:359–363
Gonorynchiformes, 4:289–290, 4:293–294
ground sharks, 4:11, 4:113, 4:119–128
Gymnotiformes, 4:369, 4:373–376
hagfishes, 4:77, 4:80–81
herrings, 4:67, 4:69, 4:277, 4:281–287
Hexanchiformes, 4:143, 4:146–148
Labroidei, 5:275, 5:284–290, 5:293, 5:301–306
labyrinth fishes, 5:427, 5:431–434
lampreys, 4:83, 4:88–89
Lampridiformes, 4:447–448, 4:452–454
lanternfishes, 4:441, 4:444–445
lungfishes, 4:201, 4:205
mackerel sharks, 4:11, 4:131–132, 4:135–141
minnows, 4:12, 4:297, 4:314–315
morays, 4:255, 4:264–266
mudminnows, 4:379, 4:383–386
mullets, 5:59, 5:62–64
Ophidiiformes, 5:15–16, 5:20–21
Orectolobiformes, 4:105, 4:108–110
Osmeriformes, 4:389–390, 4:395–401
Osteoglossiformes, 4:231, 4:237–240
Percoidei, 5:195, 5:202–208, 5:211, 5:214–216, 5:219, 5:225–232, 5:235, 5:247–253, 5:255, 5:266–268, 5:270–272
Percopsiformes, 5:5, 5:8–12
pikes, 4:379, 4:383–386
ragfishes, 5:351
Rajiformes, 4:11, 4:173–174, 4:182–188
rays, 4:10–11, 4:173–174, 4:182–188
Saccopharyngiformes, 4:271, 4:274–276
salmons, 4:405, 4:410–419
sawsharks, 4:167, 4:170
Scombroidei, 5:405, 5:409–417
Scorpaeniformes, 5:157, 5:161–162, 5:163–164, 5:169–177, 5:179, 5:186–193
Scorpaenoidei, 5:163–164
sharks, 4:10–11
skates, 4:173–174, 4:187
snakeheads, 5:437, 5:442–446
South American knifefishes, 4:369, 4:373–376
southern cod-icefishes, 5:321, 5:325–327
Stephanoberyciformes, 5:105, 5:108–109
stingrays, 4:11, 4:173
Stomiiformes, 4:421, 4:426–429
Stromateoidei, 5:421
Synbranchiformes, 5:151, 5:154–155
Tetraodontiformes, 5:467, 5:474–484
toadfishes, 5:41, 5:44–45
Trachinoidei, 5:331, 5:337–339
whale sharks, 4:11, 4:110
Zoarcoidei, 5:309, 5:315–317
insects, 3:4–5, **3:7–16**, 3:77–78
book lice, 3:243, 3:246–247
bristletails, 3:113
caddisflies, 3:375, 3:379–381
cockroaches, 3:147, 3:153–158
Coleoptera, 3:315–316, 3:327–333
diplurans, 3:107, 3:110–111
Diptera, 3:357, 3:364–373
earwigs, 3:195, 3:199–200
fleas, 3:347, 3:352–354
Hemiptera, 3:259–260, 3:267–279
Hymenoptera, 3:405, 3:411–423
lacewings, 3:305, 3:311–314
Lepidoptera, 3:383–384, 3:393–403
mantids, 3:177–178, 3:183–186
Mantophasmatodea, 3:217
mayflies, 3:125, 3:129–130
Mecoptera, 3:341, 3:345–346
Megaloptera, 3:289, 3:293–294
Odonata, 3:133, 3:137–138
Orthoptera, 3:201, 3:211–215
Phasmida, 3:221, 3:228–231
Phthiraptera, 3:249, 3:254–256
proturans, 3:93, 3:96–97
rock-crawlers, 3:189, 3:192–193
snakeflies, 3:297, 3:300–302
springtails, 3:99, 3:103–104
stoneflies, 3:141, 3:145
strepsipterans, 3:335, 3:338–339
termites, 3:161–163, 3:170–174
thrips, 3:281, 3:285–287
Thysanura, 3:119, 3:122–123
webspinners, 3:233, 3:236–237
zorapterans, 3:239, 3:241
limbed vertebrates, 6:7–11
lower metazoans and lesser deuterostomes, **1:7–14**
acoels, 1:179, 1:181–182
anoplans, 1:245, 1:249–250
arrow worms, 1:9, 1:13, 1:433, 1:437–441
box jellies, 1:10, 1:147, 1:151–152
calcareous sponges, 1:57–58, 1:61–64
cnidarians, 1:10
Anthozoa, 1:10, 1:103, 1:111–120
Hydrozoa, 1:10, 1:123, 1:134–145
comb jellies, 1:9, 1:10–11, 1:169, 1:173–176
demosponges, 1:77–78, 1:82–86
echinoderms
Crinoidea, 1:355, 1:361–364
Echinoidea, 1:401, 1:408–414
Ophiuroidea, 1:387, 1:393–398
sea cucumbers, 1:417–419, 1:425–430
sea daisies, 1:381, 1:385–386
sea stars, 1:367, 1:373–379
enoplans, 1:253, 1:256–257
entoprocts, 1:9–10, 1:13, 1:319, 1:323–325
flatworms
free-living flatworms, 1:185–186, 1:190–194
monogeneans, 1:11, 1:213, 1:218–223
tapeworms, 1:225–226, 1:234–241
gastrotrichs, 1:9, 1:12, 1:269, 1:272–273
girdle wearers, 1:9, 1:12–13, 1:343, 1:347–349
glass sponges, 1:67, 1:71–75
gnathostomulids, 1:9, 1:12, 1:331, 1:334–335
hair worms, 1:12, 1:305, 1:308–309
hemichordates, 1:13, 1:443, 1:447–449
jaw animals, 1:9, 1:12, 1:327
jellyfish, 1:10, 1:153, 1:160–167
kinorhynchs, 1:9, 1:11, 1:13, 1:275, 1:279–280
lancelets, 1:14, 1:485, 1:490–496
larvaceans, 1:473, 1:476–478
nematodes, 1:12
roundworms, 1:12, 1:283, 1:287–289, 1:291
secernenteans, 1:12, 1:293, 1:297–304

Orthonectida, 1:99, 1:101
placozoans, 1:10, 1:87
priapulans, 1:12–13, 1:337, 1:340–341
rhombozoans, 1:93, 1:96–98
rotifers, 1:9, 1:11–12, 1:259, 1:264–267
Salinella salve, 1:91
salps, 1:467, 1:471–472
sea squirts, 1:451–452, 1:458–466
sorberaceans, 1:479, 1:482–483
thorny headed worms, 1:9, 1:11–12, 1:311, 1:315–317
Trematoda, 1:11, 1:197, 1:203–205, 1:207–210
wheel wearers, 1:13, 1:351
mammals
aardvarks, 15:155
agoutis, 16:407, 16:411–414
anteaters, 13:171, 13:177–179
armadillos, 13:181–182, 13:187–190
Artiodactyla, 15:263–267
Atelidae, 14:155, 14:162–166
aye-ayes, 14:85
baijis, 15:19
bandicoots, 13:1
dry-country, 13:9, 13:14–16
rainforest, 13:9, 13:14–16
bats, 13:309
American leaf-nosed bats, 13:413, 13:423–432
bulldog bats, 13:443, 13:449
disk-winged, 13:473, 13:476
Emballonuridae, 13:355, 13:360–363
false vampire, 13:379, 13:384
funnel-eared, 13:459, 13:463–465
horseshoe, 13:387, 13:396–399
Kitti's hog-nosed, 13:367
Molossidae, 13:483, 13:490–491, 13:491–493
mouse-tailed, 13:351, 13:353
moustached, 13:435, 13:440–441
New Zealand short-tailed, 13:453, 13:457
Old World fruit, 13:319, 13:325–330, 13:333, 13:340–347
Old World leaf-nosed, 13:401, 13:407–409
Old World sucker-footed, 13:479
slit-faced, 13:371, 13:375
smoky, 13:467, 13:470
Vespertilionidae, 13:497, 13:506–514, 13:519, 13:524–525
bears, 14:256, 14:295, 14:302–305
beavers, 16:177, 16:183
bilbies, 13:19
botos, 15:27
Bovidae, 16:1–2
Antilopinae, 16:45, 16:50–55
Bovinae, 16:11, 16:18–23
Caprinae, 16:87–89, 16:98–103
duikers, 16:73, 16:80–84
Hippotraginae, 16:27–28, 16:37–41
Neotraginae, 16:59, 16:66–70
bushbabies, 14:23, 14:28–31
Camelidae, 15:313–314, 15:320–323
Canidae, 14:265–266, 14:276–282
capybaras, 16:401
Carnivora, 14:255, 14:256
cats, 14:369, 14:379–389
Caviidae, 16:389–390, 16:395–398
Cetacea, 12:14, 15:2–4, 15:133, 15:266
Balaenidae, 15:107, 15:115–117
beaked whales, 15:59, 15:64–68
dolphins, 15:41–42, 15:53–55
franciscana dolphins, 15:23
Ganges and Indus dolphins, 15:13
gray whales, 15:93
Monodontidae, 15:81, 15:90
porpoises, 15:33, 15:38–39
pygmy right, 15:103
rorquals, 15:1, 15:119, 15:127–129
sperm, 15:73, 15:79–80
Cheirogaleidae, 14:35, 14:41–44
chevrotains, 15:325–326, 15:331–333
Chinchillidae, 16:377, 16:382–383
colugos, 13:299, 13:304
coypus, 16:473
dasyurids, 12:287, 12:294–298, 12:299*t*–301*t*
Dasyuromorphia, 12:277
deer
Chinese water, 15:373
muntjacs, 15:343, 15:349–354
musk, 15:335, 15:340–341
New World, 15:379–382, 15:388–395
Old World, 15:357–358, 15:364–371*t*
Dipodidae, 16:211, 16:219–222
Diprotodontia, 13:31–32
dormice, 16:317, 16:323–326
duck-billed platypuses, 12:243
Dugongidae, 15: 199, 15:203
echidnas, 12:235
elephants, 12:*29*, 15:161–165, 15:174–175
Equidae, 15:225, 15:232–235
Erinaceidae, 13:203, 13:209–212
giant hutias, 16:469
gibbons, 14:207–208, 14:218–222
Giraffidae, 15:399, 15:408
great apes, 14:225, 14:237–239
gundis, 16:311, 16:314–315
Herpestidae, 14:347, 14:354–356
Heteromyidae, 16:199, 16:205–208
hippopotamuses, 15:301–302, 15:304, 15:310–311
humans, 14:241–243
hutias, 16:461
Hyaenidae, 14:359–360, 14:364–366
hyraxes, 15:177–178, 15:186–188
Indriidae, 14:63, 14:69–72
Insectivora, 13:193–194
koalas, 13:43
Lagomorpha, 16:479–480
lemurs, 14:47–48, 14:55–60
Cheirogaleidae, 14:35, 14:41–44
sportive, 14:3–4, 14:73–75, 14:79–83
Leporidae, 16:505, 16:511–514
Lorisidae, 14:13, 14:18–20
Macropodidae, 13:83, 13:93–100
manatees, 15: 205, 15: 211–212
marine mammals, **12:62–68**
Megalonychidae, 13:155, 13:159
moles
golden, 13:215–216, 13:220–221
marsupial, 13:25, 13:28
monitos del monte, 12:273
monkeys
Callitrichidae, 14:115, 14:127–131
Cebidae, 14:101, 14:110–112
cheek-pouched monkeys, 14:187–188, 14:197–203
leaf-monkeys, 14:4, 14:171–172, 14:178–183
night, 14:135
Pitheciidae, 14:143, 14:150–152
squirrel, 14:101, 14:108–110
monotremes, 12:11, 12:33, 12:227–228
mountain beavers, 16:131
Muridae, 16:281–283, 16:288–294
Arvicolinae, 16:225, 16:233–237, 16:238*t*
hamsters, 16:239, 16:245–246
Murinae, 16:249, 16:255–260
Sigmodontinae, 16:263–264, 16:272–275
musky rat-kangaroos, 13:69–70
Mustelidae, 14:319, 14:327–331
numbats, 12:303
octodonts, 16:433–434, 16:438–440
opossums
New World, 12:249–250, 12:257–263, 12:264*t*–265*t*
shrew, 12:267, 12:270
Otariidae, 14:256, 14:393, 14:402–406
pacaranas, 16:385–386
pacas, 16:417, 16:423
pangolins, 16:107, 16:115–119
peccaries, 15:291, 15:298–299
Perissodactyla, 15:215–216
Petauridae, 13:125, 13:131–132
Phalangeridae, 13:57, 13:*63*–65
pigs, 15:275, 15:284–288
pikas, 16:491, 16:497–501
pocket gophers, 16:185, 16:192–194
porcupines
New World, 16:365–366, 16:371–373
Old World, 16:357–363
possums
feather-tailed, 13:139–140, 13:144
honey, 13:135
pygmy, 13:105–106, 13:110–111
primates, 14:1–4
Procyonidae, 14:309, 14:313–315
pronghorns, 15:411
Pseudocheiridae, 13:113, 13:119–122
rat-kangaroos, 13:78–80
rats
African mole-rats, 16:339, 16:347–349
cane rats, 16:333, 16:337
chinchilla rats, 16:443, 16:447–448
dassie rats, 16:329
spiny, 16:449, 16:453–457
rhinoceroses, 15:249, 15:251, 15:257–262
rodents, 16:121–122
sengis, 16:517–518, 16:524–529
shrews
red-toothed, 13:247, 13:255–263
West Indian, 13:243
white-toothed, 13:265, 13:271–275
Sirenia, 15: 191–192
solenodons, 13:237, 13:240
springhares, 16:307
squirrels
flying, 16:135, 16:139–140
ground, 16:143, 16:151–157
scaly-tailed, 16:299, 16:302–305
tree, 16:163, 16:169–174
Talpidae, 13:193, 13:194, 13:279, 13:284–287
tapirs, 15:237–238, 15:245–247
tarsiers, 14:91–92, 14:97–99
tenrecs, 13:225, 13:232–233
three-toed tree sloths, 13:161, 13:167–169

tree shrews, 13:289, 13:294–296
true seals, 14:256, 14:417, 14:427–433
tuco-tucos, 16:425, 16:429
ungulates, 15:131–138
Viverridae, 14:335, 14:340–343
walruses, 14:256, 14:409
wombats, 13:51, 13:55–56
Xenartha, 13:147–149
protostomes, 2:4–5, 2:10–14
amphionids, 2:195
amphipods, 2:261, 2:265–271
anaspidaceans, 2:181, 2:183
aplacophorans, 2:379, 2:382–385
Arachnida, 2:333, 2:339–351
articulate lampshells, 2:521, 2:525–527
bathynellaceans, 2:177, 2:179
beard worms, 2:85, 2:88
bivalves, 2:451, 2:457–466
caenogastropods, 2:445, 2:449
centipedes, 2:353, 2:358–362
cephalocarids, 2:131, 2:133
Cephalopoda, 2:475, 2:482–488
chitons, 2:393, 2:397–400
clam shrimps, 2:147, 2:150–151
copepods, 2:299, 2:304–309
cumaceans, 2:229, 2:232
Decapoda, 2:197–198, 2:204–213
deep-sea limpets, 2:435, 2:437
earthworms, 2:65, 2:70–73
echiurans, 2:9, 2:103, 2:106–107
fairy shrimps, 2:135, 2:139
fish lice, 2:289, 2:292–293
freshwater bryozoans, 2:497, 2:500–501
horseshoe crabs, 2:327, 2:331–332
Isopoda, 2:249, 2:254–259
krill, 2:185, 2:189–192
leeches, 2:75, 2:80–82
leptostracans, 2:161, 2:164–165
lophogastrids, 2:225, 2:227
mantis shrimps, 2:167, 2:173–174
marine bryozoans, 2:503, 2:506–507, 2:509, 2:513–514
mictaceans, 2:241, 2:242
millipedes, 2:363, 2:367–370
monoplacophorans, 2:387, 2:390
mussel shrimps, 2:311, 2:314
mysids, 2:215, 2:219–221
mystacocarids, 2:295, 2:297
myzostomids, 2:59, 2:62
Neritopsina, 2:439, 2:442–443
nonarticulate lampshells, 2:515, 2:518–519
Onychophora, 2:109, 2:113–114
pauropods, 2:375, 2:377
peanut worms, 2:97, 2:100
phoronids, 2:491, 2:494
Polychaeta, 2:45, 2:50–56
Pulmonata, 2:411–413, 2:417–421
remipedes, 2:125, 2:128
sea slugs, 2:403, 2:407–409
sea spiders, 2:321, 2:324
spelaeogriphaceans, 2:243, 2:244
symphylans, 2:371, 2:373
tadpole shrimps, 2:141, 2:145–146
tanaids, 2:235, 2:238
tantulocaridans, 2:283, 2:286
Thecostraca, 2:273, 2:278–280
thermosbaenaceans, 2:245, 2:247
tongue worms, 2:317–318
true limpets, 2:423, 2:427
tusk shells, 2:469, 2:473
Vestimentifera, 2:91, 2:94
Vetigastropoda, 2:429, 2:433–434
water bears, 2:115, 2:120–123
water fleas, 2:153, 2:157–158
reptiles
African burrowing snakes, 7:461, 7:463
African sideneck turtles, 7:129, 7:132–133
Afro-American river turtles, 7:137, 7:140–141
Agamidae, 7:209, 7:214–221
Alligatoridae, 7:171, 7:175–177
Anguidae, 7:339, 7:342–344
Australo-American sideneck turtles, 7:77, 7:81–83
big-headed turtles, 7:135
blindsnakes, 7:379, 7:383–384
boas, 7:409, 7:413–416
Central American river turtles, 7:99
chameleons, 7:223–224, 7:235–240
colubrids, 7:465–467, 7:473–481
Cordylidae, 7:319, 7:324–325
crocodilians, 7:157, 7:*158*
Crocodylidae, 7:179, 7:184–186
early blindsnakes, 7:369
Elapidae, 7:483–484, 7:491–498
false blindsnakes, 7:387
false coral snakes, 7:399
file snakes, 7:439
Florida wormlizards, 7:283
Gekkonidae, 7:259, 7:265–269
Geoemydidae, 7:115, 7:118–119
gharials, 7:167
Iguanidae, 7:243, 7:250–256
Kinosternidae, 7:121, 7:125–126
knob-scaled lizards, 7:347, 7:351
Lacertidae, 7:297, 7:301–302
leatherback seaturtles, 7:101
Microteiids, 7:303, 7:306–307
mole-limbed wormlizards, 7:279, 7:281
Neotropical sunbeam snakes, 7:405
New World pond turtles, 7:105, 7:109–112
night lizards, 7:291, 7:294–295
pig-nose turtles, 7:75
pipe snakes, 7:395
pythons, 7:419, 7:424
seaturtles, 7:85, 7:89–91
shieldtail snakes, 7:393
skinks, 7:327–328, 7:332–337
slender blindsnakes, 7:373, 7:376
snapping turtles, 7:93, 7:95–96
softshell turtles, 7:151, 7:154–155
spade-headed wormlizards, 7:287
splitjaw snakes, 7:430
Squamata, 7:195–198, 7:*196*
sunbeam snakes, 7:401
Teiidae, 7:309, 7:314–316
Testudines, 7:*66*
tortoises, 7:143, 7:146–148
Tropidophiidae, 7:433, 7:436
tuatara, 7:189
Varanidae, 7:359, 7:365–368
Viperidae, 7:445, 7:451–459
wormlizards, 7:273, 7:276
vertebrates, 1:14
Tayassu spp., 15:267, 15:291
Tayassu pecari. *See* White-lipped peccaries
Tayassu tajacu. *See* Collared peccaries
Tayassuidae. *See* Peccaries
Taylor, E. H., 6:415, 6:419, 6:425, 6:431, 7:369
Taylor, Iain, 9:335
Taylorana spp., 6:246, 6:249, 6:251
Taylorana hascheana. *See* Penang Taylor's frogs
Tayras, 14:323, 14:*326*, 14:*327*, 14:331
Tchagra spp. *See* Bush-shrikes
Tchagra cruentus. *See* Rosy-patched shrikes
Tchagra minuta. *See* Marsh tchagras
Tchagra senegala. *See* Black-crowned tchagras
Tcharibbeena. *See* Bennett's tree kangaroos
Teacherbirds. *See* Ovenbirds
Teaching, mother chimpanzees, 12:157
Teals, 8:363, 8:364, 8:365, 8:367, 8:370, 8:371, 8:*374*, 8:*375*, 8:*384*, 8:386–*387*
Teats. *See* Nipples
Tectitethya crypta, 1:44
Tectum mesencephali, 12:6
Teeth, 4:7, 4:*10*, 4:17, 4:*337*, 7:9, 7:26, 12:9, 12:11, 12:*37*
digestion and, 12:46
domestication and, 12:174
evolution and, 12:11
feeding habits and, 12:14–15
subterranean mammals, 12:74
suckling and, 12:89
tadpoles, 6:40
See also Physical characteristics
Tegeticula moths. *See* Yucca moths
Tegeticula yuccasella. *See* Yucca moths
Tegus, 7:207, **7:309–317**, 7:*313*
Tehuantepec jackrabbits, 16:487
Teiidae, 7:204, 7:206, 7:207, **7:309–317**, 7:*313*
Teiinae, 7:309
Teinsotoma fernandesi, 2:431
Teinsotoma funiculatum, 2:431
Teira batfishes. *See* Longfin spadefishes
Teius spp., 7:309, 7:311, 7:312
Telebasis salva. *See* Damselflies
Telecanthura spp., 9:424
Teledromas spp. *See* Tapaculos
Telefomin cuscuses, 13:32, 13:60, 13:66*t*
Telemetry, 12:199–202
Telencephalon. *See* Forebrain
Teleoceros spp., 15:249
Teleogramma spp., 5:276
Teleogramma gracile. *See* Worm cichlids
Teleogryllus comodus. *See* Field crickets
Teleostei, 4:11, 4:25, 4:26, 4:34, 4:37
Telescope fishes, 4:432
Telmatherinidae. *See* Sulawesi rainbows
Telmatobiinae, 6:155–158, 6:*156*, 6:179
Telmatobius spp., 6:157, 6:158
Telmatobius culeus. *See* Titicaca water frogs
Telmatobufo spp., 6:157
Telocoprids, 3:322
Telophorus spp. *See* Bush-shrikes
Telophorus kupeensis. *See* Mount Kupé bush-shrikes
Telophorus multicolor. *See* Many-colored bush-shrikes
Telophorus olivaceus. *See* Olive bush-shrikes
Telophorus sulfureopectus. *See* Orange-breasted bush-shrikes
Temenuchus spp., 11:409
Temminck's ground pangolins. *See* Ground pangolins
Temminck's larks, 10:342
Temminck's rollers, 10:52
Temminck's stints, 9:179
Temminck's trident bats, 13:410*t*
Temnocephala chilensis, 1:*189*, 1:*193*

Temnocephalida, 1:186
Temnospondyls, 6:8–10
Temnotrogon roseigaster. See Hispaniolan trogons
Temnurus spp., 11:504
Temnurus temnurus. See Ratchet-tailed treepies
Temora longicornis, 2:*303*, 2:305–306
Temperate basses (Moronidae), 4:49, 4:68, **5:195–199**, 5:*201*, 5:*203*, **5:206**
Temperate basses (Percichthyidae), **5:219–222**
Temperature, 1:24–25, 2:26
  body, 4:37, 8:9–10
  butterflies and, 3:385
  development and, 3:44
  life histories and, 3:36–38
  moths and, 3:385
  water, 4:38–39, 4:49–50, 4:53
  See also Global warming; Habitats
Temperature-dependent sex determination (TSD), 7:7, 7:13, 7:72, 7:164
  See also Reproduction
Temple, Stanley, 9:270
Templeton, A. R., 14:244
Temporal babblers. *See* Gray-crowned babblers
1080 (Sodium monofluoroacetate), 12:186
Ten-lined mongooses. *See* Narrow-striped mongooses
Tenacigobius spp., 5:374
Tenasserim hornbills. *See* Plain-pouched hornbills
Tenches, 4:*305*, 4:*312*, 4:318–319
Tenebrio molitor. See Mealworms
Tenebrionidae, 3:320
Tengmalm's owls, 9:350
Tenniel, John, 9:270
Tenodera aridifolia sinensis. See Chinese mantids
Tenrec spp. *See* Tenrecs
Tenrec ecaudatus. See Common tenrecs
Tenrecidae. *See* Tenrecs
Tenrecinae. *See* Spiny tenrecs
Tenrecomorpha, 13:215
Tenrecs, **13:225–235**, 13:*231*
  behavior, 13:197–198, 13:227–228
  conservation status, 13:229
  distribution, 12:135, 12:136, 13:197, 13:*225*, 13:226
  echolocation, 12:86
  evolution, 13:225
  feeding ecology, 13:198, 13:228–229
  habitats, 13:197, 13:226–227
  humans and, 13:230
  physical characteristics, 13:195–196, 13:225–226
  reproduction, 13:199, 13:229
  species of, 13:232–234, 13:234*t*
  taxonomy, 13:193, 13:194, 13:225
Tent-making bats, 13:416, 13:*421*, 13:*425*, 13:431
Tentacled snakes, 7:465, 7:468, 7:*472*, 7:*473*, 7:477
Tentaculariidae, 1:226
Tenthredinidae. *See* Sawflies
Tenualosa thibaudeaui. See Laotian shads
Tenuirostritermes spp., 3:163–164
Teoidea, 7:307
Tephrinectes spp., 5:450
Tephritid flies, 2:38
Tephrodornis. See Woodshrikes
Tepui swifts, 9:422
Terapon spp., 5:220
Terapon jarbua. See Target perches
Terapontidae. *See* Grunters
Terateleotris spp., 5:373
Terathopius ecaudatus. See African bateleurs
Teratoscincus scincus. See Common plate-tailed geckos
Terebellids, 2:46, 2:47
Terebrantia, 3:281, 3:282
Terebratulina retusa. See Lampshells
Terebratulina septentrionalis. See Northern lampshells
Terecays. *See* Yellow-spotted river turtles
Teredinidae, 2:452, 2:453
Teredo navalis. See Common shipworms
Terek sandpipers, 9:176, 9:179
Terenura spodioptila. See Ash-winged antwrens
Teretrurus sanguineus, 7:391
Tergomya, 2:387
Termes fatalis. See Linnaeus's snapping termites
Termites, 2:36, 2:37, 3:49, **3:*161*–175**, 3:*162*, 3:*169*
  behavior, 3:164–166, 3:*166*
  caste in, 3:69, 3:70
  vs. cockroaches, 3:161–162
  conservation status, 3:168
  distribution, 3:163
  ecosystems and, 3:72
  evolution, 3:9, 3:161–163
  feeding ecology, 3:63, 3:64, 3:166–167
  habitats, 3:163–164
  humans and, 3:168
  as pests, 3:75, 3:168
  physical characteristics, 3:163
  queens, 3:*165*
  reproduction, 3:35–36, 3:167–168
  social structure, 3:67, 3:68, 3:69, 3:70, 3:72
  species of, 3:*170*–*175*
  taxonomy, 3:161–163
Termitidae, 3:55, 3:164, 3:168
Termopsidae. *See* Rottenwood termites
Terns, **9:203–217**, 9:*208*, 9:*210*
  behavior, 9:206
  conservation status, 9:208–209
  distribution, 9:*203*, 9:205
  evolution, 9:203–204
  feeding ecology, 9:207
  habitats, 9:205–206
  humans and, 9:209
  physical characteristics, 9:204–205
  reproduction, 9:207–208
  species of, 9:*211*–*217*
  taxonomy, 9:104, 9:203–204
Terpsiphone smithii. See Annobón paradise-flycatchers
Terpsiphone viridis. See African paradise-flycatchers
Terrace, H. S., 12:160–161
Terrapene spp. *See* Box turtles
Terrapene carolina. See Eastern box turtles
Terrapins
  diamondback, 7:107, 7:*108*, 7:109–*110*
  painted, 7:*117*, 7:*118*
  See also Helmeted turtles
Terrariums, 6:54–55
Terrestrial decapods, 2:199
Terrestrial insects, 3:11–12, 3:22, 3:82
  See also specific types of terrestrial insects
Terrestrial kraits, 7:483, 7:484, 7:486
Terrestrial leeches. *See* Tiger leeches
Terrestrial mammals
  adaptations in, 12:13–14
  field studies of, 12:200–202
  vision, 12:79
  See also specific terrestrial species
Terrestrial pangolins, 16:107–112
Terrestrial scutellerid bugs, 3:260
Terrestrial snails. *See* Lung-bearing snails
Terrestrial tree shrews, 13:*291*, 13:*292*, 13:*293*, 13:295, 13:*296*
Terrestrisuchus spp., 7:18
Terriers, 14:288, 14:292
Territoriality, 4:63, 7:43–44, 7:245, 12:*142*
  bird songs and, 8:37–41
  cichlids, 5:277
  Cyprinodontiformes, 5:92
  evolution of, 12:18
  nutritional requirements and, 12:124
  pygmy sunfishes, 5:197
  See also Behavior
Tesia olivea. See Slaty-bellied tesias
Tesias, 11:5, 11:*8*, 11:14
Tessarabrachion spp., 2:185
Tessellated darters, 5:198
Testacella spp., 2:414
Testacellids, 2:414
Testes, 12:91, 12:102–103
  See also Physical characteristics
Testosterone, 12:148
Testosterone levels, bird songs and, 8:39
Testudines. *See* Turtles
Testudinidae. *See* Tortoises
Testudo graeca. See Greek tortoises
Testudo hermanni. See Hermann's tortoises
Tetanus, subterranean mammals and, 12:77–78
Tethytragus spp., 16:1
Tetonius spp., 14:1
Tetra perez. *See* Bleeding-heart tetras
Tetrabothriidae. *See* Tetrabothriidea
Tetrabothriidea, 1:226, 1:227, 1:229
Tetracerus quadricornis. See Chousinghas
Tetraclaenodon spp., 15:132
Tetractinomorpha, 1:77
Tetradactylus eastwoodae. See Eastwood's seps
Tetragonocephalidae, 1:226
Tetragonuridae. *See* Brook trouts
Tetragonurus spp. *See* Brook trouts
Tetrakentron synaptae, 2:117
Tetramerocerata, 2:375, 2:376
Tetrao malouinus. See White-bellied seedsnipes
Tetrao namaqua. See Namaqua sandgrouse
Tetrao orientalis. See Black-bellied sandgrouse
Tetrao senegallus. See Spotted sandgrouse
Tetrao tetrix. See Black grouse
Tetrao urogallus. See Capercaillies
Tetraodontidae. *See* Pufferfishes
Tetraodontiformes, **5:467–485**, 5:*472*, 5:*473*
  behavior, 5:469–470
  conservation status, 5:471
  distribution, 5:468
  evolution, 5:467
  feeding ecology, 5:470
  habitats, 5:468–469
  humans and, 5:471
  physical characteristics, 5:467–468
  reproduction, 5:470–471
  species of, 5:*474*–484
  taxonomy, 5:467

Tetraogallus spp. *See* Snowcocks
Tetraogallus tibetanus. See Tibetan snowcocks
Tetraonchus monenteron, 1:*216,* 1:*218,* 1:221
Tetraphyllidea, 1:226, 1:227
Tetrapleurodon spp., 4:83
Tetrapods, evolution, 6:7–10
Tetrarhynchobothriidae, 1:226
Tetrarogidae. *See* Waspfishes
Tetras, 4:12, 4:66, 4:337, 4:*338,* 4:*340,* 4:*343–344*
Tetraselmis spp., 1:33
Tetrastemma spp., 1:253
Tetrax spp. *See* Bustards
Tetrax tetrax. See Little bustards
Tetrigidae, 3:201
Tetrigoidea, 3:201
Tetrix subulata, 3:203
Tetrodotoxin, 6:189, 6:325, 6:373
Tettigarctidae, 3:55
Tettigonioidea, 3:201
Tevnia spp., 2:91
Texan clam shrimps, 2:*149,* 2:*150,* 2:151
Texas alligator lizards, 7:*341,* 7:*343,* 7:344
Texas blind salamanders, 6:*394,* 6:400
Texas blindsnakes, 7:374–*376*
Texas horned lizards, 7:246, 7:*249,* 7:*251,* 7:254
Texas kangaroo rats, 16:209*t*
Texas mice, 16:278*t*
Texas Parks and Wildlife Department, on northern cat-eyed snakes, 7:481
Texas pocket gophers, 16:195*t*
Texas spotted whiptails, 7:312
Texas threadsnakes. *See* Texas blindsnakes
Texas wormsnakes. *See* Texas blindsnakes
Thalassarche spp. *See* Albatrosses
Thalasseus spp. *See* Crested terns
Thalassinidea, 2:197, 2:198, 2:201
Thalassocalyce inconstans, 1:*172,* 1:*176*–177
Thalassocalycida, 1:169
Thalassoica antarctica. See Antarctic petrels
Thalassoma spp., 5:296
Thalassoma ascensionis, 5:298
Thalassoma bifasciatum. See Blueheads
Thalassoma noronhanum. See Noronha wrasses
Thalassophryne spp., 5:41, 5:42
Thalassophryninae, 5:41, 5:42
Thalassornis leuconotus. See White-backed ducks
Thalattosuchians. *See* Marine crocodiles
Thaleichthys pacificus. See Eulachons
Thalia democratica, 1:*469,* 1:473
Thaliacea. *See* Salps
Thallomys paedulcus. See Tree rats
Thalurania spp. *See* Hummingbirds
Thamins. *See* Eld's deer
Thamnocephalidae, 2:135, 2:136
Thamnocephalus spp., 2:136
Thamnolaea cinnamomeiventris. See Mocking chats
Thamnomanes caesius. See Cinereous antshrikes
Thamnophiine natricines, 7:470
Thamnophiini, 7:466
Thamnophilidae. *See* Ant thrushes
Thamnophilus doliatus. See Barred antshrikes
Thamnophis spp. *See* North American garter snakes
Thamnophis butleri. See Garter snakes
Thamnophis radix. See Plains garter snakes
Thamnophis sauritus. See Garter snakes
Thamnophis sirtalis. See Common garter snakes
Thamnophis sirtalis infernalis. See San Francisco garter snakes
Thamnophis sirtalis parietalis. See Red-sided garter snakes
Thamnophis sirtalis tetrataenia. See San Francisco garter snakes
Thaptomys nigrita, 16:270
Thaumasioscolex didelphidis, 1:226
Thaumastocoris spp., 3:55
Thaumastodermatidae, 1:269, 1:270
Thaumastura spp. *See* Hummingbirds
Thaumastura cora. See Peruvian sheartails
Thecocodium brieni, 1:128
Thecodontia, 7:5
Thecostraca, **2:273–281,** 2:*277*
  behavior, 2:275
  conservation status, 2:276
  distribution, 2:274
  evolution, 2:273
  feeding ecology, 2:275
  habitats, 2:275
  humans and, 2:276
  physical characteristics, 2:273–275
  reproduction, 2:275–276
  species of, 2:*278–281*
  taxonomy, 2:273
Thekla larks, 10:346, 10:352
Thelenota anax, 1:419
Thelenota rubralineata. See Candy cane sea cucumbers
Theletornis spp. *See* Twigsnakes
Theloderma spp., 6:291, 6:292
Thelodonti, 4:9
Themiste spp., 2:98
Themiste lageniformes, 2:98
Thenus orientalis. See Flathead locust lobsters
Theodoxus spp., 2:440, 2:441
Theodoxus fluviatilis. See River nerites
Theonella swinhoei, 1:45
Theopropus elegans. See Boxer mantids
Theory of Mind, 12:159–160
Theragra chalcogramma. See Alaska pollocks
Therapsida, 7:5, 12:10, 12:36
Therians
  mammary nipples in, 12:5
  reproduction, 12:51
  See also Marsupials; Placentals; specific animals
Thermal imaging, in field studies, 12:198–199
Thermal isolation, 12:3
Thermals, in migration, 8:29–30
Thermokinesis, 1:39
  See also Behavior
Thermoreception, 7:32–33
Thermoregulation, 7:28, 7:44–45, 12:3, 12:36, 12:50–51, 12:59–60, 12:65–67, 12:73–74, 12:113
  Iguanidae, 7:244
  knob-scaled lizards, 7:348
  Teiidae, 7:310–311
  tunas, 5:*406*
  See also Behavior; Energy conservation; Physical characteristics
Thermosbaena mirabilis, 2:*245,* 2:*247*
Thermosbaenaceans, **2:245–247**
Thermosphaeroma thermophilum, 2:252
Thermotaxis, 1:40
  See also Behavior
Thermozodium esakii, 2:115, 2:116
Theropithecus spp. *See* Geladas
Theropithecus gelada. See Geladas
Thescelocichla leucopleura. See Swamp greenbuls
Thespidae, 3:177
Thetys vagina, 1:*469,* 1:473–474
Thick-billed cuckoos, 9:*316,* 9:318–*319*
Thick-billed flowerpeckers, 11:*192,* 11:*193*–194
Thick-billed ground pigeons, 9:244
Thick-billed larks, 10:342, 10:*345,* 10:346, 10:*347,* 10:*350*–351
Thick-billed murres, 9:104, 9:219, 9:*223,* 9:224–*225*
Thick-billed parrots, 9:279–280
Thick-billed weavers, 11:*378,* 11:*380,* 11:*391*
Thick-knees, 9:104, **9:143–149,** 9:*147–148*
Thick-lipped gouramies, 5:*427*
Thick-spined porcupines, 16:353–354, 16:*356,* 16:*357,* 16:358
Thick-spined rats. *See* Armored rats
Thick-tailed bushbabies. *See* Brown greater bushbabies
Thick-tailed opossums, 12:251, 12:*255,* 12:*260*
Thick-tailed pangolins. *See* Indian pangolins
Thick-tailed pygmy jerboas, 16:215, 16:216, 16:223*t*
Thicket-fantails, 11:83, 11:85
Thicket tinamous, 8:*60,* 8:62–63
Thiel, V., 1:57
Thigmokinesis, 1:38
  See also Behavior
Thigmotaxis, 1:40
  See also Behavior
Thimble jellys, 1:*158,* 1:*160*–161
Thin-billed murres. *See* Common murres
Thin-spined porcupines, 16:365, 16:368, 16:369, 16:374*t,* 16:450, 16:*452,* 16:*453,* 16:457
Thinhorn sheep, 12:131, 12:178, 16:87, 16:91, 16:92
Think Tank exhibit (Smithsonian), 12:151
Thinking. *See* Cognition
Thinocoridae. *See* Seedsnipes
Thinocorus spp. *See* Seedsnipes
Thinocorus orbignyianus. See Gray-breasted seedsnipes
Thinocorus rumicivorus. See Least seedsnipes
Thinopithecus avunculus. See Tonkin snub-nosed monkeys
Thinornis novaeseelandiae. See Shore plovers
Third eyes, 7:32
Thirteen-lined ground squirrels, 16:143–144
Thomasomyines, 16:263–264
Thomasomys aureus, 16:266, 16:268
Thomas's bats. *See* Shaggy-haired bats
Thomas's bushbabies, 14:33*t*
Thomas's flying squirrels, 16:141*t*
Thomasset's frogs, 6:136
Thomomys spp., 16:185
Thomomys bottae. See Valley pocket gophers
Thomomys bulbivorus. See Camas pocket gophers
Thomomys mazama. See Mazama pocket gophers
Thomomys monticola. See Mountain pocket gophers
Thomomys talpoides. See Northern pocket gophers

Thomomys umbrinus. See Southern pocket gophers
Thompson, D., 9:275
Thomson's gazelles, 12:*10*, 15:*133*, 16:*46*, 16:*49*, 16:51
coevolution and, 12:216
conservation status, 16:48
distribution, 16:*50*
in guilds, 12:119
physical characteristics, 15:269
predation and, 12:116
Thoracic excites, 3:12–13
Thoracica, 2:273
Thoraciliacus spp., 6:12
Thoracistus spp., 3:204
Thorax spp., 3:150
Thoreau, Henry David, 11:505
Thorectidae, 1:79
Thorius spp., 6:391
Thorius aureus. See Golden thorius
Thorn-tailed rayaditos, 10:*212*, 10:*219*–220
Thornbills, 9:442, 9:450, 11:55–*56*, 11:*56*, 11:60–61
Thornbirds, 10:210
Thornfishes, 5:321, 5:*322*
Thornlike treehoppers, 3:*260*
Thorny catfishes, 4:352, 4:353
Thorny devils, 7:*212*, 7:*217*, 7:218–219
Thorny headed worms, **1:311–317**
behavior, 1:312
conservation status, 1:313
distribution, 1:312
evolution, 1:311
feeding ecology, 1:312
habitats, 1:312
humans and, 1:313
physical characteristics, 1:311–312
reproduction, 1:22, 1:312–313
species of, 1:314–316
taxonomy, 1:9, 1:11–12, 1:311
Thorny tinselfishes, 5:*125*, 5:*126*, 5:127
Thorold's deer. *See* White-lipped deer
Thoropa spp., 6:156, 6:157
Thoropa miliaris. See Rock River frogs
Thoropa petropolitana, 6:35
Thrashers. *See* Mimids
Thrattidion noctivagus. See Sanaga pygmy herrings
Thraupidae. *See* Tanagers
Thraupinae, 11:286
Threadfin breams, 5:255
See also Nemipteridae
Threadfin shads, 4:*277*
Threadfins, 4:68, 5:255, 5:258–263
Threadfishes. *See* Slender snipe eels
Threadsnakes, 7:*374*, 7:*485*
See also Slender blindsnakes
Threadworms, 1:*296*, 1:297–298
See also Human whipworms
Threat displays, cranes, 9:25–26
Threatened Fishes Committee of the Australian Society for Fish Biology, 5:323
Three-banded armadillos, 13:*182*, 13:*183*, 13:184
Three-banded coursers, 9:153, 9:*154*, 9:*159*–160
Three-horned chameleons. *See* Jackson's chameleons
Three-phase lifestyle
parrotfishes and, 5:296–297
wrasses and, 5:296–297
See also Behavior
Three-pointed pangolins. *See* Tree pangolins
Three-quarters relatedness, 3:70
Three-spine horseshoe crabs. *See* Japanese horseshoe crabs
Three-stripe batagurs. *See* Painted terrapins
Three-striped dasyures, 12:300*t*
Three-striped douroucoulis. *See* Three-striped night monkeys
Three striped flatworms, 1:*3*
Three-striped night monkeys, 14:135, 14:*136*, 14:*137*, 14:*139*, 14:*140*, 14:141
Three-toed amphiumas, 6:405, 6:*408*, 6:410
Three-toed jacamars, 10:91, 10:93, 10:*94*, 10:96–*97*
Three-toed jerboas, 16:211, 16:212, 16:213, 16:214, 16:216
Three-toed toadlets, 6:6, 6:35, **6:179–182,** *6:181–182*
Three-toed tree sloths, 13:*150*, **13:161–169,** 13:*162*, 13:*164*, 13:*165*, 13:*166*
behavior, 13:*152*, 13:*153*, 13:163
conservation status, 13:153, 13:165
distribution, 13:*161*, 13:162
evolution, 13:161
feeding ecology, 13:164–165
habitats, 13:162
humans and, 13:165
physical characteristics, 13:149, 13:161–162
reproduction, 13:*147*, 13:165
species of, 13:167–169
taxonomy, 13:147
Three-toed woodpeckers, 10:149, 10:*152*, 10:*160*–161
Three-wattled bellbirds, 10:305, 10:308, 10:*311*, 10:321–*322*
Threespine sticklebacks, 5:*133*, 5:*138*, 5:*140*, 5:141
Threespot damselfishes, 5:*294*
Threespot dascyllus, 5:*293*
Threetooth puffers, 5:467–471, 5:*472*, 5:*483*, 5:484
Threnetes spp., 9:443, 9:448
Threnetes ruckeri, 9:446
Thresher sharks, 4:11, 4:132, 4:133, 4:*134*, 4:*135*–136
Threshold relationships, in ecosystems, 12:217
Threskiornis ibis. See Sacred ibises
Threskiornis melanocephalus. See Black-headed ibises
Threskiornis molucca. See Australian white ibises
Threskiornis solitarius. See Reunion flightless ibises
Thriambeutes mesembrinoides, 3:359
Thrichomys apereoides. See Punares
Thripidae, 3:281, 3:283
Thripophaga berlepschi. See Russet-mantled softtails
Thripophaga macroura. See Striated softtails
Thrips, 3:68, 3:75, **3:281–287,** 3:*284*
Thrombocytes, 12:8
Thrush-like ant-pittas, 10:*243*, 10:*255*
Thrush-like woodcreepers. *See* Plain-brown woodcreepers
Thrushes, 10:171, **10:483–491,** 10:*491*, **10:499–504**
behavior, 10:486–487
conservation status, 10:488–489
distribution, 10:*483*, 10:485
evolution, 10:483
feeding ecology, 10:487
habitats, 10:485–486
humans and, 10:489
physical characteristics, 10:483–485
reproduction, 10:487–488
species of, 10:*499–504*
taxonomy, 10:483
See also Shrike-thrushes
Thrust, for powered flight, 12:56
Thryomanes bewickii. See Bewick's wrens
Thryomanes sissonii. See Socorro wrens
Thryonomyidae. *See* Cane rats
Thryonomys spp. *See* Cane rats
Thryonomys arkelli, 16:333
Thryonomys gregorianus. See Lesser cane rats
Thryonomys logani, 16:333
Thryonomys swinderianus. See Greater cane rats
Thryothorus spp., 10:525, 10:526, 10:527, 10:528
Thryothorus ludovicianus. See Carolina wrens
Thryothorus nicefori. See Niceforo's wrens
Thryothorus nigricapillus. See Bay wrens
Thumb-splitters. *See* Mantis shrimps
Thumbed ghost frogs. *See* Rose's ghost frogs
Thumbless bats. *See* Smoky bats
Thunberg's stick insects, 3:*227*, 3:*228*, 3:229
Thunderfishes. *See* Sailfin sandfishes
Thunnus spp. *See* Tunas
Thunnus alalunga. See Albacore
Thunnus maccoyii. See Southern bluefin tunas
Thunnus obesus. See Bigeye tunas
Thunnus thynnus. See Atlantic bluefin tunas
Thurber's juncos. *See* Dark-eyed juncos
Thylacines. *See* Tasmanian wolves
Thylacinidae. *See* Tasmanian wolves
Thylacinus cynocephalus. See Tasmanian wolves
Thylacinus potens, 12:307
Thylacomyinae. *See* Bilbies
Thylacosmilus spp., 12:11
Thylamys spp. *See* Fat-tailed mouse opossums
Thylamys elegans. See Elegant fat-tailed opossums
Thylamys pusilla. See Small fat-tailed opossums
Thylogale spp. *See* Pademelons
Thylogale stigmatica. See Red-legged pademelons
Thymallus arcticus arcticus. See Arctic graylings
Thyridopteryx ephemeraeformis. See Bagworms
Thyridoropthrum spp., 3:204
Thyroid gland, lissamphibians, 6:19
Thyroptera spp. *See* Disk-winged bats
Thyroptera discifera. See Peters's disk-winged bats
Thyroptera laveli, 13:473, 13:474
Thyroptera tricolor. See Spix's disk-winged bats
Thyropteridae. *See* Disk-winged bats
Thyroxine, 12:141
Thyrsites spp., 5:406
Thyrsoidea macrura, 4:255
Thysania agrippina. See Owlet moths
Thysanoessa spp., 2:185
Thysanoessa inermis, 2:*188*, 2:*190*, 2:191
Thysanoessa raschii, 2:*188*, 2:191–*192*
Thysanoessa spinifera, 2:*188*, 2:*192*
Thysanopoda spp., 2:185
Thysanoptera. *See* Thrips
Thysanosoma actinioides, 1:230

Thysanosomatidae, 1:226
Thysanura, 3:10, 3:19, **3:119–124,** 3:*121*
See also Silverfish
Tiaris spp., 11:264
Tibetan antelopes, 12:134, 16:*9*
Tibetan argalis, 12:134
Tibetan eared-pheasants, 8:435
Tibetan foxes, 14:284*t*
Tibetan gazelles, 12:134, 16:48, 16:57*t*
Tibetan ground-jays. *See* Hume's ground-jays
Tibetan larks, 10:343
Tibetan muntjacs. *See* Tufted deer
Tibetan partridges, 8:434
Tibetan pitvipers, 7:*450*, 7:*454*, 7:455
Tibetan sand foxes. *See* Tibetan foxes
Tibetan snowcocks, 8:*439*, 8:*445*
Tibetan stream salamanders, 6:*338*, 6:*339*
Tibetan water shrews. *See* Elegant water shrews
Tibetan wild asses. *See* Kiangs
Tichodroma muraria. See Wall creepers
Tichodrominae. *See* Wall creepers
Ticinosuchus spp., 7:17
Tickell's blue flycatchers. *See* Orange-breasted blue flycatchers
Tickell's brown hornbills, 10:73
Ticks, **2:333–352,** 2:*338*
behavior, 2:335
conservation status, 2:336
distribution, 2:335
evolution, 2:333
feeding ecology, 2:335
habitats, 2:335
humans and, 2:336
physical characteristics, 2:333–334
reproduction, 2:335
taxonomy, 2:333
Tidal water bears, 2:116, 2:*119*, 2:121–122
Tide shifts, 4:60–61
Tidepool fishes, 4:47, 4:60–61, 5:181
Tidepool gunnels, 5:312
Tidepool johnnies. *See* Tidepool sculpins
Tidepool sculpins, 5:*180*, 5:181, 5:*185*, 5:*188*–189
Tiekes, 11:448, 11:449–*450*
Tiger barbs, 4:*303*, 4:315–*316*
Tiger beetles, 3:30, 3:57, 3:63, 3:82, 3:320, 3:321
Tiger botia. *See* Clown loaches
Tiger catfishes. *See* Tiger shovelnose catfishes
Tiger coral trouts. *See* Blacksaddled coral groupers
Tiger finches. *See* Red avadavats
Tiger flatworms, 2:*23*
Tiger grunters. *See* Target perches
Tiger herons, 8:233, 8:239, 8:241, 8:244
Tiger leeches, 2:75, 2:76, 2:77, 2:79, 2:*80*
Tiger-leg monkey frogs, 6:*230*
Tiger loaches, 4:*321*
Tiger moths, 3:38–39
Tiger ocelots. *See* Little spotted cats
Tiger perches. *See* Target perches
Tiger Peters frogs. *See* Indian tiger frogs
Tiger quolls. *See* Spotted-tailed quolls
Tiger ratsnakes, 7:466, 7:468
Tiger salamanders, 6:*323*, 6:*357*, 6:359–360
behavior, 6:44, 6:360
habitats, 6:356, 6:360
physical characteristics, 6:352, 6:360
reproduction, 6:*29*, 6:31, 6:360
taxonomy, 6:355, 6:359–360
Tiger sharks, 4:*4*, 4:46, 4:*114*–116, 4:*118*, 4:121–*122*
Tiger shovelnose catfishes, 4:*355*, 4:*357*, 4:364–365
Tiger snake eels, 4:*259*, 4:*268*
Tiger snakes, 7:*490*, 7:495–*496*
Tiger watersnakes. *See* Yamakagashis
Tigerfishes. *See* Grunters
Tigers, 14:*377*, 14:*380*–381
behavior, 14:371, 14:372
conservation status, 12:218, 12:220, 14:374
distribution, 12:136, 14:370
humans and, 12:139, 14:376
physical characteristics, 14:370
reproduction, 14:372–373
stripes, 12:38
in zoos, 12:210
Tiger's tail sea cucumbers, 1:*424*, 1:426, 1:*430*
Tigriopus californicus, 2:*303*, 2:308
Tijuca spp. *See* Cotingas
Tijuca condita, 10:308
Tiki tikis. *See* Blueheads
Tilamatura spp. *See* Hummingbirds
Tilapia spp., 5:277
Tilapia melanopleura, 5:280–281
Tilapia zilii, 5:280–281
Tilapiines, 5:277, 5:280–281
Tilefishes, 5:255, 5:256, 5:258–263
Tiliqua spp., 7:329
Tiliqua gerrardii, 7:329
Tiliqua rugosa. See Bobtails
Tiliqua scincoides. See Blue-tongued skinks
Tillodontia, 12:11, 15:131, 15:133, 15:135, 15:136
Timaliidae. *See* Babblers
Timaline babbers, 11:2
Timber rattlesnakes, 7:*50*, 7:448, 7:*450*, 7:*452*, 7:454
Timber wolves. *See* Gray wolves
Timbo disc frogs, 6:35, 6:*307*, 6:*314*–315
Timema spp., 3:223
Timema cristinae, 3:221
Timematidae. *See* Timematodea
Timematodea, 3:221
Timon spp., 7:297
Timor deer, 15:371*t*
Timor figbirds, 11:*430*, 11:431–*432*
Timor sparrows, 11:356
Timor wild boars, 15:290*t*
Tinamous, 8:*53*–55, **8:57–67**
Tinamus major. See Great tinamous
Tinamus osgoodi. See Black tinamous
Tinamus pentlandii. See Puna tinamous
Tinamus solitarius. See Solitary tinamous
Tinbergen, N., 8:38, 12:140–143, 12:147–148, 15:143
Tinca spp., 4:298
Tinca tinca. See Tenches
Tincinae, 4:297
Tineids, 3:387, 3:388, 3:389
Tineola bisselliella. See Webbing clothes moths
Tinselfishes, 5:124, 5:*125*, 5:127–*128*
Tipula paludosa. See European marsh crane flies
Tit-hylias, 11:147
Titanic longhorn beetles, 3:13, 3:317, 3:*318*, 3:*326*, 3:328, 3:*332*
Titanites anguiformis. See Ammonites
Titanus gigantea. See Titanic longhorn beetles
Titicaca flightless grebes, 8:*173*, 8:*174*
Titicaca water frogs, 6:157, 6:159, 6:*160*, 6:171
Titis, **14:143–154,** 14:*149*
behavior, 14:7, 14:8, 14:145, 14:146
conservation status, 14:11, 14:148
distribution, 14:144–145
evolution, 14:143
feeding ecology, 14:146–147
habitats, 14:145
humans and, 14:148
physical characteristics, 14:143–144
reproduction, 14:147
species of, 14:*150*, 14:151–153, 14:153*t*
taxonomy, 14:143
Titiscaniidae, 2:439
Titmice, **11:155–166,** 11:*159*
behavior, 11:157
conservation status, 11:158
distribution, 11:*155*, 11:156
evolution, 11:155
feeding ecology, 11:*156*, 11:157
habitats, 11:156–157
humans and, 11:158
long-tailed, 10:*169*, **11:141–146,** 11:*144*, 11:147
penduline, **11:147–153,** 11:*150*
physical characteristics, 11:155–156
reproduction, 11:157–158
taxonomy, 11:155
Tits. *See* specific types of tits
Tityra spp. *See* Tyrant flycatchers
Tityras. *See* Tyrant flycatchers
Tmetothylacus tenellus. See Golden pipits
T'o. *See* Chinese alligators
Toad head turtles, 7:*78*
Toad-headed agamas, 7:*213*, 7:*215*, 7:220–221
Toad-like treefrogs, 6:*284*, 6:288
Toadfishes, **5:41–46,** 5:*43*
Toadfrogs, Asian, 6:5, 6:61, 6:64, **6:109–117,** 6:*112*
Toadlets
Australian, **6:147–154,** 6:*150*
Madagascan, 6:6, 6:301, **6:317–321,** 6:*319–320*
three-toed, 6:6, 6:35, 6:173, **6:179–182,** 6:*181–182*
Toads, **6:61–68**
behavior, 6:46–48
chemosensory cues, 6:45–46
distribution, 6:5, 6:63–64
evolution, 6:15, 6:61–62
fire-bellied, **6:83–88,** 6:*85*
habitats, 6:26, 6:63–64
larvae, 6:39–43
medicinal uses of, 6:53
midwife, 6:*32*, 6:36, 6:47, 6:*63*, **6:89–94,** 6:*90–91*
physical characteristics, 6:25–26, 6:62–63
reproduction, 6:33, 6:35, 6:36, 6:*37*, 6:68
spadefoot, 6:4–5, 6:64, **6:119–125,** 6:*120–121*, 6:*123*
Suriname, 6:36–37, **6:99–107,** 6:*100*, 6:*101*, 6:*106*
true, **6:183–195,** 6:*186–187*
vocalizations, 6:46–48
See also specific types of toads
Tobacco doves. *See* Common ground doves
Tobacco hornworms, 3:83
Tobias's caddisflies, 3:57

Tobies, 4:70
Tockus camurus. See Red-billed dwarf hornbills
Tockus deckeni. See Von der Decken's hornbills
Tockus erythrorhynchus. See Red-billed hornbills
Tockus hemprichii. See Hemprich's hornbills
Tockus monteiri. See Monteiro's hornbills
Tockus nasutus. See African gray hornbills
Toco toucans, 10:86, 10:125, 10:126, 10:*127*, 10:*129*, 10:*134*–135
Todies, 10:1–4, **10:25–30**
Todiramphus chloris. See Collared kingfishers
Todiramphus pyrrhopygius. See Red-backed kingfishers
Todopsis kowaldi. See Blue-capped ifrits
Todus angustirostris. See Narrow-billed todies
Todus mexicanus. See Puerto Rican todies
Todus multicolor. See Cuban todies
Todus subulatus. See Broad-billed todies
Tody flycatchers. *See* Tyrant flycatchers
Tody motmots, 10:31, 10:32
Toebiters. *See* Dobsonflies; Eastern dobsonflies
Toes, Squamata, 7:198
  See also Physical characteristics
Togo mole-rats, 16:350*t*
Tok-tokkies, 3:320
Tokay geckos, 7:*261–262*, 7:263, 7:*264*, 7:*265*, 7:266
Tokyo salamanders, 6:*37*
Tolmomyias sulphurescens. See Yellow-olive flycatchers
Toltecs, reptiles and, 7:55
Tolypeutes spp. *See* Three-banded armadillos
Tolypeutes matacus. See Southern three-banded armadillos
Tolypeutes tricinctus. See Brazilian three-banded armadillos
Tolypeutinae, 13:182
Tomato bugs, 3:*265*, 3:*271–272*
  See also Southern green stink bugs
Tomato frogs, 6:55, 6:303, 6:305
Tomato fruitworms. *See* Corn earworms
Tomb bats, 13:355, 13:356, 13:357
  black-bearded, 13:*357*, 13:364*t*
  Egyptian, 13:364*t*
  Hildegarde's, 13:*357*
  Mauritian, 13:*359*, 13:*362*, 13:363
Tomba bowerbirds. *See* Archbold's bowerbirds
Tomcods, 5:26, 5:27
  See also Atlantic tomcods
Tomeurus spp., 5:93
Tomeurus gracilis, 5:93
Tomicodon spp., 5:355
Tomicodon humeralis. See Sonora clingfishes
Tomistoma schlegelii. See False gharials; Malayan gharials
Tomistominae. *See* False gharials
Tommies. *See* Thomson's gazelles
Tommy cods. *See* Kelp greenlings
Tomopeas ravus. See Peruvian crevice-dwelling bats
Tomopeatinae. *See* Peruvian crevice-dwelling bats
Tomopterna spp., 6:245, 6:248
Tomtits, 11:111
  See also Southern whitefaces
Tonatia spp. *See* Round-eared bats
Tonatia sylvicola. See White-throated round-eared bats
Tongaichthys spp., 5:406
Tongan flying foxes, 13:*323*, 13:*327*–328
Tongan fruit bats. *See* Tongan flying foxes
Tongan whistlers, 11:118
Tongue worms, **2:317–320**
  behavior, 2:33, 2:319–320
  conservation status, 2:320
  distribution, 2:319
  evolution, 2:317–318
  feeding ecology, 2:320
  habitats, 2:319
  humans and, 2:320
  physical characteristics, 2:318–319
  reproduction, 2:320
  taxonomy, 2:317–318
Tonguefishes, 5:450–453, 5:*454*, 5:*456*–457
Tongues, 7:9
  chameleons, 7:*224*, 7:*225*, 7:226–227
  crocodilians, 7:160
  Squamata, 7:198
  See also Physical characteristics
Tonicella lineata. See Lined chitons
Tonkin snub-nosed monkeys, 14:185*t*
Tonna spp., 2:447
Tool use, 11:507, 12:140–141, 12:156–158
  See also Behavior
Toolache wallabies, 13:39
Tooth-billed bowerbirds, 11:477, 11:*482*, 11:*483*–484
Tooth-billed catbirds. *See* Tooth-billed bowerbirds
Tooth-billed pigeons, 9:244, 9:247, 9:*253*, 9:*266*
Tooth-billed wrens, 10:526, 10:528
Tooth-bills, 9:438
Tooth-carps. *See* Japanese rice fishes
Tooth replacements, 12:9
Tooth shells. *See* Tusk shells
Toothed whales, 12:14, 15:1
  behavior, 15:7
  echolocation, 12:85–87
  evolution, 15:3, 15:41, 15:59
  feeding ecology, 15:8
  geomagnetic fields, 12:84
  hearing, 12:81–82
  humans and, 15:11
  physical characteristics, 12:67, 15:4, 15:81
Top minnows, 4:30
Top shells, **2:429–434**, 2:*432*
Topaza spp. *See* Hummingbirds
Topaza pella. See Crimson topaz
Topis, 15:272, 16:*2*, 16:42*t*
  behavior, 16:5, 16:30
  distribution, 16:29
  evolution, 16:27
  habitats, 16:29
  humans and, 16:34
  predation and, 12:116
  reproduction, 16:7, 16:33
  taxonomy, 16:27
Toque macaques, 14:194, 14:205*t*
Tor putitora, 4:298
Torgos tracheliotus. See Lappet-faced vultures
Tornierella spp., 6:281, 6:282
Toroas. *See* Royal albatrosses
Torontoceros spp., 15:379
Toros, 16:450–451, 16:*452*, 16:*453*, 16:454–455
Torpedinidae, 4:69
Torpedinoidei, 4:11
Torpedo bancrofti. See Lesser electric rays
Torpedo nobiliana. See Atlantic torpedos
Torpedo rays, 4:20, 4:69
Torpidity. *See* Energy conservation
Torpor, 12:113
Torrens creek rock-wallabies. *See* Allied rock-wallabies
Torrent catfishes, 4:353
Torrent ducks, 8:363, 8:365, 8:371–372, 8:*375*, 8:*382*–383
Torrent-larks, 11:453
Torrent midges, 3:360
Torrent salamanders, **6:385–388**, 6:*387*
  distribution, 6:5, 6:*385*, 6:*387*
  habitats, 6:325, 6:385, 6:387
  taxonomy, 6:3, 6:323, 6:385, 6:387
Torrentfishes, 4:57, 5:331–333, 5:335, 5:*336*, 5:337–*338*
Torreornis inexpectata. See Cuban sparrows
Torres-Mura, J. C., 16:435, 16:436
Torresian crows, 11:504, 11:*511*, 11:*522*
Torresian imperial pigeons, 9:244
Tortoise beetles, 3:30, 3:318–319, 3:322
Tortoises, **7:65–73, 7:143–149,** 7:*145*
  Aldabra, 7:47, 7:70, 7:72
  conservation status, 7:10, 7:59, 7:61–62, 7:72–73, 7:144
  distribution, 7:70, 7:*143*
  in folk medicine, 7:51
  as food, 7:47
  hearing, 7:32
  in mythology and religions, 7:55
  physical characteristics, 7:*65–70*, 7:143
  radiated, 7:*30*
  reproduction, 7:7, 7:70–72, 7:*71*, 7:143–144
  See also specific types of tortoises
Tortricidae, 3:388, 3:389
Tortuga aplanada. *See* Central American river turtles
Tortuga blanca. *See* Central American river turtles
Tortuga plana. *See* Central American river turtles
Tossunnoria spp., 16:88
Totanus tennuirostris. See Great knots
Totem poles, 8:*25*
Tou lung. *See* Chinese alligators
Toucan barbets, 10:*118*, 10:*122*–123
Toucanets, 10:126
Toucans, 10:85–88, **10:125–135,** 10:*129*
  behavior, 10:126–127
  conservation status, 10:127–128
  distribution, 10:*125*, 10:126
  evolution, 10:125
  feeding ecology, 10:127
  habitats, 10:126
  humans and, 10:128
  physical characteristics, 10:125–126
  reproduction, 10:127–128
  species of, 10:*130–135*
  taxonomy, 10:125
Touch, 4:62, 8:6, 12:82–83
  See also Physical characteristics
Toupials, 11:301
Towhees, 11:264, 11:*267*, 11:279
Townsend's cormorants. *See* Brandt's cormorants
Townsend's chipmunks, 16:*146*
Townsend's juncos. *See* Dark-eyed juncos
Townsend's solitaires, 10:487, 10:*490*, 10:*495*
Townsend's warblers, 11:286

Toxic sea urchins, 1:*404*
Toxins, 4:15, 4:70, 4:179, 4:265
  batrachotoxin, 6:53, 6:197, 6:198, 6:209
  dart poison, 6:*45*, 6:200, 6:*202*, 6:209
  diet and, 6:199
  plant, 3:46, 3:48
  salamanders, 6:325
  Taricha spp., 6:373
  tarichatoxin, 6:325
  tetrodotoxin, 6:189, 6:325, 6:373
  See also Poison frogs
Toxoderidae, 3:177
Toxostoma spp. *See* Mimids
Toxostoma crissale. See Crissal thrashers
Toxostoma guttatum. See Cozumel thrashers
Toxostoma lecontei. See Leconte's thrashers
Toxostoma redivivum. See California thrashers
Toxostoma rufum. See Brown thrashers
Toxotes jaculatrix. See Banded archerfishes
Toxotidae. *See* Archerfishes
Toy dogs, 14:288, 14:289
Trace elements, 12:120
Tracheal system, 3:21–22
Trachelochismini, 5:355, 5:356
Trachelochismus spp., 5:355
Trachelophorus giraffa. See Giraffe-necked weevils
Tracheloptychus spp. *See* Madagascan plated lizards
Trachemys scripta. See Pond sliders
Trachemys scripta elegans. See Red-eared sliders
Trachemys venusta grayi. See Gray's sliders
Trachichthyidae. *See* Roughies
Trachinidae. *See* Weeverfishes
Trachinocephalus myops. See Painted lizardfishes
Trachinoidei, **5:331–340,** 5:*336*
Trachinus draco. See Greater weevers
Trachipteridae, 4:447, 4:448, 4:449
Trachipterus spp., 4:447
Trachops cirrhosus. See Fringe-lipped bats
Trachurus trachurus. See Horse mackerels
Trachyboa spp., 7:434
Trachyboa boulengeri, 7:434
Trachyboa gularis, 7:434
Trachycephalus spp., 6:226, 6:227, 6:228–229
Trachymedusae, 1:123, 1:124
Trachypetrella spp., 3:203
Trachyphonus darnaudii. See D'Arnaud's barbets
Trachyphonus erythrocephalus. See Red-and-yellow barbets
Trachyphonus vaillantii. See Crested barbets
Trachypithecus spp., 14:172, 14:173
Trachypithecus auratus, 14:175
Trachypithecus auratus auratus, 14:*175*
Trachypithecus cristatus. See Silvery leaf-monkeys
Trachypithecus delacouri, 14:11, 14:175
Trachypithecus francoisi, 14:175
Trachypithecus geei, 14:175
Trachypithecus johnii, 14:175
Trachypithecus laotum, 14:175
Trachypithecus pileatus. See Capped langurs
Trachypithecus poliocephalus, 14:11, 14:175
Trachypithecus vetulus, 14:175
Trachypithecus villosus, 14:175
Tragelaphine antelopes. *See* Tragelaphini
Tragelaphini, 16:2, 16:3, 16:11–14
Tragelaphus angasii. See Nyalas
Tragelaphus buxtoni. See Mountain nyalas
Tragelaphus eurycerus. See Bongos
Tragelaphus imberbis. See Lesser kudus
Tragelaphus scriptus. See Bushbucks
Tragelaphus spekii. See Sitatungas
Tragelaphus strepsiceros. See Greater kudus
Tragocerinae, 15:265
Tragopan satyra. See Satyr tragopans
Tragopans, 8:434, 8:435, 8:*438*, 8:*449*
Tragulidae. *See* Chevrotains
Traguloids. *See* Chevrotains
Tragulus spp. *See* Mouse deer
Tragulus javanicus. See Lesser Malay mouse deer
Tragulus meminna. See Spotted mouse deer
Tragulus napu. See Greater Malay mouse deer
Tragus, bats, 12:54
  See also Bats
Trahiras, 4:339, 4:*341*, 4:348
Trainbearers, 9:438, 9:440, 9:447
Training, of zoo animals, 12:211
Transgenic mice, 16:128
Translocation programs, 12:224
  Chatham Islands black robins, 11:110–111
  Seychelles warblers, 11:7
Transmitters, monitoring by, 8:27
Transparent gobies, 5:375
Trap-door spiders. *See* Purse web spiders
Traps, in field studies, 12:194–197, 12:200–202
Traumatic insemination, 3:263
Trawlers (bats), 13:313–314
Tree corals, red soft, 1:114–115
Tree crickets, 3:205, 3:207
Tree dassies. *See* Tree hyraxes
Tree hyraxes, 15:177–184
Tree kangaroos, 13:34, 13:35, 13:36, 13:83–89
  Bennett's, 13:*86*, 13:*91*, 13:*99*–100
  black, 13:32
  Doria's, 13:102*t*
  forbidden, 13:32
  Goodfellow's, 13:33, 13:39
  greater, 13:36
  Lumholtz's, 13:102*t*
  Matschie's, 13:*37*, 13:*86*, 13:*91*, 13:*98*, 13:100
Tree pangolins, 16:109, 16:110, 16:112, 16:*114*, 16:*117*, 16:119–120
Tree pipits, 10:372, 10:373, 10:375
Tree porcupines, 16:365, 16:367, 16:*368*, 16:369
Tree rats, 16:*254*, 16:*257*
  Brazilian spiny, 16:451, 16:458*t*
  red-nosed, 16:268
  rufous, 16:458*t*
  tuft-tailed spiny, 16:458*t*
Tree shrews, **13:289–298,** 13:*293*
  behavior, 13:291
  conservation status, 13:292
  distribution, 12:136, 13:*289*, 13:291
  evolution, 13:289–291
  feeding ecology, 13:291–292
  habitats, 13:291
  humans and, 13:292
  Indian, 13:197, 13:292, 13:*293*, 13:*294*
  locomotion, 12:14
  physical characteristics, 13:291
  reproduction, 12:90, 12:92, 12:94, 12:95, 12:96, 13:292
  species of, 13:*294–296*, 13:297*t*–298*t*
  taxonomy, 13:289
Tree sloths, 13:147, 13:149, 13:150–151
  three-toed, **13:161–169**
  two-toed, **13:155–159**
Tree snakes, 9:59, 9:470
Tree sparrows, 8:24, 11:397, 11:*400*, 11:401–*402*
Tree squirrels, 12:115, **16:163–176,** 16:*168*
  behavior, 16:165–166
  conservation status, 16:167
  distribution, 16:*163*–164
  evolution, 16:163
  feeding ecology, 16:166
  habitats, 16:124, 16:164–165
  humans and, 16:126, 16:167
  physical characteristics, 16:123, 16:163
  reproduction, 16:166–167
  species of, 16:169–174, 16:174*t*–175*t*
  taxonomy, 16:163
Tree swallows, 10:*358*, 10:359, 10:*360*
Tree swifts, 9:415, 9:417, **9:433–436,** 9:*435–436*
Tree voles, 16:225, 16:226, 16:228, 16:229
  See also Red tree voles
Tree wagtails. *See* Forest wagtails
Tree wasps, 3:*406*
Treecreepers, 11:133, **11:177–181,** 11:*179*
Treefrogs
  African, 6:3–4, 6:6, **6:279–290,** 6:*284*
  barking, 6:*41*
  Central American, 6:40
  defense mechanisms, 6:66
  metamorphosis, 6:42
  as pets, 6:55
  reproduction, 6:28, 6:32, 6:35, 6:50
  taxonomy, 6:215
  See also Amero-Australian treefrogs; Asian treefrogs
Treehoppers, 3:259
  delphacid, 3:*265*, 3:*269*, 3:273
  thornlike, 3:*260*
Trefoil horseshoe bats, 13:389
Tregellasia capito. See Pale-yellow robins
Tremarctinae, 14:295
Tremarctos ornatus. See Spectacled bears
Trematoda, 1:11, **1:197–211,** 1:*202*
  behavior, 1:199–200
  conservation status, 1:200
  distribution, 1:198
  evolution, 1:197
  feeding ecology, 1:200
  habitats, 1:198–199
  humans and, 1:201
  physical characteristics, 1:197–198
  reproduction, 1:200
  species of, 1:*203*–211
  taxonomy, 1:197
Trematomus bernacchii. See Emerald notothens
Trematomus loennbergii, 4:49–50
Trematomus nicolai, 4:49–50
Tremblers. *See* Mimids
Treron pembaensis, 9:244
Treron sanctithomae, 9:244
Treron waalia. See Bruce's green pigeons
Treroninae. *See* Fruit doves
Tres Marias cottontails, 16:509
Tretanorhinus spp. *See* Neotropical swampsnakes
Tretioscincus bifasciatus, 7:303
Trevallys, 4:*26*, 4:48, 4:68, **5:255–263**
Triacanthidae. *See* Triplespines
Triacanthodidae. *See* Spikefishes
Triacanthus biaculeatus. See Triplespines

Triadobatrachus spp., 6:9, 6:11
Triaenodes bicolor, 3:*378,* 3:*380*
Triaenodes phalacris, 3:377
Triaenodes tridonata, 3:377
Triaenodon obesus. See Whitetip reef sharks
Triaenophoridae, 1:225
Triaenops persicus. See Persian trident bats
Triaenostreptus spp., 2:363
Triakidae. *See* Hound sharks
Triakis megalopterus, 4:116
Triakis semifasciata. See Leopard sharks
Trialeurodes vaporariorum. See Greenhouse whiteflies
Triangulins, 3:64
Triassurus spp., 6:13
Triathalassothia spp., 5:41
Triatoma spp., 3:53
Triatoma infestans. See Kissing bugs
Tribes, human, 14:251
Tribolium confusum. See Confused flour beetles
Tribolodon spp., 4:298
Tribosphenic molars, 12:9
See also Teeth
Triceratolepidophis spp., 7:445
Trichastoma rufescens. See Rufous-winged akalats
Trichechidae. *See* Manatees
Trichechinae, 15:205
Trichechus spp. *See* Manatees
Trichechus inunguis. See Amazonian manatees
Trichechus manatus. See West Indian manatees
Trichechus manatus latirostris. See Florida manatees
Trichechus manatus manatus. See Antillean manatees
Trichechus senegalensis. See West African manatees
Trichina worms, 1:35, 1:*283,* 1:*286,* 1:289–291
Trichinella spiralis. See Trichina worms
Trichiotinus spp., 3:320
Trichiuridae, 5:405
Trichiurus lepturus. See Largehead hairtails
Trichiurus nitens. See Largehead hairtails
Trichius spp., 3:320
Trichobatrachus robustus. See Hairy frogs
Trichobilharzia ocellata, 1:199
Trichobilharzia physellae, 1:200
Trichoceridae, 3:361
Trichodectidae, 3:251
Trichodontidae. *See* Sandfishes
Trichogaster spp., 5:427, 5:428
Trichogaster leeri. See Pearl gouramies
Trichoglossus haematodus. See Rainbow lorikeets
Tricholepidion gertschi. See Relic silverfish
Trichomycteridae. *See* Pencil catfishes
Trichomys spp., 16:450
Trichoniscus pusillus. See Common pygmy woodlice
Trichonotidae. *See* Sanddivers
Trichonotus setiger. See Spotted sanddivers
Trichophilopterus babakotophilus. See Lemur lice
Trichoplacidae. *See* Placozoans
Trichoplax adhaerens. See Placozoans
Trichopsis spp. *See* Croaking gouramies
Trichoptera. *See* Caddisflies
Trichosurus arnhemensis. See Northern brushtail possums
Trichosurus caninus. See Mountain brushtail possums
Trichosurus vulpecula. See Brush-tailed possums; Common brushtail possums
Trichuris trichiura. See Human whipworms
Trichys spp., 16:351
Trichys fasciculata. See Long-tailed porcupines
Tricladida, 1:186, 1:187, 1:188
Tricolioidea, 2:430
Tricolor herons, 8:*241*
Tricolored chats. *See* Crimson chats
Tridacna spp., 2:454
Tridacna gigas. See Giant clams
Tridacnidae, 2:454
Tridactylidae, 3:201
Tridactyloidea, 3:201
Trident bats, 13:*402,* 13:409*t*–410*t*
Trident leaf-nosed bats, 13:*406,* 13:*407*
Tridentata marginata, 1:30
Tridentiger trigonocephalus. See Japanese gobies
Triggerfishes, 4:14, 4:48, 4:50, 4:67, 4:70, **5:467–477,** 5:*473*
Triglidae. *See* Sea robins
Trigoniophthalmus alternatus, 3:*115,* 3:*116,* 3:117
Trigonopterygoidea, 3:201
Trillers, 10:385–387
Trilling frogs. *See* Painted frogs (Limnodynastidae)
Trilobites, 2:*12,* 2:25, 12:64
Trimeresurus spp., 7:445, 7:447
Trimeresurus popeiorum. See Pope's pitvipers
Trimmatom nanus, 5:374
Trinectes maculatus. See Hogchokers
Tringa spp. *See* Shanks
Tringa cancellata. See Tuamotu sandpipers
Tringa erythropus. See Spotted redshanks
Tringa glareola. See Marsh sandpipers
Tringa guttifer, 9:177
Tringa interpres. See Ruddy turnstones
Tringa melanoleuca. See Greater yellowlegs
Tringa nebularia. See Greenshanks
Tringa ochropus. See Green sandpipers
Tringa pugnax. See Ruffs
Tringa solitaria. See Solitary sandpipers
Tringa totanus. See Common redshanks
Tringinae, 9:175–180
Tringini. *See* Shanks
Trinidad piping-guan, 8:401
Trinidad poison frogs, 6:*201,* 6:208–209
Trinidadian funnel-eared bats. *See* White-bellied funnel-eared bats
Trioceros spp., 7:224
Triodon macropterus. See Threetooth puffers
Triodontidae. *See* Threetooth puffers
Trionychidae. *See* Softshell turtles
Trionychinae, 7:151
Trionyx spiniferus. See Eastern spiny softshell turtles
Triops spp., 2:141, 2:142, 2:143
Triops australiensis. See Shield shrimps
Triops cancriformis, 2:*141,* 2:143
Triops longicaudatus. See Longtail tadpole shrimps
Triopsidae, 2:141
Tripedalia cystophora, 1:*150,* 1:*152*
Triplespines, 5:467–471, 5:*473,* 5:*483*–484
Tripletails, 5:255, 5:258–263
Triploblastic phyla, 1:19, 2:13–14
See also specific species
Triploblasts, 1:10–11
Tripneustes gratilla. See Toxic sea urchins
Tripodfishes, 4:49, 4:433, 4:*435,* 4:*436,* 4:437–438
Triprion spp., 6:226
Triprion petasatus. See Yucatecan shovel-headed treefrogs
Triprion spatulatus, 6:230
Tripterygiidae, 5:341
Trissolcus basalis, 3:*409,* 3:*420*
Tristan albatrosses, 8:113
Tristram's red-winged starlings, 11:409
Tristram's storm-petrels, 8:135, 8:138
Triton glacialis. See Pyrenean brook salamanders
Triturus spp., 6:33, 6:45, 6:49, 6:363, 6:366, 6:367
Triturus carnifex. See Vienna newts
Triturus cristatus. See Great crested newts
Triturus marmoratus. See Marbled newts
Triturus viridescens. See Eastern newts
Triturus vulgaris. See Smooth newts
Trochilidae. *See* Hummingbirds
Trochilinae. *See* Hummingbirds
Trochilus spp. *See* Hummingbirds
Trochilus aquila. See White-tipped sicklebills
Trochilus coruscans. See Sparking violet-ears
Trochilus guy. See Green hermits
Trochilus hirsutus. See Hairy hermits
Trochilus jardini. See Velvet-purple coronets
Trochilus ludovicae. See Green-fronted lancebills
Trochilus magnificus. See Frilled coquettes
Trochilus maugaeus. See Puerto Rican emeralds
Trochilus mellivora. See White-necked jacobins
Trochilus mosquitus. See Ruby topaz
Trochilus naevius. See Saw-billed hermits
Trochilus pella. See Crimson topaz
Trochilus polytmus. See Red-billed streamertails
Trochilus ruckeri. See Band-tailed barbthroats
Trochilus tzacatl. See Rufous-tailed hummingbirds
Trochoidea, 2:429, 2:430
Trochophore larvae, 2:3–4, 2:21, 2:24
See also Reproduction
Trochozoa, 1:6
Trochus spp., 2:431
Trochus niloticus. See Top shells
Troctomorpha, 3:243
Trogiomorpha, 3:243
Troglodytes spp., 10:526
Troglodytes aedon. See House wrens
Troglodytes tanneri. See Clarion wrens
Troglodytes troglodytes. See Winter wrens
Troglodytidae. *See* Wrens
Trogon bairdii. See Baird's trogons
Trogon collaris. See Collared trogons
Trogon elegans. See Elegant trogons
Trogon massena. See Slaty-tailed trogons
Trogon narina. See Narina trogons
Trogon neoxenus. See Eared quetzals
Trogon reinwardtii. See Javan trogons
Trogon roseigaster. See Hispaniolan trogons
Trogon temnurus. See Cuban trogons
Trogon violaceus. See Violaceus trogons
Trogon viridis. See White-tailed trogons
Trogonidae. *See* Trogons
Trogoniformes. *See* Trogons
Trogonophidae. *See* Spade-headed wormlizards
Trogonophis spp., 7:287, 7:288

Trogonophis wiegmanni. See Spade-headed wormlizards
Trogons, **9:477–485,** 9:*480,* 10:1
Trogontherium spp., 16:177
Trogopterus xanthipes. See Complex-toothed flying squirrels
Troides spp., 3:389
Trondo mainties, 5:*283,* 5:*288*
Tropheus spp., 5:276
Tropheus moorii. See Blunthead cichlids
Trophon spp., 2:447
Tropical anglers, 4:*64*
Tropical brittle stars, 1:*392,* 1:*396*–397
Tropical clawed frogs, 6:99, 6:*105*
Tropical deciduous forests, 12:219
Tropical forests, 1:50, 1:53
See also Conservation status; Habitats
Tropical gars, 4:*223,* 4:*224,* 4:225
Tropical kingbirds, 10:273
Tropical plantanna. *See* Müller's plantanna
Tropical pocket gophers, 16:190, 16:195*t*
Tropical rainforests, 12:218
Tropical rat fleas. *See* Oriental rat fleas
Tropical ratsnakes. *See* Tiger snakes
Tropical savannas, 12:219
Tropical snakeflies, 3:9
Tropical walkingsticks, 3:43
Tropical whales. *See* Bryde's whales
Tropical wormlizards, 7:*280*
Tropicbirds, 8:183–186, **8:187–192**
Tropidoclonion spp. *See* Lined snakes
Tropidolaemus spp., 7:445, 7:447
Tropidophiidae, **7:433–438**
Tropidophis spp., 7:433, 7:434, 7:435
Tropidophis bucculentus. See Navassa woodsnakes
Tropidophis celiae, 7:433
Tropidophis feicki. See Banded woodsnakes
Tropidophis fuscus, 7:433
Tropidophis melanurus. See Cuban ground boas
Tropidophis paucisquamis, 7:434
Tropidophis semicinctus, 7:433
Tropidophis spiritus, 7:433
Tropidophis taczanowskyi, 7:434
Tropidophorus grayi, 7:329
Tropidosaura spp., 7:297
Tropidurinae, 7:243, 7:244, 7:246–247
Tropiese grysboks. *See* Sharpe's grysboks
Tropiometra carinata, 1:*360,* 1:*364*
Trout cichlids, 5:*282,* 5:*285*
Trout cods, 5:221, 5:222
Trout minnows, 4:391
Troutperches, 4:57, 5:*5,* **5:5–6,** 5:7, **5:11–12**
Trouts, 4:*39,* 4:50, 4:405–407, 4:*408–409*
behavior, 4:62, 4:406–407
coral, 5:255, 5:271–*272*
distribution, 4:58, 4:405
feeding ecology, 4:68, 4:407
reproduction, 4:*32,* 4:*405,* 4:*407*
species of, 4:*410*–411, 4:413–414, 4:*416*–419
True bats. *See* Microchiroptera
True bugs, 3:*36,* 3:47, 3:63–64, 3:65, 3:75, **3:259–264,** 3:*266*
See also Water bugs
True chevrotains. *See* Water chevrotains
True corals, 1:103
behavior, 1:105, 1:106
conservation status, 1:107, 1:117, 1:118, 1:119, 1:120
evolution, 1:103
feeding ecology, 1:106
humans and, 1:107, 1:108
physical characteristics, 1:104, 1:105
reproduction, 1:107
symbiosis and, 1:33
True crabs, 2:197, 2:198, 2:201
True crickets, 3:203
True elephants. *See* Elephants
True flies. *See* Diptera
True frogs, **6:245–264,** 6:*253–254*
behavior, 6:250–251, 6:255–263
conservation status, 6:251, 6:255–263
distribution, 6:4–5, 6:6, 6:*245,* 6:250, 6:*255–263*
evolution, 6:245–248
feeding ecology, 6:251, 6:255–263
as food, 6:252, 6:256
habitats, 6:250, 6:255–263
humans and, 6:252, 6:255–263
physical characteristics, 6:248–250, 6:255–263
predators and, 6:250–251
reproduction, 6:35, 6:68, 6:251, 6:255–263
species of, 6:255–264
tadpoles, 6:42, 6:249–250, 6:250–251
taxonomy, 6:245–248, 6:255–263
True grasshoppers, 3:201
True lemmings, 16:225–226
True lemurs. *See* Lemurs
True limpets, **2:423–427,** 2:*426,* 2:*427*
True mole crickets, 3:203
True owls. *See* Typical owls
True plains zebras, 15:222
True ruminants. *See* Ruminants
True sardines. *See* European pilchards
True seals, **14:417–436,** 14:*425,* 14:*426*
behavior, 12:85, 12:87, 14:418–419
conservation status, 14:262, 14:422
distribution, 12:129, 12:138, 14:*417*–418
evolution, 14:417
feeding ecology, 14:260, 14:*261,* 14:419–420
habitats, 14:257, 14:418
humans and, 14:423–424
physical characteristics, 14:256, 14:417, 14:*419*
reproduction, 12:126, 12:127, 14:261, 14:262, 14:420–422
species of, 14:*427*–434, 14:435*t*
taxonomy, 14:256, 14:417
True shrimps, 2:197, 2:198, 2:200, 2:201
True soles, 5:453
True swifts. *See* Swifts
True toads, **6:183–195,** 6:*186–187*
Trueb, L., 6:425
Truebella spp., 6:184
True's beaked whales, 15:70*t*
Trugon terrestris. See Thick-billed ground pigeons
Trumpet manucodes, 8:*39,* 11:490
Trumpeter swans, 8:363, 8:369
Trumpeters, 5:235, 5:237–240, 5:242–244, 9:1–4, 9:45, **9:77–83,** 9:*80*
Trumpetfishes, 5:131–133, 5:135
Trunk turtles. *See* Leatherback seaturtles
Truong Son muntjacs, 15:346, 15:*348,* 15:*351,* 15:353
Trupialis loyca. See Long-tailed meadowlarks
Trupialis superciliaris. See White-browed blackbirds
Tryblidiidae, 2:387
Tryblidioidea, 2:387
Tryblidium spp., 2:387
Trygon aiereba. See Freshwater stingrays
Tryngites subruficollis. See Buff-breasted sandpipers
Trypanorhyncha, 1:225–226, 1:227
Trypanosoma spp., 2:11
Trypetesa lampas, 2:277, 2:*278–279*
Tscherskia spp. *See* Greater long-tailed hamsters
Tscherskia triton. See Greater long-tailed hamsters
TSD. *See* Temperature-dependent sex determination
Tsessebes, 16:32
Tsetse flies, 3:39, 3:46, 3:47, 3:76, 3:361, 3:*362,* 3:*367–368*
Tshekardocoleidae, 3:315
Tsolov's mouse-like hamsters, 16:243
Tsushima salamanders, 6:*336*
Tuamotu sandpipers, 9:177, 9:179–180, 9:*181,* 9:*184*
Tuans. *See* Brush-tailed phascogales
Tuatara, **7:189–193,** 7:195
conservation status, 7:59, 7:192
skeleton, 7:23–24
skulls of, 7:5
third eye, 7:32
visual displays, 7:*35*
Tub gurnards, 5:*165*
Tube anemones, 1:40, 1:103, 1:104
Tube blennies, 4:70, 5:341
Tube case makers, 3:375, 3:376
Tube dwelling anemones, American, 1:*109,* 1:*111,* 1:115
Tube-eyes, 4:447, 4:448
Tube-nosed bats
brown, 13:*523,* 13:524
fruit, 13:333, 13:*334,* 13:*335,* 13:*337,* 13:*339,* 13:*346,* 13:347, 13:349*t*
Hutton's, 13:*526t*
insectivorous, 13:*520*
Tube sea anemones. *See* American tube dwelling anemones
Tubenosed seabirds. *See* Procellariiformes
Tubeshoulders, 4:390, 4:391, 4:392
Tubesnouts, 5:131, 5:132, 5:134, 5:135
Tubeworms, **2:45–57,** 2:*48,* 2:*49*
behavior, 2:46
conservation status, 2:47
distribution, 2:46
evolution, 2:45
feeding ecology, 2:46–47
habitats, 2:46
humans and, 2:47
physical characteristics, 2:45–46
reproduction, 2:47
species of, 2:*50–56*
taxonomy, 2:45
Tubiluchus corallicola, 1:*339,* 1:*340,* 1:341
Tubulanus annulatus, 1:*248,* 1:*250–251*
Tubularia spp., 1:129
Tubularia crocea, 1:30
Tubulidentata. *See* Aardvarks
Tubulifera. *See* Phlaeothripidae
Tubuliporida. *See* Stenolaemata
Tuco-tucos, 16:123, **16:425–431,** 16:*428,* 16:*430t*–431*t*
Tucunares. *See* Speckled pavons

Tucuxis, 15:6, 15:43, 15:44, 15:46, 15:48, 15:50
Tuditanus spp., 6:11
Tudor, G., 10:303
Tuft-tailed spiny tree rats, 16:458*t*
Tufted capuchins. *See* Black-capped capuchins
Tufted deer, 15:268, 15:343, 15:346, 15:*348*, 15:352–*353*
See also Muntjacs
Tufted mynas. *See* Crested mynas
Tufted puffins, 9:*222*
Tufted titmice, 8:24, 11:156
Tuis, 11:237, 11:*239*, 11:*249*–250
Tule elk, 15:*266*
Tule perches, 5:276
Tulerpeton spp., 6:7
Tumare. *See* Tiger shovelnose catfishes
Tunas, **5:405–407**, 5:*408*, **5:412–415**
conservation status, 1:51
feeding ecology, 4:68
habitats, 4:49
physical characteristics, 4:4, 4:14, 4:19
thermoregulation, 5:*406*
Tundra shrews, 13:263*t*
Tunga spp., 3:347, 3:349–350
Tunga penetrans. *See* Chigoes
Túngara frogs, 6:*47*, 6:*160*, 6:167–168
Tungidae, 3:347
Tunicates, 1:22, 1:33–34, 1:42, 1:*451*, 1:*452*, 1:*453*
See also Sea squirts; Sorberaceans
Tunney's rats, 16:*128*
Tupaia belangeri. *See* Belanger's tree shrews
Tupaia chrysogaster, 13:292
Tupaia dorsalis. *See* Golden-bellied tree shrews
Tupaia glis. *See* Common tree shrews
Tupaia gracilis. *See* Slender tree shrews
Tupaia javanica. *See* Javanese tree shrews
Tupaia longipes. *See* Bornean tree shrews
Tupaia minor. *See* Pygmy tree shrews
Tupaia montana. *See* Montane tree shrews
Tupaia nicobarica. *See* Nicobar tree shrews
Tupaia palawanensis. *See* Palawan tree shrews
Tupaia picta. *See* Painted tree shrews
Tupaia splendidula. *See* Rufous-tailed tree shrews
Tupaia tana. *See* Terrestrial tree shrews
Tupaiidae. *See* Tree shrews
Tupaiinae, 13:291, 13:292
Tupinambinae, 7:309
Tupinambis spp. *See* Tegus
Tupinambis duseni, 7:312
Tupinambis merianae, 7:312
Tupinambis rufescens, 7:310, 7:312
Tupinambis teguixin. *See* Black-and-white tegus
Turacos, **9:299–310**, 9:*303*
behavior, 9:300–301
conservation status, 9:301–302
distribution, 9:*299*, 9:300
evolution, 9:299
feeding ecology, 9:301
habitats, 9:300
humans and, 9:302
physical characteristics, 9:299–300
reproduction, 9:301
species of, 9:*304–310*
taxonomy, 9:299
Turahbuglossus cuvillieri, 5:449
Turbanellidae, 1:269
Turbaries, 15:281
Turbellarians. *See* Free-living flatworms
Turbidity, 1:24
See also Habitats
Turbot fishes, 5:*451*
Turcos. *See* Moustached turcas
Turdidae. *See* Chats; Thrushes
Turdoides squamiceps. *See* Arabian babblers
Turdoides striatus. *See* Jungle babblers
Turdus bewsheri. *See* Comoro thrushes
Turdus brachypterus. *See* Bristlebirds
Turdus carolinus. *See* Rusty blackbirds
Turdus harmonicus. *See* Gray shrike-thrushes
Turdus iliacus. *See* Redwings
Turdus leuchophrys. *See* Willie wagtails
Turdus melanophrys. *See* Bell miners
Turdus melanops. *See* Yellow-tufted honeyeaters
Turdus merula. *See* Blackbirds
Turdus migratorius. *See* American robins
Turdus morio. *See* Red-winged starlings
Turdus naumanni. *See* Dusky thrushes
Turdus olivaceus. *See* Olive thrushes
Turdus philomelos. *See* Song thrushes
Turdus pilaris. *See* Fieldfares
Turdus plumbeus. *See* Red-legged thrushes
Turdus punctatus. *See* Spotted quail-thrushes
Turdus roseus. *See* Rosy starlings
Turdus ruficollis. *See* Dark-throated thrushes
Turdus torquata. *See* Ring ouzels
Turkestan red pikas, 16:*496*, 16:*497*, 16:499
Turkey quail. *See* Plains-wanderers
Turkey vultures, 8:22, 8:235, 8:*276*, 8:278, 8:*279*, 8:*280*
Turkeyfishes. *See* Red lionfishes
Turkeys, 8:*20*, 8:400, 8:433–437
Turkish bearded reedlings, 10:507, 10:521
Turkish vans, 14:291
Turks and Caicos iguanas, 7:247
Turnagra spp., 11:116
Turnagra capensis. *See* Piopios
Turner, J. M. W., 2:425
Turner's sea daisies, 1:381, 1:382, 1:383, 1:*384*, 1:*385*, 1:386
Turnicidae. *See* Buttonquails
Turnix spp. *See* Buttonquails
Turnix castanota. *See* Chestnut-backed buttonquails
Turnix everetti. *See* Sumba buttonquails
Turnix hottentotta. *See* Black-rumped buttonquails
Turnix maculosa. *See* Red-backed buttonquails
Turnix meiffrenii. *See* Lark buttonquails
Turnix melanogaster. *See* Black-breasted buttonquails
Turnix nigricollis. *See* Madagascar buttonquails
Turnix ocellata. *See* Spotted buttonquails
Turnix olivii. *See* Buff-breasted buttonquails
Turnix pyrrhothorax. *See* Red-chested buttonquails
Turnix suscitator. *See* Barred buttonquails
Turnix sylvatica. *See* Small buttonquails
Turnix tanki. *See* Yellow-legged buttonquails
Turnix varia. *See* Painted buttonquails
Turnix velox. *See* Australian little buttonquails
Turnix worcesteri. *See* Worcester's buttonquails
Turnstones, 9:175–180
Turquoise-browed motmots, 10:31, 10:32
Turquoise cotingas, 10:308, 10:*310*, 10:*315*
Turritopsis nutricula. *See* Immortal jellyfish
Tursiops spp. *See* Bottlenosed dolphins
Tursiops aduncus. *See* Indo-Pacific bottlenosed dolphins
Tursiops truncatus. *See* Common bottlenosed dolphins
Turtle case makers, 3:375, 3:376
Turtle frogs, 6:*148*
Turtle-headed seasnakes, 7:*490*, 7:*494*–495
Turtle water bears, 2:*115*, 2:116, 2:118, 2:*119*, 2:121
Turtledoves, 9:250
Turtles, **7:65–73**
African sideneck, **7:129–134**, 7:*130*, 7:*131*
Afro-American river, **7:137–142**, 7:*139*
American mud, **7:121–127**, 7:*124*
Australo-American sideneck, **7:77–84**
big-headed, 7:32, **7:135–*136***
black-breasted leaf, 7:7, 7:*116*
black marsh, 7:72
Blanding's, 7:106
bog, 7:105
box, 7:*106*
brown roofed, 7:72
Central American river, **7:99–100**
chicken, 7:106
Chinese softshell, 7:48, 7:151, 7:152, 7:*153*, 7:*154*, 7:155
Chinese stripe-necked, 7:*117*, 7:*118*, 7:119
Chinese three-striped box, 7:119
common snakeneck, 7:*78*, 7:*80*, 7:*82*–83
conservation status, 1:51, 7:10–11, 7:59–62, 7:72–73
distribution, 7:70
eastern box, 7:*68*, 7:106, 7:*108*, 7:112–*113*
Eurasian pond, **7:115–120**
Eurasian river, **7:115–120**
evolution, 7:13–14, 7:65
farming, 7:48
in folk medicine, 7:51
as food, 7:47
food ecology, 7:67
homing, 7:43
integumentary system, 7:29
locomotion, 7:24
in mythology and religion, 7:54–55
New World pond, **7:105–113**, 7:*108*
physical characteristics, 7:*65*, 7:*65–70*
pig-nose, **7:75–76**
reproduction, 7:6–7, 7:42, 7:70–72, 7:*71*
salt glands, 7:30
shells, 7:13, 7:65–67, 7:*69*, 7:72
skeleton, 7:23, 7:*70*
skulls, 7:4–5, 7:26
softshell, 7:48, 7:65–72, **7:151–155**, 7:*153*
territoriality, 7:44
wood, **7:115–120**
See also specific types of turtles
Tusk shells, **2:469–474**, 2:*472*
Tusked frogs, 6:35, 6:140, 6:141, 6:*142*, 6:*143*
Tusked silversides. *See* Pygmy silversides
Tusks, 5:26
See also Lotidae
Tuxedo pincushion urchins, 1:*407*, 1:*414*
TV cockroaches. *See* Brownbanded cockroaches
Twain, Mark, 3:67
Twenty-rayed stars. *See* Sunflower stars
Twenty Thousand Leagues under the Sea, 2:42
Twigsnakes, 7:466, 7:467
Twilight bats. *See* Evening bats
Twinspot gobies, 5:*373*

Twinspots, 11:354, 11:*358*, 11:362–363
Two-banded coursers. *See* Double-banded coursers
The Two Crabs, 2:42
Two-humped camels. *See* Bactrian camels
Two-legged wormlizards, 7:280, 7:*281*
Two-lined caecilians, 6:*416*, 6:*417*–418
Two-lined salamanders, 6:*46*, 6:*393*, 6:399
Two-spot lizardfishes. *See* Clearfin lizardfishes
Two-spotted clingfishes, 5:*358*, 5:*361*, 5:362–363
Two-spotted suckers. *See* Two-spotted clingfishes
Two-toed amphiumas, 6:405, 6:*406*, 6:*408*, 6:*409*, 6:410
Two-toed anteaters. *See* Silky anteaters
Two-toed toadlets, 6:62
Two-toed tree sloths, 13:150, 13:*152*, 13:153, **13:155–159**, 13:*156*
Two-wattled cassowaries. *See* Southern cassowaries
Twofin flashlightfishes. *See* Splitfin flashlightfishes
Tylas eduardi. *See* Tylas vangas
Tylas vangas, 10:439, 10:440, 10:441, 10:442
Tyler, J. C., 5:405
Tyler, S., 1:11, 1:179
Tylodina corticalis, 2:*406*, 2:*407*
Tylonycteris spp., 13:311
Tylonycteris pachypus. *See* Bamboo bats
Tylopoda. *See* Camelidae
Tylos spp., 2:250, 2:251
Tylos punctatus, 2:251
Tylosurus spp., 5:80
Tylotriton verrucosus. *See* Mandarin salamanders
Tympanic bone, 12:9
See also Physical characteristics
Tympanic membrane, 12:9
See also Physical characteristics
Tympanocryptis spp., 7:210
Tympanocryptis cephalus. *See* Earless dragons
Tympanoctomys spp., 16:433, 16:434
Tympanoctomys barrerae. *See* Red viscacha rats
Tympanuchus cupido. *See* Greater prairie chickens
Typhlacontias spp., 7:328
Typhlichthys subterraneus. *See* Southern cavefishes
Typhlogobius californiensis. *See* Blind gobies
Typhlomys cinereus. *See* Chinese pygmy dormice
Typhlonectes spp., 6:436, 6:437
Typhlonectes compressicauda. *See* Cayenne caecilians
Typhlonectes natans. *See* Rio Cauca caecilians
Typhlonectidae, 6:411, 6:412, 6:435, 6:436
Typhlophis spp., 7:369–371
Typhlophis ayarzaguenai, 7:370, 7:371
Typhlophis squamosus, 7:*370*, 7:*371*
Typhlopidae. *See* Blindsnakes
Typhloplanoida, 1:186
Typhlops spp., 7:379–382
Typhlops andamanesis, 7:381
Typhlops angolensis, 7:379
Typhlops bibronii, 7:382
Typhlops cariei, 7:379
Typhlops comorensis, 7:381
Typhlops congestus, 7:380
Typhlops depressiceps, 7:380
Typhlops diardii, 7:382
Typhlops elegans, 7:381
Typhlops grivensis, 7:379
Typhlops lineolatus, 7:379
Typhlops meszoelyi, 7:381
Typhlops monensis. *See* Mona Island blindsnakes
Typhlops oatesii, 7:381
Typhlops paucisquamus, 7:380
Typhlops punctatus. *See* Spotted blindsnakes
Typhlops reticulatus, 7:380
Typhlops socotranus, 7:381
Typhlops vermicularis, 7:382
Typhlosaurus spp., 7:327–329
Typhlosaurus gariepensis, 7:329
Typhlosaurus lineatus. *See* Striped blind legless skinks
Typhlotriton spp., 6:391
Typical antbirds. *See* Ant thrushes
Typical cranes, 9:1–2
Typical owls, **9:345–365**, 9:*347*, 9:*352–353*
behavior, 9:348–349
conservation status, 9:351
distribution, 9:*345*, 9:346–347
evolution, 9:345
feeding ecology, 9:348–349
habitats, 9:347–348
humans and, 9:351
nest construction, 9:*348*, 9:350
physical characteristics, 9:345–346
reproduction, 9:350–351
species of, 9:354–365
taxonomy, 9:331, 9:345
Tyranni, 10:169–174
Tyrannidae. *See* Tyrant flycatchers
Tyranninae. *See* Tyrant flycatchers
Tyrannine flycatchers. *See* Tyrant flycatchers
Tyrannoidea, 10:170
Tyrannulets. *See* Tyrant flycatchers
Tyrannus spp. *See* Kingbirds
Tyrannus cubensis. *See* Giant kingbirds
Tyrannus dominicensis. *See* Gray kingbirds
Tyrannus forficata. *See* Scissor-tailed flycatchers
Tyrannus melancholicus. *See* Tropical kingbirds
Tyrannus savana. *See* Fork-tailed flycatchers
Tyrannus verticalis. *See* Western kingbirds
Tyrant flycatchers, **10:269–289**, 10:*274*, 10:*276–277*, 11:25
behavior, 10:271–273
conservation status, 10:274–275
distribution, 10:*269*, 10:271
evolution, 10:269–270
feeding ecology, 10:273
habitats, 10:271
humans and, 10:275
physical characteristics, 10:270
reproduction, 10:273–274
species of, 10:*278–289*
taxonomy, 10:269–270
Tyrian purple dyes, 1:44, 2:41
See also Humans
Tyrrhenian painted frogs, 6:*91*, 6:94
Tyto spp., 9:335
Tyto alba. *See* Barn owls; Common barn owls
Tyto aurantia. *See* Golden masked owls
Tyto capensis. *See* African grass owls
Tyto deroepstorffi. *See* Andaman barn owls
Tyto glaucops. *See* Ashy-faced owls
Tyto inexspectata. *See* Minahasa masked-owls
Tyto longimembris. *See* Eastern grass owls
Tyto manusi. *See* Manus masked owls
Tyto multipunctata. *See* Lesser sooty owls
Tyto nigrobrunnea. *See* Taliabu masked owls
Tyto novaehollandiae. *See* Australian masked owls
Tyto prigoginei. *See* Itombwe owls
Tyto sororcula. *See* Sulawesi barn owls
Tyto soumagnei. *See* Madagascar red owls
Tyto tenebricosa. *See* Sooty owls
Tytonidae. *See* Barn owls

U

Uakaris, **14:143–154**, 14:*149*
behavior, 14:146
conservation status, 14:148
distribution, 14:144–145
evolution, 14:143
feeding ecology, 14:146–147
habitats, 14:145
humans and, 14:148
physical characteristics, 14:143–144
reproduction, 14:147
species of, 14:*150*, 14:151, 14:153*t*
taxonomy, 14:143
Uca spp. *See* Fiddler crabs
Uca musica, 1:*254*
Uca pugilator. *See* Sand fiddler crabs
Ucayali bald uakaris, 14:148, 14:151
Udzungwa forest-partridges, 8:399, 8:433, 8:*438*, 8:*447*
Udzungwa partridges. *See* Udzungwa forest-partridges
Uganda kobs, 16:7
Ugandan squeakers, 6:*267*, 6:*269–270*
Uinta ground squirrels, 16:160*t*
Uintatheres, 15:131, 15:135, 15:136
Uintatherium spp., 15:135, 15:136
Ulendo: Travels of a Naturalist in and out of Africa, 12:218
Ulochaetes leoninus. *See* Lion beetles
Ultra-eusocial termites, 3:165–166
Ultrasound, bats and, 12:54–55
Ultraviolet B radiation, amphibians and, 6:56, 6:57
Uluguru bush-shrikes, 10:428, 10:429
Uluguru bushbabies, 14:26, 14:32*t*
Uma spp., 7:245
Uma inornata. *See* Coachella Valley fringe-toed lizards
Umbonia crassicornis. *See* Thornlike treehoppers
Umbonium spp., 2:430
Umbra spp., 4:379, 4:380
Umbra krameri. *See* European mudminnows
Umbra limi. *See* Central mudminnows
Umbra pygmaea, 5:91
Umbrella mouth gulpers. *See* Pelican eels
Umbrella species, 1:49
Umbrella wasps. *See* Golden paper wasps
Umbrellabirds
Amazonian, 10:305, 10:306, 10:308, 10:*309*, 10:*316*–317
bare-necked, 10:*311*, 10:315–*316*
long-wattled, 10:308, 10:*311*, 10:*317*–318
Umbrivaga spp., 7:469
Uncia (Panthera) uncia. *See* Snow leopards
Uncia uncia. *See* Snow leopards
Unconditioned stimuli, 1:41, 2:40
See also Behavior

Underground grass grubs, 3:389
Underleaf-sallies, 10:27
Underwoodia iuloides, *2:366*, *2:367*
Underwood's pocket gophers, 16:196*t*
Undulated antshrikes, 10:*244*, 10:*245–246*
Unequal holoblastic cleavage, 2:20
See also Reproduction
UNESCO, 8:126
Ungaliophis spp., 7:434, 7:435
Ungaliophis continentalis. *See* Dwarf boas
Ungaliophis panamensis. *See* Southern bromeliad woodsnakes
Unguis, 12:39–40
Ungulata. *See* Ungulates
Ungulates, **15:131–144**
behavior, 12:145, 15:142
coevolution and, 12:216
distribution, 12:137
domestication, 15:145–153
evolution, 15:131–138
feeding ecology, 15:141–142
field studies of, 12:200
guilds, 12:118–119
neonatal requirements, 12:126–127
physical characteristics, 12:40, 15:138–141, 15:142
predation and, 12:116
reproduction, 12:90, 12:92, 12:94, 12:95, 12:97, 12:98, 12:102, 12:107–108, 15:142–143
stomach, 12:123
See also Cattle; specific types of ungulates
Unguligrades, 12:39
Unicoloniality, 3:69
Unicolored tree kangaroos. *See* Doria's tree kangaroos
Unicorn cods. *See* Bregmacerotidae
Unicornfishes, 4:67–68, 4:447, 4:453
Uniform woodcreepers, 10:*231*, 10:*234–235*
Unionidae, 2:451
Unionoida, 2:454
Unionoidea, 2:24, 2:453–454
Unique-headed bugs, 3:*265*, 3:*272*, 3:*273–274*
United Kingdom
American mink, 12:182–183
gray squirrels, 12:183–184
United Nations Science Agency, 8:126
United States Department of the Interior
babirusas, 15:288
Heermann's kangaroo rats, 16:208
maned sloths, 13:165
San Joaquin pocket mice, 16:205
sportive lemurs, 14:76
United States Fish and Wildlife Service
Alabama cavefishes, 5:10
American alligators, 7:49
American burying beetles, 3:*332–333*
Aplodontia rufa nigra, 16:133
Arizona gray squirrels, 16:172
Atheriniformes, 5:71
babirusas, 15:280
birding and, 8:23
black-tailed prairie dogs, 16:155
Caprinae, 16:95
Central American river turtles, 7:100
Diptera, 3:361
eastern indigo snakes, 7:475
Gila monsters, 7:357
gray whales, 15:99
Heteromyidae, 16:203
Idaho ground squirrels, 16:158
Indian rhinoceroses, 12:224
introduced species, 8:24
leatherback seaturtles, 7:102
Lepidoptera, 3:388
Ozark cavefishes, 5:8
peninsular pronghorns, 15:415
pygmy hogs, 15:280, 15:287
Queen Alexandra's birdwings, 3:398–399
reptilian imports/exports, 7:53*t*
San Francisco garter snakes, 7:479
snakeheads, 5:439
Sonoran pronghorns, 15:415–416
spruce-fir moss spiders, 2:343
Vespertilioninae, 13:502
United States National Marine Fisheries Service, 4:414, 5:51, 5:56
Unken reflex, 6:*46*, 6:*62*, 6:325
Unstriped ground squirrels, 16:159*t*
Uperoleia spp., 6:147–149
Uperoleia inundata, 6:149
Uperoleia lithomoda, 6:148
Uperoleia mimula, 6:148
Upland clawed frogs. *See* Common plantanna
Upland sandpipers, 9:179, 9:180
Upper mouths, 4:*4*, 4:*304*, 4:309–*310*
See also Physical characteristics
Upside-down catfishes, 4:352
Upside-down jellyfish, 1:33, 1:*159*, 1:*161*, 1:162
Upucerthia dumetaria. *See* Scale-throated earthcreepers
Upupa epops. *See* Hoopoes
Upupa marginata. *See* Madagascar hoopoes
Upupidae. *See* Hoopoes
Uraeginthus bengalus. *See* Red-cheeked cordon-bleu
Uraeotyphlidae. *See* Kerala caecilians
Uraeotyphlinae. *See* Kerala caecilians
Uraeotyphlus spp. *See* Kerala caecilians
Uraeotyphlus narayani. *See* Kannan caecilians
Uraeotyphlus oxyurus. *See* Red caecilians
Ural owls, 9:*348*, 9:349, 9:350
Uranid moths, 3:386
Uranoscopidae. *See* Stargazers
Uranoscopus bicinctus. *See* Marbled stargazers
Urartsk mouse-like hamsters, 12:243
Uratelornis chimaera. *See* Long-tailed ground-rollers
Urchin clingfishes, 5:355, 5:356, 5:*358*, 5:*361*, 5:363
Urchins
behavior, 1:403
evolution, 1:401
feeding ecology, 1:404
habitats, 1:403
humans and, 1:405
pea, 1:*407*, 1:*411*–412
physical characteristics, 1:401, 1:402
purple-heart, 1:403
taxonomy, 1:401
tuxedo pincushion, 1:*406*, 1:*414*
See also specific types of urchins
Urea, 7:3, 12:122–123
Urechidae, 2:103
Urechis caupo. *See* Innkeeper worms
Ureters, 12:90
See also Physical characteristics
Uria aalge. *See* Common murres
Uria lomvia. *See* Thick-billed murres
Urials, 12:178, 16:6, 16:87, 16:89, 16:91, 16:92–93, 16:104*t*
Uric acid, 3:21, 7:3
Urich's tyrannulets, 10:275
Urinary tract, 12:90–91
See also Physical characteristics
Urnatella gracilis. *See* Freshwater colonial entoprocts
Urnisilla spp., 3:203
Urobatis jamaicensis. *See* Yellow stingrays
Urochordata iasis atlantic. *See* Chain salps
Urochroa spp. *See* Hummingbirds
Urocissa spp., 11:504
Urocissa ornata. *See* Sri Lanka magpies
Urocolius indicus. *See* Red-faced mousebirds
Urocolius macrourus. *See* Blue-naped mousebirds
Urocyon cinereoargenteus. *See* Gray foxes
Urocyon littoralis. *See* Island foxes
Urodeles, 6:13
Uroderma bilobatum. *See* Tent-making bats
Urodynamis taitensis. *See* Long-tailed koels
Urogale spp., 13:292
Urogale everetti. *See* Philippine tree shrews
Urogenital sinus, 12:90, 12:111
See also Physical characteristics
Urogenital system, lissamphibians, 6:17–18
Urolophidae. *See* Round rays
Uromacer spp., 7:467
Uromastyx spp. *See* Dhabb lizards
Uromastyx acanthinurus. *See* Spiny-tailed agamas
Uropeltidae. *See* Shieldtail snakes
Uropeltis spp., 7:391, 7:392
Uropeltis macrolepis macrolepis. *See* Large-scaled shieldtails
Uropeltis maculatus, 7:392
Uropeltis myhendrae, 7:391
Uropeltis ocellatus, 7:391
Uropeltis phipsonii. *See* Phipson's shieldtail snakes
Uropeltis rubromaculata, 7:392
Urophycis spp., 5:28
Urophycis blennoides, 5:*25*
Urophycis chuss. *See* Red hakes
Urophycis tenuis. *See* White hakes
Uropsila spp., 10:528
Uropsilinae, 13:279, 13:280
Uropsilus spp. *See* Shrew moles
Uropsilus investigator. *See* Yunnan shrew-moles
Uropsilus soricipes. *See* Chinese shrew-moles
Uropsylla tasmanica, 3:349
Uropygi, 2:333
Urosaurus graciosus. *See* Common sagebrush lizards
Urosticte spp. *See* Hummingbirds
Urotrichus talpoides. *See* Greater Japanese shrew moles
Ursidae. *See* Bears
Ursinae, 14:295
Ursula's sunbirds, 11:210
Ursus americanus. *See* American black bears
Ursus americanus floridanus, 14:300
Ursus arctos. *See* Brown bears
Ursus arctos middendorffi. *See* Kodiak bears
Ursus arctus. *See* Brown bears
Ursus maritimus. *See* Polar bears
Ursus thibetanus. *See* Asiatic black bears
U.S. Coral Reef Initiative, 1:107

U.S. Endangered Species Act. *See* Endangered Species Act (U.S.)
U.S. Fish and Wildlife Services. *See* United States Fish and Wildlife Service
U.S. National Marine Fisheries Service. *See* United States National Marine Fisheries Service
Usambara robin-chats, 10:488
Usambiro barbets. *See* D'Arnaud's barbets
Usazoros spp., 3:239
USDI. *See* United States Department of the Interior
Uta Hick's bearded sakis, 14:148, 14:151
Uta stansburiana. *See* Common side-blotched lizards
Utas. *See* Common side-blotched lizards
Uterine glands, 12:94
See also Physical characteristics
Uterus, 12:90–91
See also Physical characteristics
Utilitarianism, 3:84
See also Conservation status
Uvulifer ambloplitis. *See* Black-spot flatworms
Uyeno, T., 5:179
Uzbekian snakeflies, wart-headed, 3:*299*, 3:*300*, 3:302
Uzelothripidae, 3:55, 3:281
Uzungwa bushbabies, 14:33*t*

V

V-marked weavers. *See* Village weavers
Vaceletia spp., 1:78
Vaginal watering pots. *See* Watering pot shells
Vaginalis alba. *See* Pale-faced sheathbills
Vaginas, 12:91
See also Physical characteristics; Reproduction
Valais goats, 15:*268*
Valamugil spp., 5:59
Valdiviathyris spp., 2:515
Valenciennes, A., 5:427
Valenciidae, 4:57
Valley pocket gophers, 12:74, 12:*74*, 16:*188*, 16:*189*, 16:*190*, 16:*191*, 16:*192*
Vallicula multiformis, 1:*172*, 1:*174*, 1:175
Val's gundis. *See* Desert gundis
Valvifera, 2:249
Vampire bats, 8:46, 13:*308*, 13:*311*, 13:*422*
behavior, 13:419
conservation status, 13:420
distribution, 13:415
feeding ecology, 13:314, 13:418, 13:*420*
hairy-legged, 13:314, 13:433*t*
humans and, 13:316, 13:317, 13:420
infrared energy, 12:83–84
physical characteristics, 13:308, 13:414
reproduction, 13:315, 13:419, 13:420
species of, 13:*425*–426
See also False vampire bats
Vampire squids, 2:475–476, 2:477, 2:*481*, 2:*486*, 2:488–489
Vampyromorpha. *See* Vampire squids
Vampyroteuthis infernalis. *See* Vampire squids
Vampyrum spectrum. *See* Spectral bats
Van Dam's vangas, 10:439, 10:440, 10:*441*
Van Neck, Jacob Cornelius, 9:269
Van Soest, R. W. M., 1:57
Van Tuinen, M., 8:183
Van Zyl's golden moles, 13:222*t*
Vancouver Island marmots, 16:147, 16:*150*, 16:*152*, 16:156
Vandellia spp., 4:69
Vandellia cirrhosa. *See* Candiru
Vanellus spp. *See* Lapwings
Vanellus armatus. *See* Blacksmith plovers
Vanellus chilensis. *See* Southern lapwings
Vanellus gregarius. *See* Sociable plovers
Vanellus leucurus. *See* White-tailed lapwings
Vanellus macropterus. *See* Javanese lapwings
Vanellus spinosus. *See* Spur-winged lapwings
Vanga curvirostris. *See* Hook-billed vangas
Vanga shrikes, **10:439–445**
Vangidae. *See* Vanga shrikes
Vangulifer spp. *See* Shovelbills
Vanzosaura rubricauda, 7:*304*
Vaquitas, 15:9, 15:33, 15:34, 15:35–36, 15:39*t*
Varanidae, 7:9, 7:196, 7:204, 7:206, 7:207, **7:359–368,** 7:*364*
Varanines, 7:359
Varanus spp., 7:207, 7:359–362
Varanus albigularis, 7:52, 7:360, 7:362
Varanus bengalensis. *See* Bengal monitor lizards
Varanus brevicauda. *See* Short-tailed monitors
Varanus caudolineatus. *See* Stripe-tailed monitors
Varanus dumerilii, 7:363
Varanus eremius, 7:361
Varanus exanthematicus, 7:52
Varanus giganteus. *See* Perenties
Varanus gilleni, 7:361, 7:362
Varanus gouldii. *See* Australian sand goannas; Sand monitors
Varanus griseus, 7:48, 7:363
Varanus indicus, 7:360–361
Varanus komodoensis. *See* Komodo dragons
Varanus mertensi, 7:360
Varanus mitchelli, 7:360–361
Varanus niloticus. *See* Nile monitors
Varanus olivaceus, 7:361
Varanus prasinus, 7:361–363
Varanus rosenbergi, 7:362–363
Varanus salvadorii. *See* Crocodile monitors
Varanus salvator. *See* Water monitors
Varanus tristis, 7:361, 7:363
Varanus varius, 7:362–363
Varecia spp., 14:51
Varecia variegata. *See* Variegated lemurs
Vargula hilgendorfii, 2:*313*, 2:*314*
Varia jamoerensis. *See* Yamur Lake grunters
Variable bush vipers. *See* Green bush vipers
Variable coquinas. *See* Coquina clams
Variable flying foxes. *See* Island flying foxes
Variable goshawks, 8:317
Variable indigobirds. *See* Dusky indigobirds
Variable oystercatchers, 9:125, 9:*126*, 9:*127*, 9:*129*, 9:*131*
Variable pitohuis, 11:116, 11:*119*, 11:*125*
Variable pocket gophers, 16:196*t*
Variable quill-snouted snakes, 7:*463*
Variable seedeaters, 11:*267*, 11:*277–278*
Variable sunbirds, 11:208, 11:*209*
Varied buttonquails. *See* Painted buttonquails
Varied springtails, 3:*102*, 3:*103*
Varied trillers, 10:387, 10:*388*, 10:*391–392*
Variegated boobies. *See* Peruvian boobies
Variegated fairy-wrens, 11:*47*, 11:*51*
Variegated grasshoppers, 3:*210*, 3:*212*
Variegated lemurs, 14:*53*, 14:*57*, 14:58–59
Variegated pygmy perches, 5:222
Variegated spider monkeys, 14:*156*
Variegated squirrels, 16:*168*, 16:*170*, 16:172
Variegated tinamous, 8:58, 8:59, 8:60, 8:*64*
Variegated wrens. *See* Variegated fairy-wrens
Variichthys spp., 5:220
Varying hares. *See* Mountain hares
Vas deferens, 12:91
Vaught, Kay Cunningham, 2:412
Vaux's swifts, 9:417
Vectors, disease, 3:76, 3:151, 3:252, 3:350, 3:353, 3:361
Vegetarian diets. *See* Feeding ecology
Veil angelfishes. *See* Freshwater angelfishes
Veiled chameleons, 7:*224*, 7:*228*, 7:*233*, 7:*237*
Veiled chitons, 2:395, 2:*396*, 2:*397*, 2:400–401
Velcro sea stars, 1:367, 1:369, 1:*372*, 1:*373*
Veldkamp's dwarf epauletted bats. *See* Little flying cows
Veldkamp's dwarf fruit bats. *See* Little flying cows
Velds, 3:56
Velella spp., 1:29
Velella velella. *See* By the wind sailors
Veles binotatus. *See* Brown nightjars
Velifer spp., 4:447
Veliferidae, 4:448, 4:449
Velvet (antlers), 12:40
See also Physical characteristics
Velvet ants, 3:*409*, 3:*417*, 3:418
Velvet asities, 10:188, 10:*189*, 10:*190*
Velvet catfishes, 4:352
Velvet-mantled drongos, 11:438, 11:439
Velvet-purple coronets, 9:*453*, 9:*465–466*
Velvet whalefishes. *See* Red whalefishes
Velvet worms, **2:109–114,** 2:*110*
behavior, 2:37, 2:110
conservation status, 2:111
distribution, 2:110
evolution, 2:109
feeding ecology, 2:110
habitats, 2:110
humans and, 2:111
physical characteristics, 2:23, 2:109–110
reproduction, 2:17, 2:110
species of, 2:*113*–114
taxonomy, 2:109
Velvetfishes, 5:163, 5:164, 5:166
Velvety free-tailed bats. *See* Greater house bats
Venerable silverfish. *See* Relic silverfish
Veneridae, 2:451, 2:452
Venezuelan hemorrhagic fever, 16:270
Venezuelan red howler monkeys, 14:*157*, 14:*158*, 14:*161*, 14:*162*–163
Venezuelan skunk frogs, 6:200, 6:*201*, 6:*203*
Venezuelan spiny rats, 16:*450*
Venezuelan wood-quails, 8:*459*, 8:460–461
Veniliornis fumigatus. *See* Smoky-brown woodpeckers
Venison-hawks. *See* Gray jays
Venom, 7:9–10, 7:37–38
folk medicine and, 7:50–52
platypus, 12:232, 12:244
teeth and, 7:9, 7:26, 7:206
See also Elapidae; Viperidae
Venomous fishes
Scorpaenoidei, 5:165
weeverfishes, 5:335

Venomous snakes. *See* Elapidae; Viperidae
Venous hearts, bats, 12:60
Vent clams, giant, 2:*456*, 2:*462*, 2:466
Venus (Roman goddess), 7:54
Venus's flower baskets, 1:69, 1:*70*, 1:72, 1:74–75
Venus's girdles, 1:*169*, 1:*172*, 1:*174*–175
Verdins, 10:174, 11:147, 11:*150*, 11:152–*153*
Verger bruchinus, 3:377
Verheyen, R., 8:369
Vermetidae, 2:445
Vermicella spp. *See* Bandy-bandy snakes
Vermicella annulata. *See* Bandy-bandy snakes
Vermilingua, 13:171
Vermilion flycatchers, 10:270, 10:273, 10:*276*, 10:280–*281*
Vermipsylla spp., 3:350
Vermipsyllidae, 3:347, 3:348
Vermipsylloidea, 3:347
Vermivora bachmanii. *See* Bachman's warblers
Vermivora chrysoptera. *See* Golden-winged warblers
Vermivora luciae. *See* Lucy's warblers
Vermivora pinus. *See* Blue-winged warblers
Vernal pool tadpole shrimps, 2:142, 2:*144*, 2:*145*–146
Verne, Jules, 2:42
Veronavelifer spp., 4:448
Verongida, 1:77, 1:79
Veronicellidae, 2:411, 2:413
Verreaux, Jules, 8:104
Verreaux's eagle owls, 8:262
Verreaux's eagles, 8:321–322
Verreaux's sifakas, 14:66, 14:67
Vertebral column, 12:8
See also Physical characteristics
Vertebrates, 4:9
flight, 12:52–61
limbed, 6:7–11
taxonomy, 1:14
See also specific vertebrates
Vertical migration, 4:49, 4:60, 4:61, 4:442, 4:443
See also Behavior
Verticillitida, 1:77, 1:78, 1:79
Verticillitidae. *See* Verticillitida
Vervet monkeys. *See* Grivets
Vesicomyid clams. *See* Giant vent clams
Vesper mice, 16:267, 16:269, 16:270
Vespertilio murinus. *See* Parti-colored bats
Vespertilionidae, **13:497–526,** 13:*504*, 13:*505*, 13:*523*
behavior, 13:498–500, 13:521
conservation status, 13:501–502, 13:522
distribution, 13:*497*, 13:498, 13:*519*, 13:520–521
evolution, 13:497, 13:519
feeding ecology, 13:313, 13:500, 13:521
habitats, 13:498, 13:521
humans and, 13:502, 13:522
physical characteristics, 13:310, 13:497–498, 13:519–520
reproduction, 13:500–501, 13:522
species of, 13:*506*–514, 13:515*t*–516*t*, 13:524–525, 13:526*t*
taxonomy, 13:497, 13:519
Vespertilioninae, **13:497–517,** 13:*504*, 13:*505*, 13:515*t*–516*t*
Vespidae, 3:68, 3:406
Vespine wasps, 3:69
Vespula spp., 3:69
Vespula germanica. *See* Yellow jackets
Vespula sylvestris. *See* Tree wasps
Vespula vulgaris. *See* European wasps
Vestigial limbs, 7:*23*, 7:24
Vestimentifera, **2:91–95,** 2:*93*
Vestimentiferan tubeworms. *See* Hydrothermal vent worms
Vetchworms. *See* Corn earworms
Veterinary entomologists, 3:78
Veterinary medicine, in zoos, 12:210–211
Vetigastropoda, **2:429–434,** 2:*432*
Vibrations, 12:83, 12:*83*
See also Physical characteristics
Vibrissae, 12:38–39, 12:*43*, 12:72
See also Physical characteristics
Vibrissaphora spp. *See* Moustache toads
Vibrissaphora ailaonica. *See* Ailao moustache toads
Vicariance, 3:53
Victoria babblers, 11:128
Victoria crowned-pigeons, 8:*26*, 9:241, 9:250
Victoria lyrebirds. *See* Superb lyrebirds
Victoria river snappers, 7:78, 7:*80*, 7:*82*, 7:83
Victoriapithecus spp., 14:172, 14:188–189
Victoria's riflebirds, 11:490, 11:*494*, 11:*499*
Vicugna spp. *See* Vicuñas
Vicugna vicugna. *See* Vicuñas
Vicuñas, **15:313–323,** 15:*314*, 15:*317*, 15:*319*
behavior, 15:142, 15:316–317, 15:321
conservation status, 15:317–318, 15:321
distribution, 15:315, 15:*321*
domestication of, 15:151
evolution, 15:313–314
feeding ecology, 15:317, 15:321
habitats, 12:132, 15:315–316, 15:321
humans and, 15:318, 15:321
physical characteristics, 15:314–315, 15:321
reproduction, 15:321
taxonomy, 15:313–314, 15:321
Vidalamiinae, 4:229
Vidua funerea. *See* Dusky indigobirds
Vidua macroura. *See* Pin-tailed whydahs
Vidua paradisaea. *See* Paradise whydahs
Vidua purpurascens. *See* Purple indigobirds
Viduinae, 11:376, 11:378
Vieillot's black weavers, 11:386
Vienna newts, 6:*325*
Vieraella spp., 6:11
Vietnam greenfinches, 11:325
Vietnam warty pigs, 15:263, 15:280, 15:289*t*
Vietnamese pheasants, 8:433, 8:437
Vietnamese pot-bellied pigs, 15:*278*
Village weavers, 11:*380*, 11:*386*
Vinciguerria spp., 4:424
Vinciguerria lucetia, 4:424
Vine weevils, 1:*45*
Vinegar flies, 3:55, 3:61, 3:357
Vinesnakes, 7:466–468
Vinous-throated parrotbills, 10:*509*, 10:521–*522*
Violaceus trogons, 9:479
Violet-ears, 9:442, 9:447, 9:448, 9:*454*, 9:460–461
Violet gobies, 5:*378*, 5:*383*–384
Violet-headed sunbirds. *See* Orange-breasted sunbirds
Violet-throated metaltails, 9:418
Violet turacos, 9:300
Violet woodhoopoes, 10:65
Viper boas, 7:40, 7:411, 7:*412*, 7:413–*414*
Vipera spp., 7:445, 7:446, 7:448
Vipera aspis, 7:448
Vipera berus. *See* Common adders
Vipera darevskii, 7:447
Vipera dinniki, 7:447
Vipera latifii, 7:448
Vipera ursinii, 7:448
Vipera wagneri, 7:448
Viperfishes, 4:*421*, 4:423, 4:*425*–428, 4:*426*
Viperidae, **7:445–460,** 7:*449*, 7:*450*
behavior, 7:447
conservation status, 7:448
distribution, 7:*445*, 7:447
evolution, 7:445–446
feeding ecology, 7:447–448
habitats, 7:447
humans and, 7:448
physical characteristics, 7:446–447
reproduction, 7:448
species of, 7:*451*–460
taxonomy, 7:445
Viperinae. *See* Vipers
Vipers, **7:445–460,** 7:*449*
behavior, 7:205, 7:206, 7:447
conservation status, 7:448
distribution, 7:*445*, 7:447
Elapidae *vs.*, 7:483
evolution, 7:445–446
feeding ecology, 7:447–448
habitats, 7:447
humans and, 7:448
physical characteristics, 7:446–447
reproduction, 7:41, 7:42, 7:448
species of, 7:*451*–460
taxonomy, 7:445
Vir philippinensis. *See* Bubble coral shrimps
Vireo atricapillus. *See* Black-capped vireos
Vireo bellii. *See* Bell's vireos
Vireo bellii pusillus. *See* Least Bell's vireos
Vireo caribaeus. *See* Saint Andrew vireos
Vireo gilvus. *See* Warbling vireos
Vireo gracilirostris. *See* Red-eyed vireos
Vireo masteri. *See* Choco vireos
Vireo olivaceus. *See* Red-eyed vireos
Vireo osburni. *See* Blue Mountain vireos
Vireolanius leucotis. *See* Slaty-capped shrike-vireos
Vireonidae, **11:255–262,** 11:*257*
Vireos, **11:255–262,** 11:*257*
Virgin of Guadalupe, 7:55
Virginia opossums, 12:*250*–253, 12:*251*, 12:*254*, 12:*255*, 12:*257*, 12:258
Virginia rails, 9:48, 9:51
Virginia spp. *See* North American earthsnakes
Virginia striatula. *See* Rough earthsnakes
Virola surinamensis, 9:478
Viruses
in rabbit control, 12:187–188
symbiosis and, 1:31
Visayan broadbills. *See* Visayan wattled broadbills
Visayan flowerpeckers. *See* Black-belted flowerpeckers
Visayan hornbills. *See* Visayan tarictic hornbills
Visayan spotted deer, 15:360, 15:371*t*
Visayan tarictic hornbills, 10:72, 10:75, 10:77, 10:*81*
Visayan warty pigs. *See* Cebu bearded pigs

Visayan wattled broadbills, 10:178, 10:*180*, 10:*184*–185
Viscacha rats, 16:436, 16:*437*, 16:*438*, 16:439–440
Viscachas, **16:377–384**, 16:*380*, 16:*381*, 16:383*t*
Vishnu (Indian god), 7:55–56
Vision, 3:27–28, 4:21, 4:*27*, 4:60, 4:61, 7:32, 7:38–39, 8:*5*–6, 12:5–6, 12:50, 12:76–77, 12:79–80
  falcons, 8:348
  herons, 8:242
  lissamphibians, 6:*17*, 6:22
  magic eyes for, 9:396
  night-adapted, 9:332
  sunlight and, 1:24
  See also Eyes; Physical characteristics
Visual communication, 12:160
  amphibian, 6:45, 6:46
  mating and, 3:60
  See also Behavior
Visuki (Indian god), 7:56
Vitamin requirements, 12:120, 12:123
  See also Feeding ecology
Vitelline sac. *See* Yolk sac
Vitt, Laurie J., 7:304, 7:307, 7:340
Viverra spp., 14:335, 14:338
Viverra civettina. See Malabar civets
Viverra megaspila. See Oriental civets
Viverra zibetha. See Indian civets
Viverricula spp., 14:335, 14:338
Viverricula indica, 14:336
Viverridae, **14:335–345**, 14:*339*
  behavior, 14:258, 14:337
  conservation status, 14:262, 14:338
  distribution, 12:135, 12:136, 14:*335*, 14:336
  evolution, 14:335
  feeding ecology, 14:260, 14:337
  habitats, 14:337
  humans and, 14:338
  physical characteristics, 14:256, 14:335–336
  reproduction, 14:338
  species of, 14:*340–343*, 14:344*t*–345*t*
  taxonomy, 14:256, 14:335
Viverrinae, 14:335
Viviparity, 3:39, 3:150
  amphibians, 6:28, 6:38
  caecilians, 6:39
  fishes, 4:32, 4:65–66
  ground sharks, 4:115–116
  Lacerta vivipara, 7:298
  nurse sharks, 4:109
  reptiles, 7:6–7
  salamanders, 6:367
  Squamata, 7:206–207
  tasseled wobbegongs, 4:110
  Therians, 12:5, 12:7, 12:106
  See also Reproduction
Viviparous eelpouts, 5:312, 5:313
Vizcachas. *See* Plains viscachas
Vlei rats. *See* Angoni vlei rats
Vocal sac-brooding frogs, 6:6, 6:36, **6:173–177**, 6:*175–176*
Vocalizations, 12:85
  aggressive, 6:47–48
  Anura, 6:*22*, 6:26, 6:30
  choruses, 6:49
  frogs, 6:*22*, 6:26, 6:30, 6:46–48, 6:304
  reproduction and, 6:35, 6:47, 6:49
  toads, 6:46–48
  See also Behavior; Communication; Songs, bird
Vogelkops, 11:477
Voice organs, 12:8
  See also Physical characteristics
Voice prints, cranes, 9:25
Volant mammals, 12:200–201
Volatinia jacarina. See Blue-black grassquits
Volcano rabbits, 16:482, 16:483, 16:487, 16:505, 16:509, 16:*510*, 16:513–*514*
Volcano shrews, 13:277*t*
Volcano sponges. *See* Barrel sponges
Voles, **16:225–238**, 16:*231*, 16:*232*
  behavior, 12:141–142, 12:148, 16:226–227
  conservation status, 16:229–230
  distribution, 16:*225*, 16:226
  evolution, 16:225
  feeding ecology, 16:227–228
  habitats, 16:124, 16:226
  humans and, 16:127
  parasitic infestations, 12:78
  physical characteristics, 16:225–226
  predation and, 12:115
  reproduction, 16:228–229
  species of, 16:233–237*t*, 16:238*t*
  taxonomy, 16:225
Volitan lionfishes. *See* Red lionfishes
Volitinia spp., 11:264
Volutes, red, 2:*445*, 2:*446*
Volutine stoneyian tabanid flies, 3:360
Volvox spp., 1:7
Vombatidae. *See* Wombats
Vombatiformes, 13:31, 13:32
Vombatus ursinus. See Common wombats
Vomeronasal system, 6:22–23, 7:34–37, 7:195
  See also Physical characteristics
Von der Decken's hornbills, 10:72
von Gadow, H., 9:345
von Humbolt, Alexander, 9:373–374
Voodoo, snakes in, 7:57
Voracious geckos, 7:263
Vormela spp., 14:321
Vormela peregusna. See Marbled polecats
Vorontsov, N. N., 16:211
Vortex theory, of animal flight, 12:57
Voskoboinikova, O. S., 5:331
Voss, R., 16:269
Vu Quang oxen. *See* Saolas
Vulpes spp. *See* Foxes
Vulpes cana. See Blanford's foxes
Vulpes corsac. See Corsac foxes
Vulpes ferrilata. See Tibetan foxes
Vulpes pallida. See Pale foxes
Vulpes velox. See Swift foxes
Vulpes vulpes. See Red foxes
Vulpes zerda. See Fennec foxes
Vultur gryphus. See Andean condors
Vultures, 8:233–236, **8:275–285**, 8:313–316, 8:318–324, 8:*320*, 8:*325–328*, 8:*334*, 8:*335–336*
Vulturine guineafowl, 8:*401*, 8:426–428, 8:*427*, 8:*428*
Vulturine parrots. *See* Pesquet's parrots

W

Waccamaw silversides, 5:71
Waddycephalus spp., 2:318, 2:320
Wade, Christopher M., 2:411
Waders. *See* Charadriiformes
Wager, Vincent A., 7:223
Wagner's bonneted bats, 13:493*t*
Wagner's moustached bats, 13:435, 13:*436*, 13:442*t*
Wagner's sac-winged bats. *See* Chestnut sac-winged bats
Wagtail flycatchers. *See* Fantails
Wagtails, 9:313, **10:371–384**, 10:*377*
  behavior, 10:373–374
  conservation status, 10:375–376
  distribution, 10:*371*, 10:372–373
  evolution, 10:371
  feeding ecology, 10:374
  habitats, 10:373
  humans and, 10:376
  physical characteristics, 10:371–372
  reproduction, 10:374–375
  taxonomy, 10:371
  See also Willie wagtails
Wahi grosbeaks, 11:343
Wahlberg's epauletted fruit bats, 13:*313*, 13:*338*, 13:*343*
Wahlberg's screeching frogs. *See* Bush squeakers
Wahnes's parotias, 11:492
Wahoos, 4:68, 5:*408*, 5:*411*, 5:412
Waigeo cuscuses, 13:60
Waipatiidae, 15:13
Waldrapp ibises. *See* Hermit ibises
Waldrapps. *See* Hermit ibises
Walia ibex, 16:87, 16:91, 16:92, 16:94, 16:95, 16:104*t*
Walker, E. P., 16:333
Walking catfishes, 4:3
Walking snakeheads, 5:438, 5:439, 5:*440*, 5:*444*–445
Walking water-sticks. *See* Water measurers
Walkingsticks, 3:9, 3:43, 3:44, 3:59, 3:65, 3:*224*
  See also Common American walkingsticks
Wall creepers, 11:167, 11:*170*, 11:*174*–175
Wall lice. *See* Bed bugs
Wall lizards, **7:297–302**
Wallabia spp., 13:83, 13:85
Wallabia bicolor. See Swamp wallabies
Wallabies, 13:*39*, **13:83–103**, 13:*91*, 13:*92*
  behavior, 12:145, 13:36, 13:86–87
  digestive system, 12:123
  distribution, 12:137, 13:*83*, 13:84–85
  evolution, 13:31–32, 13:83
  feeding ecology, 13:87–88
  habitats, 13:34, 13:35, 13:85–86
  humans and, 13:89–90
  physical characteristics, 13:32, 13:33, 13:83–84
  reproduction, 13:88–89
  species of, 13:93–100, 13:101*t*–102*t*
  taxonomy, 13:83
Wallaby-rats. *See* Tasmanian bettongs
Wallace, Alfred R., 1:48, 3:53, 3:54, 4:56–57
Wallace's fairy-wrens. *See* Wallace's wrens
Wallace's flying frogs, 6:291
Wallace's owlet-nightjars, 9:388
Wallace's standardwings. *See* Standardwing birds of paradise
Wallace's wrens, 11:45, 11:*47*, 11:*53*
Wallaroos, 13:34, 13:101*t*
Waller's gazelles. *See* Gerenuks

Walleye pollocks. *See* Alaska pollocks
Walleyed pikes. *See* Walleyes
Walleyes, 4:68, 4:379, 4:380, 5:195–*197*, 5:199, 5:*201*, 5:*204*, 5:208
Walruses, **14:409–415**, 14:*410*, 14:*411*, 14:*413*
distribution, 12:138
feeding ecology, 14:260
habitats, 12:14
locomotion, 12:44
migrations, 12:*170*
physical characteristics, 12:46, 12:68
reproduction, 12:111
taxonomy, 14:256
Waminoa spp., 1:*179*, 1:*180*
Wandering albatrosses, 8:29, 8:*31*, 8:107, 8:108, 8:113, 8:*117*, 8:118–119
Wandering gliders, 3:*136*, 3:137–*138*
Wandering royal albatrosses, 8:113
Wandering seabird lice, 3:*253*, 3:*254*–255
Wandering violin mantids, 3:178, 3:179, 3:*182*, 3:*183*
Wapiti, 12:19, 15:271
Waptia fieldensis, 2:*11*
Warblers, 10:505
mouse, 11:*58*, 11:62–63
reed, 9:313, 9:314, 11:5, 11:*6*, 11:7
See also specific types of warblers
Warbling antbirds, 10:*243*, 10:*252*–*253*
Warbling grass-parakeets. *See* Budgerigars
Warbling honeyeaters. *See* Brown honeyeaters
Warbling silverbills. *See* African silverbills
Warbling vireos, 11:*257*, 11:*260*
Ward's trogons, 9:479
Warham, John, 8:107, 8:123
Waris. *See* White-lipped peccaries
Warnings, bird songs and, 8:38–40, 8:41
Warrens, rabbit, 12:186
Warsangli linnets, 11:324
Wart biter katydids, 3:85
Wart-headed Uzbekian snakeflies, 3:*299*, 3:*300*, 3:302
Wart snakes. *See* File snakes
Warthog lice, 3:249
Warthogs, 15:*276*, 15:277, 15:278–279, 15:280, 15:289*t*
See also Common warthogs
Warty-faced honeyeaters. *See* Regent honeyeaters
Warty frogfishes, 5:*50*
Warty newts, 6:*364*
Warty tree toads, 6:46, 6:*49*, 6:*160*, 6:*165*–166
Waryfishes, 4:432, 4:433
Washington Convention, on true frogs, 6:252
Wasmannia auropunctata. *See* Little fire ants
Wasp galls, 3:81
Waspfishes, 5:163, 5:164
Wasps, 1:*147*, 1:148, 2:22, 2:23, 2:38, **3:405–425**, 3:*409*, 3:*410*
behavior, 3:5, 3:59, 3:64, 3:65, 3:406
conservation status, 3:407
ecosystems and, 3:72
evolution, 3:405
feeding ecology, 3:63, 3:406–407
habitats, 3:406
humans and, 3:407–408
as pests, 3:75
physical characteristics, 3:43, 3:405
reproduction, 3:36, 3:39, 3:59, 3:62, 3:67, 3:407
social, 3:68, 3:69
species of, 3:*413*–424
stylopised, 3:*336*
taxonomy, 3:405
See also specific types of wasps
Watchdogs. *See* Guard dogs
Water
balance, 12:113
deep, 4:56, 4:396
depth, 1:25
exchange, 7:30
oxygen levels, 4:37, 4:38–39
pH levels, 4:37
quality, 2:43 (*See also* Conservation status)
salinity, 4:37, 4:50, 4:52
shallow, 4:53–56
stratification, 4:38–39
temperature, 4:38–39, 4:49–50, 4:53
tropical, 4:53
Water bats. *See* Daubenton's bats
Water bears, 2:14, **2:115–123**, 2:*119*
behavior, 2:117
conservation status, 2:118
distribution, 2:116
evolution, 2:115
feeding ecology, 2:117
habitats, 2:116–117
humans and, 2:118
physical characteristics, 2:115–116
reproduction, 2:23, 2:117–118
species of, 2:*120*–*122*
taxonomy, 2:115
Water bees. *See* Creeping water bugs
Water Beetle Specialist Group, 3:322
Water beetles
behavior, 3:60
as bioindicators, 3:82
conservation status, 3:322
evolution, 3:7
great, 3:*325*, 3:*329*
habitats, 3:319
humans and, 3:323
long-toed, 3:319
physical characteristics, 3:317
reproduction, 3:62
Water boatmen, 3:57, 3:60, 3:264, 3:*265*, 3:*272*–*273*
Water buffaloes, 15:145, 15:263, 16:2, 16:4, 16:11, 16:12, 16:20–21
Asian, 16:2, 16:4, 16:*17*
distribution, 16:*18*
humans and, 16:8, 16:15
physical characteristics, 16:3, 16:12, 16:13
reproduction, 16:14
short-horned, 16:1
Water bugs
behavior, 3:59, 3:261
creeping, 3:*266*, 3:*275*–*276*
evolution, 3:7, 3:259
feeding ecology, 3:63, 3:262
as food, 3:80
humans and, 3:264
physical characteristics, 3:260
reproduction, 3:62, 3:263
See also Giant water bugs
Water chevrotains, 15:325, 15:*326*, 15:328, 15:329, 15:*330*, 15:*331*
Water civets. *See* Otter civets
Water conservation
by gray treefrogs, 6:297–298
reptilian, 7:3
by water-holding frogs, 6:*227*
Water deer, 15:268
See also Chinese water deer
Water dikkops, 9:144
Water dogs. *See* Tiger salamanders
Water fleas, **2:153–159**, 2:*156*
behavior, 2:153–154
conservation status, 2:155
evolution, 2:153
feeding ecology, 2:154
habitats, 2:153
humans and, 2:155
physical characteristics, 2:153
reproduction, 2:154
species of, 2:*157*–158
taxonomy, 2:153
Water frogs, **6:147–154**, 6:*150*
See also Corrugated water frogs
Water-holding frogs, 6:*227*, 6:230, 6:*232*, 6:*240*
Water lice, 2:*253*, 2:*255*–*256*
Water measurers, 3:260, 3:*265*, 3:*274*, 3:275
Water mice, 16:*271*, 16:*272*, 16:275
Water moccasins. *See* Cottonmouths
Water mongooses. *See* Marsh mongooses
Water monitors, 7:48, 7:52, 7:360–361, 7:363
Water opossums, 12:14, 12:249–253, 12:254, 12:*255*, 12:257–*258*
Water ousels. *See* Eurasian dippers
Water ouzels. *See* American dippers
Water pythons, 7:420
Water rabbits, 16:516*t*
Water rails, 9:49
Water rats, 16:266, 16:267–268, 16:269, 16:270
false, 16:262*t*
Florida, 16:225
golden-bellied, 16:261*t*
Water scavenger beetles, 3:319
Water shrews. *See* American water shrews
Water springtails, 3:*102*, 3:103–*104*
Water striders. *See* Sea skaters
Water voles, 16:225, 16:226, 16:227, 16:229, 16:230
Waterbirds (journal), 8:27
Waterbucks, 12:116, 16:*12*, 16:27, 16:29, 16:30, 16:32, 16:*36*, 16:37–38, 16:*41*
Watercocks, 9:46, 9:52
Waterdogs. *See* Mudpuppies
Watering pot shells, 2:*456*, 2:*458*, 2:460
Waterman, J. M., 16:145
Waters, J. M., 4:59, 4:389
Watershed, 4:39
Waterthrushes. *See* American dippers
Watson, John B., 2:35
Wattle birds, New Zealand, **11:447–451**, 11:*449*–*450*
Wattle-eyed flycatchers. *See* Black-throated wattle-eyes
Wattle pigs, 3:*321*
Wattled cranes, 9:3, 9:25, 9:26, 9:27, 9:*30*, 9:*32*
Wattled crows. *See* Kokakos
Wattled curassows, 8:*417*, 8:*422*–423
Wattled guans, 8:413
Wattled honeyeaters. *See* Red wattlebirds
Watts, P., 12:57
Waugh, Evelyn, 12:183
Waved albatrosses, 8:113, 8:115, 8:*116*

Wax production, molting and, 3:32
Waxbills, 11:353–357, 11:*358*, 11:364–365, 11:378
See also Weaverfinches
Waxwings, **10:447–452**, 10:*450*
Weaning, 12:97, 12:126
See also Reproduction
Weasel lemurs. *See* Sportive lemurs
Weasel sharks, 4:113, 4:115
Weasel sportive lemurs, 14:*78*, 14:*80*, 14:82
Weasels, **14:319–328**, 14:*326*, **14:332–334**
behavior, 12:114, 14:321–323
conservation status, 14:324
distribution, 14:*319*, 14:321
evolution, 14:319
feeding ecology, 14:260, 14:323
humans and, 14:324–325
physical characteristics, 14:319–321
reproduction, 14:324
species of, 14:*327*–328, 14:332*t*, 14:333*t*
taxonomy, 14:256, 14:319
Weatherfishes, 4:*322*, 4:*325*, 4:332
Weaver ants, 3:80
Weaverfinches, 11:353–357, **11:353–373**, 11:*358–359*
behavior, 11:354–355
conservation status, 11:356–357
distribution, 11:*353*, 11:354
evolution, 11:353
feeding ecology, 11:*355*
habitats, 11:354
humans and, 11:357
physical characteristics, 11:353–354, 11:*354*, 11:*355*
reproduction, 11:355–356
species of, 11:*360–372*
taxonomy, 11:353
Weavers, 10:171, 10:173, **11:375–395**, 11:*380–381*
behavior, 11:377
conservation status, 11:378–379
distribution, 11:*375*, 11:376
evolution, 11:375–376
feeding ecology, 11:377
habitats, 11:376–377
humans and, 11:379
nest construction, 11:*376*, 11:*377*, 11:*378*, 11:*379*
physical characteristics, 11:376
reproduction, 11:377–378
species of, 11:*382–394*
taxonomy, 11:375–376
Web-foot frogs, 6:317, 6:318, 6:*319*
Web-footed geckos, 7:261, 7:*264*, 7:*267*–268
Web-spinning spiders, 2:25
Webb, J. F., 5:180
Webbing clothes moths, 3:*391*, 3:*403*–404
Weberogobious amadi. *See* Poso bungus
Webless toothed toads. *See* Schmidt's lazy toads
Webspinners, 3:9, **3:*233*–238**, 3:*235*
Webworms, 3:389
Weddell seals, 12:85, 12:87, 12:138, 14:*423*, 14:*426*, 14:*428*, 14:430–431
Wedge-capped capuchins. *See* Weeper capuchins
Wedge-rumped storm-petrels, 8:135, 8:*136*
Wedge-tailed sunbirds. *See* Orange-breasted sunbirds
Wedgebills, 11:75, 11:76
Weebills, 11:55, 11:56
Weed whitings, 4:67
Weeds, white, 1:129
Weedsuckers, 5:355, 5:357
Weedy scorpionfishes, 5:166
See also Merlet's scorpionfishes
Weedy seadragons, 5:*137*, 5:*140*, 5:147
Week's bichirs, 4:*209*
Weenie worms. *See* Green bonellias
Weeper capuchins, 14:101, 14:102, 14:103, 14:*103*, 14:*107*, 14:*110*, 14:112
Weeros. *See* Cockatiels
Weeverfishes, **5:331–335**, 5:*336*, **5:338–339**
Weevils, **3:*315*–323**, 3:*321*, **3:327**
behavior, 3:65
boll, 3:60
feeding habits, 3:63
flour, 3:57
as food, 3:80–81
giraffe-necked, 3:*325*, 3:*327*
grain, 3:57, 3:76
nut, 3:*2*
palm, 3:323
pollination by, 3:79
in religion, 3:74
reproduction, 3:44
rice, 3:76
See also Coleoptera
Wegener, Alfred, 4:58
Weidomyines, 16:263–264
Weight reduction, in bats, 12:58–59
Weitzman, Stanley H., 5:421
Welcome swallows. *See* House swallows
Welcome trogons. *See* Eared quetzals
Welsh, W. W., 5:56
Welwitch's hairy bats, 13:497
Werner, Franz, 7:223
Werneria spp., 6:184
Wessling, Bernard, 9:25
West African brush-tailed porcupines. *See* African brush-tailed porcupines
West African manatees, 15:191–193, 15:*209*, 15:*210*, 15:*211–212*
West African pygmy squirrels, 16:167, 16:174*t*
West African pythons. *See* Ball pythons
West African rousettes. *See* Egyptian rousettes
West African screeching frogs. *See* Ugandan squeakers
West African thrushes. *See* Olive thrushes
West Atlantic stalked crinoids, 1:*361*, 1:*364*, 1:365
West Atlantic trumpetfishes, 5:*138*, 5:*139*
West Australian boodies, 13:35
West Caucasian turs, 16:87, 16:91, 16:94
West Indian flamingos. *See* Greater flamingos
West Indian manatees, 15:*193*, 15:*196*, 15:*205*, 15:*210*, 15:*211–212*
behavior, 15:*207*
conservation status, 15:195
distribution, 15:192
feeding ecology, 15:193, 15:*206*
habitats, 15:192, 15:*194*
migrations, 12:87
physical characteristics, 15:192, 15:*209*
reproduction, 15:*208*
taxonomy, 15:191
West Indian monk seals, 14:422, 14:435*t*
West Indian powderpost drywood termites, 3:*169*, 3:*170*–171
West Indian sea eggs, 1:401, 1:*407*, 1:*414*–415
West Indian sea stars. *See* Cushion stars
West Indian shrews, 12:133, 13:193, 13:197, **13:243–245**, 13:*244*
West Indian sloths, **13:155–159**
West Indies. *See* Neotropical region
West Indies brotulas. *See* Key brotulas
West Nile virus, 8:25
Western, David, 12:97
Western Australian Conservation Act, 5:381
Western Australian Department of Fisheries, 4:397, 4:398
Western Australian fur seals. *See* New Zealand fur seals
Western banded geckos, 7:*264*, 7:*265*
Western barbastelles, 13:500, 13:*505*, 13:506–*507*
Western barred bandicoots, 12:*110*, 13:*1*, 13:*4*–5, 13:10, 13:12, 13:17*t*
Western bluebills, 11:*356*
Western bluebirds, 10:485
Western bonneted bats, 13:493*t*
Western bowerbirds, 11:481
Western bristlebirds, 11:57
Western chestnut mice, 16:*252*
Western chipmunks. *See* Least chipmunks
Western chuckwallas. *See* Common chuckwallas
Western collared lizards. *See* Common collared lizards
Western corella parrots, 9:277–278
Western crowned pigeons, 9:250, 9:*253*, 9:*265*–266
Western Cuban nesophontes, 13:243
Western cusimanses, 14:357*t*
Western diamondback rattlesnakes, 7:38, 7:40, 7:42, 7:44, 7:50, 7:*197*
Western earless lizards. *See* Common lesser earless lizards
Western European hedgehogs, 12:*185*, 13:*200*, 13:*204*, 13:*205*, 13:*208*, 13:*209*, 13:212*t*–213*t*
Western fat-tailed dwarf lemurs, 14:*37*, 14:*38*, 14:*40*, 14:*41*, 14:42
Western flower thrips, 3:283, 3:*284*, 3:*285*
Western galaxias. *See* Western minnows
Western gorillas, 12:222, 14:225, 14:*231*, 14:232, 14:*236*, 14:*238*
Western gray kangaroos, 13:101*t*
Western gray plantain-eaters, 9:*303*, 9:*309*–310
Western gray squirrels, 16:175*t*
Western grebes, 8:*170*, 8:171, 8:*173*, 8:*180*–181
Western ground uta. *See* Common side-blotched lizards
Western hare-wallabies. *See* Rufous hare-wallabies
Western heather voles, 16:238*t*
Western hog-nosed skunks, 14:332*t*
Western hyraxes. *See* Red-headed rock hyraxes
Western jackdaws, 11:505, 11:*511*, 11:*519*
Western jumping mice, 16:212, 16:*215*, 16:216
Western kingbirds, 10:272, 10:275, 10:*276*, 10:*279*–280
Western Malagasy broad striped mongooses, 14:*350*
Western Malagasy bushy-tailed rats, 16:*284*
Western marsh harriers, 8:320

Western mastiff bats. *See* Western bonneted bats
Western meadowlarks, 11:*307*, 11:*316*
Western minnows, 4:391, 4:*394*, 4:*396*–397
Western mountain reedbucks, 16:34
Western olive sunbirds, 11:210
Western pocket gophers. *See* Valley pocket gophers
Western pond turtles, 7:106
Western pygmy blues, 3:384
Western pygmy perches, 5:*224*, 5:*229*, 5:230
Western pygmy possums, 13:*107*
Western quolls. *See* Chuditches
Western rattlesnakes. *See* Prairie rattlesnakes
Western red colobus, 14:*174*, 14:175, 14:*176*, 14:*178*–179
Western ringtails. *See* Common ringtails
Western rock sengis, 16:522, 16:531*t*
Western rock wallabies. *See* Brush-tailed rock wallabies
Western sagebrush lizards. *See* Common sagebrush lizards
Western sand dollars, 1:405, 1:*406*, 1:*411*, 1:412
Western sandpipers, 9:176, 9:*177*
Western sandveld lizards, 7:*300*, 7:*301*
Western scrub-birds. *See* Noisy scrub-birds
Western scrub-jays, 11:506–507, 11:*510*, 11:*513*
Western shrike-thrushes. *See* Gray shrike-thrushes
Western side-blotched lizards. *See* Common side-blotched lizards
Western slender glass lizards, 7:*340*
Western spinebilled honeyeaters. *See* Western spinebills
Western spinebills, 11:*239*, 11:248–*249*
Western spotted skunks, 14:*320*, 14:324
Western swamp turtles, 7:79, 7:*80*, 7:*82*, 7:83
Western tarsiers, 12:*81*, 14:*92*, 14:93, 14:*94*, 14:95, 14:*96*, 14:*97*–98
Western thumbless bats. *See* Schnabeli's thumbless bats
Western tree hyraxes, 15:177, 15:178, 15:184, 15:*185*, 15:*186*–187
Western warbling-vireos. *See* Warbling vireos
Western wattled cuckoo-shrikes, 10:386, 10:*388*, 10:*389*
Western whipbirds, 11:76
Western wood-pewees, 10:*270*, 10:275, 10:*276*, 10:286–*287*
Western yellow robins. *See* Eastern yellow robins
Western zebra-tailed lizards. *See* Zebra-tailed lizards
Wetar figbirds, 11:428
Wetas, 3:202, 3:207
Wetlands, 4:40
  See also Habitats
Wetmore, Alexander, 10:26
Wetmorena spp., 7:339
Weyland ringtails, 13:114, 13:122*t*
Weyns's duikers, 16:85*t*
Whale fleas. *See* Gray whale lice; Sperm whale lice
Whale sharks, 4:*107*
  conservation, 4:105, 4:110
  distribution, 4:*108*
  feeding ecology, 4:67, 4:110
  physical characteristics, 4:*3*, 4:14, 4:44, 4:110
  taxonomy, 4:11, 4:110
Whalebirds. *See* Broad-billed prions
Whalefishes, **5:105–106,** 5:*107*, **5:108**
Whalelike catfishes, 4:352
Whales, 12:14, 15:1–11, 15:*84*
  beaked, **15:59–71,** 15:*63*, 15:*69t*–*70t*
  belugas, 12:*85*, 12:87, 15:9, 15:10–11, **15:81–91**
  distribution, 12:138
  gray, 12:87, 15:3, 15:4, 15:5, 15:6, 15:9, **15:93–101**
  migrations, 12:169
  neonatal milk, 12:126
  physical characteristics, 12:63, 12:67, 12:68
  reproduction, 12:91, 12:94, 12:103, 12:108
  right, **15:107–118,** 15:*114*
  smell, 12:80
  sperm, **15:73–80**
  teeth, 12:46
  See also Killer whales; Rorquals
Wheatears, 10:483–486, 10:488, 10:*490*, 10:*496*
Wheel snails, 2:412
Wheel wearers, 1:13, 1:23, **1:351–354**
Wheelbarrows. *See* Rufous treecreepers
Wheeler's shrimp gobies. *See* Gorgeous prawn-gobies
Whimbrels, 9:176
Whinchats, 10:485
Whip coral gobies, 5:*379*, 5:*380*, 5:382–383
Whip-poor-wills, 9:369, 9:401, 9:*402*, 9:*407*, 9:*410*–411
Whipbills, 11:76
Whipbirds, **11:75–81,** 11:77
Whipsnakes, 7:26, 7:57, 7:486
Whiptail lizards, **7:309–317,** 7:*313*
Whiptail wallabies, 13:*86*, 13:101*t*
Whipworms. *See* Human whipworms
Whirligig beetles, 3:81, 3:317, 3:318, 3:321, 3:*324*, 3:*328*, 3:330
Whiskered armorheads. *See* Striped boarfishes
Whiskered bats, 13:515*t*
Whiskered flowerpeckers, 11:191
Whiskered flying squirrels, 16:142*t*
Whiskered pittas, 10:195
Whiskered tree swifts, 9:417, 9:433, 9:435–*436*
Whiskered yuhinas, 10:*509*, 10:*519*
Whiskers. *See* Vibrissae
Whiskey-jacks. *See* Gray jays
Whispering bats. *See* Brown long-eared bats
Whistle pigs. *See* Woodchucks
Whistlers, **11:115–126,** 11:*119*
  behavior, 11:117
  conservation status, 11:118
  distribution, 11:*115*, 11:116
  evolution, 11:115
  feeding ecology, 11:117
  habitats, 11:116
  humans and, 11:118
  physical characteristics, 11:116
  reproduction, 11:117–118
  species of, 11:*120*–*126*
  taxonomy, 11:115–116
Whistling ducks, 8:363, 8:364, 8:366, 8:369, 8:370
Whistling hares. *See* American pikas
Whistling kites, 8:321
Whistling rats. *See* Angoni vlei rats
Whistling-thrushes, 10:484, 10:485, 10:487
Whistling warblers, 11:291
White, E. B., 2:41
White amurs. *See* Grass carps
White-backed ducks, 8:367
White-backed mousebirds, 9:*472*, 9:474–*475*
White-backed munias, 11:357
White-backed vultures. *See* White-rumped vultures
White-backed woodpeckers, 10:*153*, 10:*162*
White bald uakaris, 14:148, 14:151
White-banded sphinxlets, 9:444
White basses, 5:196
White bats, 13:*315*, 13:414, 13:416, 13:*418*, 13:*421*, 13:*427*, 13:431–432
  See also Northern ghost bats
White-bearded antshrikes, 10:242
White-bearded honeyeaters. *See* New Holland honeyeaters
White bellbirds, 10:305, 10:*311*, 10:319–*320*
White-bellied boobies. *See* Brown boobies
White-bellied cuckoo-shrikes, 10:387
White-bellied duikers, 16:84*t*
White-bellied fat-tailed mouse opossums. *See* Small fat-tailed opossums
White-bellied funnel-eared bats, 13:*462*, 13:463–*464*
White-bellied go-away-birds, 9:300, 9:*303*, 9:308–*309*
White-bellied hedgehogs. *See* Central African hedgehogs
White-bellied hummingbirds. *See* White-necked jacobins
White-bellied mangrove snakes, 7:465–466
White-bellied orioles. *See* Olive-backed orioles
White-bellied pangolins. *See* Tree pangolins
White-bellied pigeons. *See* Snow pigeons
White-bellied pitohuis, 11:118
White-bellied plantain-eaters. *See* White-bellied go-away-birds
White-bellied sea-eagles, 8:321
White-bellied seedsnipes, 9:190, 9:191, 9:*192*, 9:193–*194*
White-bellied spider monkeys, 14:160, 14:167*t*
White-bellied storm-petrels, 8:137
White-bellied wormlizards, 7:273, 7:*276*
White-billed divers. *See* Yellow-billed loons
White boobies. *See* Masked boobies
White-breasted blue mockingbirds. *See* Blue-and-white mockingbirds
White-breasted cormorants. *See* Great cormorants
White-breasted cuckoo-shrikes, 10:*386*
White-breasted dippers. *See* Eurasian dippers
White-breasted guineafowl, 8:426–429
White-breasted hedgehogs. *See* Eastern European hedgehogs
White-breasted mesites, 9:5, 9:6, 9:7, 9:*8*
White-breasted negro-finches, 11:*355*, 11:*358*, 11:360–*361*
White-breasted nuthatches, 11:*169*, 11:*170*, 11:*171*–172
White-breasted roatelos. *See* White-breasted mesites
White-breasted starlings. *See* Black-winged mynas
White-breasted thrashers, 10:466, 10:468

White-breasted waterhens, 9:62–*63*
White-breasted whistlers, 11:116
White-breasted white-eyes, 11:227
White-breasted wood wrens, 10:528
White-breasted woodswallows, 11:*460*, 11:*461*, 11:463–464
White-browed babblers, 11:127, 11:*128*
White-browed blackbirds, 11:293, 11:*308*, 11:314–*315*
White-browed crakes, 9:47
White-browed gibbons. *See* Hoolock gibbons
White-browed hawk-owls, 9:347
White-browed jungle-flycatchers, 11:27
White-browed nuthatches, 11:169
White-browed robin chats, 10:484–485, 10:*490*, 10:*493*
White-browed scrubwrens, 11:*58*, 11:*59*–60
White-browed sparrow weavers, 11:*380*, 11:382–*383*
White-browed tit-spinetails, 10:211
White-browed treecreepers, 11:134
White-browed trillers. *See* Varied trillers
White-browed woodswallows, 11:*461*, 11:*464*–465
White butterflies, 3:39
White-capped albatrosses, 8:114
White-capped dippers, 10:475, 10:*478*, 10:480–*481*
White-capped sea lions. *See* Australian sea lions
White cats. *See* Channel catfishes
White-cheeked cotingas, 10:306
White-cheeked gibbons, 14:*208*, 14:*216*, 14:*221*
White-cheeked turacos, 9:300
White-chested white-eyes, 11:229
White-chinned sapphires, 9:449
White cockatoos. *See* Sulphur-crested cockatoos
White-collared kingfishers. *See* Collared kingfishers
White-collared kites, 8:323
White-collared mangabeys. *See* Collared mangabeys
White-collared pratincoles. *See* Rock pratincoles
White-collared swifts, 9:422, 9:*425*, 9:*426*
White-crested helmet-shrikes. *See* White helmet-shrikes
White-crested jay thrushes. *See* White-crested laughing thrushes
White-crested laughing thrushes, 10:*510*, 10:514–*515*
White-crested turacos, 9:300
White-crowned forktails, 10:*490*, 10:494–*495*
White-crowned sparrows, 8:40
White-crowned tits, 11:149
White-eared bulbuls. *See* Common bulbuls
White-eared catbirds, 11:478
White-eared hummingbirds, 9:448
White-eared kobs, 16:7
White-eared myza. *See* Greater Sulawesi honeyeaters
White-eared night herons, 8:236, 8:*246*, 8:255–*256*
White-eared puffbirds, 10:102, 10:*103*, 10:*105*, 10:*107*
White-eyed honeyeaters. *See* New Holland honeyeaters
White-eyed river martins, 10:360
White-eyed starlings, 11:407, 11:410, 11:*413*, 11:*414*
White-eyes, 10:404, **11:227–234,** 11:*231*
White-faced arboreal spiny rats, 16:450, 16:*452*, 16:*455*
White-faced glossy ibises, 8:*295*, 8:296–*297*
White-faced ibises. *See* White-faced glossy ibises
White-faced nunbirds, 10:102, 10:103
White-faced owls. *See* Southern white-faced owls
White-faced sakis, 14:143–*144*, 14:147, 14:*149*, 14:*150*
White-faced scops-owls. *See* Southern white-faced owls
White-faced starlings, 11:410
White-faced storm-petrels, 8:136, 8:138
White-faced titmice. *See* Southern whitefaces
White-faced tree ducks. *See* White-faced whistling ducks
White-faced whistling ducks, 8:364, 8:*370*, 8:*374*, 8:*376*–377
White fin dolphins. *See* Baijis
White-flippered penguins. *See* Little penguins
White-footed dunnarts, 12:289, 12:301*t*
White-footed mice, 16:127, 16:128, 16:*267*, 16:*268*
White-footed sportive lemurs, 14:*74*, 14:*76*, 14:*78*, 14:*79*, 14:81–82
White-footed voles, 16:237*t*
White-footed weasel lemurs. *See* White-footed sportive lemurs
White-fronted bee-eaters, 10:*3*, 10:40, 10:41, 10:*43*, 10:*45*–46
White-fronted capuchins, 14:101, 14:102, 14:*107*, 14:*110*
White-fronted chats, 11:65, 11:*66*
White-fronted falconets, 8:351, 8:353, 8:*357*
White-fronted hairy dwarf porcupines, 16:365, 16:374*t*
White-fronted plovers, 9:163
White fruit pigeons. *See* White imperial pigeons
White grubs, 3:63
White hakes, 5:28, 5:29, 5:*31*, 5:*38*, 5:39
White-handed gibbons. *See* Lar gibbons
White-headed gulls, 9:209
White-headed hoopoes, 10:66
White-headed mousebirds, 9:*472*, 9:*474*
White-headed munias, 11:*354*
White-headed vangas, 10:439, 10:440, 10:441
White helmet-shrikes, 10:*427*, 10:*431*, 10:*432*
White herrings. *See* Alewives
White imperial pigeons, 9:*253*, 9:*265*
White-lined comb-tooth blennies, 5:*344*, 5:346
White-lined treefrogs, 6:227, 6:229, 6:*231*, 6:242
White-lipped deer, 12:134, 15:358, 15:*363*, 15:*368*, 15:370–371
White-lipped mud turtles, 7:*124*, 7:*125*, 7:126
White-lipped peccaries, 15:291–299, 15:*295*, 15:*297*
White-lipped pythons, 7:420
White-lipped tamarins, 14:*119*
White-lipped treefrogs, 6:225, 6:*229*
White-lored antpittas. *See* Fulvous-bellied ant-pittas
White mullets, 5:*59*
White-naped black tits. *See* White-naped tits
White-naped cranes, 9:3, 9:*25*, 9:*27*
White-naped swifts, 9:421
White-naped tits, 11:158, 11:*159*, 11:*164*–165
White-naped weasels, 14:333*t*
White-necked bald crows. *See* White-necked picathartes
White-necked crows, 11:508
White-necked fruit bats. *See* Tongan flying foxes
White-necked hawks, 8:321
White-necked jacobins, 9:*454*, 9:459–*460*
White-necked picathartes, 10:507, 10:*509*, 10:*522*–523
White-necked puffbirds, 10:*105*, 10:*106*
White-nosed bearded sakis, 14:144, 14:145, 14:147, 14:148, 14:153*t*
White-nosed coatis, 14:*310*, 14:*312*, 14:*314*
White-nosed monkeys, lesser, 14:*192*
White-nosed sakis. *See* White-nosed bearded sakis
White openbills. *See* Asian openbills
White owls. *See* Common barn owls
White pelicans. *See* American white pelicans
White pigs, 15:277
White-quilled bustards, 9:*95*, 9:*98*
White-quilled korhaans. *See* White-quilled bustards
White rhinoceroses, 15:215, 15:*249*, 15:*256*, 15:261
  behavior, 12:145, 15:218, 15:253
  conservation status, 15:222, 15:*254*
  distribution, 15:*260*, 15:261
  feeding ecology, 15:220, 15:253
  habitats, 15:*253*
  physical characteristics, 15:216, 15:217, 15:251, 15:*252*
  reproduction, 15:221, 15:222
  scent marking, 12:80–81
  See also Rhinoceroses
White-rumped falcons, 8:351
White-rumped green thorntails, 9:444
White-rumped kingfishers, 10:6, 10:*12*, 10:*17*
White-rumped sandpipers, 9:179
White-rumped swiftlets, 9:424, 9:*425*, 9:428–*429*
White-rumped vultures, 8:316, 8:320, 8:323, 8:*327*, 8:*335*
White salamanders. *See* Texas blind salamanders
White seals. *See* Crab-eater seals
White shads. *See* American shads
White-shafted fantails. *See* Gray fantails
White sharks, 4:*10*, 4:131–*135*, 4:*132*, 4:*134*, 4:137–138
White-shouldered capuchins. *See* White-throated capuchins
White-shouldered ibis, 8:236, 8:293
White-shouldered sea-eagles. *See* Steller's sea-eagles
White-shouldered tits, 11:156
White-sided jackrabbits, 16:516*t*
White-spotted puffers, 5:*472*, 5:*474*, 5:481
White-spotted spurdogs. *See* Piked dogfishes
White starlings. *See* Bali mynas
White storks. *See* European white storks
White-streaked tenrecs, 13:228, 13:*229*
White-striped dorcopsises, 16:102*t*
White-striped free-tailed bats, 13:*487*, 13:*489*, 13:*492*
White-striped geckos, 7:*260*, 7:*262*

White sturgeons, 4:*4*, 4:*216*, 4:218–*219*
White-tailed deer, 15:*386*, 15:*389*–390
behavior, 15:271, 15:*381*, 15:*382*
competition, 12:118
ecosystem effects, 12:217
evolution, 15:380
habitats, 15:383
humans and, 15:384–385
reproduction, 15:384
scent marking, 12:80–81, 12:*142*
translocation of, 12:224
White-tailed eagles, 8:323
White-tailed gnus. *See* Black wildebeests
White-tailed hummingbirds, 9:418, 9:450, 9:455–456
White-tailed kites, 8:317
White-tailed lapwings, 9:161, 9:162
White-tailed larks, 10:343
White-tailed mice, 16:*286*, 16:288–*289*
White-tailed monarchs. *See* White-tipped monarchs
White-tailed mongooses, 14:259, 14:350
White-tailed plantcutters. *See* Red-breasted plantcutters
White-tailed plovers, 9:163
White-tailed shrikes, 10:426, 10:427, 10:428
White-tailed sicklebills. *See* White-tipped sicklebills
White-tailed swallows, 10:360
White-tailed trogons, 9:479
White-tailed tropicbirds, 8:*190*
White-throated bee-eaters, 10:*43*, 10:*46*
White-throated bulbuls, 10:*400*, 10:409–*410*
White-throated capuchins, 14:*1*, 14:101, 14:102, 14:*103*, 14:*104*, 14:*107*, 14:*111*
White-throated dippers. *See* Eurasian dippers
White-throated fantail flycatchers. *See* White-throated fantails
White-throated fantails, 11:85, 11:86, 11:*88*, 11:*90*
See also Northern fantails
White-throated flowerpeckers, 11:191
White-throated gerygones, 11:56
White-throated greenbuls, 10:396
White-throated magpie-jays, 11:504
White-throated mountain babblers, 10:507
White-throated needletails, 9:416, 9:422, 9:424, 9:*425*, 9:*429*
White-throated nightjars, 9:*403*
White-throated rails, 9:*49*, 9:*53*, 9:*59*–60
White-throated round-eared bats, 13:433*t*
White-throated seedeaters, 9:396
White-throated sparrows, 11:*268*, 11:*270*–271, 11:304
White-throated swifts, 9:416, 9:422
White-throated tits, 11:142, 11:143
White-throated toucans, 10:125, 10:*129*, 10:*133*–134
White-throated treecreepers, 11:133, 11:*134*, 11:*136*, 11:*137*
White-throated treerunners, 10:*213*, 10:224–*225*
White-throated wallabies. *See* Parma wallabies
White-throated woodrats, 16:277*t*, 16:*282*
White throats. *See* White-throated sparrows
White tigers, 14:*372*
White-tipped monarchs, 11:*99*, 11:*102*–103
White-tipped sicklebills, 9:*454*, 9:*455*
White-toothed shrews, 13:198, 13:199, 13:247, **13:265–277**, 13:*269*, 13:*270*, 13:*276t*–*277t*
White-tufted grebes, 8:169
White-vented storm-petrels, 8:*137*
White wagtails, 10:372, 10:374
White wartyback mussels, 1:49
White weeds, 1:129
White-whiskered puffbirds, 10:*104*, 10:*105*, 10:107–*108*
White-whiskered spider monkeys, 14:160, 14:167*t*
White-winged albatrosses. *See* Wandering albatrosses
White-winged blackbirds. *See* Lark buntings
White-winged choughs, 11:453, 11:*455*, 11:456–*457*, 11:504
White-winged cotingas, 10:308, 10:321
White-winged cuckoo-shrikes, 10:387
White-winged doves, 9:*243*
White-winged ducks, 8:367
White-winged juncos, 11:271
White-winged nightjars, 9:370, 9:405
White-winged potoos, 9:396, 9:397
White-winged robins, 11:*107*, 11:*111*–112
White-winged sandpipers, 9:179–180
White-winged thicket-flycatchers. *See* White-winged robins
White-winged tits. *See* White-naped tits
White-winged trillers, 10:387
White-winged trumpeters. *See* Pale-winged trumpeters
White-winged vampires, 13:314
White-winged warblers, 11:291
Whitefaces, 11:55, 11:57, 11:*58*, 11:62
Whitefishes. *See* Atlantic menhadens; Capelins; Elephantfishes; Upper mouths
Whiteflies, 3:66, 3:*259*, 3:260, 3:261, 3:*265*, 3:*267*
Whiteheads, 11:117, 11:*119*, 11:*125*–*126*
Whitehead's broadbills, 10:178, 10:179
Whitehead's spiderhunters, 11:210
Whitehead's swiftlets, 9:418
Whitehead's trogons, 9:479
Whitelined toadfishes, 5:42
White's thrushes, 10:*491*, 10:500–*501*
White's treefrogs. *See* Green treefrogs
Whitespotted filefishes, 5:*467*
Whitespotted greenlings, 5:191
Whitespotted toadfishes, 5:42
Whitethroats, 11:*8*, 11:22–*23*
Whitetip reef sharks, 4:*67*, 4:*113*, 4:115, 4:116
Whitetip soldierfishes, 5:*114*
Whoophills, 9:28
Whooping cranes, 8:27, 9:3, 9:23–28, 9:29, 9:*35*
Whydahs, 11:356, 11:*361*, 11:375–376, 11:378, 11:*381*, 11:392–393
Whyte's barbets, 10:115
Wichety grub, 3:81
Widdowson, Elsie, 12:126
Wide-headed rottenwood termites, 3:*169*, 3:*174*–175
Wide-mouthed frogs, 6:*5*
Wide-mouthed lampreys. *See* Pouched lampreys
Wideawake terns. *See* Sooty terns
Widow monkeys. *See* Collared titis
Widows, 11:376, 11:*381*, 11:390
See also Jackson's widow-birds
Wiedomys pyrrhorhinos, 16:267
Wiesner, J. B., 1:47
Wilby, G. V., 5:351
Wilcox, Bruce, 1:48
Wild boars. *See* Eurasian wild pigs
Wild canaries. *See* American goldfinches; Yellow warblers
Wild cats, 14:257, 14:262, 14:290, 14:292, 14:375, 14:*378*, 14:*383*–384
See also Cats
Wild cattle. *See* Cattle
Wild dogs. *See* Dogs
Wild goats, 16:*88*
behavior, 16:92, 16:104*t*
conservation status, 16:104*t*
distribution, 16:91, 16:104*t*
domestication and, 15:146, 15:148
evolution, 16:87
feeding ecology, 16:93, 16:104*t*
habitats, 16:104*t*
humans and, 16:95
physical characteristics, 16:90, 16:104*t*
See also Goats
Wild junglefowl. *See* Red jungle fowl
Wild oxen. *See* Aurochs
Wild sheep, 15:269, 15:271, 15:273, 16:87, 16:*88*, 16:95
See also Sheep
Wild turkeys, 8:21, 8:399, 8:433, 8:*434*, 8:*435*, 8:*438*, 8:*441*
Wild yaks. *See* Yaks
WildBird (journal), 8:27
Wildcats. *See* Wild cats
Wildebeests, 12:*164*, **16:27–43**, 16:*35*
behavior, 16:5, 16:30–31
conservation status, 16:33–34
distribution, 16:29, 16:*38*
evolution, 16:27–28
feeding ecology, 16:31–32
in guilds, 12:119
habitats, 16:4, 16:29–30
humans and, 16:34
migrations, 12:*164*, 12:*166*
physical characteristics, 16:2, 16:28–29
reproduction, 12:101, 12:106, 16:32–33
species of, 16:37–38, 16:42*t*
taxonomy, 16:27–28
Wildlife Protection Fund (India), 7:56
Wilfredomys spp., 16:264
Wilfredomys oenax. *See* Red-nosed tree rats
Wilhelm rainforest frogs, 6:*306*, 6:311
Willets, 9:177
Williamson, D. I., 2:195
Williams's larks, 10:346
Willie wagtails, 11:83, 11:*85*, 11:86, 11:*88*, 11:*94*–95
Willow cats. *See* Channel catfishes
Willow flycatchers, 10:269, 10:*271*, 10:*272*, 10:275, 10:*276*, 10:282–*283*
Willow grouse. *See* Willow ptarmigans
Willow ptarmigans, 8:434, 8:435, 8:437, 8:*440*, 8:441–442
Willow tits, 11:155, 11:161
Willughby, F., 8:369
Wilson, D. E., 16:121, 16:389
Wilson, E. O., 1:48, 1:49, 3:71, 7:54
Wilson, M. V. H., 5:5
The Wilson Bulletin, 8:27
Wilsonia citrina. *See* Hooded warblers
Wilson's birds of paradise, 11:489, 11:492

Wilson's phalaropes, 9:*181*, 9:*187*–188
Wilson's plovers, 9:162
Wilson's storm-petrels, 8:136–138, 8:*139–140*
Wind
bird songs and, 8:39
migration and, 8:32–33, 8:35
The Wind in the Willows, 12:183, 13:200
Windowpane flounders, 5:*454*, 5:*456*, 5:462
Winemiller, Kirk O., 15:29
Wing-banded horneros, 10:*210*
Wing-barred seedeaters. *See* Variable seedeaters
Wing-beat frequencies, 3:23–24
Wing loading (WL), bats, 12:57–58
Wing sacs, 12:81
Winged insects, 3:12–13, 3:33–34, 3:38
See also specific types of winged insects
Wingertshellicus backesi, 3:11
Wingless insects, 3:10, 3:11, 3:33, 3:38
See also specific types of wingless insects
Wings, 3:5, 3:12–13, 3:22–24, 12:56–58, 12:56–60, 12:*58*
See also Physical characteristics
Wingstrandarctus spp., 2:117
Winter flounders, 5:453, 5:*454*, 5:*456*, 5:460–462, 14:228
Winter moths, 3:38
Winter skates, 4:*177*
Winter wrens, 10:526, 10:*530*, 10:*537*
Winteria telescopa, 4:392
Wire corals, 1:103
Wire-tailed manakins, 10:*297*, 10:*298*, 10:302–*303*
Wire worms. *See* Barber's pole worms
Wireworms, 3:56, 3:63
Wisents, 16:14
Witjuti, 3:81
Wiwaxia, 12:64
Wiwaxia spp., 2:45
WL. *See* Wing loading
Wobbegons, 4:11, 4:105
Woermann's bats. *See* African long-tongued fruit bats
Wolbachia spp., 3:39
Wolf, L. L., 9:444
Wolf characins. *See* Trahiras
Wolf-eels, 4:44, 5:310–313, 5:*312*, 5:*314*, 5:*315*
Wolf herrings, 4:277, 4:*280*
Wolf spiders, 2:37
Wolffishes, 4:44, 5:309–313
See also Trahiras
Wolffsohn's viscachas, 16:380, 16:383*t*
Wolfsnails, rosy. *See* Rosy wolfsnails
Wolterstorffina spp., 6:184
Wolverines, 14:*326*, 14:*328*, 14:330–331
behavior, 14:323
conservation status, 12:220
distribution, 12:131
feeding ecology, 14:323
habitats, 14:321
humans and, 14:324
reproduction, 12:105, 12:107, 12:109, 12:110, 14:324
Wolves, **14:265–278,** 14:*275*, **14:282–285**
aardwolves, 14:259, 14:260, **14:359–367**
behavior, 12:*141*, 12:*142*, 14:267–268, 14:269
conservation status, 14:262, 14:263, 14:272–273
domestic dogs and, 12:180
eurasian, 12:180
evolution, 14:265–266
Falkland Island, 14:262, 14:272, 14:284*t*
feeding ecology, 14:269, 14:270–271
habitats, 14:267
humans and, 14:273–274
physical characteristics, 14:266
predation and, 12:116
red, 12:*214*, 14:262, 14:273
reproduction, 14:272
species of, 14:*276–278*, 14:282, 14:284*t*
taxonomy, 14:265
See also Gray wolves; Tasmanian wolves
Womas, 7:420
Wombats, **13:51–56,** 13:*52*, 13:*54*
behavior, 13:52, 13:55–56
conservation status, 13:40, 13:53, 13:55–56
distribution, 13:*51*, 13:*55*–56
evolution, 13:31, 13:51
feeding ecology, 13:38, 13:52–53, 13:55–56
habitats, 13:*35*–36, 13:52, 13:55–56
humans and, 13:40, 13:53, 13:55–56
physical characteristics, 13:32, 13:33, 13:51, 13:55–56
reproduction, 13:38, 13:39, 13:53, 13:55–56
taxonomy, 13:51, 13:55–56
Women. *See* Humans
Wompoo fruit doves, 9:*253*, 9:*264*
Wonambi spp., 7:17
Wonder geckos. *See* Common plate-tailed geckos
Wonga pigeons, 9:250
Wongai ningauis, 12:300*t*
Wonsettler, A. L., 5:180
Wood ants, 3:68
Wood-boring beetles, 3:64, 3:75, 3:316, 3:320–321, 3:323
Wood-boring flies, 3:359
Wood-boring insects, 3:56
Wood boring psocids, 3:243, 3:244
Wood ducks, 8:364, 8:365, 8:369
Wood fish. *See* Silverfish
Wood frogs, 6:*35*, 6:49, 6:63
Wood ibis. *See* Wood storks
Wood larks, 10:*342*, 10:343, 10:*345*, 10:346, 10:*347*, 10:*353*
Wood lemmings, 16:228, 16:229, 16:*232*, 16:*235*–236, 16:270
Wood nuthatches. *See* Nuthatches
Wood-owls, 9:347, 9:*353*, 9:*359*–360
Wood pigeons, 9:243, 9:250
Wood quails, 8:456–457
Wood-rails, 9:46, 9:47
Wood rats, white-throated, 16:*282*
Wood sandpipers, 9:177
Wood snipes, 9:177, 9:180
Wood storks, 8:235, 8:265, 8:*266*, 8:*269*, 8:*270*
Wood thrushes, 10:484, 10:486
Wood turtles, 7:72
Wood warblers, 11:6
Wood wasps, 3:405, 3:*410*, 3:*421*–422
Woodchats, 10:427
Woodchucks, 12:*31*, 16:126, 16:*149*, 16:152–*153*
Woodcock, Martin, 10:194
Woodcocks, 9:176–180, 9:*181*, 9:182
Woodcreepers, 10:170–171, **10:229–237,** 10:*231*
Woodford's rails, 9:47–48
Woodhoopoes, 10:1–4, **10:65–69,** 10:*67*
Woodland dormice, 16:327*t*
Woodland jumping mice, 16:212, 16:213, 16:*214*, 16:215, 16:216, 16:224*t*
Woodland Park Zoo, 12:205
Woodland pipits, 10:373
Woodland voles, 16:238*t*
Woodlark Island cuscuses, 13:58, 13:60, 13:66*t*
Woodlice, **2:249–260,** 2:*253*
behavior, 2:250–251
conservation status, 2:252
distribution, 2:250
evolution, 2:249
feeding ecology, 2:251–252
habitats, 2:250
humans and, 2:252
physical characteristics, 2:249–250
reproduction, 2:252
species of, 2:*254*–260
taxonomy, 2:249
Woodpecker finches, 11:264, 11:*265*, 11:*267*, 11:277
Woodpeckers, 10:85–88, **10:147–153,** 10:*152*, **10:156–168**
Woodrats, 12:131, 16:126
Allen's, 16:277*t*
bushy-tailed, 16:*124*, 16:277*t*
dusky-footed, 16:277*t*
eastern, 16:125, 16:277*t*
white-throated, 16:277*t*
Woods, C. A., 16:436
Wood's slit-faced bats, 13:373
Woodshrikes, 10:385–387
Woodsnakes, **7:433–438**
Woodswallows, **11:459–465,** 11:*461*
Woodworker frogs, 6:140, 6:*142*, 6:144–145
Woody Woodpecker, 8:22, 10:88, 10:151
Woolly dormice, 16:327*t*
Woolly false vampire bats, 13:*414*, 13:418, 13:435*t*
Woolly flying squirrels, 16:136, 16:141*t*
Woolly horseshoe bats, 13:388, 13:389, 13:390, 13:392
Woolly lemurs, 14:63, 14:66
See also Eastern woolly lemurs
Woolly mammoths, 12:*18*, 12:*19*, 12:*21*, 12:*22*, 15:162
Woolly monkeys, 14:7, 14:11, 14:155, 14:156, 14:157, 14:158–159
Woolly mouse opossums, 12:250–253, 12:*256*, 12:*261*–*262*
Woolly-necked storks, 8:266
Woolly opossums, 12:249–253, 12:254
Woolly rhinoceroses, 15:138, 15:249
See also Rhinoceroses
Woolly spider monkeys. *See* Northern muriquis; Southern muriquis
Wooly tapirs. *See* Mountain tapirs
Worcester's buttonquails, 9:14, 9:20
Wordsworth, William, 11:509
Work dogs, 14:288, 14:289, 14:*289*
Workers, 3:67–68, 3:163
honeybee, 3:29, 3:*64*, 3:*68*
vs. queens, 3:69, 3:70, 3:71
subcastes, 3:70
termite, 3:163, 3:164, 3:165
Working memory. *See* Short-term memory
World Conservation Union (IUCN), 12:213–214

  - *See also* IUCN Red List of Threatened Species
- World Conservation Union (IUCN), Species Survival Commission, 1:48
- World Heritage Sites, 8:126
- *A World of Wounds: Ecologists and the Human Dilemma*, 12:217
- World Resources Institute, 1:107
- World Series of Birding, 8:22
- World Wildlife Fund for Nature (WWF), 12:224
- World Wildlife Fund Global 200 Terrestrial Ecoregions, 16:386–387
- Worm cichlids, 5:*282*, 5:*288*, 5:289
- Worm-eating warblers, 11:288
- Wormlizards, **7:273–277,** 7:*274*
  - Florida, **7:283–285,** 7:*284*
  - mole-limbed, **7:279–282**
  - spade-headed, **7:287–290**
- Worms
  - acorn, 1:443, 1:*444*, 1:445
  - anoplans, **1:245–251,** 1:*248*
  - arrow, **1:*433*–442,** 1:*435*, 1:*436*
  - barber's pole, 1:*296*, 1:302–303
  - beard, **2:85–89**
  - clam, 2:25, **2:45–57**
  - cod, 1:*296*, 1:304
  - cold seep, **2:91–95**
  - earth, **2:65–74**
  - enoplans, **1:*253*–257,** 1:*254*
  - gastrotrichs, **1:269–273,** 1:*271*
  - gnathostomulids, **1:331–335,** 1:*333*
  - hair, 1:9, **1:305–310,** 1:*307*
  - hydrothermal vent, **2:91–95**
  - kinorhynchs, **1:275–281,** 1:*278*
  - peanut, 2:13–14, 2:21, 2:24, **2:97–101**
  - roundworms, **1:283–292,** 1:*284*, 1:*286*
  - sand, **2:45–57**
  - tapeworms, 2:28
  - thorny headed, **1:311–317**
  - tiger flatworms, 2:*23*
  - tongue, **2:317–320**
  - tube, **2:45–57**
  - velvet, 2:37, **2:109–114**
  - *See also* Flatworms; Ribbon worms
- Wormsnakes, 7:467, 7:470
  - *See also* Slender blindsnakes
- Wourms, John, 4:192
- Woylies. *See* Brush-tailed bettongs
- Wrasses, 4:14, 4:26, 4:29, 5:239, **5:293–304,** 5:*299*, 5:*300*
  - feeding ecology, 4:50, 4:68, 4:69, 5:295
  - habitats, 4:48, 5:294
  - physical characteristics, 4:*4*, 4:*17*, 4:26, 5:293–294
  - reproduction, 4:29, 4:*33*, 4:45, 5:296–298
- Wreathed hornbills, 10:72, 10:73, 10:*74*
- Wreckfishes, 5:255, 5:258–261, 5:263
- Wrege, Peter, 10:40
- Wren-babblers, 10:506
- Wren-like rushbirds, 10:*213*, 10:220–*221*
- Wrens, **10:525–538,** 10:*530*
  - behavior, 10:527–528
  - bird songs, 8:37
  - conservation status, 10:529
  - distribution, 10:*525*, 10:526–527
  - evolution, 10:525
  - feeding ecology, 10:528
  - habitats, 10:527
  - humans and, 10:529
  - nests of, 10:174
  - physical characteristics, 10:525–526
  - reproduction, 10:528–529
  - species of, 10:*531–538*
  - taxonomy, 10:525
  - *See also* specific types of wrens
- Wrentits, 10:505, 10:506, 10:*510*, 10:*513*
- Wrinkle-faced bats, 12:79, 13:310, 13:*421*, 13:*430*, 13:432
- Wrinkle-lipped free-tailed bats, 13:493*t*
- Wrinkled bark beetles, 3:317
- Wroughton's free-tailed bats, 13:487, 13:495*t*
- Wry-billed plovers. *See* Wrybills
- Wrybills, 9:161, 9:162, 9:*165*, 9:170–*171*
- Wrymouths, 5:309–313, 5:*314*, 5:315–*316*, 6:323
- Wrynecks, **10:147–154,** 10:*153*
- Wu Wang (Emperor of China), 12:203
- *Wuchereria bancrofti*, 1:35
- Würdemann's herons. *See* Great blue herons
- Wurrups. *See* Rufous hare-wallabies
- *Wuthering Heights* (Brontë), 10:462
- Wygodzinsky, P., 3:117
- *Wyulda* spp., 13:57
- *Wyulda squamicaudata. See* Scaly-tailed possums

X

- *Xanthocephalus* spp. *See* Icteridae
- *Xanthocephalus xanthocephalus. See* Yellow-headed blackbirds
- *Xanthomyza phrygia. See* Regent honeyeaters
- *Xanthopsar flavus. See* Saffron-cowled blackbirds
- *Xanthorhysis tabrumi*, 14:91
- *Xantusia* spp., 7:291, 7:292
- *Xantusia henshawi. See* Granite night lizards
- *Xantusia riversiana. See* Island night lizards
- *Xantusia vigilis. See* Desert night lizards
- Xantusiidae. *See* Night lizards
- Xantus's murrelets, 9:222
- *Xema sabini. See* Sabine's gulls
- Xenarthra, 12:26, 12:33, 12:132, **13:147–154,** 15:134
- *Xenentodon cancila. See* Silver needlefishes
- *Xenicus gilviventris. See* Rock wrens
- *Xenicus longipes. See* Bush wrens
- *Xenicus lyalli. See* Stephens Island wrens
- Xenisthmidae, 5:373, 5:374
- *Xenocalamus* spp. *See* Quill-snouted snakes
- *Xenocalamus bicolor. See* Variable quill-snouted snakes
- Xenocyprinae, 4:297
- *Xenocypris* spp., 4:299–300
- *Xenocypris microlepis. See* Smallscale yellowfins
- Xenodermatinae, 7:465
- *Xenodon* spp. *See* False pitvipers
- Xenodontinae, 7:466–468
- *Xenoglaux loweryi. See* Long-whiskered owlets
- *Xenolepidichthys dalgleishi. See* Tinselfishes
- *Xenoligea montana. See* White-winged warblers
- *Xenomedea* spp., 5:343
- Xenomystinae, 4:232
- *Xenomystus* spp., 4:232
- *Xenomystus nigri*, 4:232, 4:233
- Xenopeltidae. *See* Sunbeam snakes
- *Xenopeltis* spp. *See* Sunbeam snakes
- *Xenopeltis unicolor. See* Common sunbeam snakes
- *Xenoperdix udzungwensis. See* Udzungwa forest-partridges
- *Xenophidion* spp., 7:433, 7:434
- *Xenophora pallidula. See* Red volutes
- *Xenophrys* spp., 6:110, 6:111
- *Xenophrys monticola*, 6:116
- *Xenophrys parva. See* Burmese spadefoot toads
- *Xenophthalmichthys* spp., 4:392
- *Xenopirostris* spp. *See* Vanga shrikes
- *Xenopirostris damii. See* Van Dam's vangas
- *Xenopirostris polleni. See* Pollen's vangas
- *Xenopirostris xenopirostris. See* Lafresnaye's vangas
- Xenopodinae. *See* Clawed frogs
- *Xenopoecilus* spp., 5:80
- *Xenopoecilus oophorus*, 5:81
- *Xenopoecilus poptae*, 5:81
- *Xenopoecilus sarasinorum*, 5:81
- *Xenops milleri. See* Rufous-tailed xenops
- *Xenopsylla* spp., 3:349
- *Xenopsylla cheopis cheopis. See* Oriental rat fleas
- *Xenopus* spp.
  - behavior, 6:47, 6:100
  - conservation status, 6:102, 6:251
  - feeding ecology, 6:100–101
  - as food, 6:54
  - physical characteristics, 6:100
  - reproduction, 6:41, 6:102
- *Xenopus arabiensis*, 6:99
- *Xenopus borealis*, 6:101
- *Xenopus gilli. See* Gill's plantanna
- *Xenopus hasaunas*, 6:99
- *Xenopus laevis. See* Common plantanna
- *Xenopus muelleri. See* Müller's plantanna
- *Xenopus romeri*, 6:99
- *Xenopus stromeri*, 6:99
- *Xenopus tropicalis. See* Tropical clawed frogs
- *Xenornis setifrons. See* Speckled antshrikes
- *Xenos* spp., 3:*335*
- Xenosauridae. *See* Knob-scaled lizards
- *Xenosaurus* spp. *See* Knob-scaled lizards
- *Xenosaurus grandis*, 7:*348*, 7:*350*, 7:*351*
- *Xenosaurus grandis grandis*, 7:347
- *Xenosaurus newmanorum. See* Newman's knob-scaled lizards
- *Xenosaurus penai*, 7:347
- *Xenosaurus phalaroantheron*, 7:347
- *Xenosaurus platyceps*, 7:347, 7:*348*
- *Xenosaurus rectocollaris*, 7:347
- *Xenospiza baileyi. See* Sierra Madre sparrows
- *Xenothrix* spp., 14:143
- *Xenothrix mcgregori. See* Jamaican monkeys
- *Xenotoca eiseni. See* Redtail splitfins
- Xenotrichulidae, 1:269
- *Xenoturbella* spp., 1:11
- *Xenoturbella bocki*, 1:10
- *Xenotyphlops* spp., 7:380, 7:381
- *Xenotyphlops grandidieri*, 7:380, 7:381
- *Xenoxybelis* spp., 7:467
- Xenungulates, 15:135, 15:136
- *Xenurobrycon polyancistrus. See* Bolivian pygmy blue characins
- *Xenus cinereus. See* Terek sandpipers
- Xerini, 16:143
- *Xeromys myoides. See* False water rats
- *Xerophila leucopsis. See* Southern whitefaces
- *Xerus* spp., 16:143, 16:144

Xerus inauris. See South African ground squirrels
Xerus rutilus. See Unstriped ground squirrels
Xestospongia muta. See Giant barrel sponges
Xestospongia testudinaria. See Barrel sponges
Xiamen Rare Marine Creatures Conservation Area, 1:490
Xiphasia spp., 5:341
Xiphasia setifer. See Hairtail blennies
Xiphias gladius. See Swordfishes
Xiphiidae. *See* Swordfishes
Xiphiopsyllidae, 3:347, 3:348–349
Xiphister atropurpureus. See Black pricklebacks
Xiphocolaptes falcirostris. See Moustached woodcreepers
Xiphocolaptes major. See Great rufous woodcreepers
Xipholena spp. *See* Cotingas
Xipholena atropurpurea. See White-winged cotingas
Xiphophorus hellerii. See Green swordtails
Xiphophorus hellerii ssp. *hellerii. See* Green swordtails
Xiphophorus maculatus. See Blue platys
Xiphosura. *See* Horseshoe crabs
Xylocopa spp., 3:406
Xylocopa virginica. See Large carpenter bees
Xyloplacidae. *See* Sea daisies
Xyloplax spp., 1:381
Xyloplax medusiformis. See Medusiform sea daisies
Xyloplax turnerae. See Turner's sea daisies
Xyrichtys virens, 5:298
Xyronotidae, 3:203

Y

Yabe, M., 5:179
Yacaré caimans, 7:171, 7:*172*
Yaks, 12:*48*, 16:*1*, 16:*17*, 16:*18*
  distribution, 16:4
  domestic, 12:177–178
  evolution, 16:11
  high-altitudes and, 12:134
  humans and, 16:15
  physical characteristics, 16:2
Yamakagashis, 7:467, 7:468, 7:*471*, 7:*473*, 7:478
Yamames, 4:412
Yamur Lake grunters, 5:222
Yangtze alligators. *See* Chinese alligators
Yankees. *See* German cockroaches
Yarra pygmy perches, 5:222
Yarrell's curassows. *See* Wattled curassows
Yarrow's spiny lizards. *See* Mountain spiny lizards
Yawlings. *See* Atlantic herrings
Yellow-and-black burrowing snakes, 7:462
Yellow-and-black mud daubers. *See* Mud dauber wasps
Yellow armadillos, 13:184, 13:*186*, 13:*188*
Yellow baboons, 14:*193*
Yellow-backed duikers, 16:*75*, 16:*78*, 16:*82*, 16:83–84
Yellow-backed orioles, 11:304s
Yellow bats, 13:311, 13:497
Yellow bear cuscuses, 13:60
Yellow-bellied capuchins. *See* Yellow-breasted capuchins
Yellow-bellied elaenias, 10:273
Yellow-bellied eremomelas, 11:*9*, 11:18–*19*
Yellow-bellied fantail flycatcher. *See* Yellow-bellied fantails
Yellow-bellied fantails, 11:83, 11:86, 11:*88*, 11:*89*
Yellow-bellied flycatchers, 10:273
Yellow-bellied gliders, 13:126, 13:*127*, 13:128–129, 13:*130*, 13:*131*–132
Yellow-bellied green pigeons. *See* Bruce's green pigeons
Yellow-bellied greenbuls, 10:398
Yellow-bellied laughing thrushes. *See* Yellow-throated laughing thrushes
Yellow-bellied marmots, 16:*147*, 16:158*t*
Yellow-bellied sapsuckers, 10:87, 10:149, 10:*153*, 10:164–*165*
Yellow-bellied seasnakes, 7:30, 7:484, 7:*489*, 7:496–*497*
Yellow-bellied sliders. *See* Pond sliders
Yellow-bellied sunbird-asities, 10:188
Yellow-bellied sunbirds. *See* Olive-backed sunbirds (*Cinnyris* spp.)
Yellow-bellied tits, 11:*159*, 11:*162*–163
Yellow-bellied toads, 6:*62*, 6:*85*, 6:87–88
Yellow-billed caciques, 11:*308*, 11:*310*
Yellow-billed cuckoos, 9:*317*, 9:*326*
Yellow-billed jacamars, 10:92, 10:*94*, 10:*98*–99
Yellow-billed kingfishers, 10:10, 10:*11*, 10:19–*20*
Yellow-billed loons, 8:159, 8:*162*, 8:*166*
Yellow-billed magpies, 11:504
Yellow-billed nuthatches, 11:169
Yellow-billed oxpeckers, 11:*409*
Yellow-billed scimitar babblers. *See* Chestnut-backed scimitar-babblers
Yellow-billed shrikes, 10:426
Yellow-billed tropicbirds. *See* White-tailed tropicbirds
Yellow-billed turacos, 9:300
Yellow-blotched palm-pitvipers, 7:*450*, 7:*453*
Yellow-blotched salamanders. *See Ensatina* spp.
Yellow boring sponges, 1:*81*, 1:*84*–85
Yellow boxfishes, 5:*472*, 5:*480*–481
Yellow-breasted apalis, 11:*9*, 11:13–*14*
Yellow-breasted birds of paradise, 11:492
Yellow-breasted bowerbirds, 11:*478*
Yellow-breasted capuchins, 14:11, 14:101, 14:102, 14:106, 14:*107*, 14:*111*, 14:112
Yellow-breasted chats, 11:287–288, 11:*292*, 11:*293*
Yellow-breasted fantails, 11:83
Yellow-breasted pipits, 10:371, 10:375, 10:*377*, 10:*381*
Yellow-breasted sunbirds. *See* Olive-backed sunbirds (*Cinnyris* spp.)
Yellow-browed bulbuls, 10:397
Yellow-browed melidectes, 11:251
Yellow-browed tits, 11:155, 11:156, 11:*159*, 11:165–*166*
Yellow-browed toucanets, 10:128, 10:*129*, 10:130–*131*
Yellow-casqued hornbills, 10:75
Yellow chats, 11:65, 11:66
Yellow-cheeked crested gibbons. *See* Golden-cheeked gibbons
Yellow-chinned sunbirds. *See* Green sunbirds
Yellow-collared ixulus. *See* Whiskered yuhinas
Yellow crazy ants, 3:72, 8:214, 8:217
Yellow-crested gardener birds. *See* Macgregor's bowerbirds
Yellow-crested helmet-shrikes, 10:429
Yellow-crowned Amazons, 9:*281*, 9:*292*–*293*
Yellow-crowned brush-tailed tree rats. *See* Toros
Yellow-crowned gonoleks, 10:*431*, 10:*434*
Yellow-crowned night herons, 8:239
Yellow-crowned parrots. *See* Yellow-crowned Amazons
Yellow-eared bulbuls, 10:396
Yellow-eared parrots, 9:279
Yellow-eared spiderhunters, 11:*212*, 11:*225*
Yellow-eyed cuckoo-shrikes, 10:386, 10:387
Yellow-eyed ensatina. *See Ensatina* spp.
Yellow-eyed flycatcher warblers. *See* Golden-spectacled warblers
Yellow-eyed penguins, 8:150, 8:151, 8:*152*, 8:*156*
Yellow-eyed starlings, 11:410
Yellow-faced pocket gophers, 16:*191*, 16:*193*, 16:194
Yellow-faced siskins, 11:325
Yellow fever, 3:76
Yellow fever mosquitos, 3:76, 3:*362*, 3:365–*366*
Yellow-flanked whistlers, 11:115, 11:117
Yellow-footed antechinuses, 12:277, 12:299*t*
Yellow-footed honeyguides, 10:138
Yellow-footed rock wallabies, 13:33, 13:39, 13:*92*, 13:*96*, 13:97
Yellow-footed snakeflies, 3:*299*, 3:301–*302*
Yellow-footed tortoises. *See* South American yellow-footed tortoises
Yellow-footed weavers, 11:379
Yellow-fronted tinker barbets. *See* Yellow-fronted tinkerbirds
Yellow-fronted tinkerbirds, 10:115, 10:*118*, 10:*119*–120
Yellow fruit bats. *See* Straw-colored fruit bats
Yellow goatfishes, 5:*241*, 5:*246*, 5:*249*, 5:250–251
Yellow golden moles, 13:198, 13:222*t*
Yellow-green vireos. *See* Red-eyed vireos
Yellow grunts. *See* Bluestriped grunts
Yellow-handed titis. *See* Collared titis
Yellow-headed Amazons. *See* Yellow-crowned Amazons
Yellow-headed blackbirds, 11:283, 11:304, 11:*308*, 11:312–*313*
Yellow-headed cisticolas. *See* Golden-headed cisticolas
Yellow-headed geckos, 7:*264*, 7:*265*, 7:266–267
Yellow-headed leafbirds. *See* Blue-winged leafbirds
Yellow-headed manakins, 10:297
Yellow-headed rockfowls. *See* White-necked picathartes
Yellow-headed sidenecks. *See* Yellow-spotted river turtles
Yellow-headed temple turtles, 7:*116*
Yellow-hooded blackbirds, 11:*308*, 11:*313*–314
Yellow-hooded wagtails, 10:372
Yellow house bats. *See* House bats
Yellow jackets, 3:60, 3:68, 3:*409*, 3:*413*, 3:423–424

Yellow-knobbed curassows, 8:414
Yellow-legged buttonquails, 9:*15*, 9:*17*
Yellow-legged tinamous, 8:59
Yellow-margined box turtles, 7:*117*, 7:*118*–119
Yellow mongooses, 14:*348*, 14:350, 14:357*t*
Yellow morays, 4:*256*
Yellow mud turtles, 7:72, 7:*124*, 7:*125*
Yellow-naped Amazons. *See* Yellow-crowned Amazons
Yellow-naped yuhinas. *See* Whiskered yuhinas
Yellow-necked francolins, 8:434
Yellow-nosed mollymawks, 8:110, 8:114, 8:115
Yellow-olive flycatchers, 10:274
Yellow orioles, 11:293
Yellow perches, 4:*15–16*, 5:196–199, 5:*197*, 5:*201*, 5:*204*, 5:207–208
Yellow pikes. *See* Walleyes
Yellow rails, 9:50
Yellow-ridged toucans, 10:125
Yellow robins. *See* Eastern yellow robins
Yellow-rumped caciques, 11:*308*, 11:*309*–310
Yellow-rumped honeyguides, 10:139, 10:*140*, 10:*144*
Yellow-rumped thornbills, 11:*58*, 11:*60–61*
Yellow-rumped tinkerbirds, 10:115
Yellow-rumped warblers, 11:286
Yellow seahorses, 4:*4*, 5:*132*
Yellow-shafted flickers. *See* Northern flickers
Yellow-shouldered blackbirds, 11:302–303
Yellow-sided opossums. *See* Southern short-tailed opossums
Yellow-spotted Amazon turtles. *See* Yellow-spotted river turtles
Yellow-spotted hyraxes, 15:177, 15:189*t*
See also Rock hyraxes
Yellow-spotted night lizards, 7:291, 7:*293*, 7:*294*
Yellow-spotted river turtles, 7:*138*, 7:*139*, 7:*140*, 7:141
Yellow-spotted rock hyraxes. *See* Rock hyraxes
Yellow stingrays, 4:*181*, 4:*185*, 4:188
Yellow-streaked greenbuls, 10:397
Yellow streaked tenrecs, 13:*199*, 13:*227*, 13:*228*, 13:234*t*
Yellow-tailed thornbills. *See* Yellow-rumped thornbills
Yellow-tailed woolly monkeys, 14:11, 14:155, 14:156, 14:159, 14:167*t*
Yellow tangs, 5:*394*, 5:*398*, 5:400
Yellow-throated bulbuls, 10:399
Yellow-throated laughing thrushes, 10:*510*, 10:*515*–516
Yellow-throated leaf-loves, 10:397, 10:398
Yellow-throated leafbirds, 10:416
Yellow-throated longclaws, 10:371, 10:372–373
Yellow-throated martens, 14:*320*
Yellow-throated miners, 11:238
Yellow-throated nicators, 10:*400*, 10:*412*–413
Yellow-throated scrubwrens, 11:70
Yellow-throated seedeaters, 11:325
Yellow-throated silvereyes, 11:229
Yellow-tipped pardalotes. *See* Striated pardalotes
Yellow tits, 11:156
Yellow-toothed cavies, 16:399*t*
See also Cuis
Yellow-tufted honeyeaters, 11:*240*, 11:*244*
Yellow-tufted malachite sunbirds. *See* Malachite sunbirds
Yellow-tufted pipits, 10:373
Yellow-vented bulbuls, 10:396, 10:*401*, 10:*404*
Yellow-vented flowerpeckers, 11:*192*, 11:*194*
Yellow wagtails, 10:372, 10:373, 10:376
Yellow warblers, 11:*292*, 11:*296*
Yellow-whiskered bulbuls. *See* Yellow-whiskered greenbuls
Yellow-whiskered greenbuls, 10:398, 10:*401*, 10:*406*
Yellow white-eyes, 11:*229*
Yellow-winged bats, 13:381, 13:*381*, 13:*383*, 13:*384*
Yellow-winged caciques, 11:302–303
Yellow-winged honeyeaters. *See* New Holland honeyeaters
Yellowbellies. *See* Golden perches
Yellowbelly cichlids. *See* Giant cichlids
Yellowbelly toads. *See* Yellow-bellied toads
Yellowfin goatfishes, 5:*236*
Yellowhammers, 11:*268*, 11:*273–274*
Yellowhead collared lizards. *See* Common collared lizards
Yellowheads, 11:57, 11:*58*, 11:*63*, 11:118
Yellowish pipits, 10:372
Yellowline arrow crabs, 2:*202*, 2:*209*–210
Yellowspotted sawtails, 5:*394*, 5:*399*–400
Yellowtail damselfishes, 5:*299*, 5:*303*, 5:305–306
Yellowtail flounders, 5:453, 5:*453*
Yellowtailed surgeonfishes, 5:*391*
Yemen accentors, 10:460
Yemen chameleons. *See* Veiled chameleons
Yemeni chameleons. *See* Veiled chameleons
Yerkes Regional Primate Research Center, 12:161
Yerkish, 12:161
Yo-yo clams, 2:*455*, 2:*461*, 2:465
Yolk sac, 12:93
See also Reproduction
York, H. F., 1:47
Yorkshire pigs, 12:*172*
Yorkshire terriers, 14:288, 14:289
Young, G. C., 4:9
Young, J. Z., 2:479, 7:19
Young fingered dragonets, 5:*365*
Yow lung. *See* Chinese alligators
Yowlers. *See* Ocean pouts
Yucatán brown brocket deer, 15:396*t*
Yucatán squirrels, 16:175*t*
Yucatan wrens, 10:529
Yucatecan shovel-headed treefrogs, 6:*231*, 6:239
Yucca moths, 3:*391*, 3:399–*400*
Yucca night lizards. *See* Desert night lizards
Yuhina flavicollis. *See* Whiskered yuhinas
Yuma antelope squirrels. *See* Harris's antelope squirrels
Yungus redbelly toads, 6:*186*, 6:193
Yunnan nuthatches, 11:169
Yunnan shrew-moles, 13:282

Z

Zabetian, C. P., 5:331
Zabrini, 3:54
Zacatuches. *See* Volcano rabbits
Zachaenus spp., 6:156
Zachaenus parvulus, 6:35
Zaedyus pichiy. *See* Pichi armadillos
Zaglossus spp. *See* Long-beaked echidnas
Zaglossus bartoni. *See* Long-beaked echidnas
Zaglossus bruijni. *See* Long-beaked echidnas
Zalambdalestids, 13:193
Zalophus californianus. *See* California sea lions
Zalophus californianus wollebaecki. *See* Galápagos sea lions
Zalophus wollebaeki. *See* Galápagos sea lions
Zambezi barbels. *See* Sharptooth catfishes
Zambian mole-rats, 12:75
Zanclidae. *See* Moorish idols
Zanclus cornutus. *See* Moorish idols
Zani, P. A., 7:304, 7:307
Zanj sengis. *See* Black-and-rufous sengis
Zanzibar bushbabies, 14:*27*, 14:*30*
Zanzibar leopards, 14:385
Zapata wrens, 10:526, 10:528, 10:529, 10:*530*, 10:*534*
Zapodidae, 16:123, 16:125, 16:283
Zapodinae, 16:212
Zappa, Frank, 1:129, 1:138
Zappa's jellyfish, 1:129, 1:*131*, 1:138
Zaprochilinae, 3:203
Zaprochilus spp., 3:206
Zaproridae. *See* Prowfishes
Zapus spp., 16:212, 16:213, 16:216
Zapus hudsonius. *See* Meadow jumping mice
Zapus hudsonius campestris, 16:216
Zapus hudsonius preblei, 16:216
Zapus princeps. *See* Western jumping mice
Zapus trinotatus, 16:212, 16:216
Zapus trinotatus orarius, 16:216
Zarhachis spp., 15:13
Zavattariornis stresemanni. *See* Stresemann's bush-crows
Zealeuctra claasseni. *See* Common needleflies
Zebra antelopes. *See* Zebra duikers
Zebra butterflies, 3:*5*
Zebra danios. *See* Zebrafishes
Zebra duikers, 16:2, 16:*75*, 16:*79*, 16:*80*, 16:81
Zebra finches, 11:354, 11:355, 11:357, 11:*359*, 11:*367*
Zebra jumping spiders. *See* Zebra spiders
Zebra mice, 16:*250*
Zebra mongooses. *See* Banded mongooses
Zebra mussels, 2:17, 2:*27*, 2:43, 2:454, 2:*456*, 2:*463*, 2:464–465
Zebra sharks, 4:105, 4:*107*, 4:*109*, 4:110–111
Zebra spiders, 2:*338*, 2:343–345, 2:*344*
Zebra-tailed lizards, 7:245, 7:*248*, 7:*253*
Zebra turkeyfishes, 5:*165*
Zebra waxbills, 11:*356*
Zebrafishes, 4:303, 4:*306*, 4:*311*
See also Logperches
Zebras, **15:225–236,** 15:*231*
behavior, 12:145, 15:142, 15:219, 15:226–228
conservation status, 15:221, 15:222, 15:229–230
distribution, 15:217, 15:*225*, 15:226
domestication and, 15:145, 15:146
evolution, 15:136, 15:137, 15:215, 15:216, 15:225
feeding ecology, 12:118–119, 15:220, 15:228
habitats, 15:218, 15:226
humans and, 15:222, 15:223, 15:230
migrations, 12:164, 12:*167*

physical characteristics, 15:139, 15:140, 15:216, 15:217, 15:225–226
reproduction, 15:221, 15:228–229
species of, 15:232–234, 15:236*t*
taxonomy, 15:225
Zebrasoma spp., 5:396
Zebrasoma flavescens. *See* Yellow tangs
Zebu cattle, 16:6
Zeidae, 5:123
Zeiformes. *See* Dories
Zeitgeber, 12:74
Zelinkaderidae, 1:275
Zeltus amasa maximianus, 1:42, 2:37–38
Zenaida asiatica. *See* White-winged doves
Zenaida auriculata. *See* Eared doves
Zenaida macroura. *See* American mourning doves
Zenion spp., 5:123, 5:124
Zenion hololepis. *See* Dwarf dories
Zeniontidae, 5:123, 5:124
Zenithoptera americana, 3:*133*
Zenkerella insignis. *See* Cameroon scaly-tails
Zenker's flying mice, 16:300, 16:*301*, 16:*304*
Zenker's fruit bats, 13:349*t*
Zenker's honeyguides, 10:138
Zenopsis conchifer. *See* Buckler dories
Zeus capensis. *See* South African Cape dories
Zeus faber. *See* John dories
Zhou, K., 15:19
Ziegler, T. E., 14:165
Zig-zag eels, 5:*152*
Zingel spp., 5:195
Zinser's pocket gophers, 16:197*t*
Ziphiidae. *See* Beaked whales
Ziphius spp., 15:59
Ziphius cavirostris. *See* Cuvier's beaked whales
Zitting cisticolas, 11:*2*, 11:4, 11:5, 11:*9*, 11:*12*
Zoantharia. *See* Hexacorallia
Zoanthids, 1:103, 1:105, 1:107
Zoanthinaria. *See* Zoanthids
Zoarces americanus. *See* Ocean pouts
Zoarces viviparus. *See* Viviparous eelpouts
Zoarcidae. *See* Eelpouts
Zoarcoidei, **5:309–319**, 5:*314*
behavior, 5:311
conservation status, 5:313
distribution, 5:310
evolution, 5:309–310
feeding ecology, 5:311–312
habitats, 5:310–311
humans and, 5:313
physical characteristics, 5:310
reproduction, 5:312–313
species of, 5:*315*–318
taxonomy, 5:309
Zohary, D., 15:147
Zokors. *See* Siberian zokors
Zolini, 3:55
Zonerodius heliosylus, 8:245
Zonocerus variegatus. *See* Variegated grasshoppers
Zonosaurus spp. *See* Madagascan plated lizards
Zonosaurus madagascariensis. *See* Madagascan plated lizards
Zonotrichia albicollis. *See* White-throated sparrows
Zonotrichia leuophrys. *See* White-crowned sparrows
Zoogeography, 4:53, 4:56–58, 12:129–139
See also Distribution
Zoological Society of London, 12:204
Zooplankton, 4:39, 4:50
Zoos, 7:62, **12:203–212**, 12:*205*, 12:*206*, 12:*207*
animal husbandry, 12:209–211
ethics in, 12:212
exhibit designs, 12:204–205
history, 12:203
inbreeding in, 12:222
management of, 12:205–212
purpose of, 12:203–204
Zootermopsis laticeps. *See* Wide-headed rottenwood termites
Zoothera spp., 10:484, 10:485, 10:487
Zoothera dauma. *See* White's thrushes
Zoothera major. *See* Amami thrushes
Zopherus chilensis. *See* Ma'kechs
Zorapterans, **3:239–241**
Zorillas, 14:321, 14:323, 14:332*t*
Zorotypidae. *See* Zorapterans
Zorotypus spp., 3:239
Zorotypus barberi, 3:240
Zorotypus gurneyi, 3:240
Zorotypus hubbardi. *See* Hubbard's zorapterans
Zorotypus palaeus, 3:239
Zorotypus swezeyi, 3:240
Zorros, 14:265
Zosteropidae. *See* White-eyes
Zosterops spp., 11:227
Zosterops abyssinicus flavilateralis. *See* White-breasted white-eyes
Zosterops albogularis. *See* White-chested white-eyes
Zosterops citrinellus, 11:228
Zosterops erythropleura, 11:229
Zosterops japonicus. *See* Japanese white-eyes
Zosterops javanicus, 11:228
Zosterops lateralis. *See* Silvereyes
Zosterops lateralis chlorocephalus. *See* Capricorn white-eyes
Zosterops lateralis lateralis. *See* Tasmanian white-eyes
Zosterops luteus, 11:228
Zosterops maderaspatanus. *See* Madagascar white-eyes
Zosterops mayottensis. *See* Chestnut-sided white-eyes
Zosterops modestus. *See* Seychelles gray white-eyes
Zosterops natalis. *See* Christmas Island white-eyes
Zosterops novaeguineae. *See* New Guinea white-eyes
Zosterops pallidus. *See* Cape white-eyes
Zosterops senegalensis. *See* Yellow white-eyes
Zosterops strenuus. *See* Robust white-eyes
Zrzavy, J., 1:57
Zu spp., 4:447
Zulu golden moles, 13:222*t*
Zweifel, Richard G., 6:301
Zygaena mokarran. *See* Great hammerhead sharks
Zygodontomys brevicauda, 16:268, 16:269, 16:270
Zygogeomys spp., 16:185
Zygogeomys trichopus. *See* Michoacán pocket gophers
Zygomatic arches, 12:8
Zygoneura, 2:4
Zygoptera. *See* Damselflies
Zygotes, 1:18, 12:92, 12:101
See also Reproduction